Spin-Crossover Materials

Spin-Crossover Materials

Properties and Applications

Edited by

MALCOLM A. HALCROW
School of Chemistry, University of Leeds, UK

WILEY
A John Wiley & Sons, Ltd., Publication

This edition first published 2013

Registered office
John Wiley & Sons Ltd, The Atrium, Southern Gate, Chichester, West Sussex, PO19 8SQ, United Kingdom

For details of our global editorial offices, for customer services and for information about how to apply for permission to reuse the copyright material in this book please see our website at www.wiley.com.

Library of Congress Cataloging-in-Publication Data applied for.

A catalogue record for this book is available from the British Library.

ISBN: 9781119998679

Typeset in 10/12pt Times by Aptara Inc., New Delhi, India

Contents

List of Contributors

Catalina Achim, Department of Chemistry, Carnegie Mellon University, USA.

Elena Bagryanskaya, International Tomography Center, Siberian Branch of the Russian Academy of Sciences, Russian Federation and N.N. Vorozhtov Novosibirsk Institute of Organic Chemistry, Siberian Branch of the Russian Academy of Science, Russia Federation.

Colette Boskovic, School of Chemistry, University of Melbourne, Australia.

Azzedine Bousseksou, Laboratoire de Chimie de Coordination, CNRS & Université de Toulouse (UPS, INPT), France.

Sally Brooker, Department of Chemistry and The MacDiarmid Institute for Advanced Materials and Nanotechnology, University of Otago, New Zealand.

Guillaume Chastanet, CNRS, Université de Bordeaux, ICMCB, France.

Majed Chergui, Ecole Polytechnique Fédérale de Lausanne, Laboratoire de Spectroscopie Ultrarapide (LSU), ISIC, Faculté des Sciences de Base, BSP, Switzerland.

Christian Chong, GEMAC, Université de Versailles, CNRS UMR 8635, France.

Eric Collet, Institut de Physique de Rennes, Université de Rennes, 1-CNRS 35042 Rennes Cedex, France.

Robert J. Deeth, Inorganic Computational Chemistry Group, Department of Chemistry, University of Warwick, UK.

Cedric Desplanches, CNRS, Université de Bordeaux, ICMCB, France.

Kim R. Dunbar, Department of Chemistry, Texas A&M University, USA.

Cristian Enachescu, Department of Physics, Faculty of Physics, Alexandru Ioan Cuza University, Romania.

Stephanie R. Fiedler, Department of Chemistry, Colorado State University, USA.

Yann Garcia, Institute of Condensed Matter and Nanosciences, MOST – Inorganic Chemistry, Université Catholique de Louvain, Belgium.

Damien Garrot, GEMAC, Université de Versailles, CNRS UMR 8635, France.

Philippe Guionneau, CNRS, Université de Bordeaux, ICMCB, F-33608 Pessac, France.

Malcolm A. Halcrow, School of Chemistry, University of Leeds, UK.

Christopher M. Handley, Inorganic Computational Chemistry Group, Department of Chemistry, University of Warwick, UK.

Shinya Hayami, Kumamoto University, Department of Chemistry, Japan.

John J. Hayward, Department of Chemistry and Biochemistry, The University of Windsor, Canada.

Benjamin J. Houghton, Inorganic Computational Chemistry Group, Department of Chemistry, University of Warwick, UK.

Shinji Kanegawa, Kyushu University, Institute for Materials Chemistry and Engineering, Japan.

Soonchul Kang, Kyushu University, Institute for Materials Chemistry and Engineering, Japan.

Christina M. Klug, Department of Chemistry, Colorado State University, USA.

Jean-François Létard, CNRS, Université de Bordeaux, ICMCB, France.

Zhao-Yang Li, Kyushu University, Institute for Materials Chemistry and Engineering, Japan.

Paulo Nuno Martinho, Institute of Nanotechnology, Karlsruhe Institute of Technology, Germany.

Seiji Miyashita, Department of Physics, Graduate School of Science, The University of Tokyo, Japan, and CREST, JST, Japan.

Gábor Molnár, Laboratoire de Chimie de Coordination, CNRS & Université de Toulouse (UPS, INPT), France.

M. Carmen Muñoz, Departamento de Física Aplicada, Universitat Politècnica de València, Spain.

Keith S. Murray, School of Chemistry, Monash University, Australia.

Anil D. Naik, Institute of Condensed Matter and Nanosciences, MOST – Inorganic Chemistry, Université Catholique de Louvain, Belgium.

Masamichi Nishino, Computational Materials Science Centre, National Institute for Materials Science, Japan.

Juan Olguín, Department of Chemistry, University of Otago, New Zealand.

Victor Ovcharenko, International Tomography Center, Siberian Branch of the Russian Academy of Sciences, Russian Federation.

Carlos M. Quintero, Laboratoire de Chimie de Coordination, CNRS & Université de Toulouse (UPS, INPT), France.

Cyril Rajnak, Institute of Nanotechnology, Karlsruhe Institute of Technology, Germany.

Jeremy M. Rawson, Department of Chemistry and Biochemistry, The University of Windsor, Canada.

José Antonio Real, Instituto de Ciencia Molecular (ICMol)/Departamento de Química Inorgánica, Universidad de Valencia, Spain.

Mario Ruben, Institute of Nanotechnology, Karlsruhe Institute of Technology, Germany.

Jean-Pascal Rueff, Synchrotron SOLEIL, France.

Lionel Salmon, Laboratoire de Chimie de Coordination, CNRS & Université de Toulouse (UPS, INPT), France.

Osamu Sato, Kyushu University, Institute for Materials Chemistry and Engineering, Japan.

Michael Shatruk, Department of Chemistry and Biochemistry, Florida State University, USA.

Helena J. Shepherd, Laboratoire de Chimie de Coordination, CNRS & Université de Toulouse (UPS, INPT), France.

Matthew P. Shores, Department of Chemistry, Colorado State University, USA.

Ahmed Slimani, GEMAC, Université de Versailles, CNRS UMR 8635, France.

François Varret, GEMAC, Université de Versailles, CNRS UMR 8635, France.

Birgit Weber, Inorganic Chemistry II, Universität Bayreuth, Germany.

Zi-Shuo Yao, Kyushu University, Institute for Materials Chemistry and Engineering, Japan.

Preface

The spin-crossover process involves the rearrangement of electrons in a metal ion, from a high spin to a low spin state. These correspond to the distributions of electrons within the metal orbital energy levels that yield the maximum and minimum number of unpaired electrons respectively. The phenomenon is particularly prevalent in iron chemistry and can occur in any phase of matter, although it is most often studied in the solid state. Spin-crossover has a large impact on the physical properties of a solid material, including its magnetic moment, colour, dielectric constant and electrical resistance. Moreover some spin-crossover materials show pronounced hysteresis, which often reflects a structural phase change during the transition. Within the hysteresis loop, the materials are genuinely bistable switches that can be either high or low spin depending on their history.

Several practical applications of spin-crossover materials have been demonstrated that make use of their switching properties. They include: display and memory devices, with pixels of a spin-transition material whose colour or dielectric constant is switched by spot-heating and cooling; electrical and electroluminescent devices, where changes in the electrical resistance of a spin-crossover thin-film can be detected, or used to quench light emission; and, using the switchable paramagnetism of a spin-crossover compound in a temperature-sensitive MRI contrast agent. Switchable liquid crystals, nanoparticles and thin films of spin-crossover materials have also been achieved, that function almost as well as the same materials in the bulk phase. Notably, most of these application studies have been carried out using just two materials, whose spin-transitions show thermal hysteresis of an appropriate width (30–50 K) that spans room temperature. The production of new switchable spin-crossover materials with technologically useful properties by design, rather than by trial and error, remains a problem of crystal engineering that is only now beginning to be addressed. This combination of technical challenge and practical application explains why an effect that was first observed in the early 1930s continues to be heavily studied by groups around the world.

For the past eight years, the "bible" in the field has been the three-book set from the *Topics in Current Chemistry* monograph series, edited by Philipp Gütlich and Harold Goodwin and published in 2004.[1] This book is intended to complement that earlier work, and concentrates on aspects of spin-crossover research that have developed since then, or are otherwise covered in less detail in the *Topics in Current Chemistry* volumes. Articles from the *Topics in Current Chemistry* series are cited in this book where appropriate, and should be referred to by the reader.

The first four chapters present an overview of the development of spin-crossover research (Murray), and more detailed surveys of the mononuclear (Weber), polynuclear (Olguín and Brooker) and polymeric (Muñoz and Real) spin-crossover complexes that have been discovered since 2004. The structures of these solid compounds are then examined, to describe the state of play in the crystal engineering of spin-transition molecular materials (Halcrow). As before, these first chapters are intended to supplement those in the *Topics in Current Chemistry* volumes,[1] which give a more comprehensive survey of the types of compounds that are known to exhibit spin-crossover.

The next chapters cover alternative types of spin state transition found in molecule-based materials, whose chemistry has developed particularly rapidly since 2004. These include two different types of charge-transfer-induced spin-transition, based on electron transfer between different metal ions (Dunbar *et al.*), and between a metal and coordinated ligand (Boskovic). Other chapters cover spin-transitions based on reversible spin pairing between organic radical centres (Rawson and Hayward), and magnetic transitions associated with Jahn–Teller switching in copper/radical coordination polymers (Ovcharenko and Bagryanskaya). The physical characteristics of these different types of transition show many similarities to metal ion spin-crossover, including examples of thermal hysteresis and excited spin-state trapping at low temperatures.

The following chapter by Shores *et al.*, updates the chemistry of spin-crossover in solution. The measurement of the thermodynamics and kinetics of spin-transitions in solution is well-established. However, there has been a recent recognition that spin-crossover is also subject to supramolecular influences in solution, and can be responsive to host–guest binding interactions.

The next topic to be discussed is the application of spin-crossover compounds, in multifunctional switchable materials and in nanotechnology. This is covered in chapters describing materials combining spin-crossover with conductivity and magnetic ordering (Sato *et al.*), with liquid crystallinity and amphiphilic behaviour (Hayami), and with fluorescence (Bousseksou *et al.*). Several of these properties have been exploited to make functional or multifunctional nanoparticles, thin films and surface patterns, or even in switchable single-molecule junctions. These aspects are brought together by Ruben *et al.*

The next set of chapters describes advances in the physical and theoretical methods for studying spin-crossover materials. Coverage is limited to methods that have grown in importance since 2004, and the reader is referred back to the *Topics in Current Chemistry* series for a more comprehensive treatment of the topic.[1] The chapter by Chergui covers ultrafast measurements of high→low-spin switching, that have deconvoluted the electron redistribution and molecular structure changes that take place during a spin-transition. Next, Varret *et al.* describe the use of optical microscopy to monitor spin-crossover in single crystals at the macroscopic level. This is followed by two chapters describing advances in the theoretical description of spin-crossover, in single molecules (Deeth *et al.*) and in bulk lattices (Enachescu *et al.*). Last are discussions of advances in the study of light- or pressure-induced spin-state trapping phenomena, in bulk materials (Létard *et al.*) and in single crystals (Guionneau and Collet).

In the final chapter, Rueff describes the importance of pressure-induced spin-crossover to geology. A large proportion of the Earth's mantle contains iron-containing oxide materials, which undergo spin-crossover at high pressures in the laboratory. This has been intensely researched during the last eight years, to determine whether these spin state changes also occur in the mantle, and whether they can explain certain anomalies in its physical properties.

Guionneau and Collet have dedicated their chapter to Andrès Goeta. Andrès was one of the leaders of a team at the University of Durham who pioneered the study of excited spin states in spin-transition materials by photo-crystallography. Andrès had been due to contribute to this book but passed away suddenly in July 2011. I would like to express my appreciation to Drs Guionneau and Collet, for stepping into the breach and providing a chapter on this important topic at short notice. But, more importantly, I also dedicate this book as a whole to Andrès' memory.

Malcolm A. Halcrow
Leeds, UK
July 2012

1. Gütlich, P., Goodwin, H. A. (Eds) (2004) Spin Crossover in Transition Metal Compounds I–III. Top. Curr. Chem., vols. 233–235. Springer Verlag, Berlin/Heidelberg, Germany.

1

The Development of Spin-Crossover Research

Keith S. Murray

School of Chemistry, Monash University, Australia

Dedicated to my good friend, the late Hans Toftlund who was a fund of knowledge on spin-crossover and many other inorganic chemistry topics.

1.1 Introduction

The approach to this chapter is a personal one and treats the topics in some depth rather than attempting to provide a compendium of all that has been published in this vast field. So the author apologises in advance to those whose contributions are not included. The subject of spin-crossover (SCO), (or spin-equilibrium or spin-transition (ST)), in *d*-block metal complexes spans some nine decades and is one of those intriguing areas of inorganic research that has had a number of quiet times and rebirths, not unlike the subject of magnetochemistry. The oft quoted work of Cambi and Szegö initiated the subject (Fig. 1.1). It was carried out in the institute of industrial chemistry of the University of Milan in 1931, and showed some 16 tris(*N*,*N*-disubstituted dithiocarbamate) iron(III) derivatives, $[Fe(R_2NCS_2)_3]$, in Table 3 of their iconic paper, with anomalous magnetic susceptibilities relative to that of the high spin (HS) d^5 value for the O-bonded $[Fe(acac)_3]$.[1] These compounds will be discussed further, later. A present day 'Googling' of spin-crossover in Wikipedia reveals a brief and useful survey of development in the subject, finishing with efforts (ongoing) at commercial applications of these molecular magnetic 'switching' materials. The Google search shows many hundreds of hits for spin-crossover. Between 1931 and 2011 there has been, in the author's view, a number of broadly distinguishable periods. Those interested in this topic and in the history of science may well disagree with the definition of such periods. But here we go.

Spin-Crossover Materials: Properties and Applications, First Edition. Edited by Malcolm A. Halcrow.

1931. B 2591

Berichte der Deutschen Chemischen Gesellschaft.

1931, Nr. 10. — Abteilung B (Abhandlungen) — 11. November.

421. L. Cambi und L. Szegö: Über die magnetische Suszeptibilität der komplexen Verbindungen.

[Aus d. Istituto di Chimica Industriale d. Universität Mailand.]
(Eingegangen am 18. Juli 1931.)

Die komplexen Verbindungen der Metalle der ersten Übergangsreihe des periodischen Systems bieten betreffs der Beziehungen zwischen dem magnetischen Verhalten und der inneren Elektronen-Struktur noch immer sehr interessante Probleme.

Tabelle III.

Substanzen	$\chi \times 10^6$ (20°)	$\chi_A \times 10^6$	p
Äthyl-xanthogenat, $(C_2H_5O.CS_2)_3Fe$	+6.93	+3100	13.3
Benzyl-xanthogenat, $(C_7H_7O.CS_2)_3Fe$	+3.68	+2523	12.1
N-Methyl-dithiocarbamat, $(CH_3.NH.CS_2)_3Fe$	+40.55	+15321	29.7
N-Äthyl-dithiocarbamat, $(C_2H_5.NH.CS_2)_3Fe$	+34.0	+14443	28.9
N-Propyl-dithiocarbamat, $(C_3H_7.NH.CS_2)_3Fe$	+25.5	+11900	26.2
N-Benzyl-dithiocarbamat, $(C_7H_7.NH.CS_2)_3Fe$	+24.3	+14954	29.4
N,N-Dimethyl-dithiocarbamat, $(C_2H_6N.CS_2)_3Fe$	+15.2	+6534	19.4
N,N-Diäthyl-dithiocarbamat, $(C_4H_{10}N.CS_2)_3Fe$	+16.7	+8635	22.3
N,N-Dipropyl-dithiocarbamat, $(C_6H_{14}N.CS_2)_3Fe$	+14.4	+8776	22.5
N,N-Di-*n*-butyl-dithiocarbamat, $(C_8H_{18}N.CS_2)_3Fe$	+18.4	+12732	27.1
N,N-Di-isobutyl-dithiocarbamat, $(C_8H_{18}N.CS_2)_3Fe$	+4.87	+3682	14.6
N,N-Methyl-phenyl-dithiocarbamat, $(C_7H_8N.CS_2)_3Fe$	+4.35	+2965	13.1
N,N-Äthyl-phenyl-dithiocarbamat, $(C_8H_{10}N.CS_2)_3Fe$	+13.7	+9143	23.0
N,N-Isoamyl-phenyl-dithiocarbamat, $(C_{11}H_{16}N.CS_2)_3Fe$	+5.5	+4672	16.4
N-Piperidyl-dithiocarbamat, $(C_5H_{10}N.CS_2)_3Fe$	+12.1	+6621	19.6
N,N-Dibenzyl-dithiocarbamat, $(C_{14}H_{14}N.CS_2)_3Fe$	+12.8	+7201	20.4
N,N-Äthyl-cyclohexyl-dithiocarbamat, $(C_8H_{16}N.CS_2)_3Fe$	+4.47	+3374	14.0
N-Piperazyl-dithiocarbamat, $^1/_2[(C_2H_4N.CS_2)_2]_3Fe_2$	+21.6	+9320	23.2
Acetyl-aceton, $(C_5H_7O_2)_3Fe$ [30]	+42.2	+15066	29.5

Figure 1.1 *Extract from the Cambi and Szegö paper on Fe^{III} tris-dithiocarbamate compounds. Note that magnetic moments (p in Table III) are in Weiss magnetons, which are ~5 x Bohr magneton values. Adapted with permission from [1]. Copyright Wiley-VCH Verlag GmbH & Co., 1931.*

Between 1931 and the 1960s the subject lay largely dormant, with the Second World War probably playing some part in the lack of activity, at least as far as publications were concerned. However, coordination chemistry, and associated magnetochemical studies of the *d*-block complexes prepared, continued during the 1940s and 1950s in Europe,[2] the USA,[3] Japan[4] and Australia – in the latter country by the likes of Burrows,[5] Nyholm[6] and Dwyer,[7] but spin-crossover did not feature. SCO was recognised by Pauling in regard to Fe^{II} heme oxygenation in 1936.[8]

The 1960–80 period can be labelled 'the renaissance in mononuclear SCO compounds' and there was great activity occurring in many research groups worldwide. Not only did Martin, Ewald and group,[9] then Figgis[10] in Australia, reinvestigate the $[Fe(R_2NCS_2)_3]$ family, including the first applied pressure work on SCO materials; on the other side of the world in Russia, Zelentsov and Gerbeleu and co-workers[11] developed bis-tridentate thiosemicarbazone Fe^{III} complexes of types $[Fe(5\text{-}X\text{-}thsa)_2]^-$ and [Fe(5-X-thsa)(5-X-tshaH)] having $FeN_2S_2O_2$ coordination spheres, the complexes often producing sharper and more hysteretic spin-transitions than the $[Fe(R_2NCS_2)_3]$ compounds. We will see later that the thsa-Fe^{III} materials have been receiving recent attention in other laboratories.[12] Iron(III) SCO monomers containing N_4O_2 or N_3O_3 ligand donor combinations, commonly from Schiff-base chelators, began to emerge from work in the USA,[13] Japan[14] and Australia,[15] with a report by Hendrickson and group catching the eye in which they found that the nature of the spin-transition (shape and $T_{1/2}$) was found to depend on the size of crystallites and how finely the crystallites were ground.[16] Such nonligand-field/noncovalent 'supramolecular' and physical effects continue to intrigue studies of cooperativity in crystalline SCO samples.

The first iron(II) d^6 SCO monomers were discovered in the mid-1960s and this led to an explosion in studying the kinds of N-donor ligand combinations that would yield SCO behaviour, a pursuit that continues today. The first examples, by König and Madeja,[17] were of the type *cis*-$[Fe(NCS)_2(1,10\text{-}phen)_2]$ and the 2,2′-bipy analogue, with a Fe$^{II}N_6$ mixed heterocyclic/pseudohalide(N) donor set providing the appropriate ligand-field. There are many such related pyridine-containing ligand systems that make up the FeN_6 chromophore, including the tetradentate tripyridylmethylamine compounds, $[Fe(NCS)_2(TPA)]$,[18] and the bis-dipyridylamine $[Fe(NCS)_2(DPA)_2]$[19] complexes or congeners thereof, such as the DPA-substituted triazines to be discussed later. Hexakis-tetrazole complexes such as $[Fe(1\text{-}propyl\text{-}tetrazole)_6]^{2+}$, discovered by Haasnoot *et al.*,[20] provided a great vehicle for the detailed study of its abrupt spin-transition by Gütlich and co-workers by use of magnetic, Mössbauer spectral, structural and thermodynamic methods.[21] It also provides a very good student demonstration of its rapid colour change, from colourless to violet, and vice versa, upon cooling in liquid nitrogen, then rewarming above its $T_{1/2}$ of 130 K. The tris-chelated picolylamine family, $[Fe(2\text{-pic})_3](Cl)_2$·solvate, was likewise much studied in this early period,[22] and has proved recently to yield detailed synchrotron X-ray structural information on intermediate phases (IPs) existing at temperatures where steps/inflections occur along the complex thermal spin-transition (see the section below on mononuclear SCO materials).[23]

Other azole N-donors, such as the ubiquitous tris-(1-pyrazolyl)-hydridoborate ('scorpionate') *facial* chelators, led Trofimenko and Jesson to study the spin-crossover properties of $[Fe(HB(pz)_3)_2]$ and substituted-pyrazole analogues.[24] Interestingly, it took another 30-plus years to see the SCO properties in the cationic tris-(1-pyrazolyl)methane analogues, $[Fe(HC(pz)_3)_2](anion)_2$.[25–27] Goodwin *et al.* made extensive studies on the synthesis, structure, Mössbauer spectra and magnetism of a variety of pyrazolyl-pyridine chelates of Fe(II)[28] and these have led to further recent advances by Halcrow,[29] Létard,[30] Ruben and co-workers.[31]

Cobalt(II) d^7 SCO complexes were some of the earliest to be investigated, by Baker *et al.* and Martin *et al.*, the $[Co(terpy)_2](anion)_2$ systems showing gradual spin-transitions that were sensitive to changes in anion.[32, 33]

1980–2012 period. Following a lessening of interest, or perhaps a slowing in the frenetic activity expended on monomeric species during 1960–1980, the subject received renewed and continuing interest primarily, but not solely, because of the challenges in polynuclear iron(II and III) and cobalt(II) SCO chemistry. The fundamental reason was to see if covalent bridging between SCO metal centres, in crystals, would influence the degree of cooperativity and thermal hysteresis loop widths when compared to monomeric analogues.[34, 35] Supramolecular bridging interactions were of similar importance. In other words, the question was posed as to whether the spin-transitions on individual metal centres would occur sequentially or simultaneously. Multistep transitions could well occur, as had already been seen in the form of 2-step transitions in some monomeric compounds. More on this will be discussed later. A second fundamental question was to investigate whether any synergy occurred between spin-crossover and spin-spin magnetic exchange, the latter originating between paramagnetic single-ion centres (e.g. HS–HS Fe^{II}; $S = 2 : S = 2$ coupling), via superexchange interactions across bridging groups, in discrete clusters, 1D chains, 2D sheets and 3D frameworks.[34, 35] The other major impetus for renewed interest in SCO compounds was the possibility of producing electronically useful 'new age' materials for use in displays, sensors and memories.[36] In Europe, a network of SCO researchers, 'TOSS' (thermally and optical spin state switching) was formed and lasted many years, to be replaced, in part, by 'MAGMANet'– the latter, ongoing network includes researchers working in all areas of molecular magnetic materials.

In the discrete cluster class, the first dinuclear Fe(II) SCO complexes studied were by Kahn, Real *et al.*[34] of type $[Fe(NCX)_2(\text{bidentate})(\mu\text{-bipyrimidine})Fe(NCX)_2(\text{N,N-bidentate})]$, bidentate = 2,2′-bipy, bis-thiazoline, etc., and the first dinuclear Co(II) complex, by Kahn and Zarembowitch,[37] was of the binucleating Schiff base fsaen type with N,O donor groups. The Leiden group of Reedijk, Haasnoot *et al.* reported trinuclear 1,2,4-triazole-bridged Fe(II) complexes in which the central ion showed the spin-transition while the terminal Fe atoms remained HS.[38, 39] Exchange coupling between neighbouring atoms was generally weak, a few cm^{-1} at most. In a $Co^{II}{}_2$ macrocyclic derivative simultaneous crossover and exchange was observed by Brooker *et al.*[40]

Lehn, Ruben *et al.*[41] reported the first tetranuclear 2×2 'grid' $Fe^{II}{}_4L_4$ SCO complex that showed rather broad spin-transitions. Other recent $Fe^{II}{}_4$ examples are described later and include crystal structures and physicochemical proof of the various spin state combinations. Dunbar and co-workers described a trigonal-pyramidally shaped $Fe^{II}{}_5$ SCO cluster.[42] The first hexanuclear SCO 'nanoball', containing six $Fe^{II}N_6$ chromophores held within ditopic scorpionate-pyridyl ligands that also had eight Cu^I centres in the hydridopyrazolylborate N_3 'inner-pockets', was made by Batten, Duriska *et al.*[43]

This chapter will now cover recent developments in polynuclear and mononuclear SCO materials, with updates on theory, then briefer sections will cover recent advances in multifunctional materials, instrumentation/measurement and, finally, applications.

1.2 Discrete Clusters of SCO Compounds

There have been a number of synthetic and physicochemical challenges in polynuclear cluster materials that contain SCO centres and many, but not all, of these have been overcome in the last decade. In the area of synthesis and design the key challenges have included:

- The design of bridging and terminal ligands and the coordination environment around constituent Fe^{II} (or Fe^{III} or Co^{II}) centres that yield a spin-transition.
- The design and isolation of tri-, tetra- and higher nuclearity SCO clusters.
- The aim of observing and understanding synergy between exchange coupling and SCO between, and within, nearest neighbour ions. Why is exchange coupling often negligible?

The coordination environment that yields the correct ligand-field and hence induce spin-crossover at Fe^{II} centres is commonly made up of six pyridyl- or azole-N donors or combinations of these with (usually) two $\underline{N}CX^-$ ligands, where X = S, Se, BH_3; or $N(CN)_2^-$ (dicyanamide, dca^-).[21, 44] Chelating ligands containing these donors are commonly used as terminal, and sometimes bridging, ligands. Recently, combinations of N,O donors, from Schiff base ligands have proved successful in Fe^{II} systems.[45, 46] The latter donors also induce spin-transitions in six-coordinate Fe^{III} compounds, both mononuclear,[12–14, 16, 47] dinuclear and trinuclear,[48] and in six- or five-coordinate Co^{II} compounds.[37, 49] However, six S-donors, such as in the abovementioned tris-dithiocarbamates are commonly employed in Fe^{III} compounds.[1, 9, 10] *N,N,N*-tridentate chelating ligands such as terpy have been long known to yield Co^{II} SCO mononuclears[50] (more recent aspects on these are given later) and one wonders if, when combined with bridging 4,4′-dipyridine type and nonbridging $\underline{N}CX^-$ ligands, they will yield new Co^{II} dinuclears as has been found to be the case for Fe^{II} and Fe^{III}.[47, 51] Care is always needed with the M:L:NCX^- stoichiometry employed in such reactions so that a dinuclear rather than a mononuclear product ML_3 is obtained.

Designing larger SCO clusters requires, for instance in $Fe^{II}{}_4$ species, the proper combination of terminal chelating groups and bridging 2-connectors. In the case of squares, a flexible tetradentate L such as trimethylpyridylamine (TPA) can be combined with linear CN^- or 4,4′-bipy to yield $[Fe^{II}{}_4L_4(bridge)_4]$ SCO species. Cubane $Fe^{II}{}_4L_4$ SCO clusters require *facial*-tridentate terminal ligands, such as the scorpionates $HB(pz)_3^-$ or $HC(pz)_3$ combined with three two-connectors of the CN^- or 4,4′-bipy types. The SCO properties are discussed later. Triangular or rectangular Fe^{II} clusters can be obtained that don't possess CN^- or 4,4′-bipy bridges by using an appropriate polypodal central linker, such as a 1,3,5-tri-tris(pyrazolyl)methane-substituted benzene, together with a $Fe^{II}(HC(X\text{-}pz)_3)$ terminal group, which yields a triangle of widely spaced $[FeN_6]$ SCO centres.[52, 53] The early, linear $Fe^{II}{}_3$-1,2,4-triazole SCO compounds (spin change at only one Fe), that could have yielded 1D $[Fe(1,2,4\text{-triazole})_3]^{2+}(anion)_2$ chains, required the correct mole ratio of reagents be used, and some luck.[38, 54] When the anion is NCS^-, $[Fe_2L_5(NCS)_4]$ dinuclears are obtained, that, apart from the first example by Reedijk and Haasnoot *et al.*,[55] have invariably remained HS–HS irrespective of the triazole used; more on this below.

Designing higher nuclearity SCO clusters, such as an octahedrally disposed $Fe^{II}{}_6$ aggregate in a 14-metal $Fe^{II}{}_6Cu^{I}{}_8$ 'nanoball' required the skills and '3D vision' of those, like Batten,[43] designing self-assembled metallosupramolecular polyhedra (Archimedean and Platonic), here of the pseudo-spherical type. A $Tp^{4\text{-}py}$ scorpionate ligand was employed to make the nanoball with tetrahedral Cu^I ions in the inner coordination

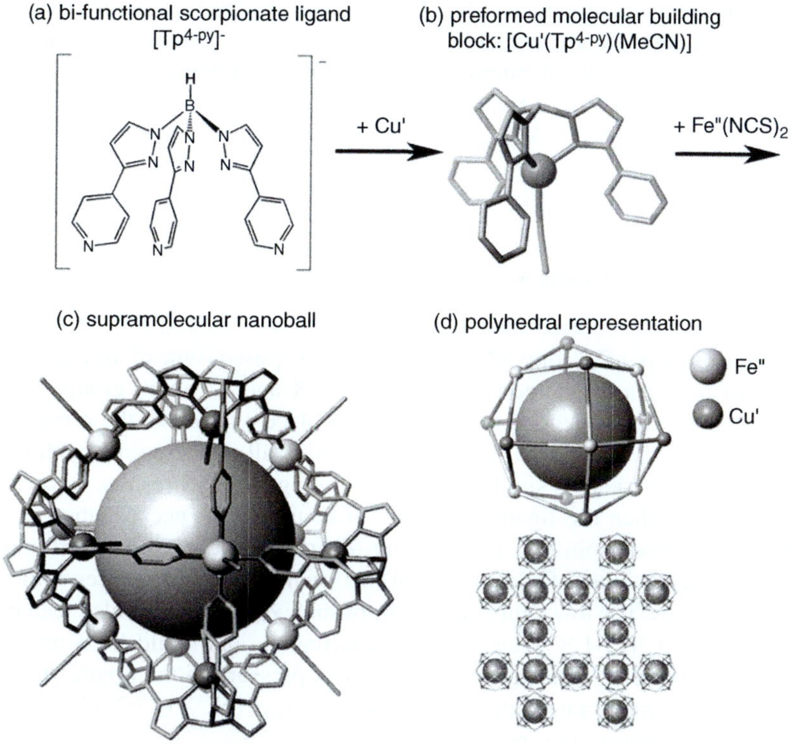

Figure 1.2 *Formation and structure of $Fe^{II}{}_6Cu^{I}{}_8$ 'nanoball' showing (a); the bifunctional $Tp^{4\text{-}py}$ ligand, (b); the $Cu^{I}(Tp^{4\text{-}py})(MeCN)$ building block, (c); the supramolecular nanoball, (d); the polyhedral representation and packing motif. Reproduced with permission from [43]. Copyright Wiley-VCH Verlag GmbH & Co., 2009.*

'pockets' and with the outer 4-pyridyl groups doing just what they were intended to do, rather than leading to polymeric alternatives (Fig. 1.2). Choice of co-ligands was, of course, important to create the crossover ligand-field at each $Fe(py)_4$ centre; viz. $\underline{N}CS^-$ and $\underline{N}CMe$ in *trans* positions. Other important features such as porosity in crystals of these nonframework nanoball materials, and the effect of guest sorption on the spin-transition, provided an added bonus. The chemistry and functions of metallosupramolecular polyhedra are receiving much current interest from a number of groups including those of Fujita,[56] MacGillivray,[57] Ward[58] and Stang.[59]

The physicochemical, electronic, magnetic and theoretical challenges in polynuclear SCO clusters included:

- Understanding 1-step, vs. 2-step, vs. multistep spin-transitions.
- Understanding cooperativity (thermal hysteresis) within and between SCO clusters and relating to structure.
- Observing and understanding photomagnetic LIESST effects in polynuclear clusters and comparing results to monomers.
- Gaining a theoretical understanding of SCO in clusters and of synergy with exchange coupling.

Answers to such challenges, or challenges that still remain, will be discussed when describing the various molecular cluster classes, below.

Recent advances in dinuclear and polynuclear SCO compounds are described in other chapters in this book, and a number of reviews on the topic are available.[35, 44, 45, 60–66] Aspects that are relevant to the development of SCO research are now given without attempting to be fully comprehensive of all such reports. The timeline dating from the *Topics in Current Chemistry*'s three volumes on 'Spin-crossover in transition metal compounds' (2004) is largely followed.[67] Many of the present subsections have related chapters within these three volumes.

1.2.1 Dinuclear Fe^{II} - Fe^{II} SCO Clusters

Kahn, Real and co-workers first reported their bipyrimidine(bpym)-bridged Fe^{II} SCO complexes, of type $[(\underline{N}CX)_2(2,2'\text{-bipy})Fe(\mu\text{-bpym})Fe(2,2'\text{-bipy})(\underline{N}CX)_2]$ in 1987 and posed some of the challenges shown above, such as is there synergy between exchange and SCO?[34] The X and 2,2′-bipy groups were systematically varied and it was noted that, even when all seemed to be in place to achieve the crossover ligand-field in such dinuclear complexes, the spin state HS–HS was stabilised at all temperatures, without SCO occurring. Many of us since then have experienced similar, rather frustrating, HS–HS results, when using other ligand combinations.[68] Effects other than the ligand-field can, of course, influence whether or not SCO occurs. There was, nevertheless, a good 'spin-off' for Kahn *et al.* in that the weak antiferromagnetic exchange coupling ($2J = -4.1\ cm^{-1}$ from $-2J\mathbf{S_1}\cdot\mathbf{S_2}$ Hamiltonian) between the HS Fe^{II} centres could be observed and quantified from $\chi_M T$ *vs.* T plots. The low temperature part of such $\chi_M T$ plots was also seen in 'half' crossover μ-bpym examples, that had a 'HS–LS' plateau above this region, and could be extrapolated to the high temperature HS–HS data, above $T_{1/2}$. Later, the application of pressure to the bpym/$\underline{N}$CS HS–HS example yielded SCO.[60, 62] One of the important properties displayed by the $[(\underline{N}CX)_2(2,2'\text{-bipy})Fe(\mu\text{-bpym})Fe(2,2'\text{-bipy})(\underline{N}CX)_2]$ family is that of reversible spin switching.[63, 69]

Some 10 years later, we[64] and others[70] began exploring other dinuclear $Fe^{II}Fe^{II}$ SCO compounds, using a variety of terminal and bridging groups, to create FeN_6 or, more recently, FeN_4O_2 donor sets at each Fe^{II}.[45] The reasons for this rekindling of interest were many, some given above as bullet points, while others included questions such as 'do other dinuclear systems exist to compare with the bipyrimidine family?'; 'do these dinuclears display 1-step HS–HS to LS–LS transitions or 2-step HS–HS to HS–LS to LS–LS transitions with decreasing temperature?'; 'is it possible to isolate the "HS–LS" molecule that gives the $\chi_M T$ plateau that occurs between the HS–HS and LS–LS states, and are these individual HS–LS molecules or 50:50 HS–HS:LS–LS molecular mixtures?'.

Bridging groups employed include pyrazolates, triazolates, triazoles, pyrimidines, 4,4′-bipy (and similar 2-connecting dipyridyls) or $N(CN)_2^-$, the first three forming part of polytopic chelating ligands (Fig. 1.3). The terminal groups are generally pyridine-derived chelates sometimes in combination with $\underline{N}CX^-$, carefully chosen to create 6-coordination at each Fe^{II}. Dinuclear helicates of stoichiometry $Fe_2L_3^{2+}$ have also been investigated where L is a ditopic N,N-chelator (Fig. 1.4).[71]

Key discoveries to emerge, usually supported by structures, magnetism, Mössbauer spectra and DFT calculations, were (i) the observation of full, 1-step HS–HS to LS–LS transitions in pyrazolate-bridged compounds,[64, 65, 70] (ii) 2-step HS–HS to 'HS–LS' to LS–LS transitions in pyrazolate,[64] triazolate,[72] pyrimidine[73] and 4,4′-bipy[51, 74] (and related) bridged compounds, where 'HS–LS' was usually found to be a HS–LS molecule but examples of 50:50 HS–HS:LS–LS were also found, (iii) 'half' SCO examples that were 'trapped' in the HS–LS form following the HS–HS to HS–LS spin-transition, in trihelicates[71] and in triazole-bridged molecules with the latter HS–LS molecules of Brooker *et al.* being the first such HS–LS form to be structurally characterised,[75, 76] (iv) the LIESST properties of dinuclear species,[60, 64, 69] (v) the DFT calculations that predicted/rationalised the 2-step transitions,[77] (vi) the very weak to zero HS–HS exchange coupling even in dinuclears that contained bridges capable of transmitting stronger exchange.[34, 35, 44, 60–64]

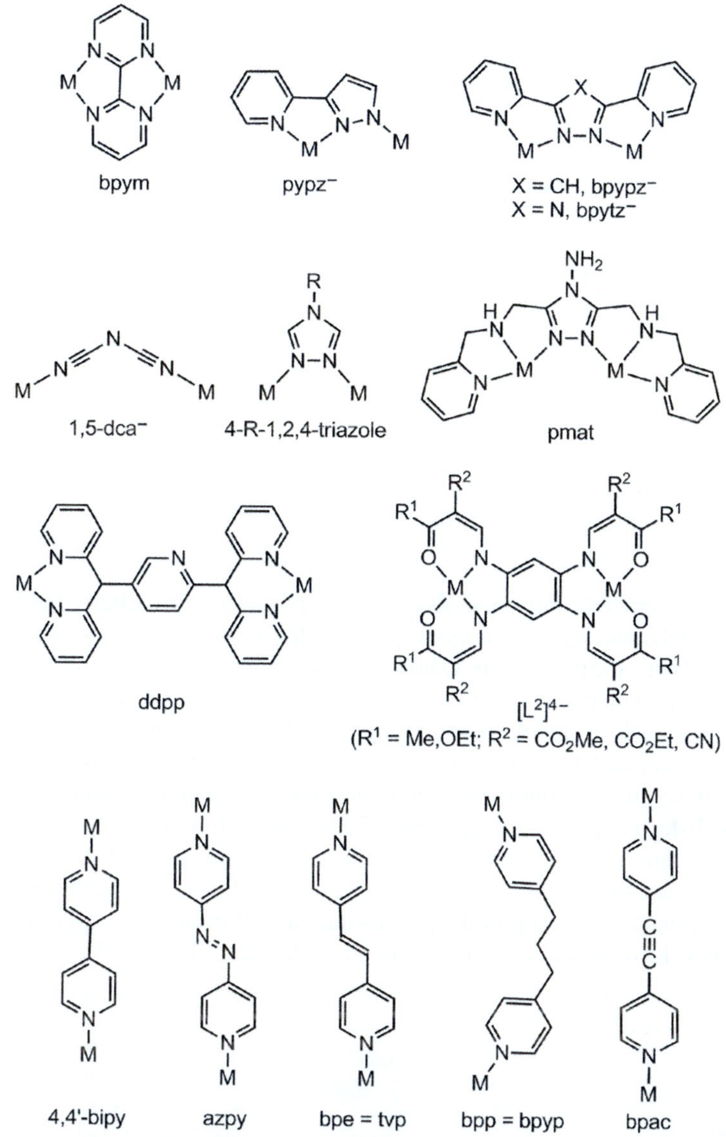

Figure 1.3 *Various bridging groups used in dinuclear* Fe^{II} *SCO compounds.*

In general, these studies of dinuclear systems that followed from the bipyrimidine-bridged work confirmed many of the findings of Kahn *et al.*[34] as well as making significant advances in our fundamental knowledge of such covalently-bridged SCO molecules.

Here, current examples containing 4,4′-bipy-type bridging, [[Fe(dpia)(*cis*-NCS)$_2$]$_2$(μ-L)] (L = 4,4′-bipy[78] or bpe,[79] Fig. 1.5) are described, first, in some detail, giving emphasis to modern developments. The synthesis required a tridentate chelating ligand, dpia (di(2-picolyl)amine), to make the Fe^{II} centres 6-coordinate with FeN_6 donor sets. Care had to be taken not to make the homoleptic *bis*-monomer, $[Fe(dpia)_2]^{2+}$. A similar compound employing L = 3-bpp, viz. [[Fe(3-bpp)(*trans*-NCS)$_2$]$_2$(μ-4,4′-bipy)]·2MeOH, with unusual

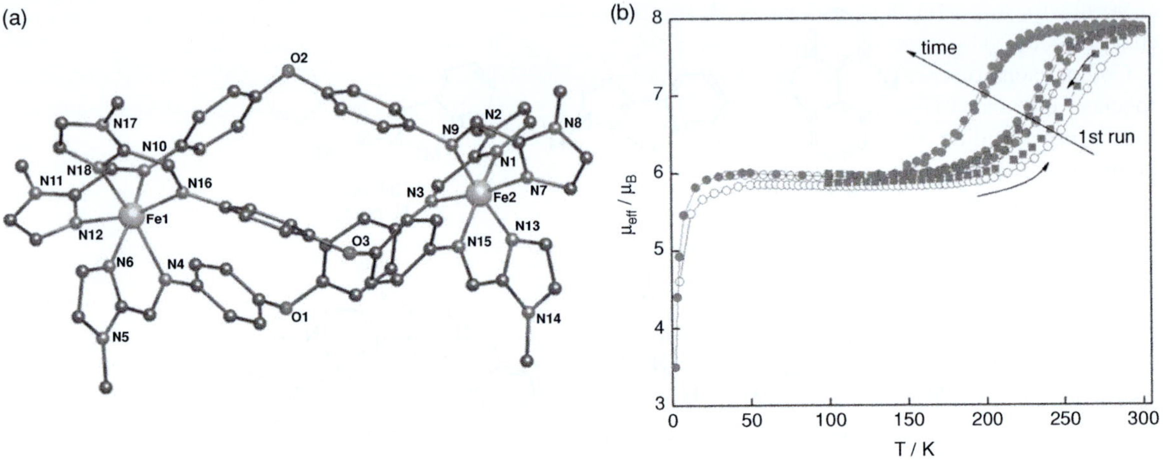

Figure 1.4 (a) Structure of an $Fe_2L_3^{2+}$ triple helicate by Kruger et al. (b) Magnetic properties as a function of time. Reproduced from [71] with permission of The Royal Society of Chemistry, 2011.

structural nuances that have been recently described,[51] was reported a little earlier. [[Fe(dpia)(*cis*-NCS)$_2$]$_2$(μ-bpe)] forms two polymorphs and a pseudopolymorph [[Fe(dpia)(*cis*-NCS)$_2$]$_2$(μ-bpe)]·2MeOH all showing quite different $\chi_M T$ vs. T plots: the polymorph **1** shows a 2-step gradual spin-transition, polymorph **2** remained HS–HS, and the methanol adduct showing an abrupt 1-step (full HS–HS to LS–LS) transition; Fig. 1.5).[79] There was no thermal hysteresis in any of the spin-transitions. Such differences in susceptibilities as these are not unusual in SCO chemistry and the authors were, of course, keen to find out why such differences occurred. The crystal structures of the dinuclear molecules, and how they packed in the crystal, were discussed in detail, as were the octahedral distortion parameters, Σ (sum of deviations of the 12 *cis* N-Fe–N angles from 90°) for the HS–HS structures, around each Fe^{II} centre, with polymorph **2** having two distinct Fe sites even at 293 K. The 2-step example **1** did not show structurally distinct Fe sites within each binuclear molecule at 183 K, the temperature (inflection point) at which [HS–LS] molecules would exist, but displayed a similar (averaged) structure to the [HS–HS] form, however with shorter Fe–N lengths. [LS–LS] Fe–N lengths were observed

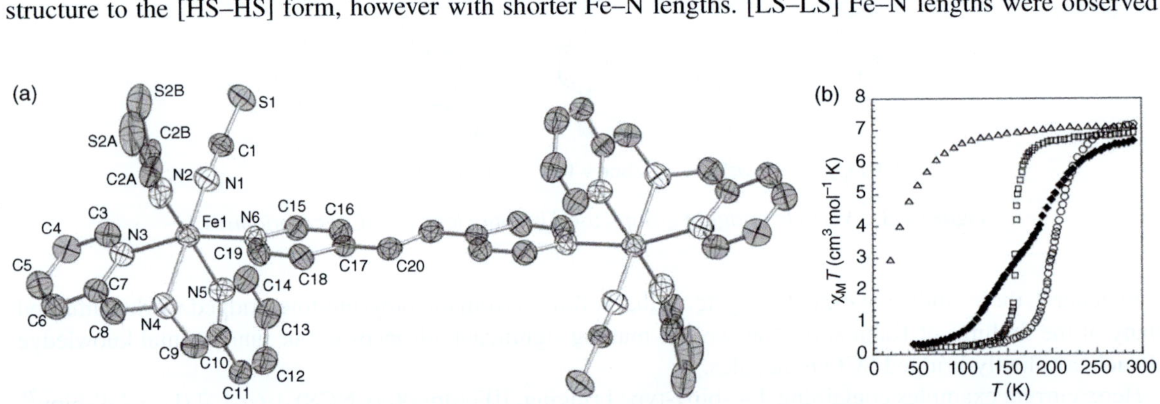

Figure 1.5 (a) Structure of [[Fe(dpia)(NCS)$_2$]$_2$(μ-bpe)]. (b) Magnetic data for various forms of [[Fe(dpia)(NCS)$_2$]$_2$(μ-bpe)]; ♦ [[Fe(dpia)(NCS)$_2$]$_2$(μ-bpe)] polymorph 1, Δ [[Fe(dpia)(NCS)$_2$]$_2$(μ-bpe)] polymorph 2, □ [[Fe(dpia)(NCS)$_2$]$_2$(μ-bpe)]·2MeOH, ○ [[Fe(dpia)(NCS)$_2$]$_2$(μ-4,4′-bipy)]. Reproduced from [79] with permission of The Royal Society of Chemistry, 2011.

at 90 K with lower octahedral distortions (lower Σ) than in the [HS–HS] form. No crystallographic phase change occurred between 300 and 90 K.

Comparisons of core geometries, Fe..Fe separations and octahedral distortions in the three bpe-bridged species were made to those in [[Fe(dpia)(*cis*-NCS)$_2$]$_2$(μ-4,4′-bipy)][78] and in other dinuclear SCO compounds having bipyrimidine, pyrazolate, triazolate and dicyanamide bridges.[64] The 'take home' message was that the nature of the spin-transition (full 1-step; full 2-step; half crossover) in dinuclear Fe^{II} SCO compounds was primarily related to the degree of the octahedral distortions at the FeN_6 cores, that is intramolecular effects, these being influenced by packing and ligand strain arising from terminal and/or bridging ligands. Inter-dinuclear interactions, viz. H-bonding, π-stacking, van der Waals, were deemed to be responsible for the differing cooperativity, highlighted by the abrupt (more cooperative) transition in [[Fe(dpia)(*cis*-NCS)$_2$]$_2$(μ-bpe)]·2MeOH. A strong distortion, having a {higher Σ/weaker ligand field} in the starting [HS–HS] form was felt to stabilise the HS state, whatever the temperature. Then the relative degree of distortion of FeN_6 sites in the [HS–HS] form was felt to be responsible for whether the 'half' transition [HS–HS] to [HS–LS] occurred, with a large distortion on the HS site preventing it going on to form LS, or whether the 2-step [HS–HS] ↔ [HS–LS] ↔ [LS–LS] transition occurred. A mild distortion was present in the HS site in the [HS–LS] form of the latter. Similar conclusions have recently been obtained for two new alkyne-linked dipyridyl bridged analogues, [[Fe(dpia)(NCS)$_2$]$_2$(bpac)]·nCH_3OH [n = 0 (**1**) and 2 (**2**), bpac = 1,2-bis(4-pyridyl)ethyne].[80]

The related compound [[Fe(3-bpp)(*trans*-NCS)$_2$]$_2$(μ-4,4′-bipy)]·2MeOH[51, 74] was not included in the comparative magnetostructural discussions given in the phia papers.[79, 80] It showed three crystallographic phases as the temperature was lowered: phase 1, 300–161 K, $P2_1/n$; phase II, 151–113 K, Cc; phase III, 115–30 K, $P\bar{1}$(and photoexcited phase III*, 30 K, $P\bar{1}$). These corresponded to the spin states [HS–HS], [HS–HS] and [HS–LS] for I to III, respectively, in agreement with magnetic data for the 'half' spin-transition, $T_{1/2}$ ~114 K (Fig. 1.6). Thermal hysteresis in the cell volume was shown in the warming mode, with $\Delta T \sim 4$ K. The phase change I to II could not be seen in the $\chi_M T$ plot but it could be clearly seen in heat capacity data. It would be interesting to see the corresponding evolution of synchrotron PXRD cell data between 300 and 30 K as described for the 3D framework SCO systems and some of our other complexes, described later. All Fe sites were identical in phase I, while there are four different Fe sites (2 per dinuclear) in phases II and III. One Fe site (type 1) underwent SCO while the other (type 2) did not. In phase III, all Fe–N lengths, volumes of Fe octahedra and Σ values pointed to two HS sites and two LS sites (2 different dinuclear molecules). Thus the partial crossover to form [HS–LS] molecules was confirmed and the LS sites were clearly identified in these [HS–LS] forms. A thermal and light-induced structural and spin state diagram is given in Figure 1.7.

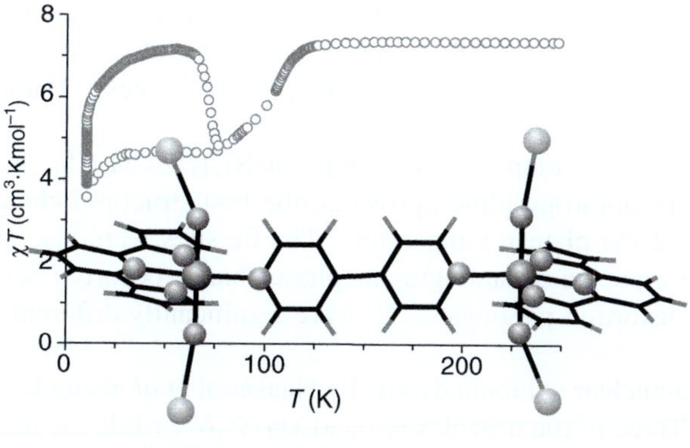

Figure 1.6 *Structure and magnetism of [[Fe(*mer-3-bpp*)(*trans-NCS*)$_2$]$_2$(μ-4,4′-bipy)]·2MeOH. Reproduced from [79] with permission of The Royal Society of Chemistry, 2011.*

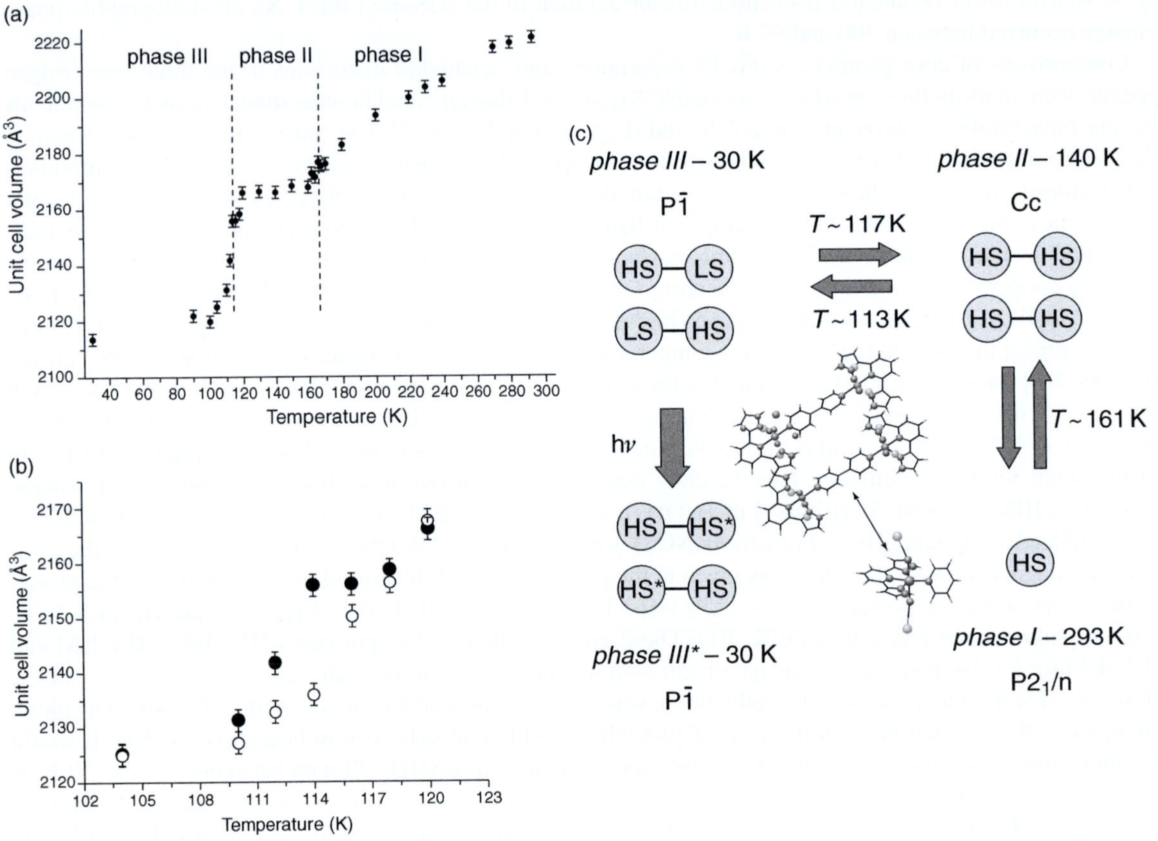

Figure 1.7 (a) Cell volume vs. temperature for phases of [[Fe(mer-3-bpp)(trans-NCS)$_2$]$_2$(μ-4,4′-bipy)]·2MeOH. (b) Thermal hysteresis in the cell volume. (c) Structural and spin state diagram. Reproduced from [74] with permission of The Royal Society of Chemistry, 2010.

The interdinuclear π–π and solvent ineractions were discussed in detail, as well as a 'switching' role being invoked for the planar to non-planar 4,4′-bipy geometric change occurring between phase I and II. Finally, this thorough study showed that the [HS–HS] forms of phases I and (photoexcited/metastable) III* were different. However, the distortion arguments given for the phia analogues[78–80] apply generally to this 3-bpp compound.

Another 2-step dinuclear SCO complex, [[Fe(ddpp)(NCS)$_2$]$_2$]·4CH$_2$Cl$_2$, with ddpp-(N-py) bridging (ddpp = 2,5-di(2′,2″-dipyridylamino)pyridine), provided the first structural characterisation of a 'ordered' [HS–LS] molecule existing at the plateau temperature. The Fe sites were also structurally inequivalent in the [HS–HS] and [LS–LS] forms. No crystallographic phase change occurred between 250 K (HS–HS) and 25 K (LS–LS). Octahedral distortion parameters, Σ, were significantly different for the two Fe sites, at all temperatures.[64, 73, 81]

One of the first reported dinuclear compounds was by Haasnoot *et al.* using L = 1,2,4(N-*p*-tolyl) triazole, viz. [Fe$^{II}_2$(L)$_5$(*cis*-<u>N</u>CS)$_4$]. Three of the triazoles bridged via N^1,N^2, while the terminal ones coordinated by one of these N atoms, with two <u>N</u>CS per Fe, *cis* disposed. Two dinuclears formed a H-bonded 'pentamer' by encompassing one [Fe(L)$_2$(NCS)$_2$(H$_2$O)$_2$ monomer.[82] The dinuclear moieties showed SCO with $T_{1/2}$ = 111 K,

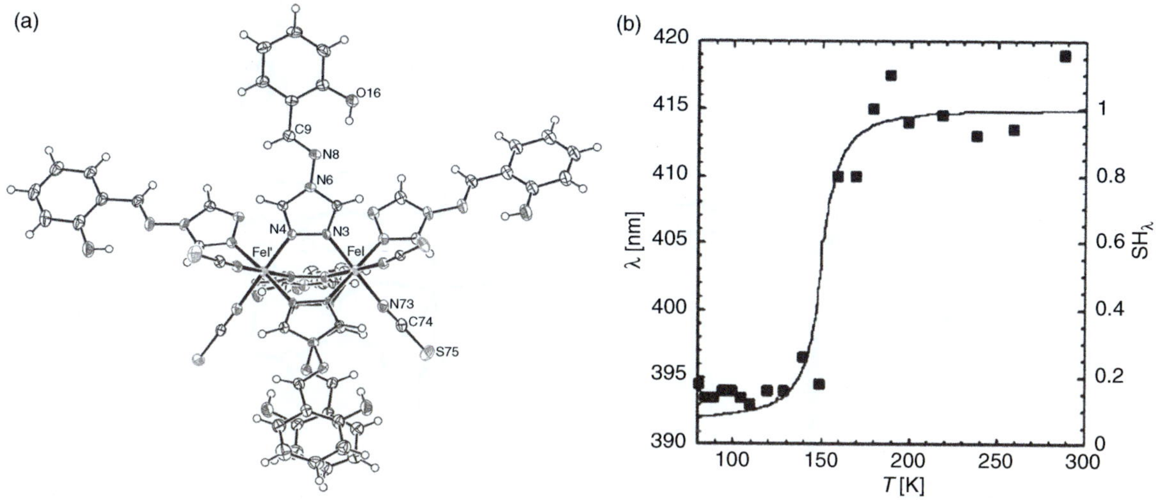

Figure 1.8 *(a) Structure of $[Fe^{II}_2(L)_5$(cis-NCS$)_4]\cdot$4MeOH where L = 4-(N=C(C_6H_4OH-2))-1,2,4-triazole. (b) Temperature dependence of λ_{max} (enol) from solid state fluorescence spectra values for $[Fe^{II}_2(L)_5$(cis-NCS$)_4]\cdot$4MeOH; the first use of fluorimetry in following spin transitions. Reprinted with permission from [84]. Copyright 2011, American Chemical Society.*

the monomer remaining HS. A $[Fe^{II}_2(L)_5($*cis*-NCS$)_4]$ derivative that we structurally characterised at around that time, with L = 1,2,4(N-picoline) remained HS–HS at all temperatures.[83] Recently, Garcia *et al.* reported another such complex, with an imino-substituted triazole, 1,2,4(N=C(C_6H_4(2-OH)-triazole).[84] The crystals were of formula $[Fe^{II}_2(L)_5($*cis*-NCS$)_4]\cdot$4MeOH, monoclinic and showed SCO with a $T_{1/2}$ of 155 K. The two Fe atoms were equivalent displaying typical LS Fe–N lengths when measured < $T_{1/2}$ and HS lengths > $T_{1/2}$ (Fig. 1.8). Intriguingly, Neville *et al.* had simultaneously used the same triazole and obtained a triclinic pseudopolymorph $[Fe^{II}_2(L)_5($*cis*-NCS$)_4]\cdot$2MeOH that had two dissimilar Fe sites, remained HS–HS and showed some differences in Fe–N–C(NCS) angles.[85] Both polymorphs displayed π–π stacking between dinuclear molecules. Such are the vagaries of SCO research!

Use of a dinucleating $[N_2O_2]_2$ Schiff base ligand (L^2; Figs 1.3 and 1.9 top), by Weber *et al.* led to a dinuclear Fe^{II} complex $[Fe_2(L^2)$(Me-imidazole$)_4]$ that displayed an abrupt [HS–HS] ↔ [LS–LS] transition, with hysteresis ($\Delta T = 21$ K; $T_{1/2} = 188$ K) when a polycrystalline sample was used and a gradual transition when a powder was used, with no hysteresis in the latter (Fig. 1.9 bottom).[86] A crystal structure at 200 K, similar to that at 125 K, showed typical [LS–LS] Fe–N lengths and O–Fe–O bite angles, thus the [HS–LS] form was not detected. Octahedral distortions such as Σ were not discussed. Miller *et al.* found a room temperature spin-transition in a dinuclear Fe^{II} complex that had a 3,6-dihydroxy-1,4-benzoquinonate bridge and N_4O_2 coordination on each Fe.[87]

1.2.1.1 Theoretical Developments in Dinuclear $Fe^{II}Fe^{II}$ SCO Complexes

The reader is referred to the detailed theoretical work, based on elastic interactions, of Spiering, Gütlich *et al.*[88] for simulating spin-transition curves for mononuclear SCO complexes, summarised nicely by Hauser *et al.*[89] and to the mean-field approach of Slichter and Drickamer (SD)[90] in determining macroscopic thermodynamic ΔH and ΔS values, although, as we see shortly, the SD approach is used in dinuclear species also. Dinuclear systems were first treated using microscopic Ising-like models.[91] Observables such as gradual vs. abrupt

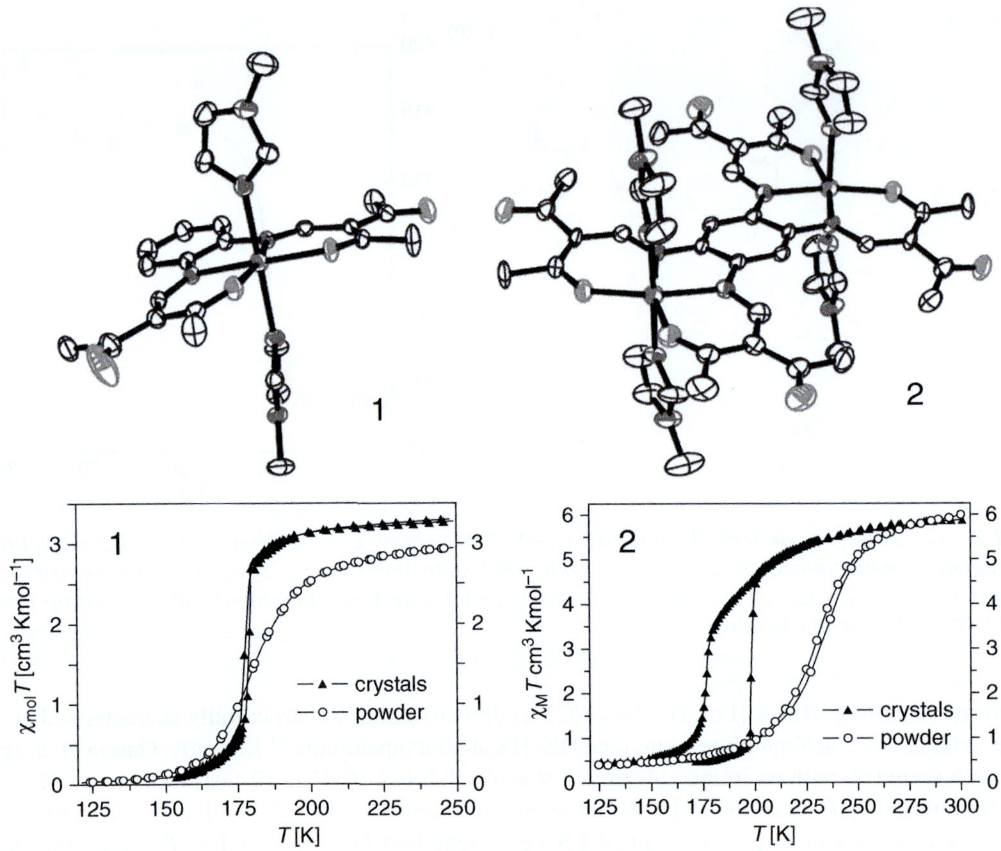

Figure 1.9 *(top) Combined figures of structures of [$Fe_2(L^2)$(Me-imidazole)$_4$],* ***2****, and mononuclear [Fe(L^1)(Me-imidazole)$_2$]· Me-imidazole,* ***1****, (where L^1 is a tetradentate analogue of L^2). (bottom) Magnetic data for crystals and powder of* ***1*** *and* ***2****. Reprinted with permission from [86]. Copyright 2008, American Chemical Society.*

spin-transitions, with or without thermal hysteresis, 2-step phenomena in mono- or dinuclear compounds or incomplete transitions were describable by these approaches. In a recent review Bousseksou *et al.*[92] also summarised recent advances in the theory of dynamical processes such as non-equilibrium photoexcitation, thermal relaxation and dynamic equilibrium processes applicable, for example, to the LIESST and LITH effects. Bousseksou *et al.*[92] went on to describe atom-phonon 1D models for real crystals, pressure induced hysteresis, vibrational densities of states and spatio-temporal development of the spin-transition. Density functional theory (DFT) for SCO systems was not mentioned in their review.

Focusing on dinuclear species and beginning with the Ising-like approach of Bousseskou, Varret, Kahn *et al.*,[93] applied first to the 2-step compound [[Fe(bt)(NCS)$_2$]$_2$(μ-bpym)], the two Fe^{II} ions can be in LS–LS, HS–LS and HS–HS states (in their paper labelled SS, SQ, QQ, where S = singlet and Q = quintet). The total enthalpy and entropy changes accompanying the LS–LS ↔ HS–HS transformation are $\Delta H = H_{HS\text{-}HS} - H_{LS\text{-}LS}$ and $\Delta S = S_{HS\text{-}HS} - S_{LS\text{-}LS}$. The dinuclear molecules were assumed to have a centre of inversion and the symmetry of each FeN_6 centre was assumed to be low enough to remove the orbital degeneracy of the HS quintet ($^5T_{2g}$) states. LS–LS gives rise to a pair state $^1\Gamma_g$, HS–LS gives two pair states $^5\Gamma_g + {}^5\Gamma_u$ while HS–HS gives five pair states $^1\Gamma_g + {}^3\Gamma_u + {}^5\Gamma_g + {}^7\Gamma_u + {}^9\Gamma_g$. The energy differences between the states arising from HS–HS are only due to intramolecular magnetic interaction, weak in the μ-bpym case, and thus ignored making HS–HS states degenerate with enthalpy ΔH, the enthalpy origin being $H_{LS\text{–}LS}$. The two pair states from HS–LS are assumed degenerate with energy $\Delta H/2 + W$ where W is a small negative or positive correction in relation to $\Delta H/2$. For $W \neq 0$ the energy of the HS–LS state is not rigorously halfway between those of HS–HS and LS–LS. W originates from electrostatic and vibronic interactions. The process LS–LS $\leftrightarrow$ HS–LS $\leftrightarrow$ HS–HS has molar fractions x, y and z for each spin isomer where $x + y + z = 1$. Following Gibbs free energy, G, calculations the authors obtained the following equation (1.1), for a single temperature:

$$\begin{aligned} G[x, y] = {} & y(\Delta H/2 + W) + (1 - x - y)\Delta H + \gamma(xy + y(1 - x - y) + 2(1 - x - y)x) \\ & - T[(y/2 + 1 - x - y)\Delta S - R(x \ln x + y \ln y + (1 - x - y)\ln(1 - x - y)] \end{aligned} \tag{1.1}$$

The state LS–LS is the enthalpy and entropy origin. It was assumed that variations $S_{HS\text{–}HS} - S_{HS\text{–}LS}$ and $S_{HS\text{–}LS} - S_{LS\text{–}LS}$ were $= \Delta S/_2$. Minima in G were calculated at single temperature points which leads to $x = f(T)$, $y = f(T)$ and $z = f(T)$ leading to the curve $c = f(T)$ where c is the molar faction of Fe^{II} in the HS state, related by $c = (y + 2z)/_2$. c is slightly different from n_{HS} (fraction HS). The crossover temperature $T_{1/2}$ is when $c = {}^1/_2$. An interaction parameter, γ, is between LS–LS and HS–LS molecules. The parameter $\rho = 2W/\Delta H$. A series of plots of c vs. T (100 to 300 K) were given for $\Delta H = 1000$ cm^{-1}, $\Delta S = 5$ cm^{-1} K^{-1}, $\rho = 0.1$, 0, 0.1 and 0.2, with γ varying between 0, 166 cm^{-1} and 332 cm^{-1}, the latter yielding the biggest and most horizontal plateau. Even with $\rho = 0$, a 2-step is calculated while a positive ρ suppresses the 2-step character. The more negative is ρ, the more pronounced the 2-step is. Thus, a negative ρ and large γ act synergistically to yield 2-step behaviour, with ρ originating from within dinuclear molecules and γ between dinuclear molecules.

The χ vs. T data for [[Fe(bt)$(NCS)_2]_2(\mu$-bpym)] gave an excellent fit for the parameter set: $\Delta H = 1100$ cm^{-1}, $\Delta S = 6.16$ cm^{-1} K^{-1}, $\gamma = 215$ cm^{-1} and $W = -40$ cm^{-1} (i.e. $\rho = -0.072$). The $T_{1/2}$ (or T_c critical temperature) $= \Delta H/\Delta S = 178.6$ K. At this temperature the molar fractions were found to be $x = z = 0.15$ and y(HS–LS) $= 0.7$. Note that the latter is different to the statistically expected value of 0.5. The different slopes noted for the two steps were also replicated. We have applied this model to one of our triazolate bridged 2-step compounds, [[Fe(bpytz)(py)$(NCBH_3)_2]_2$], as a powder, and obtained the best-fit parameter set: $\Delta H = 1883$ cm^{-1}, $\Delta S = 11.02$ cm^{-1} K^{-1}, $\gamma = 184$ cm^{-1} and $W = -119$ cm^{-1} (i.e. $\rho = -0.126$). The $T_{1/2}$ (or T_c critical temperature) $= \Delta H/\Delta S = 170.9$ K.[94]

The Slichter–Drickamer mean-field model (regular solution; Eq. (1.2)) has been used by Kaizaki *et al.*[70] to fit the 1-step HS–HS $\leftrightarrow$ LS–LS transition in the pyrazolate-bridged analogue viz *trans*-[[Fe(bpypz)(py)$(NCBH_3)_2]_2$], and in the $(NCS)_2$ analogue.

$$\ln[(1 - \gamma_{HS})/\gamma_{HS}] = (\Delta H_{HS\leftrightarrow LS} + \Gamma(1 - 2\gamma_{HS}))/RT - \Delta S_{HS\leftrightarrow LS}/R \tag{1.2}$$

where γ_{HS} is the HS fraction, Γ is an interaction parameter, ΔH and ΔS are the enthalpy and entropy changes associated with the HS $\leftrightarrow$ LS transition. A cooperativity factor $C = \Gamma/_2RT_c$. Please note the different definitions of parameters from the Ising model above. The best-fit yielded: $NCBH_3$; $\Delta H_{HS\leftrightarrow LS} = 1111.6$ cm^{-1}; $\Delta S_{HS\leftrightarrow LS} = 5.4$ cm^{-1} K^{-1}, $\Gamma = 112$ cm^{-1}; $T_c = 205$ K; $C = 0.39$; the NCS complex: $\Delta H_{HS\leftrightarrow LS} = 498$ cm^{-1}; $\Delta S_{HS\leftrightarrow LS} = 3.9$ cm^{-1} K^{-1}, $\Gamma = 154$ cm^{-1}; $T_c = 127$ K; $C = 0.87$. The bpytz/$NCBH_3$ complex has larger $\Delta H_{HS\leftrightarrow LS}$ and $\Delta S_{HS\leftrightarrow LS}$ than has the bpytz/$NCBH_3$ derivative. Parameter values obtained using this model were listed for monomeric SCO compounds by Létard *et al.*[95] and Kaizaki *et al.*[70] Cooperativity factors, C, were related, in mononuclear compounds, to features of ligands such as the length of conjugated substituents. The cooperativity in dinuclear μ-bpypz species was felt by Kaizaki to be mononuclear-like rather than due to inter-dinuclear interactions.[70]

Boukheddaden *et al.* described a general theoretical model applicable to SCO systems as well as those that showed spin changes and magnetic interactions/long range order, such as Fe/Co Prussian Blue species.[96] Boča *et al.* also reviewed aspects of theory to room temperature SCO materials[97] including a large table of ΔH and ΔS values for Fe^{II} and Fe^{III} monomers and, as we see below, the theory for Fe^{III} compounds.

1.2.1.2 DFT Calculations for Fe^{II}-Fe^{II} SCO Complexes

Zein and Borshch[77] made significant findings of energy levels, using quantum mechanical DFT calculations, for the dinuclear SCO complexes $[[Fe(bt)(NCS)_2]_2(\mu\text{-bpym})]$ and NCSe analogue, $[[Fe(bpym)(NCS)_2]_2(\mu\text{-bpym})]$ and NCSe analogue, and $[(pypzH)(NCSe)Fe(\mu\text{-pypz})_2Fe(NCSe)(pypzH)]$, the first four having bipyrimidine bridges and 2-step transitions,[60–63] the latter having pyrazolate bridges and a single HS–HS ↔ LS–LS transition.[64, 98] First, the geometries were optimised and the electronic states calculated for LS–LS, HS–LS and HS–HS states, the LS–LS and HS–HS structures assumed to have centres of symmetry. The total dimer spin for LS–LS is 0, LS–HS is 2 and HS–HS, that has exchange coupled states 4,3,2,1,0, assumed to have spin of 4. The energies of these states for all complexes is shown in Figure 1.10.

The LS–LS is lowest in energy for complexes {bt,S}, {bt,Se} and {pypz,Se} and these can have the full HS–HS ↔ LS–LS transitions. The HS–HS level was calculated as lowest for {bpym,S}, as observed experimentally while HS–LS was lowest for {bpym,Se}, thus giving the 'half' crossover observed from magnetism. The position of the HS–LS level was decisive for the appearance of a 2-step transition, in broad agreement with the predictions of the Ising model, above. Thus the energies of the HS–LS state were lower than the average energy between LS–LS and HS–HS and lead to 2-steps for {bt,S} and {bt,Se}. In contrast, the energy of HS–LS for {pypz,Se} is above the average position and just below that of HS–HS, leading to a 1-step transition, as observed. This was ascribed to differences in the bridging ligand geometry, being more

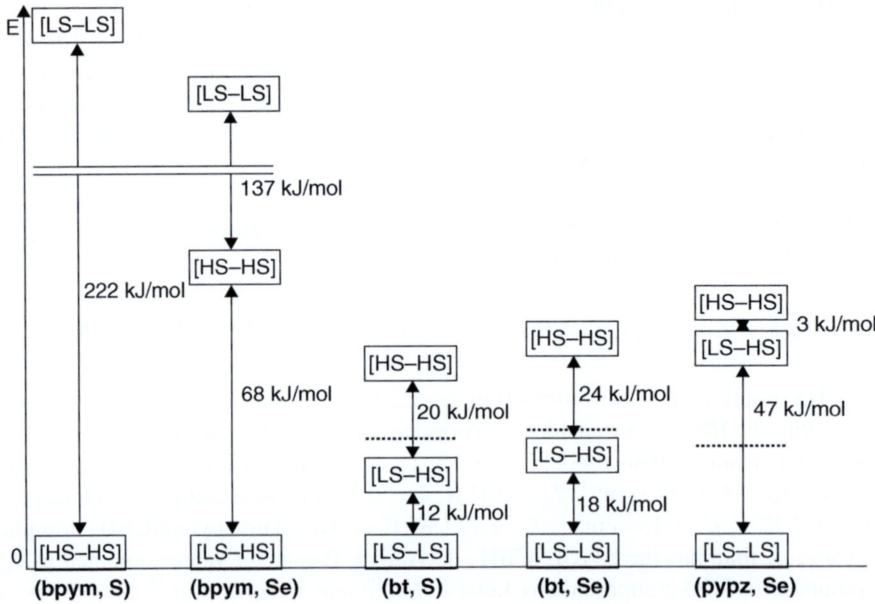

Figure 1.10 Relative energies of the LS–LS, LS–HS and HS–HS states for the dinuclear iron(II) complexes $[[Fe(bpym)(NCS)_2]_2(\mu\text{-bpym})]$, [bpym,S]; $[[Fe(bpym)(NCSe)_2]_2(\mu\text{-bpym})]$, [bpym,Se]; $[[Fe(bt)(NCS)_2]_2(\mu\text{-bpym})]$, {bt,S}; $[[Fe(bt)(NCSe)_2]_2(\mu\text{-bpym})]$, {bt,Se}; $[(pypzH)(NCSe)Fe(\mu\text{-pypz})_2Fe(NCSe)(pypzH)]$, {pypz,Se}. Reprinted with permission from [77]. Copyright 2005, American Chemical Society.

distorted in {pypz,Se} compared to {bt,S/Se}. Presumably, similar DFT calculations on the 'half' crossover compound $[Fe_2(PMAT)_2](BF_4)_4{\cdot}4DMF$[75, 76] would agree with those for {bpym,Se}. Zein and Borshch also calculated *J* coupling constants (HS–HS) and the five compounds were all weakly antiferromagnetic, with reasonable agreement with *J* measured for {bpym,S}, but too weak to affect the energy gap between LS–LS and HS–HS, thus not yielding synergy between exchange and SCO.[77]

1.2.1.3 Theoretical and Experimental Developments in Dinuclear $Fe^{III}Fe^{III}$ SCO

A series of dinuclear Fe^{III} Schiff base d^5-d^5 complexes of type $[[Fe(pentadentate\text{-}N_3O_2)]_2(\mu\text{-}4,4'\text{-bipy})]$,[99] and CN-bridged analogues, were investigated by Boča, Nemec *et al.*[48, 100] by means of experiment and theory. The former μ-4,4′-bipy type were first reported by Hayami *et al.*[101, 102]

From a theoretical and data fitting perspective, Boča *et al.*[48, 100] found that the Ising thermodynamic model discussed above for $Fe^{II}Fe^{II}$ SCO species would not fit the CN-bridged $Fe^{III}Fe^{III}$ $\chi_M T$ (μ_{eff}) vs. *T* or isothermal magnetization, *M* vs. *H* data at 2 K. So, they developed a new and more extensive model that, while it used many parameters, was capable of simultaneously fitting magnetic susceptibility, magnetisation, n_{HS} and Mössbauer spectral data. But first, we briefly discuss the experimental data and start with mononuclear 'precursors', the magnetic data of which could be fitted by use of Ising models.

Complexes $[(N_3O_2)Fe^{III}(X)]{\cdot}S$ were structurally characterised where N_3O_2 is a pentadentate R-substituted-salicylaldimine (e.g. saldptm, below) or naphthaldimine Schiff base; anionic X = CN^-, NCO^-, NCS^-, $NCSe^-$, $NCBH_3^-$, N_3^-; S = solvent.[100] Observables were $\chi_M T$ vs. *T* and *M* vs *H* at 2 and 4.6 K. The CN^- compounds were LS d^5 ($S = 1/2$) and displayed typical LS Fe^{III}–N and Fe^{III}–O distances and small octahedral distortions Σ ~25°. The HS examples had longer Fe–N/O bond lengths and higher Σ ~56° and displayed typical HS $^6A_{1g}$ zero-field split magnetism with decreases observed in $\chi_M T$ below ~5 K, the *D* parameters <1 cm^{-1}, confirmed by the magnetisation isotherms. SCO behaviour in the naphthaldimine/NCS and NCSe compounds, from magnetism and variable temperature crystallography (cell lengths and volume), followed closely our early $[Fe^{III}(salen)(imidazole)_2]^+$ data,[15] and were fitted by Ising-like theory,[100] including vibrations (ν). The theory and symbols are in the ESI of the paper and from HS mole fraction x'_{HS}, vs. temperature, the parameters for Fe^{III}-naphthaldimine/NCS were: $T_c = 150$ K, $g_{LS} = 2.25$, $g_{HS} = 2.08$, $D_{HS} = 1.8$ cm^{-1}, $\Delta_{eff} = 231$ K, $\gamma = 87$ cm^{-1}, $\nu_{LS} = 539$ cm^{-1}; calc. $\Delta H = 160$ cm^{-1} and $\Delta S = 1.04$ cm^{-1}, where Δ_{eff} = the HS–LS energy difference modified by vibrational $h\nu_{HS} - h\nu_{LS}$, where $\nu_{HS} = \nu_{LS}/1.5$.

The ΔH and ΔS values are much smaller than for the Fe^{II} ($S = 2 \leftrightarrow S = 0$) cases. Looking back at the $[Fe^{III}(salen)(imidazole)_2]ClO_4$ results,[15] that used a different model[103] but included vibrations, we saw that $\Delta_{eff} = 540$ K, with the ratio of vibrational partition functions for HS and LS forms crucial to obtain the non-Boltzmann $\chi_M T$ crossover region.[15]

In the dinuclear systems of Boča *et al.*,[99] six new complexes of type $[(saldptm)Fe^{III}(\mu\text{-}L)Fe^{III}(saldptm)](BPh_4)_2$ were structurally characterised at 90 K, all showing typical LS Fe^{III} bond distances Fe–N and Fe–O, that is LS–LS state, except for one that had a 4,4′-bi(pyridine-N oxide) bridge that remained in the HS–HS state. The Fe..Fe distances varied from 11.14 Å, for the 4,4′-bipy bridge to 13.37 Å for the bpe (conjugated ethene linker) bridge (Fig. 1.11a). These two examples showed a gradual spin-transition above ~100 K and ~150 K, respectively, the transition being close to complete HS–HS at 300 K. The plateau value of μ_{eff} of ~ 3 μ_B ($\chi_M T = 1.1$ cm^3 mol^{-1} K), per Fe_2, arises from the two magnetically isolated LS–LS ($S = 1/2$) ions, with a very small decrease below 5 K ascribed to weakly antiferromagnetically coupled LS–LS centres (Fig. 1.11b). In an ethane-linked bipyridine analogue, the LS–LS centres were felt to couple weakly ferromagnetically, which is surprising over the long Fe..Fe distance involved. The magnetic data for the 4,4′-bipy and bpe compounds were consistent with an energy level diagram having the lowest LS–LS level and isolated from the HS–HS (by $E = \Delta_1$) with the HS–LS level higher still at Δ_2 above HS–HS (Fig. 1.11c; see A/LL top right). The latter level has the potential to split, by magnetic exchange, into *S* = 5, 4, 3, 2, 1, 0 sublevels, separated by 30*J* (this HS–HS exchange could not be identified experimentally) and the LS–LS in

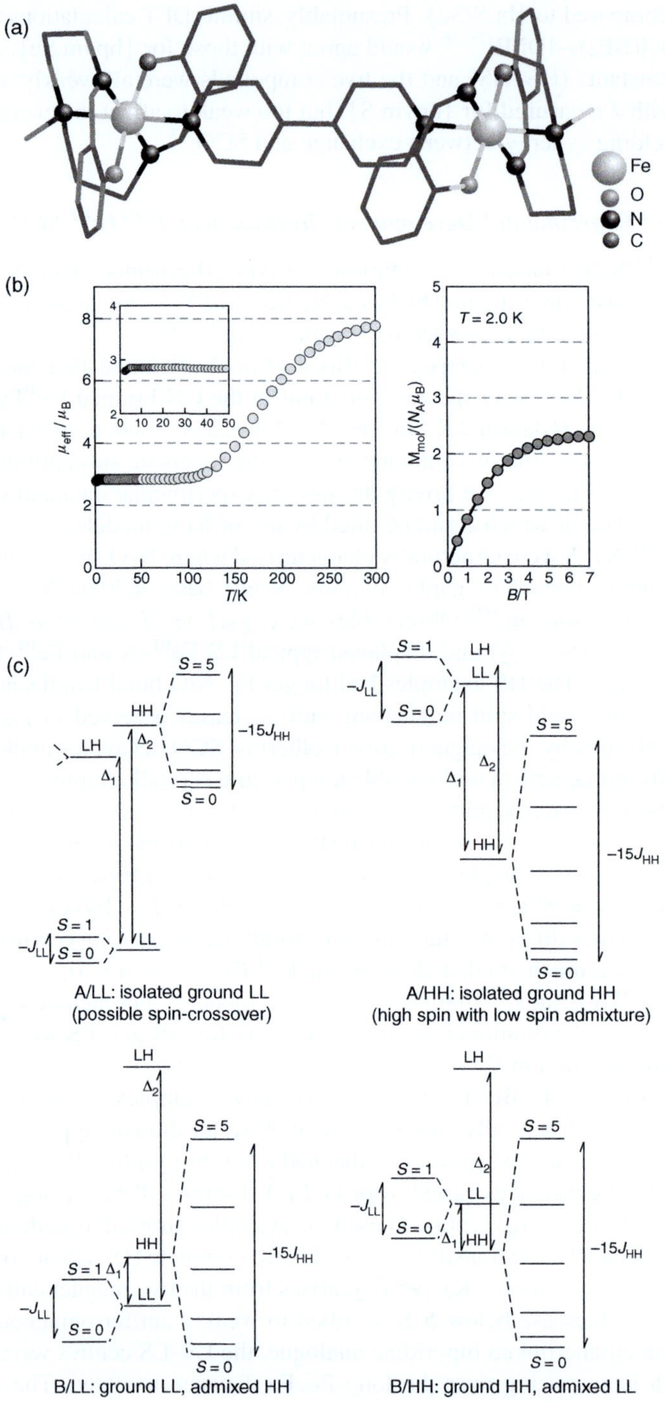

Figure 1.11 *(a) Structure of [(saldptm)Fe*^{III}*(μ-4,4′-bipy)Fe*^{III}*(saldptm)](BPh*$_4$*)*$_2$*. (b) Magnetic data (in* μ_B*) for [(saldptm)Fe*^{III}*(μ-4,4′-bipy)Fe*^{III}*(saldptm)](BPh*$_4$*)*$_2$*, (c) classes of energy level diagrams for the Fe*^{III} *dinuclear SCO materials. Reproduced from [99]. Copyright 2009. With kind permission from Springer Science and Business Media.*

to $S = 1$ and 0 sublevels, separated by $2J'$. Spin-crossover occurred between the LS–LS and HS–HS levels. Other members of this dinuclear family, with different (non-conjugated) bridges, conformed to a different energy level diagram in which the Δ_1 was small enough to lead to LS–LS/HS–HS mixing, via the exchange (band of) levels. The spin-transition for these compounds was very gradual, stretching from ~ 20 K to 300 K, with μ_{eff} values close to the HS–HS values even at 2 K, and with higher M values at 2 K and 7 T. We note that our recently reported series of 1D chain species containing tetradentate N_2O_2-Schiff base (salen, salophen, acen) Fe^{III} centres bridged by 4,4′-bipy-type ligands, described later, had similar SCO magnetism to this dinuclear series, as did one dinuclear example.[104]

The dinuclear CN-bridged Fe^{III} Schiff base family is the one that required a new model for quantitative fitting of data.[48] Diamagnetic bridges such as $Ni(CN)_4{}^{2-}$, $Pt(CN)_4{}^{2-}$ and $Ag(CN)_2{}^{-}$ were also reported and formed heterotrinuclear species. The CN-bridges lead to HS–LS combinations, the C of CN providing the strong ligand field, and also provide stronger exchange coupling. Thus, these examples pose the question as to whether SCO and exchange occur simultaneously or whether one dominates (or to quote the authors 'interferes with'[48]) the other. As in the μ-4,4′-bipy dinuclear examples above, the combination of SCO and exchange coupling can give rise to four subclasses, viz. (i) isolated ground state HS–HS, (ii) isolated ground state LS–LS, (iii) ground HS–HS, admixed LS–LS, (iv) ground LS–LS, admixed HS–HS (Fig. 1.11c). The energy gap LS–LS to HS–HS is Δ_1; HS–HS to HS–LS is Δ_2. Exchange coupling between HS–HS and LS–LS states gives multiplets that can overlap the crossover processes in (iii) and (iv). Coupling between HS–LS was assumed zero with the energy of HS–LS levels being high. Mössbauer spectra identified HS and LS Fe^{III} sites. As in the other pentadentate-blocker families reported, susceptibilities vs. temperature and magnetisation vs. field (at 2 K) were the observables, along with crystallographic bond lengths. The $M(CN)_4{}^{2-}$ bridged trinuclears gave very weak HS–HS states at all temperatures, with very weak exchange manifest at low temperatures, the M data confirming $S = 5$ ground states. The $Ag(CN)_2{}^{-}$ complex gave a very broad crossover.

The new model assumed reference states LS–LS, LS–HS, HS–LS, HS–HS each with an energy gap from zero of $\Delta_{LS\text{-}LS}$ (assumed zero), and each with exchange coupled spin multiplets. The full details of the spin Hamiltonians, partition functions, approximations, etc. are given in the paper, and cooperativity was assumed to be absent, while molecular vibrations were included in the partition functions. The minimum number of (ten) parameters was:

$$g_{LS}, g_{HS} (\sim 2.0), J_{LS-LS}, J_{LS-HS} = J_{HS-LS}, J_{HS-HS}, D_{HS} (\sim 0), D_{LS-LS}, \Delta_{LS-HS} = \Delta_{HS-LS}, \Delta_{HS-HS}, h\upsilon_{HS}$$

The $\chi_M T$ plots for five {SCO + exchange $(-JS_a \cdot S_b)$} examples gave, in general, very broad increases between ~20 K and 300 K. Taking one example, [(saldptm)Fe^{III}(CN)Fe^{III}(saldptm)]$(ClO_4)\cdot 2H_2O$ (note the paper labels saldptm = salpet), fitting of $\chi_M T$ and M vs. H isotherms simultaneously, yielded:

$$g_{LS} = 2.65, g_{HS} = 1.80, J_{LS-LS} = -3.9\ \text{cm}^{-1}, J_{LS-HS} = J_{HS-LS} = -0.16\ \text{cm}^{-1}, J_{HS-HS} = -3.67\ \text{cm}^{-1},$$
$$D_{HS} = -4.2\ \text{cm}^{-1}, D_{LS-LS} = -1.77\ \text{cm}^{-1}, \Delta_{LS-HS} = \Delta_{HS-LS} = 10\,000\ \text{K (fixed)}, \Delta_{HS-HS} = 903\ \text{K},$$
$$\upsilon_{HS} = 238\ \text{cm}^{-1}, \upsilon_{LS}/\upsilon_{HS} = 1.16.$$

The energy levels deriving from this parameter set showed well separated 'bands', at low temperatures only the LS–LS and LS–HS manifolds interplay, then the HS–HS levels become populated above ~120 K when SCO begins. The M value at 2 K/5 T is ~3.5 $N\mu_B$, not saturated and indicative of a ground (coupled) spin state of between 1 and 3. The Mössbauer spectrum of this compound shows HS and LS doublets at 20 K and 300 K, as expected and is felt to be indicative of spin-crossover occurring.[48] A comparative plot that just included exchange coupling, without SCO, to yield a $S = 2$ ground state would have been instructive.

1.2.2 Tri-, Tetra-, Penta- and Hexa-nuclear Fe^{II} SCO Clusters

1.2.2.1 Trinuclears

The first reported 1,2,4-triazole bridged species, in which only the central Fe^{II} undergoes spin-crossover, were mentioned at the beginning of this chapter.[38, 39] The Mössbauer features of one such example $[Fe_3(iptrz)_6(H_2O)_6](CF_3SO_3)_6$ (iptrz = 4(i-C_3H_7)-1,2,4-triazole) are described in a new book on Mössbauer spectroscopy.[105, 106] Synchrotron nuclear inelastic scattering methods have been used by Wolney *et al.* to obtain the vibrational ν(Fe–N) bands of the central FeN_6 ion in the 4($HOCH_2CH_2$)-1,2,4-triazole) analogue.[107] Work by Tuchagues *et al.* showed that a different kind of trinuclear complex, having a triangular rather than linear $Fe^{II}{}_3$ disposition, $[Fe_3L_2(\underline{N}CS)_4(H_2O)]$, where L is a Schiff base, showed spin-crossover, once again only occurring at the central Fe.[108] This was a somewhat surprising observation at the time because of its $Fe^{II}N_4O_2$ environment. Other examples of SCO occurring at such N,O ligand donor sites are, as we have seen earlier, now well known.[45, 86]

A mixed-valence trinuclear complex $[L^5Fe^{III}[Fe^{II}(CN)_5(NO)]Fe^{III}L^5]\cdot 0.5MeOH\cdot 3.75H_2O$ (L^5 = salpet, a pentadentate Schiff base) shows spin-crossover at the Fe^{III} d^5 centres. The metallacyanido-bridged complex $[L^5Fe^{III}[Ni(CN)_4]Fe^{III}L^5]\cdot 2MeOH$ (L^5 = MeBu-salpet) contains a high spin pair, HH, over the whole temperature range with a ferromagnetic exchange interaction postulated. A theoretical model was outlined that allowed simultaneous fitting of all available experimental data using a common set of parameters (see theory section above).[48]

Another CN-bridged trinuclear example, $[Fe^{II}{}_3(CN)_6(HB(3,5\text{-}Me\text{-}_2\text{-}pz))_2(tpa)]$, with a T-shaped geometry, possesses a central Fe(tpa)($\underline{N}$C)-$_2$ moiety that displayed a gradual crossover above 300 K, the terminal iron groups being LS.[109]

Recently we described a tritopic tris(pyrazolyl)methane bridging ligand and its $Fe^{II}{}_3$ spin-crossover derivative.[53] Each iron centre is capped by tris(3,5-dimethylpyrazolyl)methane groups, thus achieving six-coordination (Fig. 1.12a) and displaying a very gradual, incomplete spin-transition with $T_{1/2}$ ~350 K. The

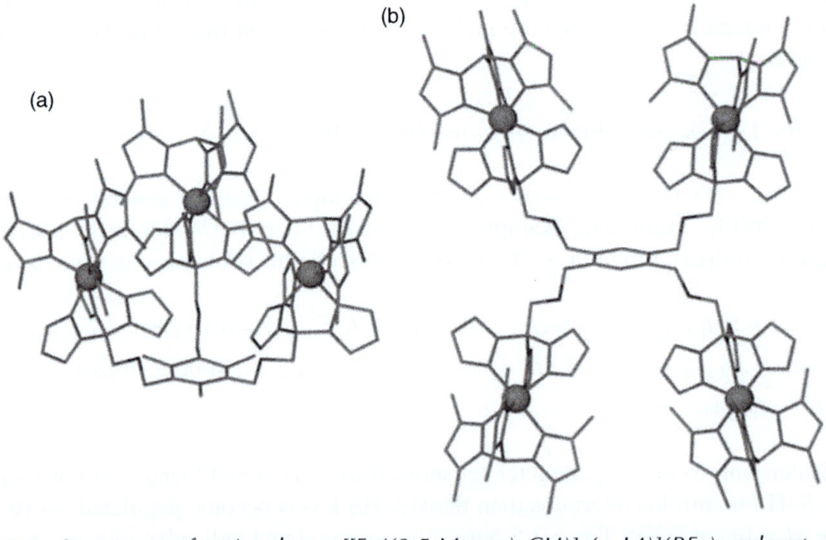

Figure 1.12 *(a) Structure of trinuclear $[[Fe((3,5\text{-}Me_2pz)_3CH)]_3(\mu\text{-}L4)](BF_4)_6\cdot$solvent, where L4 is a tritopic tris(pyrazolyl)methane bridge. (b) Structure of tetranuclear $[[Fe((3,5\text{-}Me_2pz)_3CH)]_4(\mu\text{-}L5)](BF_4)_8\cdot 8MeCN.2^tBuOMe$, where L5 is a tetratopic tris(pyrazolyl)methane bridge. Reproduced from [53], with permission of The Royal Society of Chemistry, 2011.*

three tris(pyrazolyl)methane moieties, linked via ether groups to a mesityl ring, were all on the same side of the ring in the Fe^{II}_3 complex and this led to 'capsule'-like motifs being observed in the crystal packing. As in related di- and tetra-nuclear complexes (below), the $Fe^{II}N_6$ spin centres are magnetically isolated in this trinuclear design compared to covalently-bridged compounds of the tris(μ-triazole) type.

1.2.2.2 Tetranuclears

Points of fundamental interest that emerge from assembling four Fe^{II} potential spin-crossover centres together include, apart from the synthetic/design challenges, whether the spin-transitions will occur sequentially or simultaneously, and whether cooperativity occurs within the cluster (these points are related). Two kinds of tetranuclear SCO clusters have been described, one using a tetrapodal linking ligand with tris(pyrazolyl)methane 'end' caps,[53] the other having square Fe_4 geometry formed by self-assembly of Fe^{II}–bis-terpy 2×2 grids[41] or by covalent CN-bridging, the so-called Prussian Blue fragments.[110] Recently, an analogous $N(CN)_2^-$ (dca^-) bridged square has also been reported.[111]

Historically, the 2×2 grids of Lehn, Ruben and co-workers, of formula $[Fe^{II}_4L_4]^{8+}$ where L = terpy chelators hinged at the 4,6 – positions of a pyrimidine ring (and the Co^{II} analogue), were shown, in the paramagnetic cases (some were LS–LS–LS–LS, diamagnetic at all temperatures), to display very gradual and nonhysteretic spin-transitions assigned to HS–HS–HS–LS or HS–HS–HS–HS (at 300 K) ↔ HS–HS–LS–LS or HS–LS–LS–LS (at 30 K), indicative of very weak intra-cluster cooperativity.[41] Substitution at the 2-position of the pyrimidine ring led to changes in spin states. In later work, $[Fe^{II}_4L'_4]^{8+}$ derivatives containing 3- and 4-pyridyl substituents on the central rings of the terpy moieties, were subsequently reacted with (diamagnetic) Ag^I or La^{III} and this led to further self-assembly of 1D and 2D polynuclear motifs, such as $[\text{-}Fe_4L'_4\text{-}(Ag^I)_4]_n^{12+}$, that displayed modification of the magnetism such that LS forms become stabilised, at a particular temperature, compared to the starter.[112] The very gradual and nonhysteretic spin-transitions persisted in these extended structures and their functionality as supramolecular spintronic modules was envisaged.

Oshio *et al.* have reviewed their CN-bridged molecular square work,[110] and summarised not only their SCO results, but also the control of ground spin states in mixed-metal species such as $[Fe^{II}_2Cu^{II}_2]$ or $[Fe^{III}_2Cu^{II}_2]$, with ferromagnetic exchange coupling noted across Fe^{III}-CN–Cu^{II} bridges, a feature we observed some time ago.[113] Using the flexible 4-coordinate 'end blocker' tri(pyridylmethyl)amine, tpa, in combination with bidentate 2,2′-bipy, the complex $[Fe^{II}_4(tpa)_2(2,2'\text{-bipy})_4(\mu\text{-CN})_4](PF_6)_4$ showed LS–LS–LS–LS Fe^{II}–N distances, ~1.96 Å, at 100 K while one of the four Fe centres showed HS distances, 2.1 Å, at 200 and 300 K, indicative of spin states LS–LS–LS–HS (Fig. 1.13). Magnetic susceptibility and Mössbauer studies revealed a 2-step spin-transition was occurring, involving LS–LS–LS–LS ↔ LS–LS–LS–HS ↔ LS–HS–LS–HS as the temperature was increased from 2 to 400 K, the LS centres being $Fe(2,2'\text{-bipy})_2(\underline{C}N)_2$ (*trans* – disposed across the square) with the $Fe(tpa)(\underline{N}C)_2$ centres, of weaker ligand field, undergoing SCO. Replacement of the two tpa ligands with four 2,2′-bipyrimidines, or of 2,2′-bipy with *o*-phen, gave related squares that showed only 1-step transitions and with the LS–HS–LS–HS form not stabilised at 400 K, that is only 57% of the $Fe(2,2'\text{-bpym})_2(\underline{N}C)_2$ centres were HS at 400 K. The LS–HS–LS–HS form was, however, achieved in another *o*-phen/tpa derivative for which $\chi_M T$ and crystallographic data showed no SCO between 2 and 300 K.[114]

A subtle balance of structural factors was felt to be required to generate multiple spin-transitions in these $Fe^{II}_4L_4$ cyanide-bridged squares. Surprisingly, perhaps, DFT calculations later showed that the observation of multiple steps was not intramolecular in origin, but due to asymmetrical crystal packing.[115] The $[Fe^{II}_4(tpa)_2(2,2'\text{-bipy})_4(\mu\text{-CN})_4](PF_6)_4$ 'parent' has four crystallographically different Fe sites, one of which, a $Fe(tpa)(\underline{N}C)_2$ site, undergoes π–π interactions by its tpa ring with the same site in a neighbouring tetranuclear cluster thus reducing the ligand field strength at Fe and producing the 2-step transition.[116] In contrast, there was much less distortion around Fe at the $Fe(2,2'\text{-bpym})_2(\underline{N}C)_2$ centres, and no asymmetrical

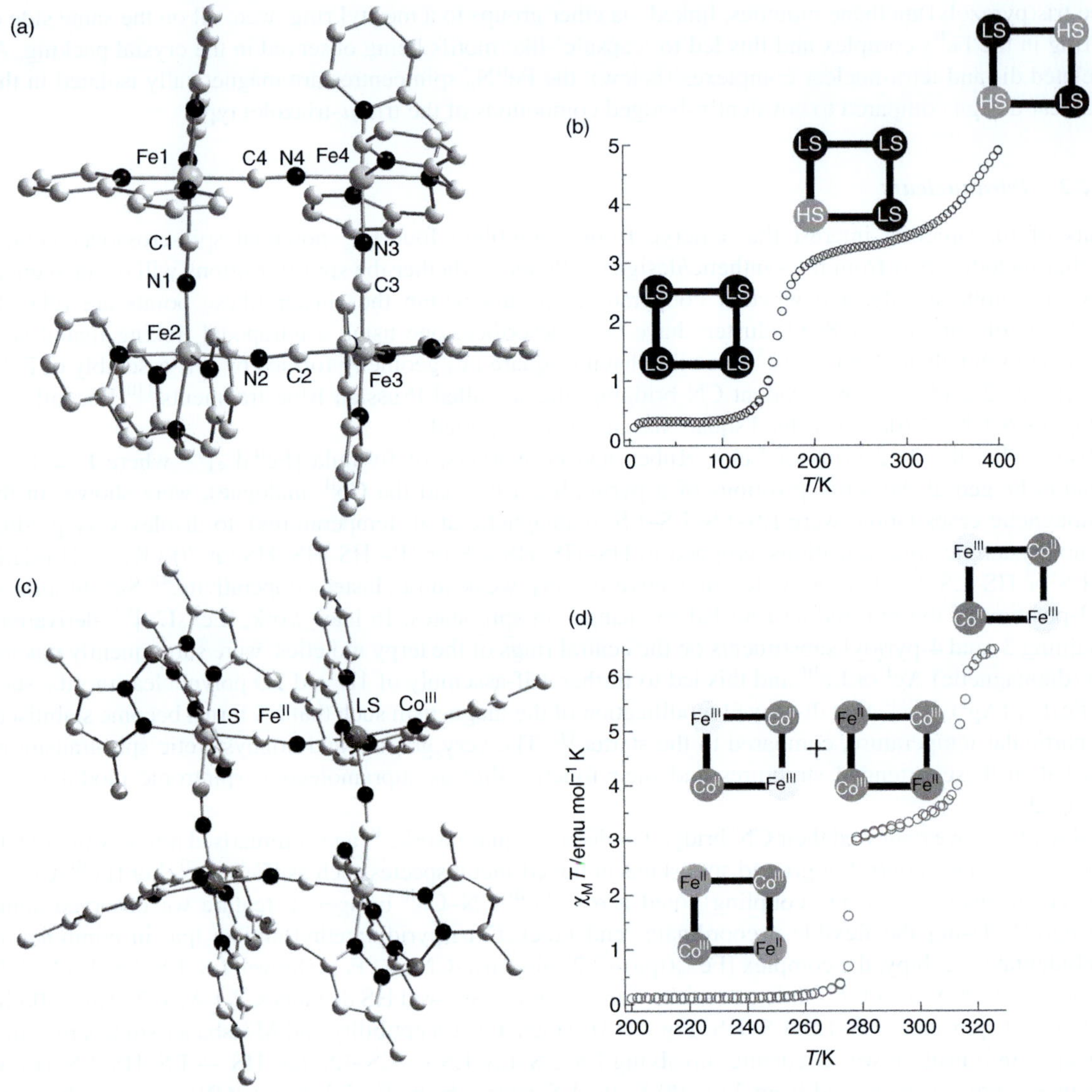

Figure 1.13 (a) Structure of $[Fe^{II}_4(tpa)_2(2,2'\text{-}bipy)_4(\mu\text{-}CN)_4](PF_6)_4$. (b) Magnetic susceptibility data for $[Fe^{II}_4(tpa)_2(2,2'\text{-}bipy)_4(\mu\text{-}CN)_4](PF_6)_4$ with appropriate spin states. (c) Structure of $[Co_2Fe_2(CN)_6(HCB(3,5\text{-}Me_2pz)_3)_2(4,4'\text{-}Bu^t_2\text{-}bipy)_4](PF_6)_2{\cdot}2MeOH$, where 4,4'-$Bu^t_2$-bipy = 4,4'-di-$Bu^t$-2,2'-bipyridine. (d) Magnetism and schematics representing the spin transition for $[Co_2Fe_2(HCB(3,5\text{-}Me_2pz)_3)_2(4,4'\text{-}Bu^t_2\text{-}bipy)_4](PF_6)_2{\cdot}2MeOH$. Reproduced with permission from [110]. Copyright Wiley-VCH Verlag GmbH & Co, 2011.

intermolecular interactions, thus yielding a 1-step transition.[109] A 2-step complete spin-transition, $LSFe^{II}_4$ to HS-Fe^{II}_4, has recently been reported for a $\mu_{1,5}$-dicyanamide-bridged square $[Fe^{II}_4(N(CN)_2)_4(tpa)_4](BF_4)_4{\cdot}2H_2O$, a molecule that showed SCO when reversibly transformed between hydrated and dehydrated forms via single crystal to single crystal transformation.[111] Metastable $HSFe^{II}_4$ states of 48% population were formed by photoexcitation using 457 nm irradiation, at 5 K, with an unusual stepwise relaxation back to $LSFe^{II}_4$ occurring on warming the metastable species back up to 50 K.

Other CN-bridged SCO systems include the bimetallic $Fe^{II}{}_2Ni^{II}{}_2$.[117] In the case of Fe_2Co_2 CN-bridged squares, the phenomenon of electron-transfer-coupled spin-transitions (ETCST), often labelled charge-transfer induced spin-transitions (CTIST) – the latter described in the iconic work of Hashimoto and Sato[118] and (later) Okhoshi[119] on Prussian Blue Co/Fe extended phases – has been investigated at the molecular level by use of thermal- or photo-stimulation. Oshio *et al.* have reviewed such studies in detail and have described Fe_2Co_2 squares,[110] $Fe^{III}{}_2Co^{II}{}_3$ trigonal bipyramidal CN-bridged clusters[120] and $Fe^{III}{}_4Co^{II}{}_4$ cubes.[121] All the squares displayed 1-step spin-transition except for an ETCST-active complex, $[Co_2Fe_2(CN)_6(HB(Me\text{-}_2pz)_3)_2(4,4'\text{-}Bu^t{}_2\text{-bipy})_4](PF_6)_2{\cdot}2MeOH$, that showed a 2-step transition (Fig. 1.13). The structure at 100 K was indicative of LS Fe^{II} or Fe^{III} and LS Co^{III}. Magnetic data obtained below the first step (275 K) was in accord with $LSFe^{II}{}_2LSCo^{III}{}_2$, the low temperature (LT) phase. Upon heating, the 2-steps were clearly observed at 275 and 310 K, indicative of ETCST from the LT to a HT phase via an intermediate (IM) state in the intervening temperature plateau. The data supported $LSFe^{III}{}_2HSCo^{II}{}_2$ as the electronic state of the HT phase, with Mössbauer spectra supporting these assignments. Half of the Fe ions had undergone ETCST in the IM phase, at 280 K. Synchrotron X-ray data at 298 K (IM phase) indicated the presence of four unique squares with states $LSFe^{II}{}_2LSCo^{III}{}_2 \times 2$ and $LSFe^{III}{}_2HSCo^{II}{}_2 \times 2$, the squares forming π-stacked layers in which HT and LT species were arranged alternately. At a chemical level, variations were made in the tri- and bidentate ligands so that the likelihood of ETCST activity in Fe_2Co_2 squares could be deduced. Light induced ETCST (LIETCST) was also noted in the LT phase of these squares to form metastable HT phases when irradiated with 808 nm light at 5 K. Light induced diamagnetic to ferromagnetic switching has also been observed in a Fe_2Co_2 square that remained in the diamagnetic $LSFe^{II}{}_2LSCo^{III}{}_2$ form at all temperatures in the absence of irradiation.[122] In view of their thermal and photoactivated properties, these molecular squares might prove valuable in device applications.

Other related, but different, transitions are found in the valence-tautomeric transitions in cobalt(II/III) catechol/semiquinone compounds where the redox and spin state changes occur at both ligand and metal centres.[123]

We have employed a different design, to make a rectangular Fe_4 SCO material, by use of a tetratopic tris(pyrazolyl)methane linking ligand, with the 6-coordination around the four Fe^{II} ions being completed by tridentate $Fe^{II}(HC(3,5\text{-}Me_2pz)_3)^{2+}$ capping groups[53] (Fig. 1.12b). A gradual, incomplete spin-transition occurred between 300 and 400 K, possibly involving the HS–LS–LS–LS state, with only small differences in magnetism noted between solvated and desolvated crystals. The SCO centres are essentially isolated (independent) from each other in an intramolecular sense, apart from mechanical/elastic interactions, when compared to the CN- or dca-bridged squares, with the 2×2 grid molecules likely to be somewhere in between. Finally, we note that Kepert *et al.* have reported a mixed spin CN-bridged square-cum-grid $Fe^{II}{}_4$ compound obtained by transformation of $NCSe^-$ to CN^-.[124]

1.2.2.3 Pentanuclears

Dunbar's compounds that contain a trigonal-bipyramidal array (Fig. 1.14) of three Fe^{II} centres (in the trigonal plane positions) and two LS Fe^{III} or Co^{III} centres (in apical positions), in the CN-bridged species $[[M^{III}(CN)_6]_2[M'^{II}(tmphen)_2]_3]$ (M/M′ = Co/Fe, Fe/Fe; tmphen = 3,4,7,8-tetramethyl-1,10-phenanthroline) led to the observation of gradual, incomplete spin-transitions above 170 K.[125] A recent theoretical treatment by Klokishner *et al.*, using a microscopic approach, has reproduced both the $\chi_M T$ vs. T behaviour and the Mössbauer spectra.[126] The Hamiltonian employed contains terms for single-ion spin-orbit coupling, Zeeman and crystal-field splitting; inter-ionic terms for isotropic exchange coupling between Fe^{III} LS ($^2T_{2g}$ state) and HS Fe^{II} ($^5T_{2g}$ state), vibrational and strain effects (both intra- and inter-cluster), short range interactions (intra-cluster). The many parameters used for best fit included Δ, axial crystal field splitting of $^5T_{2g}$ state; $\Delta_{HS\text{-}LS}$ related to Δ_o, the effective HS–LS separation in octahedral symmetry and determines the temperature

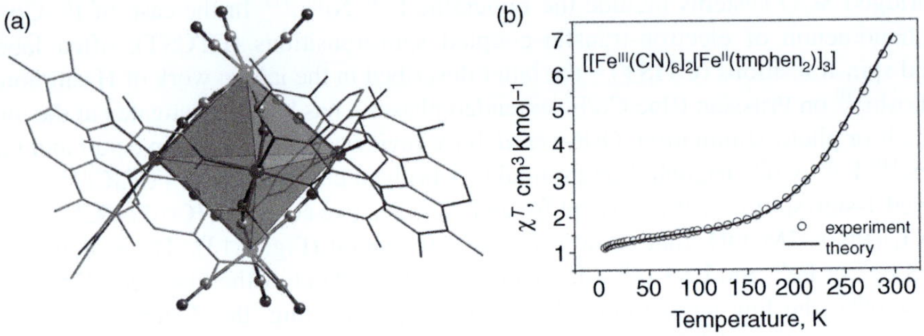

Figure 1.14 (a) Structure of $[[Fe^{III}(CN)_6]_2[Fe^{II}(tmphen)_2]_3]$. (b) Magnetism of $[[Fe^{III}(CN)_6]_2[Fe^{II}(tmphen)_2]_3]$. Reprinted with permission from [126]. Copyright 2011, American Chemical Society.

at which the $\chi_M T$ values start increasing; x, fraction of Fe^{II} ions in the HS state at all temperatures; λ and λ_1, spin-orbit coupling constants; J and J_o, exchange coupling constants, responsible for the curve steepness. The authors regarded the LS to HS transition as 'a cooperative phenomenon driven by the electronic states of Fe^{II} with totally symmetric deformation of the local coordination environment that is extended over the crystal lattice via the acoustic phonon field'.

The LIESST properties of $[[Fe^{III}(CN)_6]_2[Fe^{II}(tmphen)_2]_3]$ were quite weak, ca. 2% at the equatorial Fe^{II} sites, in comparison to the $[[Fe^{III}(CN)_6]_2[Co^{II}(tmphen)_2]_3]$ analogue. It is not clear why unless surface irradiation effects are occurring.[121]

1.2.2.4 *Hexanuclears*

As indicated at the start of this account, the 'nanoball' suprametallospecies contain $Cu^I{}_8Fe^{II}{}_6$ aggregates, with the Fe^{II} centres bonded in a *trans* arrangement to four 4-pyridine groups from the bifunctional $Tp^{4\text{-py}}$ scorpionate ligands (that have Cu^I tetrahedrally bound in the *fac* sites of the $HBpz_3^-$ inner pocket).[43] The axial positions around each Fe contain two NCS^- or one NCS^- and one MeCN ligand (solvate), the NCS^-/MeCN combination felt responsible for the observed, gradual spin-transition. In a recent review,[65] other pertinent features of these compounds have been highlighted, such as the packing of the nanoballs in the crystals (30% of volume is void space), the resulting porous properties leading to H_2- and CO_2 guest-sorption, with the $Cu^I{}_8Zn^{II}{}_6$ compound giving very strong H_2- physisorption. The efficient LIESST features of the $Cu^I{}_8Fe^{II}{}_6$ nanoball led to light-induced switching between HS and LS forms (Fig. 1.15). These spin-crossover materials, and the wide range of other $Cu^I{}_8M^{II}{}_6 - Tp^{4\text{-py}}$ analogues are being further explored by Batten *et al.* for the abovementioned properties as systematic variations are made in M^{II}, scorpionate size, counteranion and solvent (MeCN, $PhCH_2CN$). The counteranions form anion subclusters in the crystal lattice, with spherical molecular shapes resulting when the tetrahedral anions ClO_4^- or BF_4^- are used (as in the above $Cu^I{}_8Fe^{II}{}_6$ cluster). However, when octahedrally shaped PF_6^- or trigonally shaped NO_3^- anions are incorporated, the shape of the ball is distorted and the packing motifs in the crystal are significantly affected. Such supramolecular nuances are being investigated for their effects on spin-transitions in the $Cu^I{}_8Fe^{II}{}_6$ compounds.[127]

1.3 1D Chains of Fe^{II} SCO Materials

This topic was reviewed by the author up to the 2008 period.[64] A major result that fills a long-term gap is the recent crystallographic confirmation of the 1D polymeric structure of the Fe^{II} triply-bridging

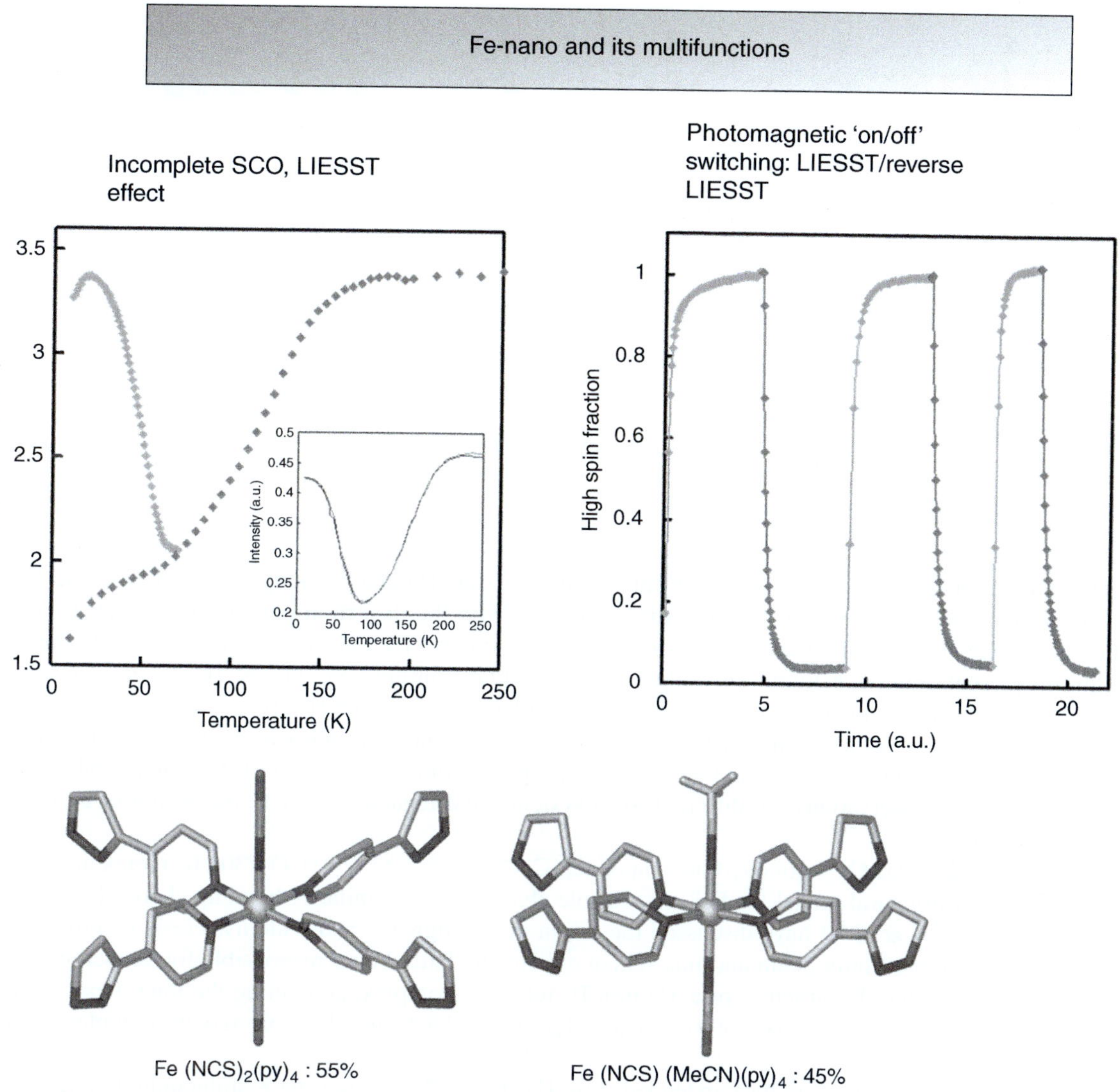

Figure 1.15 *Spin-transition magnetic data and photomagnetic switching for* $Cu^{I}_{8}M^{II}_{6}$ *- Tp*$^{4\text{-}py}$ *nanoball. Reproduced with permission from [43]. Copyright Wiley-VCH Verlag GmbH & Co., 2009.*

1,2,4(R)-triazole compounds by Guionneau and co-workers.[128] Of the various derivatives of these iconic compounds, the complex $[Fe(NH_2trz)_3](NO_3)_2{\cdot}nH_2O$; $n = 2$, proved amenable to yielding suitable, though small, crystals. The unit cell is triclinic $P\bar{1}$; R_{obs} was 9.38% at 120 K. The magnetic data showing $T_{1/2}$ above 300 K, and a concomitant thermal hysteresis, have been known since Lavrenova's work in 1986[129] and Kahn's in 1998,[130] with magnetochromic/memory applications of such materials being pursued strongly by Kahn *et al.*[131] A recent attempt to solve the 1D structure by analysing PXRD patterns was also made on $[Fe(Htrz)_2(trz)](BF_4)$, but provided limited data.[132] The chains and their mode of packing, in $[Fe(NH_2trz)_3](NO_3)_2{\cdot}2H_2O$, are shown in Figure 1.16. The Fe–N lengths show LS values as expected at 120 K. Two Fe sites are located on adjacent

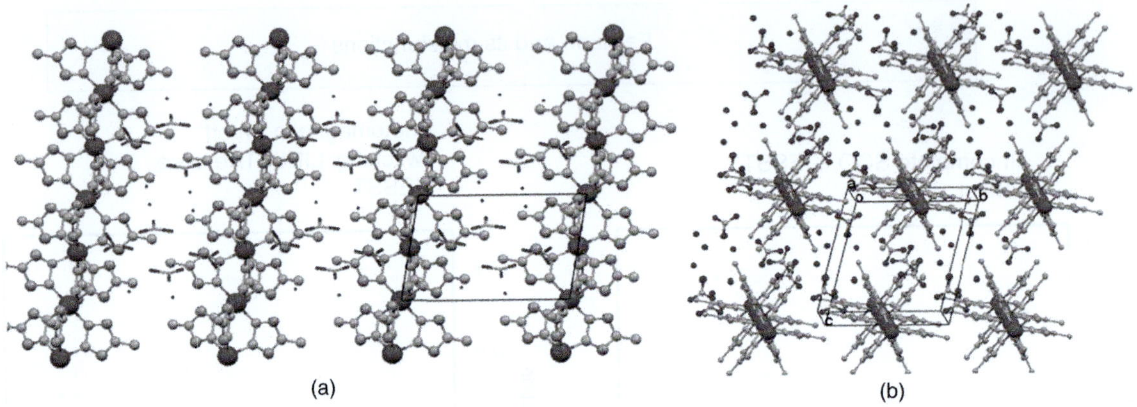

Figure 1.16 *(a) View of chains within crystal packing of [Fe(NH$_2$trz)$_3$](NO$_3$)$_2$·2H$_2$O along b direction, and (b) along a*. Reproduced from [128], with permission of The Royal Society of Chemistry.*

centres of inversion along the *a* direction thus giving zig-zag 1D chains along *a* (Fig. 1.16). The $Fe^{II}N_6$ coordination spheres have close to octahedral geometry, with the distortion parameter Σ being 15(1)° for Fe1 and 11(1)° for Fe2; these are much smaller than LS values given elsewhere in this and other articles. Thus the present chains constrain Fe to adopt rigid and regular coordination spheres. The water molecules are situated between adjacent chains and hydrogen-bond with the chains via NH_2 groups, the latter also forming H-bonds with trz of adjacent chains. Individual NO_3^- anions interact with two adjacent chains. All such (dense) interactions presumably aided the cohesive properties of crystallisation of this compound. Halcrow has made related observations, on different SCO systems, and related density to the ability of a crystal to undergo SCO.[133]

Interestingly, the 1D Cu^{II} monohydrate analogue, [Cu(NH$_2$trz)$_3$](NO$_3$)$_2$·H$_2$O, (300 K, monoclinic, *C2/c*), often used as a structural model for Fe^{II} tris-triazole chains, shows similar H-bonding chain–chain interactions[133] and Garcia *et al.*[134] have discussed the origin of the high $T_{1/2}$ and wide hysteresis loop of the Fe^{II} compound in terms of intra-chain and inter-chain effects, the latter being responsible for cooperativity. We should remember that the structure of a 1D iron(II) tetrazole complex, containing the long, flexible ligand 1,2-bis(tetrazol-1-yl)propane, viz. [Fe^{II}(btzp)$_3$](ClO$_4$)$_2$, was solved some 12 years ago, the complex showing $T_{1/2} = 130$ K.[135]

Matouzenko and co-workers have extended their 1D studies[64, 136] using the 8-aminoquinoline (aqin) terminal ligand to describe the first dca-bridged Fe^{II} example, [Fe(aqin)$_2$($\mu_{1,5}$-dca)](ClO$_4$)·MeOH].[137] The key to obtaining this material was to use a 1:1 ratio of Fe to dca, and the bidentate aqin ligand. The dicyanamide ligand, $N(CN)_2^-$, is now known to bridge in $\mu_{1,3}$ (<u>N</u>C–<u>N</u>–CN) and $\mu_{1,5}$ (<u>N</u>C–N–C<u>N</u>) modes, the former in [M(N(CN)$_2$)$_2$] extended magnets,[138] the latter in dinuclear[68] and 1D chain species.[64] A combined magnetic susceptibility, Mössbauer, Raman and structural investigation on the gradual 2-step spin-transition (complete HS to LS with no hysteresis) for [Fe(aqin)$_2$($\mu_{1,5}$-dca)](ClO$_4$)·MeOH] showed that the steps were due to inequivalent Fe sites in both the HS and LS states and that each Fe site changed spin at different temperatures. The polymeric chains are packed in sheets in the crystal. Cooperativity, estimated by fitting the $\chi_M T$ plot to the regular solution model, gave $2RT_{1/2} = 3.65$ kJmol^{-1}, leading to a lack of hysteresis, and arose via H-bonding interactions between chains and by π–π interactions between chains, with any contribution of an intra-chain type being negligible.

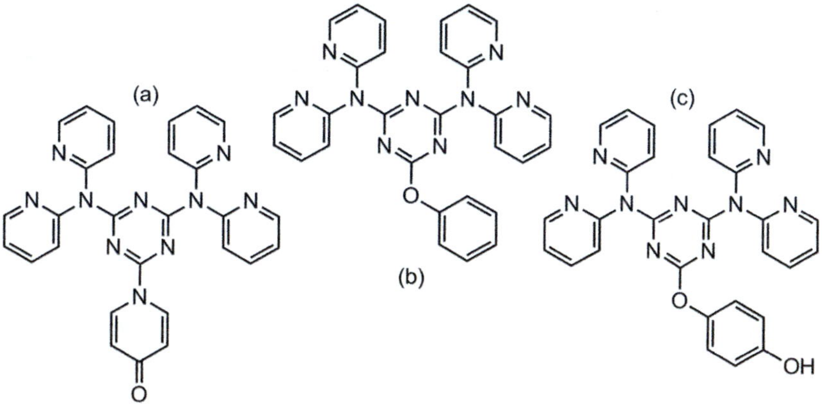

Figure 1.17 The dipyridylamine-triazine ligands: (a) DPPyT (1-(4,6-bis(dipyridin-2-ylamino)-1,3,5-triazin-2-yl)pyridin-4(1H)-one); (b) DPT (6-phenoxy-N^2,N^2,N^4,N^4-tetra-2-pyridinyl-1,3,5-triazine-2,4-diamine); and (c) DQT(4-(4,6-bis(dipyridin-2-ylamino)-1,3,5-triazin-2-yloxy)phenol). Reproduced with permission from [140]. Copyright Wiley-VCH Verlag GmbH & Co, 2011.

Our work on 2,4-dpa-based 1,3,5-triazine ligand systems[139] has led, recently, to a range of linear and zig-zag 1D SCO compounds of type *trans*-$[Fe^{II}(NCX)_2(L)]\cdot$Solvent where L is *bis*-dpa bridging ligand of the kind shown in Figure 1.17; X = S, Se, BH_3; and solvent of crystallisation can affect the spin-transition.[140] Great care had to be taken to ensure no solvent was lost during measurements. The 6-triazine position contains a variety of substituents that can produce steric, π-bonded or H-bonding effects between chains, aimed at influencing cooperativity. We have also described a crown-substituted derivative capable of showing the extra function of guest-binding.[141] Interestingly in none of these compounds was thermal hysteresis observed typical of strong cooperativity. Two such examples are described in some detail. *Trans*-$[Fe^{II}(NCS)_2(DPPyT)]\cdot 2.5CH_2Cl_2$ shows a full, gradual HS↔LS transition while *trans*-$[Fe^{II}(NCS)_2(DPT)]\cdot 2CH_3OH\cdot H_2O$ shows a 'half' SCO.

Both compounds show alternating, inequivalent Fe1 and Fe2 sites along the chain (Fig. 1.18), however the DPT compound has alternating 'trapped' HS(Fe2) and LS(Fe1) sites below the $T_{1/2}$ temperature (175 K), in agreement with the magnetism, whereas the DPPyT Fe1 and Fe2 sites are all LS, also in agreement with the magnetism. One such plot of susceptibility and LIESST data is given in Figure 1.19, showing the gradual thermal spin-transition and the partial LIESST effect, observed under green light irradiation at 10 K.[140]

The structure-magnetism-cooperativity relations, of the intra-chain and inter-chain kinds, were discussed in detail for these dpa-based 1D compounds and compared to earlier, related examples that showed trapped –HS–LS–HS–LS– sites[139] below $T_{1/2}$, or that displayed diffuse scattering as in the 2-step example *trans*-$[Fe(NCSe)_2(bdpp)]$, where bdpp = 4,6-bis(2′2″-pyridyl)pyrimidine,[142] the latter displaying averaged LS/HS character at the intermediate plateau (IP) with ordering of these sites occurring along the 1D chains but not between them. Diffuse scattering was not observed in the DPPyT and DPT systems but the trapping of –HS–LS–HS–LS– sites was in the DPT case, as indicated above. Differences in the FeN_6 octahedral distortion parameters on each Fe site (Σ^o = the sum of $|90-\theta|$ for the 12 N–Fe–N angles in the octahedron; θ^o = the sum of $|60-\theta|$ for the 24 N–Fe–N angles describing the trigonal twist angle) along the chain can affect the $T_{1/2}$ value and the abruptness of the spin-transition.[143] In this new series of chains, the octahedral distortions were not felt to be the primary determinant of spin-transition shape, rather the packing effects induced by solvate molecules played a key role.

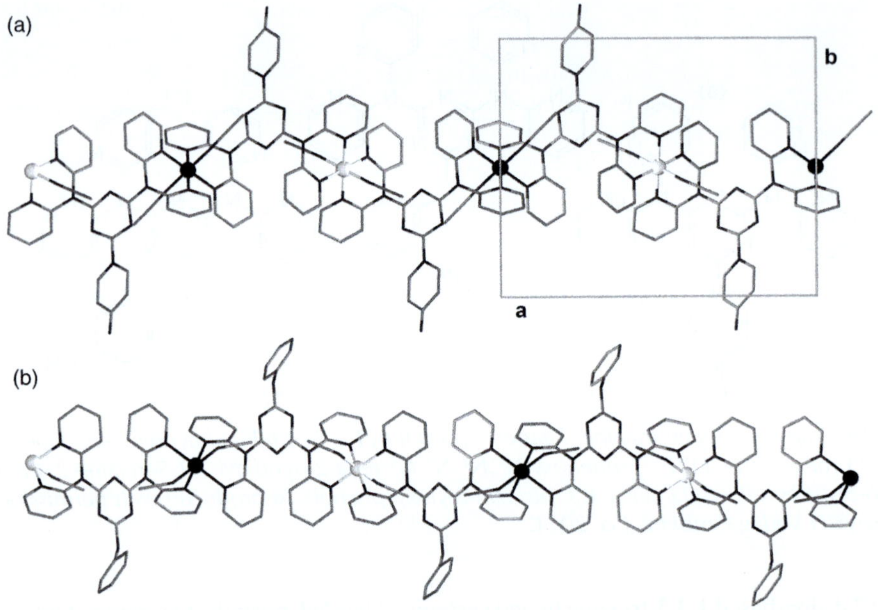

Figure 1.18 Chain structures and alternating HS and LS Fe^{II} sites of: (a) $[Fe^{II}(NCS)_2(DPPyT)]\cdot 2.5CH_2Cl_2$; and (b) $[Fe^{II}(NCS)_2(DPT)]\cdot 2CH_3OH\cdot H_2O$. Reproduced with permission from [140]. Copyright Wiley-VCH Verlag GmbH & Co, 2011.

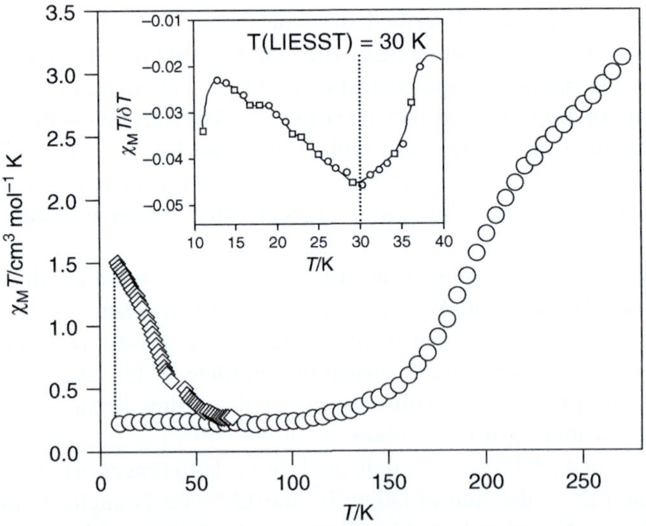

Figure 1.19 Magnetic and LIESST data for a desolvated sample of $[Fe^{II}(NCS)_2(DPPyT)]\cdot 2.5CH_2Cl_2$. Reproduced with permission from [140]. Copyright Wiley-VCH Verlag GmbH & Co, 2011.

Inter-chain interactions of various kinds were considered including S..S, from neighbouring <u>N</u>CS ligands (and Se..Se from <u>N</u>CSe congeners) and S..solvent, where, for example, the complex $[Fe^{II}(NCS)_2(DPT)]\cdot 2CH_3OH\cdot H_2O$ showed localised $(H_2O)_2(CH_3OH)_4$ ring-like H-bonded networks. Thus, the NCS sulfur atoms in neighbouring chains formed hydrogen bonds to the water O atoms in the solvent cluster and therefore allowed chain–chain interactions, a feature that was different in the compound *trans*-$[Fe^{II}(NCS)_2(DQT)]\cdot 2CH_2Cl_2$ that had a $(CH_2Cl_2)_4$ cluster in the solvent pocket, with van der Waals interactions (S..HC and S..Cl), rather than H-bonding involving the $(H_2O)_2(CH_3OH)_4$ moiety, giving a less effective way of communicating changes due to the spin-transition. Removal of the solvent molecules from $[Fe^{II}(NCS)_2(DPT)]\cdot 2CH_3OH\cdot H_2O$ caused $T_{1/2}$ to move 25 K lower with the spin-transition being still of the 'half' type; unfortunately the sample lost crystallinity during desolvation and structural comparisons could not be made in the way we were able to do in dinuclear$\cdot$xCH_2Cl_2 solvates that retained their crystal lattices for various x.[73, 81] Other inter-chain interactions in $[Fe^{II}(NCS)_2(DPT)]\cdot 2CH_3OH\cdot H_2O$, for example, were face-to-face π–π between the 6-phenoxy rings of DPT, somewhat akin to π–π effects in $[Fe(aqin)_2(\mu_{1,5}$-dca)]$(ClO_4)\cdot$MeOH, mentioned above.[137]

Clearly (*sic*), from the above discussion, there are many subtle and competing effects in crystals of 1D (and other higher dimension) Fe(II) SCO materials that affect/influence the cooperativity and, hence, the nature of the spin-transition, and it is difficult to sort out a 'one message suits all', rather each species probably needs to be treated in its own right. In contrast to the general observation of nonhysteretic, gradual transitions in the dpa-based and aqin-based Fe(II) chains,[136, 137, 139, 141] there is a marked contrast with the hysteretic (ΔT around RT), abrupt transitions of the *tris*-1,2,4-triazole linked chain systems, some recent 4-substituted examples, however, being less abrupt.[144] What can we glean from the recent crystal structure on $[Fe(NH_2trz)_3](NO_3)_2\cdot 2H_2O$? First, of course, the Fe..Fe separation, 3.65 Å, is much smaller than in the dpa-based crystals (~9.5 Å (intra-chain) and ~9.7 Å (inter-chain)) or the aqin/dca crystals of 8.9 Å, and the chains are more rigid around Fe(II), with regular coordination spheres and smaller distortions (low spin; $\Sigma \sim 11°$), than in the dpa materials (low spin; $\Sigma \sim 30$–$40°$). Importantly, the water and anions in $[Fe(NH_2trz)_3](NO_3)_2\cdot 2H_2O$ form a dense interaction network linking all of the chains and involving H-bonding of various kinds such as chain..NO_3^-..chain and (triazole)NH_2..triazole, as well as enhancing crystal cohesion. The density of these intermolecular interactions likely enhances the spin-transition and cooperativity compared to the dpa-linked systems, even though the latter possesses many of these kinds of inter-chain interactions.

Matouzenko *et al.* have discussed these and other aspects of cooperativity in their dca- bridged chain and in 4,4′-bipy type bridged chains, with theoretical arguments of the 2-sublattice Ising type given in the dca case and leading to the conclusion that inter-chain effects were dominating the 2-step, gradual transition and the associated cooperativity.[136, 137] There is always an 'outlier' compound and such a case is the 1D species of Weber *et al.* [Fe(L)(4,4′-bipy)], having a FeN_4O_2 coordination sphere (L = $N_2O_2^{2-}$ Schiff base), that displays a wide hysteresis of $\Delta T = 18$ K at $T_{1/2} = 228$ K.[145] Fitting of the susceptibility plot to the regular solution model gave a cooperativity parameter, Γ, of 5.22 kJmol^{-1} (and $C = \Gamma/2RT_{1/2}$ of 1.45)[146] typical of hysteretic behaviour. The strong cooperativity in [Fe(L)(4,4′-bipy)] was ascribed to a combination of covalent intra-chain and elastic inter-chain interactions, the latter being more dominant. Closest Fe..Fe inter-chain separations were a low ~6 Å while intra-chain Fe..Fe separations were ~11.5 Å.[145]

Weber, Garcia *et al.* also observed wide hysteresis in other $N_2O_2^{2-}$ Schiff base Fe^{II} compounds of type $[Fe^{II}L(imidazole)]$ for which hydrogen bonding played a major role in intermolecular interactions and, hence, in cooperativity.[147] Matsumoto *et al.* observed 1D packing, via NH..N hydrogen bonding, in bis-tridentate compounds $[Fe^{II}(HL^{Me})_2](anion)_2$, where L^{Me} = 2-methylimidazol-4-yl-methylideneamino-2-ethylpyridine.[148]

1.4 1D Chains of Fe^{III} SCO Materials

Using related kinds of molecular designs to those given above for the Fe^{II} species, and, specifically, extending the axial ligand directions of Fe^{III} Schiff base compounds by use of 2-connecting bi-pyridines or bi-imidazoles, has led to families of type $[[Fe^{III}(O_2N_2)(4,4'\text{-bipy})](A)]_n$. Our work on Fe^{III} acen/Fe^{III}salen/Fe^{III} salophen precursors, with common anion A = BPh_4^-, yielded species such as $[[Fe^{III}(salen)(4,4'\text{-bipy})](BPh_4)]_n$, and the salophen/acen analogues, where 4,4′-bipy could be changed for a range of linked bi-pyridines and bi-imidazoles (Fig. 1.20).[104] The $Fe^{III}N_4O_2$ coordination geometry around each Fe^{III} was similar to that employed in earlier work on SCO monomers of type $[Fe(salen)(imidazole)_2](ClO_4)$.[15] Only in the acen series was SCO observed, in the form of gradual incomplete transitions in their $\chi_M T$ plots, the salen and salophen compounds remaining HS.[105] Apart from $[[Fe^{III}(acen)(pin)](BPh_4)\cdot 3MeOH$, that had a reasonably abrupt transition at $T_{1/2}$ ~135 K (pin = *N*-4-pyridyl(isonicotinamide)), all others showed gradual, incomplete transitions measured up to 350 K, with $T_{1/2}$ of >200 K. Many crystal structures were solved for the salen series, but a 1D 'chain of H-bonded dimers' in $[(Fe(acen))_2(\mu\text{-tvp})(tvp)(tvpH)](tvpH)(BPh_4)_4\cdot 1.5MeOH$, (tvp = 4,4′-vinylpyridine; see tvp ligand in Fig. 1.3), shown in Figure 1.21, and an earlier reported analogue $[Fe(acen)(bpp)](BPh_4)$, by Imatomi *et al.*, (bpp = 1,3-bis(4-pyridyl)propane),[149] were the only structures solved for the acen series. After consideration of intra- and inter-chain interactions of the H-bonding, π-stacking and phenyl-embrace (to BPh_4^-) types, the main magnetostructural correlation to emerge was that of the ligand backbone C_2N_2 conformation (envelope/meso/planar (umbrella)) playing a key role such that the spin-transition can be allowed or inhibited.

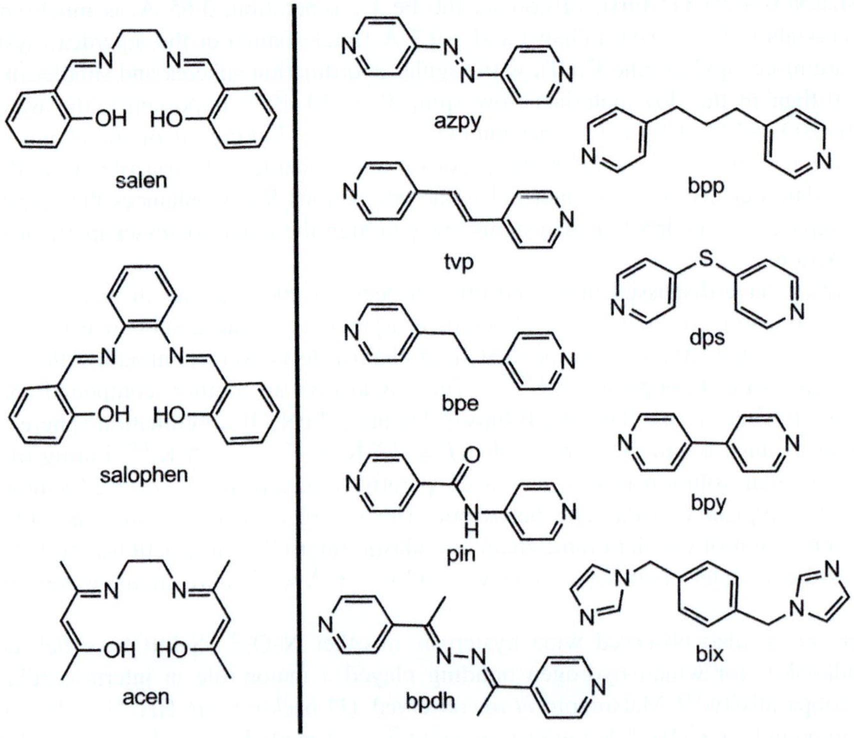

Figure 1.20 *Bridging ligands (right) and equatorial Schiff base ligands (left) used in making 1D Fe^{III} Schiff base complexes. Reproduced from [104], with permission of The Royal Society of Chemistry.*

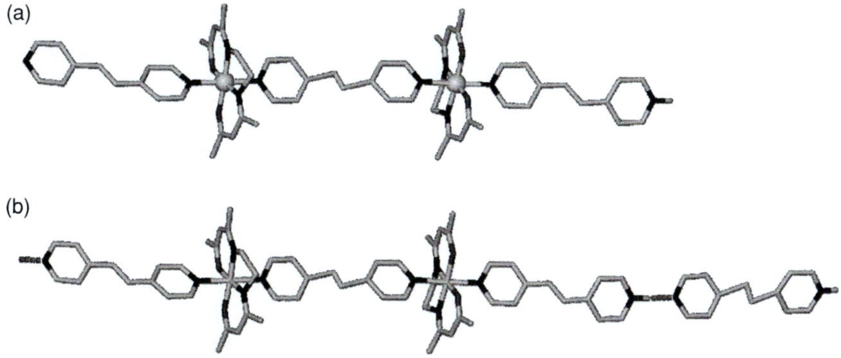

Figure 1.21 *(a) Repeat unit in the structure of [(Fe(acen))$_2$(μ-tvp)(tvp)(tvpH)](tvpH)(BPh$_4$)$_4$·1.5MeOH. (b) H-bonded chain structure of [(Fe(acen))$_2$(μ-tvp)(tvp)(tvpH)](tvpH)(BPh$_4$)$_4$·1.5MeOH. Reproduced from [104], with permission of The Royal Society of Chemistry.*

Oshio *et al.*[150] have reviewed Fe^{III} SCO compounds, prior to the above 1D Schiff base studies,[104] including their LIESST features and hybrid properties, for example SCO and conductivity in double Fe^{III} Schiff base/Ni(mnt)$_2^-$ salts.[150]

1.5 2D Sheets of Fe^{II} SCO Materials

Muňoz and Real have recently reviewed, with superb graphics, Fe^{II} metallocyanate materials (Hofmann-like phases), including 2D systems, and described their important thermo-, piezo-, photo- and chemoswitchable SCO properties.[151] They begin by contrasting these rigid and (generally) hysteretic 2D compounds with the less rigid 2D species such as [Fe(bpe)$_2$(NCS)$_2$],[152] [Fe(btr)$_2$(NCS)$_2$][153] and [Fe(azpy)$_2$(NCS)$_2$]·solvent[35, 154] (bpe = *trans*-1,2-bis(4-pyridyl)ethylene; btr = N^4,$N^{4'}$-bitriazole; azpy = 4,4′-azopyridine), most of which display interpenetration of the 2D sheets that leads to channels in the structures and resulting porosity. More is given on porosity and guest sorption, and their effects on spin-transitions, below, when discussing 3D materials. However, at this point some important facets of the 2D SCO work of Kepert *et al.* are highlighted. In the compound labelled SCOF-4(acetone), *trans*-[Fe(bpe')$_2$(NCS)$_2$]·acetone, where bpe′ = *trans*-1,2-bis(4-pyridyl)ethane, the Fe(bpe')$_2$ rhombic (4,4′) grids interpenetrate at an angle of ~90°.[155] As summarised in a recent review,[65] a very detailed crystallographic and PXRD study of SCOF-4, over a wide temperature range and in the presence and absence of solvated acetone, revealed the existence of multiple structural phases. The parent solvate showed a tetragonal to orthorhombic phase change associated with the 2-step spin-transition, that defined a checkerboard-like ordering of spin sites at the HS:LS plateau temperature. It also displayed TIESST and LIESST properties. In a related material, SCOF-3(ethanol), viz. *trans*-[Fe(bped)$_2$(NCS)$_2$]·3EtOH, where bped is *DL*-1,2-bis(4′-pyridyl)1,2-ethanediol, a potential H-bonding linker, the 1-step HS ↔ LS spin-transition changes shape with loss of ethanol.[156] The desolvation process is accompanied by a crystal-to-crystal transformation, with accompanying changes in H-bonding involving the diol groups as well as subtle changes in coordination geometries at the Fe sites. Vapour resorption of ethanol into SCOF-3 did not lead to complete solvent uptake. This study showed the delicate influence of framework structure upon the spin-transition and the relative stabilisation of HS and LS states, these, in turn, being influenced by the diol functionality.

Taking one family of CN-bridged coordination polymers (labelled SCO-CP) from the review of Muňoz and Real,[151] as an example [Fe^{II}(3-X-py)$_2$][M^{II}(CN)$_4$], these contain monodentate pyridines bonded to Fe^{II}

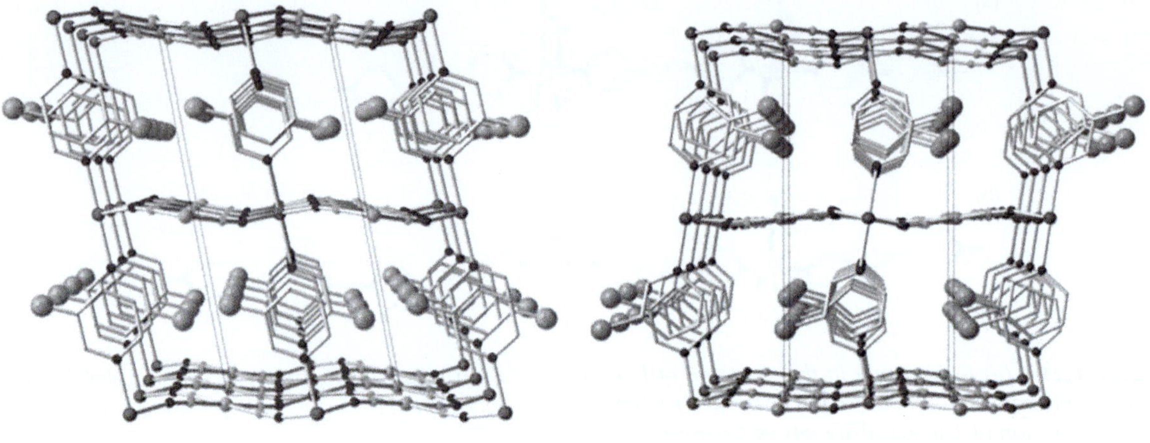

Figure 1.22 *Packing of three consecutive layers showing interdigitation of 3X-py rings and the distinct orientation of the halide X groups in the two phasesL (left) monoclinic C2/m e.g. [Fe^{II}(3-Cl-py)$_2$][Pt^{II}(CN)$_4$]; and (right) orthorhombic* Pnc2 *e.g.* Fe^{II}*(3-Cl-py)*$_2$*[*Ni^{II}*(CN)*$_4$*]. Reprinted from [151] Copyright 2011, with permission from Elsevier.*

and were first reported by Kitazawa *et al.*[157] The layers pack on top of each other such that the X-py rings interdigitate (Fig. 1.22). The cooperativity, as evidenced by the hysteresis gap, ΔT, varies with the X-substituent such that X = H, M^{II} = Pd showed $T_{1/2}$ = 210.5 K; ΔT = 5 K while X = F, M^{II} = Pd showed $T_{1/2}$ = 231.0 K; ΔT = 34.8 K. π–π effects between py rings played a part in the cooperativity differences and ΔS_{vibr} values were bigger for X = F, as was the cooperativity, and reasons were given for such differences. The $T_{1/2}$ values varied with the X group, with 3 F-being the highest.

Other 2D materials have used $M^I(CN)_2^-$ as the bridging metallocyanate, where M^I = Ag, Au, Cu and bridging pyrimidine has also been incorporated in a number of cases, often with intriguing results.[158] Linking the 2D layers in the $M^{II}(CN)_4^{2-}$ systems by 2-connecting pyrazines or bi-pyridines yields the 3D, porous Hofmann phases, and these are described below. Metallomesogen properties of 2D cyanometallate SCO materials that have long alkyl chains in the 4-position of the pyridine rings show much broader and incomplete spin-transitions compared to the crystalline, parent non-alkyl chain analogues.[159]

Interesting 2D compounds, containing triazolate-pyridyl chelating bridging ligands, have just been reported to show the highest $T_{1/2}$ values reported to date.[160] They are of type [Fe(L)$_2$] where HL = 3-(2-pyridyl)-5-(3-pyridyl)-1,2,4-triazole) and the 3-methyl-2-pyridyl analogue, the first such being an isomer of the 2-pyridyl/2-pyridyl triazolate ligand that we used to make dinuclear SCO species of type *trans*-[$Fe_2(NCX)_2(L')_2py_2$].[72] The 2D (4,4) rhombic grids in the structure pack together utilising interlayer face-to-face and edge-to-face π–π interactions. The frameworks are extremely stable up to 623 K and display 2-step, non-hysteretic transitions with $T_{1/2}$ of 329 K and 501 K, as evidenced by crystallographic, susceptibilities (using a SQUID magnetometer with an oven attachment), DSC, Raman and Mössbauer spectral studies. The stepwise behaviour was felt to arise from the two inequivalent $Fe^{II}N_6$ sites existing in the framework together with anti-cooperative supramolecular interactions occurring between 2D layers.

1.6 3D Porous SCO Materials

Some aspects have been given in the 2D SCO section above, and in recent reviews.[35, 60, 65, 151, 161] This area of bifunctional molecular materials is one of the most exciting, recent developments in SCO chemistry since it has the possibility of producing gas or solvent sensing devices that respond to guest sorption/desorption while simultaneously influencing (or being influenced by) spin switching. Real and co-workers were one of the first groups to study these spin-crossover 'molecular zeolites' and have developed, though many fine studies, the label SCO-PCPs (porous coordination polymers). Kepert and co-workers, also deeply involved, use, as indicated above, the label SCOF (spin-crossover frameworks). The more flexible frameworks, isolated when 2-connecting dipyridines bridge between $Fe^{II}(NCX)_2$ moieties, such as $[Fe(azpy)_2(NCS)_2]{\cdot}0.5EtOH$,[35, 161] have been reviewed in the Gütlich, Goodwin series[35] while an earlier example, $[Fe(tvp)_2(NCS)_2]{\cdot}MeOH$ (tvp = 1,2-di-(4-pyridyl)-ethene) was reviewed in a supramolecular monograph.[162] Use of 'bent' dipyridines, for example with S in the middle, influence the topology of the frameworks and affect the SCO properties markedly.[163]

The more rigid 3D Hofmann materials of type $[Fe^{II}(pz)[M(CN)_4]]{\cdot}$guest, where M^{II} = Ni, Pd, Pt, have produced and continue to produce fascinating results (Fig. 1.23). The reversible, hysteretic spin-transitions at the Fe^{II} centres occur simultaneously with uptake of guest molecules. Indeed, the two processes are interdependent.[164] Ohba *et al.*[165] described the Pd and Pt systems as displaying 'chemoresponsive bidirectional switching of spin state at room temperature' in which guest = 5- and 6-ring aromatics, for example benzene, favoured the HS yellow coloured state, while CS_2 (Pt) favoured the LS red coloured state. Desorption of guest produced the guest-free framework in the 'induced' spin state and the system does not recover the initial HS or LS state when releasing guest molecules while in the bistable temperature region. This memory function for the guest-free form therefore retains important information on the guest sorbed form (and vice versa). Differences in guest-binding and spin state were related to structural differences involving guest to pillar (pyrazine) interactions.

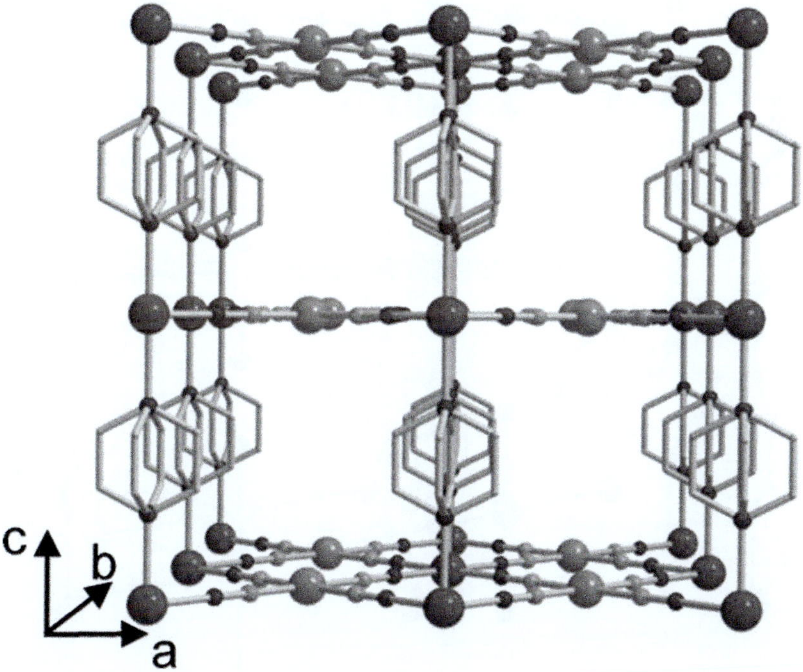

Figure 1.23 *View of a fragment of the 3D porous Hofmann SCO compounds $[Fe^{II}(pz)[M(CN)_4]]{\cdot}H_2O$, where M^{II} = Ni, Pd, Pt. Water molecules not shown. Reprinted from [151] Copyright 2011, with permission from Elsevier.*

Kepert *et al.* simultaneously reported a thorough study of guest binding to $[Fe^{II}(pz)[Ni(CN)_4]]$ in which the guests varied in size and properties between the gases N_2, O_2, CO_2 and solvents H_2O, ROH, acetone, MeCN and toluene.[164] By means of single crystal structures, synchrotron powder diffraction, adsorption and desorption isotherms (at various pressures) and magnetism, they noted such features as (i) hysteresis in the isobars for N_2, O_2 and CO_2 binding that mirrors the SCO hysteresis, (ii) the change in structural phases between $[Fe^{II}(pz)[Ni(CN)_4]]\cdot 2H_2O$ and the $.1H_2O$ and desolvated phases and concomitant change in the framework and pore structures, (iii) using the acetone solvate at 310 K, the desolvation-resolvation process $[Fe^{II}(pz)[Ni(CN)_4]]\cdot acetone_x{}^{LS} \leftrightarrow [Fe^{II}(pz)[Ni(CN)_4]]^{HS} \leftrightarrow [Fe^{II}(pz)[Ni(CN)_4]]\cdot acetone_x{}^{HS}$ shows a guest-desorption induced SCO. The key conclusion from this work was that the synergistic interplay between spin-crossover and guest exchange processes meant that guest exchange induced-SCO and SCO-induced guest exchange properties were exhibited by this Hofmann system.

Subsequent and ongoing advances in these 3D Hofmann SCO species include changes in the pz linker to longer di-pyridines such as bpac (bis(4-pyridyl)acetylene),[166] binding of halogen guests such as I_2 – with associated oxidation of Pt^{II} to Pt^{IV} and reduction of I_2 to I^- (Figs 1.24, 1.25, 1.26),[167] further variation in the nature of the guests, for pyz and other pillars, and the effects on pore size, functionality, SCO, structure and sensing/separation of guests.

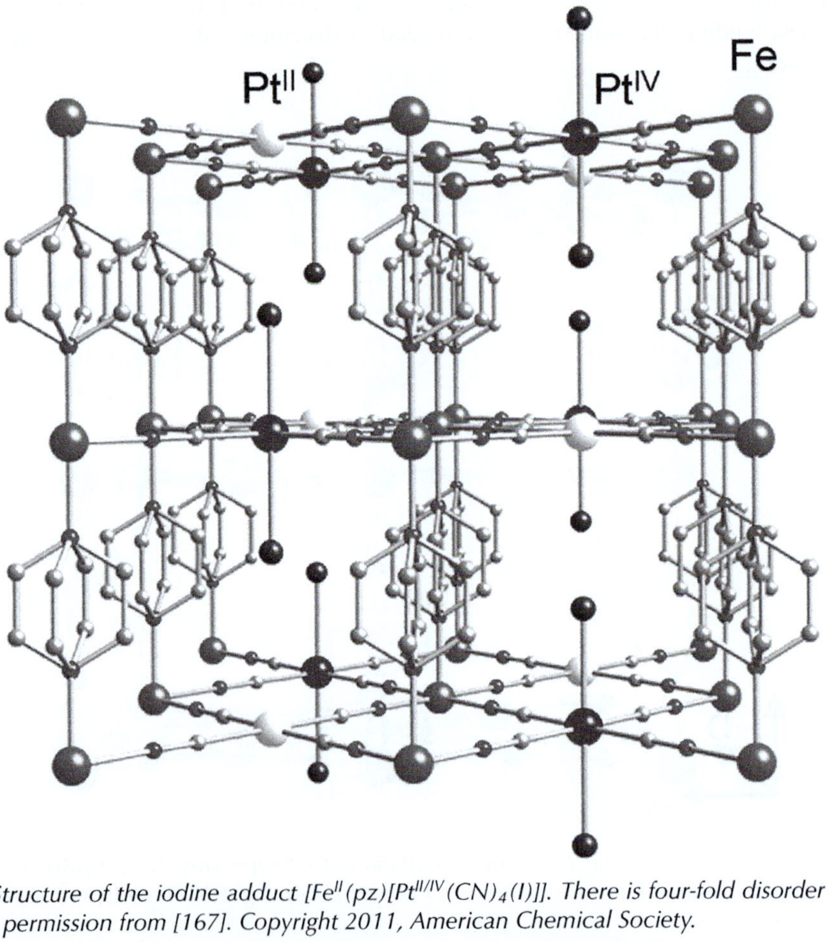

Figure 1.24 *Structure of the iodine adduct $[Fe^{II}(pz)[Pt^{II/IV}(CN)_4(I)]]$. There is four-fold disorder in the pz rings. Reprinted with permission from [167]. Copyright 2011, American Chemical Society.*

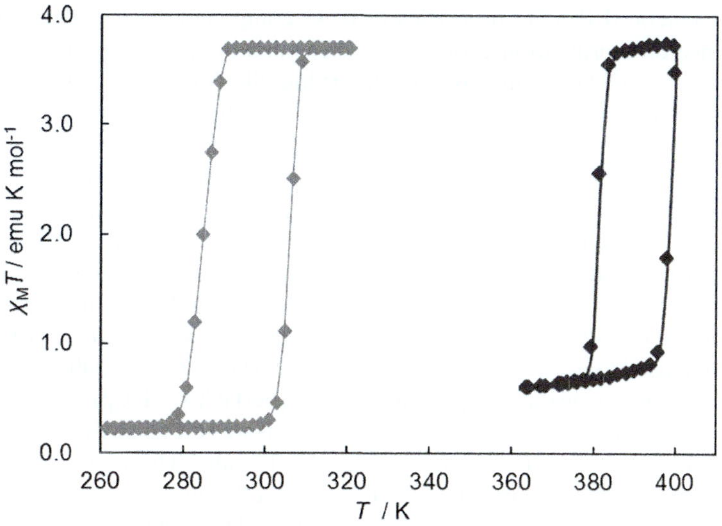

Figure 1.25 The effect of I_2 sorption on the $T_{1/2}$ and hysteresis for $[Fe^{II}(pz)[Pt^{II}(CN)_4]]$ (left) and $[Fe^{II}(pz)[Pt^{II/IV}(CN)_4(I)]]$ (right). Reprinted with permission from [167]. Copyright 2011, American Chemical Society.

1.7 Some Recent Developments in Mononuclear SCO Fe^{II}, Fe^{III} and Co^{II} Compounds

1.7.1 Iron(II) and Iron(III)

While the author has concentrated in this chapter on developments in polynuclear Fe^{II} (and Fe^{III}) SCO species, this is certainly not meant to downplay advances in mononuclear compounds, which have been extensive and important. Space only allows a short discussion of areas that have caught the author's attention. Halcrow has written two recent reviews[133,168] on Fe^{II} (and some Fe^{III}) monomers and he has a related chapter in this book. The first review[168] dealt in detail with bis-tridentate $[FeL_2]^{2+}$ complexes where L is 2,6-di(pyrazol-1-yl)pyridine (1-bpp) and pyrazole- and pyridine-substituted derivatives thereof. As well as describing the

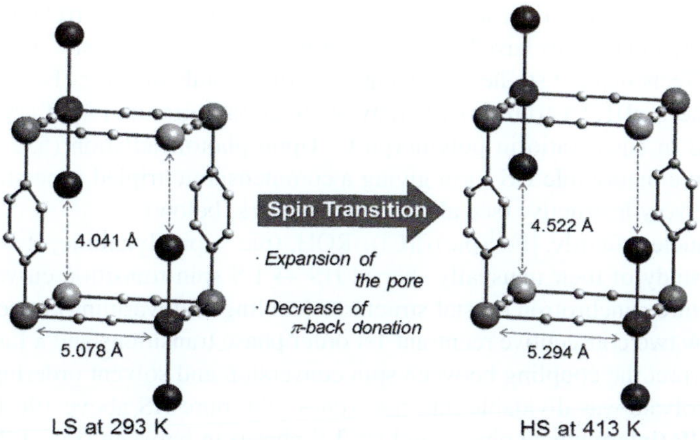

Figure 1.26 A proposed mechanism for spin-transition induced iodine migration in $[Fe^{II}(pz)[Pt^{II/IV}(CN)_4(I)]]$. Reprinted with permission from [167]. Copyright © 2011, American Chemical Society.

structural and photomagnetic (LIESST) properties of these compounds that generally display abrupt spin-transitions, the influence of coordination geometry on SCO and the influence of structure and packing on cooperativity was discussed, the latter emphasising terpyridine-like 'embraces', first described by Dance and co-workers.[169] Olguin and Brooker wrote a related review, at about the same time, on 1-bpp and 3-bpp FeL_2^{2+} species including some dinuclear derivatives.[170]

In his second review,[133] Halcrow made a valiant, and successful, attempt to analyse the structural data, above and below the $T_{1/2}$ temperatures, for various classes of Fe^{II} and Fe^{III} SCO complexes and relate structure to function, where function is the spin switching and cooperativity (with ultimate device, sensing and electronic applications in view). The octahedral distortion parameters, Σ and θ, instigated by Guionneau[143] and the angular Jahn–Teller parameters, θ vs. φ plots, (developed by Halcrow for HS forms) were given for $[Fe(1\text{-}bpp)_2]^{2+}$ families while the Fe–N1–N2–C5 torsion angle arguments of Reger *et al.*,[171] for $[Fe(HCpz_3)_2]^{2+}$ and $[Fe(HBpz_3)_2]$ salts, were used to explain the tilting of pyrazolyl rings and its relationship to SCO. We[172] also used the Σ and θ parameters for mixed scorpionate chelates $[Fe(HCpz_3)(HCpz'_3)]^{2+}$. Ligand conformation effects in $Fe^{III}(saltrien)^+$ SCO HS and LS salts were discussed, as was the salen ligand conformation (*envelope* for SCO vs. *meso* for non-SCO) for $[Fe^{III}(salen)(imidazole)_2]^+$ salts. These effects were mentioned above for 1D [Fe(salen)(4,4′-bipy)]BPh_4 type materials, with an *umbrella* conformation also observed.[105] Halcrow argues that the ligand conformation must remain similar for HS and LS forms for a spin-transition to occur, and these arguments were also applied to Weber's [Fe^{II}(Jäger Schiff base)] complexes.[45, 133] The effect of crystal density, for example in polymorphs, on SCO was discussed, with lower densities favoured for SCO to occur. A variety of structural correlations to the $T_{1/2}$ value was also given and the review finished with summaries on how cooperativity in crystalline SCO Fe^{II} complexes relates to molecular shape, crystallographic disorder, crystal packing, hydrogen-bonding and π–π interactions. Gass *et al.* have recently described supramolecular influences of di-pyridyl-appended 18-crown-6 rings, and their alkali metal derivatives, to SCO in $[Fe(3\text{-}bpp)_2]^{2+}$ salts.[173] Ross *et al.* have also discussed many of the abovementioned cooperativity aspects in mononuclear (and polynuclear) dpa-triazine Fe^{II} species in which substituent groups were incorporated into the ligand in order to enhance such intermolecular effects.[174, 175]

Létard and co-workers continue to achieve major developments, following the pioneering LIESST work of Gütlich and Hauser,[88, 89, 176] towards an understanding of photomagnetic features in large families of Fe^{II} SCO systems. Their work is described in a separate chapter of this book. Suffice to say that their linear correlations between T_{LIESST} and $T_{1/2}$ ($T_{LIESST} = T_0 - 0.3\ T_{1/2}$) from magnetism plots,[177] for families of differently coordinated Fe^{II} geometries, each giving different T_0 values, has been generally very successful, for example for the bis-*mer*-tridentate $[Fe(1\text{-}bpp)_2]^{2+}$ salts of Halcrow.[168] Interestingly, in collaborative work with Létard, our (different) bis-*mer*-tridentate Fe^{II} examples did not fit the correlations, being more hexa-monodentate in character, and reasons have been given for these 'outlier' results.[178] Other notable uses of the LIESST and TIESST effects include studies by Wang *et al.* on crystals of *trans*-$[Fe(NCS)_2(abpt)_2]$, (abpt = 4-amino-3,5-bis(pyridine-2-yl)1,2,4-triazole) whereby, without photoirradiation of crystals at 25 K, HS and LS molecules coexisted in equal ratio in polymorph C. Upon photoirradiation (532 nm) at 25 K, the LS molecule was excited to a metastable HS form giving a commensurate tripled superstructure.[179] The Bragg diffraction images were very instructive (see also Törnroos' work, below).

An important mononuclear family, $[Fe^{II}(pic)_3](Cl)_2{\cdot}ROH$, (pic = picolylamine; R = EtOH,[23] 2-PrOH[180]) has received thorough study of their unusually shaped HS $\leftrightarrow$ LS spin-transition curves by Törnroos *et al.*, using multiple temperature synchrotron crystal structures (cooling and warming) in combination with DSC data. Both solvates show two consecutive reentrant 1st order phase transitions and a Landau thermodynamic model was used to interpret the coupling between spin conversion and solvent ordering. The spin-transition curve for the 2-PrOH solvate was dividable into five zones with pure HS above 196 K and pure LS below 100 K and various HS IP (intermediate phase) and HS/LS phases in between (Fig. 1.27). The Bragg (zone) images were instructive in showing the coexistence of phases. The HS/LS phases were interpreted, at the molecular level, to be due to hydrogen-bonded arrays with a chessboard formation for the EtOH solvate and

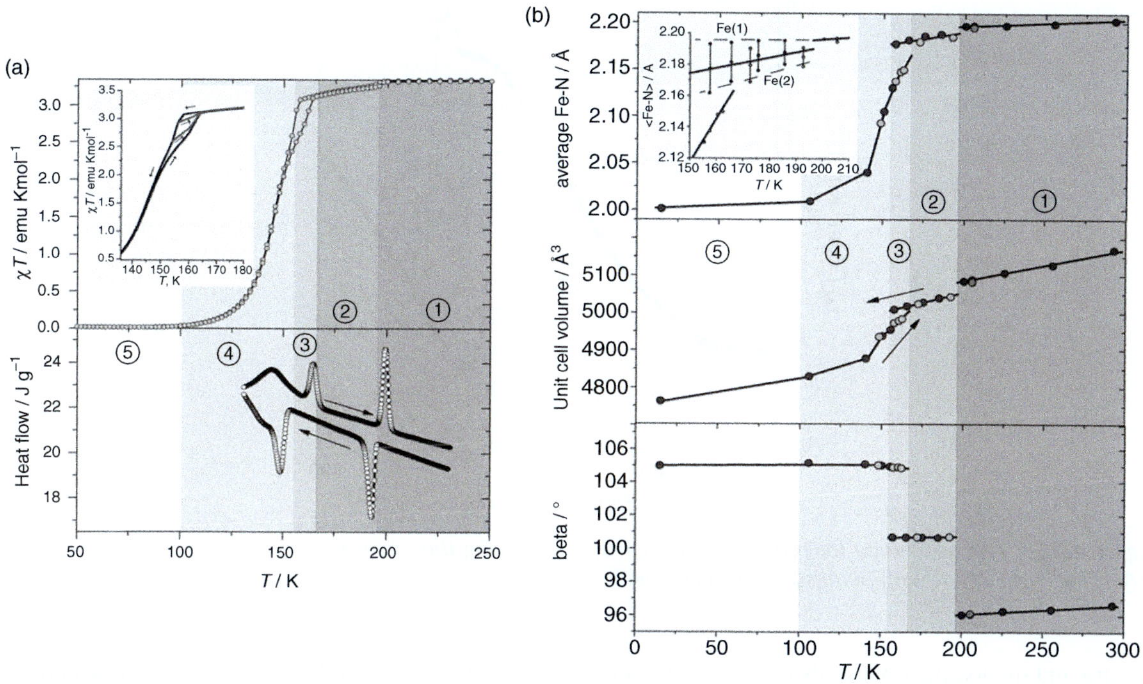

Figure 1.27 *(a) Plots of $\chi_M T$ (inset shows hysteresis cycles) and heat flow vs. temperature, with the five structural phases identified for $[Fe^{II}(pic)_3](Cl)_2 \cdot 2\text{-}PrOH$. Phase 1 is the high temperature phase, pure HS; 2 is the intermediate HS; 3 is hysteresis zone; 4 is low temperature phase, LS/HS zone; 5 is the low temperature phase, pure LS. (b) Temperature dependent cell parameters and Fe–N bond lengths. Reproduced with permission from [180]. Copyright Wiley-VCH Verlag GmbH & Co, 2006.*

an alternating zig-zag chain for the 2-PrOH solvate, one chain being all HS the other both HS and LS. These studies by Törnroos *et al.*[181, 182] set the standard for probing spin changes at the very detailed molecular level. We saw earlier the use of such approaches by Kepert *et al.* in elucidating the spin-transition and multiple phases in the acetone solvate (and desolvate) of a covalently-bridged Fe^{II} framework system, SCOF-4[155] and Guionneau *et al.* in working out the multiple phases in a 4,4′-bipyridine-bridged Fe^{II} dinuclear SCO complex.[74]

Other recent examples of stepped transitions in mononuclear Fe^{II} compounds have been described including a $[Fe^{II}(L)(NCS)_2]$ complex by Reedijk and Gamez,[181] where L is a tetradentate bis-dpa ligand, and interpreted in terms of a *symmetry breaking* spin-transition from a single HS site at high temperatures to an intermediate phase, IP, with a HS:LS ratio of 1:2. Note that these situations are different to 2-step examples in which two well defined lattice sites are present each with different $T_{1/2}$ values. Buron-Le Cointe *et al.* also described similar behaviour in a mononuclear Fe^{II} complex that was accompanied by light-induced spin state trapping.[182] Morgan *et al.* have also reported the first mononuclear Fe^{III} hexadentate Schiff base $[N_4O_2]$ example, $[Fe(L')](ClO_4)]$, with a 2-step, nonhysteretic transition and a 1:2 ratio of LS:HS states in the IP.[183]

1.7.2 Cobalt(II)

Hauser and co-workers[184] reviewed cobalt(II) $S = {}^3/_2 \leftrightarrow {}^1/_2$ SCO complexes of the $[Co(2,2'\text{-bipy})_3]^{2+}$ and $[Co(2,2':6',2''\text{-terpy})_2]^{2+}$ types with an emphasis on structure (Jahn–Teller effects), magnetism, EPR, optical

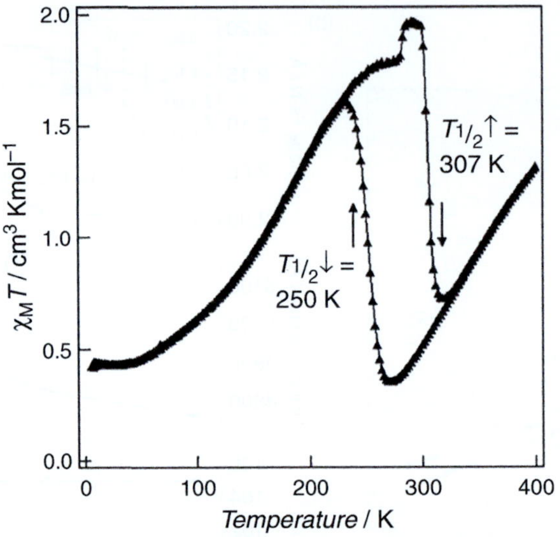

Figure 1.28 *$\chi_M T$ vs. T data for [Co^{II}(C16-terpy)$_2$](BF$_4$)$_2$ showing the 'inverse spin transition'. Reprinted from [185] Copyright 2011, with permission from Elsevier.*

spectra and theory, the review also including Co/Zn species as well as 'hybrids' formed when these cations were inserted into anionic metal-oxalate network hosts; see below. An interesting Co^{II} series containing long alkyl chains attached to the central ring of terpy chelators, has recently been reviewed by Hayami *et al.*[185] Solvated forms, such as [Co(C16-terpy)$_2$](BF$_4$)$_2$·MeOH, where C16-terpy is 4′-hexadecyloxy-2,2′:6′,2″-terpyridine, have typical Co(terpy)$_2$ coordination spheres with one of the C16 arms stuck out rod-like while the other C16 chain is twisted. There are π–π interactions between terpy pyridine rings forming tight 2D sheet packing motifs and the alkyl chains had a 'fastening' effect. This material shows a gradual, incomplete spin-transition (between 5–360 K), with a step at 150 K. The most interesting result was when it was desolvated at 400 K, the magnetism showed a 'reverse' spin-transition with a gradual (normal $^3/_2$ to $^1/_2$, though different to the parent solvate) decrease in $\chi_M T$ on cooling between 400 and 217 K, at which point an abrupt increase occurred to reach a HS value, followed by unusual S-shaped HS values down to 5 K. Rewarming then gave $\chi_M T$ values that followed the cooling curve until an abrupt decrease occurred to a LS value, at 260 K, followed by the same gradual curve between 260 and 400 K, as seen on cooling (Fig. 1.28). The large 'hysteresis' gap was reproducible on further cycling and ascribed to a phase change involving changes in the C16 chain structures, the PXRD at 50 K being different to that at 130 K.

A related, but different, situation applied to the C14 compound [Co(C14-terpy)$_2$](BF$_4$)$_2$·MeOH with changes in the C14 chain structures now observed by crystallography at 190 K, for a HS1 phase, obtained by first cooling to 100 K then warming to 190 K. A HS2 phase was obtained by cooling from 290 K to 190 K, with an LS phase solved at 10 K (Fig. 1.28). The magnetic data for this methanol solvate were interpreted in terms of two superimposed curves, an abrupt non-hysteretic one with $T_{1/2}$ at 46 K and a HS1 to HS2 transition, with hysteresis at 206 K (up) and 184 K (down), a phase change occurring at the latter temperatures. Upon desolvation at 400 K, a reverse crossover was observed and assigned as in the C16 case, now with hysteresis noted around 300 K. No structural data were presented.

The C12 example [Co(C12-terpy)$_2$](BF$_4$)$_2$·EtOH·0.5H$_2$O showed a decrease in $\chi_M T$ on warming between 5 and 70 K, HS to LS, then a gradual LS to HS plot between 70 and 300 K, assigned to reentrant SCO.[185] Not mentioned in the review[185] was a paper by Toftlund *et al.*[186] describing [Co(C12-terpy)$_2$](BF$_4$)$_2$·8.5H$_2$O, and

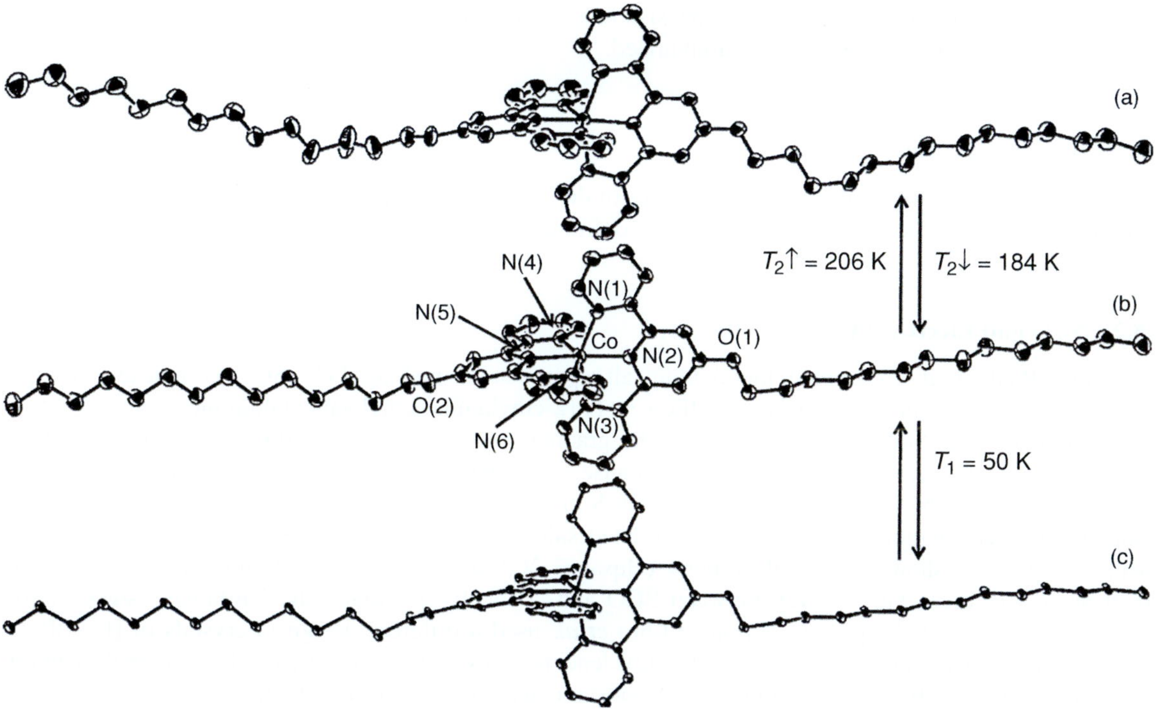

Figure 1.29 *Structures of $[Co^{II}(C14\text{-}terpy)_2](BF_4)_2{\cdot}MeOH$ showing (a) HS2 phase at 190 K, (b) HS1 phase at 190 K, (c) LS phase at 10 K. Reprinted from [185] Copyright 2011, with permission from Elsevier.*

while it was not cystallographically characterised, it showed similar magnetism between 70 and 300 K as in the EtOH/H_2O solvate but did not show the increase to a HS value below 70 K. Structures were presented for $[Co(C8\text{-terpy})_2](BF_4)_2{\cdot}H_2O$ and $[Co(C8\text{-terpy})_2](ClO_4)_2$, with disorder noted in one of the C8 chains, and for $[Co(C4\text{-terpy})_2](PF_6)_2{\cdot}3H_2O$. The $\chi_M T$ plots were of the normal, gradual SCO kind for these C8 and C4 examples. Desolvated species were, unfortunately, not studied.[188] The parent terpyridone complex that has no attached Cx chains, $[Co(terpyridone)_2](CF_3SO_3)_2{\cdot}_2O$ forms two polymorphs, one showing the normal $S = {}^1/_2$ to ${}^3/_2$ gradual spin-transition, the other, upon cooling to 217 K, shows an increase towards a HS value – less than in the C16 and C14 derivatives – then an abrupt decrease towards LS values (reaching values as for the other polymorph) at 150 K. On rewarming, the discontinuous change occurred at 188.5 K, indicative of hysteresis, with a smaller hysteresis gap also noted at 225 K. The reverse spin-transition was assigned, by Real *et al.*,[187] to a reversible reentrant structural phase transition occurring between 217 and 203 K. Following puzzling aspects at the outset,[188] there is satisfying agreement, at this time, in the interpretation of the 'reverse' transitions for the long chain and the terpyridone Co^{II} complexes.

1.8 Multifunctional/Hybrid SCO Materials

An exciting growth area is that dealing with molecular materials that have a second function to combine with the spin-crossover function and, if possible, show synergy between the functions. All such research is still at the exploratory stage and it is hoped that useful applications will emerge. Gaspar, Gütlich *et al.*[189]

reviewed this area in 2005 and here we emphasise results since then. The dual function materials that have been developed thus far are now briefly summarised.

1.8.1 SCO and Porosity

This area has been described above for frameworks of the flexible, 4,4′-bipyridine-bridged and the more rigid Hofmann-like CN-bridged phases, as well as the nonframework Fe^{II} nanoball species.[35, 43, 151–156, 161–166]

1.8.2 SCO and Electrical Conductivity

Faulmann, Real *et al.*[190] adopted the double-salt approach in making Fe^{III} SCO complexes of type $[Fe^{III}(salEen)_2]_2[Ni(dmit)_2]_5{\cdot}6MeCN$, (HsalEen = N-(2-ethylamino)ethyl-salicylaldimine; $dmit^{2-}$ = 1,3-dithia-2-thione-4,5-dithiolato) that, as well as the gradual spin-transition, also displayed semiconducting behaviour (100–300 K; non-ohmic <100 K) because of the partial oxidation inherent in this compound. Segregated stacks of $Ni(dmit)_2{}^-$ layers, with many S..S interactions were responsible for charge transport. Band structure calculations were reported. A related salt $[Fe^{III}(qsal)_2][Ni(dmit)_2]_3{\cdot}MeCN{\cdot}H_2O$, by Takahashi *et al.*,[191] also showed the dual property (Hqsal = *N*-(8-quinolyl)-salicylaldimine), and the LIESST effect at Fe^{III} centres, with synergy between SCO and conductivity. Again, the $Ni(dmit)_2{}^-$ stacks were segregated from the $[Fe^{III}(qsal)_2]^+$ cations. Oshio *et al.* used a different design in crystals of $[Fe^{II}(ttf\text{-}1\text{-}bpp)_2][Ni(mnt)_2](BF_4){\cdot}PhCN$ that had tetrathiafulvalene moieties grafted, via ethylene linkers, on the central ring (4-position) of 1-bpp, with $Ni(mnt)_2{}^-$ galvanostatically causing partial oxidation of ttf. An inflection in the resistivity plot coincided with the spin-transition, indicative of synergy between the two functions.[192] Another design, by Lemaire *et al.*, was to make complexes that contain $Fe^{III}(qsal)_2$ SCO centres that have terthienyl-alkyne groups substituted onto the phenoxo rings of qsal and then electro-oxidise these to form 'hybrid' metallo-polythiophene films having conducting and SCO functions.[193]

1.8.3 SCO and (i) Short-Range Exchange Coupling or (ii) Long-Range Magnetic Order

Dual function dinuclear materials showing simultaneous SCO and intra-cluster exchange, of the $Fe^{II}(\mu\text{-bipyrimidine})Fe^{II}$,[60–63] $Fe^{III}(\mu\text{-}4,4'\text{-bipy})Fe^{III}$[99] and $Co^{II}(\mu\text{-pyridazine})Co^{II}$ types,[40] have been described in earlier reviews and monographs and, above, in this chapter. Indeed, understanding the synergy, if any, between exchange and SCO was one of the driving forces in making dinuclear materials at the outset.[34, 35] It is still rather puzzling to observe a lack of exchange coupling, or, at best, extremely weak coupling, when bridging groups between two spin-crossover centres are employed that would normally (e.g. between Cu^{II} centres) show medium to strong coupling. Clearly, the spin-transition dominates in such cases.

An approach in which a SCO mononuclear complex is embedded within a known magnetically ordered molecular magnetic material, such as oxalate-bridged networks, has been used successfully by Coronado and co-workers.[194] They named their hybrid materials 'switching magnets'. One of their recent examples is $[Fe^{III}(5\text{-Cl-sal})_2\text{-trien})][Mn^{II}Cr^{III}(ox)_3]{\cdot}0.5(CH_3NO_2)$,[195] the anionic portion being a 3D heterometallic d^5d^3 network that is chiral, compared to earlier 2D sal_2trien analogues.[196] This trifunctional compound (SCO, order and chirality) undergoes ferromagnetic order at 5 K and shows a partial (50%) SCO in $\chi_M T$ between 300 and 60 K, with $[In^{III}(5\text{-Cl-sal})_2\text{-trien})][Mn^{II}Cr^{III}(ox)_3]$ being used to identify the cation SCO contribution. Further work using external stimuli such as light (LIESST) and pressure is planned. Decurtins *et al.* and Coronado *et al.* had earlier inserted Co^{II} and Fe^{II} SCO centres into oxalate-bridged networks.[197, 198] and Hauser *et al.* reviewed this area.[89]

1.8.4 SCO and Liquid Crystals

Gaspar, Seredyuk and Gütlich reviewed this area in great detail up to 2009.[199] Two broad approaches have been used to make liquid crystals (metallmesogens), the first using long alkyl chains on relevant ligands of the SCO centres, the approach emphasised in the review, the other using 2D CN-bridged Hofmann species that, again, had long chain substituents on the pyridine ligands.[159] In the long-chain/non-Hofmann species, interplay between spin-transitions and liquid crystal transitions were achieved in many cases. The ligand types were Fe^{II}(Cx-(tren) and Fe^{II}(Cx-tameMe) in which the Cx chains were attached to the 2-pyridyl rings via ether groups. Three classes were identified (i) systems with coupling between the electronic structure of Fe^{II} and the mesomorphic behaviour of the material, (ii) systems in which both transitions coexist in the same temperature range and (iii) systems where both transitions occur in different temperature ranges and are uncoupled. Photochromic liquid crystals are planned. Gaspar *et al.*[60] have made liquid crystals of the 1D tris-triazole Fe^{II} SCO type, with 3,5-substituted -OCx chains on 4-N-benzamide-1,2,4-triazole bridges. The short chain C4 and C6 compounds showed 'crystalline' spin-transitions while the C8, C10 and C12 compounds showed broad spin-transitions coexisting with mesomorphic behaviour in the room temperature region.[200] Columnar packing structures of the mesophases were predicted.

The 2D CN-bridged liquid crystals, by Seredyuk, Gaspar *et al.*[159] are of formula $[Fe^{II}L_2[Ag^{I}(CN)_2]]\cdot sH_2O$ or $[Fe^{II}L_2[Ni^{II}(CN)_4]]\cdot sH_2O$ and have the long alkyl chains (C6, C12, C18) substituted via ether bonds to an aromatic ring itself substituted at the 4-position of the pyridine ligand, L. Compared to the abrupt hysteretic spin-transitions exhibited in the parent crystalline (non-alkyl chain) analogues, the $\chi_M T$ plots are broad and nonhysteretic in these metallomesogens, due to poor cooperativity.

Hayami and co-workers have also reported a number of liquid crystal Fe^{II} and Co^{II} SCO compounds, including photomagnetic studies.[201] Branched chain C27-subsituted bzimpy *bis*-tridentate complexes of Fe^{II} yielded gradual transitions with an abrupt, hysteretic increase in susceptibility observed in one case.[202] Crystal to mesophase transitions were also noted in branched-chain (at C5) *bis*-terpy-Co^{II} crossover systems.[185]

1.8.5 SCO and Gels

Related to the liquid crystal 1D species by Gaspar *et al*,[200] above, Roubeau, Clérac *et al.* have used 1D Fe^{II} $[(Cx\text{-}4\text{-triazole})_3](BF_4)_2$ SCO compounds, with alkyl chains Cx = C13, C16, C18, as metal-organogelators to make gels in decane or toluene that showed colour and magnetic changes typical of these SCO materials.[203] These gels were thermo-responsive but not thermodynamically stable.

1.8.6 SCO and NLO

The functions of nonlinear optics (NLO) and ligand isomerisation in SCO species was reviewed up to 2005.[189] Developments since then include work by Lacroix *et al.*, who have coordinated the well known NLO stilbazole molecule to the vacant 6th site of the pentadentate Schiff base complex yielding $[Fe^{III}(saldpt)(DEAS)](BPh_4)$, where DEAS = 4′-diethylaminostilbazole. A gradual incomplete $S = {}^5/_2$ to ${}^1/_2$ transition was observed and NLO properties were anticipated.[204] Real *et al.* also coordinated stilbazoles to Fe^{II} in Hofmann-like SCO phases, such as in $[Fe(DEAS)_2[Ag(CN)_2]_2]$ that forms a 2D (4,4) framework. This compound showed a 33% HS to LS transition at 141 K, with a small hysteresis and with the NLO properties still to be measured.[158] DFT calculations have been reported for 2nd order NLO contributions in bis-*mer-tridentate* $Fe^{III}(N_3O_3)_2$ complexes.[205]

1.9 Developments in Instrumental Methods in Spin-Crossover Measurements

Magnetic susceptibility, Mössbauer effect spectroscopic, optical and Raman/FT-IR spectroscopic measurements, are the best known methods of following spin-transitions and these have been well described.[21, 67, 88] Photo-excitation LIESST measurements are now known using susceptibilities,[143, 151, 177] Mössbauer spectra[88, 105, 206] and single crystal structures (down to 10 K)[30, 69, 179] as the measurement probes.

We have already seen in this article that synchrotron (S) X-ray sources have enhanced, markedly, fine structural details obtainable at many temperatures (with rapid data collections) both for single crystals (e.g. to probe intermediate phases, IPs),[23, 180] but, importantly, for powders by S-PXRD vs. temperature studies. Not only can the S-PXRD data yield an alternative way of obtaining or confirming spin-transition curves, along crystal axes, or vs. cell volume, they can show the existence of phase transitions, as well as spin-transitions, by measurement of the peak evolution over particular 2θ ranges, when heating and cooling through spin-transition temperatures (Fig. 1.30).[207] Spin-crossover framework systems, including porous ones, have proved particularly amenable to this method. Phase changes in order-disorder SCO Fe^{II} monomers have also been identified this way.[174]

High pressure studies of SCO compounds have developed significantly since the early work by Ewald *et al.*[9] and reviews by Gütlich, Ksenofontov *et al.* summarised progress up to 2005.[208] In essence, application of pressure using in situ diamond anvil cells, with pressures up to 3 GPa, and following the changes in susceptibilities vs. temperature, Raman spectra[209] (e.g. of CN bands in Hofmann phases)[210] or unit cell axes/volumes or peak evolution in S-PXRD shows that SCO favours a spin-transition from HS to LS and a shift in $T_{1/2}$ to higher temperatures at higher pressures. There are a few exceptions to this, when HS forms are

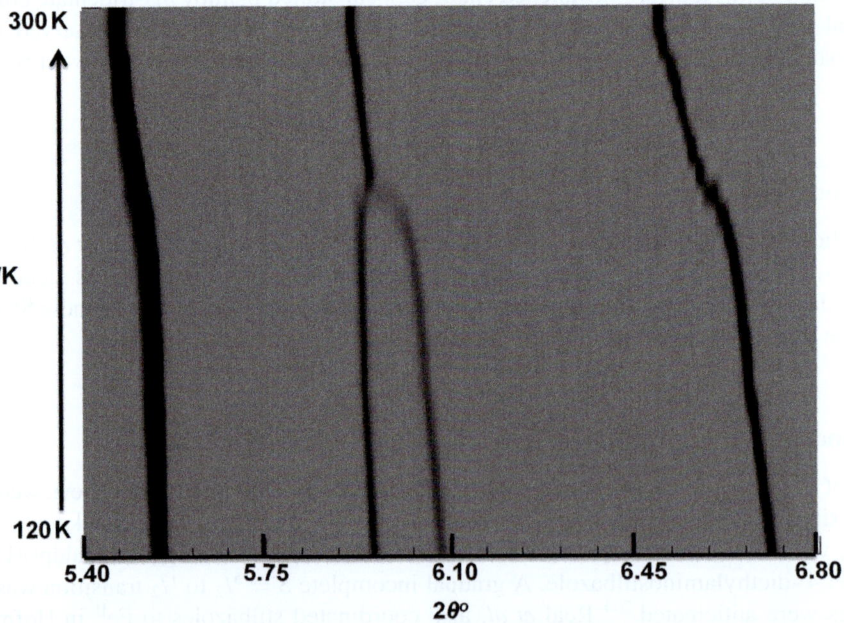

Figure 1.30 *Peak evolution in diffractograms of* cis-[$Fe^{II}(NCSe)_2(DDE)_2$], *where DDE = (N^2,N^2,N^4,N^4-tetraethyl-N^6,N^6-di(pyridin-2-yl)-1,3,5-triazine-2,4,6-triamine), between $2\theta = 5.40$–$6.80°$ over the temperature range 120–300 K, showing the [200], [111] and [002] reflections of the orthorhombic* Pbcn *phase (as seen left to right at 300 K), and the [200], [11$\bar{1}$], [111] and [002] reflections of the monoclinic P2/c phase (as seen left to right at 120 K). Reproduced from [174], with permission of The Royal Society of Chemistry.*

stabilised. Chemical systems studied under varying pressures are wide and varied and examples include (i) many 'classical' mononuclear *cis*-[Fe(phen)$_2$(NCS)$_2$] type compounds,[60, 208] (ii) Hofmann-like frameworks such as [Fe(bpac)$_2$[Ag(CN)$_2$]$_2$] for which the fully LS form could be induced under pressure at room temperature;[210] other 1D (e.g. the *tris*-triazole systems), 2D and 3D materials that have been reviewed;[208] and, a cubic Hofmann phase, [Fe(pz)[Pt(CN)$_4$]]·2H$_2$O, for which a phase change occurred at ~1 GPa during the pressure induced HS to LS spin change[211] and (iii) binuclear Fe^{II} compounds such as [Fe$_2$(μ-bpym)(bpym)$_2$(NCSe)$_4$] for which the increasing pressure had profound effects on the 2-step spin-transition and implications for HS–LS species.[60, 208] Theoretical models for pressure-induced phase transitions and hysteresis loops have been reviewed.[92]

1.10 Applications of Molecular Spin-Crossover Compounds

Létard *et al.* reviewed the application possibilities for SCO materials for the period up to 2004.[36] A very recent review by Bousseksou *et al.*[92] emphasised important aspects required to be understood before commercial applications are possible (my words), such as (i) SCO nanoparticles and thin films of Fe^{II} *tris*-triazole 1D chains and 3D Hofmann compounds [Fe(pz)[Pt(CN)$_4$]],[151, 212] (ii) nanopatterning and nanoscale organisation[213] and (iii) future possibilities of addressable nanostructures for gas sensors, displays and switchable photonic devices. Key contributions to nanoparticle SCO species, synthesis and properties, have been made by Létard *et al.*,[214] Mallah *et al.*[215] and Coronado *et al.*[216] Real *et al.* recently reviewed nanopatterning, processability and thin film growth on gold surfaces of Fe^{II} Hofmann SCO phases (Fig. 1.31).[151] Uniform thin films of an

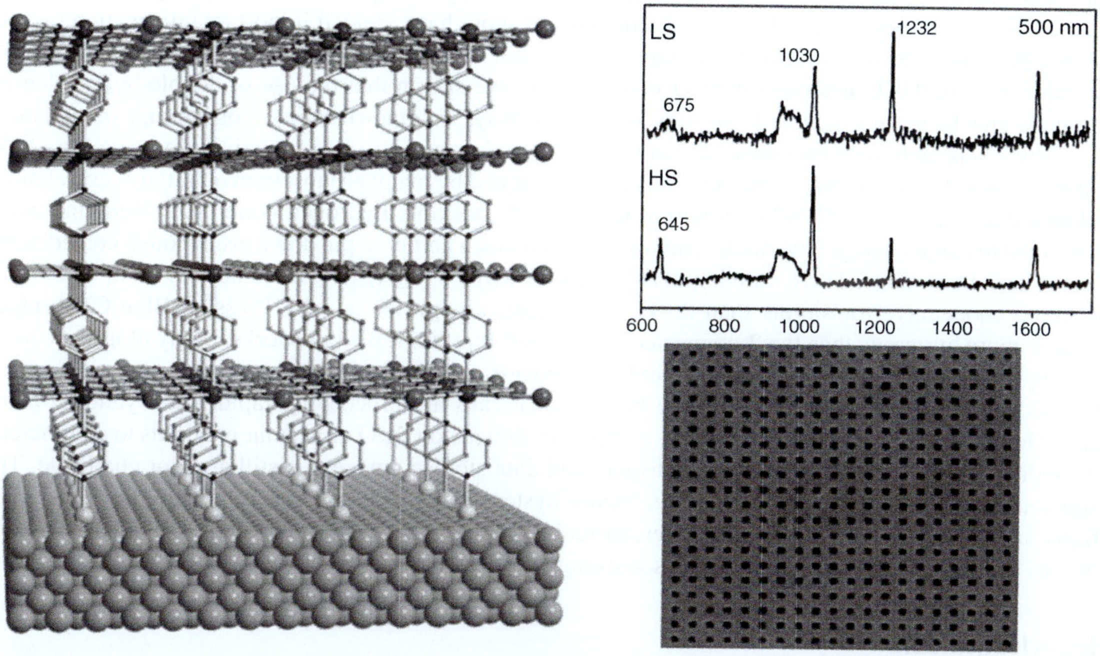

Figure 1.31 *(left) Schematic representation of the ideal epitaxial growth of [Fe(pz)[Pt(CN)$_4$]] on a 4-mercaptopyridine functionalised thin layer of gold, (right, bottom) nanopatterned thin film with 500 nm square motifs of [Fe(pz)[Pt(CN)$_4$]], (right, top) corresponding Raman spectra for the thin film in the LS and HS states. Reprinted from [151] Copyright 2011, with permission from Elsevier.*

Fe^{II} podand complex[217] and of $[Fe(phen)_2(NCS)_2]$ have been grown[218] and the latter 'iconic' compound has also been made into nanopatterns by Cavallini, Ruben *et al.*[219] using lithographic and other techniques. This group has also made nanopatterns of 1D SCO compounds.[220] Electrical field control has been achieved over single molecules of a $[Fe^{II}(py\text{-}1\text{-}bpp)_2]^{2+}$ SCO complex.[221] Cavallini has recently reviewed the status and future of thin films and patterning in SCO compounds.[222]

1.11 Summary

As indicated in this review and in subsequent chapters, the chemistry, structures, physics, theory and nano-aspects of spin-crossover compounds continue to be researched vigorously. From the chemical/design perspective of the early mononuclear $Fe^{II/III}L_3$ or $FeL_2(NCX)_2$ compounds, prepared by stoichiometric reaction of metal salt and ligand(s), we now have challenging and sophisticated designs and syntheses of ligands, both bridging and terminal, and of their discrete metal clusters or 1D, 2D and 3D polynuclear derivatives. The structures of the Hofmann Fe^{II} SCO phases, for example,[151] are aesthetically very pleasing and these materials have many and varied physical and chemical properties/functions to complement their cooperative spin-transitions. Elucidating the crystal structures of SCO compounds, at many temperatures, and sometimes under light irradiation or applied pressure, has benefitted enormously from modern X-ray diffractometers and from the development of synchrotron X-ray beamlines. Thus, intermediate phases (IPs) containing HS/LS combinations have been determined from such studies, often in combination with variable temperature powder XRD data. In turn, the mechanism or constitution of the HS to LS, or HS to HS/LS (partial) spin-transition, sometimes multistep in nature, and of any associated thermal hysteresis loops, can now be elucidated in great detail.

The theory of spin-crossover, including simulation of spin-transition curves and estimations of cooperativity parameters (memory) and thermodynamic enthalpy and entropy changes, has developed greatly since the 1960s. DFT calculations that employ crystal structures as input have yielded important information on spin state structure/energies, molecular orbital diagrams and spin density plots. Ye and Neese[223] have shown the limitations in using DFT methods for SCO systems and have shown that the use of double-hybrid density functionals and large and flexible basis sets is the best way, if somewhat time consuming, of obtaining Fe^{II} energy levels and predicting ground spin states. An understanding of cooperativity in crystalline SCO compounds, whether mono- or polynuclear, has been attempted by many groups, with qualitative correlations developed that relate intermolecular structural facets (π–π stacking, hydrogen bonding, solvent and anion interactions) to cooperativity. Quantitative answers remain much harder to pin down even though considerable advances have been made. There remain puzzles, such as why $Fe^{II}O_2N_2{\cdot}(L)_2$ mononuclear complexes often show more cooperative transitions than do $Fe^{II}(N\text{-donor})_6$ complexes or why Prussian Blue CN-bridged phases are more hysteretic than the 2-connected (e.g. μ-4,4′-bipyridine)Fe^{II} networks. Many of us have ideas about why these differences occur but they are hard to prove.

Nanotechnological developments in these bistable materials are proceeding rapidly and yielding many patents. However, bringing these, and macro- and micro- dimension SCO molecular materials to commercial applicability (memory, sensing, switching, display and data storage devices) is still a major challenge. The disappointment of the 1D room temperature, highly hysteretic $[Fe^{II}((4\text{-}N(R))\text{-}1,2,4\text{-triazole})_3]^{2+}$ materials of Kahn and co-workers not making commercial reality as high-tech displays has not tempered the skills, ambitions and intentions of those researchers working in the applied aspects of spin-crossover.

Acknowledgements

The author wishes to thank, most sincerely, his research students (Brendan Kennedy, Ben Leita, Danny Offerman, Tamsyn Ross, Caspar Schneider, Hayley Scott) and post-doctoral fellows (Boujemaa Moubaraki,

Jonathan Smith, Suzanne Neville, Ian Gass, Victor Martinez) for their great efforts in spin-crossover research. Ongoing collaborations with Professors Stuart Batten, John Cashion, Cameron Kepert and Jean-François Létard, Dr Gopalan Rajaraman and Dr John Boas, are enjoyable and productive. Grants from the Australian Research Council, the France-Australia DEST program and the Australia-India Strategic Research Fund are gratefully acknowledged. We thank the Australian Synchrotron for access to the macromolecular crystallographic beamline and to Dr David Turner (Monash University) for all his help. Nick Chilton's great skills with computers and graphics have helped enormously to assist the author in completing this chapter.

References

1. Cambi, L., Szegö, L. (1931) Über die magnetische Susceptibilität der komplexen Verbindungen. Ber. Dtsch. Chem. Ges. Teil B, 64: 2591–2598.
2. Figgis, B. N., Lewis, J. (1964) Magnetic properties in transition metal complexes. Prog. Inorg. Chem., 6: 37–239.
3. Bailar, J. C. jr. (1956) Chemistry of the Coordination Compounds, ACS Monograph no. 13, Reinhold Publishing, New York, USA. See also articles in: Kaufmann, G. B. (Ed.) (1994) Coordination Chemistry; a Century of Progress. ACS Symp. Ser., vol. 565.
4. Yamasaki, K. (1994) History of coordination chemistry in Japan during the period 1910 to the 1960s, in: Kaufmann, G. B. (Ed.), Coordination Chemistry; A Century of Progress. ACS Symp. Ser., 565: ch. 11, pp. 136–145.
5. Mellor, D. P. (1975) The Development of Coordination Chemistry in Australia. Records of Australian Academy of Science, CSIRO Publishing, 3: 29–40.
6. Livingstone, S. E. (1994) The contributions of David P. Mellor, Frank P. Dwyer, Ronald S. Nyholm to coordination chemistry, in: Kaufmann, G. B. (Ed.), Coordination Chemistry; A Century of Progress. ACS Symp. Ser., 565: ch. 10, pp. 126–135.
7. Dwyer, F. P., Mellor, D. P. (1964) Chelating Agents and Metal Chelates. Academic Press, London, UK, p. 530.
8. Pauling, L., Coryell, C. D. (1936) The magnetic properties and structure of haemoglobin, oxyhemoglobin and carbonmonoxyhemoglobin. Proc. Natl. Acad. Sci. USA, 22: 210–216.
9. Ewald, A. H., Martin, R. L., Ross, I. G., White, A. H. (1964) Anomalous behaviour of the $^6A_1$–$^2T_2$ crossover in iron(III) complexes. Proc. R. Soc. London Ser. A, 280: 235–257.
10. Figgis, B. N., Toogood, G. E. (1972) Magnetic properties of some dithicarbamate compounds of chromium(III), manganese(III) and iron(III) at very low temperatures. J. Chem. Soc. Dalton Trans., 2177–2182.
11. Ivanov, E. V., Zelentsov, V. V., Gerbeleu, N. V., Ablov, A. V. (1970) Anomalous magnetic behavior of iron(III) chelates with thiosemicarbazones of salicylaldehyde and pyroracemic acid. Dokl. Akad. Nauk SSSR, 191: 827–830.
12. Yemeli, E. W. T., Blake, G. R., Douvalis, A. P., Bakas, T., Alberda van Ekenstein, G. O. R., van Koningsbruggen, P. J. (2012) Light-induced bistability in iron(III) spin-transition compounds of 5 X-salicylaldehyde thiosemicarbazone (X = H, Cl, Br). Chem. Eur. J., in press (doi: 10.1002/chem.201002100).
13. Petty, R. H., Dose, E. V., Tweedle, M. F., Wilson, L. J. (1978) Bis(N-methylethylenediaminesalicylaldiminato)iron(III) complexes. Magnetic, Mössbauer and intersystem crossing rate studies in the solid and solution states for a new $(S = {}^1/_2) \rightleftharpoons (S = {}^5/_2)$ spin-equilibrium case. Inorg. Chem., 17: 1064–1071.
14. Oshio, H., Maeda, Y., Takashima, Y. (1983) Rapid electronic relaxation phenomenon in a $^2T_2 \rightleftharpoons {}^6A_1$ spin equilibrium system. Inorg. Chem., 22: 2684–2689.
15. Kennedy, B. J., McGrath, A. C., Murray, K. S., Skelton, B. W., White, A. H. (1987) Variable temperature magnetic, spectral and X-ray crystallographic studies of 'spin-crossover' iron(III) Schiff-base - Lewis-base adducts. Influence of noncoordinated anions on spin-state interconversion dynamics in $[Fe(salen)(imd)_2]Y$ species ($Y = ClO_4^-$, PF_6^-, BPh_4^-; imd = imidazole). Inorg. Chem., 26: 483–495.
16. Haddad, M. S., Federer, W. D., Lynch, M. W., Hendrickson, D. N. (1981) Spin-crossover ferric complexes: unusual effects of grinding and doping solids. Inorg. Chem., 20: 131–139.
17. König, E., Madeja, K. (1967) $^5T_2$–$^1A_1$ equilibriums in some iron(II)-bis(1,10-phenanthroline complexes. Inorg. Chem., 6: 48–55.

18. Matouzenko, G. S., Bousseksou, A., Lecocq, S., van Koningsbruggen, P. J., Perrin, M., Kahn, O., Collet, A.(1997) Spin transition in [Fe(DPEA)(NCS)$_2$], a compound with the new tetradentate ligand (2-aminoethyl)bis(2-pyridylmethyl)amine (DPEA): crystal structure, magnetic properties, and Mössbauer spectroscopy. Inorg. Chem., 36: 2975–2981.
19. Gaspar, A. B., Agustí, G., Martínez, V., Muñoz, M. C., Levchenko, G., Real, J. A. (2005) Spin crossover behaviour in the iron(II)-2,2′-dipyridilamine system. Synthesis, X-ray structure and magnetic studies. Inorg. Chim. Acta, 358: 4089–4094.
20. Franke, P. L., Haasnoot, J. G., Zuur, A. P. (1982) Tetrazoles as ligands. Part IV. Iron(II) complexes of monofunctional tetrazole ligands, showing high-spin ($^5T_{2g}$) $\rightleftharpoons$ low-spin transitions. Inorg. Chim. Acta, 59: 5–9.
21. Gütlich, P. (1981) Spin crossover in iron(II) complexes. Struct. Bonding (Berlin), 44: 83–195.
22. Jakobi, R., Spiering, H., Gütlich, P. (1992) Thermodynamics of the spin transition in [FexZn$_{1-}$x(2-pic)$_3$]Cl$_2$·EtOH. J. Phys. Chem. Solids, 53: 267–275.
23. Chernyshov, D., Hostettler, M., Törnroos, K. W., Bürgi, H.-B. (2003) Ordering phenomena and phase transitions in a spin crossover compound-uncovering the nature of the intermediate phase of [Fe^{II}(2-pic)$_3$]Cl$_2$·EtOH. Angew. Chem. Int. Ed., 42: 3825–3830.
24. Jesson, J. P., Trofimenko, S., Eaton, D. R. (1967) Spectra and structure of some transition metal poly(1-pyrazolyl)borates. J. Am. Chem. Soc., 89: 3148–3158.
25. McGarvey, J. J., Toftlund, H., Al-Obaidi, A. H. R., Taylor, K. P., Bell, S. E. J. (1993) Photoperturbation of the $^1A_1 \rightleftharpoons {}^5T_2$ spin equilibrium in an iron(II) complex in solution via ligand field excitation. Inorg. Chem., 32: 2469–2472.
26. Anderson, P. A., Astley, T., Hitchman, M. A., Keene, F. R., Moubaraki, B., Murray, K. S., Skelton, B. W., Tiekink, E. R. T., Toftlund, H., White, A. H. (2000) Structures and spectra of bis-tripodal iron(II) chelates, [FeL$_2$]$^{2+}$, where L = tris(pyrazol-1-yl)methane, tris(pyridin-2-yl)methane, bis(pyrazol-1-yl)(pyridin-2-yl)methane and tris(pyridin-2-yl)-phosphine oxide. Magnetism and spin crossover in the (pz)$_3$CH case. J. Chem. Soc. Dalton Trans., 3505–3512.
27. Long, G. J., Grandjean, F., Reger, D. L. (2004) Spin crossover in pyrazolylborate and pyrazolylmethane complexes, in: Gütlich, P., Goodwin, H. A. (Eds) Spin Crossover in Transition Metal Compounds I. Top. Curr. Chem., 233: 91–122.
28. Harimanow, L. S., Sugiyarto, K. H., Craig, D. C., Scudder, M. L., Goodwin, H. A. (1999) Magnetic, spectral and structural aspects of spin transitions in iron(II) complexes of 2-(pyrazol-3-yl)pyridine and 3-(thiazol-2-yl)pyrazole. Aust. J. Chem., 52: 109–122.
29. Halcrow, M. A. (2007) The spin-states and spin-transitions of mononuclear iron(II) complexes of nitrogen-donor ligands. Polyhedron, 26: 3523–3576.
30. Money, V. A., Carbonera, C., Elhaïk, J., Halcrow, M. A., Howard, J. A. K., Létard, J.-F. (2007) Interplay between kinetically slow thermal spin crossover and metastable high-spin state relaxation in an iron(II) with similar $T_{1/2}$ and T_{LIESST}. Chem. Eur. J., 13: 5503–5514.
31. Rajadurai, C., Schramm, F., Brink, S., Fuhr, O., Ghafari, M., Kruk, R., Ruben, M. (2006) Spin transition in a chainlike supramolecular iron(II) complex. Inorg. Chem., 45: 10019–10021.
32. Judge, J. S., Baker, W. A. jr. (1967) On the spin equilibrium in bis(2,2′,2″-terpyridine)cobalt(II) salts. Inorg. Chim. Acta, 1: 68–72.
33. Harris, C. M., Lockyer, T. N., Martin, R. L., Patil, H. R. H., Sinn, E., Stewart, I. M. (1969) Five- and six-coordinated complexes of cobalt(II) with 2,2′,2′-terpyridyl: unusual structure and magnetism. Aust. J. Chem., 22: 2105–2116.
34. Real, J. A., Zarembowitch, J., Kahn, O., Solans, X. (1987) Magnetic interaction and spin transition in iron(II) dinuclear compounds. Crystal structure of (μ-2,2′-bipyrimidine)bis[(2,2′-bipyrimidine)bis(thiocyanato)iron(II)]. Inorg. Chem., 26: 2939–2943.
35. Murray, K. S., Kepert, C. J. (2004) Cooperativity in spin-crossover systems. Memory, magnetism and microporosity, in: Gütlich, P., Goodwin, H. A. (Eds) Spin Crossover in Transition Metal Compounds I. Top. Curr. Chem., 233: 195–228.
36. Létard, J.-F., Guionneau, P., Goux-Capes, L. (2004) Towards spin crossover applications, in: Gütlich, P., Goodwin, H. A. (Eds) Spin Crossover in Transition Metal Compounds III. Top. Curr. Chem., 235: 221–249.
37. Zarembowitch, J., Kahn, O. Magnetic properties of some spin-crossover, high-spin and low-spin complexes with Schiff bases derived from 3-formylsalicylic acid. Inorg. Chem., 23: 589–593.

38. Vos, G., De Graaff, R. A. G., Haasnoot, J. G., Van der Kraan, A. M., de Vaal, P., Reedijk, J. (1984) Crystal structure at 300 and 105 K, magnetic properties and Moessbauer spectra of bis(triaquatris(4-ethyltriazole-N1)iron(II)-N^2,$N^{2'}$,$N^{2''}$)iron(II) hexakis(trifluoromethanesulfonate). A linear, trinuclear iron(II) compound, showing a unique high-spin-low-spin transition of the central iron atom. Inorg. Chem., 23: 2905–2910.
39. Haasnoot, J. G. (2000) Mononuclear, oligomeric and polynuclear metal coordination compounds with 1,2,4-triazole derivatives as ligands. Coord. Chem. Rev., 200–202: 131–185.
40. Brooker, S., Plieger, P. S., Moubaraki, B., Murray, K. S. (1999) First cobalt complex to exhibit both exchange coupling and spin crossover effects: the structure and unusual magnetochemistry of $[Co^{II}{}_2L(NCS)_2(SCN)_2]$ where L = macrocyclic ligand. Angew. Chem. Int. Ed., 38: 408–410.
41. Ruben, M., Breuning, E., Lehn, J.-M., Ksenofontov, V., Renz, F., Gütlich, P., Vaughan, G. B. M. (2003) Supramolecular spintronic devices: spin transitions and magnetostructural correlations in $[Fe^{II}{}_4L_4]^{8+}$ [2×2] – grid–type complexes. Chem. Eur. J., 9: 4422–4429.
42. Shatruk, M., Dragulescu-Andrasi, A., Chambers, K. E., Stoian, S. A., Bominaar, E. L., Achim, C., Dunbar, K. R. (2007) Properties of Prussian Blue materials manifested in molecular complexes: observation of cyanide linkage isomerism and spin-crossover behavior in pentanuclear cyanide clusters. J. Am. Chem. Soc., 129: 6104–6116.
43. Duriska, M. B., Neville, S. M., Moubaraki, B., Cashion, J. D., Halder, G. J., Chapman, K. W., Balde, C., Létard, J.-F., Murray, K. S., Kepert, C. J., Batten, S. R. (2009) A nano-scale molecular switch triggered by thermal-, light- and guest-perturbation. Angew. Chem. Int. Ed., 48: 2549–2552.
44. Gamez, P., Costa, J. S., Quesada, M., Aromí, G. (2009) Iron spin-crossover compounds: from fundamental studies to practical applications. Dalton Trans., 7845–7853.
45. Weber, B. (2009) Spin crossover complexes with N_4O_2 coordination sphere–the influence of covalent linkers on cooperative interactions. Coord. Chem. Rev., 253: 2432–2449.
46. Zhang, L., Xu, G.-C., Xu, H.-B., Mereacre, V., Wang, Z.-M., Powell, A. K., Gao, S. (2010) Synthesis, magnetic and photomagnetic study of new iron(II) spin-crossover complexes with N_4O_2 coordination sphere. Dalton Trans., 39: 4856–4868.
47. van Koningsbruggen, P. J., Maeda, Y., Oshio, H. (2004) Iron(III) spin crososover compounds, in: Gütlich, P., Goodwin, H. A. (Eds) Spin Crossover in Transition Metal Compounds I. Top. Curr. Chem., 233, 259–324.
48. Šalitroš, I., Boča, R., Dlháň, L., Gembický, M., Kožíšek, J., Linares, J., Moncol', J., Nemec, I., Perašínová, L., Renz, F., Svoboda, I., Fuess, H. (2009) Unconventional spin crossover in dinuclear and trinuclear iron(III) complexes with cyanido and metallacyanido bridges, Eur. J. Inorg. Chem., 3141–3154.
49. Kennedy, B. J., Fallon, G. D., Gatehouse, B. M., Murray, K. S. (1984) Spin-state differences and spin-crossovers in five-coordinate Lewis-base adducts of cobalt(II) Schiff-base complexes. Structure of high-spin cobalt(II)-saloph-2-methylimidazole adduct, Inorg. Chem., 23: 580–588.
50. Stoufer, R. C., Busch, D. H., Hadley, W. B. (1961) Unusual magnetic properties of some six-coordinate cobalt(II) complexes: electronic isomers. J. Am. Chem. Soc., 83: 3732–3734.
51. Fedaoui, D., Bouhadja, Y., Kaiba, A., Guionneau, P., Létard, J.-F., Rosa, P. (2008) Complexation of 2,6-bis(3-pyrazolyl)pyridine–bis(thiocyanato)iron(II) with a bridging 4,4′-bipyridine: a new example of a dinuclear spin crossover complex. Eur. J. Inorg. Chem., 7: 1022–1026.
52. Reger, D. L., Semeniuc, R. F., Smith, M. D. (2003) Structurally adaptive multitopic ligands containing tris(pyrazolyl)methane units as supramolecular synthons: manganese carbonyl complexes. J. Organomet. Chem., 666: 87–101.
53. Schneider, C. J., Moubaraki, B., Cashion, J. D., Turner, D. R., Leita, B. A., Batten, S. R., Murray, K. S. (2011) Spin crossover in di-, tri- and tetranuclear, mixed-ligand tris(pyrazolyl)methane iron(II) complexes. Dalton Trans., 40: 6939–6951.
54. Thomann, M., Kahn, O., Guilhem, J., Varret, F. (1994) Spin conversion versus antiferromagnetic interaction in iron(II) trinuclear species. Crystal structures and magnetic properties of $[Fe_3(p\text{-MeOptrz})_8(H_2O)_4](BF_4)_6$ and $[Fe_3(p\text{-MeOprtz})_6(H_2O)_6](tos)_6$ [p-MeOptrz = 4-(p-Methoxyphenyl)-1,2,4-triazole, tos = tosylate]. Inorg. Chem., 33: 6029–6037.
55. Kolnaar, J. J. A., de Heer, M. I., Kooijman, H., Spek, A. L., Schmitt, G., Ksenofontov, V., Gütlich, P., Haasnoot, J. G., Reedijk, J. (1999) Synthesis, structure and properties of a mixed mononuclear/dinuclear iron(II) spin-crossover compound with the ligand 4-(p-tolyl)-1,2,4-triazole. Eur. J. Inorg. Chem., 881–886.

56. Sun, Q.-F., Iwasa, J., Ogawa, D., Ishido, Y., Sato, S., Ozeki, T., Sei, Y., Yamaguchi, K., Fujita, M. (2010) Self-assembled $M_{24}L_{48}$ polyhedra and their sharp structural switch upon subtle ligand variation. Science, 328: 1144–1147.
57. MacGillivray, L. R. (2012) Design rules: a net and archimedean polyhedra score big for self-assembly. Angew. Chem. Int. Ed., 51: 1110–1112.
58. Liu, Y., Hu, C., Comotti, A., Ward, M. D. (2011) Supramolecular archimedean cages assembled with 72 hydrogen bonds. Science, 333: 436–440.
59. Stang, P. J. (1998) Molecular architecture: coordination as the motif in the rational design and assembly of discrete supramolecular species—self-assembly of metallacyclic polygons and polyhedra. Chem. Eur. J., 4: 19–27.
60. Real, J. A., Gaspar, A. B., Muñoz, M. C. (2005) Thermal, pressure and light switchable spin-crossover materials. Dalton Trans., 2062–2079.
61. Gaspar, A. B., Ksenofontov, V., Seredyuk, M., Gütlich, P. (2005) Multifunctionality in spin crossover materials. Coord. Chem. Rev., 249: 2661–2676.
62. Gaspar, A. B., Muñoz, M. C. Real, J. A., (2006) Dinuclear iron(II) spin crossover compounds: singular molecular materials for electronics. J. Mater. Chem., 16: 2522–2533.
63. Bousseksou, A., Molnár, G., Real, J. A., Tanaka, K. (2007) Spin crossover and photomagnetism in dinuclear iron(II) compounds. Coord. Chem. Rev., 251: 1822–1833.
64. Murray, K. S. (2008) Advances in polynuclear iron(II), iron(III) and cobalt(II) spin-crossover compounds. Eur. J. Inorg. Chem., 3101–3121.
65. Murray, K. S. (2009) Recent advances in molecular magnetic materials. Aust. J. Chem., 62: 1081–1101.
66. Boča, R., Nemec, I., Šalitroš, I., Pavlik, J., Herchel, R., Renz, F. (2009) Interplay between spin crossover and exchange interaction in iron (III) complexes. Pure Appl. Chem., 81: 1357–1383.
67. Gütlich, P., Goodwin, H. A. (Eds) (2004) Spin Crossover in Transition Metal Compounds I-III. Top. Curr. Chem., vols. 233–235. Springer, Berlin/Heidelberg, Germany.
68. Batten, S. R., Bjernemose, J., Jensen, P., Leita, B. A., Murray, K. S., Moubaraki, B., Smith, J. P., Toftlund, H. (2004) Designing dinuclear iron(II) spin crossover complexes. Structure and magnetism of dinitrile-, dicyanoamido-, tricyanomethanido-, bipyrimidine- and tetrazine-bridged compounds. Dalton Trans., 3370–3375.
69. Trzop, E., Buron-Le Cointe, M., Cailleau, H., Toupet, L., Molnár, G., Bousseksou, A., Gaspar, A. B., Real, J. A., Collet, E. (2007) Structural investigation of the photoinduced spin conversion in the dinuclear compound $\{[Fe(bt)(NCS)_2]_2(bpym)\}$: toward controlled multi-stepped molecular switches. J. Appl. Cryst., 40: 158–164.
70. Nakano, K., Suemura, N., Kawata, S., Fuyuhiro, A., Yagi, T., Nasu, S., Morimoto, S., Kaizaki, S. (2004) Magnetic behaviour and Mössbauer spectra of spin-crossover pyrazolate bridged dinuclear diiron(II) complexes: X-ray structures of high-spin and low-spin $[\{Fe(NCBH_3)(py)\}_2(\mu\text{-bpypz})_2]$. Dalton Trans., 982–988.
71. Archer, R. J., Hawes, C. S., Jameson, G. N. L., McKee, V., Moubaraki, B., Chilton, N. F., Murray, K. S., Schmitt, W., Kruger, P. E. (2011) Partial spin-crossover behaviour in a dinuclear iron(II) triple helicate. Dalton Trans., 40: 12368–12373 and refs therein.
72. Schneider, C. J., Cashion, J. D., Moubaraki, B., Neville, S. M., Batten, S. R., Turner, D. R., Murray, K. S. (2007) The magnetic and structural elucidation of 3,5-bis(2-pyridyl)-1,2,4-triazolate–bridged dinuclear iron(II) spin crossover compounds. Polyhedron, 26: 1764–1772.
73. Amoore, J. J. M., Kepert, C. J., Cashion, J. D., Moubaraki, B., Neville, S. M., Murray, K. S. (2006) Structural and magnetic resolution of a two-step full spin-crossover transition in a dinuclear iron(II) pyridyl-bridged compound. Chem. Eur. J., 12: 8220–8227.
74. Kaiba, A., Shepherd, H. J., Fedaoui, D., Rosa, P., Goeta, A. E., Rebbani, N., Létard, J.-F., Guionneau, P. (2010) Crystallographic elucidation of purely structural, thermal and light induced spin transitions in an iron(II) binuclear complex. Dalton Trans., 39: 2910–2918.
75. Klingele, M. H., Moubaraki, B., Cashion, J. D., Murray, K. S., Brooker, S. (2005) The first X-ray crystal structure determination of a dinuclear complex trapped in the [low spin-high spin] state: $[Fe^{II}_2(PMAT)_2](BF_4)_4{\cdot}DMF$. Chem. Commun., 987–989.
76. Kitchen, J. A., White, N. G., Jameson, G. N. L., Tallon, J. L., Brooker, S. (2011) Effect of counteranion X on the spin crossover properties of a family of diiron(II) triazole complexes $[Fe^{II}_2(PMAT)_2](X)_4$, Inorg. Chem., 50: 4586–4597.

77. Zein, S., Borshch, S. A. (2005) Energetics of binuclear spin transition complexes. J. Am. Chem. Soc., 127: 16197–16201.
78. Verat, A. Yu., Ould-Moussa, N., Jeanneau, E., Le Guennic, B., Bousseksou, A., Borshch, S. A., Marouzenko, G. S. (2009) Ligand strain and the nature of spin crossover in binuclear complexes: the two-step spin crossover in a 4,4′-bipyridine-bridged iron(II) complex [{Fe(dpia)(cis-NCS)$_2$}$_2$(4,4′-bpy)] (dpia = di(2-picolyl)amine; 4,4′-bpy = 4,4′-bipyridine). Chem. Eur. J., 15: 10070–10082.
79. Matouzenko, G. S., Jeanneau, E., Verat, A. Yu., Bousseksou, A. (20110 Spin crossover and polymorphism in a family of 1,2-bis(4-pyridyl)ethene-bridged binuclear iron(II) complexes. A key role of structural distortions. Dalton Trans., 40: 9608–9618.
80. Matouzenko, G. S., Jeanneau, E., Verat, A. Yu., De Gaetano, Y. (2012) The nature of spin crossover and coordination core distortion in a family of binuclear iron(II) complexes with bipyridyl-like bridging ligands. Eur. J. Inorg. Chem., 969–977.
81. Amoore, J. J. M., Neville, S. M., Moubaraki, B., Iremonger, S. S., Murray, K. S., Létard, J.-F., Kepert, C. J. (2010) Thermal- and light-induced spin crossover in a guest-dependent dinuclear iron(II) system. Chem. Eur. J., 16: 1973–1982.
82. Kolnaar, J. J. A., de Heer, M. I., Kooijman, H., Spck, A. L., Schmitt, G., Ksenofontov, V., Gütlich, P., Haasnoot, J. G., Reedijk, J. (1999) Synthesis, structure and properties of a mixed mononuclear/dinuclear iron(II) spin-crossover compound with the ligand 4-(p-tolyl)-1,2,4-triazole. Eur. J. Inorg. Chem., 881–886 and refs therein.
83. Smith, J. P., Leita, B. A., Moubaraki, B., Murray, K. S. unpublished data.
84. Garcia, Y., Robert, F., Naik, A. D., Zhou, G., Tinant, B., Robeyns, K., Michotte, S., Piraux, L. (2011) Spin transition charted in a fluorophore-tagged thermochromic dinuclear iron(II) complex. J. Am. Chem. Soc., 133: 15850–15853.
85. Scott, H. A., Ross, T. A., Moubaraki, B., Murray, K. S., Neville, S. M. (2013) Spin crossover in polymeric materials using schiff-base functionalized triazole ligands. Eur. J. Inorg. Chem., in press.
86. Weber, B., Kaps, E. S., Obel, J., Achterhold, K., Parak, F. G. (2008) Synthesis and characterization of a dinuclear iron(II) spin crossover complex with wide hysteresis. Inorg. Chem., 47: 10779–10787.
87. Min, K. S., DiPasquale, A., Rheingold, A. L., Miller, J. S. (2007) Room-temperature spin crossover observed for [(TPyA)Fe^{II}(DBQ^{2-})Fe^{II}(TPyA)]$^{2+}$ [TPyA = tris(2-pyridylmethyl)amine; DBQ^{2-} = 2,5-di-tert-butyl-3,6-dihydroxy-1,4-benzoquinonate]. Inorg. Chem., 46: 1048–1050.
88. Gütlich, P., Garcia, Y., Spiering, H. (2003) Spin transition phenomena, in: Miller, J. S., Drillon, M. (Eds) Magnetism Molecules to Materials IV. Wiley VCH, Weinheim, Germany, ch. 8, pp. 271–344.
89. Hauser, A., Jeftic, J., Romstedt, H., Hinek, R., Spiering, H. (1999) Cooperative phenomena and light-induced bistability in iron(II) spin-crossover compounds. Coord. Chem. Rev., 190–192: 471–491.
90. Slichter, C. P., Drickamer, H. P. (1972) Pressure-induced electronic changes in compounds of iron. J. Chem. Phys., 56: 2142–2160.
91. Bousseksou, A., Nasser, J., Linares, J., Boukheddaden, K., Varret, F. (1992) Ising-like model for the two-step spin-crossover, J. Phys. I France, 2: 1381–1403.
92. Bousseksou, A., Molnár, G., Salmon, L., Nicolazzi, W. (2011) Molecular spin crossover phenomenon: recent achievements and prospects. Chem. Soc. Rev., 40: 3313–3335.
93. Real, J. A., Bolvin, H., Bousseksou, A., Dworkin, A., Kahn, O., Varret, F., Zarembowitch, J. (1992) Two-step spin crossover in the new dinuclear compound [Fe (bt)(NCS) $_2$]$_2$bpym, with bt=2,2′-bi-2-thiazoline and bpym=2,2′-bipyrimidine: experimental investigation and theoretical approach. J. Am. Chem. Soc., 114: 4650–4658.
94. Schneider, C. J., Cashion, J. D., Chilton, N. F., Etrillard, C., Fuentealba, M., Howard, J. A. K, Létard, J.-F., Milsmann, C., Moubaraki, B., Sparkes, H. A., Batten, S. R., Murray, K. S. (2013) Spin crossover in a 3,5-bis(2-pyridyl)-1,2,4-triazolate-bridged dinuclear iron(II) complex [{Fe(NCBH$_3$)(py)}$_2$(m-L^1)$_2$]: powder vs. single crystal study. Eur. J. Inorg. Chem., in press.
95. Capes, L., Létard, J.-F., Kahn, O. (2000) Photomagnetic properties in a series of spin crossover compounds [Fe(PM-L)$_2$(NCX)$_2$] (X = S, Se) with substituted 2′-pyridylmethylene- 4-amino ligands. Chem. Eur. J., 6: 2246–2255.
96. Boukheddaden, K., Nishino, M., Miyashita, S., Varret, F. (2005) Unified theoretical description of the thermodynamical properties of spin crossover with magnetic interactions. Phys. Rev. B, 72: 014467/1–014467/11.

97. Šalitroš, I., Madhu, N. T., Boča, R., Pavlik, J., Ruben, M. (2009) Room-temperature spin-transition iron compounds. Monatsh. Chem., 140: 695–733.
98. Leita, B. A., Moubaraki, B., Murray, K. S., Smith, J. P., Cashion, J. D. (2004) Structure and magnetism of a new pyrazolate bridged iron(II) spin crossover complex displaying a single HS–HS to LS–LS transition. Chem. Commun., 2: 156–157.
99. Nemec, I., Boča, R., Herchel, R., Trávníček, Z., Gembický, M., Linert, W. (2009) Dinuclear Fe(III) complexes with spin crossover. Monatsh. Chem., 140: 815–828.
100. Nemec, I., Herchel, R., Boča, R., Trávníček, Z., Svoboda, I., Fuess, H., Linert, W. (2011) Tuning of spin crossover behaviour in iron(III) complexes involving pentadentate Schiff bases and pseudohalides. Dalton Trans., 40: 10090–10099.
101. Imatomi, S., Sato, T., Hamamatsu, T., Kitashima, R., Matsumoto, N. (2007) Spin-crossover behavior of isomorphous bi- and mononuclear iron(III) complexes. Bull. Chem. Soc. Japan, 80: 3375–3377.
102. Hayami, S., Inoue, K., Maeda, T. (1999) Structures and magnetic properties of binuclear iron(III) spin-crossover complexes. Mol. Cryst. Liq. Cryst., 335: 573–582.
103. Ewald, A. H., Martin, R. L., Sinn, E., White, A. H. (1969) Electronic equilibrium between the $^6A_1$ and $^2T_2$ states in iron(III) dithio chelates. Inorg. Chem., 8: 1837–1846.
104. Ross, T. M., Neville, S. M., Innes, D. S., Turner, D. R., Moubaraki, B., Murray, K. S. (2010) Spin crossover in iron(III) Schiff-base 1-D chain complexes. Dalton Trans., 39: 149–159.
105. Gütlich, P., Bill, E., Trautwein, A. X. (2011) Mössbauer Spectroscopy and Transition Metal Chemistry. Springer, Heidelberg, Germany, ch. 8, p. 391–464.
106. Garcia, Y., Guionneau, P., Bravic, G., Chasseau, D., Howard, J. A. K., Kahn, O., Ksenofontov, V., Reiman, S., Gütlich, P. (2000) Synthesis, crystal structure, magnetic properties and ^{57}Fe Mössbauer spectroscopy of the new trinuclear [Fe_3(4-(2′-hydroxyethyl)-1,2,4-triazole)$_6$(H_2O)$_6$](CF_3SO_3)$_6$ spin crossover compound. Eur. J. Inorg. Chem., 1531–1538.
107. Wolny, J. A., Rackwitz, S., Achterhold, K., Garcia, Y., Muffler, K., Naik, A. D., Schünemann, V. (2010) Vibrational properties of the trinuclear spin crossover complex [Fe_3(4-(2′-hydroxy-ethyl)-1,2,4-triazole)$_6$(H_2O)$_6$](CF_3SO_3)$_6$: a nuclear inelastic scattering, IR, Raman and DFT study. Phys. Chem. Chem. Phys., 12: 14782–14788.
108. Psomas, G., Bréfuel, N., Dahan, F., Tuchagues, J.-P. (2004) An unprecedented trinuclear structure involving two high-spin and one spin-crossover iron(II) centers. Inorg. Chem., 43: 4590–4594.
109. Nihei, M., Ui, M., Oshio, H. (2009) Cyanide-bridged tri- and tetra-nuclear spin crossover complexes. Polyhedron, 28: 1718–1721.
110. Newton, G. N., Nihei, M., Oshio, H. (2011) Cyanide-bridged molecular squares – the building units of Prussian Blue. Eur. J. Inorg. Chem., 3031–3042.
111. Wei, R.-J., Huo, Q., Tao, J., Huang, R.-B., Zheng, L.-S. (2011) Spin-crossover Fe^{II}_4 squares: two-step complete spin transition and reversible single-crystal-to-single-crystal transformation. Angew. Chem. Int. Ed., 50: 8940–8943.
112. Ruben, M., Ziener, U., Lehn, J.-M., Ksenofontov, V., Gütlich, P., Vaughan, G. B. M. (2005) Hierarchical self-assembly of supramolecular spintronic modules into 1D- and 2D-architectures with emergence of magnetic properties. Chem. Eur. J., 11: 94–100.
113. Gunter, M. J., Berry, K. J., Murray, K. S. (1984) A model for the cyanide form of oxidized cytochrome oxidase: an iron(III)/copper(II) porphyrin complex displaying ferromagnetic coupling. J. Am. Chem. Soc., 106: 4227–4235.
114. Wu, D.-Y., Sato, O., Duan, C.-Y. (2009) A mixed-spin Fe(II) tetranuclear cluster: preparation, structure and magnetic property. Inorg. Chem. Comm., 12: 325–327.
115. Zueva, E. M., Ryabikh, E. R., Kuznetsov, A. M., Borshch, S. A. (2011) Spin crossover in tetranuclear cyanide-bridged iron(II) square complexes: a theoretical study. Inorg. Chem., 50: 1905–1913.
116. Nihei, M., Ui, M., Yokota, M., Han, L., Maeda, A., Kishida, H., Okamoto, H., Oshio, H. (2005) Two-step spin conversion in a cyanide-bridged ferrous square. Angew. Chem. Int. Ed., 44: 6484–6487.
117. Rodríguez-Diéguez, A., Kivekäs, R., Sillanpää, R., Cano, J., Lloret, F., McKee, V., Stoeckli-Evans, H., Colacio, E. (2006) Structural and magnetic diversity in cyano-bridged bi- and trimetallic complexes assembled

from cyanometalates and [M(*rac*-CTH)]$^{n+}$ building blocks (CTH = *d,l*-5,5,7,12,12,14-hexamethyl-1,4,8,11-tetraazacyclotetradecane). Inorg. Chem., 45: 10537–10551.

118. Sato, O., Lyoda, T., Fujishima, A., Hashimoto, K. (1996) Photoinduced magnetization of a cobalt-iron cyanide. Science, 272: 704–705.
119. Shimamoto, N., Ohkoshi, S., Sato, O., Hashimoto, K. (2002) Control of charge-transfer-induced spin transition temperature on cobalt–iron prussian blue analogues. Inorg. Chem., 41: 678–684.
120. Funck, K. E., Prosvirin, A. V., Mathonière, C., Clérac, R., Dunbar, K. R. (2011) Light-induced excited spin state trapping and charge transfer in trigonal bipyramidal cyanide-bridged complexes. Inorg. Chem., 50: 2782–2789 and refs therein.
121. Li, D., Clérac, R., Roubeau, O., Harté, E., Mathonière, C., Le Bris, R., Holmes, S. M. (2008) Magnetic and optical bistability driven by thermally and photoinduced intramolecular electron transfer in a molecular cobalt–iron prussian blue analogue. J. Am. Chem. Soc., 130: 252–258.
122. Mercurol, J., Li, Y., Pardo, E., Rissit, O., Seuleiman, E., Rousselière, H., Lescouëzec, R., Julve, M. (2010) $[Fe^{II}_{LS}Co^{III}_{LS}]_2 \Leftrightarrow [Fe^{III}_{LS}Co^{II}_{HS}]_2$ photoinduced conversion in a cyanide-bridged heterobimetallic molecular square. Chem. Commun., 46: 8995–8997.
123. Sato, O., Cui, A., Matsuda, R., Tao, J., Hayami, S. (2007) Photo-induced valence tautomerism in Co complexes. Acc. Chem. Res., 40: 361–369.
124. Li, F., Clegg, J. K., Goux-Capes, L., Chastanet, G., D'Alessandro, D. M., Létard, J.-F., Kepert, C. J. (2011) A mixed-spin molecular square with a hybrid [2×2] grid/metallocyclic architecture. Angew. Chem. Int. Ed., 50: 2820–2823.
125. Funck, K. E., Hilfiger, M. G., Berlinguette, C. P., Shatruk, M., Wernsdorfer, W., Dunbar, K. R. (2009) Trigonal-bipyramidal metal cyanide complexes: a versatile platform for the systematic assessment of the magnetic properties of Prussian Blue materials. Inorg. Chem., 48: 3438–3452.
126. Klokishner, S., Ostrovsky, S., Palii, A., Shatruk, M., Funck, K., Dunbar, K. R., Tsukerblat, B. (2011) Vibronic model for cooperative spin-crossover in pentanuclear $\{[MI^{III}(CN)_6]_2[M'^{II}(tmphen)_2]_3\}$ (M/M′ = Co/Fe, Fe/Fe) compounds. J. Phys. Chem. C, 115: 21666–21677.
127. Batten, S. R., Hall, G. unpublished work.
128. Grosjean, A., Daro, N., Kauffmann, B., Kaiba, A., Létard, J.-F., Guionneau, P. (2011) The 1-D polymeric structure of the $[Fe(NH_2trz)_3](NO_3)_2{\cdot}nH_2O$ (with n = 2) spin crossover compound proven by single crystal investigations. Chem. Commun., 47: 12382–12384.
129. Lavrenova, L. G., Ikorskii, V. N., Varnek, V. A., Oglezevna, I. M., Lavrionov, S. V. (1986) High-temperature spin transition in iron(II) coordination compounds with triazoles. Koord. Khim., 12: 207–215.
130. Kahn, O., Martinez, C. J. (1998) Spin-transition polymers: from molecular materials toward memory devices. Science, 279: 44–48.
131. Kröber, J., Codjovi, E., Kahn, O., Grolière, F., Jay, C. (1993) A spin transition system with a thermal hysteresis at room temperature. J. Am. Chem. Soc., 115: 9810–9811.
132. Urakawa, A., Beek, W. V., Monrabal-Capilla, M., Galán-Mascarós, J. R., Palin, L., Milanesio, M. (2011) Combined, modulation enhanced X-ray powder diffraction and raman spectroscopic study of structural transitions in the spin crossover material $[Fe(Htrz)_2(trz)](BF_4)$. J. Phys. Chem. C, 115: 1323–1329.
133. Halcrow, M. A. (2011) Structure:function relationships in molecular spin-crossover complexes. Chem. Soc. Rev., 40: 4119–4142.
134. Dîrtu, M. M., Neuhausen, C., Naik, A. D., Rotaru, A., Spinu, L., Garcia, Y. (2010) Insights into the origin of cooperative effects in the spin transition of $[Fe(NH_2trz)_3](NO_3)_2$: the role of supramolecular interactions evidenced in the crystal structure of $[Cu(NH_2trz)_3](NO_3)_2{\cdot}H_2O$. Inorg. Chem., 49: 5723–5736.
135. van Koningsbruggen, P. J., Garcia, Y., Kahn, O., Fournès, L., Kooijman, H., Spek, A. L., Haasnoot, J. G., Moscovici, J., Provost, K., Michalowicz, A., Renz, F., Gütlich, P. (2000) Synthesis, crystal structure, EXAFS, and magnetic properties of catena [μ-tris(1,2-bis(tetrazol-1-yl)propane-*N*1,*N*1′)iron(II)] bis(perchlorate). First crystal structure of an iron(II) spin-crossover chain compound. Inorg. Chem., 39: 1891–1900.
136. Matouzenko, G. S., Perrin, M., Le Guennic, B., Genre, C., Molnár, G., Bousseksou, A., Borshch, S. A. (2007) Spin crossover behavior in a family of iron(II) zigzag chain coordination polymers. Dalton Trans., 9: 934–942.

137. Genre, C., Jeanneau, E., Bousseksou, A., Luneau, D., Borshch, S. A., Matouzenko, G. S. (2008) First dicyanamide-bridged spin-crossover coordination polymer: synthesis, structural, magnetic, and spectroscopic studies. Chem. Eur. J., 14: 697–705.
138. Batten, S. R., Murray, K. S. (2003) Structure and magnetism of coordination polymers containing dicyanamide and tricyanomethanide. Coord. Chem. Rev., 246: 103–130.
139. Neville, S. M., Leita, B. A., Offermann, D. A., Duriska, M. B., Moubaraki, B., Chapman, K. W., Halder, G. J., Murray, K. S. (2007) Spin-crossover studies on a series of 1D chain and dinuclear iron(II) triazine-dipyridylamine compounds. Eur. J. Inorg. Chem., 1073–1085.
140. Ross, T. M., Moubaraki, B., Turner, D. R., Halder, G. J., Chastanet, G., Neville, S. M., Cashion, J. D., Létard, J.-F., Batten, S. R., Murray, K. S. (2011) Spin crossover and solvate effects in 1D Fe^{II} chain compounds containing bis(dipyridylamine)-linked triazine ligands. Eur. J. Inorg. Chem., 1395–1417.
141. Ross, T. M., Moubaraki, B., Batten, S. R., Murray, K. S. (2012) Spin crossover in polymeric and heterometallic Fe^{II} species containing polytopic dipyridylamino-substituted-triazine ligands. Dalton Trans., 41: 2571–2581.
142. Neville, S. M., Leita, B. A., Halder, G. J., Kepert, C. J., Moubaraki, B., Létard, J.-F., Murray, K. S. (2008) Understanding the two-step spin-transition phenomenon in iron(II) 1D chain materials. Chem. Eur. J., 14: 10123–10133.
143. Guionneau, P., Marchivie, M., Bravic, G., Létard, J.-F., Chasseau, D. (2004) Structural aspects of spin crossover. example of the $[Fe(II)L_n(NCS)_2]$ complexes, in: Gütlich, P., Goodwin, H. A. (Eds) Spin Crossover in Transition Metal Compounds II. Top. Curr. Chem., 234: 97–128.
144. Railliet, A. P., Naik, A. D., Rotaru, A., Marchand-Brynaert, J., Garcia, Y. (2012) 1D iron(II) spin crossover complexes with 1,2,4-triazol-4-yl-propanoic acid. Hyperfine Interact., 205: 51–55.
145. Weber, B., Tandon, R., Himsl, D. (2007) Synthesis, magnetic properties and X-ray structure analysis of a 1-D chain iron(II) spin crossover complex with wide hysteresis, Z. Anorg. Allg. Chem., 633: 1159–1162.
146. Purcell, K. F., Edwards, M. P. (1984) Cooperativity in thermally induced intersystem crossing in solids: $Fe(phen)_2(NCR)_2$, R = BH_3, BPh_3, S, Se. Inorg. Chem., 23: 2620–2625.
147. Weber, B., Bauer, W., Pfaffeneder, T., Dîrtu, M. M., Naik, A. D., Rotaru, A., Garcia, Y. (2011) Influence of hydrogen bonding on the hysteresis width in iron(II) spin-crossover complexes. Eur. J. Inorg. Chem., 3193–3206.
148. Nishi, K., Arata, S., Matsumoto, N., Iijima, S., Sunatsuki, Y., Ishida, H., Kojima, M. (2010) One-dimensional spin-crossover iron(II) complexes bridged by intermolecular imidazole−pyridine NH···N hydrogen bonds, $[Fe(HL^{Me})_3]X_2$ (HL^{Me} = (2-methylimidazol-4-yl-methylideneamino-2-ethylpyridine; X = PF_6, ClO_4, BF_4). Inorg. Chem., 49: 1517–1523.
149. Imatomi, S., Kitashima, R., Hamamatsu, T., Okeda, M., Ogawa, Y., Matsumoto, N. (2006) One-dimensional polynuclear spin-crossover iron(III) complex axially bridged by 1,3-bis(4-pyridyl)propane. Chem. Lett., 35: 502–503.
150. Nihei, M., Shiga, T., Maeda, Y., Oshio, H. (2007) Spin crossover iron(III) complexes. Coord. Chem. Rev., 251: 2606–2621.
151. Muñoz, M. C., Real, J. A. (2011) Thermo-, piezo-, photo-and chemo-switchable spin crossover iron (II)-metallocyanate based coordination polymers. Coord. Chem. Rev., 255: 2068–2093.
152. Real, J. A., Andrés, E., Muñoz, M. C., Julve, M., Ganier, T., Bousseksou, A., Varret, F. (1995) Spin crossover in a catenane supramolecular system. Science, 268: 265–267.
153. Vreugdenhil, W., van Diemen, J. H., De Graaff, R. A. G., Haasnoot, J. G., Reedijk, J., van der Kraan, A. M., Kahn, O., Zarembowitch, J. (1990) High-spin to low-spin transition in $[Fe(NCS)_2(4,4'\text{-bis-}1,2,4\text{-triazole})_2](H_2O)$. X-ray crystal structure and magnetic, Mössbauer and EPR properties. Polyhedron, 9: 2971–2979.
154. Halder, G. J., Kepert, C. J., Moubaraki, B., Murray, K. S., Cashion, J. D. (2002) Guest-dependent spin crossover in a nanoporous molecular framework material. Science, 298: 1762–1765.
155. Halder, G. J., Chapman, K. W., Neville, S. M., Moubaraki, B., Murray, K. S., Létard, J.-F., Kepert, C. J. (2008) Elucidating the mechanism of a two-step spin transition in a nanoporous metal-organic framework. J. Am. Chem. Soc., 130: 17552–17562.
156. Neville, S. M., Halder, G. J., Chapman, K. W., Duriska, M. B., Southon, P. D., Cashion, J. D., Létard, J.-F., Moubaraki, B., Murray, K. S., Kepert, C. J. (2008) Single-crystal to single-crystal structural transformation and photomagnetic properties of a porous iron(II) spin-crossover framework. J. Am. Chem. Soc., 130: 2869–2876.

157. Kitazawa, T., Gomi, Y., Takahashi, M., Takeda, M., Enomoto, M., Miyazaki, A., Enoki, T. (1996) Spin-crossover behaviour of the coordination polymer $Fe^{II}(C_5H_5N)_2Ni^{II}(CN)_4$. J. Mater. Chem., 6: 119–121.
158. Agustí, G., Gaspar, A. B., Muñoz, M. C., Lacroix, P. G., Real, J. A. (2009) Spin crossover and paramagnetic behaviour in two-dimensional iron(II) coordination polymers with stilbazole push–pull ligands. Aust. J. Chem., 62: 1155–1165.
159. Seredyuk, M., Gaspar, A. B., Ksenofontov, V., Galyametdinov, Y., Verdaguer, M., Villain, F., Gütlich, P. (2010) Spin-crossover and liquid crystal properties in 2D cyanide-bridged $Fe^{II}-M^{I/II}$ metalorganic frameworks. Inorg. Chem., 49: 10022–10031.
160. Bao, X., Guo, P.-H., Liu, W., Tucek, J., Zhang, W.-X., Leng, J.-D., Chen, X.-M., Gural'skiy, I., Salmon, L., Bousseksou, A., Tong, M.-L. (2012) Remarkably high-temperature spin transition exhibited by new 2D metal–organic frameworks. Chem. Sci., 3: 1629–1633.
161. Kepert, C. J. (2006) Advanced functional properties in nanoporous coordination framework materials. Chem. Commun., 695–700.
162. Real, J. A. (1999) Bistability in iron (II) spin-crossover systems: a supramolecular function, in: Sauvage, J.-P. (ed.), Transition Metals in Supramolecular Chemistry, John Wiley and Sons, Hoboken NJ, USA, ch. 2, pp. 53–91.
163. Sciortino, N., Kepert, C. J. unpublished data.
164. Southon, P. D., Liu, L., Fellows, E. A., Price, D. J., Halder, G. J., Chapman, K. W., Moubaraki, B., Murray, K. S., Létard, J.-F., Kepert, C. J. (2009) Dynamic interplay between spin-crossover and host-guest function in a nanoporous metal-organic framework material. J. Am. Chem. Soc., 131: 10998–11009.
165. Ohba, M., Yoneda, K., Agustí, G., Muñoz, M. C., Gaspar, A. B., Real, J. A., Yamasaki, M., Ando, H., Nakao, Y., Sakaki, S., Kitagawa, S. (2009) Bidirectional chemo-switching of spin state in a microporous framework. Angew. Chem. Int. Ed., 48: 4767–4771.
166. Bartual-Murgui, C., Salmon, L., Akou, A., Ortega-Villar, N. A., Shepherd, H. J., Muñoz, M. C., Molnár, G., Real, J. A., Bousseksou, A. (2012) Synergetic effect of host-guest chemistry and spin crossover in 3D Hofmann-like metal-organic frameworks $[Fe(bpac)M(CN)_4]$ (M=Pt, Pd, Ni). Chem. Eur. J., 18: 507–516.
167. Ohtani, R., Yoneda, K., Furukawa, S., Horike, N., Kitagawa, S., Gaspar, A. B., Muñoz, M. C., Real, J. A., Ohba, M. (2011) Precise control and consecutive modulation of spin transition temperature using chemical migration in porous coordination polymers. J. Am. Chem. Soc., 133: 8600–8605.
168. Halcrow, M. A. (2009) Iron(II) complexes of 2,6-di(pyrazol-1-yl)pyridines–a versatile system for spin-crossover research. Coord. Chem. Rev., 253: 2493–2514.
169. Scudder, M. L., Goodwin, H. A., Dance, I. G. (1999) Crystal supramolecular motifs: two-dimensional grids of terpy embraces in $[ML_2]^z$ complexes (L = terpy or aromatic-N3-tridentate ligand). New J. Chem., 23: 695–705.
170. Olguín, J., Brooker, S. (2011) Spin crossover active iron (II) complexes of selected pyrazole-pyridine/pyrazine ligands. Coord. Chem. Rev., 255: 203–240.
171. Reger, D. A., Little, C. A., Smith, M. D., Rheingold, A. L., Lam, C.-K., Concolino, T. L., Long, G. J., Hermann, R. P., Grandjean, F. (2002) Synthetic, structural, magnetic and Mössbauer spectral study of $\{Fe[HC(3,5\text{-}Me_2pz)_3]_2\}I_2$ and its spin-state crossover behaviour. Eur. J. Inorg. Chem., 1190–1197.
172. Moubaraki, B., Leita, B. A., Batten, S. R., Jensen, P., Murray, K. S., Smith, J. P., Cashion, J. D., Kepert, C. J., Halder, G. J., Létard, J.-F. (2007) Structure, magnetism and photomagnetism of mixed - ligand tris(pyrazolyl)methane iron(II) spin crossover compounds. Dalton Trans., 4413–4426.
173. Gass, I. A., Forsyth, C. M., Moubaraki, B., Schneider, C. J., Murray, K. S. (2011) Supramolecular aspects of iron (II) crown-dipyridyl spin-crossover compounds. Coord. Chem. Rev., 255: 2058–2067.
174. Ross, T. M., Moubaraki, B., Wallwork, K. S., Batten, S. R., Murray, K. S. (2011) A temperature-dependent order-disorder and crystallographic phase transition in a 0D Fe^{II} spin crossover compound and its non-spin crossover Co^{II} isomorph. Dalton Trans., 40: 10147–10155.
175. Ross, T. M., Moubaraki, B., Neville, S. M., Batten, S. R., Murray, K. S. (2012) Polymorphism and spin crossover in mononuclear Fe^{II} species containing new dipyridylamino-substituted s-triazine ligands. Dalton Trans., 41: 1512–1523.
176. Hinek, R., Spiering, H., Schollmeyer, D., Gütlich, P., Hauser, A. (1996) The $[Fe(etz)_6](BF_4)_2$ spin-crossover system–part one: high-spin $\rightleftharpoons$ low-spin transition in two lattice sites. Chem. Eur. J., 2: 1427–1434.

177. Létard, J.-F. (2006) Photomagnetism of iron(II) spin crossover complexes–the T(LIESST) approach. J. Mater. Chem., 16: 2550–2557.
178. Leita, B. A., Neville, S. M., Halder, G. J., Moubaraki, B., Kepert, C. J., Létard, J.-F., Murray, K. S. (2007) Anion-solvent dependence of bistability in a family of meridional N-donor ligand containing iron(II) spin crossover complexes. Inorg. Chem., 46: 8784–8795.
179. Shih, C.-H., Sheu, C.-F., Kato, K., Sugimoto, K., Kim, J., Wang, Y., Takata, M. (2010) The photo-induced commensurate modulated structure in site-selective spin crossover complex trans-$[Fe(abpt)_2(NCS)_2]$. Dalton Trans., 39: 9794–9800.
180. Törnroos, K. W., Hostettler, M., Chernyshov, D., Vangdal, B., Bürgi, H. B. (2006) Interplay of spin conversion and structural phase transformations: re-entrant phase transitions in the 2-propanol solvate of tris(2-picolylamine)iron(II) dichloride. Chem. Eur. J., 12: 6207–6215.
181. Bonnet, S., Siegler, M. A., Costa, J. S., Molnár, G., Bousseksou, A., Spek, A. L., Gamez, P., Reedijk, J. (2008) A two-step spin crossover mononuclear iron(II) complex with a [HS–LS–LS] intermediate phase. Chem. Commun., 43: 5619–5621.
182. Buron-Le Cointe, M., Ould Moussa, N., Trzop, E., Moréac, A., Molnar, G., Toupet, L., Bousseksou, A., Létard, J.-F., Matouzenko, G. S. (2010) Symmetry breaking and light-induced spin-state trapping in a mononuclear Fe^{II} complex with the two-step thermal conversion. Phys. Rev. B, 82: 214106/1–214106/11.
183. Griffin, M., Shakespeare, S., Shepherd, H. J., Harding, C. J., Létard, J.-F., Desplanches, C., Goeta, A. E., Howard, J. A. K., Powell, A. K., Mereacre, V., Garcia, Y., Naik, A. D., Müller-Bunz, H., Morgan, G. G. (2011) A symmetry-breaking spin-state transition in iron(III). Angew. Chem. Int. Ed., 50: 896–900.
184. Krivokapic, I., Zerara, M., Daku, M. L., Vargas, A., Enachescu, C., Ambrus, C., Tregenna-Piggott, P., Amstutz, N., Krausz, E., Hauser, A. (2007) Spin-crossover in cobalt(II) imine complexes. Coord. Chem. Rev., 251: 364–378.
185. Hayami, S., Komatsu, Y., Shimizu, T., Kamihata, H., Lee, Y. H. (2011) Spin-crossover in cobalt (II) compounds containing terpyridine and its derivatives. Coord. Chem. Rev., 255: 1981–1990.
186. Nielsen, P., Toftlund, H., Bond, A. D., Boas, J. F., Pilbrow, J. R., Hanson, G. R., Noble, C., Riley, M. J., Neville, S. M., Moubaraki, B., Murray, K. S. (2009) Systematic study of spin crossover and structure in $[Co(terpyRX)_2](Y)_2$ Systems (terpyRX = 4′-alkoxy-2,2′:6′,2″-terpyridine, X = 4, 8, 12, Y = BF_4^-, ClO_4^-, PF_6^-, BPh_4^-). Inorg. Chem., 48: 7033–7047.
187. Agustí, G., Bartual, C., Martínez, V., Muñoz-Lara, F. J., Gaspar, A. B., Muñoz, M. C., Real, J. A. (2009) Polymorphism and "reverse" spin transition in the spin crossover system $[Co(4\text{-terpyridone})_2](CF_3SO_3)_2 \cdot 1H_2O$. New J. Chem., 33: 1262–1267.
188. Hayami, S., Shigeyoshi, Y., Akita, M., Inoue, K., Kato, K., Osaka, K., Takata, M., Kawajiri, R., Mitani, T., Maeda, Y. (2005) Reverse spin transition triggered by a structural phase transition. Angew. Chem. Int. Ed., 44: 4899–4903.
189. Gaspar, A. B., Ksenofontov, V., Seredyuk, M., Gütlich, P. (2005) Multifunctionality in spin crossover materials. Coord. Chem. Rev., 249: 2661–2676.
190. Faulmann, C., Jacob, K., Dorbes, S., Lampert, S., Malfant, I., Doublet, M.-L., Valade, L., Real, J. A. (2007) Electrical conductivity and spin crossover: a new achievement with a metal bis dithiolene complex. Inorg. Chem., 46: 8548–8559.
191. Takahashi, K., Cui, H.-B., Okano, Y., Kobayashi, H., Einaga, Y., Sato, O. (2006) Electrical conductivity modulation coupled to a high-spin-low-spin conversion in the molecular system $[Fe^{III}(qsal)_2][Ni(dmit)_2]_3 \cdot CH_3CN \cdot H_2O$. Inorg. Chem., 45: 5739–5741.
192. Nihei, M., Takahashi, N., Nishikawa, H., Oshio, H. (2011) Spin-crossover behavior and electrical conduction property in iron(II) complexes with tetrathiafulvalene moieties. Dalton Trans., 40: 2154–2156.
193. Djukic, B., Lemaire, M. T. (2009) Hybrid spin-crossover conductor exhibiting unusual Variable-temperature electrical conductivity. Inorg. Chem., 48: 10489–10491.
194. Clemente-León, M., Coronado, E., Martí-Gastaldo, C., Romero, F. M. (2011) Multifunctionality in hybrid magnetic materials based on bimetallic oxalate complexes, Chem. Soc. Rev., 40: 473–497.
195. Clemente-León, M., Coronado, E., López-Jordà, M., Waerenborgh, J. C. (2011) Multifunctional magnetic materials obtained by insertion of spin-crossover Fe^{III} complexes into chiral 3D bimetallic oxalate-based ferromagnets, Inorg. Chem., 50: 9122–9130.

196. Clemente-León, M., Coronado, E., López-Jordà, M., Espallargas, G. M., Soriano-Portillo, A., Waerenborgh, J. C. (2010) Multifunctional magnetic materials obtained by insertion of a spin-crossover Fe^{III} complex into bimetallic oxalate-based ferromagnets. Chem. Eur. J., 16: 2207–2219.
197. Sieber, R., Decurtins, S., Stoeckli-Evans, H., Wilson, C., Yufit, D., Howard, J. A. K., Capelli, S. C., Hauser, A. (2000) A thermal spin transition in $[Co(bpy)_3][LiCr(ox)_3]$ (ox = $C_2O_4{}^{2-}$; bpy = 2,2′-bipyridine). Chem. Eur. J., 6: 361–368.
198. Coronado, E., Galán Mascarós, J. R., Giménez-López, M. C., Almeida, M., Waerenborgh, J. C. (2007) Spin crossover Fe^{II} complexes as templates for bimetallic oxalate-based 3D magnets. Polyhedron, 26: 1838–1844.
199. Gaspar, A. B., Seredyuk, M., Gütlich, P. (2009) Spin crossover in metallomesogens. Coord. Chem. Rev., 253: 2399–2413.
200. Seredyuk, M., Gaspar, A., Ksenofontov, V., Galyametdinov, Y., Verdaguer, M., Villain, F., Gütlich, P. (47) One-dimensional iron(II) compounds exhibiting spin crossover and liquid crystalline properties in the room temperature region. Inorg. Chem., 47: 10232–10245.
201. Hayami, S., Motokawa, N., Shuto, A., Moriyama, R., Masuhara, N., Inoue, K., Maeda, Y. (2007) Spin-crossover iron(II) compounds with liquid-crystal properties. Polyhedron, 26: 2375–2380.
202. Lee, Y. H., Ohta, A., Yamamoto, Y., Komatsu, Y., Kato, K., Shimizu, T., Shinoda, H., Hayami, S. (2011) Iron(II) spin crossover complexes with branched long alkyl chain. Polyhedron, 30: 3001–3005.
203. Grondin, P., Roubeau, O., Castro, M., Saadaoui, H., Colin, A., Clérac, R. (2010) Multifunctional gels from polymeric spin-crossover metallo-gelators. Langmuir, 26: 5184–5195.
204. Lamère, J.-F., Peyrou, V., Sasaki, I., Lacroix, P. G., Vendier, L., Boillot, M.-L. (2011) Temperature effect on the optical spectra of iron(III) metal complexes exhibiting spin crossover and potential nonlinear optical properties. J. Comp. Methods Sci. Eng., 10: 447–463.
205. Zhao, H.-B., Sun, S.-L., Qiu, Y.-Q., Liu, C.-G., Su, Z.-M. (2010) Theoretical study on second-order nonlinear optical properties of spin crossover Fe(III) phenolate-pyridyl Schiff base complexes, Int. J. Quantum Chem., 110: 1863–1870.
206. Gaspar, A. B., Ksenofontov, V., Reiman, S., Gütlich, P., Thompson, A. L., Goeta, A. E., Muñoz, M. C., Real, J. A. (2006) Mössbauer Investigation of the photoexcited spin states and crystal structure analysis of the spin-crossover dinuclear complex $[\{Fe(bt)(NCS)_2\}_2bpym]$ (bt = 2,2′-bithiazoline, bpym = 2,2′-bipyrimidine. Chem. Eur. J., 12: 9289–9298.
207. Neville, S. M., Halder, G. J., Chapman, K. W., Duriska, M. B., Moubaraki, B., Murray, K. S., Kepert, C. J. (2009) Guest tunable structure and spin crossover properties in a nanoporous coordination framework material. J. Am. Chem. Soc., 131: 12106–12108.
208. Gütlich, P., Ksenofontov, V., Gaspar, A. B. (2005) Pressure effect studies on spin crossover systems. Coord. Chem. Rev., 249: 1811–1829.
209. Shepherd, H. J., Bonnet, S., Guionneau, P., Bedoui, S., Garbarino, G., Nicolazzi, W., Bousseksou, A., Molnár, G. (2011) Pressure-induced two-step spin transition with structural symmetry breaking: X-ray diffraction, magnetic, and Raman studies. Phys. Rev. B, 84: 144107/1–144107/9.
210. Shepherd, H. J., Bartual-Murgui, C., Molnár, G., Real, J. A., Muñoz, M. C., Salmon, L., Bousseksou, A. (2011) Thermal and pressure-induced spin crossover in a novel three-dimensional Hoffman-like clathrate complex. New J. Chem., 35: 1205–1210.
211. Molnár, G., Niel, V., Real, J. A., Dubrovinsky, L., Bousseksou, A., McGarvey, J. J. (2003) Raman spectroscopic dtusy of pressure effects on the spin-crossover coordination polymers $Fe(pyrazine)[M(CN)_4]{\cdot}2H_2O$ (M = Ni, Pd, Pt). First observation of a piezo-hysteresis loop at room temperature. J. Phys. Chem. B, 107: 3149–3155.
212. Cobo, S., Molnár, G., Real, J. A., Bousseksou, A. (2006) Multilayer sequential assembly of thin films that display room-temperature spin crossover with hysteresis. Angew. Chem. Int. Ed., 45: 5786–5789.
213. Bartual-Murgui, C., Akou, A., Salmon, L., Molnár, G., Thibault, C., Real, J. A., Bousseksou, A. (2011) Guest effect on nanopatterned spin-crossover thin films. Small, 7: 3385–3391.
214. Forestier, T., Mornet, S., Daro,N., Nishihara, T., Mouri, S., Tanaka, K., Fouché, O., Freysz, E., Létard, J.-F. (2008) Nanoparticles of iron(II) spin crossover. Chem. Commun., 4327–4329.
215. Volatron, F., Catala, L., Rivière, E., Gloter, A., Stéphan, O., Mallah, T. (2008) Spin-crossover coordination nanoparticles. Inorg. Chem., 47: 6584–6586.

216. Prins, F., Monrabal-Capilla, M., Osorio, E. A., Coronado, E., van der Zant, H. S. J. (2011) Room-temperature electrical addressing of a bistable spin-crossover molecular system. Adv. Mater., 23: 1545–1549.
217. Tissot, A., Bardeau, J.-F., Rivière, E., Brisset, F., Boillot, M.-L. (2010) Thermo- and photoswitchable spin-crossover nanoparticles of an iron(II) complex trapped in transparent silica thin films. Dalton Trans., 39: 7806–7812.
218. Shi, S., Schmerber, G., Arabski, J., Beaufrand, J.-B., Kim, D. J., Boukari, S., Bowen, M., Kemp, N. T., Viart, N., Rogez, G., Beaurepaire, E., Aubriet, H., Petersen, J., Becker, C., Ruch, D. (2009) Study of molecular spin-crossover complex $Fe(phen)_2(NCS)_2$ thin films. Appl. Phys. Lett., 95: 043303/1–043303/3.
219. Cavallini, M., Bergenti, I., Milita, S., Ruani, G., Salitros, I., Qu, Z.-R., Chandrasekar, R., Ruben, M. (2008) Micro- and nanopatterning of spin-transition compounds into logical structures, Angew. Chem. Int. Ed., 47: 8596–8600.
220. Cavallini, M., Bergenti, I., Milita, S., Kengne, J. C., Gentili, D., Ruani, G., Salitros, I., Meded, V., Ruben, M. (2011). Thin deposits, patterning of room-temperature-switchable one-dimensional spin-crossover compounds. Langmuir, 27: 4076–4081.
221. Meded, V., Bagrets, A., Fink, K., Chandrasekar, R., Ruben, M., Evers, F., Bernand-Mantel, A., Seldenthuis, J., Beukman, A., van der Zant, H. (2011) Electrical control over the Fe(II) spin crossover in a single molecule: theory and experiment. Phys. Rev. B, 83: 245415/1–245415/13.
222. Cavallini, M. (2012) Status and perspectives in thin films and patterning of spin crossover compounds. Phys. Chem. Chem. Phys., 14: 11867–11876.
223. Ye, S., Neese, F. (2010) Accurate modeling of spin-state energetics in spin-crossover systems with modern density functional theory. Inorg. Chem., 49: 772–774.

2

Novel Mononuclear Spin-Crossover Complexes

Birgit Weber

Inorganic Chemistry II, Universität Bayreuth, Germany

2.1 Introduction and General Considerations

Spin-crossover (SCO, or spin-transition, ST) complexes are an interesting class of switchable molecules where the switching process is induced by external perturbations as, for example, a change in temperature, pressure or by light irradiation.[1–12] The switching progress is accompanied by a change in the properties of the molecules (magnetism, structural changes, colour), some of which (e.g. colour) can be easily detected with the naked eye. Therefore possible future applications of this substance class as temperature or pressure sensors with an easy optical readout are envisaged.[13, 14] Other potential applications are as intelligent MRT contrast agents.[15]

When speaking about spin-transition one has to keep in mind that there are two principle possibilities of how a change in the spin states of a complex can be achieved – the spin-transition with or without a change of the coordination number (CN) at the metal centre. In most cases the latter possibility is thought of. However, the interest in the spin-transition with a change of the coordination number has increased, as this opens up the possibility of magnetic bistability of molecules in homogeneous solution at room temperature.[15] One classical example for this type of spin-transition is nickel(II) complexes with β–ketoenolates as bidentate ligands with the general composition [NiL_2]. In nonpolar solvents an equilibrium is observed between a monomeric species with a diamagnetic nickel(II) in a square planar coordination sphere and a trimeric species with paramagnetic nickel(II) in an octahedral coordination sphere, as illustrated on the left side of Figure 2.1.[16] The change of the splitting of the *d*-orbitals and the resulting differences in their occupation are responsible for different magnetic properties as well as a colour change from red (monomer) to green (trimer). On which side the equilibrium is situated depends on the steric demand of the substituent R, the concentration of the sample and the temperature of the system. In polar solvents (e.g. methanol, pyridine) a slightly different equilibrium is

Spin-Crossover Materials: Properties and Applications, First Edition. Edited by Malcolm A. Halcrow.

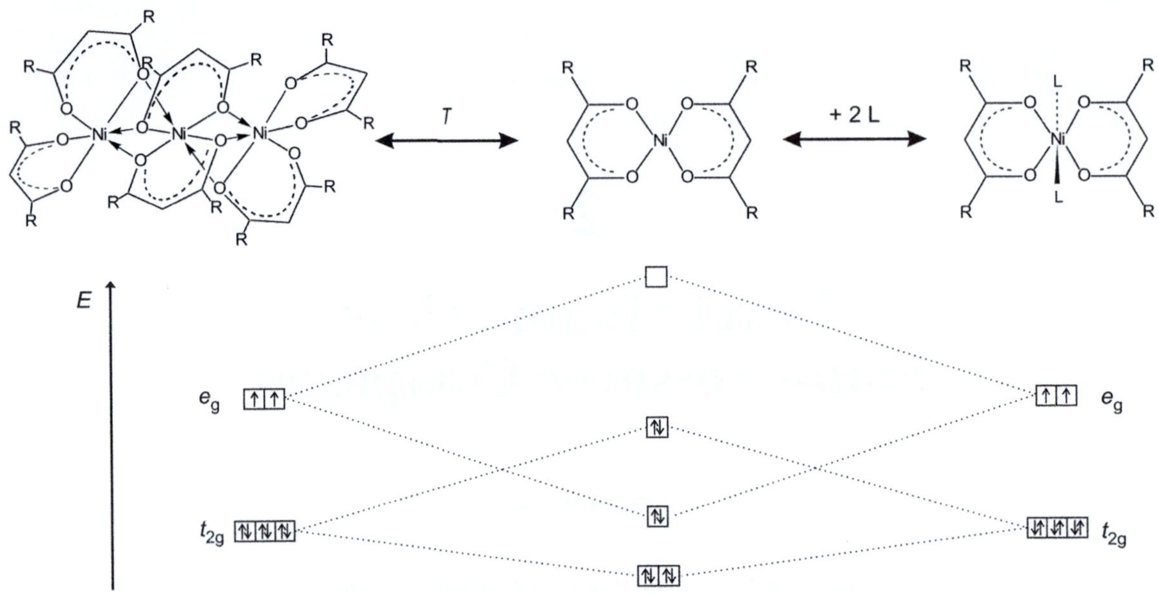

Figure 2.1 Example of an SCO complex with a change in the spin state due to a change in coordination number. In the monomeric species the nickel has the CN 4 with a square planar coordination geometry and is diamagnetic. A paramagnetic species with octahedral coordination sphere can be either obtained through trimerisation (left hand side) or by addition of two axial ligands (e.g. solvent, right hand side). Due to the change in CN and coordination geometry the splitting of the d-orbitals of the metal centre changes, resulting in different magnetic and optical properties.

observed. Here the addition of two solvent molecules in the axial positions of the monomeric complex results in the formation of a paramagnetic octahedral complex (right side of Fig. 2.1). This equilibrium is influenced by the temperature, the concentration of the sample and the solvent used.[16] Further ongoing investigations on the addition of axial ligands on square planar nickel(II) complexes showed that the stability of the octahedral complex correlates well with the basicity of axial ligand.[17]

The ST without a change of the CN is typically observed for octahedral complexes. Although any complex with d^n ($n = 4 - 7$) electron configuration could show this phenomenon, ST in octahedral iron(II) complexes is by far the most thoroughly investigated between the diamagnetic low spin (LS, $S = 0$) state and the paramagnetic high spin (HS, $S = 2$) state. Of those complexes investigated, about 90% exhibit a N_6 coordination sphere.[2] Other examples with a N_4O_2,[18–21] N_4S_2[22] or N_4C_2[23] coordination sphere are relatively rare. In contrast to this observation, there are several examples of iron(III) spin-crossover complexes with a N_4O_2 coordination sphere.[2, 24] Next to iron(II) ($n = 6$) and iron(III) ($n = 5$), cobalt(II) ($n = 7$) is also frequently used for the synthesis of octahedral SCO complexes.[2, 25, 26] For all other $3d^n$ metal centres with a $n = 4$–7 electron configuration where a ST is theoretically possible, there are only very few examples in the literature. In Figure 2.2 the basic principle of a ST under maintenance of the CN is given for a d^6 system like iron(II). The LS to HS transition leads to the occupation of the antibonding e_g orbitals in the HS state that were empty in the LS state. This results in an elongation of the metal–ligand bond lengths and a decrease of the splitting of the d-orbitals, which is responsible for the colour change of the system upon a spin-transition.

There are numerous reviews on mononuclear spin-crossover complexes in the literature. The aim of this chapter is to give an overview of some new developments since 2003, as in 2004 three volumes of *Topics in Current Chemistry* were dedicated to the spin-crossover in transition metal compounds, and presented an

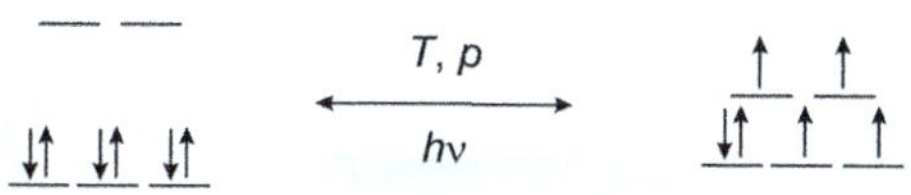

Figure 2.2 *Schematic illustration of SCO under maintenance of the CN, illustrated for a d^6 metal centre, e.g. iron(II). Upon spin-transition the occupation of the antibonding e_g orbitals changes resulting in a change of the metal–ligand bond lengths and the d-orbital splitting. This change is responsible for the colour change of the system upon spin-transition.*

excellent overview of the research up to that time.[2] As the word 'novel' in the title indicates, the focus is on ST complexes with new or unusual CNs and coordination spheres, whereas for the typical N_6-coordinated iron(II) complexes the reader may refer to one of the many review articles covering this class of substances.[5, 7, 27, 28] Such complexes are only considered in this chapter if intriguing properties are observed. Additionally, this overview is limited to mononuclear systems and the reader may refer to one of the following chapters where new developments in the field of polynuclear and polymeric spin-crossover compounds are discussed. Even if these limitations are considered, numerous new and interesting SCO systems have been introduced in the literature over the last nine years. Therefore it is only possible to present some selected examples here.

2.2 Novel Coordination Numbers (CN), Coordination Geometries and Metal Centres

In the *Topics in Current Chemistry* volumes an iron(II) SCO complex with an $N_3O_2C_2$ coordination sphere is mentioned, where the iron centre has a CN of 7.[2, 29] The following discussion starts with more detailed reflections on this example, then groups subsequent compounds by CN which is lowered step by step.

2.2.1 Coordination Number 7

The macrocyclic ligands **1** and **2** are penta-dentate ligands for a variety of metal centres (e.g. Mn(II), Fe(III), Zn(II) or Co(II)). Investigations by König *et al.* in the late 1980s on iron(II) complexes of this ligand type with two cyanide as axial ligands revealed that those complexes are either LS in the whole temperature range investigated as in the case of [Fe**2**$(CN)_2$]·H_2O or show an HS–LS transition as in the case of [Fe**1**$(CN)_2$]·H_2O.[29] The LS complexes were assumed to exhibit an approximately octahedral coordination sphere with one of the secondary amine/ether oxygens (**2** / **1**, respectively) uncoordinated.

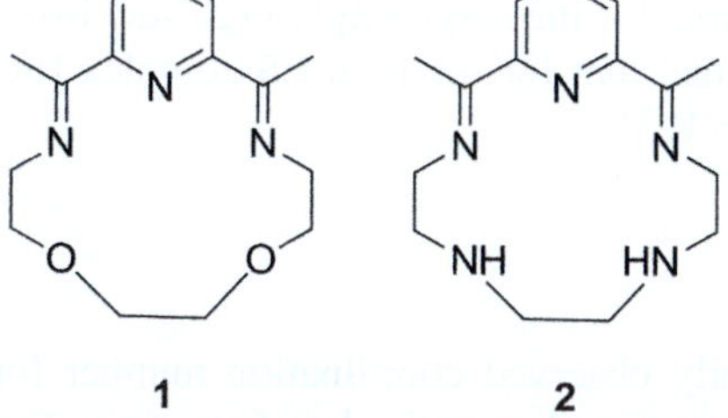

1 **2**

The unexpected structure of the complexes was attributed to a high ligand field stabilisation energy of the system in the octahedral LS state, generated by the strong field cyanide ligands. For the spin-transition complex the HS state was discussed to be either six- or seven-coordinate.[29] Detailed investigations on the

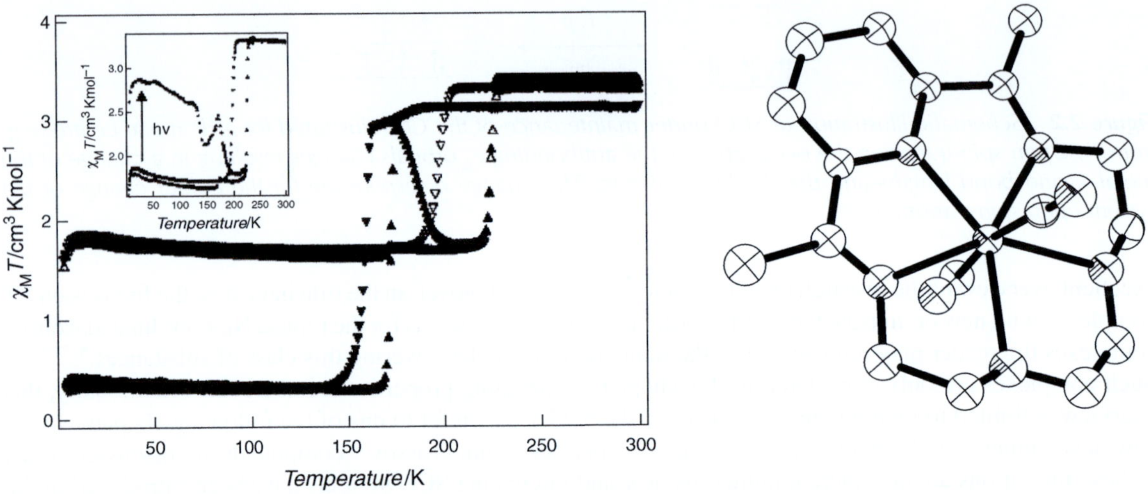

(a) (b)

Figure 2.3 *(a) $\chi_M T$ versus T plots for [Fe**1**(CN)$_2$]·H$_2$O. (▼): Cooling mode of first cycle. (▲): Warming mode of first cycle. (∇): Cooling mode of second cycle. (Δ): Warming mode of second cycle. The temperature was varied at the rate of 1 K min^{-1} without temperature overshoot. The $\chi_M T$ versus T plot in the inset was recorded in the warming mode after the sample was exposed to light illumination for 1 h. (b) Molecule structure of [Fe**1**(CN)$_2$]·H$_2$O in the HS state illustrating the CN 7. Reprinted with permission from [31]. Copyright 2001 American Chemical Society.*

magnetism of a powder sample and single crystals of this complex revealed for the single crystals a complete spin-transition with a 13 K wide thermal hysteresis loop centred around $T_{1/2} = 166$ K. After the first cycle the crystals break into a powder with an incomplete spin-transition that stops at $\gamma_{HS} = 0.5$, centred around 212 K with a 27 K wide hysteresis loop, Figure 2.3. This hysteresis loop can be repeated several times and is also observed for separately prepared powder samples. Upon rapid cooling the HS state can be quenched and upon slow warming a HS → LS transition is observed around 150 K, that is complete in both the powder sample and the single crystals.[30, 31] For the photoinduced HS state a T_{LIESST} temperature of 130 K was determined, the highest value so far obtained for a SCO compound.[31] X-ray structure analysis of the complex was first achieved in the HS state confirming the proposed CN of 7 (Fig. 2.3).[31] Six years later the LS structure was determined with the CN 6, so the reversible Fe–O bond breaking could be confirmed.[32] It appears that the reversible bond cleavage is responsible for the surprisingly high stability of the photoinduced HS state. This assumption is strengthened by the fact that for the pure LS complex [Fe**2**(CN)$_2$]·H$_2$O a photoinduced HS state is observed with $T_{LIESST} = 105$ K.[23]

2.2.2 Coordination Number 6

As this is by far the most frequently observed coordination number for SCO complexes, plenty of new interesting complexes have been introduced over the last few years. One example that meets the criteria discussed in the Introduction is a manganese(III) ($n = 4$) SCO-complex, as spin-transitions in d^4 ions are the most rare.[33] Schiff base ligands with an N_4O_2 donor atom set had already been used for the synthesis of iron(III) spin-crossover complexes[2, 24] and are also suitable for the synthesis of some iron(II) SCO complexes (see Section 3.2). In order to observe a ST for a manganese(III) complex with this donor set, care must be taken that the phenolate donors assume a *trans* configuration. This can be tuned by variation of the alkyl

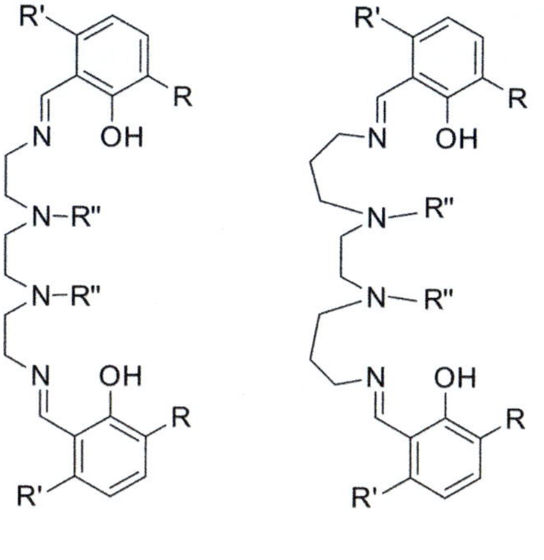

H_2sal_2trien H_2sal_2bapen

H_2sal_2trien

	R	R'	R"
3	OMe	H	H
3'	Oalkyl	H	H
16	H	H	alkyl
17	H	OMe	H
18	H	NO_2	H
19	OMe	NO_2	H
20	OEt	NO_2	H

H_2sal_2bapen

	R	R'	R"
4	OMe	H	H
4'	Oalkyl	H	H

Figure 2.4 *General structures of the ligands H_2sal_2trien and H_2sal_3bapen and their derivatives discussed in this work.*

bridges of the Schiff base ligands, leading to the basic structures of H_2sal_2trien and H_2sal_3bapen, illustrated in Figure 2.4. In the case of [Mn(**3**)]$^+$, a *cis*-configuration of the oxygen atoms is observed and the complex is HS in the entire temperature range. In the case of the H_2sal_3bapen derivative [Mn(**4**)]$^+$, a *trans*-configuration is observed and magnetic susceptibility measurements reveal a gradual spin-transition between 150–300 K.[34]

Single crystal X-ray structure analysis was possible for both spin states of [Mn(**4**)]$^+$, and for the equatorial nitrogen atoms the expected elongation of the bond length is observed when going from the LS to the HS state. In contrast to this no changes are observed for the two axial Mn–O bond lengths. This can be explained if the axial distortion of the octahedral ligand field due to the *trans*-configuration is considered, illustrated in Figure 2.5 for the d^4 HS and LS states. Upon spin-transition one electron changes from the d_{xy} to the $d_{x^2-y^2}$ orbital. Thus the occupation of only one of the antibonding e_g orbitals changes and elongation of the bond lengths is only observed for the equatorial ligands. As only a one electron spin-transition is possible, the changes of the bond length should be smaller than for iron(II) or iron(III) complexes – in agreement with the experimentally observed average value of 0.06 Å. The *trans*-phenolate configuration is a necessary precondition to obtain the axial compression of the octahedral ligand field that enables the spin-transition. This is the starting point for a new series of manganese(III) SCO complexes,[34] as shown by later work from

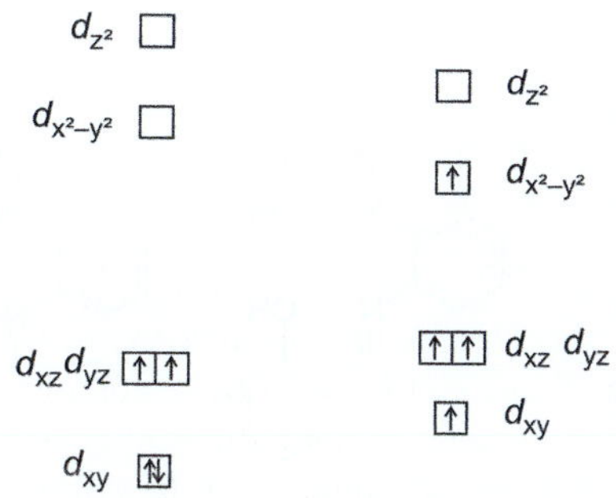

Figure 2.5 *Splitting of the d-orbitals in a compressed octahedral complex for a d^4 system in the LS (left) and the HS state (right).*

the same group. Substitution of the methoxy group by a longer chain alkoxy group (C6, C12 and C18) resulted in the ligand series **3′** and **4′**.[35] Due to the amphiphilic character of the complexes self-assembly was observed to some extent for both, in the solid state (single crystals) and at the water–air interface (Langmuir films). The longer alkyl chains stabilised the LS state, probably due to the inductive character of the alkoxy group.[35]

Another unexpected spin-transition was observed for a cobalt(II) semiquionate complex ($n = 7$).[36] In this example the combination of the tetradentate N_4 macrocycle **5** and semiquinonate leads to a N_4O_2 coordination sphere. Valence tautomerism is well known for cobalt(II) semiquinonate complexes, in which the HS cobalt(II) complex is transformed into a LS cobalt(III) complex with a catecholate ligand upon cooling.[37] For the example with a sterically demanding *tert*-butyl 'R' substituent in **5** and 3,5-di-*tert*-butylsemiquionate as co-ligand, a spin-transition of the cobalt(II) centre is observed instead of valence tautomerism.[36] At low temperatures an $S = 1$ ground state is observed due to a strong ferromagnetic coupling between the LS cobalt(II) ($S = \frac{1}{2}$) and the semiquinonate radical ($S = \frac{1}{2}$). Upon heating, the magnetic measurements reveal a gradual increase of the magnetic moment above 200 K, which can only be attributed to LS to HS spin-transition at the cobalt centre.[36]

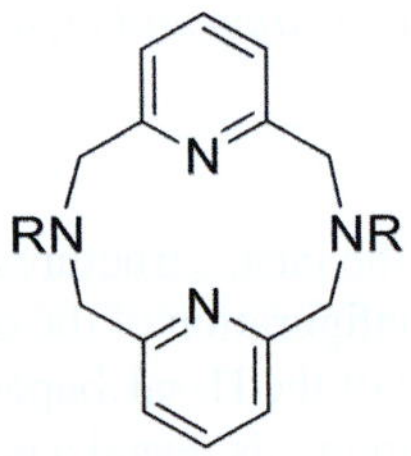

5

2.2.3 Coordination Number 5

Well-investigated examples for five-coordinated SCO complexes with an N_3O_2 coordination sphere are [Fe(salen)NO] and [Fe(salphen)NO], the first showing an abrupt $S = 1/2 \leftrightarrow S = 3/2$ spin-transition with hysteresis while the spin-transition of the latter is gradual.[38] A similar behaviour was observed for iron nitrosyl complexes of Schiff base-like ligands of type **6** with different substituents R^1 and R^2. Here the spin-transition was observed for penta-coordinated complexes with an N_3O_2 coordination sphere or six-coordinated complexes with a rather unusual N_3O_3 coordination sphere due to a methanol as sixth ligand.[39, 40] The spin state of the metal centre can be controlled by small variations of the substituents R^1 and R^2.

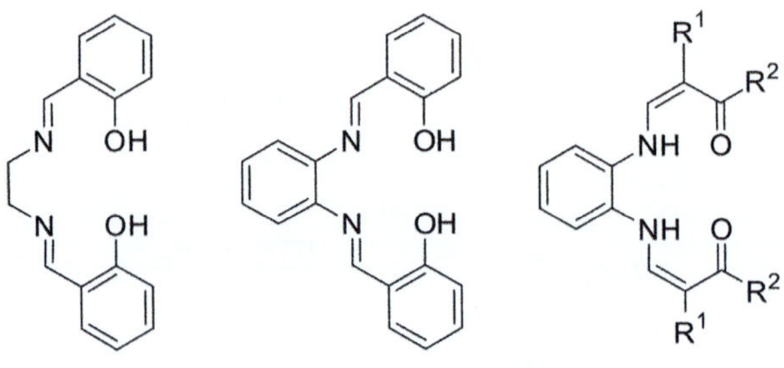

H$_2$salen H$_2$salphen **6**

In 2008 Scheidt's group reported the synthesis of a pentacoordinated iron(II) SCO complex with an N_4C coordination sphere – the first cyanidoiron(II) porphyrinato complex [K(222)][Fe(TPP)(CN)] with TPP = tetraphenylporphyrin.[41]

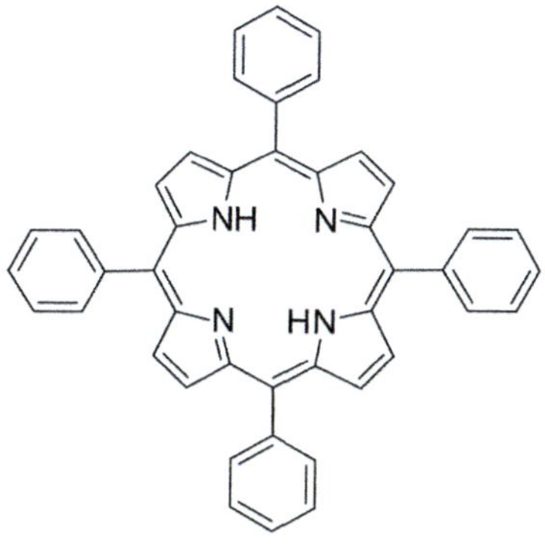

TPP

Magnetic measurements reveal that the complex performs a gradual $S = 2 \leftrightarrow S = 0$ spin-transition in the range 175–400 K. The magnetic moment at 400 K, $\chi_M T = 2.96\ cm^3 K mol^{-1}$, is in the region expected for an iron(II) complex almost fully in the HS state, therefore the spin-transition is not complete at this temperature. The $T_{1/2}$ value of the gradual spin-transition is 265 K. The spin-transition was also followed by Mössbauer spectroscopy. Here only one signal is observed in the entire temperature range with a *T*-dependent change in the quadrupole splitting and isomer shift parameters – the interconversion between the two spin states is rapid on the Mössbauer time scale and only the averaged signals of the two spin states are detected.[41] The same observation was made for most of the penta-coordinated nitrosyl iron complexes.[38, 40] As these complexes are penta-coordinated, the question arises why a $S = 2 \leftrightarrow S = 0$ spin-transition is observed, rather than a $S = 1$ intermediate spin (IS) state or a spin-transition between IS and HS or LS. Most penta-coordinated iron(II) or iron(III) porphyrinato complexes are either in the HS or in the IS state, depending on the strength of the fifth axial ligand. *T*-dependent X-ray structure analysis and DF calculations on [K(222)][Fe(TPP)(CN)] give an answer to this question.[41] As expected, an elongation of the metal–ligand bond lengths is observed upon ST. However, the change is much more pronounced for the axial Fe–C bond ($\Delta = 0.23$ Å) than for the equatorial Fe–N bonds ($\Delta = 0.10$ Å). This can be attributed to the rigid nature of the porphyrinato ligand. An even more pronounced change is observed for the displacement of the iron from the plane of the equatorial ligand, which increases from 0.17 Å in the LS state to 0.45 Å in the HS state.

In Figure 2.6 the *d*-orbital splitting for the three different possibilities in a square pyramidal ligand field is displayed. In the LS state (left) the short metal–ligand distances, especially the very short Fe–CN distance (1.88 Å), lead to a strong splitting between the e_g and the t_{2g} orbitals, while the splitting between the $d_{x^2-y^2}$ and the d_{z^2} as well as between the $d_{xz/yz}$ and the d_{xy} orbital is relatively small, leading to a stabilisation of the LS state. Following the spin-transition, an elongation of all bond lengths is observed that is especially pronounced for the Fe–CN bond. Thus the overall splitting of the *d* orbitals is reduced, but the splitting between $d_{x^2-y^2}$ and d_{z^2} increases due to the weaker influence of the cyanide ligand. For the observation of an IS state, a significantly more pronounced elongation of the Fe–CN bond length would be necessary to further increase the splitting between the $d_{x^2-y^2}$ and the d_{z^2} orbitals. At the same time the Fe–N bond lengths should be in the same order of magnitude as for the LS state. As this is not observed the structural parameters agree well with an $S = 2 \leftrightarrow S = 0$ spin-transition.[41] The analogous carbonyl complex is LS in the entire temperature range investigated, so cyanide appears to be a (slightly) weaker field ligand than carbonyl.

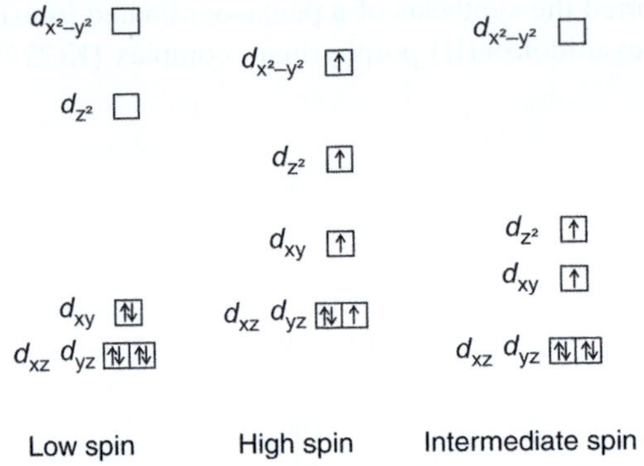

Figure 2.6 *Splitting of the d-orbitals in a square pyramidal complex for a* d^6 *system in the LS state (left), the HS state (middle) and the IS state (right).*

Spin-transitions without a change of the coordination number are not only possible for penta-coordinated iron(II) or iron(III) complexes. For a five coordinated nickel(II) complex ($n = 8$) with an N_3S_2 coordination sphere an $S = 0 \leftrightarrow S = 1$ spin-transition without a change of the coordination number was observed.[42] The trigonally distorted square pyramidal geometry at the nickel centre is obtained by a tridentate hydrotris(3-phenyl-5-methyl-1-pyrazolyl)borate (**7**) with a dithiocarbamate ($Me_2NCS_2^-$) as coligand. The complex crystallises as a green solid with two inequivalent nickel centres in the crystal lattice that are both in the HS state ($S = 1$) at room temperature.

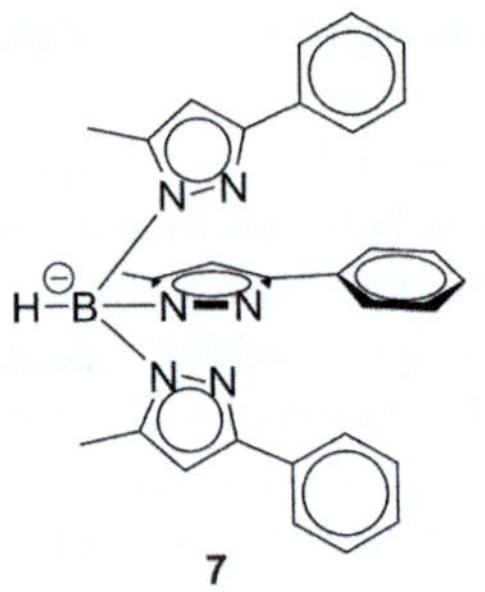

7

Susceptibility measurements and T-dependent X-ray structure analysis reveal that at 123 K Ni1 remains HS whereas Ni2 is now in the LS state. The spin-transition is accompanied by an elongation of the axial Ni–N bond and a contraction of the equatorial Ni–N/S bonds, shifting the coordination environment around the nickel centre more in the direction of a square planar system. The changes in the d-orbital splitting are similar to those displayed in Figure 2.6 for the HS and the IS state of the iron(II) complex. Further cooling below 100 K results in a further drop of the magnetic moment that can be explained by a spin-transition at Ni1.[42]

2.2.4 Coordination Number 4+1

In the Introduction, nickel(II) complexes with β-ketoenolate ligands were introduced as a typical example of a spin-transition due to a configurational equilibrium.[16] The coordination environment of the nickel centre changes from square planar to octahedral upon spin-transition. The above spin-transition in a penta-coordinated nickel(II) complex shows that sometimes much smaller changes are enough to trigger a change in the spin state. Ligand-driven spin state changes are already known for iron(II),[43, 44] where a change in the ligand field strength is responsible for a spin state change at the coordinated metal ion (light-driven ligand-induced spin change, LD-LISC). Another method to obtain spin switching would be a light-driven switch of the metal coordination number. Then it would be possible to switch the spin state of a nickel centre from LS (CN 4, square planar) to HS (CN 5, square pyramidal). The great advantage of such a light-driven switch in coordination number would be the high stability of the different spin states with regard to changes in the environment or temperature variations. Additionally the systems would be stable in solution, giving bistability in homogeneous solution at room temperature. In 2011 the group around Herges succeeded with the synthesis of such a nickel(II) complex.[15] The nickel is in the plane of a porphyrin ligand (**8**) that is functionalised with one azopyridine side chain, where a light induced switching between a *cis*- or *trans*-configuration can be obtained reversibly by irradiation with blue-green (500 nm, *trans* $\rightarrow$ *cis*) or violet-blue light (435 nm, *cis* $\rightarrow$ *trans*). In the *trans*-configuration the nickel has a square planar coordination environment within the porphyrin macrocycle and is in its diamagnetic LS state. When the azopyridine side chain assumes a *cis*-configuration, the pyridine ligand coordinates at the nickel centre leading to a square pyramidal coordination environment and a paramagnetic state for up to 75% of the molecules. The cycle can be repeated several thousand times and an application of this system in magnetic resonance imaging (MRI) is conceivable.[15]

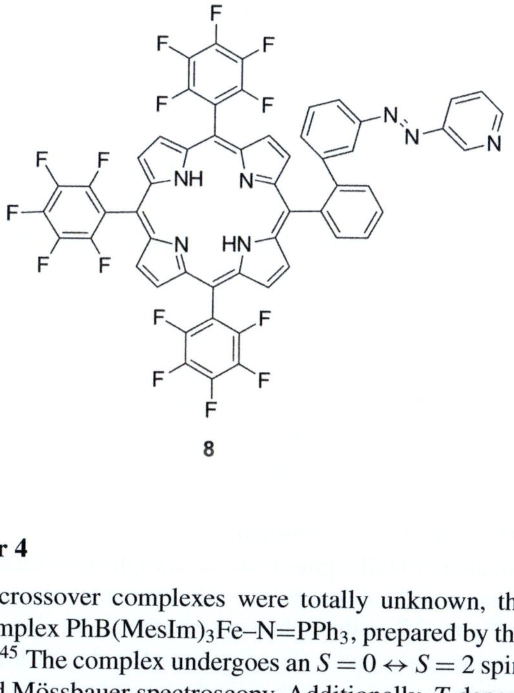

8

2.2.5 Coordination Number 4

Four-coordinate iron(II) spin-crossover complexes were totally unknown, therefore the four-coordinated iron(II) phosphoraniminato complex $PhB(MesIm)_3Fe{-}N{=}PPh_3$, prepared by the groups of Meyer and Smith, deserves special consideration.[45] The complex undergoes an $S = 0 \leftrightarrow S = 2$ spin-transition that was followed by magnetic measurements and Mössbauer spectroscopy. Additionally, *T*-dependent X-ray structure analysis and electronic structure calculations (DF) were performed.[45] On the left side of Figure 2.7 the HS structure and the *T*-dependence of the $\chi_M T$ product are given. The compound undergoes an abrupt spin-transition with $T_{1/2} = 81$ K. The magnetic moment at room temperature is $\chi_M T = 3.63\ cm^3Kmol^{-1}$, in the region expected for an iron(II) HS complex. Below 50 K the average value of $\chi_M T = 0.42\ cm^3Kmol^{-1}$ indicates that 12% of the sample remains in the HS state, which was confirmed by Mössbauer spectroscopy. Upon spin-transition a shortening of the iron–ligand bond lengths is observed that is more pronounced for the Fe–C bonds of the

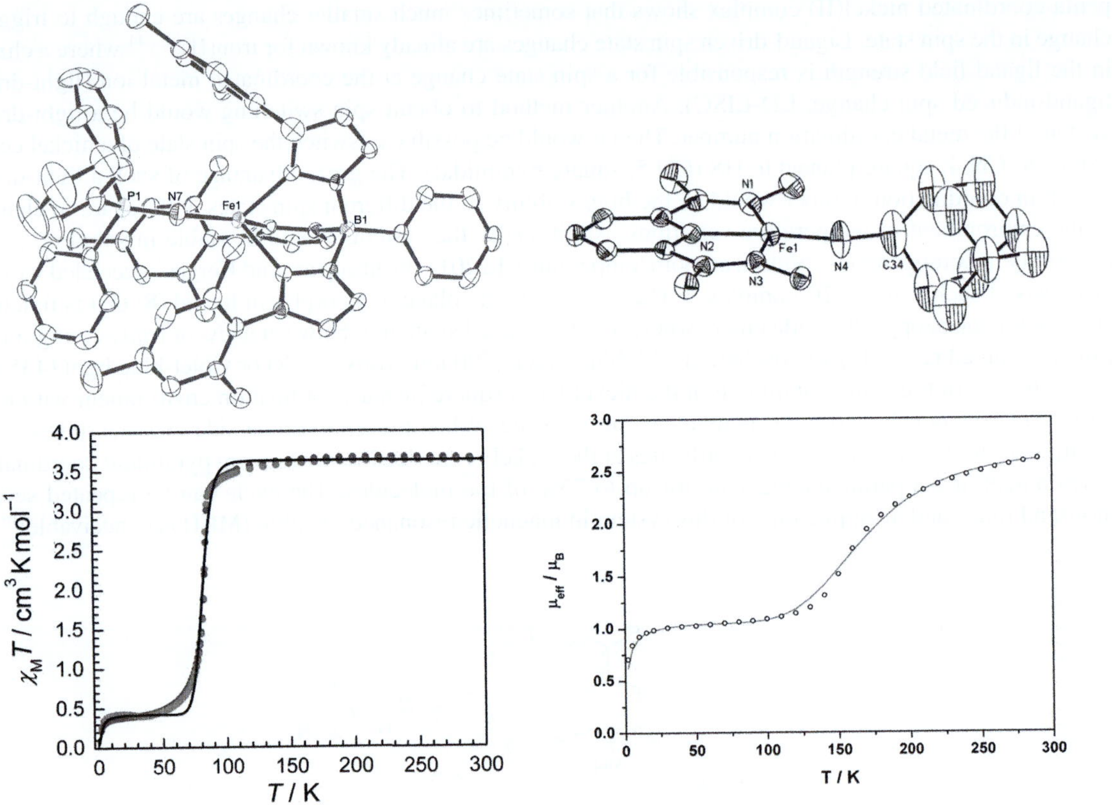

Figure 2.7 *Left side: X-ray structure (150 K) and $\chi_M T$ versus T plots for PhB(MesIm)$_3$Fe–N=PPh$_3$. The solid line corresponds to fits of the data. Right side: X-ray structure and $\chi_M T$ versus T plots for (^{iPr}PDI)FeNAd. Reprinted with permission from [45] and [46]. Copyright 2011 American Chemical Society.*

tris(carbene)borate ligand ($\Delta = 0.14$ Å) and less pronounced for the Fe–N bond to the phosphoraniminato ligand ($\Delta = 0.05$ Å). For this ligand a slight elongation of the N=P bond is observed ($\Delta = 0.03$ Å).[45] In Figure 2.8 the splitting of the *d*-orbitals in the LS and the HS states is displayed. The d_{xz} and d_{yz} orbitals are antibonding with regard to both ligands while the other three orbitals are nonbonding, thus the elongation of the bond lengths upon LS → HS transition is confirmed.

A short time later four-coordinated iron(III) spin-crossover complexes were introduced by Chirik's group.[46] Here the iron has a square planar coordination sphere with a tridentate bis(imino)pyridine and an N-alkyl substituted imide as ligands. In Figure 2.7 on the right hand side the molecular structure of (^{iPr}PDI)FeNAd and the *T*-dependence of the $\chi_M T$ product are given. The $S = 0 \leftrightarrow S = 1$ spin-transition was followed by zero- and applied-field Mössbauer spectroscopy. As both ligands are redox-active (noninnocent) ligands, X-ray absorption spectroscopy and computational studies were necessary to determine the electronic structure. The $S = 0$ state is best described as an LS ferric centre ($S = {}^1/_2$), antiferromagnetically coupled to an imidyl radical ($S = {}^1/_2$). Upon heating, an iron-centred transition to an intermediate spin ferric ion ($S = {}^3/_2$) is responsible for the total $S = 1$ state at higher temperatures.[46]

For a square planar nickel(II)(acen) complex a spin-transition from a diamagnetic LS to a paramagnetic HS state was observed upon encapsulation in a coordination cage compound. The spin-transition was followed

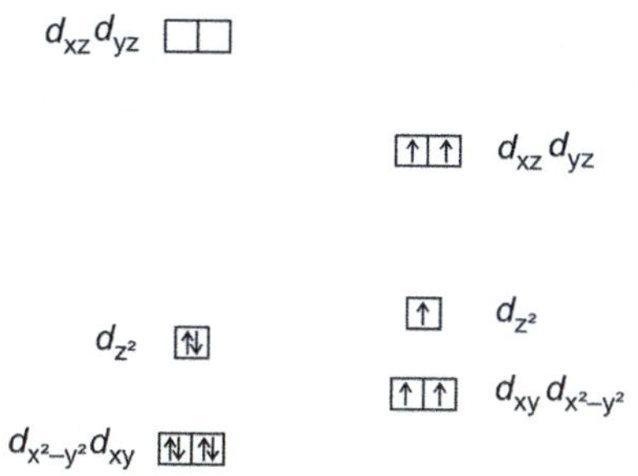

Figure 2.8 *Splitting of the d-orbitals in a tetrahedral complex PhB(MesIm)$_3$Fe–N=PPh$_3$ in the LS state (left), and the HS state (right).*

by magnetic measurements. NMR spectroscopy and results from X-ray structure analysis confirm that the square planar coordination sphere at the nickel centre is maintained and does not change upon encapsulation. A similar spin-transition was observed for a cobalt(II) complex. Interactions between the d_{z^2} orbitals of the metal centre and the π orbitals of the aromatic cage are held responsible for the observed properties.[47]

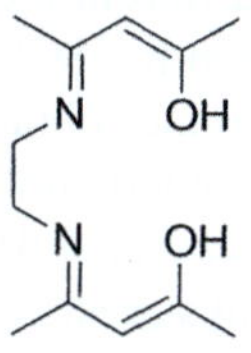

H_2acen

2.3 Iron Complexes with Novel Ligand Donor Atoms and New Ligand Systems

Next to the well-established N_6 coordination sphere for iron(II) SCO complexes,[2] other examples with a N_4O_2,[18–21] N_4S_2[22] or N_4C_2[23] coordination sphere are known, but unusual. For iron(III) spin-crossover complexes a N_4O_2 coordination sphere is observed quite frequently.[2, 24] In Section 2.2 examples of rare manganese(III)[34] and cobalt(II)[36] SCO complexes with N_4O_2 coordination spheres were given, so it should not be surprising that over the last few years several interesting iron(II) SCO complexes with the same donor atom set were reported. In the following, the donor atom set will be used as a sorting key, starting with a few selected examples with an N_6 coordination sphere, followed by a larger section on new systems with an N_4O_2 coordination sphere and some new approaches.

2.3.1 N_6 Coordination Sphere

Tris(pyrazolyl)borate ligands cannot only be used for the synthesis of five-coordinated nickel(II) SCO complexes (ligand **7**),[42] they have proved to be useful for the investigation of steric and electronic effects on the spin-transition properties of the corresponding tris(pyrazolyl)borate iron(II) complexes due to the plentiful possibilities for substituting R^1, R^2 and R' (selected examples in Fig. 2.9). It had already been demonstrated that bulky substituents in R^2-position lead to an increase of the iron–ligand distance and thus a stabilisation of the HS state (e.g. [Fe(**10**)$_2$]) whereas for the un-substituted analogue [Fe(**11**)$_2$] the LS state is stable at room temperature and the spin-transition is observed above 400 K.[48] Hamon *et al.* did show that the introduction of

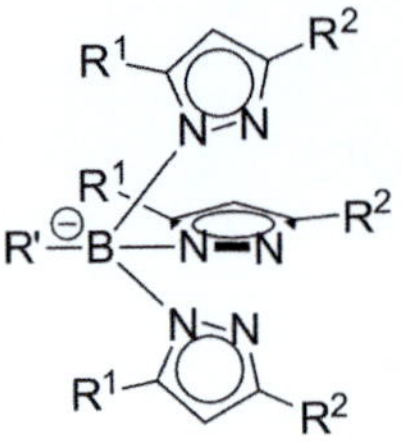

	R^1	R^2	R'
7	Me	Ph	H
9	H	Me	*t*Bu
10	H	Me	H
11	H	H	H
12	H	Me	*p*-$PhC_2C_6H_4$
13	H	Me	*p*-$Me_3SiC_2C_6H_4$
14	H	Me	*p*-$HC_2C_6H_4$

Figure 2.9 *General structure of the tris(pyrazolyl)borate ligands discussed in this work.*

a sterically demanding *t*-butyl group at the boron (R') forces the tripodal ligand **9** to lengthen. This leads to a reduction of the metal–ligand distances and a stabilisation of the smaller iron(II) LS ion. As a consequence this complex is LS at room temperature with a rare SCO above RT centered around 400 K, despite the bulky methyl group at the 3-position.[48] For less bulky substituents in the R' position (**12–14**) an analogous behaviour is observed.[49] However, these substituents allow control of the intermolecular interactions, yielding different types of spin-transition. For [Fe(**12**)$_2$] a complete and cooperative spin-transition with hysteresis (30 K) is observed whereas [Fe(**13**)$_2$] exhibits an incomplete ST involving 50% of the iron centres, although all iron centres are equivalent in the HS state. For [Fe(**14**)$_2$] the ST is stepwise, first gradual, and then abrupt. Results from X-ray structure analysis reveal that different intermolecular interactions caused by the different substituents R' are responsible for the differences in the SCO behaviour.[49]

Next to trispyrazolylborate, trispyrazolylmethanide, obtained by deprotonation of trispyrazolylmethane, was used as ligand for iron(II) to investigate the magnetic properties of the corresponding complexes. Similar bonding properties can be expected as for the tris(pyrazolyl)hydroborates (Fig. 2.9). Indeed, for the iron(II) complex of ligand **15** an abrupt SCO with a small hysteresis was observed with $T_{1/2} = 267$ K.[50] The interesting aspect of this homoleptic complex is the naked carbanion sites that point outwards and are available for further reactions.[50]

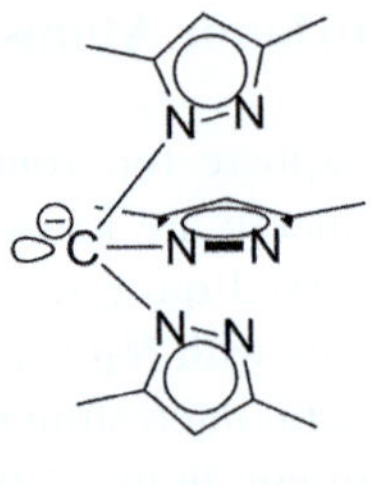

15

2.3.2 N_4O_2 Coordination Sphere

Most ligands providing a N_4O_2 coordination sphere for a spin-transition complex can be divided into two basic systems: Schiff base ligands and the very similar Schiff base-like ligands. The difference between the two ligand systems is the preferred tautomeric structure, as given in Figure 2.10. For Schiff base ligands (e.g. H_2salen, H_2salphen; R^1+R^2 = fused benzene ring) the equilibrium is on the side of the enol-imine form (left hand side). *T*-dependent investigations on various Schiff bases revealed that at lower temperatures the equilibrium can be shifted to the side of the keto-enamine form accompanied by a colour change.[51] The Schiff base-like ligands (e.g. ligand **6**) appear in the keto-enamine form with no indications of any *T*-dependent changes.

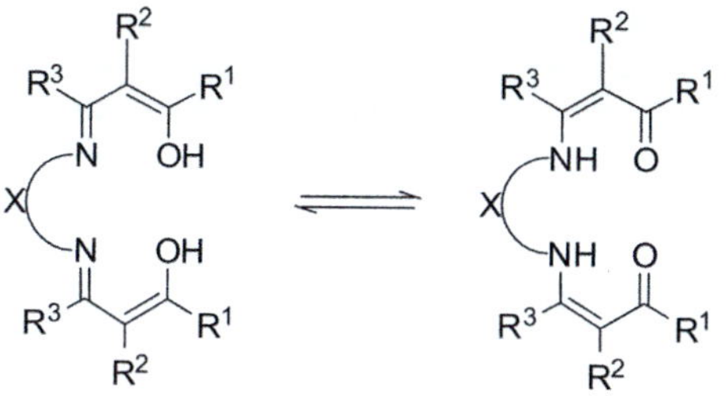

Figure 2.10 *Tautomeric structures of Schiff base and Schiff base-like ligands. For Schiff base ligands as for example the salen ligand (with R^1,R^2 a fused benzene ring, R^3 = H and X = ethylene) the enol-imine form displayed on the left hand side is observed. For Schiff base-like ligands the equilibrium is on the side of the keto-enamine form displayed on the right hand side.*

2.3.2.1 *Schiff Base Ligands*

The ligand H_2sal_2trien and derivatives (see Fig. 2.4) are not only interesting for the synthesis of manganese(III) spin-crossover complexes. The corresponding iron(III) cations $[Fe(sal_2trien)]^+$ and their derivatives are well known to show spin-crossover.[38] Due to their positive charge they can be combined with counterions bearing additional properties such as magnetism or conductivity. This was achieved, for example, by intercalation of the complex of ligand **17** (Fig. 2.4) into a layered $MnPS_3$ host lattice that is semiconducting and can acquire a spontaneous magnetisation.[52] The water-free composite material shows an abrupt spin-transition with a wide hysteresis loop (33 K) that was not observed for the pure iron(III) complex. Below 35 K a spontaneous magnetisation (ferrimagnetism) is observed.[52] Bimetallic oxalate-based ferromagnets of the type $[Mn^{II}Cr^{III}ox_3]^-$ can be used as a counterion of the $[Fe(sal_2trien)]^+$ cation leading to multifunctional magnetic composite materials.[53] Depending on the reaction conditions (solvent) either layered structures are observed or the complex cations are intercalated into a 3D anion network. In both cases the coexistence of long-range ferromagnetic ordering and SCO is observed.[53] The combination with potentially conducting redox active counterions as, for example, $[Ni(dmit)_2]^-$ (with dmit = 2-thioxo-1,3-dithiole-4,5-dithiolato) can lead to switchable molecular conductors. Here the synthesis of a composite material with a layered structure and a cooperative spin-transition with hysteresis (30 K) succeeded.[54]

Another possibility for a further functionalisation of those complexes is the addition of long alkyl chains as already reported for the corresponding manganese(III) complexes. Here different possibilities are available as the ligand can be functionalised at the aromatic ring of the salicylaldehyde (R, see Fig. 2.4) or at the amine nitrogen (R''). In an initial study the complexes $[Fe(\mathbf{3'})]^+$ with alkyl chain lengths of C_6, C_8 and C_{16} were investigated in solution. Improved cooperativity during the spin-transition was observed for the complex with the longest chains, which was explained by self-assembly of the molecules in solution due to their amphiphilic character.[55] Modification of the R'' position of the tetrafluoroborate salt of $[Fe(\mathbf{16})]^+$ with C_6 or C_{12} alkyl chains also leads to amphiphilic iron(III) complexes. While the unsubstituted complex is a pure HS complex, with increasing alkyl chain length improved SCO properties are observed (C_0 pure HS complex, C_6 incomplete SCO and C_{12} complete gradual SCO).[56]

The H_2sal_2trien ligands are not only suitable for the synthesis of iron(III) and manganese(III) SCO complexes; by introducing a nitro group in R' position (ligands **18–20** in Fig. 2.4) it is possible to obtain iron(II) SCO complexes. Small variations in the substitution pattern of R significantly influence the crystal packing and thus the spin-transition properties. For the unsubstituted complex a 2-step spin-transition with a 35 K wide plateau is observed.[19] Introduction of a methoxy group leads to a cooperative 1-step spin-transition with a 4 K wide thermal hysteresis loop. In the case of the ethoxy group the spin-transition is incomplete with only 5% of the iron centres involved.[21]

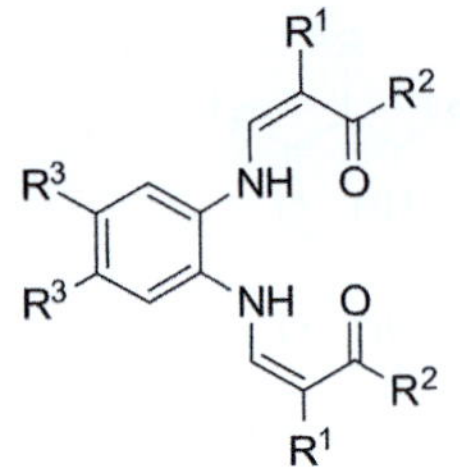

	R1	R2	R3
6a	COOEt	Me	H
6b	COOMe	Me	H
6c	COMe	Me	H
6d	COPh	Me	H
6e	COOEt	Ph	H
21a	COOEt	Me	OH
21c	COMe	Me	OH
22a	COOEt	Me	$OC_{16}H_{33}$

Figure 2.11 *General structure of the N_2O_2 coordinating Schiff base-like ligands discussed in this chapter.*

2.3.2.2 Schiff Base-Like Ligands

Schiff base-like ligands with an N_2O_2 donor set and a carbonyl group as substituent in R^1 proved to be excellent candidates for the synthesis of iron(II) SCO complexes when combined with pyridine or imidazole derivates as axial ligands.[6, 40] By varying both the substituents at the equatorial ligand (R^1, R^2 and R^3), as well as the axial ligands, a wide variety of new iron(II) SCO complexes have been synthesised and characterised over the last few years. The system is not only restricted to the synthesis of mononuclear complexes; if the monodentate axial ligand is replaced by a bridging *bis*-monodentate ligand like 4,4′-bipyridine, 1D chain iron(II) SCO complexes are obtained[57, 58] while the equatorial ligand can be modified in such a way that dinuclear complexes are received.[59, 60] Thus a highly versatile ligand system for the systematic investigation of the structure:property relationships in SCO complexes is available. Here only some of the most important results for the mononuclear complexes are presented. In Figure 2.11 the general structure of the Schiff base-like ligands discussed in the following paragraphs is displayed.

The series of new mononuclear SCO complexes of ligand **6** provided plentiful data for a systematic investigation into how small changes in the ligand sphere influence the spin-transition properties. Particular attention was given to the occurrence of wide hysteresis loops. This phenomenon is related to intermolecular interactions between the spin-transition centres that allow the ‘communication’ of the volume change associated with the spin-transition from one molecule to the next. It was demonstrated previously that an increased number of intermolecular contacts leads to a wider hysteresis loop,[61] a concept that can also be applied to the complexes of **6**.[60, 62] However, a few of these new complexes do not follow that trend, the observed hysteresis loop width being much larger than would be expected.[63] The most prominent example is a mononuclear iron(II) SCO complex with a hysteresis width of 70 K centred around room temperature, [Fe(**6a**)$(Him)_2$].[64] Results from X-ray structure analysis show a 2D network of hydrogen bonds between the spin-transition centres. In Figure 2.12 the results from the magnetic measurements and a part of the H-bond network is displayed on the left hand side.

Interestingly, a second modification of the complex can be synthesised, in which a small (5 K) hysteresis loop above room temperature is observed. Which modification precipitates depends on the reaction conditions. If the concentration of the reaction mixture is high, the complex precipitates under reflux conditions in the HS state and the wide hysteresis loop is observed (HT modification). On the other hand, if lower concentrations are used, the complex precipitates at room temperature or below in the LS state with the small hysteresis loop.[65] A network of hydrogen bonds between the complex molecules is observed for both modifications, and the number and strength of intermolecular contacts is similar for the HS and the LS crystals. Hence this concept cannot be used to explain the significant differences in the SCO properties. Isotopic labelling experiments clearly demonstrate that the hydrogen bond network significantly influences the hysteresis width.[65] A comparison of SCO complexes of different ligand systems with hydrogen bonds in the crystal packing reveals that wide hysteresis loops are observed whenever the H-bond network involves donor atoms at the iron centre, as for the complex [Fe(**6a**)$(Him)_2$].

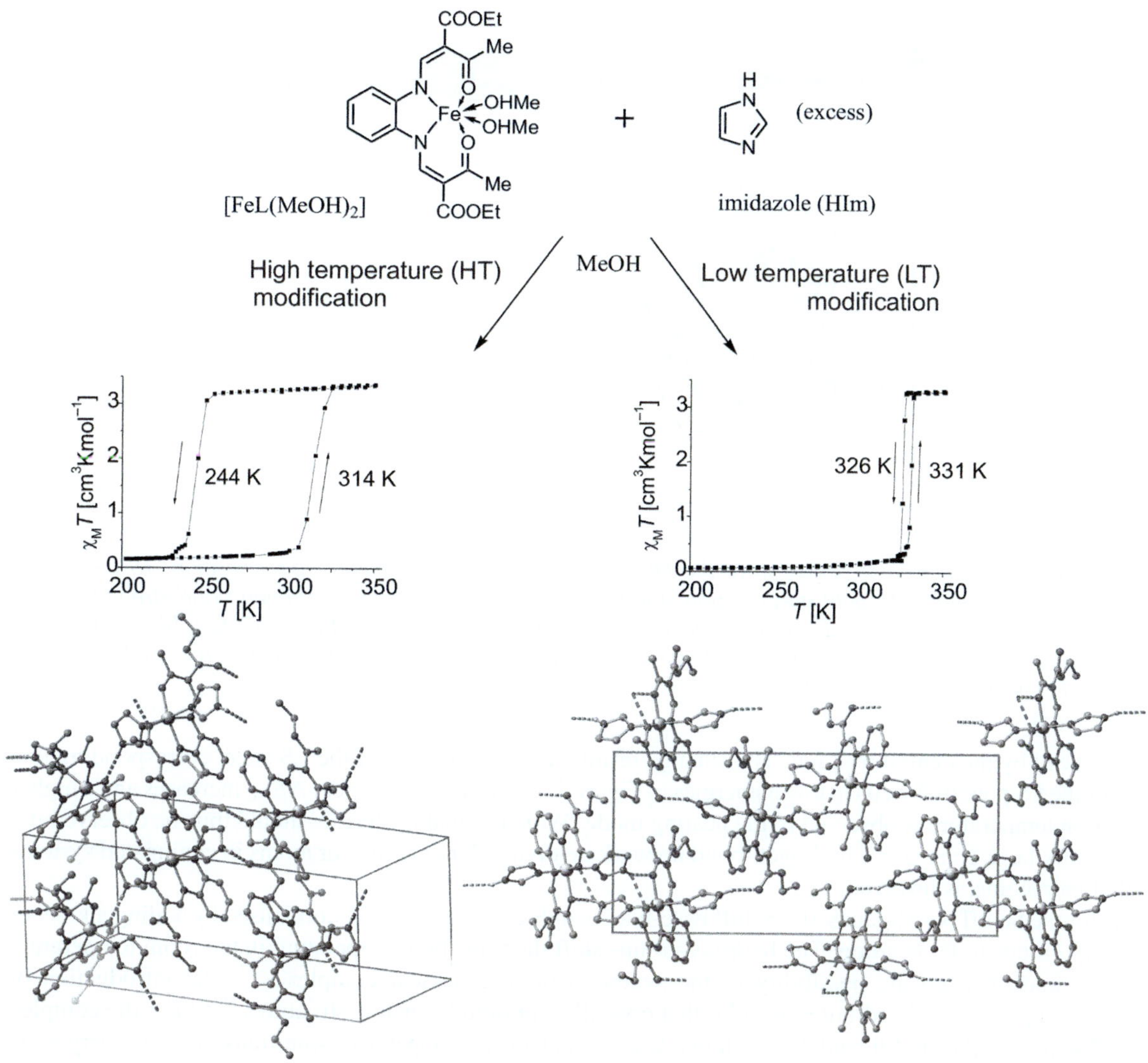

Figure 2.12 Overview over the two modifications of [Fe(**6a**)(Him)$_2$] with a $\chi_M T$ vs. T plot and an excerpt of the hydrogen-bond network of both. The high temperature modification (HT) displayed on the left hand side crystallises above 50 °C in the HS state. The low temperature modification (LT) crystallises at or below RT in the LS state. Adpated with permission from [65]. Copyright Wiley-VCH Verlag GmbH & Co, 2011.

Based on these data the following model was suggested to explain the experimental results and the influence of hydrogen bonds on the hysteresis loop width. A ST is associated with changes in the bond lengths between the iron centre and the donor atoms. In the solid state, with several intermolecular contacts, these changes can influence the length and thereby also the strength of a hydrogen bond. In the case of [Fe(**6a**)(Him)$_2$] HT, the bond length changes upon ST influence the hydrogen bond between the NH of the imidazole ligand and the carbonyl oxygen of the equatorial ligand. This influences the electron density in the oxygen that is also coordinated to the iron centre. A slight increase in electron density results in an enhanced crystal field

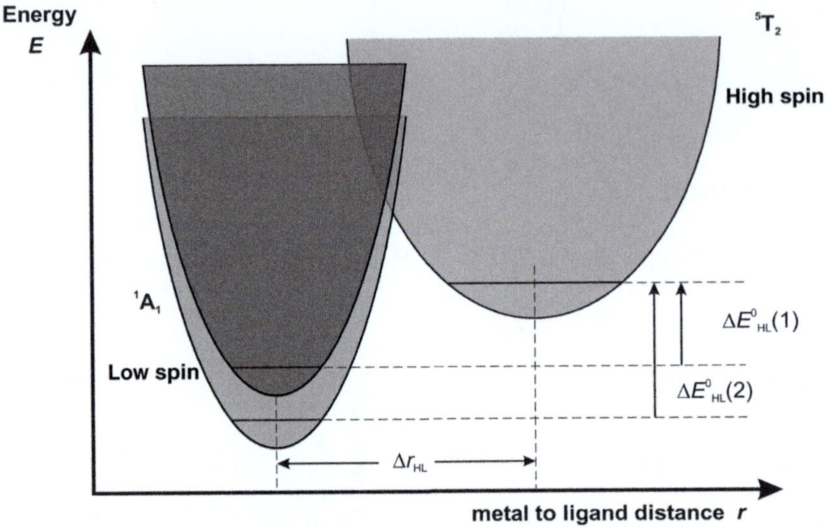

Figure 2.13 Schematic representation of the HS and LS potential wells before spin-transition (LS dark grey) and after spin-transition (LS light grey). Upon spin-transition, a change in the hydrogen bond strength results in a different ligand field strength for the LS state and therefore a higher ΔE_{HL} that leads to a higher transition temperature in the heating mode compared to the cooling mode. Reprinted with permission from [65]. Copyright Wiley-VCH Verlag GmbH & Co, 2011.

experienced by the central iron ion. This change results in a stabilisation of the LS state, corresponding to a shift of the LS potential wells to lower energies in Figure 2.13. As a consequence, ΔE_{HL} increases and a higher transition temperature is observed in the heating mode. Wide thermal hysteresis should thus be observed if a sufficiently large change in a hydrogen-bond network involving donor atoms, or atoms in conjugated systems with donor atoms, is involved.

As for the Schiff base ligands, the Schiff base-like ligands can be further functionalised by adding alkoxy or hydroxy groups in R^3 position. The hydroxy groups shift the transition temperature in the room temperature region[66] and open up the possibility of the synthesis of oligonuclear complexes.[67] The introduction of alkoxy groups leads to head-tail molecules that crystallise in lipid layer-like structures.[68, 69] For the complex [Fe(**22a**)(py)$_2$]·$^1/_4$H$_2$O of ligand **22a** with two C_{16} alkyl chains, a cooperative spin-transition with very wide thermal hysteresis (up to 50 K, Fig. 2.14) is observed. A hydrogen bond network between the iron centres is responsible for the strong cooperative interactions.[68]

2.4 Other Examples

The question of valence tautomerism vs. spin-transition is not only interesting for cobalt(II) semiquinonate complexes, as discussed in Section 2.2.[36] The iron(III) complex [Fe(TPA)(Cat)]BPh$_4$ (TPA = *tris*(2-pyridylmethyl)amine and Cat = catecholate dianion) was reported to undergo a reversible 2-step $S = {}^1/_2$ to $S = {}^5/_2$ SCO.[70] Ongoing investigations on this system revealed that two inequivalent iron sites are responsible for the 2-step nature of the spin-transition. In contrast to the initial results indicating an iron(III) catecholate complex for both spin states, DF calculations show that the electronic structure consists of a mixture of the iron(III) catecholate and the iron(II) semiquinonate form.[71]

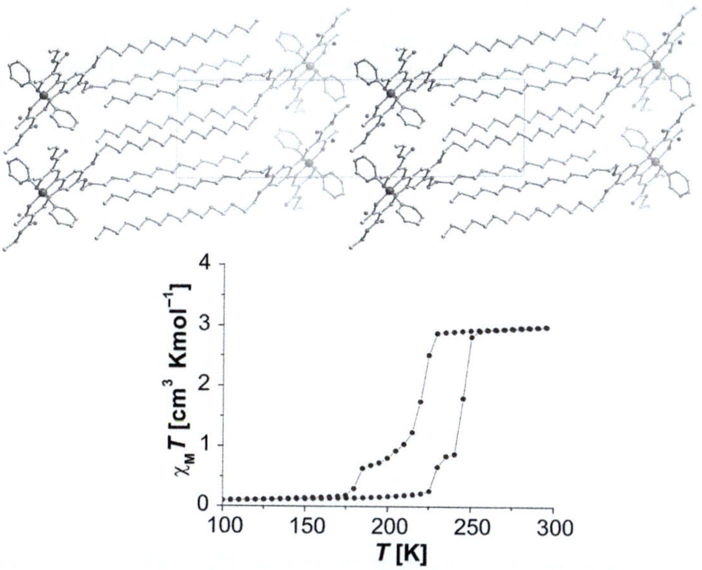

Figure 2.14 *$\chi_M T$ vs. T plot (bottom) and an excerpt of the crystal packing of [Fe(**22a**)(py)$_2$]·0.25 (top). Reproduced from [68], with permission of The Royal Society of Chemistry.*

A new ligand system for complexes with N_6 coordination sphere are noninnocent α-diimine ligands. The complex [Fe(dad)$_3$][Fe(pda)$_2$] possesses an interesting T-dependent electronic structure.[72] At low temperature both iron centres are iron(II) with the six-coordinated $[Fe(dad)_3]^{2+}$ in the LS state and the four-coordinated $[Fe(pda)_2]^{2-}$ in the HS state. Heating above 235 K resulted in a reversible increase of the magnetic moment that is associated to a one-electron transfer from the four-coordinated HS iron(II) of the anion to the dad ligand of the cation. As a consequence, the now four-coordinated iron(III) is in the $S = {}^3/_2$ intermediate spin state and the six-coordinated iron(II) cation is now HS and anti-ferromagnetically coupled to the radical $dad^{\bullet-}$ ligand. Thus, electron transfer within a cation–anion pair coupled to spin-crossover has been observed for the first time.[72]

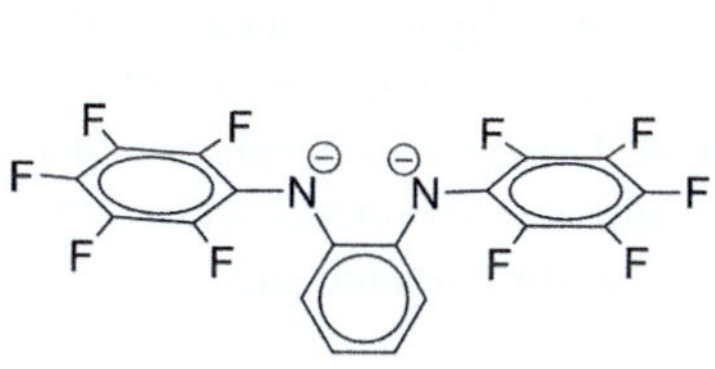

pda^{2-}

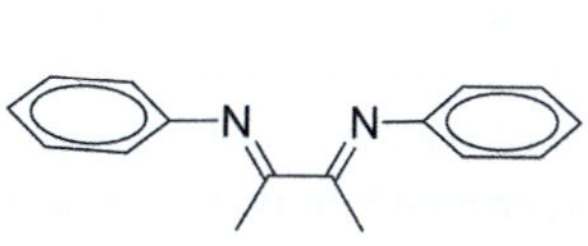

dad^{0}

The noninnocent aminophenol based ligand **23** has been used to synthesise a radical-containing Fe(III) complex [Fe(**23**)$_3$] with an unusual N_3O_3 coordination sphere. T-dependent magnetic measurements and Mössbauer spectroscopy indicate a gradual spin-crossover of the central ferric ion.[73]

Iron(III) SCO complexes with a previously unknown $N_2O_2S_2$ coordination sphere were realised with salicylaldehyde thiosemicarbazones as ligands.[74] The aim of this work was to obtain a strong molecular distortion to be able to trap the photoinduced HS state and more clearly investigate the factors required for observation of the LIESST effect in iron(III) complexes. Strong intermolecular interactions in the case of [Fe(**24**)(H**24**)] lead to a cooperative spin-transition with and up to 42 K wide thermal hysteresis loop, but no LIESST effect was detected. For the complexes [Fe(**25**)(H**25**)] and [Fe(**26**)(H**26**)] an abrupt spin-transition and the LIESST effect was observed with relatively long-lived metastable states.[74]

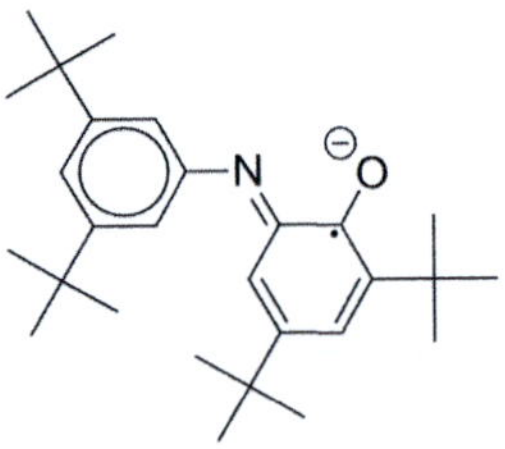

23

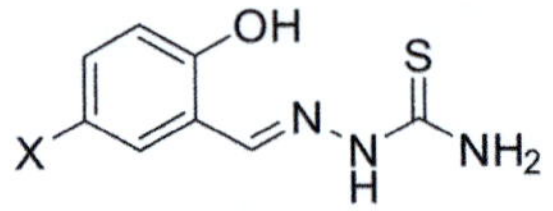

X = H H_2**24**
Cl H_2**25**
Br H_2**26**

2.5 Conclusion and Outlook

A variety of novel spin-crossover complexes has been presented. They differ strongly in coordination number and geometry, metal centres and donor atom sets. The underlying principle for the spin state change also varies, with or without maintenance of the coordination number and geometry. The examples demonstrate that purposeful ligand design can be used to control the splitting of the *d*-orbitals of the metal centre and by this the temperature dependent reversible spin state change. An interplay between electronic (ligand field strength) and steric (metal–ligand bond distances) factors influences the spin-transition temperature. Both can be controlled by substituents in the outer periphery of the ligand or modifications to the ligand structure (e.g. bridges of oligodentate ligands). This did enable the synthesis spin-crossover complexes with coordination numbers ranking between 4 and 7 and with these impressive examples in mind it is difficult to say what will come next – everything seems possible!

The new examples show that the exceptional position of N_6 coordinated iron(II) complexes for the synthesis of SCO complexes is no longer that unique. Especially for the Schiff base and Schiff base-like ligands great potential can be envisioned due to their flexibility. The synthesis of multifunctional SCO complexes is drawing more and more attention and the Schiff base and Schiff base-like ligands appear to be highly suitable for the design of such systems. Independent of the ligand system used, strong intermolecular interactions are clearly necessary for the observation of hysteresis effects in SCO. Different concepts are discussed in literature to explain the observation of wide hysteresis loops. Some of the new examples with very wide hysteresis loops show that the role of hydrogen bonds particularly needs to be investigated further.

References

1. Gütlich, P., Hauser, A., Spiering, H. (1994) Thermal and optical switching of iron(II) complexes. Angew. Chem. Int. Ed., 33: 2024–2054.
2. Gütlich, P., Goodwin, H. A. (Eds) (2004) Spin Crossover in Transition Metal Compounds I-III. Top. Curr. Chem., vols. 233–235. Springer, Berlin / Heidelberg, Germany.
3. Létard, J.-F. (2006) Photomagnetism of iron(II) spin crossover complexes–the *T*(LIESST) approach. J. Mater. Chem., 16: 2550–2559.
4. Sato, O., Tao, J., Zhang, Y.-Z. (2007) Control of magnetic properties through external stimuli, Angew. Chem. Int. Ed., 46: 2152–2187.

5. Kitchen, J. A., Brooker, S. (2008) Spin crossover in iron(II) complexes of 3,5-di(2-pyridyl)-1,2,4-triazoles and 3,5-di(2-pyridyl)-1,2,4-triazolates. Coord. Chem. Rev., 252: 2072–2092.
6. Weber, B. Spin crossover complexes with N_4O_2 coordination sphere–the influence of covalent linkers on cooperative interactions. Coord. Chem. Rev., 253: 2432–2449.
7. Halcrow, M. A. (2009) Iron(II) complexes of 2,6-di(pyrazol-1-yl)pyridines–a versatile system for spin-crossover research. Coord. Chem. Rev., 253: 2493–2514.
8. Murray, K. S. (2009) Recent advances in molecular magnetic materials. Aust. J. Chem., 62: 1081–1101.
9. Gaspar, A., Seredyuk, M., Gütlich, P. (2009) Spin crossover in iron(II) complexes: recent advances. J. Mol. Struct., 924–926: 9–19.
10. Brooker, S., Kitchen, J. A. (2009) Nano-magnetic materials: spin crossover compounds vs. single molecule magnets vs. single chain magnets. Dalton Trans., 7331–7340.
11. Koudriavtsev, A. B., Linert, W. (2010) Spin crossover — An unusual chemical equilibrium. J. Struct. Chem., 51: 335–365.
12. Halcrow, M. A. (2011) Structure:function relationships in molecular spin-crossover complexes, Chem. Soc. Rev., 40: 4119–4142.
13. Létard, J.-F., Guionneau, P., Goux-Capes, L. (2004) Towards spin crossover applications, in: Spin crossover in transition metal compounds III. Top. Curr. Chem., 235: 221–249.
14. Garcia, Y., Ksenofontov, V., Mentior, S., Dîrtu, M. M., Gieck, C., Bhatthacharjee, A., Gütlich, P. (2008) Rapid cooling experiments and use of an anionic nuclear probe to sense the spin transition of the 1D coordination polymers $[Fe(NH_2trz)_3]SnF_6{\cdot}nH_2O$ (NH_2trz=4-amino-1,2,4-triazole) Chem. Eur. J., 14: 3745–3758.
15. Venkataramani, S., Jana, U., Dommaschk, M., Sonnichsen, F. D., Tuczek, F., Herges, R. (2011) Magnetic bistability of molecules in homogeneous solution at room temperature. Science, 331: 445–448.
16. Holleman, A. F., Wiberg, E., Wiberg, N. Lehrbuch der anorganischen Chemie, de Gruyter, Berlin, 2007.
17. Thies, S., Bornholdt, C., Köhler, F., Sönnichsen, F. D., Näther, C., Tuczek, F., Herges, R. (2010) Coordination-induced spin crossover (CISCO) through axial bonding of substituted pyridines to nickel–porphyrins: σ-donor versus π-acceptor effects. Chem. Eur. J., 16: 10074–10083.
18. Petrouleas, V., Tuchagues, J. P. (1987) $Fe[5NO_2$-sal-N(1,4,7,10)]: a new iron(II) complex exhibiting an unusual two-step spin conversion afforded by a hexadentate ligand with a N_4O_2 donor set. Chem. Phys. Lett., 137: 21–25.
19. Boinnard, D., Bousseksou, A., Dworkin, A., Savariault, J. M., Varret, F., Tuchagues, J. P. (1994) Two-step spin conversion of $[Fe^{II}(5\text{-}NO_2$-sal-N(1,4,7,10))]: 292, 153, and 103 K X-ray crystal and molecular structure and infrared, magnetic, Mössbauer, calorimetric, and theoretical studies. Inorg. Chem., 33: 271–281.
20. Psomas, G., Bréfuel, N., Dahan, F., Tuchagues, J.-P. (2004) An unprecedented trinuclear structure involving two high-spin and one spin-crossover iron(II) centers. Inorg. Chem., 43: 4590–4594.
21. Salmon, L., Bousseksou, A., Donnadieu, B., Tuchagues, J.-P. (2005) Two novel iron(II) materials based on dianionic N4O2 Schiff bases: structural properties and spin-crossover characteristics in the series [Fe(3-X,5-NO_2-sal-N(1,4,7,10)] (X = H, 3-MeO, 3-EtO). Inorg. Chem., 44: 1763–1773.
22. Grillo, V. A., Gahan, L. R., Hanson, G. R., Stranger, R., Hambley, T. W., Murray, K. S., Moubaraki, B., Cashion, J. D. (1998) Iron(III) and iron(II) complexes of 1-thia-4,7-diazacyclononane ([9]aneN_2S) and 1,4-dithia-7-azacyclononane ([9]aneNS_2). X-Ray structural analyses, magnetic susceptibility, Mossbauer, EPR and electronic spectroscopy. J. Chem. Soc. Dalton Trans., 2341–2348.
23. Costa, J. S., Balde, C., Carbonera, C., Denux, D., Wattiaux, A., Desplanches, C., Ader, J.-P., Gütlich, P., Létard, J.-F. (2007) Photomagnetic properties of an iron(II) low-spin complex with an unusually long-lived metastable LIESST state. Inorg. Chem., 46: 4114–4119.
24. Nihei, M., Shiga, T., Maeda, Y., Oshio, H. (2007) Spin crossover iron(III) complexes. Coord. Chem. Rev., 251: 2606–2621.
25. Krivokapic, I., Zerara, M., Daku, M. L., Vargas, A., Enachescu, C., Ambrus, C., Tregenna-Piggott, P., Amstutz, N., Krausz, E., Hauser, A. (2007) Spin-crossover in cobalt(II) imine complexes. Coord. Chem. Rev., 251: 364–378.
26. Hayami, S., Komatsu, Y., Shimizu, T., Kamihata, H., Lee, Y. H. (2011) Spin-crossover in cobalt(II) compounds containing terpyridine and its derivatives. Coord. Chem. Rev., 255: 1981–1990.
27. Gass, I. A., Batten, S. R., Forsyth, C. M., Moubaraki, B., Schneider, C. J., Murray, K. S. (2011) Supramolecular aspects of iron(II) crown-dipyridyl spin-crossover compounds. Coord. Chem. Rev., 255: 2058–2067.

28. Olguín, J., Brooker, S. (2011) Spin crossover active iron(II) complexes of selected pyrazole-pyridine/pyrazine ligands, Coord. Chem. Rev., 255: 203–240.
29. Nelson, S. M., McIlroy, P. D. A., Stevenson, C. S., König, E., Ritter, G., Waigel, J. (1986) Quadridentate *versus* quinquedentate co-ordination of some N_5 and N_3O_2 macrocyclic ligands and an unusual thermally controlled quintet $\rightleftharpoons$ singlet spin transition in an iron(II) complex. J. Chem. Soc. Dalton Trans., 991–995.
30. König, E., Ritter, G., Dengler, J., Nelson, S. M. (1987) Detailed study of a two-step quintet $\rightleftharpoons$ singlet spin transition in an iron(II) complex with a N_3O_2 macrocyclic ligand and the kinetics of the quintet $\rightarrow$ singlet relaxation in the temperature range 115–130 K. Inorg. Chem., 26: 3582–3588.
31. Hayami, S., Gu, Z.-Z., Einaga, Y., Kobayasi, Y., Ishikawa, Y., Yamada, Y., Fujishima, A., Sato, O. (2001) A novel LIESST iron(II) complex exhibiting a high relaxation temperature. Inorg. Chem., 40: 3240–3242.
32. Guionneau, P., Le Gac, F., Kaiba, A., Costa, J. S., Chasseau, D., Létard, J.-F. (2007) A reversible metal-ligand bond break associated to a spin-crossover. Chem. Commun., 3723–3725.
33. Garcia, Y., Gütlich, P. (2004) Thermal spin crossover in Mn(II), Mn(III), Cr(II) and Co(III) coordination compounds, in: Gütlich, P., Goodwin H. A. (Eds) Spin Crossover in Transition Metal Compounds II. Top. Curr. Chem., 234: 49–62.
34. Morgan, G. G., Murnaghan, K. D., Müller-Bunz, H., McKee, V., Harding, C. J. (2006) A manganese(III) complex that exhibits spin crossover triggered by geometric tuning. Angew. Chem. Int. Ed., 45: 7192–7195.
35. Gandolfi, C., Cotting, T., Martinho, P. N., Sereda, O., Neels, A., Morgan, G. G., Albrecht, M. (2011) Synthesis and self-assembly of spin-labile and redox-active manganese(III) complexes. Dalton Trans., 40: 1855–1865.
36. Graf, M., Wolmershäuser, G., Kelm, H., Demeschko, S., Meyer, F., Krüger, H.-J. (2010) Temperature-induced spin-transition in a low-spin cobalt(II) semiquinonate complex. Angew. Chem. Int. Ed., 49: 950–953.
37. Hendrickson, D. N., Pierpont, C. G. (2004) Valence tautomeric transition metal complexes, in: Gütlich, P., Goodwin, H. A. (Eds) Spin Crossover in Transition Metal Compounds II. Top. Curr. Chem., 234: 63–95.
38. van Koningsbruggen, P. J., Maeda, Y., Oshio, H. (2004) Iron(III) spin crossover compounds, in: Gütlich, P., Goodwin, H. A. (Eds) Spin Crossover in Transition Metal Compounds I. Top. Curr. Chem., 233: 259–324.
39. Weber, B., Görls, H., Rudolph, M., Jäger, E.-G. (2002) Nitrosyliron complexes of macrocyclic [N_4^{2-}] and open chain [$N_2O_2^{2-}$] chelate ligands: influence of the equatorial ligand on the NO binding mode. Inorg. Chim. Acta, 337: 247–265.
40. Weber, B., Jäger, E.-G. (2009) Structure and magnetic properties of iron(II/III) complexes with $N_2O_2^{2-}$ coordinating Schiff base like ligands, Eur. J. Inorg. Chem., 465–477.
41. Li, J., Lord, R. L., Noll, B. C., Baik, M.-H., Schulz, C. E., Scheidt, W. R. (2008) Cyanide: a strong-field ligand for ferrohemes and hemoproteins? Angew. Chem. Int. Ed., 47: 10144–10146.
42. Ma, H., Petersen, J. L., Young, V. G., Yee, G. T., Jensen, M. P. (2011) Solid-state spin crossover of Ni(II) in a bioinspired N_3S_2 ligand field. J. Am. Chem. Soc., 133: 5644–5647.
43. Roux, C., Zarembowitch, J., Gallois, B., Granier, T., Claude, R. (1994) Toward ligand-driven light-induced spin changing influence of the configuration of 4-styrylpyridine (stpy) on the magnetic properties of $Fe^{II}(stpy)_4(NCS)_2$ Complexes. Crystal structures of the spin-crossover species Fe(*trans*-stpy)$_4$(NCS)$_2$ and of the high-spin species Fe(*cis*-stpy)$_4$(NCS)$_2$. Inorg. Chem., 33: 2273–2279.
44. Boillot, M.-L., Pillet, S., Tissot, A., Rivière, E., Claiser, N., Lecomte, C. (2009) Ligand-driven light-induced spin change activity and bidirectional photomagnetism of styrylpyridine iron(II) complexes in polymeric media. Inorg. Chem., 48: 4729–4736.
45. Scepaniak, J. J., Harris, T. D., Vogel, C. S., Sutter, J., Meyer, K., Smith, J. M. (2011) Spin crossover in a four-coordinate iron(II) complex. J. Am. Chem. Soc., 133: 3824–3827.
46. Bowman, A. C., Milsmann, C., Bill, E., Turner, Z. R., Lobkovsky, E., DeBeer, S., Wieghardt, K., Chirik, P. J. (2011) Synthesis and electronic structure determination of n-alkyl-substituted bis(imino)pyridine iron imides exhibiting spin crossover behavior. J. Am. Chem. Soc., 133: 17353–17369.
47. Ono, K., Yoshizawa, M., Akita, M., Kato, T., Tsunobuchi, Y., Ohkoshi, S., Fujita, M. (2009) Spin crossover by encapsulation. J. Am. Chem. Soc., 131: 2782–2783.
48. Hamon, P., Thépot, J.-Y., Le Floch, M., Boulon, M.-E., Cador, O., Golhen, S., Ouahab, L., Fadel, L., Saillard, J.-Y., Hamon, J.-R. (2008) Dramatic remote substitutent effects on the electronic spin state of bis(scorpionate) iron(II) complexes, Angew. Chem. Int. Ed., 47: 8687–8691.

49. Reger, D. L., Gardinier, J. R., Gemmill, W. R., Smith, M. D., Shahin, A. M., Long, G. J., Rebbouh, L., Grandjean, F. (2005) Formation of third generation poly(pyrazolyl)borate ligands from alkyne coupling reactions of Fe[(p-IC6H4)B(3-Rpz)3]2 (R = H, Me; pz = pyrazolyl): pathways toward controlling an iron(II) electronic spin-state crossover. J. Am. Chem. Soc., 127: 2303–2316.
50. Kuzu, I., Krummenacher, I., Hewitt, I. J., Lan, Y., Mereacre, V., Powell, A. K., Höfer, P., Harmer, J., Breher, F. (2009) Syntheses, structures and electronic properties of zwitterionic iron(II) and cobalt(II) complexes featuring ambidentate tris(pyrazolyl)methanide ligands, Chem. Eur. J., 15: 4350–4365.
51. Robert, F., Naik, A. D., Tinant, B., Robiette, R., Garcia, Y. (2009) Insights into the origin of solid-state photochromism and thermochromism of n-salicylideneanils: the intriguing case of aminopyridines. Chem. Eur. J., 15: 4327–4342.
52. Floquet, S., Salunke, S., Boillot, M.-L., Clément, R., Varret, F., Boukheddaden, K., Rivière, E. (2002) The spin transition of an iron(III) complex intercalated in a $MnPS_3$ layered magnet. Occurrence of a hysteresis effect on removal of lattice solvent. Chem. Mater., 14: 4164–4171.
53. Clemente-León, M., Coronado, E., López-Jordà, M., Mínguez Espallargas, G., Soriano-Portillo, A., Waerenborgh, J. C. (2010) Multifunctional magnetic materials obtained by insertion of a spin-crossover Fe^{III} complex into bimetallic oxalate-based ferromagnets. Chem. Eur. J., 16: 2207–2219.
54. Dorbes, S., Valade, L., Real, J. A., Faulmann, C. (2005) [Fe(sal$_2$-trien)][Ni(dmit)$_2$]: towards switchable spin crossover molecular conductors. Chem. Commun., 69–71.
55. Gandolfi, C., Moitzi, C., Schurtenberger, P., Morgan, G. G., Albrecht, M. (2008) Improved cooperativity of spin-labile iron(III) centers by self-assembly in solution. J. Am. Chem. Soc., 130: 14434–14435.
56. Martinho, P. N., Harding, C. J., Müller-Bunz, H., Albrecht, M., Morgan, G. G. (2010) Inducing spin crossover in amphiphilic iron(III) complexes. Eur. J. Inorg. Chem., 675–679.
57. Bauer, W., Scherer, W., Altmannshofer, S., Weber, B. (2011) Two-step versus one-step spin transitions in iron(II) 1D chain compounds. Eur. J. Inorg. Chem., 2803–2818.
58. Bauer, W., Dîrtu, M. M., Garcia, Y., Weber, B. (2012) Iron(II) spin transition coordination polymers with a zigzag structure. Cryst. Eng. Comm., 14: 1223–1231.
59. Weber, B., Kaps, E. (2005) Synthesis and magnetic properties of new dinuclear iron(II) complexes of a phenylene-bridge shiff base analogue dinucleating ligand, Heteroatom Chem., 16: 391–397.
60. Weber, B., Kaps, E. S., Obel, J., Achterhold, K., Parak, F. G. (2008) Synthesis and characterization of a dinuclear iron(II) spin crossover complex with wide hysteresis. Inorg. Chem., 47: 10779–10787.
61. Real, J. A., Gaspar, A. B., Niel, V., Muñoz, M. C. (2003) Communication between iron(II) building blocks in cooperative spin transition phenomena. Coord. Chem. Rev., 236: 121–141.
62. Weber, B., Kaps, E., Weigand, J., Carbonera, C., Létard, J.-F., Achterhold. K., Parak, F. G. (2008) Cooperative Iron(II) spin crossover complexes with N_4O_2 coordination sphere. Inorg. Chem., 47: 487–496.
63. Pfaffeneder, T. M., Thallmair, S., Bauer, W., Weber, B. (2011) Complete and incomplete spin transitions in 1D chain iron(II) compounds. New J. Chem., 35: 691–700.
64. Weber, B., Bauer, W., Obel, J. (2008) An iron(II) spin-crossover complex with a 70 K wide thermal hysteresis loop. Angew. Chem. Int. Ed., 47: 10098–10101.
65. Weber, B., Bauer, W., Pfaffeneder, T., Dîrtu, M. M., Naik, A. D., Rotaru, A., Garcia, Y. (2011) Influence of hydrogen bonding on the hysteresis width in iron(II) spin crossover complexes. Eur. J. Inorg. Chem., 3193–3296.
66. Weber, B., Obel, J., Henner-Vasquez, D., Bauer, W. (2009) Two new iron(II) spin-crossover complexes with N_4O_2 coordination sphere and spin transition around room temperature. Eur. J. Inorg. Chem., 5527–5534.
67. Weber, B., Obel, J., Lorenz, L. R., Bauer, W., Carrella, L., Rentschler, E. (2009) Control of exchange interactions in trinuclear complexes based on orthogonal magnetic orbitals, Eur. J. Inorg. Chem., 5535–5540.
68. Schlamp, S., Weber, B., Naik, A. D., Garcia, Y. (2011) Cooperative spin transition in a lipid layer like system. Chem. Commun., 47: 7152–7154.
69. Schlamp, S., Thoma, P., Weber, B. (2012) New octahedral, head-tail iron(II) complexes with spin crossover properties. Eur. J. Inorg. Chem., in press (doi: 10.1002/ejic.201101096).
70. Simaan, A. J., Boillot, M.-L., Rivière, E., Boussac, A., Girerd, J.-J. (2000) A two-step spin crossover in [(TPA)Fe^{III}(cat)]BPh_4. Angew. Chem. Int. Ed., 39: 196–198.
71. Simaan, A. J., Boillot, M.-L., Carrasco, R., Cano, J., Girerd, J.-J., Mattioli, T. A., Ensling, J., Spiering, H., Gütlich, P. (2005) Electronic, vibrational, and structural properties of a spin-crossover catecholato–iron system

in the solid state: theoretical study of the electronic nature of the doublet and sextet states. Chem. Eur. J., 11: 1779–1793.

72. Khusniyarov, M. M., Weyhermüller, T., Bill, E., Wieghardt, K. (2008) Reversible electron transfer coupled to spin crossover in an iron coordination salt in the solid state. Angew. Chem. Int. Ed., 47: 1228–1231.
73. Mukherjee, S., Weyhermüller, T., Bill, E., Wieghardt, K., Chaudhuri, P. (2005) Tuning of spin transition in radical-containing iron(III) complexes by remote ligand substituents, Inorg. Chem., 44: 7099–7108.
74. Yemeli, E. W. T., Blake, G. R., Douvalis, A. P., Bakas, T., Alberda van Ekenstein, G. O. R., van Koningsbruggen, P. J. (2012) Light-induced bistability in iron(III) spin-transition compounds of 5 X-salicylaldehyde thiosemicarbazone (X = H, Cl, Br). Chem. Eur. J., in press (doi: 10.1002/chem.201002100).

3

Spin-Crossover in Discrete Polynuclear Complexes

Juan Olguín[1] and Sally Brooker[2]

[1]Department of Chemistry, University of Otago, New Zealand
[2]Department of Chemistry and The MacDiarmid Institute for Advanced Materials and Nanotechnology, University of Otago, New Zealand

3.1 Introduction

Spin-crossover (SCO) complexes are a class of promising molecular materials that may find application in the fast-developing area of nanotechnology, for example as data storage, display devices, sensors, etc.[1–3] However, only SCO-complexes that show abrupt crossover, in which case they are identified as a spin-transition (ST), are likely to be useful, *and* for many applications room temperature (RT) thermal hysteresis loops (memory effect) will also be required.[2, 3] It has been established that strong interaction/communication between the metal centres can lead to high cooperativity which optimises the chances of observing both of these highly desired properties, that is (i) an abrupt, ST, event, and (ii) observation of a thermal hysteresis loop. This interaction/communication can be transmitted either (i) by an intramolecular pathway, usually provided by a bridge, between the metal centres, as occurs in the 'Kahn' polymers which feature triple 1,2,4-triazole N^1,N^2-bridges between the iron(II) centres in the 1-D chain,[2] and/or (ii) by intermolecular pathways between discrete complexes, involving supramolecular interactions within the crystal lattice, such as classic and nonclassic hydrogen bonds, π–π and anion/cation–π interactions.[4–11]

Polymeric complexes are therefore one target for the development of suitable SCO-materials. However, unfortunately these systems can be very difficult to crystallise and hence the structural aspects of their high cooperativity cannot always be probed in detail. The classic example of this is the family of Kahn iron(II) triazole polymers, which resisted single crystal structure analysis from the 1990s until 2011.[12] In order to learn more about the interaction between metal centres in such polymeric materials, discrete polynuclear complexes are helpful, with dinuclear complexes the simplest possible scenario. The effect on the properties

Spin-Crossover Materials: Properties and Applications, First Edition. Edited by Malcolm A. Halcrow.

of the coexistence of magnetic interactions and SCO is also a topic of considerable interest that discrete, especially dinuclear, systems offer the best chance to probe. Finally, discrete multinuclear complexes can switch between more combinations of states as they contain more than one metal centre per molecule, which could result in bigger storage capabilities and/or more complex logical operations.[13, 14] Discrete polymeric complexes can benefit from both intramolecular communication between the metal ions via a bridge(s), and from intermolecular packing interactions between the complexes.

The synthesis of discrete polynuclear complexes is generally achieved by one of two approaches. The first, 'controlled/designer-ligand', approach involves use of polydentate ligands with the desired number of metal ion binding pockets. The key advantage of this is that the nuclearity and many other features (e.g. metal ion bridging moiety) of the product are controlled by the ligand design. Also, families of closely related complexes are relatively readily generated, for example.[15] The disadvantage is that demanding design and organic synthesis is often necessary. Five- and six-membered ring heterocycles containing two or more nitrogen atoms are a widely used core moiety for this type of ligand as they can hold in close proximity at least two metal centres and commonly (i) provide a bridge(s) between them and (ii) in combination with other appropriate terminal/bridging ligands, offer about the right ligand field for SCO.

The second, 'supramolecular', approach involves the use of a combination of simpler blocking/capping and bridging ligands that are either easily accessed or commercially available. In this case the nuclearity of the resulting complex (tri-, tetra-, penta-nuclear complex, etc.) is controlled by the combination of the geometrical requirements of the metal centre (usually approximately octahedral), the angle between the donor atoms in the bridging ligand (e.g. 90 vs. 180 vs. 120°) and the denticity of the blocking/terminal/capping ligand (e.g. tridentate vs. bidentate). The big advantage of this approach is that it does not require arduous organic synthesis. The main disadvantage is that in some systems rearrangement of the ligands and/or the presence of a dynamic equilibrium between different nuclearity products can occur, for example.[16] It is also less straightforward to generate closely related families of complexes, albeit still possible in some cases, for example.[17]

As noted above, many structural factors affect the SCO-properties. A detailed analysis of these factors in mononuclear complexes was recently presented by Halcrow.[18] These include the class of ligand (e.g. imine vs. amine) and the denticity/chelate-ring(s) size, which strongly affect the ligand field imposed. Other factors include the intermolecular interactions between the SCO-molecules within the crystal lattice. These, and hence the SCO, can be significantly affected by polymorphism, crystallographic order/disorder and by the nature and quantity of solvent molecules of crystallisation (solvatomorphism), all of which are virtually impossible to predict. Crystal lattice solvent molecules can hydrogen bond to the ligand strands, significantly modifying the ligand field experienced by the metal ion. Likewise, the nature of the counter-anion (e.g. size, charge, etc) can have a big impact.

SCO between fully HS and fully LS in dinuclear complexes can occur via two different pathways:

1. From [HS–HS] → [LS–LS], a one-step transformation or
2. From [HS–HS] → '[HS–LS]' → [LS–LS], a two-step transformation.

If the enthalpy of the 'half'-SCO, '[HS–LS]', state is lower than the halfway point between the enthalpies of the [LS–LS] and [HS–HS] states, or if the difference in enthalpies is very small but strong intermolecular interactions are present, then SCO can proceed in two steps[19] (see Section 3.4.1 for the case where [HS–LS] differs from [LS–HS]). The '[HS–LS]' state can be either a localised mixed spin state [HS–LS] or a 1:1 mixture of pure [HS–HS] and [LS–LS] states (see later).

Clearly this type of analysis can be extended to higher nuclearity complexes, where the number of spin state combinations possible between the fully HS and fully LS states grows with the nuclearity of the complexes.

Polynuclear and polymeric SCO-active complexes have been reviewed before, in particular in the 2004 series of *Topics in Current Chemistry*, in articles by Garcia, Real and co-workers[20] and by Real, Gütlich and co-workers.[21] More recently Murray[22] and Real and co-workers[23] have reviewed discrete polynuclear SCO complexes. In the present work, we report the advances in discrete polynuclear SCO complexes of d^4-d^7 metal ions published from 2004–2011 inclusive. Whilst a large number of new *discrete polynuclear SCO* complexes of iron(II)[9] and iron(III)[24] have been reported (90), no new examples featuring cobalt(II),[25] cobalt(III)[26] or chromium(II)[26] have been reported although new studies on the existing complexes have been described and so are presented here. No examples of discrete polynuclear SCO complexes of manganese(II)/(III)[26] had been reported up until 2011 (but an organometallic dimanganese(II) SCO complex has was reported in 2012[27]).

3.2 Dinuclear Iron(II) Complexes

SCO-active complexes of iron(II) are the most widely studied, not least because (i) the effect of SCO is the largest (4 ↔ 0 unpaired electrons; ca. 10% change in M–N bond lengths) and (ii) an N_6 coordination environment can lead to SCO for Fe(II).[28, 29] A large number of nitrogen-based heterocycles have been used to generate such complexes. Most SCO-active iron(II) complexes are mononuclear, but the number of dinuclear examples has grown substantially in recent years. Other nuclearities, such as tetranuclear, are also growing in number, alongside the development of supramolecular chemistry; possibly due to the fact that complexes with more than three metal centres are often best synthesised by a combination of the controlled/designer-ligand and supramolecular approaches.

When SCO goes via the 'half'-SCO state '[HS–LS]' in dinuclear iron(II) complexes, both options, localised mixed spin state [HS–LS] and a 1:1 mixture of pure [HS–HS] and [LS–LS] states, have been observed, by magnetic, structural and Mössbauer spectroscopy (see later). Unfortunately, it is not possible to confidently predict which systems will undergo SCO via one- vs. two-step SCO or by which '[HS–LS]' state: more examples of dinuclear complexes are necessary to rationalise these phenomena. However, it does appear that the use of highly constrained ligands, that facilitate communication of the spin state of one iron(II) centre to the other, promotes the two step transformation occurring via a localised [HS–LS] state.[30–32] A potential disadvantage of the use of highly constrained ligands is that the [HS–LS] state is so substantially stabilised, and/or the geometrical requirements of the [LS–LS] state cannot be met, that the [LS–LS] state is never accessed and the compound, despite having two metal ions, has only two possible states available to it (see Section 3.2.4 below).

3.2.1 Supramolecular Approach

In this section the dinuclear SCO-active complexes synthesised by combining blocking/capping and bridging ligands (Fig. 3.1) are presented, along with a summary of their SCO properties (Table 3.1).

In 2004 Real, Gütlich and co-workers[33] synthesised $[Fe^{II}_2(phdia)_2(\mu\text{-}phdia)(NCS)_4]$ as a brown-green powder (Fig. 3.1). No X-ray structure determination was possible but the authors proposed, on the basis of spectroscopic data, that the two iron(II) centres are bridged by a bis-bidentate phdia ligand and that in each case the octahedral coordination sphere is completed by one terminal bidentate phdia ligand and two *cis*-bound thiocyanate anions. It undergoes incomplete 2-step ST from mostly-[HS–HS] ↔ [HS–LS] ↔ mostly-[LS–LS]. Mössbauer spectroscopy showed that at RT around 16% of the iron centres are in the LS state, so the first ST step ($T_{1/2}{}^1 = 108$ K; hysteresis $\Delta T_{1/2}{}^1 = 2$ K) is incomplete, albeit that it results in 50% of the iron centres being HS, that is a conversion to the '[HS–LS]' state, according to magnetic measurements. In the second incomplete ST step ($T_{1/2}{}^2 = 80$ K; $\Delta T_{1/2}{}^2 = 7$ K), only 33% of the iron centres undergo ST, leaving 17% in the HS state. Mössbauer spectroscopy on a quench-cooled sample under an external magnetic field revealed the presence of the metastable [HS–HS] state and localised [HS–LS] state, establishing that the ST occurs via the localised [HS–LS] state.

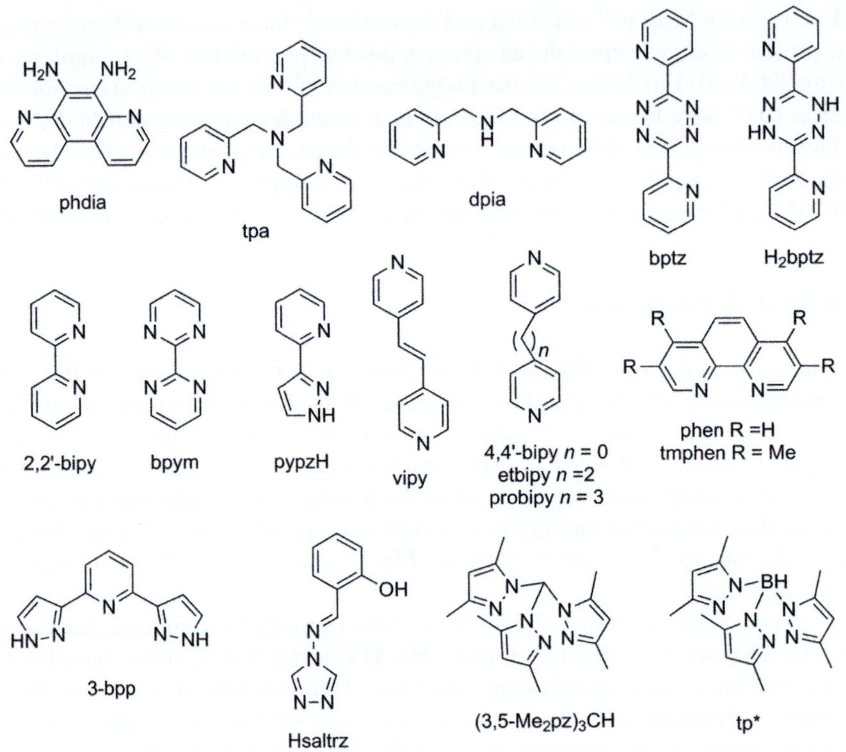

Figure 3.1 *Blocking and bridging ligands used for the 'supramolecular' approach to generation of dinuclear iron(II) SCO-active complexes.*

Even when a ligand that confers the right field to generate mononuclear SCO-active materials is used as a blocking ligand, and a bridging ligand is added to generate polynuclear systems, there is no guarantee of obtaining SCO-active polymetallic complexes, as shown in a systematic study by Murray, Toftlund and co-workers.[34] Known mononuclear SCO-active complexes were linked using different bridging ligands, but only a few of the resulting complexes were SCO-active.

The dark red complex $[Fe_2^{II}(tpa)_2(\mu\text{-}NC(CH_2)_4CN)_2](ClO_4)_4\cdot NC(CH_2)_4CN$ (Fig. 3.1) was characterised by X-ray crystallography at 173 K (Table 3.1). The two crystallographically independent iron(II) centres in the asymmetric unit are both LS (av. Fe–N 1.960 and 1.954 Å, for Fe(1) and Fe(2), respectively). Variable temperature (VT) magnetic measurements showed that it remained in the [LS–LS] state between 2 and 300 K. On warming above 300 K, SCO began to occur, but was not complete even at 350 K.

Using different bridging ligands, the blue-green complex $[Fe^{II}{}_2(tpa)_2(\mu\text{-}bptz)](ClO_4)_4$ and yellow-brown complex $[Fe^{II}{}_2(tpa)_2(\mu\text{-}H_2bptz)](ClO_4)_4$ (Fig. 3.1) were synthesised, but not structurally characterised. The blue-green complex is [LS–LS] in the 2–300 K temperature range, whilst above 300 K the beginning of a very gradual SCO is observed. In contrast the yellow-brown complex showed a reproducible and abrupt 'half'-SCO, from [HS–HS] to '[HS–LS]', from 300 to 2 K, centred at 134 K. The change from a tetrazine to a dihydrotetrazine bridge clearly has a major impact on the SCO, probably due to the ability of the latter ligand to act as a hydrogen bond donor; as it is known that such interactions can dramatically affect the SCO properties, as seen in mononuclear iron(II) SCO of 3-bpp (Fig. 3.1);[42] however, an X-ray crystal structure would be necessary to demonstrate this point.

Table 3.1 *Summary of the SCO-properties of the dinuclear iron(II) complexes synthesised following the supramolecular approach.*

Compound	Transition type[a]	$T_{1/2}$ [K][b]	Fe...Fe [Å][c]	Σ [°][d]	Reference
$[Fe^{II}_2(phdia)_2(\mu\text{-}phdia)(NCS)_4]$	1, abrupt, incomplete	$T_{1/2}{}^1 = 108$ K and $\Delta T_{1/2}{}^1 = 2$ K $T_{1/2}{}^2 = 80$ K and $\Delta T_{1/2}{}^2 = 7$ K	–	–	33
$[Fe_2{}^{II}(tpa)_2(\mu\text{-}CN\text{-}(CH_2)_4\text{-}CN)_2](ClO_4)_4\cdot CN\text{-}(CH_2)_4\text{-}CN$	Not stated, incomplete	Above RT	8.6146(8) [LS–LS]	25.1 and 51.7 [LS–LS]	34
$[Fe^{II}_2(tpa)_2(\mu\text{-}bptz)](ClO_4)_4$	Not stated, incomplete	Above RT	–	–	34
$[Fe^{II}_2(tpa)_2(\mu\text{-}H_2bptz)](ClO_4)_4$	3, abrupt	134 K	–	–	34
$[Fe^{II}_2(phen)_2(\mu\text{-}bpym)(NCS)_4]$	3, abrupt	170 K	–	–	34
$[[Fe^{II}_2(pypzH)_2(\mu\text{-}bpym)(NCSe)_4]$	3, gradual	Between 100–200 K	–	–	34
$[Fe^{II}_2(3\text{-}bpp)_2(\mu\text{-}4,4'\text{-}bpy)(SCN)_4]\cdot 2MeOH$	3, gradual	Between 70–120 K	11.413(2) [HS–HS]	87.57 [HS–HS]	35
$[Fe^{II}(dpia)_2(NCS)_4(4,4'\text{-}bipy)]$	2a, abrupt and complete	$T_{1/2}{}^1 = 218$ and $T_{1/2}{}^2 = 191$ K	11.4926(9) [HS–HS] 11.317(3) [HS–LS] 11.1078(7) [LS–LS]	62.4 [HS–HS] 50.36 [HS–LS] 31.9 [LS–LS]	36
$[Fe_2{}^{II}(dpia)_2(NCS)_4(vipy)]$(dark-red polymorph)	suggested 2a, gradual	$T_{1/2}{}^1 = 215$ and $T_{1/2}{}^2 = 132$ K	13.737(1) [HS–HS] 13.737(1) '[HS–LS]' 13.384(1) [LS–LS]	80.5 [HS–HS] 55.8 [HS–LS] 38.2 [LS–LS]	37
$[Fe_2{}^{II}(dpia)_2(NCS)_4(vipy)]\cdot 2MeOH$	1, abrupt	159 K	13.789(1) [HS–HS]	65.8 [HS–HS]	37
$[Fe_2{}^{II}(Hsaltrz)_5(NCS)_4]\cdot 4MeOH$	1, abrupt	150 K	3.966(8) [HS–HS] 3.648(7) [LS–LS]	20.72 [HS–HS] 15.23 [LS–LS]	38

[a]1 = [HS–HS] → [LS–LS], 2a = [HS–HS] → [HS–LS] → [LS–LS], 2b = [HS–HS] → [HS–HS]:[LS–LS] 1:1 → [LS–LS], 3 = [HS–HS] → '[HS–LS]'.
[b]RT = room temperature. [c]Intramolecular separation. [d]Octahedral distortion parameter = sum of the absolute deviations from 90° of the 12 *cis* angles.[39–41]

In the same study, three neutral complexes containing a bridging bipyrimidine unit (bpym, Fig. 3.1), two bidentate blocking ligands and two thio- or seleno-cyanate ligands coordinated to each iron centre were synthesised; namely, $[Fe^{II}_2(phen)_2(\mu\text{-bpym})(NCS)_4]$ and $[Fe^{II}_2(pypzH)_2(\mu\text{-bpym})(NCX)_4]$ (X = S or Se, Fig. 3.1). No colour was reported for these complexes and no X-ray structure was obtained. The phen-containing complex underwent ST from the [HS–HS] state to the '[HS–LS]' state, centred at 170 K; whereas the pypzH complex was stabilised in the [HS–HS] state, 2–300 K. Not surprisingly, the selenium analogue, $[Fe^{II}_2(pypzH)_2(\mu\text{-bpym})(NCSe)_4]$, showed a gradual SCO, [HS–HS] $\leftrightarrow$ '[HS–LS]', between 200 and 100 K, confirming the stronger ligand field conferred by the NCSe anion.

Rosa, Létard and co-workers[35] used the bridging ligand 4,4′-bipyridine (4,4′-bpy) to generate the dinuclear red complex $[Fe^{II}_2(3\text{-bpp})_2(\mu\text{-}4,4'\text{-bpy})(SCN)_4]\cdot 2MeOH$ (Fig. 3.1). The X-ray dataset acquired at 293 K showed that both of the iron atoms were coordinated to the central pyridine and to two pyrazole rings of a 3-bpp ligand strand, in a *meridional* fashion, and, in the same plane, a nitrogen donor from the 4,4′-bipyridine fragment that bridges to the second iron. Two *trans*-thiocyanate ligands complete the iron coordination sphere. Each of the two methanol molecules in the crystal lattice is hydrogen bonded to one of the pyrazole rings in each ligand strand of the complex cation. As commonly seen in this family of complexes of 3-bpp,[42] there are face-to-face π–π interactions between the pyrazole rings of neighbouring molecules and between the central pyridine ring of the ligand and a pyrazole moiety of a neighbouring molecule, forming sheets that run along the *bc* plane. There are C–H...π interactions between the H at the 5-position of the pyrazole ring and a thiocyanate anion of an adjacent molecule. Hydrogen bonding interactions are also present, between the NH of one pyrazole and a sulfur atom of a thiocyanate ion of a neighbouring complex. The bond lengths and angles, 2.108(2)–2.233(2) Å and 88.52(7)–95.73(6)°, respectively, are typical of HS iron(II) centres (Table 3.1).[35]

VT magnetic susceptibility studies on the powder sample of this complex confirmed that at room temperature the complex is stabilised in the [HS–HS] state and undergoes SCO to the '[HS–LS]' state upon cooling ($\chi_M T$ is 7.4 cm^3 Kmol^{-1} at 250 K and decreases to 4.7 cm^3 Kmol^{-1} between 70–120 K). The authors were unable to differentiate between a localised [HS–LS] or a 1:1 [HS–HS]:[HS–LS] mixture of species, either of which is consistent with the observed value of $\chi_M T$ at 70–120 K.[35] Irradiation with the 647 nm band of a Kr^+ laser at 10 K resulted in the observation of a LIESST (Light-Induced Excited Spin-State Trapping) effect, with *T*(LIESST) = 72 K, and almost quantitative photoconversion to the metastable [HS–HS] state. This sample presents a 6 K wide LITH (Light Induced Thermal Hysteresis).[35]

In 2009 Matouzenko and co-workers[36] synthesised the brown-violet dinuclear complex $[Fe^{II}_2(dpia)_2(NCS)_4(4,4'\text{-bipy})]$ (Fig. 3.1). This complex undergoes a reversible abrupt two step ST ($T_{1/2}{}^1$ = 218 and $T_{1/2}{}^2$ = 191 K) with a very narrow plateau at 204 K, which can be considered as an inflexion point, where the magnetic susceptibility indicates a 1:1 ratio of HS to LS iron(II) centres. X-ray crystal structures acquired at 295 K (above $T_{1/2}{}^1$) and at 120 K (below $T_{1/2}{}^2$), clearly showed the presence of a centrosymmetric dinuclear complex stabilised in the [HS–HS] and the [LS–LS] state respectively (Table 3.1). A dataset collected at 208 K (around the plateau region/inflexion point) revealed a similar structure to that seen at 120 and 295 K, but with Fe–N bond lengths in between those seen for the pure HS– and LS–states, that is a disordered/delocalised '[HS–LS]' state. However, for only the second time,[32] *zero-field* Mössbauer spectroscopy was used to establish that the ST actually occurs through a [HS–LS] localised state (which requires breaking of the inversion symmetry). Near the ST plateau two sets of resonances were observed for both the HS and the LS Fe(II) centres, consistent with the resolution of the three species: [HS–HS], [HS–LS] and [LS–LS]. This is a surprising finding as, in contrast to the first case where the [HS–LS] state was observed by *zero-field* Mössbauer spectroscopy (and for the first time by X-ray crystallography, see later[30]) due to the highly constrained bis-terdentate ligand PMAT (Fig. 3.2),[32] neither the bridging nor blocking ligands used here are constrained.[36]

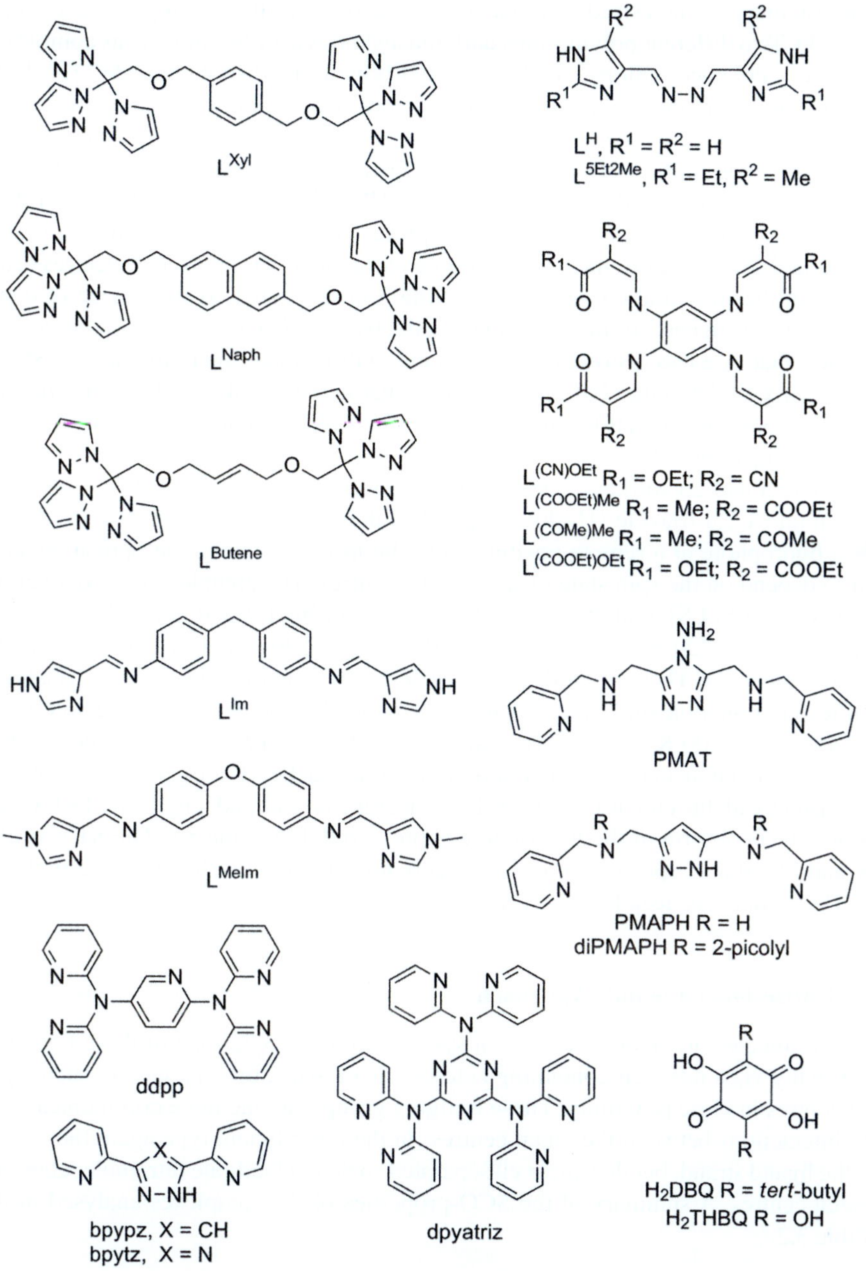

Figure 3.2 *Designed ligands used to generate dinuclear iron(II) SCO-active complexes.*

In 2011 the same authors[37] synthesised a similar system, but replaced the 4,4′-bipy bridging unit by a longer, vipy ligand (Fig. 3.1). Two different polymorphs, dark red and brown, of the anhydrous complex $[Fe_2{}^{II}(dpia)_2(NCS)_4(vipy)]$ were obtained, along with the red-brown methanol solvate $[Fe_2{}^{II}(dpia)_2(NCS)_4(vipy)]\cdot 2MeOH$. VT magnetic measurements on the anhydrous polymorphs revealed that the dark red one undergoes gradual 2-step SCO, with $T_{1/2}{}^1 = 215$ and $T_{1/2}{}^2 = 132$ K, with an inflexion point, rather than a plateau, at 182 K where the ratio of HS:LS is 1:1. The brown polymorph is locked in the [HS–HS] state. In contrast, the methanol solvate undergoes a 1-step ST, [HS–HS] to [LS–LS], at $T_{1/2} = 159$ K. A careful analysis of the X-ray crystal structures of the three complexes at different temperatures showed, as expected, that there are structural differences in the coordination environments of the iron(II) centres (Table 3.1). The authors proposed that the structural characteristics of the iron sites, such as the degree of distortion of the FeN_6 core, are caused by ligand strain and packing effects; and that these distortions correlate with the magnetic properties, that is, a less distorted HS iron centre will be more likely to undergo SCO. The degree of distortion is not easy to determine but one parameter that is commonly used to get a measure of this is Σ,[39–41] the sum of the absolute deviations of each of the 12 *cis* N–Fe–N angles from 90° (bigger value of Σ means greater distortion, Table 3.1).

Also in 2011, Garcia and co-workers[38] designed and crystallised a bright yellow dinuclear complex of *N*-salicylidene-4-amino-1,2,4-triazole (Hsaltrz, Fig. 3.1; Table 3.1), $[Fe^{II}_2(Hsaltrz)_5(NCS)_4]\cdot 4MeOH$. This ligand contains a fluorophore at a remote position from the iron(II) centres, the optical signature of which was anticipated to depend on the spin state of the iron(II) centres. The complex showed an almost complete 1-step ST, [HS–HS] ↔ [LS–LS], with $T_{1/2} = 150$ K, $\Delta H = 8.6$ kJmol^{-1} and $\Delta S = 55.5$ Jmol^{-1} K^{-1}. X-ray crystallography at 100 and 200 K was consistent with this, and also showed that the iron(II) centres are crystallographically identical. They are bridged by the N^1N^2-triazole moieties of three Hsaltrz ligands, and the remaining three coordination sites on each ion are occupied by one monodentate Hsaltrz ligand and two *cis*-bound SCN anions. At 100 K the iron(II) centres are in the [LS–LS] state (average Fe–N 1.983(2) Å), and the C–O and C=N bond distances of the fluorophore at the back of both types of triazoles, bridging and terminal, are consistent with the enol form. At 200 K the average Fe–N bond length is 2.180 Å, consistent with the [HS–HS] state; the enol form of the fluorophore is preserved during the ST. The ST was also monitored by fluorometry and Mössbauer spectroscopy; both nicely mirrored the results obtained by the VT magnetic measurements. The complex is deep burgundy at 77 K.

3.2.2 'Controlled/Designer-Ligand' Approach

Two main types of dinucleating ligands have been synthesised (Fig. 3.2). One of these ligand types bridges the metal ions, usually via a nitrogen-containing heterocycle or azine, or via a moiety such a phenolate, and holds the metal centres in close proximity. These bridging groups can facilitate communication and in some cases magnetic interactions between the metal centres. In the other ligand type again there are two binding pockets within the ligand strand, but they are well separated so do not facilitate intramolecular communication between the metal centres. A summary of the SCO-properties of the complexes analysed in this Section is presented in Table 3.2.

3.2.3 Ligands with Two Isolated Binding Pockets

In 2004 Hannon and co-workers[43] reported three orange dinuclear helical complexes, $[Fe^{II}_2(L^{Im})_3]X_4\cdot 2H_2O$ (reportedly obtained as dihydrates despite drying in vacuo), where L^{Im} is a bis-bidentate Schiff base ligand derived from the 2:1 condensation of 4(5)-imidazolecarboxaldehyde and 4,4′-methylenedianiline (Fig. 3.2) and $X^- = PF_6$, BF_4 and ClO_4. The three complexes were structurally characterised by X-ray crystallography

Table 3.2 *Summary of the SCO-properties of dinuclear iron(II) complexes synthesised from designer dinucleating ligands.*

Compound	Transition type[a]	$T_{1/2}$ [K][b]	Fe...Fe [Å]	Σ[°][c]	Reference
$[Fe^{II}_2(L^{Im})_3](PF_6)_4\cdot 2H_2O$	Not stated, gradual, incomplete	~165	11.396(2) [LS–LS]	64.0	43
$[Fe^{II}_2(L^{Im})_3](BF_4)_4\cdot 2H_2O$	Not stated, incomplete	~135	11.555(2) [HS–LS]-localised	82.8 HS & 63.1 LS	43
$[Fe^{II}_2(L^{Im})_3](ClO_4)_4\cdot 2H_2O$	Proposed 50% 2a, other 50% stays [HS–HS], gradual	$T_{1/2}{}^1 = 240$ and $T_{1/2}{}^2 = 120$	11.583(3) Between [HS–HS] & [HS–LS]-localised	76.1 LS & 93.6 HS	43, 44
$[Fe^{II}_2(L^{MeIm})_3](ClO_4)_4\cdot 2MeCN$	1, gradual	$T_{1/2}\downarrow = 140$, $T_{1/2}{}^1\uparrow = 147$ and $T_{1/2}{}^2\uparrow = 166$	11.35 [LS–LS]	55.2 & 64.0 [LS–LS]	45
$[Fe^{II}_2(L^{MeIm})_3](ClO_4)_4\cdot xH_2O$	3a, gradual	x = 3: $T_{1/2}\uparrow$ ~265 x = 1-4: 210–265	x = 1.5: 11.45(3)[HS–LS]-localised	x = 1.5: 85.3 HS and 63.0 LS	46
$[Fe^{II}_2(ddpp)_2(NCS)_4]\cdot 4CH_2Cl_2$	2a, gradual	$T_{1/2}{}^1 = 180$ and $T_{1/2}{}^2 = 80$	7.324(4) [HS–HS] 7.1585(6) [HS–LS]-localised 7.0654(15) [LS–LS]	41.8 & 46.7 [HS–HS] 27.7 & 45.9 in [HS–LS] 28.4 & 35.2 [LS–LS]	47
$[Fe^{II}_2(ddpp)_2(NCS)_4]\cdot CH_2Cl_2$	Not stated, gradual	200	7.298(13) [HS–HS]	42.4 & 40.4 [HS–HS]	47
$[Fe^{II}_2(ddpp)_2(NCS)_4]\cdot 4\mathit{CHCl_3}$	1, gradual	148	7.434(3) [HS–HS] 7.233(15) [LS–LS]	48.46 [HS–HS] 30.23 [LS–LS]	48
$[Fe^{II}_2(ddpp)_2(NCS)_4]\cdot 3\mathit{CHCl_3}$	1, gradual	200	7.373(2) [HS–HS] 7.143(9) [LS–LS]	34.2 & 51.3 [HS–HS] 27.7 & 34.4 [LS–LS]	48
$[Fe^{II}_2(ddpp)_2(NCS)_4]\cdot \mathit{CHCl_3}$	3a, gradual and incomplete	ca. 110	7.371(5) [HS–HS] 7.212(11) [HS–LS]	42.8 & 47.2 [HS–HS] 25.4 & 43.6 [HS–LS]	48
$[Fe_2{}^{II}(dpyatriz)_2(H_2O)_2(CH_3CN)_2](ClO_4)_4$	Not stated, incomplete	265	9.232(2) [HS–HS]	42.86 [HS–HS]	49
$[Fe^{II}_2[(3,5\text{-}Me_2pz)_3CH]_2(\mu\text{-}L^{Xyl})](BF_4)_4\cdot 2DME$	Not stated, abrupt	363	16.362(8) [LS–LS]	26 [LS–LS]	50
$[Fe^{II}_2[(3,5\text{-}Me_2pz)_3CH]_2(\mu\text{-}L^{Xyl})](BF_4)_4$	Not stated, abrupt	330, $\Delta T_{1/2} = 13$	–	–	50

(continued)

Table 3.2 (Continued)

Compound	Transition type[a]	$T_{1/2}$ [K][b]	Fe...Fe [Å]	Σ [°][c]	Reference
$[Fe^{II}_2[(3,5\text{-}Me_2pz)_3CH]_2(\mu\text{-}L^{Naph})](BF_4)_4 \cdot 2.5MeCN \cdot 2DME$	Not stated, gradual, incomplete	Above RT	17.944(1) & 18.311(1) [LS–LS]	24 & 26 [LS–LS]	50
$[Fe^{II}_2[(3,5\text{-}Me_2pz)_3CH]_2(\mu\text{-}L^{Butene})](BF_4)_4 \cdot nDME$ $n = 2, 0$	Not stated, gradual, incomplete	Above RT	$n = 2$: 15.4411(9)	$n = 2$: 27 [LS–LS]	50
$[Fe^{II}_2(L^H)_3](ClO_4)_4 \cdot 5MeNO_2$	3a, abrupt	240	4.0362(7) [HS–HS] 3.8507(8) [HS–LS]	115.8 & 136.3 [HS–HS] 63.8 LS & 152.6 HS in [HS–LS]	51–54
$[Fe^{II}_2(L^H)_3](BF_4)_4$ needles	3a, abrupt	254	–	–	51–53
$[Fe^{II}_2(L^H)_3](BF_4)_4$ blocks	3a, abrupt	$T_{1/2}\downarrow = 183$, $T_{1/2}\uparrow = 190$, $\Delta T_{1/2} = 7$	3.8695(7) [HS–LS]-localised	158.5 HS & 66.4 LS	51–53
$[Fe^{II}_2(L^{5Et2Me})_3](ClO_4)_4 \cdot$*solvents* plate polymorph	for $\cdot H_2O$ 3a, abrupt	for $\cdot H_2O$ 120	for $\cdot ^1/_2(H_2O) \cdot 3MeCN$ 4.1873(4) [HS–LS]	for $\cdot ^1/_2(H_2O) \cdot 3MeCN$ 137.6 HS & 64.6 LS in [HS–LS]	53
$[Fe^{II}_2(pypzH)_2(\mu\text{-}pypz)_2(NCSe)_2] \cdot 2H_2O$	1, abrupt	225	4.100(7) [HS–HS] 3.996(5) [LS–LS]	72.1 [HS–HS] 46.6 [LS–LS]	55
trans-$[Fe_2(\mu\text{-}bpypz)_2(NCBH_3)_2(4\text{-}PhPy)_2]$	2b, abrupt	200	4.2146(7) [HS–HS] 4.0605(8)-4.2000(8) [HS–HS]:[LS–LS] 4.0434(8) [LS–LS]	71.4 [HS–HS] 63.2 LS & 79.0 HS in [HS–HS]:[LS–LS] 61.3 [LS–LS]	56
trans- $[Fe_2(\mu\text{-}bpypz)_2(NCS)_2(4\text{-}PhPy)_2]$	Not stated, incomplete	150	–	–	57
$[Fe_2^{II}(bpytz)_2(NCBH_3)_2(Py)_2]$	Not stated, abrupt	$T_{1/2}{}^1 = 194$ & $T_{1/2}{}^2 = 151$	–	–	58
$[Fe^{II}_2(PMAT)_2](BF_4)_4 \cdot DMF$	3a, abrupt	224	4.2965(9) [HS–HS] 4.212(1) [HS–LS]	117.5 [HS–HS] 133.1 HS & 64.9 LS in [HS–LS]	59
$[Fe^{II}_2(PMAT)_2](Cl)_4 \cdot 2MeOH \cdot H_2O$	No magnetic or Mössbauer data	Between 90 and 298 K	4.288(1) [HS–HS] 4.229(2) [HS–LS]	110.4 [HS–HS] 123.1 HS & 69.7 LS in [HS–LS]	15

$[Fe^{II}_2(PMAT)_2](PF_6)_4$	Not stated, gradual	120	–	–	15
$[Fe^{II}_2(PMAT)_2](SbF_6)_4{\cdot}2H_2O$	Not stated, gradual	190	–	–	15
$[Fe^{II}_2(PMAT)_2](CF_3SO_3)_4{\cdot}$*solvents*	Not stated, gradual	for $\cdot H_2O$ 180	for $\cdot{}^3/_4(IPA){\cdot}H_2O$ 4.220(2) & 4.255(2) [HS–HS]	for $\cdot{}^3/_4(IPA){\cdot}H_2O$ 109.6 & 109.5 [HS–HS]	15
$[Fe^{II}_2(PMAT)_2](B(PhF)_4)_4{\cdot}{}^1/_2MeCN$	Not stated, gradual	Above RT	–	–	15
$[Fe^{II}_2(PMAT)_2](C_{16}H_{33}SO_3)_4{\cdot}$*solvents*	Not stated, gradual	for $\cdot 2H_2O$ Above RT	for $\cdot 1.6(MeOH){\cdot}0.4H_2O$ 4.232(2) [HS–HS]	$\cdot 1.6(MeOH){\cdot}0.4H_2O$ 87.5 [HS–HS]	15
$[Fe^{II}_2(L^{(CN)OEt})(Py)_4]$	Not stated, 3c gradual	175	–	–	60
$[Fe^{II}_2(L^{(COMe)Me})(Py)_4]{\cdot}x$(pyridine)	3b, relatively abrupt	for $x = 0$ ~150 for $x = 1$ no SCO	CIF not available	CIF not available	60
$[Fe^{II}_2(L^{(COOEt)Me})(Py)_4]{\cdot}x$(pyridine)	Not stated, Relatively abrupt	for $x = 0$ $T_{1/2}{}^1 \approx 260$ & $T_{1/2}{}^2 \approx 90$	CIF not available	CIF not available	60
$[Fe^{II}_2(L^{(COMe)Me})(MeIm)_4]{\cdot}x(MeIm)$	1, gradual for $x = 0$ abrupt for $x = 4$	for $x = 0$, 229 for $x = 4$ $T_{1/2}\downarrow = 178$ $T_{1/2}\uparrow = 199$ & $\Delta T_{1/2} = 21$	for $x = 4$ 7.845(2) [HS–HS] 7.802(2) [LS–LS]	for $x = 4$ 15.2 [HS–HS] 18.1 [LS–LS]	61
$[Fe^{II}_2(TPA)_2(\mu\text{-}DBQ)](BF_4)_2$	1, abrupt	~290	7.676(1) [HS–HS] 7.617(1) [LS–LS]	70.3 [HS–HS] 59.9 [LS–LS]	62
$[Fe^{II}_2(TPA)_2(\mu\text{-}THBQ)](BF_4)_2$	3b, abrupt	~250 $\Delta T_{1/2} \sim 10$	7.918(2) & 7.872(2) [HS–HS] 7.614(1) [LS–LS] &7.936(1) [HS–HS] in [HS–HS]:[LS–LS] 1:1	107.8 & 120.4 [HS–HS] 54.3 [LS–LS] & 121.7 [HS–HS] in [HSHS]:[LS–LS] 1:1	63

[a] 1 = [HS–HS] → [LS–LS], 2a = [HS–HS] → [HS–LS] → [LS–LS], 2b = [HS–HS] → [HS–HS]:[LS–LS] 1:1 → [LS–LS], 2c = [HS–HS] → '[HS–LS]' → [LS–LS], 3a = [HS–HS] → [HS–LS], 3b = [HS–HS] → [HS–HS]:[LS–LS] in a 1:1 ratio, 3c = [HS–HS] → '[HS–LS]'. [b] RT = room temperature. [c] Octahedral distortion parameter (see Table 3.1 for definition).

at 180 K (Table 3.2). As expected, three ligand strands wrap two iron(II) centres by coordination through the imine and imidazole nitrogen donors. The average Fe–N distances showed that the PF_6 complex is stabilised in the [LS–LS] state (1.999 Å, one iron(II) centre in the asymmetric unit), whilst the BF_4 analogue is stabilised in the localised [HS–LS] state (av. Fe–N 2.028 and 2.182 Å, two iron(II) centres in the asymmetric unit) and the ClO_4 analogue seems to be intermediate between [HS–HS] and [HS–LS] states (av. Fe–N 2.069 and 2.141 Å, two iron(II) centres in the asymmetric unit) at 180 K. In all three structures the imidazole NH moieties are involved in hydrogen bonding with solvent and anion molecules, and there are extensive CH...π interactions between the CH_{imine} and the imidazole and phenyl rings of neighbouring molecules, forming a 3D supramolecular network. VT magnetic measurements showed that all three complexes undergo gradual incomplete SCO (at temperatures not consistent with the structural data obtained at 180 K) and that none showed hysteresis despite the extensive supramolecular interactions found in the crystal structures. For the PF_6 and BF_4 analogues the gradual SCO is incomplete, with 20% remaining in the [HS–HS] state below 100 K. The ClO_4 analogue appeared to undergo a half-SCO: it was re-made by Gütlich, Garcia and co-workers[44] who showed by applied magnetic field ^{57}Fe Mössbauer spectroscopy, that half of the iron centres undergo a two-step [HS–HS] ↔ [LS–LS] SCO with $T_{1/2}{}^{1} = 240$ K and $T_{1/2}{}^{2} = 120$ K, whilst the other half of the iron centres remain [HS–HS]. They suggest it is likely that the gradual SCO occurs via the [HS–LS] state. These results clearly show how sensitive SCO properties can be to the nature of the anions, even when they are not coordinated.

Kruger, Schmitt, Clérac and co-workers[45] reported in 2009 the synthesis of a new helical diiron(II) complex, $[Fe^{II}_2(L^{MeIm})_3](ClO_4)_4{\cdot}2MeCN$, where L^{MeIm} is the bis-bidentate Schiff base that results from the 1:2 condensation of 4,4′-oxydianiline and 1-methyl-2-imidazolecarboxaldehyde (Fig. 3.2). This ligand is similar to L^{Im} (above), the differences being an oxygen atom linking the two phenyl groups in L^{MeIm} instead of a methylene linker in L^{Im}, and that the $NH_{imidazole}$ imidazole groups in L^{Im} are N-methylated in L^{MeIm}. At 150 K there are two iron(II) centres in the asymmetric unit with average Fe–N bond lengths of 1.996 and 1.994 Å indicative of the [LS–LS] state. As expected three ligand strands coordinate to two metal centres through the imine and imidazole nitrogen donors. VT magnetic measurements on the desolvated complex showed a $\chi_M T$ value per dinuclear unit of 6.6 cm^3 Kmol^{-1} at room temperature, as expected for the [HS–HS] state. Upon cooling, a 1-step SCO, [HS–HS] to [LS–LS], occurs with $T_{1/2} = 140$ K. Below 80 K, a residual $\chi_M T$ value per dinuclear unit of 0.35 cm^3 Kmol^{-1} remains. Interestingly, on warming the magnetic behaviour does not follow the same path, instead a two-step transition is observed with a small hysteresis loop, $T_{1/2}{}^{1}\uparrow = 147$ and $T_{1/2}{}^{2}\uparrow = 166$ K. This magnetic behaviour is reproducible. This example also highlights once again the importance of solvent effects on SCO properties, as the structure of the di-acetonitrile solvate shows it is [LS–LS] at 150 K.

A complex LIESST effect is observed for the desolvated sample. When it is irradiated at 10 K with red (650 nm) or white light a photoconversion to almost 100% the metastable [HS–HS] state occurs, after 1.1 h in the case of white light. After irradiation the magnetic behaviour of the sample was analysed in the 2–300 K temperature range (Fig. 3.3). The relaxation profile was explained by the authors as a combination of two relaxation mechanisms, each associated with one of the two iron(II) centres, that is, one of the metal centres relaxes faster than the other within the dinuclear unit, generating [HS–LS] units (below 80 K). Then there is competition between the relaxation of the photoinduced [HS–HS] state to [LS–LS] and [LS–HS] species and thermal SCO that favours the appearance of HS iron(II) centres (above 80 K).

In 2011 Kruger and co-workers reported the SCO-properties of the analogous water solvates $[Fe^{II}_2(L^{MeIm})_3](ClO_4)_4{\cdot}xH_2O$ ($x = 1$–4).[46] The X-ray crystal structure of the $x = 1.5$ hydrate acquired at 150 K reveals the same overall double helicate cation $[Fe^{II}_2(L^{MeIm})_3]^{4+}$ as for the acetonitrile solvate (see above), once again with two crystallographic independent iron(II) centres, however, the hydrate is stabilised in the localised [HS–LS] state (av. Fe–N 2.181 and 1.958 Å, Table 3.2), in contrast to the [LS–LS] state seen in the acetonitrile solvate (see above). The partial occupancy water molecules are hydrogen bonded to one another and to the anions. Both types of solvate feature intramolecular edge-to-face CH...π interactions,

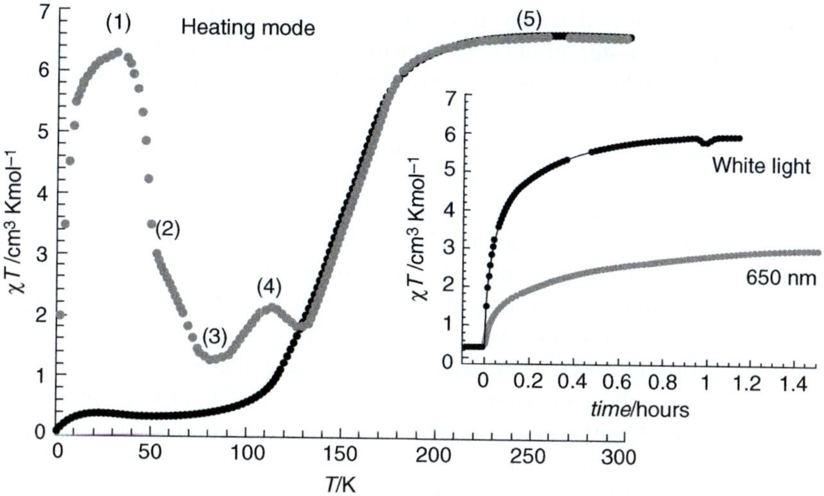

Figure 3.3 *Magnetic and photomagnetic properties of the complex $[Fe^{II}_2(L^{MeIm})_3](ClO_4)_4{\cdot}2MeCN$. Reproduced from [44] with permission of The Royal Society of Chemistry.*

however they are weaker in the hydrate which instead has intermolecular face-to-face $\pi \ldots \pi$ interactions (not seen in the acetonitrile solvate).

VT magnetic measurements (200–300 K) on a red crystalline sample of the hydrate revealed, on repeated cooling-heating cycles, a series of complex and gradual 'half' SCO profiles. The first cooling-heating cycle showed a gradual SCO from the localised [HS–LS] to mostly [HS–HS] (5.54 μ_B per Fe at 300 K) with $T_{1/2}\uparrow$ ~265 K. There is a wide [HS–LS] plateau region (4.14 μ_B per Fe), from ~200–50 K. On the second cycle, $T_{1/2}\uparrow$ drops to ~240 K with what appears to be a hysteresis loop ($\Delta T_{1/2} \approx 12$ K). However, further cycles revealed that dehydration is the cause of both the loops (not thermal hysteresis) and the shifts of the $T_{1/2}$ to lower temperatures. The sample was dried under vacuum in the SQUID at 300 K for 2 h, resulting in a dehydrated sample (TGA analysis confirms this) that gave a less gradual SCO, centred at ~200 K in both the heating and cooling modes. The dehydration process is reversible as samples retain their crystallinity, and after standing in ambient conditions for days they rehydrated and showed very similar profiles to the first cycle of freshly prepared samples. The drop in $T_{1/2}$ on desolvation indicates that the dehydrated sample can more readily go [HS–HS], which the authors tentatively suggest could be due to reduced lattice constraints.

VT ^{57}Fe Mössbauer spectroscopy confirmed the [HS–LS] to incomplete [HS–HS] SCO. This complex joins a small but growing group of 'half' SCO complexes in which the localised [LS–HS] form is stabilised.[30–32, 47, 48, 51–54] In the absence of DSC, a van't Hoff analysis using the magnetic data on the desolvated sample (cycles 2–4) gave $\Delta H = 17$–$27(2)$ kJmol^{-1} and $\Delta S = 100$–$148(10)$ Jmol^{-1} K^{-1} for this 'half' SCO event. It is noted that ΔS values above those expected for the spin state change alone ($\Delta S = R\ln[(2S+1)_{HS}/(2S+1)_{LS} = 13.4$ Jmol^{-1} K^{-1}) are commonly due to 'vibrational contributions deriving from stretching and other modes at the Fe(II) coordination centres'.[46]

The authors suggested that the dramatic difference between the magnetic behaviour of the acetonitrile and water solvates, 1-step [HS–HS]–[LS–HS] at ca. 160 K vs. half crossover [LS–HS]–[HS–HS] at ca. 260 K (Table 3.2), can be rationalised in terms of the different supramolecular interactions (see above) and in the nature and quantity of the solvent found in the crystal lattice.

Three differently solvated yellow complexes, $[Fe^{II}_2(ddpp)_2(NCS)_4]{\cdot}nCH_2Cl_2$ ($n = 0$, 1 or 4, Fig. 3.2), were reported in 2006 by Murray, Kepert and co-workers.[47] VT magnetic susceptibility studies revealed that the tetra-CH_2Cl_2 solvate undergoes a relatively gradual 2-step transition, with $T_{1/2}{}^1 = 180$ K and

$T_{1/2}{}^2 = 80$ K; no hysteresis loop was detected. In contrast, the mono-CH_2Cl_2 solvate (obtained by heating *in vacuo* at 100 °C) undergoes a 1-step gradual SCO, with $T_{1/2} = 200$ K (no hysteresis), whereas the fully desolvated complex (obtained by heating *in vacuo* at 200 °C) remains [HS–HS]. Interestingly, the mono-CH_2Cl_2 solvate can be converted back to the tetra-CH_2Cl_2 solvate by immersion in CH_2Cl_2, showing very similar magnetic properties to the freshly synthesised complex, the only difference being a residual magnetic moment below 50 K due to the formation of imperfections in the crystal lattice after the solvation/desolvation cycle. This study demonstrates once again how sensitive SCO properties are to the degree of solvation.

X-ray datasets were acquired on the mono-CH_2Cl_2 solvate (fully HS), and on the tetra-CH_2Cl_2 solvate before, during and after the SCO (at 250, 123 and 25 K). In both cases the two crystallographically independent iron(II) centres in the dinuclear unit are doubly bridged by two bis-bidentate ligand strands bound in a head-to-tail fashion; in both cases six coordination is completed by two *cis*-coordinated thiocyanate anions. There are intramolecular (between central coordinated pyridines) and intermolecular (between uncoordinated pyridines) π–π interactions as well as intermolecular S...S and S...H–C(CH_2Cl_2) interactions. In the tetra-CH_2Cl_2 solvate, the Fe–N bond lengths were entirely consistent with the SCO occurring from [HS–HS] to localised [HS–LS] to [LS–LS]; the first time all three states had been structurally characterised (the first example of [HS–LS][30] did not go [LS–LS]). The average Fe–N bond lengths were: 250 K, 2.180 and 2.169 Å; 123 K, 1.985 and 2.152 Å; 25 K, 1.974 and 1.991 Å. None of the octahedral distortion parameters, for HS or LS centres in this tetra-CH_2Cl_2 solvated ddpp complex (Table 3.2, 27.7–46.7°), are as high as those seen in the first structurally characterised [HS–LS] complex (64.9–133.1°), of the bis-terdentate PMAT ligand (Fig. 3.2).[30] These authors propose the greater flexibility of bis-bidentate ddpp is the key to the present complex being able to undergo the second SCO to [LS–LS].

In a 2010 paper the same authors[48] reported that the mono-CH_2Cl_2 solvate showed no LIESST activity whereas the tetra-CH_2Cl_2 solvate showed modest LIESST activity when irradiated with green light (λ = 530.2 nm at 5 mW cm^{-2}) at 10K. An increase in the $\chi_M T$ value to 1.56 cm^3 Kmol^{-1} was detected, indicating a 43% photoconversion to the metastable [HS–HS] species. The *T*(LIESST) value was 56 K.

In the same paper, the effect of the type and extent of solvation on the SCO properties of the $[Fe^{II}{}_2(ddpp)_2(NCS)_4]$ complex was further explored by replacement of CH_2Cl_2 by $CHCl_3$. Use of $CHCl_3$ as the crystallisation solvent gave $[Fe^{II}{}_2(ddpp)_2(NCS)_4]\cdot 4CHCl_3$.[47] Stepwise solvent removal by drying for 3 hours at 67, 100 and 200 °C, respectively, allowed the isolation of $[Fe^{II}{}_2(ddpp)_2(NCS)_4]\cdot 3CHCl_3$, $[Fe^{II}{}_2(ddpp)_2(NCS)_4]\cdot CHCl_3$ and solvent free $[Fe^{II}{}_2(ddpp)_2(NCS)_4]$.

X-ray crystallographic data were acquired, at 250 K and 123 or 110 K, for the three solvates $[Fe^{II}{}_2(ddpp)_2(NCS)_4]\cdot nCHCl_3$ (n = 4, 3 and 1; Table 3.2). As expected, the coordination environment around the metal centres is similar to the CH_2Cl_2 analogues. However, the tetra-$CHCl_3$ solvate differs in that there is only one crystallographically unique iron centre, whereas in all of the other structures in the $CHCl_3$ and CH_2Cl_2 series, both iron centres are present in the asymmetric unit. The average Fe–N bond distances reveal that the tetra-$CHCl_3$ (2.004(5) Å) and tris-$CHCl_3$ (1.977(3) and 1.995(3) Å) solvates are [LS–LS] at 123 and 110 K, respectively, whereas the mono-$CHCl_3$ solvate (2.113(4) and 1.984(4) Å) is stabilised in a localised [HS–LS] state at 110 K.

VT magnetic susceptibility studies on the tris- and tetra-$CHCl_3$ solvates showed gradual but complete 1-step SCO, centred at 200 and 148 K, respectively, while for the mono-$CHCl_3$ solvate the SCO is gradual and incomplete (~75%), occurring between 250 and 65 K. The solvent-free material remains [HS–HS]. The $T_{1/2}$ for the tris-$CHCl_3$ solvate is significantly higher than for the other $CHCl_3$ solvates, which the authors suggest is consistent with the smaller octahedral distortion parameter Σ for one of the iron centres in the [HS–HS] form in that case (34.2° vs. 42.8–48.46°), indicating that it will more readily adopt the LS-state. However, it is also interesting to note that the other iron centre has the largest Σ (51.3°), yet nevertheless the complex undergoes a 1-step SCO to the [LS–LS] form. The authors comment that there is minimal communication (sterically and electronically) between the two iron centres in this dinuclear family so it is primarily the crystallographic environment of the iron centres that determines the SCO behaviour. When the tris-$CHCl_3$ solvate was re-solvated with CH_2Cl_2, the SCO behaviour changed, from a 1-step to a 2-step process identical to that found in $[Fe^{II}_2(ddpp)_2(NCS)_4]\cdot 4CH_2Cl_2$. The sensitivity of the SCO behaviour of this complex to the choice and amount of solvent present in the lattice indicates that it should be possible to develop a solvent sensor.

In 2006 Reedijk, Gamez, Roubeau and co-workers[49] reported that pink single crystals of $[Fe_2^{II}(dpyatriz)_2(H_2O)_2(CH_3CN)_2](ClO_4)_4$ (Fig. 3.2) were obtained by diethyl ether vapour diffusion into the acetonitrile reaction solution. On exposure to air the crystals crumbled, and acetonitrile was replaced by water, resulting in the yellow compound $[Fe_2^{II}(dpyatriz)_2(H_2O)_4](ClO_4)_4$. The pink complex crystallises ($P\bar{1}$) with the asymmetric unit at RT comprising half of the dinuclear complex cation and two perchlorate anions. The symmetry related iron centres are doubly bridged, by two bis-bidentate dpyatriz ligands (binding via two bipyridylamine moieties; the third remains uncoordinated). The N_5O coordination environment is completed by a water and acetonitrile molecule. The average Fe–N (2.163 Å) and the Fe–O_w (2.140(2) Å) bond lengths are consistent with the [HS–HS] state. VT magnetic measurements showed that it undergoes an incomplete (63%) but relatively abrupt SCO with $T_{1/2} = 265$ K. This is an interesting complex as an N_5O donor set usually provides too weak a ligand field for iron(II) centres to undergo SCO, that is generally stabilising the HS state. Less surprisingly, the N_4O_2 coordinated iron(II) centres in the yellow complex remain [HS–HS] at all temperatures. However both N_5O and N_4O_2 coordination are able to generate SCO active iron(II) complexes,[64] it is simply less common than N_6 coordination.

Murray and co-workers[50] reported the synthesis of di-, tri- and tetra-nuclear mixed-ligand tris (pyrazolyl)methane iron(II) complexes in 2011. In this work three dinucleating, as well as both a tri- and a tetra-nucleating ligand (see Sections 3.3.1 and 3.3.2, respectively, for the tri and tetrametallic complexes), were prepared by reacting the appropriate number of tris(pyrazolyl)ethanol 'arms' with the chosen di-, tri- or tetra-bromomethyl linker moiety, forming ether linkages. All three of the mixed ligand dinuclear complexes, $[Fe^{II}_2[(3,5\text{-}Me_2pz)_3CH]_2(\mu\text{-}L^R)](BF_4)_4\cdot$solvents, where $(3,5\text{-}Me_2pz)_3CH$ = tris(3,5-dimethyl-pyrazolyl)methane and $L^R = L^{Xyl}$, L^{Naph} and L^{Butene} (Fig. 3.2), were structurally characterised below the SCO temperature, so were in the [LS–LS] state (1.956–1.972 Å; Σ 24–27°). Solvent loss precluded data collections at temperatures appropriate for the HS state. In all cases each iron(II) centre is coordinated to a terminal tri-dentate tris(pyrazolyl)methane ligand as well as to one end of the bis-tridentate tris(pyrazolyl)methane-based ligand (it bridges to the second metal ion). The red complex $[Fe^{II}_2[(3,5\text{-}Me_2pz)_3CH]_2(\mu\text{-}L^{Xyl})](BF_4)_4\cdot 2DME$ (DME = dimethoxyethane) showed a complete and abrupt ST with $T_{1/2} = 363$ K (no thermal hysteresis) whereas the desolvated form showed a complete ST at a lower $T_{1/2}$, 330 K, and with a 13 K wide hysteresis loop. DSC on the desolvated form confirmed the thermal hysteresis of 13 K; the first order phase transition was characterised by $\Delta H = 19.4$ kJmol^{-1} and $\Delta S = 56.3$ Jmol^{-1} K^{-1}. The authors believe that this cooperativity results from CH..π interactions between the dinuclear units. The other complexes, $[Fe^{II}_2[(3,5\text{-}Me_2pz)_3CH]_2(\mu\text{-}L^{Naph})](BF_4)_4\cdot 2.5MeCN\cdot 2DME$ and $[Fe^{II}_2[(3,5\text{-}Me_2pz)_3CH]_2(\mu\text{-}L^{Butene})](BF_4)_4\cdot n$DME ($n = 0$ or 2), were purple and displayed very broad, gradual and incomplete SCO between 200–400 K. The authors also noted that whilst SCO at or about RT is desirable, it does make studying the compound more challenging, not least due to solvent loss tending to occur in the same temperature region.

3.2.4 Ligands with Potential for Communication between Binding Pockets

3.2.4.1 Imidazole Containing Azine Ligands

From 2007 to 2010, Sunatsuki, Kojima and co-workers[51–54] synthesised a series of mono- and dinuclear complexes of four imidazole-based flexidentate azine ligands. Two dinuclear iron(II) complexes of the unsubstituted-imidazole-based ligand L^H (Fig. 3.2), namely the triple helicates $[Fe^{II}_2(L^H)_3](X)_4$ where

X = ClO_4 or BF_4, were SCO-active. VT magnetic measurements showed, and crystallography and Mössbauer spectroscopy confirmed, that the orange ClO_4 complex undergoes an abrupt ST from [HS–HS] to localised [HS–LS] with $T_{1/2}$ = 240 K (av. Fe–N & Σ, at 298 K: Fe1 2.153 Å & 115.8°, Fe2 2.197 Å & 136.3°; at 103 K: Fe1 1.975 Å & 63.8°, Fe2 2.189 Å & 152.6°). The BF_4 analogue was obtained as two polymorphs, needles and blocks. The needles showed similar magnetic behaviour to the ClO_4 analogue but with the [HS–HS] ↔ [HS–LS] ST at a slightly higher $T_{1/2}$ of 254 K. In contrast, the blocks had a more abrupt and hysteretic [HS–HS] ↔ [HS–LS] ST, with $T_{1/2}\downarrow$ = 183, $T_{1/2}\uparrow$=190 and $\Delta T_{1/2}$ = 7 K. The X-ray crystal structure of the block-polymorph at 173 K confirms a localised [HS–LS] state (Σ at 173 K: HS 158.5 and LS 66.4°). The highly constrained ligand leads to these two localised [LS–HS] state complexes exhibiting the largest Σ values seen for such species to date (152.6 and 158.5°). Hence it is not surprising that a second ST, to [LS–LS], is not observed in either case. The differences in magnetic properties between these two complexes was explained in terms of the different supramolecular interactions present in the crystal lattice: a zigzag 1D hydrogen bond chain is observed for the ClO_4 analogue whereas a 2D hydrogen bond network (in both cases between the spare $NH_{imidazol}$ and the anions) was observed in the block-polymorph of the BF_4 analogue. The 2D network is expected to result in larger cooperativity than the 1D chain, in agreement with the observation of hysteresis in the former.

Dimetallic complexes of the ligands bearing methyl groups at the 2 or 5 positions of the imidazole rings are stabilised in the [LS–LS] state even at room temperature, due to the electron donor properties of the methyl substituents increasing the ligand field experienced by the iron(II) centres. The ligand bearing both an ethyl group at the 2-position and a methyl group at the 5-position of the imidazole rings, L^{5Et2Me} (Fig. 3.2), should be similarly electron donating and provide an increased ligand field. However, none of the four solvatomorphs of $[Fe^{II}_2(L^{5Et2Me})_3](ClO_4)_4$·*solvents* are [LS–LS] at any temperature. The authors suggest that this is due to steric effects outweighing the electronic effects.[53] The yellow single crystals initially obtained, as plates$^+$ (·$0.5H_2O$·$3CH_3CN$) and blocks$^+$ (·$2CH_3CN$), on isolation and standing in air gave differently solvated plates$^\#$ (·$0.5H_2O$) and blocks$^\#$ (·$0.5CH_3CN$), respectively. VT magnetic studies on the latter samples showed that the blocks$^\#$ remain [HS–HS] down to 5 K (av. Fe–N & Σ for Fe1&2: 2.204 Å & 134.0°), whilst the plates$^\#$ showed a reversible and abrupt [HS–HS] ↔ [HS–LS] ST at 120 K. In contrast to the latter result, an X-ray crystal structure of the plates$^+$ at 163 K showed two different iron(II) centres in the asymmetric unit and revealed a localised [HS–LS] state (av. Fe–N & Σ: Fe1 2.188 Å & 137.6°; Fe2 2.009 Å & 64.6°), not the [HS–HS] state expected from the magnetic data on the plates$^\#$. Clearly the difference in the amount and nature of the solvent molecules present in the crystal lattice plays a key role in what spin state is observed: it is well known that the presence or absence of crystallisation solvent molecules affects SCO behaviour.[18, 42]

3.2.4.2 Pyrazole and Pyrazolate Containing Ligands

In 2004, Murray and co-workers prepared $[Fe^{II}_2(pypzH)_2(\mu\text{-}pypz)_2(NCSe)_2]\cdot 2H_2O$, where pypzH is 3-(2′-pyridyl)pyrazole and $pypz^-$ its anion (Figs 3.1 and 3.4), under nitrogen in aqueous methanol with a pinch of ascorbic acid, in 9% yield (C≡N 2093 cm^{-1}).[55] VT magnetic susceptibility measurements on the red-orange crystalline material revealed a complete and abrupt 1-step ST, [HS–HS] to [LS–LS], with $T_{1/2}$ = 225 K, but no hysteresis. This was confirmed by both crystal structure determinations (av. Fe–N & Σ for Fe1&2: 298 K, 2.181(7) Å & 72.1°; 123 K, 1.995(5) Å & 46.6°) and VT ^{57}Fe Mössbauer spectra (295 K, HS doublet δ_{iso} = 0.98(1) and ΔE_Q = 1.83(1) mm s^{-1}; 77 K, LS doublet δ_{iso} = 0.48(1) and ΔE_Q = 0.52(1) mm s^{-1}).[55] No phase change occurred during the ST to [HS–HS] (remains *Pbca*) but the cell volume expands 9.3%. At both temperatures the asymmetric unit consists of two half dinuclear complexes, with the other halves generated by a centre of inversion (Fig. 3.4). The crystallographically unique iron(II) centre is coordinated to one terminal NCSe anion, a terminal bidentate pypzH ligand, and to two bridging deprotonated $pypz^-$

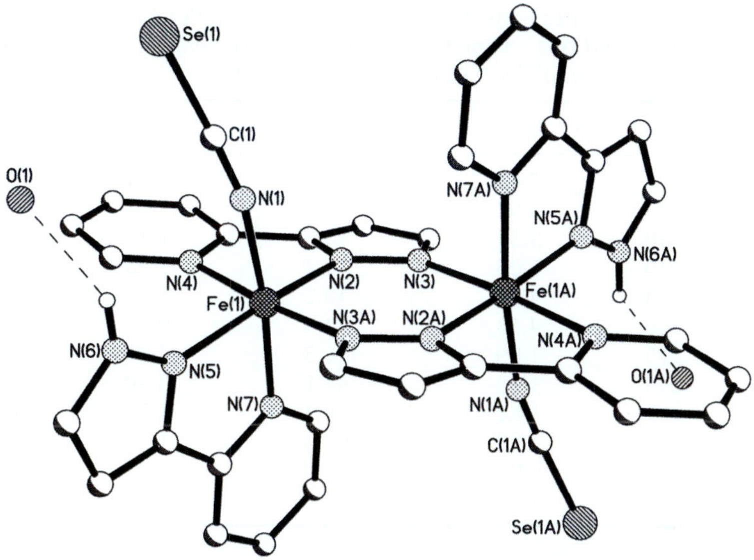

Figure 3.4 *Perspective view of complex $Fe^{II}{}_2(pypzH)_2(\mu\text{-}pypz)_2(NCSe)_2]\cdot 2H_2O$ at room temperature. The NH hydrogen atoms are shown; the OH hydrogen atoms are not as they are not present in the cif file; the CH hydrogen atoms are omitted for clarity.*[55]

ligands, one of which is coordinated to this iron centre in a bidentate mode and the other in a monodentate mode. The uncoordinated NH of the pyrazole ring is weakly hydrogen bonded to the water molecule of crystallisation. There is no π–π stacking between the dinuclear molecules. The lack of strong intermolecular interactions between the complexes is consistent with the absence of hysteresis.

A classic example of a dinucleating ligand designed to confer SCO properties on two iron(II) centres is 3,5-bis(2′-pyridyl)pyrazolate (bpypz$^-$, Fig. 3.2), which has been employed to very good effect by Kaizaki and co-workers.[17, 56, 65] The resulting family of complexes, *trans*-$[Fe_2(\mu\text{-bpypz})_2(NCE)_2(R\text{-Py})_2]$, contain two iron(II) centres doubly-bridged by two bpypz$^-$ ligands which occupy all eight of the equatorial positions, with each of the pairs of apical positions occupied by a substituted pyridine molecule (R-Py) and an NCE anion (E = S, Se or BH_3). The axial donors on adjacent iron centres are *trans* to one another.

VT magnetic studies on *trans*-$[Fe_2(\mu\text{-bpypz})_2(NCBH_3)_2(4\text{-PhPy})_2]$, where 4-PhPy = 4-phenylpyridine, revealed a relatively abrupt 2-step ST.[55] X-ray crystal structure determinations at 100, 200 and 296 K provided details of the [LS–LS], [HS–HS]:[LS–LS] 1:1 and [HS–HS] spin states, respectively. No phase change was detected during ST (remains $P\bar{1}$). One iron(II) centre is present in the asymmetric unit at 100 and 296 K (other half of the complex generated by $\bar{1}$), with average Fe–N of 1.979 and 2.167 Å respectively, consistent with the [LS–LS] and [HS–HS] states, respectively. At 200 K the cell volume doubles and the asymmetric unit comprises two iron(II) centres of different dimetallic complexes, with average Fe–N of 2.000 and 2.116 Å, consistent with a 1:1 mixture of [LS–LS] and [HS–HS].[66] This was the first structural characterisation of this type of mixed spin state. Interestingly, the yellow NCS analogue, namely *trans*-$[Fe_2(\mu\text{-bpypz})_2(NCS)_2(4\text{-PhPy})_2]$, synthesised by the same authors,[57] showed an incomplete but relatively abrupt 1-step ST, with a lower $T_{1/2}$ (150 K), as expected for the weaker NCE co-ligand NCS.[9]

Kaizaki and co-workers[16] also studied the electronic effect of pyridine ring substituent R on the SCO properties in two sets of *trans*-$[Fe_2(\mu\text{-bpypz})_2(NCE)_2(R\text{-Py})_2]$ complexes, with NCE = $NCBH_3$ or NCS. In both sets R-Py is 4-methylpyridine (4-MePy), 4-(*N,N*,-dimethylamino)pyridine (4-Me_2NPy), pyridine (Py),

***Table 3.3** SCO-parameters for the series trans-[Fe^{II}_2(μ-bpypz)$_2$(NCE)$_2$(R-Py)$_2$] and least-squared fitted thermodynamic parameters (ΔH and ΔS), interaction parameter Γ, cooperativity factor C and T(LIESST). Reproduced from [16] with permission of The Royal Society of Chemistry.*

NCE	R-Py	$T_{1/2}$ (K)	ΔH (kJmol^{-1})	ΔS (JK^{-1}mol^{-1})	Γ (kJmol^{-1})	C	T(LIESST) (K)[65]
NCS	3-BrPy	158	7.63	48.3	2.41	0.92	54
NCS	3-ClPy	148	5.97	40.4	2.45	0.97	55
NCS	3-MePy	130	5.32	40.8	1.13	0.52	51
NCS	Py	129	5.06	39.2	1.93	0.90	59
NCS	4-Me_2NPy	113	4.54	40.2	1.44	0.76	NA
NCS	4-MePy	106	4.53	42.6	0.82	0.47	59
$NCBH_3$	3-BrPy	235	14.9	56.5	3.06	0.78	NA
$NCBH_3$	3-ClPy	235	15.3	65.1	3.10	0.79	NA
$NCBH_3$	3-MePy	225	14.2	63.3	1.52	0.41	NA
$NCBH_3$	Py	207	12.9	62.6	1.54	0.46	NA
$NCBH_3$	4-Me_2NPy	188	11.9	63.0	1.47	0.47	NA
$NCBH_3$	4-MePy	180	11.6	64.6	0	0	NA

NA = not available.

3-methylpyridine (3-MePy), 3-chloropyridine (3-ClPy) or 3-bromopyridine (3-BrPy). This study clearly confirmed the expectation[9] that $NCBH_3^-$ imposes a stronger ligand field and induces a higher $T_{1/2}$ than NCS^- (Table 3.3, Fig. 3.5). Remarkably good linear correlations were obtained, for both sets of complexes, between the Hammett constant (σ^n: electron donating substituents have a negative value, electron withdrawing ones have a positive value, H is by definition zero) and the $T_{1/2}$. Specifically, for both sets the $T_{1/2}$ increased as follows: 4-MePy < 4-Me_2NPy < Py < 3-MePy, 3-ClPy ≤ 3-BrPy (Fig. 3.5). This is consistent with the more electron withdrawing substituents increasing the π-backbonding (increased π-acceptance) and decreasing the σ-bonding of the pyridine ligands, increasing the soft acidity of the iron(II) centres, stabilising the LS state and therefore increasing the $T_{1/2}$. This is an important study as it was the first, and still rare, investigation of

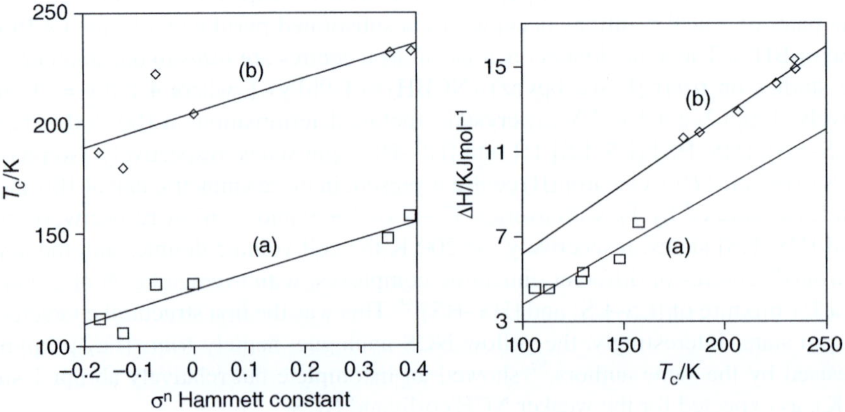

***Figure 3.5** Rare examples of correlations of $T_{1/2}$ with (left) Hammett constant of substituent R and (right) ΔH (see Table 3.3) in a family of related SCO complexes, specifically in trans-[Fe^{II}_2(μ-bpypz)$_2$(NCE)$_2$(R-Py)$_2$], where (a) E = S or (b) E = BH_3. Reproduced from [16] with permission of The Royal Society of Chemistry.*

the electronic effect of systematic variations in a ligand substituent in a family of closely related SCO-active complexes to actually identify useful correlations. Such studies open up the possibility of predicting the $T_{1/2}$ of the SCO complex by consideration of the nature of the proposed substituent on the pyridine ligand *before actually preparing it*. This is highly desirable in SCO chemistry if we are to move from good luck to good management of the properties obtained. However, the challenges of avoiding steric and/or packing effects on altering substituents mean that the observation of useful correlations is likely to remain a relatively rare occurrence for studies on discrete complexes in the solid state.

In a later publication the same authors studied the photomagnetism of the NCS family and found that the data obtained fits the established formula,[67] $T(\text{LIESST}) = T_0 - 0.3T_{1/2}$, when T_0 is 100 K (Table 3.3).[65]

3.2.4.3 Triazole and Triazolate Containing Ligands

In 2007 Murray and co-workers[58] reported a series of *triazolate* analogues of the *pyrazolate*-based *trans*-$[Fe^{II}_2(\mu\text{-bpypz})_2(NCE)_2(R\text{-Py})_2]$ complexes described in the previous Section.[17, 56, 65] Two of the three $[Fe_2^{II}(bpytz)_2(ECN)_2(Py)_2]$ complexes, where bpytz = 3,5-bis(2′-pyridyl)-1,2,4-triazolate (Fig. 3.2), were stabilised in the [HS–HS] state at all temperatures (E = S or Se, both yellow) whereas one (E = BH_3, fine orange needles) underwent a reversible 2-step ST with parameters $T_{1/2}{}^1 = 194$ K and $T_{1/2}{}^2 = 151$ K. It was not possible to determine whether the ST occurred via a localised [HS–LS] state or a mixture of [HS–HS] and [LS–LS] states.

Brooker and co-workers[59] synthesised the *bis-terdentate* triazole ligand PMAT, where PMAT = 3,5-bis{[*N*-(2-pyridinemethyl)amino]methyl}-4*H*-1,2,4-triazole (Fig. 3.2), in which the central 1,2,4-triazole ring has been functionalised at the 3- and 5-positions with pyridine-based pendant arms. The resulting pale yellow $[Fe^{II}_2(PMAT)_2](BF_4)_4\cdot DMF$ complex was reported in 2005 to undergo an abrupt and complete half ST, [HS–HS] ↔ [HS–LS], with $T_{1/2} = 224$ K.[30, 32] The structure was determined at 298 and 123 K: the latter was the first structure determination of a *localised mixed-spin state complex*, and clearly showed it was not a mixture of [HS–HS] and [LS–LS] states (Table 3.2). At 298 K half of the complex is present in the asymmetric unit (other half generated by $\bar{1}$), with both iron centres HS (av. Fe–N & Σ: 2.188 Å & 117.5°). On cooling to 123 K one axis doubles and the entire complex is in the asymmetric unit, with one iron centre clearly LS and the other HS (av. Fe–N & Σ: Fe1 1.998 Å & 64.9°; Fe2 2.202 Å & 133.1°).[30]

Also for the first time, *standard* VT ^{57}Fe Mössbauer spectra (without applied field)[32] were used to determine that the ST proceeds from [HS–HS] to localised [HS–LS]. On cooling, the ST of one HS centre to LS at 224 K is communicated, by local distortions of the two highly constrained bridging ligands that provide all 12 donors to the 2 metal centres, to the other metal centre in the dimetallic complex, giving rise to a new quadrupole doublet, for HS in [HS–LS], which is distinct from the doublet observed for HS in [HS–HS].[32] There was no evidence of a second SCO to the fully low spin [LS–LS] form, even under 10.3 Kbar at 4 K.[31]

In 2011 the same authors[15] reported that the choice of monoanion X (Cl, PF_6, SbF_6, CF_3SO_3, tetra(4-fluorophenyl)borate and hexadecylsulfonate), and the degree and nature of the solvation, had a significant effect on the SCO properties of the $[Fe^{II}_2(PMAT)_2]^{4+}$ cation. Three of the six new complexes were structurally characterised (X = Cl, CF_3SO_3 and $C_{16}H_{33}SO_3$), in all cases revealing the same overall structure as for the original BF_4 complex (see above, Table 3.2). The X = Cl complex was stabilised in the localised [HS–LS] state at 90 K (av. Fe–N 2.181 and 1.986 Å, entire molecule in the asymmetric unit) and in the [HS–HS] state at 293 K (av. Fe–N 2.155 Å, half of the complex in the asymmetric unit). Due to a lack of sufficient clean complex no magnetic or Mössbauer data was reported for that complex. The CF_3SO_3 and $C_{16}H_{33}SO_3$ analogues were both [HS–HS] at 90 K (av. Fe–N 2.119 & 2.132 Å for CF_3SO_3 with two molecules in the asymmetric unit; 2.073 Å for $C_{16}H_{33}SO_3$ with only half of the complex in the asymmetric unit). Due to the highly constrained nature of the PMAT ligand, the distortion parameter found in these PMAT complexes is consistently higher (av. Σ: HS > 109°; LS < 70°) than those found in related mononuclear $[Fe^{II}L_2(NCS)_2]$ complexes[68] (av. Σ: HS 80(5)°; LS 47(5)°). The isolated samples of these two complexes were *differently solvated* to the single crystals (Table 3.2), which led to different magnetic properties. Specifically, at 90 K magnetic and Mössbauer data showed that both are [HS–LS], not [HS–HS] as seen structurally. Further, the magnetic (4.8 BM per Fe) and Mössbauer data for the CF_3SO_3 analogue at RT is consistent the presence of 25% of the iron(II) centres in the LS–state, that is a 2:1 mixture of [HS–HS]:[HS–LS]. On further cooling a gradual almost complete [HS–HS] ↔ '[HS–LS]' SCO was observed centred at 180 K.

VT magnetic measurements on the remaining complexes showed that the PF_6 analogue undergoes a gradual and partial [HS–HS] ↔ '[HS–LS]', $T_{1/2}$ ~ 120 K, where approximately only 25% of the sample undergoes SCO. The SbF_6 analogue behaves similarly but with $T_{1/2}$ ~ 190 K. However, it is more complex as it undergoes an irreversible phase change during the first cooling cycle, but subsequent warming–cooling cycles then follow the same path. The $B(PhF)_4$ and $C_{16}H_{33}SO_3$ analogues showed an onset of gradual [HS–HS] ↔ '[HS–LS]' SCO $T_{1/2}$ > 300 K. These last two complexes showed the highest $T_{1/2}$ in this series, probably due to their size and the extensive interactions within the crystal lattice of the complexes. The $C_{16}H_{33}SO_3$ analogue packs with the long tailed anions interdigitating nicely, and separating the cations by about 31 Å.

This study of seven different salts of the $[Fe^{II}_2(PMAT)_2]^{4+}$ cation demonstrates once again that the intuitive expectation that packing effects (modified by anion and/or solvent of crystallisation) might be secondary to the primary effects of the choice of ligand bound to the metal ion(s), on the SCO behaviour of the metal ion(s), is naïve, as clearly packing effects impact strongly here, as the two PMAT ligands providing all 12 donors to the two iron centres are constant yet a wide range of magnetic behaviour is observed.

No evidence of the [LS–LS] state has been yet observed for any diiron(II) complex of any highly constrained *bis-terdentate* ligand, even when the neutral triazole unit in PMAT is replaced by a negatively charged, stronger field, pyrazolate unit. The highly air sensitive complex $[Fe_2{}^{II}(PMAP)_2](SbF_6)_2{\cdot}2.25IPA$, where PMAPH is 3,5-bis{[*N*-(2-pyridinemethyl)amino]methyl}-pyrazole (Fig. 3.2), is stabilised in the [HS–LS] state between 1.8–300 K.[69] Above 300 K VT magnetic susceptibility increases, consistent with three possible scenarios: (i) population of the [HS–HS] state, (ii) degradation of the sample and/or (iii) desolvation of the sample. No X-ray crystal structure was obtained for this complex, so it was not possible to determine the nature of this mixed spin state, either localised or 1:1 mixture of pure HS and LS states.

The same authors reported, in 2011, the extension of this approach to the *bis-tetradentate* ligand diPMAPH[70] (Fig. 3.2); adding an extra pendant arm to each side of the pyrazole ring in PMAPH. Specifically, in 1:1 MeOH/IPA, the 1:2 reaction of diPMAPH with *in situ* prepared $[Fe^{II}(MeOH)_4(NCS)_2]$, resulted in $K[Fe^{II}_2(diPMAP)(NCS)_4]{\cdot}2IPA$. The NCSe analogue could not be obtained by this route or otherwise. Rather, the use of $[Fe^{II}(Py)_4(NCSe)_2]$ gave instead the neutral complex $[Fe^{II}_2(diPMAP)(NCSe)_2(SeCN)(Py)]$. The IR spectrum is consistent with two NCSe anions being N-bound while the third is Se-bound. No crystal structures were determined. Both complexes are stabilised in a '[HS–LS]' mixed state in the 300–2 K temperature range (VT magnetic and Mössbauer spectroscopy at 4.6 K).

3.2.4.4 Jäger-Type Ligands

In 2005 Weber and co-workers[60] reported a series of dinuclear iron(II) complexes inspired by the mononuclear imine-based iron(II) SCO-active complexes of Jäger and co-workers.[71] The complexes, $[Fe^{II}_2(L^R)(Py)_4]{\cdot}xPy$, where L^R = tetra-imine ligands derived from a central 1,2,4,5-tetraaminobenzene and two substituted malonodialdehyde type units (Fig. 3.2), were made under Schlenk conditions from 1 : 2.6 L^R : iron(II) acetate in pyridine. Changing the substituents in the dicarbonyl units modified the electronic and steric properties. The two N_2O_2 ligand pockets each bind an iron centre equatorially, with the apical positions occupied by two pyridine donors. The VT magnetic properties were measured on dry pulverised samples. $[Fe^{II}_2(L^{(COOEt)OEt})(Py)_4]{\cdot}Py$, is stabilised in the [HS–HS] state; J = –1.08 cm^{-1}, g = 1.96, H = –2J

($S_1{\cdot}S_2$). In contrast, $[Fe^{II}_2(L^{(CN)OEt})(Py)_4]$ undergoes a gradual [HS–HS] ↔ '[HS–LS]' SCO, with $T_{1/2} \sim$ 175 K. $[Fe^{II}_2(L^{(COOEt)Me})(Py)_4]{\cdot}xPy$ and $[Fe^{II}_2(L^{(COMe)Me})(Py)_4]{\cdot}xPy$ can be isolated as different pyridine adducts and, as commonly occurs, the solvatomorphs present different SCO behaviour. The x=0 sample, $[Fe^{II}_2(L^{(COMe)Me})(Py)_4]$, has a relatively abrupt but incomplete [HS–HS] ↔ [HS–HS]:[LS–LS] 1:1 ($T_{1/2} \sim$ 150 K) with a small hysteresis, whereas the x = 1 solvate is stabilised in the '[HS–LS]' mixed spin state 5–295 K. In the case of $[Fe^{II}_2(L^{(COOEt)Me})(Py)_4]$ a two-step [HS–LS] ↔ '[HS–LS]' is observed ($T_{1/2}{}^1 \approx 260$ K and $T_{1/2}{}^2 \approx 90$ K). Structure determinations on the x = 7 solvate, $[Fe^{II}_2(L^{(COMe)Me})(Py)_4]{\cdot}7Py$, at 200 and 125 K, confirmed it was fully [HS–HS] at 200 K (asymmetric unit is half the complex, other half generated by $\bar{1}$; av. Fe–N/O 2.18/2.01 Å) whereas at 125 K, the cell volume quadruples and the asymmetric unit comprises two crystallographic independent complexes, one [LS–LS] (av. Fe–N/O 1.92½.007 Å) and one [HS–HS] (av. Fe–N/O 2.166/2.026 Å). It is likely that the SCOs observed for the other complexes of this type also occur from [HS–HS] to [LS–LS] through a 1:1 mixture of the two states. Solution NMR studies revealed different but gradual partial SCO.[72]

In 2008 the same research group reported $[Fe^{II}_2(L^{(COMe)Me})(MeIm)_4]{\cdot}xMeIm$,[61] where two *N*-methylimidazole ligands (MeIm) occupy the axial positions on both iron(II) centres. The x = 0 powder (dark brown) and x = 4 crystalline (black) samples present different magnetic properties. The powder had a gradual SCO, $T_{1/2} = 229$ K, with no thermal hysteresis, whereas the crystalline sample showed an abrupt, asymmetric, ST, with parameters $T_{1/2}{\downarrow} = 178$, $T_{1/2}{\uparrow} = 199$ and $\Delta T_{1/2} = 21$ K. The hysteresis and asymmetric nature of the ST was confirmed by Mössbauer spectroscopy. VT X-ray crystallography on the crystalline sample confirmed that the ST occurred from [HS–HS] → [LS–LS], and revealed that changes in the 3D packing interactions during the change in spin state (including disorder/order of the MeIm of crystallisation) led to the asymmetric and hysteretic ST. This was only the third report of a dinuclear complex with hysteresis,[33, 63] and it currently holds the record loop width for a diiron(II) complex (21 K, scan rate unknown; NB far wider loop widths are known for both mononuclear[10] and polynuclear[73] complexes).

3.2.4.5 Dihydroxy-Benzoquinone-Type Ligands

In 2007 Miller and co-workers[62] reported that the dark-brown complex $[Fe^{II}_2(TPA)_2(\mu\text{-}DBQ)](BF_4)_2$, where TPA = tris(2-pyridylmethyl)amine and DBQ^{2-} = 2,5-di*tert*-butyl-3,6-dihydroxy-1,4-benzoquinone dianion (Fig. 3.2), undergoes a relatively abrupt, and almost complete, ST, with $T_{1/2} \sim 290$ K. X-ray crystal structures at 208 and 298 K show that the bisbidentate DBQ^{2-} moiety bridges the two iron centres, whilst a capping tetradentate TPA ligand completes each of the N_4O_2-coordination spheres. The average Fe–N and Fe–O distances on either side of the ST do not vary as much as usual: 1.930(2) and 1.955(1) Å at 208 K vs. 1.955(1) and 2.006(1) at 298 K, but are consistent with increased HS character with increased temperature. The DBQ^{2-} fragment is potentially noninnocent,[74] but the bond lengths within it are in agreement with the dianion form, and are nearly identical at the two temperatures. A VT Mössbauer study was not performed but would be useful to confirm that this is not a case of valence tautomerism. Despite CH–π and offset face-to-face π interactions between the pyridyl moieties and the benzene ring of the DBQ^{2-} fragment, no hysteresis is observed. The complex also undergoes three reversible 1e redox processes.

The same authors[63] reported that dark brown $[Fe^{II}_2(TPA)_2(\mu\text{-}THBQ)](BF_4)_2$, where $THBQ^{2-}$ = 2,3,5,6-tetrahydroxy-1,4-benzoquinonate (Fig. 3.2), undergoes an abrupt [HS–HS] ↔ [HS–HS]:[LS–LS] 1:1 ST centred at ~250 K with a ~10 K hysteresis loop (VT magnetic measurements). Below 50 K the magnetic moment increases to a maximum of 6.25 BM at 9 K, consistent with ferromagnetic interactions between the HS centres. Dark brown crystals suitable for X-ray crystallography were obtained from an undisturbed methanol solution. At both 208 and 295 K the asymmetric unit consists of two halves of two independent iron(II) complexes, the other two halves are generated by $\bar{1}$). The coordination is the same as for the previous complex, but here the bridging ligand is $THBQ^{2-}$ instead of DBQ^{2-}. Unlike the previous complex, the usual change in bond lengths is observed on ST: at 208 K the average Fe–N and Fe–O bond lengths are 1.953(2) and 1.958(2) Å for one iron(II) centre compared to 2.089(2) and 2.180(2) Å for the other. This clearly shows that there is a 1:1 mixture of [HS–HS] and [LS–LS] complexes at 208 K. At 295 K the average Fe–N and Fe–O bond lengths for both iron centres are consistent with the presence of two [HS–HS] complexes (2.087(3) and 2.163(2) Å). Whilst the parameters for the bridge differ, they are considered consistent with it being in the $THBQ^{2-}$ form and not in the $THBQ^{3-}$ radical. Mössbauer spectroscopy confirmed the presence of 1:1 HS:LS at 77.5 K ($\delta_{LS} = 0.37$ mm s^{-1}, $\Delta E_{Q(LS)} = 0.52$ mm s^{-1}, relative area = 51% and $\delta_{HS} = 0.98$ mm s^{-1}, $\Delta E_{Q(HS)} = 3.21$ mm s^{-1}, relative area = 49%).

3.3 Higher Nuclearity Iron(II) Compounds

3.3.1 Trinuclear Iron(II) Complexes

The three trinuclear systems presented in this section were prepared using the two approaches, controlled/designer-ligand and supramolecular, introduced above. Each has a unique motif, namely a bent alkoxo-bridged system, a well-spaced cluster system and a linear triazole-bridged system. In the former and latter cases only one of the iron(II) centres undergoes SCO, whereas in the cluster system a very gradual and incomplete SCO from the fully LS-state is observed upon heating.

Tuchagues and co-workers[75] synthesised the *bent* trinuclear complex $[Fe^{II}_3(\mu\text{-}L^{DAP})_2(SCN)_4(H_2O)]$ where $(L^{DAP})^-$ is the deprotonated version of the ditopic Schiff base ligand that results from the condensation of two equivalents of 2-acetylpyridine and one equivalent of 1,3-diaminopropan-2-ol (Fig. 3.6). Single crystals of

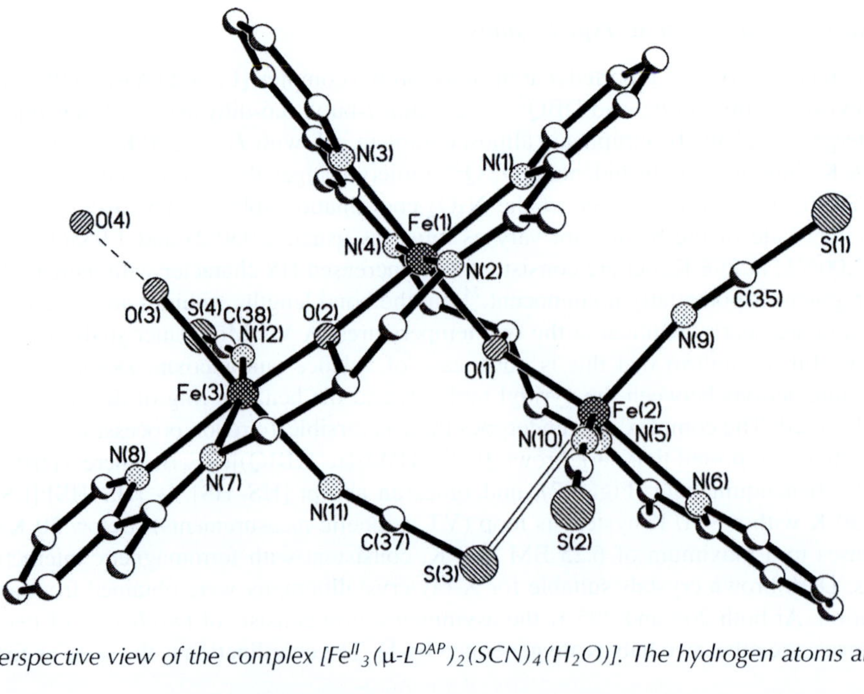

Figure 3.6 Perspective view of the complex $[Fe^{II}_3(\mu\text{-}L^{DAP})_2(SCN)_4(H_2O)]$. The hydrogen atoms are omitted for clarity.[75]

this deep blue complex, obtained in 39% yield by slow evaporation of the reaction solution, were characterised by X-ray crystallography at 180 K. The asymmetric unit comprises the entire trinuclear complex. The three crystallographically independent iron(II) centres are arranged in a triangular geometry and have differing coordination spheres (Fig. 3.6). The central iron centre, Fe(1), is coordinated to two N_2O-tridentate ligand strands (N-donors are pyridine and imine; O-donor is bridging alkoxide), so is in a N_4O_2 environment. The Fe(1)–N/O bond distances (1.977(4)–2.054(4) Å; $\Sigma = 92.7°$) are consistent with it being LS at 180 K. The Fe(3) centre is also found in a N_4O_2 coordination sphere, but it is made up of the second N_2O-pocket of one of the ligand strands already coordinated to Fe(1), plus two N-bound NCS anions and one water molecule. This iron centre is HS (Fe(3)–N/O 2.044(3)–2.244(4) Å; $\Sigma = 72.6°$, lower distortion than in LS Fe(1) is presumably due to 3 monodentate ligands in place of 1 tridentate ligand). The Fe(2) centre is in an unusual $N_4O(S)$ environment and is also HS at 180 K (Fe(2)–N/O 2.022(3) to 2.215(3) Å; Fe(2)...S 3.063(2) Å; $\Sigma = 115.8°$). It is coordinated to the second N_2O-pocket of the other ligand strand already coordinated to Fe(1), two N-bound NCS anions and a weakly interacting S-bound NCS anion that forms a 1,3-bridge between Fe(2) and Fe(3). VT magnetic measurements and Mössbauer spectroscopy show that *only one* of the iron(II) centres undergoes a very gradual and incomplete SCO, between fully LS at low temperature and mostly HS at RT. This is expected to be Fe(1), which is in a unique $(N_{pyr})_2(N_{imine})_2(O_{alkoxo})_2$ environment for SCO.

Murray and co-workers[50] synthesised the tritopic ligand L^{Mes} (Fig. 3.7) in which a central mesityl core has three ether-linked pendant arms, each terminating in tris-pyrazolymethane moieties. The solvent content of the resulting pale pink trinuclear complex $[Fe^{II}_3[(3,5\text{-}Me_2pz)_3CH]_3(\mu\text{-}L^{Mes})](BF_4)_6$·solvent could not be established by X-ray crystallography, microanalysis or TGA. X-ray crystal structures determined at 100 and 220 K revealed, as expected, three well separated iron(II) centres coordinated in the three N_3-donor tris-pyrazolylmethane pockets, and an N_3-donor $(3,5\text{-}Me_2Pz)_3CH$ capping co-ligand. At both temperatures the space group is $P\bar{1}$ and the asymmetric unit comprises two crystallographically independent trinuclear complexes: in both the three ether-linked arms are oriented towards the same side of the plane formed by the mesitylene group forming a cup. The arms of pairs of inversion related 'cups' interdigitate resulting in a capsule-like motif which accommodates a BF_4 anion (disordered across the $\bar{1}$ site; hydrogen bonded to a number of C–H of the pyrazolyl groups). The average Fe–N bond lengths (1.973(8) and 1.972(6) Å) and distortion parameters ($\Sigma = 24°$ and $26°$) are consistent with the complex being in the [LS–LS–LS] state at both 100 and 220 K. VT magnetic measurements showed a very broad gradual and incomplete SCO starting

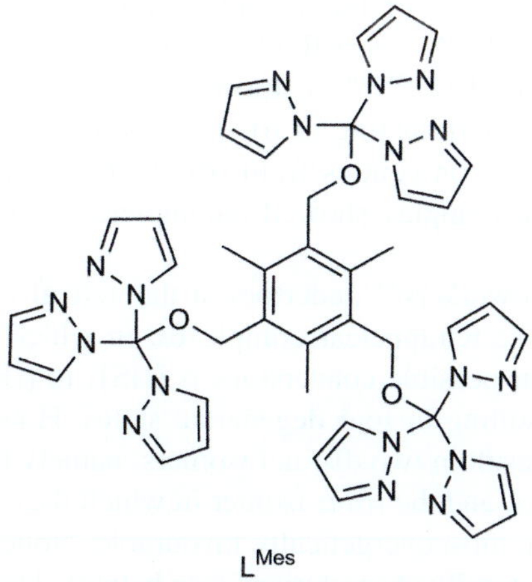

Figure 3.7 *Structure of the trinucleating ligand L^{Mes}.*

above 200 K from the [LS–LS–LS] state. Even at 350 K only 55% of the iron centres had converted to the HS state; the authors estimate $T_{1/2} \sim 347$ K. interestingly, a reproducible abnormality in the VT magnetic profile, a small rapid rise in moment, was observed at ~200 K. The authors suggested this could be due to the abrupt crossover of one of the iron(II) centres, but also point out that the X-ray crystallography results are inconsistent with this as they show the complex to be fully LS at both 100 and 220 K.

The light green *linear* trinuclear complex reported in 2011 by Murugesu and co-workers,[76] $[Fe^{II}_3(npt)_6(EtOH)_4(H_2O)_2](ptol)_6{\cdot}4EtOH$ where npt = 4-(4′-nitrophenyl)-1,2,4-triazole and ptol = *p*-toluensulfonate, was anticipated to undergo SCO at the central iron(II) ion only, given earlier reports on closely related complexes.[77, 78] As expected, VT magnetic measurements showed reversible [HS–HS–HS] ↔ [HS–LS–HS] SCO, at $T_{1/2} = 148$ K.[75] This was confirmed by X-ray crystal structures determined at 181 and 100 K. At both temperatures this revealed the expected linear arrangement of the iron centres, where the terminal iron centres are N_3O_3 coordinated, to three bridging triazole moieties and three oxygen donors (two ethanol and one water), whereas the central iron(II) centre is N_6-coordinated, to all six bridging triazole ligands. The Fe–N bond lengths of the central iron(II) centre change from typical for HS at 181 K (2.152 Å) to typical for LS at 100 K (2.009 Å), confirming the transition to the [HS–LS–HS] state. The octahedral distortion parameter, whilst remaining small, unexpectedly increased on going LS (Σ: HS 9.0°; LS 17.1°).

3.3.2 Tetranuclear Iron(II) Complexes

With only one exception, the tetranuclear complexes presented in this section can be classified into two types, molecular squares and [2×2] grids. In the molecular square complexes capping ligands block some of the coordination sites and the other coordination sites are occupied by bridging ligands which form the square. Most of these feature bridging cyanide ions, coordinated through both the C and N donors (like in Prussian Blue). As a result the two C-bound iron(II) centres, located in opposite corners of the square, remain in the LS-state throughout the experimental temperature range, whilst the two N-bound centres can undergo SCO. In contrast, when the bridging moieties are dicyanamide ions the iron centres are equivalent so in principle each of them can undergo SCO. Likewise, in the [2×2] grid complexes, in which the four iron(II) centres are coordinated to four ligand strands, all of the metal centres are equivalent, and each one can potentially undergo SCO.

The exception to the above two types of tetrametallic iron(II) complexes is the cluster $[Fe^{II}_4[(3,5\text{-}Me_2pz)_3CH]_4(\mu\text{-}L^{TetraMeBn})](BF_4)_8{\cdot}8MeCN{\cdot}2$-*tert*-BuOMe (Fig. 3.8), reported by Murray and co-workers.[50] In this case the large tetratopic ligand $L^{TetraMeBn}$, featuring a central 1,2,4,5-tetramethylbenzene core with four ether-linked pendant arms, each terminating in tris-pyrazolymethane moieties (similar to the tritopic ligand L^{Mes}, Fig. 3.7), 'insulates' the iron centres from one another. The magnetic properties of both the *solvated* and *desolvated* tetranuclear complex showed incomplete and gradual SCO, with $T_{1/2} \sim 350$ and 380 K, respectively.

Recently Borshch, Zueva and co-workers[79] undertook a theoretical analysis of the four possible spin-transitions that can occur in grid-like tetranuclear complexes, in which the four metal ions are sitting in equivalent positions. One of the four possible conversions is $[HS]_4$ to $[(HS)_3\text{–}LS]$ for which there are four possible, but identical, isomers, resulting in four degenerate states. However, the next possible transition, $[(HS)_3\text{–}LS]$ to $[(HS)_2\text{–}(LS)_2]$, can result in two distinct isomers, namely the *cis* isomer in which the two LS centres are located next to each other, and the *trans* isomer in which they are at opposite corners. According to the authors the *cis*-complex is the most energetically favourable. Nonetheless, examples of both types of isomers have been isolated and structurally characterised (see below). The next possible transitions are from $[(HS)_2\text{–}(LS)_2]$ to $[(LS)_3\text{–}HS]$ and finally to the fully LS state $[LS]_4$.

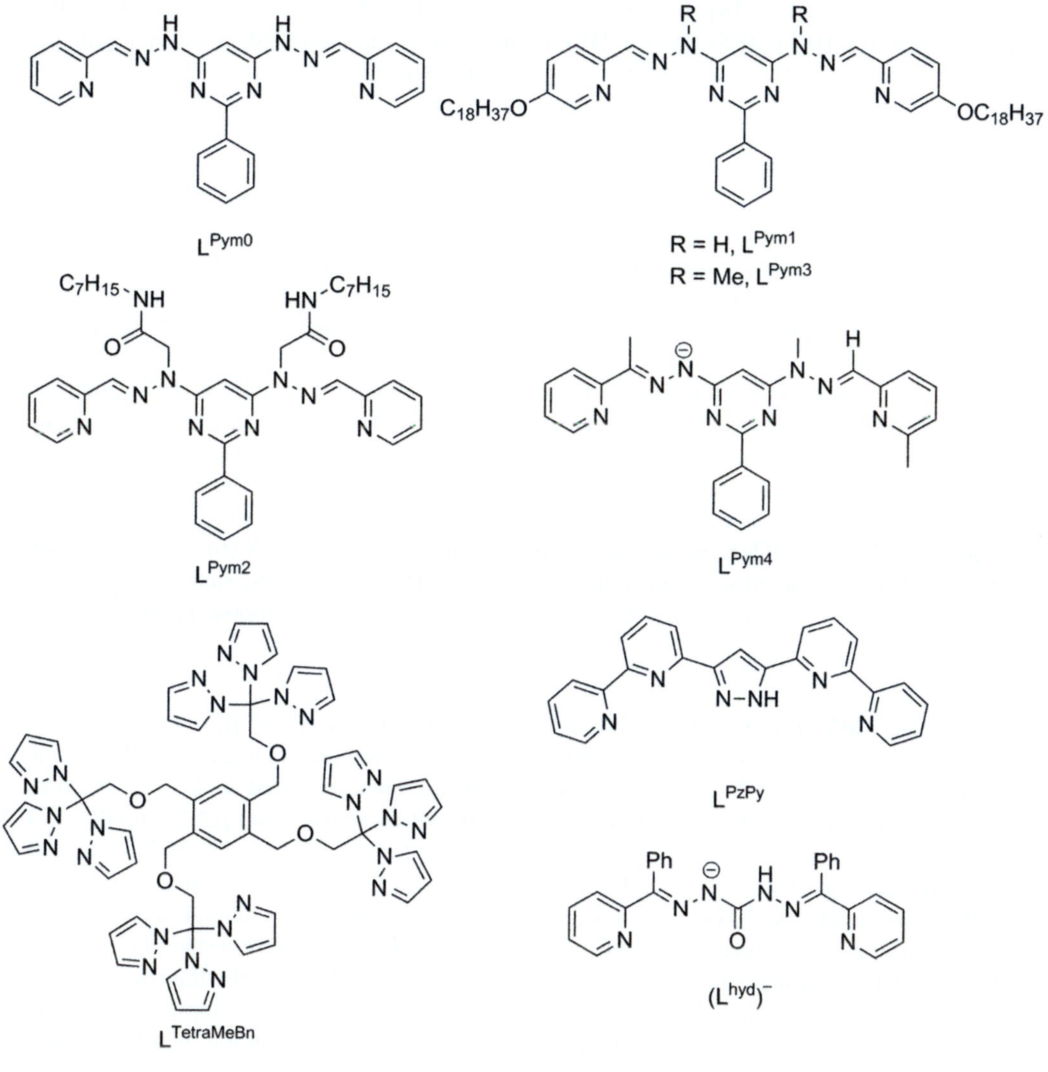

Figure 3.8 *Ligands used to generate tetranuclear iron(II) SCO complexes.*

3.3.2.1 *Cyanide Based Molecular Square Complexes*

Oshio and co-workers reported the dark red molecular square complex $[Fe^{II}(2,2'\text{-bipy})_4(tpa)_2(\mu\text{-CN})_4](PF_6)_4$ in 2005 (Fig. 3.1).[80] The structure determination showed that all four iron(II) centres have an octahedral geometry, where two $[Fe^{II}(2,2'\text{-bipy})_2]^{2+}$ units and two $[Fe^{II}(tpa)]^{2+}$ fragments are alternately arranged in the square and are bridged by four CN ligands which are C-bound to the $[Fe^{II}(2,2'\text{-bipy})_2]^{2+}$ moieties and N-bound to the $[Fe^{II}(tpa)]^{2+}$ moieties. The Fe–N bond lengths vary with temperature. At 100 K, all four iron centres are LS (Fe–N: $[Fe^{II}(2,2'\text{-bipy})_2]^{2+}$ 1.958 Å; $[Fe^{II}(tpa)]^{2+}$ 1.976 Å) whereas at 200 and 300 K one iron centre, in one of the $[Fe^{II}(tpa)]^{2+}$ fragments, is HS (Fe–N 2.154–2.165 Å). VT magnetic measurements were consistent with this, showing that on warming the complex undergoes a 2-step SCO with an abrupt

[LS–LS–LS–LS] to [LS–LS–LS–HS] SCO at ~160 K followed by the second, more gradual, SCO above 300 K, from [LS–LS–LS–HS] to [LS–HS–LS–HS]. Even at 400 K the second SCO was not quite complete. No hysteresis was observed. Mössbauer spectra confirmed these results.

In 2009 they reported $[Fe^{II}_4(2,2'\text{-bipy})_4(bpym)_4(\mu\text{-}CN)_4](PF_6)_4{\cdot}6MeOH{\cdot}4H_2O$ (Fig. 3.1), a dark red complex, and structurally characterised it at 200 K.[81] In this case the square is formed by two $[Fe^{II}(2,2'\text{-bipy})_2]^{2+}$ moieties and two $[Fe(bpym)]^{2+}$ units bridged by four CN anions, which are C-bound to the iron(II) centres in the former fragment and N-bound to the latter fragment, completing octahedral coordination for each metal centre. VT magnetic susceptibility showed a reversible, gradual and complete one-step SCO ($T_{1/2} \approx 350$ K), from $[LS]_4$ to *trans*-[LS–HS–LS–HS] for which $\chi_M T$ at 400 K is 3.42 emu mol^{-1}. The two $[Fe^{II}(2,2'\text{-bipy})_2]^{2+}$ fragments are coordinated to two strong field C-bound CN ligands and so stay LS. It is the *trans*-disposed $[Fe(bpym)]^{2+}$ moieties containing two N-bound CN ligands at which SCO occurs.

A related molecular square, containing phen and tpa (Fig. 3.1) as capping ligands, is $[Fe^{II}_4(phen)_4(tpa)_2(\mu\text{-}CN)_4](PF_6)_4$, was reported by Real and co-workers.[82] The C-bound CN ligands coordinate to the $[Fe^{II}(phen)_2]^{2+}$ fragment which remains LS. VT magnetic measurements showed that on warming the deep red complex undergoes a gradual and almost complete one-step SCO, from the pure LS-state to the *trans*-[HS–LS–HS–LS]-state, again centred above room temperature. Both spin states were characterised by X-ray crystallography (at 100 and 370 K). Interestingly, introduction of a 2-methyl-substituent onto one or two of the three pyridine rings in tpa led to stabilisation of the *trans*-[HS–LS–HS–LS]-state; no SCO to the fully LS state was observed for those two tetranuclear iron(II) squares, which was proposed to be due to increased steric hindrance.

Dicyanamide-Based Molecular Squares Tao and co-workers[83] also used the $[Fe^{II}(tpa)]^{2+}$ fragment, but linked four (not two) of them together using dicyanamide (not cyanide) linkers to generate the red molecular square complex $[Fe^{II}_4(tpa)_4(\mu\text{-}(CN)_2N)_4](BF_4)_4{\cdot}2H_2O$. This complex, reported in 2011, is the first square Fe^{II}_4 complex to undergo complete SCO of all four iron(II) centres, that is from $[LS]_4$ to $[HS]_4$. VT X-ray crystallography showed that: at 150 K it is fully LS (av. Fe–N: Fe1 1.973 Å; Fe2 1.967 Å); at 250 K it is *trans*-[HS–LS–HS–LS] (av. Fe–N: Fe1 2.130 Å; Fe2 1.988 Å); at 350 K desolvation occurs (lose $\cdot 2H_2O$) giving the fully HS-state (av. Fe–N: Fe1 2.151 Å; Fe2 2.138 Å). On cooling this anhydrous complex to 250 K then 150 K the resulting X-ray structures varied only slightly from the dihydrate. On standing for a week in air the crystals regain $2H_2O$. VT magnetic susceptibility studies confirmed that the dihydrate undergoes a 2-step, reversible and complete SCO with $T_{1/2}{}^1 = 302$ and $T_{1/2}{}^2 = 194$ K. The anhydrous sample is very similar, with parameters $T_{1/2}{}^1 = 294$ and $T_{1/2}{}^2 = 211$ K, and a small hysteresis loop (6 K) in the high temperature SCO. The complex is also LIESST active.

3.3.2.2 Grid-Like Complexes

Since Lehn and co-workers reported the first tetranuclear SCO complexes, Fe^{II}_4 grids, in 2000[14] they have since synthesised many further examples of bis-tridentate pyrimidine-based ligands and tetrametallic grids with a wide range of interesting properties, including tuning SCO by varying the ligand substituents.[84] In 2005 they showed, by NMR spectroscopy in acetonitrile, that $[Fe^{II}_4L^{Pym4}{}_4](BF_4)_8$ has two LS and two SCO-active centres.[85] Then, in 2009, they reported an extended family of ditopic bis-hydrazone pyrimidine-based ligands and iron(II) grids, made with the aim of tuning the SCO properties, namely $[Fe^{II}_4(L^{Pym0})_4](OTf)_8$ and $[Fe^{II}_4(L^{Pymn})_4](BF_4)_8$, where $n = 1$–4 (Fig. 3.8, OTf = triflate).[86] All five grids showed gradual SCO around RT, but complexes $[Fe^{II}_4(L^{Pym0})_4](OTf)_8$ and $[Fe^{II}_4(L^{Pym2})_4](BF_4)_8$ went from [HS–HS–HS–LS] to

[HS–LS–LS–LS], whereas the other complexes went from a mixture of [HS–HS–HS–HS] and [HS–HS–HS–LS] to a $[(HS)_2–(LS)_2]$ state. These complexes were not structurally characterised so it was not possible to determine whether the mixed spin state is localised (either *cis* or *trans*) or a mixture of pure states.

The grid-type complex $[Fe^{II}{}_4L^{PzPy}{}_4](BF_4)_4$[87] of the bis-terpyridine-like pyrazole-based ligand, L^{PzPy} (Fig. 3.8), reported by Meyer and co-workers, showed gradual stepwise SCO: from fully HS to the $[(HS)_3–LS]$-state with $T_{1/2} \approx 200$ K, followed by an incomplete transition to the [HS–LS–HS–LS] state, $T_{1/2} \approx 100$ K. X-ray crystallography at 133 K allowed the characterisation of the $[(HS)_3–LS]$-state. Along with SCO, this complex shows four reversible redox processes, resulting in mixed-valence grids. The mixed-valence complex $[Fe^{II}{}_2Fe^{III}{}_2L^{PzPy}{}_4](BF_4)_6$ was structurally characterised by X-ray crystallography but is SCO inactive (both Fe^{II} HS; both Fe^{III} LS). This is a nice example of a multiply-switchable complex. Such systems may be required in the future to access materials capable of more complex logic operations.

More recently, Sato and co-workers reported a grid derived from a bis-hydrazide Schiff base ligand HL^{hyd} (Fig. 3.8), namely $[Fe^{II}{}_4(\mu\text{-}L^{hyd})_4](BF_4)_4$.[88] The iron(II) centres are in octahedral N_4O_2 environments. VT magnetic susceptibility studies showed a relatively abrupt and complete SCO from $[(HS)_4]$ to the *cis*-[HS–HS–LS–LS] state, at ca. 175 K. X-ray crystallography at 123 K showed two *neighbouring* iron(II) centres were HS (av. Fe–N 1.985 Å) whilst the other two were LS (2.136 Å), confirming this is a localised *cis*-isomer spin state, not the *trans* isomer usually observed. At 283 K all four iron(II) centres are HS (Fe–N 2.139-2.160 Å). This complex also showed a LIESST effect, with *T*(LIESST) = 77 K.

3.3.3 Higher Nuclearity Mixed Metal/Valent Iron(II) Complexes

In 2007 Dunbar, Achim and co-workers[89] reported a series of pentanuclear mixed metal and mixed valence cyanide-based complexes of the type $[Fe^{II}(tmphen)_2]_3[M^{III}(CN)_6]_2$ (Fig. 3.1), where M^{III} = Fe, Co and Cr. X-ray crystallography revealed that the five metal ions in these complexes are arranged in a trigonal-bipyramidal motif. The three $[Fe^{II}(tmphen)_2]^{2+}$ moieties occupy the equatorial positions and are capped by two apical $[M^{III}(CN)_6]^{3-}$ fragments each of which bridges via a single CN ion to each of the three equatorial moieties, binding to the iron(II) centres by the N of the CN ion. VT X-ray structure, magnetic and Mössbauer studies show that the Fe^{II} centres in $[Fe^{II}(tmphen)_2]_3[Cr^{III}(CN)_6]_2$ remain fully LS, whereas in the Fe^{III} and Co^{III} analogues both Fe^{II} centres undergo gradual SCO from fully LS to fully HS, centred at ~300 K for both complexes (SCO in the Co^{III} analogue is not complete; ca. 10% of Fe^{II} remains HS at low temperature). Such complexes have a wide range of interesting properties besides SCO, for example some are Single Molecule Magnets.[90] More recently, Dunbar, Mathonière, Clérac and co-workers[91] used optical reflectivity measurements to test for a photomagnetic response in the Co^{III} analogue: this indicated that photoexcitation does occur, and that at the *surface* of this intense dark red solid it is nearly quantitative. Subsequent photomagnetic measurements showed 20% photoconversion of the bulk solid to the metastable HS–state. The Fe^{III} analogue showed only 2% photoconversion.

In 2009, Batten and co-workers reported the largest discrete SCO-system known to date.[92] Crystals of this yellow complex were obtained in almost quantitative yield (Fig. 3.9). The structure determination reveals that this complex forms a 3 nm 'nanoball', $[Fe^{II}{}_6\,[Cu^{I}(tp^{4\text{-}py})]_8]^{14+}$, comprising six octahedrally arranged iron(II) ions, plus eight cuboidally arranged copper(I) ions each of which is capped by a $(tp^{4\text{-}py})^-$ anion. The nanoballs pack inefficiently so, in addition to the internal volume of the nanoball, this is where the considerable amount of disordered solvent, and perchlorate anions, are located. There are two distinct iron(II) environments $[Fe^{II}(NCS)_2(py)_4]$ (Fe-A) and $[Fe^{II}(NCS)(CH_3CN)(py)_4]$ (Fe-B), present in ca. 56:44 ratio. The $\chi_M T$ value per iron(II) drops from 3.4 to 1.8 cm^3 mol^{-1} K on cooling from 200 to 80 K, corresponding to ca. 50% of the iron(II) centres undergoing SCO to LS. Mössbauer data is consistent with this. The others suggest that it is the Fe-B site that undergoes SCO. SCO is triggered not only thermally, but also by light (LIESST) and by removal and replacement of guest acetonitrile molecules.

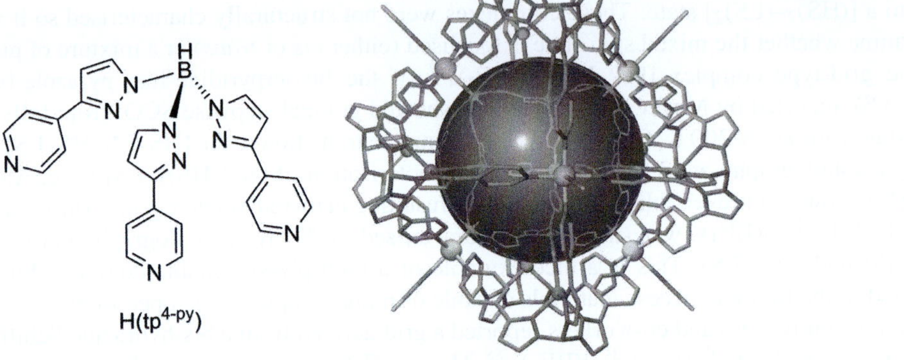

Figure 3.9 The 4-pyridyl substituted tris-pyrazolylborate ligand (left) used to cap each of the copper(I) centres in the largest known discrete SCO system, the $Fe^{II}{}_6Cu^{I}{}_8$ nanoball (right). Adpated with permission from [92]. Copyright Wiley-VCH Verlag GmbH & Co, 2009.

3.4 Iron(III)

Whilst the first SCO reported in the literature was an iron(III) complex,[93] and mononuclear SCO-active iron(III) complexes are not uncommon,[24] there are far fewer examples than for iron(II). Octahedral iron(III) complexes generally undergo gradual SCO between the LS state, $(t_{2g})^5(e_g)^0$ $S = {}^1/_2$, and the HS state, $(t_{2g})^3(e_g)^2$ $S = {}^5/_2$.[24,94] However, in some cases complexes that adopt other geometries, for example pentacoordinate, around the iron(III) centre can access the intermediate spin state (IS), $S = {}^3/_2$. In those complexes the SCO can in principle occur between the LS, IS and HS states.

In contrast to iron(II) SCO complexes, the LS state in iron(III) complexes is paramagnetic so the change observed in Fe-donor bond lengths during a crossover is not as dramatic as for iron(II). Nevertheless it is usually still possible to differentiate between the LS and HS states by X-ray crystallography.[24]

A wide variety of six-coordination environments promote SCO in iron(III),[24] the most common being N_4O_2, which is often provided by salicylaldehyde-derived Schiff bases. Not surprisingly these, usually N_3O_2-donor, ligands have been used as blocking ligands to cap metal ions, leaving one position for a ditopic N-donor ligand to bridge to a second such moiety, generating dinuclear complexes.

3.4.1 Dinuclear Iron(III) Complexes

In dinuclear complexes featuring metal ions in differing environments (e.g. CN bridged), there are four, not three, possible spin states: [HS–HS], [HS–LS], [LS–HS] and [LS–LS] (ignoring IS states). In dinuclear *iron(II)* complexes only the HS state, $S = 2$, is active (LS is $S = 0$) so intramolecular exchange only occurs in the [HS–HS] state. In contrast, in *iron(III)* complexes intramolecular exchange magnetic interactions (J) can occur between the two iron centres in all of these spin states. These interactions give rise to energy bands for all of the possible reference states in dinuclear iron(III) complexes (Fig. 3.10). The interplay between SCO and magnetic exchange produces a number of different types of SCO systems: the reader is referred to the work of Boča and co-workers for more details.[95, 96]

In 2005 Matsumoto and co-workers[97] reported $[Fe_2{}^{III}(L^{MeOsaldpt})_2(\mu\text{-vipy})](BPh_4)_4$ (Figs 3.1 and 3.11), a black complex. VT magnetic measurements showed that it undergoes a gradual 2-step SCO, from fully HS to [HS–LS] with $T_{1/2} \approx 200$ K, then a more gradual and incomplete SCO towards the fully LS state,

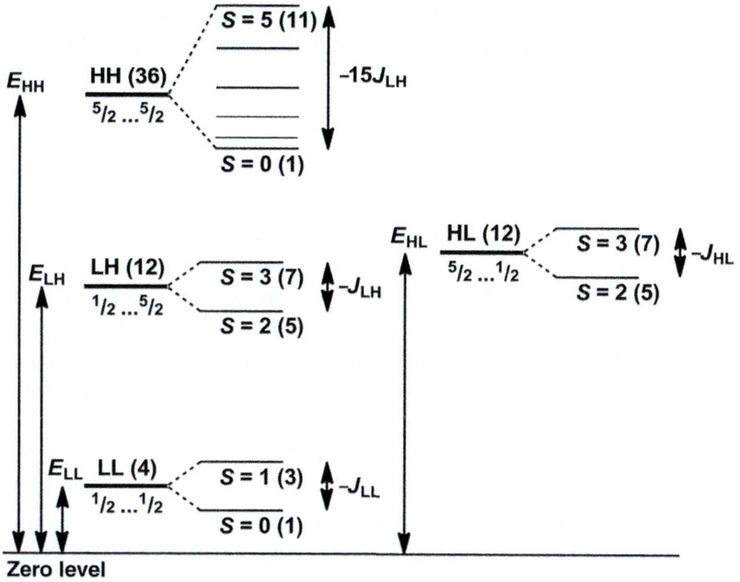

Figure 3.10 *Energy diagram of the possible reference states and corresponding spin multiplets (multiplicities in brackets), arising from the combination of magnetic exchange interactions and SCO, for dinuclear iron(III) complexes. Adpated with permission from [65]. Copyright Wiley-VCH Verlag GmbH & Co, 2009.*

centred around 70 K. X-ray crystallography data at 296 and 125 K were consistent with the fully HS and a localised [HS–LS] state, respectively. At both temperatures the iron(III) centres are crystallographically independent and are coordinated to one end-capping N_3O_2-donor Schiff base ligand with the octahedral N_4O_2 coordination environment completed by the bridging vipy ligand. At 296 K the Fe^{III}–N/O bond lengths (av. Fe–N/O: Fe(1)–N/O, 2.146/1.916 Å; Fe(2)–N/O, 2.128/1.911 Å) are clearly consistent with [HS–HS] whereas at 125 K only those around the Fe(1) centre are characteristic of HS (av. Fe(1)–N/O: 2.142/1.911 Å), whilst those around Fe(2) are significantly reduced (av. Fe(2)–N/O: 2.001/1.873 Å) clearly showing it is LS. This was the first report of a structurally characterised localised mixed spin state dinuclear iron(III) complex.

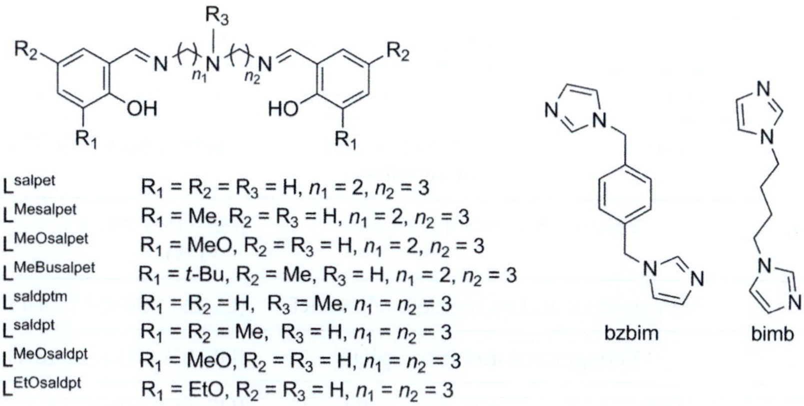

Figure 3.11 *L^R 'capping' ligands and two examples of ditopic bridging ligands L^B (bzbim and bimb; others shown in Fig. 3.1), used for the generation of discrete polynuclear iron(III) SCO-active complexes.*

They went on to instead use 1,4-bisimidazolylbutane (bimb, Fig. 3.11) as the bridging ligand, giving black blocks of $[Fe_2^{III}(L^{MeOsaldpt})_2(\mu\text{-bimb})](BPh_4)_4$.[98] VT magnetic measurements showed that it undergoes a relatively abrupt, closely-spaced two-step, reversible SCO with $T_{1/2}^{\ 1} = 90$ and $T_{1/2}^{\ 2} = 60$ K. X-ray crystallography at both 296 and 100 K revealed only one crystallographically independent iron(III) centre. At 296 K the Fe–N/O bond lengths (av. Fe–N/O 2.116/1.914 Å) are consistent with the [HS–HS] state, whereas at 100 K this is significantly reduced (av. Fe–N/O 1.981/1.869 Å), consistent with SCO to the fully [LS–LS] state. The complex was thermochromic in the solid state, from purple at RT to green at liquid nitrogen temperature.

In a similar manner, Boča and co-workers[95] prepared a series of six dinuclear complexes, bridging two $[Fe^{III}(L^R)]^+$ fragments (L^R in Fig. 3.11) by a range of bridging ligands, generating complexes of the type $[Fe_2^{III}(L^R)_2(\mu\text{-}L^B)](BPh_4)_4$, where L^B = 4,4′-bipy, vipy, etbipy, probipy and bzbim (Figs 3.1 and 3.11). All of the complexes, except for the bzbim analogue, were characterised by X-ray crystallography at 90 K (Table 3.4). Two complexes, $[Fe_2^{III}(L^{saldptm})_2(\mu\text{-}L^B)](BPh_4)_4$ where L^B = 4,4′-bipy and vipy, undergo a complete but gradual SCO from fully HS to fully LS state, centred at ca. 205 and 260 K, respectively: the average Fe–N/O bond lengths of 2.002/1.869 and 1.989/1.868 Å, respectively, clearly showed that at 90 K both these complexes are [LS–LS]. The other three structurally characterised complexes, $[Fe_2^{III}(L^{saldpt})_2(\mu\text{-}L^B)](BPh_4)_4$, where L^B = etbipy and probipy, plus $[Fe_2^{III}(L^{EtOsaldpt})_2(\mu\text{-4,4'-bipy})](BPh_4)_4$, showed very gradual and incomplete SCO: the average Fe–N/O distances for $[Fe_2^{III}(L^{saldpt})_2(\mu\text{-etbipy})](BPh_4)_4$ and $[Fe_2^{III}(L^{EtOsaldpt})_2(\mu\text{-4,4'-bipy})](BPh_4)_4$ complexes (1.984/1.877 and 1.985/1.864 Å, respectively) indicate

Table 3.4 *Summary of the SCO properties and Fe...Fe separations of some dinuclear complexes $[Fe_2^{III}(L^R)_2(L^B)](BPh_4)$. See Figures 3.1 and 3.11 for diagrams of L^R and L^B.*

L^R	L^B	$T_{1/2}$ (K)	Transition	X-ray Fe...Fe (Å)	References
$L^{MeOsaldpt}$	vipy	Two steps $T_{1/2}^{\ 1} \approx 200$ $T_{1/2}^{\ 2} \approx 70$	[HS–HS] → [HS–LS] → [LS–LS]	[HS–HS] / 296 K / 13.683(3) [LS–LS] / 100 K / 13.528(3)	97
$L^{MeOsaldpt}$	bimb	Two steps $T_{1/2}^{\ 1} \approx 90$ $T_{1/2}^{\ 2} \approx 60$	[HS–HS] → [LS–LS]	[LS–LS] / 100 K / 13.080(4) [HS–LS]^a / 296 K /13.307(4)	98
$L^{saldptm}$	4,4′-bipy	205	[HS–HS] → [LS–LS]	[LS–LS] / 90 K / 11.145^b	95
$L^{saldptm}$	vipy	260	[HS–HS] → [LS–LS]	[LS–LS] / 90 K / 13.348^b	95
$L^{saldptm}$	etbipy	~150	Very gradual and incomplete	[LS–LS] / 90 K / 13.374^b	95
$L^{saldptm}$	probipy	Very gradual and incomplete		[HS–HS] / 293 K / 11.310^b	95
$L^{EtOsaldptm}$	vipy	Very gradual and incomplete (above RT)		[LS–LS] / 90 K / 11.143^b	95
L^{salpet}	CN	Very gradual and incomplete		[HS–LS] / RT / 5.256(1)	96
$L^{MeBusalpet}$	CN	~110	[HS–HS] → [HS–LS]	[HS–LS]^a / 90 K / 5.1485(6)	96

^{a}= disordered delocalised [HS–LS]; ^{b}CIF not available.

both are in the [LS–LS] state at 90 K, whereas for complex $[Fe_2^{III}(L^{saldpt})_2(\mu\text{-probipy})](BPh_4)_4$ the average Fe–N/O of 2.155/1.917 Å showed this complex is stabilised in the [HS–HS] state at RT. The only complex that was not structurally characterised, $[Fe_2^{III}(L^{saldptm})_2(\mu\text{-bzbim})](BPh_4)_4$, shows a very gradual and incomplete spin-crossover.

They went on to employ CN anions as the bridge in such complexes,[96] thereby generating dinuclear complexes in which one of the iron(III) centres is coordinated to the N-donor of the CN bridge whereas the other one is C-bound, namely $[Fe_2^{III}(L^{salpet})(\mu\text{-CN})]ClO_4{\cdot}2H_2O$ and $[Fe_2^{III}(L^{MeBusalpet})(\mu\text{-CN})](BPh_4){\cdot}2MeCN$ (Fig. 3.11). The inherent asymmetry of the iron(III) centres breaks the degeneracy of the mixed-spin [HS–LS] pair, so the [HS–LS] and [LS–HS] states are no longer energetically equivalent (Fig. 3.10). Both complexes are stabilised as a mixture of the [LS–LS] and [HS–LS] states at low temperature: in the latter the N-bound iron(III) is HS whereas the C-bound iron(III) is LS. Both complexes were characterised by X-ray crystallography. At RT $[Fe_2^{III}(L^{salpet})(\mu\text{-CN})]ClO_4{\cdot}2H_2O$ has two crystallographically independent iron(III) centres in the asymmetric unit and is in a localised [HS–LS] state: Fe2 is N-bound to the CN bridge (av. Fe2–N/O 2.091/1.920 Å) and is HS, whereas Fe1 is C-bound to the CN bridge (av. Fe1–N/O/C 1.981/1.896/1.998 Å) and LS, as expected. At 90 K $[Fe_2^{III}(L^{MeBusalpet})(\mu\text{-CN})](BPh_4){\cdot}2MeCN$ has one unique iron(III) centre, with an inversion centre in between the two metal ions and hence in the middle of the CN bridge, resulting in an inherently disordered CN bridge. The average Fe–N/O bond lengths (1.893/2.011 Å; disordered groups not taken into account) are an average of the pure HS and LS states, so at 90 K this complex is in a delocalised [HS–LS] state. VT magnetic measurements on the complex of L^{salpet} showed a very gradual and incomplete SCO, whereas for the complex of $L^{MeBu\text{-}salpet}$ a complete SCO, [LS–LS] to [LS–HS], was observed, centred around 110 K. Mössbauer spectroscopy results were in agreement with these findings.

Lemaire and co-workers[99] synthesised a series of complexes $[Fe_2^{III}(QsalH)_2(\mu\text{-bisQsalH})]X_2$ (Fig. 3.12) where X = Cl, SCN, PF_6 and ClO_4. VT magnetic measurements showed that for the Cl, SCN and PF_6 analogues the $\chi_M T$ values at RT are lower than the expected; Mössbauer confirmed that this is due to the presence of some iron(III) centres in the LS state even at RT. Upon cooling $\chi_M T$ decreases gradually between 350 down to 20 K, then from 20 to 2 K a more rapid decrease is observed. At 2 K the $\chi_M T$ of *ca.* 3.5 $cm^3\ Kmol^{-1}$ is consistent with the presence of a 35:65 ratio of HS:LS. VT Mössbauer spectroscopy studies were generally in good agreement with the magnetic measurements, except that at 6 K the Mössbauer results suggest that the HS:LS ratio was 50:50. The ground state for these complexes appears to be the '[HS–LS]' state, however it was not possible to assign this a localised mixed spin state or a 1:1 mixture of [HS–HS]:[LS–LS]. The VT magnetic measurements on the ClO_4 analogue showed that at 350 K a large number of the iron(III) centres are stabilised in the LS state ($\chi_M T = 5.1\ cm^3\ Kmol^{-1}$). Mössbauer spectroscopy at 293 K confirmed 55% of the iron centres are LS, and that at 6 K this rises to 80% LS (20% HS).

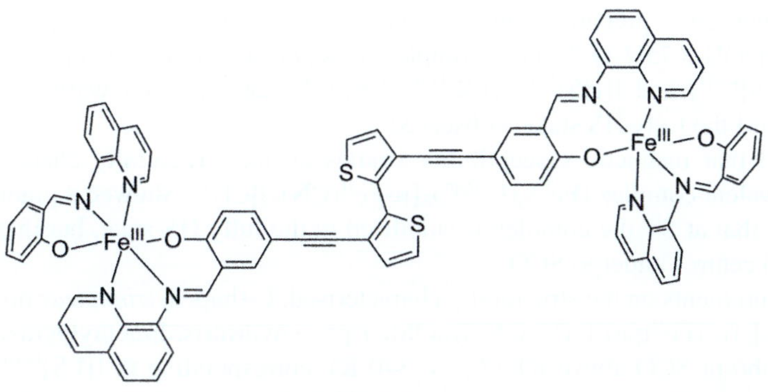

Figure 3.12 *A dinuclear iron(III) SCO-active complex formed using an elaborate ditopic bridging ligand.*

3.4.2 Mixed Metal Iron(III) Complexes

By replacing the ditopic bridges used to generate dimetallic complexes (see above) by a $[Cr(CN)_6]^{3-}$ bridging moiety, Herchel, Renz and co-workers[100] were able to connect *three* N_3O_2 capped iron(III) complexes, forming a tetranuclear T-shaped complex $[Fe^{III}{}_3(L^{saldpt})_3[\mu\text{-}Cr^{III}(CN)_6]]$ (Fig. 3.11) in 2004. VT magnetic measurements and Mössbauer spectroscopy show that it undergoes a very gradual and incomplete SCO from fully HS at RT to a mixture comprising ca. 50% HS at 4.5 K. The authors proposed that either (i) there are two sublattices, one staying fully HS and the other undergoing SCO from fully HS to fully LS, or (ii) [HS–LS–LS] and [HS–HS–LS] coexist at low temperature and undergo SCO to the fully HS state on warming.

Renz and co-workers went on to report three *pentanuclear* complexes $[Fe^{III}{}_4(L^R)_4[\mu\text{-}Sn^{IV}(CN)_4]]Cl_4$, where L^R = salpet,[101] Mesalpet and MeOsalpet[102] (Fig. 3.11), were studied by ^{119m}Sn and ^{57}Fe Mössbauer spectroscopy. The ^{57}Fe Mössbauer data for the complexes of $L^{Mesalpet}$ and L^{salpet} showed that at RT the iron(III) centres are present in a 22:78 ratio of HS:LS states for the former complex, whereas for the latter complex there are two types of HS signals and a LS signal present in a 45:47:8 ratio. At 78 K the percentage of LS iron(III) increased to 40 and 34% respectively. The ^{119m}Sn Mössbauer spectra of both complexes showed only one tin(IV) signal at RT, and likewise at low temperature for the $L^{Mesalpet}$ complex. In contrast, the L^{salpet} complex showed two signals at low temperature, corresponding to a 89:11 ratio of tin(IV) to tin(II) and consistent with some valence tautomerism. Taken together these results indicate that the L^{salpet} complex undergoes SCO at the iron(III) centres and valence tautomerism at the tin centres, with the concomitant reduction of 5.5% iron(III) to iron(II), which is not seen in ^{57}Fe Mössbauer as it is buried by the large iron(III) signal. The related systems $[Fe^{III}{}_4(L^{salpet})_4[\mu\text{-}Sn^{IV}(SCN)_4]]Cl_4$ and $[Fe^{III}{}_4(L^{MeOsalpet})_4[\mu\text{-}Sn^{IV}(SCN)_4]]Cl_4$ do not undergo SCO or valence tautomerism.

They used ^{57}Fe Mössbauer spectroscopy to show that the *hepta-* and *nona-nuclear* mixed metal complexes[103–105] $[Fe^{III}{}_6(L^{saldpt})_6[\mu\text{-}Co^{III}(CN)_6]]Cl_3$[103] and $[Fe^{III}{}_8(L^R)_8[\mu\text{-}Mo^{IV}(CN)_8]]Cl_4$, where L^R = saldpt[103] and EtOsaldpt[104] (Fig. 3.11), are stabilised in the fully HS state at RT. On cooling SCO was detected, with a HS:LS ratio at 20 K of 66:34 for the heptanuclear complex and 68:32 and 67:33 for the nonanuclear saldpt and EtOsaldpt complexes, respectively. Likewise they showed that the *dodecanuclear* mixed-metal complex $[[Fe^{III}{}_5(L^{salpet})_5][Fe^{II}(CN)_5](\mu\text{-}CN)[Fe^{III}{}_5(L^{salpet})_5][Co^{III}(CN)_5]]Cl_4$ (Fig. 3.11)[106] was fully HS at RT and underwent SCO to a HS:LS ratio of 73:18 at 20 K. None of these large clusters were structurally characterised.

3.4.3 Mixed Valence Iron(II)/(III) Complexes

For details of *pentanuclear* mixed-valent cyanide-based complexes of the type $[Fe^{II}(tmphen)_2]_3[Fe^{III}(CN)_6]_2$, reported by Dunbar and co-workers, see Section 3.3.3.

Boča and co-workers prepared the *trinuclear* mixed-valent complex $[Fe_2{}^{III}(L^{salpet})_2[\mu\text{-}Fe^{II}(CN)_5(NO)]]\cdot{}^1/_2MeOH\cdot 3.75H_2O$.[96] This complex is stabilised at low temperature as a mixture of $[[LS]^{FeIII}\text{–}[LS]^{FeII}\text{–}[LS]^{FeIII}]$ and $[[HS]^{FeIII}\text{–}[LS]^{FeII}\text{–}[LS]^{FeIII}]$ states and on warming a very gradual and incomplete SCO toward the fully HS state is observed.

They also reported that magnetic susceptibility studies on the structurally characterised, star shaped, *heptanuclear* mixed-valent complex $[Fe^{III}{}_6(L^{salpet})_6[\mu\text{-}Fe^{II}(CN)_6]]Cl_2$[107] showed a gradual SCO. Mössbauer spectroscopy showed that at RT the complex is stabilised in the fully HS-state, but that on cooling to 20 K ~22% of the iron(III) centres undergo SCO.

VT magnetic measurements on the structurally characterised, L-shaped, *trinuclear* mixed valent iron complex $[[Fe^{III}(tp^*)(CN)_3]_2[\mu\text{-}(Fe^{II}tpa)]]\cdot 4MeCN\cdot{}^tBuOMe$ (tp* = hydrotris(dimethylpyrazolyl)borate, Fig. 3.1) showed a relatively abrupt SCO above RT ($T_{1/2}$ = 340 K), corresponding to $[[LS]^{FeIII}[LS]^{FeII}[LS]^{FeIII}] \rightarrow$

$[[LS]^{FeIII}[HS]^{FeII}[LS]^{FeIII}]$.[81] X-ray crystallography data at 200 K is in agreement with a fully LS state, in which each of the two iron(III) centres is coordinated to a terminal/capping tp* ligand, and to three C-bound CN ligands one of which is N-bound to the central iron(II) moiety. Mössbauer spectroscopy at 20 K showed two quadrupole doublets in a 2:1 ratio, with parameters consistent with LS iron(III) and HS iron(II), respectively.

3.5 Cobalt(II)

Octahedral cobalt(II) (d^7) complexes can undergo thermal spin-crossover between the HS state, $S = {}^3/_2$, $(t_{2g})^5(e_g)^2$ (4.7–5.2 μ_B), and the LS state, $S = {}^1/_2$, $(t_{2g})^6(e_g)^1$ (1.8–2.2 μ_B).[24] There are several key differences between cobalt(II) and iron(II) SCO complexes: (i) in the LS state cobalt(II) is Jahn–Teller active; (ii) cobalt(II) requires a greater field strength than iron(II) to go LS and (iii) SCO-active cobalt(II) complexes are often readily oxidised to cobalt(III). SCO-active cobalt(II) complexes are far less common than those of iron(II) or iron(III).[25, 108] For SCO-active cobalt(II) complexes an N_6 environment is particularly well known.[109] Others include N_4O_2[110, 111] and N_4S_2,[112] as well as pentacoordinate N_3O_2, N_2O_3, N_2PX_2 and NP_2X_2 systems.[25]

The structural changes, driven by changes in M-donor distances on SCO, are smaller for cobalt(II) than for iron(III) or iron(II), typically ca. 0.10 vs. 0.15 vs. 0.20 Å, respectively, consistent with the smaller change in spin state.[108] Hence cobalt(II) thermal SCO is generally gradual and incomplete, with only a few examples of complete and hysteretic ST, all of which correspond to octahedral, *mononuclear* cobalt(II) complexes containing either, terpyridine-,[113] 4-terpyridone-,[114] 3-formylsalicylic acid-derived,[115–117] or pyrazine-derived imide[11] ligands.

Only three families of SCO-active discrete polymetallic cobalt(II) complexes are known, two dinuclear and one trinuclear (Fig. 3.13), all of which were reported before 2004.[25] No new systems were reported during 2004–2011, however new studies on some of the existing systems have been described so these are presented here after a brief summary of the key features of each of these three families.

3.5.1 Dinuclear Cobalt(II) Complexes

In 1998, Brooker and co-workers reported the first dicobalt(II) complex of MC^{Pz}, $[Co^{II}_2(MC^{Pz})(CH_3CN)_4]$, where MC^{Pz} is the Schiff base macrocycle derived from [2+2] condensation of 3,6-diformylpyridazine and 1,3-diaminopropane (Fig. 3.13), and showed that it was stable in five different redox states.[118] They went on to prepare the first dinuclear cobalt complex to be shown to undergo both SCO and exchange

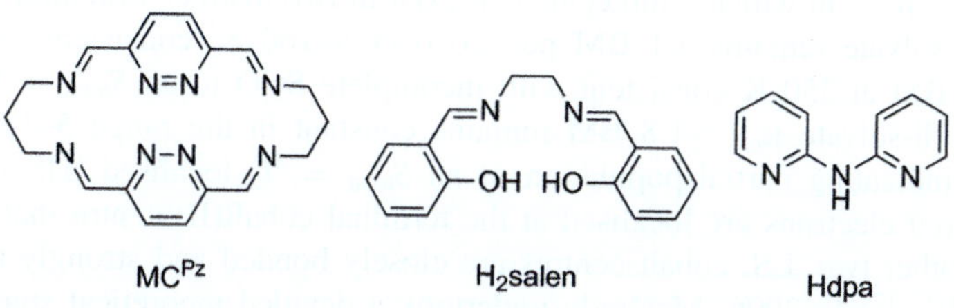

Figure 3.13 The three ligand types used for the generation of polynuclear cobalt(II) SCO-active complexes to date.

interactions, namely $[Co^{II}_2(MC^{Pz})(SCN)_2(NCS)_2]$.[112] The X-ray structure determination at 160 K revealed one unique octahedral cobalt(II) centre in a N_5S coordination environment, with the other half of the complex generated by inversion. The equatorial positions are occupied by the N_4 pocket provided by the macrocycle (av. Co^{II}–N_{macro} 1.951(4) Å) and the axial positions by two NCS anions, one N-bound then other S-bound. The Co–N_{NCS} distance (2.115(3) Å) is 0.164(7) Å longer than the average equatorial Co–N distance, consistent with the expected Jahn–Teller distortion for a LS d^7 ion. VT magnetic measurements showed that the μ_{eff} of 3.15 BM per cobalt(II) at 350 K, drops to 0.70 BM at 4.5 K. A plateau region between 40 K and 224 K is observed, corresponding to the [LS–LS] state: fitting of the low temperature data show that antiferromagnetic interactions are mediated by the pair of pyridazine bridges between the two LS $S = {}^1/_2$ ions ($g = 2.13$, $2J = -11.7$ cm^{-1}). From 225 to 350 K the magnetic moment slowly increases as a very gradual and incomplete SCO towards the [HS–HS] state occurs. In 2002 they reported that modifying the axial donors tuned the spin state,[119] and in 2009 the redox properties of the $[Co^{II}_2(MC^{Pz})(CH_3CN)_4]$ were further probed by Peters and co-workers.[120]

The classic example of an *SCO-active dinuclear cobalt(II)* complex is the *dimer* $[Co^{II}(salen)]_2$ where $salen^{2-}$ = *N,N′*-ethylenebis(salicylideniminato) (Fig. 3.13). The cobalt(II) centres are *five coordinate* due to the dimerization, as shown by X-ray structure determinations in 1969 and 1971.[121, 122] The equatorial positions of the square pyramid are occupied by the N_2O_2 donor set of the salen ligand (av. Co–N/O 1.92/1.90 Å) and the apical position is fulfilled by the bridging oxygen atom (Co–O 2.25 Å) of one of the phenolate groups of the salen ligand bound to the other cobalt(II) in the dimer. The magnetic properties of this complex, and related complexes, have been exhaustively studied.[25] In a 2009 study Miller and co-workers re-examined this complex and showed it also undergoes both SCO *and* antiferromagnetic exchange interactions, the latter somewhat stronger than in the above example, leading to a diamagnetic ground state at 5 K (in the [LS–LS] state $g = 2.26$ and $J = -21$ cm^{-1}). At 200 K, χT per Co^{II} is 0.43 emu $Kmol^{-1}$, whereas at 5 K it is almost zero.[123] Above 200 K the magnetic moment increases, consistent with gradual SCO, however it is not complete even at 525 K, where only 71% of centres are HS. Beyond 525 K the complex decomposes.

3.5.2 Trinuclear Cobalt(II) Complexes

Cotton, Murillo, Dunbar and co-workers followed up on earlier studies, by themselves[124] and others,[125] and showed that on deprotonation the ligand dipyridylamine (Hdpa, Fig. 3.13) reacts with cobalt(II) chloride to give a linear metal-metal bridged tricobalt complex that can be isolated as two different solvates, namely $[Co^{II}_3(dpa)_4Cl_2]\cdot nCH_2Cl_2$, $n = 1$ or 2.[124] Both solvates have been characterised by X-ray crystallography revealing differing Co–Co bonds, the mono-solvate is symmetrical (2.3369(4) Å at 296 K) whereas the di-solvate is asymmetric (2.299(1) and 2.4719(1) Å at 298 K). This complex was proposed to be an example of bond-stretch isomerism,[126] in which a molecule can exist in two distinct structural forms. The magnetic moment of the mono-solvate remains 2.1 BM per Co over 8–160 K, consistent with a fully LS state, then increases, to 3.8 BM at 350 K consistent with incomplete SCO to an $S_{total} = 5/2$ delocalised [LS–HS–LS] state.[124] The di-solvate μ_{eff} ~1.8 BM remains constant in the range 5–150 K, then increases, to 2.6 BM at 300 K indicating partial population of an $S_{total} = {}^3/_2$ localised $[HS–(LS–LS)_{anti}]$ state, in which the three unpaired electrons are localised at the terminal cobalt(II) centre that has the longest Co–Co distance, and the other two, LS, cobalt centres are closely bonded and strongly antiferromagnetically coupled (diamagnetic[125]).[124] In 2006, McGrady undertook a detailed theoretical study of this fascinating system.[127]

The same team reported in 2000 that this tricobalt(II) complex can be oxidised by one electron to give the mixed-valent complex $[Co^{III}Co^{II}_2(dpa)_4Cl_2](BF_4)\cdot 2CH_2Cl_2$.[128] At low temperatures, 1.8–50 K, the magnetic moment is virtually zero indicating a $S = 0$ ground state, $[(LS)]^{Co(III)}$–$(LS–LS)^{Co(II)}{}_{antiferro}$]. Upon warming, a reversible and gradual two-step SCO is observed, with $T_{1/2}{}^1 = 315$ and $T_{1/2}{}^2 \approx 330$ K. The 2-step nature of the conversion was explained in terms of the presence of an intermediate spin state: the first step was ascribed to $S = 0 \rightarrow S = 1$, and the second to $S = 1 \rightarrow S = 2$. The 2-step transition was also observed in solution with parameters $T_{1/2}{}^1 = 201$ and $T_{1/2}{}^2 = 281$ K.

In 2004 they prepared the complex $[Co^{II}{}_3(dpa)_4(CCPh)_2]$ where CCPh is phenylacetylide.[129] The X-ray crystal structure acquired at 213 K showed two different Co–Co distances, that is an asymmetric structure similar to $[Co^{II}{}_3(dpa)_4Cl_2]{\cdot}2CH_2Cl_2$. As expected, from the molecular structure, this complex also undergoes a localised SCO to the [HS–(LS–LS)$_{anti}$] state, at around ~250 K. The complex is more readily oxidised than the halide analogue.

3.6 Dinuclear Chromium(II) Complex

Octahedral chromium(II) complexes can undergo SCO from the paramagnetic HS state ($S = 2$) to the paramagnetic LS state ($S = 1$).[26]

In 1994, O'Hare and co-workers reported the synthesis and characterisation of the first examples of dinuclear d^4 SCO complexes, specifically the dark green triple-decker organometallic complexes $[Cr^{II}{}_2(\mu^2{:}\eta^5\text{-}P_5)(\eta^5\text{-}C_5Me_5)_2]X$, where $X^- = PF_6$ or SbF_6 (Fig. 3.14). Upon cooling, below about 150 K the moment starts to drop slowly, then at at low temperature both undergo abrupt and hysteretic ST (PF_6 $T_{1/2} \approx 33$ and $\Delta T_{1/2} = 2$ K; SbF_6 $T_{1/2} \approx 23$ and $\Delta T_{1/2} = 2$ K) below which they are essentially diamagnetic.[130]

In 2007, a very detailed VT X-ray crystallographic study of the SbF_6 salt was carried out,[131] the results of which are in good agreement with the magnetic data. On cooling from 290 to 170 K (during which there is one independent complex cation in the asymmetric unit), the Cr...Cr distance reduces from 3.1928(5) to 3.1588(4) Å. Below this a phase change occurs, from F_{ddd} at RT, to *I2/a* (nonstandard setting of *C2/c*) below 160 K and the asymmetric unit now comprises two independent complex cations. On further cooling, between 150 and 25 K, one of these two cations (cation A) undergoes dramatic changes in geometry (mainly involving Cr and P atoms). Compared to 290 K, at 25 K the Cr...Cr distance is reduced by 0.31 Å (2.886(2) Å) for cation A, but only by 0.08 Å (3.108(2) Å) for cation B. A dramatic decrease in the unit cell volume occurs between 20 and 18 K. At 15 K, the Cr...Cr distances had decreased further, and are similar to one another (cation A 2.782(2) Å; cation B 2.798(2) Å). All of these structural changes correlate well with the magnetic profile ($\mu_{eff} = 4.1$ μ_B above 160 K drops to ~2.8 μ_B at 25 K then ST occurs at $T_{1/2} \approx 23$ K); the slight difference in $T_{1/2}$ value is likely due to temperature calibration differences. The authors suggested that the origin of the two transitions (at ca. 160 and 23 K) could be due to the effects of changes to the extent of antiferromagnetic coupling between Cr^{II} centres mediated by the P_5 ring or two-step SCO [LS–LS] → [HS–HS]:[LS–LS] → [HS–HS].[131]

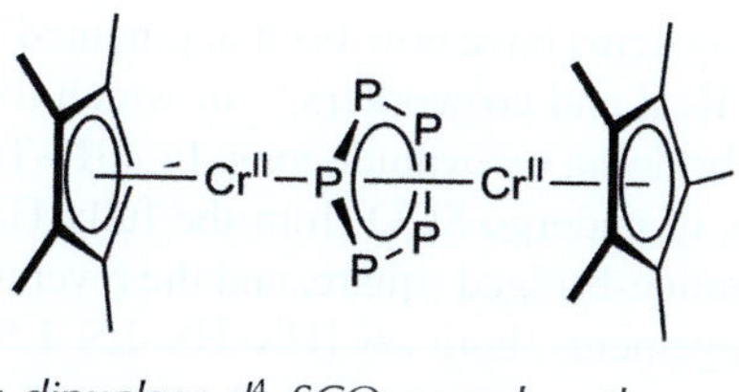

Figure 3.14 *The cation of the only dinuclear* d^4 *SCO complex: the organometallic triple-decker* $[Cr^{II}(\mu^2{:}\eta^5\text{-}P_5)(\eta^5\text{-}C_5Me_5)_2]X$*, where* $X^- = PF_6{}^-$ *or* $SbF_6{}^-$.

3.7 Concluding Remarks

In this chapter we have reviewed the recent advances, 2004–2011, in the development of discrete polynuclear SCO-active complexes. All of the 90 new examples reported are of iron, with the majority of iron(II) (68) and the remainder of iron(III) (22), hence VT ^{57}Fe Mössbauer spectra frequently complement the VT magnetic studies. Of the 90, 63 are dinuclear, 3 trinuclear, 14 tetranuclear and the remaining 10 are an assortment of higher nuclearities including mixed metal clusters.

In dinuclear complexes the two possible SCO routes from fully HS to fully LS, 1-step [HS–HS] to [LS–LS] and 2-step that passes through a '[HS–LS]' half-SCO state (localised or delocalised), are well established. Key advances in the review period were the first reports of structure determinations of each type of half-SCO state, a 1:1 mixture of [HS–HS]:[LS–LS] by Kaizaki and co-workers late in 2004,[56] then a localised [HS–LS] state early in 2005 by Brooker and co-workers.[30] Both were dinuclear iron(II) complexes. In the latter case, due to the constrained nature of the bis-terdentate PMAT ligands it was also possible to differentiate between the HS state in [HS–HS] and in [HS–LS] by zero-field ^{57}Fe Mössbauer spectroscopy for the first time.[32] Given that such spectrometers are far more numerous and hence accessible than applied field instruments this is a key finding. A few years later, Matouzenko and co-workers[36] used zero-field ^{57}Fe Mössbauer spectroscopy to distinguish the quadrupole doublets (four of them) characteristic of each of the LS and HS centres present in a mixture of the three species, [HS–HS], [HS–LS] and [LS–LS]. Interestingly in this case the ligand used is quite flexible so it appears that crystal packing interactions are sufficient to promote the formation, and identification, of localised [HS–LS] pairs. It is well known that crystal packing is a major factor that can promote or deactivate SCO, for example different polymorphs often show different behaviour. Analysis of the distortion of the metal centres from a perfect geometry as a function of spin state is informative and such analysis has become established as the norm over the review period.

In 2005 the first dinuclear iron(III) complex to be structurally characterised in the localised [HS–LS] state was reported by Matsumoto and co-workers.[97]

In an important study Kaizaki and co-workers demonstrated a beautiful relationship between $T_{1/2}$ and pyridine substituent (R) in the two families of dinuclear iron(II) complexes *trans*-$[Fe^{II}_2(\mu\text{-bpypz})_2(NCE)_2(R\text{-}Py)_2]$, where E = S or BH_3. This was the first, and is still a very rare, example of an investigation of the electronic effect of systematic variations in a ligand substituent in a family of closely related SCO-active complexes to actually identify useful correlations. Such findings are rare due to the impact that such variations can have on the solvent content and/or crystal packing interactions, as these often have as important a role in determining magnetic behaviour.

The report by Garcia and co-workers in 2011 of an SCO-active dinuclear iron(II) complex featuring an appended fluorophore is exciting as this provides not only dual functionality but also an additional method of monitoring SCO.[38]

Recent advances in trinuclear systems by Tuchagues and co-workers[75] have shown a triangular complex, in which only one metal centre undergoes SCO. However, there are no examples of linear or triangular systems in which a complete [HS–HS–HS] $\leftrightarrow$ [LS–LS–LS] SCO occurs. Tetranuclear systems are more numerous, with the first example of a SCO-active tetranuclear complex of any sort reported by Lehn and co-workers in 2000.[14] Their grid-like systems have now been augmented by squares developed by Oshio and co-workers[80] and expanded upon by Real and co-workers,[82] in which there are usually two differing metal ion environments due to asymmetric-bridging by cyanide ions. In 2011 Tao and co-workers reported the first example of a square iron(II) complex to undergo SCO from the fully HS state, $[HS]_4$ to the fully LS state, $[LS]_4$.[83] It is a symmetrically-dicyanamide-bridged square, and the reversible SCO occurs in two steps. Of the possible 50:50 mixed spin state arrangements, both *cis*-[HS–HS–LS–LS] and *trans*-[HS–LS–HS–LS] have been observed.

Ten examples of even larger, mixed-metal or mixed-valent systems have been reported. In most cases these exhibit more than one property, for example in addition to SCO they may be redox active or exhibit valence tautomerism. The standout amongst these large systems is the report by Batten and co-workers of the largest known discrete SCO complex, $[Fe^{II}_6[Cu^{I}(tp^{4\text{-}py})]_8]^{14+}$, which forms a 3 nm nanoball that is switched by thermal, light and guest perturbation.[92]

Discrete polynuclear SCO complexes of metal ions other than iron remain very uncommon, with no new examples reported in the review period (2004–2011). Prior to this only three such cobalt systems (all Co^{II}: two dinuclear,[112, 119, 123] and one trinuclear[124, 125]) and one chromium system (Cr^{II}: dinuclear[130]) had been reported. The first manganese example (Mn^{II} dinuclear) was reported in 2012.[27] Clearly there is considerable room for development of discrete polynuclear SCO complexes of metal ions other than iron.

Hysteresis loops of a few Kelvin are more common now, however wide thermal hysteresis loops remain uncommon and unpredictable in discrete polynuclear complexes. Control of this feature in the solid state would require control of crystal packing, something which is not yet attainable in the general case, as intermolecular interactions are critically important to hysteresis. The widest hysteresis loop observed to date for a discrete dinuclear iron(II) complex is 21 K reported by Weber and co-workers in 2008;[61] they have reported far wider loop widths (e.g. 70 K) in hydrogen bonded monometallic analogues.[10] So whilst discrete polynuclear SCO complexes offer the advantage of more possible spin states and hence denser information storage, to date hysteresis loop widths for such complexes are disappointingly small. It should be noted that authors need to include the temperature sweep rates employed as otherwise the loop width quoted cannot be compared with others (the faster you sweep the bigger the lag in response, and the wider the loop).

LIESST studies are increasingly common, and this represents a potentially powerful switching method for practical applications. However, the light-induced metastable HS state is not accessible for all complexes. More examples of polynuclear LIESST-active complexes are necessary if useful trends are to be identified, as has been done by Létard and co-workers for mononuclear and polymeric LIESST-active complexes (correlations found between the type of ligands and *T*(LIESST)).[132] DSC measurements are also increasingly presented, providing a growing database of ΔH and ΔS values associated with SCO events, and providing another method of confirming the $T_{1/2}$.

The bottom line in this field is that gaining predictability of SCO properties is highly desirable if this type of system is to be systematically developed as potential nano-components for future molecular computers, sensors and so on. However, this will likely only be possible in the general case when full, predictable, manipulation of the crystal packing via crystal engineering is possible or larger assemblies (e.g. polymers,[2, 133] nanoparticles[134–137] or amphiphilic assemblies[138]). Meantime, we will continue to use informed minds, intelligence and creativity to develop this field in exciting new directions.

References

1. Létard, J.-F., Guionneau, P., Goux-Capes, L. (2004) Towards spin crossover applications, in: Gütlich, P., Goodwin, H. A. (Eds) Spin Crossover in Transition Metal Compounds III. Top. Curr. Chem., 235: 221–249.
2. Kahn, O., Martinez, C. J. (1998) Spin-transition polymers: from molecular materials toward memory devices. Science, 279: 44–48.
3. Brooker, S., Kitchen, J. A. (2009) Nano-magnetic materials: spin crossover compounds *vs.* single molecule magnets *vs.* single chain magnets. Dalton Trans., 7331–7340.
4. Martinez, C. R., Iverson, B. L. (2012) Rethinking the term "pi-stacking". Chem. Sci., 3: 2191–2201.
5. Salonen, L. M., Ellermann, M., Diederich, F. (2011) Aromatic rings in chemical and biological recognition: energetics and structures. Angew. Chem. Int. Ed., 50: 4808–4842.
6. Hunter, C. A., Sanders, J. K. M. (1990) The nature of π–π interactions. J. Am. Chem. Soc., 112: 5525–5534.
7. Arunan, E., Desiraju, G. R., Klein, R. A., Sadlej, J., Scheiner, S., Alkorta, I., Clary, D. C., Crabtree, R. H., Dannenberg, J. J., Hobza, P., Kjaergaard, H. G., Legon, A. C., Mennucci, B., Nesbitt, D. J. (2011) Definition of the hydrogen bond (IUPAC recommendations 2011). Pure Appl. Chem., 83: 1637–1641.

8. Desiraju, G. R., Steiner, T. (1999) The Weak Hydrogen Bond in Structural Chemistry and Biology. Oxford University Press, Oxford, UK, p. 521.
9. Murray, K. S., Kepert, C. J. (2004) Cooperativity in spin crossover systems: memory, magnetism and microporosity, in: Gütlich, P., Goodwin, H. A. (Eds) Spin Crossover in Transition Metal Compounds I. Top. Curr. Chem., 233: 195–228.
10. Weber, B., Bauer, W., Obel, J. (2008) An iron(II) spin-crossover complex with a 70 K wide thermal hysteresis loop. Angew. Chem. Int. Ed. 47: 10098–10101.
11. Cowan, M. G., Olguín, J., Narayanaswamy, S., Tallon, J. L., Brooker, S. (2012). Reversible switching of a cobalt complex by thermal, pressure, and electrochemical stimuli: abrupt, complete, hysteretic spin crossover. J. Am. Chem. Soc., 134: 2892–2894.
12. Grosjean, A., Daro, N., Kauffmann, B., Kaiba, A., Létard, J.-F., Guionneau, P. (2011). The 1-D polymeric structure of the $[Fe(NH_2trz)_3](NO_3)_2{\cdot}nH_2O$ (with $n = 2$) spin crossover compound proven by single crystal investigations. Chem. Commun., 47: 12382–12384.
13. Ruben, M., Breuning, E., Gisselbrecht, J. P., Lehn, J.-M. (2000) Multilevel molecular electronic species: electrochemical reduction of a [2×2] Co grid-type complex by 11 electrons in 10 reversible steps. Angew. Chem. Int. Ed., 39: 4139–4142.
14. Breuning, E., Ruben, M., Lehn, J.-M., Renz, F., Garcia, Y., Ksenofontov, V., Gütlich, P., Wegelius, E., Rissanen, K. (2000) Spin crossover in a supramolecular Fe^{II}_4 [2×2] grid triggered by temperature, pressure, and light. Angew. Chem. Int. Ed., 39: 2504–2507.
15. Kitchen, J. A., White, N. G., Jameson, G. N. L., Tallon, J. L., Brooker, S. (2011) Effect of counteranion X on the spin crossover properties of a family of diiron(II) triazole complexes $[Fe^{II}_2(PMAT)_2](X)_4$. Inorg. Chem., 50: 4586–4597.
16. Kitchen, J. A., Noble, A., Brandt, C. D., Moubaraki, B., Murray, K. S., Brooker, S. (2008) High and low spin mononuclear and dinuclear iron(II) complexes of 4-amino and 4-pyrrolyl-3,5-di(2-pyridyl)-4*H*-1,2,4-triazoles. Inorg. Chem., 47, 9450–9458.
17. Nakano, K., Suemura, N., Yoneda, K., Kawata, S., Kaizaki, S. (2005) Substituent effect of the coordinated pyridine in a series of pyrazolato bridged dinuclear diiron(II) complexes on the spin-crossover behavior. Dalton Trans., 740–743.
18. Halcrow, M. A. (2011) Structure : function relationships in molecular spin-crossover complexes. Chem. Soc. Rev., 40: 4119–4142.
19. Zein, S., Borshch, S. A. (2005) Energetics of binuclear spin transition complexes. J. Am. Chem. Soc., 127: 16197–16201.
20. Real, J. A., Gaspar, A. B., Muñoz, M. C., Gütlich, P., Ksenofontov, V., Spiering, H. (2004) Bipyrimidine-bridged dinuclear iron(II) spin crossover compounds, in: Gütlich, P., Goodwin, H. A. (Eds) Spin Crossover in Transition Metal Compounds I. Top. Curr. Chem., 233: 167–193.
21. Garcia, Y., Niel, V., Muñoz, M. C., Real, J. A. (2004) Spin crossover in 1D, 2D and 3D polymeric Fe(II) networks, in: Gütlich, P., Goodwin, H. A. (Eds) Spin Crossover in Transition Metal Compounds I. Top. Curr. Chem., 233: 229–257.
22. Murray, K. S. (2008) Advances in polynuclear iron(II), iron(III) and cobalt(II) spin-crossover compounds. Eur. J. Inorg. Chem., 3101–3121.
23. Gaspar, A. B., Muñoz, M. C., Real, J. A. (2006) Dinuclear iron(II) spin crossover compounds: singular molecular materials for electronics. J. Mater. Chem., 16: 2522–2533.
24. van Koningsbruggen, P. J., Maeda, Y., Oshio, H. (2004) Iron(III) spin crossover compounds, in: Gütlich, P., Goodwin, H. A. (Eds) Spin Crossover in Transition Metal Compounds I. Top. Curr. Chem., 233: 259–324.
25. Goodwin, H. A. (2004) Spin crossover in cobalt(II) systems, in: Gütlich, P., Goodwin, H. A. (Eds) Spin Crossover in Transition Metal Compounds II. Top. Curr. Chem., 234: 23–47.
26. Garcia, Y., Gutlich, P. (2004) Thermal spin crossover in Mn(II), Mn(III), Cr(II) and Co(III) coordination compounds, in: Gütlich, P., Goodwin, H. A. (Eds) Spin Crossover in Transition Metal Compounds II. Top. Curr. Chem., 234: 49–62.
27. Scheuermayer, S., Tuna, F., Bodensteiner, M., Scheer, M., Layfield, R. A. (2012) Spin crossover in phosphorus- and arsenic-bridged cyclopentadienyl-manganese(II) dimers. Chem. Commun., 48: 8087–8089.

28. Gütlich, P., Goodwin, H. A. (2004) Spin crossover–an overall perspective in: Gütlich, P., Goodwin, H. A. (Eds) Spin Crossover in Transition Metal Compounds I. Top. Curr. Chem., 233: 1–47.
29. Goodwin, H. A. (2004) Spin crossover in iron(II) tris(diimine) and bis(terimine) systems, in: Gütlich, P., Goodwin, H. A. (Eds) Spin Crossover in Transition Metal Compounds I. Top. Curr. Chem., 233: 59–90.
30. Klingele, M. H., Moubaraki, B., Cashion, J. D., Murray, K. S., Brooker, S. (2005) The first X-ray crystal structure determination of a dinuclear complex trapped in the [low spin–high spin] state: $[Fe^{II}_2(PMAT)_2](BF_4)_4{\cdot}DMF$. Chem. Commun., 987–989.
31. Bhattacharjee, A., Ksenofontov, V., Kitchen, J. A., White, N. G., Brooker, S., Gütlich, P. (2008) Effect of pressure and light on the spin transition behavior of the dinuclear iron(II) compound $[Fe^{II}_2(PMAT)_2](BF_4)_4{\cdot}DMF$. Appl. Phys. Lett., 92: 174104/1–174104/3.
32. Grunert, C. M., Reiman, S., Spiering, H., Kitchen, J. A., Brooker, S., Gütlich, P. (2008) Mixed spin-state [HS–LS] pairs in a dinuclear spin-transition complex: confirmation by variable-temperature ^{57}Fe Mössbauer spectroscopy. Angew. Chem. Int. Ed., 47: 2997–2999.
33. Ksenofontov, V., Gaspar, A. B., Niel, V., Reiman, S., Real, J. A., Gütlich, P. (2004) On the nature of the plateau in two-step dinuclear spin-crossover complexes. Chem. Eur. J., 10: 1291–1298.
34. Batten, S. R., Bjernemose, J., Jensen, P., Leita, B. A., Murray, K. S., Moubaraki, B., Smith, J. P., Toftlund, H. (2004) Designing dinuclear iron(II) spin crossover complexes. Structure and magnetism of dinitrile-, dicyanamido-, tricyanomethanide-, bipyrimidine- and tetrazine-bridged compounds. Dalton Trans., 3370–3375.
35. Fedaoui, D., Bouhadja, Y., Kaiba, A., Guionneau, P., Létard, J.-.F, Rosa, P. (2008) Complexation of 2,6-bis(3-pyrazolyl)pyridine-bis(thiocyanato)iron(II) with a bridging 4,4′-bipyridine: a new example of a dinuclear spin crossover complex. Eur. J. Inorg. Chem., 1022–1026.
36. Verat, A. Y., Ould-Moussa, N., Jeanneau, E., Boris, L. G., Borshch, S. A., Matouzenko, G. S. (2009) Ligand strain and the nature of spin crossover in binuclear complexes: two-step spin crossover in a 4,4′-bipyridine-bridged iron(II) complex $[\{Fe(dpia)(NCS)_2\}_2(4,4'\text{-bpy})]$ (dpia = di(2-picolyl)amine; 4,4′-bpy = 4,4′-bipyridine) Chem. Eur. J., 15: 10070–10082.
37. Matouzenko, G. S., Jeanneau, E., Verat, A. Y., Boussekou, A. (2011) Spin crossover and polymorphism in a family of 1,2-bis(4-pyridyl)ethene-bridged binuclear iron(II) complexes. A key role of structural distortions. Dalton Trans., 40: 9608–9618.
38. Garcia, Y., Robert, F., Naik, A. D., Zhou, G., Tinant, B., Robeyns, K., Michotte, S., Piraux, L. (2011) Spin transition charted in a fluorophore-tagged thermochromic dinuclear iron(II) complex. J. Am. Chem. Soc., 133: 15850–15853.
39. Guionneau, P., Brigouleix, C., Barrans, Y., Goeta, A. E., Létard, J.-F., Howard, J. A. K., Gaultier, J., Chasseau, D. (2001) High pressure and very low temperature effects on the crystal structures of some iron(II) complexes. C. R. Acad. Sci. Paris Ser. IIc, Chimie, 4: 161–171.
40. Deeney, F. A., Charles, J. H., Morgan, G. G., McKee, V., Nelson, J., Teat, S. J., Clegg, W. (1998) Response to steric constraint in azacryptate and related complexes of iron-(II) and -(III). J. Chem. Soc. Dalton Trans., 1837–1844.
41. Drew, M. G. B., Harding, C. J., McKee, V., Morgan, G. G., Nelson, J. (1995) Geometric control of manganese redox state. J. Chem. Soc. Chem. Commun., 1035–1038.
42. Olguín, J., Brooker, S. (2011) Spin crossover active iron(II) complexes of selected pyrazole-pyridine/pyrazine ligands. Coord. Chem. Rev., 255: 203–240.
43. Tuna, F., Lees, M. R., Clarkson, G. J., Hannon, M. J. (2004) Readily prepared metallo-supramolecular triple helicates designed to exhibit spin-crossover behaviour. Chem. Eur. J., 10: 5737–5750.
44. Garcia, Y., Grunert, M., Reiman, S., van Campenhoudt, O., Gütlich, P. (2006) The two-step spin conversion in a supramolecular triple helicate dinuclear iron(II) complex studied by Mössbauer spectroscopy. Eur. J. Inorg. Chem., 3333–3339.
45. Pelleteret, D., Clérac, R., Mathonière, C., Harte, E., Schmitt, W., Kruger, P. E. (2009) Asymmetric spin crossover behaviour and evidence of light-induced excited spin state trapping in a dinuclear iron(II) helicate. Chem. Commun., 221–223.
46. Archer, R. J., Hawes, C. S., Jameson, G. N. L., McKee, V., Moubaraki, B., Chilton, N. F., Murray, K. S., Schmitt, W., Kruger, P. E. (2011) Partial spin crossover behaviour in a dinuclear iron(II) triple helicate. Dalton Trans., 40: 12368–12373.

47. Amoore, J. J. M., Kepert, C. J., Cashion, J. D., Moubaraki, B., Neville, S. M., Murray, K. S. (2006) Structural and magnetic resolution of a two-step full spin-crossover transition in a dinuclear iron(II) pyridyl-bridged compound. Chem. Eur. J., 12: 8220–8227.
48. Amoore, J. J. M., Neville, S. M., Moubaraki, B., Iremonger, S. S., Murray, K. S., Létard, J.-F., Kepert, C. J. (2010) Thermal- and light-induced spin crossover in a guest-dependent dinuclear iron(II) system. Chem. Eur. J., 16: 1973–1982.
49. Reedijk, J., Quesada, M., de Hoog, P., Gamez, P., Roubeau, O., Aromí, G., Donnadieu, B., Massera, C., Lutz, M., Spek, A. L. (2006) Coordination dependence of magnetic properties within a family of related $[Fe^{II}_2]$ complexes of a triazine-based ligand. Eur. J. Inorg. Chem., 1353–1361.
50. Schneider, C. J., Moubaraki, B., Cashion, J. D., Turner, D. R., Leita, B. A., Batten, S. R., Murray, K. S. (2011) Spin crossover in di-, tri- and tetranuclear, mixed-ligand tris(pyrazolyl)methane iron(II) complexes. Dalton Trans., 40: 6939–6951.
51. Sunatsuki, Y., Kawamoto, R., Fujita, K., Maruyama, H., Suzuki, T., Ishida, H., Kojima, M., Iijima, S., Matsumoto, N. (2009) Structures and spin states of bis(tridentate)-type mononuclear and triple helicate dinuclear iron(II) complexes of imidazole-4-carbaldehyde azine. Inorg. Chem., 48: 8784–8795.
52. Sunatsuki, Y., Kawamoto, R., Fujita, K., Maruyama, H., Suzuki, T., Ishida, H., Kojima, M., Iijima, S., Matsumoto, N. (2010) Structures and spin states of mono- and dinuclear iron(II) complexes of imidazole-4-carbaldehyde azine and its derivatives. Coord. Chem. Rev., 254: 1871–1881.
53. Sunatsuki, Y., Maruyama, H., Fujita, K., Suzuki, T., Kojima, M., Matsumoto, N. (2009) Mononuclear bis(tridentate)-type and dinuclear triple helicate iron(II) complexes containing 2-ethyl-5-methylimidazole-4-carbaldehyde azine. Bull. Chem. Soc. Jpn., 82: 1497–1505.
54. Fujita, K., Kawamoto, R., Tsubouchi, R., Sunatsuki, Y., Kojima, M., Iijima, S., Matsumoto, N. (2007) Spin states of mono- and dinuclear iron(II) complexes with bis(imidazolylimine) Ligands. Chem. Lett., 36: 1284–1285.
55. Leita, B. A., Moubaraki, B., Murray, K. S., Smith, J. P., Cashion, J. D. (2004) Structure and magnetism of a new pyrazolate bridged iron(II) spin crossover complex displaying a single HS–HS to LS–LS transition. Chem. Commun., 156–157.
56. Nakano, K., Kawata, S., Yoneda, K., Fuyuhiro, A., Yagi, T., Nasu, S., Morimoto, S., Kaizaki, S. (2004) Direct two-step spin-crossover through [HS–HS]...[LS–LS] at the plateau in dinuclear diiron(II) complex $[\{Fe(NCBH_3)(4phpy)\}_2(\mu\text{-bpypz})_2]$. Chem. Commun., 2892–2893.
57. Yoneda, K., Nakano, K., Fujioka, J., Yamada, K., Suzuki, T., Fuyuhiro, A., Kawata, S., Kaizaki, S. (2005) Synthesis, structure and magnetic properties of pyrazolate-bridged dinuclear complexes $[\{M(NCS)(4\text{-Phpy})\}_2(\mu\text{-bpypz})_2]$ ($M = Co^{2+}$ and Fe^{2+}). Polyhedron, 24: 2437–2442.
58. Schneider, C. J., Cashion, J. D., Moubaraki, B., Neville, S. M., Batten, S. R., Turner, D. R., Murray, K. S. (2007) The magnetic and structural elucidation of 3,5-bis(2-pyridyl)-1,2,4-triazolate-bridged dinuclear iron(II) spin crossover compounds. Polyhedron, 26: 1764–1772.
59. Klingele, M. H., Moubaraki, B., Murray, K. S., Brooker, S. (2005) Synthesis and some first-row transition-metal complexes of the 1,2,4-triazole-based bis(terdentate) ligands TsPMAT and PMAT. Chem. Eur. J., 11: 6962–6973.
60. Weber, B., Kaps, E. (2005) Synthesis and magnetic properties of new dinuclear iron(II) complexes of a phenylene-bridge Schiff base analogue dinucleating ligand. Heteroatom Chem., 16: 391–397.
61. Weber, B., Kaps, E. S., Obel, J., Achterhold, K., Parak, F. G. (2008) Synthesis and characterization of a dinuclear iron(ii) spin crossover complex with wide hysteresis. Inorg. Chem., 47: 10779–10787.
62. Min, K. S., DiPasquale, A., Rheingold, A. L., Miller, J. S. (2007) Room-temperature spin crossover observed for $[(TPyA)Fe^{II}(THBQ^{2-})Fe^{II}(TPyA)](BF_4)_2$ [TPyA = tris(2-pyridylmethyl)amine; $THBQ^{2-}$ = 2,3,5,6-tetrahydroxy-1,4-benzoquinonate]. Inorg. Chem., 46: 1048–1050.
63. Min, K. S., Swierczek, K., DiPasquale, A. G., Rheingold, A. L., Reiff, W. M., Arif, A. M., Miller, J. S. (2008) A dinuclear iron(II) complex, $[(TPyA)Fe^{II}(THBQ^{2-})Fe^{II}(TPyA)](BF_4)_2$ [TPyA = tris(2-pyridylmethyl)amine; $THBQ^{2-}$ = 2,3,5,6-tetrahydroxy-1,4-benzoquinonate] exhibiting both spin crossover with hysteresis and ferromagnetic exchange, Chem. Commun., 317–319.
64. Weber, B. (2009) Spin crossover complexes with N_4O_2 coordination sphere–the influence of covalent linkers on cooperative interactions. Coord. Chem. Rev., 253: 2432–2449.

65. Létard, J.-F., Carbonera, C., Real, J. A., Kawata, S., Kaizaki, S. (2009) Photomagnetism of a series of dinuclear iron(II) complexes. Chem. Eur. J., 15: 4146–4155.
66. Nakano, K., Suemura, N., Kawata, S., Fuyuhiro, A., Yagi, T., Nasu, S., Morimoto, S., Kaizaki, S. (2004) Magnetic behavior and Mössbauer spectra of spin-crossover pyrazolate bridged dinuclear diiron(II) complexes: X-ray structures of high-spin and low-spin $[\{Fe(NCBH_3)(py)\}_2(\mu\text{-bpypz})_2]$. Dalton Trans., 982–988.
67. Létard, J.-F., Capes, L., Chastanet, G., Moliner, N., Létard, S., Real, J. A., Kahn, O. (1999) Critical temperature of the LIESST effect in iron(II) spin crossover compounds. Chem. Phys. Lett., 313: 115–120.
68. Guionneau, P., Marchivie, M., Bravic, G., Létard, J.-F., Chasseau, D. (2004) Structural aspects of spin crossover. Example of the $[Fe^{II}L_n(NCS)_2]$ complexes, in: Gütlich, P., Goodwin, H. A. (Eds.) Spin crossover in transition metal compounds II. Top. Curr. Chem., 234: 97–128.
69. Noble, A., Olguín, J., Clérac, R., Brooker, S. (2010) Doubly pyrazolate-bridged dinuclear complexes of a highly constrained bis-terdentate ligand: observation of a [high spin-low spin] state for $[Fe^{II}_2(PMAP)_2][SbF6]_2{\cdot}2.25(C_3H_8O)$ (PMAP = 3,5-bis{[*N*-(2-pyridylmethyl)amino]methyl}-1*H*-pyrazolate). Inorg. Chem., 49: 4560–4569.
70. Olguín, J., Jameson, G. N. L., Brooker, S. (2011) Two dinuclear iron(II) complexes $K[Fe_2(L^1)(SCN)_4]{\cdot}2(C_3H_8O)$ and $[Fe_2(L^1)(SeCN)_3(C_5H_5N)]{\cdot}H_2O$ are stabilised in the '[HS–LS]' state by a bis-tetradentate pyrazolate-based ligand. Dalton Trans., 40: 5086–5089.
71. Jäger, E.-G., Häussler, E., Rudolph, M., Schneider, A. (1985) Synthese, Adduktbildung, Redoxpotentiale und photochrome Iodderivate von Eisen(II)-Komplexen Schiffscher Basen mit elektronenziehenden Substituenten. Z. Anorg. Allg. Chem., 525: 67–85.
72. Weber, B., Walker, F. A. (2007) Solution NMR studies of iron(II) spin-crossover complexes. Inorg. Chem., 46: 6794–6803.
73. Garcia, Y., van Koningsbruggen, P. J., Codjovi, E., Lapouyade, R., Kahn, O., Rabardel, L. (1997) Non-classical Fe^{II} spin-crossover behaviour leading to an unprecedented extremely large apparent thermal hysteresis of 270 K: application for displays. J. Mater. Chem., 7: 857–858.
74. Min, K. S., DiPasquale, A. G., Rheingold, A. L., White, H. S., Miller, J. S. (2009) Observation of redox-induced electron transfer and spin crossover for dinuclear cobalt and iron complexes with the 2,5-di-*tert*-butyl-3,6-dihydroxy-1,4-benzoquinonate bridging ligand. J. Am. Chem. Soc., 131: 6229–6236.
75. Psomas, G., Brefuel, N., Dahan, F., Tuchagues, J.-P. (2004) An unprecedented trinuclear structure involving two high-spin and one spin-crossover iron(II) centers. Inorg. Chem., 43: 4590–4594.
76. Savard, D., Cook, C., Enright, G. D., Korobkov, I., Burchell, T. J., Murugesu, M. (2011) Gradual spin crossover behaviour in a linear trinuclear Fe^{II} complex. Cryst. Eng. Comm, 13: 5190–5197.
77. Vos, G., le Fèbre, R. A., de Graaff, R. A. G., Haasnoot, J. G., Reedijk, J. (1983) Unique high-spin-low-spin transition of the central ion in a linear, trinuclear iron(II) triazole compound. J. Am. Chem. Soc., 105: 1682–1683.
78. Vos, G., de Graaff, R. A. G., Haasnoot, J. G., van der Kraan, A. M., de Vaal, P., Reedijk, J. (1984) Crystal structure at 300 and 105 K, magnetic properties and Mössbauer spectra of bis(triaquatris(4-ethyltriazole-N^1)iron(II)-N^2, $N^{2'}$,$N^{2''}$)iron(II) hexakis(trifluoromethanesulfonate). A linear, trinuclear iron(II) compound, showing a unique high-spin-low-spin transition of the central iron atom. Inorg. Chem., 23: 2905–2910.
79. Zueva, E. M., Ryabikh, E. R., Borshch, S. A. (2011) Theoretical analysis of spin crossover in iron(II) [2 × 2] molecular grids. Inorg. Chem., 50: 11143–11151.
80. Nihei, M., Ui, M., Yokota, M., Han, L., Maeda, A., Kishida, H., Okamoto, H., Oshio, H. (2005) Two-step spin conversion in a cyanide-bridged ferrous square. Angew. Chem. Int. Ed., 44: 6484–6487.
81. Nihei, M., Ui, M., Oshio, H. (2009) Cyanide-bridged tri- and tetra-nuclear spin crossover complexes. Polyhedron, 28: 1718–1721.
82. Boldog, I., Muñoz-Lara, F. J., Gaspar, A. B., Muñoz, M. C., Seredyuk, M., Real, J. A. (2009) Polynuclear spin crossover complexes: synthesis, structure, and magnetic behavior of $[Fe_4(\mu\text{-CN})_4(phen)_4(L)_2]^{4+}$ squares. Inorg. Chem., 48: 3710–3719.
83. Wei, R.-J., Huo, Q., Tao, J., Huang, R.-B., Zheng, L.-S. (2011) Spin-crossover Fe^{II}_4 squares: two-step complete spin transition and reversible single-crystal-to-single-crystal transformation. Angew. Chem. Int. Ed., 50: 8940–8943.

84. Ruben, M., Breuning, E., Lehn, J.-M., Ksenofontov, V., Renz, F., Gütlich, P., Vaughan, G. B. M. (2003) Supramolecular spintronic devices: spin transitions and magnetostructural correlations in $[Fe_4{}^{II}L_4]^{8+}$ [2×2]-grid-type complexes. Chem. Eur. J., 9: 4422–4429.
85. Uppadine, L. H., Gisselbrecht, J.-P., Kyritsakas, N., Nättinen, K., Rissanen, K., Lehn, J.-M. (2005) Mixed-valence, mixed-spin-state, and heterometallic [2×2] grid-type arrays based on heteroditopic hydrazone ligands: synthesis and electrochemical features. Chem. Eur. J., 11: 2549–2565.
86. Stefankiewicz, A. R., Rogez, G., Harrowfield, J., Drillon, M., Lehn, J.-M. (2009) Structural features directing the specificity and functionality of metallo-supramolecular grid-type architectures. Dalton Trans., 5787–5802.
87. Schneider, B., Demeshko, S., Dechert, S., Meyer, F. (2010) Double-switching multistable Fe_4 grid complex with stepwise spin-crossover and redox transitions. Angew. Chem. Int. Ed, 49: 9274–9277.
88. Wu, D.-Y., Sato, O., Einaga, Y., Duan, C.-Y. (2009) A spin-crossover cluster of iron(II) exhibiting a mixed-spin structure and synergy between spin transition and magnetic interaction. Angew. Chem. Int. Ed., 48: 1475–1478.
89. Shatruk, M., Dragulescu-Andrasi, A., Chambers, K. E., Stoian, S. A., Bominaar, E. L., Achim, C., Dunbar, K. R. (2007) Properties of Prussian blue materials manifested in molecular complexes: observation of cyanide linkage isomerism and spin-crossover behavior in pentanuclear cyanide clusters. J. Am. Chem. Soc., 129: 6104–6116.
90. Funck, K. E., Hilfiger, M. G., Berlinguette, C. P., Shatruk, M., Wernsdorfer, W., Dunbar, K. R. (2009) Trigonal-bipyramidal metal cyanide complexes: a versatile platform for the systematic assessment of the magnetic properties of Prussian blue materials. Inorg. Chem., 48: 3438–3452.
91. Funck, K. E., Prosvirin, A. V., Mathonière, C., Clérac, R., Dunbar, K. R. (2011) Light-induced excited spin state trapping and charge transfer in trigonal bipyramidal cyanide-bridged complexes. Inorg. Chem., 50, 2782–2789.
92. Duriska, M. B., Neville, S. M., Moubaraki, B., Cashion, J. D., Halder, G. J., Chapman, K. W., Balde, C., Létard, J. F., Murray, K. S., Kepert, C. J., Batten, S. R. (2009) A nano-scale molecular switch triggered by thermal-, light- and guest-perturbation. Angew. Chem. Int. Ed., 48: 2549–2552.
93. Cambi, L., Szego, L. (1931) Über die magnetische susceptibilität der komplexen verbindungen. Ber. Dtsch. Chem. Ges. Teil B, 64: 2591–2598.
94. Nihei, M., Shiga, T., Maeda, Y., Oshio, H. (2007) Spin crossover iron(III) complexes. Coord. Chem. Rev., 251: 2606–2621.
95. Nemec, I., Boča, R., Herchel, R., Trávníček, Z., Gembický, M., Linert, W. (2009) Dinuclear Fe(III) complexes with spin crossover. Monatsh. Chem., 140: 815–828.
96. Šalitroš, I., Boča, R., Dlháň, L., Gembický, M., Kožíšek, J., Linares, J., Moncol', J., Nemec, I., Perašínová, L., Renz, F., Svoboda, I., Fuess, H. (2009) Unconventional spin crossover in dinuclear and trinuclear iron(III) complexes with cyanido and metallacyanido bridges, Eur. J. Inorg. Chem., 3141–3154.
97. Kitashima, R., Imatomi, S., Yamada, M., Matsumoto, N., Maeda, Y. (2005) Gradual two-step spin crossover behavior of binuclear iron(III) complex bridged by trans-1,2-bis(4-pyridyl)ethylene. Chem. Lett., 34: 1388–1389.
98. Imatomi, S., Sato, T., Hamamatsu, T., Kitashima, R., Matsumoto, N. (2007) Spin-crossover behavior of isomorphous bi- and mononuclear iron(III) complexes. Bull. Chem. Soc. Jpn., 80: 2375–2377.
99. Djukic, B., Poddutoori, P. K., Dube, P. A., Seda, T., Jenkins, H. A., Lemaire, M. T. (2009) Bimetallic iron(3+) spin-crossover complexes containing a 2,2′-bithienyl bridging bis-qsalH ligand. Inorg. Chem., 48: 6109–6116.
100. Herchel, R., Boča, R., Gembický, M., Kožísek, J., Renz, F. (2004) Spin crossover in a tetranuclear $Cr(III){-}Fe(III)_3$ complex. Inorg. Chem., 43: 4103–4105.
101. Renz, F., Jung, S., Klein, M., Menzel, M., Thünemann, A. F. (2009) Molecular switching complexes with iron and tin as central atom. Polyhedron, 28: 1818–1821.
102. Jung, S., Renz, F., Klein, M., Menzel, M., Boča, R., Stößer, R. (2010) Molecular switching in iron complexes bridged via tin-cyanides observed by Mössbauer and ESR spectroscopy. J. Phys. Conf. Ser., 217: 012027/1–012027/6.
103. Renz, F., Kerep, P. (2005) Unprecedented multiple electronic spin transition in hepta- and nonanuclear complex compounds observed by Mössbauer spectroscopy. Polyhedron, 24: 2849–2851.
104. Renz, F., Kerep, P. (2004) The nonanuclear $[Mo(IV)\{(CN)Fe(III)(3\text{-ethoxy-saldptn})\}_8]Cl_4$ complex compound exhibits multiple spin transitions observed by Mössbauer spectroscopy. Hyperfine Interact., 156/157: 371–377.
105. Renz, F., Kerep, P., Hill, D., Klein, M. (2006) Complexes based on ethylene- and propylene-bridged-pentadentate-Fe(III)-units allow interplay between magnetic centers and multistability investigated by Mössbauer spectroscopy. Hyperfine Interact., 168: 981–987.

106. Renz, F., Hill, D., Klein, M., Hefner, J. (2007) Unprecedented multistability in dodecanuclear complex compound observed by Mössbauer spectroscopy. Polyhedron, 26: 2325–2329.
107. Boča, R., Šalitroš, I., Kožíšek, J., Linares, J., Moncoľ, J., Renz, F. (2010) Spin crossover in a heptanuclear mixed-valence iron complex. Dalton Trans., 39: 2198–2200.
108. Hayami, S., Komatsu, Y., Shimizu, T., Kamihata, H., Lee, Y. H. (2011) Spin-crossover in cobalt(II) compounds containing terpyridine and its derivatives. Coord. Chem. Rev., 255: 1981–1990.
109. Krivokapic, I., Zerara, M., Daku, M. L., Vargas, A., Enachescu, C., Ambrus, C., Tregenna-Piggott, P., Amstutz, N., Krausz, E., Hauser, A. (2007) Spin-crossover in cobalt(II) imine complexes. Coord. Chem. Rev., 251: 364–378.
110. Krüger, H.-J. (2009) Spin transition in octahedral metal complexes containing tetraazamacrocyclic ligands. Coord. Chem. Rev., 253: 2450–2459.
111. Graf, M., Wolmershäuser, G., Kelm, H., Demeschko, S., Meyer, F., Krüger, H.-J. (2010) Temperature-induced spin-transition in a low-spin cobalt(II) semiquinonate complex. Angew. Chem. Int. Ed., 49: 950–953.
112. Brooker, S., Plieger, P. G., Moubaraki, B., Murray, K. S. (1999) $[Co^{II}_2L(NCS)_2(SCN)_2]$: the first cobalt complex to exhibit both exchange coupling and spin crossover effects. Angew. Chem. Int. Ed., 38: 408–410.
113. Hayami, S., Murata, K., Urakami, D., Kojima, Y., Akita, M., Inoue, K. (2008) Dynamic structural conversion in a spin-crossover cobalt(II) compound with long alkyl chains. Chem. Commun., 6510–6512.
114. Agustí, G., Bartual, C., Martínez, V., Muñoz-Lara, F. J., Gaspar, A. B., Muñoz, M. C., Real, J. A. (2009) Polymorphism and "reverse" spin transition in the spin crossover system $[Co(4\text{-terpyridone})_2](CF_3SO_3)_2{\cdot}H_2O$. New J. Chem., 33: 1262–1267.
115. Zarembowitch, J., Kahn, O. (1984) Magnetic properties of some spin-crossover, high-spin, and low-spin cobalt(II) complexes with Schiff bases derived from 3-formylsalicylic acid. Inorg. Chem., 23: 589–593.
116. J. Zarembowitch, R. Claude, O. Kahn (1985) High-spin ↔. low-spin transition in solid $Co(H_2fsa_2en)(H_2O)_2$ $[H_4fsa_2en =$ *N,N'*-ethylenebis(3-carboxysalicylaldimine)]: a magnetic investigation. Inorg. Chem., 24: 1576–1580.
117. Thuery, P., Zarembowitch, J. (1986) Spin state of cobalt(II) in five- and six-coordinate lewis base adducts of (*N,N'*-ethylenebis(3-carboxysalicylaldiminato))cobalt(II). New spin-crossover complexes. Inorg. Chem., 25: 2001–2008.
118. Brooker, S., Kelly, R. J., Plieger, P. G. (1998) A pyridazine Schiff-base macrocycle hosts a dicobalt centre in five different redox states: evidence for a mixed valent Co^ICo^{II} species. Chem. Commun., 1079–1080.
119. Brooker, S., de Geest, D. J., Kelly, R. J., Plieger, P. G., Moubaraki, B., Murray, K. S., Jameson, G. B. (2002) Exchange-coupled high-spin, low-spin and spin-crossover dicobalt(II) complexes of a pyridazine-containing Schiff-base macrocycle: control of cobalt(II) spin state by choice of axial ligands. J. Chem. Soc. Dalton Trans., 2080–2087.
120. Szymczak, N. K., Berben, L. A., Peters, J. C. (2009) Redox rich dicobalt macrocycles as templates for multi-electron transformations. Chem. Commun., 6729–6731.
121. Holt, S. L., DeIasi, R., Post, B. (1971) Crystal structure of the oxygen-inactive form of bis(salicylaldehyde)ethylenediiminecobalt(II). Inorg. Chem., 10: 1498–1500.
122. Brückner, S., Calligaris, M., Nardin, G., Randaccio, L. (1969) The crystal structure of the form of *N,N'*-ethylenebis(salicylaldehydeiminato)cobalt(II) inactive towards oxygenation. Acta Cryst. Sect. B, 25: 1671–1674.
123. Min, K. S., Arthur, J., Shum, W. W., Bharathy, M., zur Loye, H.-C., Miller, J. S. (2009) Tristability arising from singlet-triplet and quartet spin states for dimeric Co^{II}salen. Inorg. Chem., 48: 4593–4594.
124. Clérac, R., Cotton, F. A., Daniels, L. M., Dunbar, K. R., Kirschbaum, K., Murillo, C. A., Pinkerton, A., Schultz, A. J., Wang, X. (2000) Linear tricobalt compounds with di(2-pyridyl)amide (dpa) ligands: temperature dependence of the structural and magnetic properties of symmetrical and unsymmetrical forms of $Co_3(dpa)_4Cl_2$ in the solid state. J. Am. Chem. Soc., 122: 6226–6236 and refs therein.
125. Yang, E.-C., Cheng, M.-C., Tsai, M.-S., Peng, S.-M. (1994) Structure of a linear unsymmetrical trinuclear cobalt(II) complex with a localized $Co^{II}–Co^{II}$ bond : dichlorotetrakis[μ3-bis(2-pyridyl)amido]tricobalt(II). J. Chem. Soc. Chem. Commun., 2377–2378.
126. Rohmer, M.-M., Strich, A., Bénard, M., Malrieu, J.-P. (2001) Metal—metal bond length variability in $Co_3(dipyridylamide)_4Cl_2$: bond-stretch isomerism, crystal field effects, or spin transition process? A DFT study. J. Am. Chem. Soc., 123: 9126–9134.
127. Pantazis, D. A., McGrady, J. E. (2006) A three-state model for the polymorphism in linear tricobalt compounds. J. Am. Chem. Soc., 128: 4128–4135.

128. Clérac, R., Cotton, F. A., Dunbar, K. R., Lu, T., Murillo, C. A., Wang, X. (2000) A new linear tricobalt compound with di(2-pyridyl)amide (dpa) ligands: two-step spin crossover of $[Co_3(dpa)_4Cl_2][BF_4]$. J. Am. Chem. Soc., 122: 2272–2278.
129. Berry, J. F., Cotton, F. A., Murillo, C. A., Roberts, B. K. (2004) An efficient synthesis of acetylide/trimetal/acetylide molecular wires. Inorg. Chem., 43: 2277–2283.
130. Hughes, A. K., Murphy, V. J., O'Hare, D. (1994) Synthesis, X-ray structure and spin crossover in the triple-decker complex $[(\eta^5\text{-}C_5Me_5)Cr(\mu^2{:}\eta^5\text{-}P_5)Cr(\eta^5\text{-}C_5Me_5)]^+[A]^-$ ($A = PF_6$, SbF_6). J. Chem. Soc. Chem. Commun., 163–164.
131. Goeta, A. E., Howard, J. A. K., Hughes, A. K., O'Hare, D., Copley, R. C. B. (2007) Structural-magnetic correlations on the first dinuclear spin crossover d^4 system. J. Mater. Chem., 17: 485–492.
132. Létard, J.-F., Guionneau, P., Nguyen, O., Costa, J. S., Marcén, S., Chastanet, G., Marchivie, M., Goux-Capes, L. (2005) A guideline to the design of molecular-based materials with long-lived photomagnetic lifetimes. Chem. Eur. J., 11: 4582–4589.
133. Grondin, P., Siretanu, D., Roubeau, O., Achard, M.-F., Clérac, R. (2012) Liquid-crystalline zinc(II) and iron(II) alkyltriazoles one-dimensional coordination polymers. Inorg. Chem., 51: 5417–5426.
134. Titos-Padilla, S., Herrera, J. M., Chen, X.-W., Delgado, J. J., Colacio, E. (2011) Bifunctional hybrid SiO_2 nanoparticles showing synergy between core spin crossover and shell luminescence properties. Angew. Chem. Int. Ed., 50: 3290–3293.
135. Raza, Y., Volatron, F., Moldovan, S., Ersen, O., Huc, V., Martini, C., Brisset, F., Gloter, A., Stephan, O., Bousseksou, A., Catala, L., Mallah, T. (2011) Matrix-dependent cooperativity in spin crossover Fe(pyrazine)$Pt(CN)_4$ nanoparticles. Chem. Commun., 47: 11501–11503.
136. Galan-Mascaros, J. R., Coronado, E., Forment-Aliaga, A., Monrabal-Capilla, M., Pinilla-Cienfuegos, E., Ceolin, M. (2010) Tuning size and thermal hysteresis in bistable spin crossover nanoparticles. Inorg. Chem., 49: 5706–5714.
137. Forestier, T., Mornet, S., Daro, N., Nishihara, T., Mouri, S.-i., Tanaka, K., Fouché, O., Freysz, E., Létard, J. F. (2008) Nanoparticles of iron(II) spin-crossover. Chem. Commun., 4327–4329.
138. Gandolfi, C., Moitzi, C., Schurtenberger, P., Morgan, G. G., Albrecht, M. (2008) Improved cooperativity of spin-labile iron(III) centres by self-assembly in solution. J. Am. Chem. Soc., 130: 14434–14435.

4

Polymeric Spin-Crossover Materials

M. Carmen Muñoz[1] and José Antonio Real[2]

[1] *Departamento de Física Aplicada, Universitat Politècnica de València, Spain*
[2] *Instituto de Ciencia Molecular (ICMol)/ Departamento de Química Inorgánica, Universidad de Valencia, Spain*

4.1 Introduction

The design of coordination polymers (CPs) with switching and memory functions is an important subject of current interest in the search for new advanced materials with potential applications. Implementation of CPs with electronically labile metal ions able to display spin-crossover (SCO) behaviour is a singular approach to this end. In this chapter we provide a survey of 1-, 2- and 3-dimensional (1–3D) spin-crossover coordination polymers (nD-SCO-CPs) reported in the period 2004–2011. This chapter contains three independent parts devoted to 1D-, 2D- and 3D-SCO-CPs, respectively. Each part has been organised according to the nature of the bridging ligands, that is triazoles, bis-tetrazoles, bis- and poly-pyridine-like groups; and anionic ligands essentially derived from metallocyanate complexes ($[M^{I}(CN)_2]^-$ (M^{I} = Cu, Ag, Au); $[M^{II}(CN)_4]^{2-}$ M^{II} = Ni, Pd, Pt; $[M^{III}(CN)_6]^{3-}$ (M^{III} = Cr) and $[M^{IV}(CN)_6]^{4-}$ (M^{IV} = Nb). In these sections we shall briefly describe essential structural features and spin-crossover properties of relevant SCO compounds. A number of reviews devoted to specific families of SCO-CPs have appeared during this period[1–8]

4.2 One-Dimensional SCO-CPs

4.2.1 Triazole Based Bridges

1,2,4-Triazole based 1D-CPs have attracted much interest due to their relevant intrinsic SCO properties, simple preparation, chemical stability and easy processability. They typically display abrupt thermo-induced magnetic and chromatic spin-transitions accompanied by large hysteresis loops characterised by SCO temperatures, T_{SCO}, near room temperature, which have created prospects of application in memory and sensory devices.

Spin-Crossover Materials: Properties and Applications, First Edition. Edited by Malcolm A. Halcrow.

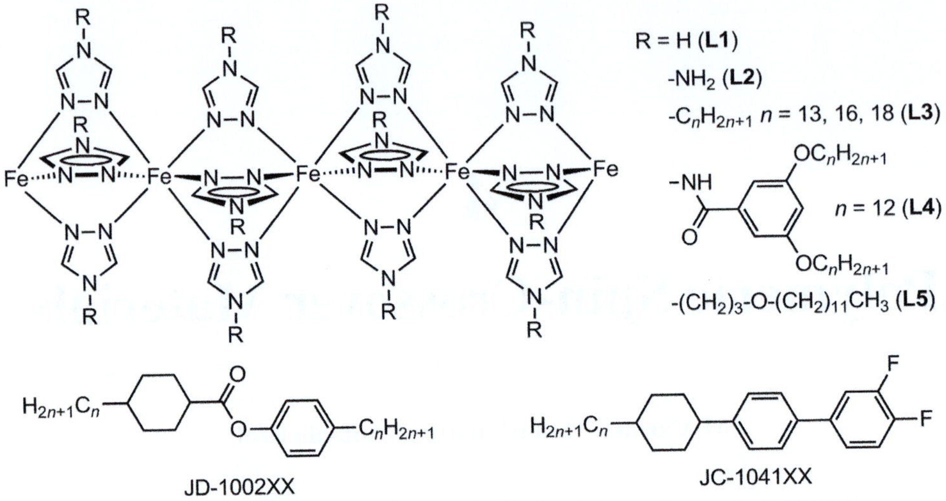

Scheme 4.1

The linear 1D structure of the paradigmatic [Fe(H**L1**)$_2$(**L1**)](BF$_4$) compound has recently been confirmed from synchrotron X-ray powder diffraction analysis (Scheme 4.1).[9] The Fe(II) ions are octahedrally surrounded by six triazole ligands, which act as bridges through the N1 and N2 atoms, being the separation between two consecutive triple bridged iron atoms in the range 3.6–3.7 Å. The strong cooperative behaviour shown by this compound has essentially been attributed to the effective propagation of short-range elastic interactions between neighbouring Fe(II) ions through the triazole bridges. However, there is experimental evidence indicating that for related [Fe(4-R-**L1**)$_3$](A)$_2$ complexes the cooperative behaviour of the SCO strongly depends on the R substituent attached to the 4N position of the H**L1** ligand, the nature of the counterions and the presence of solvent molecules in the crystal. Indeed, the cooperative behaviour of the SCO significantly decreases when R contains bulky groups and/or long alkyl groups (*vide infra*). This suggests that inter-chain interactions, supported by counterions and solvent molecules, must play an important role in cooperativity. Indeed, occurrence of extended hydrogen bonding has been inferred for [Fe(**L2**)$_3$](A)$_x$ on the basis of muon spin relaxation experiments (A = NO_3^-, x = 2)[10, 11] and crystallographic data ([A = $M^{IV}F_6]^{2-}$ (M^{IV} = Ti, Zr, Sn, Ge) and $[TaF_7]^{2-}$, x = 1).[12, 13] A general discussion about cooperativity in triazole-based SCO materials has recently been reported on the basis of thermodynamic parameters derived from calorimetric data.[14] Furthermore, the mechanism of photoinduction of SCO irradiating with pulsed laser at temperatures near or within the hysteresis loop has also been investigated for [Fe(**L2**)$_3$](A)$_2$·nH$_2$O (A = NO_3^-, n = 1; Br^-, n = 3).[15, 16]

Transferring the SCO properties to technologically convenient platforms such as composites and gels,[17–28] liquid-crystalline materials,[29, 30] dendrimers,[31] Langmuir–Blodgett films,[32] and nanoparticles[33–43] has been the most important activity involving 1,2,4-triazole based 1D-SCO-CPs. In this respect, lipophilic functionalisation of the triazole ring has represented an important strategy to confer soft texture and solubility in organic solvents to these materials. For example, the polymer [Fe(**L3**)$_3$](ptol)$_2$·xH$_2$O (ptol$^-$ = para-toluensulfonate) displays SCO with characteristic temperature T_{SCO} = 320 or 280 K (x = 2 or 0, n =18). The solid is purple below T_{SCO}, and becomes white above T_{SCO}. It dissolves in paraffinic solvents and toluene at temperatures above 413 K to give transparent paramagnetic colourless monophasic viscous liquids.

Upon cooling to room temperature these liquids typically form purple diamagnetic gels. SCO occurs in less than 10 K in the temperature interval 320–329 K for these gels independently of the solvent and complex concentration.[23, 24]

Studies of gels generated from solutions of $[Fe(\mathbf{L4_n})_3][C_{n'}H_{2n'+1}\text{-}SO_3]$ (n = 12; n′ = 8, 12, 16) in dodecane have suggested that the sol-gel phase transition can reversibly assist their cooperative SCO behaviour. These transparent purple diamagnetic gels collapse to give pale-yellow paramagnetic solutions at 353 K and recover the initial purple colour upon cooling ($T_{SCO} \approx$ 328–330 K). Furthermore, they display a typical liquid-crystalline mesophase that disappears completely upon heating at 335 K during the spin-state change, indicating that the gel phase stabilises the LS state while the sol phase stabilises the HS state. Indeed, stabilisation of the gel state occurs concomitantly with generation of a hydrogen-bonding network (involving RSO_3^-, the triazole moiety and H_2O molecules), which locks the Fe–N bond length at a distance characteristic of the LS state. This network collapses during the transformation gel-to-sol stabilising the HS state.[25] In a similar way, the complex $[Fe[\mathbf{L5}]_3]Cl_2$ forms gels when dispersed in the organic liquid crystals JC-1041XX and JD-10002XX (Scheme 4.1). These gels are purple at 293 K and pale-yellow at temperatures higher than 363 K. Interestingly, the SCO in the gel phase is complete, steeper and occurs at higher temperatures than for the powder material, and also manifests hysteretic behaviour: $T_{SCO}{}^{up} \approx 334$ K and $T_{SCO}{}^{down} \approx 324$ K (in JC-1041XX) and $T_{SCO}{}^{up} \approx 324$ K and $T_{SCO}{}^{down} \approx 319$ K (in JD-10002XX). This hysteretic behaviour seems to be triggered by a phase transition from a nematic liquid crystalline phase to an isotropic phase observed for these gels in the temperature interval 361–363 K.[26]

A different strategy to explore synergies between SCO and liquid crystal phenomena is based on the possibility of engineering lipophilic triazole ligands capable of transferring intrinsic liquid crystal properties to the resulting iron(II) 1D-SCO-CPs. In this respect, simultaneous thermally induced crystal-to-liquid-crystal and SCO transitions occur near room temperature in the series $[Fe(\mathbf{L4_n})_3]A_2{\cdot}xH_2O$ (n = 4, 6, 8, 10, 12; A = ptol⁻, $CF_3SO_3^-$ and BF_4^- and x = 0-1), which represent a new class of thermochromic liquid crystal materials.[29, 30]

Miniaturization of SCO systems at nanometric scale has been an important motivating topic in the processing of these materials. First attempts at producing SCO nanoparticles were reported for 60–200 nm nanoparticles (NPs) of $[Fe(\mathbf{L2})_3]Br_2{\cdot}3H_2O$ prepared from water in oil microemulsions.[33] Successive studies on this material have demonstrated that the hysteretic behaviour remains for NPs larger than 30 nm but disappears for smaller sizes.[34–36] A new synthetic approach has enabled to obtain mono-dispersed colloidal suspensions of 3–4 nm spherical NPs of $[Fe(\mathbf{L2})_3](ptol)_2$ showing cooperative SCO without hysteresis. Interestingly, this method avoids the problem of aggregation, often observed in the precipitation method, and enables to correlate particle size and SCO properties without ambiguity.[37] Microtransfer moulding techniques have been used to fabricate homogeneous micro- and nano-patterns of NPs of the latter compound, dispersed in *n*-octane.[38]

Contrarily to expectations, 10–15 nm NPs of $[Fe(H\mathbf{L1})_2(\mathbf{L1})]BF_4$ dispersed in octane show SCO properties comparable with those found for the bulk material.[39, 40] Even smaller NPs (2–5 nm) of the same material well dispersed in silica films show stable hysteresis loops ca. 21 K width. The incomplete character of the SCO at high temperatures and the significantly high critical temperatures compared to those of the bulk material, have been ascribed to positive chemical pressure induced by the rigid SiO_2 matrix.[41]

Synergy between SCO and fluorescence has been demonstrated for $[Fe(\mathbf{L2})_3](ptol)_2$ NPs doped with the fluorescent agent rhodamine. At room temperature the emission spectrum of rhodamine overlaps the ${}^1A_1 \rightarrow {}^1T_1$ absorption band characteristic of Fe(II) in the LS state, consequently the emission is partially quenched. Bleaching of this band in the HS state induces and increases the NPs' fluorescence.[42] Similar results have been observed for large rod-like $[Fe(H\mathbf{L1})_2(\mathbf{L1})]BF_4$ NPs coated with a silica shell functionalised by grafting fluorophore species.[43]

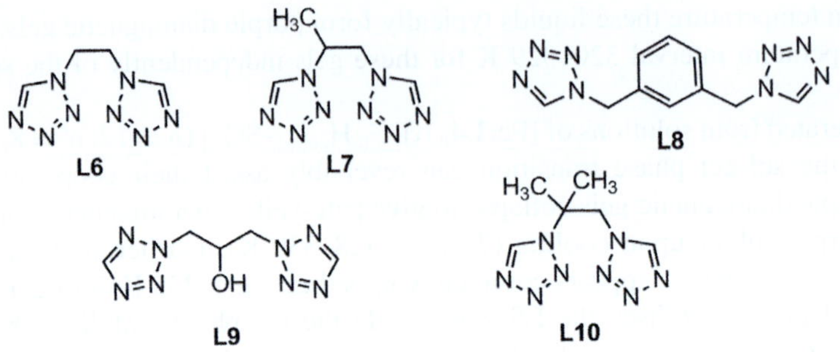

Scheme 4.2

4.2.2 Tetrazole Based Bridges

Unlike the triazole ring, the tetrazole ring invariably acts as a monodentate ligand. Thus, a number of mononuclear $[FeL_6]A_2$ SCO complexes containing simple tetrazole ligands have been reported since long ago, and reviewed.[44] The first examples of 1D-SCO-CPs generically formulated $[Fe(\mathbf{L})_3]A_2$ were isolated in the early 2000s from bis-tetrazole ligands, more specifically **L6** and **L7** (Scheme 4.2). These bridging ligands adopt *syn* conformation with almost eclipsed disposition of the tetrazole rings when coordinated to the Fe(II) ion. Three bridging ligands connect two adjacent Fe(II) in the chain. The weak cooperativity observed in these complexes was ascribed to the flexible alkyl spacers.[45, 46] Similar structure and SCO behaviour has been recently reported for $[Fe(\mathbf{L8})_3]A_2{\cdot}S$; $A = PF_6^-$ ($S = CH_3OH$), $CF_3SO_3^-$ ($S = CH_3CN$) and ClO_4^- ($S = CH_3CN$). The PF_6^- derivative displays an almost complete SCO ($T_{SCO} = 160$ K), the $CF_3SO_3^-$ shows ca. 50% conversion ($T_{SCO} = 110$ K) and the ClO_4^- derivative is mainly a paramagnetic system, which shows quite incomplete SCO (ca. 20%) in the interval 50–150 K. Interestingly, the $CF_3SO_3^-$ derivative shows two different crystallographic phases which account for observation of 50% SCO in this compound. At low temperature the centres are regularly disposed, alternating the LS and HS states in the chain.[47]

Two double-bridged chains containing the bridging ligands **L9** and **L10** have been reported. In the complex $[Fe(\mathbf{L9})_2(CH_3CN)_2](ClO_4)_2{\cdot}CH_3CN$ the equatorial positions of Fe(II) are occupied by two **L9** bridging ligands, while the axial positions are occupied by two acetonitrile molecules. This complex shows poor cooperative SCO with $T_{SCO} = 110$ K.[48] A similar situation occurs for $[Fe(\mathbf{L10})_4](ClO_4)_2$ where the Fe(II) ions are double bridged by two **L10** ligands while two additional **L10** ligands act as terminal axial ligands. In contrast, this complex displays strong cooperative SCO ($T_{SCO}^{down} = 133$ K, $T_{SCO}^{up} = 136$ K) concomitantly with an order–disorder transition involving the perchlorate anions.[49]

Thermal- and photoinduced SCO have been reported for complexes derived from related bistetrazole ligands with longer alkyl spacers (pentyl, hexyl, etc.), however, to our knowledge no crystal structures have been determined yet.[50, 51]

4.2.3 Bis-Monodentate Pyridine-Like Bridges

The synthesis of 1D-SCO-CPs based on bis-monodentate pyridine-like bridges necessarily requires participation of ancillary ligands. This is the case for the positively charged zig-zag chains $[Fe(\mathbf{L11})_2(\mathbf{L12\text{-}14})](ClO_4)_2{\cdot}nEtOH$, where **L11** is a bidentate ligand and **L12-14** are bis-monodentate bridging ligands (Scheme 4.3). A linear chain has also been isolated for the particular case of $[Fe(\mathbf{L15})_2(\mathbf{L12})](ClO_4)_2{\cdot}2EtOH$. The common feature of this series of compounds is the occurrence of complete and relatively gradual SCO with T_{SCO} in the interval 176–220 K.[52–54]

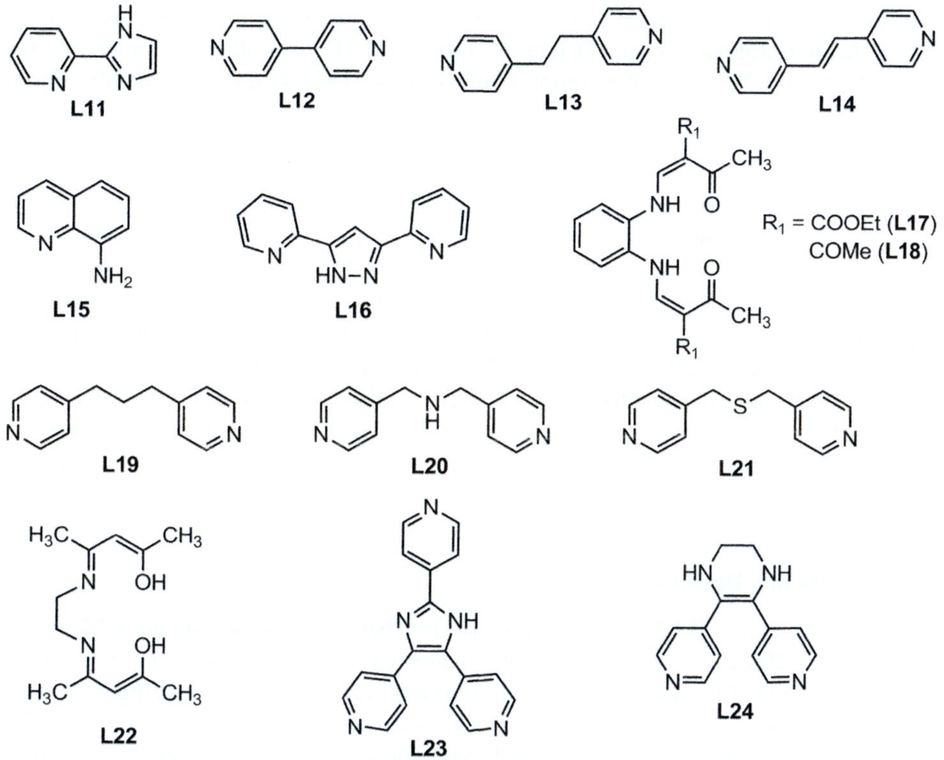

Scheme 4.3

Combination of the deprotonated bis-bidentate bridging ligand **L16** with the bis-monodentate ligand **L12** has afforded an unusual infinite chain of dinuclear iron(II) SCO units formulated $[[Fe^{II}(\mathbf{L16})(NCS)]_2(\mathbf{L12})]\cdot$MeOH. The two Fe(II) ions in the dinuclear unit are bridged by two equatorial **L16** ligands while consecutive dimeric units are connected through **L12**, which occupy the remaining axial positions of the iron(II) ions. Interestingly, this compound displays a cooperative transition with a hysteresis loop 2 K wide centred at ca. 160 K. The photo-generated HS–HS species ($T < 50$ K) clearly denote the occurrence of intradimer antiferromagnetic coupling.[55]

An important family of iron(II) compounds based on the tetradentate N_2O_2 Schiff base-like ligands **L17** and **L18** has been reported.[4] The deprotonated form of **L17** and **L18** occupies the four equatorial positions of the Fe(II) ion while the remaining axial positions are occupied by bis-monodentate bridging ligands **L-L** (**L12**, **L13**, **L19**, **L20** and **L21**) affording a rich variety of neutral 1D-SCO-CPs formulated $[Fe^{II}(\mathbf{L17/18})(\mu\text{-}(L\text{-}L))]\cdot n$S. In particular, $[Fe^{II}(\mathbf{L18})(\mathbf{L12})]$ displays strong cooperativity with a 18 K wide hysteresis loop ($T_{SCO}{}^{up} = 237$ K and $T_{SCO}{}^{down} = 219$ K). However, the presence of solvent molecules included in the crystal completely blocks the HS state (MeOH, n = 0.25; DMF, n = 1).[56, 57] When the bridging ligand is **L13** a cooperative 2-step transition is observed, $T_{SCO1} = 123$ K and $T_{SCO2} = 95$ K with hysteresis widths $\Delta T =$ 11 K and 9 K for n = 0, respectively, and $T_{SCO1} = 106$ K ($\Delta T = 22$ K) and $T_{SCO2} = 98$ K ($\Delta T = 7$ K) for nS = 0.5MeOH. The less rigid bridge **L19** also originates a 2-step behaviour for the MeOH solvate (n = 1), however, as expected, the SCO is gradual without hysteresis ($T_{SCO1} = 166$ K and $T_{SCO2} = 90$ K). The crystal structure of this compound, determined for the pure HS and mixed spin HS–LS phases, denotes the occurrence of a crystallographic phase transition between the *C*2/*c* space group (HS state) and the *P*2$_1$/*c* space group in the plateau where the HS and LS ions orderly alternate along the chain. In contrast to [Fe^{II}(**L18**)(**L12**)], the homologous [Fe^{II}(**L17**)(**L12**)] compound is HS over all temperatures. Replacement of **L12** by **L13**, **L19**, **L20** and **L21** in the latter complex, affords new 1D-SCO-CPs with mainly continuous SCO behaviours characterised by T_{SCO} in the interval 170–200 K.[4] Following a similar synthetic strategy a series of Fe(III) 1D-SCO-CPs, based on the more basic Schiff base acacen (**L22**), has been reported to display poor cooperative behaviour.[58–60]

The nitrile-rich [$Fe(CH_3CN)_4(\mu\text{-pz})](ClO_4)_2$ (pz = pyrazine) constituted of four equatorially coordinated CH_3CN ligands and two axial pyrazine bridges displays a 3-step continuous SCO with ca. 50, 38 and 12% conversions characterised by two inflexion points, one at 127 K and the other at 102 K (T_{SCO1} = 146 K, T_{SCO2} = 115 and T_{SCO3} = 85 K). This unusual behaviour seems to be correlated with order–disorder transitions experienced by the pz rings and the perchlorate ions when cooling from 146 K to 115 K, which involve reorganisation of intermolecular contacts and the appearance of novel ones.[61]

The ligands **L23** and **L24** have afforded double bridged chains formulated [$Fe(L)_2(NCS)_2$]·nH_2O. Four pyridine rings occupy the equatorial positions of the [FeN_6] octahedron while the NCS^- groups fill the axial positions. Only two of the three pyridine rings of **L23** coordinate the Fe(II) ion. Both compounds display poor cooperative 50% thermal induced SCO in the temperature range 100–300 K.[62, 63]

4.2.4 Polydentate Chelate Bridges

The unsymmetrical bridging ligand **L25** (Scheme 4.4), a hybrid between 4,4′-bipyridine (**L12**) and terpyridine, forms the complex [Co(**L25**)Cl_2]·2S which represents an unusual example of Co(II) 1D-SCO-CPs. The Co(II) ion is equatorially surrounded by one terpy and one pyridine moieties of two **L25** ligands, the remaining axial coordination positions are completed with two Cl^- anions. The methanol solvate (S = CH_3OH) is a typical HS Co(II) complex, however, the dihydrate homologue exhibits a cooperative $S = {}^3/_2$ (HS) ↔ $S = {}^1/_2$ (LS) spin-transition centred at 222.5 K featuring a 1 K wide thermal hysteresis loop.[64] A similar synthetic strategy has afforded the symmetrical rigid bisterdentate bridging ligand **L26**. The resulting [Fe(**L26**)]$(BF_4)_2$ complex is a rigid 1D-SCO-CPs in which the Fe(II) ion is coordinated by two 2,6-di(pyrazol-1-yl)pyridine

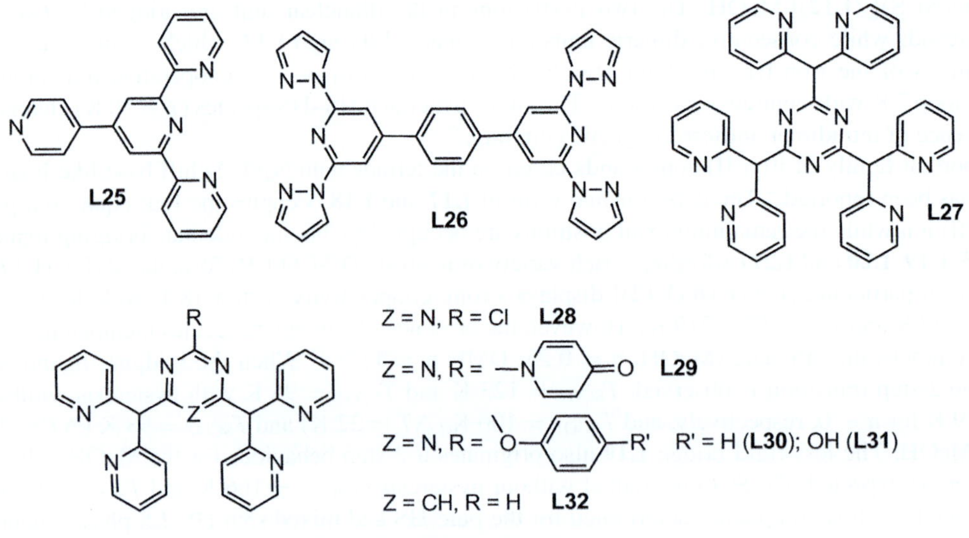

Scheme 4.4

moieties belonging to two distinct ligands. This salt undergoes a cooperative SCO with T_{SCO} centred at 323 K characterised by a 10 K wide hysteresis loop.[65]

The more sophisticated polypyridine ligand **L27** affords a rather complicated 1D-polymer consisting of $[Fe_3(\mathbf{L27})_2(C_2H_5CN)_4(BF_4)_2]_\infty{}^{4+}$ units. Each Fe(II) ion is coordinated by two dipyridylamine moieties belonging to adjacent **L27** ligands. Two of the three Fe(II) complete their coordination sphere with two *cis*-coordinated propionitrile molecules while the third Fe(II) completes the axial positions with two F atoms of adjacent BF_4^- anions. These voluminous repeating units define pores filled by four uncoordinated BF_4^- anions and four propionitrile solvent molecules. The two $[FeN_6]$ centres display gradual SCO at T_{SCO} = 300 K while the $[FeN_4F_2]$ centre remains HS over all temperatures. The coordinated propionitrile molecule can be reversibly replaced by acetonitrile or hydroxilic solvents (water, propanol). The former shifts T_{SCO} ca. 27 K towards low temperatures while the latter stabilises the HS state at any temperature.[66]

Closely related *trans*-[Fe(**L28-31**)(NCX)$_2$]·nS (X = S, Se) complexes have been synthesised from the bis(dipyridylamine) triazine ligands **L28-31**. The structure is essentially the same for all members of this family. Two dipyridylamine moieties bridge two consecutive Fe(II) ions, which complete their coordination with two NCX^- groups in *trans* conformation. These compounds undergo gradual SCO with T_{SCO} in the temperature range 175–200 K.[67, 68] Full 2-step SCO has been observed for the unsolvated compounds derived from the bis(dipyridylamine)pyrimidine ligand **L32** characterised by T_{SCO1} = 135 K (200 K) and T_{SCO2} = 90 K (125 K) for X = S (Se). A HS polymorph has also been characterised for the S derivative, which is isostructural to the Se derivative. The main structural difference found between both polymorphs is the presence and lack of π–π stacking between the chains for the SCO and HS species, respectively. A structural analysis of the SCO polymorph reveals the presence of two crystallographically distinct Fe(II) sites in the temperature interval 25–250 K. At 123 K, just in the middle of the plateau, the chains show an arrangement of alternating HS and LS sites. In contrast, the X = Se derivative shows the presence of one Fe(II) site in the temperature interval 90–250 K with an averaged LS/HS character of the Fe(II) site in the plateau (150 K). Interestingly, the diffraction images of the latter compound show the occurrence of diffuse scattering planes in the diffraction images, at 90 K and 150 K, suggesting that long-ordering –LS–HS–LS–HS– occurs along the direction of the chains but at a random distribution between adjacent chains.[69]

4.2.5 Anionic Bridging Ligands

As far as we know, only four 1D-SCO-CPs based on anionic bridging ligands have been reported so far. These ligands are $[N(CN)_2]^-$, $[C[C(CN)_2]_3]^{2-}$, $[Ni^{II}(CN)_4]^{2-}$ and $[Cu^I(3CNpy)(CN)_2]^-$ (3CNpy = 3-cyanopyridine). The chain compounds [Fe^{II}(**L15**)$_2$[N(CN)$_2$]](ClO$_4$)·MeOH,[70] [Fe(**L33**)$_2$([C[C(CN)$_2$]$_3$])][71] and [Fe(**L33**)$_2$(Ni(CN)$_4$]·nH$_2$O[72] can be described in a similar way (Scheme 4.5). The equatorial positions of the Fe(II) are coordinated by two identical chelate bidentate ligands, L-L, defining $[Fe(L\text{-}L)_2]^{2+}$ moieties axially linked by the anionic bridging ligand. These compounds undergo gradual SCO with T_{SCO} = 201 K, ca. 95 K and 360 K for $[N(CN)_2]^-$, $[C[C(CN)_2]_3]^{2-}$ and $[Ni^{II}(CN)_4]^{2-}$ respectively. Detailed LIESST effect studies have also been reported for the $[C[C(CN)_2]_3]^{2-}$ derivative.[71]

In contrast to the precedent systems, double bridges are formed in [Fe(3CNpy)$_2$[Cu(3CNpy)(CN)$_2$]. The trigonal metalloligand [Cu^I(3CNpy)(CN)$_2$] predisposes the formation chains since the NC–Cu–CN moiety defines an angle of 119.1°. The Fe(II) ion, located at an inversion centre, defines a pseudoctahedral $[FeN_6]$ coordination site with two 3CNpy groups occupying the axial positions while the equatorial positions are occupied by the N atoms of four $[Cu(3CNpy)(CN)_2]^-$ bridges. The chains assemble in such a way that favours short Cu·····Cu metallophilic contacts (2.6358 Å). This compound undergoes a complete and relatively cooperative SCO without hysteresis characterised by $T_{SCO} = 172$ K, $\Delta H = 7.9$ kJmol^{-1}; $\Delta S = 45$ J K^{-1} mol^{-1}. In addition the unit cell volume experiences noticeable variation upon SCO, $\Delta V_{HL} = 46.2$ Å^3, concomitantly with a large variation of inter-chain Cu····Cu distances (0.045 Å) upon SCO.[73] The structurally related 1D-CP containing 3-X-pyridine ligands [X = Cl (polymorph α), Br, I] remain HS at any temperature.[74]

4.3 Two Dimensional SCO-CPs

4.3.1 Neutral Organic Bridging Ligands

A number of physical studies have been reported in recent years on the paradigmatic 2D-SCO-CP [Fe(**L34**)$_2$(NCS)$_2$]·H_2O (Scheme 4.5). They include detailed analysis of the LS and HS state single crystal structures,[75] observation of SCO behaviour using longitudinal field muon spin relaxation measurements,[76] investigation of the hysteresis loop of the solid solutions [Fe_xZn_{1-x}(btr)$_2$(NCS)$_2$]·H_2O by means of the FORC method,[77] experimental electron density analysis of the SCO,[78] kinetics of the light-induced first-order phase transition,[79] optical, magnetic and structural properties in the light-induced and thermally quenched metastable states[80] and calorimetric measurements of solid solutions.[81] Unexpectedly, the closely related complex [Fe(**L35**)$_2$(NCS)$_2$] is HS at any temperature even at pressures as high as 12 kbar. This situation has been discussed in terms of pure crystallographic analysis. The [FeN_6] coordination sphere for the **L35** derivative is significantly more distorted ($\Sigma = 21.2°$) and features larger average Fe–N bond distances (2.186(3) Å) than [Fe(**L34**)$_2$(NCS)$_2$]·H_2O in the HS state (14.8° and 2.164(3) Å). Furthermore, the inter-layer spacing for **L35** derivative is the same as that of the **L34** derivative in the LS state, a fact that probably enhances stabilisation of the HS state.[82]

Strong cooperative thermal- and light-induced SCO behaviour with a hysteresis loop 13 K wide centred around 107 K has been observed for [Fe(**L36**)$_3$](ClO_4)$_2$. This compound displays a singular (3,6) 2D network topology characterised by sharing [Fe_3(**L36**)$_3$] triangular motives. Detailed single crystal absorption measurements have shown that: (i) the hysteresis width depends on the temperature sweep rate indicating that kinetics of the phase transition are quite slow, and (ii) the apparent transition temperature for the HS-to-LS transition at a given temperature sweep rate depends upon the history of the crystal. These observations

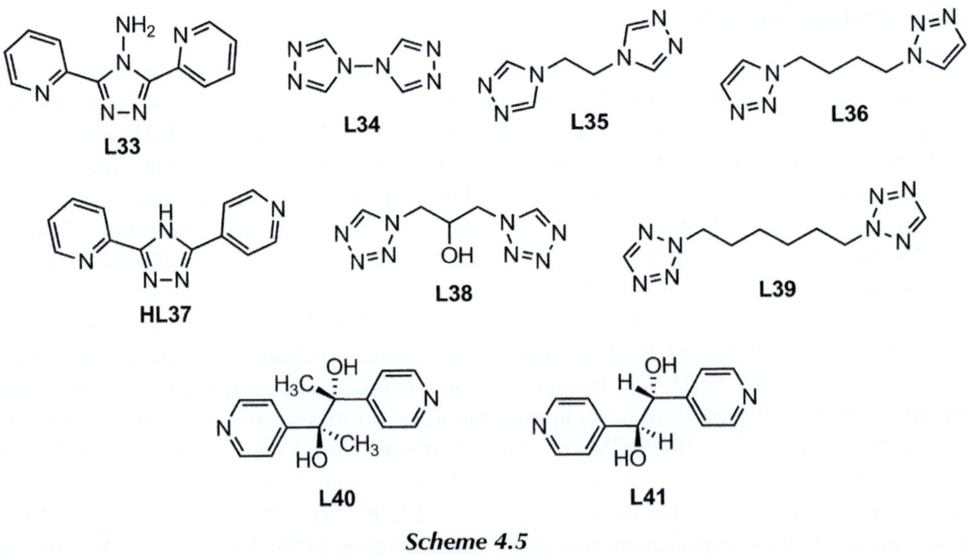

Scheme 4.5

strongly support the occurrence of a crystallographic phase transition below T_{SCO}^{down}. Indeed, at temperatures close below T_{SCO}^{down} the relaxation kinetics of the thermally trapped HS state are governed by the kinetics of the crystallographic phase transition.[83–85] In the highly diluted solid solutions $[Fe_xZn_{1-x}(\mathbf{L36})_3](ClO_4)_2$ (x = 0.02 and 0.1) the system displays a very incomplete (20%) and gradual SCO. These highly diluted crystals show quantitative LIESST and reverse-LIESST effects.[86, 87]

Self-assembly of Fe(II) and H**L37** has afforded four architectural isomers formulated $[Fe(\mathbf{L37})_2]\cdot nS$ where **L37** is the deprotonated form of H**L37**. This asymmetric ligand has the ability to act simultaneously as bidentate and monodentate ligand. Four **L37** groups coordinate the Fe(II) in *trans* conformation, two of them occupy four equatorial positions and other two coordinate the remaining axial positions. Three of these isomers (with n = $2.17H_2O$, $2.5H_2O$ and 4dioxane·$4H_2O$) are low spin 3D-CPs with the NbO topology while the remaining unsolvated isomer (n = 0) forms a planar 2D-CP with (4,4) topology. The latter compound shows a cooperative SCO with T_{SCO}^{up} = 278.8 K and T_{SCO}^{down} = 273.4 K.[88] A fifth architectural isomer with n = $4H_2O$ has also been reported. In this hydrate the **L37** ligands coordinate the Fe(II) centres in *cis* conformation giving corrugated layers with (4,4) topology, which stack in a face-to-face manner. This *cis* form shows a rather gradual 50% SCO covering the temperature interval 100–300 K. Interestingly, a sharp structural transformation occurs at considerable high temperature (633 K) involving a change of conformation of the L37 ligands from *cis*- to *trans*- disposition to give the planar unsolvated form referred above. This happens without loss of crystallinity.[89]

Two bis(tetrazole)-type ligands, **L38** and **L39**, have been specifically designed to synthesise new 2D-SCO-CPs. Self-assembly of $Fe(BF_4)_2\cdot 6H_2O$ and **L38** affords the complex $[Fe(\mathbf{L38})_{1.8}(\mathbf{L38}\text{-}OBF_3)_{1.2}](BF_4)_{0.8}\cdot(H_2O)_{0.8}(CH_3CN)$ where **L38**-OBF_3 is formed spontaneously during the synthesis. It is made up of square grids with the Fe(II) ions connected through single **L38** bridges in the *b* direction and double bridges, **L38** and **L38**-OBF_3, in the *c* direction. This compound shows a gradual SCO with T_{SCO} = 112 K.[90] Compound $[Fe(\mathbf{L39})_4(CH_3CN)_2](ClO_4)_2$ forms a topologically similar (4,4) 2D network in which four **L39** bridging ligands coordinate the equatorial positions of Fe(II) while the two axial coordination sites are occupied by two CH_3CN molecules. This compound displays a relatively cooperative SCO characterised by T_{SCO} = 128 K.[91]

Detailed synthetic and structural studies dealing with the influence of host–guest interactions on the SCO behaviour of the porous 2D-CP $[Fe(\mathbf{L})_2(NCS)_2]\cdot$nSolvent constituted of perpendicular interpenetration of equivalent 2D layers have been carried out using the bis-monodentate bridging ligands L = **L40**, **L41**, and **L13**. The framework $[Fe(\mathbf{L40})_2(NCS)_2]\cdot$Solvent was synthesised to investigate the effect on the SCO of potential host–host and host–guest hydrogen bonding interactions provided by the diol groups of the ligand. Indeed, the interpenetrated frameworks are stabilised by hydrogen bonding interactions between the diol groups and the sulfur atoms of the NCS^- groups. It undergoes a spin conversion at T_{SCO} = 162 K that practically remains unchanged after removing an included molecule of acetone, which does not interact with the host.[92] Similar structural characteristics were observed for the acetonitrile solvate, which shows less cooperative SCO at lower temperatures (T_{SCO} = 137 K). In contrast, protic solvents like MeOH, EtOH and 1-PrOH, which show host–guest NCS····OH hydrogen bonds, undergo considerably more cooperative SCO with T_{SCO}^{down} = 142 K T_{SCO}^{up} = 145 K for MeOH, T_{SCO}^{down} = 156 K T_{SCO}^{up} = 166 K for EtOH and T_{SCO} = 187 K for 1-PrOH.[93] Similar host–host hydrogen bonding stabilisation of the frameworks and host–guest interactions were observed for the closely related compound $[Fe(\mathbf{L41})_2(NCS)_2]\cdot$nEtOH. Nevertheless, it displays poor cooperativity and incomplete SCO for n = 3 and 0.[94] When the bridging ligand is **L13**, the more flexible framework $[Fe(\mathbf{L13})_2(NCS)_2]\cdot$3Acetone is obtained. It shows a gradual 2-step SCO behaviour with characteristic temperatures T_{SCO1} = 133 K and T_{SCO2} = 80 K accompanied by a structural phase transition, from tetragonal ($P\bar{4}2_1c$) to orthorhombic ($P2_12_12$), with checkerboard-like ordering of spin sites at the intermediate HS:LS plateau.[95] Photoexcitation of the HS state at 10 K (LIESST effect) has been studied on nanoparticles of the latter complex.[96] The longer and flexible ligand **L19** facilitates parallel interpenetration of 2D-CP in the complexes [Fe(**L19**)$_2$(NCX)$_2$] (X = S, BH$_3$). Only the X = BH$_3$ derivative undergoes SCO (T_{SCO} = 123 K).[97]

The compound [Fe$_2$(2,2′-bipyridine)(imidazolate)$_4$] forms extended 2D double-layer sheets of alternating tetrahedral and octahedral Fe(II) singly bridged by imidazolate ligands, with the octahedral centres additionally coordinated by 2,2′-bipyridine ligands that occupy positions between the sheets. This uncommon compound undergoes two structural phase transitions accompanied by two quite incomplete cooperative SCO events with critical temperatures centred at ca. 135 K and 151 K.[98]

4.3.2 Dicyanometalate [M^I(CN)$_2$]$^-$ Bridging Ligands (M^I = Cu, Ag, Au)

These 2D-SCO-CPs share the structural principle by which the equatorial positions of the [FeN$_6$] octahedron are invariably occupied by four CN groups of the metallocyanate bridging ligands while the axial positions are occupied by monodentate pyridine-like ligands. The Cu(I) atom shows a marked tendency to expand the coordination number of the linear [Cu(CN)$_2$]$^-$ species to adopt distorted trigonal [CuL(CN)$_2$]$^-$ or tetrahedral [CuL$_2$(CN)$_2$]$^-$ geometries. For example, self-assembly of Fe(II), 3-fluoropyridine (3Fpy) and [Cu(CN)$_2$]$^-$ affords the compound [Fe(3Fpy)$_2$[Cu(3Fpy)$_{1.5}$(CN)$_2$]$_2$] consisting of strongly corrugated 2D grids made up of edge-sharing [Fe$_4$] squares whose edges are defined by two distorted pseudo-trigonal [Cu(3Fpy)(CN)$_2$]$^-$ and two pseudo-tetrahedral [Cu(3Fpy)$_2$(CN)$_2$]$^-$ anions. The compound is essentially LS at 300 K, but undergoes an irreversible structural and magnetic phase transition at $T \approx 350$ K. The stable new phase shows two SCO regimes, a smooth one in the interval 225–400 K and a steep one in the temperature range 150–250 K. Interestingly, the coordination bond lengths and angles of the Cu(I) centres show noticeable changes upon SCO suggesting strong coupling between the Fe(II) ions and the [Cu(CN)$_2$]$^-$ bridging moieties. This flexibility combined with the normal SCO structural changes associated to Fe(II) cause a large change of the unit cell volume ΔV_{HL} = 50.7 Å^3 per Fe(II).[74] In contrast, the use of the ditopic axial ligand pyrimidine (pmd) affords a rigid 2D-SCO-CP formulated [Fe(pmd)$_2$[Cu(CN)$_2$]$_2$]. The Fe(II) ion lies at the inversion centre of an elongated octahedron with the equatorial positions occupied by four [Cu(CN)$_2$]$^-$ ions, which define [Fe[Cu(CN)$_2$]$_2$]$_\infty$ linear chains. The axial positions are occupied by the pmd groups that also coordinate the Cu(I) ions of the adjacent chains to generate the 2D-CP. Short interlayer Cu····Cu distances, 3.5329 Å (293 K), indicate the presence of metallophilic interactions connecting the layers. This compound displays cooperative SCO characterised by $T_{SCO}{}^{down}$ = 132 K and $T_{SCO}{}^{up}$ = 142 K. Thermal, pressure and light induction including photo-crystallography and photo-calorimetric studies have also been reported for this compound.[99]

The mentioned tendency to expand the coordination sphere is strongly mitigated for the [Ag(CN)$_2$]$^-$ ion, and no examples are known for [Au(CN)$_2$]$^-$. This is, in part, due to the persistent presence of efficient metallophilic interactions in the resulting polymers and the more inert character of M^I. A large family of compounds based on 3-substituted pyridines (3Xpy) of general formula [Fe(3Xpy)$_2$[M(CN)$_2$]] (X = H, CH$_3$, F, Cl, Br and I for M^I = Ag; X = CN, F, Cl, Br, I for M^I = Au) (hereafter abbreviated **M3Xpy**) have been reported. The lineal [M(CN)$_2$]$^-$ bridges connect pairs of pseudoctahedral Fe(II) sites defining 2D grids made up of edge-sharing [Fe[M(CN)$_2$]]$_4$ motifs, which are organised in such a way that the iron atoms of one layer are below/above the centre of the windows defined by the other layers. The layers are arranged in pairs with short M^I...M^I contacts (3.0–3.2 Å) indicating the occurrence of strong metallophilic interactions.[100, 101] Only four members of this family show SCO: **Ag3Fpy**, **Ag3Clpy**, **Ag3Hpy** and **Au3Fpy**. The remaining compounds [M = Ag; X = Br, I) and (M = Au; X = H, Cl, Br, I)] are HS. Compound **Ag3Fpy**, displays a gradual cooperative 2-step transition with characteristic temperatures T_{SCO1} = 162 K and T_{SCO2} = 96 K, reflecting the occurrence of two crystallographically distinct Fe(II) sites. This SCO behaviour compares well with that reported for the homologue **Ag3Hpy** characterised by T_{SCO1} = 146.3 K and T_{SCO2} = 91 K.[102] A more steep 50% ST taking place at relatively low temperatures, T_{SCO} = 106 K, has been observed for **Ag3Clpy** in agreement with the average Fe–N bond distance and the crystal volume change at 100 K. For this compound only one type of Fe(II) site, averaged between the HS and LS states, has been observed.[100]

Two different SCO behaviours have been described for the system **Au3Fpy**. One displays a cooperative half-spin-transition (T_{SCO}^{down} = 140 K and T_{SCO}^{up} = 145 K). At room temperature the crystals display the monoclinic $P2_1/c$ space group characterised by one crystallographically Fe(II) site in the HS state. At 120 K, the crystals change to the triclinic $P\bar{1}$ space group where two crystallographically different and equally populated Fe(II) sites, one LS and the other HS, are observed.[101] This compound displays an almost complete two-step transition (T_{SCO1} = 200 K and T_{SCO2}^{down} = 95.6 K and T_{SCO2}^{up} = 122 K) at a pressure of 0.18 GPa. The second reported SCO behaviour shows a complete 2-step spin-transition at ambient pressure with critical temperatures T_{SCO1} = 147.9 K, T_{SCO2}^{down} = 98.2 K and T_{SCO2}^{up} = 118.6 K.[103, 104] The fully LS crystal structure of this 'second' form, solved at 80 K, shows that this system recovers the monoclinic $P2_1/c$ space group. Although the magnetic behaviour suggests the existence of two polymorphs, it is difficult to justify them from a crystallographic viewpoint.

This series of **M3Xpy** (M^I = Ag, Au) halogeno-pyridine complexes suggests a downward shift of the T_{SCO} temperature when moving from silver to gold. Furthermore, the opposite trend from the expected decrease of the ligand field (and therefore T_{SCO}) as the electronegativity of X increases is observed. The former trend can be related to the more electronegative character of Au and consequently to the weaker donor capability of the CN groups in $[Au(CN)_2]^-$. The latter trend has tentatively been justified considering attenuation of the lattice pressure as the polarisability of X increases.

Three architectural isomers have been isolated from slow evaporation of Fe-3CNpy-$[Au(CN)_2]^-$ solutions in MeOH/H_2O, two of them called **Au3CNpy** (polymorphs α and β) are 2D-CP and the third is a 3D-CP (*vide infra*). The α form displays a double layer structure sustained by strong Au····Au contacts (3.1060 Å) while in the β form the layers are distributed homogeneously and no short Au····Au contacts are observed. The α form displays a cooperative 50% SCO (T_{SCO}^{down} = 107 K and T_{SCO}^{up} = 120 K), a fact consistent with the presence of two crystallographically distinct Fe(II) ions in the temperature range 105–293 K; in contrast polymorph β is HS.[105] Formation of the 2D-SCO-CP $[Fe(3CNpy)_2(CH_3OH)_{2/3}[Au(CN)_2]_2]$ is favoured from solutions essentially constituted of MeOH.[106] The 2D grids contain three independent Fe(II) centres: two $[Fe(1,2)N_6]$ and one $[Fe(3)N_4O_2]$. An extended 3D hydrogen-bonding network exhibiting the NbO topology defines the crystal packing. This compound undergoes an incomplete gradual SCO involving ca. 33% of the Fe(II) centres (T_{SCO} = 169 K). Reversible loss of the coordinated methanol molecules and the concomitant coordination of two terminal CN groups at the Fe(3) sites involves the change of space group from triclinic $P\bar{1}$ to trigonal $P\bar{3}$. Consequently, the 2D layers connected via hydrogen bonds convert into a rigid polymer made up of triple interpenetrated 3D coordination networks preserving the NbO topology. These reversible structural changes noticeably affect the critical temperature of the spin-transition observed for this compound (see Section 4.4.2).

In situ generation of $[Ag(3Xpy)(CN)_2]^-$ species formed in solution during the diffusion process was reported for $[Fe(3Xpy)_2[Ag(CN)_2][Ag(3Xpy)(CN)_2]]\cdot 3Xpy$ (**Ag3Xpy***) (X = Br and I). These isostructural compounds combine two different bridging $[Ag(CN)_2]^-$ and $[Ag(3Xpy)(CN)_2]^-$ units, which define slightly corrugated layers. Unlike **Ag3Xpy**, the layers are regularly separated by ca. 8 Å. The space between the layers is partially occupied by 3Brpy or 3Ipy guest molecules. Consequently, no argentophilic interactions are observed. The guest and coordinated 3-Xpy molecules interact via π-stacking defining infinite chains. Furthermore, there are weak X····X contacts between the three types of 3-Xpy molecules.[100] No homologous **Au3Xpy*** compounds have been reported, however, the presence of an excess of 3-Ipy favours its inclusion during the diffusion process to form $[Fe(3Ipy)_2[Au(CN)_2]]\cdot(3Ipy)_{0.5}$ (**Au3Ipy·(3Ipy)$_{0.5}$**). Interestingly, the double layer structure remains intact in this compound, most likely due to the greater efficiency of Au(I) to give aurophilic interactions. Furthermore, strong N→I donor–acceptor interactions occur between the nitrogen and iodine atoms of the guest and the coordinated 3-Ipy, respectively. The N····I distance is 2.994(7) Å at 293 K, which shortens by 0.087(7) Å at 80 K.[101]

It is worth noting that the near room temperature SCO found for **Ag3Xpy*** (T_{SCO} = 306 K for X = Br and T_{SCO} = 275 for X = I) strongly contrasts with the paramagnetic behaviour observed for the corresponding **Ag3Xpy** compounds. A similar situation is observed for the paramagnetic compound **Au3Ipy** and its related 3Ipy clathrate derivative **Au3Ipy·(3Ipy)$_{0.5}$**. The latter displays a 2-step transition (T_{SCO1} = 155 K; $T_{SCO2}{}^{down}$ = 97 K and $T_{SCO2}{}^{up}$ = 110 K). These results suggest that the 3-Xpy guest molecules play an important role in the stabilisation of the LS spin state. This is particularly true for the gold derivative since the composition and the double layer structure of the host framework remains essentially unchanged despite the presence of the guest molecules.

The use of pyridine ligands bearing bulky substituents has afforded the compounds generically formulated as $[Fe(L)_2[M(CN)_2]_2]\cdot nH_2O$ (**ML**) (M = Ag or Au) where L is 4-phenylpyridine (4phpy),[107, 108] **L42** or **L43** (Scheme 4.6).[109] Compounds **Ag4phpy·H$_2$O** and **Au4phpy·1/2H$_2$O** undergo relatively poorly cooperative SCO behaviour with T_{SCO1} = 213 K and T_{SCO2} = 162 K, and T_{SCO} = 225 K, respectively. The 2-step transition observed for the Ag derivative transforms to a 1-step transition (T_{SCO} = 226 K) after dehydration. Compound **AgL43** represents a new example of a 33% cooperative SCO characterised by $T_{SCO}{}^{down}$ = 138 K and $T_{SCO}{}^{up}$ = 144 K. A reversible crystallographic phase transition featuring a near-doubling of the unit cell volume in the low temperature phase accompanies this SCO. Compound **AuL42** displays a complete gradual 2-step SCO behaviour with T_{SCO1} = 226 K and T_{SCO2} = 132 K, in contrast to **AuL43** which is essentially HS. Self-assembly of **L42**, $[Ag(CN)_2]^-$ and Fe(II) affords the only member of this series in which the Ag(I) ion expands its coordination sphere to give pseudo-trigonal bridging units [Ag**L42**$(CN)_2]^-$. This fact together with the length of **L42** causes marked congestion between two consecutive parallel $[Fe[M(CN)_2]_2]_\infty$ grids. This compound is HS over all temperatures.

4.3.3 Tetracyanometalate $[M^{II}(CN)_4]^{2-}$ Bridging Ligands (M^{II} = Ni, Pd, Pt)

Self-assembly of Fe(II), monodentate ligands derived from pyridine and tetracyanometallate complexes $[M(CN)_4]^{2-}$ have provided a series of CPs generically formulated $[Fe(L)_2[M(CN)_4]]$ where M = Ni, Pd, Pt and L = 3-Xpy (X = F, Cl, Br, I) (**M3Xpy**),[110] 4-phenylpyridine (4phpy) (**M4phpy**)[107, 108] and **L44** (ttfadpy) (**ML44**).[111] Furthermore, the ditopic ligand 5-bromopyrimidine (5Brpmd) acts as a monodentate ligand in presence of $[M(CN)_4]^{2-}$ to give $[Fe(5Brpmd)_2[M(CN)_4]]$ (**M5Brpmd**).[112] The complex [Fe(3-methylpyridine)$_2$[Ni(CN)$_4$]] has also been recently investigated.[113] Although detailed structural studies have only been carried out for the **M3Xpy** compounds, their essential structural characteristics can be safely extended to the other members of this series. The structural principles concerning the octahedral $[FeN_6]$ site remain unaltered with respect to those discussed for the dicyanometallates. However, the connectivity mediated by $[M(CN)_4]^{2-}$ is twice as many that of $[M(CN)_2]^-$ giving more dense and rigid 2D-CPs. This enables more effective transmission of the structural changes associated with the SCO and, hence, occurrence of larger thermal hysteresis loops. The spin-crossover properties deduced from magnetic and calorimetric measurements for the SCO members of these series of compounds are summarised in Table 4.1. Except for the poorly cooperative 2-step SCO observed in the **L44** family, the remaining compounds display strong cooperative behaviour with large hysteresis. Regardless of M, the average T_{SCO} temperatures follow approximately the trend: 3-Mepy $\leq$ 3-Clpy $\ll$ **L44** $\approx$ 4-phpy $\approx$ 5-Brpmd $<$ 3-Fpy. The compounds **Ni3Mepy** and **Ni3Clpy** are surprisingly similar since they have comparable structures, undergo 50% cooperative SCO characterised by the lowest critical temperatures in the series and display 2-step SCO under pressure. The 2-step SCO observed for **Ni3Clpy** is consistent with that observed for its isostructuctural Pd counterpart. The crystal structure data of the latter, obtained in the middle of the plateau where 50% of spin conversion takes place, reveals the occurrence of a crystallographic phase transition.[110]

Table 4.1 *Hofmann-like two-dimensional spin-crossover coordination polymers (2D-SCO-CPs).*

Compound	SCO*	T_{SCO}^{up} (K)	T_{SCO}^{down} (K)	ΔT_{SCO} (K)	ΔH_{av} (kJ/mol)	ΔS_{av} (J / Kmol)	Ref.
$Fe(3\text{-}Fpy)_2[Ni(CN)_4]$	c	205.8	234.4	28.6	19.1	87.4	110
$Fe(3\text{-}Fpy)_2[Pd(CN)_4]$	c	213.6	248.4	34.8	21.4	93.4	110
$Fe(3\text{-}Fpy)_2[Pt(CN)_4]$	c	214.0	239.5	25.5	20.6	91.7	110
$Fe(3\text{-}Clpy)_2[Ni(CN)_4]$	50%	105.8	123.4	17.6	—	—	110
$Fe(3\text{-}Clpy)_2[Pd(CN)_4]$	2-step	1) 169.6 2) 141.4	1) 164.5 2) 148.4	1) 4.9 2) 7	1) 8.8 2) —	1) 48.7 2) —	110
$Fe(3\text{-}Clpy)_2[Pt(CN)_4]$	c	141.1	161.4	19.3	10.6	70.4	110
$Fe(5\text{-}Brpmd)_2[Ni(CN)_4]$	c	170	180	10	11	62	112
$Fe(5Brpmd)_2[Pd(CN)_4]$	c	204	214	10	16	76	112
$Fe(5\text{-}Brpmd)_2[Pt(CN)_4]$	c	197	223	26	16	81	112
$Fe(3\text{-}Mepy)_2[Ni(CN)_4]$	50%	87.5	105	ca 10	—	—	113
$Fe(4\text{-}phpy)_2[Ni(CN)_4]$	i	135	158	23	7.8$	55.7$	107
$Fe(4\text{-}phpy)_2[Pd(CN)_4]$	c	163	203	40	11.8	64.1	107
$Fe(4\text{-}phpy)_2[Pt(CN)_4]$	c	185	225	40	12.8	64	107
$Fe(ttfadpy)_2[Ni(CN)_4]$#	two-step	1) 172 2) 111		—	—	—	111
$Fe(ttfadpy)_2[Pd(CN)_4]$#	two-step	1) 228 2) 197		—	1) 5.7 2) 6.5	1) 24 2) 30.4	111
$Fe(ttfadpy)_2[Pt(CN)_4]$#	two-step	1) 228 2) 202		—	1) 6.2 2) 7.2	1) 26.6 2) 32.8	111

*i = incomplete; c = complete; #These compounds contain 1.5–2 molecules of H_2O; $This compound displays ca. 14.6% of HS impurities in the LS state. Extrapolation to 100% gives $\Delta H = 9.1$ kJmol^{-1} and $\Delta S = 65$ J K^{-1} mol^{-1}.

The thermal hysteresis width observed for **M3Fpy** is essentially unaffected by pressure (in the range 10^5 HPa–0.5 GPa), while T_{SCO} may be tuned at will in the interval 200–400 K. Taking advantage of the sharp difference of colour between the HS and LS states, visible spectroscopic studies at 298 K have demonstrated that these materials display strongly cooperative pressure-induced SCO characterised by a 0.1 GPa wide piezo-hysteresis.[110,114] This is a remarkable result because room temperature switches and memories are of singular relevance in materials science. The influence of the nanoparticle size on the T_{SCO} and hysteresis width has also been investigated for the **M3Fpy** series.[115]

4.4 Three-Dimensional SCO-CPs

4.4.1 Neutral Organic Bridging Ligands

A series of 1,n-bis(tetrazol-1-yl)alkane (nditrz) ligands have been specifically designed to investigate thermal- and light-induced SCO in new Fe(II) 3D-CPs (Scheme 4.6). The complex $[Fe(\mathbf{L45})_3](PF_6)_2$ is the only member of this series whose single crystal structure was investigated. It is made up of triple interpenetrated 3D networks with the α-polonium topology. Each Fe(II) is octahedrally coordinated by six tetrazole rings with the butylene spacer stretched in a zig-zag *trans* configuration spanning Fe(II) centres. Each **L45** ligand connects two Fe(II) atoms to form a 3D network. Given the inherently flexible nature of the ligand, a gradual SCO behaviour could be anticipated for this compound. However, it features a cooperative two-step SCO with characteristic temperatures $T_{SCO1} = 184$ K, $T_{SCO2}^{down} = 168$ K and $T_{SCO2}^{up} = 174$ K. Apparently the packing

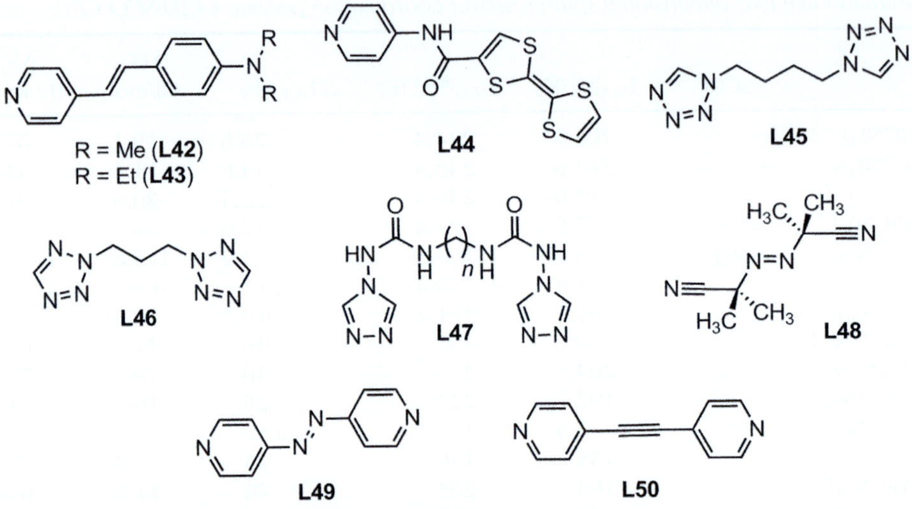

Scheme 4.6

density of the compound due to the 3-fold interlocked system, and the perfect size ratio of the PF_6^- filling the cavities, suppresses distortion and twisting of the ligand in the structure.[116–118] Replacement of PF_6^- by ClO_4^- does not affect the 2-step character of the SCO but provokes a significant decrease of cooperativity and the critical temperatures ($T_{SCO1} = 84$ K, $T_{SCO2} = 134$ K). In the latter compound, the HS-to-LS relaxation of the photo-excited HS state mirrors the thermal 2-step transition. The complex [Fe(**L46**)$_3$(ClO_4)$_2$]·2EtOH exhibits a similar 3D structure, however, the shorter propylene spacer connecting the terminal tetrazole moieties in **L46** prevents interpenetration of the network. This compound displays a gradual one-step SCO with $T_{SCO} \approx$ 160 K.[119] In a similar way, the synthesis of n-alkyl *N,N′*-bis(4-*N′*-1,2,4-triazol)diurea (n-**L47**) ligands has afforded the series of complexes [Fe(n-**L47**)$_3$](Tos)$_2$ (n = 4, 6, 8 or 12), which display rather incomplete SCO behaviour centred at ca. 350 K.[120]

A 3D-SCO-CP with the NiAs topology has been prepared from tetrazole-1H-acetic acid (trzCOOH). This compound, [[(Fe^{II})$_2$$Fe^{III}$(trzCOO)$_2$($H_2O$)]$_3$O]($ClO_4$)$_2$($NO_3$)(EtOH)($H_2O$), contains triangular [[Fe^{III}(trzCOO)$_2$(H_2O)]$_3$$\mu_3$O]$^+$ units supported by double carboxylate bridges. The tetrazole groups point outward coordinating Fe(II) centres forming [Fe^{II}(trz)$_6$] SCO moieties, which connect six different triangular units. This singular compound undergoes a rather gradual SCO in the temperature interval 100–300 K.[121]

Two SCO compounds with the α-polonium 3D network formulated [Fe(**L48**)$_3$][MCl_4] (M = Fe(III), In(III)) with **L48** being the new organodinitrile bridging ligand 2,2′-azobisisobutanenitrile have been synthesised. Both derivatives undergo a gradual SCO with T_{SCO} = 118 K and 170 K for M = In^{III} and Fe^{III}, respectively.[122]

4.4.2 Dicyanometalate [$M^I(CN)_2$]$^-$ Bridging Ligands

In Section 4.2.5 we indicated that self-assembly of Fe(II), [$Cu(CN)_2$]$^-$ and 3Xpy ligands predominantly affords 1D-CPs. However, in the specific case of X = Cl two architectural isomers [Fe(3Clpy)$_2$[Cu(3Clpy)(CN)$_2$] were formed, namely polymorphs α and β. The former is a HS 1D-CP while the latter is a 3D-SCO-CP. In this β-polymorph, deviation of the NC–Cu–CN angle from 180°, resulting from expansion of the coordination sphere from 2 to 3, is noticeably smaller (144–148°) than in polymorph α, thus facilitating the formation of a 3D four-connected network with the expanded structure of the open-framework $CdSO_4$. This polymorph displays a 2-step SCO (T_{SCO1} = 210 K and T_{SCO2} = 169 K), a fact that agrees with the presence of two slightly different $[FeN_6]$ sites. As in other examples involving $[Cu3Xpy(CN)_2]^-$ bridges the variation of the unit cell volume per Fe(II) ion is unusually large ($\Delta V_{SCO} = 59$ Å^3).[74]

Three architectural isomers were observed for the system $[Fe(3CNpy)_2[Au(CN)_2]_2]\cdot nH_2O$ (cf. Section 4.3.2). One of them (n = $^2/3$) forms a triple interpenetrated four-connected 3D network with the prototypal NbO structure,[105] which is isostructural with the Ag counterpart.[123] Viewed along the [001] direction, the crystal packing displays trigonal and hexagonal channels partially filled by the ligands and water molecules. The vertices and edges of the trigonal channels are defined by the $[FeN_6]$ sites and the $[Au(CN)_2]^-$ groups, respectively. Moderate $Au^I\cdots\cdot Au^I$ interactions are present in the triangular motifs (3.4212–3.3952 Å). It undergoes a cooperative SCO characterised by a ca. 6 K hysteresis width involving only ca. 33% of the Fe(II) ions ($T_{SCO}{}^{down}$ = 122 K and $T_{SCO}{}^{up}$ = 128 K). Application of pressure revealed two additional ca. 33% SCO steps not observed at atmospheric pressure.[124] The previously mentioned 2D-SCO-CP $[Fe(3CNpy)_2(CH_3OH)_{2/3}[Au(CN)_2]_2]$ is formally the precursor of this 3D-SCO-CP (cf. Section 4.3.2).

The system Fe(II)-$[Ag(CN)_2]$-pmd (pmd = pyrimidine) has generated an exceptionally rich variety of 3D-SCO-CPs with interesting structural physical and chemical properties. The complex $[Fe(pmd)[Ag_3(CN)_2][Ag(CN)_2]]$ is rather singular as it contains three different bridging ligands, namely pmd and the two linear anions $[Ag(CN)_2]^-$ and $[Ag_3(CN)_2]^-$ (generated *in situ*) characterised by their markedly different length, which determine the formation of an intricate self-interpenetrated 3D network with five crystallographically distinct $[FeN_6]$ sites. The pmd bridges define infinite ···Fe–pmd–Fe··· chains running in the *c* direction while the other two unequivalent bridges radiating from these chains connect other equivalent chains facilitating the occurrence of strong argentophilic interactions with very short Ag···Ag distances, 2.98–3.02 Å. The thermodynamic and kinetic SCO properties of this singular system have been exhaustively investigated from its thermal and light-induced SCO properties combining magnetic, calorimetric measurements, single crystal absorption spectroscopy, detailed variable temperature crystal determinations and photo-crystallographic studies. This compound undergoes a moderately cooperative 2-step SCO, T_{SCO1} = 185 K and T_{SCO2} = 147 K, without a crystallographic phase change. At the middle of the plateau two crystallographically unequivalent ···Fe–pmd–Fe··· chains show the occurrence of ···HS–LS··· order while a third chain shows averaged LS-HS Fe–N bond lengths. Interestingly, strong synergy between SCO behaviour and argentophilic interactions is observed. About 60% of the Ag····Ag short contacts change significantly during the low temperature step, a fact that seems to be correlated with the larger variations of entropy and unit cell volume observed in this step. Interestingly, photomagnetic, photooptical and photo-crystallographic LIESST studies show a 2-step relaxation from the HS to the LS mirroring the thermal induced SCO.[125]

Likewise, $[Fe(3,5Mepy)_2[Ag_2(CN)_3][Ag(CN)_2]]$ (3,5Mepy = 3,5-dimethylpyridine) contains the $[Ag_2(CN)_3]^-$ anion. However, the nonbridging nature of the 3,5Mepy ligand induces generation of a triple-interpenetrated 3D framework with the topology of $CdSO_4$. Inter-framework argentophilic interactions are observed, which strongly depend on the spin state of the iron (i.e. Ag····Ag = 3.224 Å (HS) to 3.073 Å (LS)). The magnetic behaviour denotes the occurrence of moderately cooperative SCO without hysteresis at T_{SCO} = 235 K.[126]

The $[Fe(pmd)_2[Ag(CN)_2]$ system represents another example of architectural isomerism. The α isomer is made up of an stack of 2D $[Fe_4[Ag(CN)_2]_4]_\infty$ grids with the axial sites of the $[FeN_6]$ occupied by pmd ligands. The pmd ligands also interact with one of two crystallographically distinct Ag(I) atoms belonging to adjacent grids to form a 3D-CP. The resulting pseudo-tetrahedral coordination of this Ag(I) atom provokes strong deviation from the linearity of the $[Ag(CN)_2]^-$ bridges, thereby inducing parallel interpenetration of the resulting strongly undulating grids, which favours strong argentophillic interactions (Ag····Ag = 2.9972 Å). In the β isomer, the Ag(I) centres adopt pseudo-trigonal coordination as a result of the interaction with the pmd ligands of adjacent grids. Consequently, the bending of the $[Ag(CN)_2]^-$ moieties is markedly smaller than that of isomer α and no parallel interpenetration occurs. Despite their similar $[FeN_6]$ coordination centres, isomer α is HS over all temperatures while isomer β undergoes cooperative SCO ($T_{SCO}^{down} = 181.1$ K and $T_{SCO}^{up} = 186.3$ K; $\Delta H_{av} = 11.2 \pm 0.5$ kJmol^{-1}; $\Delta S_{av} = 62 \pm 3$ J K^{-1}mol^{-1}).[127]

The ligand 5-Br-pyrimidine (5Brpmd) affords the 3D-SCO-CP [Fe(5Brpmd)$_2$[Ag(CN)$_2$]] in which the 5Brpmd bridges the Fe(II) ions defining the same intricate framework described for [Fe(pmd)[Ag(CN)$_2$]$_2$.[128] Furthermore, as the latter compound the 5Brpmd derivative also undergoes a strongly cooperative half transition at similar temperatures ($T_{SCO}^{down} = 149$ and $T_{SCO}^{up} = 167$ K; $\Delta H_{av} = 12 \pm 0.5$ kJmol^{-1}; $\Delta S_{av} = 76 \pm 3$ J K^{-1}mol^{-1}).[112]

The effect of pressure on the hysteretic behaviour of the 3D-SCO-CP [Fe(pmd)(H_2O)[Ag(CN)$_2$]$_2$][128] has been recently investigated from magnetic and visible absorption spectroscopy. The critical temperatures increase about 100 K in the pressure interval 10^5 Pa–0.34 GPa without decreasing the hysteresis loop width. This dependence of T_{SCO}^{up} and T_{SCO}^{down} on pressure is sigmoidal and defines a piezo-hysteresis ca. 0.1 GPa wide. These properties enable to tune the critical temperatures of the thermo-hysteresis in a large range of temperatures including room temperature.[129]

4.4.3 Tetracyanometalate [M^{II}(CN)$_4$]$^{2-}$ Bridging Ligands

As described in Section 4.3.3, self-assembly of Fe(II) ions and [M^{II}(CN)$_4$]$^{2-}$ (M^{II} = Ni, Pd, Pt) anions affords parallel stacks of rigid 2D (4,4) planar grids [Fe[M^{II}(CN)$_4$]]$_\infty$ with the Fe(II) axial positions occupied by monodentate ligands. Formal replacement of the monodentate axial ligands by bis-monodentate bridges has originated a new kind of Hofmann-like 3D-SCO-CPs. Three series of compounds based on the bridges pz, **L49** and **L50** have been reported so far. These compounds combine two complementary functions, namely porosity and strong cooperative SCO behaviour. These attributes have enabled investigation of synergies between host–guest chemistry and SCO phenomena. Particularly interesting is the system [Fe(pz)[M^{II}(CN)$_4$]] which, in addition, exhibits a hysteresis loop 25 K wide at room temperature. Indeed, the guest-free framework adsorbs various guest molecules in the gas phase or solution giving the correspondent clathrates (M = Ni,[130, 131] Pd[130] and Pt[130]). A bimodal reversible change of spin state at the iron(II) sites is observed concomitantly with the uptake of guest molecules. The HS clathrate, characterised by a yellow colour, is stabilised by hydroxylic solvents, five- and six-membered aromatic molecules, while the LS clathrate, characterised by a red colour, is stabilised by CS_2 at 298 K. Stabilisation of the LS state was also observed for the CH_3CN clathrate of the Fe(II)-Ni(II) framework.[131] Interestingly, at 298 K within the LS branch of the thermal hysteresis these frameworks adsorb benzene or pyrazine when exposed to vapours with simultaneous change from the LS (diamagnetic, red) state to the HS (paramagnetic, yellow) state. At same temperature, the HS form of the framework adsorbs CS_2 with a simultaneous change to the red colour characteristic of the LS state. These sensory sorption–desorption processes take place rapidly (within 2 min.) with a memory effect since the system retains the HS or the LS state after desorption of benzene or CS_2, respectively.[130] Furthermore, photoinduced bidirectional switching between the HS and LS states occurs when irradiating inside the hysteresis region, with a nanosecond pulsed laser ($\lambda = 532$ nm), on bulk and single-crystal samples.[132, 133]

The presence of coordinatively unsaturated Pt(II) centres in the framework favours adsorption of dihalogen molecules involving associative oxidation with partial transformation of Pt(II) to Pt(IV) and reduction of the dihalogen to the corresponding halide, affording [Fe(pz)[Pt^{II,IV}(CN)$_4$(X)]] [X = Cl^-, Br^-, I^-]. All Pt atoms are crystallographically equivalent with the axial positions occupied by two X^- anions with half occupancy. Thus, there are twice as many available positions as X atoms to form the compound; consequently, the holes and X atoms are uniformly and randomly distributed in the crystal. The three derivatives undergo strongly cooperative SCO with critical temperatures $T_{SCO}^{up} \approx 270$ K (Cl), 324 K (Br) and 392 K (I) and $T_{SCO}^{down} \approx 258$ K (Cl), 293 K (Br) and 372 K (I). It is worth noting that the T_{SCO} values decrease as the electronegativity of X^- increases, a fact that suggests some degree of electron withdrawal at the nitrogen atom of the Fe–NC–Pt–X moiety with concomitant reduction of the ligand field at the $[FeN_6]$ site.[134] Fully iodine-loaded crystals strongly stabilises the LS state showing an increase in T_{SCO} by ca. 100 K with respect to the guest-free $[Fe(pz)[Pt^{II}(CN)_4]]$ compound. A further relevant property is the observation of iodine mobility from concentrated to diluted regions of the $[Fe(pz)[Pt^{II,IV}(CN)_4(I)_x]]$ crystals, a consequence of the presence of additional available equivalent crystallographic positions mentioned above. Taking advantage of this, continuous modulation of T_{SCO}, in the interval 300–400 K, has been achieved by heating appropriate homogeneous mixtures of guest-free and fully loaded iodine derivatives at temperatures just above 400 K.[135]

Substitution of pz by the longer bridging ligand **L49** has afforded a new series Hofmann-like 3D-SCO-CPs with enhanced porosity. However, compared to the pz series, the monohydrate [Fe(**L49**)$[M^{II}(CN)_4]]\cdot H_2O$ derivatives show much less cooperative SCO behaviour with average critical temperatures, $T_{SCO}{}^{av}$, centred at 245 K (M^{II} = Ni), 291 K (M^{II} = Pd) and 280 K (M^{II} = Pt). The two latter transitions are characterised by a hysteresis loop 9 K and 10 K wide, respectively. Although the molecule of H_2O included in the channels seems not to interact appreciably with the framework, $T_{SCO}{}^{av}$ is shifted ca. 100 K to low temperatures upon dehydration (191 and 182 K for the Pd and Pt derivatives, respectively) with concomitant increase of the hysteresis width to ca. 20 K. It is worth noting that, despite the observed strong cooperativity, the spin conversions are quite smooth for the precipitated samples, most probably due to the high insolubility of these compounds, which induce crystallite size in-homogeneities and consequently a relatively wide distribution of critical temperatures.[136] Similar phenomenology has been observed when using the rigid, and even longer, ligand **L50** as a bridging ligand. The resulting compound [Fe(**L50**)$[M^{II}(CN)_4]]\cdot{}^1/2$**L50**$\cdot H_2O$ crystallises with non-coordinated molecules of **L50** installed in the pores, a fact which does not preclude this system to store other invited molecules.[137]

Using combined top-down/bottom-up approaches micro- and nano-sized patterns constituted of square motives of $[Fe(pz)[Pt^{II}(CN)_4]]$ have been grown on gold surfaces. Assembled continuous and micro-patterned (size of the squares $\geq$ 2 μm) thin films ca. 150 nm thick display similar SCO properties as the bulk material. Using Raman spectroscopy it is even possible to detect the spin-state change of single objects 500 nm^2 or arrays of 200 nm^2 objects.[138, 139] Similar studies have been extended to the **L49**[136] and **L50**[140] homologues.

Water-in-oil micro-emulsion techniques have been employed to obtain surfactant-free square shaped nanocrystals of $[Fe(pz)[Pt^{II}(CN)_4]]$.[141, 142] Three main consequences appear as the size of guest-free crystals decreases from the bulk to the nanometer size: (i) increase of the residual HS fraction at low temperatures; (ii) decrease of the critical temperatures and (iii) decrease of the hysteresis width. These consequences seem to be closely related to substantial increase of the specific surface area as the size of the nanocrystals decreases. Residual paramagnetic impurities may be generated at the surface of the nanocrystals due to coordinative defects. This in turn will decrease the chemical pressure on the iron(II) SCO centres near the HS Fe(II) impurities (defects) thereby reducing T_c as well as cooperativity. Nevertheless, ultra-small nanoparticles (3.8 nm) of $[Fe(pz)[Ni^{II}(CN)_4]]$ grown in porous matrices of chitosan display cooperative ST with a hysteresis loop 10 K wide ($T_{SCO}{}^{down}$ = 280 K and $T_{SCO}{}^{up}$ = 290 K) despite only 33% of iron(II) ions undergoing SCO.[143]

4.4.4 Hexa- and Octacyano-metallate Bridging Ligands

Hexa- and octa-cyanometallate building blocks have been extensively used for the construction of CPs exhibiting magnetic ordering. A relatively small fraction of these undergo thermal- pressure- and light-induced valence tautomeric phenomena, namely spin state change associated with the metal-to-metal electron transfer process. However, as far as we know only two examples have been reported to exhibit SCO phenomenon.

The Prussian Blue analogue $[CsFe[Cr(CN)_6]]$ crystallises in the cubic $F\bar{4}3m$ space group. The $[Cr(CN)_6]^{3-}$ building blocks connect six iron(II) atoms which are coordinated by six $[Cr(CN)_6]^{3-}$ ligands. This compound undergoes a cooperative ST characterised by $T_{SCO}{}^{down} = 211$ K and $T_{SCO}{}^{up} = 238$ K and a hysteresis loop $\Delta T = 27$ K wide. The high temperature phase (HT) contains 6% of LS Fe(II) centres while the low temperature phase contains 11% of Fe(II) ions in the HS state. Due to these HS impurities the solid shows spontaneous magnetisation with magnetic ordering at 9 K.[144] In a subsequent study it was found that this compound displays an abrupt isosymmetric phase transition, accompanied by dramatic unit cell volume collapse, in the ascending branch of the ST, namely in the range 245–265 K (note that for the new sample the hysteresis loop was found to be $\Delta T = 37$ K wide, $T_{SCO}{}^{down} = 208$ K and $T_{SCO}{}^{up} = 245$ K). This phase transition, accompanied by sudden transformation of the HS Fe(II) ions to the LS state, is induced by absorption of X-ray photons which generates photo-excited LS iron(II) domains whose size rapidly grows with time until the percolation threshold is reached and the structure collapse is triggered.[145] Pressure experiments have shown that the cooperative SCO is extremely sensitive to compression which shifts very rapidly to higher temperatures with the shape of the accompanying thermal hysteresis loop not affected by pressurisation.[146]

In the octacyano-bridged Nb-Fe(II) bimetallic assembly $[Fe_2[M(CN)_8](3\text{-methanolpyridyl})_8]\cdot nH_2O$ (M = Mo, Nb), the two axial positions of the Fe(II) ion are occupied by cyanide nitrogen atoms of $[Nb(CN)_8]^{4-}$ whereas the equatorial positions are occupied by four nitrogen atoms from (3-pyridyl)methanol ligands.Four CN groups of $[Nb(CN)_8]^{4-}$ bridge four Fe centres while remaining CN groups are free. This CP undergoes a continuous poorly cooperative SCO in the temperature range 150–350 K centred about $T_{SCO} = 250$ K. As in the previous example, spontaneous magnetisation ($T_{SCO} = 12$ K) reveals antiferromagnetic interactions between the residual HS iron(II) ions and Nb(IV) ($S = {}^1/_2$).[147] Replacement of (3-pyridyl)methanol by 4-pyridinealdoxime lowers T_{SCO} to ca. 120 K, a fact that has enabled observation of quantitative LIESST effect and consequently reversible photoswitching between paramagnetic and ferrimagnet states at around 5 K.[148]

4.5 Conclusion

The period 2004–2011 has witnessed an impressive increase of activity in the field of SCO-CPs. The paradigmatic triazole-based 1D-SCO-CPs have inspired much work focused on transferring their important physical properties to more convenient technological platforms such as preparation of composites, gels, liquid crystalline materials, dendrimers, Langmuir–Blodget films and nanoparticles. A large series of bridging ligands with triazole, tetrazole and/or pyridine functions have specifically been designed to study new 1-3D-SCO-CPs. In this respect, particularly significant has been the synthesis and characterisation of new functional porous SCO compounds made up of orthogonal interpenetration of 2D-CPs. Furthermore, development of a new class of 1-3D-SCO-CPs based on metallocyanate bridging ligands has contributed to the discovery of new important aspects of the SCO phenomenon. These new materials feature structural diversity, supramolecular isomerism, interpenetrating frameworks, structure flexibility, reversible solid-state chemical reactions, metallophilic interactions, porosity, physi- and chemisorption, or processability at nanoscale level, in addition to inherent SCO properties.

Acknowledgement

The authors are grateful for financial support from the Spanish Ministerio de Ciencia e Innovacián (MICINN) and FEDER funds (CTQ2010-18414), the Generalitat Valenciana through PROMETEO/2012/049 and ACOMP2012/233 projects.

References

1. van Koningsbruggen, P. J. (2004) Special classes of iron(II) azole spin crossover compounds, in: Gütlich, P., Goodwin, H. A. (Eds) Spin Crossover in Transition Metal Compounds I. Top. Curr. Chem., 233: 123–149.
2. Garcia, Y., Niel, V., Muñoz, M. C., Real, J. A. (2004) Spin crossover in 1D, 2D and 3D polymeric Fe(II) networks, in: Gütlich, P., Goodwin, H. A. (Eds) Spin Crossover in Transition Metal Compounds I. Top. Curr. Chem., 233: 229–257.
3. Muñoz, M. C., Real, J. A. (2011) Thermo-, piezo-, photo- and chemo-switchable spin crossover iron(II)-metallocyanate based coordination polymers. Coord. Chem. Rev., 255: 2068–2093.
4. Weber, B. (2009) Spin crossover complexes with N_4O_2 coordination sphere - the influence of covalent linkers on cooperative interactions. Coord. Chem. Rev., 253: 2432–2449.
5. Bodenthin, Y., Schwarz, G., Tomkowicz, Z., Lommel, M., Geue, T., Haase, W., Mohwald, H., Pietsch, U., Kurth, D. G. (2009) Spin-crossover phenomena in extended multi-component metallo-supramolecular assemblies. Coord. Chem. Rev., 253: 2414–2422.
6. Benmansour, S., Atmani, C., Setifi, F., Triki, S., Marchivie, M., Gómez-García, C. J. (2010) Polynitrile anions as ligands: From magnetic polymeric architectures to spin crossover materials, Coord. Chem. Rev., 254: 1468–1478.
7. Real, J. A., Gaspar, A. B., Muñoz, M. C. (2005) Thermal, pressure and light switchable spin-crossover materials. Dalton Trans., 2062–2079.
8. Aromí, G., Barrios, L. A., Roubeau, O., Gamez, P. (2011) Triazoles and tetrazoles: Prime ligands to generate remarkable coordination materials. Coord. Chem. Rev., 255: 485–546.
9. Urakawa, A., Van Beek, W., Monrabal-Capilla, M., Galán-Mascarós, J. R., Palin, L., Milanesio, M. (2011) Combined, modulation enhanced X-ray powder diffraction and raman spectroscopic study of structural transitions in the spin crossover material $[Fe(Htrz)_2(trz)](BF_4)$. J. Phys. Chem. C, 115: 1323–1329.
10. Garcia, Y., Campbell, S. J., Lord, J. S., Boland, Y., Ksenofontov, V., Gütlich, P. (2007) Dynamics and supramolecular organization of the 1D spin transition polymeric chain compound $[Fe(NH_2trz)_3](NO_3)_2$. Muon spin relaxation. J. Phys. Chem. B, 111: 11111–11119.
11. Garcia, Y., Campbell, S. J., Lord, J. S., Gütlich, P. (2008) Muon spin relaxation studies of iron(II) spin crossover complexes. Inorg. Chim. Acta, 361: 3577–3585.
12. Dîrtu, M. M., Garcia, Y., Nica, M., Rotaru, A., Linares, J., Varret, F. (2007) Iron(II) spin transition 1,2,4-triazole chain compounds with novel inorganic fluorinated counteranions. Polyhedron, 26: 2259–2263.
13. Dîrtu, M. M., Rotaru, A., Gillard, D., Linares, J., Codjovi, E., Tinant, B., Garcia, Y. (2009) Prediction of the spin transition temperature in Fe^{II} one-dimensional coordination polymers: an anion based database, Inorg. Chem., 48: 7838–7852.
14. Roubeau, O., Castro, M., Burriel, R., Haasnoot, J. G., Reedijk, J. (2011) Calorimetric investigation of triazole-bridged Fe(II) spin-crossover one-dimensional materials: measuring the cooperativity. J. Phys. Chem. B, 115: 3003–3012.
15. Fouché, O., Degert, J., Jonusauskas, G., Daro, N., Létard, J.-F., Freysz, E. (2010) Mechanism for optical switching of the spin crossover $[Fe(NH_2\text{-}trz)_3](Br)_2{\cdot}3H_2O$ compound at room temperature. Phys. Chem. Chem. Phys., 12: 3044–3052.
16. Galle, G., Deldicque, D., Degert, J., Forestier, T., Létard, J.-F., Freysz, E. (2010) Room temperature study of the optical switching of a spin crossover compound inside its thermal hysteresis loop. Appl. Phys. Lett., 96: 041907/1–041907/3.
17. Kuroiwa, K., Shibata, T., Sasaki, S., Ohba, M., Takahara, A., Kunitake, T., Kimizuka, N. (2006) Supramolecular control of spin-crossover phenomena in lipophilic Fe(II)-1,2,4-triazole complexes. J. Polym. Sci. Part A - Polym. Chem., 44: 5192–5202.
18. Lee, S. W., Lee, J. W., Jeong, S. H., Park, I. W., Kim, Y. M., Jin, J. I. (2004) Processable magnetic plastics composites – spin crossover of PMMA/Fe(II)-complexes composites. Synth. Met., 142: 243–249.
19. Suzuki, A., Fujiwara, M., Nishijima, M. (2008) High spin/low spin phase transitions of a spin-crossover complex in the emulsion polymerization of trifluoroethylmethacrylate (TFEMA) using PVA as a protective colloid. Colloid Polym. Sci., 286: 525–534.

20. Suzuki, A., Iguchi, M., Oku, T., Fujiwara, M. (2010) Magnetic properties of the Fe^{II} spin crossover complex in emulsion polymerization of trifluoroethylmethacrylate using poly(vinyl alcohol). J. Solid State Chem., 183: 951–956.
21. Rubio, M., Hernández, R., Nogales, A., Roig, A., López, D. (2011) Structure of a spin-crossover Fe(II)-1,2,4-triazole polymer complex dispersed in an isotactic polystyrene matrix. Eur. Polym. J., 47: 52–60.
22. Chen, Y., Ma, J. G., Zhang, J. J., Shi, W., Cheng, P., Liao, D. Z., Yan, S. P. (2010) Spin crossover-macromolecule composite nano film material. Chem. Commun., 46: 5073–5075.
23. Roubeau, O., Colin, A., Schmitt, V., Clérac, R. (2004) Thermoreversible gels as magneto-optical switches. Angew. Chem. Int. Ed., 43: 3283–3286.
24. Grondin, P., Roubeau, O., Castro, M., Saadaoui, H., Colin, A., Clérac, R. (2010) Multifunctional gels from polymeric spin-crossover metallo-gelators. Langmuir, 26: 5184–5195.
25. Fujigaya, T., Jiang, D. L., Aida, T. (2007) Spin-crossover physical gels: a quick thermoreversible response assisted by dynamic self-organization. Chem. Asian J., 2: 106–113.
26. Kuroiwa, K., Kikuchi, H., Kimizuka, N. (2010) Spin crossover characteristics of nanofibrous Fe^{II}-1,2,4-triazole complexes in liquid crystals. Chem. Commun., 46: 1229–1231.
27. Matsukizono, H., Kuroiwa, K., Kimizuka, N. (2008) Lipid-packaged linear iron(II) triazole complexes in solution: controlled spin conversion via solvophobic self-assembly. J. Am. Chem. Soc., 130: 5622–5623.
28. Rubio, M., López, D. (2009) Effect of solvent on the gelation properties of a metallo-organic polymer of [Fe(II)(4-octadecyl-1,2,4-triazole)$_3$(ClO$_4$)$_2$]. Eur. Polym. J., 45: 3339–3346.
29. Seredyuk, M., Gaspar, A. B., Ksenofontov, V., Reiman, S., Galyametdinov, Y., Haase, W., Rentschler, E., Gütlich, P. (2006) Room temperature operational thermochromic liquid crystals. Chem. Mater., 18: 2513–2519.
30. Seredyuk, M., Gaspar, A. B., Ksenofontov, V., Galyametdinov, Y., Verdaguer, M., Villain, F., Gütlich, P. (2008) One-dimensional iron(II) compounds exhibiting spin crossover and liquid crystalline properties in the room temperature region. Inorg. Chem., 47: 10232–10245.
31. Fujigaya, T., Jiang, D. L., Aida, T. (2005) Spin-crossover dendrimers: generation number-dependent cooperativity for thermal spin transition. J. Am. Chem. Soc., 127: 5484–5489.
32. Roubeau, O., Natividad, E., Agricole, B., Ravaine, S. (2007) Formation, structure, and morphology of triazole-based Langmuir-Blodgett films. Langmuir, 23: 3110–3117.
33. Letard, J.-F., Guionneau, P., Goux-Capes, L. (2004) Towards spin crossover applications, in: Gütlich, P., Goodwin, H. A. (Eds) Spin Crossover in Transition Metal Compounds III. Top. Curr. Chem., 235: 221–249.
34. Forestier, T., Mornet, S., Daro, N., Nishihara, T., Mouri, S., Tanaka, K., Fouché, O., Freysz, E., Létard, J.-F. (2008) Nanoparticles of iron(II) spin-crossover. Chem. Commun., 4327–4329.
35. Forestier, T., Kaiba, A., Pechev, S., Denux, D., Guionneau, P., Etrillard, C., Daro, N., Freysz, E., Létard, J.-F. (2009) Nanoparticles of [Fe(NH$_2$-trz)$_3$]Br$_2$·3H$_2$O (NH$_2$-trz = 2-amino-1,2,4-triazole) prepared by the reverse micelle technique: influence of particle and coherent domain sizes on spin-crossover properties. Chem. Eur. J., 15: 6122–6130.
36. Arnaud, C., Forestier, T., Daro, N., Freysz, E., Létard, J.-F., Pauliat, G., Roosen, G. (2009) Observation of an asymmetry in the thermal hysteresis loop at the scale of a single spin-crossover particle. Chem. Phys. Lett., 470: 131–135.
37. Tokarev, A., Salmon, L., Guari, Y., Nicolazzi, W., Molnár, G., Bousseksou, A. (2010) Cooperative spin crossover phenomena in [Fe(NH$_2$trz)$_3$](tosylate)$_2$ nanoparticles. Chem. Commun., 46: 8011–8013.
38. Thibault, C., Molnár, G., Salmon, L., Bousseksou, A., Vieu, C. (2010) Soft lithographic patterning of spin crossover nanoparticles. Langmuir, 26: 1557–1560.
39. Coronado, E., Galán-Mascarós, J. R., Monrabal-Capilla, M., García-Martínez, J., Pardo-Ibáñez, P. (2007) Bistable spin-crossover nanoparticles showing magnetic thermal hysteresis near room temperature. Adv. Mater., 19: 1359–1361.
40. Galán-Mascarós, J. R., Coronado, E., Forment-Aliaga, A., Monrabal-Capilla, M., Pinilla-Cienfuegos, E., Ceolin, M. (2010) Tuning size and thermal hysteresis in bistable spin crossover nanoparticles. Inorg. Chem., 49: 5706–5714.
41. Faulmann, C., Chahine, J., Malfant, I., de Caro, D., Cormary, B., Valade, L. (2011) A facile route for the preparation of nanoparticles of the spin-crossover complex Fe(Htrz)$_2$(trz)(BF$_4$) in xerogel transparent composite films. Dalton Trans., 40: 2480–2485.

42. Salmon, L., Molnár, G., Zitouni, D., Quintero, C., Bergaud, C., Micheau, J. C., Bousseksou, A. (2010) A novel approach for fluorescent thermometry and thermal imaging purposes using spin crossover nanoparticles. J. Mater. Chem., 20: 5499–5503.
43. Titos-Padilla, S., Herrera, J. M., Chen, X. W., Delgado, J. J., Colacio, E. (2011) Bifunctional hybrid SiO_2 nanoparticles showing synergy between core spin crossover and shell luminescence properties. Angew. Chem. Int. Ed., 50: 3290–3293.
44. Kusz, J., Gütlich, P., Spiering, H. (2004) Structural investigations of tetrazole complexes of iron(II), in: Gütlich, P., Goodwin, H. A. (Eds.) Spin Crossover in Transition Metal Compounds II. Top. Curr. Chem., 234: 129–153.
45. van Koningsbruggen, P. J., Garcia, Y., Kahn, O., Fournès, L., Kooijman, H., Spek, A. L., Haasnoot, J. G., Moscovici, J., Provost, K., Michalowicz, A., Renz, F., Gütlich, P. (2000) Synthesis, crystal structure, EXAFS, and magnetic properties of catena [μ-tris(1,2-bis(tetrazol-1-yl)propane-N^1,$N^{1'}$)iron(II)] bis(perchlorate). First crystal structure of an iron(II) spin-crossover chain compound. Inorg. Chem., 39: 1891–1900.
46. Schweifer, J., Weinberger, P., Mereiter, K., Boca, M., Reichl, C., Wiesinger, G., Hilscher, G., van Koningsbruggen, P.J., Kooijinan, H., Grunert, M., Linert, W. (2002) catena-[μ-Tris(1,2-bis(tetrazol-1-yl)ethane-*N4*,*N4′*)iron(II)] bis(tetrafluoroborate): synthesis, structure, spectroscopic and magnetic characterization of a chain-type coordination polymer spin-crossover compound. Inorg. Chim. Acta, 339: 297–306.
47. Quesada, M., Prins, F., Bill, E., Kooijman, H., Gamez, P., Roubeau, O., Spek, A. L., Haasnoot, J. G., Reedijk, J. (2008) Counterion effect on the spin-transition properties of the cation $[Fe(btzx)_3]^{2+}$ (btzx = *m*-xylylenebis(tetrazole)). Chem. Eur. J., 14: 8486–8499.
48. Białońska, A., Bronisz, R., Weselski, M. (2008) A new family of spin-crossover complexes based on a $Fe^{II}(tetrazolyl)_4(MeCN)_2$-type core. Inorg. Chem., 47: 4436–4438.
49. Quesada, M., Kooijman, H., Gamez, P., Costa, J. S., van Koningsbruggen, P. J., Weinberger, P., Reissner, M., Spek, A. L., Haasnoot, J. G., Reedijk, J. (2007) $[Fe(\mu\text{-btzmp})_2(btzmp)_2](ClO_4)_2$: a doubly-bridged 1D spin-transition bistetrazole-based polymer showing thermal hysteresis behaviour. Dalton Trans., 5434–5440.
50. Absmeier, A., Bartel, M., Carbonera, C., Jameson, G. N. L., Weinberger, P., Caneschi, A., Mereiter, K., Létard, J.-F., Linert, W. (2006) Both spacer length and parity influence the thermal and light-induced properties of iron(II) α,ω-bis(tetrazole-1-yl)alkane coordination polymers. Chem. Eur. J., 12: 2235–2243.
51. Absmeier, A., Bartel, M., Carbonera, C., Jameson, G. N. L., Werner, F., Reissner, M., Caneschi, A., Létard, J.-F., Linert, W. (2007) Mutual influence of spacer length and noncoordinating anions on thermal and light-induced spin-crossover properties of iron(II)-α,ω-bis(tetrazol-1-yl)alkane coordination polymers. Eur. J. Inorg. Chem., 3047–3054.
52. Matouzenko, G. S., Molnár, G., Brefuel, N., Perrin, M., Bousseksou, A., Borshch, S. A. (2003) Spin-crossover iron(II) coordination polymer with zigzag chain structure. Chem. Mater., 15: 550–556.
53. Genre, C., Matouzenko, G. S., Jeanneau, E., Luneau, D. (2006) A spin-crossover iron(II) coordination polymer with the 8-aminoquinoline ligand: synthesis, crystal structure and magnetic properties of $[Fe(aqin)_2(4,4'\text{-}bpy)](ClO_4)_2{\cdot}2EtOH$ (aqin = 8-aminoquinoline, 4,4′-bpy = 4,4′-bipyridyl). New. J. Chem., 30: 1669–1674.
54. Matouzenko, G. S., Perrin, M., Le Guennic, B., Genre, C., Molnár, G., Bousseksou, A., Borshch, S. A. (2007) Spin crossover behavior in a family of iron(II) zigzag chain coordination polymers. Dalton Trans., 934–942.
55. Yoneda, K., Adachi, K., Hayami, S., Maeda, Y., Katada, M., Fuyuhiro, A., Kawata, S., Kaizaki, S. (2006) A steep one-step [HS-HS] to [LS-LS] spin transition in a 4,4′-bipyridine linked one-dimensional coordination polymer constructed from a pyrazolato bridged Fe(II) dimer. Chem. Commun., 45–47.
56. Weber, B., Tandon, R., Himsl, D. (2007) Synthesis, magnetic properties and X-ray structure analysis of a 1-D chain iron(II) spin crossover complex with wide hysteresis. Z. Anorg. Allg. Chem., 633: 1159–1162.
57. Weber, B., Kaps, E. S., Desplanches, C., Létard, J.-F. (2008) Quenching the hysteresis in single crystals of a 1D chain iron(II) spin crossover complex. Eur. J. Inorg. Chem., 2963–2966.
58. Imatomi, S., Kitashima, R., Hamamastu, T., Okeda, M., Ogawa, Y., Matsumoto, N. (2006) One-dimensional polynuclear spin-crossover iron(III) complex axially bridged by 1,3-bis(4-pyridyl)propane. Chem. Lett., 35: 502–503.
59. Imatomi, S., Hashimoto, S., Matsumoto, N. (2009) Inter- and intrachain spin-transition processes in one-dimensional polynuclear iron(III) complexes of *N*,*N′*-ethylenebis(acetylacetonylideneimine) bridged by 1,3-bis(4-pyridyl)propane and 1,4-bis(imidazolyl)butane. Eur. J. Inorg. Chem., 721–726.

60. Ross, T. M., Neville, S. M., Innes, D. S., Turner, D. R., Moubaraki, B., Murray, K. S. (2010) Spin crossover in iron(III) Schiff-base 1-D chain complexes. Dalton Trans., 39: 149–159.
61. Białońska, A., Bronisz, R., Darowska, K., Drabent, K., Kusz, J., Siczek, M., Weselski, M., Zubko, M., Zarowski, A. (2010) Nitrile-rich coordination polymer $^{1}_{\infty}\{Fe(CH_3CN)_4(pyrazine)(ClO_4)_2\}$ exhibiting a HS $\rightleftharpoons$ LS transition. Inorg. Chem., 49: 11267–11269.
62. Yang, F. L., Li, B., Hanajima, T., Einaga, Y., Huang, R. B., Zheng, L. S., Tao, J. (2010) An iron(II) incomplete spin-crossover compound: pressure effects and Mossbauer spectroscopy study. Dalton Trans., 39: 2288–2292.
63. Yang, F. L., Tao, J., Huang, R. B., Zheng, L. S. (2011) Temperature-dependent in situ ligand cyclization via C=C coupling and formation of a spin-crossover iron(II) coordination polymer. Inorg. Chem., 50: 911–917.
64. Hayami, S., Hashiguchi, K., Juhasz, G., Ohba, M., Okawa, H., Maeda, Y., Kato, K., Osaka, K., Takata, M., Inoue, K. (2004) 1-D cobalt(II) spin transition compound with strong interchain interaction: $[Co(pyterpy)Cl_2]\cdot X$. Inorg. Chem., 43: 4124–4126.
65. Rajadurai, C., Fuhr, O., Kruk, R., Ghafari, M., Hahn, H., Ruben, M. (2007) Above room temperature spin transition in a metallo-supramolecular coordination oligomer/polymer. Chem. Commun., 2636–2638.
66. Quesada, M., de la Peña-O'Shea, V. A., Aromí, G., Geremia, S., Massera, C., Roubeau, O., Gamez, P., Reedijk, J. (2007) A molecule-based nanoporous material showing tuneable spin-crossover behavior near room temperature. Adv. Mater., 19: 1397–1402.
67. Neville, S. M., Leita, B. A., Offermann, D. A., Duriska, M. B., Moubaraki, B., Chapman, K. W., Halder, G. J., Murray, K. S. (2007) Spin-crossover studies on a series of 1D chain and dinuclear iron(II) triazine-dipyridylamine compounds. Eur. J. Inorg. Chem., 1073–1085.
68. Ross, T. M., Moubaraki, B., Turner, D. R., Halder, G. J., Chastanet, G., Neville, S. M., Cashion, J. D., Létard, J.-F., Batten, S.R., Murray, K. S. (2011) Spin crossover and solvate effects in 1D Fe^{II} chain compounds containing bis(dipyridylamine)-linked triazine ligands. Eur. J. Inorg. Chem., 1395–1417.
69. Neville, S. M., Leita, B. A., Halder, G. J., Kepert, C. J., Moubaraki, B., Létard, J.-F., Murray, K. S. (2008) Understanding the two-step spin-transition phenomenon in iron(II) 1D chain materials. Chem.-Eur. J., 14: 10123–10133.
70. Genre, C., Jeanneau, E., Bousseksou, A., Luneau, D., Borshch, S. A., Matouzenko, G. S. (2008) First dicyanamide-bridged spin-crossover coordination polymer: synthesis, structural, magnetic, and spectroscopic studies. Chem. Eur. J., 14: 697–705.
71. Dupouy, G., Triki, S., Marchivie, M., Cosquer, N., Gómez-García, C. J., Pillet, S., Bendeif, E. E., Lecomte, C., Asthana, S., Létard, J.-F. (2010) Cyanocarbanion-based spin-crossover materials: photocrystallographic and photomagnetic studies of a new iron(II) neutral chain. Inorg. Chem., 49: 9358–9368.
72. Herchel, R., Trávníček, Z., Zbořil, R. (2011) Spin crossover behavior of a one-dimensional polymeric-chain compound $\{Fe(abpt)_2[\mu\text{-}Ni(CN)_4]\cdot xH_2O\}_n$ ($x = 0.5 > 0$): synthesis, spectral, thermal and magnetic properties. Inorg. Chim. Acta, 365: 458–461.
73. Galet, A., Muñoz, M. C., Real, J. A. (2006) $\{Fe(3CNpy)_2[Cu(3CNpy)(\mu\text{-}CN)_2]_2\}$: a one-dimensional cyanide-based spin-crossover coordination polymer. Inorg. Chem., 45: 4583–4585.
74. Agustí, G., Muñoz, M. C., Gaspar, A. B., Real, J. A. (2009) Spin-crossover behavior in cyanide-bridged iron(II)-copper(I) bimetallic 1-3D metal-organic frameworks. Inorg. Chem., 48: 3371–3381.
75. Pillet, S., Hubsch, J., Lecomte, C. (2004) Single crystal diffraction analysis of the thermal spin conversion in $Fe(btr)_2(NCS)_2\cdot H_2O$: evidence for spin-like domain formation. Eur. Phys. J. B, 38: 541–552.
76. Roubeau, O., Gubbens, P. C. M., Visser, D., Blaauw, M., de Réotier, P. D., Yaouanc, A., Haasnoot, J. G., Reedijk, J., Sakarya, S., Jayasooriya, U. A., Cottrell, S. P., King, P. J. C. (2004) Observation of the spin-crossover in $[Fe(btr)_2(NCS)_2]\cdot H_2O$ (btr=4,4′-bis-1,2,4-triazole) with μ-SR. Chem. Phys. Lett., 395: 177–181.
77. Enachescu, C., Tanasa, R., Stancu, A., Codjovi, E., Linares, J., Varret, F. (2004) FORC method applied to the thermal hysteresis of spin transition solids: first approach of static and kinetic properties. Physica B, 343: 15–19.
78. Legrand, V., Pillet, S., Souhassou, M., Lugan, N., Lecomte, C. (2006) Extension of the experimental electron density analysis to metastable states: a case example of the spin crossover complex $Fe(btr)_2(NCS)_2\cdot H_2O$, J. Am. Chem. Soc., 128: 13921–13931.

79. Pillet, S., Legrand, V., Souhassou, M., Lecomte, C. (2006) Kinetics of light-induced first-order phase transformation in molecular solids: $Fe(btr)_2(NCS)_2{\cdot}H_2O$. Phys. Rev. B, 74: 140101/1 –140101/4.
80. Legrand, V., Pillet, S., Carbonera, C., Souhassou, M., Létard, J.-F., Guionneau, P., Lecomte, C. (2007) Optical, magnetic and structural properties of the spin-crossover complex $Fe(btr)_2(NCS)_2{\cdot}H_2O$ in the light-induced and thermally quenched metastable states. Eur. J. Inorg. Chem., 5693–5706.
81. Rotaru, A., Dîrtu, M. M., Enachescu, C., Tanasa, R., Linares, J., Stancu, A., Garcia, Y. (2009) Calorimetric measurements of diluted spin crossover complexes $Fe_xM_{1\text{-}x}(btr)_2(NCS)_2{\cdot}H_2O$ with M^{II} = Zn and Ni. Polyhedron, 28: 2531–2536.
82. Garcia, Y., Bravic, G., Gieck, C., Chasseau, D., Tremel, W., Gütlich, P. (2005) Crystal structure, magnetic properties, and ^{57}Fe Mossbauer spectroscopy of the two-dimensional coordination polymers [M(1,2-bis(1,2,4-triazol-4-yl)ethane)$_2$(NCS)$_2$] (M^{II} = Fe, Co). Inorg. Chem., 44: 9723–9730.
83. Bronisz, R. (2005) 1,4-Di(1,2,3-triazol-1-yl)butane as building block for the preparation of the iron(II) spin-crossover 2D coordination polymer. Inorg. Chem., 44: 4463–4465.
84. Krivokapic, I., Enachescu, C., Bronisz, R., Hauser, A. (2008) The interaction between the spin transition and a crystallographic phase transition in the spin-crossover compound $[Fe(bbtr)_3](ClO_4)_2$: nucleation, formation of domains and fluctuations, Inorg. Chim. Acta, 361: 3616–3622.
85. Krivokapic, I., Enachescu, C., Bronisz, R., Hauser, A. (2008) Spin transition and relaxation dynamics coupled to a crystallographic phase transition in a polymeric iron(II) spin-crossover system. Chem. Phys. Lett., 455: 192–196.
86. Krivokapic, I., Chakraborty, P., Enachescu, C., Bronisz, R., Hauser, A. (2011) Low-spin → high-spin relaxation dynamics in the highly diluted spin-crossover system $[Fe_xZn_{1-x}(bbtr)_3](ClO_4)_2$. Inorg. Chem., 50: 1856–1861.
87. Krivokapic, I., Chakraborty, P., Bronisz, R., Enachescu, C., Hauser, A. (2010) Significant variation of the singlet-quintet intersystem crossing rate constant in an iron(II) high-spin complex as a function of temperature. Angew. Chem. Int. Ed., 49: 8509–8512.
88. Bao, X., Liu, J. L., Leng, J. D., Lin, Z. J., Tong, M. L., Nihei, M., Oshio, H. (2010) Spin crossover versus low-spin behaviour exhibited in 2D and 3D supramolecular isomers of $[Fe^{II}(2,4\text{-bpt})_2]$·Guest. Chem.-Eur. J., 16: 7973–7978.
89. Bao, X., Guo, P. H., Liu, J. L., Leng, J. D., Tong, M. L. (2011) Crystalline-state *cis*-to-*trans* transformation of a two-dimensional spin-crossover system. Chem.-Eur. J., 17: 2335–2339.
90. Quesada, M., Prins, F., Roubeau, O., Gamez, P., Teat, S. J., van Koningsbruggen, P. J., Haasnoot, J. G., Reedijk, J. (2007) A 2D Fe^{II}-bistetrazole coordination polymer exhibiting spin-crossover properties. Inorg. Chim. Acta, 360: 3787–3796.
91. Białońska, A., Bronisz, R. (2010) High spin and spin-crossover two-dimensional coordination polymers containing Fe^{II}(tetrazol-2-yl)$_4$(solv)$_2$ (solv = ethanol, acetonitrile) cores linked by flexible/elastic spacers. Inorg. Chem., 49: 4534–4542.
92. Neville, S. M., Moubaraki, B., Murray, K. S., Kepert, C. J. (2007) A thermal spin transition in a nanoporous iron(II) coordination framework material. Angew. Chem. Int. Ed., 46: 2059–2062.
93. Neville, S. M., Halder, G. J., Chapman, K. W., Duriska, M. B., Moubaraki, B., Murray, K. S., Kepert, C. J. (2009) Guest tunable structure and spin crossover properties in a nanoporous coordination framework material. J. Am. Chem. Soc., 131: 12106–12108.
94. Neville, S. M., Halder, G. J., Chapman, K. W., Duriska, M. B., Southon, P. D., Cashion, J. D., Létard, J.-F., Moubaraki, B., Murray, K. S., Kepert, C. J. (2008) Single-crystal to single-crystal structural transformation and photomagnetic properties of a porous iron(II) spin-crossover framework. J. Am. Chem. Soc., 130: 2869–2876.
95. Halder, G. J., Chapman, K. W., Neville, S. M., Moubaraki, B., Murray, K. S., Létard, J.-F., Kepert, C. J. (2008) Elucidating the mechanism of a two-step spin transition in a nanoporous metal-organic framework. J. Am. Chem. Soc., 130: 17552–17562.
96. Neville, S. M., Etrillard, C., Asthana, S., Létard, J.-F. (2010) Light-induced stored information in nanoparticles. Eur. J. Inorg. Chem., 282–288.
97. Atsuchi, M., Higashikawa, H., Yoshida, Y., Nakashima, S., Inoue, K. (2007) Novel 2D interpenetrated structure and occurrence of the spin-crossover phenomena of assembled complexes, $Fe(NCX)_2(bpp)_2$ (X = S, Se, BH_3; bpp=1,3-bis(4-pyridyl)propane). Chem. Lett., 36: 1064–1065.

98. Patrick, B. O., Reiff, W. M., Sanchez, V., Storr, A., Thompson, R. C. (2004) Poly(2,2′-bipyridine) tetrakis(imidazolato)diiron(II): structural and spin-state phase transitions and low-temperature magnetic ordering in a unique 2-dimensional material. Inorg. Chem., 43: 2330–2339.
99. Agustí, G., Thompson, A. L., Gaspar, A. B., Muñoz, M. C., Goeta, A. E., Rodríguez-Velamazán, J. A., Castro, M., Burriel, R., Real, J. A. (2008) Thermal, pressure and light induced spin transition in the two-dimensional coordination polymer $\{Fe(pmd)_2[Cu(CN)_2]_2\}$, Dalton Trans., 642–649.
100. Muñoz, M. C., Gaspar, A. B., Galet, A., Real, J. A. (2007) Spin-crossover behavior in cyanide-bridged iron(II)-silver(I) bimetallic 2D Hofmann-like metal-organic frameworks. Inorg. Chem., 46: 8182–8192.
101. Agustí, G., Muñoz, M. C., Gaspar, A. B., Real, J. A.(2008) Spin-crossover behavior in cyanide-bridged iron(II)-gold(I) bimetallic 2D Hofmann-like metal-organic frameworks. Inorg. Chem., 47: 2552–2561.
102. Rodríguez-Velamazán, J. A., Carbonera, C., Castro, M., Palacios, E., Kitazawa, T., Létard, J.-F., Burriel, R. (2010) Two-step thermal spin transition and LIESST relaxation of the polymeric spin-crossover compounds Fe(X-py)$_2$[Ag(CN)$_2$]$_2$ (X = H, 3-methyl, 4-methyl, 3,4-dimethyl, 3-Cl). Chem. Eur. J., 16: 8785–8796.
103. Kosone, T., Kachi-Terajima, C., Kanadani, C., Saito, T., Kitazawa, T. (2008) A two-step and hysteretic spin-crossover transition in new cyano-bridged hetero-metal $Fe^{II}Au^{I}$ 2-dimensional assemblage. Chem. Lett., 37: 422–423.
104. Kosone, T., Kanadani, C., Saito, T., Kitazawa, T. (2009) Synthesis, crystal structures, magnetic properties and fluorescent emissions of two-dimensional bimetallic coordination frameworks Fe^{II}(3-fluoropyridine)$_2$[$Au^{I}(CN)_2$]$_2$ and Mn^{II}(3-fluoropyridine)$_2$[$Au^{I}(CN)_2$]$_2$. Polyhedron, 28: 1930–1934.
105. Galet, A., Muñoz, M. C., Martínez, V., Real, J. A. (2004) Supramolecular isomerism in spin crossover networks with aurophilic interactions. Chem. Commun., 2268–2269.
106. Galet, A., Muñoz, M. C., Real, J. A. (2006) Coordination polymers undergoing spin crossover and reversible ligand exchange in the solid. Chem. Commun., 4321–4323.
107. Seredyuk, M., Gaspar, A. B., Ksenofontov, V., Verdaguer, M., Villain, F., Gütlich, P. (2009) Thermal- and light-induced spin crossover in novel 2D Fe(II) metalorganic frameworks $\{Fe(4\text{-}PhPy)_2[M(CN)_x]_y\}\cdot sH_2O$: spectroscopic, structural, and magnetic studies. Inorg. Chem., 48: 6130–6141.
108. Seredyuk, M., Gaspar, A. B., Ksenofontov, V., Galyametdinov, Y., Verdaguer, M., Villain, F., Gütlich, P. (2010) Spin-crossover and liquid crystal properties in 2D cyanide-bridged Fe^{II}-$M^{(I/II)}$ metalorganic frameworks. Inorg. Chem., 49: 10022–10031.
109. Agustí, G., Gaspar, A. B., Muñoz, M. C., Lacroix, P. G., Real, J. A. (2009) Spin crossover and paramagnetic behaviour in two-dimensional iron(II) coordination polymers with stilbazole push-pull ligands. Aust. J. Chem., 62: 1155–1165.
110. Martínez, V., Gaspar, A. B., Muñoz, M. C., Bukin, G. V., Levchenko, G., Real, J.A. (2009) Synthesis and characterisation of a new series of bistable iron(II) spin-crossover 2D metal-organic frameworks. Chem. Eur. J., 15: 10960–10971.
111. Martínez, V., Gaspar, A. B., Muñoz, M. C., Ballesteros, R., Ortega-Villar, N., Ugalde-Saldívar, V. M., Moreno-Esparza, R., Real, J. A. (2007) Spin-crossover 2D metal-organic frameworks with a redox-active ligand: [Fe(ttf-adpy)$_2$M(CN)$_4$]·nH$_2$O (ttf-adpy = 4-tetrathiafulvalenylcarboxamidopyridine; M^{II} = Ni, Pd, Pt). Eur. J. Inorg. Chem., 303–310.
112. Agustí, G., Gaspar, A. B., Muñoz, M. C., Real, J. A. (2007) Thermal- and pressure-induced cooperative spin transition in the 2D and 3D coordination polymers $\{Fe(5\text{-}Br\text{-}pmd)_2[M(CN)_x]_y\}$ (M = Ag^{I}, Au^{I}, Ni^{II}, Pd^{II}, Pt^{II}). Inorg. Chem., 46: 9646–9654.
113. Molnár, G., Guillon, T., Moussa, N. O., Rechignat, L., Kitazawa, T., Nardone, M., Bousseksou, A. (2006) Two-step spin-crossover phenomenon under high pressure in the coordination polymer Fe(3-methylpyridine)$_2$[Ni(CN)$_4$]. Chem. Phys. Lett., 423: 152–156.
114. Levchenko, G., Bukin, G. V., Terekhov, S. A., Gaspar, A. B., Martínez, V., Muñoz, M. C., Real, J. A. (2011) Pressure-induced cooperative spin transition in iron(II) 2D coordination polymers: room-temperature visible spectroscopic study. J. Phys. Chem. B, 115: 8176–8182.
115. Martínez, V., Boldog, I., Gaspar, A. B., Ksenofontov, V., Bhattacharjee, A., Gütlich, P., Real, J. A. (2010) Spin crossover phenomenon in nanocrystals and nanoparticles of Fe(3-Fpy)$_2$[M(CN)$_4$] (M^{II} = Ni, Pd, Pt) two-dimensional coordination polymers. Chem. Mater., 22: 4271–4281.

116. Grunert, C. M., Schweifer, J., Weinberger, P., Linert, W., Mereiter, K., Hilscher, G., Müller, M., Wiesinger, G., van Koningsbruggen, P. J. (2004) Structure and physical properties of [μ-tris(1,4-bis(tetrazol-1-yl)butane-N^4,$N^{4'}$)iron(II)] bis(hexafluorophosphate), a new Fe(II) spin-crossover compound with a three-dimensional threefold interlocked crystal lattice. Inorg. Chem., 43: 155–165.
117. Bartel, M., Absmeier, A., Jameson, G. N. L., Werner, F., Kato, K., Takata, M., Boca, R., Hasegawa, M., Mereiter, K., Caneschi, A., Linert, W. (2007) Modification of spin crossover behavior through solvent assisted formation and solvent inclusion in a triply interpenetrating three-dimensional network. Inorg. Chem., 46: 4220–4229.
118. Jameson, G. N. L., Werner, F., Bartel, M., Absmeier, A., Reissner, M., Kitchen, J. A., Brooker, S., Caneschi, A., Carbonera, C., Létard, J.-F., Linert, W. (2009) Anion, solvent and time dependence of high-spin-low-spin interactions in a 3D coordination polymer. Eur. J. Inorg. Chem., 3948–3959.
119. Bronisz, R. (2007) Tetrazol-2-yl as a donor group for incorporation of a spin-crossover function based on Fe(II) ions into a coordination network. Inorg. Chem., 46: 6733–6739.
120. Rentschler, E., von Malotki, C. (2008) Spin transition in three-dimensional bridged coordination polymers of iron (II)-urea-triazoles. Inorg. Chim. Acta, 361: 3646–3653.
121. Bai, Y. L., Tao, J., Huang, R. B., Zheng, L. S., Zheng, S. L., Oshida, K., Einaga, Y. (2008) Pressure effects and Mössbauer spectroscopic studies on a 3D mixed-valence iron spin-crossover complex with NiAs topology. Chem. Commun., 1753–1755.
122. Chainok, K., Neville, S. M., Moubaraki, B., Batten, S. R., Murray, K. S., Forsyth, C. M., Cashion, J. D. (2010) Synthesis, structures and spin crossover properties of infinite 3D frameworks of iron(II) containing organodinitrile bridging ligands. Dalton Trans., 39: 10900–10909.
123. Galet, A., Niel, V., Muñoz, M. C., Real, J. A. (2003) Synergy between spin crossover and metallophilicity in triple interpenetrated 3D nets with the NbO structure type. J. Am. Chem. Soc., 125: 14224–14225.
124. Galet, A., Gaspar, A. B., Agustí, G., Muñoz, M. C., Real, J. A. (2007) Pressure effect studies on the 3D spin crossover system: $\{Fe(3CN\text{-}py)_2[M(CN)_2]_2\}\cdot nH_2O$ ($n \leq 2/3$, M = Ag(I), Au(I)). Chem. Phys. Lett., 434: 68–72.
125. Niel, V., Thompson, A. L., Goeta, A. E., Enachescu, C., Hauser, A., Galet, A., Muñoz, M. C., Real, J. A. (2005) Thermal- and photoinduced spin-state switching in an unprecedented three-dimensional bimetallic coordination polymer. Chem. Eur. J., 11: 2047–2060.
126. Kosone, T., Suzuki, Y., Ono, S., Kanadani, C., Saito, T., Kitazawa, T. (2010) A new spin crossover heterometallic $Fe^{II}Ag^{I}$ coordination polymer with the $[Ag_2(CN)_3]^-$ unit: crystallographic and magnetic study. Dalton Trans., 39: 1786–1790.
127. Galet, A., Muñoz, M. C., Gaspar, A. B., Real, J. A. (2005) Architectural isomerism in the three-dimensional polymeric spin crossover system $\{Fe(pmd)_2[Ag(CN)_2]_2\}$: synthesis, structure, magnetic properties, and calorimetric studies. Inorg. Chem., 44: 8749–8755.
128. Niel, V., Thompson, A. L., Muñoz, M. C., Galet, A., Goeta, A. E., Real, J. A. (2003) Crystalline-state reaction with allosteric effect in spin-crossover, interpenetrated networks with magnetic and optical bistability. Angew. Chem. Int. Ed., 42: 3760–3763.
129. Galet, A., Gaspar, A. B., Muñoz, M. C., Bukin, G. V., Levchenko, G., Real, J. A. (2005) Tunable bistability in a three-dimensional spin-crossover sensory- and memory-functional material. Adv. Mater., 17: 2949–2953.
130. Ohba, M., Yoneda, K., Agustí, G., Muñoz, M. C., Gaspar, A. B., Real, J. A., Yamasaki, M., Ando, H., Nakao, Y., Sakaki, S., Kitagawa, S. (2009) Bidirectional chemo-switching of spin state in a microporous framework. Angew. Chem. Int. Ed., 48: 4767–4771.
131. Southon, P. D., Liu, L., Fellows, E. A., Price, D. J., Halder, G. J., Chapman, K. W., Moubaraki, B., Murray, K. S., Létard, J.-F., Kepert, C. J. (2009) Dynamic interplay between spin-crossover and host-guest function in a nanoporous metal-organic framework material. J. Am. Chem. Soc., 131: 10998–11009.
132. Bonhommeau, S., Molnár, G., Galet, A., Zwick, A., Real, J. A., McGarvey, J. J., Bousseksou, A. (2005) One shot laser pulse induced reversible spin transition in the spin-crossover complex $[Fe(C_4H_4N_2)\{Pt(CN)_4\}]$ at room temperature. Angew. Chem. Int. Ed., 44: 4069–4073.
133. Cobo, S., Ostrovskii, D., Bonhommeau, S., Vendier, L., Molnár, G., Salmon, L., Tanaka, K., Bousseksou, A. (2008) Single-laser-shot-induced complete bidirectional spin transition at room temperature in single crystals of $(Fe^{II}(pyrazine)(Pt(CN)_4))$. J. Am. Chem. Soc., 130: 9019–9024.

134. Agustí, G., Ohtani, R., Yoneda, K., Gaspar, A. B., Ohba, M., Sánchez-Royo, J. F., Muñoz, M. C., Kitagawa, S., Real, J. A. (2009) Oxidative addition of halogens on open metal sites in a microporous spin-crossover coordination polymer. Angew. Chem. Int. Ed., 48: 8944–8947.
135. Ohtani, R., Yoneda, K., Furukawa, S., Horike, N., Kitagawa, S., Gaspar, A. B., Muñoz, M. C., Real, J. A., Ohba, M. (2011) Precise control and consecutive modulation of spin transition temperature using chemical migration in porous coordination polymers. J. Am. Chem. Soc., 133: 8600–8605.
136. Agustí, G., Cobo, S., Gaspar, A. B., Molnár, G., Moussa, N.O., Szilágyi, P. A., Pálfi, V., Vieu, C., Muñoz, M. C., Real, J. A., Bousseksou, A. (2008) Thermal and light-induced spin crossover phenomena in new 3D Hofmann-like microporous metalorganic frameworks produced as bulk materials and nanopatterned thin films. Chem. Mater., 20: 6721–6732.
137. Bartual-Murgui, C., Ortega-Villar, N. A., Shepherd, H. J., Muñoz, M. C., Salmon, L., Molnár, G., Bousseksou, A., Real, J. A. (2011) Enhanced porosity in a new 3D Hofmann-like network exhibiting humidity sensitive cooperative spin transitions at room temperature. J. Mater. Chem., 21: 7217–7222.
138. Cobo, S., Molnár, G., Real, J. A., Bousseksou, A. (2006) Multilayer sequential assembly of thin films that display room-temperature spin crossover with hysteresis. Angew. Chem. Int. Ed., 45: 5786–5789.
139. Molnár, G., Cobo, S., Real, J. A., Carcenac, F., Daran, E., Vieu, C., Bousseksou, A. (2007) A combined top-down/bottom-up approach for the nanoscale patterning of spin-crossover coordination polymers. Adv. Mater., 19: 2163–2167.
140. Bartual-Murgui, C., Salmon, L., Akou, A., Thibault, C., Molnár, G., Mahfoud, T., Sekkat, Z., Real, J. A., Bousseksou, A. (2011) High quality nano-patterned thin films of the coordination compound {Fe(pyrazine)[Pt(CN)$_4$]} deposited layer-by-layer. New J. Chem., 35: 2089–2094.
141. Boldog, I., Gaspar, A. B., Martínez, V., Pardo-Ibañez, P., Ksenofontov, V., Bhattacharjee, A., Gütlich, P., Real, J. A. (2008) Spin-crossover nanocrystals with magnetic, optical, and structural bistability near room temperature. Angew. Chem. Int. Ed., 47: 6433–6437.
142. Volatron, F., Catala, L., Rivière, E., Gloter, A., Stephan, O., Mallah, T. (2008) Spin-crossover coordination nanoparticles. Inorg. Chem., 47: 6584–6586.
143. Larionova, J., Salmon, L., Guari, Y., Tokarev, A., Molvinger, K., Molnár, G., Bousseksou, A. (2008) Towards the ultimate size limit of the memory effect in spin-crossover solids. Angew. Chem. Int. Ed., 47: 8236–8240.
144. Kosaka, W., Nomura, K., Hashimoto, K., Ohkoshi, S. (2005) Observation of an Fe(II) spin-crossover in a cesium iron hexacyanochromate. J. Am. Chem. Soc., 127: 8590–8591.
145. Papanikolaou, D., Margadonna, S., Kosaka, W., Ohkoshi, S., Brunelli, M., Prassides, K. (2006) X-ray illumination induced Fe(II) spin crossover in the Prussian blue analogue cesium iron hexacyanochromate. J. Am. Chem. Soc., 128: 8358–8363.
146. Papanikolaou, D., Kosaka, W., Margadonna, S., Kagi, H., Ohkoshi, S., Prassides, K. (2007) Piezomagnetic behavior of the spin crossover Prussian blue analogue CsFe[Cr(CN)]$_6$. J. Phys. Chem. C, 111: 8086–8091.
147. Arai, M., Kosaka, W., Matsuda, T., Ohkoshi, S. (2008) Observation of an iron(II) spin-crossover in an iron octacyanoniobate-based magnet. Angew. Chem. Int. Ed., 47: 6885–6887.
148. Ohkoshi, S., Imoto, K., Tsunobuchi, Y., Takano, S., Tokoro, H. (2011) Light-induced spin-crossover magnet. Nature Chem., 3: 564–569.

5

Structure:Function Relationships in Molecular Spin-Crossover Materials

Malcolm A. Halcrow

School of Chemistry, University of Leeds, UK

5.1 Introduction

The most challenging aspect of making a new spin-crossover material with technologically useful properties is controlling the form of the transition. That is, whether it occurs gradually or abruptly; with or without hysteresis; in a single step or in stepwise fashion; and, whether or not it occurs to full completeness (Fig. 5.1).[1] These aspects are determined by the cooperativity in the solid lattice – the efficiency with which structural changes at individual spin-crossover metal sites are transmitted through the bulk material.[2] To that extent, a spin-crossover transition can be considered a model for the propagation of any first-order phase transition through a crystalline material.

On one level, differing degrees of spin-crossover cooperativity in a solid lattice can be described using theoretical descriptions of elastic interactions between the functional sites in the material.[2] This can reproduce the common types of spin-transition curve, if the interactions between molecules in their different spin states are chosen appropriately in the mathematical model (this is described further in Chapter 18). On the other hand, such descriptions do not shed light on the origins of these nearest neighbour interactions, which correspond to bonding between the spin centres in the solid material. That is, they tell you what combination of interaction enthalpies are required to give a particular type of cooperative transition, but not how to achieve them.

The spin centres may be linked by covalent bonding, if they are contained in a polynuclear complex or coordination polymer.[3] Alternatively, in a molecular crystal, they arise from weaker hydrogen bonding, π–π interactions or simple mechanical contacts between molecules.[4, 5] Importantly, it is unsafe to assume that the strongest bonding between spin sites will simply yield the most cooperative transition. As describe below, the

Spin-Crossover Materials: Properties and Applications, First Edition. Edited by Malcolm A. Halcrow.

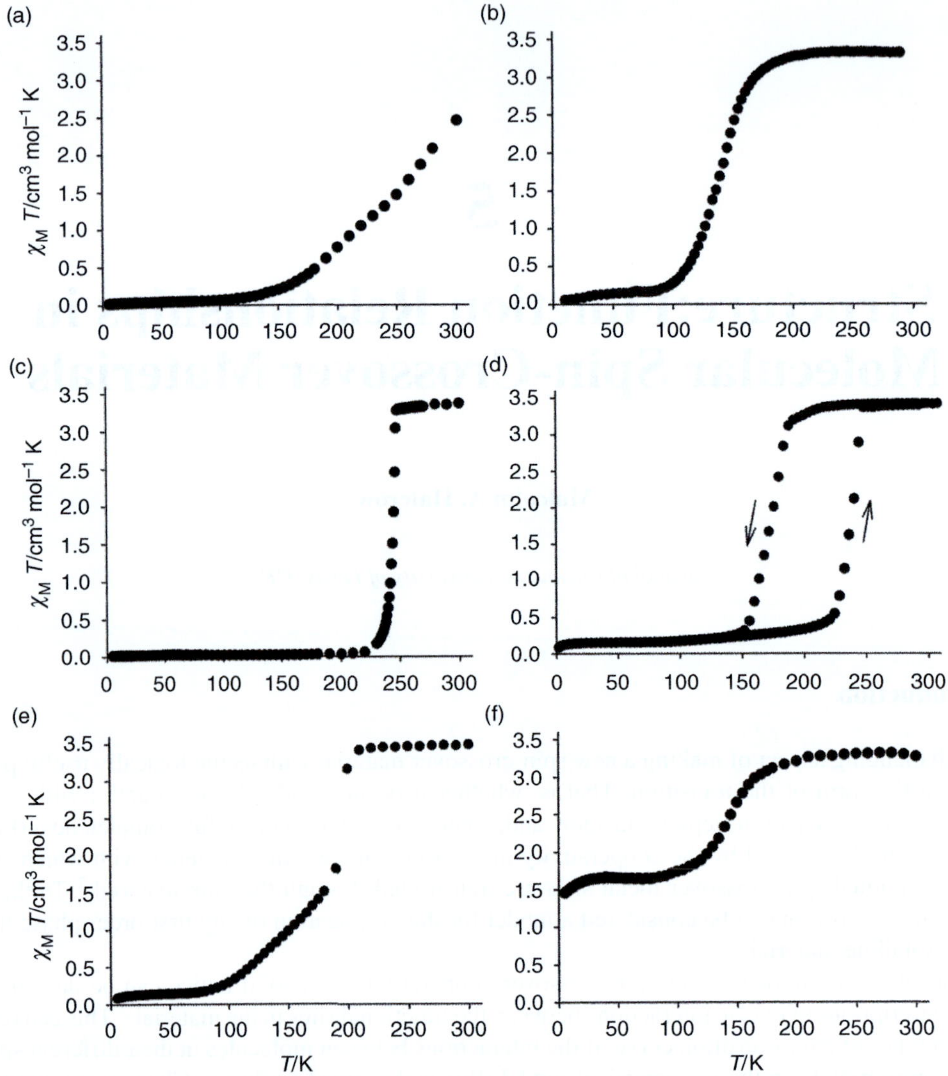

Figure 5.1 *Six representative spin-crossover transitions of molecular iron(II) complexes, illustrating the different types of cooperativity that can be observed. (a)→(c) increasing degrees of abruptness. (d) thermal hysteresis. (e) discontinuous transition. (f) incomplete transition. Data taken from refs*[7–10].

most cooperative known spin-crossover materials are all molecular crystals, whose transitions are mediated by secondary interactions, not coordination polymers. This apparently arises because solid state cooperativity reflects changes in the solid lattice energy between the spin states, which cannot be well understood by considering each bonding interaction in isolation. Rather, it is a sum of all the different intermolecular interactions and steric contacts in the crystal lattice.

Thus, a full understanding of the cooperativity of a spin-transition requires detailed knowledge of the structure of the sample and of the structural differences between the different spin states. This chapter surveys the available structural data from different classes of complex, with a view to correlating cooperativity and structure changes associated with spin-crossover. The survey is necessarily incomplete, because structural data from spin-crossover compounds are often unavailable, or only available in one spin state. That is particularly true for the most cooperative materials like Figure 5.1(d), where the large structure changes causing the cooperativity often lead to crystal decomposition during spin-crossover. None-the-less, enough structure:function data are available to allow some conclusions to be drawn.[5, 6] The compounds referred to in the following discussions are drawn in Scheme 5.1.

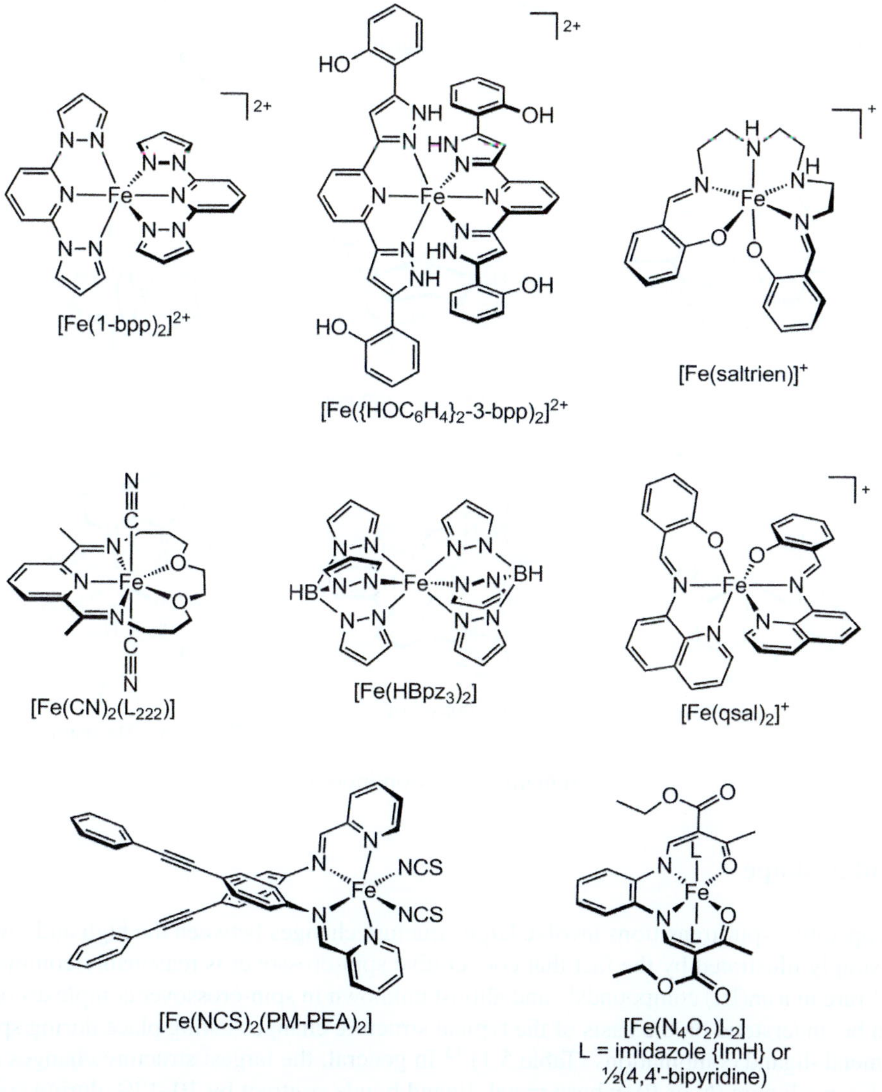

Scheme 5.1 *Complexes referred to in this chapter. The compounds are shown in the order in which they appear in the text.*

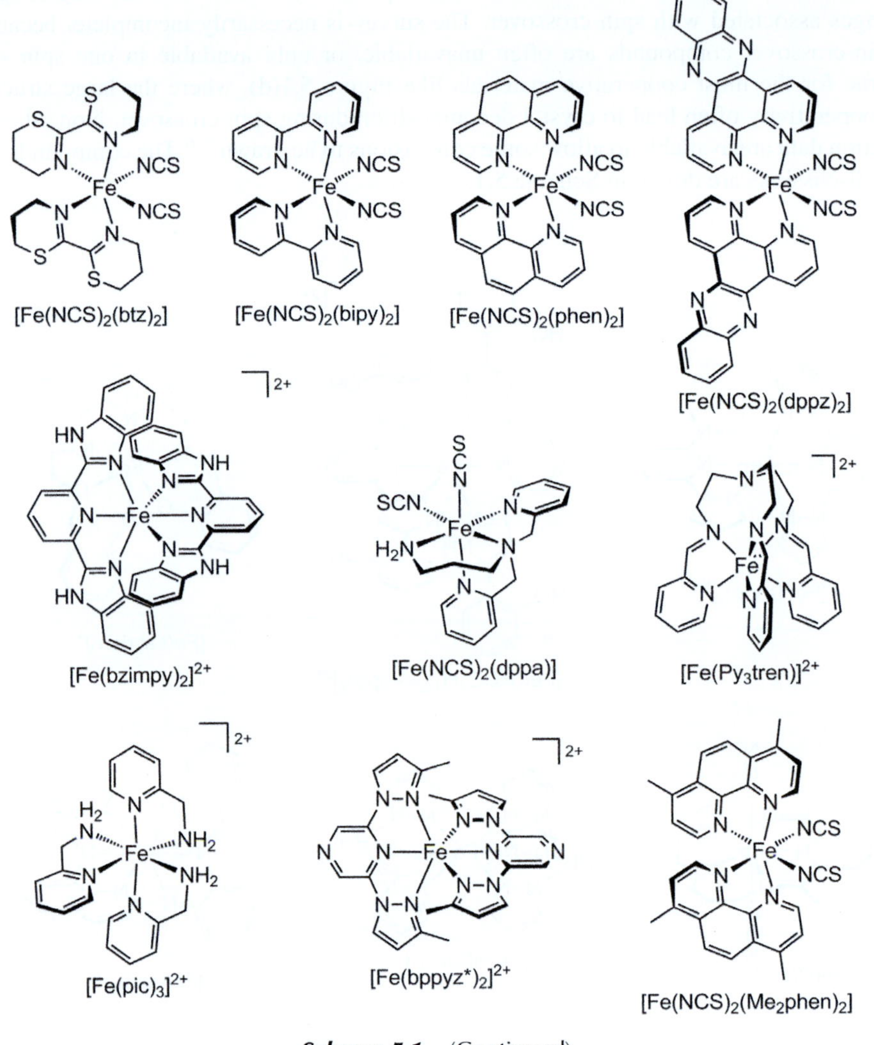

Scheme 5.1 (Continued)

5.2 Molecular Shape

In general, cooperative spin-transitions involve large structure changes between the high and low spin states. This is most simply illustrated by the fact that cooperative spin-crossover is reasonably common in iron(II) compounds,[11] rare in iron(III) compounds[12] and almost unknown in spin-crossover complexes of other metal ions. This can be understood on the basis of the typical structure changes taking place during spin-crossover for different metal–ligand combinations (Table 5.1).[13] In general, the largest structure changes are observed for group 15 ligand donors (N or P), whose metal–ligand bonds contract by 10–13% during spin-crossover. Spin-crossover in O-, S- and Cl-donor complexes gives rise to much smaller structure changes in comparison. Moreover, spin-crossover in octahedral d^6 ($S = 2 \rightarrow 0$) and d^5 ($^5/_2 \rightarrow {}^3/_2$) metal ions generally involves greater structural changes than for d^4 ($S = 2 \rightarrow 1$) or d^7 ($^3/_2 \rightarrow {}^1/_2$), since it involves a greater depopulation of the

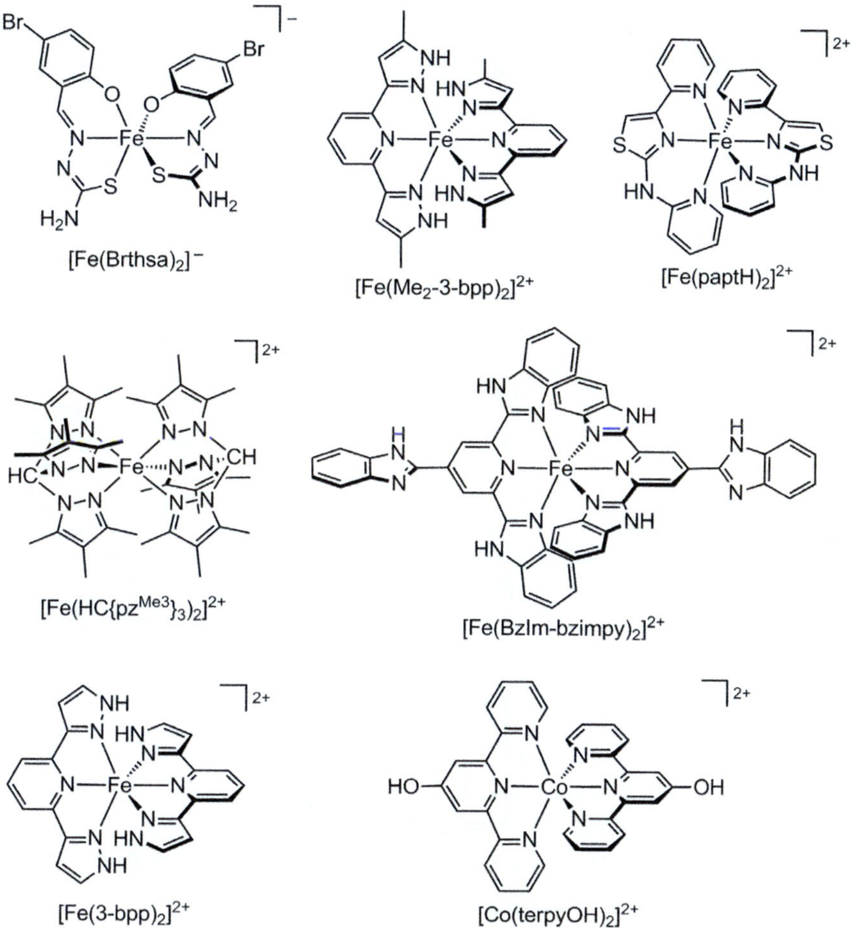

Scheme 5.1 (Continued)

antibonding e_g *d*-orbital manifold. Of the metal/ligand combinations that commonly afford spin-crossover, these criteria imply that the $[FeN_6]^{2+}$ moiety should give the most cooperative spin-transitions, as observed.[11]

The above generalisation takes account only of the change in size of the spin centre during the transition, which contracts and expands as transition is cycled. An isotropic 'breathing' transformation can be accommodated by a solid lattice relatively easily. However, more pronounced changes in the *angular* shape of a molecule between the spin-states also play an important role in controlling spin-crossover between them. The high spin state of any metal centre always shows more structural variation than the low spin, because of the weaker metal–ligand bonding in the high spin state. Therefore, the form of a spin-transition can be related to the shape of the compound in its high spin state, and the degree to which this differs from the low spin state. Changes in shape of the spin centres during spin-crossover result in large rearrangements in the crystal lattice at the molecular level, resulting in anisotropic expansions or contractions of the bulk material which may also lead to crystallographic phase changes. Such modifications are harder to achieve in a rigid lattice than a simple isotropic contraction, leading to higher activation barriers for the transition and larger changes in lattice energy.

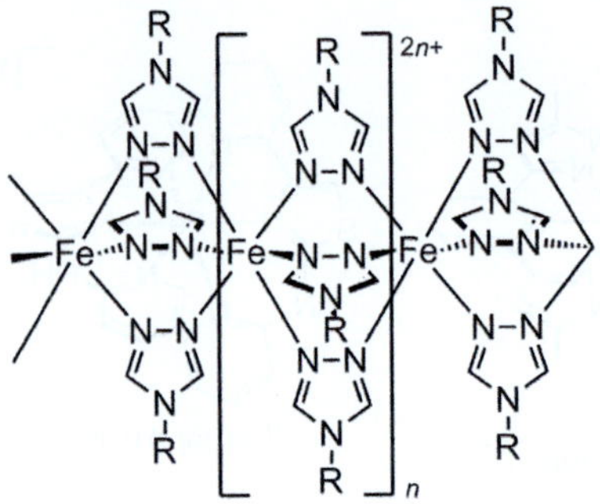

$[Fe(Rtz)_3]^{2+}$

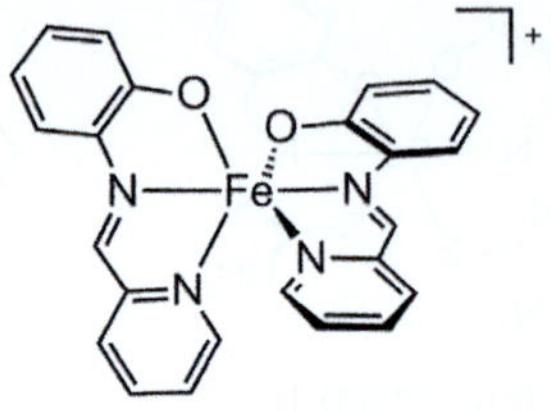

$[Fe(pap)_2]^+$

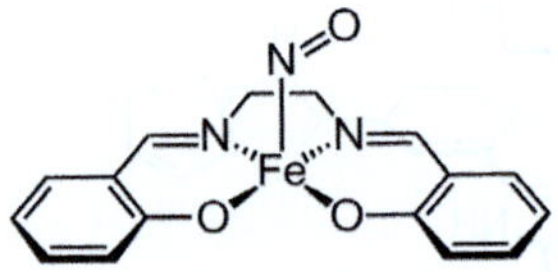

[Fe(salen)(NO)]

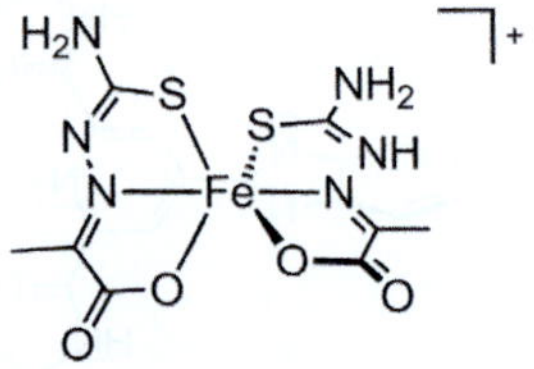

[Fe(thpu)(Hthpu)]

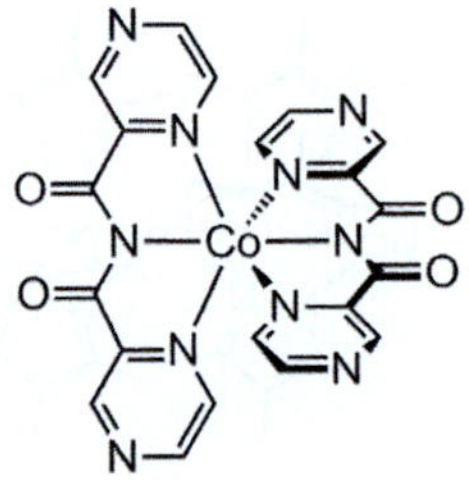

$[Co(dpzca)_2]$

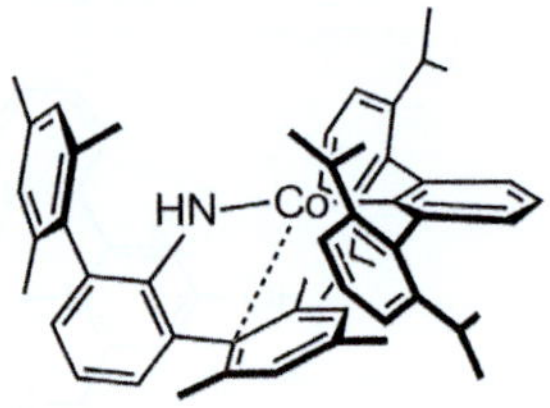

$[Co(NHC_6H_3\text{-}2,6\text{-}\{C_6H_2\text{-}2,4,6\text{-}Me_3\}_2)(C_6H_3\text{-}2,6\text{-}\{C_6H_3\text{-}2,6\text{-}iPr_2\}_2)]$

Scheme 5.1 (Continued)

Table 5.1 *Typical metal–ligand bond lengths (Å) in the different spin states of six-coordinate spin-crossover complexes, with different combinations of metal ion and ligand donor atoms.*

		N	O	P	S	Cl
Iron(II)	High spin	2.2–2.3	2.0–2.1	2.6	–	2.4
	Low spin	1.9–2.0	2.0	2.3	2.3	2.3
Iron(III)	High spin	2.1–2.2	1.9–2.0	–	2.4	–
	Low spin	1.9–2.0	1.9	–	2.2–2.3	–
Cobalt(II)	High spin	2.0–2.2	–	–	2.4–2.6	–
	Low spin	1.9–2.1	–	–	–	–
Manganese(III)[a]	High spin	2.0–2.3	1.9–2.1	–	–	–
	Low spin	2.0–2.2	1.9–2.0	–	–	–

[a]Comparison is complicated by the pronounced Jahn–Teller elongations shown by high-spin complexes of manganese(III).

There are several examples of compounds where strong cooperativity appears to be related to a change in shape of the molecule, caused by a change in the metal coordination geometry, a ligand conformational rearrangement, or both. Conversely, however, if the difference in shape of the high and low spin states is too great, spin-crossover in the solid state is kinetically inhibited and does not occur at all.[5]

5.2.1 Molecular Shape Inducing Cooperativity

Two examples of this come from the $[Fe(1\text{-bpp})_2]^{2+}$ series of complexes:[14] $[Fe(1\text{-bpp})_2][Ni(mnt)_2]_2{\cdot}MeNO_2$[15] and $[Fe(Me_2\text{-}1\text{-bpp})_2][BF_4]_2{\cdot}xH_2O$.[16] Both these compounds exhibit hysteretic spin-transitions on cooling, which involve multiple crystallographic phase-changes. While several other complexes of this type also exhibit spin-crossover, these are the only examples to exhibit noteworthy cooperativity.[14] A likely explanation for their behaviour is seen in Figure 5.2, which compares the high and low spin structures of $[Fe(Me_2\text{-}1\text{-bpp})_2][BF_4]_2{\cdot}xH_2O$. The high spin state of the compound has a somewhat twisted ligand conformation, so that the dihedral angle between the least squares planes of the two tridentate ligands (θ) is only 76.4° (as opposed to 90° in an undistorted six-coordinate geometry). In its fully low spin form, this dihedral angle θ has expanded to 81.9°, a difference of $\Delta\theta = 5.5°$. A comparable $\Delta\theta = 4.5°$ is also seen in $[Fe(1\text{-bpp})_2][Ni(mnt)_2]_2{\cdot}MeNO_2$. In contrast, however, less cooperative spin-crossover complexes of the $[Fe(1\text{-bpp})_2]^{2+}$ type exhibit $\Delta\theta \leq 0.6°$, showing that essentially no ligand conformational rearrangements take place between their spin states.[14] The spin-crossover cooperativity and complicated structural chemistry in both complexes is probably directly related to these ligand conformational changes, and the major lattice rearrangements required to accommodate them.

Another, closely related complex is a solvated phase of $[Fe([HOC_6H_4]_2\text{-}3\text{-bpp})_2][ClO_4]_2$, which undergoes spin-crossover at 153 K in one step, with a 40 K thermal hysteresis loop.[17] There are no crystallographic phase changes involved in this transition, which does however proceed with large changes in ligand conformation. This time it is the pendant phenoxy groups, rather than the heterocyclic core of the ligand, that rearrange during the transition, resulting in atomic displacements of up to 1.5 Å between the two spin states (Fig. 5.3). Although changes to anion disorder between the spin states may also be relevant,[17] this large change in molecular structure should be a major contributor to the cooperativity of the transition.[5] Similar structure changes are also exhibited by the spin-crossover phase of $[Fe(saltrien)][Ni(dmit)_2]$ ($dmitH_2$ = 1,3-dithiole-2-thione-4,5-dithiol),[18, 19] whose spin-transition involves large rearrangements of the phenoxo groups in the

Figure 5.2 *Overlay of the high spin (white) and low spin (grey) structures of the complex dication in* $[Fe(Me_2\text{-}1\text{-bpp})_2][BF_4]_2$*, emphasising the changes in ligand conformation during spin-crossover.*[16] *All H atoms have been omitted for clarity.*

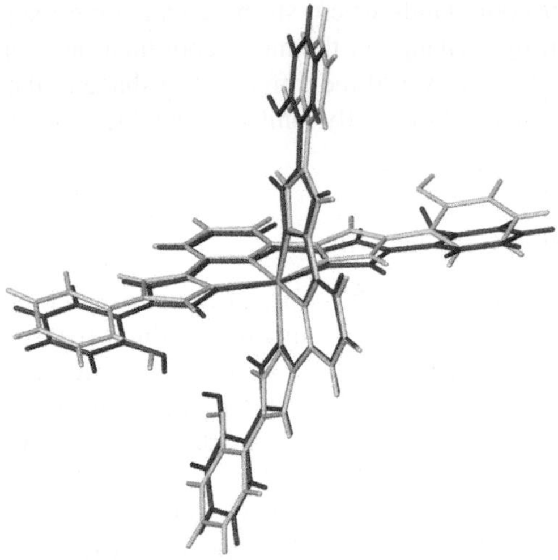

Figure 5.3 Overlay of the high spin (white) and low spin (grey) structures of the complex dication in $[Fe([HOC_6H_4]_2\text{-}3\text{-}bpp)_2][ClO_4]_2\cdot H_2O\cdot 2(CH_3)_2CO$, showing the ligand conformational rearrangement during spin-crossover that may give rise to its thermal hysteresis.[16]

$[saltrien]^{2-}$ ligand.[18] The 30 K thermal hysteresis loop shown by this phase is one of the widest known for an iron(III) complex.[12]

A related phenomenon involves cooperativity caused by changes in disorder of a coordinated ligand. This is an established phenomenon in some iron(II) complexes of conformationally flexible poly(alkylamine) chelates, although the maximum cooperativity that has been observed in such cases is a hysteresis loop of 10 K.[20–22]

A final, rather different example is $[Fe(CN)_2(L_{222})]\cdot H_2O$,[23] whose thermodynamic high and low spin forms are seven- and six-coordinate, respectively.[24, 25] This change in coordination number and the associated ligand rearrangement occur slowly in the solid lattice. Hence, the low→high spin transition in this material occurs in two steps, involving metastable seven-coordinate low spin, and six-coordinate high spin, species. The latter is initially formed upon heating to room temperature, as the kinetic product of the transition, and only reverts to the thermodynamic seven-coordinate high spin phase over a period of hours.[23, 26, 27]

5.2.2 Molecular Shape Inhibiting Spin-Crossover

Relatively small differences in shape seem to favour cooperative spin-crossover. However, if the difference in shape between the high and low spin forms of a material is too large, it will freeze the compound in its high spin state. This is because a solid lattice cannot accommodate the structural rearrangement that would be required for the transition to proceed.[5] This is a common phenomenon in iron(II) complexes of 1-bpp,[14] whose high spin states are subject to a strong angular Jahn–Teller distortion that results in an unusual, twisted six-coordinate geometry (Fig. 5.4).[28–30] This distortion is not adopted by the low spin state of the compound, so any spin-crossover would require a rotation of one ligand around the metal ion, by up to 30°. That is impossible in the solid state. The few known examples of high-spin $[Fe(terpy)_2]^{2+}$ complexes also adopt this Jahn–Teller distorted structure in the crystal, and do not show spin-crossover.[31, 32] Importantly, though, this is purely a solid state effect; both sets of compounds undergo spin-crossover in solution, where large changes to molecular geometry can take place.[28, 30, 31]

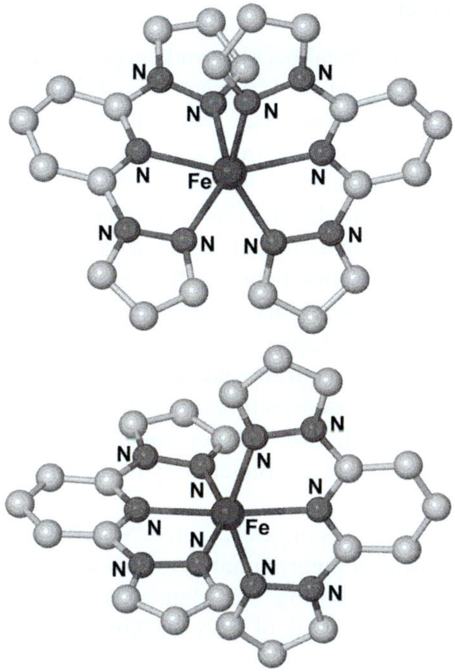

Figure 5.4 *Structures of the high spin forms of the PF_6^- (top) and BF_4^- (bottom) salts of $[Fe(1\text{-}bpp)_2]^{2+}$. All H atoms have been omitted for clarity. Only the BF_4^- salt exhibits spin-crossover.*[28]

Inhibition of spin-crossover by a structural distortion in the high spin complex also occurs in iron(II) complexes of pyrazolylborate and pyrazolylmethane 'scorpionate' ligands like $[Fe(HBpz_3)_2]$ (Scheme 5.1);[33, 34] in some salts of the iron(III) complex $[Fe(qsal)_2]^+$;[5] and, in iron(III) complexes of tetradentate $[salen]^{2-}$ and hexadentate $[saltrien]^{2-}$ Schiff bases.[35–38] In contrast to the 1-bpp and terpy systems, the structural distortions that prohibit spin-crossover in these examples involve changes to ligand conformation, rather than the metal coordination geometry. The effect of the distortion on the overall shape of the molecule is the same in each case, however.

5.3 Crystal Packing

The most cooperative spin-transitions will be observed when the structural changes occurring at the molecular level are transmitted most efficiently between spin centres in the bulk material. In a molecular crystal, this requires strong intermolecular contacts between nearest neighbour spin centres. As mentioned above, a full description of this cooperativity should take account of all intermolecular contacts.[39] However, in at least some compounds, one particularly short or strong lattice interaction can be identified that is a major contributor to cooperative behaviour.

While hydrogen bonding and π–π interactions can easily be identified in a lattice, it is unclear to what extent these directional, but weak, secondary interactions contribute to cooperative behaviour in themselves. More likely is that hydrogen bonds and π–π interactions lead to short interatomic contacts between neighbouring molecules, which transmit changes in structure mechanically through the material.

5.3.1 Short Intermolecular Contacts

The first unambiguous relation between crystal packing and cooperativity was demonstrated by Guionneau et al., in a series of complexes $[Fe(NCS)_2(PM\text{-}R)_2]$.[40] When R = aryl, these usually crystallise in one of two, closely related crystal forms showing the same basic crystal packing motif. Both structure types undergo spin-crossover if R is a phenyl or biphenyl-type group, although the cooperativity of the transitions varies strongly between the compounds. There are no π–π interactions between the aromatic arms of the molecules in these structures. Rather, the closest intermolecular interactions are C–H...S contacts from the aromatic 'R' groups of one molecule and the isothiocyanate ligands of its neighbour (Fig. 5.5). Experimentally[40, 41] and theoretically,[42] there is a good linear correlation between the length of these contacts in the crystal, and the width of the transition from magnetic susceptibility measurements. A similar observation was made about a complex from the $[Fe(N_4O_2)L_2]$ series (Chapter 2), that crystallises in two spin-crossover polymorphs. These both undergo the transition abruptly, with thermal hysteresis widths of 5 and 70 K respectively; the latter is one of the widest spin-crossover hysteresis loops known.[43–45] The cooperative form exhibits a short intermolecular hydrogen bond to one of the ligand O-donor atoms that is not present in the gradual polymorph (see Section 5.5.1, below). That could provide an efficient way of transmitting cooperativity directly between the iron centres.[45]

A different relationship was noted for the complexes $[Fe(NCS)_2L_2]$, where L is a *bis*-heterocyclic chelate. Several of these compounds also form an (approximately) isostructural series, where the molecules pack by an interdigitation of the aromatic ligands (Fig. 5.5). Importantly, the cooperativity of the transition for different L ligands perfectly parallels the strength of these stacking interactions:[4]

$$L = \text{btz} < \text{bipy} \approx \text{phen} < \text{dppz}$$

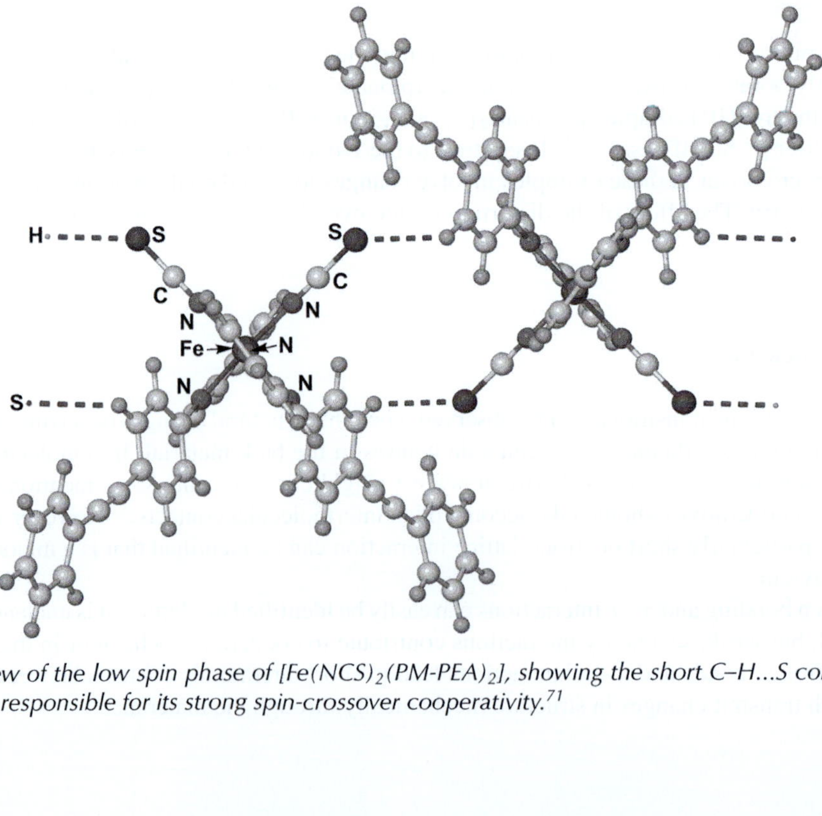

Figure 5.5 *View of the low spin phase of $[Fe(NCS)_2(PM\text{-}PEA)_2]$, showing the short C–H...S contacts that were proposed to be responsible for its strong spin-crossover cooperativity.*[71]

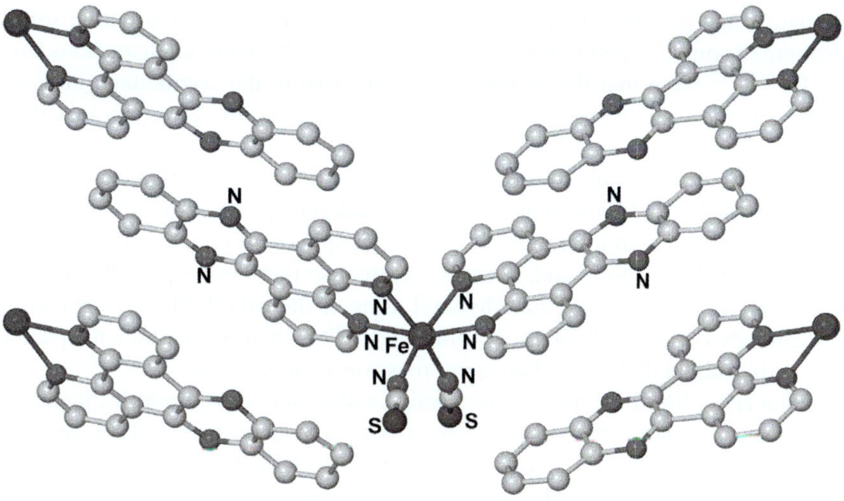

Figure 5.6 View of the high spin form of $[Fe(NCS)_2(dppz)_2]{\cdot}py$, showing the interdigitation of the dppz ligands that gives rise to a 55 K spin-crossover hysteresis loop, without the involvement of a crystallographic phase transition.[46–48]

The most cooperative complex $[Fe(NCS)_2(dppz)_2]$, undergoes its spin-transition around 128 K, with a hysteresis width of up to 55 K depending on the sample history.[46, 47] Notably this transition does not involve a crystallographic phase change, implying that the extensive overlap between the polycyclic ligand backbone is sufficient in itself to afford this wide hysteresis (Fig. 5.6).[48] A comparable relationship has been noted in salts of $[Fe(qsal)_2]^+$, some of which exhibit similarly wide thermal hysteresis. The cations in all these salts associate into co-parallel 1D chains through strong π–π overlap. Spin-crossover cooperativity in these materials appears to correlate with the number of secondary contacts between these chains in the second and third dimension.[49]

An alternative way of expressing a relationship between crystal packing and cooperativity is to identify similarities between structurally related materials. One case where that has been achieved is in the 'terpyridine embrace' lattice type, adopted by several compounds from the $[Fe(1\text{-}bpp)_2]^{2+}$ series as well as some other spin-crossover complexes. This structure involves 2D layers of intimately interdigitated complex cations, with adjacent layers being more weakly linked by van der Waals contacts. Several compounds with this packing motif exhibit essentially identical spin-transitions that occur abruptly with a narrow 2–3 K hysteresis loop.[50] Notably the salt $[Fe(bzimpy)_2][ClO_4]_2$ (bzimpy = 2,6-*bis*{benzimidazol-2-yl}pyridine) also adopts this lattice type, but with strengthened intra-layer nearest-neighbour interactions, and exhibits a wider hysteresis width of 12 K.[51] Hence, the consistent cooperativity in these compounds probably arises from mechanical coupling of structure changes between the molecules in the layers.

5.3.2 Inhibition of Spin-Crossover by Steric Congestion

While strong intermolecular interactions are required to afford a strongly cooperative spin-transition, if the intermolecular environment about a spin centre is too congested it can prevent a spin-transition from taking place. An established example of this is $[Fe(NCS)_2(dppa)_2]$, which crystallises in three polymorphs, two of which exhibit spin-crossover on cooling while the third remains high spin.[52] The inactive polymorph contains a distorted ligand conformation that is enforced by an intermolecular steric contact. This steric crowding inhibits the atomic movements in the crystal that would be associated with a spin-transition.[52, 53] Comparable quenching of spin-crossover by sterically-imposed conformational distortions has also been seen in other compounds.[37, 54, 55] Note that this situation is distinct from the compounds in Section 2.2, where spin-crossover is quenched by a molecular distortion that is intrinsic to the compound itself, not imposed by the lattice.

A more subtle example is seen in complexes of hexadentate podand ligands based on the *tris*-(2-aminoethyl)amine (tren) skeleton. Many such compounds undergo spin-crossover below room temperature, mostly in a gradual manner, but a small number remain high spin at all temperatures. There is no aspect of the molecular structures of these compounds that can rationalise that observation. However, in some cases it can be related to steric crowding about the ethylenyl arms of the tripodal ligands, which undergo a substantial rearrangement between the high and low spin states.[5, 56] If there is insufficient space in the lattice to accommodate this conformational change, then spin-crossover cannot occur.

In other systems that crystallise in spin-crossover-active and -inactive polymorphs, it can be more difficult to identify specific structural features that prevent spin-crossover from occurring. There are several examples, however, where the inactive form has a significantly higher crystallographic density than the active polymorph(s). It may be that the more restricted molecular environment produced by a more efficient, and dense, crystal packing is sufficient to inhibit spin-crossover in these cases.[5]

5.4 Cooperativity Mediated by Disorder

The influence of ligand conformational disorder on cooperativity has already been noted (Section 2.1). One system where cooperativity is known to be controlled by disorder in the lattice solvent is $[Fe(pic)_3]Cl_2{\cdot}ROH$ (pic = 2-{aminomethyl}pyridine), where ROH is one of six different alkyl alcohols.[57] These all adopt the same basic lattice type, involving a 2D network of hydrogen bonding. However, their spin-state behaviour is rather variable. Four of them undergo spin-crossover at approximately the same temperature, gradually (R = Me), in two abrupt steps (R = Et) or with more complicated stepped transition curves (R = *i*Pr or allyl). The other two solvates, with R = *n*Pr and *t*Bu, remain high spin above 5 K. Subsequent studies have implied that this variability is related to the lattice solvent, which is strongly disordered at room temperature. Spin-crossover is coupled to changes in this disorder, and/or to movement of the solvent within the lattice,[58, 59] sometimes leading to crystallographic phase changes affecting the form of the transition.[60, 61] In the *n*PrOH and *t*BuOH solvates, the bulkier alcohols may inhibit these structural rearrangements in steric grounds.

Anion disorder has also been shown to control the form of a spin-transition. An example is $[Fe(bppyz^*)_2]X_2$ ($X^- = BF_4^-$ or ClO_4^-), which exhibit a clear discontinuity in their spin-transitions at 50% conversion (the ClO_4^- salt is compound (e) in Fig. 5.1).[62] This is associated with an abrupt order:disorder transition in one of the two anion sites in the crystal, with the larger perchlorate ion having a more pronounced effect on the spin-transition.[59, 63]

5.5 Compounds Showing Wide Thermal Hysteresis

Molecular compounds exhibiting unusually wide thermal hysteresis, defined arbitrarily as $\geq$30 K, are listed in Table 5.2. The majority of the compounds in the Table exhibit symmetric hysteresis loops between the high and low spin states. However, there are also some examples with more complicated behaviour. These may involve step-wise transitions involving intermediate mixed-spin solid phases, which can occur on only one side of the hysteresis loop (usually the low$\rightarrow$high spin transition when re-warming the sample). Alternatively, hysteresis may only be observed when the spin state transition is scanned on a slow temperature ramp. Much of

Table 5.2 *Molecular spin-crossover complexes showing thermal hysteresis widths ≥30 K that can be repeatedly scanned. The data quoted were obtained from powder samples of the materials. Data from crystalline samples are also given in square brackets where these are appreciably different.*

	$T_{1/2}\downarrow$ [K]	$T_{1/2}\uparrow$ [K]	ΔT [K]	Ref.
$[Fe(3\text{-bpp})_2][CF_3SO_3]_2\cdot H_2O$	147	170, 285[a]	138[a]	83–85
$[Fe(NCS)_2(PM\text{-}PEA)_2]$ at 2.6 kbar	–	–	100	73, 74
$[Fe(pic)_3]Cl_2\cdot H_2O$	199	290[b]	91[b]	78, 79
$[Fe(N_4O_2)(\mu\text{-}PyN{=}NPy)]\cdot CH_3OH$	153	235	82	45
$[Fe(N_4O_2)(ImH)_2]$, HT polymorph	244	314	70	43, 45
$[Fe(qsal)_2]NCSe$	212	215, 282[a]	70[a]	64, 87
$[Fe(1\text{-bpp})_2][Ni(dmit)_2]_2\cdot CH_3NO_2$	175	176, 187, 194, 244[a]	70[a]	15
$[Fe(Me_2\text{-}3\text{-bpp})_2][BF_4]_2$, *in vacuo*	190 [172]	227 [237]	37 [65]	77
$[Fe(NCS)_2(PM\text{-}PEA)_2]$ at 1 bar	195 [194]	255 [231]	60 [37]	71, 72
$[Fe(N_4O_2)(\mu\text{-}4,4'\text{-bipy})]$	205, 245[a]	220, 304[a]	59[a]	88
$[Fe(NCS)_2(Me_2Phen)_2]\cdot 2\text{-}MePy$	146 [153]	202 [195]	56 [42]	75
$[Fe(NCS)_2(dppz)_2]\cdot Py$	100	155	55	46–48
$[Fe(qsal)_2]NCS$	~240, 290[a]	210, 290[a]	~50[a]	64, 87
$[Fe([HOC_6H_4]_2\text{-}3\text{-bpp})_2][ClO_4]_2\cdot H_2O\cdot 2(CH_3)_2CO$	133	173	40	17
$Li[Fe(Brthsa)_2]\cdot H_2O$	294	~260, 333[a]	39[a]	76
$[Fe(HC[pz^{Me3}]_3)_2][BF_4]_2$	105	143	38	82
$[Fe(BzIm\text{-}bzimpy)_2][ClO_4]_2$	306	341	35	65
$[Fe(paptH)_2][NO_3]_2$	263	229[b]	34[b]	80
$[Co(terpyOH)_2][CF_3SO_3]_2\cdot H_2O$, form 2	156	189	33	86
$[Fe(saltrien)][Ni(dmit)_2]$, form *t*-1	228	258	30	18, 19

[a]Transition takes place in more than one step. The hysteresis width is quoted at the widest point of the loop.
[b]This thermal hysteresis is only observed on slow thermal scanning of the material.

this behaviour arises from the coexistence of spin-crossover and one or more crystallographic phase changes, occurring concomitantly or separately during thermal cycling. Indeed, it is often thought that spin-transitions accompanied by changes in crystallographic symmetry are most likely to show hysteresis, because of the activation energy associated with the additional rearrangement of the crystal lattice. However, at least three of the compounds in the Table show wide hysteresis without such a crystallographic phase change.

Importantly, the Table only lists compounds showing wide hysteresis that is reproduced upon multiple thermal scanning. So-called apparent hysteresis is also well-known, where an initial thermal scan causes an irreversible change in the material that is reflected in changes to its spin-state behaviour. This is usually induced by desolvation or decomposition of a sample upon heating, which stabilises its high spin state.[56, 64, 66–69] More rarely, it can also reflect an irreversible crystallographic phase change above room temperature in a solvent free material.[70]

5.5.1 Compounds with Symmetric Hysteresis Loops

Materials with abrupt and complete hysteretic switching are the most desirable for application purposes. The most cooperative materials of this type are two iron(II) complexes of Jäger-type Schiff base ligands (Chapter 2). The complex $[Fe(N_4O_2)(ImH)_2]$ can be obtained in two different polymorphs depending on the temperature of crystallisation. One of these (the 'high temperature' or HT form) exhibits a spin-transition with a 70 K hysteresis loop that, unusually, spans room temperature.[43] The 'low temperature' (LT) polymorph, in

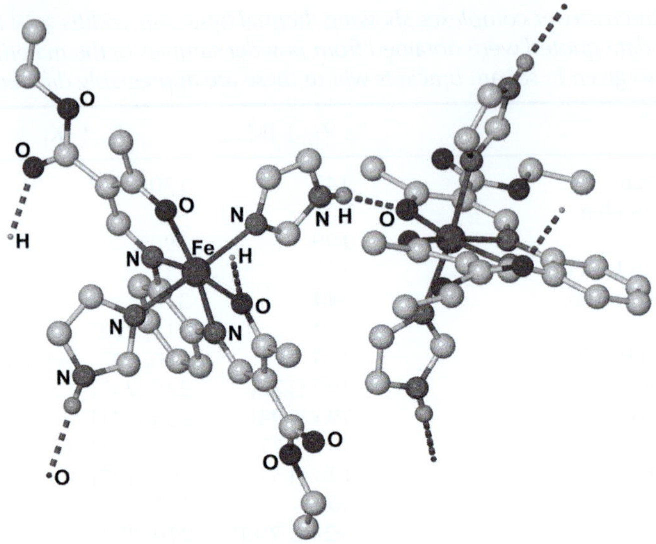

Figure 5.7 View of the HT polymorph of [Fe(N_4O_2)(ImH)$_2$] in its high spin state, showing the short intermolecular N–H...O hydrogen bond that may be the origin of its highly spin-crossover cooperativity.[45] All C-bound H atoms have been omitted for clarity.

contrast, undergoes spin-crossover at a higher temperature, and with a much narrower hysteresis, of 4 K.[44] Crystal structures of both materials are only available in one spin-state, the high spin-state of the HT form and the low spin-state of the LT form. Hence it is unclear whether a crystallographic phase change is the origin of the cooperativity in either material, although this seems likely in the HT form whose crystals crack when they become low spin.

Comparison of the structures of these two phases is instructive, however. Both crystals show the same pattern of intermolecular N–H...O hydrogen bonding involving the imidazole ligands. One of these hydrogen bonds is accepted by a peripheral carboxy group on the Schiff base chelate, and the other by a metal-coordinated O atom (Fig. 5.7). The latter interaction should be particularly effective for transmission of intermolecular cooperativity through the crystal, and is somewhat shorter in the HT crystal [N...O = 2.832(5) Å] than in the LT [2.950(3) Å]. Although the comparison is complicated by the different spin-states in the two crystal structures, this provides a plausible rationale for the increased spin-crossover cooperativity in the HT polymorph.[39, 45] Notably deuteration of the imidazole NH groups in the HT material leads to a small, but measurable, narrowing of the hysteresis loop, implying that hydrogen bonding involving these NH groups does contribute to their cooperative behaviour.[45] No such change is observed upon partial deuteration of the LT structure.

A 1D coordination polymer material from the same family, *catena*-[Fe(N_4O_2)(μ-PyN=NPy)]·CH_3OH, exhibits an even wider spin-transition hysteresis, with $T_{1/2}\downarrow = 153$ and $T_{1/2}\uparrow = 235$ K. There are no direct inter-chain hydrogen bonds in this material, meaning the origin of its cooperativity is less clear cut. However, the high cooperativity is dependent on the presence of the methanol solvent; desolvation of the material affords a powder with rather gradual spin-crossover and no hysteresis, centred at 314 K. The partly deuterated compound *catena*-[Fe(N_4O_2)(μ-PyN=NPy)]·CD_3OD also shows a reduced spin hysteresis width, comparable to that seen for the partially deuterated HT [Fe(N_4O_2)(ImH)$_2$] material. All these observations suggest that the inter-chain cooperativity in the coordination polymer may be mediated by the solvent molecule.[45]

The most cooperative spin-crossover complex that has been structurally characterised in both spin-states, is $[Fe(NCS)_2(PM\text{-}PEA)_2]$.[71] Its spin-crossover depends strongly on the form of the sample. Unusually, the material is more strongly cooperative as a powder than in the single crystal (the opposite trend is more common), with powder samples showing a hysteresis loop of up to 60 K.[72] The crystalline material undergoes a change in space group between its spin-states, associated with the introduction of an additional crystallographic C_2 axis bisecting the molecule in the low spin phase. There are no π–π interactions between the aryl substituents in the crystal. Rather, the cooperativity may be transmitted by close intermolecular C–H...S contacts of 2.8 Å between the phenylene substituents on one molecule and the thiocyanate ligands of its neighbours (Fig. 5.5). There is a good linear relationship between the length of this contact and the spin-crossover cooperativity in other compounds related to this one (Section 5.3.1). Intriguingly, placing $[Fe(NCS)_2(PM\text{-}PEA)_2]$ under 2.6 kbar of hydrostatic pressure increases its hysteresis loop to around 100 K width,[73] which may reflect a pressure-induced phase change in the material (as has been observed in a closely related complex[74]).

Although they have not been crystallographically characterised, other complexes where symmetrical, hysteretic spin-crossover is known to be accompanied by a crystallographic phase change are $[Fe(NCS)_2(Me_2Phen)_2]\cdot 2\text{-}MePy$,[75] $Li[Fe(Brthsa)_2]\cdot H_2O$[76] and the low pressure polymorph of $[Fe(Me_2\text{-}3\text{-}bpp)_2][BF_4]_2$.[77] The latter compound is noteworthy, in that the spin-crossover material is only obtained upon *in vacuo* dehydration of a hydrate phase of the complex. There are also two compounds where wide hysteresis is only observed upon slow thermal scanning of the sample, $[Fe(pic)_3]Cl_2\cdot H_2O$[78, 79] and $[Fe(paptH)_2][NO_3]_2$.[80] A careful calorimetry study of $[Fe(pic)_3]Cl_2\cdot H_2O$ showed that this behaviour reflects a kinetically slow crystallographic phase change in the sample, occurring around 200 K. If the temperature of the sample is scanned too quickly, there is insufficient time for this phase change to occur and the hysteresis loop collapses.[79] It is unknown whether a similar phenomenon underlies the behaviour of $[Fe(paptH)_2][NO_3]_2$, but it seems likely. A comparable, but less pronounced, dependence of spin-crossover on a slow crystallographic phase change has also been seen in at least one other material.[81]

In contrast, the complexes $[Fe(NCS)_2(dppz)_2]\cdot Py$ (Fig. 5.6),[48] $[Fe([HOC_6H_4]_2\text{-}3\text{-}bpp)_2][ClO_4]_2\cdot H_2O\cdot 2(CH_3)_2CO$ (Fig. 5.3)[17] and one polymorph of $[Fe(saltrien)][Ni(dmit)_2]$[18] show 30–60 K hysteresis without a phase change between the spin states. All three compounds exhibit a significant migration of their aryl substituents around the metal ion between the high and low spin phases, because of a change in ligand conformation or a change in coordination geometry (Section 5.2.1). Although they do not affect the crystal symmetry, such molecular rearrangements lead to atomic displacements of up to 1.5 Å within the crystal lattice that are comparable to a full crystallographic phase change. Thus, the rationale of hysteresis associated with large structural changes in the lattice also applies in these cases. In addition to this conformational rearrangement, changes to perchlorate anion disorder between the spin states of $[Fe([HOC_6H_4]_2\text{-}3\text{-}bpp)_2][ClO_4]_2\cdot H_2O\cdot 2(CH_3)_2CO$ may also be relevant to its thermal hysteresis.[17]

Finally, there are two more compounds in Table 5.2 showing symmetric spin-crossover hysteresis loops of 35–40 K by susceptibility and/or Mössbauer measurements, but with no available structural data of any type. These are $[Fe(HC[pz^{Me3}]_3)_2][BF_4]_2$[82] and $[Fe(BzIm\text{-}bzimpy)_2][ClO_4]_2$.[65]

5.5.2 Compounds with Structured Hysteresis Loops

A few compounds in Table 5.2 show more complicated, stepped hysteresis loops, which may be symmetric (steps occurring equally in both cooling and warming mode) or unsymmetric (steps only observed on cooling or on warming the sample). One such example is $[Fe(3\text{-}bpp)_2][CF_3SO_3]_2\cdot H_2O$, whose high→low spin transition occurs abruptly in one step. However, the reverse low→high spin transition takes place in two steps separated by a wide thermal plateau, where ca. one-third of the molecules are high spin.[83, 84] At its widest point this hysteresis loop spans 138 K, which is the largest known hysteresis for a molecular material. No single crystal or powder X-ray diffraction data are available, but EPR linewidth measurements on a doped $[Fe,Mn(3\text{-bpp})_2][CF_3SO_3]_2\cdot H_2O$ sample concluded that a crystallographic phase change takes place during the initial high→low spin transition.[85]

More structural detail is available from $[Fe(1\text{-bpp})_2][Ni(dmit)_2]_2\cdot CH_3NO_2$, whose hysteresis loop is even more structured.[15] This complex undergoes an abrupt and complete high→low spin transition on cooling, as before, but the low→high spin process now occurs in four discrete steps leading to a maximum hysteresis width of 70 K. Crystal structure determinations were achieved on the high spin and low spin crystals, and on two of the three intermediate mixed spin phases. These show that the initial high→low spin transition of the iron centre is accompanied by dimerisation of the $[Ni(dmit)_2]^{\bullet -}$ radical anions through intermolecular S...S interactions (this type of dimerisation is commonly found in $[Ni(dmit)_2]^{\bullet -}$ salts). In warming mode these anion dimers dissociate in step-wise fashion, via intermediate phases containing a combination of monomer and dimer anion assemblies and mixed high and low spin iron sites. Unusually large changes in the molecular conformation of the $[Fe(1\text{-bpp})_2]^{2+}$ cation during spin-crossover may also play a role in this unusual cooperativity (Section 5.2.1).[15]

A rather different susceptibility curve is presented by polymorph 2 of $[Co(terpyOH)_2][CF_3SO_3]_2\cdot H_2O$, whose spin-crossover has two abrupt components overlaid onto a more gradual transition baseline curve.[86] Both the abrupt steps show thermal hysteresis and, very unusually, one of them involves an 'inverse' low→high spin event on cooling. Although a structure determination of the low spin phase was not achieved, the inverse transition step involves a crystallographic phase change, with no change in space group but a doubling of the unit cell volume. The increased entropy associated with the expansion of the unit cell, together with any change in anion or solvent disorder, is evidently sufficient to overcome the unfavourable electronic entropy of the inverse spin-crossover at the cobalt centres.[86] Such entropy-driven inverse spin-transitions have also been observed in other $[Co(terpy)_2]^{2+}$ derivatives, some of which also show wide hysteresis. These are discussed in Chapter 12.

The iron(III) salts $[Fe(qsal)_2]X$ ($X^- = NCS^-$ or $NCSe^-$) are isostructural at room temperature. Their thermal spin-transitions are also broadly similar, with hysteresis loops resembling that in $[Fe(3\text{-bpp})_2][CF_3SO_3]_2\cdot H_2O$ (see above) in showing a pronounced step on the warming curve only.[64, 87] Although this stepped hysteresis was simulated for the selenocyanate salt using a simple lattice elasticity model,[64] no structural data were reported to relate the results of this simulation to the structural chemistry of the material. Finally, a polymorph of the 1D coordination polymer *catena*-$[Fe(N_4O_2)(\mu\text{-}4,4'\text{-bipy})]$, obtained by desolvating a crystalline methanol solvate of the compound, undergoes spin-crossover in two steps with respective hysteresis widths of 15 K and 59 K.[88] A crystal structure of the precursor methanol solvate phase was reported, but showed no unusual features that would explain this observation.

5.6 Other Noteworthy Compounds

5.6.1 Iron(II) Triazole Coordination Polymers

Complexation of iron(II) salts with 1,2,4-triazoles yields linear coordination polymers of type $[Fe(Rtz)_3]X_2\cdot nH_2O$ (R = H, NH_2, C_2H_4OH or another substituent; X^- = anion) or, when R = H, $[Fe(tz)(Htz)_2]X\cdot nH_2O$ (Chapter 4).[89, 90] These are the most important class of spin-crossover material for application purposes. They routinely undergo spin-crossover with thermal hysteresis loops of 20–40 K[89–93] (although examples with ΔT up to 50 K are known[94]), whose temperature can be controlled by variation of the triazole substituent,[92] the anion[93] and the water content. These were the first spin-crossover materials to be prepared showing room temperature hysteresis, which has made them very important for device applications.[91]

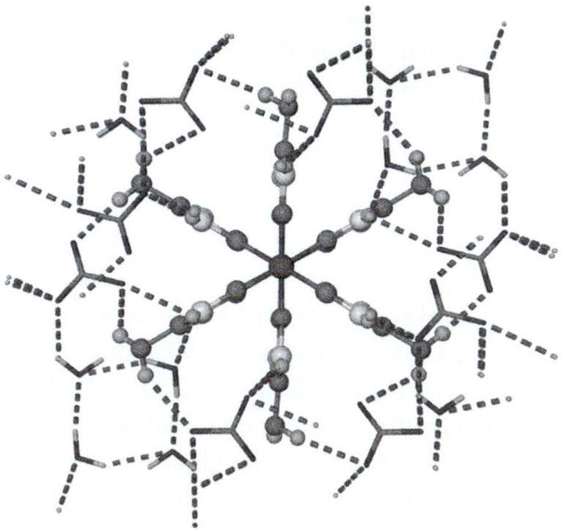

Figure 5.8 *View of the crystal structure of [Fe(atz)$_3$][NO$_3$]$_2$·2H$_2$O, showing the hydrogen bond network around the polymer coordination chain.[98] The view is along the Fe...Fe vector. The nitrate anions and water molecules are de-emphasised for clarity.*

Although their technological importance was first recognised 20 years ago,[91] their poorly crystalline nature meant that few structural data on these materials were available. The first structure determinations were of analogous copper(II) coordination polymers $[[Cu(Rtz)_3]]_nX_{2n}$,[93–96] which showed the expected structural features but were not necessarily isostructural with their iron(II) congeners. However, very recently two structure determinations of the iron materials have been published, from single crystal or powder diffraction data.[97, 98] A common feature of all these iron and copper structures is the absence of any direct contact between the coordination polymer chains in the material. Rather, there is extensive hydrogen bonding between the chains and the anions and water, involving the acidic triazole C–H groups as well as the peripheral substituents (Fig. 5.8). Since hysteretic spin-transitions require cooperativity between the chains, as well as within them, this hydrogen bonding must be an important mediator of inter-chain interactions. Mechanical coupling of structure changes on neighbouring chains via the hydrogen bonded triazole C–H groups is likely to be particularly efficient, since these are adjacent to the ligand N-donor atoms (c.f. $[Fe(N_4O_2)(Him_2)]$, Section 5.5.1).

5.6.2 Cooperative Complexes of Other Metal Ions

Cooperative behaviour is much rarer in complexes of metal ions other than iron(II), because of the smaller structural changes between their high and low spin states (Table 5.1). Individual examples of highly cooperative iron(III) and cobalt(II) complexes have been mentioned in Sections 5.2.1, 5.5.1 and 5.5.2. Other examples with weaker cooperativity include the iron(III) complexes [Fe(salen)NO],[99, 100] $[Fe(pap)_2]X$ ($X^- = BF_4^-$ and ClO_4^-)[87] and [Fe(thpu)(Hthpu)],[101] and the cobalt(II) compound $[Co(dpzca)_2]$,[102] which all exhibit thermal hysteresis loops of ca. 10 K. This weaker cooperativity is still noteworthy, in the context of the chemistry of these metal ions. Importantly the above compounds, and the examples in Table 5.2, all contain approximately planar polydentate ligands, with predominantly aromatic donor groups. That is consistent with the discussion in Sections 5.2 and 5.3.1, where cooperativity can be promoted by changes in conformation of rigid aromatic ligands between the spin states. A last example is the unusual, two-coordinate organometallic complex $[Co(NHC_6H_3\text{-}2,6\text{-}[C_6H_2\text{-}2,4,6\text{-}Me_3]_2)(C_6H_3\text{-}2,6\text{-}[C_6H_3\text{-}2,6\text{-}iPr_2]_2)]$, which exhibits an 8 K hysteresis loop.[103] The origin of this cooperativity may be the steric consequences of slippage of an aromatic substituent on the amido ligand during the transition. This forms an additional weak Co...C contact to the cobalt ion that has a different hapticity in the two spin states.

5.7 Conclusions

The compounds in Table 5.2 with hysteresis widths > 50 K are the most cooperative spin-transition materials known. Importantly, these molecular complexes show wider spin-crossover hysteresis than any known polymer or framework material (Chapter 4), showing that covalently linking the spin centres is not always the most effective way to transmit cooperativity through a lattice. Rather, the examples discussed in this chapter imply that two structural features are particularly effective at mediating cooperativity in a molecular crystal. First is a strongly aspherical molecular shape that allows neighbouring molecules in the lattice to interdigitate effectively. Planar or wedge shaped aromatic ligand donor groups associating through intermolecular π–π interactions are particularly efficient in this regard (Fig. 5.6), not because of the π–π interactions (which are weak), but because they present a large contact surface area between neighbouring molecules. This allows changes in molecular shape during spin-crossover (Figs 5.2 and 5.3) to be propagated between molecules more efficiently. Another structural feature that appears to work equally well is intermolecular hydrogen bonding, or other close steric contacts, involving ligand donor atoms (Fig. 5.7), or groups immediately adjacent to them (Figs 5.5 and 5.8). These will transmit changes between the iron spin centres more effectively than hydrogen bonds at the periphery of the molecule, which have little impact on the steric environment of the metal ion.

References

1. Real, J. A., Gaspar, A. B., Muñoz, M. C. (2005). Thermal, pressure and light switchable spin-crossover materials. Dalton Trans., 2062–2079.
2. Hauser, A., Jeftić, J., Romstedt, H., Hinek, R., Spiering, H. (1999) Cooperative phenomena and light-induced bistability in iron(II) spin-crossover compounds. Coord. Chem. Rev., 190–192: 471–491.
3. Muñoz, M. C., Real, J. A. (2011) Thermo-, piezo-, photo- and chemo-switchable spin crossover iron(II)-metallocyanate based coordination polymers. Coord. Chem. Rev., 255: 2068–2093.
4. Real, J. A., Gaspar, A. B., Niel, V., Muñoz, M. C. (2003) Communication between iron(II) building blocks in cooperative spin transition phenomena. Coord. Chem. Rev., 236: 121–141.
5. Halcrow, M. A. (2011) Structure : function relationships in molecular spin-crossover complexes. Chem. Soc. Rev., 40: 4119–4142.
6. Tao, J., Wei, R.-J., Huang, R.-B., Zheng, L.-S. (2012) Polymorphism in spin-crossover systems. Chem. Soc. Rev., 41: 703–737.
7. Lazar, H. Z., Forestier, T., Barrett, S. A., Kilner, C. A., Létard, J.-F., Halcrow, M. A. (2007) Thermal and light-induced spin-crossover in salts of the heptadentate complex [tris(4-{pyrazol-3-yl}-3-aza-3-butenyl)amine]iron(II). Dalton Trans., 4276–4285.
8. Roberts, T. D., Tuna, F., Malkin, T. L., Kilner, C. A., Halcrow, M. A. (2012) An iron(II) complex exhibiting five anhydrous phases, two of which interconvert by spin-crossover with wide hysteresis. Chem. Sci., 3: 349–354.
9. Money, V. A. , Elhaïk, J., Evans, I. R., Halcrow, M. A., Howard, J. A. K. (2004) A study of the thermal and light-induced spin transition in $[FeL_2](BF_4)_2$ and $[FeL_2](ClO_4)_2$ L = 2,6-di(3-methylpyrazol-1-yl)pyrazine. Dalton Trans., 65–69.
10. Elhaïk, J., Mohammed, R., Halcrow, M. A. (2012). Unpublished data.

11. Halcrow, M. A. (2007) The spin-states and spin-transitions of mononuclear iron(II) complexes of nitrogen-donor ligands. Polyhedron, 26: 3523–3576.
12. Nihei, M., Shiga, T., Maeda, Y., Oshio, H. (2007) Spin crossover iron(III) complexes. Coord. Chem. Rev., 251: 2606–2621.
13. König, E. (1987) Structural changes accompanying continuous and discontinuous spin-state transitions. Prog. Inorg. Chem., 35: 527–622.
14. Halcrow, M. A. (2009) Iron(II) complexes of 2,6-di(pyrazolyl)pyridines– a versatile system for spin-crossover research. Coord. Chem. Rev., 253: 2493–2514.
15. Nihei, M., Tahira, H., Takahashi, N., Otake, Y., Yamamura, Y., Saito, K., Oshio, H. (2010) Multiple bistability and tristability with dual spin-state conversions in $[Fe(dpp)_2][Ni(mnt)_2]_2 \cdot MeNO_2$. J. Am. Chem. Soc., 132: 3553–3560.
16. Money, V. A., Carbonera, C., Elhaïk, J., Halcrow, M. A., Howard, J. A. K., Létard, J.-F. (2007) Interplay between kinetically slow thermal spin-crossover and metastable high-spin state relaxation in an iron(II) complex with similar $T_{1/2}$ and *T*(LIESST). Chem. Eur. J., 13: 5503–5514.
17. Craig, G. A., Costa, J. S., Roubeau, O., Teat, S. J., Aromí, G. (2011). Coupled crystallographic order–disorder and spin state in a bistable molecule: multiple transition dynamics. Chem. Eur. J., 17: 3120–3127.
18. Dorbes, S., Valade, L., Real, J. A., Faulmann, C. (2005). $[Fe(sal_2\text{-trien})][Ni(dmit)_2]$: towards switchable spin crossover molecular conductors. Chem. Commun., 69–71.
19. Faulmann, C., Szilágyi, P. A., Jacob, K., Chahine, J., Valade, L. (2009) Polymorphism and its effects on the magnetic behaviour of the $[Fe(sal_2\text{-trien})][Ni(dmit)_2]$ spin-crossover complex. New J. Chem., 33: 1268–1276.
20. Matouzenko, G. S., Luneau, D., Molnár, G., Ould-Moussa, N., Zein, S., Borshch, S. A., Bousseksou, A. and Averseng, F. (2006) A two-step spin transition and order–disorder phenomena in the mononuclear compound $[Fe(Hpy\text{-}DAPP)](BF_4)_2$. Eur. J. Inorg. Chem., 2671–2682.
21. Matouzenko, G. S., Bousseksou, A., Borshch, S. A., Perrin, M., Zein, S., Salmon, L., Molnár, G., Lecocq, S. (2004) Inorg. Chem., 43: 227–236.
22. Brefuel, N., Shova, S., Lipkowski, J., Tuchagues, J.-P. (2006) Fe^{II} bi-stable materials based on dissymmetrical ligands: N_4 Schiff bases including 2-pyridyl and 5-methylimidazol-4-yl rings yield various Fe^{II} spin-crossover phenomena around 300 K. Chem. Mater., 18: 5467–5479.
23. Nelson, S. M., Mcllroy, P. D. A., Stevenson, C. S., König, E., Ritter, G., Waigel, J. (1986) Quadridentate *versus* quinquedentate co-ordination of some N_5 and N_3O_2 macrocyclic ligands and an unusual thermally controlled quintet $\rightleftharpoons$ singlet spin transition in an iron(II) complex. J. Chem. Soc. Dalton Trans., 991–995.
24. Guionneau, P., Costa, J. S., Létard, J.-F. (2004) Revisited crystal symmetry of the high-spin form of the iron(II) spin-crossover complex dicyano[2,13-dimethyl-6,9-dioxa-3,12,18-triazabicyclo[12.3.1]-octadeca-1(18),2,12,14,16-pentaene]iron(II) monohydrate. Acta Cryst. Sect. C, 60: m587–m589.
25. Guionneau, P., Le Gac, F., Kaiba, A., Costa, J. S., Chasseau, D., Létard, J.-F. (2007) A reversible metal–ligand bond break associated to a spin-crossover. Chem. Commun., 3723–3725.
26. König, E., Ritter, G., Dengler, J., Nelson, S. M. (1987) Detailed study of a two-step quintet $\rightleftharpoons$ singlet spin transition in an iron(II) complex with a N_3O_2 macrocyclic ligand and the kinetics of the quintet $\rightarrow$ singlet relaxation in the temperature range 115–130 K. Inorg. Chem., 26: 3582–3588.
27. Costa, J. S., Guionneau, P., Létard, J.-F. (2005) Photomagnetic properties of the $[Fe(L_{222}(N_3O_2))(CN)_2]\cdot H_2O$ complex : a fascinating example of multi-metastability. J. Phys. Conf. Ser., 21: 67–72.
28. Holland, J. M., McAllister, J. A., Kilner, C. A., Thornton-Pett, M., Bridgeman, A. J., Halcrow, M. A. (2002) Stereochemical effects on the spin-state transition shown by salts of $[FeL_2]^{2+}$ [L = 2,6-di(pyrazol-1-yl)pyridine]. J. Chem. Soc. Dalton Trans., 548–554.
29. Elhaïk, J., Evans, D. J., Kilner, C. A., Halcrow, M. A. (2005) A structural, magnetic and Mössbauer spectroscopic study of an unusual angular Jahn-Teller distortion in a series of high-spin iron(II) complexes. Dalton Trans., 1693–1700.
30. Haryono, M., Heinemann, F. W., Petukhov, K., Gieb, K., Müller, P., Grohmann, A. (2009). Parallel crystallization of a "static" and a spin-crossover polymorph of an iron(II) complex from the same solution. Eur. J. Inorg. Chem., 2136–2143.

31. Constable, E. C., Baum, G., Bill, E., Dyson, R., van Eldik, R., Fenske, D., Kaderli, S., Morris, D., Neubrand, A., Neuburger, M., Smith, D. R., Wieghardt, K., Zehnder, M., Zuberbühler, A. D. (1999) Control of iron(II) spin states in 2,2′:6′,2″-terpyridine complexes through ligand substitution. Chem. Eur. J., 5: 498–508.
32. Brauchli, S. Y., Constable, E. C., Harris, K., Häussinger, D., Housecroft, C. E., Rösel, P. J., Zampese, J. A. (2010) Towards catenanes using π-stacking interactions and their influence on the spin-state of a bis(2,2′:6′,2″-terpyridine)iron(II) domain. Dalton Trans., 39: 10739–10748.
33. Reger, D. L., Little, C. A., Smith, M. D., Rheingold, A. L., Lam, K. C., Concolino, T. L., Long, G. J. , Hermann, R. P., Grandjean, F. (2002) Synthetic, structural, magnetic, and Mössbauer spectral study of $\{Fe[HC(3,5\text{-}Me_2pz)_3]_2\}I_2$ and its spin-state crossover behavior. Eur. J. Inorg. Chem., 1190–1197.
34. Reger, D. L., Gardinier, J. R., Elgin, J. D., Smith, M. D., Hautot, D., Long, G. J., Grandjean, F. (2006) Structure-function correlations in iron(II) tris(pyrazolyl)borate spin-state crossover complexes. Inorg. Chem., 45: 8862–8875.
35. Kennedy, B. J., McGrath, A. C., Murray, K. S., Skelton, B. W., White, A. H. (1987) Variable-temperature magnetic, spectral, and x-ray crystallographic studies of "spin-crossover" iron(III) Schiff-base-Lewis-base adducts. Influence of noncoordinated anions on spin-state interconversion dynamics in [Fe(salen)(imd)2]Y species ($Y = ClO_4^-$, BF_4^-, PF_6^-, BPh_4^-; imd = imidazole). Inorg. Chem., 26: 483–495.
36. Ross, T. M., Neville, S. M., Innes, D. S., Turner, D. R., Moubaraki, B., Murray, K. S. (2010) Spin crossover in iron(III) Schiff-base 1-D chain complexes. Dalton Trans., 39: 149–159.
37. Pritchard, R., Barrett, S. A., Kilner, C. A., Halcrow, M. A. (2008) The influence of ligand conformation on the thermal spin transitions in iron(III) saltrien complexes. Dalton Trans., 3159–3168.
38. Pritchard, R., Barrett, S. A., Kilner, C. A., Halcrow, M. A. (2009) The influence of ligand conformation on the thermal spin transitions in iron(III) saltrien complexes–correction. Dalton Trans., 10622.
39. Pfaffeneder, T. M., Thallmair, S., Bauer, W., Weber, B. (2011) Complete and incomplete spin transitions in 1D chain iron(II) compounds. New J. Chem., 35: 691–700.
40. Guionneau, P., Marchivie, M., Bravic, G., Létard, J.-F., Chasseau, D. (2004) Structural aspects of spin crossover. Example of the $[Fe^{II}L_n(NCS)_2]$ complexes, in: Gütlich, P., Goodwin, H. A. (Eds) Spin Crossover in Transition Metal Compounds II. Top. Curr. Chem., 234: 97–128.
41. Marchivie, M., Guionneau, P., Létard, J.-F., Chasseau, D. (2003) Towards direct correlations between spin-crossover and structural features in iron(II) complexes. Acta Cryst. Sect. B, 59: 479–486.
42. Kepenekian, M., Le Guennic, B., Robert, V. (2009) Primary role of the electrostatic contributions in a rational growth of hysteresis loop in spin-crossover Fe(II) complexes. J. Am. Chem. Soc., 131: 11498–11502.
43. Weber, B., Bauer, W., Obel, J. (2008) An iron(II) spin-crossover complex with a 70 K wide thermal hysteresis loop. Angew. Chem. Int. Ed. 47: 10098–10101.
44. Müller, B. R., Leibeling, G., Jäger, E.-G. (2000) Cooperative magnetic behavior of self-assembled iron(II) chelate chain compounds. Chem. Phys. Lett., 319: 368–374.
45. Weber, B., Bauer, W., Pfaffeneder, T., Dîrtu, M. M., Naik, A. D., Rotaru, A., Garcia, Y. (2011) Influence of hydrogen bonding on the hysteresis width in iron(II) spin-crossover complexes. Eur. J. Inorg. Chem., 3193–3206.
46. Zhong, Z. J., Tao, J.-Q., Yu, Z., Dun, C.-Y., Liu, Y.-J., You, X.-Z. (1998) A stacking spin-crossover iron(II) compound with a large hysteresis. Dalton Trans., 327–328.
47. Yu, Z., Liu, K., Tao, J. Q., Zhong, Z. J., You, X. Z., Siu, G. G. (1999) Rapid cooling and magnetic field-induced cooperative effect for the metastable quintet state in a spin crossover complex. Appl. Phys. Lett., 74: 4029–4031.
48. Shepherd, H. J., Palamarciuc, T., Rosa, P., Guionneau, P., Molnár, G., Létard, J.-F., Bousseksou, A. (2012). Antagonism between extreme negative linear compression and spin crossover in $[Fe(dpp)_2(NCS)_2]\cdot py$. Angew. Chem. Int. Ed., 51: 3910–3914.
49. Takahashi, K., Sato, T., Mori, H., Tajima, H., Sato, O. (2010) Correlation between the magnetic behaviors and dimensionality of intermolecular interactions in Fe(III) spincrossover compounds. Physica B, 405: S65–S68.
50. Pritchard, R., Kilner, C. A., Halcrow, M. A. (2007) Iron(II) complexes with a terpyridine embrace packing motif show remarkably consistent cooperative spin-transitions. Chem. Commun., 577–579.
51. Boča, R., Boča, M., Dlháň, L., Falk, K., Fuess, H., Haase, W., Jaroščiak, R., Papánková, B., Renz, F., Vrbová, M., Werner, R. (2001) Strong cooperativeness in the mononuclear iron(II) derivative exhibiting an abrupt spin transition above 400 K. Inorg. Chem., 40: 3025–3033.

52. Matouzenko, G. S., Bousseksou, A., Lecocq, S., van Koningsbruggen, P. J., Perrin, M., Kahn, O., Collet, A. (1997) Polymorphism in spin transition systems. Crystal structure, magnetic properties, and Mössbauer spectroscopy of three polymorphic modifications of [Fe(DPPA)(NCS)$_2$] [DPPA = (3-aminopropyl)bis(2-pyridylmethyl)amine]. Inorg. Chem., 36: 5869–5879.
53. Zein, S., Matouzenko, G. S., Borshch, S. A. (2005) Quantum chemical study of three polymorphs of the mononuclear spin-transition complex [Fe(DPPA)(NCS)$_2$]. J. Phys. Chem. A, 109: 8568–8571.
54. Elhaïk, J., Kilner, C. A., Halcrow, M. A. (2006) Structural diversity in iron(II) complexes of 2,6-di(pyrazol-1-yl)pyridine and 2,6-di(3-methylpyrazol-1-yl)pyridine. Dalton Trans., 823–830.
55. Dias, J. C., Soriano-Portillo, A., Clemente-Léon, M., Giménez-Saiz, C., Galán-Mascarós, J. R., Gómez-García, C. J., Coronado, E. (2007). Inorg. Chim. Acta, 360: 3843–3847.
56. Lazar, H. Z., Forestier, T., Barrett, S. A., Kilner, C. A., Létard, J.-F., Halcrow, M. A. (2007) Thermal and light-induced spin-crossover in salts of the heptadentate complex [tris(4-{pyrazol-3-yl}-3-aza-3-butenyl)amine]iron(II). Dalton Trans., 4276–4285.
57. Hostettler, M., Törnroos, K. W., Chernyshov, D., Vangdal, B., Bürgi, H.-B. (2004) Challenges in engineering spin crossover: structures and magnetic properties of six alcohol solvates of iron(II) tris(2-picolylamine) dichloride. Angew. Chem. Int. Ed., 43: 4589–4594.
58. Chernyshov, D., Vangdal, B., Törnroos, K. W., Bürgi, H.-B. (2009) Chemical disorder and spin crossover in a mixed ethanol–2-propanol solvate of FeII tris(2-picolylamine) dichloride. New J. Chem., 33: 1277–1282.
59. Chernyshov, D., Klinduhov, N., Törnroos, K. W., Hostettler, M., Vangdal, B., Bürgi, H.-B. (2007) Coupling between spin conversion and solvent disorder in spin crossover solids. Phys. Rev. B, 76: 014406/1–014406/7.
60. Chernyshov, D., Hostettler, M., Törnroos, K. W., Bürgi, H.-B. (2003) Ordering phenomena and phase transitions in a spin-crossover compound–uncovering the nature of the intermediate phase of [Fe(2-pic)$_3$]Cl$_2$·EtOH. Angew. Chem. Int. Ed., 42: 3825–3830.
61. Törnroos, K. W., Hostettler, M., Chernyshov, D., Vangdal, B., Bürgi, H.-B. (2006) Interplay of spin conversion and structural phase transformations: re-entrant phase transitions in the 2-propanol solvate of tris(2-picolylamine)iron(II) dichloride. Chem. Eur. J., 12: 6207–6215.
62. Elhaïk, J., Money, V. A., Barrett, S. A., Kilner, C. A., Evans, I. R., Halcrow, M. A. (2003) The spin-states and spin-crossover behaviour of iron(II) complexes of 2,6-dipyrazol-1-ylpyrazine derivatives. Dalton Trans., 2053–2060.
63. Money, V. A., Elhaïk, J., Evans, I. R., Halcrow, M. A., Howard, J. A. K. (2004) A study of the thermal and light-induced spin transition in [FeL$_2$](BF$_4$)$_2$ and [FeL$_2$](ClO$_4$)$_2$ L = 2,6-di(3-methylpyrazol-1-yl)pyrazine. Dalton Trans., 65–69.
64. Hayami, S., Gu, Z.-z., Yoshiki, H., Fujishima, A., Sato, O. (2001) Iron(III) spin-crossover compounds with a wide apparent thermal hysteresis around room temperature. J. Am. Chem. Soc., 123: 11644–11650.
65. Boča, R., Renz, F., Boča, M., Fuess, H., Haase, W., Kickelbick, G., Linert, W., Vrbová-Schikora, M. (2005) Tuning the spin crossover above room temperature: iron(II) complexes of substituted and deprotonated 2,6-bis(benzimidazol-2-yl)pyridine. Inorg. Chem. Commun., 8: 227–230.
66. Clemente-León, M., Coronado, E., Giménez-López, M. C., Romero, F. M., Asthana, S., Desplanches, C., Létard, J.-F. (2009) Structural, thermal and photomagnetic properties of spin crossover [Fe(bpp)$_2$]$^{2+}$ salts bearing [Cr(L)(ox)$_2$]$^-$ anions. Dalton Trans., 8087–8095.
67. Šalitroš, I., Pavlik, J., Boča, R., Fuhr, O., Rajadurai, C., Ruben, M. (2010) Supramolecular lattice-solvent control of iron(II) spin transition parameters. Cryst. Eng. Comm., 12: 2361–2368.
68. Zhang, L., Xu, G.-C., Xu, H.-B., Zhang, T., Wang, Z.-M., Yuan, M., Gao, S. (2010) Abrupt spin transition around room temperature and light induced properties in Fe^{II} complexes with N_4O_2 coordination sphere. Chem. Commun., 46: 2554–2556.
69. Ostermeier, M., Berlin, M.-A., Meudtner, R. M., Demeshko, S., Meyer, F., Limberg, C., Hecht, S. (2010) Complexes of click-derived bistriazolylpyridines: remarkable electronic influence of remote substituents on thermodynamic stability as well as electronic and magnetic properties. Chem. Eur. J., 16: 10202–10213.
70. Salmon, L., Molnár, G., Cobo, S., Oulié, P., Etienne, M., Mahfoud, T., Demont, P., Eguchi, A., Watanabe, H., Tanaka, K., Bousseksou, A. (2009). Re-investigation of the spin crossover phenomenon in the ferrous complex [Fe(HB(pz)$_3$)$_2$]. New J. Chem., 33: 1283–1289.

71. Létard, J.-F., Guionneau, P., Codjovi, E., Lavastre, O., Bravic, G., Chasseau, D., Kahn, O. (1997). Wide thermal hysteresis for the mononuclear spin-crossover compound cis-bis(thiocyanato)bis[*N*-(2′-pyridylmethylene)-4-(phenylethynyl)anilino]iron(II). J. Am. Chem. Soc., 119: 10861–10862.
72. Daubric, H., Cantin, C., Thomas, C., Kliava, J., Létard, J.-F., Kahn, O. (1999) Spin transition with a very large thermal hysteresis in a molecular crystal: an EPR study of $Fe(PM\text{-}PEA)_2(NCS)_2$. Chem. Phys., 244: 75–88.
73. Ksenofontov, V., Levchenko, G., Spiering, H., Gütlich, P., Létard, J.-F., Bouhedja, Y., Kahn, O. (1998) Spin crossover behaviour under pressure of $Fe(PM\text{-}L)_2(NCS)_2$ compounds with substituted 2′-pyridylmethylene-4-anilino ligands. Chem. Phys. Lett., 294: 545–553.
74. Rotaru, A., Varret, F., Codjovi, E., Boukheddaden, K., Linares, J., Stancu, A., Guionneau, P., Létard, J.-F. (2009) Hydrostatic pressure investigation of the spin crossover compound $[Fe(PM-BiA)_2(NCS)_2]$ polymorph I using reflectance detection. J. Appl. Phys., 106: 053515/1–053515/5.
75. König, E., Ritter, G., Kulshreshtha, S. K., Csatary, N. (1984) Mössbauer effect and X-ray diffraction at the high-spin ($^5T_2$) $\rightleftharpoons$ low-spin ($^1A_1$) transition in bis(thiocyanato)bis(4,7-dimethyl-l,l0-phenanthroline)iron(II)-α-picoline: thermal hysteresis, associated crystallographic phase change, time dependence of $^5T_2 \rightarrow {}^1A_1$ transformation, particle size effects, and related phenomena. Inorg. Chem., 23: 1903–1910.
76. Floquet, S., Boillot, M.-L., Rivière, E., Varret, F., Boukheddaden, K., Morineauc, D., Négrierd, P. (2003) Spin transition with a large thermal hysteresis near room temperature in a water solvate of an iron(III) thiosemicarbazone complex. New J. Chem., 27: 341–348.
77. Roberts, T. D., Tuna, F., Malkin, T. L., Kilner, C. A., Halcrow, M. A. (2012) An iron(II) complex exhibiting five anhydrous phases, two of which interconvert by spin-crossover with wide hysteresis. Chem. Sci., 3: 349–354.
78. Sorai, M., Ensling, J., Hasselbach, R. M., Gütlich, P. (1977). Mössbauer effect study on low-spin $^1A_1 \rightleftharpoons$ high-spin $^5T_2$ transition in $[Fe(2\text{-}pic)_3]Cl_2$. II. Influence of non-coordinating solvent molecule in $[Fe(2\text{-}pic)_3]Cl_2{\cdot}X$, X = C_2H_5OH, CH_3OH, H_2O and $2H_2O$. Chem. Phys., 20: 197–208.
79. Nakamoto, T., Bhattacharjee, A., Sorai, M. (2004) Cause for unusually large thermal hysteresis of spin crossover in $[Fe(2\text{-}pic)_3]Cl_2{\cdot}H_2O$. Bull. Chem. Soc. Jpn., 77: 921–932.
80. Ritter, G., König, E., Irler, W., Goodwin, H. A (1978) The high-spin ($^5T_2$) $\rightleftharpoons$ low-spin ($^1A_1$) transition in solid bis[2-(2-pyridylamino)-4-(2-pyridyl)thiazole]iron(II) dinitrate. its dependence on time land on the previous history of the specimen. Inorg. Chem., 17: 224–228.
81. Kusz, J., Zubko, M., Neder, R. B., Gütlich, P. (2012) Structural phase transition to disorder low-temperature phase in $[Fe(ptz)_6](BF_4)_2$ spin-crossover compounds. Acta Cryst. Sect. B, 68: 40–56.
82. Reger, D. L., Elgin, J. D., Smith, M. D., Grandjean, F., Rebbouh, L., Long, G. J. (2004) A Study of the Electronic Spin-State Crossover in $\{Fe[HC(3,4,5\text{-}Me_3pz)_3]_2\}(BF_4)_2$. Eur. J. Inorg. Chem., 3345–3352.
83. Buchen, T., Gütlich, P., Sugiyarto, K. H., Goodwin, H. A. (1996) High-spin → low-spin relaxation in $[Fe(bpp)_2](CF_3SO_3)_2{\cdot}H_2O$ after LIESST and thermal spin-state trapping–dynamics of spin transition versus dynamics of phase transition. Chem. Eur. J., 2: 1134–1138.
84. Sugiyarto, K. H., Weitzner, K., Craig, D. C., Goodwin, H. A. (1997) Structural, magnetic and Mössbauer studies of bis(2,6-bis(pyrazol-3-yl)pyridine)iron(II) triflate and its hydrates. Aust. J. Chem., 50: 869–873.
85. Sung, R. C. W., McGarvey, B. R. (1999) A study of the spin-state transition and phase transformation in $[Fe(bpp)_2][CF_3SO_3]_2{\cdot}H_2O$ and $[Fe(bpp)_2][BF_4]_2$ using Mn^{2+} electron spin resonance. Inorg. Chem., 38: 3644–3650.
86. Agustí, G., Bartual, C., Martínez, V., Muñoz-Lara, F. J., Gaspar, A. B., Muñoz, M. C., Real, J. A. (2009) Polymorphism and "reverse" spin transition in the spin crossover system $[Co(4\text{-}terpyridone)_2](CF_3SO_3)_2{\cdot}H_2O$. New J. Chem., 33: 1262–1267.
87. Hayami, S., Hiki, K., Kawahara, T., Maeda, Y., Urakami, D., Inoue, K., Ohama, M., Kawata, S., Sato, O. (2009) Photo-induced spin transition of iron(III) compounds with π–π intermolecular interactions. Chem. Eur. J., 15: 3497–3508.
88. Bauer, W., Scherer, W., Altmannshofer, S., Weber, B. (2011) Two-step versus one-step spin transitions in iron(II) 1D chain compounds. Eur. J. Inorg. Chem., 2803–2818.
89. Garcia, Y., Niel, V., Muñoz, M. C., Real, J. A. (2004). Spin crossover in 1D, 2D and 3D polymeric Fe(II) networks in: Gütlich, P., Goodwin, H. A. (Eds.) Spin crossover in transition metal compounds I. Top. Curr. Chem., 233: 229–257.

90. Aromí, G., Barriosa, L. A., Roubeau, O., Gamez, P. (2011). Triazoles and tetrazoles: prime ligands to generate remarkable coordination materials. Coord. Chem. Rev., 255: 485–546.
91. Kröber, J., Codjovi, E., Kahn, O., Grolibre, F., Jay, C. (1993). A spin transition system with a thermal hysteresis at room temperature. J. Am. Chem. Soc.,115: 9810–9811.
92. Kahn, O., Sommier, L., Codjovi, E. (1997). Spin transition molecular alloys: an attempt of fine tuning of the transition temperatures. Chem. Mater., 9: 3199–3205.
93. Dîrtu, M. M., Rotaru, A., Gillard, D., Linares, J., Codjovi, E., Tinant, B., Garcia, Y. (2009). Prediction of the spin transition temperature in Fe^{II} one-dimensional coordination polymers: an anion based database. Inorg. Chem., 48: 7838–7852.
94. Garcia, Y., Moscovici, J., Michalowicz, A., Ksenofontov, V., Levchenko, G., Bravic, G., Chasseau, D., Gütlich, P. (2002). A spin transition molecular material with a wide bistability domain. Chem. Eur. J., 8: 4992–5000.
95. Garcia, Y., van Koningsbruggen, P. J., Bravic, G., Guionneau, P., Chasseau, D., Cascarano, G. L., Moscovici, J., Lambert, K., Michalowicz, A., Kahn, O. (1997). Synthesis, crystal structure, EXAFS, and magnetic properties of *catena*-poly[μ-tris(4-(2-hydroxyethyl)-1,2,4-triazole-N^1,N^2)copper(II)] diperchlorate trihydrate: relevance with the structure of the iron(II) 1,2,4-triazole spin transition molecular materials. Inorg. Chem., 36: 6357–6365.
96. Dîrtu, M. M., Neuhausen, C., Naik, A. D., Rotaru, A., Spinu, L., Garcia, Y. (2010). Insights into the origin of cooperative effects in the spin transition of $[Fe(NH_2trz)_3](NO_3)_2$: the role of supramolecular interactions evidenced in the crystal structure of $[Cu(NH_2trz)_3](NO_3)_2{\cdot}3H_2O$. Inorg. Chem., 49: 5723–5736.
97. Urakawa, A., Van Beek, W., Monrabal-Capilla, M., Galán-Mascarós, J. R., Palin, L., Milanesio, M. (2011). Combined, modulation enhanced X-ray powder diffraction and raman spectroscopic study of structural transitions in the spin crossover material $[Fe(Htrz)_2(trz)](BF_4)$. J. Phys. Chem. C, 115: 1323–1329.
98. Grosjean, A., Daro, N., Kauffmann, B., Kaiba, A., Létard, J.-F., Guionneau, P. (2011). The 1-D polymeric structure of the $[Fe(NH_2trz)_3](NO_3)_2{\cdot}nH_2O$ (with $n = 2$) spin crossover compound proven by single crystal investigations. Chem. Commun., 47: 12382–12384.
99. Haller, K. J., Johnson, P. L., Feltham, R. D., Enemark, J. H., Ferraro, J. R., Basile, L. J. (1979). Effects of temperature and pressure on the molecular and electronic structure of *N,N′*-ethylenebis(salicylideneiminato)nitrosyliron, Fe(NO)(salen). Inorg. Chim. Acta, 33: 119–130.
100. Wells, F. V., McCann, S. W., Wickman, H. H., Kessel, S. L., Hendrickson, D. N., Feltham, R. D. (1982). Mössbauer effect and magnetic investigation of the $S = {}^3/_2 \leftrightarrow S = {}^1/_2$ spin crossover in [Fe(5-Cl-salen)NO] and the $S = {}^3/_2$ state in [Fe(5-Cl-salen)NO]. Inorg. Chem., 21: 2306–2311.
101. Hayami, S., Hashiguchi, K., Inoue, K., Maeda, Y. (2004). Photo-induced spin transition for an iron(III) pyruvic acid thiosemicarbazone compound. J. Nucl. Radiochem. Sci., 5: N1–N3.
102. Cowan, M. G., Olguín, J., Narayanaswamy, S., Tallon, J. L., Brooker, S. (2012). Reversible switching of a cobalt complex by thermal, pressure, and electrochemical stimuli: abrupt, complete, hysteretic spin crossover. J. Am. Chem. Soc., 134: 2892–2894.
103. Ni, C., Fettinger, J. C., Long, G. J., Power, P. P. (2009). Spin-state crossover with structural changes in a cobalt(II) organometallic species: low-coordinate, first row, heteroleptic amido transition metal aryls. synthesis and characterization of $Ar'MN(H)Ar^{\#}$ (M = Mn, Fe, Co) ($Ar' = C_6H_3$-2,6-$(C_6H_3$-2,6-$iPr_2)_2$, $Ar^{\#} = C_6H_3$-2,6-$(C_6H_2$-2,4,6-$Me_3)_2$). Inorg. Chem, 48: 2443–2448.

6

Charge Transfer-Induced Spin-Transitions in Cyanometallate Materials

Kim R. Dunbar[1], Catalina Achim[2] and Michael Shatruk[3]

[1] *Department of Chemistry, Texas A&M University, USA*
[2] *Department of Chemistry, Carnegie Mellon University, USA*
[3] *Department of Chemistry and Biochemistry, Florida State University, USA*

6.1 Introduction

Complexes of transition metals with *o*-quinones have long been recognized for their ability to exhibit valence tautomeric transitions associated with a redox equilibrium between the *o*-semiquinone radical ($S = {}^1/_2$) and *o*-catecholate dianion ($S = 0$) forms of the ligand.[1] About 30 years ago, Buchanan and Pierpont were the first to show that in the complex $[Co^{II}(dbSq^{\bullet -})_2(bpy)]$ (dbSq = 3,5-di-*tert*-butylsemiquinonate; bpy = 2,2′-bipyridine) the change in temperature induces charge transfer (CT) from the Co^{II} ion to one of the monoanionic $dbSq^{\bullet -}$ ligands, to afford the mixed semiquinonato-catecholato form of the complex, $[Co^{III}(dbSq^{\bullet -})(dbCat^{2-})(bpy)]$ (dbCat = 3,5-di-*tert*-butylcatecholate).[2] The change in oxidation state of cobalt in the complex is accompanied by a change in spin state, specifically from high spin paramagnetic Co^{II} (HS, $S = {}^3/_2$) to low spin diamagnetic Co^{III} ion (LS, $S = 0$). Thus, the transition of the complex between the two states could be classified as a Charge Transfer-Induced Spin-Transition (CTIST).

A CTIST process can be induced by photoexcitation or changes in temperature, pressure, or solvent content. The former is arguably the most appealing approach from a practical point of view. Similar to the traditional spin-crossover,[3,4] which occurs without a CT event, the CTIST phenomenon is accompanied by pronounced changes in the magnetic and optical properties of the compounds, as well as by significant alteration of the metal–ligand bond lengths, which are more pronounced for the ion that undergoes the transition between the HS and LS states. Requisite conditions for the occurrence of CTIST include: (i) the existence of a CT excited state that can be accessed under the influence of one of the aforementioned external stimuli, (ii) the ability of the metal ion to change oxidation state in the given coordination environment, and

Spin-Crossover Materials: Properties and Applications, First Edition. Edited by Malcolm A. Halcrow.

Table 6.1 *Examples of cyanometallate-based compounds that exhibit CTIST.*

Chemical formula	Structure	Characterization[a]	Ref.
$Fe[Fe(CN)_6]_{0.75}\cdot 3.5H_2O$[b]	Prussian Blue (PB), 3D	SCXRD, Magn, IR	14
Extended Structures			
$K_{0.14}Co[Fe(CN)_6]_{0.71}\cdot 4.9H_2O$	PB	Magn, IR	5
$Na_{0.43}Co[Fe(CN)_6]_{0.76}\cdot 3.3H_2O$	PB	PXRD, Magn, IR	15
$Rb_{0.53}Co[Fe(CN)_6]_{0.80}\cdot 3.4H_2O$	PB	PXRD, Magn, IR, Möss	16
$Co[Os(CN)_6]_{0.67}\cdot 4H_2O$	PB	PXRD, Magn, IR	17
$Co[Fe(CN)_5(NH_3)]\cdot 6H_2O$	3D	PXRD, Magn, IR, Möss	18
$[[(Tp)Fe(CN)_3]_2[Co(bpe)]]\cdot 5H_2O$	2D	SCXRD, Magn, IR	19
$Cs[Co(3\text{-}CNpy)_2][W(CN)_8]\cdot H_2O$	2D	SCXRD, Magn, IR	20
$[[Co(H_2O)_2][Co(pym)_2]_2[W(CN)_8]_2]\cdot 4H_2O$	3D	SCXRD, PXRD, Magn, IR	21
Nanomaterials			
$Na_{0.31}Co[Fe(CN)_6]_{0.77}\cdot 3.8H_2O$	Thin film	PXRD, Magn, IR	22
$Rb_{0.18}Co[Fe(CN)_6]_{0.75}\cdot zH_2O$	Thin film	Magn, IR, AFM, SEM	23
$Rb_xCo[Fe(CN)_6]_y\cdot zH_2O$	Nanoparticles	Magn, TEM	24
Molecular Structures			
$[[Co(tmphen)_2]_3[Fe(CN)_6]_2]\cdot n$Solv	Trig. bipyramid	SCXRD, Magn, IR, Möss	8
$[[Co(tmphen)_2]_3[Fe(CN)_6]_2]\cdot n$Solv	Trig. bipyramid	SCXRD, Magn, IR, Möss	25
$[[Co(^tBu\text{-}bpy)_2]_2[(Tp^*)Fe(CN)_3]_2](PF_6)_2\cdot 2MeOH$	Square	SCXRD, Magn, IR	26
$[(PY5Me_2)_4Co_4Re(CN)_7](PF_6)_5$	Star-like	SCXRD, Magn, IR	27
$[[(pzTp)Fe(CN)_3]_4[Co(pz)_3CCH_2OH]_4](ClO_4)_4$	Cube	SCXRD, Magn, IR	28

[a]XRD = X-ray diffraction, Magn = magnetometry, IR = infra-red spectroscopy, Möss = Mössbauer spectroscopy, AFM = atomic force microscopy, SEM = scanning electron microscopy, TEM = transmission electron microscopy.
[b]The original Prussian Blue is shown for comparison.

(iii) the ability of the metal ion involved in CT to exist in the LS and HS configurations in one of the oxidation states. The last condition limits the observation of CTIST to compounds of d^4-d^7 metal ions of the first transition series (Table 6.1).

The valence-tautomeric transitions accompanied by changes in spin state are made possible by the similarity in the energies of the frontier orbitals of the metal ion and the ligand. One could anticipate similar transitions that are exclusively metal based to occur in polynuclear metal complexes, in which different metal ions have *d*-orbitals of comparable energy and in which there is an efficient CT pathway. In this vein, Hashimoto and his collaborators first demonstrated this behavior in coordination polymers based on cyanometallates. In 1996, they reported that $K_{0.2}Co_{1.4}[Fe(CN)_6]\cdot 6.9H_2O$, a solid with a Prussian blue type structure, exhibited a remarkable enhancement of the ferrimagnetic ordering temperature, saturation magnetization, and coercivity upon irradiation with $\lambda = 660$ nm at 5 K.[5] The observed behavior was explained by a CTIST that converts the material with diamagnetic LS–Co^{III}/LS–Fe^{II} ions to the material with paramagnetic (and antiferromagnetically coupled!) HS–Co^{II}/LS–Fe^{III} ions. Thus, besides changing the spin state of the Co sites, the CTIST also "turned on" magnetic superexchange across CN^- bridges and led to the emergence of long-range ferrimagnetic ordering. The photoinduced, magnetically-ordered state persisted for several days at 5 K.

The discovery of photoinduced CTIST and associated changes in optical and magnetic properties of $K_{0.2}Co_{1.4}[Fe(CN)_6]\cdot 6.9H_2O$ spurred interest in the examination of cyanometallates as candidates for materials with CTIST. These research activities have been growing in parallel with a veritable explosion of interest in the assembly of various architectures originating from cyanometallate building blocks,[6] in particular, paramagnetic discrete molecules that can act as single-molecule magnets.[7]

Shortly after the discovery of the CTIST in the extended Prussian Blue type phases, our groups were the first to demonstrate in 2004 that this phenomenon can also be achieved in materials based on polynuclear metal complexes.[8, 9] Some excellent reviews have been published on the topic, but each of these reviews has mainly focused on one class of materials.[10–13] This account is devoted to CTIST in both extended molecule-based structures and molecular complexes, as well as in nanoscale materials. After summarizing some general methods used for the observation and characterization of the CTIST phenomenon, we address separately each type of cyanometallate-based materials that exhibit CTIST, that is extended solids, nanoparticles, and mixed-metal polynuclear complexes (Table 6.1).

6.2 Characterization of CTIST Compounds

Crystallographic, magnetic, and spectroscopic methods have been used in the identification of CTIST. We present these methods using as an example two Co-Fe Prussian Blue analogs (PBAs), $Co[Fe(CN)_6]_{0.67}\cdot 4H_2O$ and $Rb_{0.53}Co[Fe(CN)_6]_{0.8}\cdot 3.4H_2O$.[16] The compounds exist predominantly in the HS–Co^{II}/LS–Fe^{III} and LS–Co^{III}/LS–Fe^{II} states, respectively, between 2 K and 300 K. $Rb_{0.53}Co[Fe(CN)_6]_{0.8}\cdot 3.4H_2O$ but not $Co[Fe(CN)_6]_{0.67}\cdot 4H_2O$ undergoes a light-induced CTIST from the LS–Co^{III}/LS–Fe^{II} to the HS–Co^{II}/LS–Fe^{III} state.

The difference in the oxidation and spin states of the metal ions is revealed by examining the crystal structure, magnetism, IR and Mössbauer spectra of these compounds as a function of temperature and light irradiation. The χT values for the two compounds (χ – molar magnetic susceptibility) in the 2–300 K temperature range indicate that they are predominantly in HS–Co^{II}/LS–Fe^{III} ($xT \sim 4.5$ emu mol^{-1} K^{-1}) and LS–Co^{III}/LS–Fe^{II} ($xT \sim 0$ emu mol^{-1} K^{-1}) states, respectively. $Co[Fe(CN)_6]_{0.67}\cdot 4H_2O$ exhibits ferrimagnetic ordering with $T_C = 16$ K but $Rb_{0.53}Co[Fe(CN)_6]_{0.8}\cdot 3.4H_2O$ does not. After irradiation with $\lambda = 500$–750 nm for 40 min at 5 K, the value of χT for $Co[Fe(CN)_6]_{0.67}\cdot 4H_2O$ remains unchanged but that for $Rb_{0.53}Co[Fe(CN)_6]_{0.8}\cdot 3.4H_2O$ increases significantly indicating the occurrence of photoinduced CTIST from the LS–Co^{III}/LS–Fe^{II} to the HS–Co^{II}/LS–Fe^{III} state. The photogenerated HS–Co^{II}/LS–Fe^{III} state shows magnetic ordering at $T_C = 22$ K.

The Fe–C bond lengths for the LS–Fe^{II} and LS–Fe^{III} ions that have the same coordination environment are close in value. In contrast, the Co–N bond lengths for LS–Co^{III} and HS–Co^{II} ions with the same set of ligands differ by more than 0.1 Å.[1] Hence, structural information on the Co sites is useful to identify the CTIST. Unfortunately, single crystals of the two PBAs discussed in this section, $Co[Fe(CN)_6]_{0.67}\cdot 4H_2O$ and $Rb_{0.53}Co[Fe(CN)_6]_{0.8}\cdot 3.4H_2O$, could not be obtained. Nevertheless, structural information for these compounds was obtained from powder X-ray diffraction patterns, which are consistent with a face-centered cubic structure of PB. The cubic unit cell parameter a is 10.32 Å in $Co[Fe(CN)_6]_{0.67}\cdot 4H_2O$ and 9.96 Å in $Rb_{0.53}Co[Fe(CN)_6]_{0.8}\cdot 3.4H_2O$.[16] Assuming similar Fe–C and C≡N bond lengths in these two PBAs and the original PB structure, one obtains Co–N bond lengths of 2.11 Å in $Co[Fe(CN)_6]_{0.67}\cdot 4H_2O$ and 1.93 Å in $Rb_{0.53}Co[Fe(CN)_6]_{0.8}\cdot 3.4H_2O$, which is consistent with the presence of HS–Co^{II} and LS–Co^{III}, respectively, in these materials.

The cyanide stretching frequency, ν_{CN}, is sensitive to the oxidation state of the metal ion(s) to which the cyanide is coordinated. The ν_{CN} values for bridging cyanides in Fe-containing PBAs of 3d transition metals are in the ranges of 2145–2185 cm^{-1} and 2065–2100 cm^{-1} for M^{II}–NC–Fe^{III} and M^{II}–NC–Fe^{II}, respectively.[29, 30] These values have been used extensively to determine the oxidation states of transition metal ions in PBAs. For example, the ν_{CN} stretch at 2163 cm^{-1} observed for $Co[Fe(CN)_6]_{0.67}\cdot 4H_2O$ indicates that the compound is in the HS–Co^{II}/LS–Fe^{III} state.[16] The IR spectrum of the compound did not show any changes under illumination at 15 K. $Rb_{0.53}Co[Fe(CN)_6]_{0.8}\cdot 3.4H_2O$, however, had an IR peak at 2133 cm^{-1}, with a small shoulder on the low-energy side. The 2133 cm^{-1} peak was assigned to the cyanide in the LS–Co^{III}/LS–Fe^{II} state based on its similarity to the ν_{CN} stretch observed for the dinuclear complex $[(NC)_5Co^{III}–NC–Fe^{II}(CN)_5]^{6-}$ (2130 cm^{-1}).[31] The low-energy shoulder was attributed to the cyanide in HS–Co^{II}-NC-LS–Fe^{II} units, which exists because some of the Co ions in $Co[Fe(CN)_6]_{0.67}\cdot 4H_2O$ are in the Co^{II} oxidation state. Upon illumination at 15 K, the ν_{CN} stretch at 2133 cm^{-1} gradually decreased in intensity and two new stretches appeared at 2169 cm^{-1} and 2103 cm^{-1}. These new peaks have been assigned to the bridging cyanide in HS–Co^{II}/LS–Fe^{III} and HS–Co^{II}/LS–Fe^{II} states, respectively.

Since the majority of known CTIST compounds contain Fe centers, ^{57}Fe Mössbauer spectroscopy provides important information on the individual oxidation and spin state of these centers. The room-temperature Mössbauer spectra of $Co[Fe(CN)_6]_{0.67}\cdot 4H_2O$ and $Rb_{0.53}Co[Fe(CN)_6]_{0.8}\cdot 3.4H_2O$ show quadrupole doublets with similar isomer shifts (δ) of –0.17 mm/s and –0.16 mm/s, respectively, but very different quadrupole splitting (ΔE_Q) of 0.61 mm/s and 0.16 mm/s, respectively. Based on these parameters, one can conclude that $Co[Fe(CN)_6]_{0.67}\cdot 4H_2O$ and $Rb_{0.53}Co[Fe(CN)_6]_{0.8}\cdot 3.4H_2O$ contain LS–Fe^{III} and LS–Fe^{II} ions, respectively. At 25 K and in the absence of an applied magnetic field, the Mössbauer spectrum of $Rb_{0.53}Co[Fe(CN)_6]_{0.8}\cdot 3.4H_2O$ still shows only one quadrupole doublet with $\delta = 0.02$ mm/s and $\Delta E_Q = 0.15$ mm/s, which is indicative of LS–Fe^{II} ions. After light irradiation, the intensity of this doublet decreases and a new doublet appears with $\delta = 0.09$ mm/s, $\Delta E_Q = 1.13$ mm/s. The parameters of the new doublet are indicative of LS Fe^{III} ions.

As shown by the foregoing discussion, the combination of magnetic measurements, magnetism, X-ray crystallography, and IR and Mössbauer spectroscopy data clearly prove the occurrence of photoinduced CTIST in $Rb_{0.53}Co[Fe(CN)_6]_{0.8}\cdot 3.4H_2O$. These methods have also been used to characterize molecular $[Fe_2Co_2]$ squares that exhibit CTIST (see Section 6.4). The charge distribution in those clusters was identified by (i) the Co–N bond lengths, which for LS–Co^{III} and HS–Co^{II} are ~1.93 Å and ~2.10 Å, respectively, (ii) the ^{57}Fe Mössbauer isomer shift, which for Fe ions with the same coordination typically is lower for LS–Fe^{III} than for LS–Fe^{II}, and quadrupole splitting, which typically is lower for LS–Fe^{II} than for LS–Fe^{III}, (iii) the IR stretching frequencies for bridging cyanides, which are below 2150 cm^{-1} in Fe^{II}–CN–Co^{III} and above 2150 cm^{-1} in Fe^{III}–CN–Co^{II}, and (iv) the χT values, which are significantly lower for the clusters that contain LS–Fe^{II}/LS–Co^{III} pairs than for those that contain LS–Fe^{III}/HS–Co^{II} pairs. We also note that in some cases the high temperature form of the complexes that undergo temperature-induced CTIST can be trapped and studied at low temperature by fast cooling. This made possible the isolation of clusters containing LS–Fe^{III}/HS–Co^{II} pairs, and the low temperature magnetization of this form of the clusters revealed that the $S = {}^1/_2$ and $S = {}^3/_2$ spins of the metal sites are weakly ferromagnetically coupled. Metal-to-metal charge transfer (MM′CT) transitions of the type Co^{III}-to-Fe^{II} and Co^{II}-to-Fe^{III} have been also observed for the clusters in the ranges of 750–850 nm and 550–560 nm, respectively.

The physical characterization methods identified in this section constitute a basic suite of tools for interrogation of CTIST phenomena. In addition, a number of other techniques have been used for the study of CTIST compounds, and they will be mentioned in the text whenever appropriate.

6.3 CTIST in Coordination Polymers

6.3.1 Co-Fe Prussian Blue Analogs

As mentioned above, the CTIST phenomenon was first reported for a heterometallic cyanide, viz., $K_{0.2}Co_{1.4}[Fe(CN)_6]\cdot 6.9H_2O$.[5] Subsequently, the study of CTIST in extended cyanide-bridged structures focused predominantly on Co-Fe PBAs, which explains why the most extensive information on the CTIST phenomenon has been obtained for these materials. For this reason we discuss first CTIST in Co-Fe PBAs and then CTIST in extended solids. Prior to considering specific examples, we briefly describe the Prussian blue structure type and introduce the conventions used in the rest of the chapter.

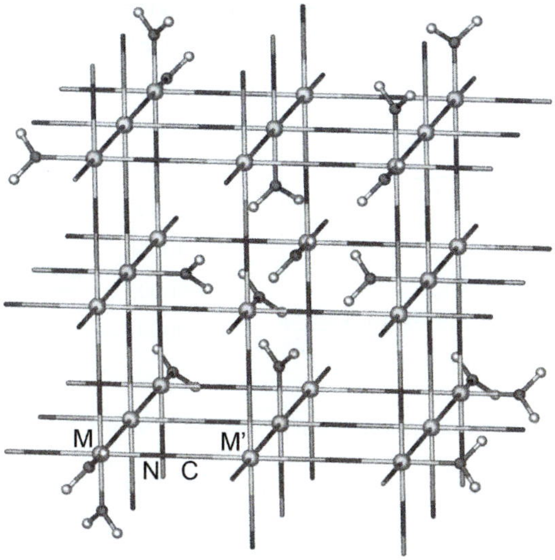

Figure 6.1 *The Prussian Blue structure type. Note that the hexacyanometallate unit in the middle is absent, which causes coordination of water molecules to the adjacent metal sites.*

Prussian Blue (PB) is the most ubiquitous structure type adopted by coordination polymers based on cyanide-bridged metal ions.[14] PB and its analogs are typically synthesized by combining an aqueous solution of a transition metal ion M^{n+} with an aqueous solution of a hexacyanometallate anion $[M'(CN)_6]^{m-}$.[32] The ensuing compound has a structure that consists of metal ions connected into a three-dimensional (3D) cubic array through CN^- anions (Fig. 6.1). The voids of the structure contain counterions (usually alkali metal ions A^+) and water molecules. In an ideal structure, in which all the M and M′ sites are occupied, the composition of the material is $A_xM[M'(CN)_6]\cdot yH_2O$, where x is the number of counterions necessary for charge balance. In many cases, including that of the original PB structure,[14] there are vacancies in the $[M'(CN)_6]^{m-}$ sites, a fact that has two important consequences for the composition and structure of the compound. First, the number of counterions x is smaller than in the ideal structure. Second, the M^{n+} sites experience a deficit of CN^- ligands and coordinate water molecules to complete the six-coordinate coordination sphere. Thus, on the extreme opposite to the $A_xM[M'(CN)_6]\cdot zH_2O$ composition, the compound has a $M[M'(CN)_6]_{1-(x/m)}\cdot zH_2O$ composition, in which the concentration of $[M'(CN)_6]^{m-}$ vacancies is so high that there are no charge-compensating A^+ counterions and the M^{n+} ions are coordinated by the greatest possible number of water molecules. In the following discussion, we write the composition of PB-type compounds as $A_xM[M'(CN)_6]_y\cdot zH_2O$, because the M sites are always completely occupied but the A and $[M'(CN)_6]$ sites exhibit variable degrees of occupancy. In this convention, the concentration of $[M'(CN)_6]$ vacancies is given by $(1-y)$. The composition of PB, usually written as $Fe^{III}{}_4[Fe^{II}(CN)_6]_3\cdot 14H_2O$, is herein expressed as $Fe^{III}[Fe^{II}(CN)_6]_{0.75}\cdot 3.5H_2O$. In PB, 25% of $[Fe^{II}(CN)_6]$ sites are vacant and the average coordination environment of the Fe^{III} sites is $N_{4.5}O_{1.5}$. We now proceed to the discussion of CTIST in PB-type compounds.

The synthesis of the material $K_{0.14}Co[Fe(CN)_6]_{0.71}\cdot 4.9H_2O$, for which Hashimoto reported CTIST in 1996, bears similarities to the synthesis of PB from a ferrous salt and ferricyanide. In both cases, the oxidation and spin state of the metal ions changes during synthesis. $K_{0.14}Co[Fe(CN)_6]_{0.71}\cdot 4.9H_2O$ is prepared from a solution that contains HS Co^{II} ($S = {}^3/_2$) cations and LS $[Fe^{III}(CN)_6]^{3-}$ ($S = {}^1/_2$) anions. The resulting material contains predominantly diamagnetic ($S = 0$) LS–Co^{III}/LS–Fe^{II} units. Similarly, the PB $Fe^{III}[Fe^{II}(CN)_6]_{0.75}\cdot 3.5H_2O$ is prepared from a solution that contains HS Fe^{II} ($S = 2$) and LS $[Fe^{III}(CN)_6]^{3-}$ ($S = {}^1/_2$) ions. The resulting

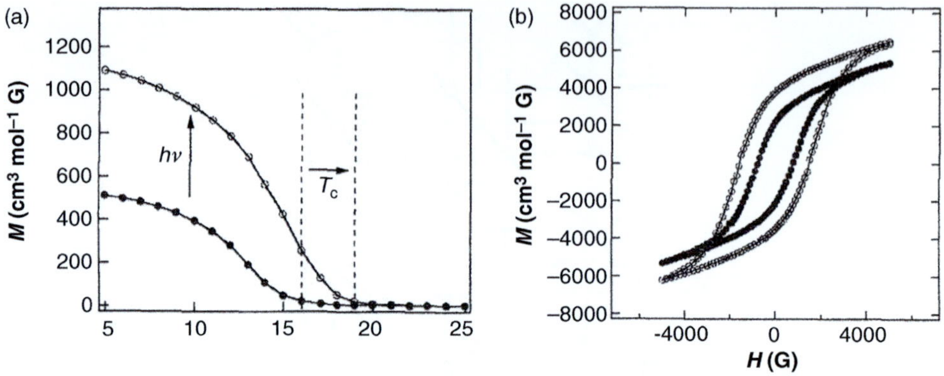

Figure 6.2 *Photoinduced magnetization enhancement in $K_{0.14}Co[Fe(CN)_6]_{0.71}\cdot4.9H_2O$ as observed in the increase of the ferrimagnetic ordering temperature T_C (a) and the broader hysteresis loop (b). The data before and after irradiation are shown with closed and open circles, respectively. Reproduced from [5]. Copyright American Association for the Advancement of Science, 1996.*

PB contains HS Fe^{III} ($S = {}^5/_2$) and LS $[Fe^{II}(CN)_6]^{3-}$ ($S = 0$) ions.[33] In both compounds, the LS Fe^{II} centers, obtained as the result of CT during synthesis, are coordinated by carbon-bound CN^- ligands. The Co-Fe PBA is very sensitive to external conditions, such as temperature and pressure, while the original PB material is not.

$K_{0.14}Co[Fe(CN)_6]_{0.71}\cdot4.9H_2O$ exhibits ferrimagnetic ordering at $T_C = 16$ K.[5] Owing to the presence of vacancies in the $[Fe(CN)_6]$ crystallographic sites (Fig. 6.1), the compound contains a mixture of diamagnetic LS–Co^{III} ($S = 0$) and paramagnetic HS–Co^{II} ($S = {}^3/_2$) sites, with the valence description being $K_{0.14}Co^{II}{}_{0.29}Co^{III}{}_{0.71}[Fe^{II}(CN)_6]_{0.71}\cdot4.9H_2O$. (It is also possible that a small residual fraction of LS–Fe^{III} sites is present in the material, which would explain the observation of ferrimagnetic ordering at 16 K even in the absence of photoexcitation.) Upon irradiation with red light ($\lambda = 660$ nm) at 5 K, the T_C increases to 19 K, with a concomitant increase in the remnant magnetization and coercivity of the material (Fig. 6.2). These increases are attributed to photoinduced electron transfer that converts diamagnetic LS–Co^{III}/LS–Fe^{II} units to paramagnetic HS–Co^{II}/LS–Fe^{III} ones (Scheme 6.1). LS–Fe^{III} sites are generated under irradiation conditions for the similar compound, $K_{0.4}Co[Fe(CN)_6]_{0.8}\cdot2.9H_2O$.[34]

Hashimoto and co-workers demonstrated the importance of the A^+ cation for the occurrence of CTIST by comparing the behavior of $Co[Fe(CN)_6]_{0.67}\cdot4H_2O$ and $Rb_{0.53}Co[Fe(CN)_6]_{0.8}\cdot3.4H_2O$, which can be formulated as $Co^{II}[Fe^{III}(CN)_6]_{0.67}\cdot4H_2O$ and $Rb_{0.53}Co^{II}{}_{0.33}Co^{III}{}_{0.67}[Fe^{II}(CN)_6]_{0.8}$, respectively.[16] (See also

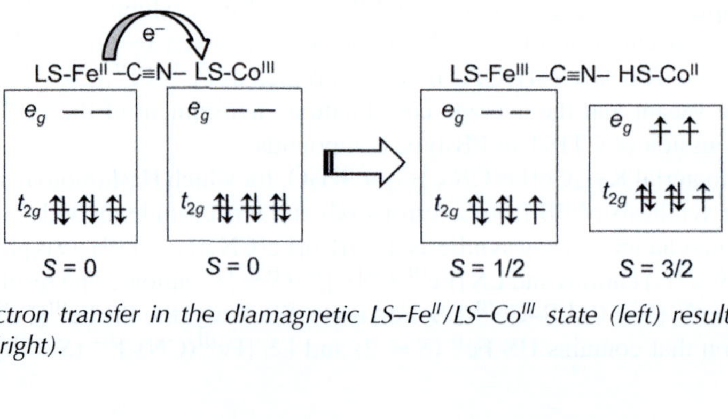

Scheme 6.1 *The electron transfer in the diamagnetic LS–Fe^{II}/LS–Co^{III} state (left) results in the paramagnetic LS–Fe^{III}/HS–Co^{II} state (right).*

Table 6.2 *Properties of compounds $A_xCo[Fe(CN)_6]_y{\cdot}zH_2O$ (A = Cs, Na).*

Composition	% composition of Co and Fe sites				CTIST (thermal)	CTIST (photoinduced)
	HS–Co^{II}	LS–Fe^{III}	LS–Co^{III}	LS–Fe^{II}		
$Co[Fe(CN)_6]_{0.67}{\cdot}4H_2O$	100	100	0	0	No	No
$Cs_{0.08}Co[Fe(CN)_6]_{0.70}{\cdot}4.5H_2O$	97	96	3	4	No	No
$Cs_{0.18}Co[Fe(CN)_6]_{0.73}{\cdot}4H_2O$	88	86	12	14	Yes	Yes
$Cs_{0.30}Co[Fe(CN)_6]_{0.80}{\cdot}4H_2O$	57	47	43	53	Yes	Yes
$Cs_{0.58}Co[Fe(CN)_6]_{0.85}{\cdot}3.5H_2O$	40	30	60	70	No	No
$Cs_{0.98}Co[Fe(CN)_6]_{0.98}{\cdot}3.3H_2O$	30	28	70	72	No	No
$Na_{0.05}Co[Fe(CN)_6]_{0.67}{\cdot}4.2H_2O$	100	93	0	7	No	No
$Na_{0.27}Co[Fe(CN)_6]_{0.73}{\cdot}3.5H_2O$	100	89	0	11	Yes	Yes
$Na_{0.43}Co[Fe(CN)_6]_{0.76}{\cdot}3.3H_2O$	100	83	0	17	Yes	Yes
$Na_{0.48}Co[Fe(CN)_6]_{0.79}{\cdot}3.1H_2O$	85	70	15	30	Yes	Yes
$Na_{0.82}Co[Fe(CN)_6]_{0.87}{\cdot}2.6H_2O$	34	0	66	100	No	No

Section 6.2.) $Co[Fe(CN)_6]_{0.67}{\cdot}4H_2O$ exhibits ferrimagnetic ordering at $T_C = 16$ K and does not respond to illumination, as the constituent metal ions are already in the HS–Co^{II}/LS–Fe^{III} state. Note that in this case, no electron transfer from Co^{2+} to $[Fe(CN)_6]^{3-}$ takes place upon formation of the compound. $Rb_{0.53}Co[Fe(CN)_6]_{0.8}{\cdot}3.4H_2O$ shows a much lower magnetic response over the entire temperature range and does not order magnetically even at 2 K. Nevertheless, after irradiation with $\lambda = 500$-750 nm for 40 min at 5 K, the compound exhibits ferrimagnetic ordering with $T_C = 22$ K. The $Rb_{0.45}Co[Fe(CN)_6]_{0.83}{\cdot}3.2H_2O$ compound studied by Bleuzen *et al.* showed properties similar to those of $Rb_{0.53}Co[Fe(CN)_6]_{0.8}{\cdot}3.4H_2O$.[35]

The presence of the alkali metal ion in the structure is not a necessary and sufficient cause for the manifestation of CTIST, but it appears that the presence of $[Fe(CN)_6]$ vacancies is. To demonstrate the importance of the hexacyanometallate vacancies in stabilizing the CTIST effect, a series of $Cs_xCo[Fe(CN)_6]_y{\cdot}zH_2O$ compounds was investigated (Table 6.2).[36] The Cs-rich phase, $Cs_{0.98}Co[Fe(CN)_6]_{0.98}{\cdot}3.3H_2O$, which consists primarily of LS–Co^{III}/LS–Fe^{II} states but has very low concentration of $[Fe(CN)_6]$ vacancies, shows a negligible CTIST effect. A detailed analysis of the relationship between the concentration of vacancies and the oxidation states of Co and Fe was performed by XANES spectroscopy, using as reference points the most disparate members of the series, $Co[Fe(CN)_6]_{0.67}{\cdot}4H_2O$ and $Cs_{0.98}Co[Fe(CN)_6]_{0.98}{\cdot}3.3H_2O$. This study showed that (i) compounds with $y = 0.70$ and 0.73 mainly contain HS–Co^{II} sites, as is the case in $Co[Fe(CN)_6]_{0.67}{\cdot}4H_2O$; (ii) the compound with $y = 0.85$ contains LS–Co^{III} sites, as is the case in $Cs_{0.98}Co[Fe(CN)_6]_{0.98}{\cdot}3.3H_2O$; (iii) the compound with $y = 0.80$ contained a mixture of HS–Co^{II} and LS–Co^{III} .

The compounds with larger y values have lower concentrations of $[Fe(CN)_6]$ vacancies ($1-y$). Consequently, they contain more diamagnetic LS–Co^{III}/LS–Fe^{II} pairs and have lower values of χT at room temperature. As the temperature decreases, the χT values of the samples with high ($y = 0.67$, 0.7) and low ($y = 0.85$, 0.98) vacancy concentrations decrease only slightly (Fig. 6.3), indicating the persistence of mainly HS–Co^{II}/LS–Fe^{III} and LS–Co^{III}/LS–Fe^{II} states, respectively. The samples with intermediate y values, that is 0.73 and 0.80, exhibit a temperature-induced CTIST between 280 and 170 K, from LS–Co^{III}/LS–Fe^{II} at low temperature to LS–Co^{II}/LS–Fe^{III} at high temperature. Photoinduced CTIST at 10 K is also observed for all the $Cs_xCo[Fe(CN)_6]y{\cdot}zH_2O$ compounds that exhibit temperature-induced CTIST.

Similar observations were made for a $Na_xCo[Fe(CN)_6]_y{\cdot}zH_2O$ series.[15] In "as-prepared" samples, the HS–Co^{II}/LS–Fe^{III} prevailed in compounds with low Na^+ ion content and a large number of $[Fe(CN)_6]$ vacancies while the LS–Co^{III}/LS–Fe^{II} prevailed in compounds with high Na^+ content. A thermal CTIST was observed

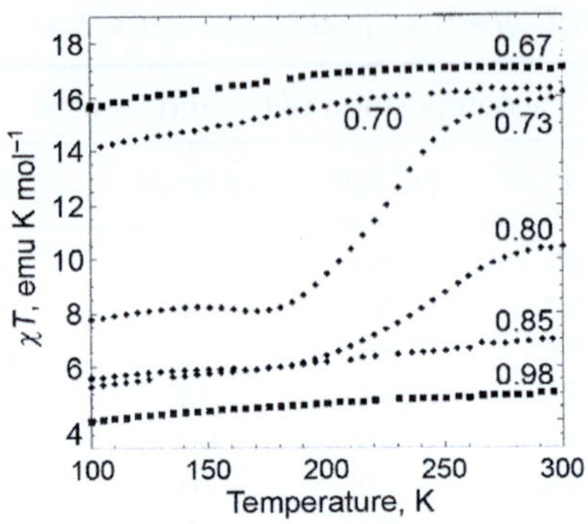

Figure 6.3 *Temperature dependence of χT in the $Cs_xCo[Fe(CN)_6]_y \cdot zH_2O$ series, with the value of y indicated for each curve. Reprinted with permission from [36]. Copyright 2001 American Chemical Society.*

only for compounds in the middle of the series (Table 6.2). The temperature of the CTIST gradually increases with lower concentrations of $[Fe(CN)_6]$ vacancies (Fig. 6.4a).

The intermediate compounds in the $Na_xCo[Fe(CN)_6]_y \cdot zH_2O$ series,[15] with $y = 0.73$, 0.76, and 0.79, show hysteresis in the temperature-induced CTIST (Fig. 6.4a). They also demonstrate a photoinduced CTIST under irradiation at 5 K, which leads to a ferrimagnetically ordered state with $T_C = 26$ K. In contrast, the extreme compositions, that is $Na_{0.05}Co[Fe(CN)_6]_{0.67} \cdot 4.2H_2O$ and $Na_{0.82}Co[Fe(CN)_6]_{0.87} \cdot 2.6H_2O$, do not show photoinduced CTIST. The former compound, however, contains mainly HS–Co^{II}/LS–Fe^{III} pairs and exhibits ferrimagnetic ordering at 16 K.

According to the information obtained in these studies, it is obvious that the occurrence of CTIST in the Co-Fe PBAs requires that electron transfer between the Co^{2+} and $[Fe(CN)_6]^{3-}$ ions occurs during the synthesis and leads to diamagnetic LS–Co^{III}/LS–Fe^{II}, CN-bridged pairs. The likelihood of electron transfer increases as the Fe/Co ratio in $A_xCo[Fe(CN)_6]_y \cdot zH_2O$ (A = Cs, Na) approaches 1. This fact can be explained by the evolution of the average coordination environment of Co ions from N_4O_2 in $Co[Fe(CN)_6]_{0.67} \cdot 4H_2O$ to N_6 in $ACo[Fe(CN)_6] \cdot zH_2O$. The higher number of N-bound ligands increases the ligand field strength at the Co sites, thus stabilizing the LS–Co^{III} state and preventing the CTIST. On the other hand, when the concentration of vacancies becomes sufficiently low, the extended structure is not sufficiently flexible to accommodate the significant change in the Co–N bond lengths that accompanies the transition and the CTIST also cannot occur. The combination of these effects translates to an optimal range of compositions for which the average ligand field at Co sites and the concentration of vacancies in the $[Fe(CN)_6]$ sites are such that thermal or photoinduced CTIST can take place.[36]

Interestingly, the photoinduced HS–Co^{II}/LS–Fe^{III} state of $Na_xCo[Fe(CN)_6]_y \cdot zH_2O$ persists at temperatures far above the magnetic ordering temperature (Fig. 6.4b), which is also the case for similar compounds that exhibit the CTIST. Thus, after photoexcitation and heating above the T_C, the χT value for $Na_{0.27}Co[Fe(CN)_6]_{0.73} \cdot 3.5H_2O$ remains similar to that observed for the high temperature (HT) form of the compound. The relaxation to the low temperature (LT) form (with prevailing LS–Co^{III}/LS–Fe^{II} states) is observed only at temperatures above 150 K. Upon further heating, the compound undergoes the thermal CTIST and the HT state is restored. As the concentration of $[Fe(CN)_6]$ vacancies decreases as y increases from 0.73 to 0.76 and 0.79, the relaxation temperature of the photoinduced HS–Co^{II}/LS–Fe^{III} state also decreases. The increased stability range of the diamagnetic LS–Co^{III}/LS–Fe^{II} state is also reflected in the larger difference between the relaxation temperature and the thermal CTIST (Fig. 6.4b).

The HT state of $Na_{0.32}Co[Fe(CN)_6]_{0.74} \cdot 3.4H_2O$ can be thermally quenched by fast cooling the material from room temperature to 5 K.[37] In a manner akin to the photoinduced magnetically ordered state, the

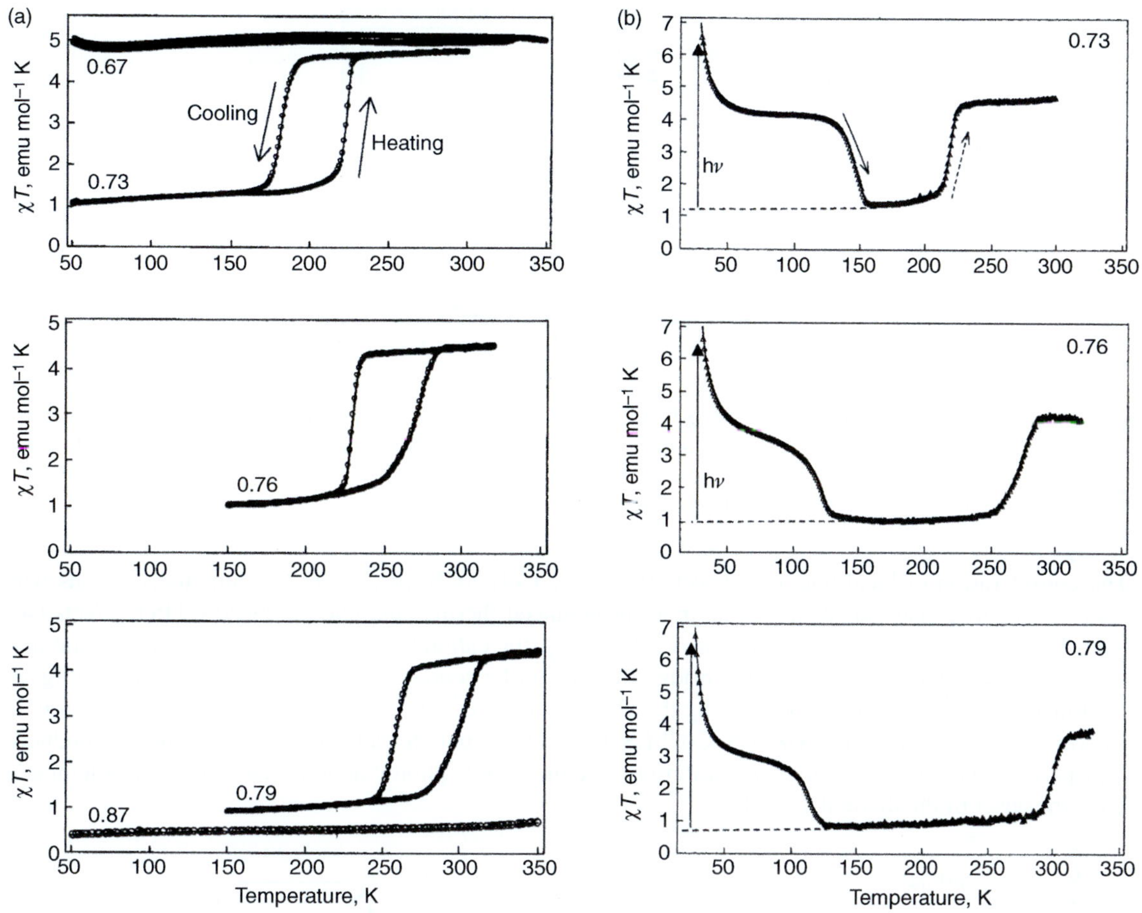

***Figure 6.4** Magnetic and photomagnetic behavior of $Na_xCo[Fe(CN)_6]_y{\cdot}zH_2O$ (the value of y is indicated for each curve): (a) Temperature dependence of χT. The cooling and heating curves coincide for y = 0.67 and y = 0.87. (b) Relaxation of the photoinduced magnetically ordered state upon heating, with the subsequent thermal CTIST at higher temperatures. Adapted with permission from [15]. Copyright 2002, American Chemical Society.*

metastable form of the compound exhibits a ferrimagnetic transition at $T_C = 20$ K and irreversible relaxation to the diamagnetic LT state above 160 K (Fig. 6.5). Nevertheless, due to the thermal CTIST, the HT state is fully recovered upon heating above 240 K.

To obtain more insight into the influence of the alkali cation on the thermal and photoinduced CTIST properties, Bleuzen *et al.* studied the series of compounds $A_{0.5}Co[Fe(CN)_6]_{0.83}{\cdot}2.8H_2O$ (A = Na, Rb, Cs).[38] In spite of the fact that they contain the same concentration of $[Fe(CN)_6]$ vacancies, the Rb^+- and Cs^+-containing phases exist predominantly in the diamagnetic LS–Co^{III}/LS–Fe^{II} state, whereas the Na^+ containing compound exhibits a thermal CTIST with hysteresis (Fig. 6.6a). Under illumination at 10 K, all compounds exhibit a photoinduced increase in magnetization, with the magnetic ordering occurring at $T_C = 21$ K (Fig. 6.6b). The strongest photomagnetic effect is observed for the Na^+ analog. The stability of the photoinduced state also depends on the nature of the A^+ ion, with the relaxation temperature increasing from 95 K to 110 K to 135 K for A = Cs, Rb, and Na, respectively.

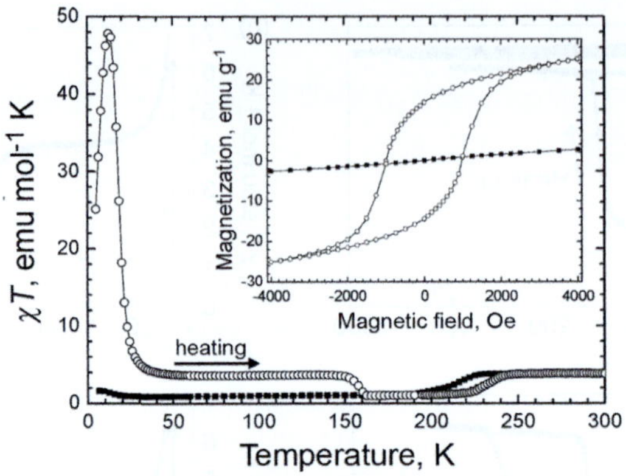

Figure 6.5 Temperature dependence of χT and field dependence of magnetization at 5 K (inset) for $Na_{0.32}Co[Fe(CN)_6]_{0.74}\cdot 3.4H_2O$. The closed squares represent the behavior of a slowly cooled sample (0.9 K/min); the open circles represent the behavior of a sample that was rapidly quenched to 5 K. Reprinted with permission from [28]. Copyright 2001 American Chemical Society.

The consideration of local bonding aspects in the $A_{0.5}Co[Fe(CN)_6]_{0.83}\cdot 2.8H_2O$ compounds reveals why the Na^+-containing materials demonstrate more pronounced thermal and photoinduced CTIST effects (see Table 6.1, Figs. 6.4 and 6.6). As the size of A^+ decreases, the cation becomes more polarizing and interacts more strongly with the CN^- ligands that bridge the Co and Fe centers. This interaction weakens the ligand field around the Co^{II} ion, thereby stabilizing the HS–Co^{II}/LS–Fe^{III} state, which becomes closer in energy to the thermodynamic ground state, viz., LS–Co^{III}/LS–Fe^{III}. This situation leads to a more efficient photoinduced CTIST process, a higher relaxation temperature for the metastable photomagnetic state and a lower temperature for the onset of the thermal CTIST.

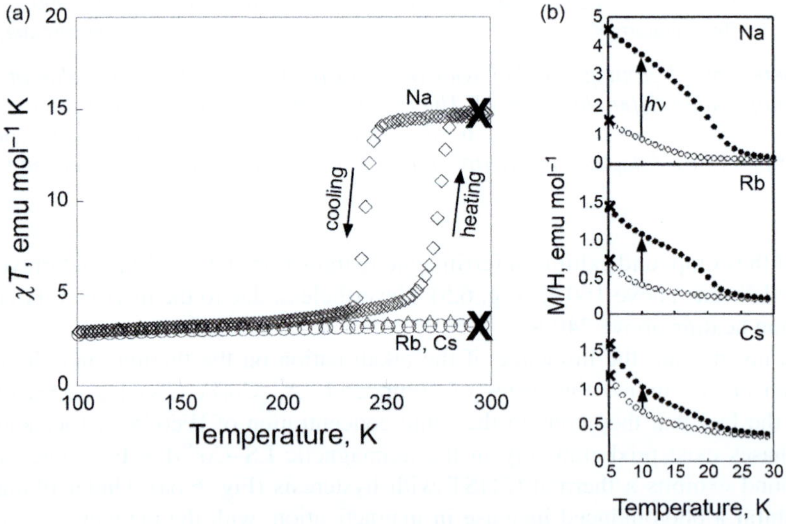

Figure 6.6 Temperature dependence of χT (a) and photoinduced increase in magnetization (b) of $A_{0.5}Co[Fe(CN)_6]_{0.83}\cdot 2.8H_2O$ (A = Na, Rb, Cs). The crosses indicate the temperatures at which X-ray absorption spectra were collected. Reprinted with permission from [28]. Copyright 2001 American Chemical Society.

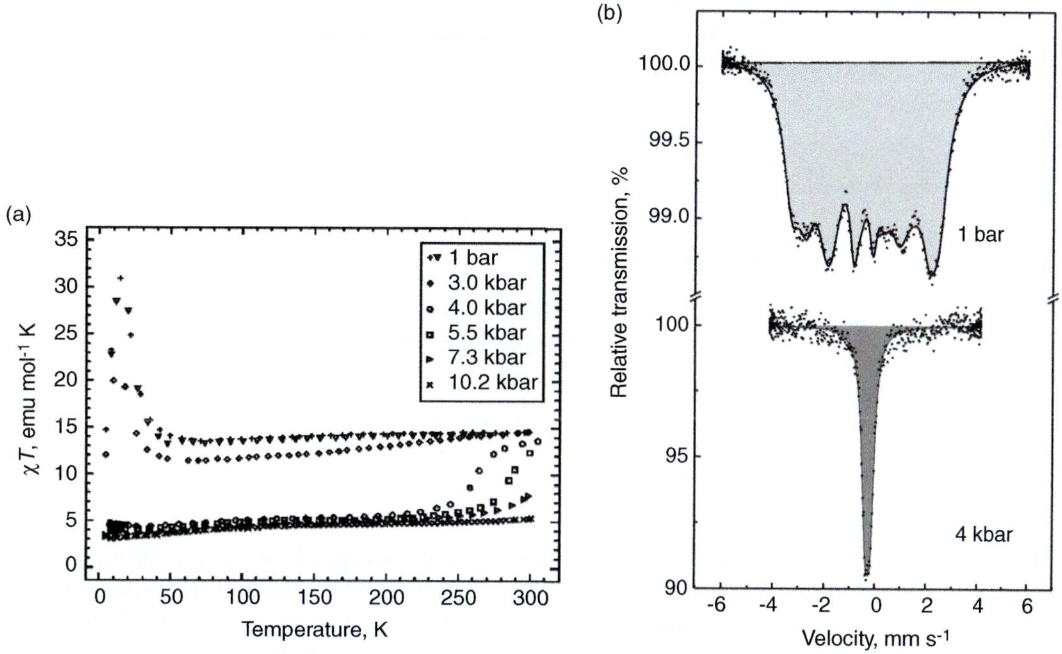

Figure 6.7 *Magnetic properties of $K_{0.03}Co[Fe(CN)_6]_{0.68}\cdot 4.5H_2O$: (a) temperature dependence of χT at different pressures; (b)^{57}Fe Mössbauer spectra recorded at 4.2 K and at 1 or 4 kbar.*

The CTIST behavior is strongly dependent on the external pressure. This sensitivity originates in the appreciable difference in the volume of the coordination sphere of the HS–Co^{II} and LS–Co^{III} ions. The LS–Co^{III}/LS–Fe^{II} pairs are stabilized under applied pressure because of the decrease in the unit cell volume upon conversion from the HS–Co^{II}/LS–Fe^{III} to LS–Co^{III}/LS–Fe^{II} pairs.[39, 40] For example, under ambient conditions, $K_{0.03}Co[Fe(CN)_6]_{0.68}\cdot 4.5H_2O$ exists exclusively in the HS–Co^{II}/LS–Fe^{III} state and undergoes a ferrimagnetic transition at 16 K (Fig. 6.7a).[39] The magnetic ordering is suppressed by application of pressure over 4.0 kbar, as conclusively demonstrated by the disappearance of the hyperfine splitting in the ^{57}Fe Mössbauer spectrum (Fig. 6.7b). Furthermore, a thermal CTIST is observed above this pressure and the transition temperature increases with increasing pressure. Both of these observations confirm the pressure-induced stabilization of the diamagnetic LS–Co^{III}/LS–Fe^{II} state.

One of the most appealing practical applications of compounds that exhibit the CTIST phenomenon is photoinduced switching between two magnetic states. By ramping the temperature down and up during the cooling and heating phases of the hysteresis loop, metastability is imparted to HS–Co^{II}/LS–Fe^{III} or LS–Co^{III}/LS–Fe^{II} state on the LT and HT side of the hysteresis region, respectively (Fig. 6.8). Consequently, a transition between the two states in $Na_{0.27}Co[Fe(CN)_6]_{0.76}\cdot 4.2H_2O$ is achieved by a single-shot laser pulse excitation at 2.38 eV (522 nm).[41] Irradiation of a sample cooled to 205 K (point 3 in Fig. 6.8) resulted in the transition HS–Co^{II}/LS–Fe^{III} $\rightarrow$ LS–Co^{III}/LS–Fe^{II} (3$\rightarrow$5). The change was followed by comparing absorption spectra before and after the excitation. The reverse transition was achieved by irradiating the sample at 225 K (6$\rightarrow$2). Each transition takes place within ~1 ns if the incident photon density overcomes a certain threshold value to achieve full propagation of the excitation throughout the sample.

A number of studies have been aimed at the elucidation of the CT dynamics in these systems. Transient absorption measurements taken on $Na_{0.35}Co[Fe(CN)_6]_{0.79}\cdot 3.7H_2O$ revealed two transient bands at 2.5 eV and

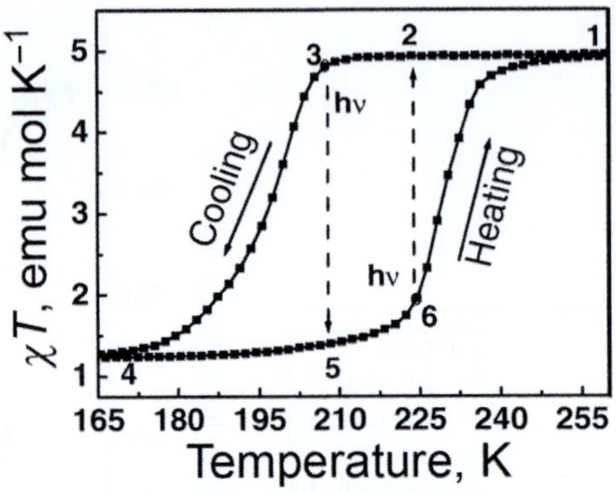

Figure 6.8 Switching between the two states of $Na_{0.27}Co[Fe(CN)_6]_{0.76}\cdot 4.2H_2O$ can be accomplished within the hysteresis loop, which was recorded by changing the temperature from point 1 to point 4 and back, as indicated with the consecutive numbers and arrows. The switching was accomplished by irradiating the sample at point 3 or 6. Reprinted with permission from [41]. Copyright © 2003, American Physical Society.

1.9 eV. Both bands had a biexponential decay. The fast component corresponds to the spin-allowed excitation to the intermediate Frank–Condon CT state LS–Co^{II}/LS–Fe^{III} (I state in Scheme 6.2). This state is populated within ~150 fs of the photoexcitation, and then decays via intersystem crossing to the final photoinduced HS–Co^{II}/LS–Fe^{III} state (PI state in Scheme 6.2), which corresponds to the LS–Co^{II} ($S = ^1/_2$) → HS–Co^{II} ($S = ^3/_2$) spin-transition. The slow component is due to intersystem crossing. The lifetime of the I state ($\tau_I = 300$ ns) is temperature-independent. The lifetime of the PI state is significantly longer and decreases with increasing temperature ($\tau_{PI} = 20$ μs at 215 K and 8 μs at 250 K), indicating a thermally activated process.

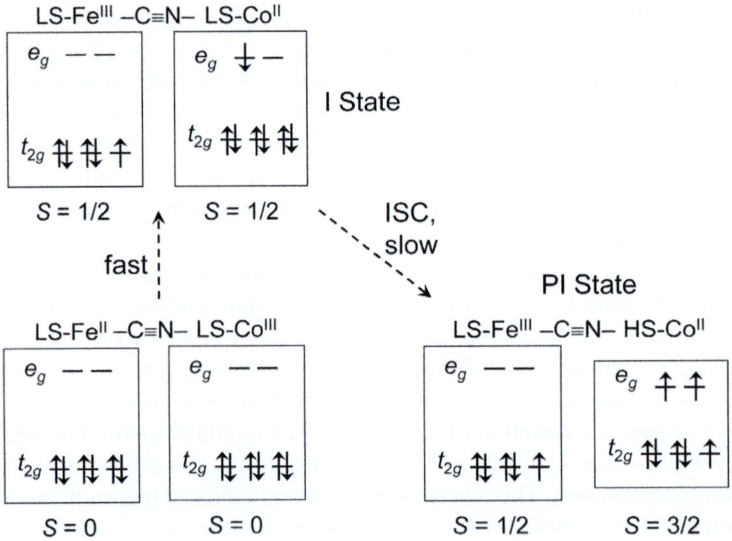

Scheme 6.2 The dynamics of CTIST in the Co-Fe PBA. The initial fast excitation from the diamagnetic LS–Fe^{II}/LS–Co^{III} state to the intermediate (I) LS–Fe^{III}/LS–Co^{II} state is followed by the slower inter-system crossing (ISC) to the PI LS–Fe^{III}/HS–Co^{II} state.

6.3.2 Other Prussian Blue Analogs

An interesting variant of the Prussian Blue structure is obtained when one of the CN^- ligands of hexacyanoferrate is replaced by an ammonia molecule. The reaction between Co^{II} precursors and $[Fe(CN)_5(NH_3)]^{2-}$ anions results in the formation of $Co[Fe(CN)_5(NH_3)]\cdot 6H_2O$,[18] whose 3D CN-bridged structure closely resembles that of Prussian Blue. The difference is that the sixth CN^- linker is missing and the corresponding coordination site on the Co center, that is the site facing the NH_3 molecule coordinated to the Fe center, is occupied by a water molecule. The coordination environment of each Co center is N_5O, which is similar to the average coordination environment in the Co-Fe PBAs that exhibit CTIST behavior. The IR spectrum of $Co[Fe(CN)_5(NH_3)]\cdot 6H_2O$ reveals two ν_{CN} stretches at 2162 and 2117 cm^{-1}, assigned to the bridging CN in Fe^{III}–C≡N–Co^{II} and Fe^{II}–C≡N–Co^{III} sites, respectively. The coexistence of LS Fe centers in the two different oxidation states was also confirmed by Mössbauer spectral data which revealed that the Fe^{III}/Fe^{II} ratio is 3:1 at room temperature, thus indicating that the magnetically coupled HS–Co^{II}/LS–Fe^{III} pairs dominate in the "as-prepared" material.

Similar to the Co-Fe PBAs, $Co[Fe(CN)_5(NH_3)]\cdot 6H_2O$ exhibits ferrimagnetic ordering at 11 K (Fig. 6.9). After irradiation with red light (620–750 nm) at 18 K, the intensity of the IR band at 2162 cm^{-1} increases, that of the 2117 cm^{-1} feature decreases, and the 2-K magnetization increases by ~25%. These changes indicate that a photoinduced electron transfer from the Co^{II} to Fe^{III} centers takes place.

Out of six water molecules in $Co[Fe(CN)_5(NH_3)]\cdot 6H_2O$, one is coordinated to the Co center and the other five are interstitial as determined by heating to 380 K, which results in the loss of only five H_2O molecules per formula unit of the compound. Upon dehydration, the lattice constant changes from 10.20 Å below 340 K to 9.93 Å above this temperature. This change is accompanied by an increase in the intensity of the 2117 cm^{-1} peak and decay of the 2162 cm^{-1} stretch in the IR spectrum, and decrease in the effective magnetic moment upon heating to 340 K. Therefore, dehydration of $Co[Fe(CN)_5(NH_3)]\cdot 6H_2O$ appears to favor the diamagnetic LS–Co^{III}/LS–Fe^{II} state.

Lower dimensionality CN-bridged structures can be obtained by employing capping (blocking) ligands whose presence prevents the growth of the three-dimensional PB-type structure.[6] By using this approach, Sato *et al.* obtained the layered compound $[[(Tp)Fe(CN)_3]_2[Co(bpe)]]\cdot 5H_2O$ (Tp = hydrotris(pyrazolyl)borate, bpe = 1,2-bis(4-pyridyl)ethane).[19] The capping Tp ligand blocks three coordination sites at the Fe^{III} ion

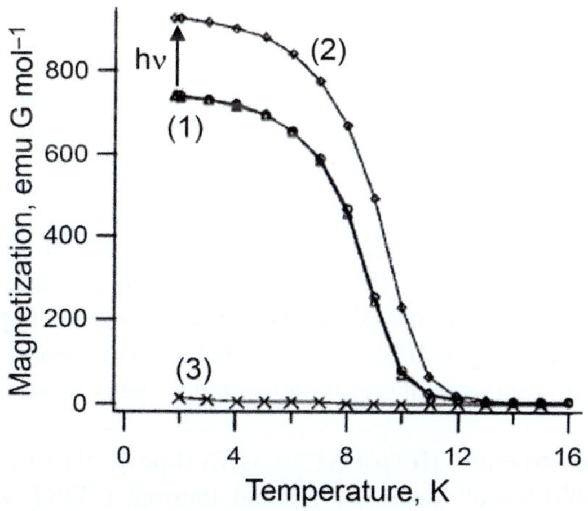

Figure 6.9 *Temperature dependence of magnetization for $Co[Fe(CN)_5(NH_3)]\cdot 6H_2O$ before irradiation (curve 1), after irradiation (curve 2), and after dehydration at 380 K. Reprinted from [18] Copyright 2001, with permission from Elsevier.*

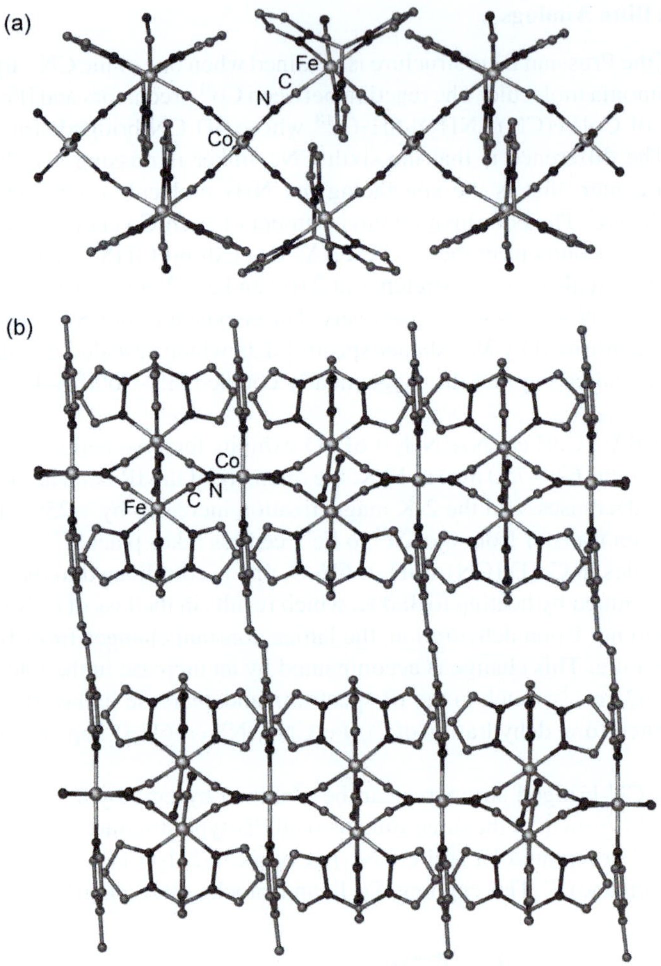

Figure 6.10 *The crystal structure of [[(Tp)Fe(CN)₃]₂[Co(bpe)]]·5H₂O: (a) a fragment of the infinite "[(Tp)Fe(CN)₃]₂Co" chain; (b) a top view of the layer formed by chains connected via ditopic bpe linkers. Hydrogen atoms and interstitial water molecules are omitted for the sake of clarity. Adpated with permission from [19]. Copyright Wiley-VCH Verlag GmbH & Co, 2010.*

and leads to the formation of a "$[(Tp)Fe(CN)_3]_2Co$" chain (Fig. 6.10a). In this structural fragment, two of three CN^- ligands at each Fe^{III} center bridge the Fe^{III} and Co^{II} ions. In turn, each Co^{II} ion is surrounded by four N-bound CN^- ligands in the equatorial plane. The axial sites are occupied by bidentate bpe linkers, which connect the chains into layers (Fig. 6.10b). The space between the layers is filled with interstitial water molecules. Thus, this structure is composed of one-dimensional stripes carved of the PB structure, connected into layers by bpe linkers.

Similar to its parent PB-type structure, $[[(Tp)Fe(CN)_3]_2[Co(bpe)]]\cdot 5H_2O$ exhibits a CTIST between the HS–Co^{II}/LS–Fe^{III} and LS–Co^{III}/LS–Fe^{II} states. A gradual thermal CTIST with a narrow (<10 K) hysteresis is observed between 210 and 150 K (Fig. 6.11a). The transition is accompanied by a pronounced change in the Co–N bond lengths. A photoinduced CTIST is also observed as an increase in the χT value when the sample is irradiated at 5 K for more than 12 h (Fig. 6.11b). The photoinduced state

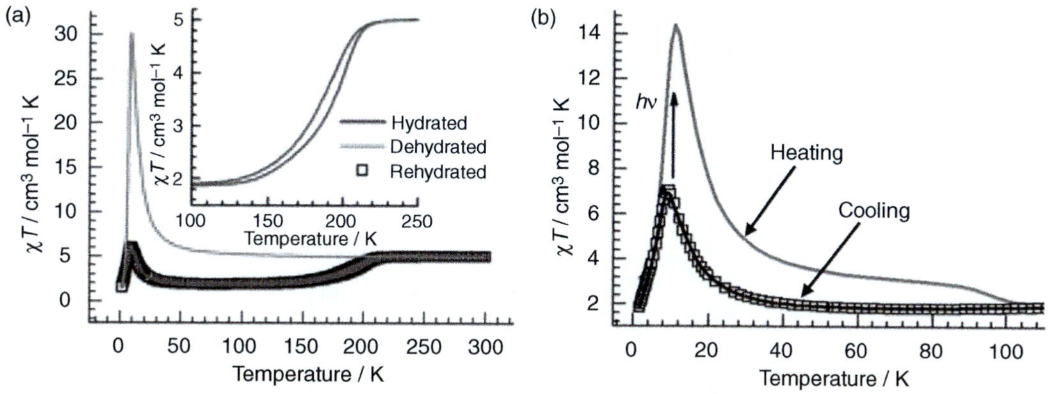

Figure 6.11 *The temperature dependence of χT for [[(Tp)Fe(CN)$_3$]$_2$[Co(bpe)]]·5H$_2$O: (a) hydrated, dehydrated, and rehydrated samples (inset shows a magnified temperature region that corresponds to the thermal CTIST); (b) hydrated sample before and after irradiation, measured in cooling and heating modes, respectively. Adpated with permission from [19]. Copyright Wiley-VCH Verlag GmbH & Co, 2010.*

relaxes back to the ground LS–Co^{III}/LS–Fe^{II} state after heating above 100 K. Interestingly, single crystals of [[(Tp)Fe(CN)$_3$]$_2$[Co(bpe)]]·5H$_2$O can be completely dehydrated without loss of crystallinity to afford [[(Tp)Fe(CN)$_3$]$_2$[Co(bpe)]]. The resulting material does not exhibit CTIST and exists only in the HS–Co^{II}/LS–Fe^{III} state (Fig. 6.11a). Nevertheless, when placed in water overnight or exposed to water vapor for more than one week, the hydrated form of the compound is restored, and the CTIST behavior is observed again.

In recent years, the study of CTIST in Prussian Blue-type materials has been extended to compositions that include 5*d* metal ions. Co[Os(CN)$_6$]$_{0.67}$·4H$_2$O exhibits behavior very similar to its Co-Fe analogs.[17] A thermal CTIST is observed with a 20 K wide hysteresis loop centered at 245 K (Fig. 6.12a). The HT form dominated by the HS–Co^{II}/LS–Os^{III} pairs can be quenched by rapid cooling. The quenched state exhibits ferrimagnetic ordering at ~17 K and reverts to the LT (LS–Co^{III}/LS–Os^{II}) state at 170 K. A photoinduced CTIST is achieved by white-light irradiation of the LT state obtained by slow cooling from room temperature. Similar to the quenched state, the photoinduced form of the compound exhibits an enhanced magnetic moment and ferrimagnetic ordering at ~17 K (Fig. 6.12b). In a strict sense, it is important to note that the exact Fe-containing analog, Co[Fe(CN)$_6$]$_{0.67}$·4H$_2$O,[16, 35] exists as the HS–Co^{II}/LS–Fe^{III} form over the entire temperature range, and does not exhibit neither thermal nor photoinduced CTIST behavior. Thus, the use of 4*d* and 5*d* metal ions for the construction of the CT frameworks should lead to interesting behavior that does not exactly track the properties of the corresponding 3*d* based materials.

It also should be mentioned that another PBA, RbMn[Fe(CN)$_6$], exhibits fascinating photomagnetic properties, but since the behavior is related only to charge transfer and not associated with spin-transitions, we will not consider this compound in the present account. Interested readers are referred to an excellent recent review by Tokoro and Ohkoshi.[42]

6.3.3 Coordination Polymers Based on Octacyanometallates

Octacyanometallates [M(CN)$_8$]$^{n-}$ (M = Mo, W) exist in two common oxidation states, M^{IV} and M^{V} with $n = 4$ and 3, respectively. Therefore, a CTIST can be expected in assemblies that combine these anions with Co^{II} cations. Indeed, in 2003, Ohkoshi, Hashimoto, *et al.* reported a photomagnetic coordination polymer Cs[Co(3-CNpy)$_2$][W(CN)$_8$]·H$_2$O (3-CNpy = 3-cyanopyridine),[20] whose structure consists of Co and W

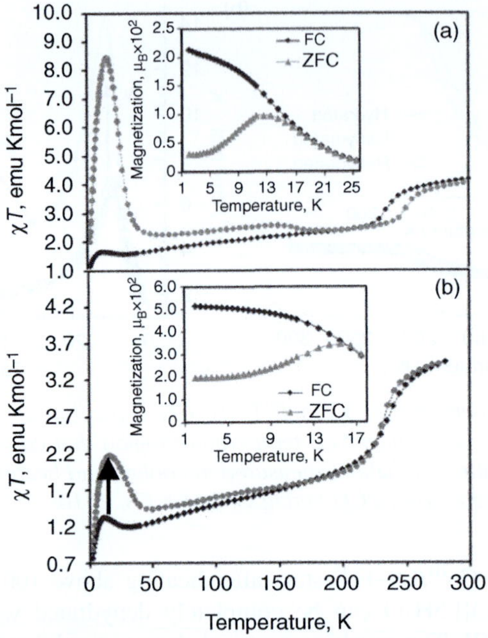

Figure 6.12 *The temperature dependence of χT and low temperature field cooled (FC) vs. zero-field cooled (ZFC) magnetization (insets) for $Co[Os(CN)_6]_{0.67}{\cdot}4H_2O$: (a) a slowly cooled sample (dark) vs. a thermally quenched sample (pale); (b) a slowly cooled sample (dark) vs. an irradiated sample (pale). Reprinted with permission from [17]. Copyright 2010, American Chemical Society.*

centers bridged by CN^- ligands to form a two-dimensional layered architecture (Fig. 6.13a). Each Co site is octahedral with four N-bound CN^- ligands from octacyanotungstate ions and two pyridyl N atoms of 3-CNpy molecules which are coordinated above and below the layer (Fig. 6.13b). The compound exhibits an abrupt thermal CTIST that occurs at $T_{1/2\downarrow} = 167$ K upon cooling and at $T_{1/2\uparrow} = 216$ K upon heating* (Fig. 6.13c). The transition corresponds to a change from the paramagnetic HT form which contains HS–Co^{II} ($S = 3/2$) and LS–W^V ($S = 1/2$) ions to the diamagnetic LT form which contains LS–Co^{III} and LS–W^{IV} ions (both $S = 0$). Similar to the Co-Fe and Co-Os PBAs, the HS–Co^{II}/LS–W^V state can be quenched by rapid cooling (>10 K/min) of the HT form[43, 44] or generated from the LS–Co^{III}/LS–W^{IV} state by photoexcitation of the LT form at lower temperatures.[20, 44] The photoinduced form of the compound exhibits magnetic ordering at 30 K (Fig. 6.13d). Interestingly, the CTIST can be induced not only by visible light, but also by X-ray irradiation, as was revealed during studies of this material by XANES and EXAFS spectroscopies.[43]

The significant decrease in the Co–N distances observed upon conversion from the HS–Co^{II} to the LS–Co^{III} ion is reflected in an abrupt change in the unit cell volume associated with the CTIST.[45] This change causes a strong dependence of CTIST in $Cs[Co(3\text{-}CNpy)_2][W(CN)_8]{\cdot}H_2O$ on external pressure.[46] With increasing applied pressure, the thermal CTIST hysteresis loop is shifted to progressively higher temperatures, owing to the pressure-induced stabilization of the smaller-volume diamagnetic LT form. Finally, above 0.44 GPa only the LT form was observed in the 200–300 K temperature range. Interestingly, the pressure-induced change in magnetic behavior is also accompanied by hysteresis; as the pressure was released, the HT form re-appeared only below 0.30 GPa.

* $T_{1/2}$ indicates the temperature at which the transition is half complete.

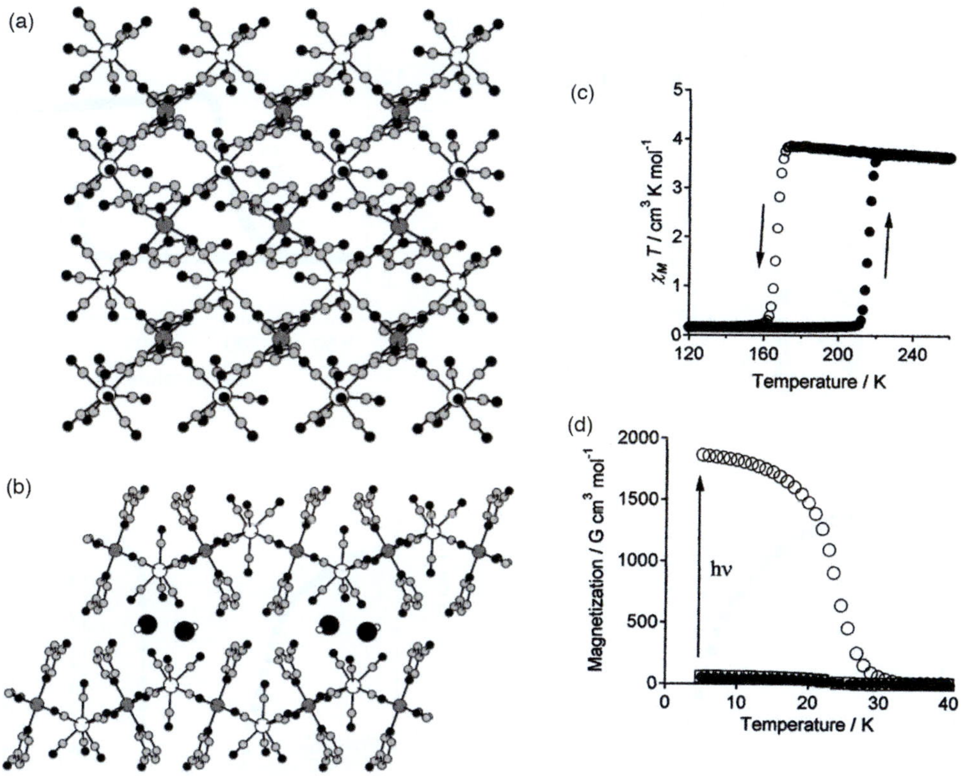

Figure 6.13 *(a) The top view of a CN-bridged layer in the crystal structure of Cs[Co(3-CNpy)$_2$][W(CN)$_8$]·H_2O. (b) A side view of two adjacent layers separated by interstitial water molecules. (c) The temperature dependence of χT. (d) The emergence of the photoinduced magnetically ordered state upon irradiation at low temperature. Reprinted with permission from [20]. Copyright 2003 American Chemical Society.*

By replacing 3-CNpy with pyrimidine (pym), the Ohkoshi group isolated the three-dimensional coordination polymer [[Co(H_2O)$_2$][Co(pym)$_2$]$_2$[W(CN)$_8$]$_2$]·4H_2O.[21] The formation of this structure is favored due to the improved bridging ability of the pym ligand as compared to 3-CNpy. Two of the three Co sites (Co1) in the formula unit form a CN-bridged layer with W sites, similar to the situation in the structure of Cs[Co(3-CNpy)$_2$][W(CN)$_8$]·H_2O. The octahedral coordination environment of these Co sites includes four N-bound CN^- ligands in the equatorial plane and two pym ligands in the axial positions (Fig. 6.14a). The layers are "stitched" into the three-dimensional framework by means of the remaining Co site (Co2), which is positioned between them and coordinated by one CN^- and one pym ligand from each layer. Two water molecules complete the octahedral coordination environment of this ion. Thus, the pym ligands participate in bridging Co1 and Co2 sites. The [W(CN)$_8$] fragment shares four of its CN^- ligands to create the layers and another CN^- ligand is used to connect the layers via the Co2 sites.

The material [[Co(H_2O)$_2$][Co(pym)$_2$]$_2$[W(CN)$_8$]$_2$]·4H_2O exhibits a thermal CTIST with a remarkably large hysteresis. The transition from the HT HS–Co^{II}/LS–W^V state to the LT LS–Co^{III}/LS–W^{IV} state occurs at $T_{1/2\downarrow} = 208$ K, while the reverse transition (in the heating regime) takes place at $T_{1/2\uparrow} = 298$ K (Fig. 6.14b). Only two of the three Co^{II} ions undergo the CTIST, which can be explained by the availability of only two W^V centers for the MM′CT, as well as by the weaker ligand field at the Co2 site. The latter remains in the HS–Co^{II} state at all temperatures, thus contributing to the rather large residual χT value in the LT form of

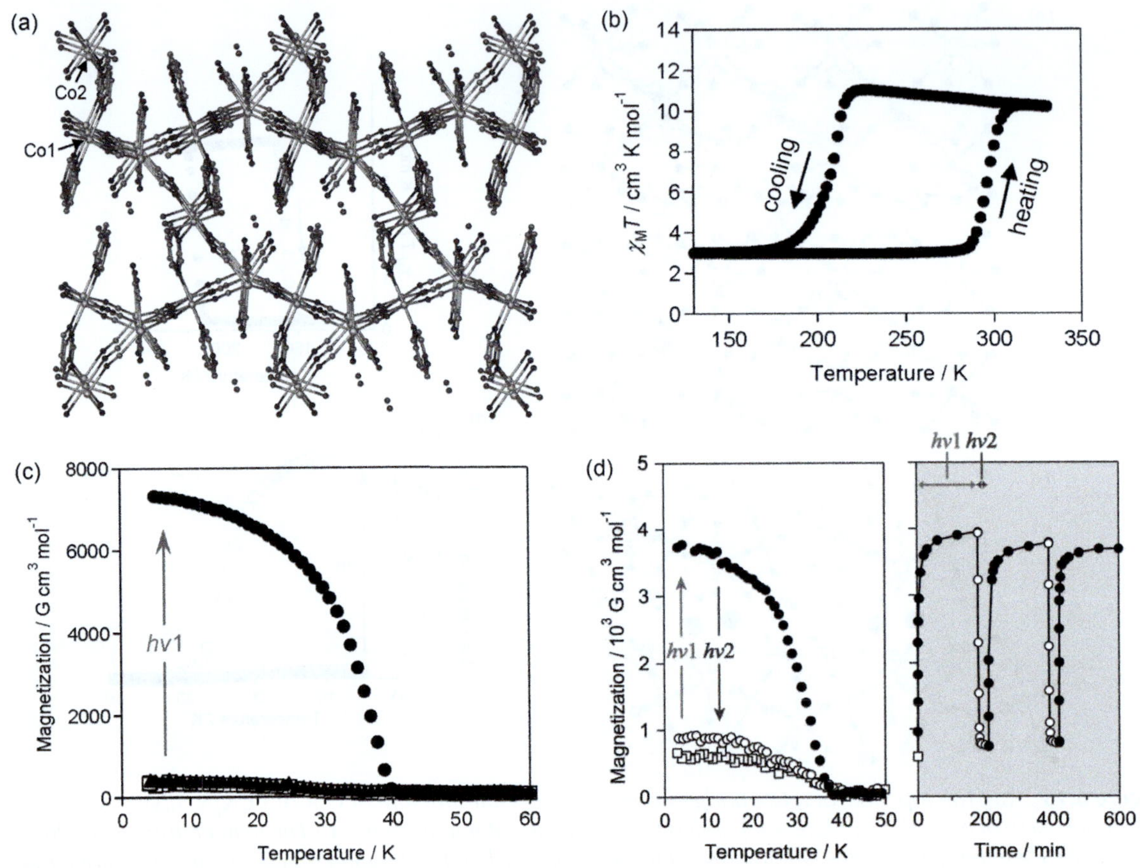

Figure 6.14 (a) The crystal structure of $[[Co(H_2O)_2][Co(pym)_2]_2[W(CN)_8]_2]\cdot 4H_2O$. The Co1 and Co2 sites are indicated. The H atoms are omitted for the sake of clarity. (b) The temperature dependence χT in the cooling and heating modes. (c) Photogeneration of the magnetically ordered state. (d) The reversible photoswitching between the "dark" and photomagnetic states by means of two different irradiation wavelengths ($\lambda_1 = 840$ nm, $\lambda_2 = 532$ nm). Reprinted with permission from [21]. Copyright 2008, American Chemical Society.

the compound. The LT state does not exhibit any signatures of magnetic ordering, but, after irradiation with a 840-nm laser at 5 K, the material becomes ferromagnetic with $T_C = 40$ K and $H_c = 1.2$ T at 2 K (Fig. 6.14c). As in previous cases, the emergence of magnetic ordering is due to a photoinduced MM′CT from the LS–Co^{III} to LS–W^{IV} ion which results in the magnetically coupled HS–Co^{II}/LS–W^{V} state. The photoinduced state relaxes back to the LS–Co^{III}/LS–W^{IV} state above 150 K. The irradiation wavelength of 840 nm was chosen because of its overlap with the broad MM′CT transition of the LT form ($\lambda_{max} = 772$ nm). Irradiation into the band corresponding to the M′MCT transition of the photoinduced state ($\lambda_{max} = 473$ nm) with a 532-nm laser results in the recovery of the LT form. This reversible switching between the photoinduced and LT forms of the compound can be repeated multiple times (Fig. 6.14d).

Upon standing under mother liquor, $[[Co(H_2O)_2][Co(pym)_2]_2[W(CN)_8]_2]\cdot 4H_2O$ slowly converts to a different isomer, namely $[[Co(H_2O)_2][Co(pym)(H_2O)]_2[W(CN)_8]_2]\cdot 2H_2O\cdot 2pym$.[21] The structure of the new phase is very similar to the parent compound, but one of the pym ligands in the axial positions of each Co1 (in-layer) site has been replaced by a water molecule. Consequently, the Co1 site experiences a weaker ligand field and does not exhibit CTIST behavior. The structure, therefore, contains only HS–Co^{II} and LS–W^{V} ions, which are magnetically coupled via the CN^- bridges, and the compound exhibits ferromagnetic ordering at 32 K, with $H_c = 1.2$ T at 2 K.

Compounds designed on a similar principle were reported by Sieklucka *et al.*,[47] who used pyrazine (pyz) or 4,4′-bipyridine (bpy) instead of pym to connect the CN-bridged Co/W layers into a three-dimensional framework. The formulae of the resulting solids, $[[Co_3(H_2O)_6(pyz)_3][W(CN)_8]_2]\cdot 3.5H_2O$ and $[[Co_3(H_2O)_4(bpy)_{4.5}][W(CN)_8]_2]\cdot 6H_2O$, are reminiscent of $[[Co(H_2O)_2][Co(pym)_2]_2[W(CN)_8]_2]\cdot 4H_2O$, but their exact structures remain unknown because of the lack of single crystal samples. Given the fact that the bpy-containing compound exhibits an abrupt increase in magnetization under irradiation with green light at 4.2 K,[48] it would be interesting to further pursue the properties of these materials.

6.4 CTIST in Nanoscale Materials

6.4.1 Thin Films

After the initial discovery of CTIST in bulk solid Co-Fe PBAs, the Hashimoto group demonstrated that this effect also could be observed in thin films.[22] Moreover, their studies revealed a strong dependence of the observed behavior on the identity of the alkali metal ions present in the structure, similar to the studies discussed above for the bulk materials.[16, 35] Thin-film $Na_{0.31}Co[Fe(CN)_6]_{0.77}\cdot 3.8H_2O$ was prepared on the surface of a Pt electrode. Powder X-ray diffraction revealed a unit cell parameter of 10.33 Å and the IR spectrum showed a ν_{CN} stretch at 2160 cm^{-1}, suggesting the presence of HS–Co^{II}/LS–Fe^{III} pairs. The LS–Fe^{III} state was also confirmed by Mössbauer spectroscopy. The material exhibits thermal CTIST when cooled below 280 K, with the LS–Co^{III}/LS–Fe^{II} state dominating below 230 K. Interestingly, conversion to the diamagnetic state was also achieved by cation exchange. After immersing the film in 1 M solution of KCl only for 5 s at 313 K, the ν_{CN} stretch was shifted to 2135 cm^{-1} and the unit cell parameter decreased to 9.96 Å. Elemental analysis indicated essentially complete substitution of K^+ for Na^+ ions. The HS–Co^{II}/LS–Fe^{III} state was partially restored after immersing the film of $K_{0.31}Co[Fe(CN)_6]_{0.77}\cdot 3.8H_2O$ into 1 M NaCl solution. The behavior observed for the thin-film samples is also in accord with the demonstrated stabilization of the HS–Co^{II}/LS–Fe^{III} state in the bulk materials with smaller A^+ ions (see pp. 179–180 and Fig. 6.6).[38]

An interesting phase-separation phenomenon was reported for electrochemically synthesized films that contain a variable ratio of Na^+ and K^+ ions, $(Na_{1-\delta}K_\delta)_{0.35}Co[Fe(CN)_6]_{0.74}\cdot zH_2O$.[49] At 300 K, a sample with $\delta = 0.07$ showed preference for the HS–Co^{II}/LS–Fe^{III} pairs, while a sample with $\delta = 0.85$ was dominated by the LS–Co^{III}/LS–Fe^{II} pairs. Samples with intermediate Na^+/K^+ ratios ($\delta = 0.37, 0.50, 0.58, 0.71$) exhibited two sets of X-ray powder diffraction peaks (Fig. 6.15) that were well reproduced by the Rietveld refinement assuming the co-existence of two phases, namely HS–Co^{II}/LS–Fe^{III} (Na^+-rich) and LS–Co^{III}/LS–Fe^{II} (K^+-rich). The phase separation was possible due to the facile migration of cations in the film.

Thermal CTIST was observed in all the materials that contained HS–Co^{II}/LS–Fe^{III} pairs at room temperature. In the material with $\delta = 0.07$, which had predominantly HS–Co^{II}/LS–Fe^{III} pairs, the CTIST was identified by a change in the fcc unit cell parameter from 10.34 Å above 230 K to 10.00 Å below 230 K. In the biphasic materials with $\delta = 0.37, 0.50, 0.58, 0.71$, the CTIST was identified by the coalescence of the peaks of the HS–Co^{II}/LS–Fe^{III} (Na^+-rich) phase with those of the LS–Co^{III}/LS–Fe^{II} (K^+-rich) phase below 230 K.

The thin-film and bulk samples of Co-Fe PBAs exhibit similar CTIST features in their powder diffraction patterns as well as IR and Mössbauer spectra, that is in the properties related to the local bonding characteristics of the metal sites undergoing the CTIST. On the other hand, the long-range magnetic ordering for the samples with a large abundance of HS–Co^{II}/LS–Fe^{III} pairs reveals distinct anisotropic effects for

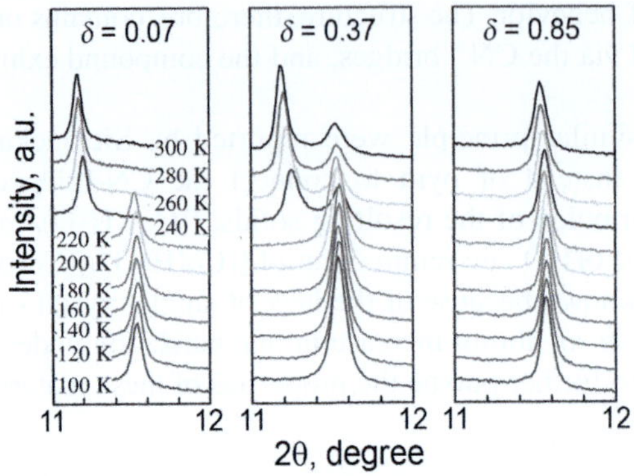

Figure 6.15 Temperature dependence of the (400) reflection in the powder X-ray diffraction patterns of thin-film samples of $(Na_{1-\delta}K_\delta)_{0.35}Co[Fe(CN)_6]_{0.74}\cdot zH_2O$. Reproduced with permission from [49]. Copyright © 2009, American Physical Society.

lower-dimensional structures. A lamellar mesostructure containing monolayers of CN-bridged Co^{II} and Fe^{III} ions separated by clay-amphiphile nanosheets (Fig. 6.16a) was prepared by Langmuir–Blodgett templating.[50] The mesostructure contains 100 repeat units, with the regular spacing of ~22 Å. The ν_{CN} stretches appeared at 2159 and 2094 cm^{-1} at 300 K, but at 8 K their intensities were decreased at the expense of a growing ν_{CN} stretch at 2112 cm^{-1}, suggesting an increase in the fraction of LS–Co^{III}/LS–Fe^{II} pairs due to gradual CTIST. Due to the large separation between the monolayers, this mesostructure represents an excellent model for the evaluation of quantum confinement effects in the magnetism of the Co-Fe PBA. The sample showed magnetic ordering at 15 K, but the saturation magnetization was significantly higher for the case when the monolayers were oriented parallel to the magnetic field than for the perpendicular alignment (Fig. 6.16b). The hysteresis loop also showed higher coercivity for the former orientation. A photoinduced increase in magnetization was observed for each orientation of the sample, but again, it was more pronounced when the field was parallel to the film (Fig. 6.16c). Thus, the monolayers clearly exhibit easy in-plane magnetization and the direction normal to the plane can be considered a hard axis.

The research groups of Meisel and Talham prepared thin films of $Rb_{0.18}Co[Fe(CN)_6]_{0.75}$ by sequentially dipping a Melinex substrate into aqueous solutions of Co^{II} and $[Fe(CN)_6]^{3-}/Rb^+$ ions.[23] It was shown that the

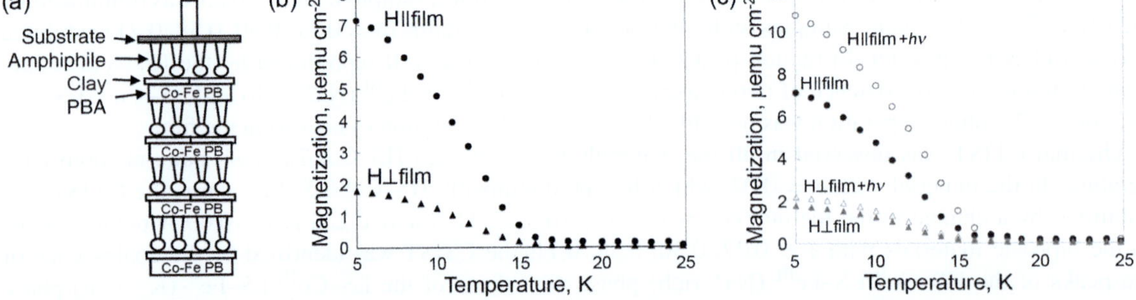

Figure 6.16 (a) A schematic representation of the lamellar mesostructure, in which clay-amphiphile nanosheets alternate with Co-Fe PBA monolayers. (b) Temperature-dependent magnetization with the field parallel and perpendicular to the surface of the film. (c) Photoinduced increase of magnetization for two different orientations of magnetic field relative to the film surface. Reprinted with permission from [50]. Copyright 2005, American Chemical Society.

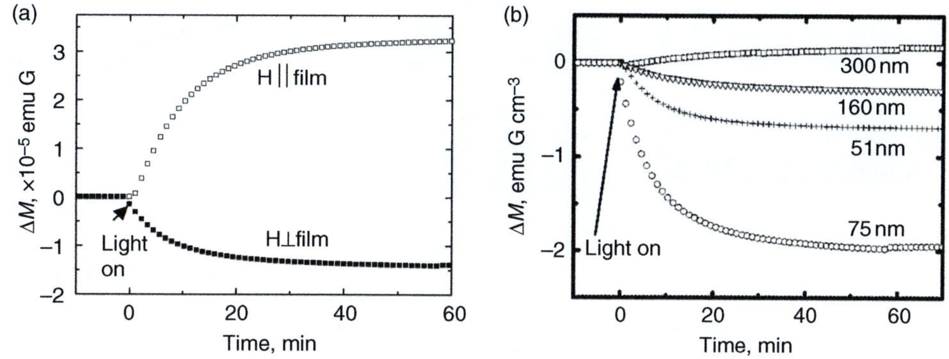

Figure 6.17 *The dependence of light-induced changes in the magnetization of $Rb_{0.18}Co[Fe(CN)_6]_{0.75}$ thin films on the orientation of the applied magnetic field (a) and the thickness of the film (b). Reprinted with permission from [23] Elsevier and with permission from [51] Copyright 2007, American Chemical Society.*

films prepared at short immersion times (~3 s) and without thorough rinsing between changes in solutions were thicker and appeared powder-like when examined by scanning electron microscopy. In contrast, the films prepared with longer immersion times (~60 s) and with careful rinsing after each dipping were about 5-times thinner and showed a smooth, quasi-2D texture. Unlike the monolayers described above, these films did not exhibit a notable anisotropy for the magnetic susceptibility. Nevertheless, an interesting anisotropic effect was observed in their photomagnetic behavior. When the applied magnetic field is parallel to the film, a photoinduced increase in magnetization was observed, but the field applied perpendicular to the film surface led to a photoinduced decrease in magnetization (Fig. 6.17a). This effect is observed for the films thinner than 200 nm (Fig. 6.17b).[51] For thicker films, the photoinduced magnetic behavior becomes similar to that observed for bulk samples. Initially, this anisotropic magnetic response was explained by a dipolar (stray) field emanating from primordial ferrimagnetic regions that exist in the material prior to irradiation.[52] The dipolar field acts on the remaining diamagnetic regions that undergo photoinduced magnetic ordering in the direction opposite to the applied magnetic field when the latter is oriented perpendicular to the film surface, thus causing a decrease in total magnetization. This explanation, however, was later questioned by the same authors as being inconsistent with some of the changes associated with the photoinduced CTIST in the thin film, and the anisotropic magnetization was attributed to an anisotropy field generated at the film–substrate interface.[51]

The ability to control the thickness of the PB-type film offers opportunities not feasible in the case of bulk synthesis. Thus, some recent efforts have been directed at the preparation of thin-film heterostructures combining alternating layers of the photomagnetic Co-Fe PBA with another PB-type material (Fig. 6.18a). In such heterostructures, generated by the aforementioned sequential adsorption method, a layer of photomagnetic $Rb_{0.18}Co[Fe(CN)_6]_{0.75}$ can be sandwiched between two layers of ferromagnetic $Rb_{0.2}Ni[Cr(CN)_6]_{0.73}$ (T_C ~ 70 K).[53, 54] Remarkably, the multilayer architecture exhibits a photoinduced decrease of magnetization below the Curie temperature of the Ni-Cr PBA film (Fig. 6.18b), despite the fact that the latter phase is not, itself, photomagnetic! (Note that in these studies all the measurements were performed with the magnetic field parallel to the film surface in order to avoid possible demagnetization effects observed for thin films of $Rb_{0.18}Co[Fe(CN)_6]_{0.75}$.[52]) Thus, one can assume that the photoinduced changes observed at temperatures below 70 K stem from the CTIST in the middle Co-Fe layer, and the decrease in the total magnetization is explained by the influence of this layer on the adjacent Ni-Cr layers.

Indeed, the bulk phase of $Ni[Cr(CN)_6]_{0.67}\cdot 4H_2O$ was shown to exhibit lower magnetization under applied pressure.[55] This phenomenon was attributed to canting of ferromagnetically aligned magnetic moments of Ni^{II} and Cr^{III} ions under external stress. A similar mechanism was proposed for the thin-film heterostructures. The

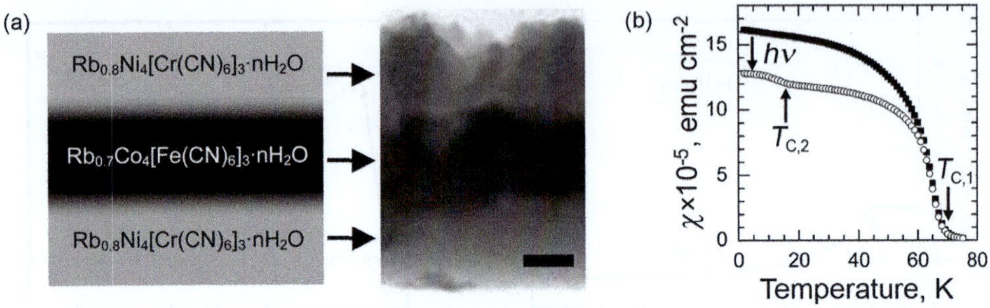

Figure 6.18 *(a) A schematic representation of the thin-film heterostructure composed of two different PBAs (left) and a transmission electron micrograph showing the cross-section of the real heterostructure (right). (b) The temperature dependence of magnetic susceptibility for the trilayer heterostructure before and after irradiation (black and white circles, respectively). Reprinted with permission from [53]. Copyright 2010, American Chemical Society.*

photoinduced CTIST results in the increase of the unit cell parameter for the Co-Fe layer due to the transition from the low-volume LS–Co^{III}/LS–Fe^{II} state to the high-volume HS–Co^{II}/LS–Fe^{III} state. This expansion introduces a substantial strain in the interface region between the Co-Fe and Ni-Cr films which is transmitted to the Ni-Cr part of the heterostructure, causing canting of the ordered magnetic moments. Importantly, the underlying mechanism of the photoinduced changes in these heterostructures remains the same as in the pure Co-Fe PBAs, which implies that the magnetism of the Co-Fe layer should increase under irradiation conditions. A careful examination of the low-temperature region of the magnetic susceptibility data for the photoinduced phase indicates another magnetic transition around 15 K, which was not observed for the dark state (Fig. 6.18b). This transition stems from the photoinduced ferrimagnetic ordering in the Co-Fe layer, which increases the magnetization in this part of the heterostructure, thus contributing to the smaller decrease in the total magnetization. Nevertheless, the total magnetization is still lower in the photoinduced state as compared to the dark state, owing to the larger relative contribution from the Ni-Cr layers. It is also important to note that an appreciable photoinduced effect was observed only for heterostructures, in which each layer was at least 200 nm thick, thus suggesting that certain long-range propagation of the strain generated in the interface region is necessary to achieve significant photoinduced changes in magnetization.

6.4.2 Nanoparticles

Nanoscaled PB-type materials have been actively investigated since the seminal report by Mann and co-workers on the use of "water-in-oil" microemulsions (reverse micelles) for the preparation of PB nanoparticles (NPs).[56] The same authors were the first to report the synthesis of the Co-Fe PBA in the NP form (Fig. 6.19).[57] The size of these cube-shaped NPs can be controlled by varying the water-to-surfactant ratio. By changing the $[H_2O]/[AOT]$ ratio from 10 to 30 (AOT = bis(2-ethylhexyl)sulfosuccinate), that is by increasing the volume of the water droplet in the microemulsion, the particle size was varied from 17 ± 3 to 22 ± 2.5 nm. The IR spectra of NPs suspended in isooctane showed ν_{CN} stretches at 2160, 2120, and 2090 cm^{-1}, corresponding to Co^{II}–NC–Fe^{III}, Co^{III}–NC–Fe^{II}, and Co^{II}–NC–Fe^{II} linkages, respectively. Thus, the CT between the Co and Fe sites also takes place in the nanomaterial, similar to the CT in the bulk and thin-film samples.

The evolution of magnetic and photomagnetic properties with the size of NPs was studied by Meisel, Talham, and co-workers[24] who prepared NPs of $Rb_xCo[Fe(CN)_6]_y{\cdot}zH_2O$ protected by polyvinylpyrrolidone (PVP). The size of the particles was varied from 3.3(8) to 13(3) nm by changing the concentration of PVP.

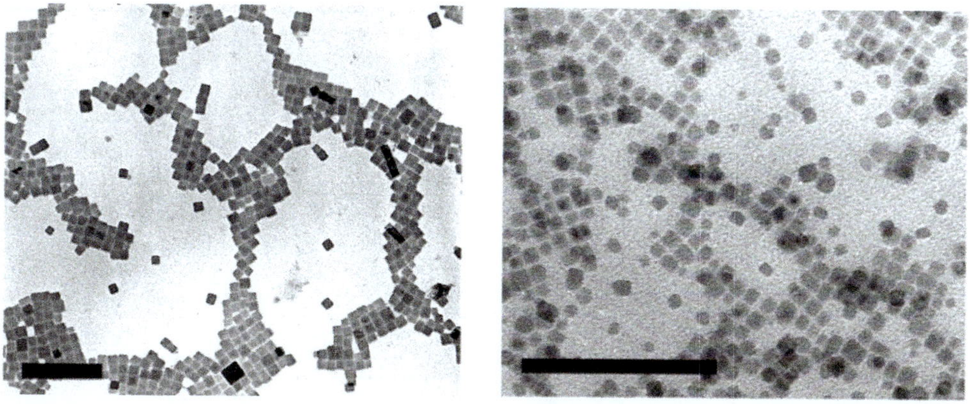

Figure 6.19 *Cubic nanoparticles of the Co-Fe PBA at various magnifications. The length of the black bar corresponds to 200 nm. Reprinted with permission from [57]. Copyright 2002, American Chemical Society.*

The ν_{CN} stretch observed at 2124 cm^{-1} indicated the dominant presence of diamagnetic LS–Co^{III}/LS–Fe^{II} pairs. Under irradiation at 5 K, NPs of all sizes exhibited an increase in magnetization due to photoinduced CTIST, but magnetic ordering was observed only for NPs of ~10 nm and larger (Fig. 6.20), suggesting that smaller NPs are below the single-domain limit. The Curie temperature increases with larger particle sizes, as well as under irradiation. The photoinduced CTIST also resulted in larger coercivities for the magnetically ordered samples.

Steigman and co-workers embedded Co-Fe PBA NPs into an optically transparent silica xerogel.[58] The NP size established by transmission electron microscopy was 8–10 nm. Magnetic measurements indicated magnetic ordering below 13 K, and a photoinduced increase in magnetization was observed under irradiation with $\lambda \geq 550$ nm at 5 K. Similar xerogel nanocomposites were reported by Bleuzen and co-workers,[59, 60] but, despite photoinduced increase in magnetization, no magnetic ordering was observed. This might be explained by the smaller sizes of the NPs obtained in the latter studies ($\leq$8 nm) and appears to agree with the single-domain size limit established in the studies of surfactant-protected NPs of the Co-Fe PBA.[24] Superparamagnetism was reported below the single-domain limit.[58]

Shortly after the first reports on the preparation of nanosized PB and its analogs in reverse micelles, it was demonstrated that the NPs of these materials could be stabilized by means of electrostatic effects, even in the absence of surfactants.[61] When the synthesis is performed with an access of cyanometallate, the particles become negatively charged, which prevents aggregation and stabilizes the suspension of NPs. In addition, the surface of the particles remains reactive, allowing the growth of successive shells around the NP core.[62] Using this strategy, the research team of Talham and Meisel investigated core-shell photomagnetic NPs[63] following on their earlier studies of similar thin-film heterostructres.[53, 54] The growth of NPs containing various combinations of Co-Fe and Ni-Cr PBAs in core-shell or core-shell-shell configuration, with well-defined interface separations (Fig. 6.21), was achieved by carefully controlling the concentration and the addition rate of the reactants. Cognizant of the fact that the photoinduced effects were most evident in films thicker than 200 nm,[54] the authors targeted 100–200 nm diameters for each core and shell of the NP. The obtained heterostructured NPs behaved similarly to the heterostructured films, that is the photoinduced decrease in magnetization was observed below the magnetic ordering point of the Ni-Cr substructure (~70 K). The structural expansion of the Co-Fe substructure caused by the photoinduced CTIST from low-volume diamagnetic LS–Co^{III}/LS–Fe^{II} pairs to high-volume magnetically coupled HS–Co^{II}/LS–Fe^{III} pairs exerts strain on the Ni-Cr substructure. This "internal pressure" leads to canting of the ferromagnetically aligned

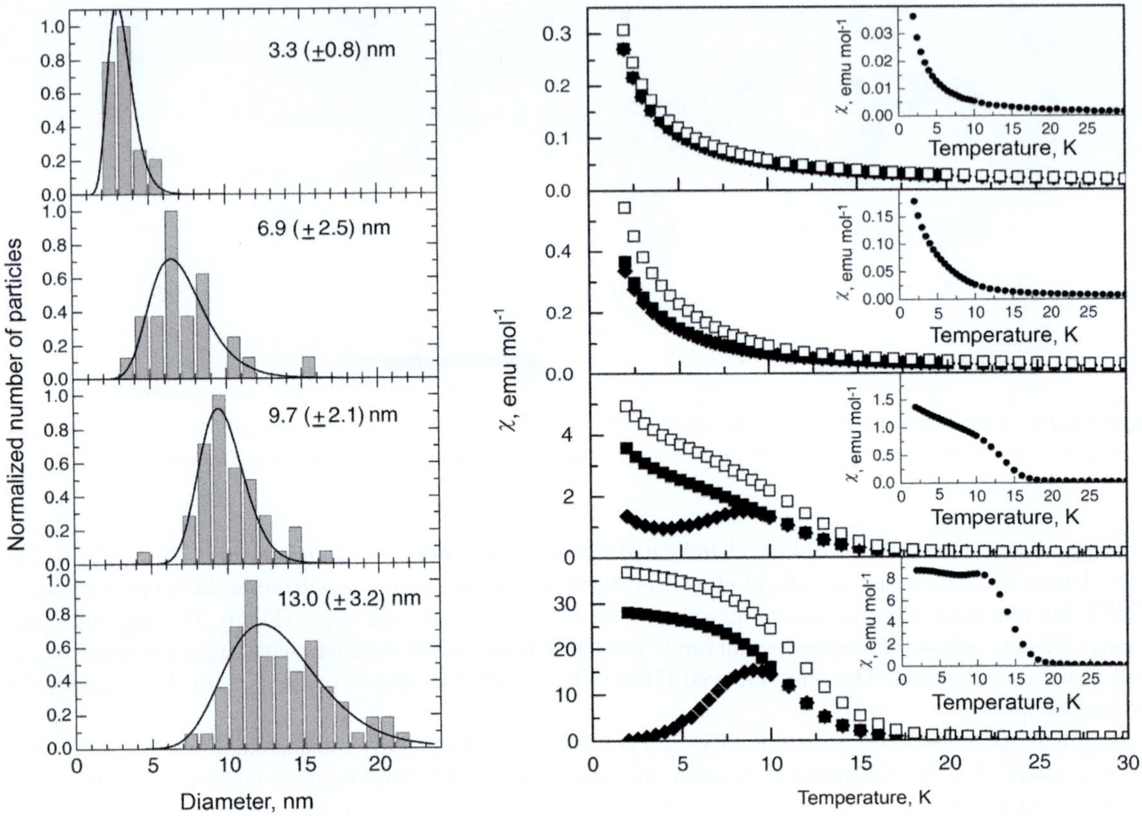

Figure 6.20 Size distribution and corresponding magnetic behavior of four different batches of Co-Fe PBA nanoparticles. The temperature dependence of magnetic susceptibility was measured in the field cooled (■) and zero-field cooled (♦) modes, as well as in the field cooled mode after irradiation (□). The insets show the difference in the field cooled susceptibility before and after irradiation. Reproduced with permission from [55]. Copyright Institute of Physics Publishing.

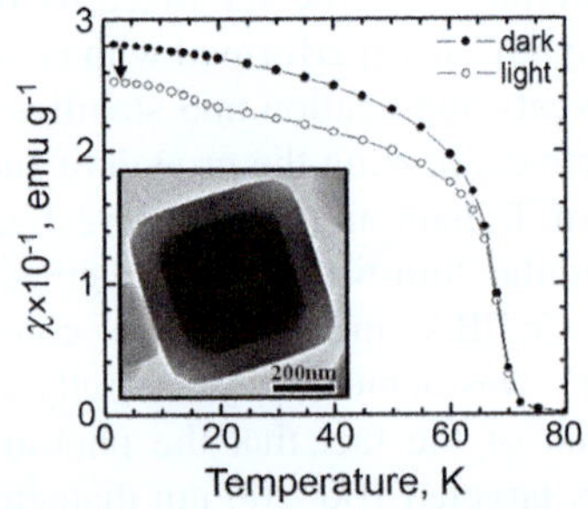

Figure 6.21 Temperature dependence of magnetic susceptibility for the core-shell nanocomposite of Co-Fe and Ni-Cr PBAs before and after photoexcitation (black and white circles, respectively). The arrow indicates the photoinduced decrease in total magnetization. Inset: a high-resolution TEM micrograph of the corresponding core-shell nanoparticle. Reprinted with permission from [63]. Copyright 2011, American Chemical Society.

Ni^{II} and Cr^{III} moments and decrease in the total magnetization, provided the relative change in magnetization is higher in the Ni-Cr substructure than in the Co-Fe one. This explanation is supported by two additional observations: (i) the decrease in magnetization is less pronounced below 20 K, the ferrimagnetic ordering temperature for the Co-Fe substructure (Fig. 6.21) and (ii) when the relative volume of the Co-Fe part is made sufficiently large, a photoinduced increase in magnetization is observed.

6.5 CTIST in Polynuclear Transition Metal Complexes

In 2004, Dunbar, Achim, and their collaborators reported the observation of a CTIST in $[[Co(tmphen)_2]_3[Fe(CN)_6]_2]\cdot n$Solv.[8] This complex belongs to a general family of trigonal bipyramid complexes of the type $[[M(LL)_2]_3[M'(CN)_6]_2]$, in which LL represents a bidentate ligand such as bipyridine (bpy) or o-phenantroline (phen) and M and M′ are transition metal ions in oxidation state +2 or +3. In the last 15 years, Dunbar and her collaborators have demonstrated that complexes in this family exhibit a remarkably broad range of magnetic properties including single molecule magnetism and spin crossover as well as CTIST (Scheme 6.3).[64]

The crystal structure of $[[Co(tmphen)_2]_3[Fe(CN)_3]_2(\mu\text{-}CN)_6]\cdot n$Solv revealed that it is a molecular species[8, 9] related to the PBAs for which CTIST was originally reported.[5] While the latter PBAs exhibit an infinite 3D arrangement of metal cations and cyanide ligands, the $[[Co(tmphen)_2]_3[Fe(CN)_3]_2(\mu\text{-}CN)_6]\cdot n$Solv is constrained to the molecular regime owing to the presence of tmphen blocking ligands coordinated to Co, which prevent the growth of extended structures.[6] Such blocking ligands have been used since 2004 to prepare several other molecule-based materials that exhibit CTIST. These materials are based on heterometallic, cyanide-bridged clusters with square (Table 6.3),[26, 65–68] star,[27] cube,[28] and crown[69] core geometries (Scheme 6.4).

The majority of the complexes that exhibit CTIST behavior are based on a combination of cobalt and iron ions. In these clusters, the CTIST relates HS–Co^{II}/LS–Fe^{III} and LS–Co^{III}/LS–Fe^{II} pairs, in a manner akin to the extended structures of PBAs discussed in the previous two sections. The relative energy of the two pairs is important for the manifestation of CTIST behavior. On the basis of electrochemical data, Oshio and coworkers determined that the difference in free energy between the two pairs (expressed as the corresponding potential difference) is 0.41 V for a $[Fe_2Co_2]$ square cluster which undergoes CTIST.[26] When the absolute difference in free energy is greater than 0.57 V, the molecular squares cannot undergo CTIST and contain

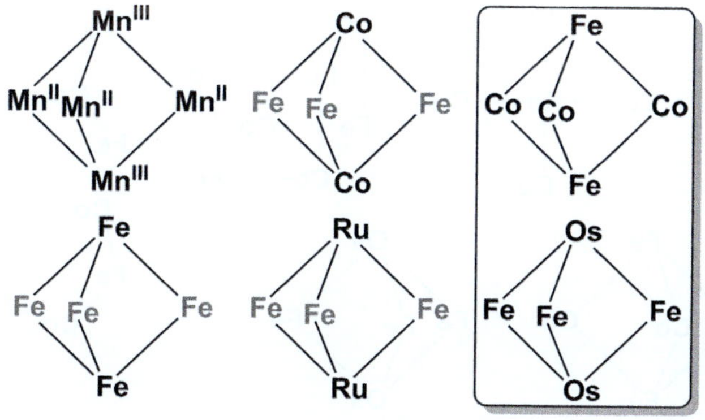

Scheme 6.3 *Core of the heterometallic trigonal bipyramid clusters. The Mn cluster is a single molecule magnet. The highlighted molecules display CTIST, Fe sites shown in gray display spin crossover between LS Fe^{II} and HS Fe^{II}.*

Table 6.3 *Chemical and physical properties of $[[FeL^3]_2[Co(L^2)_2]_2(\mu\text{-}CN)_4]]X_2$ squares.*[a]

(Fe)L^3	(Co)L^2	X	CTIST $h\nu/\Delta$	d(Co-N) (Temp/K)	λ_{MMCT} Fe^{II}-Co^{III}/Fe^{III}-Co^{II}	References
Tp	dmphen	ClO_4^-	HS–Co^{II}/LS–Fe^{III}			66, 70
$B(pz)_4^-$	dmphen	ClO_4^-	HS–Co^{II}/LS–Fe^{III}			
$B(pz)_4^-$	bik	ClO_4^-	Yes / No	1.92–1.95 (200)	NR	
Tp*	DMF_4	OTf^-	No			28, 67
Tp*	Me-bpy	OTf^-	Yes / Yes	2.09(250) 1.93(120)	770/560	
Tp*	bpy	OTf^-	Yes / Yes	2.11(230)	845/520	
Tp*	bpy	PF_6^-	HS–Co^{II}/LS–Fe^{III}	2.09–2.15(100)	–/560 (solution)	26, 65
Tp*	^{t}Bu-bpy	PF_6^-	Yes / Yes	2.11(330) 1.89–1.94(100)	770/560 (solution)	
Tp	^{t}Bu-bpy	PF_6^-	LS–Co^{III}/LS–Fe^{II}	1.91(100)	740/– (solution)	

[a]Ligand abbreviations: Tp = hydrotris(pyrazol-1-yl)borate; Tp*= hydrotris(3,5-dimethylpyrazol-1-yl)borate; $B(pz)_4$ = tetrapyrazolylborate; bik = bis(1-methylimidazol-2-yl)ketone; Me-bpy = 4,4′-dimethyl-2,2′-bipyridine; Bu-bpy = 4,4′-di-*tert*-butyl-2,2′-bipyridine; dmphen = 2,9-dimethyl-1,10-phenantroline.

either HS–Co^{II}/LS–Fe^{III} or LS–Co^{III}/LS–Fe^{II} pairs. The reduction potential of the metal ions in the clusters, as well as the difference in free energy between the two pairs, depend on the chemical nature of the ligands coordinated to the Fe and Co ions, which are typically a combination of cyanide and bidentate or tridentate ligands. This relationship is similar to that in which the coordination environment of Fe^{II} complexes affects the ligand field splitting at the Fe^{II} center and, implicitly, the ability of an Fe^{II} complex to display spin-crossover. This platform also makes it possible to (i) adjust the electronic structure of the clusters to achieve CTIST behavior, and (ii) synthesize "model complexes" for comparative studies of structures with closely related coordination environments, some of which exhibit CTIST and others of which do not (see below).

An excellent illustration of the relationship between the cobalt ion coordination in related molecular square and the observation of CTIST comes from the work of Holmes[67, 68] and Oshio[26] and their collaborators. The two groups have reported four different Fe_2Co_2 squares in which the Fe ions have the same coordination environment, viz. $[Fe(Tp^*)(CN)_3]$. The Co ions in all four squares have two bridging cyanide ligands with the four other coordination sites being occupied by different ligands, specifically $[Co(CN)_2(DMF)_4]$,

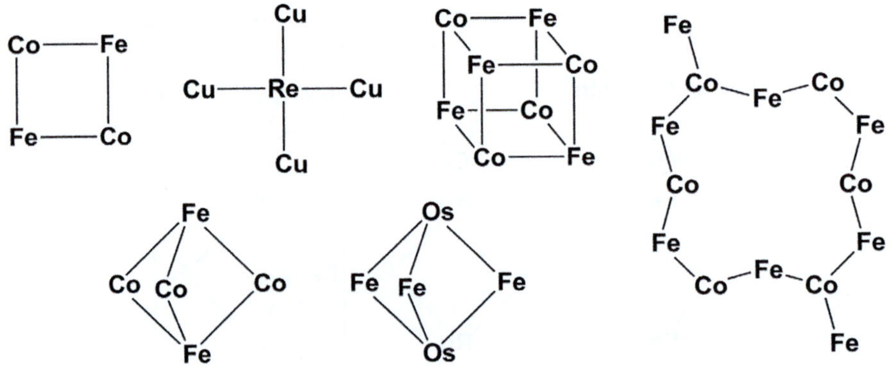

Scheme 6.4 *The core of clusters in molecule-based materials that display CTIST. The lines represent bridging μ_2-CN^- ligands.*

$[Co(CN)_2(bpy)_2]$, $[Co(CN)_2(Me\text{-}bpy)_2]$, $[Co(CN)_2(^tBu\text{-}bpy)_2]$. The complex that contains DMF adopts a HS–Co^{II}/LS–Fe^{III} configuration and does not undergo CTIST. The other three complexes that have bipyridine or its alkylated derivative coordinated to the Co centers can undergo CTIST both in solution (in certain solvents) and in the solid state. The three complexes undergo CTIST in methyl- or butyro-nitrile solvents at 198 K or 230 K. These data indicate that (i) the CTIST is an intrinsic property of the clusters, and (ii) the ligand substituents are important for dictating the temperature at which CTIST takes place in solution. Notably, in the solid state, the bpy- and Me-bpy-containing squares undergo CTIST at temperatures 20–55 K lower than in solution, whereas the ^{t}Bu-bpy-containing square undergoes CTIST at a temperature approximately 40 K higher than in solution. This difference may be due to the ligand effects, counterions (OTf^- and PF_6^-), or crystallization solvent, all of which affect the packing of the molecules within the crystal.

The important role of the counterion and the solvent on the ability of a complex to undergo CTIST is underscored by the fact that $[[Co(bpy)_2]_2[Fe(Tp^*)(CN)]_2(\mu\text{-}CN)_4]^{2+}$ undergoes a CTIST when crystallized with OTf^-, DMF, and H_2O,[67] but not with PF_6^- and MeOH.[26] In another example of the role of the interstitial solvent, crystals of the charge neutral $[[Co(tmphen)_2]_3[Fe(CN)_6]_2]$ molecule undergo a temperature-induced CTIST when directly isolated from the acetonitrile mother liquor but if the solvent is removed from the crystals under vacuum, the molecules exist in the $(HS\text{–}Co^{II})_3(LS\text{–}Fe^{III})_2$ state up to room temperature. When crystals of the same complex are exposed to a humid atmosphere for several days, the complex exists in the $(HS\text{–}Co^{II})(LS\text{–}Co^{III})_2(LS\text{–}Fe^{II})_2$ form at all temperatures up to room temperature.[9] These observations support the conclusion that the CTIST is affected by subtle factors such as relatively weak interactions between the solvent and/or the counterions and the clusters.

The complex $[[Co(^tBu\text{-}bpy)_2]_2[Fe(Tp^*)(CN)]_2(\mu\text{-}CN)_4](PF_6)_2 \cdot 2MeOH$ synthesized by Oshio and coworkers is remarkable in two respects.[26] First, it is the only compound that displays a two-step CTIST process ($T_{1/2} = 275$ K and $T_{1/2} = 310$ K) reported to date. Based on a combination of data obtained from synchrotron X-ray crystallography, magnetometry, and IR spectroscopy, Oshio established that the intermediate state consists of a 1:1 mixture of $[(LS\text{–}Fe^{II})_2(LS\text{–}Co^{III})_2]$ and $[(LS\text{–}Fe^{III})_2(HS\text{–}Co^{II})_2]$ clusters. The two types of clusters are positionally ordered into a quadrupled unit cell in which every cluster of one type is surrounded by clusters of the other type. Secondly, the same compound was shown to undergo a CTIST in butyronitrile solution when protonated. The transition from the LS–Fe^{III}/HS–Co^{II} to the LS–Fe^{II}/LS–Co^{III} state was attributed to the protonation of the terminal cyanide ligands on the Fe^{III} site, which causes a positive shift in the reduction potential of the Fe^{III} ion and favors electron transfer from the Co^{II} to the Fe^{III} ion.

Three cyanide-bridged pentanuclear complexes that exhibit CTIST have been identified to date. As mentioned above, the first molecular solid for which a CTIST was observed was the trigonal bipyramid cluster $[[Co(tmphen)_2]_3[Fe(CN)_3]_2(\mu\text{-}CN)_6] \cdot n$Solv.[8, 9, 73] This complex undergoes CTIST as a function of temperature[8, 9] and also under photoexcitation at low temperature.[71] The manifestation of CTIST depends critically on the crystallization solvent (see above), with differences leading to the isolation and characterization of materials in which the cluster is in either the LS–Fe^{III}/HS–Co^{II} form or the LS–Fe^{II}/LS–Co^{III} form at all temperatures or in, the form that undergoes CTIST. Given the stoichiometry of the cluster, only two of the three Fe ions are involved in the CTIST. Although the three Fe ions in the complex have identical ligands, the two that undergo the CTIST have tmphen ligands that are involved in π–π interactions different from the third ion.

Another member of the same family of trigonal bipyramid complexes, namely the heterobimetallic cluster $[[Fe(tmphen)_2]_3[Os(CN)_6]_2]$ also undergoes a reversible CTIST between the LS–Fe^{II}/Os^{III} and the HS–Fe^{III}/Os^{II} forms as a function of temperature.[25] The transition is unique in that it relates LS Fe^{II} and HS Fe^{III} ions whose spin is different by $^5/_2$, the largest possible change for an octahedral d metal ion. While HS–Fe^{II} ions with a coordination environment of four imine nitrogens and two N-coordinated cyanide ligands are very well known in Fe^{II} complexes with spin crossover, a HS–Fe^{III} complex with the same coordination was not known until it was generated by the CTIST in the $[[Fe(tmphen)_2]_3[Os(CN)_6]_2]$ cluster.

A pentanuclear cluster with a star-like geometry, $[(PY5Me_2)_4Co_4Re(CN)_7](PF_6)_5$, was recently reported by Long and collaborators.[27] This complex has thermochromiac and photochromic properties. Based on observed changes in the solid-state reflectivity of the complex with (i) temperature changes from 290 K to 10 K, and (ii) irradiation with white light of a solid state sample at 5 K, it was suggested that the compound may undergo a transition between LS–Co^{III}/LS–Re^{III} and HS–Co^{II}/LS–Re^{IV} states. If one approximates the difference in free energy between these states as the difference between the reduction potentials for the $Co^{III/II}$ and $Re^{IV/III}$ couples, one obtains a $|\Delta E| = 0.71$ V, which is large when compared to the $|\Delta E|$ value for square complexes for which CTISTs between the LS–Fe^{III}/HS–Co^{II} and LS–Fe^{III}/HS–Co^{II} states have been observed.

In terms of higher nuclearity molecules that exhibit CTIST, Holmes and collaborators reported a CTIST in an octanuclear Fe_4Co_4 cube,[28] and Oshio *et al.* observed CTIST in a tetradecanuclear Fe_8Co_6 cluster that contains a 12-member crown of alternating Fe and Co ions.[69] Notably, the Fe ions in these clusters have a coordination similar to that of the Fe ions in molecular squares, namely a tridentate, facially-coordinated Tp or $B(pz)_4$ ligand and three cyanide ligands. The Co ions possess a combination of one or two bidentate, imine-nitrogen coordinating ligands (LL) and cyanide ligands to achieve a coordination number of six. Both compounds undergo solid state CTIST between the paramagnetic LS–Fe^{III}/HS–Co^{II} state and the diamagnetic LS–Fe^{II}/LS–Co^{III} state induced by temperature or by light, when irradiated at low temperatures. For the cube compound, the CTIST is complete, that is only the LS–Fe^{III}/HS–Co^{II} form is present at high temperature and only the LS–Fe^{II}/LS–Co^{III} form is present at low temperature. In contrast, the degree to which CTIST generates the high-temperature LS–Fe^{III}/HS–Co^{II} form of the crown-like cluster depends significantly on the degree of solvation of the sample. Solid samples of this cluster which contain some solvent exist in the $[(LS–Fe^{III})_8(HS–Co^{II})_6]$ form at room temperature but, at most, three LS–Fe^{II}/LS–Co^{III} pairs have been observed at low temperature. We note that the cluster contains four Co sites with $[Co(LL)_2(NC)_2]$ and $[Co(LL)(NC)_4]$ coordination; the CTIST may not be possible for the latter Co sites. This question could be addressed by studying the crown complex in solution by UV-visible spectroscopy and electrochemistry if the complex is sufficiently soluble.

6.6 Summary and Outlook

The spin-crossover phenomenon was reported over 80 years ago and has intrigued chemists and physicists ever since its discovery.[3] In contrast, charge transfer induced spin-transition (CTIST) behavior was realized only 16 years ago. The results of research on CTIST carried out over this relatively short period of time and presented in this chapter demonstrate that this property manifests itself in a broad range of materials, ranging from infinite 2- and 3D structures to thin films, nanoparticles, and molecule-based materials. Magnetic measurements, electronic, infrared and Mössbauer spectroscopies, and X-ray crystallography have been used to study CTIST in these diverse materials. These investigations have revealed that CTIST can be induced by a broad range of perturbations including temperature, pressure, photoexcitation, desolvation, protonation, and electric bias. Importantly, the results also underscore the fact that the manifestation of CTIST depends on a subtle balance between the electronic structures of the transition metal ions in the material. This balance is often achieved in extended, quantum-confined, and molecular materials that contain Fe-CN-Co units.

Several challenges remain largely unexplored in this vibrant and quickly emerging area of research. A detailed understanding of the mechanism of CTIST would facilitate the rational design of new CTIST materials with specific transition temperatures and photomagnetic responses. In this vein, it would be useful to understand the sequence and timescales of the spin change and charge transfer both in solution and in solid state, the relationship between the electronic structure of the metal ions that undergo CTIST, and the role of intermolecular interactions with guest molecules or between the CTIST complexes themselves. Theoretical studies conducted in conjunction with experimental methods mentioned in this chapter and ultrafast spectroscopy may provide insight into the microscopic origin and dynamics of CTIST-related processes. A combination of the powerful tools available today for both experimental and computational studies is expected to reveal fundamental relationships between charge transfer induced spin-transitions, spin-crossover, and simple charge transfer in inorganic materials. The future of this field is just over the horizon.

Acknowledgement

KRD thanks the Department of Energy for generous support of her project in cyanide magnetism including results highlighted in this chapter. The authors thank their co-workers who contributed to their collaborative research on the topics presented herein.

References

1. Hendrickson, D. N., Pierpont, C. G. (2004) Valence tautomeric transition metal complexes, in: Gütlich, P., Goodwin, H. A. (Eds) Spin Crossover in Transition Metal Compounds II. Top. Curr. Chem., 234: 63–95.
2. Buchanan, R. M., Pierpont, C. G. (1980) Tautomeric catecholate–semiquinone interconversion via metal–ligand electron transfer. Structural, spectral, and magnetic properties of (3,5-di-tert-butylcatecholato)(3,5-di-tert-butylsemiquinone)(bipyridyl)cobalt(III), a complex containing mixed-valence organic ligands. J. Am. Chem. Soc., 102: 4951–4957.
3. Gütlich, P., Goodwin, H. A. (2004) Spin crossover–an overall perspective, in: Gütlich, P., Goodwin, H. A. (Eds) Spin Crossover in Transition Metal Compounds I. Top. Curr. Chem., 233: 1–47.
4. Hauser, A. (2004) Ligand field theoretical considerations, in: Gütlich, P., Goodwin, H. A. (Eds) Spin Crossover in Transition Metal Compounds I. Top. Curr. Chem., 233: 49–58.
5. Sato, O., Iyoda, T., Fujishima, A., Hashimoto, K. (1996) Photoinduced magnetization of a cobalt-iron cyanide. Science, 272: 704–705.
6. Shatruk, M., Avendano, C., Dunbar, K. R. (2009) Cyanide-bridged complexes of transition metals: a molecular magnetism perspective. Prog. Inorg. Chem., 56: 155–334.
7. Atanasov, M., Comba, P., Hausberg, S., Martin, B. (2009) Cyanometalate-bridged oligonuclear transition metal complexes–possibilities for a rational design of SMMs. Coord. Chem. Rev., 253: 2306–2314.
8. Berlinguette, C. P., Dragulescu-Andrasi, A., Sieber, A., Galán-Mascarós, J. R., Güdel, H. U., Achim, C., Dunbar, K. R. (2004) A charge-transfer-induced spin transition in the discrete cyanide-bridged complex $\{[Co(tmphen)_2]_3[Fe(CN)_6]_2\}$. J. Am. Chem. Soc., 126: 6222–6223.
9. Berlinguette, C. P., Dragulescu-Andrasi, A., Sieber, A., Güdel, H. U., Achim, C., Dunbar, K. R. (2005) Charge-transfer-induced spin transition in a discrete complex: the role of extrinsic factors in stabilizing three electronic isomeric forms of a cyanide-bridged co/fe cluster. J. Am. Chem. Soc., 127: 6766–6779.
10. Sato, O. (2003) Optically switchable molecular solids: photoinduced spin-crossover, photochromism, and photoinduced magnetization. Acc. Chem. Res., 36: 692–700.
11. Dei, A. (2005) Photomagnetic effects in polycyanometallate compounds: an intriguing future chemically based technology? Angew. Chem. Int. Ed., 44: 1160–1163.
12. Sato, O., Tao, J., Zhang, Y.-Z. (2007) Control of magnetic properties through external stimuli. Angew. Chem. Int. Ed., 46: 2152–2187.
13. Halcrow, M. A. (2008) Trapping and manipulating excited spin states of transition metal compounds. Chem. Soc. Rev., 37: 278–289.
14. Buser, H. J., Schwarzenbach, D., Petter, W., Ludi, A. (1977) The crystal structure of Prussian blue: $Fe_4[Fe(CN)_6]_3{\cdot}xH_2O$. Inorg. Chem., 16: 2704–2710.
15. Shimamoto, N., Ohkoshi, S., Sato, O., Hashimoto, K. (2002) Control of charge-transfer-induced spin transition temperature on cobalt–iron Prussian blue analogues. Inorg. Chem., 41: 678–684.

16. Sato, O., Einaga, Y., Fujishima, A., Hashimoto, K. (1999) Photoinduced long-range magnetic ordering of a cobalt–iron cyanide. Inorg. Chem., 38: 4405–4412.
17. Avendano, C., Hilfiger, M. G., Prosvirin, A., Sanders, C., Stepien, D., Dunbar, K. R. (2010) Temperature and light induced bistability in a $Co_3[Os(CN)_6]_2{\cdot}6H_2O$ Prussian blue analog. J. Am. Chem. Soc., 132: 13123–13125.
18. Gu, Z.-Z., Einaga, Y., Sato, O., Fujishima, A., Hashimoto, K. (2001) Photo- and dehydration-induced charge transfer processes accompanied with spin transition on CoFe(CN)5NH3·6H2O. J. Solid State Chem., 159: 336–342.
19. Liu, T., Zhang, Y.-J., Kanegawa, S., Sato, O. (2010) Water-switching of spin transitions induced by metal-to-metal charge transfer in a microporous framework. Angew. Chem. Int. Ed., 49: 8645–8648.
20. Arimoto, Y., Ohkoshi, S., Zhong, Z. J., Seino, H., Mizobe, Y., Hashimoto, K. (2003) Photoinduced magnetization in a two-dimensional cobalt octacyanotungstate. J. Am. Chem. Soc., 125, 9240–9241.
21. Ohkoshi, S., Hamada, Y., Matsuda, T., Tsunobuchi, Y., Tokoro, H. (2008) Crystal structure, charge-transfer-induced spin transition, and photoreversible magnetism in a cyano-bridged cobalt–tungstate bimetallic assembly. Chem. Mater., 20: 3048–3054.
22. Sato, O., Einaga, Y., Iyoda, T., Fujishima, A., Hashimoto, K. (1997) Cation-driven electron transfer involving a spin transition at room temperature in a cobalt iron cyanide thin film. J. Phys. Chem., 101: 3903–3905.
23. Frye, F. A., Pajerowski, D. M., Lane, S. M., Anderson, N. E., Park, J. H., Meisel, M. W., Talham, D. R. (2007) Effect of film thickness on the photoinduced decrease in magnetism for thin films of the cobalt iron Prussian blue analogue Rb0.7Co4[Fe(CN)6]3.0. Polyhedron, 26: 2281–2286.
24. Pajerowski, D. M., Frye, F. A., Talham, D. R., Meisel, M. W. (2007) Size dependence of the photoinduced magnetism and long-range ordering in Prussian blue analogue nanoparticles of rubidium cobalt hexacyanoferrate. New J. Phys., 9: 222 (11 pp.).
25. Hilfiger, M. G., Chen, M., Brinzari, T. V., Nocera, T. M., Shatruk, M., Petasis, D. T., Musfeldt, J. L., Achim, C., Dunbar, K. R. (2010) An unprecedented charge transfer induced spin transition in an Fe–Os cluster. Angew. Chem. Int. Ed., 49: 1410–1413.
26. Nihei, M., Sekine, Y., Suganami, N., Nakazawa, K., Nakao, A., Nakao, H., Murakami, Y., Oshio, H. (2011) Controlled intramolecular electron transfers in cyanide-bridged molecular squares by chemical modifications and external stimuli. J. Am. Chem. Soc., 133: 3592–3600.
27. Zadrozny, J. M., Freedman, D. E., Jenkins, D. M., Harris, T. D., Iavarone, A. T., Mathonière, C., Clérac, R., Long, J. R. (2010) Slow magnetic relaxation and charge-transfer in cyano-bridged coordination clusters incorporating $[Re(CN)_7]^{3-/4-}$. Inorg. Chem., 49: 8886–8896.
28. Li, D., Clérac, R., Roubeau, O., Harté, E., Mathonière, C., Le Bris, R., Holmes, S. M. (2008) Magnetic and optical bistability driven by thermally and photoinduced intramolecular electron transfer in a molecular cobalt–iron Prussian blue analogue. J. Am. Chem. Soc., 130: 252–258.
29. Bertrán, J. F., Blanco Pascual, J., Hernandez, M., Rodriguez, R. (1988) Tribochemical oxidation of potassium bromide by transition metal ferricyanides. React. Solids, 5: 95–100.
30. Reguera, E., Bertrán, J. F., Diaz, C., Blanco, J., Rondón, S. (1990) Mössbauer and infrared spectroscopic studies of novel mixed valence states in cobaltous ferrocyanides and ferricyanides. Hyperfine Interact., 53, 391–395.
31. Hester, R. E., Nour, E. M. (1981) Resonance-Raman and infrared studies of cyanide-bridged dimetal complexes. J. Chem. Soc. Dalton Trans., 939–941.
32. Dunbar, K. R., Heintz, R. A. (1997) Chemistry of transition metal cyanide compounds: modern perspectives. Prog. Inorg. Chem., 45: 283–391.
33. Robin, M. B. (1962) The color and electronic configurations of Prussian blue. Inorg. Chem., 1:, 337–342.
34. Einaga, Y., Sato, O., Iyoda, T., Kobashi, Y., Ambe, F., Hashimoto, K., Fujishima, A. (1997) Electronic states of cobalt iron cyanides studied by ^{57}Fe Mössbauer spectroscopy. Chem. Lett., 26: 289–290.
35. Bleuzen, A., Lomenech, C., Escax, V., Villain, F., Varret, F., Cartier dit Moulin, C., Verdaguer, M. (2000) Photoinduced ferrimagnetic systems in Prussian blue analogues $C^I{}_xCo_4[Fe(CN)_6]_y$ (C^I = Alkali Cation). 1. Conditions to observe the phenomenon. J. Am. Chem. Soc., 122: 6648–6652.
36. Escax, V., Bleuzen, A., Cartier dit Moulin, C., Villain, F., Goujon, A., Varret, F., Verdaguer, M. (2001) Photoinduced ferrimagnetic systems in Prussian blue analogues $C^I{}_xCo_4[Fe(CN)_6]_y$ (C^I = Alkali Cation). 3. Control of the photo- and thermally induced electron transfer by the $[Fe(CN)_6]$ vacancies in cesium derivatives. J. Am. Chem. Soc., 123: 12536–12543.

37. Gawali-Salunke, S., Varret, F., Maurin, I., Enachescu, C., Malarova, M., Boukheddaden, K., Codjovi, E., Tokoro, H., Ohkoshi, S., Hashimoto, K. (2005) Magnetic and Mössbauer investigation of the photomagnetic prussian blue analogue $Na_{0.32}Co[Fe(CN)_6]_{0.74}\cdot 3.4H_2O$: cooperative relaxation of the thermally quenched state. J. Phys. Chem. B, 109: 8251–8256.
38. Cafun, J. D., Champion, G., Arrio, M. A., Cartier dit Moulin, C., Bleuzen, A. (2010) Photomagnetic CoFe Prussian blue analogues: role of the cyanide ions as active electron transfer bridges modulated by cyanide–alkali metal ion interactions. J. Am. Chem. Soc., 132: 11552–11559.
39. Ksenofontov, V., Levchenko, G., Reiman, S., Gütlich, P., Bleuzen, A., Escax, V., Verdaguer, M. (2003) Pressure-induced electron transfer in ferrimagnetic Prussian blue analogs. Phys. Rev. B, 68: 024415/1–024415/6.
40. Hanawa, H., Moritomo, Y., Tateishi, J., Ohishi, Y., Kato, K. (2004) Pressure-induced spin state transition in Co-Fe cyanide. J. Phys. Soc. Jpn., 73: 2759–2762.
41. Liu, H. W., Matsuda, K., Gu, Z.-Z., Takahashi, K., Cui, A. L., Nakajima, R., Fujishima, A., Sato, O. (2003) Reversible valence tautomerism induced by a single-shot laser pulse in a cobalt-iron Prussian blue analog. Phys. Rev. Lett., 90: 167403/1–167403/4.
42. Tokoro, H., Ohkoshi, S. (2011) Novel magnetic functionalities of Prussian blue analogs. Dalton Trans., 40: 6825–6833.
43. Yokoyama, T., Okamoto, K., Ohta, T., Ohkoshi, S., Hashimoto, K. (2002) Local structure and electronic state of a photomagnetic material of CoW cyanide studied by x-ray-absorption fine structure spectroscopy. Phys. Rev. B, 65: 064438/1–064438/8.
44. Le Bris, R., Mathonière, C., Létard, J.-F. (2006) Cooperative relaxation of the metastable states in the photomagnetic octacyanotungstate $Cs^{I}[\{Co^{II}(3\text{-}CN\text{-}py)_2\}\{W^{V}(CN)_8\}]\cdot H_2O$. Chem. Phys. Lett., 426: 380–386.
45. Kim, J. E., Ohishi, Y., Moritomo, Y., Kato, K., Takata, M., Ohkoshi, S. (2007) Charge-transfer transition in two-dimensional $Cs[Co(3\text{-}cyanopyridine)_2][W(CN)_8]\cdot H_2O$ as investigated by angle-resolved x-ray diffraction. Phys. Rev. B, 76: 014106/1–014106/6.
46. Kim, J. E., Ohishi, Y., Moritomo, Y., Kato, K., Takata, M., Ohkoshi, S. (2008) Pressure-temperature phase diagram for charge-transfer transition in $Cs[Co(3\text{-}CNpy)_2][W(CN)_8]H_2O$. Phys. Rev. B, 77: 012101/1–012101/4.
47. Podgajny, R., Bałanda, M., Sikora, M., Borowiec, M., Spałek, L., Kapusta, C., Sieklucka, B. (2006) Cobalt(II) octacyanotungstate(V) organic-inorganic hybrid ferromagnetic materials with pyrazine and 4,4′-bipyridine. Dalton Trans., 2801–2809.
48. Zieliński, P. M., Tracz, P., Podgajny, R., Pełka, R., Bałanda, M., Wasiutyński, T., Sieklucka, B. (2007) Photo-induced relaxation of magnetization in molecular magnet. Acta Phys. Pol. A, 112: S/183–S/188.
49. Abe, Y., Kim, J. E., Matsuda, T., Shibata, T., Tanida, H., Uruga, T., Moritomo, Y. (2009) Phase separation driven by mobile cations in $(Na_{1-x}K_x)_{0.35}Co[Fe(CN)_6]_{0.74}\cdot zH_2O$ Phys. Rev. B, 80: 134101/1–134101/6.
50. Yamamoto, T., Umemura, Y., Sato, O., Einaga, Y. (2005) Observation of the anisotropic photoinduced magnetization effect in Co–Fe Prussian blue thin films fabricated by using clay Langmuir-Blodgett films as a template. J. Am. Chem. Soc., 127: 16065–16073.
51. Frye, F. A., Pajerowski, D. M., Park, J. H., Meisel, M. W., Talham, D. R. (2008) Anisotropic photoinduced magnetism in thin films of the Prussian blue analogue $A_jCo_k[Fe(CN)_6]_l\cdot nH_2O$. Chem. Mater., 20: 5706–5713.
52. Park, J. H., Cizmar, E., Meisel, M. W., Huh, Y. D., Frye, F., Lane, S., Talham, D. R. (2004) Anistropic photoinduced magnetism of a $Rb_jCo_k[Fe(CN)_6]_l\cdot nH_2O$ thin film. Appl. Phys. Lett., 85: 3797–3799.
53. Pajerowski, D. M., Andrus, M. J., Gardner, J. E., Knowles, E. S., Meisel, M. W., Talham, D. R. (2010) Persistent photoinduced magnetism in heterostructures of Prussian blue analogues. J. Am. Chem. Soc., 132: 4058–4059.
54. Pajerowski, D. M., Gardner, J. E., Frye, F. A., Andrus, M. J., Dumont, M. F., Knowles, E. S., Meisel, M. W., Talham, D. R. (2011) Photoinduced magnetism in a series of Prussian blue analogue heterostructures. Chem. Mater., 23: 3045–3053.
55. Zentková, M., Arnold, Z., Kamarád, J., Kavečansky, V., Lukáčova, M., Mat'aš, S., Mihalik, M., Mitróová, Z., Zentko, A. (2007) Effect of pressure on the magnetic properties of $TM_3[Cr(CN)_6]_2\cdot 12H_2O$. J. Phys. Condens. Matter, 19: 266217 (10 pp.).
56. Vaucher, S., Li, M., Mann, S. (2000) Synthesis of Prussian blue nanoparticles and nanocrystal superlattices in reverse microemulsions. Angew. Chem. Int. Ed., 39: 1793–1796.

57. Vaucher, S., Fielden, J., Li, M., Dujardin, E., Mann, S. (2002) Molecule-based magnetic nanoparticles: synthesis of cobalt hexacyanoferrate, cobalt pentacyanonitrosylferrate, and chromium hexacyanochromate coordination polymers in water-in-oil microemulsions. Nano Lett., 2: 225–229.
58. Moore, J. G., Lochner, E. J., Ramsey, C., Dalal, N. S., Stiegman, A. E. (2003) Transparent, superparamagnetic $K^{I}_{x}Co^{II}_{y}[Fe^{III}(CN)_6]$–silica nanocomposites with tunable photomagnetism. Angew. Chem. Int. Ed., 42: 2741–2743.
59. Fornasieri, G., Bleuzen, A. (2008) Controlled synthesis of photomagnetic nanoparticles of a prussian blue analogue in a silica xerogel. Angew. Chem. Int. Ed., 47: 7750–7752.
60. Fornasieri, G., Aouadi, M., Durand, P., Beaunier, P., Rivière, E., Bleuzen, A. (2010) Fully controlled precipitation of photomagnetic CoFe Prussian blue analogue nanoparticles within the ordered mesoporosity of silica monoliths. Chem. Commun., 46: 8061–8063.
61. Brinzei, D., Catala, L., Louvain, N., Rogez, G., Stéphan, O., Gloter, A., Mallah, T. (2006) Spontaneous stabilization and isolation of dispersible bimetallic coordination nanoparticles of $Cs_xNi[Cr(CN)_6]_y$. J. Mater. Chem., 16: 2593–2599.
62. Catala, L., Brinzei, D., Prado, Y., Gloter, A., Stéphan, O., Rogez, G., Mallah, T. (2009) Core–multishell magnetic coordination nanoparticles: toward multifunctionality on the nanoscale. Angew. Chem. Int. Ed., 48: 183–187.
63. Dumont, M. F., Knowles, E. S., Guiet, A., Pajerowski, D. M., Gomez, A., Kycia, S. W., Meisel, M. W., Talham, D. R. (2011) photoinduced magnetism in core/shell Prussian blue analogue heterostructures of $K_jNi_k[Cr(CN)_6]_l{\cdot}nH_2O$ with $Rb_aCo_b[Fe(CN)_6]_c{\cdot}mH_2O$. Inorg. Chem., 50: 4295–4300.
64. Funck, K. E., Hilfiger, M. G., Berlinguette, C. P., Shatruk, M., Wernsdorfer, W., Dunbar, K. R. (2009) Trigonal-bipyramidal metal cyanide complexes: a versatile platform for the systematic assessment of the magnetic properties of Prussian blue materials. Inorg. Chem., 48: 3438–3452.
65. Nihei, M., Sekine, Y., Suganami, N., Oshio, H. (2010) Thermally two-stepped spin transitions induced by intramolecular electron transfers in a cyanide-bridged molecular square. Chem. Lett., 39: 978–979.
66. Mercurol, J., Li, Y., Pardo, E., Risset, O., Seuleiman, M., Rousselière, H., Lescouëzec, R., Julve, M. (2010) 2024A>$[Fe^{II}_{LS}Co^{III}_{LS}]_2 \Leftrightarrow [Fe^{III}_{LS}Co^{II}_{HS}]_2$ photoinduced conversion in a cyanide-bridged heterobimetallic molecular square. Chem. Commun., 46: 8995–8997.
67. Siretanu, D., Li, D., Buisson, L., Bassani, D. M., Holmes, S. M., Mathonière, C., Clérac, R. (2011) Thermally induced electron transfer in cyano-bridged molecular squares: from solid state to solution. Chem. Eur. J., 17: 11704–11708.
68. Zhang, Y., Li, D., Clérac, R., Kalisz, M., Mathonière, C., Holmes, S. M. (2010) Reversible thermally and photoinduced electron transfer in a cyano-bridged $\{Fe_2Co_2\}$ square complex. Angew. Chem. Int. Ed., 49: 3752–3756.
69. Mitsumoto, K., Oshiro, E., Nishikawa, H., Shiga, T., Yamamura, Y., Saito, K., Oshio, H. (2011) Cyanide-bridged $[Fe_8M_6]$ clusters displaying single-molecule magnetism (M = Ni) and electron-transfer-coupled spin transitions (M = Co). Chem. Eur. J., 17: 9612–9618.
70. Pardo, E., Verdaguer, M., Herson, P., Rousseliere, H., Cano, J., Julve, M., Lloret, F., Lescouëzec, R. (2011) Synthesis, crystal structures, and magnetic properties of a new family of heterometallic cyanide-bridged $Fe^{III}_2M^{II}_2$ (M = Mn, Ni, and Co) square complexes. Inorg. Chem., 50: 6250–6262.
71. Funck, K. E., Prosvirin, A. V., Mathonière, C., Clérac, R., Dunbar, K. R. (2011) Light-induced excited spin state trapping and charge transfer in trigonal bipyramidal cyanide-bridged complexes. Inorg. Chem., 50, 2782–2789.
72. Klokishner, S., Ostrovsky, S., Palii, A., Shatruk, M., Funck, K., Dunbar, K. R., Tsukerblat, B. (2011) Vibronic model for cooperative spin-crossover in pentanuclear $\{[M^{III}(CN)_6]_2[M'^{II}(tmphen)_2]_3\}$ (M/M′ = Co/Fe, Fe/Fe) compounds. J. Phys. Chem. C, 115: 21666–21677.
73. Ostrovsky, S., Palii, A., Klokishner, S., Shatruk, M., Funck, K., Achim, C., Dunbar, K. R., Tsukerblat, B. (2012) Vibronic approach to the cooperative spin transitions in crystals based on cyano-bridged pentanudear M_2Fe_3 (M = Co, Os) clusters, in: Atanasov, M., Daul, C. (Eds) Vibronic Iteractions and the Jahn-Teller Effect Prog. Theor. Chem. Phy., 23: 379–395.

7

Valence Tautomeric Transitions in Cobalt-dioxolene Complexes

Colette Boskovic

School of Chemistry, University of Melbourne, Australia

7.1 Introduction

Valence tautomeric (VT) transitions involve a stimulated intramolecular electron transfer between a redox-active metal center and a redox-active ligand.[1–7] VT transitions have been observed in solution and the solid state for complexes with metals including Mn, Fe, Co, Ni, Cu, Rh and Ir; and with a variety of redox-active ligand types, including *o*-dioxolenes, phenoxyl ligands, ferrocenes, crown ethers, tetrathiafulvalene and polychlorotriphenylmethyl radicals. However, the family of compounds that have provided by far the most examples of VT transitions have been octahedral cobalt complexes with *o*-dioxolene ligands. The most commonly investigated of these have been the 1:1 and 1:2 cobalt:dioxolene mononuclear complexes $[Co(N_2L)(diox)_2]$ and $[Co(N_4L)(diox)]^+$ (N_2L and N_4L are bi- and tetradentate nitrogen donor ancillary ligands, respectively; diox = *o*-dioxolene in the catecholate or semiquinonate form) and related polynuclear complexes. In these species a spin transition at the metal center accompanies electron transfer, such that an electron is transferred from a diamagnetic catecholate (cat) ligand to a low spin (LS)-Co^{III} center, yielding a high spin (HS)-Co^{II} ion coordinated to a paramagnetic semiquinonate (sq) ligand with antiferromagnetic coupling (Fig. 7.1). This concerted intramolecular electron transfer and metal-based spin-transition renders VT compounds something of a combination of conventional spin-crossover (SCO) systems and heterometallic charge transfer induced spin transition (CTIST) systems, with all three classes of compound characterized by electronic lability. Like SCO transitions, VT transitions are stimulated by the variation of an external parameter and can be induced thermally or by application of pressure or a magnetic field, or by irradiation with visible light or soft X-rays. In order for a VT transition to occur, two requirements must be met in the metal complex: (i) the degree of covalency in the interaction between metal ion and redox-active ligand must be small; and (ii) their frontier orbitals must be of similar energy. Fulfilment of these criteria has often been

Spin-Crossover Materials: Properties and Applications, First Edition. Edited by Malcolm A. Halcrow.

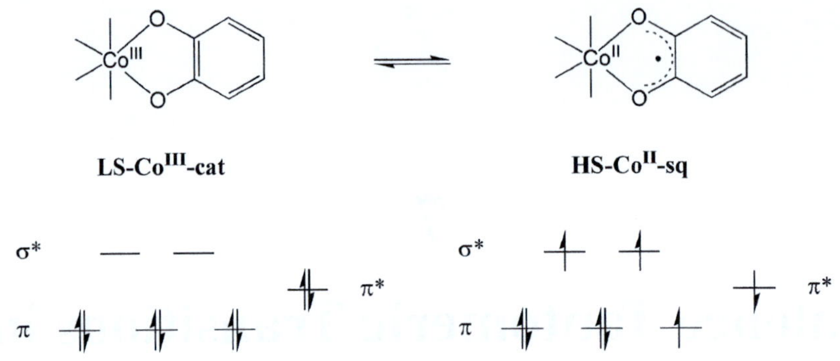

Figure 7.1 Representation of the electronic states associated with a VT transition in cobalt-dioxolene complexes.

achieved by the employment of the substituted dioxolene ligands 3,5 and 3,6 di-*tert*-butyl dioxolene (3,5-dbdiox and 3,6-dbdiox, respectively) and 9,10-phenanthrenequinone (phendiox), with the electron donating substituents on these ligands helping to stabilize oxidation from the catecholate to the semiquinonate form.

Transition metal complexes that display VT transitions are important for several reasons. First, they are invaluable model systems for fundamental studies of electron transfer between metal centers and organic ligands and coupled electron transfer and spin-transitions. Second, as coordination complexes with redox-active ligands, their study can provide insights into other situations where both metal and ligand are redox-active, even though stimulated VT transitions do not necessarily occur. For example, redox-active phenoxyl radical (tyrosyl) ligands are important to the function of the active sites of a number of metalloenzymes including a number of copper oxidases and the water oxidation complex in photosystem II.[8] In addition, the quest for new catalysts for organic transformations[9–11] and for energy conversion and storage (for example in the form of water oxidation catalysts for artificial photosynthesis)[12, 13] has led to the deliberate incorporation of redox-active ligands in metal complexes. Finally, as is true for SCO transitions, VT transitions are typically accompanied by distinct and reversible changes in structural, optical and magnetic properties. This feature may afford future applications for species that undergo VT transitions in molecule-based materials for display devices, data storage, molecular switches or sensors. Of particular interest in this context are bistable materials that display a hysteretic VT transition around room temperature, with a reasonably wide thermal hysteresis loop. Also potentially important are systems in which a relatively long-lived metastable state can be produced (e.g. photogenerated) at accessible temperatures.[7]

The topic of VT transition metal complexes was reviewed by Hendrickson and Pierpont in the 2004 issue of *Topics in Current Chemistry* dedicated to spin-transitions.[1] This review discussed the thermodynamic origins of VT transitions, mainly focusing on the $[Co(N_2L)(diox)_2]$ family of complexes, which were the complexes for which VT transitions were first observed and which have since been widely studied. Also discussed was the utility of magnetic susceptibility and electronic absorption spectroscopic measurements in elucidating VT behavior. Subsequently, several other articles have reviewed complexes that display VT transitions. These include a review discussing the potential of $[Co(N_4L)(diox)]^+$ mononuclear complexes and dinuclear complexes with bis-dioxolene ligands as molecular switches;[2] a microreview focused largely on alternative redox-active ligands to dioxolenes;[3] a review of the effect of factors additional to the redox-active metal and ligand on the VT transition, including the effect of ancillary ligands, counterions, redox-activity and the environment;[4] a more general review discussing VT complexes in the context of controlling magnetic properties through external stimuli;[5] a review specifically dedicated to photoinduced valence tautomerism;[6] and a discussion of the potential of VT cobalt-dioxolene complexes as devices for molecular spintronics.[7]

The scope of the present work is to review recent advances that specifically relate to VT transitions in cobalt-dioxolene complexes.

7.2 Induction of Valence Tautomeric Transitions

7.2.1 Thermally Induced Valence Tautomerism

It was upon heating and cooling [Co(2,2′-bpy)(3,5-dbdiox)$_2$] (2,2′-bpy = 2,2′-bipyridine) that VT transitions were first observed[14] and because of its experimental simplicity, thermal cycling remains the most common method of stimulating a VT transition. The thermal induction of VT transitions has been comprehensively discussed in previous reviews.[1–5, 7] The two tautomeric forms, LS-Co^{III}-cat and HS-Co^{II}-sq have different charge distributions (Fig. 7.1), with an enhanced population of antibonding e_g* orbitals in HS-Co^{II} versus LS-Co^{III}, which gives rise to larger Co-O/N distances of 0.16–0.22 Å in the HS-Co^{II}-sq tautomer. The thermal induction of VT transitions in Co-dioxolene complexes is entropically driven, with the large entropy gain arising from (i) the higher density of vibrational states of the HS-Co^{II}-sq complex, due to the longer Co O/N bond lengths, and (ii) the higher spin state degeneracy of HS-Co^{II}-sq versus LS-Co^{III}-cat. Typical values of the entropy contribution to this process are 80–150 JK^{-1}mol^{-1}, which require enthalpic contributions in the range of 20–60 kJmol^{-1}.

7.2.2 Pressure Induced Valence Tautomerism

In contrast to thermally-induced VT transitions, few studies of pressure-induced VT transitions have been reported, with this method of stimulation relatively little explored owing to the practical difficulties associated with the experimental measurements. Room temperature cobalt K-edge extended X-ray absorption fine structure (EXAFS) and X-ray Absorption Near Edge Structure (XANES) studies of [Co(phen)(3,5-dbdiox)$_2$] (phen = 9,10-phenanthroline) at pressures up to 2.5 GPa (25 kbar) showed that increasing the pressure induced the transition from the HS-Co^{II}-sq tautomer to the LS-Co^{III}-cat one.[16] Variable temperature magnetic susceptibility measurements on [Co(cth)(phendiox)]PF$_6$·H$_2$O (cth = *dl*-5,7,7,12,14,14-hexamethyl-1,4,8,11-tetraazacyclotetradecane) at pressures up to 7.4 kbar indicated that as the pressure increases, the transition temperature increases and the transition became more gradual.[17] Both of these studies suggest that, as for SCO transitions in iron(II) complexes,[18] the LS-Co^{III}-cat tautomer is favored at higher pressures because of its smaller molecular volume.

However, interesting results were reported by Sato *et al.* from a more recent study of the application of pressure on two dinuclear cobalt complexes with a bridging tetraoxolene ligand.[19] Thermally-induced VT transitions (Section 7.4.1) are observed at one cobalt center for both dinuclear complexes in [[Co(tpa)]$_2$(dhbq)](PF$_6$)$_3$ and [[Co(dpqa)]$_2$(dhbq)](PF$_6$)$_3$ (tpa = tris(2-pyridylmethyl)amine; dpqa = di(2-pyridylmethyl)-N-(quinolin-2-ylmethyl)amine and dhbq = bis-bidentate tetraoxolene ligand derived from 2,5-dihydroxy-1,4-benzoquinone; Fig. 7.2). Upon the application of pressure, magnetic susceptibility measurements indicated that the transition became more gradual, with larger values of the magnetic susceptibility obtained at elevated pressure at all temperatures. Curiously, for pressures up to 1.3 kbar the transition temperature *decreased with pressure*, while as the pressure was further increased from 1.3 kbar to nearly 5 kbar the transition temperature increased, consistent with the previous observation for mononuclear VT complexes.[16, 17] This behavior was interpreted in terms of a shut-down of the electron transfer step of the VT transition following a pressure-induced SCO transition. The authors suggested that the pressure induced an increase in the enthalpic separation between the two valence tautomers, thereby stabilizing the LS-Co^{II}-sq state over the LS-Co^{III}-cat state at low temperature.

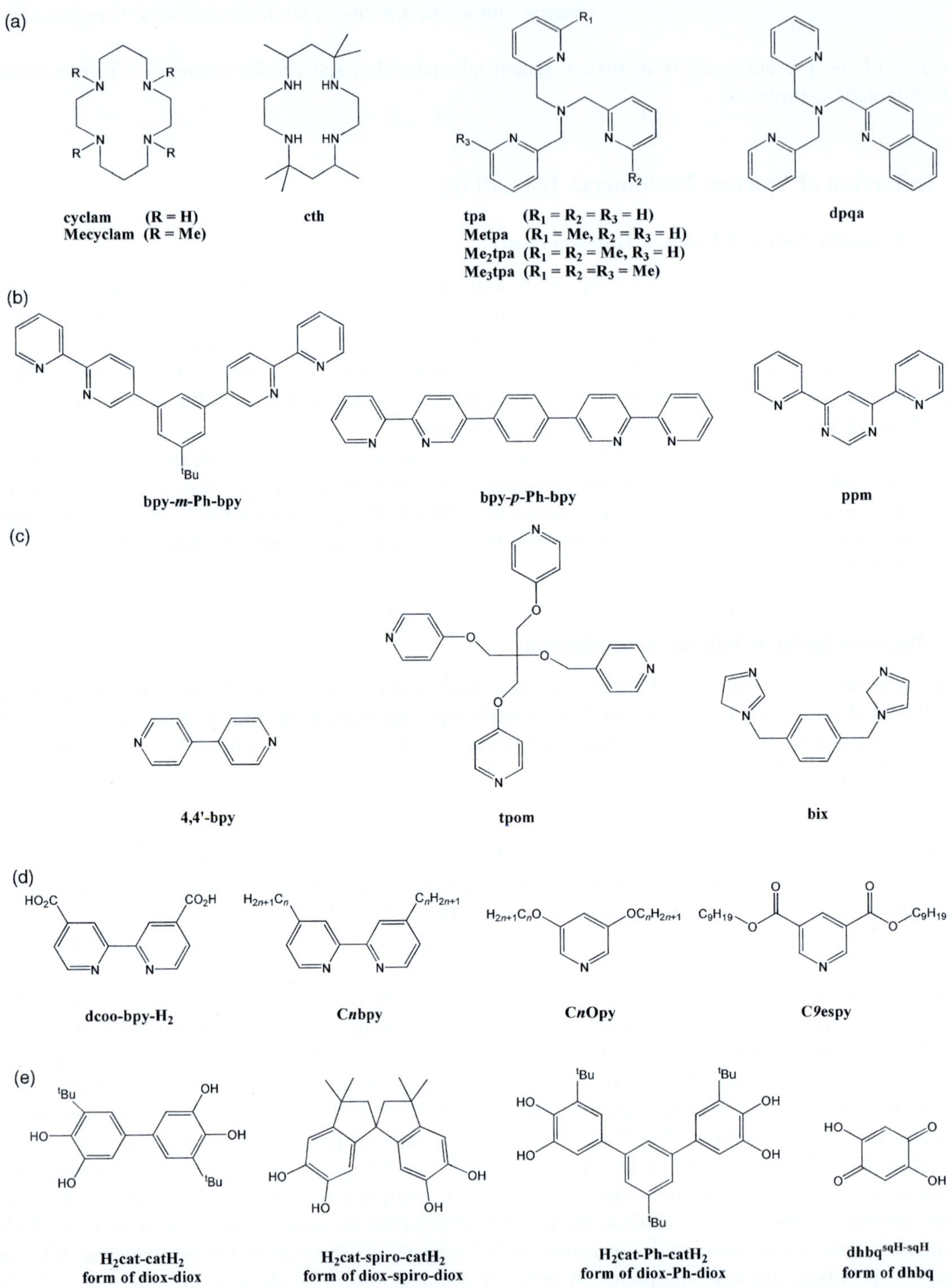

Figure 7.2 *(a) Tetradentate nitrogen donor (N_4L) ancillary ligands; (b) bridging bis-bidentate (N_2LN_2) ancillary ligands; (c) divergent bridging ancillary ligands for VT coordination polymers and networks; (d) functionalized ancillary ligands for bifunctional VT complexes; and (e) bridging bis-bidentate dioxolene and tetraoxolene redox-active ligands for polynuclear VT complexes.*

7.2.3 Light Induced Valence Tautomerism

Previous reviews have discussed the earliest investigations of light induced VT transitions, which were carried out at relatively high temperatures using pulsed laser photolysis on the nano-, pico- and femtosecond timescale to investigate VT transitions in solution.[1, 3, 5] Very recently Lapini and co-workers have demonstrated that transient infra-red spectroscopy is also a powerful technique for probing light induced VT transitions with sub-picosecond time resolution.[20] For photoinduced VT transitions generally, excitation of the ligand to metal charge transfer (LMCT) band of a LS-Co^{III}-cat tautomer affords a LMCT excited state, which relaxes to the transient HS-Co^{II}-sq tautomer. Alternatively, excitation of the metal to ligand charge transfer (MLCT) band of a HS-Co^{II}-sq tautomer gives rise to a transient LS-Co^{III}-cat tautomer. In both cases the photoinduced tautomer displays lifetimes in the nanosecond range before relaxing back to the original tautomeric form following Arrhenius-like behavior, due to the presence of an activation energy barrier.

In response to the discovery of, and now considerable body of work investigating, the LIESST (light-induced excited spin state trapping) effect in iron(II) and iron(III) SCO complexes, there have been numerous reports of analogous photoinduced VT transitions on solid samples at low temperature that give rise to long-lived metastable species.[6] This is the case for a number of mononuclear, dinuclear and polymeric complexes that also show thermally induced VT transitions. These transitions are generally monitored by irradiation with a laser coupled to a SQUID magnetometer, but have also been monitored by infra-red, electronic absorption, electron paramagnetic resonance (EPR) and cobalt K-edge XANES spectroscopy.[5–7, 21–39] With a photomagnetic set-up, often only a fraction of the bulk solid sample undergoes a photoinduced VT transition, which is generally attributed to opacity of the sample preventing penetration of the light beyond the surface. Indeed reflectivity measurements on $[[Co(cth)]_2(dhbq)](PF_6)_3$ confirmed almost quantitative photoinduced interconversion on the surface of the sample versus only 40% interconversion of the bulk sample measured by magnetic susceptibility.[22] Incomplete photoconversion may also be due to overlap of the transitions of the two valence tautomers giving rise to a photostationary state.[26] Notably, pulsed laser irradiation can allow greater penetration of light into the sample and has been shown to photoinduce the VT transition much more efficiently.[25]

Relaxation of the photo-induced HS-Co^{II}-sq tautomer back to the original LS-Co^{III}-cat tautomer is typically investigated following the experimental protocol developed by Létard *et al.*, originally for SC complexes.[40] The sample is held at low temperature (10 K) and irradiated until it reaches the photostationary limit. The excitation is switched off and the system is slowly heated (0.3 K min^{-1}). The temperature at which the thermal susceptibility profile measured after irradiation rejoins the non-irradiated curve is referred to as an effective T_{LIEEST}, or T_{LIVT} (LIVT = light induced valence tautomerism). For VT complexes effective T_{LIEEST} values have generally been reported in the range 38–80 K. Multiple cycles of irradiation and heating have been performed for complexes such as $[Co(tmeda)(3,5\text{-dbdiox})_2]$ (tmeda = tetramethylethylenediamine), demonstrating the reversibility of the VT transition and the robustness of the system.[21] Relaxation of the metastable photoinduced tautomer back to the LS-Co^{III}-cat species can also be achieved by irradiating the MLCT band of this tautomer with red light.[28] This is described as a "reverse" photoinduced VT transition and corresponds to the reverse LIESST effect in SCO systems. It is possible, for instance with $[Co(phen)(3,5\text{-dbdiox})_2]$, to achieve multiple cycles by alternately irradiating the LMCT (532 nm) and MLCT (830 nm) bands.[29]

The relaxation of the system can also be probed by measuring relaxation of the magnetization following photoirradiation at different temperatures. Fitting the relaxation curves obtained in this way allows a plot of $\ln(\tau)$ versus $(1/T)$ (τ is the relaxation time, T is the temperature), which generally shows that the relaxation rate does not follow a simple Arrhenius law as expected for a single thermally activated process, Rather, two distinct relaxation regimes are typically observed. At low temperatures (typically below $T \approx 25$ K) little or no temperature dependence is observed, with pseudo-Arrhenius behavior sometimes evident for

1:1 Co:dioxolene complexes with an activation energy between 5 and 15 cm^{-1}. This is interpreted in terms of a tunneling mechanism assisted by low-energy photons. In the higher temperature regime the behavior follows an Arrhenius law with activation energies in the ranges 242–295 and 110–135 cm^{-1} for 1:1 and 1:2 Co:dioxolene complexes, respectively.[19, 23, 28, 30, 31, 37] These results indicate a thermally assisted relaxation mechanism at higher temperature, with the disparity in activation energies for the 1:1 and 1:2 complexes in agreement with a dominant contribution from the totally symmetric Co–O vibrational breathing mode (300–400 cm^{-1}) in promoting the relaxation process from the metastable state.[7, 31] In the low temperature, essentially temperature-independent, relaxation regime, the lifetimes of the photo-induced tautomers at cryogenic temperatures are typically 10^4-10^5 s,[32] with no apparent correlation between the cryogenic relaxation rates of the photoinduced tautomer and the critical temperature associated with the thermally-induced VT transition.[31]

7.2.4 Magnetic Field Induced Valence Tautomerism

Following reports of magnetic field induced SCO transitions in iron(II) complexes, critical magnetic fields that would induce the VT transition were predicted at between 140 and 220 T for several [Co(N_2L)($diox)_2$] complexes.[41] These were based on the energy separation between the two tautomers derived from the critical temperature values for the thermally-induced transitions. A subsequent experiment performed by Mischenko and co-workers involved exposure of [Co(bpy)(3,5-dbdiox$)_2$] to an explosion magnetic field at 4.2 K, giving rise to anomalous behavior associated with a magnetic field-induced VT transition at around 100 T.[42] Subtraction of the intrinsic "parasitic" signal from the imbalanced signal associated with the transition affords a smooth and sharp transition in the magnetization versus magnetic field plot (Fig. 7.3). The value of the entropy change associated with the transition and the energy difference between the two tautomers

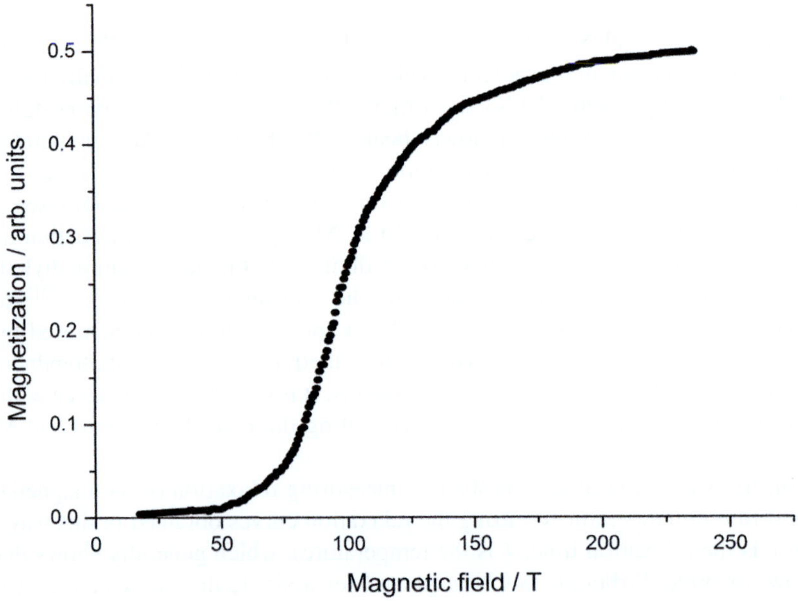

Figure 7.3 Magnetization of [Co^{III}(bpy)(3,5-dbcat)(3,5-dbsq)] at 4.2 K versus the applied magnetic field. Adpated with permission from [42]. Copyright Wiley-VCH Verlag GmbH & Co, 2006.

derived from the measured critical field are consistent with the values derived from the thermally-induced VT transition.

7.2.5 X-Ray Induced Valence Tautomerism

Synchrotron X-ray absorption spectroscopy (XAS) has proved to be very useful for elucidating VT behavior in cobalt complexes, making use of both the cobalt K- and L-edges.[16, 24, 33, 38] In the process of using XAS to probe the light-induced VT transition in [Co(Me$_2$tpa)(3,5-dbdiox)](PF$_6$)·toluene, Dei and co-workers found that the soft X-rays themselves could induce the VT transition, without irradiation with visible light.[44] While a long exposure to the X-ray beam with a flux of 10^8 photons s^{-1} yielded a weak conversion of the LS-Co^{III}-cat tautomer to the HS-Co^{II}-sq tautomer, raising the flux to 10^{10} photons s^{-1} gave rise to XAS spectra qualitatively identical to those resulting from irradiation with 904 nm laser light under the lower X-ray flux density (Fig. 7.4). Thermal cycling between 10 and 300 K demonstrates that the process is reversible, with photoinduced degradation only apparent after the acquisition of 600 spectra on the same spot of the sample at high flux. Although the mechanism of the X-ray induced VT transition is not certain, it has been attributed to the production by inelastic scattering of secondary electrons with energy in the range of the LMCT transition that is used to stimulate the light-induced VT transition. The authors suggest that this method of inducing VT transitions might be particularly useful for studying bifunctional photochromic VT complexes.

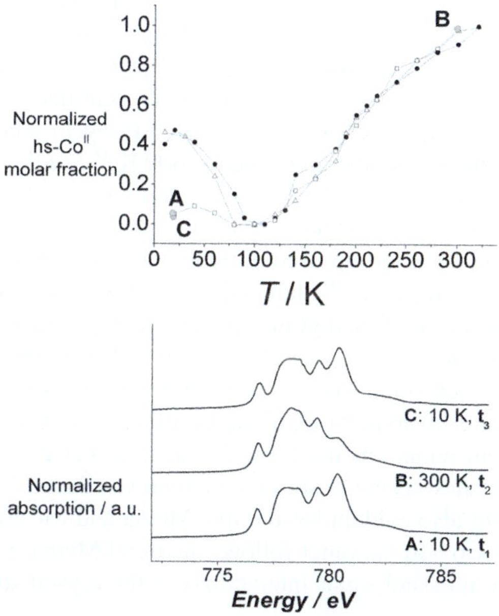

Figure 7.4 *Upper panel: Temperature dependence of the HS-Co^{II}-sq fraction undergoing VT equilibrium for a sample of [Co(Me$_2$tpa)(3,5-dbdiox)](PF$_6$)·toluene measured under a 10^8 X-rays photons s^{-1} flux in the dark (squares), and after 1 h of 904 nm laser light irradiation (triangles). The filled gray circles labeled A, B, and C are extracted from the spectra in the lower panel. The filled black circles represent the fraction of HS-Co^{II}-sq obtained in the dark under a 10^{10} X-ray photons s^{-1} flux. Lower panel: Evidence of the reversibility of the soft-X-ray-induced VT inter-conversion: L_3-edge XAS spectra at 10 K before (A) and after (C) a complete thermal cycle up to 300 K (B) (time scale $t_1 < t_2 < t_3$). Reproduced from [43], with permission from John Wiley and Sons, 2010.*

7.3 Other Factors that Contribute to the Valence Tautomeric Transition

7.3.1 Ancillary Ligand Effects

It has become evident that the ancillary ligands do far more than simply occupy the additional metal coordination sites, with the electronic and steric properties of these ligands proving to play a key role in modulating the VT transitions. Early studies largely focused on the 1:2 cobalt:dioxolene complexes [Co(N_2L)(diox)$_2$] showed a correlation between the donor strength of the bidentate diimine ancillary ligand and the transition temperature of the thermally-induced VT transition.[1, 4] This is due to the ligand field of the ancillary ligand essentially tuning the energy of the frontier orbitals of the dioxolene ligand relative to the cobalt 3d orbitals. Other studies have shown that 1:2 cobalt:dioxolene mononuclear complexes with monodentate pyridyl derivatives as ancillary ligands can also display VT transitions.[35–37, 44–48] In these complexes of formula [Co(NL)$_2$(diox)$_2$] (NL = monodentate nitrogen donor ligand) the nitrogen donors of the non-chelating pyridyl ligands are *trans* to each other, rather than *cis*, as is the case for the chelating bidentate ligands. In itself, this seems to have little effect on the VT behavior, with thermal and light-induced VT transitions evident in these complexes comparable to those in the [Co(N_2L)(diox)$_2$]. However, quite remarkable behavior has been observed for several of these [Co(NL)$_2$(diox)$_2$] complexes, which appears to be due to cooperative effects in the system due to ancillary-ligand mediated intermolecular interactions (Section 7.3.3).[35, 36]

A move to investigating 1:1 cobalt:dioxolene complexes [Co(N_4L)(diox)]$^+$ saw the incorporation of tetradentate nitrogen donor ligands (Fig. 7.2), initially by the research group of Dei. These have included the tetrazamacrocylic ligands cth, 1,4,8,11-tetraazacyclotetradecane (cyclam) and 1,4,8,11-tetraaza-1,4,8,11-tetramethylcyclotetradecane (Me$_4$cyclam), each of which has four amino nitrogen donor atoms.[7, 48] This is in contrast to the other widely used family of N_4L ancillary ligands based on tpa, each of which has one amino and three imino nitrogen donor atoms.[7, 26] Of the macrocylic ligands, studies have shown that while Me$_4$cyclam, with its (weaker) tertiary amine donor atoms, stabilizes the HS-Co^{II}-sq tautomer in the temperature range 4–300 K, the cth and cyclam ligands with (stronger) secondary amine donors stabilize transition to the LS-Co^{III}-cat tautomer at temperatures below 300 K.[48] Studies of these complexes by the Dei group emphasized the utility of electrochemical measurements to access the redox potentials of the ligand and metal-based processes, which correlate directly with the energies of the frontier orbitals that are so important to the VT transfer. The same group used similar principles in an elegant demonstration of the tuning of the VT transitions in [Co(Me$_n$tpa)(3,5-dbdiox)]$^+$ ($n = 0$-3) complexes by derivatization of the tpa ancillary ligand.[7, 26] Sequential addition of methyl groups into the 6-position of the pyridine rings increases the steric hindrance at the cobalt center, thereby modulating its redox properties and controlling the charge distribution across the cobalt-dioxolene unit. Magnetic susceptibility measurements in the temperature range 0–320 K indicate that a temperature-invariant HS-Co^{II}-sq tautomer is evident for the Me$_3$tpa complex, while the tpa and Metpa complexes both remain as the LS-Co^{III}-cat tautomer throughout this temperature range. In contrast, the Me$_2$tpa complex showed the onset of a thermally-induced VT transition at around 280 K. Photoinduced VT transitions were also evident for the tpa, Metpa and Me$_2$tpa complexes. The lifetimes of the photoinduced metastable HS-Co^{II}-sq tautomer follows the trend Metpa $\geq$ Me$_2$tpa > tpa and the authors correlate this to the extent of the intermolecular interactions in the crystal structures which increase in the order tpa < Me$_2$tpa < Metpa.

7.3.2 Counterion and Solvation Effects

Studies involving systematic variation of the counterions of Co-dioxolene complexes are hampered by the near impossibility of isolating the effect of the counterion from the effects of solvation and crystal packing generally. However in an early study that attempted to investigate this, the compounds [Co(cth)(phendiox)]Y ($Y = PF_6^-$, I^- and BPh_4^-) showed an increase in the VT transition temperature in the order $PF_6^- < I^- < BPh_4^-$.[17] The transition also appears to be more gradual in the case of BPh_4^-. These results were interpreted in terms of the increased molecular volume of I^- versus PF_6^- giving rise to steric hindrance that inhibits the molecular expansion associated with the VT transition. In an analogy with the effect of physical pressure, this "chemical pressure" has the effect of increasing the transition temperature for the thermally-induced VT transition. On the other hand, the increased transition temperature observed for the BPh_4^- salt was interpreted in terms of a decreased degree of localization of electrons in BPh_4^- versus PF_6^-, which increases the coulombic interaction between cation and anion, thereby stabilizing the LS-Co^{III}-cat tautomer.

It has also become clear that the molecules of solvation present in the crystal lattice can have a significant effect on VT transitions in the solid state. One of the first systems for which such a comparison could be made is comprised of the ethanol and toluene solvates $[Co(Me_2tpa)(3,5\text{-dbdiox})]PF_6{\cdot}C_2H_5OH$ and $[Co(Me_2tpa)(3,5\text{-dbdiox})]PF_6{\cdot}$toluene.[26, 28] Upon heating from low temperature, the magnetic susceptibility data for the ethanol solvate shows the onset of a VT transition from the LS-Co^{III}-cat to the HS-Co^{II}-sq tautomer at about 280 K. In contrast the transition for the toluene solvate is essentially complete by 300 K, with two clearly separated steps with transition temperatures of 125 and 185 K (Fig. 7.5). The crystal structure of the toluene solvate is in fact comprised of two independent molecules, which undergo thermally-induced VT transitions at two different temperatures, giving rise to the well defined steps in the magnetic susceptibility profile. Although a detailed analysis was not provided, the differing crystal packing and therefore intermolecular interactions are presumably responsible for the much higher transition temperature for the ethanol solvate than for the toluene one.

In a similar study by Ruiz-Molina *et al.* the complex $[Co(phen)(3,5\text{-dbdiox})_2]$ was recrystallized from a variety of solvents, with the microcrystalline samples showing an increase in the VT transition temperature in the order toluene < acetonitrile < acetone < dichloromethane < ethanol.[49] The absence of single crystal structures precluded any correlation with the structures in this case.

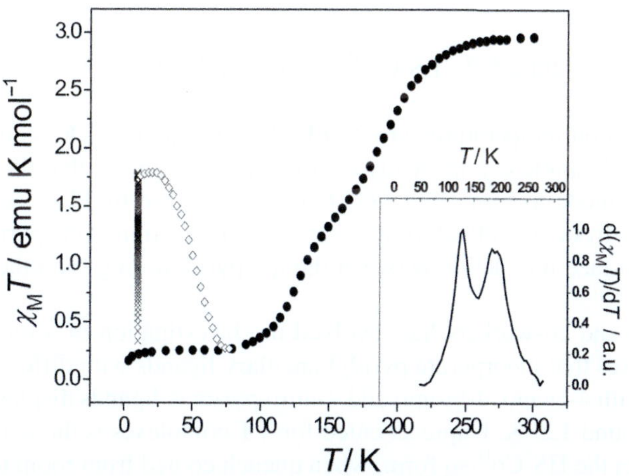

Figure 7.5 *Plot of $\chi_M T$ vs T for $[Co(Me_2tpa)(3,5\text{-dbdiox})]PF_6{\cdot}$toluene measured before (•) and after (◊) irradiation at 904 nm. The crosses indicate the increase in $\chi_M T$ with time upon irradiation of the sample. The inset shows the first derivative $\chi_M T$ with respect to T, evidencing the 2-step thermally induced transition. Reproduced from [28], with permission from John Wiley and Sons, 2008.*

A more detailed investigation was reported by Boskovic *et al.* for three different solvates of the complex *trans*-[Co(py)$_2$(3,5-dbdiox)$_2$] (py = pyridine).[37] In this case while transition temperatures for thermally-induced VT transitions are around 250 K for the pyridine and acetonitrile solvates, the hexane solvate does not show the onset of a transition until around 350 K. In addition, desolvation of the acetonitrile solvate increases the transition temperature by around 50 K and gives rise to a much more abrupt transition. Of most interest in this work is the pyridine solvate, in the crystal structure of which there are two independent molecules. Variable temperature crystallography indicates that the VT transition only occurs in one of these two molecules up to 300 K, consistent with magnetic susceptibility data. Inspection of the structure suggests that "chemical pressure" associated with the proximity of a pyridine molecule of solvation to one of the independent [Co(py)$_2$(3,5-dbdiox)$_2$] molecules inhibits the VT transition for this molecule, while in the absence of similar steric hindrance the other molecule is able to undergo the expansion in molecular volume that accompanies the VT transition.

7.3.3 Cooperativity

An early report of a complex that displays significant cooperativity associated with a VT transition involved [Co(py$_2$O)(3,6-dbdiox)$_2$] (py$_2$O = bis(pyridyl) ether), which displays a hysteretic VT transition with transition temperatures of 330 K (heating) and 100 K (cooling), affording a 130 K wide hysteresis loop.[50] The origin of the hysteresis is attributed to a structural change of the py$_2$O ancillary ligand from a non-planar conformation in the low temperature LS-Co^{III}-cat tautomer to a planar conformation in the high temperature HS-Co^{II}-sq tautomer, which imposes an energy barrier to the intramolecular electron transfer in either direction. The ability of the compound to undergo the conformational changes associated with the VT transition depends on the packing in the crystal lattice. A sample recrystallized from acetone displays a sufficiently open structure to allow the conformational change of the py$_2$O ligand, while a sample recrystallized from toluene is comprised of a layered structure that prevents this structural change and therefore the VT transition.

A large hysteresis at room temperature was reported by Sato *et al.* for a PF_6^- salt of the dinuclear tetraoxalene complex [[Co(tpa)]$_2$(dhbq)]$^{3+}$ (Fig. 7.6a), which displays a thermally induced VT transition at one cobalt center according to:[51]

$$\text{LS-Co}^{III}\text{—dhbq}^{\text{sq-cat}}\text{—LS-Co}^{III} \rightleftharpoons \text{LS-Co}^{III}\text{—dhbq}^{\text{sq-sq}}\text{—HS-Co}^{II} \quad (7.1)$$

For this complex the transition temperatures are 310 K (heating) and 297 K (cooling), giving a hysteresis width of 13 K (Fig. 7.6c). A single crystal X-ray diffraction study shows that the dinuclear complexes are stacked in a head-to-waist mode to form 2D layers that are separated by PF_6^- counteranions. The authors attribute the presence of the hysteresis effect to an offset $\pi \cdots \pi$ interaction of the pyridine rings of the ancillary tpa ligands and an edge-to-face interaction between the tpa pyridine rings and the phenyl ring of the dhbq ligands (Fig. 7.6b).

Recent work by Shultz and co-workers has involved the investigation of a remarkable family of *trans*-[Co(NL)$_2$(diox)$_2$] complexes that incorporate pyridyl ancillary ligands with different substituents (NL).[35, 36] Of these, the complexes with 4-cyano, 4-bromo and 4-nitro pyridine ligands display thermal hysteresis loops of width 5 K, centered around 125 K. Unprecedented for VT complexes is the feature that these complexes can be thermally trapped in the HS-Co^{II}-sq form, when quench cooled from room temperature down to 10 K (Fig. 7.7). The HS-Co^{II}-sq tautomer can also be photogenerated and the photogenerated tautomer persists upon heating up to around 90 K, higher than for any other VT complex. In addition, thermal cycling while maintaining irradiation also gives rise to a light-induced thermal hysteresis loop, while below 50 K the light-induced tautomer relaxes substantially more slowly than has been reported previously for any other VT complex. A detailed comparison of the three structures of the [Co(NL)$_2$(diox)$_2$] complexes that display these

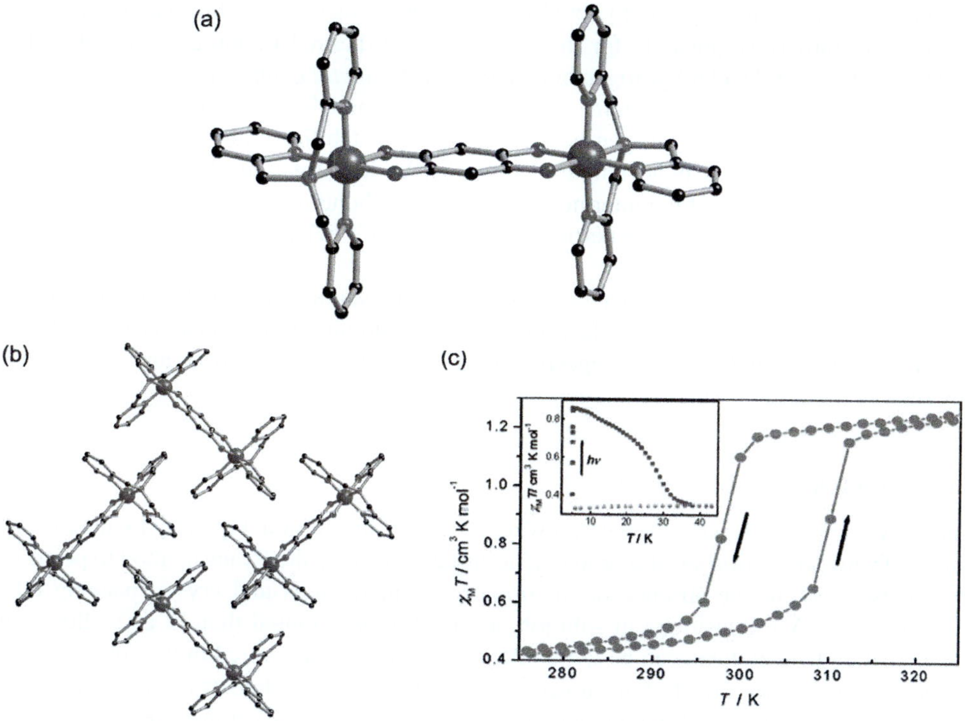

Figure 7.6 *(a) Molecular structure; (b) packing diagram; and (c) plot of* $\chi_M T$ *vs. T for [[Co(tpa)]*$_2$*(dhbq)](PF*$_6$*)*$_3$ *with the inset showing the photoinduced changes after irradiation at 532 nm. Reprinted with permission from [51], Copyright 2006, American Chemical Society.*

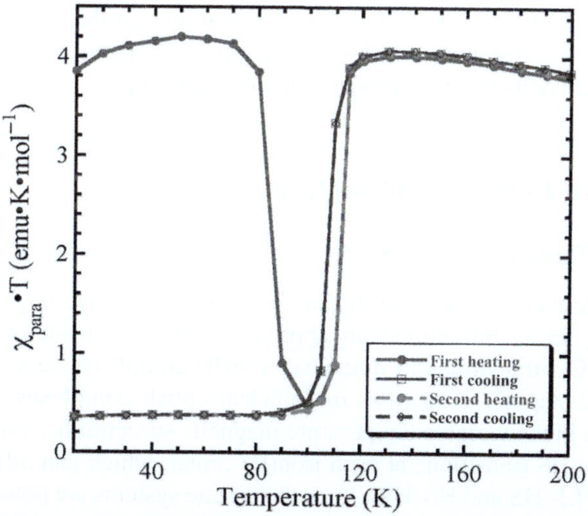

Figure 7.7 *Variable temperature* $\chi_M T$ *cycles for trans-[Co(4-CNpy)*$_2$*(3,5-dbdiox)*$_2$*] (4-CNpy = 4-cyanopyridine), the first heating cycle demonstrates thermal trapping of the high temperature stable form upon rapid quench from 300 to 10 K. Reprinted with permission from [35], Copyright 2010 American Chemical Society.*

unusual features, versus the analogs that do not (NL = pyridine, 4-methyl pyridine and 4-methoxy pyridine) reveal that the cooperativity responsible for the hysteresis and thermal trapping is strongly reliant on an extensive network of intermolecular hydrogen bonds *between the metal complexes directly* (i.e. not mediated by solvent molecules), which is present in the 4-cyano, 4-bromo and 4-nitro pyridine derivatives but absent in the other three complexes. The hydrogen bonding appears to stabilize both tautomers and likely lowers the activation barrier for Co–O/N bond length elongation. An investigation of different solvates of these complexes reveals that disruption of the intermolecular hydrogen bonding between metal complexes through "lattice softening" by solvent intercalation perturbs the observed VT behavior, diminishing or eliminating the cooperative effects.

Another recent report from the Dei group concerns solid state studies of [Co(Me_2tpa)(3,5-dbcat)] diluted with its zinc analog. These indicate that the relaxation rate of the photoinduced metastable HS-Co^{II}-sq tautomer is independent of the degree of dilution,[34] suggesting that intermolecular interactions and cooperativity effects are not important in this case.

7.3.4 Valence Tautomerism in Solution

Much of the early work on valence tautomerism was carried out in solution, typically by measuring variable temperature electronic absorption spectra or magnetic susceptibility by the Evans method to probe thermally-induced VT transitions.[1] In the absence of intermolecular interactions and crystal packing effects, it is generally observed that VT transitions in solution are much more gradual than in crystalline samples. In addition the transition temperatures can be quite different from those measured in the solid state, which has been interpreted in terms of the effect of the molecular environment on the density of vibrational states, which relates to the variations in the Co–O/N bond lengths that accompany the VT transition.[52]

Ruiz-Molina and co-workers have reported a study of the variable temperature electronic absorption spectra of dilute solutions of [Co(N_2L)(3,5-dbdiox)$_2$] (N_2L = 2,2′-bpy and phen) in five different solvents, which revealed a shift in the critical temperature of about 70 K in the order: acetonitrile > ethanol > dichloromethane > acetone > toluene.[49] Interestingly, this ordering does not correspond to the ordering of the dielectric constants of the solvent. The authors suggest that instead the endoergic term for each solvent may play an important role. This term measures the work required to separate the solvent molecules to provide an enclosure of suitable dimensions to accommodate the metal complex.

7.4 Polynuclear Valence Tautomeric Complexes

7.4.1 Dinuclear Valence Tautomeric Complexes

Efforts to synthesize dinuclear complexes that display VT transitions have in part been driven by a desire to incorporate cooperativity in an intramolecular approach to achieve hysteretic transitions. In addition, the observation of 2-step SCO transitions in dinuclear iron(II) complexes serves as an inspiration for the achievement of analogous 2-step VT transitions in dinuclear cobalt complexes. Two-step SCO transitions can be observed as a plateau in the variable temperature magnetic susceptiblity profile.[53, 54] The origin of the steps is often separate LS to HS transitions at each iron(II) center, which can afford a species that exists in three distinct states (LS-LS, LS-HS and HS-HS). Such three-state systems are potentially important for future applications as three-state switches or molecular ternary memory. For these applications to be realized three metastable states are required, which are accessible at reasonable temperatures. Also, molecular materials with three states based on VT transitions will require that the degree of electronic communication between the separate VT units is small, but non-zero, so that each state can be accessed separately. Furthermore, the ability to interconvert these states by different stimuli (for instance, heat and light) may be important for molecular electronics applications in processes such as write-lock-read-unlock-erase cycles.[55]

The conceptually simplest approach to dinuclear cobalt-dioxolene VT complexes is to use bridging bis-bidentate nitrogen donor ancillary ligands, such as bis(diimine) ligands to link two [Co(diox)$_2$] units. This approach has been employed by Jung with the complex [[Co(3,6-dbdiox)$_2$]$_2$(N$_2$LN$_2$)], where N$_2$LN$_2$ is 2,2′-bipyrimidine,[56] and by Shultz, Preuss, Kirk and co-workers with the complexes [[Co(3,5-dbdiox)$_2$]$_2$(N$_2$LN$_2$)], where N$_2$LN$_2$ is 4,6-di-2′-pyridylpyrimidine (ppm) or ligands based on *p*- and *m*-phenylene bridged 2,2′-bipyridine (bpy-*m*-Ph-bpy and bpy-*p*-Ph-bpy)[57, 58] (Fig. 7.2). The first such complex synthesized, the bipyrimidine complex, did not showed any thermally-induced VT transition in the range 100–400 K, with magnetic susceptibility and NIR data indicating that the complex stayed in the [HS-Co^{II}-sq]$_2$ tautomeric form.[56] In contrast magnetic susceptibility, X-ray crystallography and XAS and electronic spectroscopy measurements on the phenylene bridged bpy complexes showed evidence for all three tautomeric forms: [LS-Co^{III}-cat]$_2$, [LS-Co^{III}-cat][HS-Co^{II}-sq] and [HS-Co^{II}-sq]$_2$ in the temperature range 200–400 K.[57] However the observed behavior was complicated and showed considerable dependence on the environment, with different behavior observed for solvated and desolvated microcrystalline samples and thin films. In any case there was no clear evidence of either hysteresis or a 2-step transition. In contrast, a 2-step thermally-induced VT transition was identified from the variable temperature (88–428 K) NIR spectra of a thin amorphous film of [[Co(3,5-dbdiox)$_2$]$_2$(ppm)] according to:[58]

$$[\text{LS-Co}^{III}\text{-cat}]_2 \rightleftharpoons [\text{LS-Co}^{III}\text{-cat}][\text{HS-Co}^{II}\text{-sq}] \rightleftharpoons [\text{HS-Co}^{II}\text{-sq}]_2 \quad (7.2)$$

This transition is incomplete though, with the [LS-Co^{III}-cat]$_2$ state not fully realized down to 88 K. Powder and solution samples of this complex were also investigated, with magnetic susceptibility studies on an amorphous powder showing evidence for the onset of a transition from the [LS-Co^{III}-cat][HS-Co^{II}-sq] to the [HS-Co^{II}-sq]$_2$ tautomer in the range 2–300 K. Spectroscopic studies on a dilute toluene solution suggested the [HS-Co^{II}-sq]$_2$ tautomer at 345 K showing evidence for inter-conversion to the [LS-Co^{III}-cat][HS-Co^{II}-sq] tautomer upon cooling to 194 K. Although at least a partial VT transition occurs for all three phases of the complex, the observed behavior is different in each case and is clearly critically dependent on the environment of the cobalt complex.

An alternative approach to dinuclear cobalt-dioxolene VT complexes is for the redox-active ligand to also serve as a bridge between [Co(N$_4$L)] units. Efforts in this directions have employed bis-bidentate bis-dioxolene ligands (Fig. 7.2) derived from bis-5,5′-*tert*-butyl-3,3′,4,4′-tetrahydroxybiphenyl (diox-diox) and 3,3,3′,3′-tetramethyl-1,1′-spirobisindane-5,5′,6,6′-tetrol (diox-spiro-diox).[2, 30, 59] The first such complexes employed the diox-diox ligand with two complexes formulated as [[Co(cth)]$_2$(diox-diox)]$^{n+}$ (n = 2 and 3) reported as the LS-Co^{III}-cat-cat-LS-Co^{III} and LS-Co^{III}-cat-sq-LS-Co^{III} species.[59] No VT transition is reported for either complex and EPR spectra of the LS-Co^{III}-cat-sq-LS-Co^{III} species suggest complete delocalization (Robin and Day class III behavior)[60] of the unpaired electron across both dioxolene units of the ligand. Although not reported directly, a review article mentions the synthesis of complexes formulated as [[Co(cth)]$_2$(diox-spiro-diox)]$^{n+}$ (n = 3 and 4),[7] which are the LS-Co^{III}-cat-spiro-sq-LS-Co^{III} and LS-Co^{III}-sq-spiro-sq-LS-Co^{III} forms. While EPR studies on the LS-Co^{III}-cat-spiro-sq-LS-Co^{III} species suggest that the unpaired electron is fully localized on one dioxolene unit of the bis-dioxolene ligand (Robin and Day class I behavior),[60] a weak antiferromagnetic coupling is observed between the two radical ($S = {}^1/_2$) semiquinonate halves of the ligand for the LS-Co^{III}-sq-spiro-sq-LS-Co^{III} species.[7]

Subsequent work from Boskovic and co-workers has demonstrated that use of the tpa family of ligands allows control of the redox state of the dinuclear complexes of formula [[Co(Me$_n$tpa)]$_2$(diox-spiro-diox)]$^{2+}$ (Fig. 7.8),[38] as was observed previously for the mononuclear complexes [Co(Me$_n$tpa)(3,5-dbdiox)]$^+$.[26] In

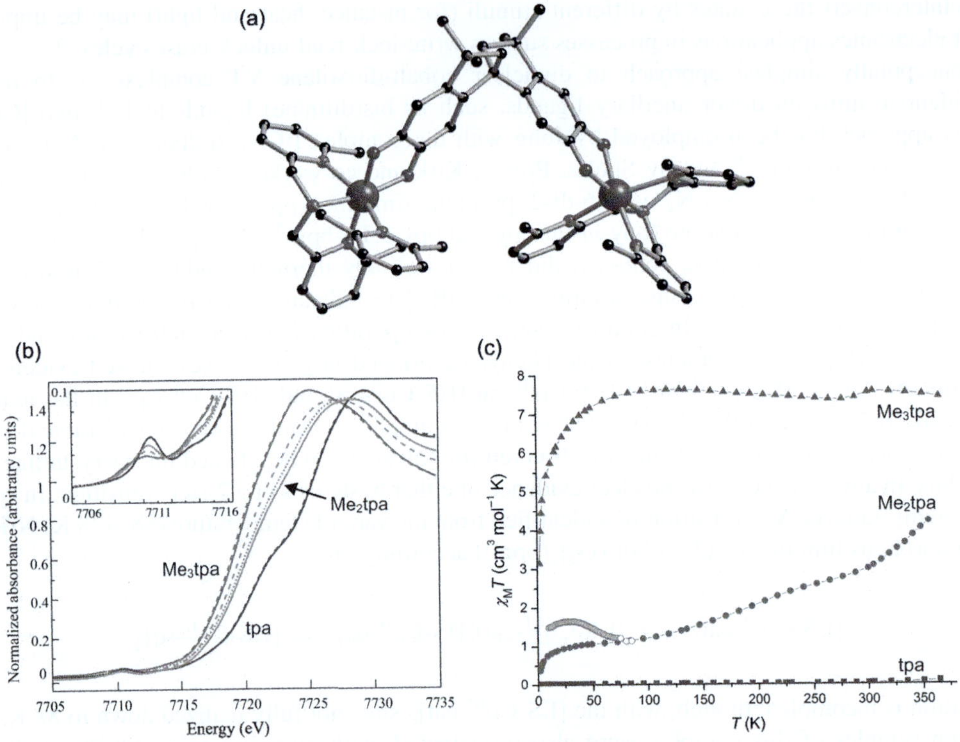

Figure 7.8 *(a) Structural representation of [[Co(Me$_3$tpa)]$_2$(diox-spiro-diox)]$^{2+}$. Variable temperature (b) Co K-edge XANES spectra for ClO_4^- salts of [[Co(Me$_n$tpa)]$_2$(diox-spiro-diox)]$^{2+}$ (n = 0, 2 and 3), at 16 (solid), 100 (dotted) and 298 K (dashed); (c) $\chi_M T$ data for ClO_4^- salts of [[Co(tpa)]$_2$(diox-spiro-diox)]$^{2+}$ (■), [[Co(Me$_2$tpa)]$_2$(diox-spiro-diox)]$^{2+}$ (•) and [[Co(Me$_3$tpa)]$_2$(diox-spiro-diox)]$^{2+}$ (▲) and photoinduced $\chi_M T$ data for [[Co(Me$_2$tpa)]$_2$(diox-spiro-diox)]$^{2+}$ (∘). Reprinted with permission from [38], Copyright 2012 American Chemical Society.*

this case the employment of the tpa and Me$_3$tpa ancillary ligands afford the temperature-invariant LS-Co^{III}-cat-spiro-cat-LS-Co^{III} and HS-Co^{II}-sq-spiro-sq-HS-Co^{II}, respectively across the temperature range 2–360 K. However variable temperature magnetic susceptibility and Co K-edge XANES studies on the Me$_2$tpa analog show evidence of the two-step thermally-induced VT transition:

$$\begin{aligned} \text{LS-Co}^{\text{III}}\text{–cat-spiro-cat–LS-Co}^{\text{III}} &\rightleftharpoons \text{LS-Co}^{\text{III}}\text{–cat-spiro-sq–HS-Co}^{\text{II}} \\ &\rightleftharpoons \text{HS-Co}^{\text{II}}\text{–sq-spiro-sq–HS-Co}^{\text{II}} \end{aligned} \tag{7.3}$$

in the temperature range 100–360 K, with both steps incomplete. This VT transition can also be partially photoinduced.

In related systems, the bis-bidentate tetraoxolene ligand dhbq (Fig. 7.2) has been used to bridge [Co(N$_4$L)] units in the dinuclear complexes [[Co(N$_4$L)]$_2$(dhbq)]$^{3+}$. Early work by Dei, Létard and co-workers on a complex, which could not be crystallographically characterized, but was formulated as [[Co(cth)]$_2$(dhbq)]$^{3+}$,[22, 23] revealed both thermal- and photoinduced VT transitions involving one cobalt center according to Eq. (7.1). Subsequently Sato and co-workers were able to obtain an X-ray crystal structure of a PF_6^- salt of

$[[Co(tpa)]_2(dhbq)]^{3+}$. As mentioned above (Section 7.3.3), the thermally induced VT transition at one cobalt center is accompanied by hysteresis centered at 303 K, which is attributed to the cooperative effect associated with intermolecular interactions between adjacent dinuclear complexes.[51] Replacement of the PF_6^- counteranions with BF_4^- and ClO_4^- increases the transition temperature by more than 50 K in both cases, with hysteresis (width 20 K) evident for the ClO_4^- salt, but not for the BF_4^-.[61] In contrast, changing the ancillary ligand to dpqa leads to a decrease in the transition temperature and a significant diminution in the hysteresis width.[19, 25] Again small changes in the ancillary ligands and counteranions give rise to changes in crystal packing and intermolecular interactions that have a profound effect on the VT transition. These VT transitions can also be photoinduced (Section 7.2.3) and pressure-induced (Section 7.2.2).

7.4.2 Polymeric Valence Tautomeric Complexes

Other efforts to achieve intramolecular cooperativity have focused on polymeric VT complexes. The simplest approach to one-dimensional coordination polymers that display VT behavior has involved the use of divergent nitrogen donor ligands such as pyrazine (pyz) and 4,4′-bipyridine (4,4′-bpy) to link cobalt-dioxolene units into a chain. The chain complexes $[Co(pyz)(3,6\text{-dbdiox})_2]_\infty$ and $[Co(4,4'\text{-bpy})(3,5\text{-dbdiox})_2]_\infty$ have both been characterized by single crystal X-ray diffraction.[62, 63] Magnetic susceptibility and near infrared spectroscopy studies of these chain complexes indicate thermally-induced VT transitions and close to complete inter-conversion between the $[LS\text{-}Co^{III}\text{-cat}]_\infty$ and $[HS\text{-}Co^{II-}sq]_\infty$ tautomers, although no hysteresis was observed. Interestingly $[Co(pyz)(3,6\text{-dbdiox})_2]_\infty$ recrystallized from hexane crystallizes as thin tetragonal plates, with the polymer chain direction perpendicular to the plane of the plate. These crystals show a reversible photomechanical effect in the form of a physical distortion upon light irradiation, with the change in Co–N bond length associated with the VT transition corresponding to a change of 0.06 mm per mm of polymer length.[62]

A second approach has employed the redox-active ligand as the bridging unit, making use of bis-bidentate bis-dioxolene ligands with N_2L ancillary ligands. Although they could not be characterized crystallographically, complexes formulated as $[Co(phen)(diox\text{-}Ph\text{-}diox)]_\infty$ (diox-Ph-diox = bis-bidentate bis-dioxolene ligand derived from 3,5-bis(3′,4′-dihydroxy-5′-*tert*-butylphenyl)-1-*tert*-butylbenzene; Fig. 7.2) and $[Co(N_2L)(diox\text{-}spiro\text{-}diox)]_\infty$ (N_2L = 2,2′-bpy and phen), and believed to have one dimensional chain structures, have been reported to display thermally and light-induced VT transitions.[30, 32, 64] This family of complexes allow for comparison between nominally Robin and Day class I (diox-spiro-diox) and class II (diox-Ph-diox) bridging-ligands,[60] however the unfortunate lack of structural data precludes a definitive elucidation of the VT behavior.

There is only a single report of a higher dimensional coordination network, or metal-organic-framework, that incorporates VT building blocks, which is the species $[Co(tpom)(3,5\text{-dbdiox})_2]_\infty$ reported by Tao, Zheng *et al.* (Fig. 7.9), where tpom is the tetrapyridyl bridging ligand tetrakis(4-pyridyloxymethylene)methane (Fig. 7.2).[39] This complex has a two-dimensional layer structure with the tpom ligands acting as tetrahedral nodes linking two-connecting $[Co(3,5\text{-dbdiox})_2]$ units. The three dimensional structure is assembled from two layers of cobalt atoms on either side of a "sandwich" layer of bridging tpom ligands (Fig. 7.9). The layers are stacked on top of each other to give a porous structure with solvent molecules between the repeating layers. The material displays thermally- and light-induced VT transitions, with the thermal transition complicated by solvent loss. Interestingly, there is little cooperativity between cobalt centers, particularly following solvent loss, with the thermally induced transition showing a very gradual profile. The authors attribute this to the relative flexibility of the ether units in the linking ancillary ligand. Nevertheless this species serves as proof of concept for the organization of VT units into framework-type materials.

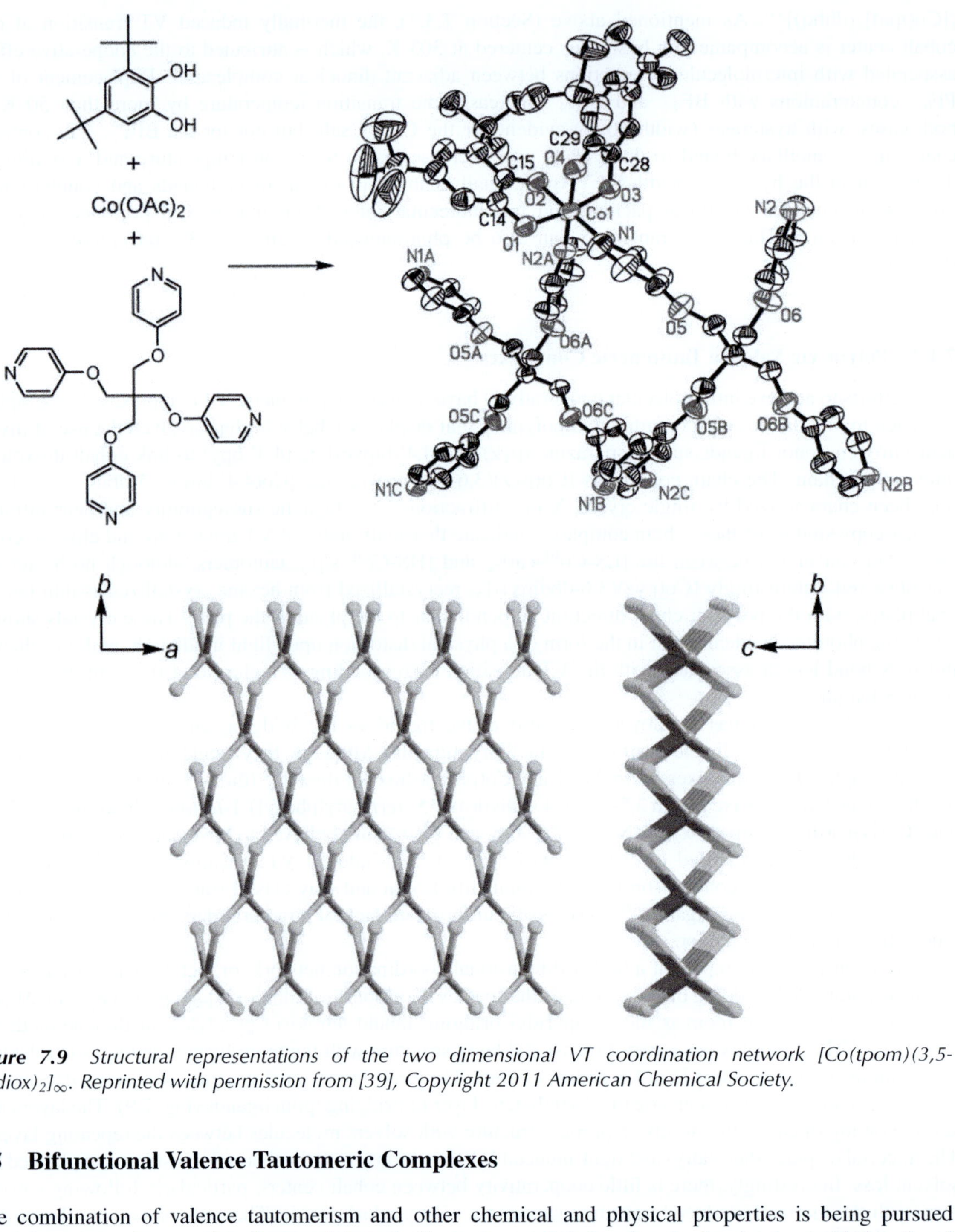

Figure 7.9 *Structural representations of the two dimensional VT coordination network* $[Co(tpom)(3,5\text{-}dbdiox)_2]_\infty$. *Reprinted with permission from [39], Copyright 2011 American Chemical Society.*

7.5 Bifunctional Valence Tautomeric Complexes

The combination of valence tautomerism and other chemical and physical properties is being pursued through a range of approaches that are ultimately directed towards informing the development of bifunctional molecule-based materials for future applications. An example of this is the synthesis by Ruiz-Molina *et al.*

of amorphous spherical nanoparticles based on the one dimensional VT coordination polymer [Co(bix)(3,5-dbdiox)$_2$]$_\infty$ (bix = 1,4-bis(imidazol-1-ylmethyl)benzene; Fig. 7.2).[63] It was possible to tune the size of the nanoparticles to obtain batches with diameters in the range 76 ± 9 to 204 ± 13 nm. Magnetic susceptibility studies on aqueous colloidal suspensions of the different sized nanoparticles indicated a gradual thermally-induced VT transition, which is incomplete up to 350 K, with the magnetic behavior essentially independent of the size of the nanoparticles.

Efforts by Frank's group have afforded molecules that combine valence tautomerism with photochromism. These species were obtained by incorporating the photoswitchable photchromic ligand spiro(indoline-phenanthrolinoxazine) (ipso) as an ancillary ligand attached to a [Co(3,5-dbdiox)$_2$] unit.[65] The ipso ligand photoisomerizes between a UV-induced ring-opened photomercocyanine form (pmc, $\lambda_{max} \approx 600$ nm) and a ring-closed spirooxazine form (so, $\lambda_{max} \approx 350$ nm). In principle this molecular system can exist in four different electronic states, which can be inter-converted by a combination of thermal and optical stimuli (Fig. 7.10). Spectroscopic and magnetic studies in solution and the solid state revealed complex behavior and low photoresponsivities, however the system represents a prototype for bifunctional multistate molecular materials.

Another clever modification of the ancillary ligands of the prototype VT complex, this time by Yoshida, Saito and co-workers, afforded the anionic complex [Co(dcoo-bpy)(3,5-dbdiox)$_2$]$^{2-}$ (dcoo-bpyH$_2$ = 2,2′-bipyridine-4,4′-dicarboxylic acid; Fig. 7.2), and combination with the bulky trihexyl(tetradecyl)phosphonium cation afforded the first VT ionic liquid.[66] This compound is a highly viscous liquid at room temperature

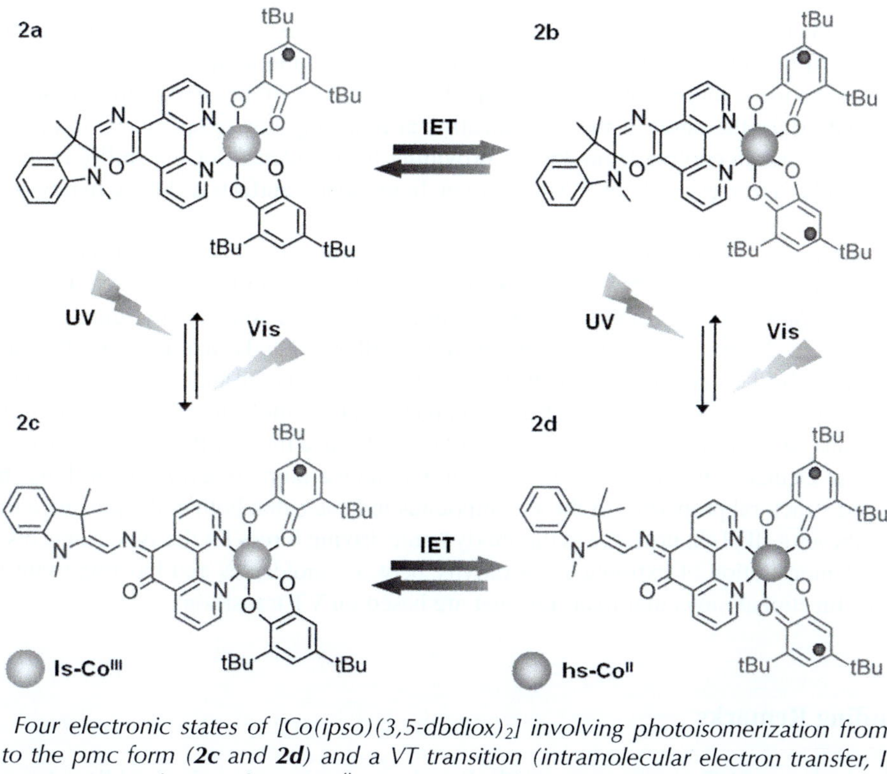

Figure 7.10 *Four electronic states of [Co(ipso)(3,5-dbdiox)$_2$] involving photoisomerization from the so form (**2a** and **2b**) to the pmc form (**2c** and **2d**) and a VT transition (intramolecular electron transfer, IET) from the LS-Co^{III}-cat tautomer (**2a** and **2c**) to the HS-Co^{II}-sq tautomer (**2b** and **2d**). Reproduced from [65] with permission of The Royal Society of Chemistry.*

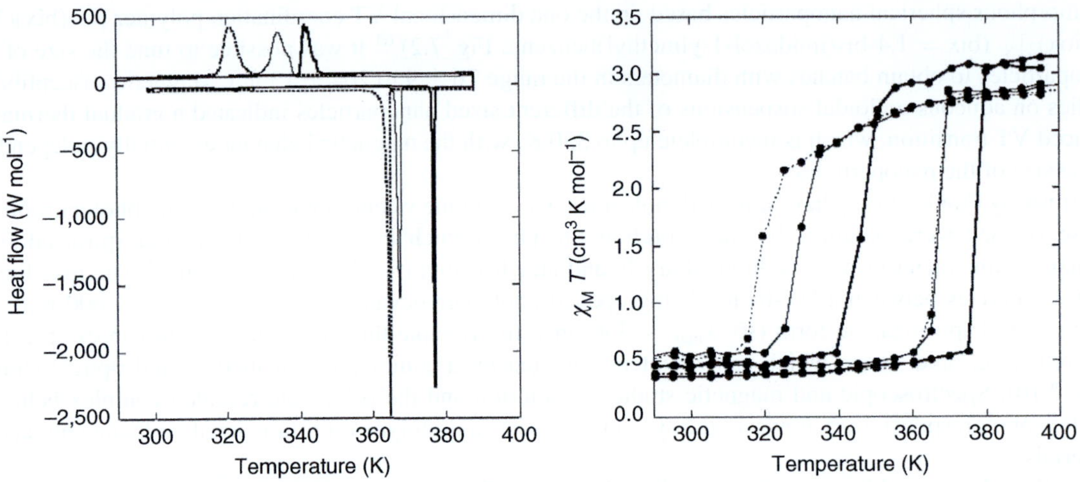

Figure 7.11 Plots of the temperature dependent DSC curves (left) and magnetic susceptibility (right) observed for [Co(CnOpy)$_2$(3,6-dbdiox)], with n = 9 (bold), 12 (solid) and 17 (dotted), on the first heating processes at a scan rate of 1 K min^{-1}. Reprinted with permission from [45], Copyright 2008 American Chemical Society.

with a glass transition temperature of 262 K. Variable temperature magnetic susceptibility and spectroscopic measurements reveal a gradual thermally-induced VT transition in the temperature range 2 to 350 K. The authors argue that the loose molecular packing in the ionic liquid facilitates expansion to the larger Co^{II}-sq tautomers, accounting for the onset of the VT transition at relatively low temperature. In addition, thermal hysteresis is observed when cooling from 350 K, providing firm evidence for bistability in the liquid state. Thus this compound is a rare example of a "smart" ionic liquid that manifests a change in physical properties with temperature.

A series of studies by Chang, Kitagawa *et al.* on the carefully designed cobalt-dioxolene complexes [Co(C*n*Opy)$_2$(3,6-dbdiox)], [Co(C*n*bpy)$_2$(3,5-dbdiox)] and [Co(C*n*espy)$_2$(3,5-dbdiox)]] (C*n*Opy = 3,5-dialkoxy-pyridine; C*n*bpy = 4,4′-dialkyl-2,2′-bipyridine; C9espy = dinonyl-pyridine-3,5-dicarboxylate; Fig. 7.2), functionalized with long alkyl chains on the ancillary ligands, has revealed that the VT interconversions exhibited by these species are synchronized with crystal–crystal and crystal–melt phase transitions (Fig. 7.11).[45–47, 67] In these compounds the structural changes at the molecular level (resulting from the VT transition) are propagated throughout the ensemble of molecules in a manner that gives rise to a synchronous macroscopic phase change. An integrated series of variable temperature experiments probing the magnetic, spectroscopic and structural properties of these compounds indicates that both enthalpic and entropic effects associated with the long alkyl chains serve as thermodynamic driving forces for the synchronic bistability. This is essentially a demonstration of extreme cooperativity between molecules and has important implications for the design of functional molecular materials that are based on VT transitions.

7.6 Concluding Remarks

The fundamental origins of VT transitions in cobalt-dioxolene complexes have been well-understood for some time. However the effect of both the ancillary ligands and the molecular environment on the manifestation of the VT transition is becoming increasingly apparent. In some instances careful chemical design of both ancillary and redox-active ligands has allowed the deliberate modulation of the VT transitions or the combination of VT transitions with other physical properties to generate bifunctional species. Although crystal packing, and therefore intermolecular interactions, remain essentially beyond the control of the synthetic chemist, serendipity has afforded several compounds where cooperativity resulting from intermolecular interactions has given rise to the two properties that have been identified as essential for functional molecular materials based on VT transitions. These are room temperature bistability with wide hysteresis on the one hand, and very long-lived accessible metastable states on the other. While these two important properties have been observed in crystalline samples, their extension to environments suitable for practical applications in future devices, such as on surfaces or in thin films, remains a challenge.

Acknowledgements

Ho-Chol Chang, Natia Frank, David Shultz and Lorenzo Sorace and are thanked for their generous provision of figures. Kerwyn Alley is thanked for his assistance with thc preparation of this manuscript.

References

1. Hendrickson, D. N., Pierpont, C. G. (2004) Valence tautomeric transition metal complexes, in: Gütlich, P., Goodwin, H. A. (Eds) Spin Crossover in Transition Metal Compounds II. Top. Curr. Chem., 234: 63–95.
2. Dei, A., Gatteschi, D., Sangregorio, C., Sorace, L. (2004). Quinonoid metal complexes: toward molecular switches. Acc. Chem. Res., 37: 827–835.
3. Evangelio, E., Ruiz-Molina, D. (2005) Valence tautomerism: new challenges for electroactive ligands. Eur. J. Inorg. Chem., 2957–2971.
4. Evangelio, E., Ruiz-Molina, D. (2008) Valence tautomerism: more actors than just electroactive ligands and metal ions. C. R. Chimie, 11: 1137–1154.
5. Sato, O., Tao, J., Zhang, Y.-Z. (2007) Control of magnetic properties through external stimuli. Angew. Chem. Int. Ed., 46: 2152–2187.
6. Sato, O., Cui, A., Matsuda, R., Tao, J., Hayami, S. (2007). Photo-induced valence tautomerism in Co complexes. Acc. Chem. Res., 40: 361–369.
7. Dei, A., Sorace, L. (2010) Cobalt-dioxolene redox isomers: potential spintronic devices. Appl. Magn. Reson., 38: 139–153.
8. Kaim, W., Schwederski, B. (2010) Non-innocent ligands in bioinorganic chemistry–an overview. Coord. Chem. Rev., 254: 1580–1588.
9. Dzik, W. I., van der Vlugt, J. I., Reek, J. N. H., de Bruin, B. (2011) Ligands that store and release electrons during catalysis. Angew. Chem. Int. Ed., 50: 3356–3358.
10. Heyduk, A. F., Zarkesh, R. A., Nguyen, A. I. (2011) Designing catalysts for nitrene transfer using early transition metals and redox-active ligands. Inorg. Chem., 50: 9849–9863.
11. Smith, A. L., Hardcastle, K. I., Soper, J. D. (2010) Redox-active ligand-mediated oxidative addition and reductive elimination at square planar cobalt(III): multielectron reactions for cross-coupling. J. Am. Chem. Soc., 132: 14358–14360.
12. Blanchard, S., Derat, E., Desage-El Murr, M., Fensterbank, L., Malacria, M., Mouriès-Mansuy, V. (2012) Non-innocent ligands: new opportunities in iron catalysis. Eur. J. Inorg. Chem., 376–389.
13. Wada, T., Tsuge, K., Tanaka, K. (2000) Electrochemical oxidation of water to dioxygen catalyzed by the oxidized form of the bis(rutheniumhydroxo) complex in H_2O. Angew. Chem. Int. Ed., 39: 1479–1482.
14. Ghosh, S., Baik, M.-H. (2012) The mechanism of O–O bond formation in Tanaka's water oxidation catalyst. Angew. Chem. Int. Ed., 51:1221–1224.

15. Buchanan, R. M., Pierpont, C. G. (1980) Tautomeric catecholate–semiquinone interconversion via metal–ligand electron transfer. Structural, spectral, and magnetic properties of (3,5-di-tert-butylcatecholato)(3,5-di-tert-butylsemiquinone)(bipyridyl)cobalt(III), a complex containing mixed-valence organic ligands. J. Am. Chem. Soc., 102: 4951–4957.
16. Roux, C., Adams, D. M., Itié, J. P., Polian, A., Hendrickson, D. N., Verdaguer, M. (1996) Pressure-induced valence tautomerism in cobalt o-quinone complexes: an X-ray absorption study of the low-spin [Co^{III}(3,5-DTBSQ)(3,5-DTBCat)(phen)] to high-spin [Co^{II}(3,5-DTBSQ)$_2$(phen)] interconversion. Inorg. Chem., 35: 2846–2852.
17. Caneschi. A., Dei, A., De Biani, F. F., Gütlich, P., Ksenofontov, V., Levchenko, A., Hoefer, A., Renz, F. (2001) Pressure- and temperature-induced valence tautomeric interconversion in a *o*-dioxolene adduct of a cobalt-tetraazamacrocycle complex. Chem. Eur. J., 7: 3926–3930.
18. Gütlich, P., Ksenofontov, V., Gaspar, A. B. (2005) Pressure effect studies on spin crossover systems. Coord. Chem. Rev., 249: 1811–1829.
19. Li, B., Yang, F.-L., Tao, J., Sato, O., Huangand, R.-B., Zheng, L.-S. (2008) The effects of pressure on valence tautomeric transitions of dinuclear cobalt complexes, Chem. Commun. 6019–6021.
20. Tourón Touceda, P., Mosquera Vázquez, S., Lima, M., Lapini, A., Foggi, P., Dei, A., Righini, R. (2012) Transient infrared spectroscopy: a new approach to investigate valence tautomerism. Phys. Chem. Chem. Phys., 14: 1038–1047.
21. Cui, A., Takahashi, K., Fujishima, A., Sato, O. (2004). Mechanism and relaxation kinetics of photo-induced valence tautomerism of [Co(phen)(3,5-DBSQ)$_2$]·C_6H_5Cl, J. Photochem. Photobiol. A Chem., 167: 69–73.
22. Carbonera, C., Dei, A., Létard, J.-F., Sangregorio, C., Sorace, L. (2004) Thermally and light-induced valence tautomeric transition in a dinuclear cobalt-tetraoxolene complex. Angew. Chem. Int. Ed., 43: 3136–3138.
23. Carbonera, C., Dei, A., Létard, J.-F., Sangregorio, C., Sorace, L. (2007). Relaxation dynamics of a photoinduced di-cobalt-tetraoxolene valence tautomers. Inorg. Chim. Acta, 360: 3825–3828.
24. Yokoyama, T., Okamoto, K., Nagai, K., Ohta, T., Hayami, S., Gu, Z.-Z., Nakajima, R., Sato, O. (2001) Photo-induced magnetized state of Co(DTBSQ)(DTBCat)(phen).$C_6H_5CH_3$ studied by X-ray absorption spectroscopy, Chem. Phys. Lett. 345: 272–276.
25. Teki, Y., Shirokoshi, M., Kanegawa, S., Sato, O. (2011) ESR study of light-induced valence tautomerism of a dinuclear co complex. Eur. J. Inorg. Chem., 3761–3767.
26. Beni, A., Dei, A., Laschi, S., Rizzitano, M., Sorace, L. (2008) Tuning the charge distribution and photoswitchable properties of cobalt-dioxolene complexes by using molecular techniques. Chem. Eur. J., 14: 1804–1813.
27. Li, B., Tao, J., Sun, H.-L., Sato, O., Huang, R.-B., Zheng, L.-S. (2008) Side-effect of ancillary ligand on electron transfer and photodynamics of a dinuclear valence tautomeric complex. Chem. Commun., 2269–2271.
28. Dapporto, P., Dei, A., Poneti, G., Sorace, L. (2008) Complete direct and reverse optically induced valence tautomeric interconversion in a cobalt-dioxolene complex. Chem. Eur. J., 14: 10915–10918.
29. Sato, O., Hayami, S., Gu, Z.-Z., Takahashi, K., Nakajima, R., Seki, K., Fujishima, A. (2002) Photo-induced valence tautomerism in a Co compound, J. Photochem. Photobiol. A Chem., 149: 111–114.
30. Beni, A., Dei, A., Shultz, D., Sorace, L. (2006) Ligand design modulates photoinduced properties of cobalt-dioxolene valence tautomers. Chem. Phys. Lett., 428: 400–404.
31. Beni, A., Dei, A., Rizzitano, M., Sorace, L. (2007) Unprecedented optically induced long-lived intramolecular electron transfer in cobalt-dioxolene complexes. Chem. Commun., 2160–2162.
32. Affronte, M., Beni, A., Dei, A., Sorace, L. (2007) Valence tautomerism interconversion triggers transition to stable charge distribution in solid polymeric cobalt–polyoxolene complexes. Dalton Trans., 5253–5259.
33. Poneti, G., Mannini, M., Sorace, L., Sainctavit, P., Arrio, M.-A., Rogalev, A., Wilhelm, F., Dei, A. (2009) X-ray absorption spectroscopy as a probe of photo- and thermally induced valence tautomeric transition in a 1:1 cobalt-dioxolene complex. Chem. Phys. Chem., 10: 2090–2095.
34. Dei, A., Poneti, G., Sorace, L. (2010) Metal dilution effects on entropy and light-induced valence tautomeric interconversion in a 1:1 cobalt−dioxolene complex. Inorg. Chem., 49: 3271–3277.
35. Schmidt, R. D., Shultz, D. A., Martin, J. D. (2010) Magnetic bistability in a cobalt bis(dioxolene) complex: long-lived photoinduced valence tautomerism. Inorg. Chem., 49: 3162–3168.
36. Schmidt, R. D., Shultz, D. A., Martin, J. D., Boyle, P. D. (2010) Goldilocks effect in magnetic bistability: remote substituent modulation and lattice control of photoinduced valence tautomerism and light-induced thermal hysteresis. J. Am. Chem. Soc., 132: 6261–6273.

37. Mulyana, Y., Poneti, G., Moubaraki, B., Murray, K. S., Abrahams, B. F., Sorace, L., Boskovic, C. (2010) Solvation effects on the valence tautomeric transition of a cobalt complex in the solid state. Dalton Trans., 39: 4757–4767.
38. Alley, K. G., Poneti, G., Aitken, J. B., Hocking, R. K., Moubaraki, B., Murray, K. S., Abrahams, B. F., Harris, H. H., Sorace, L., Boskovic, C. (2012) A two-step valence tautomeric transition in a dinuclear cobalt complex. Inorg. Chem., 51: 3944–3946.
39. Li, B., Chen, L.-Q., Wei, R.-J., Tao, J., Huang, R.-B., Zheng, L.-S., Zheng, Z. (2011) Thermally induced and photoinduced valence tautomerism in a two-dimensional coordination polymer. Inorg. Chem., 50: 424–426.
40. Létard, J.-F. (2006) Photomagnetism of iron(II) spin crossover complexes–the T(LIESST) approach. J. Mater. Chem., 16, 2550–2559.
41. Zvezdin, A. K., Mischenko, A. S. (2002) Field-induced spin phase transitions in the cobalt valence tautomers. arXiv:cond-mat/0209337, p. 17.
42. Markevtsev, I. N., Monakhov, M. P., Platonov, V. V., Mischenko, A. S., Zvezdin, A. K., Bubnov, M. P., Abakumov, G. A., Cherkasov, V. K. (2006). Field-induced spin phase transition in a Co complex. J. Magn. Mag. Mater., 300: e407–e410.
43. Poneti, G., Mannini, M., Sorace, L., Sainctavit, P., Arrio, M.-A., Otero, E., Cezar, J. C., Dei, A. (2010) Soft-X-ray-induced redox isomerism in a cobalt dioxolene complex. Angew. Chem. Int. Ed., 49: 1954–1957.
44. Liang, H., Cha, M. S., Lee, Y.-A., Lee, S. S., Jung, O.-S. (2007) First bistable monocobalt complex with six free nitrogen donors: 3,5-di-tert-butylbenzoquinonatocobalt complex containing triethanolaminetriisonicotinate. Inorg. Chem. Commun., 10: 71–73.
45. Kiriya, D., Chang, H.-C., Kitagawa, S. (2008) Molecule-based valence tautomeric bistability synchronized with a macroscopic crystal-melt phase transition. J. Am. Chem. Soc., 130: 5515–5522.
46. Kiriya, D., Chang, H.-C., Nakamura, K., Tanaka, D., Yoneda, K., Kitagawa, S. (2009) Polymorph-dependent molecular valence tautomerism synchronized with crystal-melt phase transitions. Chem. Mater., 21: 1980–1988.
47. Kiriya, D., Nakamura, K., Kitagawa, S. Chang, H.-C. (2010) Modulable cooperativity in a valence tautomeric complex functionalized with branched alkyl chains. Chem. Commun., 46: 3729–3731.
48. Bencini, A., Caneschi, A., Carbonera, C., Dei, A., Gatteschi, D., Righini, R., Sangregorio, C., Van Slageren, J. (2003) Tuning the physical properties of a metal complex by molecular techniques: the design and the synthesis of the simplest cobalt-o-dioxolene complex undergoing valence tautomerism. J. Mol. Struct., 656: 141–154.
49. Evangelio, E., Rodriguez-Blanco, C., Coppel, Y., Hendrickson, D. N., Sutter, J. P., Campo, J., Ruiz-Molina, D. (2009) Solvent effects on valence tautomerism: a comparison between the interconversion in solution and solid state. Solid State Sci., 11: 793–800.
50. Jung, O.-S., Jo, D. H., Lee, Y.-A., Conklin, B. J., Pierpont, C. G. (1997) Bistability and molecular switching for semiquinone and catechol complexes of cobalt. Studies on redox isomerism for the bis(pyridine) ether series $Co(py_2X)(3,6\text{-}DBQ)_2$, X = O, S, Se, and Te. Inorg. Chem., 36: 19–24.
51. Tao, J., Maruyama, H., Sato, O. (2006) Valence tautomeric transitions with thermal hysteresis around room temperature and photoinduced effects observed in a cobalt-tetraoxolene complex. J. Am. Chem. Soc., 128: 1790–1791.
52. Caneschi, A., Cornia, A., Dei, A. (1998) Valence Tautomerism in a Cobalt Complex of a Schiff base diquinone ligand. Inorg. Chem., 37: 3419–3421.
53. Amoore, J. J. M., Neville, S. M., Moubaraki, B., Iremonger, S. S., Murray, K. S., Létard, J.-F., Kepert, C. J. (2010) Thermal- and light-induced spin crossover in a guest-dependent dinuclear iron(II) system. Chem. Eur. J., 16: 1973–1982.
54. Matouzenko, G. S., Jeanneau, E., Verat, A. Yu., Bousseksou, A. (2011) Spin crossover and polymorphism in a family of 1,2-bis(4-pyridyl)ethene-bridged binuclear iron(II) complexes. A key role of structural distortions. Dalton Trans., 40: 9608–9618.
55. Balzani, V., Credi, A., Venturi, M. (2008) Molecular Devices and Machines: Concepts and Perspectives for the Nanoworld. Wiley-VCH, Weinheim, p. 588.
56. Jung, O.-S., Jo, D. H., Lee, Y.-A., Chae, H. K., Sohn, Y. S. (1996) Charge distribution dependent on the chelating ratio. $[Co(\mu\text{-}bpym)(dbbq)_2]$ vs. $[\{Co(dbbq)_2\}(\mu\text{-}bpym)]$ (bpym=2,2′-bipyrimidine; dbbq=3,6-di-t-butyl-1,2-benzoquinone). Bull. Chem. Soc. Jpn., 69: 2211–2214.

57. Bin-Salamon, S., Brewer, S. H., Depperman, E. C., Franzen, S., Kampf, J. W., Kirk, M. L., Kumar, R. K., Lappi, S., Peariso, K., Preuss, K. E., Shultz, D. (2006) Testing bridge-mediated differences in dinuclear valence tautomeric behaviour. Inorg. Chem., 45: 4461–4467.
58. Hearns, N. G. R., Korĉok, J. L., Paquette, M. M., Preuss, K. E. (2006) Dinuclear cobalt bis(dioxolene) complex exhibiting two sequential thermally induced valence tautomeric transitions. Inorg. Chem., 45: 8817–8819.
59. Bencini, A., Daul, C. A., Dei, A., Mariotti, F., Lee, H., Shultz, D. A., Sorace, L. (2001) Charge distribution in bis-dioxolene radical metal complexes. Synthesis and DFT characterization of dinuclear Co(III) and Cr(III) complexes with a mixed-valent, $S = {}^{1}/_{2}$ semiquinone-catecholate ligand. Inorg. Chem., 40: 1582–1590.
60. Robin, M. B., Day, P. (1967). Mixed valence chemistry. A survey and classification. Adv. Inorg. Chem. Radiochem., 10: 247–403.
61. Sato, O., Miura, S., Maruyama, H., Zhang, Y., Wu, D., Zhang, W., Xu, H., Matsuda, R., Sun, H., Tao, J. (2008) Magnetic and photo-magnetic properties of Co dinuclear complexes. Inorg. Chim. Acta, 361: 3659–3662.
62. Jung, O.-S., Pierpont, C. G. (1994). Photomechanical polymers. J. Am. Chem. Soc., 116: 2229–2230.
63. Imaz, I., Maspoch, D., Rodríguez-Blanco, C., Pérez-Falcón, J. M., Campo, J., Ruiz-Molina, D. (2008) Valence-tautomeric metal-organic nanoparticles, Angew. Chem. Int. Ed., 47: 1857–1860.
64. Bodnar, S. H., Caneschi, A., Dei, A., Shultz, D. A., Sorace, L. (2001) Thermal hysteresis in molecule-based magnetic materials can be induced by appropriate dioxolene ligand design. Chem. Commun., 2150–2151.
65. Paquette, M. M., Kopelman, R. A., Beitler, E., Frank, N. L. (2009) Incorporating optical bistability into a magnetically bistable system: a photochromic redox isomeric complex. Chem. Commun., 5424–5426.
66. Yoshida, Y., Tanaka, H., Saito, G., Ouahab, L., Yoshida, H., Sato, N. (2009) Valence-tautomeric ionic liquid composed of a cobalt bis(dioxolene) complex dianion. Inorg. Chem., 48: 9989–9991.
67. Kiriya, D., Chang, H.-C., Kamata, A., Kitagawa, S. (2006) Polytypic phase transition in alkyl chain-functionalized valence tautomeric complexes. Dalton Trans., 1377–1383.

8

Reversible Spin Pairing in Crystalline Organic Radicals

Jeremy M. Rawson and John J. Hayward

Department of Chemistry and Biochemistry, The University of Windsor, Canada

8.1 Introduction

The concept of high spin and low spin configurations in *d*-block complexes is taught in every undergraduate introduction to the coordination chemistry of the transition metals. In this situation the interplay between inter-electron repulsion or 'pairing energy' (P_E) and crystal (ligand) field splitting (Δ) determines the electronic structure and provides an elegant example of how chemical tuning and modification of the ligand donor set can manipulate electronic structure. When Δ is of a similar magnitude to P_E then there is a fine balance between crystal field stabilisation energy (an enthalpic term which favours the low spin configuration) and maximising the number of microstates (an entropic term which favours the high spin configuration). In these cases spin-transitions typically occur between a low temperature enthalpically stabilised low spin configuration and a high temperature entropically favoured high spin configuration. The ability to drive spin-transitions thermally, through light-irradiation or pressure-induced transitions are the focus of the rest of this book. In this chapter we consider organic 'spin-transition' materials whose electronic structures can be manipulated in a conceptually similar but substantially different chemical fashion, that is by examining the fine interplay between inter-electron repulsion (P_E) and promotion energy to a low-lying vacant orbital (Δ) within the context of organic chemistry.

In order to understand the behaviour of such systems we shall begin with a discussion (Section 8.2) of stable free-radicals and their tendency to associate to form dimers, focusing on computational and experimental studies of the electronic structures of these dimers in the gas phase and in solution. In Section 8.3 we extend these discussions to the solid state and investigate examples in which we observe (i) thermal population of electronic excited states leading to a gradual thermal evolution of paramagnetism upon warming and (ii) first-order solid state phase transitions in which bond cleavage can lead to abrupt diamagnetic-paramagnetic phase

Spin-Crossover Materials: Properties and Applications, First Edition. Edited by Malcolm A. Halcrow.

transitions. These have direct parallels to inorganic spin-transition chemistry where we observe incomplete or complete spin transition behaviour with or without thermal hysteresis.[1] In some cases these phase transitions occur with significant windows of thermal hysteresis and the structural origins of the thermal hysteresis is discussed through selected examples. We will see that the origin of the spin-transition behaviour is closely linked to solid-state structure and we discuss methodologies which can be employed to tailor the structures of molecules at the molecular level in order to adopt macroscopic structures which exhibit this spin-transition behaviour. This section concludes with a case study on TTTA which is arguably the most comprehensively studied member of this family of spin-transition compounds in which both light-induced and pressure-induced phase transitions have been determined. Section 8.4 concludes with a summary of the current level of understanding of these organic spin-transition molecules and focuses on future perspectives for the development of this area of research.

8.2 Radical Pairs: Solution and Gas Phase Studies

8.2.1 Radical Dimerisation in Solution

In the absence of electronic and/or steric stabilisation, most σ-radicals will undergo (i) irreversible σ-bond forming reactions or (ii) disproportionation reactions for example via H-abstraction (Scheme 8.1).[2]

Electronic and/or steric stabilisation of the radical can make the dimerisation enthalpy (ΔH_{dim}) less favourable. In particular extensive π-delocalisation disfavours dimerisation since the electron density in the overlap region between two radicals is low, disfavouring a localised $2c,2e^-$ bond. Alternatively steric protection of regions of high spin density also inhibits dimerisation. Variable temperature UV/vis and EPR spectroscopies have been used to probe the enthalpy and entropy for dimerisation processes for a number of radicals and selected values are presented in Table 8.1.[3–8] The temperature dependence of their spectroscopic properties can be quite marked, as illustrated in Figure 8.1 for PhCNSSN•. The λ_{max} for isolated monomers falls in the UV region of the spectrum leading to a pale yellow colour at ambient temperature (due to a tail of λ_{max} encroaching into the visible). Conversely a lower energy absorption band becomes apparent when the radicals associate to form dimer pairs at low temperature and is reflected in an intense red solution.

In most cases the EPR and UV/vis data (Table 8.1) exhibit some solvent dependency but clearly reflect general qualitative trends; specifically (i) the typically low enthalpies of dimerisation and (ii) the favourable entropic term for dissociation for stable π-radicals.

8.2.2 Computational Studies on Dimerisation

A number of computational studies have been undertaken on π^*-π^* dimers in the gas phase. At short intermolecular distances, when the overlap integral is large, then these dimers typically exhibit closed-shell

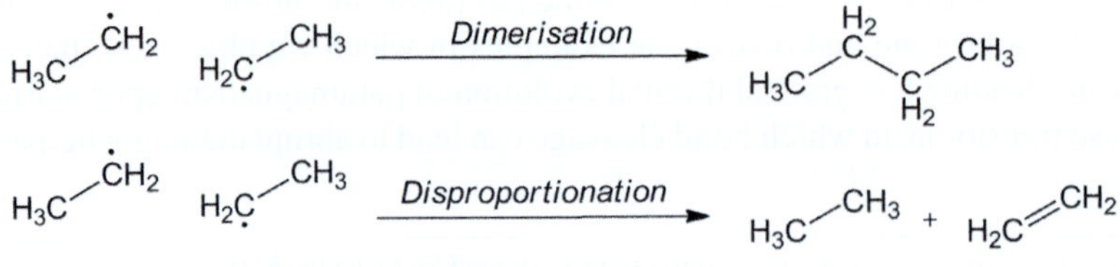

Scheme 8.1

Table 8.1 *Thermodynamic parameters for solution dimerisation of a cross-section of free radicals.*

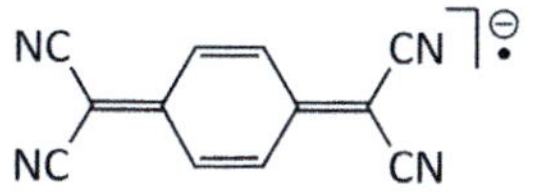	$\Delta H = -44\ kJmol^{-1}$	$\Delta S = -176\ JK^{-1}mol^{-1}$	Reference 3
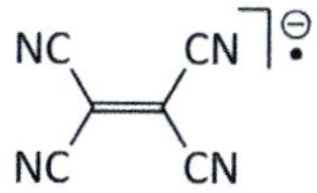	$\Delta H = -37\ kJmol^{-1}$	$\Delta S = -151\ JK^{-1}mol^{-1}$	Reference 4
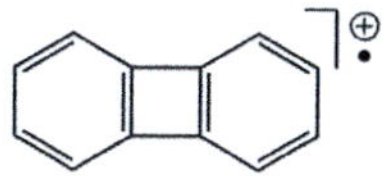	$\Delta H = -37\ kJmol^{-1}$	$\Delta S = -134.4\ JK^{-1}mol^{-1}$	Reference 4
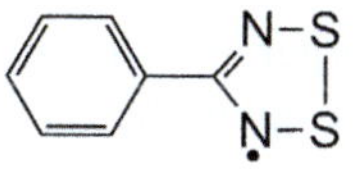	$\Delta H = -35\ kJmol^{-1}$	$\Delta S = -121\ JK^{-1}mol^{-1}$	Reference 5
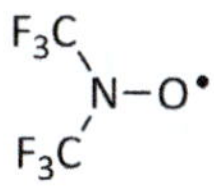	$\Delta H = -10\ kJmol^{-1}$	$\Delta S = -29\ JK^{-1}mol^{-1}$	Reference 7
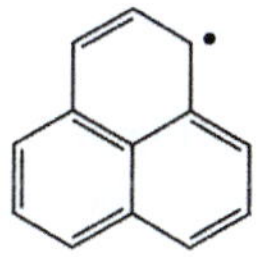	$\Delta H = -40\ kJmol^{-1}$	$\Delta S = -151\ JK^{-1}mol^{-1}$	Reference 4
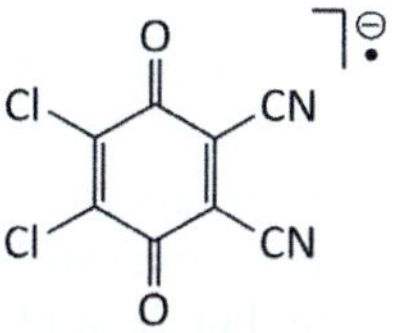	$\Delta H = -38\ kJmol^{-1}$	$\Delta S = -138\ JK^{-1}mol^{-1}$	Reference 4
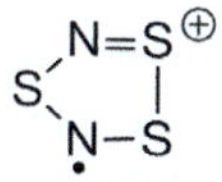	$\Delta H = -47\ kJmol^{-1}$		Reference 5
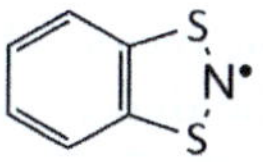	ΔH » 0 $kJmol^{-1}$		Reference 6
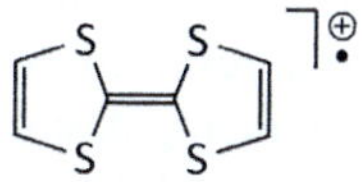	$\Delta H = -16\ kJmol^{-1}$	$\Delta S = -75\ JK^{-1}mol^{-1}$	Reference 8

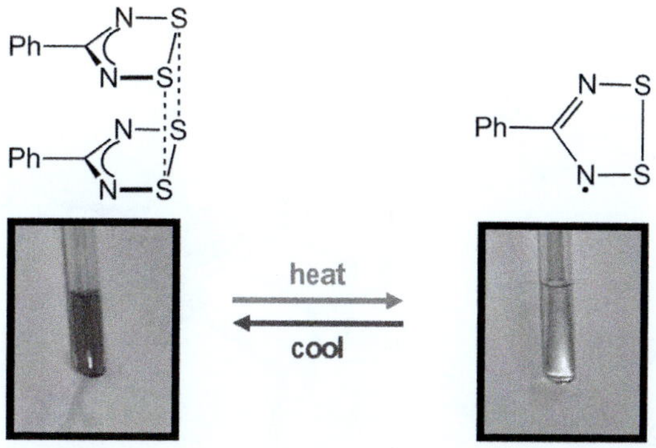

Figure 8.1 *Solution of PhCNSSN• in THF at (left) 100 K and (right) 298 K.*

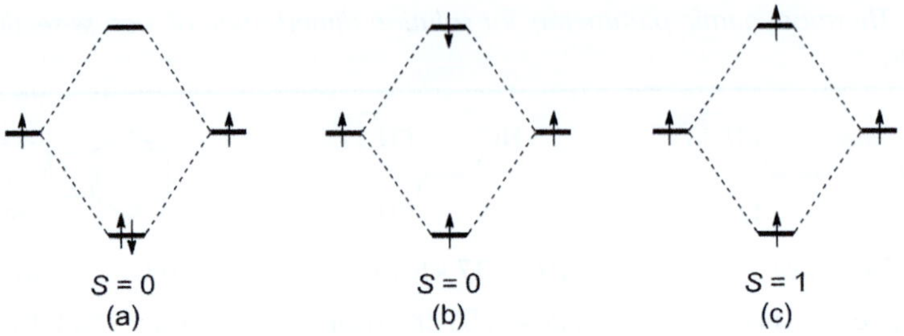

Figure 8.2 *(a) singlet ground state; (b) open-shell singlet; (c) triplet configuration.*

singlet ground states (Fig. 8.2a). However when the overlap integral is low, at larger intermolecular separations, then the orbital energy between bonding and antibonding MOs is low and open shell singlet and triplet configurations are also possible. The presence of the closed shell singlet as an excited state configuration assists stabilisation of the open shell singlet ground state in these cases. Thus the open shell singlet states are found to be more stable than the closed shell singlets at larger intermolecular separations.[9]

For charged radicals such as $TTF^{\bullet+}$, $S_3N_2{}^{\bullet+}$, $TCNE^{\bullet-}$ and $TCNQ^{\bullet-}$ the presence of unfavourable electrostatic, charge-charge, repulsion would appear to mitigate against dimerisation and dimers are not always stable minima on the potential energy surface. Yet the solution measurements clearly reveal favourable enthalpies of dimerisation (see Table 8.1). Calculations indicate that in the case of $TCNE^{\bullet-}$, for example, solvation has an important role to play in stabilising the dimer pair.[10] Similarly solvent dependent studies on $TTF^{\bullet+}$ have revealed ΔH_{dim} in the region –16 to –38 $kJmol^{-1}$ reflecting the important role of solvation effects on stabilising/destabilising the dimer. In other cases it may be likely that ion pairing can play an important role in stabilising such dimers. This is certainly the case in the solid state. For example the crystal structure of $[S_3N_2]_2Cl_2$ comprises a π^*-π^* dimer with the Cl^- anion forming close S...Cl contacts to S atoms in both $S_3N_2{}^{\bullet+}$ rings[11] whilst the structure of Na[TCNE]·$MeOCH_2CH_2OMe$ reveals that Na^+ ions adopt bridging positions within $[TCNE]_2{}^{2-}$ ions (Fig. 8.3).[12] The absence of significant counterion effects on dimerisation enthalpies in solution would indicate that solvation is probably the dominant stabilising factor in many instances.[9]

Figure 8.3 *Cation...anion interactions stabilise (left) radical cation pairs in $[S_3N_2]_2Cl_2$ and (right) radical anion pairs in $Na_2[TCNE]_2 \cdot 2MeOCH_2CH_2OMe$.*

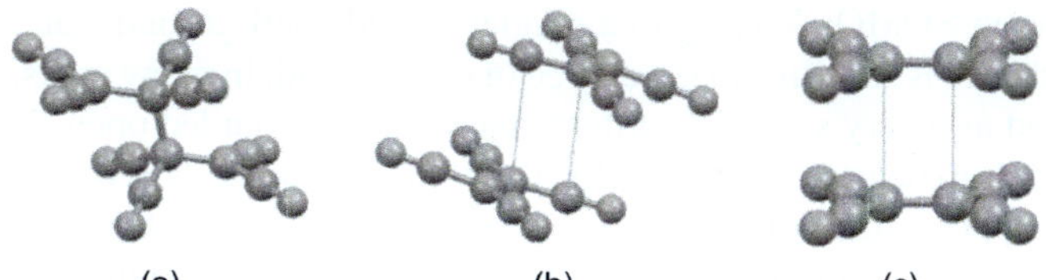

Figure 8.4 *Different modes of association of dimers of the radical anion, $TCNE^{\bullet -}$; (a) σ-bond dimerisation; (b) weak π^*-π^* dimer; (c) the common π^*-π^* dimer.*

8.3 Dimerisation in the Solid State

8.3.1 Structural Studies

The dimerisation enthalpies measured in solution for stable π-radicals (typically <50 kJmol^{-1}, Table 8.1) are consistent with formation of a weak multicentre bonding interaction rather than a localised 2c,2e$^-$ bonding interaction (E—E single bond enthalpies typically being in the order of 150–350 kJmol^{-1}). Conversely these low dimerisation enthalpies are comparable with a range of intermolecular forces such as hydrogen bonding motifs which are often considered as structure-directing in the solid state.[13] As a consequence it is not surprising that many of these radicals are found to exhibit dimeric solid-state structures (Figs 8.4 and 8.5) in which there are close intra-dimer contacts between radicals. These contacts are typically longer than conventional covalent E—E bonds, yet significantly less than the sum of the van der Waals radii.

The delocalised nature of the radical spin density means that there is typically more than one way for radicals to dimerise to afford a net bonding interaction between them; whilst one mode of association may be energetically preferred, the low dimerisation enthalpies indicate that packing effects play an important role. For example $TCNE^{\bullet -}$ typically adopts a face-to-face dimer motif, $(TCNE)_2^{2-}$ (Fig. 8.4c), which maximises π^*-π^* overlap ($d_{C...C}$ = 2.8–3.1 Å) but slipped dimers have also been reported (Fig. 8.4a) associated *via* σ-bond formation ($d_{C\text{–}C}$ = 1.6 Å) or via a weak π^*-π^* interaction ($d_{C...C}$ = 3.4–3.5 Å, Fig. 8.4b).[12, 14–16] Conversely the dithiadiazolyl radicals, $RCN_2S_2^{\bullet}$ typically adopt any one of the four dimeric motifs (Fig. 8.5), all of which exhibit a net bonding interaction between singly-occupied molecular orbitals with inter-planar separations of ca. 2.9–3.2 Å which are significantly less than the sum of the van der Waals' radii of S (3.6 Å).[17]

8.3.2 Electronic Structure and Bonding

Magnetic studies on many of these dimers typically reveal that they are essentially diamagnetic (with a small Curie-tail due to radical defects in the lattice) and this is often ascribed to a weak multicentre π^*-π^*

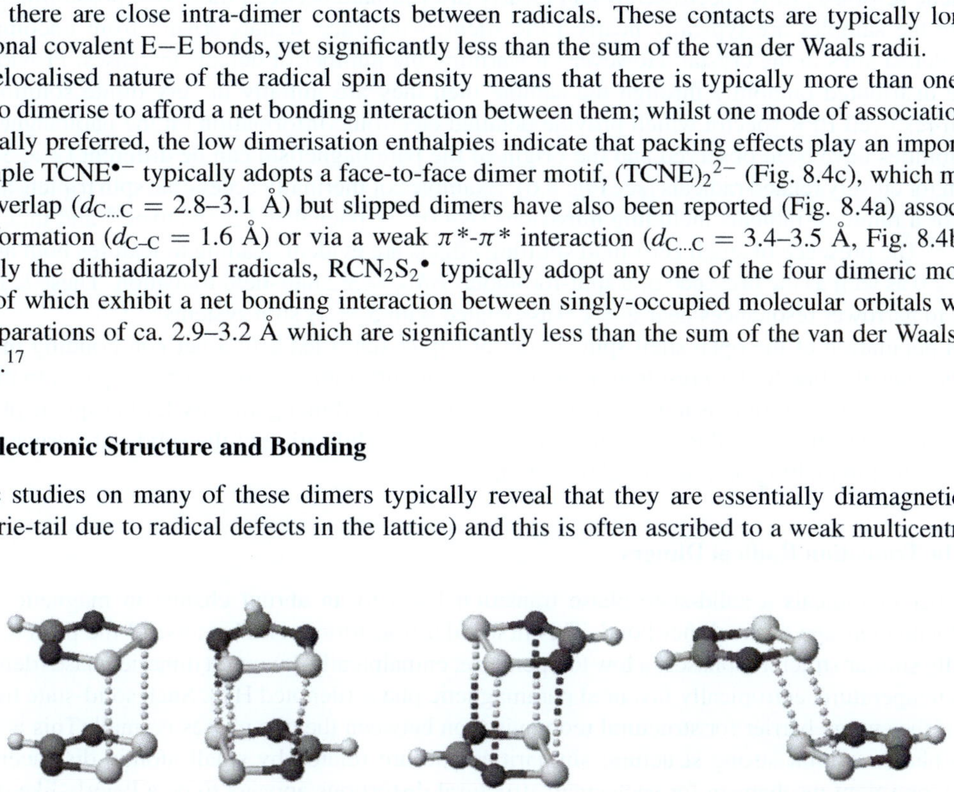

Figure 8.5 *Different modes of π^*-π^* dimerisation of dithiadiazolyl radicals, $[RCN_2S_2]_2$.*

bonding interaction between SOMOs leading to a singlet closed shell ground state (Fig. 8.2a). As already discussed the spin ground state is sensitive to the degree of overlap and, in many cases, the electronic structure could equally be described as a very strongly antiferromagnetically-coupled open shell singlet ground state. Experimentally it is difficult to determine directly the difference between closed-shell and open-shell singlet ground states since both appear diamagnetic. However it is clear from spectroscopic studies (Fig. 8.1) that association to form dimers leads to marked changes in electronic structure. One approach to probe whether there is a formal bonding interaction between the two centres is to investigate whether there is a (3, –1) bond critical point in the electronic charge density.[18] Computational studies have revealed the presence of such bond critical points in several dimers indicating a closed shell singlet ground state configuration.[14, 19–21] To the best of our knowledge, however, direct experimental evidence for such (3, −1) bond critical points from X-ray diffraction studies of radical dimers has not been reported.

8.3.3 Thermally Accessible Triplet States

Whilst these π^*-π^* dimers exhibit spin singlet ($S = 0$) characteristics (Fig. 8.2a or b), the temperature dependence of the magnetic susceptibility (χ) of these species in the solid state reveals an increase in paramagnetism upon warming. Two potential mechanisms can give rise to this behaviour; either rupture of the weak intra-dimer bond to regenerate two independent $S = {}^1\!/_2$ radicals or thermal population of a low-lying spin triplet (Fig. 8.2c). Whilst SQUID magnetometry does not clearly distinguish between the origins of the bulk magnetic response, EPR spectroscopy can clearly discriminate between these two possibilities. At low temperature the samples are typically nearly EPR silent, with trace signals arising from uncompensated radicals at defect sites in the crystal. However on warming the number of defects increases. Provided these defects are distributed randomly through the sample then they are initially in very dilute solution giving rise to well-resolved EPR spectra which provide detailed electronic information on the paramagnetic sites. Both possibilities have been observed and the origin of the paramagnetism can be different ($2 \times S = {}^1\!/_2$ or $S = 1$) even for closely related radicals (see Fig. 8.6). Examples of thermally accessible spin triplets have been observed for a range of radicals including nitroxides (N/O)[22, 23] phenalenyl derivatives[24] and thiazyl (N/S) radicals.[25–27] The presence of small zero-field splitting effects are evident, leading to features both above and below $g = 2.0$ as well as the presence of a spin-forbidden $\Delta M_S = \pm 2$ half-field transition. These features are in contrast to narrower resonances near $g = 2.0$ associated with $S = {}^1\!/_2$ spin systems.[25–27]

Thermal population of the open shell spin singlet or triplet states leads to a net non-bonding interaction between the radicals (Fig. 8.2b/c) resulting in an increase in intra-radical separation compared to the closed shell singlet; at low concentrations the crystal lattice constrains the dimer geometry for the spin triplet but, as the number of defects increase, this can eventually lead to a destabilisation of the solid state structure which might be reflected in melting or a structural transition.

8.3.4 Spin-Transition Radical Dimers

For a number of radicals a solid-state phase transition leads to an abrupt change in magnetic response associated with cleavage of the dimer bond. This first order transition arises because of the presence of two energetically similar structural phases; a low temperature, enthalpically-favoured dimeric phase (denoted LT) and a high temperature, entropically favoured paramagnetic phase (denoted HT). Such solid-state transitions can occur if the energy barrier for structural reorganisation between the two phases is small. This is favoured if the two phases exhibit strong structural similarities and are related by small atomic displacements. A particularly prevalent mechanism for such small structural distortions appears to be a Peierls-like distortion of a regular π-stack of radicals (paramagnetic) to form a distorted diamagnetic π-stack (Fig. 8.7). Notably the enthalpic stabilisation of the dimeric phase associated with the distorted π-stack appears strongly correlated

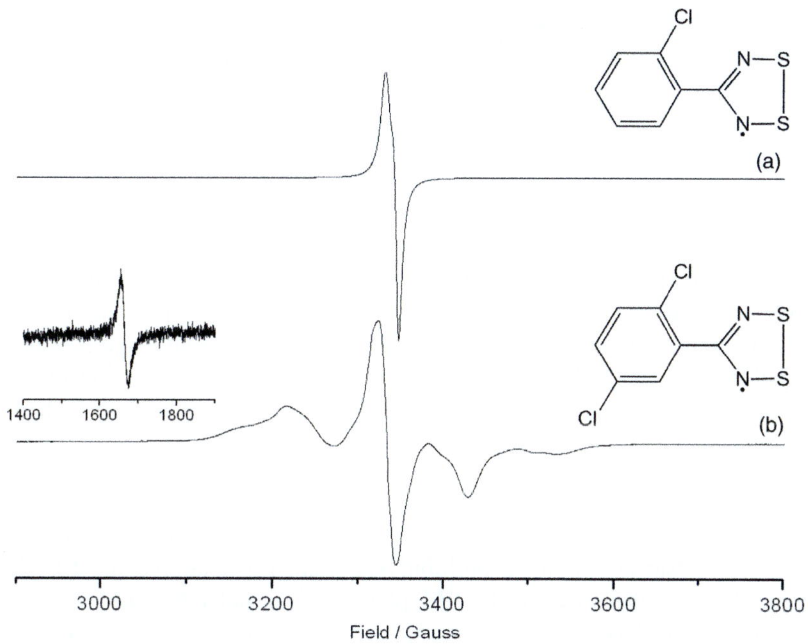

Figure 8.6 *Solid state X-band EPR spectra of: (a) 2-ClC$_6$H$_4$CNSSN exhibiting $S = {}^1/_2$ defects, and (b) 2,5-Cl$_2$C$_6$H$_3$CNSSN exhibiting an $S = 1$ spectrum with spin-forbidden $\Delta M_S = 2$ transition (inset).*

with the solution dimerisation enthalpy. Since the energetics of LT and HT phases should be near equivalent to exhibit such phase transitions, those radicals with low dimerisation enthalpies appear to exhibit a more marked tendency to undergo such solid state transitions. In this chapter we focus on the dithiazolyl (DTA, Scheme 8.2) family of radicals which have attracted particular attention as spin-transition materials and exhibit $\Delta H_{dim} \sim 0$ (Table 8.1). By way of comparison, none of the closely related dithiadiazolyl (DTDA, Scheme 8.2) radicals ($\Delta H_{dim} \sim 35$ kJmol^{-1}, Table 8.1) exhibit such spin-transition behaviour, presumably because of significant enthalpic stabilisation of the dimer phase.

When there is no significant energy barrier for interconversion, these spin transitions appear as abrupt discontinuities in the magnetism as the enthalpic preference for the dimeric diamagnetic phase is overcome

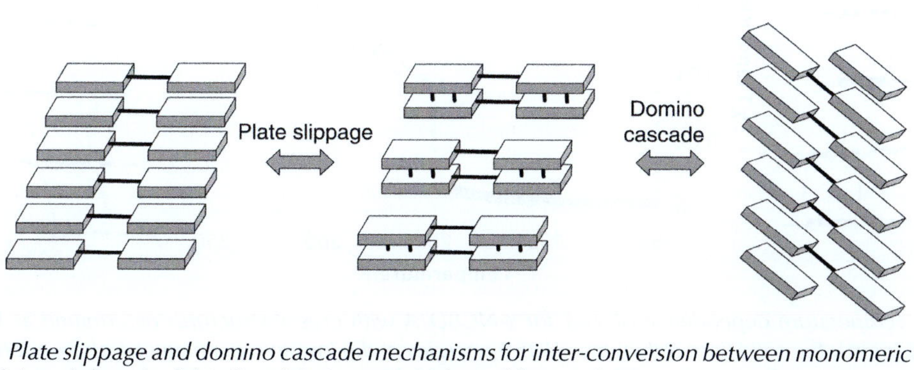

Figure 8.7 *Plate slippage and domino cascade mechanisms for inter-conversion between monomeric dithiazolyls (left and right) and dimeric dithiadiazolyls (centre) (Adapted from ref. 21).*

Scheme 8.2

by the entropic contributions, for example 2-NCBDTA undergoes an abrupt but fully reversible phase transition at 250 K ($\Delta H = +0.46$ kJmol^{-1}) from a diamagnetic low temperature phase ($d_{S...S} = 3.3$ Å) to a paramagnetic high temperature phase ($d_{S...S} = 3.7$ Å Fig. 8.8) via a plate slippage mechanism (Fig. 8.7).[28] Notably the inter-molecular contacts between radicals along the stacking direction in the HT phase lead to strong antiferromagnetic exchange interactions and thus the $\chi_M T$ values are substantially lower than the 0.375 emu·K·mol^{-1} expected for $S = {}^1/_2$ paramagnets.

In cases where there is a marked energy barrier to structural reorganisation (where strong ***inter***-stack interactions need to be cleaved) then the phase transition may be more complex and give rise to thermal hysteresis and a metastable region. Whilst the transition may be thermodynamically spontaneous at a lower temperature, the activation barrier might be prohibitively high. For example the LT phase of PDTA is metastable above ca. 295 K with respect to the HT phase;[29, 30] initially the activation energy barrier for interconversion from LT to HT phases is too prohibitive to permit rearrangement. Only once the temperature exceeds 340 K does the system undergo a LT–HT phase transition (via a 'domino cascade' Fig. 8.7). Conversely the sample stays in the HT phase on cooling down to 295 K at which point the LT phase becomes more stable and the HT–LT transition occurs. This leads to a region of bistability (Fig. 8.9).

A compilation of transition temperatures for dithiazolyls is given in Table 8.2.[28–36] Several other DTA derivatives exhibit spin-Peierls like behaviour.[37, 38] Amongst these spin-transition dithiazolyls, TTTA has been most comprehensively studied and is presented as a case study in the next section.

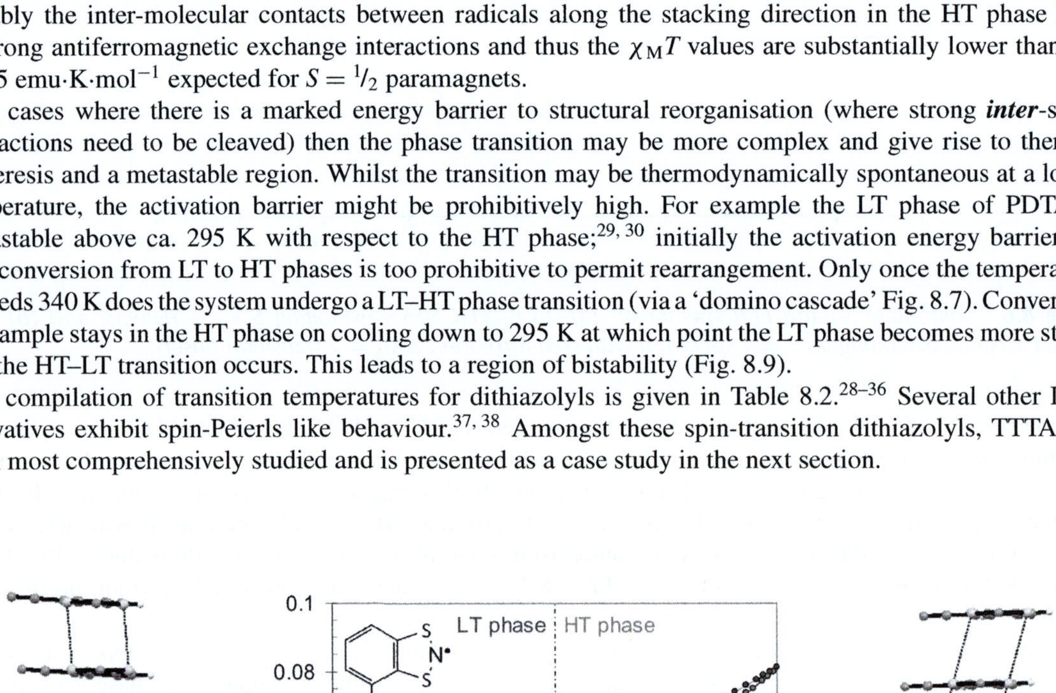

Figure 8.8 *Temperature dependence of $\chi_M T$ for 2-NCBDTA with crystal structures determined at 180 K (left) and at 300 K (right); the solid lines reflect the relative energies of the low spin diamagnetic phase (dark grey) and high spin paramagnetic phase (pale grey).*

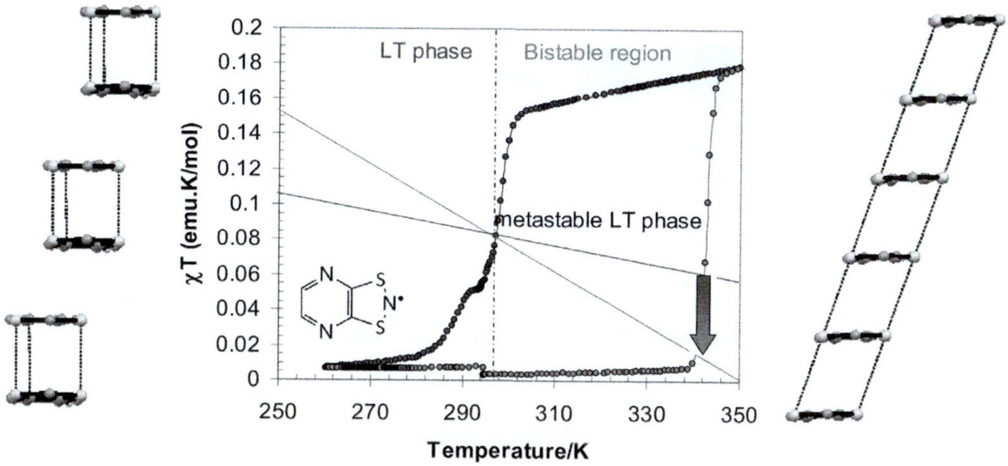

Figure 8.9 *Temperature dependence of $\chi_M T$ for PDTA with crystal structures determined at 180 K (left) and at 300 K (right); the solid lines reflect the relative energies of the low spin diamagnetic phase (dark grey) and high spin paramagnetic phase (pale grey).*

8.3.5 Trithiatriazinyl, TTTA: A Case Study

The synthesis and room temperature crystal structure of TTTA was first reported by Wolmershauser in 1989.[39] However it was not until 1999 that Awaga reported that TTTA exhibited bistability between 230 and 320 K;[36] an exceptionally large window of hysteresis which encompasses room temperature (Fig. 8.10). Vacuum coating onto gold leads to deposition of the dimeric LT phase onto the Au surface,[40] though subsequent studies of its physical properties are constrained to bulk samples. Detailed structural studies[41, 42] of TTTA reveal the LT phase adopts a distorted π-stacked motif ($d_{S...S} = 3.2$ Å, space group $P\bar{1}$) whereas the HT phase

Table 8.2 *Spin-transition temperatures for 1,3,2-DTA radicals.*

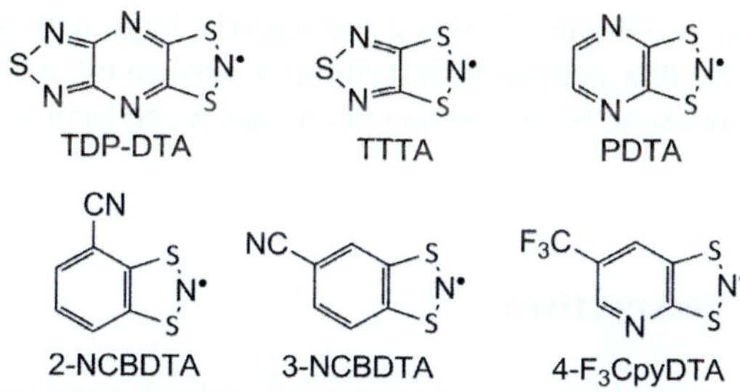

Radical	$T_{C(\uparrow)}$/K	$T_{C(\downarrow)}$/K	ΔT/K
TDP-DTA	200	150	50
TTTA	317	225	92
PDTA	343	297	46
2-NCBDTA	250	250	0
3-NCBDTA	304	293	7
4-F_3CpyDTA	50	50	0

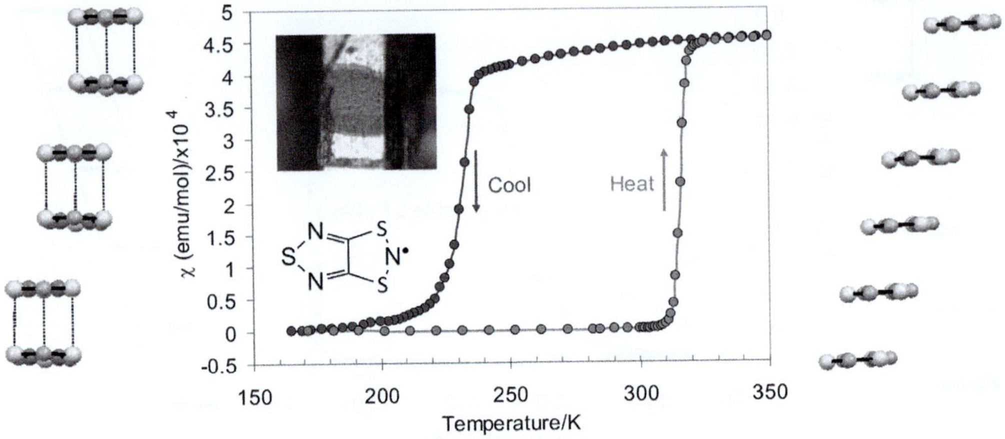

Figure 8.10 *Temperature dependence of χ for TTTA with crystal structures determined at 180 K (left) and at 300 K (right). Inset: A single crystal of TTTA after laser-irradiation (2.64 eV).*

adopts a regular π-stack ($d_{S...S}$ = 3.7 Å space group $P2_1/c$). The heat change associated with the transition is just 2 kJmol^{-1} consistent with the energetic similarity of the two phases and the low ΔH_{dim} which contributes to the stability of the LT phase. Computational studies by Novoa indicate that the diamagnetism associated with the LT phase can be attributed to an open-shell singlet rather than closed-shell singlet configuration.[43] The magnetism of the paramagnetic HT phase can be modelled as a one-dimensional chain (J/k = –320 K) consistent with strong antiferromagnetic interactions along the stacking direction with weaker inter-chain interactions.

The distortion of the π-stacked structure leads to marked changes in colour between the two phases; the HT phase is purple whereas the LT phase is gold-green dichroic. Light irradiation (2.64 eV) at room temperature leads to a photoinduced phase transition from LT to HT phases.[44] Further studies revealed that the position of the thermal hysteresis loop is sensitive to pressure with the region of bistability moving to higher temperature under pressure, consistent with stabilisation of the LT phase under pressure (Fig. 8.11).[45]

The large region of bistability is associated with a particularly large activation energy for distortion of the π-stack. As with other DTAs this activation is primarily associated with the cleavage of inter-stack interactions (predominantly electrostatic S...N and van der Waals S...S) hypothesised as a 'domino-cascade' mechanism (Fig. 8.7).[29]

8.4 Summary and Future Perspectives

Solution UV/vis and EPR studies reveal a propensity for stable radicals to undergo reversible dimerisation in solution giving rise to singlet ground state dimers. In many cases computational studies reveal the presence of (3, −1) bond critical points consistent with bond formation and a closed-shell singlet ground state. However at longer intermolecular separations the closed-shell singlet configuration is destabilised with respect to the open-shell singlet due to poorer spatial overlap. Increases in paramagnetism on warming can be attributed to one of two mechanisms; thermal population of a triplet (S = 1) excited state, or cleavage of the weak bond (generating two S = $^1/_2$ radicals). For many π-stacked DTA radicals, an abrupt spin-transition is observed which is associated with a first order phase transition between distorted π-stacked and regular π-stacked states. The cleavage of strong intermolecular inter-stack interactions appears central to the activation energy

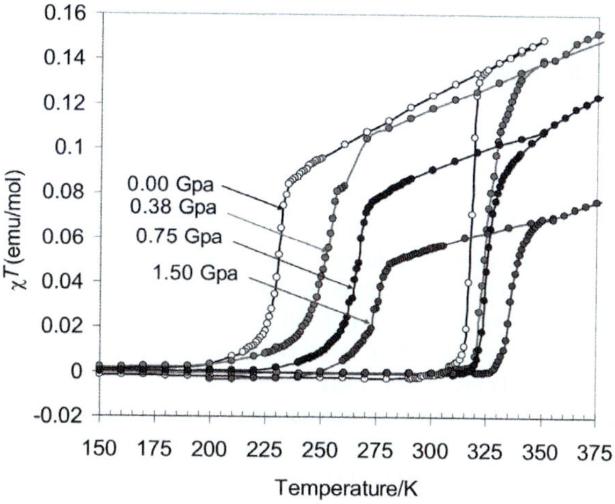

Figure 8.11 *Pressure dependence of χT for TTTA illustrating the increase in transition temperatures with increasing pressure.*

for the structural reorganisation which is the origin of the structural bistability. Whilst the control of molecular solid state structure still poses considerable challenges for the synthetic chemist, it is clear that a range of organic spin-transition compounds can now be prepared with large windows of thermal hysteresis which auger well for smart device applications. Whilst neutral organic molecules tend to favour herringbone motifs,[46] the inclusion of electronegative atoms in the molecular backbone of DTAs appears to favour the layer-like motifs desirable for spin-transition properties.[31–35]

Acknowledgements

We would like to thank Profs. Awaga, Oakley and Preuss for providing data for Figures 8.11, 8.9 and 8.1 respectively.

References

1. Kahn, O. (1993) Molecular Magnetism. VCH, Weinheim, Germany, p. 380.
2. Clayden, J., Greeves, N., Warren, S., Wothers, P. (2001) Organic Chemistry. Oxford University Press, Oxford, UK, p. 1512.
3. Boyd, R. H., Phillips, W. D. (1965) Solution dimerization of the tetracyanoquinodimethane ion radical. J. Chem. Phys., 43: 2927–2929.
4. Lu, J.-M., Rosokha, S. V., Kochi, J. K. (2003) Stable (long-bonded) dimers via the quantitative self-association of different cationic, anionic, and uncharged π-radicals: structures, energetics, and optical transitions. J. Am. Chem. Soc., 125: 12161–12171.
5. Fairhurst, S. A., Johnson, K. M., Sutcliffe, L. H., Preston, K. F., Banister, A. J., Hauptman, Z. V., Passmore, J. (1986) Electron spin resonance study of $CH_3CNSSN^•$, $C_6H_5CNSSN^•$, and $SNSSN^{•+}$ free radicals. Dalton Trans., 1465–1472.
6. Awere, E. G., Burford, N., Mailer, C., Passmore, J., Schriver, M. J., White, P. S., Banister, A. J., Oberhammer, H., Sutcliffe, L. H. (1987) The high yield preparation, characterisation, and gas phase structure of the thermally

stable $CF_3CSNSCCF_3^{\bullet}$, 4,5-bis(trifluoromethyl)-1,3,2-dithiazolyl and the X-ray crystal structure of benzo-1,3,2-dithiazolyl. Chem. Commun., 66–69.
7. Blackley, W. D., Reinhard, R. R. (1965) A new stable radical, bis(trifluoromethyl) nitroxide. J. Am. Chem. Soc., 87: 802–805.
8. Rosokha, S. V., Kochi, J. K. (2007) Molecular and electronic structures of the long-bonded π-dimers of tetrathiafulvalene cation-radical in intermolecular electron transfer and in (solid-state) conductivity. J. Am. Chem. Soc., 129: 828–838.
9. Garcia-Yoldi, I., Miller, J. S., Novoa, J. J. (2009) Theoretical study of the electronic structure of $[\text{tetrathiafulvalene}]_2^{2+}$ dimers and their long, intradimer multicenter bonding in solution and the solid state. J. Phys. Chem. A, 113: 484–492.
10. Garcia-Yoldi, I., Miller, J. S., Novoa, J. J. (2007) Structure and Stability of the $[TCNE]_2^{2-}$ dimers in dichloromethane solution: a computational study. J. Phys. Chem. A, 111: 8020–8027.
11. Small, R. W. H., Banister, A. J., Hauptman, Z. V. (1984) Hexasulphur tetranitrogen dichloride, $S_6N_4Cl_2$: crystal growth and crystal structure. Dalton Trans., 1377–1381.
12. Bock, H., Nather, C., Ruppert, K. (1992) Strukturen ladungsgestörter oder räumlich überfüllter Moleküle. 15. Tetracyanethen-Natrium-Dimethoxyethan. Z. Anorg. Allg. Chem., 614: 109–114.
13. Desiraju, G. R. (1989) Crystal Engineering: The Design of Organic Solids. Elsevier, Amsterdam, Netherlands, p. 312.
14. Miller, J. S., Novoa, J. J. (2007) Four-center carbon−carbon bonding. Acc. Chem. Res., 40: 189–198.
15. Zhang, J., Liable-Sands, L. M., Rheingold, A. L., Del Sesto, R. E., Gordon, D. C., Burkhart, B. M., Miller, J. S. (1998) Isolation and structural determination of octacyanobutanediide, $[C_4(CN)_8]^{2-}$; precursors to M(TCNE)x magnets. Chem. Commun., 1998, 1385–1386.
16. Clemente, D. A., Marzotto, A. (1996) Structure of two polymorphs of the TTF-TCNE charge-transfer complex and the degree of iconicity. J. Mater. Chem., 6: 941–946.
17. Rawson, J. M., Alberola, A., Whalley, A. (2006) Thiazyl radicals: old materials for new molecular devices. J. Mater. Chem., 16: 2560–2575.
18. Bader, R. F. W. (1994) Atoms in Molecules - A Quantum Theory. Clarendon Press, Oxford, UK, ch. 2, pp. 13–52.
19. Garcia-Yoldi, I., Miller, J. S., Novoa, J. J. (2009) Long, multicenter bonding in π-$[\text{terthiophene}]_2^{2+}$ dimers. Theor. Chem. Acc., 123: 137–143.
20. Garcia-Yoldi, I., Miller, J. S., Novoa, J. J. (2008) $[\text{Cyanil}]_2^{2-}$ dimers possess long, two-electron ten-center ($2e^-/10c$) multicenter bonding. Phys. Chem. Chem. Phys., 10: 4106–4109.
21. Capdevila-Cortada, M., Novoa, J. J. (2012) The nature of the $[TTF]^{\bullet+}\cdots[TTF]^{\bullet+}$ interactions in the $[TTF]_2^{2+}$ dimers embedded in charged [3]catenanes: room-temperature multicenter long bonds. Chem. Eur. J., 18: 5335–5344.
22. Genoug, F., Decorps, M. (1977) Pair exchange interaction in a crystalline nitroxide radical. ESR study of dimers in the triplet state. Mol. Phys., 34: 1583–1594.
23. Hatfield, W. E., Hoffmann, S. K., Corvan, P. J., Singh, P., Sethulekshmi, C. N., Metzger, R. M. (1983) Excited triplet state EPR of sigma-bonded TCNQ dimers in $[Cu(DMP)_2]_2[TCNQ]_2$. J. Phys. Coll., C3-1377–C3-1380.
24. Morita, Y., Aoki, T., Fukui, K., Nakazawa, S., Tamaki, K., Suzuki, S., Fuyuhiro, A., Yamamoto, K., Sato, K., Shiomi, D., Naito, A., Takui, T., Nakasuji, K. (2002) A new trend in phenalenyl chemistry: a persistent neutral radical, 2,5,8-tri-*tert*butyl-1,3-diazaphenalenyl, and the excited triplet state of the gable *syn*-dimer in the crystal of column motif. Angew. Chem. Int. Ed., 41: 1793–1796.
25. Cameron, T. S., Decken, A., Kowalczyk, R. M., McInnes, E. J. L., Passmore, J., Rawson, J. M., Shuvaev, K. V., Thompson, L. K. (2006) Characterisation of the thermally accessible spin triplet state in dimers of the 7π $ClCNSSS^{+\bullet}$ in the solid state. Chem. Commun., 2277–2279.
26. Shuvaev, K. V., Decken, A., Grein, F., Abedin, T. S. M., Thompson, L. K., Passmore, J. (2008) NC–$(CF_2)_4$–$CNSSN^{\bullet}$ containing 1,2,3,5-dithiadiazolyl radical dimer exhibiting triplet excited states at low temperature and thermal hysteresis on melting–solidification: structural, spectroscopic, and magnetic characterization. Dalton Trans., 4029–4037.
27. Alberola, A., Carter, E., Constantinides, C. P., Eisler, D. J., Murphy, D. M., Rawson, J. M. (2011) Crystal structures, EPR and magnetic properties of 2-$ClC_6H_4CNSSN^{\bullet}$ and 2,5-$Cl_2C_6H_3CNSSN^{\bullet}$. Chem. Commun., 47: 2532–2534.
28. Alberola, A., Collis, R. J., Humphrey, S. M., Less, R. J., Rawson, J. M. (2006) Spin transitions in a dithiazolyl radical: preparation, crystal structures, and magnetic properties of 3-cyanobenzo-1,3,2-dithiazolyl, $C_7H_3S_2N_2^{\bullet}$. Inorg. Chem., 45: 1903–1905.

29. Brusso, J. L., Clements, O. P., Haddon, R. C., Itkis, M. E., Leitch, A. A., Oakley, R. T., Reed, R. W., Richardson, J. F. (2004) Bistabilities in 1,3,2-dithiazolyl radicals. J. Am. Chem. Soc., 126: 8256–8265.
30. Brusso, J. L., Clements, O. P., Haddon, R. C., Itkis, M. E., Leitch, A. A., Oakley, R. T., Reed, R. W., Richardson, J. F. (2004) Bistability and the phase transition in 1,3,2-dithiazolo[4,5-b]pyrazin-2-yl. J. Am. Chem. Soc., 126: 14692–14693.
31. Barclay, T. M., Cordes, A. W., deLaat, R. H., Goddard, J. D., Haddon, R. C., Jeter, D. Y., Mawhinney, R. C., Oakley, R. T., Palstra, T. T. M., Patenaude, G. W., Reed, R. W., Westwood, N. P. C. (1997) The heterocyclic diradical benzo-1,2:4,5-bis(1,3,2-dithiazolyl). Electronic, molecular and solid state structure. J. Am. Chem. Soc., 119: 2633–2641.
32. Alberola, A., Collis, R. J., Less, R. J., Rawson, J. M. (2007) New synthetic pathways into dithiazolyl radicals: preparation and characterisation of 3′-methyl-benzo-1,3,2-dithiazolyl, M'BDTA. J. Organomet. Chem., 692: 2743–2749.
33. Alberola, A., Burley, J., Collis, R. J., Less, R. J., Rawson, J. M. (2007) Structural control of dithiazolyl radicals: case studies on 3′- and 4′-cyano-benzo-1,3,2-dithiazolyl, $NCC_6H_3S_2N$. J. Organomet. Chem., 692: 2750–2760.
34. Alberola, A., Clements, O. P., Collis, R. J., Cubbitt, L., Grant, C. M., Less, R. J., Oakley, R. T., Rawson, J. M., Reed, R. W., Robertson, C. M. (2008) Polymorphism in a π-stacked 1,3,2-dithiazolyl radical: pyridyl-1,3,2-dithiazolyl. Cryst. Growth. Des., 8:155–161.
35. Alberola, A., Eisler, D. J., Harvey, L., Rawson, J. M. (2011) Molecular tailoring of spin-transition materials: preparation, crystal structure and magnetism of trifluoromethyl-pyridyl-1,3,2-dithiazolyl. CrystEngComm, 13: 1794–1796.
36. Fujita, W., Awaga, K. (1999) Room-temperature magnetic bistability in organic radical crystals. Science, 286: 261–262.
37. Fujita, W., Awaga, K., Kondo, R., Kagoshima, S. (2006) A one-dimensional coordination polymer, BBDTA·$InCl_4$: possible spin-Peierls transition with high critical temperature of 108 K. J. Am. Chem. Soc., 128: 6016–6017.
38. Fujita, W., Kikuchi, K., Awaga, K. (2008) Structural study of a dimerization process in an organic radical magnet, BBDTA·$InBr_4$. Angew. Chem. Int. Ed., 47: 9480–9483.
39. Wolmershäuser, G., Johann, R. (1989) 1,3,5-Trithia-2,4,6-triazapentalenyl–a stable sulfur-nitrogen radical. Angew. Chem. Int. Ed. Engl., 28: 920–921.
40. Iketaki, K., Kanai, K., Fujita, W., Awaga, K., Tsutsumi, J., Yoshida, H., Sato, N., Knupfer, M., Ouchi, Y., Seki, K. (2008) Electronic structure of 1,3,5-trithia-2,4,6-triazapentalenyl on gold. Chem. Phys. Lett., 451: 58–62.
41. Fujita, W., Awaga, K., Matsuzaki, H., Okamoto, H. (2002) Room-temperature magnetic bistability in organic radical crystals: Paramagnetic-diamagnetic phase transition in 1,3,5-trithia-2,4,6-triazapentalenyl. Phys. Rev. B, 65: 064434/1–064434/9.
42. McManus, G. D., Rawson, J. M., Feeder, N., McInnes, E. J. L., Novoa, J. J., Burriel, R., Palacio, F., Oliete, P. (2001) Synthesis, crystal structures, electronic structure and magnetic behaviour of the trithiatriazapentalenyl radical, $C_2S_3N_3$. J. Mater. Chem., 11: 1992–2003.
43. Clarke, C. S., Jornet-Somoza, J., Mota, F., Novoa, J. J., Deumal, M. (2010) Origin of the magnetic bistability in molecule-based magnets: a first-principles bottom-up study of the TTTA crystal. J. Am. Chem. Soc., 132: 17817–17830.
44. Matsuzaki, H., Fujita, W., Awaga, K., Okamoto, H. (2003) Photoinduced phase transition in an organic radical crystal with room-temperature optical and magnetic bistability. Phys. Rev. Lett., 91: 017403/1–017403/4.
45. Tanaka, T., Fujita, W., Awaga, K. (2004) Pressure effects on magnetic bistability in a heterocyclic thiazyl radical TTTA. Chem. Phys. Lett., 393: 150–152.
46. Wright, J. D. (1994) Molecular Crystals, 2nd edn. Cambridge University Press, Cambridge, UK, p. 236.

9

Breathing Crystals from Copper Nitroxyl Complexes

Victor Ovcharenko[1] and Elena Bagryanskaya[1,2]

[1]*International Tomography Center, Siberian Branch of the Russian Academy of Sciences, Russian Federation*
[2]*N.N. Vorozhtov Novosibirsk Institute of Organic Chemistry, Siberian Branch of the Russian Academy of Sciences, Russia Federation*

9.1 Introduction

Coordination chemistry of transition metals with stable nitroxides is one of the actively developed branches of modern chemistry which is an important tool in solving the problems of molecular magnetism. The results of studies of metal complexes with nitroxides and their synthesis have been reported from various viewpoints in monographs and reviews.[1–19] Studies of heterospin compounds have led to the creation of new classes of compounds with new combinations of physical characteristics.[19] These include breathing crystals based on Cu(II) coordination compounds with stable nitroxides. Interest in these compounds has recently developed enormously and their number has increased dramatically. This prompted us to summarize data on their magnetostructural correlations. The discussion concentrated on Cu(II) hexafluoroacetylacetonate complexes $[Cu(hfac)_2]$ with pyrazolyl-substituted nitronyl nitroxides. These compounds have been studied for more than 10 years[20, 21] and are well defined. The first part of the chapter summarizes the main features of breathing crystals, obtained using X-ray and magnetic measurements, and the second part overviews the results of EPR. EPR spectroscopy as applied to breathing crystals has a number of peculiarities which demands the development of new approaches. EPR study of breathing crystals is very informative, even though the exchange interaction in nitroxide-copper–nitroxide clusters significantly exceeds the value of Zeeman interaction.

A very interesting characteristic of the complexes is the similarity of their magnetic properties to spin-transitions in classical systems based on the temperature dependence of the effective magnetic moment (μ_{eff}). Note that the classical spin-crossover[22–35] is absolutely impossible in Cu(II) complexes with diamagnetic ligands, which contain only d^9 paramagnetic centers. The main distinction of spin-transitions in Cu(II)

Spin-Crossover Materials: Properties and Applications, First Edition. Edited by Malcolm A. Halcrow.

complexes with stable organic radicals from classical spin-crossover lies in *the possibility of a change in the net spin of a heterospin exchange cluster* with *two or more* paramagnetic centers, resulting from a structural rearrangement of the metal environment induced by external treatment (change of temperature or pressure, or irradiation). In other words, for the compound to have a magnetic anomaly, it should contain at least two exchange-coupled paramagnetic centers in the solid state in which the exchange coupling parameter can change in the course of the phase transformation induced by the external treatment.

Rey *et al.* described[36] a tetranuclear complex of $[Cu(hfac)_2]$ and a 3-pyridyl-substituted nitronyl nitroxide, whose magnetic behavior was similar to spin-crossover in the temperature range 70–140 K. This effect was attributed to the transition of the coordinated O atoms of the nitroxyl fragments from axial to equatorial coordination sites. Later, a modification of a tetranuclear $[Cu(hfac)_2]$ complex with 3-pyridyl-substituted imino nitroxide was described, with a similar structure, and a 2-step transition (at 140–152 and 70–80 K) was recorded for it.[37] The high-temperature transition was assumed to be a pseudospin transition, and the magnetic effect was correlated to the ordinary change of the equatorial coordination of the imine N atom at the exocyclic metal atom to the axial coordination in a square pyramid. Quantum-chemical calculations confirmed the main ideas used to explain these spin-crossover-like effects in Cu(II) complexes with nitronyl nitroxides.[38] A spin-crossover-like effect was also recorded at 44–50 K for a trinuclear complex of $[Cu(hfac)_2]$ with dinitroxide.[39] For all the compounds described above, the structure was determined at room temperature, that is above the spin-transition temperature. Below the transition temperature, the structure of the compound was determined based on a number of model approximations for only one complex.[36, 38] The reason is quite evident: structural rearrangements as profound as these generally lead to crystal decay.[36, 37] For this compound, χT decreased slowly when the temperature decreased from 100 to 30 K and then started to increase. This unusual effect certainly deserved a special study, but the crystal structure of the compound was not studied below the transition point because the single crystals exploded below 30 K.[40] Later it was found, however, that all the magnetic effects described in,[36, 38–40] had the same nature and were associated with a structural phase transition, which was induced by the external treatment and changed the distances and angles between the paramagnetic centers in heterospin exchange clusters and hence the exchange-coupling energy.[19–21, 38] Based on the accumulated data on the magnetochemical behavior of these heterospin complexes and their properties, it is clear that this is a widespread phenomenon and the spin-transition-like behavior is now recorded even in the case of insignificant structural rearrangements of heterospin solid phases during polymorphic transformations[41, 42].

While investigating the complexes of $[Cu(hfac)_2]$ with nitronyl nitroxides containing a pyrazole substituent in the 2nd position of the imidazoline ring (L^R), we found a family of heterospin complexes capable of spin-crossover-like effects at relatively high temperatures (100–250 K).[19, 38, 43, 44] They had a valuable property: the single crystals of the complexes did not crack in the temperature range of the structural phase transition and the ensuing magnetic transition. Due to the mechanical plasticity of these compounds, these processes were regarded as single-crystal-to-single-crystal phase transitions in the majority of cases. For all heterospin complexes and solid solutions on their basis,[20, 45] therefore, the structure was determined at temperatures both before and after the magnetic transition, which offers a unique opportunity for observing the structural dynamics in the range of phase transition in these systems.

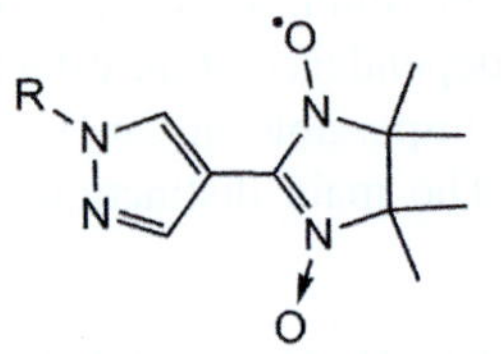

L^R (R = Me, Et, *n*-Pr, *i*-Pr, *n*-Bu)

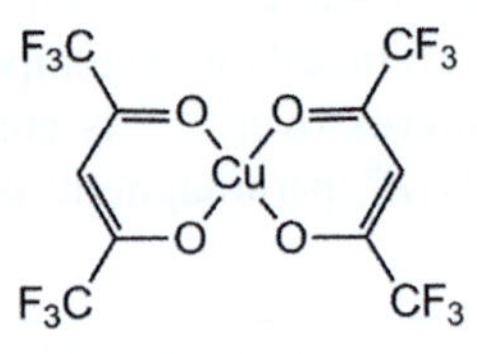

$Cu(hfac)_2$

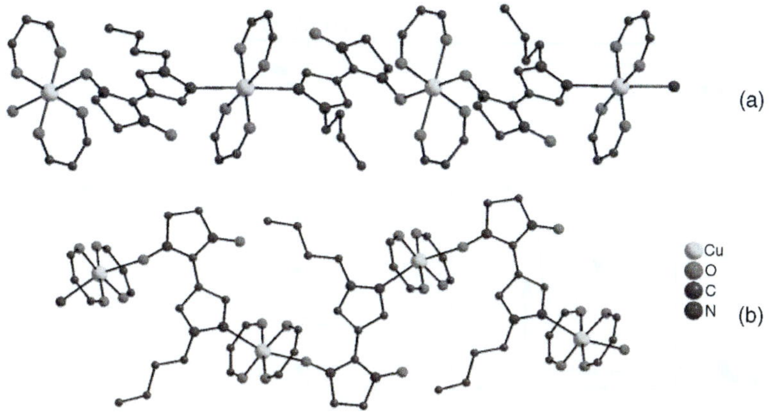

Figure 9.1 Formation of polymer chains with "head-to-head" (a) and "head-to-tail" (b) motifs due to the N,O-bridging environment of $L^{n\text{-}Bu}$ in two polymorphic modifications of $[Cu(hfac)_2L^{n\text{-}Bu}]$. The hydrogen atoms, the geminal methyl groups of $L^{n\text{-}Bu}$, and the trifluoromethyl groups of the hfac ligands are omitted for clarity.

Note that all solids characterized by spin-crossover-like magnetic effects and studied by us have a polymer-chain (rather than molecular) structure, which appears as a result of the bridging bidentate coordination of the paramagnetic ligands (Fig. 9.1). Phase transitions without crystal decomposition are extremely rare in these structures because a single-crystal-to-single-crystal polymorphic transformation requires a large number of coherent motions of polymer chains inside the crystal, but not only motions in chains. The repeated cooling–heating cycles cause a considerable and reversible change in the cell volume at the microlevel and in the volume of the entire crystal at the macrolevel. Since crystal compression and expansion of this kind resembles a periodic change in the volume of the thorax, the crystals were called breathing crystals.[19]

9.2 Structural and Magnetic Anomalies

One of the components of breathing crystals is the metal-containing matrix $[Cu(hfac)_2]$. It is characterized by stereochemical nonrigidity, which is a challenge for the development of synthesis of compounds containing organic radicals and exhibiting a spin-transition effect. For example, in studies of the reaction of $[Cu(hfac)_2]$ with the spin-labeled pyrazole L^{Me}, 12 compounds were isolated which differed in the composition and/or the structure of the solid.[27] These compounds can form simultaneously during the synthesis, and the only method of purification is mechanical separation of a mixture of single crystals after their isolation from the mother solution. The separation is absolutely necessary because insignificant changes in the structure of the solid heterospin complex can substantially change the magnetic properties of the compound. Special attention should be paid to cases when the crystals of the separated complexes are very similar in color and shape. In this case, each (!) crystal of the compound should be analyzed by XRD before magnetic measurements.

Studies of breathing crystals revealed a wide variety of magnetic anomalies on the $\mu_{eff}(T)$ (or $\chi T(T)$) curves even when changes in the structure of the organic radical were insignificant. Let us compare, for example, the data for $[Cu(hfac)_2L^R]$ where R is Me, Et, *n*-Pr, or *n*-Bu. The solid phases of all these compounds are formed by polymer chains, in which the nitroxide performs the bridging function; the chains contain Cu^{2+}–O$^\bullet$–N< or >N–$^\bullet$O–Cu^{2+}–O$^\bullet$–N< heterospin exchange clusters.

For $[Cu(hfac)_2L^{Me}]$, the abrupt change in μ_{eff} takes place in a narrow temperature range and is typical (in shape) for spin-transitions in the classical compounds that exhibit this effect, passing from high to low

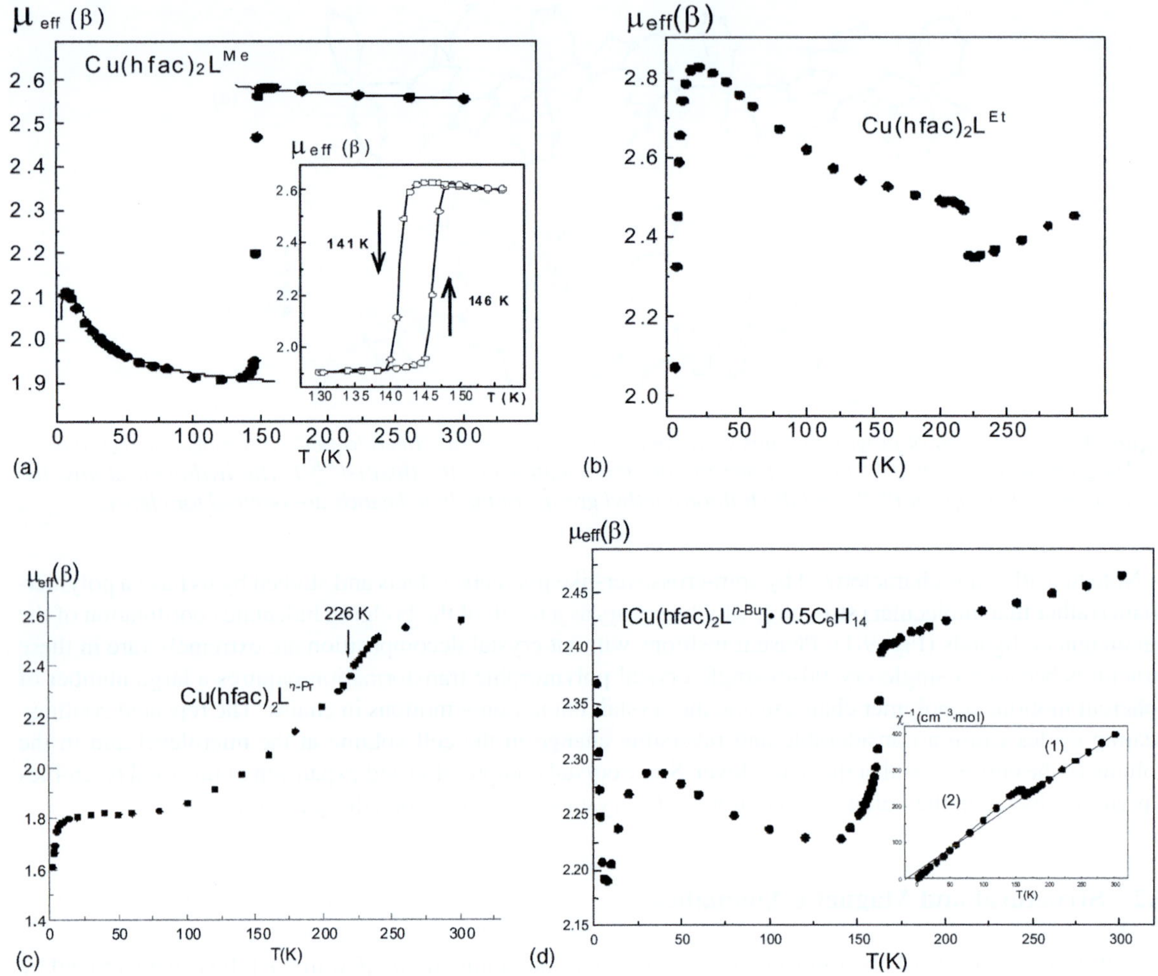

Figure 9.2 *Temperature dependence of* μ_{eff} *for [Cu(hfac)*$_2$*L*^{R}*] (R = Me (a), Et (b),* n-Pr *(c)) and [Cu(hfac)*$_2$*L*$^{n\text{-}Bu}$*]·0.5C*$_6$*H*$_{14}$ *(d).*

spin state on cooling (Fig. 9.2a). The abrupt decrease in μ_{eff} after the phase transition region corresponds to the $\sqrt{2}$-fold decrease in the effective magnetic moment, indicating that half of all spins with $S = {}^1/_2$ vanished from the sample. This is caused by the fact that in the course of the phase transformation inside the chains with the "head-to-tail" motif (Fig. 9.1b), the nitroxyl O atoms and the pyrazole N atoms, which initially occupied the axial positions, moved to the equatorial positions below 150 K in half of all CuO_5N coordination units. At 295 K the distance from the Cu atom to the axial nitroxyl O atom (Cu–O_L) was 2.484(5) Å and the distance from the Cu atom to the axial N atom of the pyrazole ring (Cu–N_L) was 2.329(5) Å. At 140 K the distances decreased to 1.992(9) and 2.014(10) Å, respectively, in *half* of all coordination units. Two O atoms of hfac ligands (O_{hfac}) shifted to the axial positions. The Cu–O_L distance, which is long enough at room temperature (2.484 Å), is the reason for the predominance of ferromagnetic exchange interaction in Cu^{2+}–$O^{\bullet}$–N< exchange clusters.[20, 37, 46–48] The shift of the coordinated nitroxyl O_L atoms from axial to equatorial position below 150 K leads to a change of ferromagnetic exchange to a strong antiferromagnetic interaction in half of all Cu^{2+}–O•–N< exchange clusters. As a result, spins are compensated in these units, which leads to the $\sqrt{2}$-fold decrease in μ_{eff}.

In solid [Cu(hfac)$_2$L$^{n\text{-Pr}}$], the polymer chains have a "head-to-head" motif, which leads to the formation of alternating CuO_6 and CuO_4N_2 coordination units (Fig. 9.1a). The CuO_6 units contain heterospin-triads of >N–•O–Cu^{2+}–O•–N< paramagnetic centers, in which the change in the bond lengths at lowered temperatures determines the character of the $\mu_{eff}(T)$ dependence. The significant shortening of the distances between the Cu atoms and the coordinated oxygen atoms (Cu–O_L) in CuO_6 units is accompanied by a simultaneous increase in their Cu–O_{hfac} distances along the O_{hfac}–Cu–O_{hfac} "axis." In distorted octahedral CuO_6 Jahn–Teller units, we actually have a replacement of the O_L–Cu–O_L "octahedron axis," which is lengthened at room temperature (shortened by 0.598 Å on cooling), by the O_{hfac}–Cu–O_{hfac} "axis," which is lengthened at low temperature (lengthened by 0.576 Å on cooling).[20] As a result of the phase transition, the crystal symmetry also changes from *C2/c* at 293 K to $P2_1/c$ at 203 K and 115 K.

Therefore, for [Cu(hfac)$_2$L$^{n\text{-Pr}}$] below 230 K, the $\mu_{eff}(T)$ curve quickly decreases and goes to a plateau with μ_{eff} ~1.8 β after a smooth decrease in μ_{eff} at lowered temperature (Fig. 9.1c). This points to a dramatic strengthening of antiferromagnetic interactions in >N–•O–Cu^{2+}–O• N< exchange clusters after the structural phase transition at 226 K. In Figure 9.2c, this temperature is marked by the arrow because the abrupt change on the $\mu_{eff}(T)$ curve is moderate. An XRD analysis above and below this temperature showed that the structural characteristics abruptly changed exactly in regions adjacent to 226 K. The decrease to 1.8 β in μ_{eff} at 50 K indicates that half of all spins vanished in [Cu(hfac)$_2$L$^{n\text{-Pr}}$] on cooling. This agrees well with the decrease in the spin of the >N–•O–Cu^{2+}–O•–N< exchange cluster to $S = {}^1/_2$. At $T < 50$ K the contribution to the magnetic moment of [Cu(hfac)$_2$L$^{n\text{-Pr}}$] is only from the residual spin of the given exchange cluster and Cu^{2+} ions of CuO_4N_2 units.

The motif of polymer chains in solid [Cu(hfac)$_2$L$^{n\text{-Bu}}$]·0.5C_6H_{14} is the same as in [Cu(hfac)$_2$L$^{n\text{-Pr}}$]. The structural transformation of coordination units in [Cu(hfac)$_2$L$^{n\text{-Bu}}$]·0.5C_6H_{14} at lowered temperatures, however, differs considerably from that in [Cu(hfac)$_2$L$^{n\text{-Pr}}$]. At room temperature, all CuO_6 units are lengthened octahedra with axial Cu–O_L distances of 2.320 Å and equatorial Cu–O_{hfac} distances of 1.970-1.983 Å. The shape of the coordination polyhedron is similar in CuO_4N_2 units, in which the axial Cu–N_L distances are long enough (2.514 Å), while the markedly shorter equatorial Cu–O_{hfac} distances are 1.937 and 1.977 Å. When the temperature is lowered to 167 K, the Cu–O_L distances are shortened from 2.320 to 2.250 Å. Further decrease to 123 K in the temperature actually doubles the cell volume as a consequence of the doubling of the independent part of the structure. The CuO_4N_2 units in [Cu(hfac)$_2$L$^{n\text{-Bu}}$]·0.5C_6H_{14} become pseudosymmetric with Cu–N_L distances of 2.457 and 2.468 Å. The CuO_6 units remain centrosymmetric but are now divided into two types. For one type, the axial Cu–O_L distances (2.250–2.348 Å) drastically increase. For the other, the Cu–O_L distances (2.250–2.007 Å) drastically decrease, leading to a transition of O_L atoms to the equatorial positions and of two O_{hfac} atoms to the axial positions (2.023–2.240 Å) in these units.

The structural phase transition in [Cu(hfac)$_2$L$^{n\text{-Bu}}$]·0.5C_6H_{14} at 163 K, as a result of which the Cu–O_L distances are shortened from 2.32 Å to ~2.00 Å in half of all >N–•O–Cu^{2+}–O•–N< exchange clusters, leads to a strong antiferromagnetic interaction between the odd electrons of Cu^{2+} and N–•O groups in these spin-triads. As a result, the effective spin of this group of triads decreases to ${}^1/_2$, causing a decrease in μ_{eff} at 163 ± 2 K (Fig. 9.2d). In the other half of >N–•O–Cu^{2+}–O•–N< exchange triads, the Cu–O_L distances increase from 2.32 to 2.35 Å, changing the sign of exchange interaction from negative to positive. As a result, μ_{eff} ceases to decrease after the structural transition when the temperature is lowered from 300 to 165 K and starts to gradually increase at $T < 150$ K. The insert in Figure 9.2d shows the temperature curve of the reversible magnetic susceptibility. The parameters of the Curie–Weiss equation for regions (1) and (2) are $C_1 = 0.805$, $\theta_1 = -19$ K and $C_2 = 0.625$, $\theta_2 = 0.3$ K, respectively. Since the C_1 and C_2 parameters are proportional to the number of

Table 9.1 *Optimum parameters of the spin Hamiltonian* for describing the magnetic properties of [M(hfac)$_2$L^{Et}] complexes.*

Compound	g_M	J, K	$J'z$, K
[Cu(hfac)$_2$L^{Et}] ($T < 220$ K)	2.0	41.7	−0.6
[Cu(hfac)$_2$L^{Et}] ($T > 230$ K)	2.52	−121	−0.34
[Mn(hfac)$_2$L^{Et}]	2.0	−110	−0.017
[Ni(hfac)$_2$L^{Et}]	2.33	−368	−1.0

*$\hat{H} = -2J(\hat{s}_{NO}\hat{s}_{Cu} + \hat{s}_{Cu}\hat{s}_{NO}) - \beta(2g_{Cu}\hat{s}_{Cu} + 2g_{NO}\hat{s}_{NO})H - 2J'z\hat{S}\langle\hat{S}\rangle$; the spin Hamiltonian defines the spin coupling in the exchange cluster, Zeeman interaction, and weak inter-chain and inter-cluster interactions.[49]

paramagnetic centers, the decrease in C_2 compared with C_1 is direct evidence for the effect of "vanishing" of some of the paramagnetic centers in >N–•O–Cu^{2+}–O•–N< exchange clusters with strong antiferromagnetic interactions. The ratio $C_2/C_1 = 0.78$ is close to 0.75, which corresponds to complete coupling of two spins in the clusters. The observed character of the $\mu_{eff}(T)$ dependence for [Cu(hfac)$_2$L$^{n\text{-Bu}}$]·0.5C_6H_{14} thus completely correlates with the structural dynamics of the heterospin complexes at lower temperatures.

Solid [Cu(hfac)$_2$L^{Et}], as well as [Cu(hfac)$_2$L^R] (R = *n*-Pr and *n*-Bu), is formed from polymer chains with a "head-to-head" motif. The most important structural feature of [Cu(hfac)$_2$L^{Et}] at 293 K is short axial distances Cu–O$_L$ and Cu–N$_L$ (2.237 and 2.375 Å) in the crystallographically independent centrosymmetric CuO_6 and CuO_4N_2 units, respectively.

These data explain the $\mu_{eff}(T)$ dependence shown in Figure 9.2b for [Cu(hfac)$_2$L^{Et}]. The μ_{eff} value is close to the theoretical value 2.45 β for a system of almost noninteracting spins of Cu^{2+} and nitroxide per [Cu(hfac)$_2$L^{Et}] fragment at room temperature. When the sample is cooled to ~225 K, μ_{eff} gradually decreases to ~2.3 β. The short axial Cu–O$_L$ distance (2.237 Å) in the octahedral CuO_6 units is the reason for the predominance of the antiferromagnetic contribution in >N–•O–Cu^{2+}–O•–N< exchange clusters in the temperature range 225–300 K.[37, 46] The results of the structural studies of [Cu(hfac)$_2$L^{Et}] after the phase transition show that at 188 K the Cu–O$_L$ distance in CuO_6 units increases by only ~0.02 Å. This bond lengthening, however, leads to a drastic decrease in the antiferromagnetic contribution and the dominance of ferromagnetic interactions in >N–•O–Cu^{2+}–O•–N< exchange clusters. The character of exchange changes abruptly at 220 K (Fig. 9.2b). This jump of μ_{eff} at lowered temperatures can often be observed for breathing crystals. The results of simulation of experimental data for two phase regions (before and after the transition) are given in Table 9.1.

Thus we have different situations when the net spin of the exchange cluster is formed with participation of *two or more* paramagnetic centers:

(a) As a result of a structural phase transition, the ferromagnetic exchange interaction such as the one in [Cu(hfac)$_2$L^{Me}] changes to the antiferromagnetic one, or the antiferromagnetic interaction increases sharply in energy as it does in [Cu(hfac)$_2$L$^{n\text{-Pr}}$]. In this case, μ_{eff} drops at the transition point. This magnetic behavior of the compound is similar in character to the $\mu_{eff}(T)$ dependence of spin-transitions in classical systems.
(b) As a result of the structural phase transition, the antiferromagnetic interaction of odd electrons in the exchange cluster becomes ferromagnetic as it does in [Cu(hfac)$_2$L^{Et}]. In this case, μ_{eff} jumps at the point of the phase transition.
(c) As a result of the structural phase transition, situation (a) takes place in some of the exchange clusters and (b) in others, as in [Cu(hfac)$_2$L$^{n\text{-Bu}}$]. In this case, μ_{eff} drops at the phase transition point and subsequently increases.

These data demonstrate the diversity (in the character of the $\mu_{eff}(T)$ dependence) of the magnetic anomalies produced by multicenter exchange clusters that undergo structural rearrangements. It appeared that the structural dynamics of coordination polyhedra containing exchange clusters was considerably affected by the structure of the R substituent (Me, Et, *n*-Pr, *n*-Bu) in the paramagnetic L^R ligand not involved in the coordination unit. That is, the magnetic anomaly is sensitive to minor variations in the structure of the solid. On the one hand, this is not surprising because the spin-crossover-like effects are cooperative, with each crystal of the compound working as an entity in the course of the phase transformation. For this reason, compounds with different compositions and hence packings behave differently in respect of the $\mu_{eff}(T)$ functional dependence. On the other hand, the question arises: What scale of structural changes in heterospin crystals is admissible for the magnetic anomaly to be reproduced in a controlled way?

9.3 Relationship between the Chemical Step and the Physical Property

Our studies showed that the shape of the magnetic anomaly was preserved during isotopic replacement, for example during the replacement of the CH_3 substituent in the pyrazole ring L^{Me} by CD_3 (L^{CD3}).[20] The shape of the magnetic anomaly also changed smoothly during the formation of limited solid solutions $[M_xCu_{1-x}(hfac)_2L^R]$ (M = Mn, Ni, Co) or (rarely) during the formation of solid solutions with a variable ratio of paramagnetic organic ligands, $[Cu(hfac)_2L^{Me}{}_xL^{Et}{}_{1-x}]$.[19, 20] However, neither the metal-containing matrix nor the paramagnetic L^R ligands were modified. Then, what can be changed in the composition of a solid breathing crystal in such a way that the main features of the magnetic anomaly could be preserved? It would be even better to know how to adjust the form of the magnetic anomaly in a controllable way. What kind of chemical step is admissible in order to change the structure of some part of a molecule to have a reproducible form of the magnetic anomaly? This is not a trivial question, but it is very important for developing methods for controllable chemical effort to change the physical property of a heterospin compound.

Certain progress has been made in studies of a series of isostructural compounds $[Cu(hfac)_2L^{n\text{-}Bu}]\cdot 0.5solv$ capable of accepting solvent molecules at the same positions of the inter-chain space. Let us first consider a series of solvates of normal hydrocarbons. We studied the relationship between the chemical step and the physical property using a specially chosen series of crystals. It was shown that a mild modification of T_c for $[Cu(hfac)_2L^{n\text{-}Bu}]\cdot 0.5solv$ required a *considerably smaller structural step than one methylene unit in the homologous series, typical in organic chemistry.*[50]

Figure 9.3 illustrates the important circumstance that the packing mode of polymer units is the same in solid solvates with hexane, heptane, octane, and octene, which is confirmed by the form of chains in both perpendicular and axial directions as shown in Figure 9.3 for packings of $[Cu(hfac)_2L^{n\text{-}Bu}]\cdot 0.5C_6H_{14}$ (a) and $[Cu(hfac)_2L^{n\text{-}Bu}]\cdot 0.5C_8H_{18}$ (b) as an example. The regions where solvate molecules lie are also the same. The only difference lies in the volume occupied by solvent molecules and the character of twisting of hydrocarbon chains in voids.

However, the form of the magnetic anomaly and its temperature for $[Cu(hfac)_2L^{n\text{-}Bu}]\cdot 0.5C_6H_{14}$ (a), $[Cu(hfac)_2L^{n\text{-}Bu}]\cdot 0.5C_7H_{16}$ (b), $[Cu(hfac)_2L^{n\text{-}Bu}]\cdot 0.5C_8H_{18}$ (c), and $[Cu(hfac)_2L^{n\text{-}Bu}]\cdot 0.5C_8H_{16}$ (d) differ. This can readily be seen, for example, if we compare the $\mu_{eff}(T)$ dependences for the solvate with hexane (Fig. 9.2d) and the solvate with octane (Fig. 9.4). This result is quite unexpected because, in this case, the temperature and form of the magnetic anomaly can only be affected by van der Waals interactions of heterospin polymer chains with included solvent molecules. Before studies of magnetostructural correlations for the solvates, it was assumed that variation of solvent molecules in the inter-chain space away from the structurally rearranged heterospin exchange clusters had an insignificant effect on the temperature of the magnetic anomaly. This was confirmed by the known experimental data on the effects of noncoordinated solvent molecules on the spin-transition temperature (T_c) and the form of the $\mu_{eff}(T)$

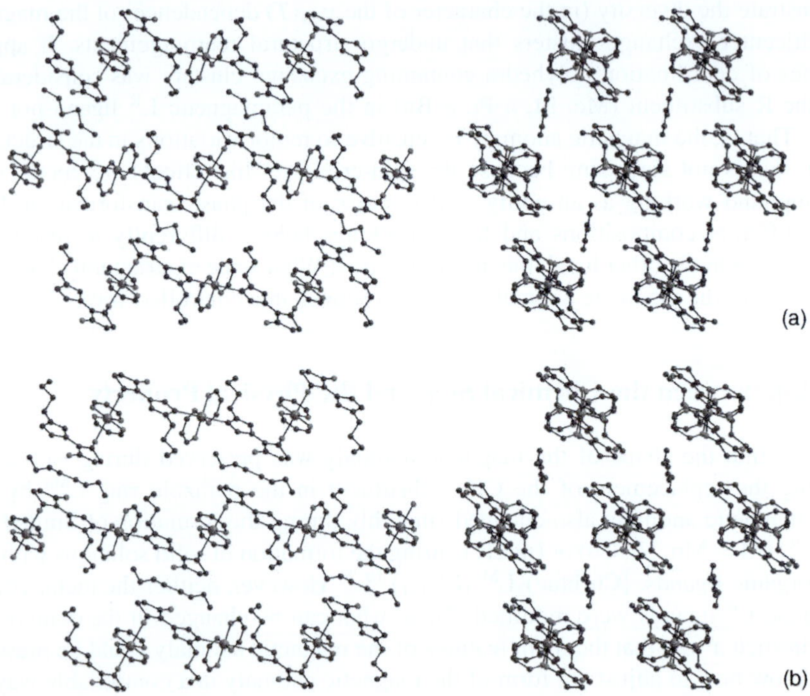

Figure 9.3 Similarity of packings for $[Cu(hfac)_2L^{n\text{-}Bu}]{\cdot}0.5C_6H_{14}$ (a) and $[Cu(hfac)_2L^{n\text{-}Bu}]{\cdot}0.5C_8H_{18}$ (b) solvates in high temperature phases. The left part is a view in one of the directions perpendicular to the chains; the right part shows the same along the chains. Reprinted with permission from [50]. Copyright 2008, American Chemical Society.

curve for solvates of classical Fe^{II} complexes with diamagnetic ligands.[22, 51–53] It is known that solvents that can hydrogen bond to anions or coordinated ligands in solid iron(II) spin-crossover complexes can change the cell symmetry and crystal packing, and hence the T_c and the shape of the $\mu_{eff}(T)$ curve.[53, 54] Low-polarity solvents, however, had an insignificant effect on these properties.[22, 55–58] This proved not to be the case for heterospin breathing crystals, which are very sensitive paramagnetic sensors for hydrocarbon solvate molecules.

This prompted us to investigate an even smaller variation of the structure of the incorporated solvent molecule. As spatial analogs of hexane, we used amyl chloride, bromide, and iodide. Figure 9.5 shows a comparative scale of difference between the radii of the terminal groups or atoms linked to the amyl fragment; the scale was constructed on the basis of reference data of covalent atomic radii.[59] If the methyl group is taken as a zero reference point, its replacement by a Cl atom leads to a certain decrease in the size of the terminal fragment. In contrast, its replacement by a Br atom should lead to an increase, though very small, in the size of the terminal fragment. The terminal group is lengthened significantly after replacement by an I atom and still more significantly after replacement by an ethyl group. As mentioned above, the packing of chains and the motif of the arrangement of solvent molecules in the solvates is the same (Fig. 9.3). If electronic and polarization effects make a small contribution, solvates with amyl bromide and chloride would be expected to be closest to $[Cu(hfac)_2L^{n\text{-}Bu}]{\cdot}0.5C_6H_{14}$ in their structural dynamics and hence their magnetic behavior.

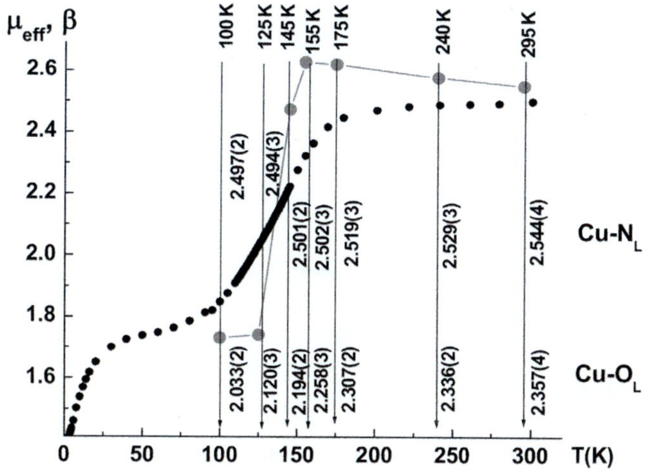

Figure 9.4 Temperature dependence of the effective magnetic moment of $[Cu(hfac)_2L^{n\text{-}Bu}]\cdot 0.5C_8H_{18}$ (•). The vertical arrows indicate the temperature of the X-ray diffraction experiment; the upper and lower rows of figures correspond to the Cu–N$_L$ and Cu–O$_L$ bond lengths in the CuN_2O_4 and CuO_6 units, respectively. The results of quantum-chemical calculations are shown by full gray circles connected by gray lines to demonstrate good agreement between the experimental and calculated dependences $\mu_{eff}(T)$. Reprinted with permission from [50]. Copyright 2008, American Chemical Society.

Indeed, our experiments showed that the experimental $\mu_{eff}(T)$ dependences for solvates with amyl chloride and bromide were very similar to the corresponding dependence for the solvate with hexane (Fig. 9.6).

The similarity of changes in the magnetic properties in the solvates is explained by the similarity of their structural transformations at different temperatures.[50] For $[Cu(hfac)_2L^{n\text{-}Bu}]\cdot 0.5C_5H_{11}I$, the structural characteristics at lowered temperatures are quite different, the dominant tendency being the lengthening of the Cu–O$_L$

Figure 9.5 Comparative scale of difference between the radii of the terminal groups or atoms linked to the amyl fragment.

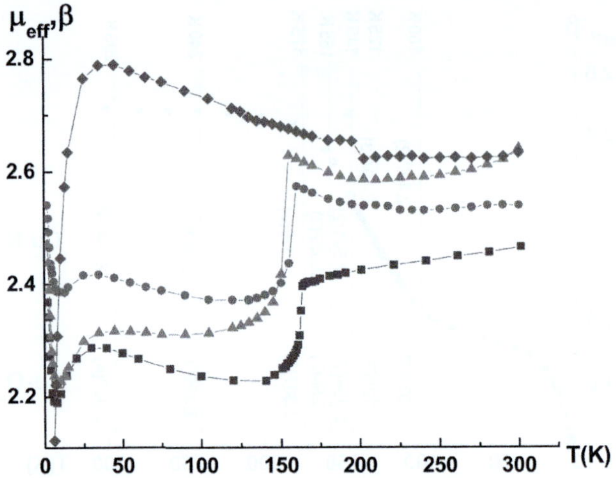

Figure 9.6 *Temperature dependences of the effective magnetic moment of [Cu(hfac)$_2$L$^{n\text{-}Bu}$]·0.5C$_6$H$_{14}$ (squares), [Cu(hfac)$_2$L$^{n\text{-}Bu}$]·0.5C$_5$H$_{11}$Br (triangles), [Cu(hfac)$_2$L$^{n\text{-}Bu}$]·0.5C$_5$H$_{11}$Cl (circles), and [Cu(hfac)$_2$L$^{n\text{-}Bu}$]·0.5C$_5$H$_{11}$I (diamonds). Reprinted with permission from [50]. Copyright 2008, American Chemical Society.*

bond after the transition. Therefore, the abrupt change in μ_{eff} and the character of $\mu_{eff}(T)$ are different for this compound (Fig. 9.6). When the temperature was lowered from 295 K to 100 K for [Cu(hfac)$_2$L$^{n\text{-}Bu}$]·0.5C$_5$H$_{11}$I, the Cu–N$_L$ distances smoothly decreased by ~0.09 Å, while the moderate shortening of Cu–O$_L$ distances changed to lengthening after the transition at 197 ± 4 K. Since the ferromagnetic exchange interactions became dominant after the structural transition (<197 K) in [Cu(hfac)$_2$L$^{n\text{-}Bu}$]·0.5C$_5$H$_{11}$I, this led to a moderate increase in μ_{eff} at that point.

Thus our data show that the tendency toward the preservation of the form of the $\mu_{eff}(T)$ dependence and transition temperature does not vanish when the hexane solvate molecules are replaced by amyl chloride and bromide molecules. The lengthening of the solvent molecule on passing to amyl iodide and especially to heptane is then excessive and drastically changes the form of the $\mu_{eff}(T)$ dependence.

Studies of magnetostructural correlations inherent in breathing crystals are important from the viewpoint of the classical magnetochemical approach, widely used for describing exchange interactions in multispin systems using the Heisenberg–Dirac–Van Vleck model. These studies use data on the character of changes in the magnetic susceptibility of the sample at different temperatures, which are analyzed using the corresponding spin Hamiltonian containing one or more exchange parameters (J_i). It was assumed that J_i was almost independent of temperature.[6] In rare cases when a change in the exchange parameter was recorded, it did not exceed 1 cm^{-1} in the order of magnitude.[60] Breathing crystals, however, are specific compounds in which the variation of temperature substantially affects the geometrical characteristics of heterospin exchange clusters. It is reasonable to assume that the considerable spatial evolution of [Cu^{II}–O$^•$–N>] or [>N–$^•$O–Cu^{II}–O$^•$–N<] exchange clusters in the course of the structural phase transition can cause a change in J_i. This effect was studied by EPR for [Cu(hfac)$_2$L$^{n\text{-}Bu}$]·0.5C$_8$H$_{18}$ crystals[61]. Indeed, J_i was found to depend significantly on temperature. The dependence, however, was reliably recorded only in the region of the phase transition. Near this region, the experimental error became large, which distorted the $J_i(T)$ dependence and made it impossible to understand the mechanism of the generation of a new (low- or high-temperature) phase. It was important to solve this problem in a quantum-chemical study of the $J_i(T)$ dependence for a real change in the structure of the compound.

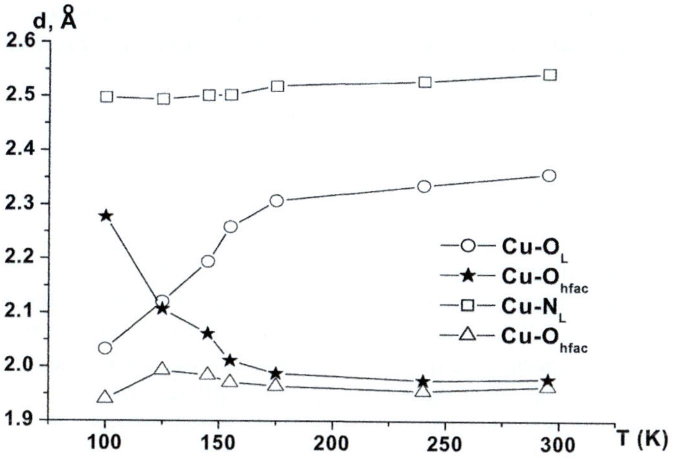

Figure 9.7 *Variation of Cu–O$_L$, Cu–N$_L$, and Cu–O$_{hfac}$ bond lengths in CuO$_6$ and CuN$_2$O$_4$ coordination units in [Cu(hfac)$_2$L$^{n\text{-}Bu}$]·0.5C$_8$H$_{18}$ at different temperatures.*

For $[Cu(hfac)_2L^{n\text{-}Bu}]\cdot 0.5C_8H_{18}$ whose structural rearrangement of exchange clusters at different temperatures is shown in Figure 9.7, it can be seen that when the temperature was lowered from 295 K to 175 K, in the range above the transition temperature, the Cu–O$_L$ distances slightly decreased. According to quantum-chemical calculations, the exchange interaction became slightly more efficient without sign reversal in the copper-radical channel (Fig. 9.8). The abrupt shortening of the Cu–O$_L$ distance in the range of the phase transition is accompanied (according to the calculations) not only by a significant change in the efficiency of the corresponding superexchange channel, but also by a reversal in sign.

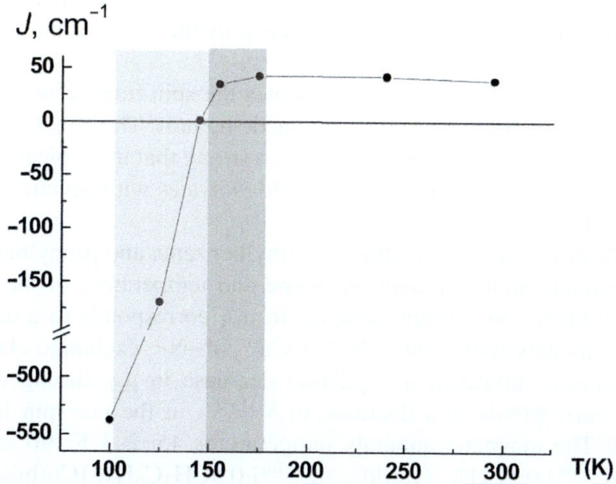

Figure 9.8 *Temperature dependence of the intra-cluster exchange coupling parameter J (Cu–nitroxide) of [Cu(hfac)$_2$L$^{n\text{-}Bu}$]·0.5C$_8$H$_{18}$ (calculated data). Reprinted with permission from [50]. Copyright 2008, American Chemical Society.*

The calculated J(Cu–nitroxide) values at seven different temperatures were used to reconstruct the experimental $\mu_{eff}(T)$ dependence for $[Cu(hfac)_2L^{n\text{-}Bu}]\cdot 0.5C_8H_{18}$.[50] The resulting $\mu_{eff}(T)$ curve is shown in Figure 9.4 as gray lines connecting the calculated magnetic moments. Good agreement is observed between the experimental and calculated $\mu_{eff}(T)$ curves. Note that the observed decrease in the magnetic moment after the lowering of temperature from 295 to 175 K cannot be explained by X-ray diffraction data, but resulted from a gradual increase in the fraction of the low-temperature phase in the high-temperature phase. The XRD data for the range ~175–150 K (in Fig. 9.8, this region is shown as a dark gray band) in Figure 9.7 correspond to the result of averaging of the long $Cu–O_L$ bonds of the high-temperature phase and the short $Cu–O_L$ bonds of the low-temperature phase, but not to a real decrease in the distances. In fact, below this temperature range, regions of the low-temperature phase (characterized by short $Cu–O_L$ distances) are gradually generated in the high-temperature phase. That is, a solid solution of the low-temperature phase is formed in the high-temperature phase. On the experimental $\mu_{eff}(T)$ curve, this is reflected as a decreased effective magnetic moment. After a certain critical content of the low-temperature phase is reached, a phase transition of the rest of the high-temperature phase to the low-temperature phase is initiated. This is the region at ~150–100 K, marked by a pale gray band in Figure 9.8. Here, indeed, all of the remaining long $Cu–O_L$ distances decrease, naturally leading to enormous changes in the exchange parameter (Fig. 9.8). The theoretical curve in Figure 9.4, therefore, represents a more drastic transformation of $\mu_{eff}(T)$, whereas the experimental $\mu_{eff}(T)$ dependence is smoother than the calculated curve obtained by theoretical analysis of the evolution of the exchange parameter on passing from one phase to the other for reasons stated above.

Thus XRD and magnetochemical studies of a series of $[Cu(hfac)_2L^{n\text{-}Bu}]\cdot 0.5$solv polymer chain complexes with included solvent molecules revealed high intra-crystalline mobility due to temperature variation in these compounds. The motion is concentrated in $>N–{}^{\bullet}O–Cu^{II}–O^{\bullet}–N<$ heterospin exchange clusters containing copper(II) Jahn–Teller ions. The lengthening or shortening of $Cu–O_{NO}$ bonds is accompanied by a corresponding shortening or lengthening of two $Cu–O_{hfac}$ bonds along one of the orthogonal axes in CuO_6 units. While experiencing changes in their structure, the heterospin polymer chains move along solvent molecules as if gliding on inner joints, which, in turn, make coherent motions concerted with the deformation of CuO_6 units. This ensures the mechanical stability of crystals, which do not decay during repeated cooling–heating cycles. A consistent structural representation of the compound using the temperature as the fourth coordinate makes it possible to visualize the complex concerted intra-crystalline motions, actually representing it as a movie. These motion pictures are direct evidence of the cooperative character of the observed phenomenon.

The strong effect of the incorporated solvent differentiates the spin transitions in question from classical spin-crossover in transition metal complexes with diamagnetic ligands. The effect of the incorporated solvent on the magnetic anomaly of $[Cu(hfac)_2L^{n\text{-}Bu}]\cdot 0.5$solv was so strong that the typical modification (addition of a methylene unit) here was excessive. This was discussed for solvates with acyclic solvents. The same holds for solvates with aromatic solvents.

Figure 9.9 shows that solvates with benzene, toluene, ethylbenzene, and propylbenzene, whose molecules formally differ by one methylene unit, differ in the scope and temperature of the magnetic anomaly. For solvates with benzene and toluene, the abrupt decrease in μ_{eff} corresponds to a decrease from $S = {}^3/_2$ to ${}^1/_2$ in the net spin in approximately half of all $>N–{}^{\bullet}O–Cu^{II}–O^{\bullet}–N<$ exchange clusters. For solvates with ethylbenzene and propylbenzene, however, the $\sqrt{2}$-fold decrease in μ_{eff} during the transition from high to low temperature phase corresponds to a decrease to $S = {}^1/_2$ in the net spin in all the $>N–{}^{\bullet}O–Cu^{II}–O^{\bullet}–N<$ exchange clusters. The magnetic anomaly temperatures 45 ± 3 K, 90 ± 3 K, 103 ± 3 K, and 205 ± 3 K for $[Cu(hfac)_2L^{n\text{-}Bu}]\cdot 0.5C_6H_6$, $[Cu(hfac)_2L^{n\text{-}Bu}]\cdot 0.5CH_3C_6H_5$, $[Cu(hfac)_2L^{n\text{-}Bu}]\cdot 0.5C_3H_7C_6H_5$, and $[Cu(hfac)_2L^{n\text{-}Bu}]\cdot 0.5C_2H_5C_6H_5$, respectively, also differ substantially. This series of T_c values cannot be justifiably correlated with the change in the composition of the solvate molecule. The reasons for the difference

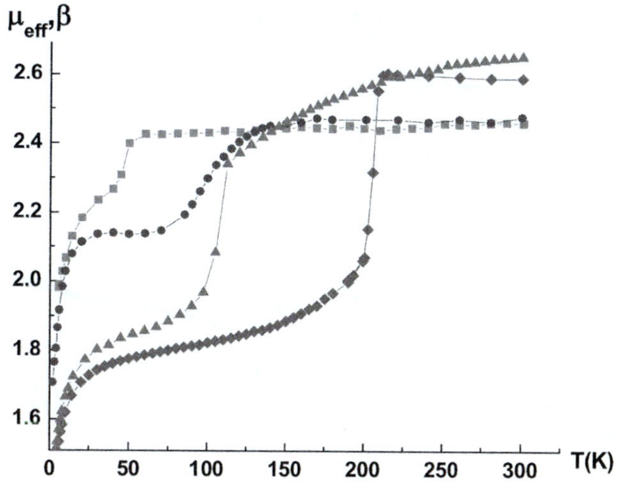

Figure 9.9 *Temperature dependences of the effective magnetic moment of [Cu(hfac)$_2$L$^{n\text{-}Bu}$]·0.5C$_6$H$_6$ (squares), [Cu(hfac)$_2$L$^{n\text{-}Bu}$]·0.5CH$_3$C$_6$H$_5$ (circles), [Cu(hfac)$_2$L$^{n\text{-}Bu}$]·0.5C$_2$H$_5$C$_6$H$_5$ (diamonds) and [Cu(hfac)$_2$L$^{n\text{-}Bu}$]·0.5C$_3$H$_7$C$_6$H$_5$ (triangles). Reprinted with permission from [42]. Copyright 2011, American Chemical Society.*

in the high temperature asymptotics of μ_{eff} are also not clear (Fig. 9.9). Our study of aromatic solvates showed that the magnetic characteristics of breathing crystals were sensitive not only to slight changes in the structure and composition of incorporated solvent molecules, but also to changes in their orientation relative to the paramagnetic fragments.[42]

9.4 Relationship between the Thermally Induced Reorientation of Aromatic Solvate Molecules and the Character of the Magnetic Anomaly

In studies of solvates with aromatic solvent molecules, the product with styrene [Cu(hfac)$_2$L$^{n\text{-}Bu}$]·0.5C$_2$H$_3$C$_6$H$_5$ proved a valuable object of investigation. Figure 9.10a shows that the $\mu_{eff}(T)$ curve for this compound has two abrupt features, with a very narrow hysteresis loop at low temperature ($T^1{}_c\uparrow = 185$ K, $T^1{}_c\downarrow = 183$ K) and a wider one at high temperature ($T^2{}_c\uparrow = 260$ K, $T^2{}_c\downarrow = 251$ K). The figure also gives the Cu–N$_L$ and Cu–O$_L$ bond lengths for [Cu(hfac)$_2$L$^{n\text{-}Bu}$]·0.5C$_2$H$_3$C$_6$H$_5$ at 105, 180, 240, and 293 K. The dashed lines connect the bond lengths between the Cu atoms and the coordinated imine N atoms of the pyrazole rings (*d*Cu–N$_L$, full blue circle in Fig. 9.10a), demonstrating the dynamics of variation of inter-atomic distances due to the variation of temperature. The dashed lines connect the points corresponding to the bond lengths between the Cu atoms and the coordinated nitroxyl O atoms (*d*Cu–O$_L$, full circles) and hexafluoroacetylacetonate anions (*d*Cu–O$_{hfac}$, open squares and triangles). According to the plots, each stage of the magnetic anomaly corresponds to drastic changes in CuO$_6$ units. When the temperature is lowered from 293 to 240 K, the Cu–O$_L$ distances decrease from 2.367(4) to 2.246(4) Å. The stage corresponding to the decrease from 180 to 105 K in temperature is accompanied by a more drastic decrease in dCu–O$_L$ from 2.263(4) Å to 2.020(6) Å. In the CuO$_6$ units, the Cu–O$_{hfac}$ bonds considerably increased from 1.999(3) Å to 2.245(6) Å in one of the axial directions. Thus [Cu(hfac)$_2$L$^{n\text{-}Bu}$]·0.5C$_2$H$_3$C$_6$H$_5$ undergoes two phase transitions at lowered temperatures, resulting in a change in the Jahn–Teller axis in CuO$_6$ coordination units of the low temperature phase of this compound. All nitroxyl O atoms pass from axial to equatorial positions, which is favorable

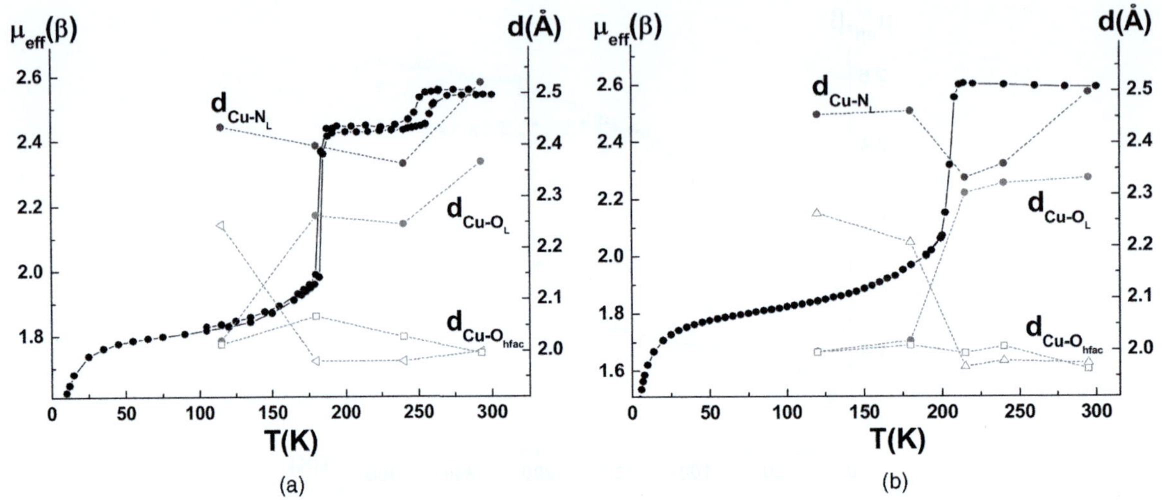

Figure 9.10 *Temperature dependences of the effective magnetic moment and Cu–N_L and Cu–O_L bond lengths at different temperatures for $[Cu(hfac)_2L^{n\text{-}Bu}]\cdot 0.5C_2H_3C_6H_5$ (a) and $[Cu(hfac)_2L^{n\text{-}Bu}]\cdot 0.5C_2H_5C_6H_5$ (b). Reprinted with permission from [42]. Copyright 2011, American Chemical Society.*

for the low spin state in >N–•O–Cu(II)–O•–N< spin-triads. This leads to complete coupling of half of all spins and a $\sqrt{2}$-fold decrease in μ_{eff}. The minor change in μ_{eff} in the range 240–260 K, corresponding to a hysteresis loop with a width of the order of 9 K, recorded during repeated cooling–heating cycles, is caused by a structural rearrangement in the exchange triads. As is known[37, 50, 58–65] the shortening of distances between the paramagnetic centers (dCu–O_L) leads to a decrease in the ferromagnetic contribution and an increase in the antiferromagnetic contribution to the exchange interaction.

An analysis of a large set of data revealed that each plateau on the $\mu_{eff}(T)$ curve for $[Cu(hfac)_2L^{n\text{-}Bu}]\cdot 0.5C_2H_3C_6H_5$ corresponds to a certain orientation of styrene molecules in the crystal denoted as A, B, or C in Figure 9.11a. This is clearly demonstrated by two projections of the structure of $[Cu(hfac)_2L^{n\text{-}Bu}]\cdot 0.5C_2H_3C_6H_5$ at 105, 180, 240, and 293 K corresponding to μ_{eff} above and below the anomaly regions.

The generalized orientations of solvate molecules A, B, and C in solid $[Cu(hfac)_2L^{n\text{-}Bu}]\cdot 0.5$solv served as a key for comparing and understanding the magnetochemical behavior of solvates with different included solvent molecules. In Figure 9.12, the dependence $\mu_{eff}(T)$ for $[Cu(hfac)_2L^{n\text{-}Bu}]\cdot 0.5C_2H_3C_6H_5$ is given as a reference for comparison with μ_{eff}(T) for $[Cu(hfac)_2L^{n\text{-}Bu}]\cdot 0.5C_2H_5C_6H_5$ and $[Cu(hfac)_2L^{n\text{-}Bu}]\cdot 0.5(1,3\text{-}(CH_3)_2C_6H_4)$. In Figure 9.12, these curves are shown in dark gray.

Let us compare the dependences $\mu_{eff}(T)$ for $[Cu(hfac)_2L^{n\text{-}Bu}]\cdot 0.5C_2H_3\text{-}C_6H_5$ and $[Cu(hfac)_2L^{n\text{-}Bu}]\cdot 0.5C_2H_5C_6H_5$. The data of Figure 9.12 (in Fig. 9.12, these curves are shown in green) show that the high (>260 K) and low temperature (<170 K) asymptotics of $\mu_{eff}(T)$ for these solvates virtually coincide. This means that the solvent molecules in these solvates are orientated according to type A at room temperature and type C at low temperatures. The same conclusion is reached if we compare the structural solutions for $[Cu(hfac)_2L^{n\text{-}Bu}]\cdot 0.5C_2H_3C_6H_5$ and $[Cu(hfac)_2L^{n\text{-}Bu}]\cdot 0.5C_2H_5C_6H_5$ at different temperatures (Fig. 9.11a, b). Thus the orientation of the aromatic solvent in solid $[Cu(hfac)_2L^{n\text{-}Bu}]\cdot 0.5C_2H_3C_6H_5$ and $[Cu(hfac)_2L^{n\text{-}Bu}]\cdot 0.5C_2H_5C_6H_5$ is identical. It is quite logical, therefore, that the high- (>260 K) and low-temperature (<170 K) asymptotics of $\mu_{eff}(T)$ for these solvates are close.

This approach allowed us to summarize all data obtained for solvates with aromatic solvents (ethylbenzene, propylbenzene, butylbenzene, styrene, *o*-xylene, *m*-xylene, *p*-xylene, 1,4-bis(trifluoromethyl)benzene,

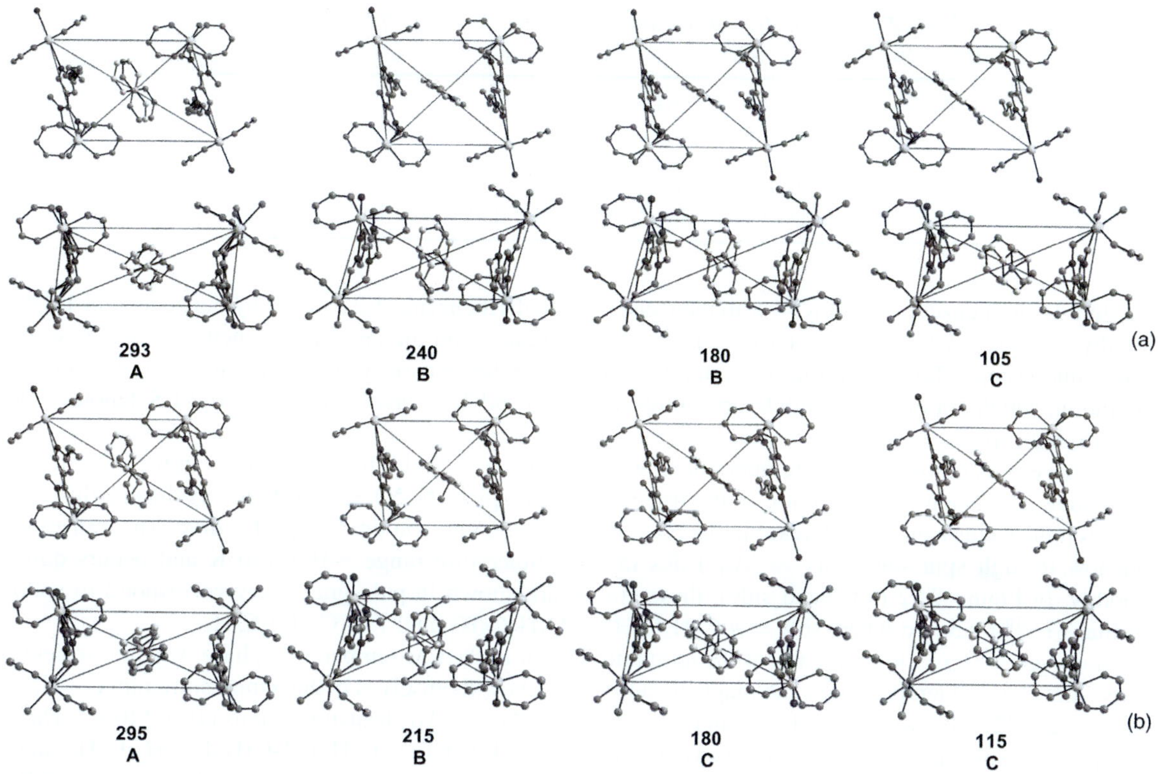

Figure 9.11 *Variation of the orientation of the styrene (a) and ethylbenzene (b) solvate molecules due to the variation of temperature. Reprinted with permission from [42]. Copyright 2011, American Chemical Society.*

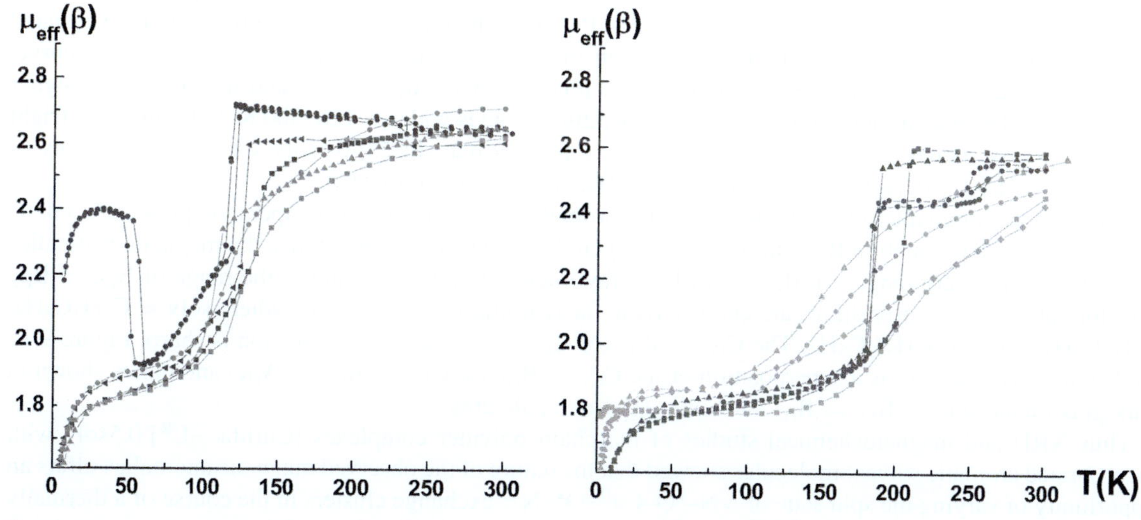

Figure 9.12 *Left: superposition of the sharp $C_{LS} \rightleftharpoons A_{HS}$ anomalies (dark) on the smooth $A_{LS} \rightleftharpoons A_{HS}$ anomalies (pale). Right: superposition of the sharp $C_{LS} \rightleftharpoons B_{HS}$ or $C_{LS} \rightleftharpoons B_{HS} \rightleftharpoons A_{HS}$ anomalies (dark) on the smooth $B_{LS} \rightleftharpoons B_{HS}$ anomalies (pale). Reprinted with permission from [42]. Copyright 2011, American Chemical Society.*

Table 9.2 *Types of transition, and the corresponding graphs in Figure 9.12.*

1) $A_{LS} \rightleftharpoons A_{HS}$ (left hand side, pale symbols)	A
2) $C_{LS} \rightleftharpoons A_{HS}$ (left, dark symbols)	
3) $B_{LS} \rightleftharpoons B_{HS}$ (right, pale symbols)	B
4) $C_{LS} \rightleftharpoons B_{HS}$ and/or $C_{LS} \rightleftharpoons B_{HS} \rightleftharpoons A_{HS}$ (right, dark symbols)	

1-methyl-4-ethylbenzene, 1-methyl-4-vinylbenzene, 1,4-diethylbenzene, 1,2,3-trimethylbenzene, and 1,2,4-trimethylbenzene). All of the structural transformations studied and the ensuing magnetic anomalies were divided into groups (Table 9.2). Entries 1 and 2 in Table 9.2 refer to case A of the structure with inter-chain orientation, which dominates at high temperatures; entries 3 and 4, to case B. The subscript LS denotes low spin; HS, high spin.

If we have orientation A at low temperatures, which is preserved at high temperatures, an increase in the temperature causes a gradual increase in the fraction of high spin >N–•O–Cu^{II}–O•–N< exchange clusters, which is reflected as a gradual increase in the effective magnetic moment on the μ_{eff}(T) curve. The transition from low to high spin state ($A_{LS} \rightleftharpoons A_{HS}$) lies in the temperature range ~100–140 K and occurs quite smoothly. In Figure 9.12 (left hand side), these curves are shown in pale gray. They correspond to solid [Cu(hfac)$_2$L$^{n\text{-Bu}}$]·0.5solv, where solv = n-$C_3H_7C_6H_5$, n-$C_4H_9C_6H_5$, and 1,4-$(C_2H_5)_2C_6H_4$.

If, however, we have orientation C of the aromatic solvent in the low temperature phase, which re-orients to type A at elevated temperatures, the magnetic anomalies show themselves in the temperature range ~115–135 K (Fig. 9.12, left), which also lies in in the range of the $A_{LS} \rightleftharpoons A_{HS}$ transformations (100–140 K). They correspond to solid [Cu(hfac)$_2$L$^{n\text{-Bu}}$]·0.5solv, where solv = 1,4-$(CF_3)_2$-C_6H_4), 1-CH_3-4-C_2H_3-C_6H_4, and 1-CH_3-4-C_2H_5-C_6H_4. The $C_{LS} \rightleftharpoons A_{HS}$ transition is quite distinct, however, in being more abrupt. Figure 9.12 vividly demonstrates the superposition of the $C_{LS} \rightleftharpoons A_{HS}$ anomalies shown in blue on the smooth anomalies corresponding to the $A_{LS} \rightleftharpoons A_{HS}$ transformations shown in pale grey.

As in the case of the $A_{LS} \rightleftharpoons A_{HS}$ transformation, if we have orientation B which is preserved at high temperatures, the fraction of high spin >N–•O–Cu^{II}–O•–N< exchange clusters gradually increases at elevated temperatures, as reflected by the gradual increase in the effective magnetic moment on the μ_{eff}(T) curve. The $B_{LS} \rightleftharpoons B_{HS}$ transition, however, is much more hindered than $A_{LS} \rightleftharpoons A_{HS}$ and thus occurs at higher temperatures. The transition is smooth and its temperature range is wider, ~120–300 K. In Figure 9.12 (right hand side), it is shown as pale curves corresponding to [Cu(hfac)$_2$L$^{n\text{-Bu}}$]·0.5solv solvates, where solv = 1,2-$(CH_3)_2C_6H_4$, 1,4-$(CH_3)_2C_6H_4$, 1,2,3-$(CH_3)_3C_6H_3$, and 1,2,4-$(CH_3)_3C_6H_3$.

Finally, if the initial inter-chain orientation of the aromatic solvent in the low temperature phase is C, but its molecules pass to orientation B (or initially B and then A) at elevated temperatures, the magnetic anomalies fall in the temperature range ~170–200 K (dark gray lines in Fig. 9.13), lying in the range of $B_{LS} \rightleftharpoons B_{HS}$ transformations. These transitions are characteristic for [Cu(hfac)$_2$L$^{n\text{-Bu}}$]·0.5solv, where solv = $C_2H_3C_6H_5$, $C_2H_5C_6H_5$, and 1,3-$(CH_3)_2C_6H_4$. The $C_{LS} \rightleftharpoons B_{HS}$ (or $C_{LS} \rightleftharpoons B_{HS} \rightleftharpoons A_{HS}$) transition is sharp. Figure 9.12 (right) demonstrates this as a superposition of the $C_{LS} \rightleftharpoons B_{HS}$ (or $C_{LS} \rightleftharpoons B_{HS} \rightleftharpoons A_{HS}$) anomalies shown in dark gray on the smooth $B_{LS} \rightleftharpoons B_{HS}$ anomalies shown in pale gray.

Thus XRD and magnetochemical studies of the chain polymer complexes [Cu(hfac)$_2$L^R]·0.5solv with incorporated aromatic solvent molecules showed that the nature of the observed magnetic anomalies offers an opportunity of varying the spin state of >N–•O–Cu^{II}–O•–N< exchange clusters in the course of a thermally induced structural rearrangement of Jahn–Teller coordination units. The form of the magnetic anomaly on the $\mu_{eff}(T)$ curve and the temperature of the effect depend on the type of orientation of solvate molecules in crystal and on changes in this orientation in the course of the structural transformation.

In addition to [Cu(hfac)$_2$], the other copper(II) acceptors shown below were investigated as components in breathing crystals. All of them form complexes with nitronyl nitroxides, including L^R. Spin transitions, however, were not recorded in any of these compounds.[66–69]

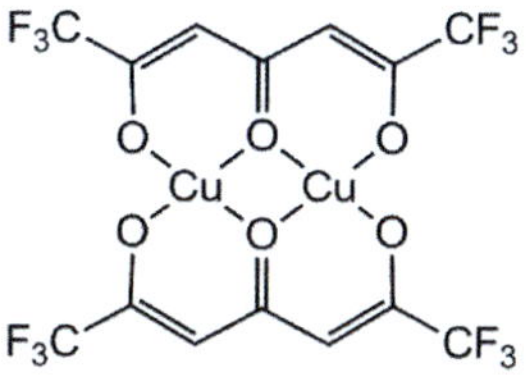

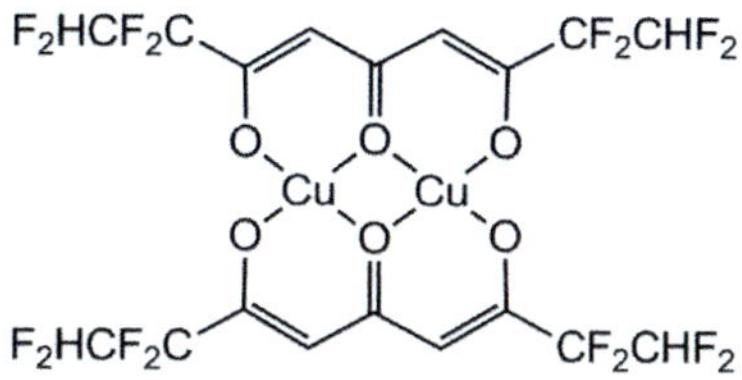

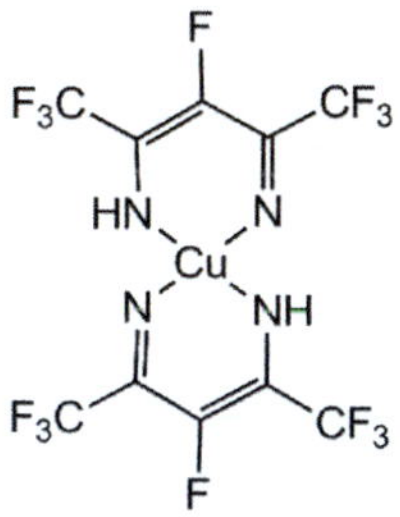

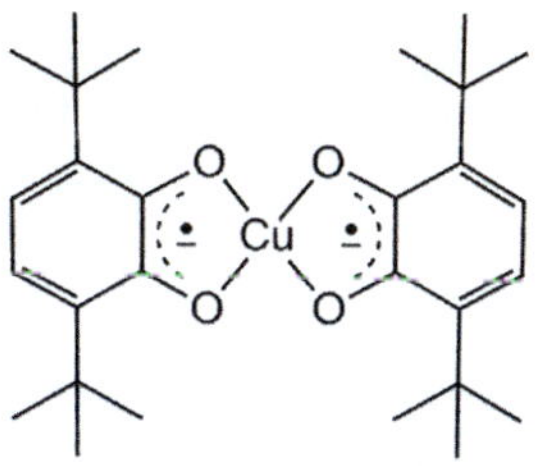

The reaction of [Cu(hfac)$_2$] with nitronyl nitroxide biradicals L–(CH$_2$)$_n$–L afforded a large group of heterospin complexes. The cross-linking of two spin-labeled fragments with –(CH$_2$)$_n$– polymethylene groups yielded layered-polymer and framework complexes with a complex spatial structure. The complexes did not exhibit spin-transitions either. The stereochemical nonrigidity effect in the solids showed itself as a deformation of –(CH$_2$)$_n$– polymethylene fragments, leaving coordination units with Cu^{2+}–O$^\bullet$–N< or >N–$^\bullet$O–Cu^{2+}–O$^\bullet$–N< exchange clusters intact.[70, 71]

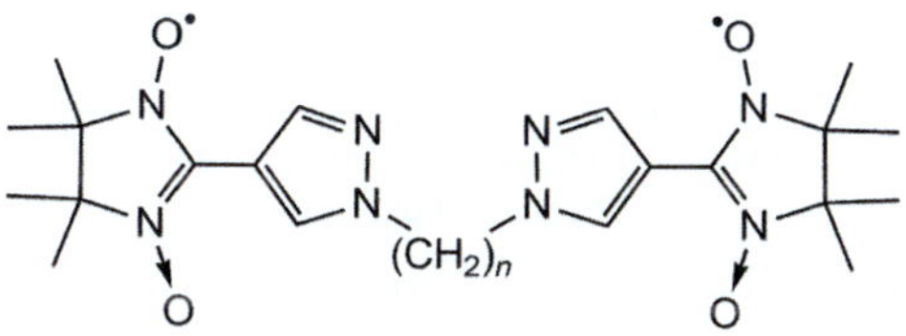

L-(CH$_2$)$_n$-L (*n* = 4, 6, 8, 10, 12)

9.5 EPR Study of Breathing Crystals

EPR studies of breathing crystals [Cu(hfac)$_2$L^R] reveal interesting and specific features, which allow us to characterize spin transitions and obtain qualitative information concerning exchange interaction in these systems.[21, 61, 72–83] The specific characteristics originate from predominant population of ground spin multiplet[72] and dynamic mixing (electron spin exchange) processes.[73] EPR of these compounds complements the magnetic susceptibility and X-ray techniques, and represents an efficient method for studying exchange interactions and coordination modes. At the same time, EPR has a much faster timescale, and does not require synthesis of single crystals. Below we consider general trends in EPR spectroscopy of strongly coupled spin-triads, and overview different approaches to studying spin-transitions and intra-cluster and inter-cluster exchange interactions,[75, 79, 81–83] as well as light-induced spin-transition in breathing crystals.[76, 80, 84]

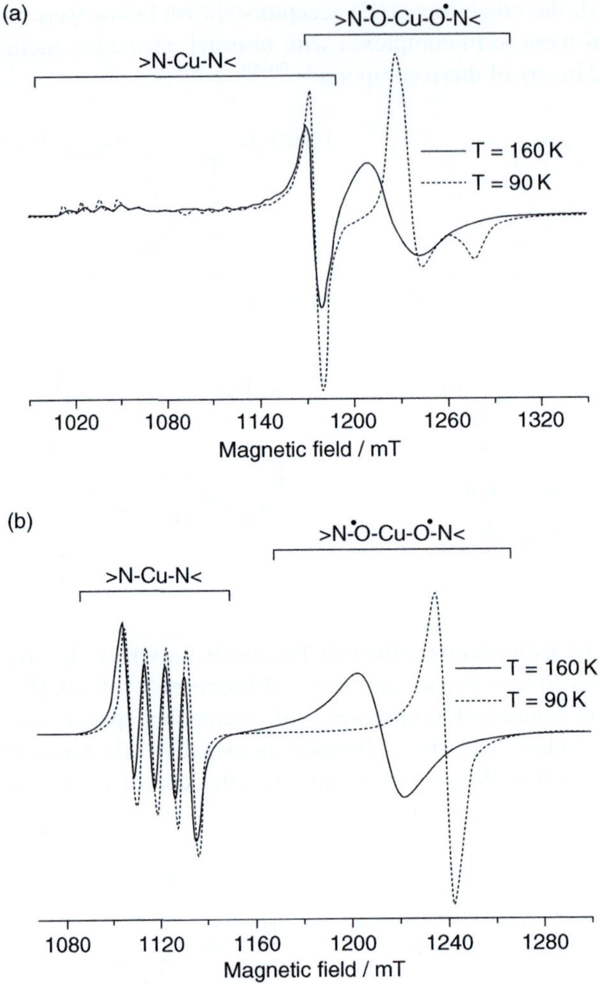

Figure 9.13 *EPR spectra of Cu(hfac)$_2$L^{Pr} at temperatures 90 and 160 K measured in Q-band ($\nu_{mw} \approx 34$ GHz): (a) polycrystalline powder and (b) single crystal (arbitrary orientation). Reprinted with permission from [83]. Copyright 2011, American Chemical Society.*

9.5.1 General Trends of EPR of Strongly Exchange-coupled Spin Triads

The theory of EPR of two- and three-spin exchange-coupled clusters has been well developed and comprehensive reviews of theoretical and experimental research have been reported in monographs.[86, 87] When the exchange interaction characterized by the exchange interaction constant J is weak or comparable in magnitude with the external magnetic field ($|J| \ll B$ or $|J| \approx B$), the EPR spectra exhibit splitting of spectral lines of individual paramagnetic centers and/or additional transitions whose positions depend on the magnitude of J. If the exchange interaction is much stronger than the external magnetic field ($|J| \gg B$), the EPR spectra of two-spin systems correspond to only a triplet state with the effective average g-factor $g = (g_1 + g_2)/2$, where g_1 and g_2 are the individual g-factors of the exchange-coupled spins. The temperature dependence of the intensity of this spectral line characterizes the relative populations of the triplet and singlet terms, and allows one to estimate the exchange constant and determine its sign. For three-spin systems, the EPR spectrum consists of three EPR lines corresponding to the three terms of the system (a quartet term and two doublet terms) if their populations are comparable with one another, or is a single line otherwise. EPR spectroscopy is usually inappropriate for the determination of the constants of the exchange interactions in the cluster if the exchange interaction is strong ($|J| \gg B$).[88–90]

The first experimental EPR study of breathing crystals revealed a strong temperature dependence of the spectra, reflecting the structural and magnetic changes in the copper clusters.[72] Figure 9.13 shows typical continuous wave EPR spectra of a single crystal and polycrystalline powder of [Cu(hfac)$_2$L$^{n\text{-Pr}}$] measured at *X*-band (9.72 GHz) at 90 and 160 K. The magnetic susceptibility behavior for this compound is shown in Figure 9.2c. Both powder and single crystal EPR spectra show signals from the isolated copper(II) ions, and the strongly temperature-dependent signals from the spin-triads. The spectrum at 160 K shows the usual pattern of Cu^{2+} ions in an octahedral environment ($g_{\perp} = 2.075$, $g_{\parallel} = 2.371$) superimposed onto a broad and weak line with a negative wing which is visible at $g = 2$. The former spectrum was attributed to the one-spin system >N–Cu^{2+}–N< (the CuO_4N_2 units), and the latter one to the three-spin cluster >N–O–Cu^{2+}–O–N< (the CuO_6 units). The most pronounced change at lower temperatures is the appearance of a new feature in the high-field region of the spectrum. Because no significant structural changes occur in the CuO_4N_2 units, this new signal was unambiguously assigned to the >N–$^{\bullet}$O–Cu^{II}–O$^{\bullet}$–N< three-spin system (in the CuO_6 unit), in the spin state $S = {}^1/_2$.

9.5.2 Predominant Population of the Ground Multiplet

To understand these EPR transformations, consider the spin Hamiltonian for a three-spin system, which can be written as

$$\hat{H} = \beta \mathbf{B} \mathbf{g}^R (\mathbf{S}^{R1} + \mathbf{S}^{R2}) + \beta \mathbf{B} \mathbf{g}^{Cu} \mathbf{S}^{Cu} - 2J(\mathbf{S}^{R1} + \mathbf{S}^{R2})\, \mathbf{S}^{Cu} \tag{9.1}$$

where *B* is the static magnetic field and β is the Bohr magneton. The superscripts R1 and R2 correspond to the spins of the two nitroxides, and superscript Cu corresponds to the spin of the copper. The nitroxides are assumed to be equivalent with isotropic values g^R and the same isotropic exchange coupling constant *J* with the copper ion. For simplicity the isotropic g^{Cu} value for the Cu^{2+} complex is assumed. The assumption of zero exchange between the two nitroxides is justified by the linear geometry of the $^{\bullet}$O–Cu^{2+}–O$^{\bullet}$ chain.[20, 86]

Figure 9.14 shows the energy level scheme for the case $J \gg B$ and the energies considering only terms linear on B/J. The eigenfunctions |1>–|4> describe the quartet $S = {}^3/_2$ state (multiplet C), the eigenfunctions |5>,

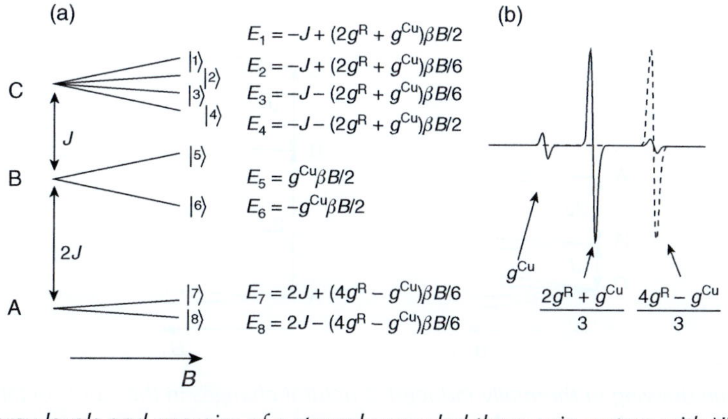

Figure 9.14 *(a) Energy levels and energies of a strongly coupled three-spin system with $|J| \gg B$, $J < 0$. (b) Calculated EPR spectra for $|J| \ll kT$ (solid line) and $J \gg kT$ (dashed line).*

Table 9.3 *EPR transitions to be observed from a >N–•O–Cu^{II}–O•–N< spin-triad when $J \gg B$.*

Transitions	Effective g-factor
$\|1\rangle \leftrightarrow \|2\rangle$, $\|2\rangle \leftrightarrow \|3\rangle$, $\|3\rangle \leftrightarrow \|4\rangle$	$g = (E_i - E_j)/\beta B = (2g^R + g^{Cu})/3$, $(i, j = 1, 2; 2, 3; 3, 4)$
$\|5> \leftrightarrow \|6>$	$g = (E_5 - E_6)/\beta B = g^{Cu}$
$\|7\rangle \leftrightarrow \|8\rangle$	$g = (E_7 - E_8)/\beta B = (4g^R - g^{Cu})/3$

$|6>$ and $|7>$, $|8>$ describe the two doublet $S = {}^1/_2$ states (multiplets A and B). For $|J| \gg B$ five transitions with three different frequencies will be observed (Fig. 9.14 and Table 9.3).[89, 96] Note that for breathing crystals the transition $|7\rangle \leftrightarrow |8\rangle$ will have $g < 2$, because $g^R \approx 2$ and $g^{Cu} > 2$.

At high temperatures where the thermal energy kT exceeds the value of the exchange interaction ($kT > |J|$), all of the spin states $S = {}^1/_2$ and $S = {}^3/_2$ are nearly equally populated, and the Boltzman factor $|J|/kT$ is small. The multiplets A, B, and C are all populated nearly equally and all resonance lines in the EPR spectra should be detected. At low temperature, $kT \gg |J|$, only the lowest doublet spin levels should be populated and consequently only the transition $|7\rangle \leftrightarrow |8\rangle$ with $g < 2$, should be observed. This prediction is nicely confirmed by experimental results. Despite g-factors of copper and nitroxides exceeding 2, at low temperatures the experimentally observed EPR line of spin-triad has $g < 2$ and corresponds to the ground-state multiplet, while other multiplets are not populated due to the high value of exchange interaction in spin-triad $J \sim 100$ cm^{-1}.

The main features of the EPR spectra of nitroxide–copper–nitroxide three-spin systems exchange-coupled with $|J| > kT$ were studied in[72, 74, 80] and are summarized below. As has been mentioned above, structural rearrangements in octahedral units CuO_6 at lower temperatures lead to a significant shortening or lengthening of the Cu–O distances[11] and, consequently, to the changes of the exchange-interaction magnitude (Fig. 9.15).

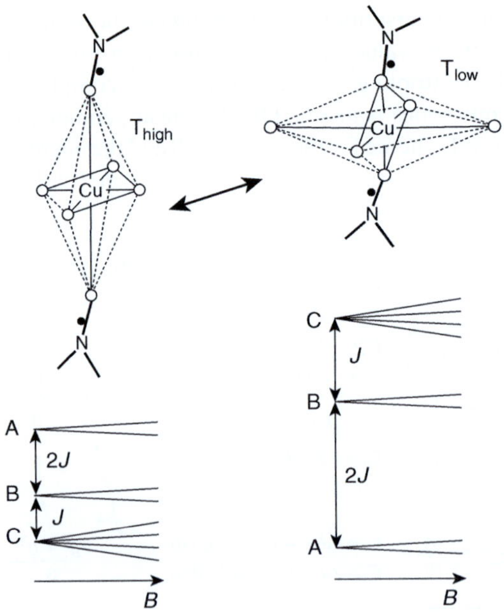

Figure 9.15 *Schematic drawing of thermally induced structural changes in the CuO_6 octahedron and theoretically expected energy levels of the spin-triad.*

As a result, the factor $|J|/kT$ changes significantly because of the simultaneous increase of $|J|$ and decrease of T, resulting in the predominant population of the ground state (multiplet A) ($S = {}^1\!/_2$). A strong signal may be observed at $g_{\text{eff}} < 2$ when the exchange coupling is antiferromagnetic (Fig. 9.14b, dashed line). The true g values can be obtained by taking into account the correct population of the energy levels in the simulation of the spectra. In cases where essentially only "$g < 2$" signals are observed (predominant population of multiplet A in accordance with the Boltzman factor $|J|/kT$), the exchange-coupled spin system has an effective $g < 2$, whereas all the individual components have $g > 2$. In the opposite case of a ferromagnetic exchange coupling, no signals at $g < 2$ would be observed. Thus, EPR can determine the sign of J for a three-spin system with $|J| > kT$. This is not the case for two-spin systems, or for the case when $|J| < kT$. When the "$g < 2$" signals are well separated from the rest of the spectrum (at high magnetic field), the absolute value of J can be obtained via the measurement of temperature dependence of intensity of this line. It should be noted that quantum chemical calculations of the g-factors of breathing crystals are in good agreement with experimental data and also suggest the presence of signals with $g < 2$ at low temperatures.[92]

9.5.3 Dynamic Spin Exchange Processes

At higher temperatures at $T > 100$ K the EPR spectra of breathing crystals (Fig. 9.13) consist of only one wide single line, instead of the expected spectra corresponding to the multiplets A, B, and C (Fig. 9.14b). The origin of this phenomenon was understood after a detailed experimental study of a series of breathing crystals using EPR in Q- and W-bands (35 and 94 GHz, respectively) at a wide range of temperatures.[73, 78] It was found that the linewidth and position of this line are strongly temperature dependent (Fig. 9.16). This effect was explained by spin (or frequency) exchange between different EPR lines using the following model.

Each spin-triad can be found in state *A*, *B*, or *C* (Fig. 9.15) with the corresponding probabilities specified by the Boltzmann distribution and effective g-factors. It was assumed that each triad continuously undergoes stochastic transitions between different multiplet states. As a result, the Larmor frequency of the electron spin randomly changes between the corresponding frequencies of the multiplets *A*, *B*, and *C*. The process is similar to fast changes in molecular geometry leading to changes in magnetic parameters, or the well known spin exchange between different spectral lines in EPR and NMR.[93] If the rate of transitions between different multiplets of the spin-triad is high compared to the Larmor frequency difference between different multiplets (fast exchange), a single EPR line at the "center of gravity" of the spectrum should be observed. Temperature change leads to changes in population of the EPR lines of different multiplets, and as a result shifts the position of the observed averaged EPR line. The dynamic spin exchange was taken into account in simulation of EPR spectra[78] which allows reproduction of the temperature dependence of EPR spectra of the spin-triads as is shown on Figure 9.16.

It was assumed that the mechanisms of the dynamic spin exchange are a modulation of exchange interactions between copper and nitroxide spins by lattice vibrations.[73] The transitions between doublets A and B can be induced by an isotropic exchange interaction, whereas the other two transitions (A↔C and B↔C) are only allowed if the exchange interaction is anisotropic (Fig. 9.16). Estimates show[73] that, for large values of $|J| \sim 100\ \text{cm}^{-1}$, the rates of these transitions can indeed be as high as 10^{10}–$10^{12}\ \text{s}^{-1}$. However, the exchange interactions in most breathing crystals significantly weaken to $J \approx 1$–$10\ \text{cm}^{-1}$ at high temperatures, as a result of a phase spin-transition (Fig. 9.8, Figure 9.15a, and Table 9.1). Thus in this case the modulation of exchange interaction by local vibrations of the crystal lattice could not explain the observed high mixing rates. At the same time, at high temperatures, it is reasonable to expect the dynamic Jahn–Teller effect in copper ions; which can lead to much higher rates of spin exchange.

To elucidate the possible mechanisms of the dynamic spin exchange, a further EPR study of spin dynamic processes in breathing crystals was recently performed using high field EPR.[78] The measurement of EPR

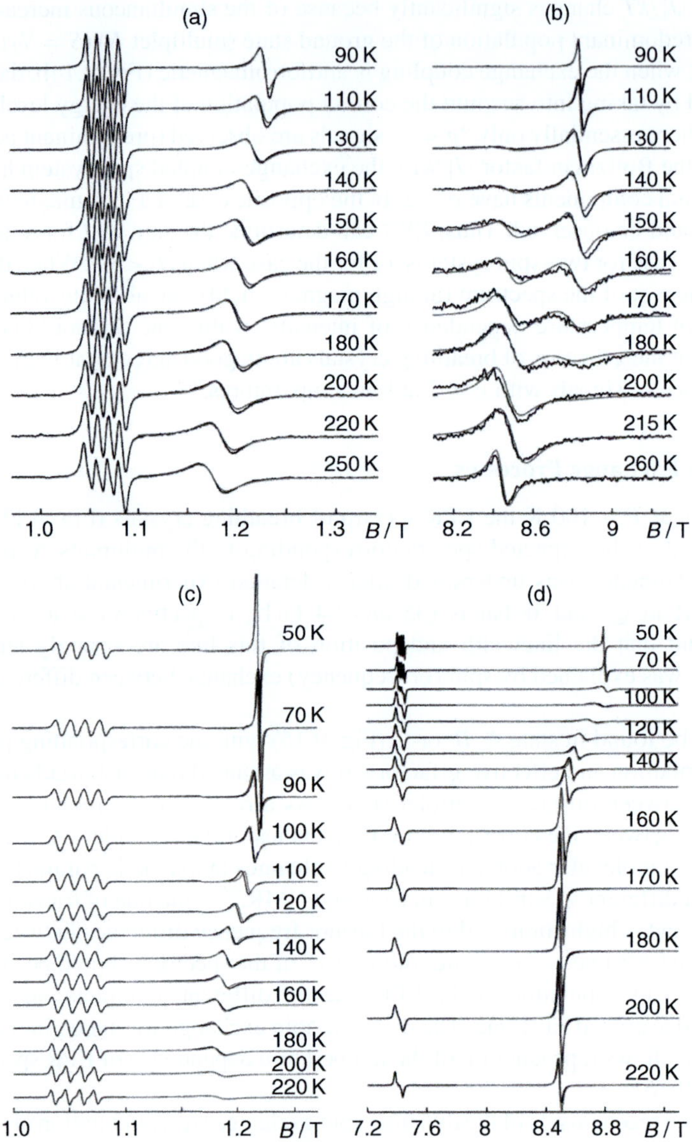

Figure 9.16 *Temperature-dependent EPR spectra of monocrystals of: [Cu(hfac)$_2$L$^{n\text{-}Pr}$] at (a) $\nu_{mw} = 33.97$ GHz and (b) $\nu_{mw} = 243.10$ GHz; and, of [Cu(hfac)$_2$L$^{n\text{-}Bu}$]·0.5C$_8$H$_{18}$ at (c) $\nu_{mw} = 33.97$ GHz and (d) $\nu_{mw} = 243.80$ GHz. Gray lines show the simulations. Reproduced from [78] with permission of The Royal Society of Chemistry.*

spectra at high different magnetic fields using different frequencies of the EPR spectrometer, allows changing of the mode of dynamic spin exchange, from fast to intermediate, or slow. This approach was used for detailed study of the dynamic mixing processes in strongly-coupled spin-triads of breathing crystals, [Cu(hfac)$_2$L^R], using high-field multifrequency EPR at 34, 122, and 244 GHz.[78] Using simulation of the multifrequency EPR data the characteristic rates of the exchange processes were estimated and it was shown that the typical range of exchange processes extend from 10^9 to 10^{12} s^{-1} and higher. Based on these values and the comparison of simulation of experiment data, it was concluded that the contribution of the dynamic Jahn–Teller effect in breathing crystals is substantial. Note that the most important consequence of the dynamic spin exchange study is the applicability of the method of measuring the temperature dependence of the exchange interaction[75, 83] by EPR, which is considered in detail in the next section of this review.

9.6 Classification of Spin-Transitions in Breathing Crystals and Correlations with Magnetic Susceptibility

The efficiency of EPR for the study of thermally induced spin-transitions and exchange interactions in spin-triads of nitroxide-copper(II)–nitroxide was demonstrated using numerous examples of compounds $[Cu(hfac)_2L^R]$ (R = Et, *n*-Pr, *n*-Bu) and $[Cu(hfac)_2L^{n\text{-}Bu}]\cdot 0.5solv$, with various organic solvents (solv) included in a crystal structure.[74] The manifestations of spin-transitions in both magnetic susceptibility and EPR data can be understood using a simple phenomenological model,[74] where the dependencies of the effective magnetic moment μ_{eff} and the effective g factor of a spin-triad g_{eff} are considered as functions of the parameter $|J(T)|/kT$.

The effective magnetic moment of a spin-triad can be calculated as:

$$\mu_{eff}^2(T) = \sum_{I=A,B,C} (g^I)^2 S^I(S^I+1) \cdot K_B^I \tag{9.2}$$

The factors K_B^I take account of the Boltzman populations of the corresponding multiplets (I = A, B or C; Figs 9.14 and 9.15):

$$K_B^I = \frac{(2S^I+1)\cdot e^{-E^I/kT}}{\sum_{P=A,B,C}(2S^P+1)\cdot e^{-E^P/kT}} \tag{9.3}$$

E^I are the energies of the corresponding multiplets. For the case $B < |J|$, Zeeman energy differences between $(2S+1)$ states of each multiplet can be neglected. Then, for antiferromagnetic exchange ($J < 0$) the temperature dependence of the effective magnetic moment of a spin-triad can be written using Eqs 9.2 and 9.3 as:

$$\mu_{tr,eff}^2(T) = \frac{3(g^A)^2 + 3(g^B)^2 \cdot e^{2J/kT} + 30(g^C)^2 \cdot e^{3J/kT}}{4(1+e^{2J/kT}+2e^{3J/kT})} \tag{9.4}$$

where g_I and S_I are the average (isotropic) g value and the total spin of the multiplets A, B, and C.

Due to the presence of two types of paramagnetic centers in breathing crystals, both EPR and magnetic susceptibility data are described as the superposition of two contributions. In magnetic susceptibility measurements, an integral susceptibility is obtained, and then an effective magnetic moment corresponding to the single copper center is calculated. Thus, the experimentally measured value follows:

$$\mu_{eff}^2 = 0.5\mu_{tr,eff}^2 + 0.5\mu_{is,eff}^2 = \frac{3(g^A)^2 + 3(g^B)^2 \cdot e^{2J/kT} + 30(g^C)^2 \cdot e^{3J/kT}}{8(1+e^{2J/kT}+2e^{3J/kT})} + 0.5\mu_{is,eff}^2 \tag{9.5}$$

where subscripts "tr" and "is" correspond to the spin-triad and the isolated copper ion, respectively. Figure 9.17a shows the calculated dependence of μ_{eff} versus $|J|/kT$ for this case.

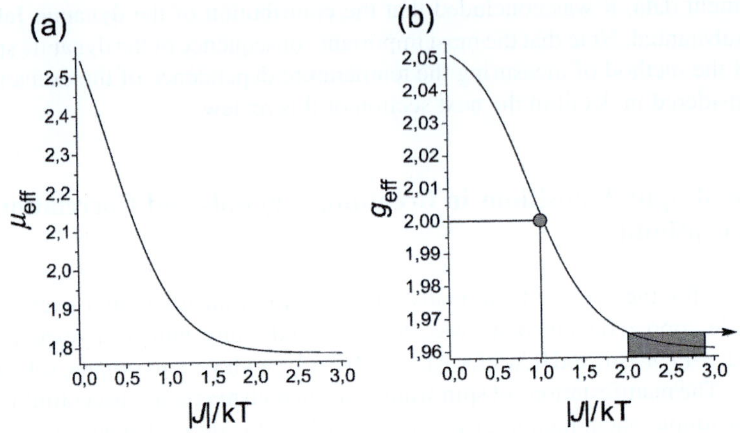

Figure 9.17 *(a) The dependence μ_{eff} vs. $|J|/kT$ ($J < 0$) calculated for [Cu(hfac)$_2$L^R] using Eq. (9.7) and assuming $\mu_{is,eff} \approx 1.86$. (b) The dependence g_{eff} vs. $|J|/kT$ ($J < 0$) calculated for [Cu(hfac)$_2$L^R] using Eq. (9.5). The typical average g-values $g^A = 1.96$, $g^B = 2.15$ and $g^C = 2.05$ are used. Reprinted with permission from [74]. Copyright 2007, American Chemical Society.*

In the case of fast spin exchange the expression for $g_{eff}(T)$, determining "the center of gravity" of the EPR spectrum, is as follows:

$$g_{eff}(T) = \sum_{\mathrm{I=A,B,C}} g^{\mathrm{I}} P_B^{\mathrm{I}} \Big/ \sum_{\mathrm{I=A,B,C}} P_B^{\mathrm{I}} = \frac{g^{\mathrm{A}} P_B^{\mathrm{A}} + g^{\mathrm{B}} P_B^{\mathrm{B}} + g^{\mathrm{C}} P_B^{\mathrm{C}}}{P_B^{\mathrm{A}} + P_B^{\mathrm{B}} + P_B^{\mathrm{C}}}, \tag{9.6}$$

where the coefficients P_B^{I} describe the probability of absorption of microwave energy by an electron in the multiplets A, B, or C, respectively and include the Boltzmann populations of the levels and the number of EPR transitions for each multiplet and probability of EPR transitions.

$$P_B^{\mathrm{I}} = e^{-E^{\mathrm{I}}/kT} \sum_i p_i^{\mathrm{I}} \tag{9.7}$$

where p_i^{I} is the intensity of EPR transitions in multiplet I. Taking into account the intensities of the corresponding transitions, which are proportional to the square of the matrix elements $m_{ij}^2 = |\langle i|\hat{S}_x|j\rangle|^2$ and are equal to: $m_{12}^2 = m_{34}^2 = {}^3/_4$, $m_{23}^2 = 1$ and $m_{56}^2 = m_{78}^2 = {}^1/_4$, the following expression for effective g-factor was obtained:

$$g_{\mathrm{eff}}(T) = \frac{g^{\mathrm{A}} + g^{\mathrm{B}} \cdot e^{2J/kT} + 10 g^{\mathrm{C}} \cdot e^{3J/kT}}{1 + e^{2J/kT} + 10 e^{3J/kT}} \tag{9.8}$$

At high temperature, $|J|/kT \approx 0$ when all multiplets are equally populated and $g_{\mathrm{eff}} = g^{\mathrm{C}}$, as $g^{\mathrm{A}} + g^{\mathrm{B}} = 2g^{\mathrm{C}}$ in accordance with expression for g-factors of different multiplets. At low temperatures ($|J|/kT \gg 1$, $J < 0$) when only the lowest multiplet is populated $g_{\mathrm{eff}} = g^{\mathrm{A}}$. Thus for the typical isotropic values $g^{\mathrm{A}} = 1.96$ and $g^{\mathrm{C}} = 2.05$, the changes of effective g-factor of the spin-triad are $\Delta g_{\mathrm{eff}} \sim 0.1$. This value could be easily detected using EPR in Q-band or W-band.

The effective g value reflects the dependence $J(T)$ and may change abruptly or smoothly depending on this function. It is useful to note that the relation $g_{eff} \approx g^R \approx 2$ holds for $|J| \approx kT$, $J < 0$, which can be derived from Eq. 9.8. Thus, the temperature at which the EPR line of a spin-triad crosses the value $g \approx 2$ can be used for the simple estimation of the exchange interaction value (at this certain temperature). Another useful prediction is that when the parameter $|J|/kT$ ($J < 0$) reaches the value ≈ 2, the effective g factor g_{eff} almost reaches its minimum value g^A: about 90% of the transformation $g_{eff} = g^C \rightarrow g_{eff} = g^A$ is complete, which can also be used for the estimations of J. These two estimations are shown on Figure 9.17b, which shows the dependence of g_{eff} on the parameter $|J|/kT$ calculated using Eq. 9.8 for the typical average isotropic g values $g^A = 1.96$, $g^B = 2.15$, and $g^C = 2.05$. The shape of the curve for $g_{eff} = |J|/kT$ (Fig. 9.17b) is somewhat different compared to that of μ_{eff} vs. ($|J|/kT$) calculated using Eq. 9.5 (Fig. 9.17a), which also is an important point for the interpretation of experimental results.

If the exchange interaction is a function of temperature $J = J(T)$, which takes place for the compounds Cu–$(hfac)_2L^R$, the factor $|J|/kT$ may change significantly within a narrow temperature range, especially if J changes abruptly during structural rearrangements. An additional factor influencing the amplitude of change of magnetic susceptibility is the significant difference in effective g values of multiplets A, B, and C for spin-triads of nitroxide-copper(II)–nitroxide. Reasonable values $g_{Cu} = 2.15$ (isotropic value) and $g^R = 2.007$ lead to $g^A = 1.96$, $g^B = 2.15$, and $g^C = 2.05$, and thus $\mu_{eff} = 3.077$ β for $|J|/kT \sim 0$ and $\mu_{tr,eff} = 1.697$ β for $|J|/kT = 1$, $J < 0$.[74]

If $|J|/kT < 1$, all of the total spin states of a triad (two doublets and one quartet) are almost equally populated, but if $|J|/kT > 1$, the ground state becomes predominately populated. Thus, the passage between situations $|J|/kT < 1$ and $|J|/kT > 1$ leads to the changes in the effective magnetic moment value and in the EPR spectra. Very informative EPR spectra are observed if the transition leads to a strong increase in $|J|$ and brings the spin system into condition $|J| > kT$. In this case, it is straightforward to determine the sign of J and to estimate its magnitude $|J|$ by monitoring the temperature dependent position of the EPR line of a triad. As already mentioned, an intense signal with $g_{eff} < 2$ denotes the existence of the strong antiferromagnetic exchange interaction $|J| > kT$. The absence of this signal indentifies that the exchange interaction is ferromagnetic or weakly antiferromagnetic ($|J| < kT$). EPR can be employed as a useful technique for the studies of strongly coupled spin-triads, especially when the exchange interaction in a triad is antiferromagnetic. Thus, EPR provides more accurate information than magnetic susceptibility on the dependence $J(T)$ due to the direct determination of g tensors and due to the possibility of the selective study of the exchange-interaction within a spin-triad.

The different character of temperature-dependent structural rearrangements in [$Cu(hfac)_2L^R$] results in different dependencies $J(T)$ and different manifestations of spin-transitions. Spin-transitions can be characterized using three characteristics: the abruptness of the transition (ΔT_c, the typical temperature width); the characteristic mean temperature where it occurs (T_c); and its amplitude ($\Delta\mu_{eff}$ in magnetic susceptibility or Δg_{eff} in EPR). The high temperature transition implies a larger magnitude of J compared to the low temperature transition of the same abruptness and amplitude. This is obvious because smaller changes of J are required at low temperatures to vary $|J|/kT$ to the same extent as at high temperatures (Eqs 9.5 and 9.8). The abruptness of the spin-transition characterizes the contribution of the $J(T)$ dependence into the spin dynamics of a triad. A very abrupt transition implies a jump in J, whereas a very gradual one implies a slow evolution with temperature or a nearly constant J. Note that the amplitude of the spin transition is directly linked to the scale of the exchange-interaction and its change, as is clear from Eqs 9.5 and 9.8.[74]

This simple phenomenological model and the approach described above were successfully used for the analysis of the different types of spin-transitions in breathing crystals.[74] Let's consider three examples with different characters of spin-transitions. Compound [$Cu(hfac)_2L^{Bu}$]·$0.5C_7H_{16}$ undergoes an abrupt spin-transition at $T_c \approx 125$ K, which causes the effective magnetic moment to decrease by $\Delta\mu_{eff} \approx 0.35$ β within the temperature range $\Delta T_c \approx 10$ K (see Fig. 9.18). The overall decrease in the magnetic moment between

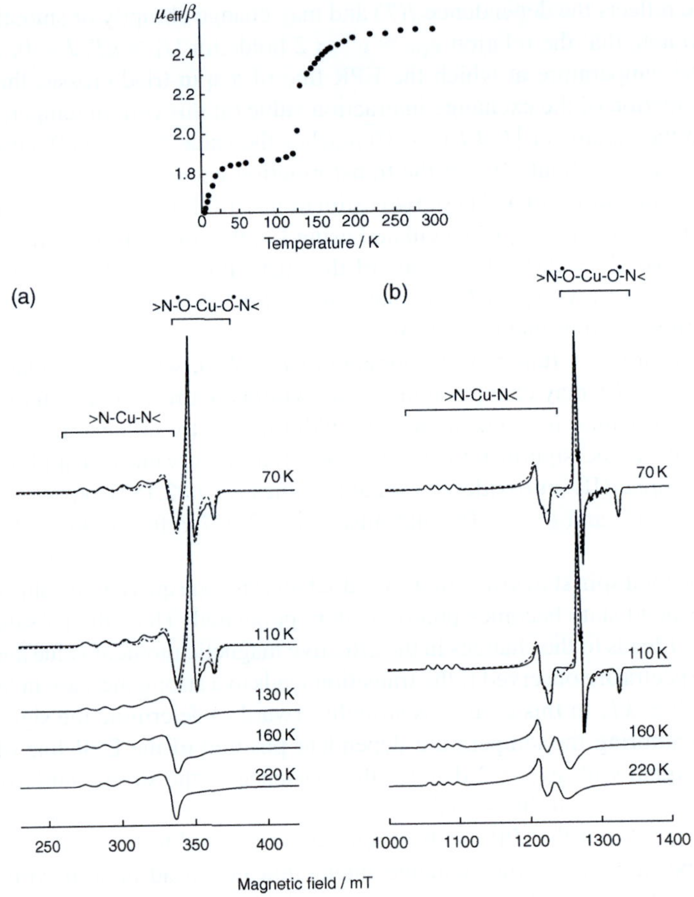

Figure 9.18 *CW EPR spectra at different T (a,b) and dependence $\mu_{eff}(T)$ (inset) of $[Cu(hfac)_2L^{Bu}]\cdot 0.5C_7H_{16}$. (a) X-band ($\nu_{mw} = 9.714$ GHz, $B = 347$ mT at g_{eff}) 2). (b) Q-band ($\nu_{mw} = 35.395$ GHz, $B = 1264$ mT at $g_{eff} = 2$). The temperatures are indicated on the right; the curves are scaled. Calculation parameters for $T = 110$ K: $g^{Cu} = [2.048, 2.078, 2.314]$, $g^R = 2.007$, $J = -200$ cm^{-1}; $g^{isolated\ Cu} = [2.060, 2.082, 2.346]$. Calculation parameters for $T = 70$ K: $g^{Cu} = [2.054, 2.084, 2.320]$, $g^R = 2.007$, $J = -200$ cm^{-1}; $g^{isolated\ Cu} = [2.060, 2.082, 2.346]$. Reprinted with permission from [74]. Copyright 2007, American Chemical Society.*

$T = 300$ and $T = 50$ K is $\Delta\mu_{eff} \approx 2.53 - 1.85 = 0.68\ \beta$. The magnetic behavior at temperatures below ~25 K is governed by inter-cluster exchange-interactions between different paramagnetic centers of the polymeric chains. Using Figure 9.17, one can establish a simple correlation between the manifestation of spin-transitions in the magnetic susceptibility and EPR measurements for $[Cu(hfac)_2L^{Bu}]\cdot 0.5C_7H_{16}$. On the one hand, a decrease in the magnetic moment by 0.7 β to $\mu_{eff} \approx 1.85$ corresponds to the transition of the system from the state $|J|/kT \ll 1$ to the state $|J|/kT > 1$ (see. Fig. 9.18a). On the other hand, at $|J|/kT > 1$ one should expect $g_{eff} < 2$ and, therefore, intense EPR signals should be observed in a high field (see Fig. 9.18b).

$Cu(hfac)_2L^{Bu}\cdot 0.5C_8H_{18}$ (octane) represents an intermediate case between abrupt and gradual spin transitions (Fig. 9.19). The dependence of the effective magnetic moment displays a nearly linear decrease by $\Delta\mu_{eff} = 0.6\ \beta$ within $T = 100$–170 K, and the overall decrease between $T = 60$–240 K is $\Delta\mu_{eff} = 0.74\ \beta$.

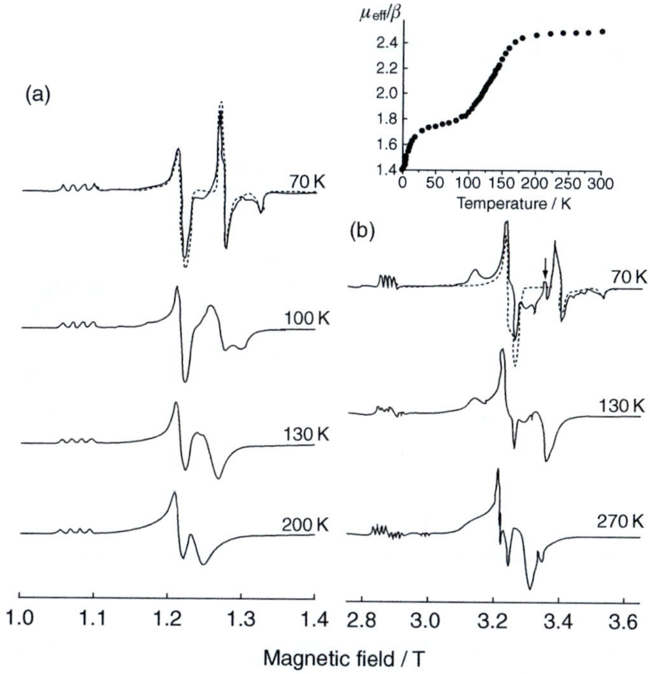

Figure 9.19 *CW EPR spectra at different T (a,b) and dependence $\mu_{eff}(T)$ (inset) of [Cu(hfac)$_2$L^{Bu}]·0.5C$_8$H$_{18}$. (a) Q-band (ν_{mw} =35.320 GHz, B = 1.262 T at $g_{eff} = 2$). (b) W-band (ν_{mw} = 94.200 GHz, B = 3.365 T at $g_{eff} = 2$). The temperatures are indicated on the right, the curves are scaled, and the arrow on the top trace (b) indicates the position of the signal from nitroxide defects. Calculation parameters for T at 70 K: g^{Cu} = [2.058, 2.090, 2.304], $g^R = 2.007$, $J = -115$ cm^{-1}; $g^{\text{isolated Cu}}$ = [2.062, 2.078, 2.343]. Reprinted with permission from [74]. Copyright 2007, American Chemical Society.*

The value $\Delta\mu_{eff} < 1.81\ \beta$ at $T < 90$ K unambiguously implies that the condition $|J| > kT$ is achieved. The transformations of the *Q*-band EPR spectra with temperature are fully consistent with this dependence. The gradual transformation of the single line into a line with resolved *g* anisotropy is observed upon lowering the temperature between 200 and 70 K, which is accompanied by a gradual shift of the line position toward lower *g* values (higher fields). The position of the EPR line of a spin-triad at $T \sim 130$ K is close to 2, allowing for the estimation $|J| \approx kT \approx 90$ cm^{-1}. The *Q*- and *W*-band EPR spectra at 70 K can be simulated using g_{Cu} = [2.058, 2.090, 2.304], and thus $g_{eff} = g^A$ = [1.990, 1.979, 1.908]. Because the major transformation has already occurred at $T \sim 70$ K, one can estimate $|J| > 2kT \approx 100$ cm^{-1} at $T < 70$ K.

The last example is the complex [Cu(hfac)$_2$L^{Et}] ($T_c \approx 226$ K, $\Delta T_c \approx 5$ K, $\Delta\mu_{eff} \approx 0.15\ \beta$) which shows a completely different dependence $\mu_{eff}(T)$ than the two previous cases (Fig. 9.20). Upon lowering the temperature in the range ~226–300 K, the magnetic moment decreases, manifesting a strong antiferromagnetic coupling within a spin-triad; however, the spin transition leads to an increase of μ_{eff}. This can be explained by an abrupt decrease in the antiferromagnetic coupling or even a change of its sign during the spin-transition, which is consistent with the abrupt lengthening of the Cu–O distances below 226 K.[38, 43] This kind of transition should lead to an increase in g_{eff}, in contrast to previous cases. As one can see from Figure 9.20 where *W*-band EPR spectra at 240 and 200 K are shown, exactly this is observed in experiments. The EPR lines of the isolated copper ion and the spin-triad are not resolved, but the disappearance of the broad high-field part

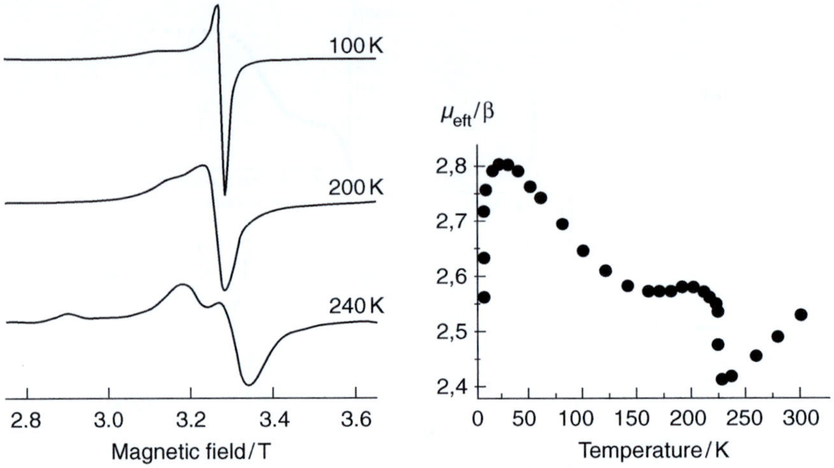

Figure 9.20 *W-band CW EPR spectra at different T and dependence μ_{eff} (T) (inset) of [Cu(hfac)$_2$L^{Et}] (ν_{mw} = 94.300 GHz, B = 3.369 T at g_{eff} = 2). The temperatures are indicated on the right; the curves are scaled. Reprinted with permission from [74]. Copyright 2007, American Chemical Society.*

of this overlapped line at 200 K is obvious. Thus, the EPR study of [Cu(hfac)$_2$L^{Et}] nicely illustrates that this model works for transitions when $|J|/kT$ also decreases. For all cases of the [Cu(hfac)$_2$L^R] family, the EPR behavior agrees completely with the magnetic susceptibility data, and the use of simple rules for the positions of the EPR signals allows for estimations of J and characterization of spin-transitions.[74]

9.7 The Detailed Magnetic Structure of Breathing Crystals

9.7.1 EPR Measurements of Temperature Dependence of Intra-cluster Exchange Interaction

In the overwhelming majority of the exchange-coupled clusters investigated in the field of molecular magnetism, the exchange interaction is constant with temperature. Very unusual and highly important characteristics were observed in [Cu(hfac)$_2$L^R] breathing crystals by means of *W*-band EPR (95 GHz).[61, 83] In [Cu(hfac)$_2$L^R] breathing crystals, the exchange interaction is strongly temperature dependent as a consequence of the structural rearrangements in spin-triads. Unambiguous experimental evidence for the explicit dependence $J(T)$ was obtained for the first time by using the high-field EPR on single crystals of [Cu(hfac)$_2$L^{Bu}]·0.5C$_8$H$_{18}$. The $\mu_{eff}(T)$ dependence cannot be satisfactorily described in a model of constant exchange interaction using Eq. (9.5), which is clearly demonstrated in Figure 9.21a by attempts at achieving agreement separately in the high and low temperature regions of the curve. However, this discrepancy alone cannot prove the temperature dependence of the exchange interaction in the spin-triad since the compound contains different paramagnetic centers contributing to $\mu_{eff}(T)$, meaning alternative explanations are possible. The signals from the spin-triads and one-spin units in *W*-band EPR spectra of breathing crystals are resolved which allows definite conclusions to be drawn.

Figure 9.21b shows the experimental dependence $g_{eff}(T)$ of a single crystal of [Cu(hfac)$_2$L$^{n\text{-}Bu}$]·0.5C$_8$H$_{18}$ that was obtained from the temperature dependent *W*-band spectra. Similar to the temperature dependence of $\mu_{eff}(T)$ in Figure 9.21a, it is obvious that $g_{eff}(T)$ cannot be simulated in a model of constant exchange interaction: the experimental slope is much steeper than the theoretical ones in the high or low temperature

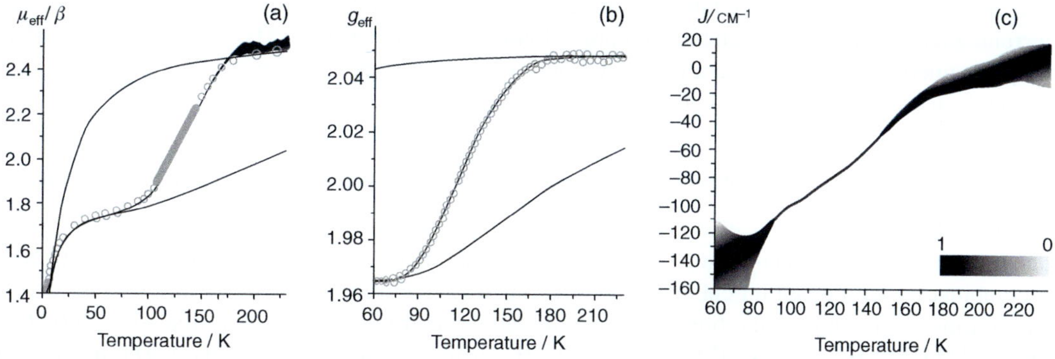

Figure 9.21 *Experimental (gray circles) $\mu_{eff}(T)$ (a) and $g_{eff}(T)$ (b) dependence of a single crystal of $[Cu(hfac)_2L^{n\text{-}Bu}]\cdot 0.5C_8H_{18}$ and related calculations (black) using J(T) dependence. Solid lines show the result obtained for a constant $J = -10$ cm^{-1} (dark line) and -120 cm^{-1} (pale line). (c) Temperature dependence of exchange interaction J(T) of $[Cu(hfac)_2L^{n\text{-}Bu}]\cdot 0.5C_8H_{18}$ calculated using the experimental function $g_{eff}(T)$ and Eq. (9.8). The normalized probability of the exchange interaction value J in the range 1 to 0 is shown. The error of the experimental determination of the g-factor (0.0005) was taken into account during the fitting procedure and determines the width of the calculated J(T) distribution at intermediate temperatures. Reprinted with permission from [61]. Copyright 2008, American Chemical Society.*

regions. Numerical fitting of the dependence of $g_{eff}(T)$ using Eq. (9.8) was used to obtain the distribution of functions $J(T)$ shown in Figure 9.21c. The width of the distribution characterizes the ambiguity of $J(T)$ determination. The broad distributions at high and low temperature regions correspond to the plateaus of the dependence $g_{eff}(T)$, where the EPR spectrum does not depend on temperature anymore. However, in the middle region of temperatures, the precision of $J(T)$ measurement is high and, overall, it is remarkable that the exchange interaction is changed by about an order of magnitude during the reversible structural rearrangements.

Further development of EPR study of the dependences of $g_{eff}(T)$ and $J(T)$ for several compounds and various experimental settings (e.g. microwave frequency band, single crystal vs. powder sample etc.) was done by Veber *et al.*[83] using two compounds with gradual spin-transitions, $[Cu(hfac)_2L^{n\text{-}Pr}]$ and $[Cu(hfac)_2L^{n\text{-}Bu}]\cdot 0.5C_8H_{18}$. Two methodologies of $J(T)$ measurement were used, using polycrystalline and single crystal samples. Although the $g_{eff}(T)$ function can in principle be measured in powders, the best results were obtained using single crystals properly oriented in the magnetic field. The origin of the gradual $g_{eff}(T)$ and $J(T)$ dependence in the two studied complexes was explained using three models: gradual structural changes in the CuO_6 octahedral units; vibrational averaging due to the dynamic Jahn–Teller effect; and, static spin-triads coupled by strong inter-cluster exchange. On the basis of the available EPR, X-ray, and magnetometry data it is not possible to distinguish which model is most valid to describe the phase transitions in breathing crystals on a molecular level. The interpretation of the observed exchange-interaction is different in these models, whether it is an exact or effective quantity. However, in any case the developed approaches for studying and interpreting $g_{eff}(T)$ and $J(T)$ dependences are useful and instructive for the characterization of the magnetic properties in the family of breathing crystals $[Cu(hfac)_2L^R]$.

The approach considered above was used recently for the X-band EPR study of series of *bis*(o-semiquinonato)copper(II) complexes with different ligands.[79] EPR data reveal that exchange-interactions in spin-triads of bis(o-semiquinonato)copper(II) complexes are very sensitive to the structure of the ligand(s) coordinated to the central copper(II) ion. The ligands moderate the copper–radical and radical–radical exchange interactions and strongly change the magnetic properties of the compound. Depending on the ligands, ferromagnetic or antiferromagnetic exchange dominates in the system. The EPR results complement previously obtained spectroscopic data on these compounds and suggest the pronounced plasticity of the clusters manifested in the broad distributions of their rhombicity parameters.

9.7.2 EPR Measurement of Dipole–Dipole Interaction and Inter-cluster Exchange Interaction

Application of EPR to breathing crystals allows one to obtain information about not only intra-cluster exchange interaction, but on inter-cluster exchange and dipole–dipole interaction as well. The application of EPR in combination with quantum chemical calculations was used recently to study inter-cluster exchange pathways in four compounds of the breathing crystals family, $[Cu(hfac)_2L^{n\text{-}Pr}]$, $[Cu(hfac)_2L^{n\text{-}Bu}]{\cdot}0.5C_7H_{16}$, $[Cu(hfac)_2L^{n\text{-}Bu}]{\cdot}0.5C_8H_{18}$, and $[Cu(hfac)_2L^{Bu}]{\cdot}0.5C_8H_{10}$ using *Q*-band EPR (34 GHz). The magnetic susceptibility data (Fig. 9.22, left column) show a different character of magnetic anomalies for each compound.[20, 42–44] Typically, at $T < 50$ K, all spin-triads are already in the strongly coupled state $S = {}^1\!/_2$ ($\mu_{eff} \approx 1.8\ \beta$), and the magnetic behavior at lower temperatures is governed by inter-cluster exchange interactions. Two of the compounds, $[Cu(hfac)_2L^{n\text{-}Bu}]{\cdot}0.5C_7H_{16}$ and $[Cu(hfac)_2L^{n\text{-}Bu}]{\cdot}0.5C_8H_{18}$ (Fig. 9.22a,b left column), demonstrate pronounced decreases in their magnetic susceptibility starting at $T < 50$ K, which implies relatively strong inter-cluster exchange interaction of about 10 cm^{-1}. The other two, $[Cu(hfac)_2L^{n\text{-}Pr}]$ and $[Cu(hfac)_2L^{n\text{-}Bu}]{\cdot}0.5C_8H_{10}$ (Fig. 9.22c,d left column), display nearly constant values of μ_{eff} at T from 50 to 4 K, with a decrease only at $T < 4$ K, which indicates relatively weak inter-cluster exchange on the order of a few cm^{-1}. However, the origin of this inter-cluster exchange interaction was not clear. The decrease of μ_{eff} at low temperatures can be due to antiferromagnetic coupling between neighboring centers of the same polymer chain. Alternatively, it might originate from copper(II)–copper(II) or triad–triad interactions between the polymer chains (or a combination of both). Magnetic susceptibility data do not allow one to distinguish between these possible exchange pathways, and exact quantum chemical computation of this complicated system is at the moment not possible.

As was shown above at low temperatures, the EPR signal of spin-triads in the strongly coupled state (with $g < 2$) and the signal of one-spin copper(II) ions (with $g > 2$) are spectrally well separated (Fig. 9.13). It is known that the double integral (I) over the EPR spectrum of paramagnetic species is proportional to its magnetic susceptibility.[94, 95] The temperature dependence of γ – the ratio of the second integral over the spectrum of spin-triads to the second integral over the spectrum of one-spin copper(II) ions – for all four compounds is shown on Figure 9.22. It is evident that the shapes of the $\gamma(T)$ curves essentially reproduce the corresponding shapes of $\mu_{eff}(T)$ dependencies for each compound. The decrease of $\gamma(T)$ at $T < 50$ K is observed for breathing crystals $[[Cu(hfac)_2L^{n\text{-}Bu}]{\cdot}0.5C_7H_{16}$ and $[Cu(hfac)_2L^{n\text{-}Bu}]{\cdot}0.5C_8H_{18}$ (Fig. 9.22 a,b left column), meaning that the magnetic susceptibility of spin-triads decreases. This is exact evidence that the inter-cluster antiferromagnetic exchange responsible for the low temperature magnetic behavior of the whole compound occurs between spin-triads. In contrast, for breathing crystals $[Cu(hfac)_2L^{n\text{-}Pr}]$ and $[Cu(hfac)_2L^{n\text{-}Bu}]{\cdot}0.5C_8H_{10}$ (Fig. 9.22 c,d), no decrease of $\gamma(T)$ is observed down to 4 K, which is in perfect agreement with the dependence of $\mu_{eff}(T)$, where the magnetic moment decreases at $T < 4$ K. Thus, in two different cases of relatively strong and relatively weak inter-cluster exchange, the shapes of the $\mu_{eff}(T)$ and $\gamma(T)$ dependencies agree in the low temperature region $4 < T < 50$ K, meaning that all observed effects of inter-cluster antiferromagnetic exchange should be attributed to the coupling between spin-triads.

Quantum chemical calculations nicely supported the experimental conclusions (Fig. 9.22, right column). A simple model accounting for the exchange integrals between selected pairs of radicals only were calculated for the X-ray diffraction crystal structure of the above-mentioned compounds.[81] Calculated inter-cluster exchange interactions between nitronyl nitroxides for breathing crystals $[Cu(hfac)_2L^{n\text{-}Pr}]$ and $[Cu(hfac)_2L^{n\text{-}Bu}]{\cdot}0.5C_8H_{10}$ were a factor of 10 weaker than those for $[Cu(hfac)_2L^{n\text{-}Bu}]{\cdot}0.5C_7H_{16}$ and

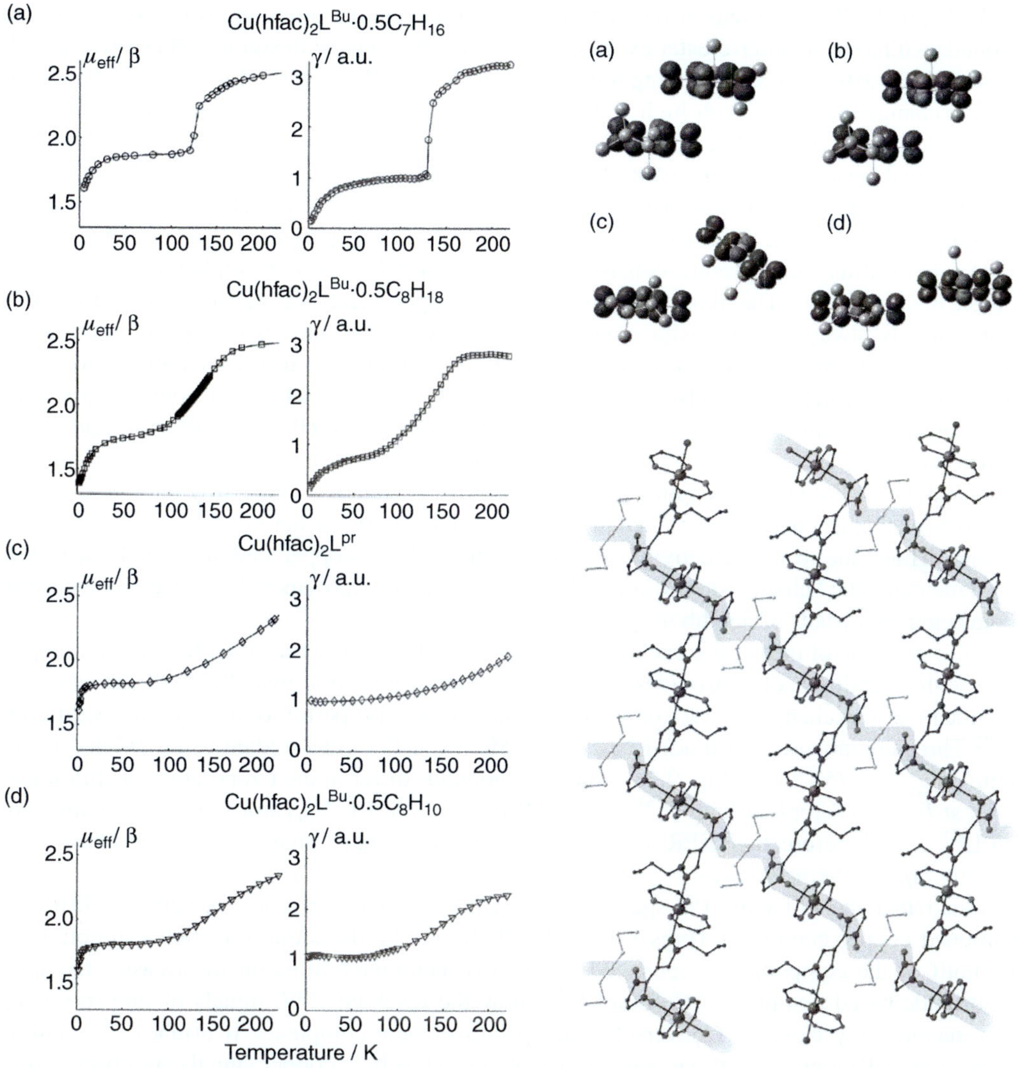

Figure 9.22 *Left column: Temperature dependence of the effective magnetic moment for the four studied compounds and corresponding temperature dependence of the ratio of second integrals* $\gamma = I_{triad}/I_{one} \sim \chi_{triad}/\chi_{one}$ *(in absolute units) measured by EPR. Solid lines show the simulated functions* $\gamma(T)$ *in the low temperature region. Right column spin-density distribution in the high spin state of the nearest nitronyl nitroxide radicals of* $[Cu(hfac)_2L^{n\text{-}Bu}]\cdot 0.5C_7H_{16}$ *(a),* $[Cu(hfac)_2L^{n\text{-}Bu}]\cdot 0.5C_8H_{18}$ *(b),* $[Cu(hfac)_2L^{n\text{-}Pr}]$ *(c), and* $[Cu(hfac)_2L^{n\text{-}Bu}]\cdot 0.5C_8H_{10}$ *(d) from the UB3LYP/6-31+G(d) calculations. Hydrogen atoms are omitted. Below in the right column: Structure of polymer chains and exchange interaction pathways (highlighted by the gray band) in breathing crystals exemplified using* $[Cu(hfac)_2L^{n\text{-}Bu}]\cdot 0.5C_7H_{16}$*. Note that, in the 3D structure, the solvent molecules (heptane) do not occupy positions between NO groups of neighboring chains as it appears in the picture. Reprinted with permission from [81]. Copyright 2010, American Chemical Society.*

$[Cu(hfac)_2L^{n\text{-Bu}}]{\cdot}0.5C_8H_{18}$. The reason for that is the different relative orientations of interacting radicals. For compounds with a strong inter-cluster exchange, the O and N atoms with significant positive spin density are situated against each other, encouraging antiferromagnetic coupling, while for compounds with a weak inter-cluster exchange the atoms with high spin density (N and O) of one radical are shifted from the corresponding atoms of the partner. Thus, in agreement with experiment, the 10-fold difference in inter-cluster exchange was found between two pairs of studied compounds.

Note that the EPR spectrum of spin-triads at $T < 50$ K does not show any characteristics of the triplet-state spectrum that would arise from exchange-coupled pairs of two spins $S = {}^1\!/_2$ (Fig. 9.13). In fact, the spectrum of a triad at any temperature consists of a single EPR line, which implies exchange narrowing between more than two spins. The hyperfine structure from copper nuclei was never observed for spin-triads, although it is nearly always observed for copper(II) ions in CuO_4N_2 units with typical splitting constant $A_{zz} \approx 14$ mT ≈ 0.014 cm^{-1}.[72–75] In the case of interaction between spin-triads, this allows for estimation of exchange-interaction between triads $|J_{\text{triad-triad}}| > A_{zz} \approx 0.014$ cm^{-1}, which, of course, agrees with the obtained values of 1–10 cm^{-1}. In the case of interaction between copper(II) ions and spin-triads, this leads to an estimate of $|J_{\text{triad–isolated Cu}}| < A_{zz} \approx 0.014$ cm^{-1}. Both estimations support exchange coupling between triads and magnetic isolation of one-spin units. Therefore, in agreement with X-ray data, it is evident that spin-triads form exchange coupled infinite chains that spread across the structural polymer chains (Fig. 9.22, right column). Thus, due to the selectivity of EPR for the isolated copper and spin-triads, it was proved reliably that the magnetic "chains" do not coincide with the structural polymer chains but are instead formed by interconnected spin-triads of neighboring polymer chains.

Another nice example of measuring exchange interaction pathways is a recent EPR study of complexes of $[Cu(hfac)_2]$ with *t*-butylpyrazolylnitroxides – a different type of nitroxide ligand (Fig. 9.23) which showed strong modification of exchange-interaction pathways and physical properties of the crystals by different type of ligand.[82] Three compounds $[Cu(hfac)_2L^{R}{}_{tert}]$ (R = Me, Et, *n*-Pr) were studied using *Q*-band EPR in the temperature range $T = 75$–293 K (Fig. 9.23). These compounds differ in the temperature of spin-transitions, which occur at $T_C \approx 90$ K for $[Cu(hfac)_2L^{Me}{}_{tert}]$, 190 K for $[Cu(hfac)_2L^{Et}{}_{tert}]$ and above room temperature for $[Cu(hfac)_2L^{n\text{-Pr}}{}_{tert}]$. Therefore, their EPR spectra should experience major changes due to the phase transitions at significantly different temperatures.

It was found that the *Q*-band EPR spectra of these novel compounds are principally different from abovementioned copper complexes (Figs 9.13, 9.16, 9.18, and 9.20), supporting the assumption that the magnetic motif of the compound has changed. As it was mentioned above in most cases, the spectra of nitronyl–nitroxide-based complexes $[Cu(hfac)_2L^R]$ show the resolved EPR signals of one-spin copper(II) ion that is magnetically isolated. Conversely, the spectra of new complexes $[Cu(hfac)_2L^{R}{}_{tert}]$ show that the one-spin copper(II) ions are not magnetically isolated (no resolved copper signals are observed), but are exchange coupled with the signals of spin-triads. This means that the inter-cluster exchange coupling between spin-triads and one-spin copper(II) ions is large on the EPR scale and thus exchange narrowing occurs.

As a result, the observed EPR signals are broad and their shape seems to be uninformative. Two well-distinguishable types of spectra were detected depending on temperature: a broad line with completely unresolved structure above the phase transition, and a more complicated pattern with clearly resolved spectral features below the phase transition (Fig. 9.23). The following simple theoretical model was used to describe these results. The spin-triad was described as an effective paramagnetic center with g^A in the strongly coupled (SS) state and g^C in the weakly coupled (WS) state. To describe the line shape of EPR spectrum produced by exchange narrowing between spin-triad and one-spin copper(II) ion, the approach of the modified Bloch equations similar to ref. 74 was used. Theoretical spectra shown in Figure 9.23 describe rather well the positions and shapes of experimentally observed spectra in the WS and SS states for $[Cu(hfac)_2L^{Me}{}_{tert}]$ and $[Cu(hfac)_2L^{Et}{}_{tert}]$. Although additional peaks (at $B \approx 1.05$ and 1.35 T) were not reproduced in theoretical spectrum, these signals are caused by weaker inter-chain exchange or dipolar interactions neglected in this

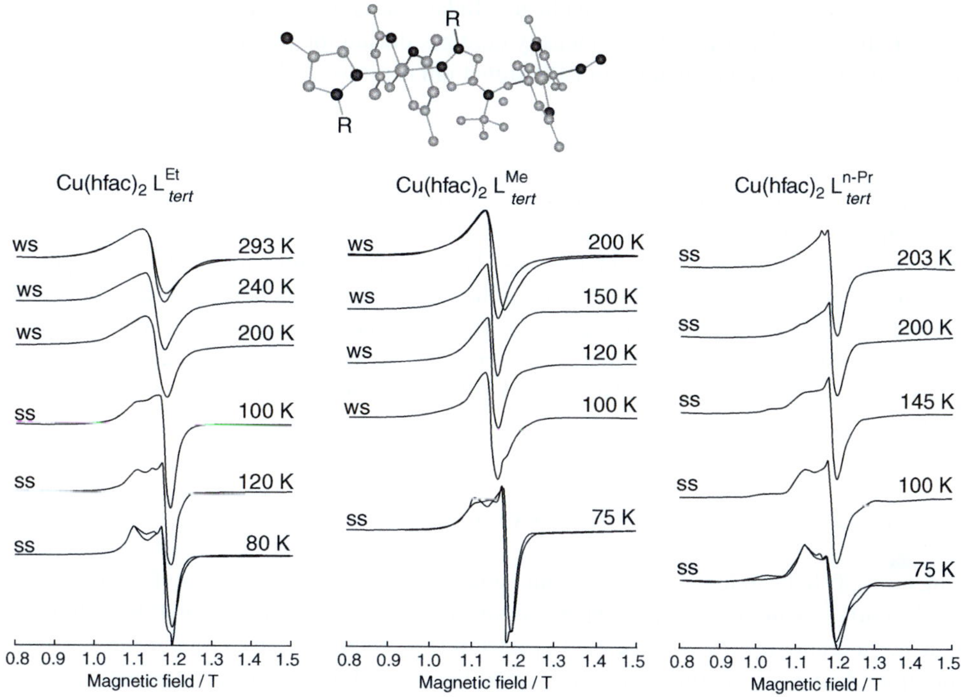

Figure 9.23 *Experimental Q-band ($\nu_{mw} \approx 33.52$ GHz) EPR spectra of [Cu(hfac)$_2$L$^{Et}{}_{tert}$], [Cu(hfac)$_2$L$^{Me}{}_{tert}$], [Cu(hfac)$_2$L$^{n\text{-}Pr}{}_{tert}$] (polycrystalline powder sample) at temperatures indicated on the right. The assignments to the WS or SS states are indicated on the left. Calculated spectra at 293 and 80 K are shown by thin dashed lines. Reprinted with permission from [82]. Copyright 2011 Springer-Verlag.*

simplified model. Otherwise, qualitative agreement in line shapes and quantitative agreement in their positions was found for the main spectral features of all three compounds. The magnitudes of inter-cluster exchange interactions based on the observed line shapes and theoretical modeling were estimated as $|J_{inter}| > 2.5 \times 10^9$ $s^{-1} \sim 0.1$ cm^{-1}. The dominant inter-cluster exchange interaction was found to be along the structural polymer chains in these type of compounds in contrast to nitronyl–nitroxide-based complexes [Cu(hfac)$_2$L^R].

The family [Cu(hfac)$_2$L^R] also includes complexes with the "head-to-tail" structure motif, whose polymeric chains include two-spin copper(II)–nitroxide clusters. One of a few representatives of this type of compound is the complex [Cu(hfac)$_2$L^{Me}] whose polymer chain structure is shown in Figure 9.1b. In spite of the apparently regular character of the chain, lowering the temperature to 230 K is accompanied by irreversible doubling of its unit cell and formation of alternating nonidentical two-spin clusters. Structural rearrangements in the polymeric chain of [Cu(hfac)$_2$L^{Me}] mainly occur on cooling to the structural phase transition temperature ($Tc{\downarrow} = 141$ K) and lead to an abrupt decrease in μ_{eff} by a factor of $\sqrt{2}$, which indicates a "disappearance" of 50% of the $S = {}^1\!/_2$ spins in the sample. As a result, in 50% of coordination sites the inter-spin distances Cu—O$_L$ are shortened, and antiferromagnetic exchange interaction is abruptly increased. The other 50% of exchange clusters are characterized by insignificant variations of their geometries and weak ferromagnetic exchange interaction still dominates. At temperatures above 141 K, the EPR spectrum of polycrystalline [Cu(hfac)$_2$L^{Me}] represents a broad singlet line ($g \approx 2.1$, $\Delta\omega \approx 30$ mT) with unresolved structure. After the phase transition (at $T < Tc{\downarrow}$), the EPR spectrum changes significantly and exhibits a number of characteristic features typical of the triplet states, including the "forbidden" double-quantum half-field transitions. The observed triplet state was assigned to the two-spin copper–nitroxide clusters linked through a weak ferromagnetic interaction (the "second type" of clusters). The EPR signal intensity changes during the phase transition are quite similar to the value of μ_{eff}, and the observed changes in EPR signal intensity were about 50%, in good agreement with magnetic susceptibility data.

At $T < T_C$, distances between neighboring clusters of the same polymer chain are doubled, because the spins of every second spin pair copper(II)–nitroxide become antiferromagnetically coupled to a singlet state. Thus, the reversible structural phase transition leads to an effective diamagnetic dilution of magnetically concentrated sample and opens up the possibility of observing previously unresolved structure, which is mainly determined by intra-cluster dipole–dipole interactions. The satisfactory agreement between experimental and simulated spectra can only be achieved by assuming the temperature dependence of the zero-field splitting parameter D and inter-cluster exchange interaction J. Based on the analysis of X-ray data it was concluded that below phase transition all $S = 1$ exchange clusters located within one slice can be divided into pairs with the distances between noncoordinated oxygens 3.898 Å, and the magnetic "chains" do not coincide with polymer chains.[75]

9.8 EPR-detected LIESST on Breathing Crystals

The phenomenon of light-induced excited spin state trapping (LIESST) was found recently in the $[Cu(hfac)_2L^R]$ family,[76] similar to iron(II) spin crossover compounds. It was found that illumination with light at at 7 K has a dramatic effect on the EPR spectrum (Fig. 9.24a–f). The signal of the strongly coupled state SS (multiplet A, $S = {}^1/_2$) strongly decreases, and the broad line assigned to the weakly coupled state (WS), $S = {}^3/_2$ appears instead. Simulations of experimental spectra showed that the observed signal contains ≈82% WS and ≈18% SS states, so the light-induced SS→WS conversion is 82%. Most interesting is the fact that the excited spin state relaxes to the ground state on the timescale of hours at cryogenic temperatures $T <$ 15 K. The mechanism of the observed LIESST-like effect in exchange-coupled clusters of breathing crystals is expected to be generally similar to that in iron compounds. The excitation at the wavelength of 900 nm most likely corresponds to a *d*-*d* absorption band of copper. After the light-induced promotion of an electron to the vacancy in the $d_{x2\text{-}y2}$ orbital, the system can relax from this excited state back to the unperturbed SS or, alternatively, undergo structural rearrangement and convert to the WS. The latter process involves the rotation of the Jahn–Teller axis and thus lowers the energy of the system due to the interchange of equatorial and axial *d*-orbitals, so that the electronically excited state becomes a ground state. Similarly to classical LIESST, the potential barrier between weakly and strongly coupled states is mainly due to their geometrical differences. It provides for a slow WS→SS relaxation and an efficient trapping of the WS state on a timescale of hours.

A detailed study of the character and relaxation mechanism of the light-induced excited spin state in three selected compounds of breathing crystals ($[Cu(hfac)_2L^{Pr}]$ (I), $[Cu(hfac)_2L^{Bu}]\cdot C_3H_7\cdot C_5H_6$ (II), and $[Cu(hfac)_2L^{Bu}]\cdot mC_8H_{10}$ (III)) was performed recently.[80] These compounds represent the cases of relatively strong or weak cooperativity, and different temperatures of thermal spin-transition. It was observed that the relaxation shows a pronounced self-decelerating character on a timescale of hours at cryogenic temperatures (<20 K) for all the compounds studied. As an example Figure 9.24 shows the relaxation curves for the compound I at $T = 5$, 10, and 13 K measured over 4 hours after photoswitching. The value $\gamma_n(t)$ is the fraction of light-switched triads and is normalized to the maximum of the conversion depth. The comparison of $\gamma_n(t)$ dependences obtained at $T =$ 5–16 K for compounds I–III reveal significant dependence of relaxation rate within this relatively narrow temperature range, and noticeable differences between compounds. Regardless of the degree of cooperativity, breathing crystals demonstrate an unusual self-decelerating relaxation trend,

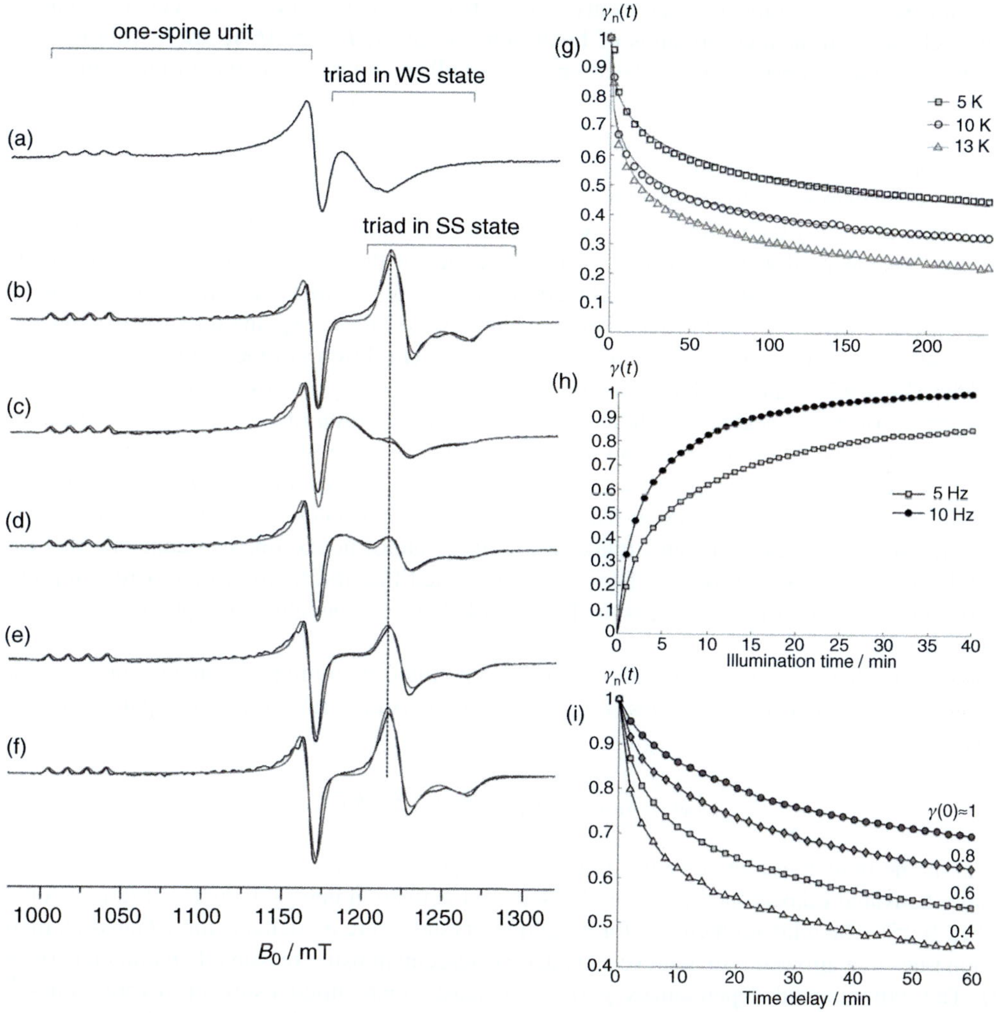

Figure 9.24 *CW EPR spectra of pure polycrystalline powder of [Cu(hfac)$_2$L$^{n\text{-}Pr}$] at T = 210 K, ν_{mw} =33.98 GHz (a), and of [Cu(hfac)$_2$L^{Pr}] mixed with glycerol at T = 7 K, ν_{mw} = 33.94 GHz (b–f). (b) dark spectrum; (c) immediately after illumination (900 nm); (d) 5 min and (e) 210 min after illumination; (f) after warming the illuminated sample up to 20 K and then cooling back to 7 K. Simulations (gray lines) yield an estimation of the SS→WS state conversion depth ≈ 82% (c), 65% (d) and 40% (e). Reproduced with permission from ref. 76. (g) Normalized relaxation dependences γ_n(t) (the fraction of light-switched triads in the weakly coupled state) for compound I measured at 5, 10, and 13 K (indicated in the legend). Solid lines show the simulations. (h) Conversion depth vs. illumination time measured for the compound I at 5 K using two frequencies of laser shots (5 and 10 Hz as indicated in the legend). (i) Normalized relaxation dependences γ_n(t) measured for the compound I after illumination of different duration (initial conversion depth γ(t = 0) is indicated on the right). Reprinted with permission from [76]. Copyright 2008 WILEY-VCH Verlag GmbH & Co. KGaA, Weinheim.*

opposite to the self-acceleration in neat iron(II) spin-crossover compounds. The general trends of self-decelerating relaxation in neat compounds of breathing crystals at $T = 5$–16 K are as follows: (i) at high temperatures $T > 16$ K the observed effect becomes very small and relaxation to the ground state is fast for the studied compounds; (ii) the relaxation has a self-decelerating character for all three studied compounds and even at lowest temperature (5 K) the shapes of the relaxation curves are pronouncedly self-decelerating; (iii) the relaxation rate increases drastically with temperature, and the character of this temperature dependence varies strongly from compound to compound.

Similar to spin-crossover compounds[94–97] two different explanations for the deviation of relaxation character from monoexponential were proposed: (i) a dependence of the relaxation rate on the concentration of excited states; and (ii) a distribution of the relaxation rates for all excited states in a certain range (usually described by a Gaussian).[76, 80, 84] In order to distinguish between these explanations the dependence of the conversion depth (γ) on time for the same irradiation power, but different repetition rate of the laser shots, was measured (Figs 9.24h and i). The conversion curve was clearly not exponential (as to be expected from the Lambert–Beer law neglecting photo-bleaching) and the initial rapid rise of the conversion during first several minutes was followed by a long and slow tail, so that the plateau was not completely reached even after 40 minutes of illumination. Based on these results the distribution model was confirmed as the main explanation. As a noticeable fraction (ca. 20%) of the excited states relax on the subsecond timescale and a larger fraction of the excited states relax on the timescale of hours, one can conclude that the distribution of relaxation rates is very broad. Thus it was assumed that the fraction of fast-relaxing clusters is excited at short illumination times, whereas the fraction of slow-relaxing clusters is photoswitched only after longer illumination.

The temperature dependence of relaxation kinetics was fitted assuming a tunneling process and broad distribution of effective activation energies (Fig. 9.25). The relaxation rate constant of particular cluster can be written as:

$$k_{\text{rel}} = k_{\text{tunnel}} + k^0_{\text{therm}} \exp(-E_A/kT), \tag{9.9}$$

where k_{tunnel} is the tunneling rate constant in the low-temperature limit, k^0_{therm} is the pre-exponential factor for thermally-activated relaxation, and E_{A} is the apparent activation energy which is different for different clusters (Fig. 9.25). The relaxation curves for breathing crystals were modeled using a Gaussian distribution of relaxation rates that allowed it to achieve satisfactory agreement using reasonable parameters (Figs 9.24g, h, and i). The experimental dependences $\gamma_{\text{n}}(t)$ were least-square fitted using the expression $k^{\text{observed}}_{\text{rel}} = k_{\text{rel}} \exp(-\delta E_A/kT)$, where δE_{A} was normally distributed around zero with the standard deviation parameter σ, and the values k_{rel} and σ were varied. The temperature dependences of k_{rel} were analyzed using the pseudo-Arrhenius plot to obtain the estimated values of activation energy E_{A} and pre-exponential factor k^0_{therm} for all three compounds. At $T > 7$ K the description using the Arrhenius law is reasonably good, whereas noticeable deviation from linearity at 5 K probably occurs because the tunneling relaxation rate k_{tunnel} becomes comparable with the thermally-activated relaxation rate.

The broad distribution of effective activation energies was explained by following causes: (i) the clusters located closer to the surface of crystal relax faster, as structural rearrangements deep inside the crystal may be more sterically-hindered compared to the crystal surface; (ii) some conformational disorder in the crystal packing (orientations of solvent molecules and alkyl substituents of the nitroxides) was previously found by X-ray analysis; (iii) the existence of thermally-populated low-lying spin levels of spin-triads also gives rise to a distribution of activation energies and relaxation parameters. The mechanism of LIESST is shown schematically in Figure 9.25. The broader distribution of electronic structure parameters in breathing crystals is not totally unexpected, because these materials have principally a 1D structure, that implies a larger flexibility of polymer chains compared to 2D and 3D structures. In addition, the changes of spin configuration

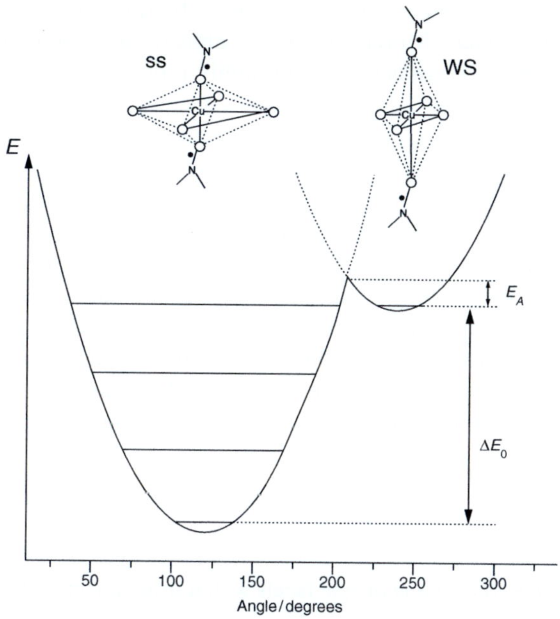

Figure 9.25 *Schematic circular section of the potential energy surface associated with the two Jahn–Teller valleys in breathing crystals. Structures corresponding to SS and WS states are sketched on top. Spin levels of the triads are not shown for simplicity. Reprinted with permission from [80]. Copyright 2012, American Chemical Society.*

in breathing crystals occur in the exchange-coupled clusters, not inside the *d*-shell of spin-transition ion as is in iron(II) spin-crossover compounds. This principal difference allows one to expect that breathing crystals are much more sensitive to external chemical and physical influences compared than iron(II) spin-crossover compounds. In perspective, this may be used to fine-tune the properties of breathing crystals for various applications.

9.9 Conclusion

To summarize, the study of stereochemically nonrigid [Cu(hfac)$_2$] nitroxide adducts capable of spin-transitions is a rather nontrivial problem because of their numerous polymorphic modifications, whose magnetic properties can markedly vary after insignificant changes in structure. The diversity of structural motifs, however, determines the diversity of magnetic effects. Of particular interest is the fact that all compounds with magnetic anomalies under study have a polymeric structure in the solid state. The motions of polymer chains relative to one another should therefore be highly concerted on passing from one polymorphic modification to another without crystal decay.

Heterospin systems based on Cu^{2+} complexes with nitroxides are very interesting for detailed studies of structural phase transitions that cause magnetic anomalies similar to classical spin-crossover. While compounds with these effects are yet not numerous, the magnetic transitions cover a wide temperature range of 30–250 K for known compounds. The highest potential transition temperature for such complexes is limited only by their decomposition temperature, which is preliminarily estimated at 350–400 K. A valuable feature of the heterospin complexes under study is their existence as high-quality, mechanically stable (over a wide temperature range) single crystals. Studies of the relationship between structural dynamics and the variation of magnetic properties will provide unique information for subsequent quantum-chemical analysis of the electronic structure of exchange clusters. Moreover, these complexes have numerous readily functionalizable paramagnetic ligands. This opens up prospects for thoroughly rationalizing all spatial motions of atoms in single crystals (i.e. the structural dynamics of a compound in the region of the phase transition), revealing the factors governing the form of the $\mu_{eff}(T)$ dependence, and developing methods for controllable modification of the magnetic transition characteristics by means of the *de novo* design of the paramagnetic ligand with a desired structure, and the introduction of dopant molecules in crystal.

We also would like to note the high utility of EPR spectroscopy as applied to breathing crystals $[Cu(hfac)_2L^R]$. Most magnetically concentrated compounds reveal strongly broadened or exchange-narrowed unresolved EPR lines and EPR spectroscopy thus becomes uninformative. In contrast, EPR as applied to breathing crystals allows us to obtain very important information on exchange interactions. The technique can measure the temperature dependence of intra-cluster exchange interaction $J(T)$, even though this is a few orders of magnitude higher than the energy of a microwave field quantum; it can determine the origin of inter-cluster exchange interaction and measure its value and observe the effect of light on the spin-transitions in breathing crystals.

References

1. Eaton, S. S., Eaton, G. R. (1978) Interaction of spin labels with transition metals, Coord. Chem. Rev., 26: 207–262 (1978).
2. Drago, R. S. (1980) Free radical reactions of transition metal systems. Coord. Chem. Rev., 32: 97–110.
3. Zolotov, Yu. Petrukhin, O. M., Nagy, V. Yu., Volodarsky, L. B. (1980) Stable free-radical complexing reagents in application of electron spin resonance to the determination of metals. Anal. Chim. Acta, 115: 1–23.
4. Larionov, S. V. (1988) Imidazoline nitroxides in coordination chemistry, in: Volodarsky, L. B. (Ed.) Imidazoline Nitroxides. Synthesis, Properties and Applications, vol 2. CRC Press Inc., Boca Raton, FL, USA, pp. 81–113.
5. Eaton, S. S., Eaton, G. R. (1988) Interaction of spin labels with transition metals, part 2. Coord. Chem. Rev., 83: 29–72.
6. Caneschi., A., Gatteschi, D., Rey, P. (1991) The chemistry and magnetic properties of metal nitronyl nitroxide complexes. Prog. Inorg. Chem., 39: 331–429.
7. Kahn, O. (1993) Molecular Magnetism. VCH, Weinheim, Germany, p. 380.
8. Volodarsky, L. B., Reznikov, V. A., Ovcharenko, V. I. (1993) Synthetic Chemistry of Stable Nitroxides. CRC Press Inc., Boca Raton, FL, USA, p. 240.
9. Caneschi, A., Gatteschi, D., Sessoli, R., Rey, P. (1989) Toward molecular magnets: the metal-radical approach. Acc. Chem. Res., 22: 392–398.
10. Iwamura, H., Inoue, K., Hayamizu, T. (1996) High-spin polynitroxide radicals as versatile briging ligands for transition metal complexes with high ferri/ferromagnetic T_C. Pure Appl. Chem., 68: 243–252.
11. Ovcharenko, V. I., Sagdeev, R. Z. (1999) Molecular ferromagnets. Russ. Chem. Rev., 68: 345–363.
12. Luneau, D., Rey, P. (2005) Magnetism of metal-nitroxide compounds involving bis-chelating imidazole and benz-imidazole substituted nitronyl nitroxide free radicals. Coord. Chem. Rev., 249: 2591–2611.
13. Luneau, D., Borta, A., Chumakov, Y., Jacquot, J.-F., Jeanneau, E., Lescop, C., Rey, P. (2008) Molecular magnets based on two-dimensional Mn(II)(-nitronyl nitroxide frameworks in layered structures. Inorg. Chim. Acta, 361: 3669–3676.
14. Vostrikova, K. E. (2008) High-spin molecules based on metal complexes of organic free radicals. Coord. Chem. Rev., 252: 1409–1419.
15. Iwamura, H., Inoue, K. (2001) Magnetic ordering in metal coordination complexes with aminoxyl radicals, in: Miller, J. S., Drillon, M. (Eds) Magnetism: Molecules to Materials II. Molecule-Based Materials. Wiley–VCH, Weinheim, Germany, pp. 61–108.
16. Oshio, H., Ito, T. (2000) Assembly of imino nitroxides with Ag(I) and Cu(I) ions. Coord. Chem. Rev., 198: 329–346.

17. Ouahab, L. (1998) Coordination complexes in conducting and magnetic molecular materials. Coord. Chem. Rev., 178–180: 1501–1531.
18. Sato, O., Tao, J., Zhang, Y-Z. (2007) Control of magnetic properties through external stimuli. Angew. Chem. Int. Ed., 46: 2152–2187.
19. Ovcharenko, V. I. (2010) Metal nitroxide complexes: synthesis and magnetostructural correlations, in: Hicks, R. (Ed.) Stable Radicals: Fundamentals and Applied Aspects of Odd-Electron Compounds. John Wiley and Sons, Chichester, UK, ch.13, pp. 461–506.
20. Ovcharenko, V. I., Maryunina, K. Yu., Fokin, S. V., Tretyakov, E. V., Romanenko, G. V., Ikorskii, V. N. (2004) Spin transition in non-classical systems. Russ. Chem. Bull., 53: 2406–2427.
21. Fedin, M. V., Veber, S. L., Sagdeev, R. Z., Ovcharenko, V. I., Bagryanskaya, E. G. (2010) Spectroscopy of thermally induced and light induced spin transitions in heterospin exchange clusters of compounds $Cu(hfac)_2L^R$. Russ. Chem. Bull., 59: 1065–1079.
22. Gütlich, P., Hauser, A., Spiering, H. (1994) Thermal and optical switching of iron (II) complexes. Angew. Chem. Int. Ed. Engl., 33: 2024–2054.
23. Bacci, M. (1988) Static and dynamic effects in spin equilibrium systems. Coord. Chem. Rev., 86: 245–271.
24. Kahn, O., Krober, J., Jay, C. (1992) Spin transition molecular materials for displays and data rcording. Adv. Mater, 4: 718–728.
25. König, E., Ritter, G., Kulshreshthsa, S. K. (1985) The nature of spin-state transitions in solid complexes of iron (II) and the interpretation of some associated phenomema. Chem. Rev., 85: 219–234.
26. König, E. (1987) Structural changes accompanying continuous and discontinuous spin-state transitions. Prog. Inorg. Chem., 35: 527–622.
27. König, E. (1991) Nature and dynamics of the spin-state interconversion in metal-complexes. Struct. Bonding (Berlin), 76: 51–152.
28. Toftlund, H. (1989) Spin equlibria in iron(II) complexes. Coord. Chem. Rev., 94: 67–108.
29. Zelentsov, V. V. (1992) Role of cooperative effects in coordinated compounds of 3D-metals. Koord. Khim., 18: 787–795.
30. Zelentsov, V. V. (1996) Magnetochemistry of iron chelates with spin transitions. Russ. Chem. J., 40: 86–92.
31. Lavrenova, L. G., Larionov, S. V. (1998) Spin transitions in coordination compounds of Fe(II) with 1,2,4-triazoles and tetrazoles. Russ. J. Coord. Chem., 24: 379–395.
32. Larionov, S. V. (2008) Spin transition in iron(III) and iron(II) complexes. Russ. J. Coord. Chem., 34: 237–250.
33. Gütlich, P., Goodwin, H. A. (Eds) (2004) Spin Crossover in Transition Metal Compounds I. Top. Curr. Chem., 233: p. 341.
34. Gütlich, P., Goodwin, H. A. (Eds) (2004) Spin Crossover in Transition Metal Compounds II. Top. Curr. Chem., 234: p. 294.
35. Gütlich, P., Goodwin, H. A. (Eds) (2004) Spin Crossover in Transition Metal Compounds III. Top. Curr. Chem., 235: p. 268.
36. Lanfranc de Panthou, F., Belorizky, E., Calemczuk, R., Luneau, D., Marcenat, C., Ressouche, E., Turek, P., Rey, P. (1995) A new-type of thermally-induced spin transition associated with an equatorial↔axial conversion in a copper(II)-nitroxide cluster, J. Am. Chem. Soc., 117: 11247–11253.
37. Lanfranc de Panthou, F., Luneau, D., Musin, R., Öhrström, L., Grand, A., Turek, P., Rey, P. (1996) Spin-transition and ferromagnetic interactions in copper(II) complexes of a 3-pyridyl-substituted imino nitroxide. Dependence of the magnetic properties upon crystal packing, Inorg. Chem., 35: 3484–3491.
38. Rey, P., Ovcharenko, V. I. (2003) Copper(II) nitroxide molecular spin transition complexes, in: Miller, J. S., Drillon, M. (Eds) Magnetism: Molecules to Materials IV. Wiley-VCH, Weinheim, Germany, ch. 2, pp. 41–64.
39. Iwahory, F., Inoue, K., Iwamura, H. (1999) Synthesis and magnetic properties of bis(hexafluoroacetylacetonato) copper(II) complex with 5-bromo-a,3-phenylenebis(N-tert-butylaminoxyl) as a bridging ligand. Mol. Cryst. Liq. Cryst. Sci. Tech., Sect. A, 334: 533–538.
40. Caneschi, A., Chiesi, P., David, L., Ferraro, F., Gatteschi, D., Sessoli, R. (1993) Crystal-structure and magnetic-properties of two nitronyl nitroxide biradicals and of their copper (II) complexes. Inorg. Chem., 32: 1445–1453.
41. Okazawa, A., Hashizume, D., Ishida, T. (2010) Ferro- and antiferromagnetic coupling switch accompanied by twist deformation around the copper(ii) and nitroxide coordination bond. J. Am. Chem. Soc., 132: 11516–11524.

42. Romanenko, G. V., Maryunina, K. Yu., Bogomyakov, A. S., Sagdeev, R. Z., Ovcharenko, V. I. (2011) Relationship between the thermally induced reorientations of aromatic solvate molecules in $Cu(hfac)_2$-nitroxide breathing crystals and the character of the magnetic anomaly. Inorg. Chem., 50: 6597–6609.
43. Fokin, S. V., Romanenko, G. V., Shvedenkov, Yu. G., Ikorskii, V. N., Tretyakov, E. V., Vasilevskii, S. F., Ovcharenko, V. I. (2002) Crystal structure and magnetic properties of Cu(II) complexes with pyrozole-substituted iminonitroxyl radicals: new trinuclear heterospin complexes. J. Struct. Chem., 43: 828–834.
44. Ovcharenko, V. I., Fokin, S. V., Romanenko, G. V., Ikorskii, V. N., Tretyakov, E. V., Vasilevskii, S. F., Sagdeev, R. Z. (2002) Unsual spin transitions. Mol. Phys., 100: 1107–1115.
45. Maryunina, K., Fokin, S., Ovcharenko, V., Romanenko, G., Ikorskii, V. (2005) Solid solutions: an efficient way to control the temperature of spin transition in heterospin crystals $M_xCu_{1-x}(hfac)_2L$ (M = Mn, Ni, Co; L = nitronyl nitroxide). Polyhedron, 24: 2094–2101.
46. Musin, R. N., Schastnev, P. V., Malinovskaya, S. A. (1992) Delocalization mechanism of ferromagnetic exchange interactions in complexes of copper (II) with nitroxyl radicals. Inorg. Chem., 31: 4118–4121.
47. Musin, R. N., Ovcharenko,V. I., Öhrström, L., Rey, P. (1997) Magnetic properties of Cu(II) bischelate complexes with 3-imidazoline nitroxides. 1. An *ab initio* analysis of spin density delocalization channels. J. Struct. Chem., 38: 703–710.
48. Musin, R. N., Ovcharenko,V. I., Öhrström, L., Rey, P. (1997) Magnetic properties of Cu(II) bischelate complexes with 3-imidazoline nitroxides. 2. An *ab initio* analysis of exchange interaction mechanisms. J. Struct. Chem., 38: 711–715.
49. Ovcharenko, V. I., Shvedenkov, Yu. G., Musin, R. N., Ikorskii, V. N. (1990) Determination of exchange interaction parameters in heterospin exchange clusters. J. Struct. Chem., 40: 29–34.
50. Ovcharenko, V. I., Romanenko, G. V., Maryunina, K. Yu., Bogomyakov, A. S., Gorelik, E. V. (2008) Thermally induced magnetic anomalies in solvates of the bis(hexafluoroacetylacetonate)copper(II) complex with pyrazolyl-substituted nitronyl nitroxide Inorg. Chem., 47: 9537–9552.
51. Gütlich, P., Köppen, H., Steinhäuser, H. G. (1980) Deuterium isotope effect on the high-spin $\rightleftharpoons$ low-spin transition in deuterated solvates of tris(2-picolylamine) iron(II) chloride. Chem. Phys. Lett., 74: 475–480.
52. Murray, K. S., Kepert, C. J. (2004) Cooperativity in spin crossover systems: memory, magnetism and microporosity, in: Gütlich, P., Goodwin, H. A. (Eds) Spin Crossover in Transition Metal Compounds I. Top. Curr. Chem., 233: 195–228.
53. Sorai, M. (2004) Heat capacity studies of spin crossover systems, in: Gütlich, P., Goodwin, H. A. (Eds.) Spin Crossover in Transition Metal Compounds III. Top. Curr. Chem., 235: 153–170.
54. Wiehl, L., Kiel, G., Köhler, C. P., Spiering, H., Gütlich, P. (1986) Structure determination and investigation of the high-spin $\leftrightarrow$ low-spin transition of $[Fe(2\text{-}pic)_3]Br_2.EtOH$. Inorg. Chem., 25: 1565–1571.
55. Cecconi, F., Di Vaira, M., Midollini, S., Orlandini, A., Sacconi, L. (1981) Singlet $\rightleftharpoons$ quintet spin transition of iron (II) complexes with a P4Cl2 door set – X-ray structures of the compound $FeCl_2(Ph_2PCH{=}CHPPh_2)_2$ and of its acetone solvate at 130 K and 295 K. Inorg. Chem., 20: 3423–3430.
56. Hostettler, M., Törnroos, K. W., Chernyshov, D., Vangdal, B., Bürgi, H.-B. (2004) Challenges in engineering spin crossover: structures and magnetic properties of six alcohol solvates of iron(II) tris(2-picolylamine) dichloride. Angew. Chem. Int. Ed., 2004, 43: 4589–4594.
57. Halder, G. J., Kepert, C. J., Moubaraki, B., Murray, K. S., Cashion, J. D. (2002) Guest-dependent spin crossover in a nanoporous molecular framework material. Science, 298: 1762–1765.
58. Törnroos, K. W., Hostettler, M., Chernyshov, D., Vangdal, B., Bürgi, H.-B. (2006) Interplay of spin conversion and structural phase transformations: re-entrant phase transitions in the 2-propanol solvate of tris(2picolylamine)iron(II) dichloride. Chem. Eur. J., 12: 6207–6215.
59. Gordon, A. J., Ford, R. A. (1972) The Chemist's Companion. A Handbook of Practical Data, Techniques, and References. John Wiley and Sons, New York, NY, USA, pp 95–101.
60. Kennedy, T. A., Choh, S. H., Seidel, G. (1970) Temperature dependence of the exchange interaction in $K_2CuCl_4 \cdot 2H_2O$. Phys. Rev. B, 2: 3645–3651.
61. Veber, S. L., Fedin, M. V., Potapov, A. I., Maryunina, K. Yu., Romanenko, G. V., Sagdeev, R. Z., Ovcharenko, V. I., Goldfarb, D., Bagryanskaya, E. G. (2008) High-field EPR reveals the strongly temperature-dependent exchange interaction in "breathing" crystals $Cu(hfac)_2L^R$. J. Am. Chem. Soc., 130: 2444–2445.

62. Osip'yan, Yu. A., Morgunov, R. B., Baskakov, A. A., Ovcharenko, V. I., Fokin, S. V. (2003) Structural defects in molecular crystals based on heterospin copper complexes. Phys. Solid State, 45: 1465–1470.
63. Zueva, E. M., Ryabykh, E. R., Kuznetsov, An. M. (2009) Microscopic mechanisms of magnetic transitions in chain polymeric copper(II) complexes with nitronyl nitroxide radicals. Russ. Chem. Bull., 8: 1654–1662.
64. Postnikov, A. V., Galakhov, A. V., Blügel, S. (2005) Magnetic interactions in a Cu–containing heterospin polymer. Phase Transitions, 78: 689–699.
65. Ovcharenko, V., Fursova, E., Romanenko, G., Eremenko, I., Tretyakov, E., Ikorskii, V. (2006) Synthesis, structure, and magnetic properties of (6-9)-nuclear Ni(II) trimethylacetates and their heterospin complexes with nitroxides. Inorg. Chem., 45: 5338–5350.
66. Ovcharenko, V. I., Fokin, S. V., Romanenko, G. V., Tretyakov, E. V., Boltacheva, N. S., Filyakova, V. I., Charushin, V. N. (2006) New metal-containing matrix in the design of heterospin systems: bis(1,1,1,5,5,5-hexafluoro-4-iminopent-2-en-2-olato)copper(II) complex with nitroxide. Russ. Chem. Bull., 55: 2122–2124.
67. Ovcharenko, V. I., Fokin, S. V., Romanenko, G. V., Bogomyakov,A. S., Yachevskii, D. S., Chizhov, D. L., Charushin, V. N., Chupakhin, O. N. (2010) Heterospin complexes of fluorinated dinuclear Cu^{II} and Mn^{II} triketonates with nitroxides. Russ. Chem. Bull., 59: 1162–1174.
68. Fokin, S. V., Ovcharenko, V. I., Romanenko, G. V., Tretyakov,E. V., Bogomyakov, A. S., Saloutina, L. V., Filyakova, T. I., Saloutin, V. I., Charushin, V. N., Chupakhin, O. N. (2011) Bis(1,1,1,3,5,5,5-heptafluoro-4-iminopent-2-ene-2-aminato)copper(II) - a new metal-containing matrix in the design of heterospin systems. Russ. Chem. Bull., 60: 816–823.
69. Ovcharenko, V. I., Gorelik, E. V., Fokin, S. V., Romanenko, G. V., Ikorskii, V. N., Krashilina, A. V., Cherkasov, V. K., Abakumov, G. A. (2007) Ligand effects on the ferro- to antiferromagnetic exchange ratio in bis(o-semiquinonato)copper(II). J. Am. Chem. Soc., 129: 10512–10521
70. Tretyakov, E., Fokin, S., Romanenko, G., Ikorskii, V., Vasilevsky, S., Ovcharenko, V. (2006) 2D and 3D $Cu(hfac)_2$ complexes with nitronyl nitroxide biradicals. Inorg. Chem., 45: 3671–3678.
71. Romanenko, G. V., Tolstikov, S. E., Tretyakov, E. V., Fokin, S. V., Ikorskii, V. N., Ovcharenko, V. I. (2007) Shift of stereochemical nonrigidity from coordination units to polymethylene fragments in heterospin copper(II) hexafluoroacetylacetonate complexes with nitronyl nitroxide biradicals. Russ. Chem. Bull., 56: 1795–1804.
72. Fedin, M. V., Veber, S. L., Gromov, I. A., Ovcharenko, V. I., Sagdeev, R. Z., Schweiger, A., Bagryanskaya, E. G. (2006) EPR of three-spin nitroxide-copper(II)-nitroxide clusters coupled by a strong exchange interaction. J. Phys. Chem. A, 110: 2315–2317.
73. Fedin, M. V., Veber, S. L., Gromov, I. A., Ovcharenko, V. I., Sagdeev, R. Z., Bagryanskaya, E. G. (2007) Electron spin exchange processes in strongly coupled spin triads. J. Phys. Chem. A, 111: 4449–4455.
74. Fedin, M., Veber, S., Gromov, I., Maryunina, K., Fokin, S., Romanenko, G., Ovcharenko, V., Sagdeev, R., Bagryanskaya, E. (2007) Thermally induced spin transitions in nitroxide-copper(II)-nitroxide spin triads studied by EPR. Inorg. Chem., 46: 11405–11415.
75. Veber, S. L., Fedin, M. V., Maryunina, K. Yu., Romanenko, G. V., Sagdeev, R. Z., Bagryanskaya, E. G., Ovcharenko, V. I. (2008) Diamagnetic dilution due to the phase spin transition – an opportunity for the EPR study of intercluster exchange in "breathing" crystals of copper(II) hexafluoroacetylacetonate with pyrazole-substituted nitronyl nitroxide. Inorg. Chim. Acta, 361: 4148–4152.
76. Fedin, M., Sagdeev, R., Ovcharenko, V., Reijerse, E., Lubitz, W., Bagryanskaya, E. (2008) Light-induced excited spin state trapping in an exchange-coupled nitroxide-copper(II)-nitroxide cluster. Angew. Chem. Int. Ed., 47: 6897–6899.
77. Fedin, M., Ovcharenko, V., Bagryanskaya, E. (2009) "EPR aiding the development of molecule-based magnetic devices", in: "Electron Paramagnetic Resonance: From Fundamental Research to Pioneering Applications & Zavoisky Award", ISBN 978-1-877524-01-1.
78. Fedin, M. V., Veber, S. L., Romanenko, G. V., Ovcharenko, V. I., Sagdeev, R. Z., Klihm, G., Reijerse, E., Lubitz, W., Bagryanskaya, E. G. (2009) Dynamic mixing processes in spin triads of "breathing crystals" $Cu(hfac)_2L^R$: a multifrequency EPR study at 34, 122 and 244 GHz. Phys. Chem. Chem. Phys., 11: 6654–6663.
79. Veber, S. L., Fedin, M. V., Fokin, S. V., Sagdeev, R. Z., Ovcharenko, V. I., Bagryanskaya, E. G. (2010) EPR study of ligand effects in spin triads of bis(o-semiquinonato)copper(II) complexes, Appl. Magn. Reson., 37: 693–701.
80. Fedin, M. V., Maryunina, K. Yu., Sagdeev, R. Z., Ovcharenko, V. I., Bagryanskaya, E. G. (2012) Self-decelerating relaxation of the light-induced spin states in molecular Magnets $Cu(hfac)_2L^R$ studied by electron paramagnetic resonance, Inorg. Chem., 51: 709–717.

81. Fedin, M. V., Veber, S. L., Maryunina, K. Yu., Romanenko, G. V., Suturina, E. A., Gritsan, N. P., Sagdeev, R. Z., Ovcharenko, V. I., Bagryanskaya, E. G. (2010) Intercluster exchange pathways in polymer-chain molecular magnets $Cu(hfac)_2L^R$ unveiled by electron paramagnetic resonance. J. Am. Chem. Soc., 132: 13886–13891.
82. Fedin, M. V., Drozdyuk, I. Yu., Tretyakov, E. V., Tolstikov, S. E., Ovcharenko, V. I., Bagryanskaya, E. G. (2011), EPR of Spin Transitions in Complexes of $Cu(hfac)_2$ with tert-butylpyrazolylnitroxides. Appl. Magn. Reson., 41: 383–392.
83. Veber, S. L., Fedin, M. V., Maryunina, K. Yu., Potapov, A., Goldfarb, D., Reijerse, E., Lubitz, W., Sagdeev, R. Z., Ovcharenko, V. I., Bagryanskaya, E. G. (2011) Temperature-dependent exchange interaction in molecular magnets $Cu(hfac)_2L^R$ studied by EPR: methodology and interpretations. Inorg. Chem., 50: 10204–10212.
84. Fedin, M. V., Bagryanskaya, E. G., Matsuoka, H., Yamauchi, S., Veber, S. L., Maryunina , K. Y., Tretyakov, E. V., Ovcharenko, V. I., Sagdeev, R. Z. (2012) W-band time-resolved electron paramagnetic resonance study of light-induced spin dynamics in copper-nitroxide-based switchable molecular magnets. J. Am. Chem. Soc., 134: 16319–16326.
85. Bencini, A., Gatteschi, D. (1990) Electron Paramagnetic Resonance of Exchange-Coupled Systems. Springer-Verlag, Berlin, Germany, p. 297.
86. Yablokov, Yu. V., Voronkova, V. K., Mosina, L. V. (1988) Paramagnetic resonance in exchange clusters. Nauka, Moscow, Russia, p. 181.
87. Banci, L., Bencini, A., Gatteschi, D. (1983a) EPR spectra of trinuclear complexes. octachloro diadeninium tricopper(II). Inorg. Chem., 22: 2681–2683.
88. Banci, L., Bencini, A., Dei, A., Gatteschi, D. (1983b) EPR Spectra of and exchange interactions in trinuclear complexes. 2. Metal(II) adducts of tetradentate Schiff base copper(II) complexes. Inorg. Chem., 22: 4018–4021.
89. Boguslavskii, E. G., Shklyaev, A. A., Yudanov, V. F., Ovcharenko, V. I., Larionov, S. V. (1984) Magnetic-interactions in copper (II) complexes with derivatives of nitroxyl radical of imidazoline. Bull. Acad. Sci. USSR, Div. Chem. Sci., 33: 1394–1400.
90. Benelli, C., Gatteschi, D., Zanchini, C., Latour, J. M., Rey, P. (1986) Weak exchange interactions between nitroxides and copper(II) ions monitored by EPR Spectroscopy, Inorg. Chem., 25: 4242–4244.
91. Vancoillie, S., Rulíšek, L., Neese, F., Pierloot, K. (2009) Theoretical description of the structure and magnetic properties of nitroxide–Cu(II)-nitroxide spin triads by means of multiconfigurational ab initio calculations. J. Phys. Chem. A, 113: 6149–6157.
92. Carrington, A., McLachlan, A. D. (1979) Introduction to magnetic resonance with applications to chemistry and chemical physics. Chapman and Hall, New York, NY, USA, p. 266.
93. Coronado, E., Delhaës, P., Gatteschi, D., Miller, J. S. (Eds.) (1996) Molecular Magnetism: From Molecular Assemblies to the Devices. Nato ASI Ser. E: Appl. Sci., 321. Kluwer Academic Publisher: Dordrecht, Netherlands, p. 608.
94. Poole, C. P. jr. (1987) Electron Spin Resonance: A Comprehensive Treatise of Experimental Techniques. Dover Publications, Mineola, NY, USA, p. 780.
95. Mishra, V., Mishra, H., Mukherjee, R., Codjovi, E., Linarès, J., Létard, J.-F., Desplanches, C., Baldé, C., Enachescu, C., Varret, F. (2009) Spin-transition in $[Fe^{II}(L^5)_2][ClO_4]_2$ [L^5 = 2-[3-(2′-pyridyl)pyrazol-1-ylmethyl](1-methylimidazole)]: a further example of coexistence of features typical for disorder and cooperativity. Dalton Trans., 7462–7472.
96. Enachescu, C., Linares, J., Varret, F., Boukheddaden, K., Codjovi, E., Salunke, S. G., Mukherjee, R. (2004) Nonexponential relaxation of the metastable state of the spin-crossover system $[Fe(L)_2](ClO_4)_2{\cdot}H_2O$ [L = 2,6-bis(pyrazol-1-ylmethyl)pyridine]. Inorg. Chem., 43, 4880–4888.
97. Neville, S. M., Leita, B. A., Halder, G. J., Kepert, C. J., Moubaraki, B., Letard, J.-F., Murray, K. S. (2008) Understanding the two-step spin-transition phenomenon in iron(II) 1D chain materials. Chem. Eur. J., 14: 10123–10133.
98. Dupouy, G., Triki, S., Marchivie, M., Cosquer, N., Gómez-García, C. J., Pillet, S., Bendeif, E.-E., Lecomte, C., Asthana, S., Létard, J.-F. (2010) Cyanocarbanion-based spin-crossover materials: photocrystallographic and photomagnetic studies of a new iron(II) neutral chain. Inorg. Chem., 49: 9358–9368.

10

Spin-State Switching in Solution

Matthew P. Shores, Christina M. Klug and Stephanie R. Fiedler

Department of Chemistry, Colorado State University, USA

10.1 Introduction and Scope

Although most examinations of spin-crossover (SC) properties focus on the solid state, the solution phase provides complementary chemical insights. In addition, the sensitivity of SC to small perturbations of ligand environments offers the intriguing possibility of using spin-state switching as a reporter for chemical changes, one which may be exploitable in solution-based chemosensing schemes. The topic of spin equilibria for Fe(II) complexes in solution has been reviewed by Toftlund[1,2] and others;[3,4] the last review was in 2004. The application of isokinetic and isoequilibrium relationships to SC systems in solution has been addressed in some detail by Linert and co-workers.[5] More recently, noninnocent ligand effects on metal spin centers, including redox-induced spin-state switching,[6] and charge-transfer induced spin-state changes, have been developed. Similarly, dispersions of SC nanoparticles and liquid-crystalline materials represent important recent efforts to control or exploit spin in dilute liquid media. However, because these topics are covered in other chapters, we will focus our attention on solution studies of systems where the spin events are largely localized to single, isolated metal centers. Most of the species studied are based on Fe(II) complexes, but Co(II)- and Ni(II)-containing species also feature in the most recent reports.

Here, we will summarize the kinds of information (Section 10.2) and methods (Section 10.3) available for solution-phase spin switching investigations. Then we will survey the recent literature, focusing on the effects of solvent and ions (Section 10.4), ligand distortions and complex isomerizations (Section 10.5), and noncoordinating anion–cation interactions (Section 10.6) on spin-state switching behavior. We will conclude by highlighting several challenges and opportunities in this area (Section 10.7). Our treatment of solution investigations will emphasize chemical-interaction-based spin-state switching.

Spin-Crossover Materials: Properties and Applications, First Edition. Edited by Malcolm A. Halcrow.

10.2 Spin-Crossover: Solid State Versus Solution

The positive and negative attributes to studying SC in solution are complementary to those found for the solid state. The main advantage for solution studies is that intermolecular interactions for pure compounds are not limited to the rigid (and sometimes capricious) environment imparted by the lattice, but can be perturbed by judicious choice of solvent and additives. Solution studies can provide insight into ligand field effects on the SC complex, as well as probe interactions with charge balancing ions, added salts, and other solvent/guest molecules. A major challenge in solution studies is ensuring that the complex is resistant to ligand substitution when dissolved. Unfortunately, simply using multidentate ligands does not assure solution stability; we caution against use of overall binding constants as the sole measure of stability because temporary loss of even one part of the ligand can impact spin. Second, cooperativity is restricted significantly by dilution: species which show sharp SC transitions in the solid state often show gradual spin equilibria in solution. Note that this is not insurmountable: some local interactions can overcome dilution effects.[5] Third, the solution phase offers much smaller operative temperature ranges than the solid state, due to phase limitations (Table 10.1) and reduced solubility of complexes at lower temperatures. Thus, fitting gradual equilibria often requires extrapolation. Finally, ion pairing can be a complicating factor if noncovalent influences on SC are being probed. On balance, we would encourage more solution studies be undertaken on SC compounds because of the complementary information and environmental flexibility offered. The main caveat is that the solution and solid state structures of a complex can be sufficiently dissimilar – especially in terms of intermolecular interactions – that property comparisons in the two states should be treated carefully.

Before moving further, we find it useful to formally differentiate *spin-crossover in solution* from *spin state sensitivity to chemical environment*. The former is affected primarily by perturbations (usually thermal) to a complex's intrinsic ligand field, while the latter exploits spin state differences between subtly different

Table 10.1 *Liquid ranges for common solvents.*[a]

Solvent	T_{min} (K)[7]	T_{max} (K)[7]	ρ_{min} (g/cm^3)[8]	ρ_{max} (g/cm^3)[8]	$\rho_{295\ K}$ (g/cm^3)[b]	$\rho_{deuterated}$ (g/cm^3)[b]
dimethyl sulfoxide	291	462.15	1.102	0.9272	1.099[8]	1.190
acetonitrile	229.33	354.8	0.8488	0.7166	0.786	0.844
nitromethane	244.77	374.34	1.198	1.023	1.127[8]	1.183
acetone	178.45	329.2	0.9132	0.7484	0.791	0.872
water	273.15	373.15	0.9998	0.9584	0.997[8]	1.107
ethyl acetate	189.35	350.26	1.012	0.8284	0.902	0.984
tetrahydrofuran	164.71	338.15	1.010	0.8362	0.884[8]	0.985
methanol	175.0	337.75	0.8865	0.7494	0.791	0.888
ethanol	159.0	351.44	0.9006	0.7342	0.790[8]	0.892
dichloromethane	178.1	313.15	1.527	1.290	1.325	1.362
toluene	178.2	383.78	0.9694	0.7796	0.865	0.943
benzene	278.64	353.24	0.8926	0.8156	0.874	0.95
hexane	177.8	341.88	0.7553	0.6146	0.659	0.767
pentane	143.48	309.21	0.7565	0.6099	0.626	0.731

[a] Arranged by decreasing dipole moment.
[b] Unless noted otherwise, densities at 295 K obtained from MSDS.

chemical compositions; for example, a host species resides in a different spin state than a related host–guest complex. In practice, it is difficult to separate them experimentally. For example, the simple act of decreasing the solution temperature often increases the density, which is felt at the local level as closer intermolecular contacts. Most potential applications in chemical sensing are based on spin state sensitivity, but the small energies involved in host–guest interactions often mean that complex spin states must be poised in the thermally-accessible SC regime. Thus we will use the term "spin-state switching" to capture both aspects of spin behavior in solution.

10.3 Practical Considerations

Here, we will briefly summarize techniques and resources that we have found to be useful. For SC characterization, we are often concerned with measuring magnetic susceptibilities versus temperature to determine γ_{HS}, from which thermodynamic quantities (ΔS, ΔH) can be derived. When host–guest or other intermolecular interactions are important, careful titration studies are critical; finding orthogonal signals to independently characterize intermolecular interactions and spin states is challenging. Some powerful solid state techniques are much less practical in solution: for example Mössbauer spectroscopy requires relatively expensive ^{57}Fe enrichment to overcome solvent dilution; and dissolution removes some important tensor orientation information from EPR measurements. Others, like ultrafast spectroscopic experiments, are treated in accompanying chapters.

A key consideration is the stability of the complex in solution. With the exception of LS Co(III), SC complexes are labile in solution, the HS species more so. Temperature ranges and densities for commonly used solvents in SC and/or host–guest chemistry are provided in Table 10.1, arranged in order of decreasing polarity. Ligand dissociation and substitution are enhanced in more polar solvents, whereas ion pairing effects are significant in less polar solvents.

10.3.1 NMR Characterization

(^1H) NMR spectroscopy provides both structural and susceptibility information, and is relatively accessible and inexpensive. Temperature ranges of >200 K (173–393 K) are commonly available for instruments equipped with a variable temperature apparatus. For the Fe(II) complexes with which we are most familiar, we find that analyte concentrations of ~5 mM are necessary to provide acceptable intensities for the paramagnetically-shifted/broadened signals (acquired in reasonably short experiments). Since both paramagnetic and diamagnetic species may be present,* it is important to probe the "normal" ^1H window as well as a much larger frequency range; signals ranges vary for SC complexes but a 200 ppm range for HS Fe(II) is not uncommon. Accurate $\Delta\delta$ determinations for Evans method experiments (*vide infra*) are also aided by probing the "normal" ^1H NMR window.

The Evans method determines $\chi_M T$ of a paramagnetic species from the frequency shift ($\Delta\delta$) of diamagnetic (solvent) molecules which interact with the paramagnetic species, relative to an external reference (in practice, the same solvent trapped in an inner capillary).[9] Examples of experiments and calculations are available.[10] With the development and widespread use of superconducting NMR spectrometers, a correction to the first

* We are referring to possible multiple species in solution, not the LS component of an SC complex; as explained below, HS–LS switching is fast compared to the NMR timescale so only an average signal is observed.

term of Evans' original formula was developed.[11, 12] Converting to more common units and measurements and condensing the constants gives:

$$\chi_M = \frac{477\Delta\delta}{2 \times 10^6 M} \tag{10.1}$$

where χ_M is molar susceptibility in cm^3/mol, $\Delta\delta$ is the frequency shift in ppm and M is the concentration in mol/dm^3. The second and third terms of Evans' susceptibility equation (not shown above) are corrections for the diamagnetic susceptibility of the solvent as well as the density difference between the pure solvent and the solution under investigation. In most cases, these corrections can be ignored.[13] We have observed (in dichloromethane) that diamagnetic species (e.g. excess anion salts, ligand exchange products) also have negligible effects on susceptibility values.[14] On the other hand, several reports have shown the importance of correcting for changes in solvent density during variable temperature experiments (Table 10.1); otherwise artificially low concentrations give artificially high $\chi_M T$ values.[15]

Errors of no less than 5% in the measurements can be expected. Accurate weights, particularly of air-sensitive compounds, can be difficult to obtain and introduce large error in the measurements.[16, 17] Moreover, it is a bulk rather than local measurement; errors arising from other paramagnetic or diamagnetic impurities are possible.

Tracking δ changes during variable temperature experiments can provide a powerful complement to Evans method $\chi_M T$ values. Weber and Walker have described this nicely for Fe(II) SC.[17, 18] Because Fe(II) spin-state switching is faster than the NMR timescale, observed resonances represent the population-weighted average of the HS and LS states of the molecule. Peak shifts can be related to γ_{HS} and thus compared to Evans method $\chi_M T$ values;[17, 18] both experiments can be done simultaneously. Tracking multiple peak shifts allows characterization of multiple HS configurations should they be present in solution. Assignment of the peaks is a necessary, though complicated, condition for successfully tracking the resonances as they shift; determining δ values for a purely HS (or LS) species is another requirement, one which is not always available in the solvent's temperature range. When successful, identifying the peaks provides evidence for complex stability in solution.

Thermodynamic parameters can be gleaned from fitting solution state studies from:[19]

$$\chi T = \frac{\left[\frac{g_{HS}^2}{4} C_{HS} + (TIP_{HS})\,T\right] - \left[\frac{g_{LS}^2}{4} C_{LS} + (TIP_{LS})\,T\right]}{1 + exp\left[\frac{\Delta H}{R}\left(\frac{1}{T} - \frac{1}{T_{1/2}}\right)\right]} + \left[\frac{g_{LS}^2}{4} C_{LS} + (TIP_{LS})\,T\right] \tag{10.2}$$

where g_{HS} and g_{LS} are the Landé factors of HS and LS states, C_{HS} and C_{LS} are the Curie constants for HS and LS states, R is the gas constant, ΔH is the enthalpy and $T_{1/2}$ is the critical temperature where equimolar amounts of HS and LS states are present ($\Delta S = \Delta H/T_{1/2}$; for Fe(II) compounds we set $g_{LS} = 2.00$ and $g_{HS} = 2.11$[20]). The thermodynamic parameters obtained can help explain critical temperature differences between metal complexes of related ligands. Entropy plays a significant role in spin-state switching preferences, and is often the driving force for transitions to the HS state.[21] Entropically favorable transitions are more likely to occur at lower temperatures. In cases with similar entropy, a lower energy HS state (enthalpically favorable) may be more likely to undergo a spin-transition at a lower temperature.[19]

10.3.2 SQUID Magnetometry

Compared to the Evans method, the SQUID offers higher sensitivity and accuracy, allowing smaller sample sizes and more dilute solutions. Like the Evans method, SQUID measurements probe bulk sample properties and are susceptible to impurities; sample analysis also requires that the mass be known accurately. A wider temperature regime is available in the SQUID magnetometer, however, aggregation problems related to low solubility (crystallization or precipitation) are possible.[22] In cases where glasses (frozen solutions) are used, the molecular environment will approximate solution conditions just above the glass transition temperature.[23] Recent reports illustrate the use of SQUID measurements collected on frozen solutions,[22, 24, 25] including a study of the effect of solvent variations on the spin-transition of one complex.[22]

10.3.3 Electronic Absorption Spectroscopy

Thermochromism is well-recognized for Fe(II) SC complexes: the HS state is much less absorbing than LS; MLCT in LS Fe(II) tends to lead to dark violet colors. Spin-dependent color changes for other SC ions can be less drastic. Many solvents do not have a 150 K liquid range, so quantitatively mapping a gradual-yet-full LS–HS conversion for Fe(II) is not often done. On the other hand, in cases where SC behavior is modulated by host–guest interactions, color changes can characterize noncovalent interactions independently from a spin state change.[26]

UV-visible investigations of SC behavior are potentially more sensitive than Evans method. More dilute solutions can (or sometimes must) be used to track spin state properties, depending on the ligand sets involved. In our experience, solutions of ~0.5 mM are useful for investigating spin state changes as well as ion binding ratios (via Job plot titrations).

10.4 Spin-Crossover in Solution

In this part of our survey, we describe recent examples of SC complexes characterized in solution, highlighting instances where solvent and/or anion influences on SC properties have been observed.

10.4.1 Solution Characterization

Compounds where solution-phase SC has been observed and thermodynamic parameters assessed are collected in Table 10.2. As expected, Fe(II) is the most commonly studied ion in solution, along with a few Co(II) complexes. The first coordination environments are generally FeN_6 with the exceptions of CoP_3O_2 in [Co(**1**)(**2a-c**)]$^{1+}$,[27] and FeN_4O_2 in [Fe(**4**)$(X)_2$] and [Fe_2(**5**)$(X)_4$] (R_1 = Me, R_2 = COOEt or COMe; X = py or meim).[18, 28] The most common solvents employed for these studies are acetone, acetonitrile, methanol and dichloromethane, while the most common anions are BF_4^- and PF_6^-. There is a wide range of SC temperatures and thermodynamic parameters observed in solution, consistent with the wide variety of ligands studied. Some general trends can be seen from this data. Higher spin-transition temperatures yield larger ΔH_{SC} and ΔS_{SC} values, as expected. For a species without signs of ligand dissociation/exchange (*vide infra*), ΔH_{SC} does not exceed 30 kJ/mol and ΔS_{SC} is less than 130 J/mol K. In addition, there seems to be little correlation between ligand denticity or the first coordination sphere and thermodynamic parameters.

Table 10.2 *Solution characterization of SC compounds.*

Complex	Anion	Solvent	$T_{1/2}$ (K)	ΔH_{SC} (kJ/mol)	ΔS_{SC} (J/mol·K)	Ref.
Co(II)						
$[Co(tbta)_2]^{2+}$	ClO_4^-	CH_3CN	~325	13.9	43	44
[Co(**1**)(**2a**)]	–	CD_3OD	*b*	6.82	34.8	27
[Co(**1**)(**2b**)]	–	CD_3OD	*b*	5.83	35.6	27
[Co(**1**)(**2c**)]	–	CD_3OD	*b*	5.36	31.4	27
$[Co_3(dpa)_4Cl_2]^{1+,a}$	BF_4^-	CH_2Cl_2	201	8.7	43	45
$[Co_3(dpa)_4Cl_2]^{1+,a}$	BF_4^-	CH_2Cl_2	281	22.8	81	45
Fe(II)						
$[Fe(1\text{-}bpp)_2]^{2+}$	BF_4^-	$(CD_3)_2CO$	248	24.1	101(1)	46
$[Fe(\mathbf{3})]^{2+}$	OTf^-	$(CD_3)CO$	200	19.0(5)	95(2)	47
$[(Fe(bztpen))_2(N(CN)_2)]^+ \cdot H_2O$	PF_6^-	$(CD_3)_2CO$	258	18.1	70	48
$[(Fe(bztpen))_2(N(CN)_2)]^+$	PF_6^-	$(CD_3)_2CO$	252	12.6	50	48
$[Fe(\mathbf{8a})]^{2+}$	BF_4^-	$(CD_3)_2CO$	279	27	98	49
$[Fe(py5)(N_3)]^+$	OTf^-	CD_3OD	207	6.8	33	50
$[Fe(py5)(MeOH)]^{2+}$	OTf^-	CD_3OD	192	4.3	22	50
$[Fe(btpa)]^{2+}$	PF_6^-	CH_3OH	~310	27.5	88.5	51
$[Fe(H_2bip)_3]^{2+}$	BPh_4^-	CD_2Cl_2	201	11	55	14
$[Fe(pipi)(H_2bip)_2]^{2+}$	BPh_4^-	CD_2Cl_2	367	26	71	14
$[Fe(bpy)(H_2bip)_2]^{2+}$	BPh_4^-	CD_2Cl_2	402	28	70	14
$[Fe(phen)(H_2bip)_2]^{2+}$	BPh_4^-	CD_2Cl_2	386	25	65	14
$[Fe(6\text{-}Mepy)(H_2bip)_2]^{2+}$	Br^-	CD_2Cl_2	242	8.8	36.4	52
$[Fe(6\text{-}Mepy)(H_2bip)_2]^{2+}$	BPh_4^-	CD_2Cl_2	133	3.1	23.3	52
$[Fe(\mathbf{4})(Me/COOEt)(py)_2]$	–	*c*	207	18.2	88	18
$[Fe(\mathbf{4})(Me/COMe)(py)_2]$	–	*c*	211	25.2	121	18
$[Fe_2(\mathbf{5})(Me/COOEt)(py)_4]$	–	*c*	218	28.3	130	18
$[Fe_2(\mathbf{5})(Me/COMe)(py)_4]$	–	*c*	209	22.4	107	18
$[Fe(\mathbf{4})(Me/COMe)(meim)_2]$	–	toluene-d_8/	177	8.0	43.9	28
$[Fe_2(\mathbf{5})(Me/COMe)(meim)_4]$	–	*d*	191	14.9	77.9	28
		d				
$[Fe(\mathbf{6a})]^{2+}$	PF_6^-	CD_3CN	403	22.5	56	53
$[Fe(\mathbf{6b})]^{2+}$	PF_6^-	CD_3CN	375	20.6	55	53
$[Fe(\mathbf{6c})]^{2+}$	PF_6^-	CD_3CN	341	19.1	56	53
$[Fe(3\text{-}bpp)_2]^{2+}$	BF_4^-	CD_3CN	244	26	105	54
$[Fe(3\text{-}bpp)_2]^{2+}$	BF_4^-	$(CD_3)_2CO$	247	25	100	54
$[Fe(3\text{-}bpp)_2]^{2+}$	BF_4^-	CD_3OD	255	23	89	54
$[Fe(3\text{-}bpp)_2]^{2+}$	BF_4^-	$(CD_3)_2NCDO$	257	26	101	54
$[Fe(3\text{-}bpp)_2]^{2+}$	BF_4^-	D_2O	317	21	67	54
$[Fe(tacn)_2]^{2+}$	Br^-	D_2O	343	23	67	55
$[Fe(tacn)_2]^{2+}$	Br^-	DMF-d_7	344	21	61	55
$[Fe(tacn)_2]^{2+}$	Br^-	DMSO-d_6	324	22	68	55
$[Fe(tacn)_2]^{2+}$	Br^-	CD_3CN	318	21	66	55
$[Fe(tacn)_2]^{2+}$	Br^-	$(CD_3)_2CO$	328	24	73	55

[a]Two step process.
[b]Not reported.
[c]toluene-d8/pyridine-d5 (50/50, v/v).
[d]toluene-d8/methanol-d4 (50/50, v/v).

There are numerous other examples of compounds that undergo SC in solution, but their SC parameters are not calculated[29–37] or they will be discussed in other chapters; among these include liquid crystals[38–40] and nanoparticles.[41–43]

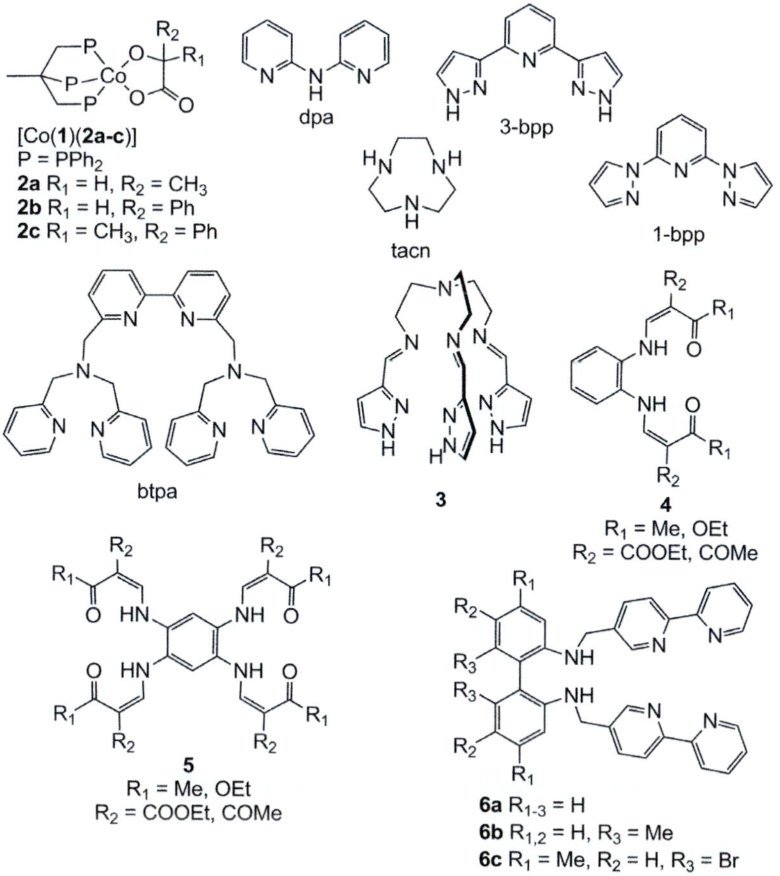

10.4.2 Solvent Effects

Investigations of a compound in multiple solutions are relatively rare. Studies of $[Fe(3\text{-bpp})_2](PF_6)_2$ in aprotic and protic solvents shows that the spin-crossover temperature depends on the hydrogen bonding ability of the solvent: in D_2O, $T_{1/2}$ is 60 K higher than in polar aprotic solvents such as CD_3NO_2 and CD_3CN; in mixed $(CD_3)_2CO/D_2O$ media, $T_{1/2}$ steadily increases with higher D_2O concentrations. In the related HS compound $[Fe(1\text{-bpp})_2](BF_4)_2$, which lacks any hydrogen bonding functionality, isotropic shifts of the paramagnetic compound are not affected by solvent, indicating that hydrogen bonding is operative in stabilizing the low spin state.[54]

SC properties for $[Fe(tacn)_2]^{2+}$ have been examined in a series of polar solvents (D_2O, DMF-d_7, DMSO-d_6, acetonitrile-d_3, acetone-d_6). Solvent choice has little effect on ΔH and ΔS values, which are comparable to those determined by variable temperature UV-vis experiments in D_2O. However, the $T_{1/2}$ values vary by solvent:

those able to participate in hydrogen bonding interactions with the ligand undergo SC at higher temperatures.[55] DFT calculations (B3LYP*) performed on this complex show that the spin state and redox potentials correlate with experimentally obtained data, bridging a gap between computational and experimental determination of spin state properties. Comparison of tridentate $[Fe(tacn)_2]^{2+}$ and monodentate $[Fe(NH_3)_6]^{2+}$ show that two ligand parameters are pivotal for SC: the chemical nature of the coordinating atom, and the ligand framework; for which theory can contribute to future ligand design strategies.[56]

10.4.3 Substituent Effects

Altering the substituents on **7** in $[Fe(\mathbf{7a,b})(cat)]^{+}$ complex cations yields SC that is dependent on the electron-withdrawing group. More electron-withdrawing substituents on **7** weaken Fe-**7** interactions, leading to stronger Fe-catecholate interactions. This in turn stabilizes the LS state. Cooling the **7b** analog in acetonitrile affords partial SC between 303 and 243 K; the behavior is more gradual when the monosubstituted ligand **7a** is used.[57]

A rare example of SC in Cr(II) complexes (d^4) has been investigated. Solid state and solution (toluene) studies of a series of $[Cr(Ind^{xMe})_2]$ sandwich-type complexes show that variations in the position and number of methyl substituents affect the spin states observed. Substitution on the six-membered ring drives SC, while modification of the five-membered ring tends to stabilize the HS state.[58]

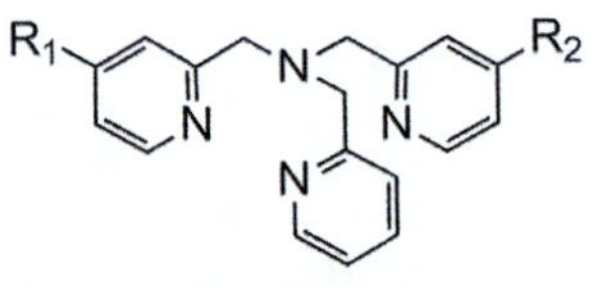

7a $R_1 = H, R_2 = NO_2$
7b $R_1 = NO_2, R_2 = NO_2$

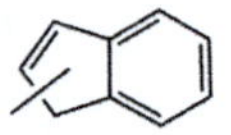

$HInd^{xMe}$

10.5 Ligation Changes Driving Spin-State Switching in Solution

Here, we focus on examples from the literature where controlled changes to the first coordination sphere of complexes results in interesting spin state changes, sometimes in non-SC-capable ions.

10.5.1 Solvent Exchange/Loss

Ligand exchange with coordinating solvents can lead to interesting chemical equilibria and spin state properties. For example, SC is observed in acetonitrile for $[Fe(bpmen)(CH_3CN)_2](ClO_4)_2$, but large ΔS and ΔH values relative to literature precedents suggest additional chemical processes are operative. It is proposed that the HS–LS equilibrium depends on the binding or loss of one acetonitrile molecule: studies in mixed solvents with varying $[CD_3CN]$ show an increase in the HS species $[Fe(bpmen)(CH_3CN)_{2-x}]^{2+}$ as $[CD_3CN]$ decreases. Desolvation of the cation destabilizes the LS state, leading to a decrease in SC temperature. Studies in pure acetone produce a HS complex and no apparent signs of SC.[59]

Co(II) complexes of tripodal ligands based on 1,2,3-triazoles, prepared via "click" chemistry, also show chemical equilibrium-dependent spin state properties. In acetonitrile, the HS complex $[Co(tbta)(CH_3CN)_3](ClO_4)_2$ can be transformed to the SC species $[Co(tbta)_2](ClO_4)_2$ by addition of free ligand; this can be reversed by subsequent addition of metal salt.[44]

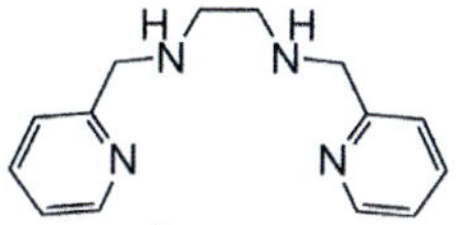

bpmen

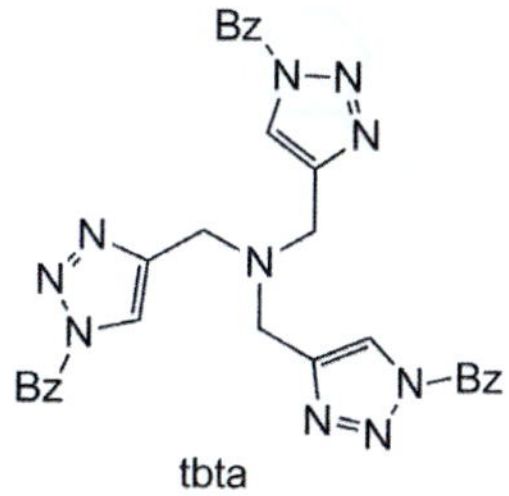

tbta

10.5.2 Anion Exchange/Loss

Tuning the ligand field through direct anion coordination and exchange produces interesting spin behavior in solution. Complexation of the neutral pentadentate ligand py5 with Fe(II) affords a family of $[Fe(py5)(X)]^{n+}$ complexes (X = Cl^-, CH_3CN, CN^-, pyridine, N_3^-, CH_3OH). Coordination of the weak field Cl^- anion generates a HS complex, while CH_3CN, CN^- and pyridine produce purely LS complexes, and N_3^- and CH_3OH produce complexes that undergo SC in CD_3OD. In the case of N_3^- the transition is not complete by 203 K, but is more complete than the CH_3OH derivative.[50] Note that solvent exchange is operative,[60] but the different spin behaviors show influence of the sixth ligand.

Similarly, changing the coordination environment of $[Fe(bztpen)(X)](PF_6)_n$ (X = Cl^-, Br^-, I^-, NCO^-, NCS^-, $N(CN)_2^-$, CH_3CN, CN^-, $n = 1,2$) systematically varies spin states depending on the ligand field strength imparted by the axial species.[61] Monitoring by electronic absorption spectroscopy, HS Fe(II) complexes with Cl^-, Br^-, I^-, NCO^-, NCS^- and $N(CN)_2^-$ display a *d-d* transition at longer wavelengths (850–960 nm), while this transition is not apparent in LS complexes with CH_3CN and CN^-. Cyclic voltammograms probing complex stability show some ligand dissociation operative in CH_3CN solutions as various amounts of $[Fe(bztpen)(CH_3CN)]^{2+}$ are present, with the highest level of exchange occurring for the charge-diffuse iodide, and strongest binding for cyanide.

Complexes of Fe(II) and Co(II) with the bidentate ligands Rdpt (adpt or pldpt) produce HS $[M(Rdpt)_2(C_{16}SO_3)_2]$ in the solid state. In solution, the Co(II) complex maintains its stoichiometry, while the Fe(II) complex equilibrates between the neutral HS complex $[Fe(Rdpt)_2(C_{16}SO_3)_2]$ and cationic LS species $[Fe(Rdpt)_3](C_{16}SO_3)_2$. Addition of excess ligand favors the LS tris complex as monitored by the Evans method and visible/near-IR spectroscopy.[62]

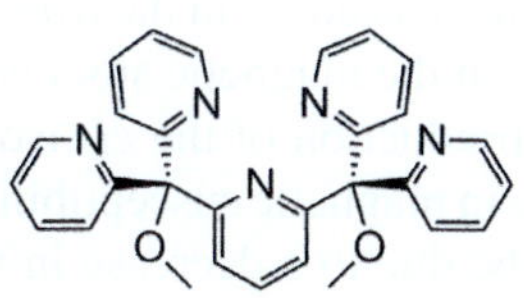

py5

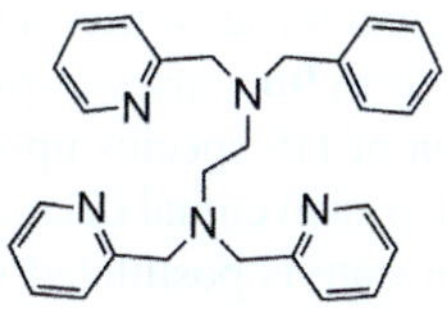

bztpen

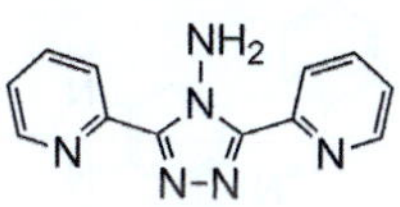

adpt

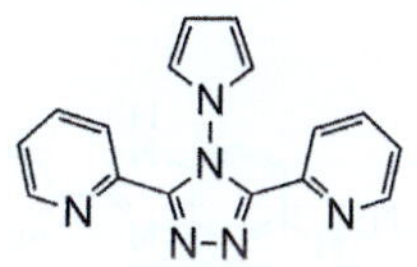

pldpt

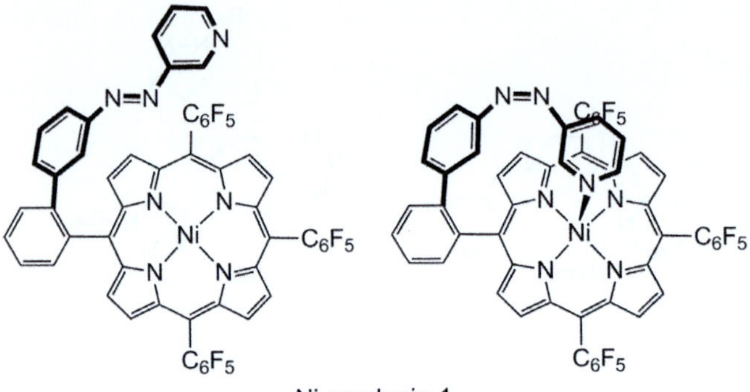

Ni-porphyrin-1

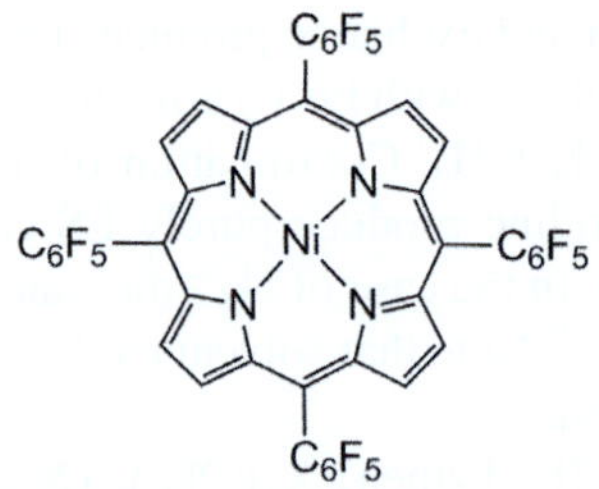

tpfpp

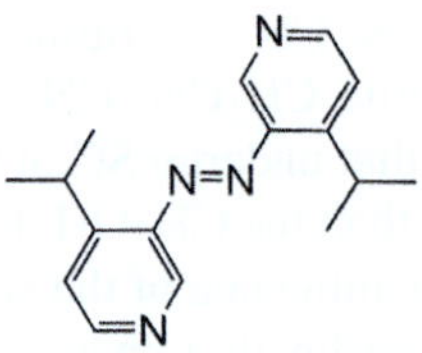

8

Figure 10.1 *Assemblies for combining photo- and spin switching.*

10.5.3 (Photo)Isomerization

There are several examples of altered coordination environments of Ni(II) porphyrins leading to spin state changes.[63–65] In some cases, photoisomerization of coordinating ligands drives spin state changes in solution by association and disassociation of those ligands with the metal center. For Ni-porphyrin-1[65] and [Ni(tpfpp)(**8**)$_n$] (n = 0, 1),[64] isomerization of a photoswitchable ligand changes the Ni(II) metal complex geometry from square-planar to square-pyramidal, causing a diamagnetic-to-paramagnetic switch (Fig. 10.1). Meanwhile, the spin-states of related Fe(III) salten complexes are not affected by photoisomerization of the coordinated azopyridine,[66] indicating that ligand isomerization alone does not impart sufficient energy to alter the spin state.

In contrast, *cis/trans* isomerism appears to be sufficient to tune Fe(II) spin state properties in at least one case. Complexes with the bidentate ligand **9a** show thermal SC in acetone in the *trans* isomer.[49] Photoirradiation at room temperature in acetone to form **9b** causes a spike in the magnetic susceptibility, detected by the Evans method, indicating the formation of HS species upon production of the *cis* isomer. Repeated irradiations to switch between the isomers lead to an eventual decrease in magnetic susceptibility, potentially due to complex degradation. The change in spin state is postulated to be due to a decrease in the π-accepting ability of the *cis* isomer compared to the *trans*.

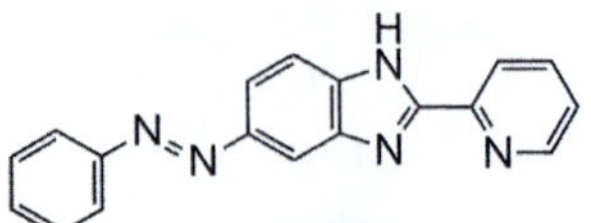

9a

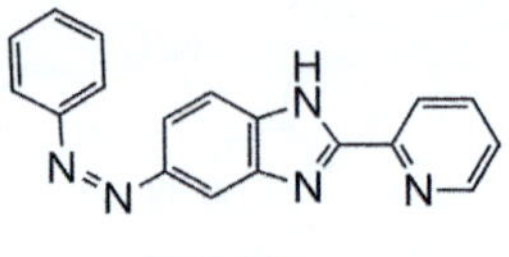

9b

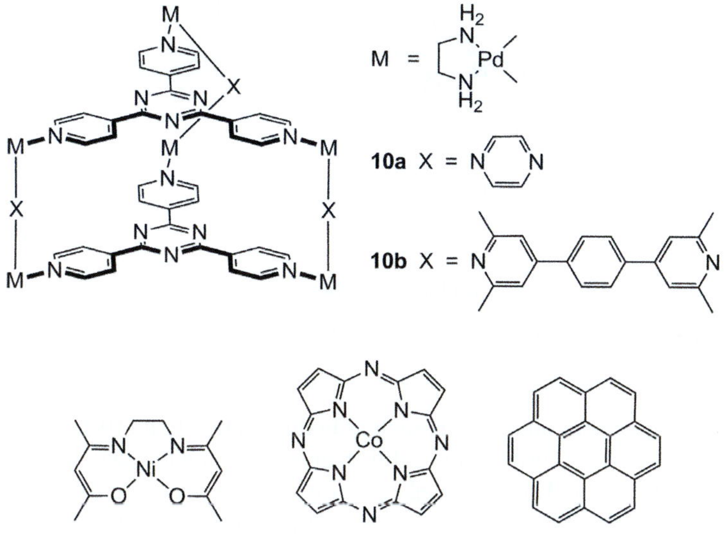

11 **12** **13**

Figure 10.2 Assembly for spin-state switching via encapsulation.

10.5.4 Encapsulation

Well designed host–guest chemistry can effectively alter the first coordination sphere of Co(II) and Ni(II) ions, with concomitant spin state changes (Fig. 10.2).[67] Introduction of **11** into **10a** to produce **10a⊃11** induces a spin state change from diamagnetic to paramagnetic for the Ni(II) complex. To induce SC in a square-planar Co(II) center, sandwiching LS [Co(tap)] (**12**) between two coronene molecules (**13**) and encapsulating within the C_3-symmetric coordination cage **10b** produces the Co(II) SC assembly **10b⊃(13·12·13)**, as detected by variable temperature ESR. The proximity of the coronenes to the metal center alters the ligand field of Co(II) and forms a "pseudo octahedral" ligand field with π-$d_z{}^2$ interactions from coronene and Co(II). Interestingly, encapsulating only [Co(tap)] within the shorter **10a** cage (without coronene) does not induce SC.

10.6 Second Coordination Sphere Triggers for Spin-State Switching

The sensitivity of SC and spin-sensitive coordination complexes to small environmental changes extends beyond the inner coordination sphere. In the solid state, the groups of Reedijk[68] and Kepert[69] first studied the effect of systematically changing guest molecules on SC in Fe(II) coordination network solids. Since then, the subject of chemical triggers for SC in the solid state has been examined by several groups, most notably Kepert and Murray and Real.[69–72] Since 2000, there have been >200 papers studying host–guest effects on complex spin states compared to ~100 papers in the 35 years prior.[†]

Robustly incorporating noncovalent host–guest interactions into spin-state switching schemes is difficult to achieve synthetically: HS–LS transition barriers for Fe(II) (2200–2900 cm^{-1}, 6.3–8.3 kcal/mol),[73] are

[†] For illustration, SciFinder searches for the terms "spin crossover," "valence tautomer" or "metal to metal charge transfer" result in 2177 publications in English between the years 1965–2010. Refinement for the terms "solvent or anion or guest" results in 328 publications over the same time period.

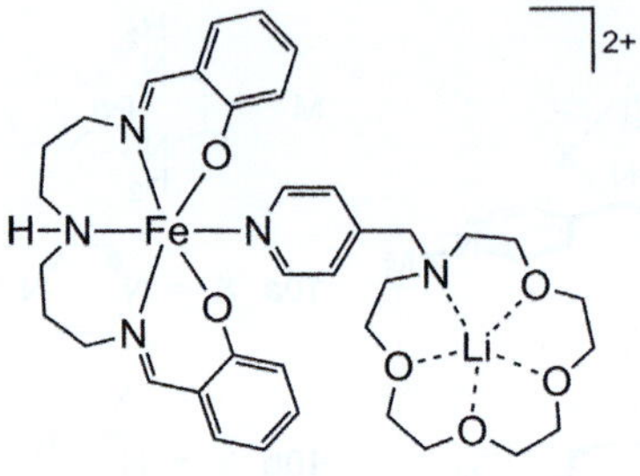

Figure 10.3 *Proposed cation binding spin-switching assembly.*

comparable to common noncovalent interactions such as hydrogen bonding (10.1–13.8 kcal/mol) and anion-π interactions (7.2–8.2 kcal/mol).[74] Further, for useful chemical sensing the switching event must be isolated at the level of host–guest interaction. Efforts to make precise correlations in the solid state through systematic structural variation can be thwarted by capricious crystal chemistry.[75] To our knowledge, the only prior art in solution is one report on alkali metal salts co-precipitated with the Fe(III) complex [Fe(salten)(**14**)]ClO_4 in which a cation binding aza-crown ether is poised to transmit binding as a change in ligand field (Fig. 10.3).[76] Solid state SC is observed for a mixture of the complex salt and $LiClO_4$, compared to HS behavior for the other combinations. In solution, titrations with Li^+ or Na^+ give slight color changes, but these do not correlate with the temperature-dependent spectral changes of the parent complex.

Because cooperativity is weakened or nonexistent in solution, ligand field effects should dominate. Although it is common to denote a spin-transition *point* as a function of ligand field, in reality there is a *range* of Δ/B values where SC-capable complexes will distort to favor the HS or LS configuration.[1, 77] If the ligands position the ground spin state of the complex on either side of the spin-transition, then the small energies associated with host–guest interactions can be amplified by transition through the "forbidden zone" to achieve guest-triggered spin-state switching events. Expected effects on spin state are illustrated for Fe(II) in Figure 10.4.

Since most Fe(II) SC compounds are cationic,[1, 2] and anions play important roles in biological, industrial and environmental processes, focusing on anionic targets represents an attractive option for spin-switching-based sensing architectures. Aside from solvent/side chain influences[78] and anion sterics,[79, 80] it appears that

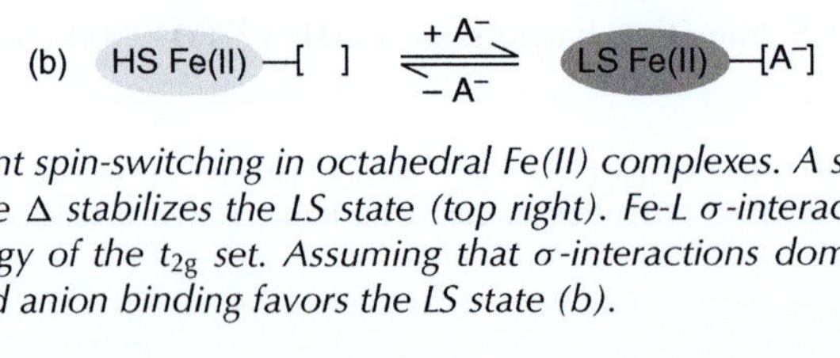

Figure 10.4 *Guest dependent spin-switching in octahedral Fe(II) complexes. A small ligand field (Δ) stabilizes the HS state (top left); a large Δ stabilizes the LS state (top right). Fe-L σ-interactions shift the e_g orbitals, and π-interactions move the energy of the t_{2g} set. Assuming that σ-interactions dominate, external cation binding stabilizes the HS state (a), and anion binding favors the LS state (b).*

hydrogen bonding and anion–ligand interactions primarily increase the σ-donating ability of the ligands bound to Fe(II) ions (Figure 10.1b).[81, 82]

10.6.1 External Anion-Dependent Spin Switching

Our initial efforts have focused on Fe(II) complexes containing the H_2bip ligand, homoleptic complex salts of which were recognized some 30 years ago as showing anion-dependent SC properties in the solid state.[83, 84] The solution spin state of the homoleptic complex $[Fe(H_2bip)_3]^{2+}$ can be correlated to the extent of hydrogen bonding interactions with charge balancing anions.[26] The tetraphenylborate and bromide salts are both SC species in dichloromethane, but at 233 K (–40 °C) they are essentially HS and LS, respectively. At room temperature, titration of the tetraphenylborate salt with Bu_4NBr shows $\delta_{N\text{-}H}$ shifts and small changes in UV-visible spectra, indicating N-H···Br^- interactions. At 233 K, anions capable of chelation by H_2bip displace bulky tetraphenylborate anions and cause HS→LS switching, as monitored by ^{1}H NMR. The affinities track in the order $Br^- > Cl^- > I^- > NO_3^- > ClO_4^-$, consistent with the fitting (or lack thereof) of the anion in the H_2bip pocket, and consistent with binding constants derived from the room-temperature electronic absorption measurements. The results demonstrate that significant spin state changes can be achieved in solution via secondary, noncovalent host–guest interactions.

10.6.2 Using Ligand Fields to Tune Anion Triggered Spin-State Switching in Solution

To establish guest-dependent spin-state switching as a viable chemosensing technique, it is necessary to increase operating temperatures and improve environmental robustness. Strengthening the ligand field may address both concerns simultaneously, since stronger fields stabilize the LS state, which in turn increases SC temperatures and reduces complex lability. Such optimization requires synthetic tunability in the form of heteroleptic complexes, where some ligand(s) are responsible for anion binding while others adjust the ligand field into the near-SC regime. The preparation of $[(H_2bip)_2FeBr_2]$[14] offers a path to a family of heteroleptic Fe(II) anion-binding complexes $[(H_2bip)_2Fe(NN)]^{2+}$; the chelating N-donor (NN) ligands can raise or lower "operating" temperatures (change in ligand field based on $[Fe(NN)_3]^{2+}$,[81] and approximation from Ni(II) analogs[85]), and can include substituents which may interact with anions/solvent (Fig. 10.5).

Strong field ligands (NN = phen, bpy, pipi) promote anion dependent spin-state switching at ambient temperatures with improved complex air stability.[14] Bromide salts are diamagnetic in dichloromethane at room temperature, while the tetraphenylborate salts are ~15% HS. The ^{1}H NMR spectral analyses show direct links between anion binding and spin-state switching for the minority HS species in solution. Importantly, whereas susceptibility changes upon titration are modest ($\Delta\chi_M T < 0.5$ cm^3·K·mol^{-1}), proton signals *not* involved in hydrogen bonding show a steady upfield shift as bromide is introduced, consistent with HS→LS switching. Reversible color changes are observed when $[(H_2bip)_2Fe(phen)]^{2+}$ is mixed with various TBA^+X^- ($X = Cl^-, Br^-, I^-, NO_3^-, ClO_4^-$) salts in dichloromethane.

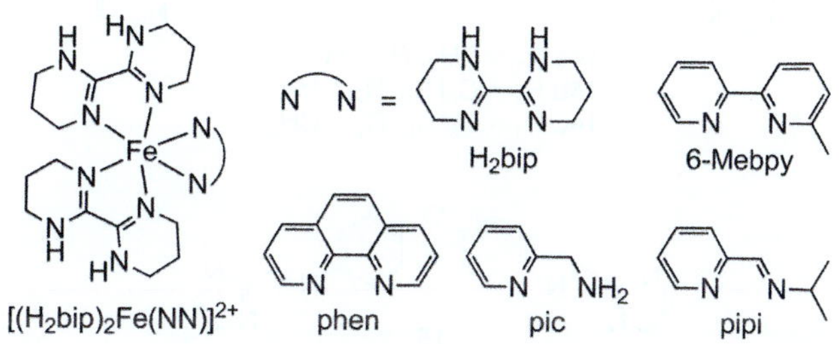

Figure 10.5 *Complexes for external anion triggered spin-state switching.*

Heteroleptic complexes containing weaker field ancillary ligands also show anion triggered spin-state switching, albeit at much lower temperatures than $[Fe(H_2bip)_3]^{2+}$, in accordance with the hypothesis of ligand field matching. One example is $[(H_2bip)_2Fe(6\text{-}Mebpy)]X_2$: the steric bulk of the methyl group adjacent to the N-donor atom stabilizes the HS state significantly.[52]

Additional supramolecular interactions impact spin state sensitivity to anions.[86] The anion-sensitive working temperature for the picolylamine complex $[(H_2bip)_2Fe(pic)]^{2+}$ is much lower than expected given that pic exerts a stronger field than H_2bip. Combining solution and solid state studies implicate ancillary hydrogen bonding between the NH_2 group of pic and solvent and/or anions as a drain on the σ-donation properties of the $H_2bip\cdots Br^-$ assembly, resulting in the suppression of spin-state switching. Note that the solution studies are carried out in the presence of background ligand dissociation processes.

10.7 Challenges and Opportunities

10.7.1 New Opportunities for Anion Reporting in Solution

Less developed than cation sensing, the area of water-stable anion sensors is small but growing,[87–89] and offers many opportunities for new research directions. One target is the development of environmentally relevant anion sensors that exploit spin state reporting. Corralling HS ions in polar solvents poses a significant challenge, and will require novel ligands.

Di- and tripodal ligands offer one potential route to complexes that are stable in polar solvents. Metal complexes with tren-based iminopyridine ligands have been studied in some depth; steric tuning at positions adjacent to metal binding donor atoms can alter the spin state of Fe(II) complexes. Numerous related ligand sets (py substituted for other N-donor groups) have been studied;[1, 47, 90, 91] many of these investigations have been summarized by Halcrow.[92] On the host–guest side, tripodal systems developed by Fabbrizzi and coworkers use cationic imidazolium groups to attract anions even in water.[93] Conceptually, it is not difficult to imagine combining both SC and anion binding into one assembly by relatively small ligand modifications. In principle, these types of ligands offer hexa- or even heptadentate coordination modes and large stability constants, even in polar solvents such as methanol. However, their stability must be probed in detail: both Toftlund[1] and we[94] have observed rapid exchange of solvent with ligand "arms" which could affect host–guest and spin state behavior in unpredictable ways. For stable species, appending hydrogen- or other anion-binding groups would create chelating pockets capable of selective interaction with anions. For hydrogen bonding in polar protic media, encapsulation of the H-bonding groups is likely necessary.[95]

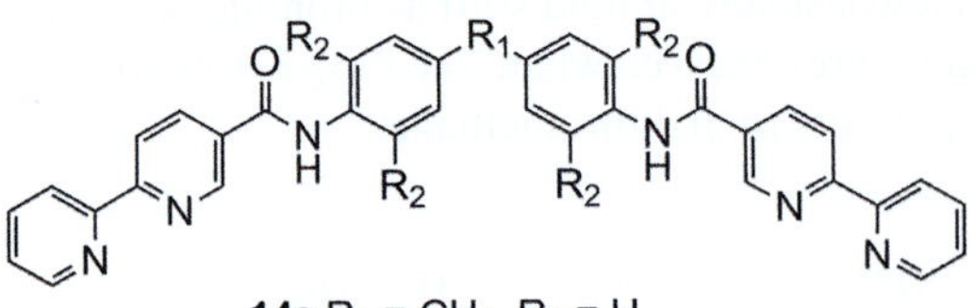

14a $R_1 = CH_2$, $R_2 = H$
14b $R_1 = C_5H_{10}$, $R_2 = H$
14c $R_1 = C_5H_{10}$, $R_2 = CH_3$

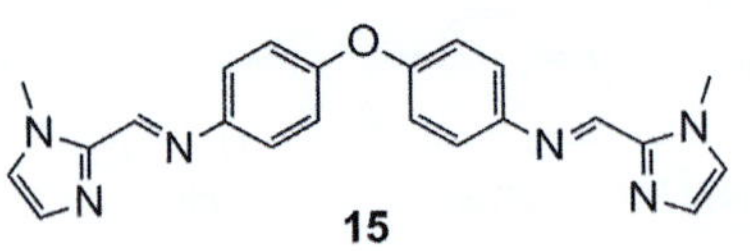

15

Multinuclear species represent another promising route. Kruger and co-workers have examined complexes with potential anion sensor applications. For example, the helicate complex salts $[Fe_2(\mathbf{15a\text{-}c})_3](BF_4)_4$ are selective for Cl^- in DMSO over F^-, Br^- or HSO_4^- (^1H NMR).[96] Although these complexes do not show changes in spin state, the related complex $[Fe_2(\mathbf{16})_3](ClO_4)_4$ shows a SC transition at 150 K in the solid state.[97] Furthermore, an encapsulating ligand similar to **10a** has been shown to fully encapsulate guanine and cytosine base pairs in water, protecting them from interrogation of water or competing base pairs.[95] A complex which combined these properties would exhibit anion sensing in a relatively polar solvent. More generally, the development of multinuclear SC systems is a promising way to target sharp spin-transitions in solution media, toward a solution variant of cooperativity.

10.7.2 MRI Contrast

An alternative spin reporting mechanism would rely on access to the first coordination sphere. Magnetic resonances for water molecules involved in ligand substitution on paramagnetic metal ions shift and relax differently from bulk water, leading to contrast. Recent investigations of iron-based contrast agents illustrate the differences between diamagnetic LS and paramagnetic HS Fe(II) species.[98] Here, solution spin state properties become more important than the solid state properties. For example, colorless HS $[Fe(Me_3tacn)(OTf)_2]$ becomes SC in solutions when the anions can be replaced by polar solvent molecules.[99] Solvent- and anion-dependent spin state properties surveyed above could be applied to "turn on" sensors where a LS→HS conversion would activate MR contrast.

10.8 Conclusions/Outlook

Spin-state switching in solution is highly sensitive to small environmental changes. Extensive efforts have been devoted to exploring the influence of solvent, anions and/or salts on SC properties, and some useful structure–property correlations have been obtained. Due to the complexity of complex behavior in solution, systematic studies in solution are still relatively scarce; nevertheless such experiments offer an alternative method for assessing the effects of secondary noncovalent interactions on spin state properties. Overall, this represents a promising yet underdeveloped area in inorganic chemical research.

Acknowledgement

We thank the NSF (CHE-1058889) and Colorado State University for support. MPS thanks Z. Ni and A. McDaniel for their efforts toward the results presented in Chapter 10.6.

Abbreviations

γ_{HS}	high spin mole fraction
MLCT	metal-to-ligand charge transfer
1-bpp	2,6-di(pyrazol-1-yl)pyridine
3-bpp	2,6-di(pyrazol-3-yl)pyridine
6-Mebpy	6-methyl-2,2′-bipyridine
adpt	4-amino-3,5-bis(2-pyridyl)-1,2,4-triazole
bpmen	*N,N*′-dimethyl-*N,N*′-bis(2-pyridylmethyl)-1,2-diaminoethane
bpy	2,2′-bipyridine

btpa	*N,N,N′,N′*-tetrakis(2-pyridylmethyl)-6,6′-bis(aminomethyl)-2,2′-bipyridine
bztpen	tris(4-[pyrazol-3-yl]-3-aza-3-butenyl)amine)
cat	catecholate
dpa	anion of di-2-pyridylamine
DMF	dimethylformamide
DMSO	dimethylsulfoxide
H_2bip	2,2′-bi-1,4,5,6-tetrahydropyrimidine
HS	high spin
ind	indene
LS	low spin
meim	*N*-methylimidazole
phen	1,10-phenanthroline
pic	picolylamine
pipi	2-pyridinalisopropylimine
pldpt	4-pyrrolyl-3,5-bis(2-pyridyl)-1,2,4-triazole
py	pyridine
pybim	2-(2′-pyridyl)benzimidazole
py5	2,6-bis(bis(2-pyridyl)methoxymethane)pyridine
salten	4-aza-heptamethlene-1,7-bis(salicylideneiminate)
SC	spin crossover
$T_{1/2}$	critical temperature where equimolar amounts of HS and LS states are present
tacn	1,4,7-triazacyclononane
tap	tetraazaporphinato
tbta	tris[(1-benzyl-1H-1,2,3-triazol-4-yl)methyl]amine
tpa	tris(pyridin-2-ylmethyl)amine
tpfpp	tetrakis(pentafluorophenyl)porphyrin

References

1. Toftlund, H. (1989). Spin equilibria in iron(II) complexes. Coord. Chem. Rev., 94: 67–108.
2. Toftlund, H. (2001). Spin equilibrium in solutions. Monatsh. Chem., 132: 1269–1277.
3. Brady, C., McGarvey, J. J., McCusker, J. K., Toftlund, H., Hendrickson, D. N. (2004). Time-resolved relaxation studies of spin crossover systems in solution, in: Gütlich, P., Goodwin, H. A. (Eds) Spin Crossover in Transition Metal Compounds III. Top. Curr. Chem., 235: 1–22.
4. Toftlund, H., McGarvey, J. J. (2004). Iron(II) spin crossover systems with multidentate ligands, in: Gütlich, P., Goodwin, H. A. (Eds) Spin Crossover in Transition Metal Compounds I. Top. Curr. Chem., 233: 151–166.
5. Hassan, N., Koudriavtsev, A. B., Linert, W. (2008). Isoequilibrium relationships and cooperative effects in spin-state transitions in solution. Pure Appl. Chem., 80: 1281–1292.
6. Miller, J. S., Min, K. S. (2009). Oxidation leading to reduction: redox-induced electron transfer (RIET). Angew. Chem. Int. Ed., 48: 262–272.
7. Haynes, W. M. (Ed.) CRC Handbook of Chemistry and Physics, 92nd edn. CRC Press, Boca Raton, FL., USA.
8. Yaws, C. L. (2003) Yaws' handbook of thermodynamic and physical properties of chemical compounds. Knovel Corp., http://www.knovel.com.
9. Evans, D. F. (1959) The determination of the paramagnetic susceptibility of substances in solution by nuclear magnetic resonance. J. Chem. Soc., 2003–2005.
10. Crawford, T. H., Swanson, J. (1971) Temperature dependent magnetic measurements and structural equilibria in solution. J. Chem. Educ., 48: 382–386.

11. Live, D. H., Chan, S. I. (1970). Bulk susceptibility corrections in nuclear magnetic resonance experiments using superconducting solenoids. Anal. Chem., 42: 791–792.
12. Schubert, E. M. (1992) Utilizing the Evans method with a superconducting NMR spectrometer in the undergraduate laboratory. J. Chem. Educ., 69: 62.
13. Grant, D. H. (1995). Paramagnetic susceptibility by NMR: the "solvent correction" reexamined. J. Chem. Educ., 72: 39–40.
14. Ni, Z., McDaniel, A. M., Shores, M. P. (2010). Ambient temperature anion-dependent spin state switching observed in "mostly low spin" heteroleptic iron(II) diimine complexes. Chem. Sci., 1: 615–621.
15. Ostfeld, D., Cohen, I. A. (1972). A cautionary note on the use of the Evans method for magnetic moments. J. Chem. Educ., 49: 829.
16. Yatsunyk, L. A., Walker, F. A. (2004). Structural, NMR, and EPR studies of $S = {}^1/_2$ and $S = {}^3/_2$ Fe(III) bis(4-cyanopyridine) complexes of dodecasubstituted porphyrins. Inorg. Chem., 43: 757–777.
17. Weber, B. (2009). Spin crossover complexes with N_4O_2 coordination sphere - the influence of covalent linkers on cooperative interactions. Coord. Chem. Rev., 253: 2432–2449.
18. Weber, B., Walker, F. A. (2007). Solution NMR studies of iron(II) spin-crossover complexes. Inorg. Chem., 46: 6794–6803.
19. Berry, J. F., Cotton, F. A., Lu, T., Murillo, C. A. (2003). Increasing solubility and stability of linear tricobalt(II) chains with depa (diethyldipyridylamide) ligands. Inorg. Chem., 42: 4425–4430.
20. Ozarowski, A., Zvyagin, S. A., Reiff, W. M., Telser, J., Brunel, L.-C., Krzystek, J. (2004). High-frequency and -field epr of a pseudo-octahedral complex of high-spin Fe(II): bis(2,2′-bi-2-thiazoline)bis(isothiocyanato)iron(II). J. Am. Chem. Soc., 126: 6574–6575.
21. Gütlich, P., Hauser, A., Spiering, H. (1994). Thermal and optical switching of iron(II) complexes. Angew. Chem. Int. Ed., 33: 2024–2054.
22. Siretanu, D., Li, D., Buisson, L., Bassani, D. M., Holmes, S. M., Mathonière, C., Clérac, R. (2011). Controlling thermally induced electron transfer in cyano-bridged molecular squares: from solid state to solution. Chem. Eur. J., 17: 11704–11708.
23. König, E., Ritter, G., Goodwin, H. A. (1977). Mössbauer effect studies of high-spin($^5T_2$) $\rightleftharpoons$ low-spin($^1A_1$) transitions in transitions in frozen solutions of iron(II) complexes - 1 bis[2-(2-pyridylamino)-4-(2-pyridyl)thiazolato]iron(II). J. Inorg. Nucl. Chem., 39: 1131–1135.
24. Inglis, R., Bendix, J., Brock-Nannestad, T., Weihe, H., Brechin, E. K., Piligkos, S. (2010). Frozen-solution magnetisation dynamics of hexanuclear oxime-based Mn^{III} single-molecule magnets. Chem. Sci., 1: 631–636.
25. Tobinaga, H., Suehiro, M., Ito, T., Zhou, G., Karasawa, S., Koga, N. (2007). Magnetic property of 1:2 mixture of Co(*p*-tolsal)$_2$; *p*-tolsal = *N*-*p*-tolylsalicylideniminato, and cyclic pentacarbene-pyridine with $S = {}^{10}/_2$ in dilute frozen solution. Polyhedron, 26: 1905–1911.
26. Ni, Z., Shores, M. P. (2009). Magnetic observation of anion binding in iron coordination complexes: toward spin-switching chemosensors. J. Am. Chem. Soc., 131: 32–33.
27. Enamullah, M., Hasegawa, M., Fukuda, Y., Linert, W., Hoshi, T. (2002). Synthesis, characterization and spin-crossover behaviors of [Co(hydroxycarboxylato)(triphos)] complexes. Bull. Chem. Soc. Jpn., 75: 2449–2453.
28. Weber, B., Kaps, E. S., Obel, J., Achterhold, K., Parak, F. G. (2008). Synthesis and characterization of a dinuclear iron(II) spin crossover complex with wide hysteresis. Inorg. Chem., 47: 10779–10787.
29. Charbonnière, L. J., Williams, A. F., Piguet, C., Bernardinelli, G., Rivara-Minten, E. (1998). Structural, magnetic, and electrochemical properties of dinuclear triple helices: comparison with their mononuclear analogues. Chem. Eur. J., 4: 485–493.
30. Doukov, T., Li, H., Sharma, A., Martell, J. D., Soltis, S. M., Silverman, R. B., Poulos, T. L. (2011). Temperature-dependent spin crossover in neuronal nitric oxide synthase bound with the heme-coordinating thioether inhibitors. J. Am. Chem. Soc., 133:, 8326–8334.
31. Draksharapu, A., Li, Q., Logtenberg, H., van den Berg, T. A., Meetsma, A., Killeen, J. S., Feringa, B. L., Hage, R., Roelfes, G., Browne, W. R. (2012). Ligand exchange and spin state equilibria of Fe(II)(N4Py) and related complexes in aqueous media. Inorg. Chem., 51: 900–913.
32. Goodman, M. A., Nazarenko, A. Y., Casavant, B. J., Li, Z., Brennessel, W. W., DeMarco, M. J., Long, G., Goodman, M. S. (2012). Tris(5-methylpyrazolyl)methane: synthesis and properties of its iron(II) complex. Inorg. Chem., 51: 1084–1093.

33. Hain, S. K., Heinemann, F. W., Gieb, K., Mueller, P., Hoerner, G., Grohmann, A. (2010). On the spin behaviour of iron(II)–dipyridyltriazine complexes and their performance as thermal and photonic spin switches. Eur. J. Inorg. Chem., 221–232.
34. Lawson, D. L. M., Hauser, A. (2010). Ab initio molecular dynamics study of an aqueous solution of $[Fe(bpy)_3](Cl)_2$ in the low-spin and in the high-spin states. J. Phys. Chem. Lett., 1: 1830–1835.
35. Min, K. S., DiPasquale, A. G., Rheingold, A. L., White, H. S., Miller, J. S. (2009). Observation of redox-induced electron transfer and spin crossover for dinuclear cobalt and iron complexes with the 2,5-di-*tert*-butyl-3,6-dihydroxy-1,4-benzoquinonate bridging ligand. J. Am. Chem. Soc., 131: 6229–6236.
36. Pritchard, R., Barrett, S. A., Kilner, C. A., Halcrow, M. A. (2008). The influence of ligand conformation on the thermal spin transitions in iron(III) saltrien complexes. Dalton Trans., 3159–3168.
37. Telfer, S. G., Bocquet, B., Williams, A. F. (2001). Thermal spin crossover in binuclear iron(II) helicates: negative cooperativity and a mixed spin state in solution. Inorg. Chem., 40: 4818–4820.
38. Gandolfi, C., Moitzi, C., Schurtenberger, P., Morgan, G. G., Albrecht, M. (2008). Improved cooperativity of spin-labile iron(III) centers by self-assembly in solution. J. Am. Chem. Soc., 130: 14434–14435.
39. Martinho, P. N., Ortin, Y., Gildea, B., Gandolfi, C., McKerr, G., O'Hagan, B., Albrecht, M., Morgan, G. G. (2012). Inducing hysteretic spin crossover in solution. Dalton Trans., 41: 7461–7463.
40. Schlamp, S., Thoma, P., Weber, B. (2012). New octahedral, head–tail iron(II) complexes with spin crossover properties. Eur. J. Inorg. Chem., 2759–2768.
41. Coronado, E., Galán-Mascarós, J. R., Monrabal-Capilla, M., García-Martínez, J., Pardo-Ibáñez, P. (2007). Bistable spin-crossover nanoparticles showing magnetic thermal hysteresis near room temperature. Adv. Mater., 19: 1359–1361.
42. Galán-Mascarós, J. R., Coronado, E., Forment-Aliaga, A., Monrabal-Capilla, M., Pinilla-Cienfuegos, E., Ceolin, M. (2010). Tuning size and thermal hysteresis in bistable spin crossover nanoparticles. Inorg. Chem., 49: 5706–5714.
43. Titos-Padilla, S., Herrera, J. M., Chen, X.-W., Delgado, J. J., Colacio, E. (2011). Bifunctional hybrid SiO_2 nanoparticles showing synergy between core spin crossover and shell luminescence properties. Angew. Chem. Int. Ed., 50: 3290–3293.
44. Schweinfurth, D., Weisser, F., Bubrin, D., Bogani, L., Sarkar, B. (2011). Cobalt complexes with "click"-derived functional tripodal ligands: spin crossover and coordination ambivalence. Inorg. Chem., 50: 6114–6121.
45. Clérac, R., Cotton, F. A., Dunbar, K. R., Lu, T., Murillo, C. A., Wang, X. (2000). A new linear tricobalt compound with di(2-pyridyl)amide (dpa) ligands: two-step spin crossover of $[Co_3(dpa)_4Cl_2][BF_4]$. J. Am. Chem. Soc., 122: 2272–2278.
46. Holland, J. M., McAllister, J. A., Kilner, C. A., Thornton-Pett, M., Bridgeman, A. J., Halcrow, M. A. (2002). Stereochemical effects on the spin-state transition shown by salts of $[FeL_2]^{2+}$ [L = 2,6-di(pyrazol-1-yl)pyridine]. J. Chem. Soc. Dalton Trans., 548–554.
47. Lazar, H. Z., Forestier, T., Barrett, S. A., Kilner, C. A., Létard, J.-F., Halcrow, M. A. (2007). Thermal and light-induced spin-crossover in salts of the heptadentate complex [*tris*(4-{pyrazol-3-yl}-3-aza-3-butenyl)amine]iron(II). Dalton Trans., 4276–4285.
48. Ortega-Villar, N., Thompson, A. L., Muñoz, M. C., Ugalde-Saldívar, V. M., Goeta, A. E., Moreno-Esparza, R., Real, J. A. (2005). Solid- and solution-state studies of the novel μ-dicyanamide-bridged dinuclear spin-crossover system $\{[(Fe(bztpen)]_2[\mu\text{-}N(CN)_2]\}(PF_6)_3 \cdot nH_2O$. Chem. Eur. J., 11: 5721–5734.
49. Hasegawa, Y., Kume, S., Nishihara, H. (2009). Reversible light-induced magnetization change in an azobenzene-attached pyridylbenzimidazole complex of iron(II) at room temperature. Dalton Trans., 280–284.
50. Rudd, D. J., Goldsmith, C. R., Cole, A. P., Stack, T. D. P., Hodgson, K. O., Hedman, B. (2005). X-ray absorption spectroscopic investigation of the spin-transition character in a series of single-site perturbed iron(II) complexes. Inorg. Chem., 44: 1221–1229.
51. Schenker, S., Stein, P. C., Wolny, J. A., Brady, C., McGarvey, J. J., Toftlund, H., Hauser, A. (2001). Biphasic behavior of the high-spin → low-spin relaxation of $[Fe(btpa)](PF_6)_2$ in solution (btpa = *N,N,N′,N′*-Tetrakis(2-pyridylmethyl)-6,6′-bis(aminomethyl)-2,2′-bipyridine). Inorg. Chem., 40: 134–139.
52. Ni, Z., Fiedler, S. R., Shores, M. P. (2011). Investigation of anion-dependence in the spin-state switching properties of $[(H_2bip)_2Fe(6\text{-}Mebpy)]X_2$. Dalton Trans. 40: 944–950.

53. Petzold, H., Heider, S. (2011). *N,N′*-Bis(2,2′-bipyridine-6-ylmethyl)-2,2′-biphenylenediamines: a tuneable ligand scaffold for room temperature Fe^{2+} SCO complexes. Eur. J. Inorg. Chem., 1249–1254.
54. Barrett, S. A., Kilner, C. A., Halcrow, M. A. (2011). Spin-crossover in $[Fe(3\text{-bpp})_2][BF_4]_2$ in different solvents–a dramatic stabilisation of the low-spin state in water. Dalton Trans., 40: 12021–12024.
55. Turner, J. W., Schultz, F. A. (2001). Solution characterization of the iron(II) bis(1,4,7-triazacyclononane) spin-equilibrium reaction. Inorg. Chem. 40: 5296–5298.
56. Lord, R. L., Schultz, F. A., Baik, M.-H. (2009). Spin crossover-coupled electron transfer of $[M(tacn)_2]^{3+/2+}$ complexes (tacn = 1,4,7-triazacyclononane; M = Cr, Mn, Fe, Co, Ni). J. Am. Chem. Soc., 131: 6189–6197.
57. Hitomi, Y., Higuchi, M., Minami, H., Tanaka, T., Funabiki, T. (2005). Tuning of spin crossover equilibrium in catecholatoiron(III) complexes by supporting ligands. Chem. Commun., 1758–1760.
58. Meredith, M. B., Crisp, J. A., Brady, E. D., Hanusa, T. P., Yee, G. T., Pink, M., Brennessel, W. W., Young, V. G. jr. (2008). Tunable spin-crossover behavior in polymethylated bis(indenyl)chromium(II) complexes: the significance of benzo-ring substitution. Organometallics, 27: 5464–5473.
59. Bryliakov, K. P., Duban, E. A., Talsi, E. P. (2005). The nature of the spin-state variation of $[Fe^{II}(BPMEN)(CH_3CN)_2](ClO_4)_2$ in solution. Eur. J. Inorg. Chem., 72–76.
60. Goldsmith, C. R., Jonas, R. T., Cole, A. P., Stack, T. D. P. (2002). A spectrochemical walk: single-site perturbation within a series of six-coordinate ferrous complexes. Inorg. Chem., 41: 4642–4652.
61. Ortega-Villar, N., Ugalde-Saldívar, V. M., Muñoz, M. C., Ortiz-Frade, L. A., Alvarado-Rodríguez, J. G., Real, J. A., Moreno-Esparza, R. (2007). Synthesis and relative stability of a series of compounds of type $[Fe(II)(bztpen)X]^+$, where bztpen = pentadentate ligand, N_5, and X^- = monodentate anion. Inorg. Chem., 46: 7285–7293.
62. White, N. G., Feltham, H. L. C., Gandolfi, C., Albrecht, M., Brooker, S. (2010). Towards Langmuir–Blodgett films of magnetically interesting materials: solution equilibria in amphiphilic iron(II) complexes of a triazole-containing ligand. Dalton Trans., 39: 3751–3758.
63. Thies, S., Bornholdt, C., Köhler, F., Sönnichsen, F. D., Näther, C., Tuczek, F., Herges, R. (2010). Coordination-induced spin crossover (CISCO) through axial bonding of substituted pyridines to nickel–porphyrins: σ-donor versus π-acceptor effects. Chem. Eur. J., 16: 10074–10083.
64. Thies, S., Sell, H., Schuett, C., Bornholdt, C., Näther, C., Tuczek, F., Herges, R. (2011). Light-induced spin change by photodissociable external ligands: a new principle for magnetic switching of molecules. J. Am. Chem. Soc., 133: 16243–16250.
65. Venkataramani, S., Jana, U., Dommaschk, M., Sönnichsen, F. D., Tuczek, F., Herges, R. (2011). Magnetic bistability of molecules in homogeneous solution at room temperature. Science, 331: 445–448.
66. Bannwarth, A., Schmidt, S. O., Peters, G., Sönnichsen, F. D., Thimm, W., Herges, R., Tuczek, F. (2012). Fe^{III} spin-crossover complexes with photoisomerizable ligands: experimental and theoretical studies on the ligand-driven light-induced spin change effect. Eur. J. Inorg. Chem., 2776–2783.
67. Ono, K., Yoshizawa, M., Akita, M., Kato, T., Tsunobuchi, Y., Ohkoshi, S., Fujita, M. (2009). Spin crossover by encapsulation. J. Am. Chem. Soc., 131: 2782–2783.
68. Roubeau, O., Gomez, J. M. A., Balskus, E., Kolnaar, J. J. A., Haasnoot, J. G., Reedijk, J. (2001). Spin-transition behaviour in chains of Fe^{II} bridged by 4-substituted 1,2,4-triazoles carrying alkyl tails. New J. Chem., 25: 144–150.
69. Halder, G. J., Kepert, C. J., Moubaraki, B., Murray, K. S., Cashion, J. D. (2002). Guest-dependent spin crossover in a nanoporous molecular framework material. Science, 298: 1762–1765.
70. Neville, S. M., Halder, G. J., Chapman, K. W., Duriska, M. B., Southon, P. D., Cashion, J. D., Létard, J.-F., Moubaraki, B., Murray, K. S., Kepert, C. J. (2008). Single-crystal to single-crystal structural transformation and photomagnetic properties of a porous iron(II) spin-crossover framework. J. Am. Chem. Soc., 130: 2869–2876.
71. Nihei, M., Han, L. Q., Oshio, H. (2007). Magnetic bistability and single-crystal-to-single-crystal transformation induced by guest desorption. J. Am. Chem. Soc., 129: 5312–5313.
72. Quesada, M., de la Peña-O'Shea V. A., Aromí, G., Geremia, S., Massera, C., Roubeau, O., Gamez, P., Reedijk, J. (2007). A molecule-based nanoporous material showing tuneable spin-crossover behavior near room temperature. Adv. Mater., 19: 1397–1402.
73. König, E. (1991). Nature and dynamics of the spin-state interconversion in metal complexes. Struct. Bonding, 76: 51–152.

74. Alberto, M. E., Mazzone, G., Russo, N., Sicilia, E. (2010). The mutual influence of non-covalent interactions in π-electron deficient cavities: the case of anion recognition by tetraoxacalix[2]arene[2]triazine. Chem. Commun., 46: 5894–5896.
75. Hostettler, M., Törnroos, K. W., Chernyshov, D., Vangdal, B., Bürgi, H.-B. (2004) Challenges in engineering spin crossover: structures and magnetic properties of six alcohol solvates of iron(II) tris(2-picolylamine) dichloride. Angew. Chem. Int. Ed., 43: 4589–4594.
76. Maeda, Y., Suzuki, M., Hirose, S., Hayami, S., Oniki, T., Sugihara, S. (1998). Effects of the encapsulation of alkali ions into iron(III) complexes with a macrocyclic ring on magnetic properties. Bull. Chem. Soc. Jpn., 71: 2837–2843.
77. Hauser, A. (2004). Ligand field theoretical considerations, in: Gütlich, P., Goodwin, H. A. (Eds) Spin Crossover in Transition Metal Compounds I. Top. Curr. Chem., 233: 49–58.
78. Zhang, W., Zhao, F., Liu, T., Yuan, M., Wang, Z. M., Gao, S. (2007). Spin crossover in a series of iron(II) complexes of 2-(2-alkyl-2*H*-tetrazol-5-yl)-1,10-phenanthroline: effects of alkyl side chain, solvent, and anion. Inorg. Chem. 2007, 46, 2541–2555.
79. Costa, J. S., Lappalainen, K., de Ruiter, G., Quesada, M., Tang, J. K., Mutikainen, I., Turpeinen, U., Grunert, C. M., Gütlich, P., Lazar, H. Z., Létard, J.-F., Gamez, P., Reedijk, J. (2007). Remarkable steric effects and influence of monodentate axial ligands L on the spin-crossover properties of *trans*-$[Fe^{II}(N_4$ Ligand)L] complexes. Inorg. Chem., 46: 4079–4089.
80. Yamada, M., Hagiwara, H., Torigoe, H., Matsumoto, N., Kojima, M., Dahan, F., Tuchagues, J.-P., Re, N., Iijima, S. (2006). A variety of spin-crossover behaviors depending on the counter anion: two-dimensional complexes constructed by $NH\cdots Cl^-$ hydrogen bonds, $[Fe^{II}H_3L^{Me}]Cl\cdot X$ ($X = PF_6^-$, AsF_6^-, SbF_6^-, $CF_3SO_3^-$; H_3L^{Me} = tris[2-{[(2-methylimidazol-4-yl)methylidene]amino}ethyl]amine). Chem. Eur. J., 12: 4536–4549.
81. Goodwin, H. A. (2004). Spin crossover in iron(II) tris(diimine) and bis(terimine) systems, in: Gütlich, P., Goodwin, H. A. (Eds) Spin Crossover in Transition Metal Compounds I. Top. Curr. Chem., 233: 59–90.
82. Lemercier, G., Bréfuel, N., Shova, S., Wolny, J. A., Dahan, F., Verelst, M., Paulsen, H., Trautwein, A. X., Tuchagues, J.-P. (2006). A range of spin-crossover temperature $T_{1/2} > 300$ K results from out-of-sphere anion exchange in a series of ferrous materials based on the 4-(4-imidazolylmethyl)-2-(2-imidazolylmethyl)imidazole (trim) ligand, $[Fe(trim)_2]X_2$ (X = F, Cl, Br, I): comparison of experimental results with those derived from density functional theory calculations. Chem. Eur. J., 12: 7421–7432.
83. Burnett, M. G., McKee, V., Nelson, S. M. (1981). $^5T_2 \rightleftharpoons {}^1A_1$ and $^6A_1 \rightleftharpoons {}^2T_2$ spin transitions in iron(II) and iron(III) complexes of 2,2′-bi-2-imidazoline and related ligands. J. Chem. Soc. Dalton Trans., 7: 1492–1497.
84. Yoder, J. C., Roth, J. P., Gussenhoven, E. M., Larsen, A. S., Mayer, J. M. (2003). Electron and hydrogen-atom self-exchange reactions of iron and cobalt coordination complexes. J. Am. Chem. Soc., 125: 2629–2640.
85. Robinson, M. A., Busch, D. H., Curry, J. D. (1963). Complexes derived from strong field ligands. XVII. Electronic spectra of octahedral nickel(II) complexes with ligands of the α-diimine and closely related classes. Inorg. Chem., 2: 1178–1181.
86. Ni, Z., Shores, M. P. (2010). Supramolecular effects on anion-dependent spin-state switching properties in heteroleptic iron(II) complexes. Inorg. Chem., 49: 10727–10735.
87. Kubik, S. (2010). Anion recognition in water. Chem. Soc. Rev., 39: 3648–3663.
88. Mannel-Croise, C., Meister, C., Zelder, F. (2010). "Naked-eye" screening of metal-based chemosensors for biologically important anions. Inorg. Chem., 49: 10220–10222.
89. Vallejos, S., Estévez, P., García, F. C., Serna, F., de la Peña, J. L., García, J. M. (2010). Putting to work organic sensing molecules in aqueous media: fluorene derivative-containing polymers as sensory materials for the colorimetric sensing of cyanide in water. Chem. Commun., 46: 7951–7953.
90. Hoselton, M. A., Wilson, L. J., Drago, R. S. (1975). Substituent effects on the spin equilibrium observed with hexadentate ligands on iron(II). J. Am. Chem. Soc., 97: 1722–1729.
91. Brewer, G., Butcher, R. J., Viragh, C., White, G. (2007). Supramolecular assemblies prepared from an iron(II) tripodal imidazole complex. A molecular scaffolding for the self assembly of icosahedral complexes of K^+, Rb^+, Cs^+ and NH_4^+ cations. Dalton Trans., 4132–4142.
92. Halcrow, M. A. (2007). The spin-states and spin-transitions of mononuclear iron(II) complexes of nitrogen-donor ligands. Polyhedron, 26: 3523–3576.

93. Amendola, V., Boiocchi, M., Colasson, B., Fabbrizzi, L., Douton, M. J. R., Ugozzoli, F. (2006). A metal-based *tris*imidazolium cage that provides six C–H hydrogen-bond-donor fragments and includes anions. Angew. Chem. Int. Ed., 45: 6920–6924.
94. Klug, C. M., McDaniel, A. M., Fiedler, S. R., Schulte, K. A., Newell, B. S., Shores, M. P. (2012). Anion dependence in the spin-crossover properties of a Fe(II) podand complex. Dalton Trans., 41: 12577–12585.
95. Sawada, T., Fujita, M. (2010). A single Watson-Crick G·C base pair in water: aqueous hydrogen bonds in hydrophobic cavities. J. Am. Chem. Soc., 132: 7194–7201.
96. Goetz, S., Kruger, P. E. (2006). A new twist in anion binding: metallo-helicate hosts for anionic guests. Dalton Trans., 1277–1284.
97. Pelleteret, D., Clérac, R., Mathonière, C., Harté, E., Schmitt, W., Kruger, P. E. (2009). Asymmetric spin crossover behaviour and evidence of light-induced excited spin state trapping in a dinuclear iron(II) helicate. Chem. Commun., 221–223.
98. Dorazio, S. J., Morrow, J. R. (2012). The development of iron(II) complexes as paraCEST MRI contrast agents. Eur. J. Inorg. Chem., 2006–2014.
99. Blakesley, D. W., Payne, S. C., Hagen, K. S. (2000). Spin-state variation in solid state and solution of mononuclear iron(II) 1,4,7-trimethyl-1,4,7-triazacyclonane complexes. Inorg. Chem., 39: 1979–1989.

11

Multifunctional Materials Combining Spin-Crossover with Conductivity and Magnetic Ordering

Osamu Sato, Zhao-Yang Li, Zi-Shuo Yao, Soonchul Kang and Shinji Kanegawa

Kyushu University, Institute for Materials Chemistry and Engineering, Japan

11.1 Introduction

The development of multifunctional materials, in which more than two properties co-exist or interact in a compound, has recently attracted considerable attention,[1–5] and many examples have been reported. Multifunctional materials range from materials that couple magnetic properties and conductivity,[6, 7] or magnetic properties and optical properties,[8–11] to those that combine dielectric properties and magnetic properties.[12] An interesting example is the multifunctional material that exhibits spin-crossover behavior and other properties.[3, 4] In this chapter, we first introduce compounds synthesized to develop new spin-crossover conductors, in which spin-crossover phenomena and conductivity coexist or interact. Next, we introduce spin-crossover magnets, in which spin-crossover phenomena and magnetic interaction coexist or interact.

11.2 Spin-Crossover and Conductivity: Spin-Crossover Conductors

A possible route for the synthesis of multifunctional materials exhibiting spin-crossover and conductivity is the preparation of double salts, by combining spin-crossover cations/anions and conducting molecular anions/cations through electrostatic interactions. Another route is the preparation of a spin-crossover complex with a potentially conducting ligand. Note that "conducting molecule" refers to one of those typically used to develop highly conductive materials such as metallic and superconducting molecular compounds.

Spin-Crossover Materials: Properties and Applications, First Edition. Edited by Malcolm A. Halcrow.

A spin-crossover complex has different spin multiplicities and various physical properties depending on whether it is in the high spin (HS) or low spin (LS) state.[13] Furthermore, the molecular volume of the HS state is generally larger than that of the LS state. In Fe^{III} spin-crossover complexes, the electronic structure of the HS state is $t_{2g}{}^3e_g{}^2$ with $S = {}^5/_2$, whereas that of the LS state is $t_{2g}{}^5e_g{}^0$ with $S = {}^1/_2$. Because the e_g orbitals are antibonding, the coordination bond between Fe^{III} and a ligand in the HS state with two electrons in the e_g orbital is weaker than that in the LS state with no electrons in the e_g orbital. In other words, the ligand–Fe^{III} bond length is longer in the HS state than that in the LS state; the difference is about 0.15 Å. Similarly, the ligand–Fe^{II} bond length in the HS state is longer than that in the LS state by ca. 0.18 Å. Hence, the molecular volume of the HS state is larger than that of the LS state. Spin-crossover between the HS and LS states involves a change in the molecular volume, which generates chemical pressure in the crystal. The chemical pressure modulates the transfer integral between conducting molecules in the conducting layer. This changes the conductivity of hybrid materials consisting of spin-crossover complexes and conducting molecules. Note that conducting molecules should form segregated columns within the compound. Furthermore, to achieve high conductivity, the conducting molecules should have a fractional oxidation state.

Several salts composed of a spin-crossover complex and conducting molecules have been reported;[14–22] all are semiconductors or insulators. The structural formulae of the spin-crossover complex and the conducting molecules used to synthesize spin-crossover conductors are shown in Scheme 11.1. Some examples are introduced below. Note that synergy between spin-crossover and conductivity has been successfully observed in three salts: $[Fe^{III}(qsal)_2][Ni(dmit)_2]_3\cdot CH_3CN\cdot H_2O$, $[Fe^{III}(qnal)_2][Pd(dmit)_2]_5\cdot$acetone and $[Fe^{II}(dppTTF)_2][Ni(mnt)_2]_2(BF_4)\cdot$PhCN (see below).

$[Fe^{III}(acpa)_2](TCNQ)_n$ ($n = 1, 2$): one of the earliest studies on the preparation of spin-crossover conductors was conducted by Nakano *et al.*[14] They synthesized two charge-transfer salts containing the organic acceptor molecule TCNQ and an Fe^{III} spin-crossover complex: $[Fe(acpa)_2](TCNQ)$ and $[Fe(acpa)_2](TCNQ)_2$ [Hacpa = *N*-(1-acetyl-2-propylidene)-2-pyridyl-methylamine; TCNQ = 7,7,8,8-Tetracyanoquinodimethane]. $[Fe(acpa)_2](TCNQ)$ shows an abrupt spin-transition at 159.5 K, whereas $[Fe(acpa)_2](TCNQ)_2$ shows a gradual spin-transition at around 370 K. The electrical conductivity of $[Fe(acpa)_2](TCNQ)_2$ with a fractional oxidation state in TCNQ is as high as 2.8×10^{-3} S cm^{-1} at room temperature.

$[Fe^{III}(sal_2\text{-trien})][Ni(dmit)_2]$: a cooperative spin-transition with hysteresis was observed in $[Fe(sal_2\text{-trien})][Ni(dmit)_2]$ (H_2sal_2-trien = the Schiff base derived from triethylenetetramine and salicylaldehyde; $dmit^{2-}$ = 1,3-dithiole-2-thione-4,5-dithiolato), in which $[Fe(sal_2\text{-trien})]$ is a spin-crossover complex and $[Ni(dmit)_2]$ is a conducting molecule (Fig. 11.1).[15] The magnetic properties indicate that spin-transitions occur abruptly at around 228 K on cooling and 258 K on warming. The hysteresis width is ca. 30 K (Fig. 11.1). This is a suitable precursor for the synthesis of switchable spin-crossover conductors.[15]

$[Fe^{III}(salten)Mepepy][M(dmit)_2]\cdot CH_3CN$ (M = Ni, Pd, Pt) and $[Fe^{III}(salten)Mepepy][Ni(dmit)_2]_3$: a gradual spin-transition was observed in $[Fe(salten)Mepepy][M(dmit)_2]\cdot CH_3CN$ (M = Ni, Pd, Pt; H_2salten = 4-azaheptamethylene-1,7-bis(salicylideneiminate); Mepepy = 1-(pyridin-4-yl)-2-(*N*-methylpyrrol-2-yl) ethane).[16] The temperature-dependent spin-crossover behavior of the three samples was similar to that of the starting material, $[Fe(salten)Mepepy](BPh_4)$.[23] $[Fe(salten)Mepepy][Ni(dmit)_2]_3$, with a fractional dithiolene anion oxidation state, was synthesized by the electro-oxidation of $[Fe(salten)Mepepy][Ni(dmit)_2]$. The $\chi_M T$ value of this material decreased almost linearly with decreasing temperature, while its room temperature electrical conductivity measured by the two-probe method was 0.1 S cm^{-1}.

$[Fe^{III}(salEen)_2]_2[Ni(dmit)_2]_5\cdot 6CH_3CN$ [salEen = *N*-(2-ethylamino)ethyl)salicylaldimine]: this compound has a fractional anion oxidation state, and exhibits a gradual spin-transition between 350 and 100 K.[17] The electrical conductivity is ca. 0.12 S cm^{-1} at 295 K, which decreases with decreasing temperature, indicating semiconducting behavior. The conductivity at 100 K is 1×10^{-4} S cm^{-1}.

$[Fe^{III}(qsal)_2][Ni(dmit)_2]\cdot xCH_3CN$ ($x = 0, 2$) and $[Fe^{III}(qsal)_2][Ni(dmit)_2]_3\cdot CH_3CN\cdot H_2O$ [qsalH = *N*-(8-quinolyl)salicylaldimine]: an unsolvated sample, $[Fe^{III}(qsal)_2][Ni(dmit)_2]$, exhibits an extremely gradual

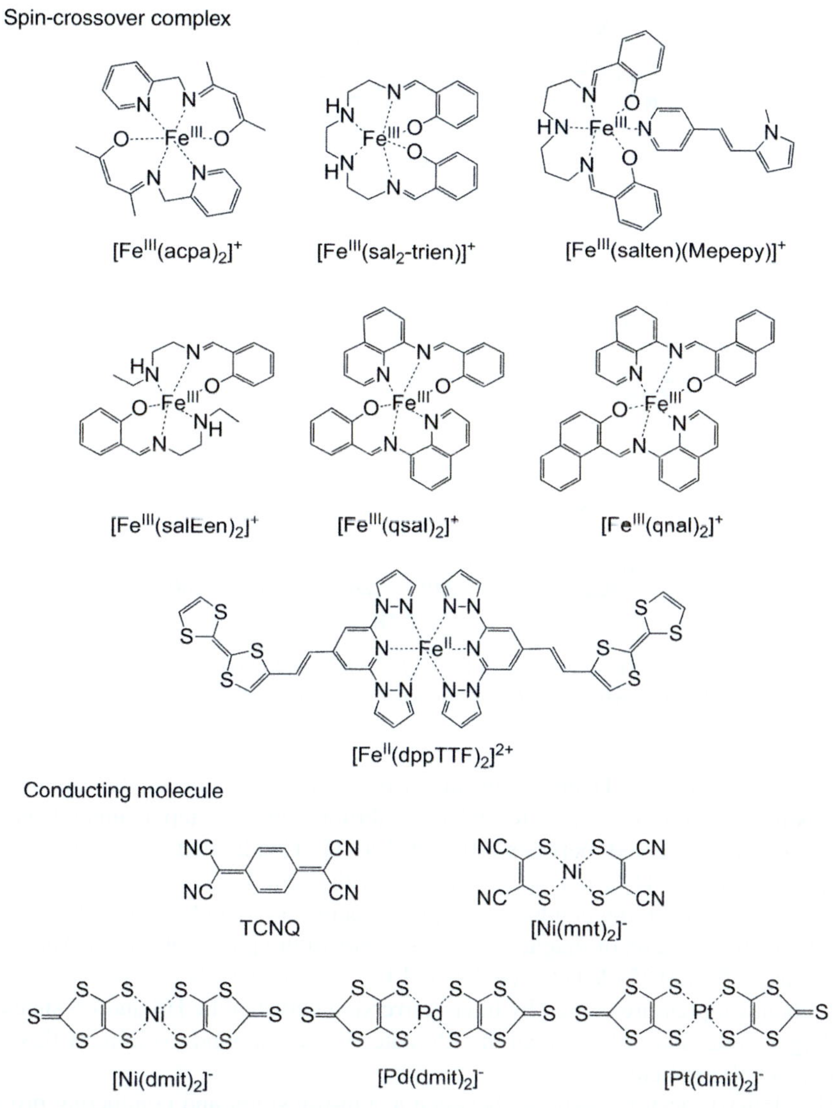

Scheme 11.1 *Molecules used for the development of spin-crossover conductors.*

spin-transition.[18] On the other hand, the solvated complex $[Fe^{III}(qsal)_2][Ni(dmit)_2]\cdot 2CH_3CN$ exhibits an abrupt spin-transition with a hysteresis loop ($T_{1/2}\downarrow = 194$ K and $T_{1/2}\uparrow = 231$ K).[19] Photomagnetic studies indicated that the solvated complex exhibits light-induced excited spin state trapping (LIESST) effects and a thermally induced spin-transition.

Synergy between the spin-transition and the conducting properties has been observed in partially oxidized species, for example $[Fe(qsal)_2][Ni(dmit)_2]_3\cdot CH_3CN\cdot H_2O$.[20] To achieve fractional oxidation, the complex was synthesized via an electrochemical route from a solution, in which $[Fe(qsal)_2][Ni(dmit)_2]\cdot 2CH_3CN$ was dissolved. $[Fe(qsal)_2][Ni(dmit)_2]_3\cdot CH_3CN\cdot H_2O$ has a segregated column structure. $[Fe(qsal)_2]^+$ molecules are dimerized by π–π interactions in the one-dimensional (1D) chain, while the $[Ni(dmit)_2]$ molecules

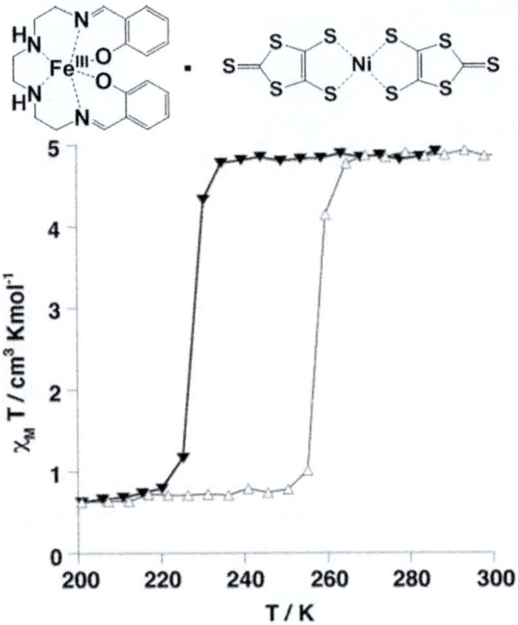

Figure 11.1 *Magnetic susceptibility measurements of [Fe^{III}(sal$_2$-trien)][Ni(dmit)$_2$] showing thermal hysteresis loop. Reprinted from [15] with permission of The Royal Society of Chemistry.*

associate into trimers within the 1D chain. Its magnetic properties indicate that spin-crossover occurs gradually from room temperature to 120 K and then abruptly in a 2-step manner from 120 to 60 K. A narrow hysteresis was observed between 90 and 120 K. Conductivity measurements showed that [Fe(qsal)$_2$][Ni(dmit)$_2$]$_3$·CH$_3$CN·H$_2$O is a semiconductor with a conductivity of 2.0 S cm^{-1} at room temperature. A hysteretic behavior in the conductivity appears between 90 and 120 K, which is consistent with the temperature at which hysteresis of magnetization vs. temperature plot is observed. The activation energy for conductivity changes from 0.25 eV between 250 and 125 K to 0.17 eV between 90 and 70 K. The change is induced by the chemical pressure caused by inter-conversion between the HS and LS states, where the HS molecule has a larger molecular volume than the LS state. The compound exhibited LIESST effects when irradiated at 5 K with 830 nm light.

[Fe^{III}(qnal)$_2$][Pd(dmit)$_2$]$_5$·acetone: synergy between a spin-transition and conducting properties has also been observed in [Fe^{III}(qnal)$_2$][Pd(dmit)$_2$]$_5$·acetone [qnalH = *N*-(8-quinolyl)-2-hydroxy-1-naphthaldimine] (Fig. 11.2).[21] This was electrochemically crystallized in a solution of [Fe^{III}(qnal)$_2$][Pd(dmit)$_2$]·acetone. The salt formed a segregated column in the crystal. The [Fe(qnal)$_2$]$^+$ molecules have a uniform 1D structure through π–π interactions, while the [Pd(dmit)$_2$]$^{n-}$ molecules form a 1D chain. Neighboring anion chains interact, forming a two-dimensional (2D) conducting layer. The [Pd(dmit)$_2$]$^{n-}$ chains consist of two crystallographically independent dimers, [[Pd(dmit)$_2$]$_2$]^0 and [[Pd(dmit)$_2$]$_2$]$^-$, and one neutral monomer, [Pd(dmit)$_2$]^0. The magnetic properties indicate that spin-crossover occurs around 220 K (Fig. 11.2), and conductivity measurements revealed semiconducting characteristics with room temperature conductivity of 0.016 S cm^{-1}. An anomaly in the conductivity appears at the spin-crossover temperature (Fig. 11.2). The activation energies for conductivity above and below the temperature of the anomaly are 0.37 and 0.24 eV, respectively. Careful investigation of the structure revealed that the lattice parameter along the *a* axis decreases in this temperature range; the [Fe(qnal)$_2$]$^+$ molecules align one-dimensionally along this axis through π–π interactions.

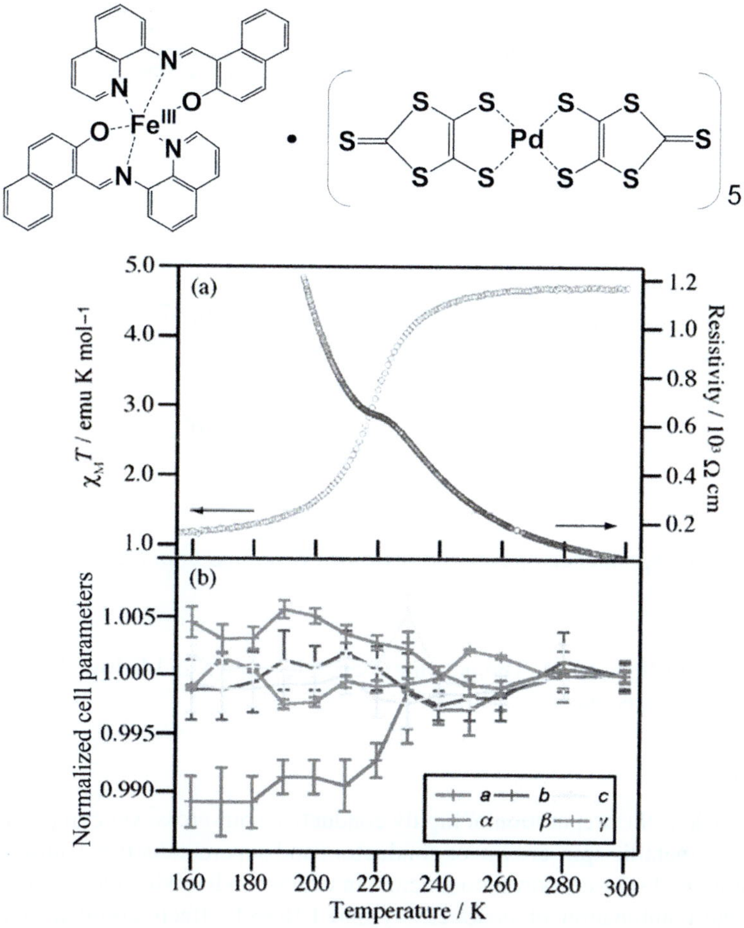

Figure 11.2 *Temperature dependence of (a) $\chi_M T$ and resistivity and (b) the cell parameters normalized for 300 K. Reprinted with permission from [21]. Copyright © 2008, American Chemical Society.*

The qnal$^-$ ligand in the LS state has a more planar structure than that in the HS state, which increases the intermolecular π–π overlap on going from the HS state to the LS state. The increase in the overlap in turn decreases the distance between the qnal$^-$ ligands, causing contraction along the *a* axis. The conductivity of the salt is modulated by the chemical uniaxial strain. Thus, the decrease in the volume at a molecular level between the HS and LS states, and π–π interactions between $[Fe(qnal)_2]^+$ molecule, play important roles in the synergy between spin-crossover and conductivity in $[Fe^{III}(qnal)_2][Pd(dmit)_2]_5$·acetone.

$[Fe^{II}(dppTTF)_2][Ni(mnt)_2]_2(BF_4)$·PhCN: another example exhibiting synergistic spin-transition and conducting properties is $[Fe(dppTTF)_2][Ni(mnt)_2]_2(BF_4)$·PhCN [dppTTF = 1-[2-(1,3-dithiol-2ylinene)-1,3-dithiolyl]-2-[2,6-bis(1-pyrazolyl)pyridyl]ethylene; mnt = maleonitriledithiolate), which was synthesized via galvanostatic oxidation (Fig. 11.3). Unlike the previous examples, this compound is an Fe^{II} complex.[22] The TTF and $[Ni(mnt)_2]^-$ form 2D layers that are linked through the $[Fe(dppTTF)_2]^{3+}$ centers. The Fe^{II} compound exhibits a gradual spin-transition between 170 and 300 K (Fig. 11.3). It is a semiconductor with room temperature conductivity of 2.6×10^{-3} S cm^{-1}. An anomaly in conductivity was clearly observed at the

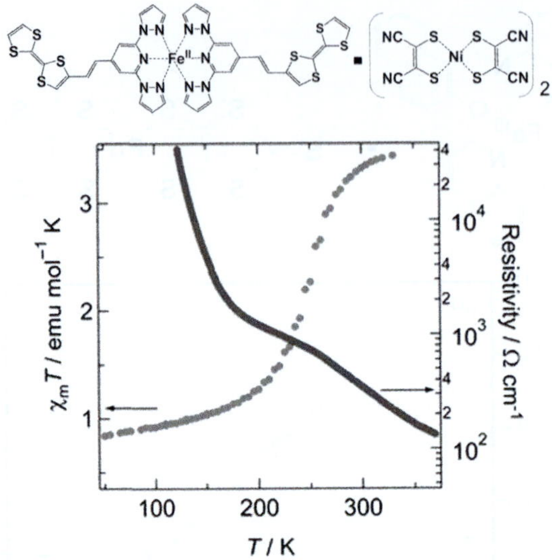

Figure 11.3 Temperature dependence of $\chi_m T$ and resistivity. Reprinted from [22] with permission of The Royal Society of Chemistry.

spin-transition temperature (Fig. 11.3). The activation energies above and below the conductivity anomaly were 0.119 and 0.129 eV, respectively.

11.2.1 Conclusions

The challenge in this field is the preparation of highly conductive spin-crossover complexes. In particular, it is important to synthesize metallic spin-crossover conductors and superconducting spin-crossover conductors. Moreover, the preparation of spin-crossover conductors in which conductivity and LIESST exhibit synergistic effects is desired. The combination of conductivity and LIESST effects could lead to the development of new compounds having LIESST-induced metallic conductivity or LIESST-induced superconductivity. Furthermore, it is interesting to develop new systems in which hybrid effects are induced through electrical interactions between spin-crossover complexes and conducting molecules, not through chemical pressure arising from the difference in the molecular volume between the HS and LS states. The achievement of strong electrical interactions may lead to the production of new spintronic molecular compounds whose conductivity can be controlled by manipulating the spin state in spin-crossover complexes.

11.3 Spin-Crossover and Magnetic Interaction: Spin-Crossover Magnets

Multifunctional materials exhibiting both spin-crossover and magnetic interaction can be achieved by combining a spin-crossover cation/anion with an anionic/cationic magnetic framework. Another route is the incorporation of spin-crossover sites into magnetic frameworks.

11.3.1 Hybrid Spin-Crossover Cation and Anionic Magnetic Framework

When spin-crossover and a magnetic framework are combined, oxalate-based magnets have often been used as the magnetic framework. Several molecular magnets with the chemical formula $A[MM'(ox)_3]$ (A = cation;

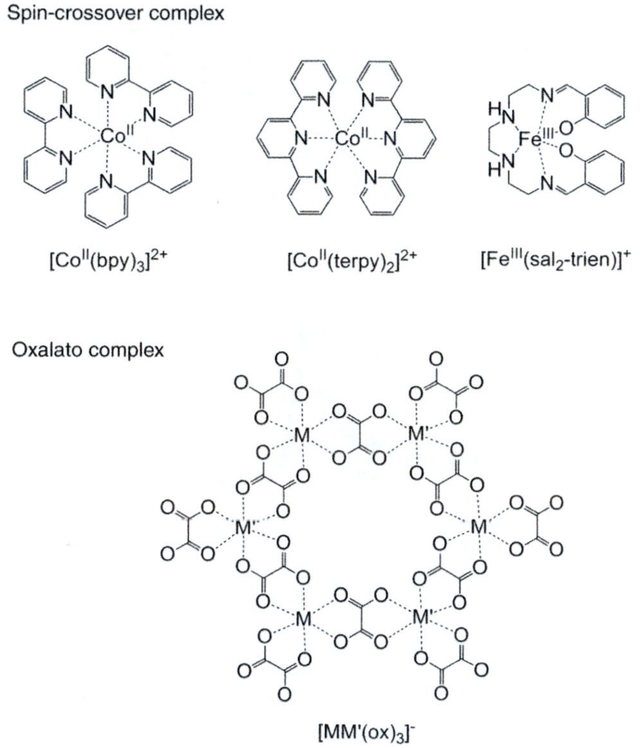

Scheme 11.2 *Molecules used for the development of hybrid spin-crossover materials and magnets.*

M and M′ = metal; ox = $C_2O_4^{2-}$) have been reported.[24] When A is replaced by a spin-crossover cation, a hybrid system having spin-crossover and magnetic coupling can be developed (Scheme 11.2).[25–31] Some examples are introduced below. In particular, $[Fe^{III}(sal_2\text{-trien})][Mn^{II}Cr^{III}(ox)_3]\cdot CH_2Cl_2$ exhibited the coexistence of spin-crossover, magnetic ordering, and photomagnetic (LIESST) effects (see below).

$[Co^{II}(bpy)_3][ACr^{III}(ox)_3]$ (A = Na or Li): two complexes, $[Co(bpy)_3][NaCr(ox)_3]$ and $[Co(bpy)_3][LiCr(ox)_3]$, have been synthesized (bpy = 2,2′-bipyridine).[25] $[Co(bpy)_3]$ in $[Co(bpy)_3][NaCr(ox)_3]$ is in the HS state, whereas $[Co(bpy)_3]$ in $[Co(bpy)_3][LiCr(ox)_3]$ exhibits spin-crossover behavior. The cavity size of the oxalate network of $[LiCr(ox)_3]^{2-}$ is smaller than that of $[NaCr(ox)_3]^{2-}$. Because the resulting chemical pressure stabilizes the LS state, spin-crossover behavior was observed in $[Co(bpy)_3][LiCr(ox)_3]$. Note that the anionic frameworks of $[LiCr(ox)_3]$ and $[NaCr(ox)_3]$ are paramagnetic, not ferromagnetic.

$[Co^{II}(terpy)_2][Mn^{II}(H_2O)ClCr^{III}(ox)_3]\cdot H_2O\cdot {}^1/_2CH_3OH$: the title complex consists of a 1D $[Mn(H_2O)Cl(ox)Cr(ox)_2]^{2-}$ anionic chain and a $[Co(terpy)_2]^{2+}$ cation (terpy = 2,2′:6,2″-terpyridine).[26] The anionic chain contains alternating $[Mn(H_2O)Cl]^{+}$ and $[Cr(ox)_3]^{3-}$ centers connected by two of the oxalate ligands from $[Cr(ox)_3]^{3-}$. Magnetization measurements showed that on heating the $\chi_m T$ value increases slightly from 260 to 300 K because of the induction of spin-crossover in the incorporated $[Co(terpy)_2]^{2+}$. The anionic framework does not show long-range magnetic ordering.

$[Fe^{III}(sal_2\text{-trien})]_2[Mn^{II}{}_2(ox)_3]\cdot 4H_2O\cdot C_3H_7NO$: the coexistence of spin-crossover and magnetic ordering has been observed in $[Fe^{III}(sal_2\text{-trien})]_2[Mn^{II}{}_2(ox)_3]\cdot 4H_2O\cdot C_3H_7NO$ (H_2sal_2-trien = *N,N′*-disalicylidenetriethylenetetramine) (Fig. 11.4).[27] The magnetic properties indicate that gradual spin-crossover of the intercalated $[Fe^{III}(sal_2\text{-trien})]^{+}$ occurs between 300 and 80 K. The homometallic Mn^{II} oxalate layers exhibit weak ferromagnetic behavior below a critical temperature of ca. 8.1 K.

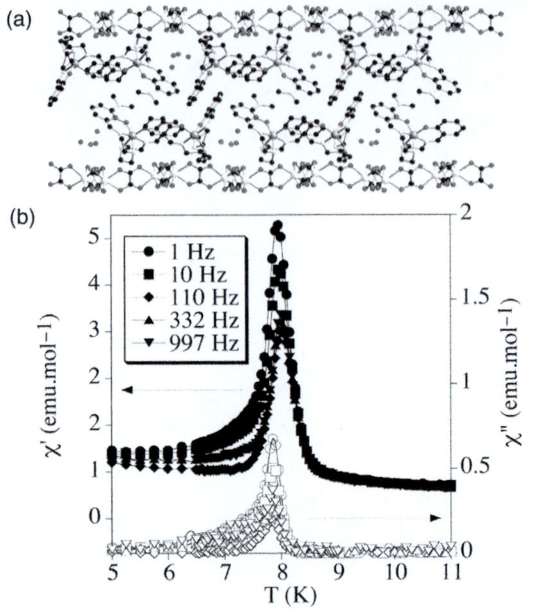

Figure 11.4 (a) Homometallic 2D honeycomb anionic layer formed by Mn^{II} ions linked through oxalate ligands and a cationic double layer of $[Fe(sal_2\text{-trien})]^+$ complexes intercalated between the 2D oxalate network. (b) Temperature dependence of the in-phase ac susceptibility (χ'; filled symbols) and the out-of-phase ac susceptibility (χ''; open symbols). Reprinted with permission from [27]. Copyright 2008, American Chemical Society.

$[Fe^{III}(sal_2\text{-trien})][Mn^{II}Cr^{III}(ox)_3]\cdot CH_2Cl_2$: this complex also exhibits the coexistence of spin-crossover and magnetic ordering (Fig. 11.5).[28] Single crystal analysis showed that $[Mn^{II}Cr^{III}(ox)_3]^-$ presents a 2D honeycomb structure. Spin-crossover occurs at around 250 K. Ferromagnetic ordering is observed below ca. 5.4 K.[28] Furthermore, a photomagnetic study showed that this salt exhibits LIESST effects.[31] The LIESST effects in $[Fe^{III}(sal_2\text{-trien})]$ have a negligible effect on the ferromagnetic behavior of the 2D oxalate network.

The challenge in the field of hybrid materials consisting of spin-crossover cations and an anionic magnetic framework is the preparation of spin-crossover magnets, in which magnetic ordering is influenced by spin-crossover and/or LIESST effects.

11.3.2 Incorporation of Spin-Crossover Sites in a Magnetic Framework

Spin-crossover complexes are typically paramagnetic. Because a metal center with high spin density is usually surrounded by a large diamagnetic ligand, magnetic interactions between molecules are almost negligible. However, several examples of magnetic interactions coupled to a spin-transition have been recently reported. The choice of the ligand is important in the development of the spin-crossover complexes that exhibit magnetic interactions. Magnetic interaction through space is weak for complexes with large ligands. Hence, bridging ligands such as 2,2′-bithiazoline,[32] have been used to obtain complexes in which magnetic interactions operate. A spin-crossover complex in which spin-crossover Fe ions are bridged by a single oxygen atom has also been reported.[33] Moreover, the CN^- ligand, which has frequently been used to construct molecular magnets, has been used to develop hybrid materials showing spin-crossover and magnetic interaction.

Here, we first introduce two spin-crossover clusters that exhibit magnetic superexchange interactions; bulk magnets showing spin-crossover; and finally a molecular nanomagnet exhibiting electron transfer coupled

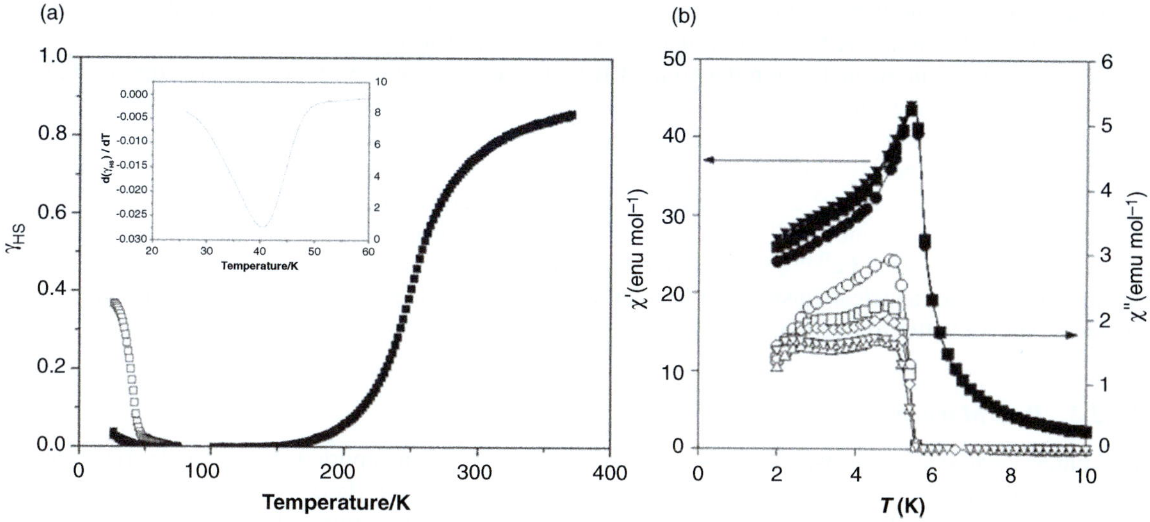

Figure 11.5 (a) Fraction of HS Fe^{III} (γ_{HS}) before (■) and after (□) irradiation. Inset: χT derivative for photomagnetism. Reprinted with permission from ref. 31. Copyright 2011, Royal Society of Chemistry. (b) Temperature dependence of the in-phase ac susceptibility (χ') (filled symbols) and the out-of-phase ac susceptibility (χ'') (open symbols): • 997 Hz, ■ 332 Hz, □ 110 Hz, ▲ 10 Hz, ▼ 1 Hz. Reprinted with permission from [28]. Copyright Wiley VCH, 2010.

spin transition. Note that electron transfer coupled spin transition is also called charge transfer induced spin-transition (CTIST).

11.3.2.1 Clusters

$[Fe^{II}(bt)(NCS)_2]\cdot bpym\cdot[Fe^{II}(bt)(NCS)_2]$: synergy between spin-crossover and magnetic interactions has been observed, particularly in the LIESST effects, in $[Fe^{II}(bt)(NCS)_2]\cdot bpym\cdot[Fe^{II}(bt)(NCS)_2]$ (bt = 2,2′-bithiazoline; bpym = 2,2′-bipyrimidine).[32] When the compound was illuminated at 10 K, the magnetization value increased. However, the $\chi_m T$ value after illumination corresponds to only 24% of that in the HS state at room temperature. Furthermore, when the sample was heated after irradiation, $\chi_m T$ increased with increasing temperature up to 44 K. These phenomena can be explained by the presence of antiferromagnetic interactions between two Fe^{II} centers in the dinuclear complex. The ground state of the complex with two LS states is $S = 0$. On the other hand, two $Fe^{II\text{-}HS}$ ions are generated in the dinuclear complex after irradiation. However, because of the presence of antiferromagnetic interaction, the ground state after irradiation is also $S = 0$. Hence, the $\chi_m T$ value with two HS states approaches 0 with decreasing temperature. Hence, the increase in the $\chi_m T$ value of the dinuclear complex by irradiation is smaller than that of the typical spin-crossover complex with no appreciable magnetic interaction. Moreover, because the low-lying magnetic spin states are progressively populated with increasing temperature, the $\chi_m T$ value after irradiation increases on heating.

$[Fe^{II}_4(HL^1)_4](BF_4)_4\cdot 2H_2O\cdot CH_3OH$: a similar synergistic effect is observed in a tetranuclear Fe^{II} molecular square complex, $[Fe^{II}_4(HL^1)_4](BF_4)_4\cdot 2H_2O\cdot CH_3OH$ [H_2L^1 = 2,2′-bis(phenyl-2-pyridinylmethylene) carbonic dihydrazide].[33] The Fe^{II} complex exhibits a 2-step decrease in magnetization due to a spin-transition and antiferromagnetic interaction. The first step is a decrease at ca. 175 K on cooling caused by the spin-transition from the $Fe^{II\text{-}HS}{}_4$ state to a $Fe^{II\text{-}HS}{}_2Fe^{II\text{-}LS}{}_2$ state, with the HS centers occupying adjacent vertices in the molecular square framework. The second step, a decrease below 50 K, is due to antiferromagnetic interactions between the two neighboring $Fe^{II\text{-}HS}$ ions in the cluster, and to zero-field splitting. When this complex was irradiated at 5 K by 785 nm light to excite the metal-to-ligand charge transfer (MLCT) band, only a small increase in magnetization due to the LIESST effect was observed. On heating after irradiation, the difference in the magnetization value before and after irradiation increased to 50 K and then decreased, finally relaxing back to the original value at around 80 K. As in the case of the above dinuclear complex, these phenomena can be explained by the presence of intramolecular antiferromagnetic interactions. The compound has a *cis*-type $Fe^{II\text{-}HS}{}_2Fe^{II\text{-}LS}{}_2$ state with antiferromagnetic interaction between the $Fe^{II\text{-}HS}$ ions. Hence, the ground state of the $Fe^{II\text{-}HS}{}_2Fe^{II\text{-}LS}{}_2$ state is $S = 0$. On the other hand, two HS sites are generated after irradiation, forming a $Fe^{II\text{-}HS}{}_4$ state. Because the four $Fe^{II\text{-}HS}$ ions interact antiferromagnetically, the ground state of the tetranuclear cluster after irradiation is also $S = 0$. Hence, only a small increase in the $\chi_m T$ value was observed after irradiation.

11.3.2.2 Bulk Magnets

Multidimensional systems in which spin-crossover occurs have been reported, for example 1D cyanide-bridged systems with a –NC–Fe–CN–Mn– heterometallic structure, such as $[Fe(L)(CN)_2][Mn(hfac)_2]$ (L is the Schiff base macrocyclic ligand derived from the condensation of 2,6-diacetylpyridine with 3,6-dioxaoctane-1,8-diamine, and Hhfac is hexafluoroacetylacetone). This system shows thermally and photoinduced spin-crossover.[34] Furthermore, various bulk magnets exhibiting spin-crossover behavior have been reported. Typical examples of these are CN^--bridged coordination polymers such as Prussian Blue analogs. Note that many Prussian Blue analogs have been synthesized, and their properties have been extensively investigated (Fig. 11.6). These analogs have been used as pigments. More recently, they have attracted considerable attention as electrochromic compounds.[35] Furthermore, they have been extensively studied as typical molecule-based magnets.[36] Room temperature magnets and tunable magnets have been synthesized recently.[9, 37]

Spin crossover systems

(a) CrFe Complexes
$CsFe^{II}[Cr^{III}(CN)_6]\cdot 1.3H_2O$: an interesting phenomenon in Prussian Blue analogs is thermally induced spin-crossover.[38] The title CrFe Prussian Blue compound, $CsFe^{II}[Cr^{III}(CN)_6]\cdot 1.3H_2O$,

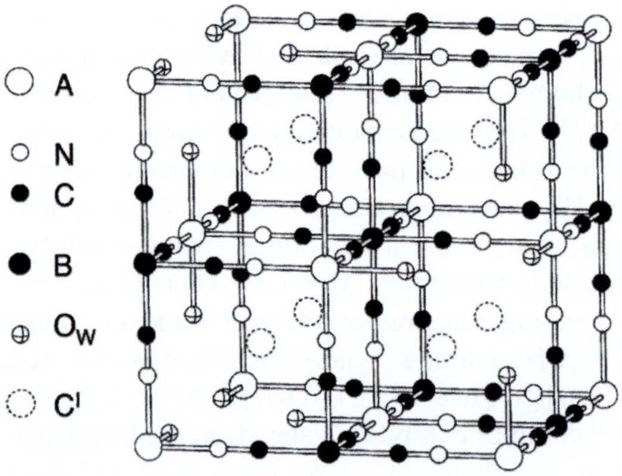

Figure 11.6 View of the unit cell of a Prussian Blue analog showing random distribution of $[B(CN)_6]^{3-}$ vacancies. Reprinted with permission from [41]. Copyright 2008, American Chemical Society.

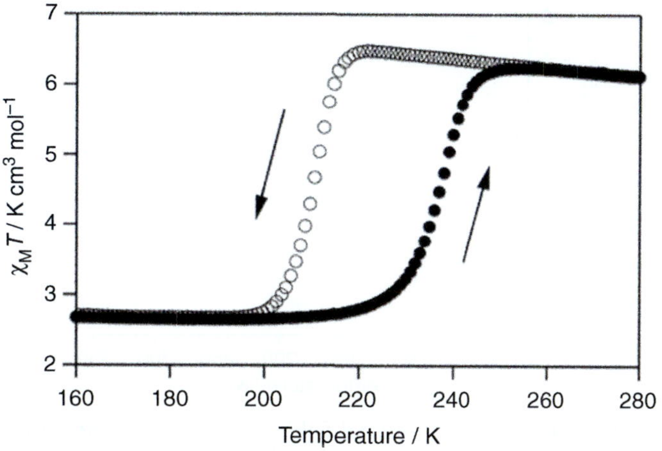

Figure 11.7 *Temperature dependence of $\chi_M T$ in cesium iron hexacyanochromate in an external magnetic field of 5000 G measured during cooling (○) and warming (•). Reprinted with permission from [38]. Copyright 2005, American Chemical Society.*

shows spin-crossover; $Cs^{I}[Fe^{II\text{-}HS}[Cr^{III}(CN)_6]]_{0.94}[Fe^{II\text{-}LS}[Cr^{III}(NC)_6]]_{0.06}\cdot 1.3H_2O \leftrightarrow Cs^{I}[Fe^{II\text{-}HS}{}_{0.12}Fe^{II\text{-}LS}{}_{0.88}[Cr^{III}(CN)_6]]_{0.94}[Fe^{II\text{-}LS}[Cr^{III}(NC)_6]]_{0.06}\cdot 1.3H_2O$. It exhibits a hysteresis loop in its $\chi_m T$ vs. T plot (Fig. 11.7). The transition temperatures are 211 K on cooling and 238 K on heating. The low temperature (LT) phase is a ferromagnet having a Curie temperature of 9 K. Furthermore, $CsFe[Cr(CN)_6]$ exhibits X-ray-illumination-induced Fe^{II} spin-crossover.[39] That is, when the complex was studied by time-dependent high-resolution synchrotron X-ray diffraction, conversion from $Fe^{II\text{-}HS}$ to $Fe^{II\text{-}LS}$ was observed.

$K_{0.4}Fe^{II}{}_4[Cr^{III}(CN)_6]_{2.8}\cdot 16H_2O$: pressure-induced linkage isomerization between Cr^{III}-CN-Fe^{II} and Cr^{III}-NC-Fe^{II} was observed in a CrFe Prussian Blue compound, $K_{0.4}Fe_4[Cr(CN)_6]_{2.8}\cdot 16H_2O$.[40, 41] The CN flip induced a spin-transition in Fe^{II} from HS to LS because the C site in CN^- exerts a stronger ligand field than the N site. The pressure-induced CN flip involving a spin-transition changes the Curie temperature from 18.5 to 7.5 K at 1200 MPa, because the number of diamagnetic $Fe^{II\text{-}LS}$ sites increases with the application of pressure.

(b) NbFe Complexes

$Fe^{II}{}_2[Nb^{IV}(CN)_8](3\text{-}pyCH_2OH)_8\cdot 4.6H_2O$ (3-py = 3-pyridyl): this octacyanide-bridged NbFe bimetallic assembly displays thermally induced spin-crossover in Fe^{II} sites (Fig. 11.8).[42] The electronic states of the high temperature (HT) and LT phases are close to the formulae $(Fe^{II\text{-}HS})_2[Nb^{IV}(CN)_8](3\text{-}pyCH_2OH)_8\cdot 4.6H_2O$ and $Fe^{II\text{-}HS}Fe^{II\text{-}LS}[Nb^{IV}(CN)_8](3\text{-}pyCH_2OH)_8\cdot 4.6H_2O$, respectively. The LT phase shows ferrimagnetic ordering with a T_c of 12 K.

$Fe^{II}{}_2[Nb^{IV}(CN)_8](4\text{-}pyridinealdoxime)_8\cdot 2H_2O$: this NbFe complex also shows thermally induced spin-crossover.[43] Furthermore, when it is irradiated, it changes from paramagnetic ($Fe^{II\text{-}LS}$–NC–Nb^{IV}–CN–$Fe^{II\text{-}LS}$) to ferromagnetic ($Fe^{II\text{-}HS}$–NC–Nb^{IV}–CN–$Fe^{II\text{-}HS}$) through the occurrence of LIESST effects in Fe^{II}. The critical temperature after irradiation is 20 K (Fig. 11.9).

(c) MoFe Complex

$Fe^{II}{}_2[Mo^{IV}(CN)_8]\cdot(3\text{-}pyCH_2OH)_8\cdot 3H_2O$: the MoFe complex $Fe^{II}{}_2[Mo^{IV}(CN)_8]\cdot(3\text{-}pyCH_2OH)_8\cdot 3H_2O$ (3-py = 3-pyridyl) exhibits an extremely gradual Fe^{II} spin-crossover over a wide temperature range.[44] However, long-range magnetic ordering is not observed.

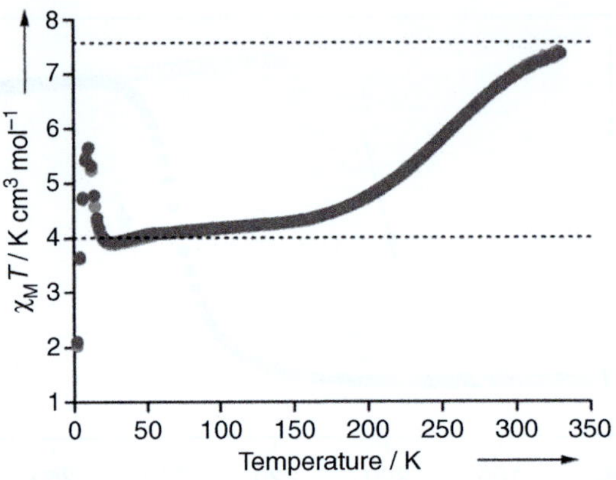

Figure 11.8 $\chi_M T$ *vs. T plots measured in an external field of 5000 Oe. Reprinted with permission from [42]. Copyright 2008 WILEY-VCH Verlag GmbH & Co. KGaA, Weinheim.*

Electron transfer coupled spin transition systems

(d) FeCo Complexes
$Na_{0.37}[Co_{1.37}Fe(CN)_6]\cdot 4.8H_2O$: the Prussian blue analog $A_xCo_y[Fe(CN)_6]\cdot zH_2O$ (A = alkali cation such as Na, K, Rb, Cs; Fig. 11.6) shows thermally and photoinduced electron transfer coupled spin transition. Many groups have studied FeCo systems.[9–11] $Na_{0.37}Co_{1.37}Fe(CN)_6\cdot 4.8H_2O$ has an $Fe^{III\text{-}LS}$–CN–$Co^{II\text{-}HS}$ state in the HT phase and an $Fe^{II\text{-}LS}$–CN–$Co^{III\text{-}LS}$ state in the LT phase.[45] The phase transition temperatures are $T_{1/2}\downarrow = 180$ K and $T_{1/2}\uparrow = 220$ K with a hysteresis width of ca. 40 K (Fig. 11.10). Irradiation with 532 nm light induces electron transfer coupled spin transition. The magnetic properties change from paramagnetic to ferrimagnetic at a T_c of 26 K (Fig. 11.10).

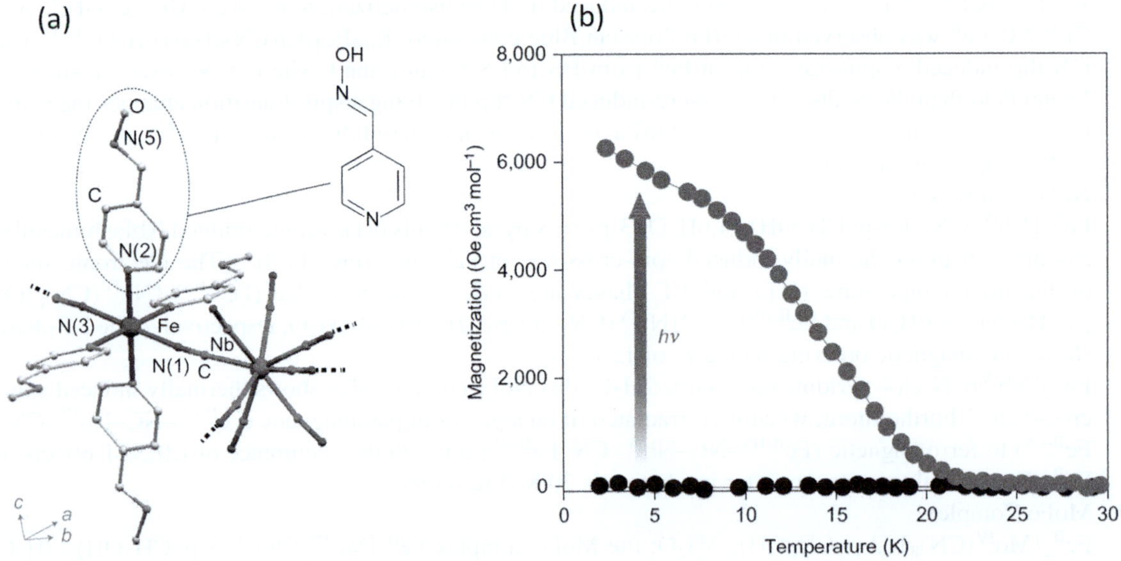

Figure 11.9 *(a) Coordination environments around Fe and Nb in* $Fe_2[Nb(CN)_8]\cdot(4\text{-pyridinealdoxime})_8\cdot 2H_2O$*. (b) Magnetization versus temperature at 100 Oe. Light irradiation induces spontaneous magnetization with a Curie temperature of 20 K. Reprinted by permission from Macmillan Publishers Ltd [43], Copyright 2011.*

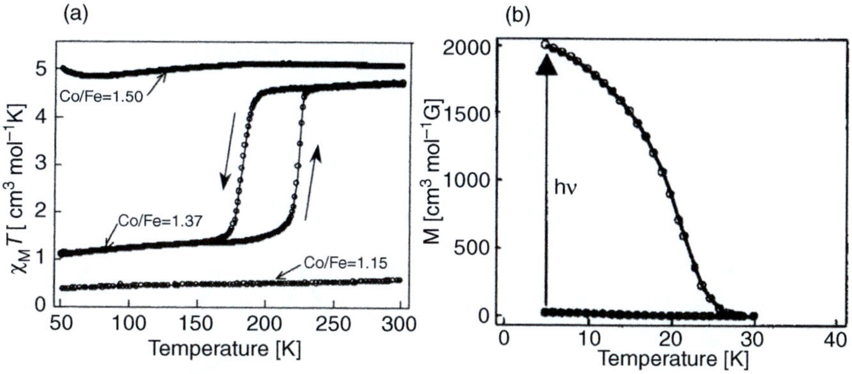

Figure 11.10 *(a) $\chi_M T$ vs. T for $Na_{0.07}Co_{1.50}Fe(CN)_6 \cdot 6.3H_2O$ (Co/Fe = 1.5), $Na_{0.37}Co_{1.37}Fe(CN)_6 \cdot 4.8H_2O$ (Co/Fe = 1.37), and $Na_{0.94}Co_{1.15}Fe(CN)_6 \cdot 3.0H_2O$ (Co/Fe = 1.15) during cooling and warming at H = 5000 G. (b) Field-cooled magnetization curves for $Na_{0.37}Co_{1.37}Fe(CN)_6 \cdot 4.8H_2O$ (Co/Fe = 1.37) before (•) and after (○) light irradiation at H = 5 G. Reprinted with permission from [45]. Copyright 2002, American Chemical Society.*

$[Fe^{III}Tp(CN)_3]_2Co^{II}(bpe) \cdot 5H_2O$: an FeCo complex with a double zigzag chain structure, $[FeTp(CN)_3]_2Co(bpe) \cdot 5H_2O$ [Tp^- = hydrotris(pyrazoly)borate; bpe = 1,2-bis(4-pyridyl)ethane], is a metamagnet showing long-range antiferromagnetic ordering below 9.6 K without detectable magnetic hysteresis. Magnetic investigation revealed that it exhibited reversible light- and temperature-induced charge transfer between $Fe^{III\text{-}LS}$–CN–$Co^{II\text{-}HS}$ and $Fe^{II\text{-}LS}$–CN–$Co^{III\text{-}LS}$ redox pairs.[46] Furthermore, through de- and rehydration, the light- and heat-induced charge transfer can be reversibly switched, providing water-switchable metal to metal charge transfer in a microporous framework.

(e) WCo Complexes

$CsCo^{II}(3\text{-cyanopyridine})_2[W^V(CN)_8] \cdot H_2O$: electron transfer coupled spin transition has been observed in WCo systems, which consist of W octacyanide. The first example of such systems is the $CsCo^{II}(3\text{-cyanopyridine})_2[W^V(CN)_8] \cdot H_2O$ complex.[47] The WCo complex exhibits thermally and photoinduced electron transfer coupled spin transition. It is expressed by $W^V(S = {}^1/_2)$–CN–$Co^{II\text{-}HS} \leftrightarrow W^{IV}(S = 0)$–CN–$Co^{III\text{-}LS}$. The hysteresis width is 49 K. It changes from the paramagnetic to the ferromagnetic state at $T_c = 30$ K by irradiation.

$[[Co^{II}(pyrimidine)(H_2O)]_2[Co^{II}(H_2O)_2][W^V(CN)_8]_2](pyrimidine)_2 \cdot 2H_2O$: this is another example of the WCo system that exhibits electron transfer coupled spin transition. The phase transition temperatures are $T_{1/2}\downarrow = 208$ K and $T_{1/2}\uparrow = 298$ K with a hysteresis width of 90 K. The critical temperature after irradiation (wavelength = 840 nm) is 40 K.[48]

(f) OsCo Complex

$Co^{II}_3[Os^{III}(CN)_6]_2 \cdot 6H_2O$: a new Prussian Blue analog containing Os^{III} ions exhibits thermally and photoinduced charge transfer coupled to a spin-transition.[49] The light- and temperature-induced charge transfer process can be expressed by $Os^{III\text{-}LS}$–CN–$Co^{II\text{-}HS} \leftrightarrow Os^{II\text{-}LS}$–CN–$Co^{III\text{-}LS}$. Photomagnetic measurements revealed a ferrimagnetic ordering temperature of 16.5 K after irradiation.

11.3.2.3 Nanomagnets

$[[Fe^{III}(bpy)(CN)_4]_2Co^{II}(4,4'\text{-bipyridine})] \cdot 4H_2O$: photofunctional molecular nanomagnets (single-chain magnets; SCMs) have recently been developed.[50] A photoswitchable SCM is a CN-bridged $Fe^{III}{}_2Co^{II}$ double zigzag chain system (Fig. 11.11). Magnetic measurements indicate that it exhibits charge transfer coupled to a spin-transition that involves a transformation between an HT phase with $Fe^{III\text{-}LS}$ ($S = {}^1/_2$) and $Co^{II\text{-}HS}$

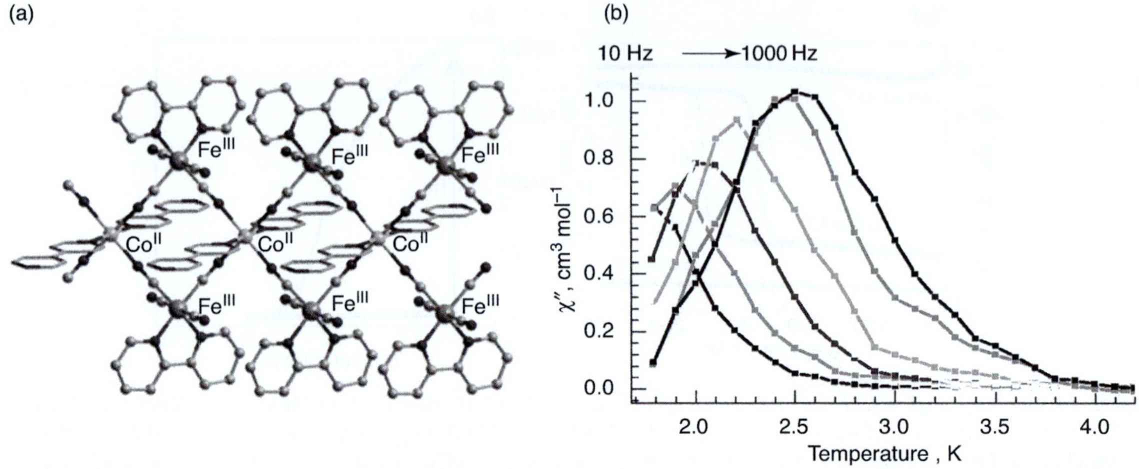

Figure 11.11 *(a) Side view of the 1D double zigzag chain of FeCo complex. (b) Temperature dependence of the imaginary part of the ac susceptibility after irradiation in zero dc-field at varying ac frequency and with a 3 Oe ac field. Reprinted with permission from [50]. Copyright 2010, American Chemical Society.*

($S = {}^3/_2$) and an LT phase with $Fe^{II\text{-}LS}$ ($S = 0$) and $Co^{III\text{-}LS}$ ($S = 0$). When it was irradiated with 532 nm laser light, the valence state changed from the $Fe^{II\text{-}LS}$–CN–$Co^{III\text{-}LS}$ state to the $Fe^{III\text{-}LS}$–CN–$Co^{II\text{-}HS}$ state. Furthermore, the spin state in Co changed from the LS to the HS state. Variable-temperature ac susceptibility measurements revealed a strong frequency dependence of both in-phase (χ') and out-of-phase (χ'') components (Fig. 11.11). The 1D chain undergoes weak antiferromagnetic interaction. Hence, the state after irradiation was an antiferromagnetic ordered phase of SCMs.

More recently, a novel phototunable single molecule magnet and SCMs have been reported.[51–54]

11.3.3 Conclusion

The challenge in this field is the preparation of room temperature photoswitchable magnets and nanomagnets. Room temperature molecular magnets have already been developed.[37] Furthermore, photofunctional MoCu systems, in which a photoinduced metastable state can be trapped even at approximately room temperature, have been reported although this is not a spin-crossover phenomenon.[55] Hence, it is expected that room temperature photomagnets can be developed by combining the knowledge in these two fields.

References

1. Clemente-León, M., Coronado, E., Martí-Gastaldoz, C., Romero, F. M. (2011) Multifunctionality in hybrid magnetic materials based on bimetallic oxalate complexes. Chem. Soc. Rev., 40: 473–497.
2. Ouahab, L. (1997) Organic/inorganic supramolecular assemblies and synergy between physical properties. Chem. Mater, 9: 1909–1926.
3. Gaspar, A. B., Ksenofontov, V., Seredyuk, M., Gütlich, P. (2005) Multifunctionality in spin crossover materials. Coord. Chem. Rev., 249: 2661–2676.
4. Muñoz, M. C., Real, J. A. (2011) Thermo-, piezo-, photo- and chemo-switchable spin crossover iron(II)-metallocyanate based coordination polymers. Coord. Chem. Rev., 255: 2068–2093.

5. Sato, O., Tao, J., Zhang, Y. Z. (2007) Control of magnetic properties through external stimuli. Angew. Chem. Int. Ed., 46: 2152–2187.
6. Coronado, E., Galán-Mascarós, J. R., Gómez-García, C. J., Laukhin, V. (2000) Coexistence of ferromagnetism and metallic conductivity in a molecule-based layered compound. Nature, 408: 447–449.
7. Uji, S., Shinagawa, H., Terashima, T., Yakabe, T., Terai, Y., Tokumoto, M., Kobayashi, A., Tanaka, H., Kobayashi, H. (2001) Magnetic-field-induced superconductivity in a two-dimensional organic conductor. Nature, 410: 908–910.
8. Decurtins, S., Gütlich, P., Kohler, C. P., Spiering, H., Hauser, A. (1984) Light-induced excited spin state trapping in a transition-metal complex - the hexa-1-propyltetrazole-iron(II) tetrafluoroborate spin-crossover system. Chem. Phys. Lett., 105: 1–4.
9. Sato, O., Iyoda, T., Fujishima, A., Hashimoto, K. (1996) Photoinduced magnetization of a cobalt iron cyanide. Science, 272: 704–705.
10. Bleuzen, A., Lomenech, C., Escax, V., Villain, F., Varret, F., Cartier dit Moulin, C., Verdaguer, M. (2000) Photoinduced ferrimagnetic systems in Prussian blue analogues $C_x{}^ICo_4[Fe(CN)_6]_y$ (C^I = alkali cation). 1. Conditions to observe the phenomenon. J. Am. Chem. Soc., 122: 6648–6652.
11. Verdaguer, M. (1996) Molecular electronics emerges from molecular magnetism. Science, 272: 698–699.
12. Cui, H.-B., Wang, Z., Takahashi, K., Okano, Y., Kobayashi, H., Kobayashi, A. (2006) Ferroelectric porous molecular crystal, $Mn_3(HCOO)_6(C_2H_5OH)$, exhibiting ferrimagnetic transition. J. Am. Chem. Soc., 128: 15074–15075.
13. Gütlich, P., Hauser, A., Spiering, H. (1994) Thermal and optical switching of iron(II) complexes. Angew. Chem. Int. Ed. Engl., 33: 2024–2054.
14. Nakano, M., Fujita, N., Matsubayashi, G. E., Mori, W. (2002) Modified chesnut model for spin-crossover semiconductors $Fe(acpa)_2(TCNQ)_n$. Mol. Cryst. Liquid Cryst., 379: 365–370.
15. Dorbes, S., Valade, L., Real, J. A., Faulmann, C. (2005) $[Fe(sal_2\text{-trien})][Ni(dmit)_2]$: towards switchable spin crossover molecular conductors. Chem. Commun., 69–70.
16. Faulmann, C., Dorbes, S., Garreau de Bonneval, B., Molnár, G., Bousseksou, A., Gomez-Garcia, C. J., Coronado, E., Valade, L. (2005) Towards molecular conductors with a spin-crossover phenomenon: Crystal structures, magnetic properties and Mössbauer spectra of $[Fe(salten)Mepepy][M(dmit)_2]$ complexes. Eur. J. Inorg. Chem., 3261–3270.
17. Faulmann, C., Jacob, K., Dorbes, S., Lampert, S., Malfant, I., Doublet, M.-L., Valade, L., Real, J. A. (2007) Electrical conductivity and spin crossover: a new achievement with a metal bis dithiolene complex. Inorg. Chem., 46: 8548–8559.
18. Faulmann, C., Dorbes, S., Lampert, S., Jacob, K., Garreau de Bonneval, B., Molnár, G., Bousseksou, A., Real, J. A., Valade, L. (2007) Crystal structure, magnetic properties and Mössbauer studies of $Fe(qsal)_2Ni(dmit)_2$. Inorg. Chim. Acta, 360: 3870–3878.
19. Takahashi, K., Cui, H.-B., Kobayashi, H., Einaga, Y., Sato, O. (2005) The light-induced excited spin state trapping effect on $Ni(dmit)_2$ salt with an Fe(III) spin-crossover cation: $[Fe(qsal)_2][Ni(dmit)_2]\cdot 2CH_3CN$. Chem. Lett., 34: 1240–1241.
20. Takahashi, K., Cui, H.-B., Okano, Y., Kobayashi, H., Einaga, Y., Sato, O. (2006) Electrical conductivity modulation coupled to a high-spin-low-spin conversion in the molecular system $[Fe^{III}(qsal)_2][Ni(dmit)_2]_3\cdot CH_3CN\cdot H_2O$. Inorg. Chem., 45: 5739–5741.
21. Takahashi, K., Cui, H.-B., Okano, Y., Kobayashi, H., Mori, H., Tajima, H., Einaga, Y., Sato, O. (2008) Evidence of the chemical uniaxial strain effect on electrical conductivity in the spin-crossover conducting molecular system: $[Fe^{III}(qnal)_2][Pd(dmit)_2]_5\cdot$acetone. J. Am. Chem. Soc., 130: 6688–6689.
22. Nihei, M., Takahashi, N., Nishikawa, H., Oshio, H. (2011) Spin-crossover behavior and electrical conduction property in iron(II) complexes with tetrathiafulvalene moieties. Dalton Trans., 40: 2154–2156.
23. Sour, A., Boillot, M.-L., Rivière, E., Lesot, P. (1999) First evidence of a photoinduced spin change in an Fe^{III} complex using visible light at room temperature. Eur. J. Inorg. Chem., 2117–2119.
24. Tamaki, H., Zhong, Z. J., Matsumoto, N., Kida, S., Koikawa, M., Achiwa, N., Hashimoto, Y., Okawa, H. (1992) Design of metal-complex magnets. Syntheses and magnetic-properties of mixed-metal assemblies $\{NBu_4MCr(OX)_3\}X$ ($NBu_4{}^+$= tetra(n-butyl)ammonium ion, ox^{2-} = oxalate ion M = Mn^{2+}, Fe^{2+}, Co^{2+}, Ni^{2+}, Cu^{2+}, Zn^{2+}) J. Am. Chem. Soc., 114: 6974–6979.

25. Sieber, R., Decurtins, S., Stoeckli-Evans, H., Wilson, C., Yufit, D., Howard, J. A. K., Capelli, S. C., Hauser, A. (2000) A thermal spin transition in $[Co(bpy)_3][LiCr(ox)_3]$ (ox = $C_2O_4^{2-}$; bpy = 2,2′-bipyridine). Chem. Eur. J., 6: 361–368.
26. Kou, H. Z., Sato, O. (2007) New oxalate-bridged Cr(III)-Mn(II) polymeric network incorporating a spin-crossover $Co(terpy)_2^{2+}$ cation. Inorg. Chem., 46: 9513–9515.
27. Clemente-León, M., Coronado, E., Giménez-López, M. C., Soriano-Portillo, A., Waerenborgh, J. C., Delgado, F. S., Ruiz-Pérez, C. (2008). Insertion of a spin crossover Fe^{III} complex into an oxalate-based layered material: Coexistence of spin canting and spin crossover in a hybrid magnet. Inorg. Chem., 47: 9111–9120.
28. Clemente-León, M., Coronado, E., López-Jordà, M., Mínguez Espallargas, G., Soriano-Portillo, A., Waerenborgh, J. C. (2010) Multifunctional magnetic materials obtained by insertion of a spin-crossover Fe(III) complex into bimetallic oxalate-based ferromagnets. Chem. Eur. J., 16: 2207–2219.
29. Clemente-León, M., Coronado, E., López-Jordà, M., Waerenborgh, J. C. (2011) Multifunctional magnetic materials obtained by insertion of spin-crossover Fe(III) complexes into chiral 3d bimetallic oxalate-based ferromagnets. Inorg. Chem., 50: 9122–9130.
30. Clemente-León, M., Coronado, E., López-Jordà, M. (2010) 2D and 3D bimetallic oxalate-based ferromagnets prepared by insertion of different Fe(III) spin crossover complexes. Dalton Trans., 39: 4903–4910.
31. Clemente-León, M., Coronado, E., López-Jordà, M., Desplanches, C., Asthana, S., Wang, H., Létard, J.-F. (2011) A hybrid magnet with coexistence of ferromagnetism and photoinduced Fe(III) spin-crossover. Chem. Sci., 2: 1121–1127.
32. Létard, J.-F., Real, J. A., Moliner, N., Gaspar, A. B., Capes, L., Cador, O., Kahn, O. (1999) Light induced excited pair spin state in an iron(II) binuclear spin-crossover compound. J. Am. Chem. Soc., 121: 10630–10631.
33. Wu, D. Y., Sato, O., Einaga, Y., Duan, C. Y. (2009) A spin-crossover cluster of iron(II) exhibiting a mixed-spin structure and synergy between spin transition and magnetic interaction. Angew. Chem. Int. Ed., 48: 1475–1478.
34. Hayami, S., Juhasz, G., Maeda, Y., Yokoyama, T., Sato, O. (2005) Novel structural and magnetic properties of a 1-D iron(II)-manganese(II) LIESST compound bridged by cyanide. Inorg. Chem., 44: 7289–7291.
35. Itaya, K., Uchida, I., Neff, V. D. (1986) Electrochemistry of polynuclear transition-metal cyanides–prussian blue and its analogs. Acc. Chem. Res., 19: 162–168.
36. Verdaguer, M., Bleuzen, A., Marvaud, V., Vaissermann, J., Seuleiman, M., Desplanches, C., Scuiller, A., Train, C., Garde, R., Gelly, G., Lomenech, C., Rosenman, I., Veillet, P., Cartier, C., Villain, F. (1999) Molecules to build solids: high T_c molecule-based magnets by design and recent revival of cyano complexes chemistry Coord. Chem. Rev., 190–192: 1023–1047.
37. Holmes, S. M., Girolami, G. S. (1999) Sol-gel synthesis of $KV^{II}[Cr^{III}(CN)_6]\cdot 2H_2O$: a crystalline molecule-based magnet with a magnetic ordering temperature above 100° C. J. Am. Chem. Soc., 121: 5593–5594.
38. Kosaka, W., Nomura, K., Hashimoto, K., Ohkoshi, S. (2005) Observation of an Fe(II) spin-crossover in a cesium iron hexacyanochromate. J. Am. Chem. Soc., 127: 8590–8591.
39. Papanikolaou, D., Margadonna, S., Kosaka, W., Ohkoshi, S., Brunelli, M., Prassides, K. (2006) X-ray illumination induced Fe(II) spin crossover in the Prussian blue analogue cesium iron hexacyanochromate. J. Am. Chem. Soc., 128: 8358–8363.
40. Coronado, E., Giménez-López, M. C., Levchenko, G., Romero, F. M., García-Baonza, V., Milner, A., Paz-Pasternak, M. (2005) Pressure-tuning of magnetism and linkage isomerism in iron(II) hexacyanochromate. J. Am. Chem. Soc., 127: 4580–4581.
41. Coronado, E., Giménez-López, M. C., Korzeniak, T., Levchenko, G., Romero, F. M., Segura, A., García-Baonza, V., Cezar, J. C., de Groot, F. M. F., Milner, A., Paz-Pasternak, M. (2008) Pressure-induced magnetic switching and linkage isomerism in $K_{0.4}Fe_4[Cr(CN)_6]_{2.8}\cdot 16H_2O$: X-ray absorption and magnetic circular dichroism studies. J. Am. Chem. Soc., 130: 15519–15532.
42. Arai, M., Kosaka, W., Matsuda, T., Ohkoshi, S. I. (2008) Observation of an iron(II) spin-crossover in an iron octacyanoniobate-based magnet. Angew. Chem. Int. Ed., 47: 6885–6887.
43. Ohkoshi, S., Imoto, K., Tsunobuchi, Y., Takano, S., Tokoro, H. (2011) Light-induced spin-crossover magnet. Nat. Chem., 3: 564–569.
44. Kosaka, W., Tokoro, H., Matsuda, T., Hashimoto, K., Ohkoshi, S. (2009) Extremely gradual spin-crossover phenomenon in a cyano-bridged Fe-Mo bimetallic assembly. J. Phys. Chem. C, 113: 15751–15755.

45. Shimamoto, N., Ohkoshi, S., Sato, O., Hashimoto, K. (2002) Control of charge-transfer induced spin transition temperature on cobalt-iron prussian blue analogues. Inorg. Chem., 41: 678–684.
46. Liu, T., Zhang, Y. J., Kanegawa, S., Sato, O. (2010) Water-switching of spin transitions induced by metal-to-metal charge transfer in a microporous framework. Angew. Chem. Int. Ed., 49: 8645–8648.
47. Arimoto, Y., Ohkoshi, S., Zhong, Z. J., Seino, H., Mizobe, Y., Hashimoto, K. (2003) Photoinduced magnetization in a two-dimensional cobalt octacyanotungstate. J. Am. Chem. Soc. 125: 9240–9241.
48. Ohkoshi, S.-I., Ikeda, S., Hozumi, T., Kashiwagi, T., Hashimoto, K. (2006) Photoinduced magnetization with a high curie temperature and a large coercive field in a cyano-bridged cobalt-tungstate bimetallic assembly. J. Am. Chem. Soc., 128: 5320–5321.
49. Avendano, C., Hilfiger, M. G., Prosvirin, A., Sanders, C., Stepien, D., Dunbar, K. R. (2010) Temperature and light induced bistability in a $Co_3[Os(CN)_6]_2 6H_2O$ Prussian Blue analog. J. Am. Chem. Soc., 132: 13123–13125.
50. Liu, T., Zhang, Y. J., Kanegawa, S., Sato, O. (2010) Photoinduced metal-to-metal charge transfer toward single-chain magnet. J. Am. Chem. Soc., 132: 8250–8251.
51. Nihei, M., Okamoto, Y., Sekine, Y., Hoshino, N., Shiga, T., Liu, I. P. C., Oshio, H. (2012) A light-induced phase exhibiting slow magnetic relaxation in a cyanide-bridged $[Fe_4Co_2]$ complex. Angew. Chem. Int. Ed., 51: 6361–6364.
52. Dong, D. P., Liu, T., Kanegawa, S., Kang, S., Sato, O., He, C., Duan, C. Y. (2012) Photoswitchable dynamic magnetic relaxation in a well-isolated $\{Fe_2Co\}$ double-zigzag chain. Angew. Chem. Int. Ed., 51: 5119–5123.
53. Hoshino, N., Iijima, F., Newton, G. N., Yoshida, N., Shiga, T., Nojiri, H., Nakao, A., Kumai, R., Murakami, Y., Oshio, H. (2012) Three-way switching in a cyanide-bridged [CoFe] chain. Nature Chem., 4: 921–926.
54. Verdaguer, M. (2012) Molecular magnetism: A three-headed Janus material. Nature Chem., 4: 871–872.
55. Herrera, J. M., Marvaud, V., Verdaguer, M., Marrot, J., Kalisz, M., Mathonière, C. (2004) Reversible photoinduced magnetic properties in the heptanuclear complex $[Mo^{IV}(CN)_2(CN\text{-}CuL)_6]^{8+}$: a photomagnetic high-spin molecule Angew. Chem. Int. Ed., 43: 5468–5471.

12

Amphiphilic and Liquid Crystalline Spin-Crossover Complexes

Shinya Hayami

Kumamoto University, Department of Chemistry, Japan

12.1 Introduction

Flexible and soft molecules are at the focus of materials research for the construction of functional devices from polymers, Langmuir–Blodgett (LB) films, gels, liquid crystals, and a number of biological materials.[1] For example, metal proteins and enzymes consist of inner flexible space formed by connections of amino acids and active centers in the biological system, and the flexible space plays an important role for operation of inherent catalytic functions.[2] This flexible space plays a very important role in the properties of soft matter, which stimulates us to construct metal complexes with long alkyl chains to incorporate such a flexible space about a central metal complex. These flexible and soft metal complexes can easily be produced by adding long alkyl chains to their periphery. Soft metal complexes are very interesting from the point of view not only of functional materials, but also for phase transitions involving a synchronicity between the central metal complex and the long alkyl chains.[3–5]

On the other hand, a variety of d^n ($n = 4$–7) transition metal compounds exhibiting spin-crossover (SCO) between their high spin (HS) and low spin (LS) states have been reported.[6] Gradual or abrupt SCO may be observed in the solid state, depending on cooperativity. In the solid state, cooperativity is worthy of consideration if intermolecular interactions are sufficiently strong. Cooperativity induces abrupt spin-transitions and hysteresis loops in SCO compounds, which is the reason why the role of intermolecular interactions in spin-transitions has been extensively studied. In this regard, one of the main challenges in the design of SCO compounds has become the production of large cooperativity. It has been theoretically suggested and experimentally confirmed that cooperativity can be increased by designing polymeric structures in which the active sites are linked to each other by chemical bridges.[7–12] Indeed, SCO compounds forming strong intermolecular interactions, such as π-π stacking, hydrogen bonding or inter-chain interactions, exhibit

Spin-Crossover Materials: Properties and Applications, First Edition. Edited by Malcolm A. Halcrow.

abrupt transitions and hysteresis loops.[13–30] So far, investigations of SCO compounds with cooperative effects have been focused on rigid systems. In contrast, the synthesis of molecular assemblies based on flexible molecular units is also a very important factor in achieving a synergy of various interesting and novel physical properties among advanced materials.[31–35] Placing a metal complex exhibiting SCO in the flexible space formed by long alkyl chains is expected to lead to novel physical properties, based on a synergy between SCO and the response of the flexible field toward external stimuli.

In the present review, we focus our attention on flexible and soft SCO iron(II), iron(III) and cobalt(II) compounds bearing long alkyl chains. Flexible and soft SCO molecules can produce soft materials, such as LB films, liquid crystal, and self-assembled molecules. Here we attempt to illustrate four topics: (i) unique magnetic properties; (ii) liquid crystals; (iii) LB films; and (iv) self-assembled molecules from SCO compounds with long alkyl chains.

12.2 Unique Magnetic Properties of SCO Cobalt(II) Compounds with Long Alkyl Chains

Cobalt(II) SCO compounds have been reviewed by Goodwin.[6] The SCO phenomenon for cobalt(II) complexes involves spin conversion between $S = {}^3/_2$ ($t_{2g}{}^5e_g{}^2$) and $S = {}^1/_2$ ($t_{2g}{}^6e_g{}^1$) spin states. The magnetic behavior is characterized by temperature-dependent magnetic susceptibility, which is measured in the form of the $\chi_M T$ versus T plots where χ_M is the molar magnetic susceptibility and T is the temperature. For SCO cobalt(II) complexes, the $\chi_M T$ value in the LS state is around 0.5 cm^3 Kmol^{-1}, whereas that in the HS state appears in the range of 1.9–3.5 cm^3 Kmol^{-1} due to the contribution of orbital angular momentum. Such a SCO involves one electron transfer between the t_{2g} and e_g orbitals, in contrast to the similar process in iron(II) or iron(III) SCO complexes where transfer of two electrons occurs. This is the origin of several important differences in the features between SCO in cobalt(II) complexes and in iron(II) or iron(III) complexes. The significant difference is in the changes of distance between the metal and the coordinated donor atom (Δr_{HL}) accompanying SCO. The change of approximately ~0.10 Å, ~0.15 Å and ~0.20 Å in Δr_{HL} takes place concomitantly with the SCO in cobalt(II), iron(III) and iron(II) complexes, respectively. Since the molecular volume change with respect to SCO in cobalt(II) is less pronounced than in iron(II) or iron(III), it leads to gradual SCO in the vast majority of cobalt(II) complexes. The entropy difference based on SCO (ΔS_{spin}) can be expressed by the equation:

$$\Delta S_{spin} = R\ln[(2S+1)_{HS}/(2S+1)_{LS}]$$

Thus the value for SCO cobalt(II) complexes ($\Delta S_{spin} = 5.8$ JK^{-1}mol^{-1}) is smaller than those for SCO iron(II) ($\Delta S_{spin} = 13.38$ JK^{-1}mol^{-1}) or iron(III) ($\Delta S_{spin} = 9.13$ JK^{-1}mol^{-1}) complexes. Therefore, SCO cobalt(II) complexes require smaller external stimuli to show spin conversion between their HS and LS states compared with iron(II) or iron(III) SCO complexes. Although SCO iron(II) or iron(III) compounds with long alkyl chains have been reported, novel and unique properties for SCO cobalt(II) compounds with long alkyl chains are described in this section.

12.2.1 Reverse Spin-Transition for Cobalt(II) Compounds

It is known that cobalt(II) compounds with 2,2′:6′,2″-terpyridine (terpy) ligands [Co(terpy)$_2$]X$_2$·nH$_2$O (X^- = halide, pseudohalide, NO_3^- or ClO_4^-; and, n = 0–5) exhibit gradual SCO behavior.[36–46] SCO cobalt(II) compounds exhibit $S = {}^1/_2 \leftrightarrow S = {}^3/_2$ spin conversion. Hayami *et al.* have reported reverse spin-transitions triggered by structural phase transitions for [Co(Cn-terpy)$_2$](BF$_4$)$_2$ (n = 14, 16) with long alkyl chains.[47–49] The single crystal X-ray analysis for [Co(C16-terpy)$_2$](BF$_4$)$_2$ revealed that the Co–N distances for the central

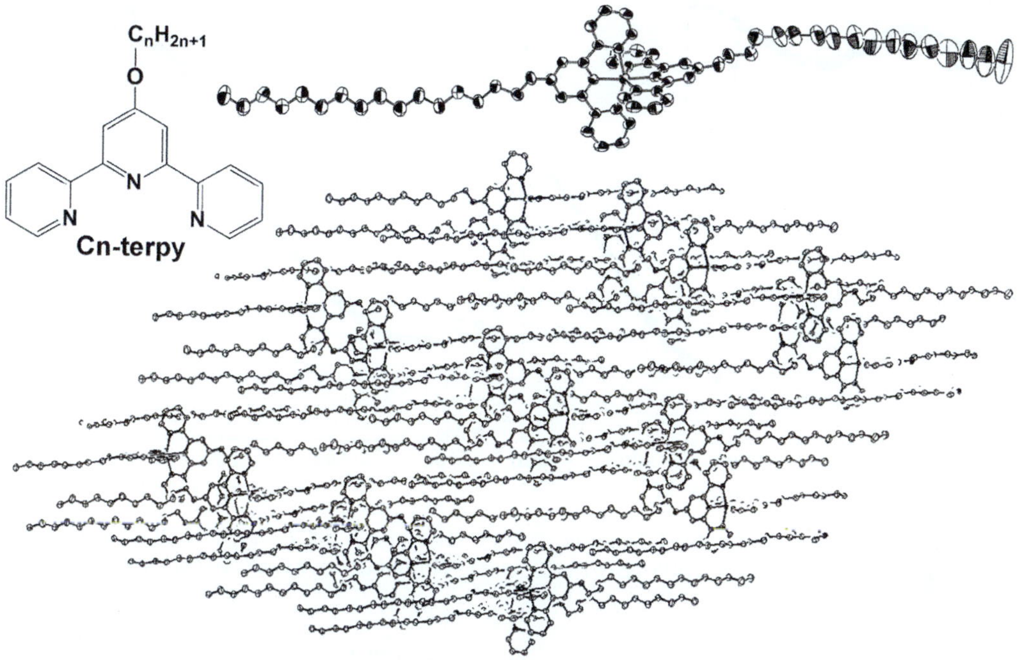

Figure 12.1 *Molecular structure and crystal packing of the [Co(C16-terpy)^2]$^{2+}$ cation. Reprinted with permission from [47]. Copyright 2005 WILEY-VCH Verlag GmbH & Co. KGaA, Weinheim.*

pyridine donor in the terpyridine unit (1.99 Å) are shorter than the Co–N distances for the terminal pyridines (2.13 Å), which induces a pronounced distortion of the CoN_6 octahedron (Fig. 12.1). The long alkyl chains $C_{16}H_{33}O$ which form a rod like structure protrude from the 4′-position of the terpyridine moiety. The two tridentate C16-terpy ligands are found to be perpendicular to one another. The $[Co(C16\text{-}terpy)_2]^{2+}$ cations form short contacts (3.49 Å) through the terminal pyridine and nearest neighbor pyridine rings in the ligands in the *ab* crystal plane, producing a 2D sheet extended by π–π interactions. Further there is fastener effect between the chains which face each other in the complexes. The molecular packing of the compound is very tight. This suggests that there are strong intermolecular interactions in the molecular assembly.

The $\chi_M T$ value for $[Co(C16\text{-}terpy)_2](BF_4)_2$ gradually decreases from 1.58 cm^3 Kmol^{-1} at 400 K to 0.41 cm^3 Kmol^{-1} at 226 K, this gradual decrease of $\chi_M T$ represents normal thermal SCO behavior (Fig. 12.2, left). On further cooling, the $\chi_M T$ value increases abruptly at around $T_{1/2}\downarrow = 217$ K, to 2.01 cm^3 Kmol^{-1} at 206 K. On further heating, the $\chi_M T$ values abruptly dropped at around $T_{1/2}\uparrow = 260$ K, showing the transition from a HS to a LS state. Then the $\chi_M T$ value gradually increases in the temperature range from 267 K to 400 K. The wide thermal hysteresis loop ($\Delta T = 43$ K) remained unchanged in successive thermal cycles. The magnetic behavior shows that the compound exhibits a "reverse spin-transition" phenomenon between LS and HS states.

The compound $[Co(C14\text{-}terpy)_2](BF_4)_2$ was also reported,[47, 48] whose structure is similar to that of $[Co(C16\text{-}terpy)_2](BF_4)_2$. The $\chi_M T$ value for this compound gradually decreases from 1.57 cm^3 Kmol^{-1} at 400 K to 0.52 cm^3 Kmol^{-1} at 268 K, then abruptly increases at $T_{1/2}\downarrow = 250$ K to 1.73 cm^3 Kmol^{-1} at 230 K (Fig. 12.2, right). On further cooling, $\chi_M T$ gradually decreases to 0.44 cm^3 Kmol^{-1} at 5 K, which follows normal SCO behavior. On heating, $\chi_M T$ gradually increases to 1.96 cm^3 Kmol^{-1} at 280 K, until a jump in $\chi_M T$ from 1.96 cm^3 Kmol^{-1} to 2.15 cm^3 Kmol^{-1} was observed at 298 K. On further heating, the

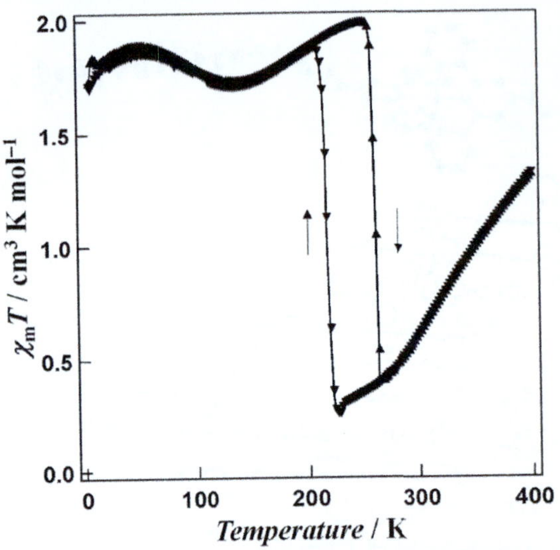

Figure 12.2 $\chi_m T$ versus T plots for $[Co(C16\text{-}terpy)_2](BF_4)_2$ (left) and $[Co(C14\text{-}terpy)_2](BF_4)_2$ (right). Reprinted with permission from [47]. Copyright 2005 WILEY-VCH Verlag GmbH & Co. KGaA, Weinheim.

$\chi_M T$ value abruptly decreases from HS to LS state at $T_{1/2}\uparrow = 307$ K to 0.95 cm^3 Kmol^{-1} at 320 K. Then the $\chi_M T$ value gradually increases up to 400 K, as before. The thermal hysteresis loop ($\Delta T = 57$ K) also remained unchanged in successive thermal cycles, and the magnetic behavior also shows that the compound $[Co(C14\text{-}terpy)_2](BF_4)_2$ also exhibited "reverse spin-transition." Furthermore, it should be emphasized that room temperature is located within the hysteresis loop for this material.

12.2.2 Re-Entrant Spin-Transition for Cobalt(II) Compounds

Re-entrant phenomena can be observed in soft materials, and re-entrant SCO behavior was observed for the solvated compound $[Co(C12\text{-}terpy)_2](BF_4)_2{\cdot}EtOH{\cdot}0.5H_2O$ and nonsolvated compound $[Co(C12\text{-}terpy)_2](BF_4)_2$. Their structures are similar to those of $[Co(C16\text{-}terpy)_2](BF_4)_2$ and $[Co(C14\text{-}terpy)_2](BF_4)_2$.[50] The $\chi_M T$ value for the solvated compound is 1.90 cm^3 Kmol^{-1} at 5 K, which decreases on heating at $T_{1/2} = 49$ K to reach 0.93 cm^3 Kmol^{-1} at 69 K. On further heating, $\chi_M T$ gradually increases to 2.03 cm^3 Kmol^{-1} at 300 K (Fig. 12.3, left). On the other hand, the $\chi_M T$ value gradually decreases on cooling from 2.03 cm^3 Kmol^{-1} at 300 K to 0.93 cm^3 Kmol^{-1} at 69 K, but then increases at $T_{1/2} = 49$ K. This magnetic behavior shows "re-entrant SCO."

The $\chi_M T$ value for the nonsolvated compound is 1.33 cm^3 Kmol^{-1} at 5 K, which abruptly decreases at $T_{1/2} = 49$ K to 0.64 cm^3 Kmol^{-1} at 59 K on heating (Fig. 12.3, right). On further heating, the $\chi_M T$ value gradually increases to 2.02 cm^3 Kmol^{-1} at 265 K, then two $\chi_M T$ jumps at 269 K and 330 K were observed. On further heating, the $\chi_M T$ value is constant (2.14 cm^3 Kmol^{-1}) between 334 K and 382 K, and abruptly drops at around 400 K. On the other hand, the $\chi_M T$ value gradually decreases from 1.89 cm^3 Kmol^{-1} at 400 K to 1.15 cm^3 Kmol^{-1} at 276 K, and abruptly increases at $T_{1/2}\downarrow = 258$ K on cooling. On further cooling, the $\chi_M T$ value gradually decreases from 1.88 cm^3 Kmol^{-1} at 248 K to 0.64 cm^3 Kmol^{-1} at 59 K, and abruptly increases at $T_{1/2} = 49$ K. This unusual magnetic curve remained unchanged in successive thermal cycles. The $\chi_m T$ change in the temperature range of 60–250 K can be assigned to normal SCO behavior. On the other hand, the $\chi_m T$ changes at $T_{1/2} = 49$ K, and at $T_{1/2}\uparrow \geq 400$ K and $T_{1/2}\downarrow = 258$ K, can be assigned to a reverse

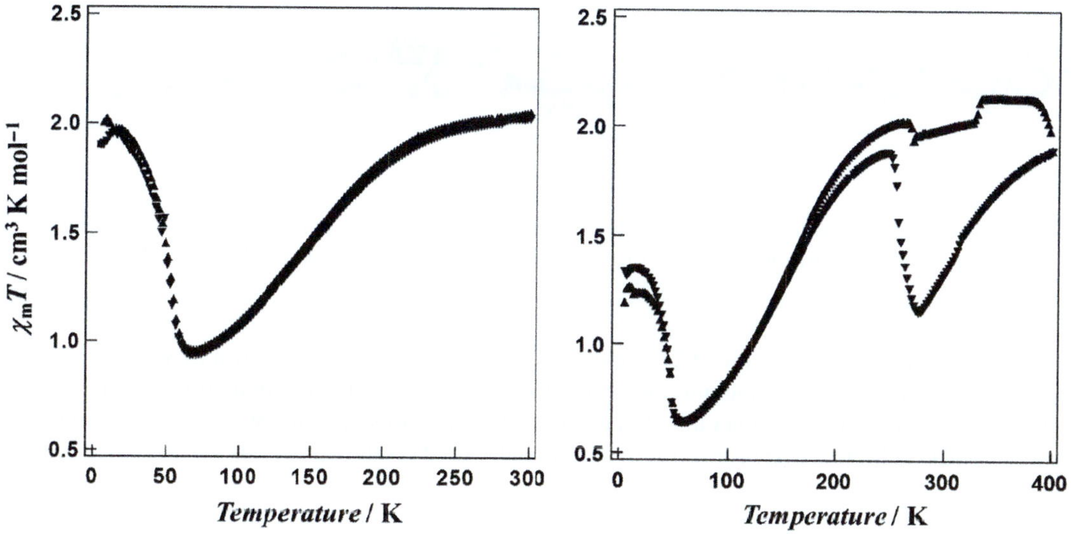

Figure 12.3 $\chi_m T$ versus T plots for $[Co(C12\text{-}terpy)_2](BF_4)_2{\cdot}EtOH{\cdot}0.5H_2O$ (left) and $[Co(C12\text{-}terpy)_2](BF_4)_2$ (right). Reproduced from [50] with permission of The Royal Society of Chemistry.

spin-transition. The reverse spin-transition at higher temperature shows very wide thermal hysteresis loop ($\Delta T \geq 142$ K), and room temperature is located within the hysteresis loop.

In conclusion, cobalt(II) compounds with long alkylated terpyridine ligands show unique magnetic behavior because of structural phase transitions associated with thermal motion of the long alkyl chains.

12.3 Liquid Crystalline SCO Compounds

Liquid crystals are fascinating functional materials that are important in fields of advanced materials such as electrooptic devices. Organic liquid crystals are diamagnetic and can be easily oriented by an electric field. It is possible to align liquid crystals by a magnetic field in a manner analogous to alignment by an electric field. The construction of metal-containing liquid crystals (metallomesogens) has recently attracted a great deal of attention.[51–55] Strong magnetic fields are required for the alignment of diamagnetic liquid crystals. On the other hand, metallomesogens containing unpaired electrons are paramagnetic, and the magnetic field strength required to align paramagnetic liquid crystals is much smaller. The investigation of magnetic liquid crystals containing rare-earth ions has been reported, because these ions often have a large magnetic anisotropy.[51] Furthermore, polymeric compounds bearing nickel(II) side chains have been synthesized, which exhibited cooperative magnetic properties toward construction of ferromagnets with liquid crystal properties.[53] Therefore, liquid crystals that can be controlled by several types of external field have been developed and anticipated.

The idea of combining SCO behavior and liquid crystal in a single material could produce fascinating peculiarities, and it has not yet been achieved. Interplay or synergy between a spin-transition and liquid crystal transition has been observed for SCO metallomesogens. These include systems with coupled phase transitions, subdivided into three groups *a*, *b*, and *c* (in *a* the structural changes associated with the crystal–liquid crystal transition drive the spin-transition; in *b* these structural changes influence the spin state of the metallic centers, but are not the driving force of the spin state transition; and *c* concerns with systems where

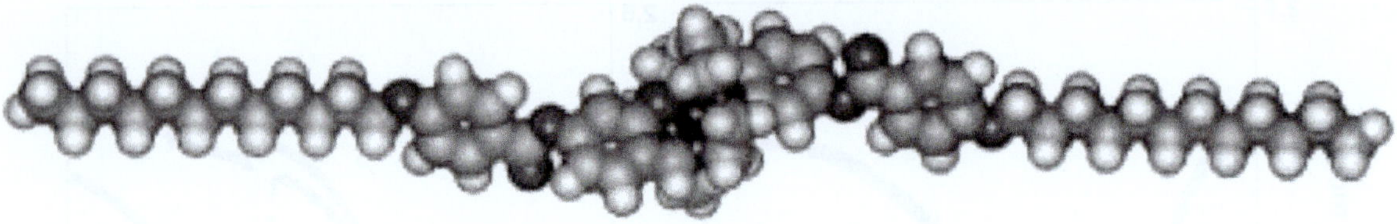

Figure 12.4 Computer simulation of the molecular structure of the liquid crystalline iron(III) complex (counterion omitted). Reprinted with permission from [56]. Copyright 2001 WILEY-VCH Verlag GmbH & Co. KGaA, Weinheim.

vitrification of the material inhibits the SCO properties). Alternative behavior is shown by systems where both transitions coexist in the same temperature region but are not coupled due to competition with dehydration of the material and systems with uncoupled phase transitions. In this section, SCO metallomesogens with independent, and synchronized, spin-transitions and crystal–liquid crystal transitions are described separately.

12.3.1 Metallomesogens with SCO Property

Galyametdinov *et al.* reported the first SCO iron(III) compound ($S = {}^1/_2 \leftrightarrow S = {}^5/_2$) with liquid crystal properties in 2001.[56] Liquid crystal properties were confirmed in the crystal state and mesophase by polarizing optical microscopy, differential scanning calorimetry (DSC), and powder X-ray scattering. The SCO iron(III) compound forms a rod-like molecular structure (Fig. 12.4), and exhibits the fan-shaped texture usually attributed to the smectic A (SmA) mesophase in the range 388–419 K. Its magnetic properties were investigated in the temperature range 4.5–460 K; clearly, the compound exhibits gradual SCO behavior (Fig. 12.5). The value of the effective magnetic moment per iron(III) ion, μ_{eff}, is 5.14 μ_B at 400 K, and a LS fraction of about 88% follows from the magnetic susceptibility measured at $T = 4.6$ K. A specific feature in the magnetic behavior of the sample is the jump in the magnetic moment at $T = 380$ K associated with the crystal–liquid crystal phase transition. Due to the presence of paramagnetic ions, the SCO iron(III) metallomesogen exhibits enhanced magnetic anisotropy and can be aligned by a magnetic field in the mesophase of a liquid crystal.

Figure 12.5 Temperature dependence of the effective magnetic moment μ_{eff} [μ_B] of the iron(III) complex. Reprinted with permission from [56]. Copyright 2001 WILEY-VCH Verlag GmbH & Co. KGaA, Weinheim.

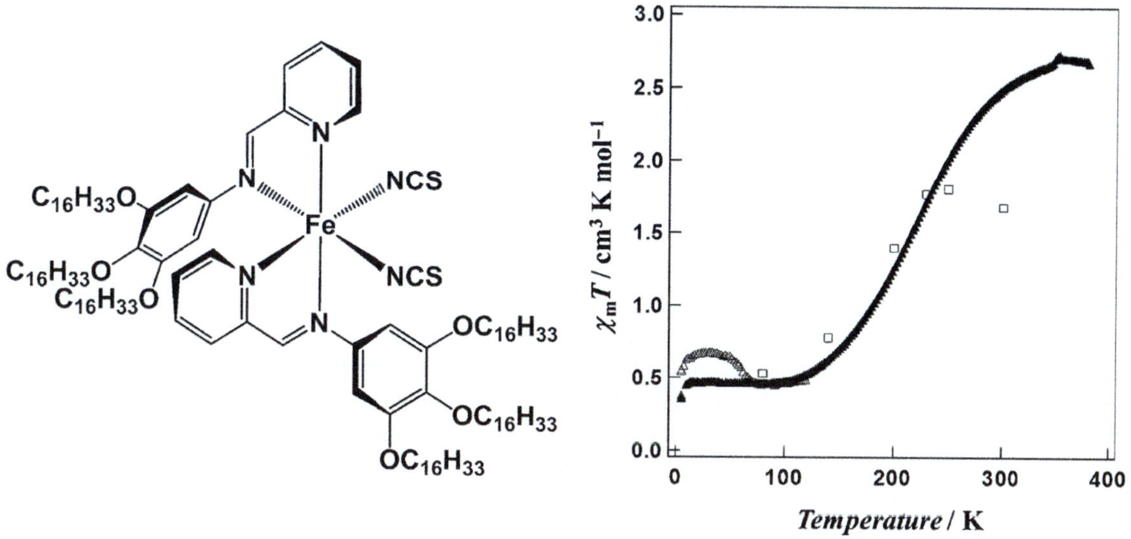

Figure 12.6 Chemical structure and $\chi_m T$ versus T plots for the iron(II) complex [Fe(3C16-L)$_2$(NCS)$_2$]. The magnetic susceptibilities of the sample were measured on warming from 5 K to 380 K (▲). Magnetic data estimated from the area ratios of Mössbauer absorption between HS and LS species also are (□). A LIESST experiment for the sample was carried out, and $\chi_m T$ versus T plots were recorded in the warming mode after the sample was exposed to light illumination for 1 hour at 5 K (△). Reprinted with permission from [57]. Copyright 2004 WILEY-VCH Verlag GmbH & Co. KGaA, Weinheim.

Subsequently, Hayami *et al.* reported an iron(II) metallomesogen [Fe(3C16-L)$_2$(NCS)$_2$] (Fig. 12.6).[57, 58] Liquid–crystalline property for this iron(II) compound was also confirmed by DSC, texture, and powder X-ray patterns. The results show that the compound exhibits a crystal–mesophase transition at 345 K, and a melting point at 400 K. SCO in the iron(II) compound was followed by magnetic susceptibility measurements and Mössbauer spectra, and the compound showed gradual SCO behavior at $T_{1/2} = 217$ K (Fig. 12.6). A jump in the $\chi_m T$ value associated with the crystal–mesophase transition was also observed at around 350 K, and the jump indicates the crystal–mesophase transition may be due to liquid crystal alignment. When the sample was illuminated ($\lambda = 532$ nm) at 5 K, an increase in the susceptibility upon irradiation was observed. This suggests that the compound shows light-induced excited spin state trapping (LIESST). Therefore, the iron(II) compound has three coexisting physical properties – SCO, LIESST ,and liquid crystallinity – in a single compound.

Recently, Gaspar *et al.* also reported a novel family of 2D iron(II) metallomesogens exhibiting SCO properties (Fig. 12.7).[59] In the 2D SCO parent compounds without alkyl chains, the observed spin-transitions are of the first order type and accompanied by large hysteresis loops. In contrast, the 2D iron(II) metallomesogens exhibit incomplete and continuous spin-transitions. The apparent loss of cooperativity can be related to the lack of intermolecular contacts between the [FeM(CN)$_4$]$_\infty$ sheets. In the reported metallomesogens, the photo- and thermally induced spin-transition and the crystal–mesophase transition each take place at different temperature intervals.

12.3.2 Synchronization of SCO and Liquid Crystal Transition

Ever since SCO metallomesogens were reported,[56] the synchronization of spin-transition and liquid crystal transitions has been a focus since this can lead to a number of advantages in practical applications.

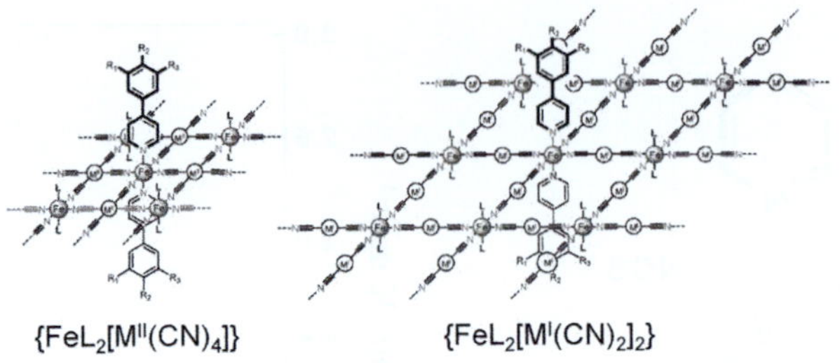

Figure 12.7 Structures of iron(II) metallomesogens. Reprinted with permission from [59]. Copyright 2010, American Chemical Society.

Fujigaya *et al.* succeeded in the first demonstration of synchronization SCO and phase transition by use of self-assembling triazole ligands with long alkyl chains for the polynucleation of iron(II) species (Figs 12.8 and 12.9).[60, 61] This approach allowed mesoscopic cooperativity among the magnetic species through inter-digitation of the alkyl chains and enabled a "lock-and-release" feature of the spin state. Gaspar *et al.* have also investigated the possibility of synchronizing both transitions in iron(II) complexes.[3, 62, 63] They have suggested that the first step in tailoring these hybrid materials has been the choice of SCO systems showing abrupt spin-transitions near or above room temperature. The second step of this approach then concerns the incorporation of a liquid crystalline moiety onto the SCO system, which implies the attachment of an aromatic core with alkyl chain substituents to the triazole or pyridine ligands. The 1D metallomesogens $[Fe(Cn\text{-}trz)_3](p\text{-}tol)_2$ were reported (Fig. 12.10) to be in the LS state at room temperature. Upon heating they show an abrupt and complete spin-transition accompanied by a pronounced change of color. The complex's

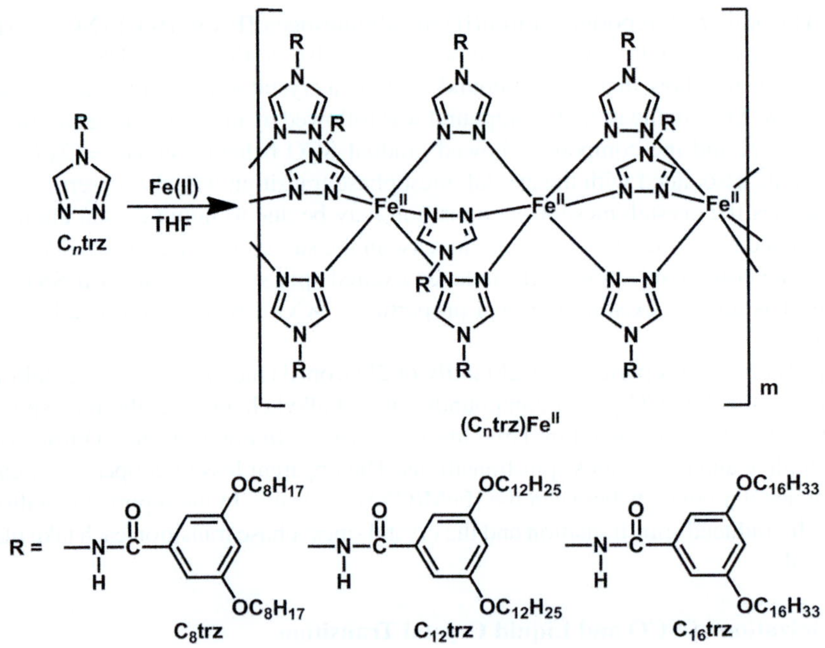

Figure 12.8 Structures of self-assembled iron(II) complexes having triazole ligands with two long alkyl chains $(C_ntrz)Fe^{II}$.[60]

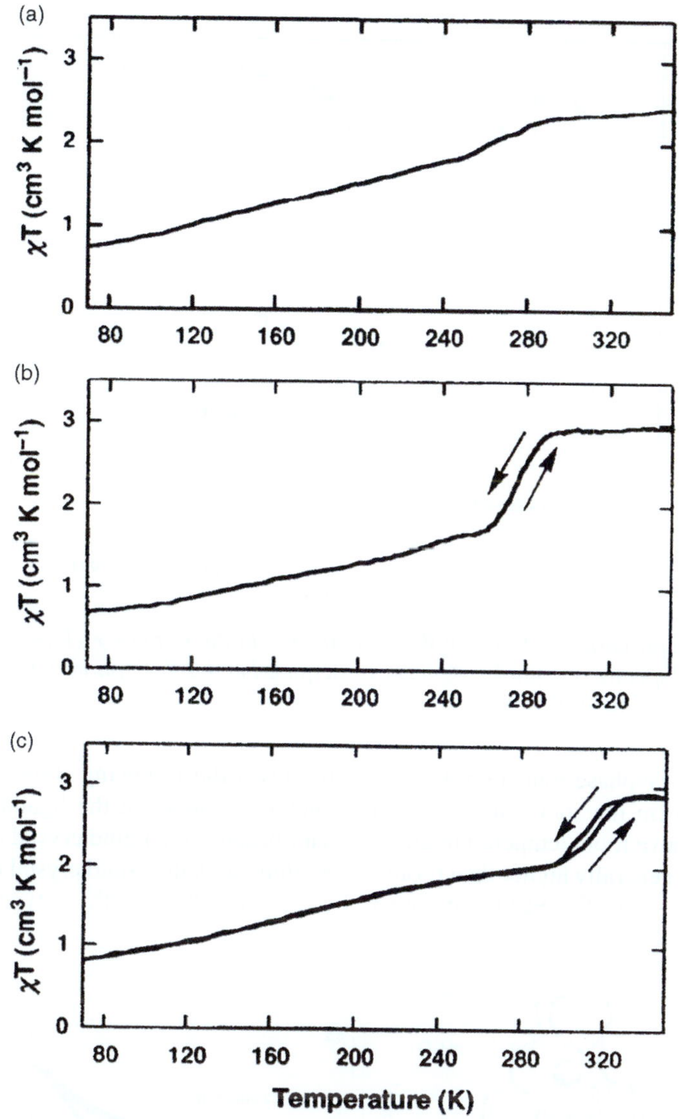

Figure 12.9 *Magnetic susceptibility profiles of (A) (C_8trz)Fe^{II}, (B) (C_{12}trz)Fe^{II}, and (C) (C_{16}trz)Fe^{II}. Reprinted with permission from [60]. Copyright 2003, American Chemical Society.*

DSC measurements, X-ray powder diffraction analysis, and optical microscopy studies have proven that the transition from the crystalline state to the mesophase (discotic columnar) takes place synchronously with the spin-transition. Furthermore iron(II) metallomesogens [Fe(C*n*-trenH)]$(A)_2\cdot sH_2O$ ($A^- = F^-$, Cl^-, Br^- and I^-) were also reported (Fig. 12.11).[3, 4, 64] The spin states for the compounds depend strongly on their degree of hydration, and not on the state of the matter. In partially dehydrated samples, however, the structural changes associated with the crystal–liquid crystal phase transition drive the spin state transition (Fig. 12.11).

Hayami *et al.* have also reported synchronizing the spin-transition and crystal–liquid crystal phase transition in a cobalt(II) compound.[65] In order to develop a variety of switchable molecules that exhibit spin-transition

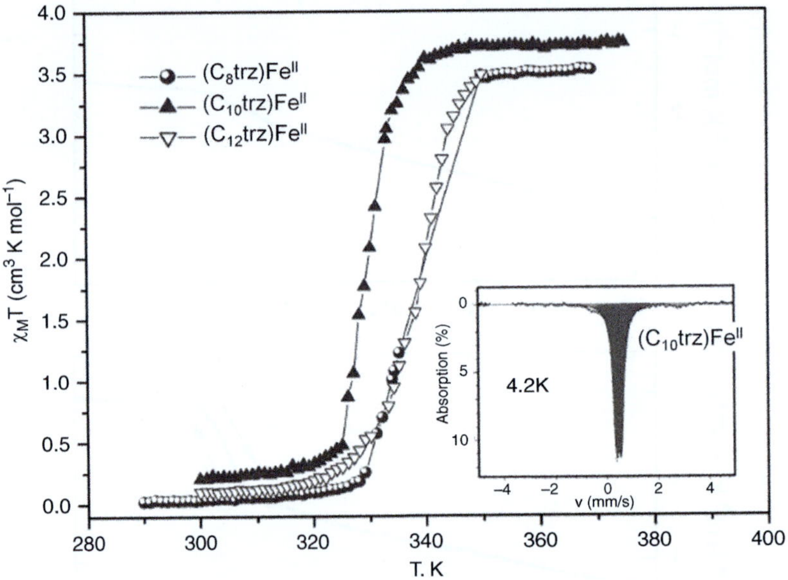

Figure 12.10 Magnetic properties of the complexes $(C_ntrz)Fe^{II}$ in the form of χ_MT vs T recorded at a rate of 1 K/min. Inset: Mössbauer spectra of complex $(C_{10}trz)Fe^{II}$ acquired at 4.2 K. Reprinted with permission from [62]. Copyright 2006, American Chemical Society.

induced by a crystal–mesophase transition, they have proposed the following strategies: SCO compounds are necessary to transform the spin states by subtle structural changes in the ligands or molecules; SCO compounds with $T_{1/2}$ above room temperature are necessary because the liquid crystal transition temperature for metallomesogens is generally higher than room temperature; and, the liquid crystal transition temperature should be adjusted to $T_{1/2}$ in the SCO compounds by changing the length or type of the alkyl chains.

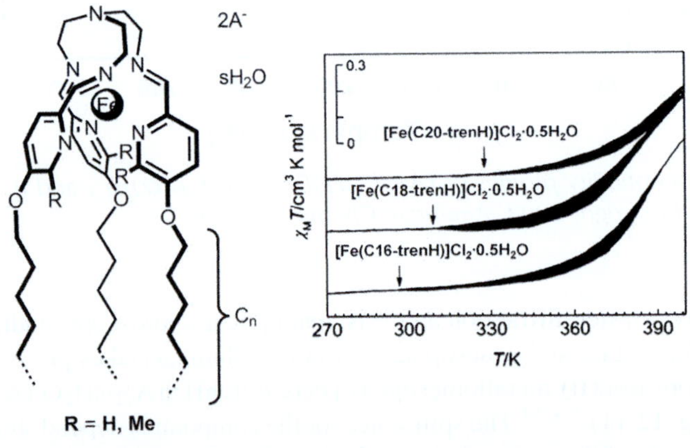

Figure 12.11 The family of complexes with formula $[Fe(Cn\text{-}trenH)](A)_2{\cdot}sH_2O$ ($A = F^-$, Cl^-, Br^- and I^-), and χ_MT versus T for $[Fe(Cn\text{-}trenH)]Cl_2{\cdot}0.5H_2O$ ($n = 16, 18, 20$). Arrows indicate the temperatures at which heating and cooling curves diverge. The black areas indicate the hysteresis loops. Reprinted with permission from [64]. Copyright 2008, American Chemical Society.

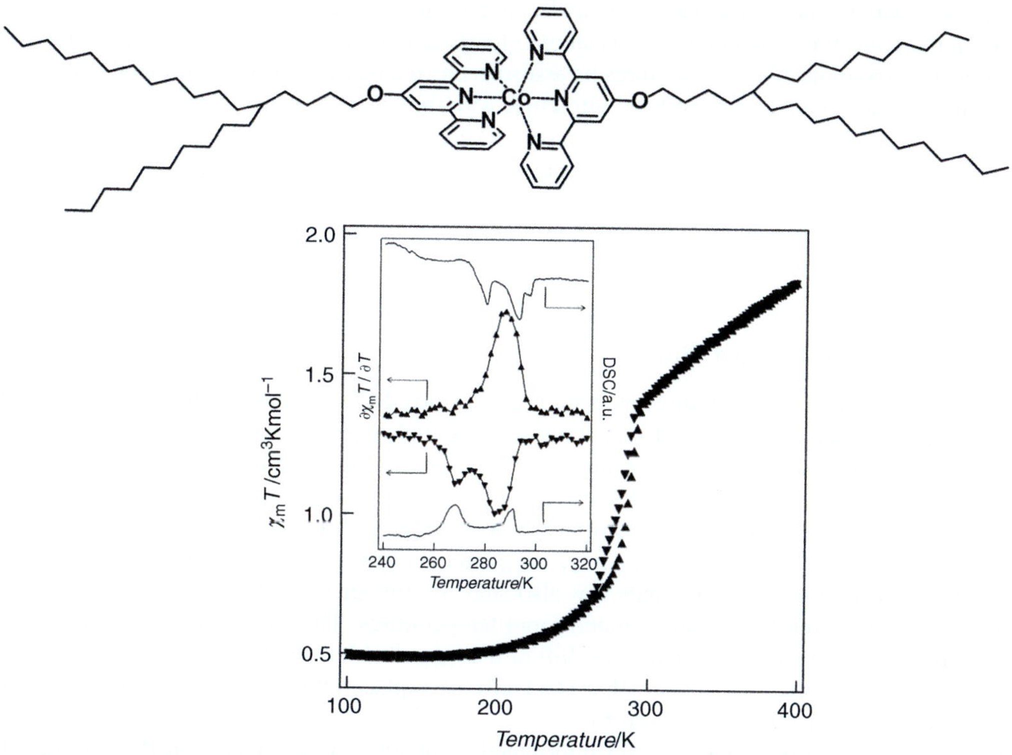

Figure 12.12 Chemical structure and $\chi_m T$ vs. T plot for [Co(C5C12C10-terpy)$_2$](BF$_4$)$_2$. The inset graph shows the derivative $\partial\chi_m T/\partial T$ plot as a function of the temperature. Reprinted with permission from [65]. Copyright 2007, American Chemical Society.

Therefore, the cobalt(II) metallomesogen [Co(C5C12C10-terpy)$_2$](BF$_4$)$_2$ with branched alkyl chains was reported (Fig. 12.12).[65] The branched alkyl chains were introduced in order to decrease the mesophase transition temperature and synchronize the liquid crystal and spin-transition temperatures for the compound (Fig. 12.12). Thus, the cobalt(II) compound also shows spin-transition and crystal–liquid crystal phase transitions in the same temperature range.

12.4 Langmuir–Blodgett Films and Amphiphilic SCO Compounds

The LB technique is a method of preparing organic monolayer and multilayer films.[66] An inert aliphatic matrix is alternated with polar planes in which specific properties are confined. LB films are usually based on organic molecules or polymers, and they have been used in investigations of nonlinear optics, energy transfer or electron transfer, organic conductors, molecular rectifiers, and ferroelectrics.[67] The LB method has been developed to incorporate inorganic species, coordination compounds, clusters, and so on into the films. The magnetism in LB films is interesting because the films can be compared to solid state materials that are studied as 2D systems. 2D magnets have been obtained using LB films of manganese stearate by Pomerantz *et al.*,[68–70] Melville *et al.*[71], and Haseda *et al.*[72] On the other hand, self-assembly of amphiphilic molecules has great advantages for the creation of materials for biological applications and environmentally friendly processability.[73–76] Depending on the external environment, on the molecular structure and shape, and the relative volume of their hydrophilic and hydrophobic parts, these molecules self-assemble into diverse supramolecular architectures, including spherical or cylindrical micelles, vesicles, ribbons, and tubules.[77] To date, a variety of supramolecular architectures have also been created by the use of the metal complexes.[78–80] These supramolecular complexes have been attracting widespread interest. However, since they consist of discrete metal complexes, consequently their applications are largely limited to shape-related functions such as host–guest inclusion. It was only recently that interactions among the connecting metal ions at low temperatures have been investigated.[81] At the same time, control of the magnetism and electronic states in the solid metal complexes represent other important issues.[82] In this section, SCO LB films and amphiphilic SCO compounds are described.

12.4.1 SCO Langmuir–Blodgett Films

The Langmuir–Blodgett (LB) technique has been used to study SCO because of the interest in controlling between HS and LS states in organized thin films for potential applications. Recently, some groups have investigated SCO compounds in LB films, using iron(II)[83, 84] and manganese(III)[85] compounds having transition temperatures near room temperature in solid states. The first attempt was carried out by Ruaudel-Teixier *et al.* in 1988 with the amphiphilic iron(II) complex $[Fe(OP_3)_2(NCS)_2]$ (Fig. 12.13).[86, 87] Three long aliphatic chains were grafted on the phenanthroline moiety, and good LB films for the compounds were obtained on the substrates. Thermal SCO in the compounds was also studied in magnetic LB films. A temperature-induced SCO was observed between liquid nitrogen and room temperatures. This SCO was characterized in the IR spectral absorption by a 40 cm^{-1} frequency up shift of the ν_{NCS} doublet when changing from HS to LS states (Fig. 12.13). From these results it appears that SCO still takes place in a 2D system, as it does in the same nonamphilic complex in a 3D single crystal form.

Soyer *et al.* also reported that a semi-fluorinated derivative of a SCO iron(II) complex $[Fe(Bi\text{-}DFOD)_2(NCS)_2]$ shows astonishing stability in the LB film, allowing the formation of perfectly defined LB films of this molecule (Fig. 12.14).[88–91] They succeeded in measuring the SCO behavior of these multilayers by using a SQUID susceptometer for the first time (Fig. 12.15). A powdered sample of the iron(II) complex shows SCO behavior at around 230 K, over an interval of ca. 150 K. In the LB film, the SCO behavior was somewhat different. Initially, at 300 K, the $\chi_m T$ product of the LB film is rather small and close

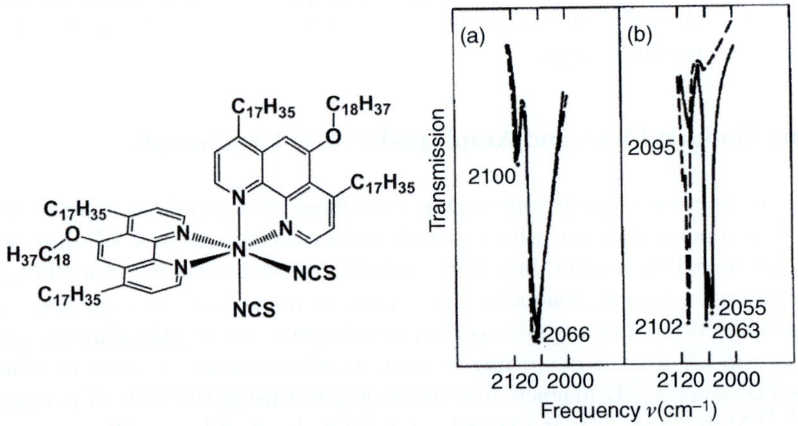

Figure 12.13 Chemical structure of $[Fe(OP_3)_2(NCS)_2]$, and its IR spectra: (a) powder (solid line: 293 K; dot line: 77 K); (b) LB film (solid line: 293 K; dot line: 77 K). Reprinted from [86], with permission from Elsevier.

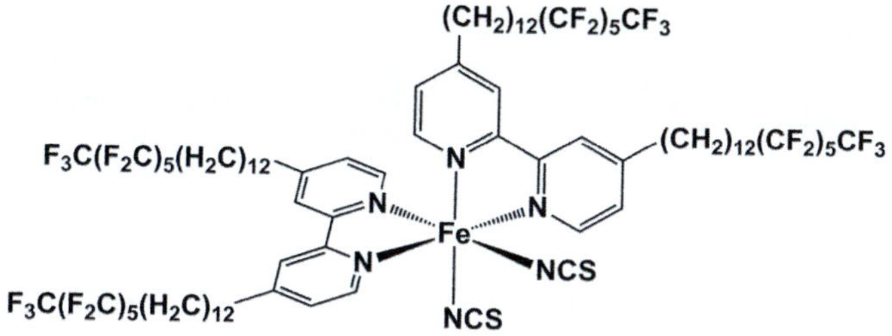

Figure 12.14 *Structure of $[Fe(Bi\text{-}DFOD)_2(NCS)_2]$.*[90]

to 0.9 emu $Kmol^{-1}$. When the sample was cooled down, the $\chi_m T$ decreases very slightly to reach a value of ca. 0.5 emu $Kmol^{-1}$. These results seem to indicate the almost total absence of a spin conversion in the LB film. When the sample was annealed at 370 K, the successive thermal cycles show a behavior similar to that of the complex in powder. The resultant curve was easily fitted with a standard thermodynamic spin conversion model, indicating a very weak cooperativity of the phenomenon in powder as well as in LB films. Therefore, due to the packing and the organization imposed by the LB technique, the molecules are mainly in a LS state in the pristine LB film. These results demonstrated the strong modification of SCO phenomenon induced by the lamellar organization imposed by the LB technique. Furthermore Létard *et al.* reported on the first evidence of LIESST effect in a LB film incorporating the iron(II) SCO complex $[Fe(Bi\text{-}DFOD)_2(NCS)_2]$ (Fig. 12.16).[92] The lower standard deviation and the faster LIESST population observed in the LB film as compared to the powder were correlated to a higher structural organization in the film. The faster dynamics of relaxation in the LB film was attributed to a pressure effect.

12.4.2 Amphiphilic SCO Compounds

The lipophilic modification of 1D coordination chains is a useful method for producing soluble inorganic polymers through the integration of solid-state inorganic chemistry with polymer and supramolecular chemistry. Kimizuka *et al.* developed the application of host–guest interactions at the surface of linear triazole

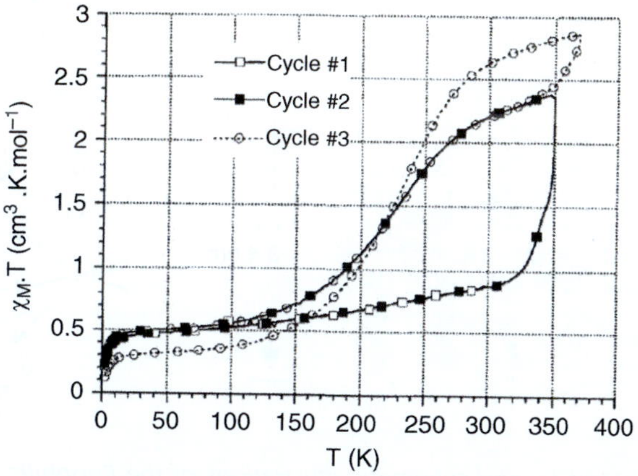

Figure 12.15 *$\chi_M T$ versus T plots for a LB film of 1200 layers of $[Fe(Bi\text{-}DFOD)_2(NCS)_2]$ deposited onto Mylar in a magnetic field of 2 T. Thermal cycles: (1) 300 K → 2 K → 300 K → 2 K; (2) 2 K → 350 K → 2 K; (3) 2 K → 370 K → 2 K. Reprinted with permission from [90]. Copyright 1999 WILEY-VCH Verlag GmbH & Co. KGaA, Weinheim.*

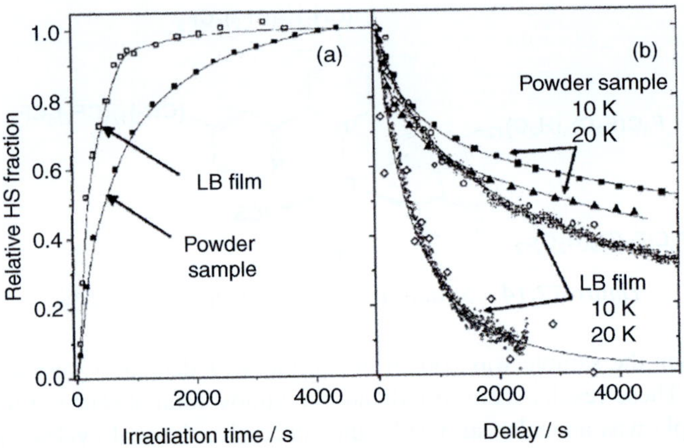

Figure 12.16 LIESST experiment on [Fe(Bi-DFOD)$_2$(NCS)$_2$]. (a) Relative HS molar fraction deduced from magnetic measurements under light irradiation as a function of time at 10 K. (b) Decay of the metastable HS state. Data obtained for a powder sample at (■) 10 K and (▲) 20 K from magnetic measurements. Data obtained for a LB film at (O) 10 K and (◇) 20 K from magnetic measurements, and (•) from UV-visible spectroscopy. Reprinted with permission from [92]. Copyright 1999, American Chemical Society.

complexes to control SCO, in the iron(II) complex Fe(**1**)$_3$Cl$_2$ (Fig. 12.17).[93] This iron(II) complex is in its purple LS state in powder form, while it is transformed into pale-yellow HS organogels when dissolved in chloroform. These gels are formed by nanofibrous aggregates, as confirmed by transmission electron microscopy (TEM). The casting of chloroform solutions on solid supports produced purple transparent films (LS state). The cast films exhibit gradual SCO behavior without thermal hysteresis. At room temperature, the cast films exhibit regular lamellar structures before and after the doping of alcohol, as confirmed by

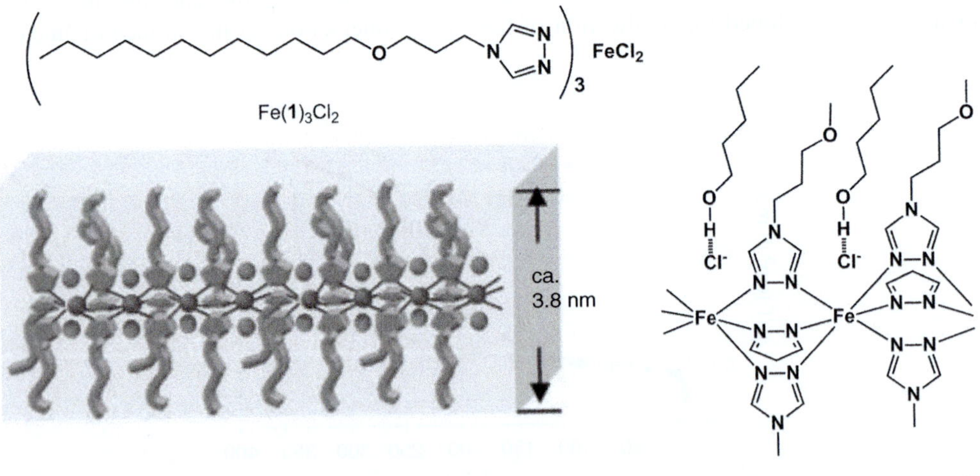

Figure 12.17 Structure of Fe(**1**)$_3$Cl$_2$, and schematic illustrations of the lipophilic molecular wire of Fe(**1**)$_3$Cl$_2$ (left), and of ionic hydrogen bonding (right). The recognition of the guest chain length is determined by (i) van der Waals interactions between the alkyl group of ligand **1** and alcohol molecules and (ii) ionic hydrogen bonding (ROH···Cl$^-$). Reprinted with permission from [93]. Copyright 2006 WILEY-VCH Verlag GmbH & Co. KGaA, Weinheim.

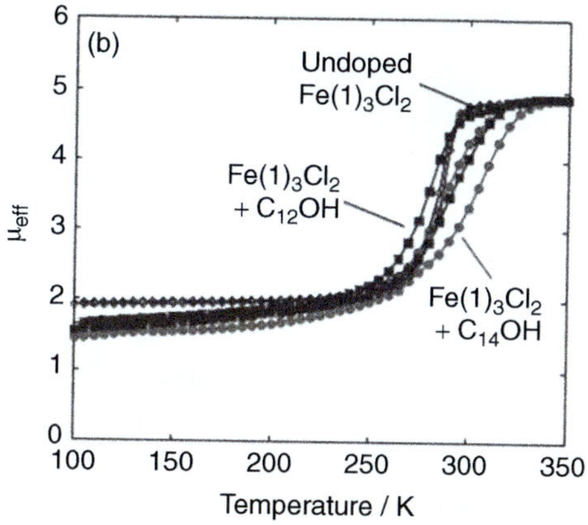

Figure 12.18 *Temperature dependence of the effective magnetic moment (μ_{eff}) of Fe(**1**)$_3$Cl$_2$/C$_n$OH samples. Reprinted with permission from [93]. Copyright 2006 WILEY-VCH Verlag GmbH & Co. KGaA, Weinheim.*

wide-angle X-ray diffraction (WAXD) measurements. Interestingly, the films Fe(**1**)$_3$Cl$_2$/C$_n$H$_{2n+1}$OH (n = 12 and 14) showed reversible abrupt SCO with thermal hysteresis (Fig. 12.18). The observed bistability is related to dynamic structural transformations between lamellar and hexagonal structures (Fig. 12.17). This provides a novel supramolecular strategy for controlling the bistability of SCO phenomena.

To convert 1D inorganic chains into soluble nanowires, amphiphilic packaging strategies have been developed in response to temperature changes and interactions with guest or solvent molecules. Kimizuka *et al.* also reported photodependent morphological changes for an azobenzene-containing iron(II) triazole complex [Fe(**2**)$_3$](BF$_4$)$_2$ (Fig. 12.19), which was soluble in common organic solvents.[94] Atomic force microscopy (AFM) of the sample in dilute chloroform solutions showed flexible molecular wires (length: 10–100 nm,

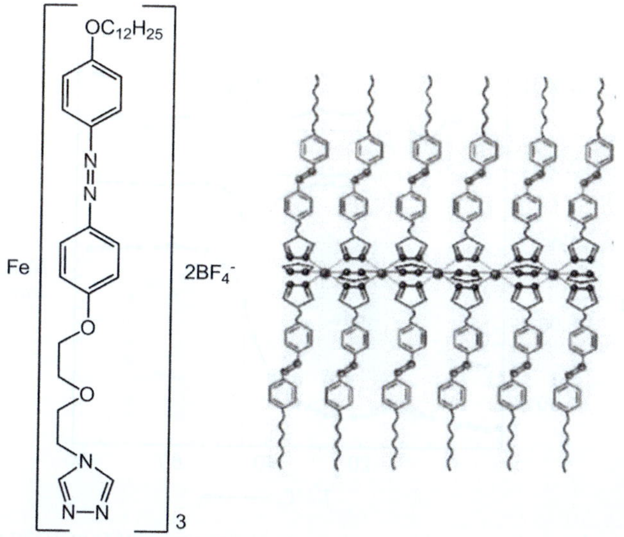

Figure 12.19 *Structure of [Fe(**2**)$_3$](BF$_4$)$_2$ and schematic illustration of a molecular wire of [Fe(**2**)$_3$](BF$_4$)$_2$. Reproduced from [94] with permission of The Royal Society of Chemistry.*

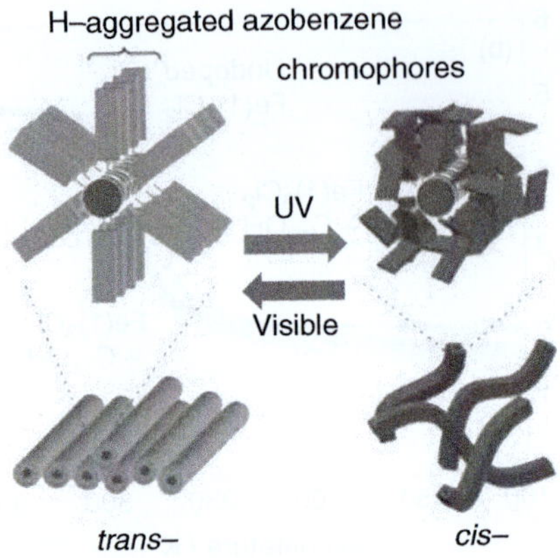

Figure 12.20 Schematic illustration of light-induced morphological changes to $[Fe(\mathbf{2})_3](BF_4)_2$. Reproduced from [94] with permission of The Royal Society of Chemistry.

width: ca. 7 nm). $[Fe(\mathbf{2})_3](BF_4)_2$ also formed organogels when dispersed in chlorocyclohexane. UV-light irradiation of [Fe(*trans*-**2**)$_3$]$(BF_4)_2$ as a yellow chlorocyclohexane gel caused its fluidization into an orange-colored solution of [Fe(*cis*-**2**)$_3$]$(BF_4)_2$. Visible-light irradiation of [Fe(*cis*-**2**)$_3$]$(BF_4)_2$ reversibly regenerated [Fe(*trans*-**2**)$_3$]$(BF_4)_2$, which is accompanied by physical gelation (Fig. 12.20).

Clérac *et al.* reported that $[Fe(\mathbf{3})_3](ptol)_2{\cdot}2H_2O$ gelates with decane and shows SCO (Fig. 12.21).[95] Aida *et al.* showed that an iron(II) complex $[Fe(C_{12}trz)_3](\mathbf{4})_2$ of double-chained 1,2,4-triazole ligands shows SCO, which was accompanied by dissolution of the dodecane gel (Fig. 12.22).[96] In these systems, however, LS complexes are maintained only in the solid state or in organogels composed of crystalline aggregates. In addition, covalent modification of triazole ligands with bulky substituents generally causes lengthening of Fe(II)–ligand bonds, resulting in the destabilization of the LS states.[61] It is thought that the application of the

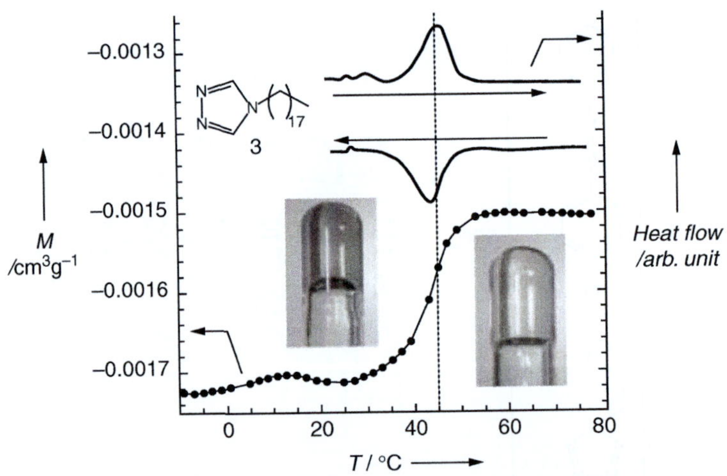

Figure 12.21 Temperature dependence of the massic magnetic moment (*M*) at 1000 Oe and DSC thermograms (inset; 5 °C min^{-1}) of a 5 wt% $[Fe(\mathbf{3})_3](ptol)_2{\cdot}2H_2O$/decane gel. The dotted line corresponds to the spin-crossover temperature (T_{SCO} = 44 °C) deduced from DSC and magnetic measurements. Reprinted with permission from [95]. Copyright 2004 WILEY-VCH Verlag GmbH & Co. KGaA, Weinheim.

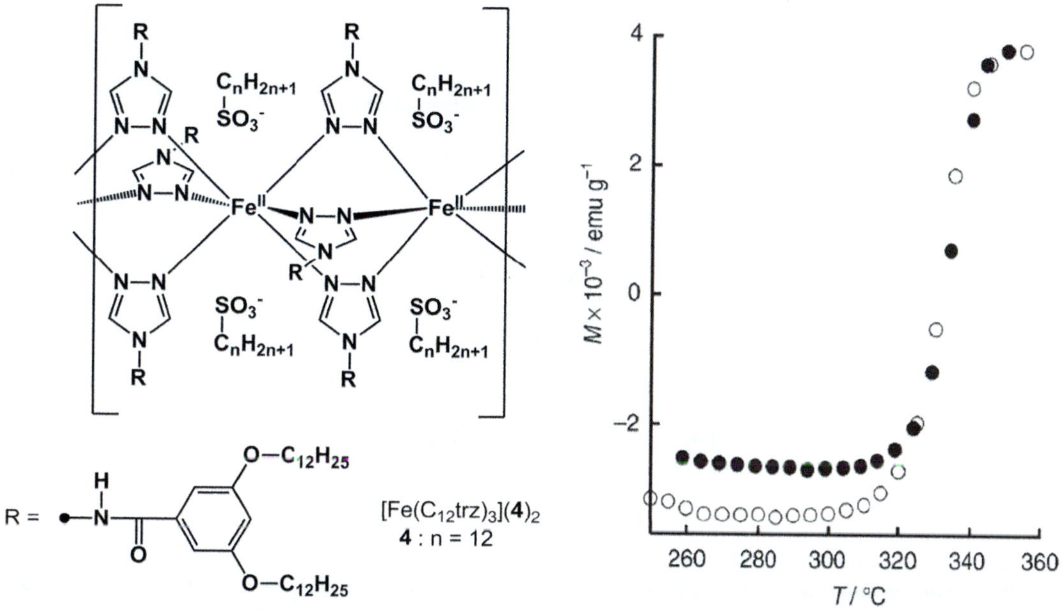

Figure 12.22 Structure of iron(II) triazolate coordination polymers with sulfonate counterions, and the magnetic susceptibility of $[Fe(C_{12}trz)_3](4)_2$ (40 wt%) upon heating (○) and cooling (•). Reprinted with permission from [96]. Copyright 2007 WILEY-VCH Verlag GmbH & Co. KGaA, Weinheim.

lipid packaging technique leads to stabilization of the LS complexes in solution, and consequent control of spin conversion via dynamic self-assembly of linear coordination chains.

Two principal strategies have been adopted in ligand design, to affect the spin state of the coordinating metal ion. The strength of the ligand field can be reduced by either steric strain or steric bulk in the coordinating ligands, thus making the HS configuration more favorable. For instance, the terpy ligand induces a strong ligand field; therefore, the resulting complexes with iron(II) of the form $[Fe(terpy)_2]X_2$ are exclusively LS irrespectively of the counteranion X^-. While substitution in the 4′-position of the central pyridine ring has no effect, it has been recognized that bulky substituents in the 6- and/or 6′-positions do affect the spin state. Kurth *et al.* reported another approach to affecting SCO, by introducing mechanical distortion through an amphiphilic phase transition in a lamellar superstructure. This was constructed with a quasi 1D rigid rod-type metallo-supramolecular polyelectrolyte (MEPE) based on a ditopic bis-terpy ligand (Fig. 12.23).[97–100] Self-assembly of the bis-terpy ligand and iron(II) acetate in aqueous solution results in formation of the MEPE. Subsequent self-assembly of the MEPE and dihexadecyl phosphate (DHP) affords the polyelectrolyte-amphiphile complex (PAC), whose magnetic properties were studied in a LB multilayer. The mechanical distortion is strong enough to reduce the strength of the ligand field at the iron ion, giving rise to a spin-transition from the LS to the HS state (Fig. 12.24). The observed high-temperature molecular magnetism differs from those of other magnetic materials. Instead of competing spin ordering versus thermal disorder, the present material is characterized by spin-transition induced by gap reduction.

The lipid packing strategy was also applied to $[Fe(NH_2\text{-trz})]^{2+}$ and $[Fe(HOC_2\text{-trz})]^{2+}$ by Kimizuka *et al.* (Fig. 12.25).[101] The lipid complexes showed SCO as pure powders, but the SCO temperatures observed for toluene dispersions were significantly higher than observed in the crystalline state (Fig. 12.25). The remarkable stabilization of the LS configuration in solution indicates an increase in the ligand field strength, which occurs as a consequence of contraction in Fe–N and nearest neighbor Fe···Fe distances. AFM measurements showed nanofibers (widths: 20–30 nm, heights: ca. 7 nm) in the LS state. In contrast, fragmented nanofibers or dots

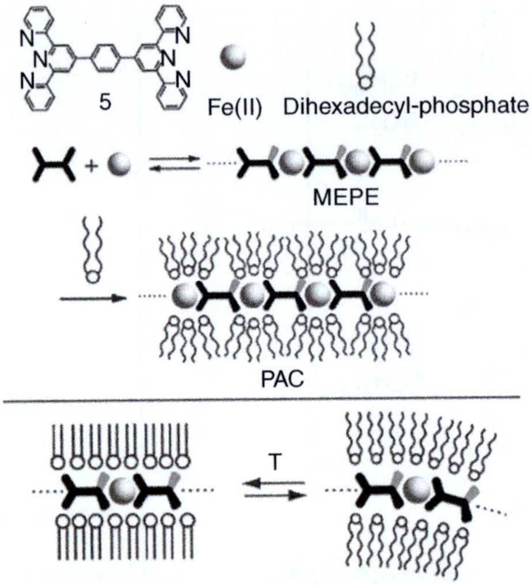

Figure 12.23 (Top) Self-assembly of ditopic ligand **5** and Fe(II) results in formation of the linear, rigid rod-type metallo-supramolecular polyelectrolyte (MEPE). In a consecutive step, MEPE is assembled with dihexadecyl phosphate, resulting in formation of the corresponding polyelectrolyteamphiphile complex (PAC). (Bottom) Upon heating of the multilayer, the alkyl chains of the mesophase melt, resulting in a distortion of the metal ion coordination geometry. Reprinted with permission from [97]. Copyright 2005, American Chemical Society.

are observed for HS complexes. These observations clearly indicate that nanofibers of LS complexes undergo thermal SCO to labile HS complexes, which dissociate into fragments. Lipids with enhanced thermal mobility would force the fragments to adopt the HS state, as indicated by the abrupt spin-transition. The observation of developed nanofibers after recooling the heat-dispersed samples, and the recovery of the LS complexes, clearly indicate reversible self-assembly of HS fragments into the LS nanofibers.

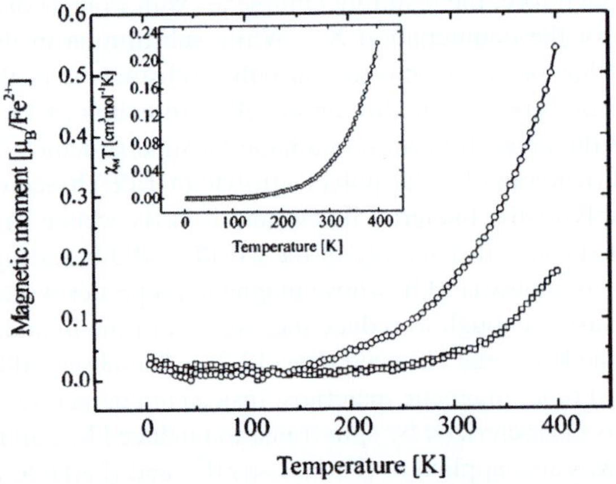

Figure 12.24 The magnetic moment for two samples consisting of 11 and 15 Langmuir monolayers in the temperature range from 5 to 400 K and H =10 kOe. The material behaves diamagnetically at low temperatures and becomes paramagnetic above room temperature. (Inset) Measurement of the molar susceptibility as $\chi_M T$ vs. T. Reprinted with permission from [97]. Copyright 2005, American Chemical Society.

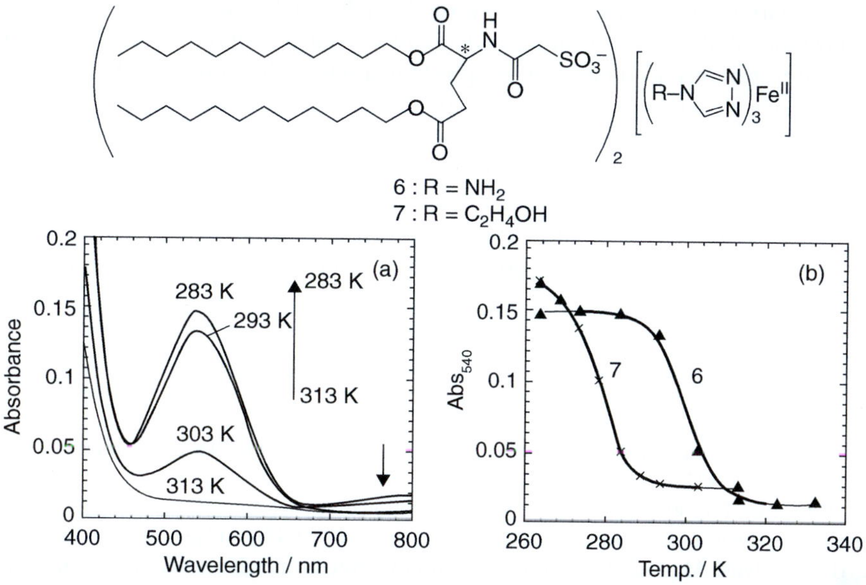

Figure 12.25 Structure of the iron(II) compounds **6** and **7**, and: (a) Temperature dependence of UV/vis absorption spectra for complex **6** (5 unit mM in toluene, cooling process). (b) Temperature dependences of peak absorption intensities at 540 nm (LS complex) for **6** and **7**. Absorbance changes in heating and cooling cycles traced the same curves. Reprinted with permission from [101]. Copyright 2008, American Chemical Society.

12.5 Conclusion and Outlook

The phenomenon of SCO has been known for 80 years. To date, there are many approaches for applications of SCO because of belief in a switching molecular device in the near future. One approach is developing in soft materials. There are many methods for the developments of SCO soft materials, and the simplest one is attaching the long alkyl chains in SCO compounds. The long alkylated SCO compounds are expected to show not only unique magnetic properties but also liquid crystalline properties, LB films, and amphiphilic self-assembly. Synergistic functions between SCO and soft material transitions are very important.

The magnetic properties and the structural characteristics of several SCO cobalt(II) complexes with terpy derivatives possessing various substituents at the 4-position have been described, where thermal motion of the long alkyl chains induces a spin state transition. These ligands have aroused special interest in SCO materials to enhance cooperative effects, because they can generate abrupt SCO with thermal hysteresis which is important to further application. This is controlled by employing a terminal functionality linked covalently or by introducing other molecules from the outside. The cobalt(II) complexes show diverse SCO behavior caused by intermolecular interactions. Recently some papers about the relationship between SCO and long alkyl chains have been also published.[102–104] From the aforementioned results, we believe that novel physical properties may be discovered in these metal complexes through coupling between the motion of flexible alkyl chains and the electronic states.

Second, achievements of the interplay and synergy between spin-transitions and liquid crystal transitions are described. Future activity in the area of SCO metallomesogens can be developed towards the creation of new systems on the basis of the present results. Gaspar *et al.* describe a promising next step in this research line with the development of photochromic liquid crystals, by inducing spin-state switching by ligand driven isomerization, analogous to the work of Zarembowitch and Boillot.[6, 105–107] We also aim to achieve synchronization of the spin-transition and liquid crystal transition, to develop photo-switchable crystal–liquid crystal transitions accompanied by spin-transitions within a thermal hysteresis loop. Finally, spin-transitions involving disintegration of coordination bonds are described, as spin conversion directed by self-assembly. The lipid-packaging of charged coordination chains not only provides solubility to linear metal coordination polymers, but also improves their potential functionality. The solvophobic contraction of charged metal complexes in organic media is a new concept, which would be widely applicable to developing functional nanofibers that show improved performance compared to those in the solid state.

These soft materials with bistability are important as a means to develop novel switchable molecular devices in future.

References

1. Hamley, I. W. (2000) Introduction to Soft Matter: Polymers, Colloids, Amphiles and Liquid Crystals. J. Wiley and Sons, Chichester, UK, p. 342.
2. Mizutani, Y., Kitagawa, T. (2001) Ultrafast structural relaxation of myoglobin following photodissociation of carbon monoxide probed by time-resolved resonance Raman spectroscopy. J. Phys. Chem. B, 105: 10992–10999.
3. Gaspar, A. B., Seredyuk, M., Gütlich, P. (2009) Spin crossover in metallomesogens. Coord. Chem. Rev., 253: 2399–2413.
4. Gaspar, A. B., Seredyuk, M., Gütlich, P. Spin crossover in iron(II) complexes: recent advances. J. Mol. Struct., 924–926, 9–19 (2009)
5. Hayami, S., Komatsu, Y., Shimizu, T., Kamihata, H., Lee, Y. H. (2011) Spin-crossover in cobalt(II) compounds containing terpyridine and its derivatives. Coord. Chem. Rev., 255: 1981–1990.
6. Gütlich, P., Goodwin, H. A. (Eds) (2004) Spin Crossover in Transition Metal Compounds I-III. Top. Curr. Chem., vols. 233–235. Springer, Berlin / Heidelberg, Germany.
7. Vreugdenhil, W., van Diemen, J. H., de Graaf, R. A, Haasnoot, J. G., van der Kraan, A. M., Kahn, O., Zarembowitch, J. (1990) High-spin ↔ low-spin transition in $[Fe(NCS)_2(4,4\text{-bis-}1,2,4\text{-triazole})_2](H_2O)$ X-ray crystal structure and magnetic, Mössbauer and EPR properties. Polyhedron, 9: 2971–2979.
8. Kröber, J., Codjovi, E., Kahn, O., Grolière, F., Jay, C. (1993) A spin transition system with a thermal hysteresis at room temperature. J. Am. Chem. Soc., 115: 9810–9811.
9. Sato, O., Iyoda, T., Fujishima, A., Hashimoto, K. (1996) Photoinduced magnetization of a cobalt-iron cyanide. Science, 272: 704–705.
10. Niel, V., Martínez-Agudo, J. M., Muñoz, M. C., Gaspar, A. B., Real, J. A. (2001) Cooperative spin crossover behavior in cyanide-bridged Fe(II)–M(II) bimetallic 3D Hofmann-like networks (M = Ni, Pd, and Pt). Inorg. Chem., 40: 3838–3839.
11. Létard, J.-F., Guionneau, P., Nguyen, O., Costa, J. S., Marcen, S., Chastanet, G., Marchivie, M., Goux-Capes, L. (2005) A guideline to the design of molecular-based materials with long-lived photomagnetic lifetimes. Chem. Eur. J., 11: 4582–4589.
12. Kosaka, W., Nomura, K., Hashimoto, K., Ohkoshi, S. (2005) Observation of an Fe(II) spin-crossover in a cesium iron hexacyanochromate. J. Am. Chem. Soc., 127: 8590–8591.
13. Real, J. A., Gallois, B., Granier, T., Suez-Panama, F., Zarembowitch, J. (1992) Comparative investigation of the spin-crossover compounds $Fe(btz)_2(NCS)_2$ and $Fe(phen)_2(NCS)_2$ (where btz = 2,2′-bi-4,5-dihydrothiazine and phen = 1,10-phenanthroline). Magnetic properties and thermal dilatation behavior and crystal structure of $Fe(btz)_2(NCS)_2$ at 293 and 130 K. Inorg. Chem., 31: 4972–4979.
14. Gallois, B., Real, J. A., Hauw, C., Zarembowitch, J. (1990) Structural changes associated with the spin transition in bis(isothiocyanato)bis(1,10-phenanthroline)iron: a single-crystal X-ray investigation. Inorg. Chem., 29: 1152–1158.
15. Zhong, Z. J., Tao, J., Yu, Z., Dun, C., Liu, Y., You, X. (1988) A stacking spin-crossover iron(II) compound with a large hysteresis. J. Chem. Soc. Dalton Trans., 327–328.

16. Vreugdenhil, W., Haasnoot, J. G., Kahn, O., Thuéry, P., Reedijk, J. (1987) A copper(II) dope as a detector for the high-spin ↔ low-spin transition in the two-dimensional compound [trans-bis(thiocyanato)bis(4,4′-bi-1,2,4-triazole)iron] hydrate. J. Am. Chem. Soc., 109: 5272–5273.
17. Real, J. A., Andrés, E., Muñoz, M. C., Julve, M., Granier, T., Bousseksou, A., Varret, F. (1995) Spin crossover in a catenane supramolecular system. Science, 268: 265–267.
18. Moliner, N., Muñoz, M. C., Létard, S., Solans, X., Menéndez, N., Goujon, A., Varret, F., Real, J. A. (2000) Spin crossover bistability in three mutually perpendicular interpenetrated (4,4) nets. Inorg. Chem., 39: 5390–5393.
19. Halder, G. J., Kepert, C. J., Moubaraki, B., Murray, K. S., Cashion, J. D. (2002) Guest-dependent spin crossover in a nanoporous molecular framework material Science, 298: 1762–1765.
20. Garcia, Y., Kahn, O., Rabardel, L., Chansou, B., Salmon, L., Tuchagues, J. P. (1999) Two-step spin conversion for the three-dimensional compound tris(4,4′-bis-1,2,4-triazole)iron(II) diperchlorate, Inorg. Chem., 38: 4663–4670.
21. Chernyshov, D., Hostettler, M., Törnroos, K. W., Bürgi, H.-B. (2003) Ordering phenomena and phase transitions in a spin-crossover compound–uncovering the nature of the intermediate phase of $[Fe(2\text{-}pic)_3]Cl_2{\cdot}EtOH$. Angew. Chem. Int. Ed., 42: 3825–3830.
22. Kitazawa, T., Gomi, Y., Takahashi, M., Takeda, M., Enomoto, M., Miyazaki, A., Enoki, T. (1996) Spin-crossover behaviour of the coordination polymer $Fe^{II}(C_5H_5N)_2Ni^{II}(CN)_4$. J. Mater. Chem., 6: 119–121.
23. Niel, V., Martínez-Agudo, J. M., Muñoz, M. C., Gaspar, A. B., Real, J. A. (2001) Cooperative spin crossover behavior in cyanide-bridged Fe(II)−M(II) bimetallic 3D Hofmann-like networks (M = Ni, Pd, and Pt). Inorg. Chem., 40: 3838–3839.
24. Galet, A., Niel, V., Muñoz, M. C., Real, J. A. (2003) Synergy between spin crossover and metallophilicity in triple interpenetrated 3D nets with the NbO structure type. J. Am. Chem. Soc., 125: 14224–14225.
25. Hayami, S., Gu, Z. Z., Shiro, M., Einaga, Y., Fujishima, A., Sato, O. (2000) First observation of light-induced excited spin state trapping for an iron(III) complex. J. Am. Chem. Soc., 122: 7126–7127.
26. Hayami, S., Gu, Z. Z., Yoshiki, H., Fujishima, A., Sato, O. (2001) Iron(III) spin-crossover compounds with a wide apparent thermal hysteresis around room temperature, J. Am. Chem. Soc., 123: 11644–11650.
27. Létard, J.-F., Guionneau, P., Codjovi, E., Lavastre, O., Bravic, G., Chasseau, D., Kahn, O. (1997) Wide thermal hysteresis for the mononuclear spin-crossover compound cis-bis(thiocyanato)bis[*N*-(2′-pyridylmethylene)-4-(phenylethynyl)anilino]iron(II). J. Am. Chem. Soc., 119: 10861–10862.
28. Xie, C. L., Hendrickson, D. N. (1987) Mechanism of spin-state interconversion in ferrous spin-crossover complexes: direct evidence for quantum mechanical tunneling. J. Am. Chem. Soc., 109: 6981–6988.
29. Conti, A. J., Xie, C. L., Hendrickson, D. N. (1989) Tunneling in spin-state interconversion of ferrous spin-crossover complexes. Concentration dependence of apparent activation energy determined in solution by laser-flash photolysis, J. Am. Chem. Soc., 111: 1171–1180.
30. Hauser, A., Adler, J., Gütlich, P. (1988) Light-induced excited spin state trapping (LIESST) in $[Fe(2\text{-}mephen)_3]^{2+}$ embedded in polymer matrices, Chem. Phys. Lett., 152: 468–472.
31. Kiriya, D., Chang, H.-C., Nakamura, K., Tanaka, D., Yoneda, K., Kitagawa, S. (2009) Polymorph-dependent molecular valence tautomerism synchronized with crystal-melt phase transitions. Chem. Mater., 21: 1980–1988.
32. Kiriya, D., Nakamura, K. Chang, H.-C., Kitagawa, S. (2009) Bimodal three-membered valence tautomerism of an alkyl chain-functionalized manganese complex, Chem. Commun., 2085–2087.
33. Kiriya, D., Nakamura, K., Kitagawa, S., Chang, H.-C. (2010) Modulable cooperativity in a valence tautomeric complex functionalized with branched alkyl chains, Chem. Commun., 46: 3729–3731.
34. Chang, H.-C., Komasaka, K., Kishida, K., Shiozaki, T., Ohmori, T., Matsumoto, T., Kobayashi, A., Kato, M., Kitagawa, S. (2011) Chromic behaviors of hexagonal columnar liquid crystalline platinum complexes with catecholato, 2-thiophenolato, and benzenedithiolato. Inorg. Chem., 50: 4279–4288.
35. Takaishi, S., Takamura, M., Kajiwara, T., Miyasaka, H., Yamashita, M., Iwata, M., Matsuzaki, H., Okamoto, H., Tanaka, H., Kuroda, S., Nishikawa, H., Oshio, H., Kato, K., Takata, M. (2008) Charge-density-wave to Mott−Hubbard phase transition in quasi-one-dimensional bromo-bridged Pd compounds, J. Am. Chem. Soc., 130: 12080–12084.
36. Hogg, R., Wilkins, R. G. (1962) Exchange studies of certain chelate compounds of the transitional metals. Part VIII. 2,2′,2″-terpyridine complexes. J. Chem. Soc., 341–350.

37. Schmidt, J. G., Brey, W. S., Stoufer, R. C. (1967) Complexes of cobalt(II). IV. Electron paramagnetic resonance spectra of some magnetically anomalous complexes of cobalt(II). Inorg. Chem., 6: 268–271.
38. Kremer, S., Henke, W., Reinen, D. (1982) High-spin-low-spin equilibriums of cobalt(2+) in the terpyridine complexes $Co(terpy)_2X_2.nH_2O$. Inorg. Chem., 21: 3013–3022.
39. Harris, C. M., Lockyer, T. N., Martin, R. L., Patil, H. R. H., Sinn, E., Stewart, I. M. (1969) Five- and six-coordinated complexes of cobalt(II) with 2,2′,2′-terpyridyl: Unusual structure and magnetism. Aust. J. Chem., 22: 2105–2116.
40. Judge, J. S., Baker, W. A. (1967) On the spin equilibrium in bis(2,2′,2″-terpyridine) cobalt(II) salts. Inorg. Chim. Acta, 1: 68–72.
41. Maslen, E. N., Raston, C. L., White, A. H. (1974) Crystal structure of bis-(2,2′:6′,2″-terpyridyl)cobalt(II) bromide trihydrate. J. Chem. Soc. Dalton Trans., 1803–1807.
42. Figgis, B. N., Kucharski, E. S., White, A. H. (1983) Crystal structure of Bis(2,2′:6′,2″-terpyridyl)cobalt(II) iodide dihydrate at 295 K and at 120 K. Aust. J. Chem., 36: 1527–1535.
43. Stoufer, R. C., Smith, D. W., Clevenger, E. A., Norris, T. E. (1966) Complexes of cobalt(II). I. On the anomalous magnetic behavior of some six-coordinate cobalt(II) complexes, Inorg. Chem., 5: 1167–1171.
44. Oshio, H., Spiering, H., Ksenofontov, V., Renz, F., Gütlich, P. (2001) Electronic relaxation phenomena following $^{57}Co(EC)^{57}Fe$ nuclear decay in $[Mn^{II}(terpy)_2](ClO_4)_2\cdot\frac{1}{2}H_2O$ and in the spin crossover complexes $[Co^{II}(terpy)_2]X_2\cdot nH_2O$ (X = Cl and ClO_4): a Mössbauer emission spectroscopic study. Inorg. Chem., 40: 1143–1150.
45. Figgis, B. N., Kucharski, E. S., White, A. H. (1983) Crystal structure of bis(2,2′:6′,2″-terpyridyl)cobalt(II) perchlorate c. 1.3 hydrate. Aust. J. Chem., 36: 1537–1561.
46. Kilner, C. A., Halcrow, M. A. (2010) An unusual discontinuity in the thermal spin transition in $[Co(terpy)_2][BF_4]_2$. Dalton Trans., 39: 9008–9012.
47. Hayami, S., Shigeyoshi, Y., Akita, M., Inoue, K., Kato, K., Osaka, K., Takata, M., Kawajiri, R., Mitani, T., Maeda, Y. (2005) Reverse spin transition triggered by a structural phase transition, Angew. Chem. Int. Ed., 44: 4899–4903.
48. Hayami, S., Murata, K., Urakami, D., Kojima, Y., Akita, M., Inoue, K. (2008) Dynamic structural conversion in a spin-crossover cobalt(II) compound with long alkyl chains. Chem. Commun., 6510–6512.
49. Hayami, S., Urakami, D., Kojima, Y., Yoshizaki, H., Yamamoto, Y., Kato, K., Fuyuhiro, A., Kawata, S., Inoue, K. (2010) Stabilization of long-lived metastable state in long alkylated spin-crossover cobalt(II) compound. Inorg. Chem., 49: 1428–1432.
50. Hayami, S., Kato, K., Komatsu, Y., Fuyuhiro, A., Ohba, M. (2011) Unique spin transition and wide thermal hysteresis loop for a cobalt(II) compound with long alkyl chain. Dalton Trans., 40: 2167–2169.
51. Binnemans, K., Galyametdinov, Y. G., Deun, R. V., Bruce, D. W., Collinson, S. R., Polishchuk, A. P., Bikchantaev, I., Haase, W., Prosvirin, A. V., Tinchurina, L., Litvinov, I., Gubajdullin, A., Rakhmatullin, A., Uytterhoeven, K., Meervelt, L. V. (2000) Rare-earth-containing magnetic liquid crystals. J. Am. Chem. Soc., 122: 4335–4344.
52. Ohta, K., Yamaguchi, N., Yamamoto, I. (1998) Discotic liquid crystals of transition metal complexes. Part 24 Synthesis and mesomorphism of porphyrin derivatives substituted with two or four bulky groups. J. Mater. Chem., 8: 2637–2650.
53. Werner, R., Falk, K., Ostrovsky, S., Haase, W. (2001) Metallopolymers with schiff base side chains. Synthesis and characterization of some nickel(II) containing polymers with unexpected cooperative magnetic properties. Macromol. Chem. Phys., 202: 2813–2823.
54. Piechocki, C., Simon, J., Skoulios, A., Guillon, D., Weber, P. (1982) Annelides. 7. Discotic mesophases obtained from substituted metallophthalocyanines. Toward liquid crystalline one-dimensional conductors. J. Am. Chem. Soc., 104: 5245–5247.
55. Giroud-Godquin, A. M., Marchon, J. C. (1984) Discotic mesophase of copper(II) laurate. J. Phys. Lett., 45: 681–684.
56. Galyametdinov, Y., Ksenofontov, V., Prosvirin, A., Ovchinnikov, I., Ivanova, G., Gütlich, P., Haase, W. (2001) First example of coexistence of thermal spin transition and liquid-crystal properties, Angew. Chem. Int. Ed., 40: 4269–4271.
57. Hayami, S., Danjobara, K., Inoue, K., Ogawa, Y., Matsumoto, N., Maeda, Y. (2004) A photoinduced spin transition iron(II) complex with liquid-crystal properties. Adv. Mater., 16: 869–872.

58. Hayami, S., Motokawa, N., Shuto, A., Masuhara, N., Someya, T., Ogawa, Y., Inoue, K., Maeda, Y. (2007) Photo-induced spin transition for iron(II) compounds with liquid-crystal properties. Inorg. Chem., 46: 1789–1794.
59. Seredyuk, M., Gaspar, A. B., Ksenofontov, V., Galyametdinov, Y., Verdaguer, M., Villain, F., Gütlich, P. (2010) Spin-crossover and liquid crystal properties in 2D cyanide-bridged Fe^{II}-$M^{I/II}$ metalorganic frameworks. Inorg. Chem., 49: 10022–10031.
60. Fujigaya, T., Jiang, D.-L., Aida, T. (2003) Switching of spin states triggered by a phase transition: spin-crossover properties of self-assembled iron(II) complexes with alkyl-tethered triazole ligands. J. Am. Chem. Soc., 125: 14690–14691.
61. Fujigaya, T., Jiang, D.-L., Aida, T. (2005) Spin-crossover dendrimers: generation number-dependent cooperativity for thermal spin transition. J. Am. Chem. Soc., 127: 5484–5489.
62. Seredyuk, M., Gaspar, A. B., Ksenofontov, V., Reiman, S., Galyametdinov, Y., Haase, W., Rentschler, E., Gütlich, P. (2006) Room temperature operational thermochromic liquid crystals. Chem. Mater., 18: 2513–2519.
63. Seredyuk, M., Gaspar, A. B., Ksenofontov, V., Galyametdinov, Y., Verdaguer, M., Villain, F., Gütlich, P. (2008) One-dimensional iron(II) compounds exhibiting spin crossover and liquid crystalline properties in the room temperature region, Inorg. Chem., 47: 10232–10245.
64. Seredyuk, M., Gaspar, A. B., Ksenofontov, V., Galyametdinov, Y., Kusz, J., Gütlich, P. (2008) Does the solid-liquid crystal phase transition provoke the spin-state change in spin-crossover metallomesogens? J. Am. Chem. Soc., 130: 1431–1439.
65. Hayami, S., Moriyama, R., Shuto, A., Maeda, Y., Ohta, K., Inoue, K. (2007) Spin transition at the mesophase transition temperature in a cobalt(II) compound with branched alkyl chains, Inorg. Chem., 46: 7692–7694.
66. Ulman, A. (1991) An Introduction to Ultrathin Organic Films: From Langmuir-Blodgett to Self-Assembly. Academic Press, Boston, USA, p. 442.
67. Roberts, G. (1990) Langmuir-Blodgett Films. Plenum Press, New York, USA, p. 444.
68. Pomerantz, M., Pollak, R. A. (1975) Spin state of manganese in monolayer films of manganese arachidate. Chem. Phys. Lett., 31: 602–604.
69. Pomerantz, M., Dacol, F. H., Segmüller, A. (1978) Preparation of literally two-dimensional magnets. Phys. Rev. Lett., 40: 246–249.
70. Dash, J. G., Ruvalds, J. (Eds) (1980) Phase Transitions in Surface Films. Plenum Press, New York, USA, p. 367.
71. Asaolu, I. A., Blott, B. H., Khan, W. I., Melville, D. (1983) Two-dimensional magnetism in Langmuir-Blodgett films. Thin Solid Films, 99: 263–264.
72. Haseda, T., Yamakawa, H., Ishizuka, M., Okuda, Y., Kubota, T., Hata, M., Amaya, K. (1977) Magnetic two-dimensional system: manganese stearate. Solid State Commun., 24: 599–602.
73. Shimizu, T., Masuda, M., Minamikawa, H. (2005) Supramolecular nanotube architectures, based on amphiphilic molecules. Chem. Rev., 105: 1401–1443.
74. Rosi, N. L., Mirkin, C. A. (2005) Nanostructures in biodiagnostics. Chem. Rev., 105: 1547–1562.
75. Lim, Y.-b., Moon, K.-S., Lee, M. (2009) Recent advances in functional supramolecular nanostructures assembled from bioactive building blocks. Chem. Soc. Rev., 38: 925–934.
76. Cavalli, S., Albericio, F., Kros, A. (2010) Amphiphilic peptides and their cross-disciplinary role as building blocks for nanoscience. Chem. Soc. Rev., 39: 241–263.
77. Kim, H.-J., Kim, T., Lee, M. (2011) Responsive Nanostructures from aqueous assembly of rigid-flexible block molecules. Acc. Chem. Res., 44: 72–82.
78. Mobian, P., Kern, J. M., Sauvage, J. P. (2003) A [2]catenane constructed around a $Ru(diimine)_3{}^{2+}$ complex used as a template. J. Am. Chem. Soc., 125: 2016–2017.
79. Hansenknopf, B., Lehn, J.-M., Boumediene, N., Gervais, A. D., Dorsselaer, A. V., Kneisel, B., Fenske, D. (1997) Self-assembly of tetra- and hexanuclear circular helicates. J. Am. Chem. Soc., 119: 10956–10962.
80. Fujita, M., Tominaga, M., Hori, A., Therrien, B. (2005) Coordination assemblies from a Pd(II)-cornered square complex. Acc. Chem. Res., 38: 369–378.
81. Ruben, M., Breuning, E., Lehn, J.-M., Ksenofontov, V., Renz, F., Gütlich, P., Vaughan, G. B. M. (2003) Supramolecular spintronic devices: spin transitions and magnetostructural correlations in $[Fe_4{}^{II}L_4]^{8+}$ [2×2]-grid-type complexes, Chem. Eur. J., 9: 4422–4429.
82. Gatteschi, D. (1994) Molecular Magnetism: a basis for new materials. Adv. Mater., 6: 635–645.

83. Kitchen, J. A., White, N. G., Gandolfi, C., Albrecht, M., Jameson, G. N. L., Tallond, J. L., Brooker, S. (2010) Room-temperature spin crossover and Langmuir-Blodgett film formation of an iron(II) triazole complex featuring a long alkyl chain substituent: the tail that wags the dog. Chem. Commun., 46: 6464–6466.
84. White, N. G., Feltham, H. L. C., Gandolfi, C., Albrecht, M., Brooker, S. (2010) Towards Langmuir-Blodgett films of magnetically interesting materials: solution equilibria in amphiphilic iron(II) complexes of a triazole-containing ligand. Dalton Trans., 39: 3751–3758.
85. Gandolfi, C., Cotting, T., Martinho, P. N., Sereda, O., Neels, A., Morgan, G. G., Albrecht, M. (2011) Synthesis and self-assembly of spin-labile and redox-active manganese(III) complexes. Dalton Trans., 40: 1855–1865.
86. Ruaudel-Teixier, A., Barraud, A., Coronel, P., Kahn, O. (1988) Spin transition in a magnetic Langmuir-Blodgett film. Thin Solid Films, 160: 107–115.
87. Armand, F., Badoux, C., Bonville, P., Ruaudel-Teixier, A., Kahn, O. (1996) Langmuir-Blodgett films of spin transition iron(II) metalloorganic polymers. 1. Iron(II) complexes of octadecyl-1,2,4-triazole. Langmuir, 11: 3467–3472.
88. Soyer, H., Mingotaud, C., Boillot, M.-L., Delhaès, P. (1998) Spin crossover of a Langmuir-Blodgett film based on an amphiphilic iron(II) complex. Langmuir, 14: 5890–5895.
89. Soyer, H., Mingotaud, C., Boillot, M.-L., Delhaes, P. (1998) Spin-crossover complex stabilized on a formamide/water subphase. Thin Solid Films, 327–329: 435–438.
90. Soyer, H., Dupart, E., Gómez-García, C. J., Mingotaud, C., Delhaès, P. (1999) First magnetic observation of a spin crossover in a Langmuir-Blodgett film. Adv. Mater., 11: 382–384.
91. Soyer, H., Dupart, E., Mingotaud, C., Gómez-García, C. J., Delhaès, P. (2000) Spin crossover phenomenon of a semi-fluorinated iron(II) complex organized in a Langmuir-Blodgett film. Colloids Surf. A, 171: 275–282.
92. Létard, J.-F., Nguyen, O., Soyer, H., Mingotaud, C., Delhaès, P., Kahn, O. (1999) First evidence of the LIESST effect in a Langmuir-Blodgett film. Inorg. Chem., 38: 3020–3021.
93. Kuroiwa, K., Shibata, T., Sasaki, S., Ohba, M., Takahara, A., Kunitake, T., Kimizuka, N. (2006) Supramolecular control of spin-crossover phenomena in lipophilic Fe(II)-1,2,4-triazole complexes. J. Polym. Sci. Part A Polym. Chem., 44: 5192–5202.
94. Kume, S., Kuroiwa, K., Kimizuka, N. (2006) Photoresponsive molecular wires of Fe^{II} triazole complexes in organic media and light-induced morphological transformations. Chem. Commun., 2442–2444.
95. Roubeau, O., Colin, A., Schmitt, V., Clérac, R. (2004) Thermoreversible gels as magneto-optical switches. Angew. Chem. Int. Ed., 43: 3283–3286.
96. Fujigaya, T., Jiang, D.-L., Aida, T. (2007) Spin-crossover physical gels: a quick thermoreversible response assisted by dynamic self-organization. Chem. Asian J., 2: 106–113.
97. Bodenthin, Y., Pietsch, U., Möhwald, H., Kurth, D. G. (2005) Inducing spin crossover in metallo-supramolecular polyelectrolytes through an amphiphilic phase transition. J. Am. Chem. Soc., 127: 3110–3114.
98. Bodenthin, Y., Schwarz, G., Tomkowicz, Z., Nefedov, A., Lommel, M., Möhwald, H., Haase, W., Kurth, D. G., Pietsch, U. (2007) Structure-driven remanent high-spin state in metallosupramolecular assemblies. Phys. Rev. B, 76: 064422/1–064422/6.
99. Bodenthin, Y., Schwarz, G., Tomkowicz, Z., Geue, T., Haase, W., Pietsch, U., Kurth, D. G. (2009) Liquid crystalline phase transition induces spin crossover in a polyelectrolyte amphiphile complex. J. Am. Chem. Soc., 131: 2934–2941.
100. Schwarz, G., Bodenthin, Y., Tomkowicz, Z., Haase, W., Geue, T., Kohlbrecher, J., Pietsch, U., Kurth, D. G. (2011) Tuning the structure and the magnetic properties of metallo-supramolecular polyelectrolyte-amphiphile complexes. J. Am. Chem. Soc., 133: 547–558.
101. Matsukizono, H., Kuroiwa, K., Kimizuka, N. (2008) Lipid-packaged linear iron(II) triazole complexes in solution: controlled spin conversion via solvophobic self-assembly. J. Am. Chem. Soc., 130: 5622–5623.
102. Martinho, P. N., Harding, C. J., Müller-Bunz, H., Albrecht, M., Morgan, G. G. (2010) Inducing spin crossover in amphiphilic iron(III) complexes. Eur. J. Inorg. Chem., 675–679.
103. Rubio, M., Hernandez, R., Nogales, A., Roig, A., Lopez, D. (2011) Structure of a spin-crossover Fe(II)–1,2,4-triazole polymer complex dispersed in an isotactic polystyrene matrix. Eur. Polym. J., 47: 52–60.
104. Schlamp, S., Weber, B., Naik, A. D., Garcia, Y. (2011) Cooperative spin transition in a lipid layer like system. Chem. Commun., 47: 7152–7154.

105. Zarembowitch, J., Roux, C., Boillot, M. L., Claude, R., Itie, J. P., Polian, A., Bolte, M. (1993) Temperature-, pressure- and light-induced electronic spin conversions in transition metal complexes. Mol. Cryst. Liq. Cryst., 234: 247–254.
106. Roux, C., Zarembowitch, J., Gallois, B., Granier, T., Claude, R. (1994) Toward ligand-driven light-induced spin changing. Influence of the configuration of 4 styrylpyridine (stpy) on the magnetic properties of $Fe^{II}(stpy)_4(NCS)_2$ complexes. Crystal structures of the spin-crossover species Fe(*trans*-stpy)$_4$(NCS)$_2$ and of the high-spin species Fe(*cis*-stpy)$_4$(NCS)$_2$. Inorg. Chem., 33: 2273–2279.
107. Boillot, M. L., Sour, A., Delhaes, P., Mingotaud, C., Soyer, H. (1999) A photomagnetic effect for controlling spin states of iron(II) complexes in molecular materials. Coord. Chem. Rev., 192: 47–59.

13

Luminescent Spin-Crossover Materials

Helena J. Shepherd, Carlos M. Quintero, Gábor Molnár, Lionel Salmon and Azzedine Bousseksou

Laboratoire de Chimie de Coordination, CNRS & Université de Toulouse, France

13.1 General Introduction

In the past few years, a number of multifunctional materials combining luminescence and spin-crossover (SCO) properties have been reported with the aim of achieving luminescence signal modulation resulting from the spin state switching phenomenon. These efforts have been motivated by a variety of different goals. First, the change of the energy transfer mechanism between the SCO complex and the luminophore upon the spin state change is a conceptually interesting phenomenon in its own right. Perhaps more importantly, a luminophore may also be used to report upon the spin state of SCO materials. Indeed, in many cases the detection of a luminescence signal is more efficient or easier to implement than other techniques commonly used in the spin-crossover field (magnetometry, optical absorption and so on). In general, luminescence can provide superior contrast and sensitivity for noncontact (remote) signal detection with fairly high spatial and temporal resolution. These assets can be used advantageously in various photonic applications. The combination with SCO will add the possibility of tuning the luminescence signal or detecting physicochemical changes in the environment of the SCO centre. Examples include photonic switches, thermometers, gas sensors, etc. Finally, one shall note that recent interest in SCO systems for fundamental studies of size reduction effects[1] combined with the possibility of detecting changes in luminescent signal, even down to the scale of a single molecule,[2] also open up very exciting perspectives for luminescent SCO nanomaterials.

Transition metal ions such as Fe(III), Fe(II), Co(II) and Ni(II) are reputed luminescence 'killers' and consequently, the desire to combine luminescence and spin-crossover materials may at first sight appear rather counterintuitive. However, the spin-crossover phenomenon involves such drastic changes in the electronic and crystal structure of the material that a change in spin state may perturb the environment of the luminophore enough to allow a luminescent response upon the SCO. While of course both electronic and structural effects are intrinsically linked in SCO materials, their respective impact on the interaction between SCO and luminescent materials is subtly different. The luminescence signal may be modulated by the structural

Spin-Crossover Materials: Properties and Applications, First Edition. Edited by Malcolm A. Halcrow.

changes that occur on SCO, for example changes in the density (and hence rigidity) of the lattice can affect both the wavelength and intensity of the light emitted by the luminophore. In this way the probe could detect the elastic changes due to the spin-transition propagating through the material, independently of the spectral properties of the SCO centres. Alternatively, the luminescence can be modulated directly due to the change in the electronic configuration of the SCO material during the transition. If the separation between a pair of energy levels associated with the luminophore closely matches an energy level spacing of the SCO centre (in a given spin state) the excited state energy can be transferred to the latter. For example, if the emission and/or excitation bands of the luminophore show significant overlap with the absorption bands of the SCO material, the luminescent response can be quenched via different energy transfer processes. Hence, through the judicious choice of a luminophore with suitable spectral overlap, it should be possible to modulate the luminescent response as a function of spin state of the SCO centres through changes in the optical absorption of the latter material. Key factors in rationalising and optimising the energy transfer mechanisms in these mixed materials include the nature (radiative or nonradiative) of the quenching process, the relative proportion of luminophores with respect to the spin-active centres as well as their separation. However, if the two entities are too close, nonradiative energy transfer mechanisms might result in severe quenching at all temperatures and loss of the luminescent signal. It is important to remember also that when combining SCO and luminescent systems, there is no guarantee that both properties will be preserved in a single hybrid material. This stems largely from the very high sensitivity of both the luminescent and SCO materials to minor chemical and structural changes caused simply by the presence of the other constituent. In addition, it is important that the luminophore is itself not sensitive to the stimulus used to switch the spin state. For example, if one considers the most usual *thermal* spin state switching, it is preferable to choose a luminophore which exhibits negligible thermal quenching in the temperature range of interest.

Since articles dealing with luminescence are relatively rare in the spin-crossover literature, we begin this chapter with a rather general introduction to luminescent materials, followed by a brief overview of electronic transitions and optical properties of spin-crossover complexes. It should be noted that several detailed reviews are available on both topics.[3–9] Here we try to bring together only those ingredients which are the most important for the design and application of multifunctional luminescent SCO entities. In the second part of the chapter we review the existing literature concerning materials that combine SCO and luminescence properties and conclude with discussion on the perspectives in this area of research.

13.2 Introduction to Luminescent Materials and Luminescence Energy Transfer

From lasers, lamps, computer displays, biological markers and security inks to even basic things such as the piece of paper that we use to write on, or a glow in the dark sticker, photoluminescent materials are ubiquitous in our everyday life. All of these materials absorb electromagnetic radiation (light) and then re-emit a part of it at a different wavelength. Photoluminescence is a particular case of luminescence. In general, luminescence leads to the conversion of specific types of energy into radiation over and above thermal radiation.[4] It is currently defined as: ‘Spontaneous emission of radiation from an electronically excited species or from a vibrationally excited species not in thermal equilibrium with its environment’.[10] Depending on the excitation source, additional types of luminescence can considered: cathodoluminescence (where the excitation source is a beam of electrons rather than light), electroluminescence (electrical current), chemiluminescence (chemical energy), triboluminescence (mechanical energy) and so on. Photoluminescence processes can be divided into two main steps that are typically represented in a Jablonski diagram as shown in Figure 13.1. The first is excitation, where energy is absorbed by an absorbing centre or ‘sensitiser’ (S) that is initially in its ground state S_0 and it is promoted to an excited state S_1^*. The second step is the return to the ground state. Here, the excited centre can either emit a photon (fluorescence, phosphorescence, delayed fluorescence) following – in

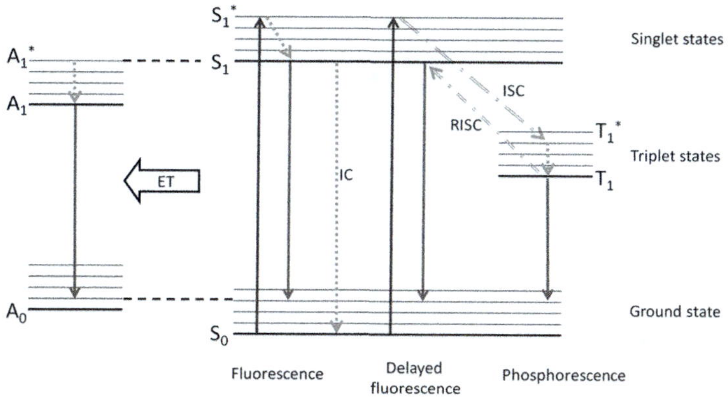

Figure 13.1 *Jablonski diagram showing excitation, internal conversion (IC), fluorescence, intersystem crossing (ISC), reverse intersystem crossing (RISC), phosphorescence, delayed fluorescence and energy transfer processes (ET).*

some cases – internal conversion or intersystem crossing(s) processes, it can dissipate the energy in the form of heat, or it can transfer the energy to an 'Acceptor' (A) nearby. These steps are considered in more detail in the following sections.

13.2.1 Photoexcitation of Luminescent Materials

To observe a luminescent response, it is necessary to consider which conditions are appropriate for elevating the system into an excited state, that is how to give energy to the system efficiently. In the particular case of photoluminescence, it is important to consider the optical absorption spectrum of the material under study. The *configurational coordinate model* is often employed to explain optical properties of a material. This model recreates the potential energy curves of an absorbing centre as a function of a configurational coordinate.[3, 4] In the simplest case, an absorbing centre in its ground state, $\boldsymbol{g}$, and its surrounding ligands are oscillating around an equilibrium distance R_0 in the so called 'breathing mode' (see Fig. 13.2). Thus, this system can be modelled as a harmonic oscillator and as a result its energy can be resolved in the form of a parabolic potential well as a function of the centre–ligand distance, R. Here, the potential well is composed of several discrete energy or vibrational levels, each of them with a characteristic wave function. It is important to note that in the lowest vibrational level, the highest probability of finding the system is at R_{g0}. By contrast, for the other levels the highest probability is located at the intersections with the parabola that describes the well. Once the centre has absorbed light, it passes from its ground state $\boldsymbol{g}$ to an excited state $\boldsymbol{e}$. Excited states can also be described as parabolic wells in the configurational coordinate diagram. However, the excited chemical bond between the absorbing centre and the ligand is altered (usually becoming weaker); as a result, the shape of the parabola and the equilibrium distance R_{e0} might be different from those of the ground state (Fig. 13.2a).

In an optical absorption event, an electron is promoted into a higher energetic level and as a result, the configuration of the system will correspond to a higher vibrational level in the excited state, $\boldsymbol{e_n}$. According to the *Franck–Condon principle*, an electronic transition happens without change in the position of the absorbing centre or its surroundings. In fact, the promotion of an electron to an excited state can take around 10^{-15} s; this is too fast for rearrangement of the heavy nuclei of the atoms involved in the process.[5] Thus, the absorption of a photon represents as a vertical transition in the configurational coordinate diagram. Assuming that the absorbing centre is initially at the lowest vibrational level of $\boldsymbol{g}$ and taking into account the overlap between

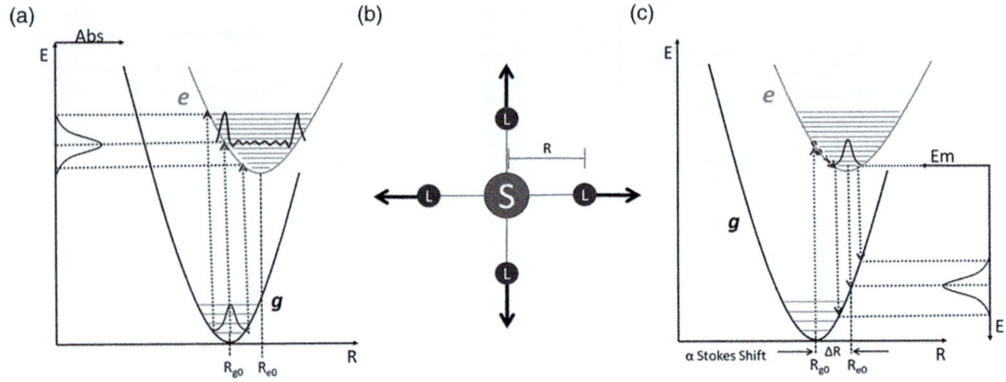

Figure 13.2 *(a) Configurational coordinate diagrams for the ground state and an excited state displaying the potential energy as a function of the parameter R for the system shown in (b). The difference between the minimum of the parabolas ($\Delta R \neq 0$) gives rise to a broad absorption band. The ground state* g *and the excited state* e *have an equilibrium distance at R_{g0} and R_{e0}, respectively. The probability density functions for the lowest vibrational level of* g *and the level corresponding to the transition of maximum absorption in* e *are also displayed. (b) Diagram illustrating an adsorbing centre and its four ligands oscillating in the symmetrical stretching or 'breathing' mode. (c) Once the system is excited, it relaxes by internal conversion to the lowest vibrational level of the excited state. The overlap of the wave function at this point with those functions of the vibronic levels of the ground state produces an emission band or ray depending on the difference of the equilibrium distances (ΔR) of the two potential wells. Adapted with permission from [3] © 1979, Springer-Verlag and with permission from [4] © 2004, Springer-Verlag.*

its probability density function with those of the vibrational levels in ***e***, the optical absorption transition is more likely to happen at R_{g0}, towards the border of the parabolic well of ***e*** (Fig. 13.2a). It is precisely at this point that the maximum of the absorption band can be found (maximum overlap of both wave functions).[3] However, it is also possible (although less likely), that the initial transition begins at $R > R_{g0}$ or $R < R_{g0}$. This situation will lead to transitions that will need a little more or a little less energy and, as a result, these variations give rise to a certain width of the absorption band. This configurational coordinate model allows us to explain the different features of band or line structure in the absorption and emission (*vide infra*) spectra of various materials and also their temperature and pressure dependence.

So far, we have described the origins of the optical absorption transitions and their width. Another feature that we will have to consider is the intensity of these transitions. Not every possible transition observed in the configurational coordinate diagram from **g** to ***e*** can happen as an optical transition. There are selection rules that make some of the transitions highly probable or *allowed* and very intense, some others less probable or *forbidden* and very weak in intensity. The principal selection rules are:

- *Laporte* or *parity selection rule*: Transitions that do not involve a change in parity (such as *d–d* and *f–f*) are forbidden ($\Delta l = \pm 1$).
- *Spin selection rule*: Transitions involving a change in spin multiplicity are forbidden ($\Delta S = 0$).

It should be stressed that these rules should not be considered as absolute. They can be relaxed due to wave function admixtures. For example, phenomena such as spin-orbit coupling, electron-vibration coupling and uneven crystal-field terms can partially allow forbidden transitions.[3, 6] An important consequence is that the absorption spectrum of a luminescent centre often differs depending on whether it is a free ion/molecule or embedded in a lattice surrounded by different types of ligands. One parameter often employed to compare the intensities of the absorption bands of a given material is the molar absorptivity (ε). This quantity reflects

Table 13.1 *Typical orders of magnitude for oscillator strengths and extinction coefficients for different absorption bands present in various transition metal complexes.*[11]

Type of transition	Approximate f	Approximate ε
Spin-forbidden, Laporte forbidden	10^{-7}	0.1
Spin-allowed, Laporte forbidden	10^{-5}	10
Spin-allowed, Laporte allowed	10^{-1}	10 000

the ability of its active centres to absorb light at a given wavelength. It is usually expressed in units of $L{\cdot}mol^{-1}{\cdot}cm^{-1}$. Additionally, a quantity known as the oscillator strength (*f*) is also used for expressing the strength of a given transition. This dimensionless parameter can be calculated as shown in Eq. (13.1) (where n is the refractive index) from the area under the absorption band when its extinction coefficient is plotted as a function of the frequency ($\bar{\upsilon}$, expressed in cm^{-1}) and not of the wavelength.[5, 11]

$$f = \frac{4.32 \times 10^{-9}}{n} \int \varepsilon(\bar{\upsilon}) d\bar{\upsilon}. \tag{13.1}$$

Table 13.1 gives approximate orders of magnitude for the extinction coefficients and the oscillator strengths of different absorption bands depending on their nature and the extent to which they are allowed for various transition metal complexes.

Finally, one should not forget that in addition to the transition probability, the absorption of a given material is also governed by other factors, such as the population of states, the path length (l) and the molar concentration of the active centres (c). This dependence is condensed in the *Lambert–Beer law* which states that the transmission (T) of light through a material is an exponential function of the product of ε, c, l as it is shown in Eq. (13.2).

$$T = \frac{I}{I_0} = e^{-\varepsilon l c}, \tag{13.2}$$

where I and I_0 are the transmitted and incident light intensities, respectively. Usually, Eq. 13.2 is employed in an alternative form in terms of optical density (OD) (or absorbance, A) instead of transmission, as displayed in Eq. (13.3).[5]

$$OD = -\log_{10} \frac{I}{I_0} 0.434 \varepsilon l c = 0.434\ A \tag{13.3}$$

13.2.2 Return to the Ground State

Once a luminescent system has absorbed energy, it has different options for returning to the ground state. These include radiative return to the ground state (emission), nonradiative return to the ground state and energy transfer between active sites. Obviously, this latter process is of central interest for the design of luminescent SCO materials.

13.2.2.1 Radiative Return

Once the system is promoted into a higher vibrational level of the excited state $\boldsymbol{e_n}$, it will rapidly relax to the lowest vibrational level of $\boldsymbol{e}$ by returning the excess energy to the surroundings. This can also be interpreted in the configurational coordinate diagram as the rearrangement of the positions between the optical centre and its ligands so that the new centre–ligand distance (R) is equal to the equilibrium distance R_{e0} of the excited state (see Fig. 13.2c). This process of disipating energy in the form of vibrations or heat is known as internal conversion and it is a significantly slower process (10^{-13}–10^{-12} s) than the electronic transitions.[3, 4] As soon as the system is at $\boldsymbol{e}_0$, it will be capable of returning to the ground state via spontaneous emission of a photon with an energy that matches the energy gap between $\boldsymbol{e}_0$ and the energetic level of the ground state to which the system is restored, $\boldsymbol{g}_n$. The maximum of the emission will be found at the vertical transition obtained when the system is at R_{e0} and the crossing point with the ground state. Similarly to the absorption processes, an offset between the centres of the two potential wells will cause the overlap of the wave function associated with $\boldsymbol{e}_0$, with energy levels of the ground state at different values of R. As a result, when $\Delta R \neq 0$ the emission will be a band composed of several wavelengths and if $\Delta R \approx 0$ the emission will be close to a monochromatic ray.[3] After emission, the system will be in the ground state, but not necessarily in the lowest vibrational level. Thus, another process of internal conversion will take place in order to return the system once again to R_{g0}. From Figure 13.2c, we can infer that due to these nonradiative relaxations, the emitted photons will be of the same or lower energy than the excitation, the difference being the so-called *Stokes shift*. The Stokes shift gives us direct information about the separation (ΔR) between $\boldsymbol{g}$ and $\boldsymbol{e}$. From the configurational coordinate diagram we are able to observe that the larger the Stokes shift is, the bigger ΔR will be and therefore the broader the emission band of the system. The significant consequence is that the emission of a luminescent centre will be sensitive to the rigidity of its environment and the interactions with surrounding species. Additionally, it is also possible to confirm that the wavelengths where emission is observed (although due to different nonradiative transitions not necessarily their intensities) will be virtually independent of the excitation energy, provided it is sufficient to populate the excited state.

In contrast to the absorption, where the key parameters for studying the phenomenon are the extinction coefficient and the oscillator strength, for luminescence the factors for characterising the emission of a given system are the *luminescence lifetime* and the *quantum yield*. The former yields kinetic information about the emission events and the latter about the efficiency of the process. So far, the energy diagrams used to describe the luminescence have not taken into account the duration of the excited states or their possible decay and, as a result, the concept of lifetime is necessary. We can define the lifetime as the average duration of time a luminophore remains in the excited state following excitation.[5] A luminescent centre in its lowest excited state ($\boldsymbol{e}_0$) can relax to the ground state essentially in two ways: luminescent (radiative) pathways (L) and nonluminescent (nonradiative) pathways (NL). Depending on the amount of available deactivation options, the system will be prone to rapidly emit a photon (very high number of Ls and low number of NLs), rapidly return to the ground state without emission (very low number of Ls and high number of NLs) or remain at the excited state due to lack of options (very low number of Ls and NLs). In other words, the higher the number of exit paths of a particular type (L or NL), the more likely the system is to take it. In this case it is possible to infer that there will be a certain probability per unit of time (rate) that the system will *exit the excited state* via either an emissive or a nonemissive transition derived from the number of relaxation pathways. Furthermore, since the luminescent centre is in an excited 'equilibrium' state ($\boldsymbol{e}_0$), this probability will not change during the time the system remains excited (memoryless). Hence, the time 'T' at which the system leaves the excited state $\boldsymbol{e}_0$ is a continuous random variable with a probability density function $p_T(t)$, and it can be demonstrated that it is distributed exponentially with a known expected value calculated as:[14]

$$< T > = \int_0^\infty t \cdot p_T(t) = \int_0^\infty t \cdot [(k_L + k_{NL})e^{-(k_L+k_{NL})t}]dt = \frac{1}{k_L + k_{NL}} \tag{13.4}$$

Hence the expected value for the time at which a luminophore leaves the excited state or the lifetime of the excited state $\boldsymbol{e}$ (τ_e), is the inverse of the sum of its spontaneous emission rate (k_L) and nonluminescent decay rate (k_{NL}). If we think of a material with a number, N_0, of identical excited centres that behave in the same manner, the population of the excited state of a material becomes:

$$N(t) = N_0 \cdot P(T > t) = N_0 e^{-(k_L + k_{NL})t} = N_0 e^{-t/\tau_e}, \tag{13.5}$$

where $P(T > t)$ is the probability of leaving the excited state after a given time t.

In order to estimate the efficiency of the luminescence for a given system, we can define the quantum yield (Q) as the ratio between the spontaneous emission rate of the active centre and its total exit rate from the excited state (Eq. (13.6)). Here, we will be able to measure how often the system will relax via a radiative pathway compared to its total relaxation processes to the ground state.

$$Q = \frac{k_L}{k_L + k_{NL}} \tag{13.6}$$

The selection rules for absorption transitions also apply for emission processes. Even though populating an excited level that has different multiplicity to that of the ground state is unlikely to happen (spin selection rule), electrons in the excited state are susceptible to undergoing spin conversion and passing to a lower excited level with different multiplicity. This process is known as *intersystem crossing* (ISC). As a result, those electrons will be trapped in an excited state with a forbidden transition for returning to the ground state. As a consequence, there are two main emission processes that can be distinguished:

- *Fluorescence*: luminescence obtained from the radiative return to the ground state without a change of the initial spin state of the system. Due to the allowed nature of the transition, it is a fast process and its lifetime is $\approx 10^{-7}$–10^{-5} s. This type of emission is usually illustrated by employing the particular case of an organic luminophore emitting a photon from an excited singlet state, S_1, to a singlet ground state, S_0.
- *Phosphorescence*: luminescence obtained from the radiative return to the ground state from an excited energetic level of different spin multiplicity. Due to the forbidden nature of the transition (spin selection rule), it is a much slower process than fluorescence and its lifetime goes from $\approx 10^{-3}$ to 1 s and more.[3, 4] Additionally, since the levels involved in the emission process are closer than in the case of fluorescence, the emitted photons are of lower energy (see Fig. 13.1). Phosphorescence is often exemplified in the literature with molecules that emit a photon from an excited triplet state, T_1, to a singlet ground state, S_0. It is worth stating that due to the low emission rates of this process, phosphorescence can be mistaken for another emission mechanism denominated delayed fluorescence. In this case, following ISC, it is possible to return to the original excited state by thermal activation or by collision with another excited molecule (in solution) (*reverse intersystem crossing*, RISC) and then emit a photon in the same manner as in conventional fluorescence.[7]

13.2.2.2 Nonradiative Return

In a luminescent material, all energy that has been absorbed and has not been emitted in the form of radiation, must be dissipated in the form of heat to the surroundings via internal conversion. So far, we have seen that after the absorption event, the system relaxes to the lowest available excited state (e_0). However, if we consider two potential wells with $\Delta R \neq 0$ and an excited state with sufficiently high energy, it is possible for the system to pass through a crossing point of the two potential wells and so transit nonradiatively to the ground state (Fig. 13.3a). With this model is possible to explain the typical quenching of the luminescence with temperature or so called *thermal quenching*. At higher temperatures, higher energetic excited levels

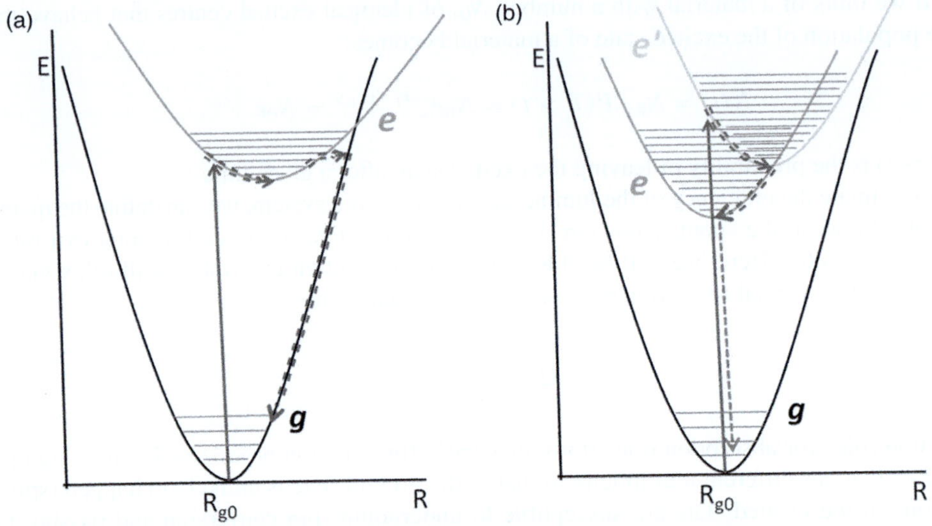

Figure 13.3 Configurational coordinate diagram of nonradiative processes: (a) Nonradiative return from the excited state e to the ground state g via a crossing point of the two potential wells. (b) Nonradiative transition from e′ to e that effectively populates the emitting excited state e. Adapted with permission from [3] © 1979, Springer-Verlag and with permission from [4] © 2004, Springer-Verlag.

can be reached and the probability of finding a crossing point with the ground state (or another excited state) increases. It is also possible to observe that the greater the offset between the two parabolas, the more nonradiative transitions can be obtained. Conversely, if the two potential wells are centred and their shapes are alike (as often observed for strongly forbidden transitions), nonradiative transitions are unlikely to happen. As a result, luminescent centres with stiff surroundings that restrict expansion after optical excitation ($\Delta R \approx 0$), usually present less nonradiative processes, higher quantum yields and low thermal quenching.[3]

It appears that nonradiative transitions are the primary competing processes for luminescence. However, we should also consider all those nonradiative transitions that may serve to feed an emitting level, as shown in Figure 13.3c. Supposing a strongly forbidden transition occurs between ***e*** and ***g***, the excited level will be difficult to populate. However, if there is also an additional excited level ***e***′ above ***e*** with an allowed optical transition from ***g***, it is possible for the system to find a crossing point between ***e***′ and ***e*** during the internal conversion. Thus, we will be able to effectively populate the excited state ***e***, circumventing its forbidden transition from the ground state.[3] From this point of view, it is clear that some sections of the bands observed in the absorption spectrum of a material may not display a luminescent response as the energy may be dissipated through nonradiative transitions. For this reason, while considering the luminescent response of a material, the *excitation spectrum* provides more information. In contrast to the absorption spectrum, where one observes how much light of a particular wavelength is absorbed by the material, in the excitation spectrum one measures the emission of the material at a fixed wavelength while varying the excitation source wavelength. In this way, the excitation spectrum will take into account not only where the system absorbs energy but also which wavelengths produce an efficient luminescent response.

An additional phenomenon to consider within the context of nonradiative pathways for returning to the ground state is the photodesctruction of the system during the excited state or *photobleaching*. Usually in the case of molecular luminophores, each time they enter into an excited state, there is a probability per unit of time to be destroyed via a photochemical process (very often photooxidation).[2] As a result, the luminophore may change irreversibly to a different molecular form that does not absorb or emit light. Additionally, the photobleaching processes happen on a timescale of seconds or minutes (depending on the power of the excitation), making them practically 10 orders of magnitude slower than fluorescence. Consequently, careful interpretation of the emission spectrum as a function of time is required, particularly in the case of experiments with long exposure times and/or high power excitation.

13.2.2.3 Energy Transfer

In the previous discussions of an excited state returning to the ground state by either radiative or nonradiative transitions, the excitation energy was dissipated to the surroundings by the same entity that absorbed it. However, it is important to also consider how the energy can flow from one place to another inside a given medium and the aspects that influence these processes. Indeed, these processes will actually constitute the mechanism by which a luminescent SCO material will function. In an energy transfer scheme, there are two entities: a sensitiser (S) responsible for absorbing the excitation energy and an acceptor (A) that will receive the energy of the excited sensitiser. The interaction between the two is described below.

13.2.2.3.1 Radiative Energy Transfer Radiative energy transfer, also known as *emission–reabsorption* process is the 'trivial' case where the sensitiser (S) absorbs energy, emits a photon and then the acceptor (A) reabsorbs the emitted photon. If the excited acceptor is a luminophore, it will emit a second photon and we will have cascade luminescence.[3] However, if the acceptor dissipates the energy a quenching effect (suppression of luminescence) will be observed. In a radiative energy transfer, both entities (sensitiser and acceptor) are independent of each other and do not interact directly. As a result, their lifetimes will not be affected by the presence of the other.

The primary condition for radiative energy transfer to occur is that there should be a spectral overlap between the emission of S and the absorption of A; only photons emitted within this spectral window will participate in the transfer. The efficiency of the process (starting from the excitation event) will be influenced by the quantum yield of S (Q_S), and by the oscillator strength of A at the spectral overlap.

For radiative energy transfer, the distance that separates the sensitiser and the acceptor does not influence the event of absorbing the emitted photon. However, this regime will be ruled by the Lambert–Beer law. Even if the above mentioned criteria for the efficiency of the process are fulfilled, the emission of S will be absorbed by an A (there could be more than one acceptor) depending on the concentration of A and also on the geometry of the sample in consideration. In order to have reabsorption, it is necessary to have an acceptor in the pathway of the emitted photon. For example, at a fixed concentration, if the size of the sample is very small, there will be little chance for the emitted photon to find and acceptor.

13.2.2.3.2 Nonradiative Energy Transfer We consider a sensitiser–acceptor pair fixed in a material, separated by a distance, r_{SA}. If r_{SA} is sufficiently small that there is a nonvanishing interaction between them (such as a wave function overlap or an electric or magnetic multipolar interaction) and there is a spectral overlap between the two (resonance condition), there is a probability per unit of time that the energy is transferred from the sensitiser to the acceptor by a nonradiative process, that is without an emission and reabsorption of a photon.[3] In 1946, Theodor Förster introduced for the first time his theory of nonradiative resonant energy transfer between two interacting dipoles.[12–14] Later Dexter expanded the theory for higher order electromagnetic as well as for exchange interactions.[15, 16] One shall note that the energy transfer can

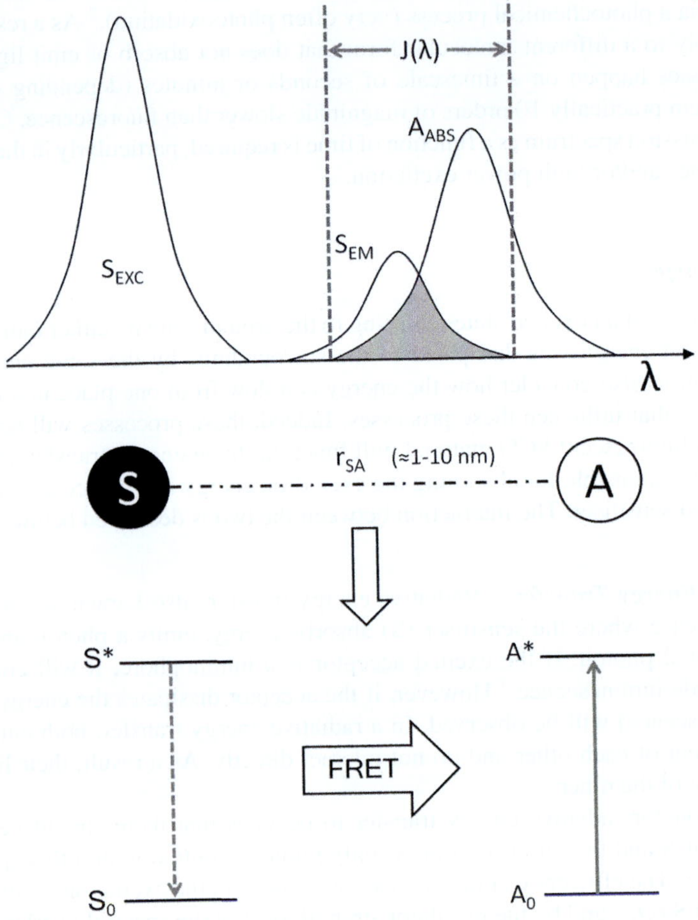

Figure 13.4 *Schematic illustration of resonant energy transfer between the centres S and A. Adapted with permission from [4] © 2004, Springer-Verlag.*

also be a nonresonant (phonon-assisted) and multistep process (energy migration). In the following we will consider only the case of a single-step, Förster-type (dipolar) resonant energy transfer process and for further details on this topic we recommend the excellent review of Powell and Blasse.[8]

In contrast to a radiative energy transfer, in a FRET (Förster resonance energy transfer) scheme the sensitiser and the acceptor are interacting with each other and they behave as if they were one whole system (see Fig. 13.4). In the case of an electric dipole–dipole interaction, Eq. (13.7) defines the transfer rate (probability per unit of time) k_T for a given sensitiser–acceptor pair separated by a distance r_{SA} as:

$$k_T = \left(\frac{1}{\tau_S}\right)\left(\frac{R_0}{r_{SA}}\right)^6, \tag{13.7}$$

where R_0 is the so-called Förster distance characteristic of the sensitiser–acceptor pair under consideration.[5, 8, 17] It represents the separation of the pair at which k_t equals the exit rate (k_s) from the excited state for S in absence of A; typical values for this distance range between 0.5 and 10 nm.[17] The rate k_s is the inverse of the lifetime of S (τ_s) and it includes all deactivation pathways of S other than the energy transfer. The value of R_0 for a pair S–A is given by Eq. (13.8):[5]

$$(R_0)^6 = \frac{9000(\ln 10)k^2 Q_S}{N_A 128\pi^5 \eta^4}\left[\frac{\int_0^\infty \varepsilon_A(\lambda)F_S(\lambda)\lambda^4 d\lambda}{\int_0^\infty F_S(\lambda)d\lambda}\right] \tag{13.8}$$

Here, ε_A is the extinction coefficient of A expressed in units of L·mol^{-1}·cm^{-1}, F_S is the luminescence intensity of S in arbitrary units, λ is the wavelength expressed in cm, η is the refractive index of the medium, Q_s is the quantum yield of S, N_A is Avogadro's number and k is a factor that describes the spatial orientation of the electric fields of the interacting dipoles of S and A. The value of k ranges between 0 and 4; however, if the orientation between the sensitisers and the acceptors in a given material is assumed to be random, its value is usually set to $^2/_3$.[17] The ratio of the integrals inside the brackets is known as the overlap integral ($J(\lambda)$) and employing the units mentioned above, it will be expressed in L·mol^{-1}·cm^3.

If a nonradiative energy transfer can take place between a given S–A pair, by exciting exclusively S and approaching A, the luminescence intensity and lifetime of the first decreases. The energy transfer towards A without an emission can be understood as an energy leak for S. It constitutes a relaxation pathway for the excited state of S, in addition to those which are inherent to it (luminescent and nonluminescent) and under no circumstance does it modify them. As a result, whenever A is close enough, the lifetime of the excited state for S will be reduced and also its radiative response, not because the intrinsic radiative rate of S has been modified, but due to the additional exit from the excited state that FRET introduces. In presence of FRET the modified life time τ_s' of the sensitiser will be:

$$\tau_S' = (k_S + k_T)^{-1} = \left((k_{L_s} + k_{NL_S}) + k_T\right)^{-1}, \tag{13.9}$$

$$\tau_S' = \left(^1/_{\tau_S} + k_T\right)^{-1}. \tag{13.10}$$

As seen in Eqs (13.7) and (13.8), the Förster resonance energy transfer is a phenomenon extremely sensitive to the distance between the sensitiser and the activator ($\alpha \cdot 1/R^6$). As a result, it is often used to measure distances between two moieties at particular places within a molecule before and after a configurational change. It may constitute a powerful tool for inferring spatial information below the diffraction limit by employing visible light. Additionally, if both the sensitiser and the acceptor are luminescent, the luminescence of A due to FRET can be employed as a marker to indicate the specific points where the two luminescent species, which were independently spread into a medium, come into contact. Another interesting application of FRET is the analysis of the local environment of the S–A pair. The nonradiative energy transfer is sensitive to fluctuations of the optical properties of the S–A pair; any change to their local environment could lead to a change of the orientation of the dipoles, emission and absorption spectra or quantum yields involved in the process and, as a result, a drastic variation of k_T.

13.2.2.3.3 Self-Quenching So far energy transfer has been considered only for different active centres. However, it is possible to have an energy transfer (radiative and/or nonradiative) between identical centres and this should be taken into consideration while characterising the luminescent response of a particular system. This phenomenon is observed when the concentration of luminescent centres is very high and it is termed *inner filter*, *self-quenching* or *concentration quenching* by different authors.[3,5] *A priori*, if the active centre has a large Stokes shift, it is unlikely to have an energy transfer (radiative or nonradiative) with another centre of the same type. However, if we consider luminophores with smaller Stokes shift, its excitation and emission bands can be broad enough to overlap. In a highly concentrated system, the separation between centres is small and the chance of having energy transfer increases dramatically. As a consequence, the excitation energy might be displaced more than once between different centres and travel long distances without an emission (energy migration). Thus, the probability of finding a quenching moiety, a defect in the material or a nonradiative transition increases; accordingly, the luminescent response measured under such saturated conditions will be of lower intensity than the response obtained from a diluted sample. Furthermore, this quenching effect will be restricted only to the wavelengths involved in the spectral overlap of the excitation and emission spectra, and as a result, the measured emission spectrum will be deformed.[5]

13.3 Electronic Transitions and Optical Properties of Spin-Crossover Complexes

The molecular spin-crossover (SCO) phenomenon, observed for octahedral transition metal complexes with a $3d^4$-$3d^7$ electronic configurations, can be understood on the basis of ligand field theory (LFT). On the other hand, LFT is also an important tool to rationalise the optical properties of SCO compounds. A detailed account of these topics can be found in the excellent review of Hauser[18] and thus we will focus here primarily on the concepts that will be important to understand the possible interplay between luminescence and spin-crossover properties.

The change of the spin state of the molecule naturally leads to a complete change of its electronic absorption spectrum. For example, in the case of an octahedral complex with $3d^6$ electronic configuration (Fe^{II}, Co^{III}) the ground state of the molecule may change from the singlet ($^1A_{1g}$) LS to the quintet ($^5T_{2g}$) HS electronic configuration (Fig. 13.5). From the Tanabe–Sugano diagram (Fig. 13.6) it is clear that in the HS state there is only one spin-allowed metal-centred *d–d* transition ($^5T_2 \rightarrow {}^5E$), while in the LS state there are two transitions of this type ($^1A_1 \rightarrow {}^1T_1$, $^1A_1 \rightarrow {}^1T_2$). In addition, weak spin-forbidden transitions ($^1A_1 \rightarrow {}^3T_1$, $^1A_1 \rightarrow {}^3T_2$) can be also detected in the LS spectra.

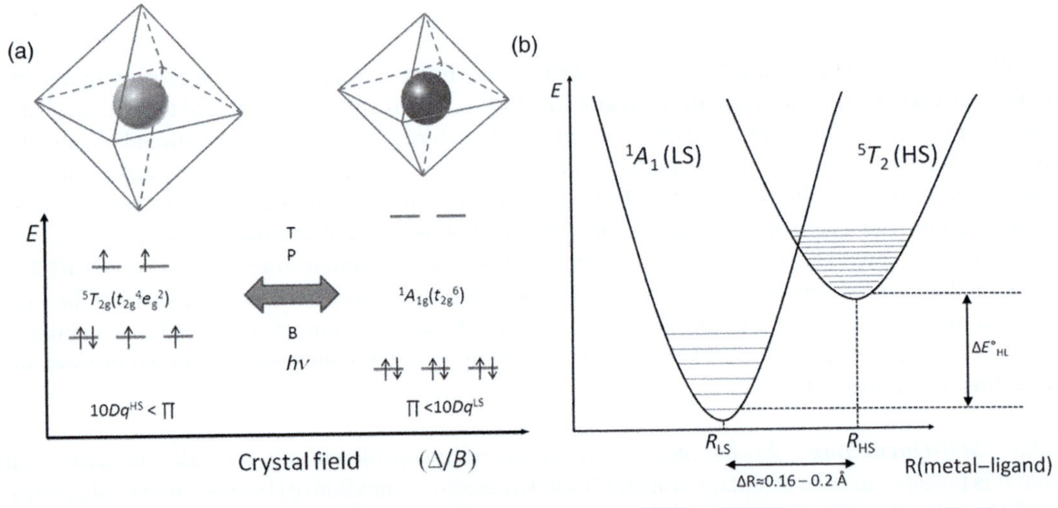

Figure 13.5 (a) High spin (weak ligand field) and low spin (strong ligand field) electronic configurations of a $3d^6$ ion (Fe^{II}, Co^{III}) in an octahedral complex. (b) Potential energy of the high spin ($^5T_{2g}$) and low spin ($^1A_{1g}$) states along the totally symmetric metal – ligand stretching coordinate in the harmonic approximation. Reproduced from [15]. Copyright 2004. With kind permission from Springer Science and Business Media.

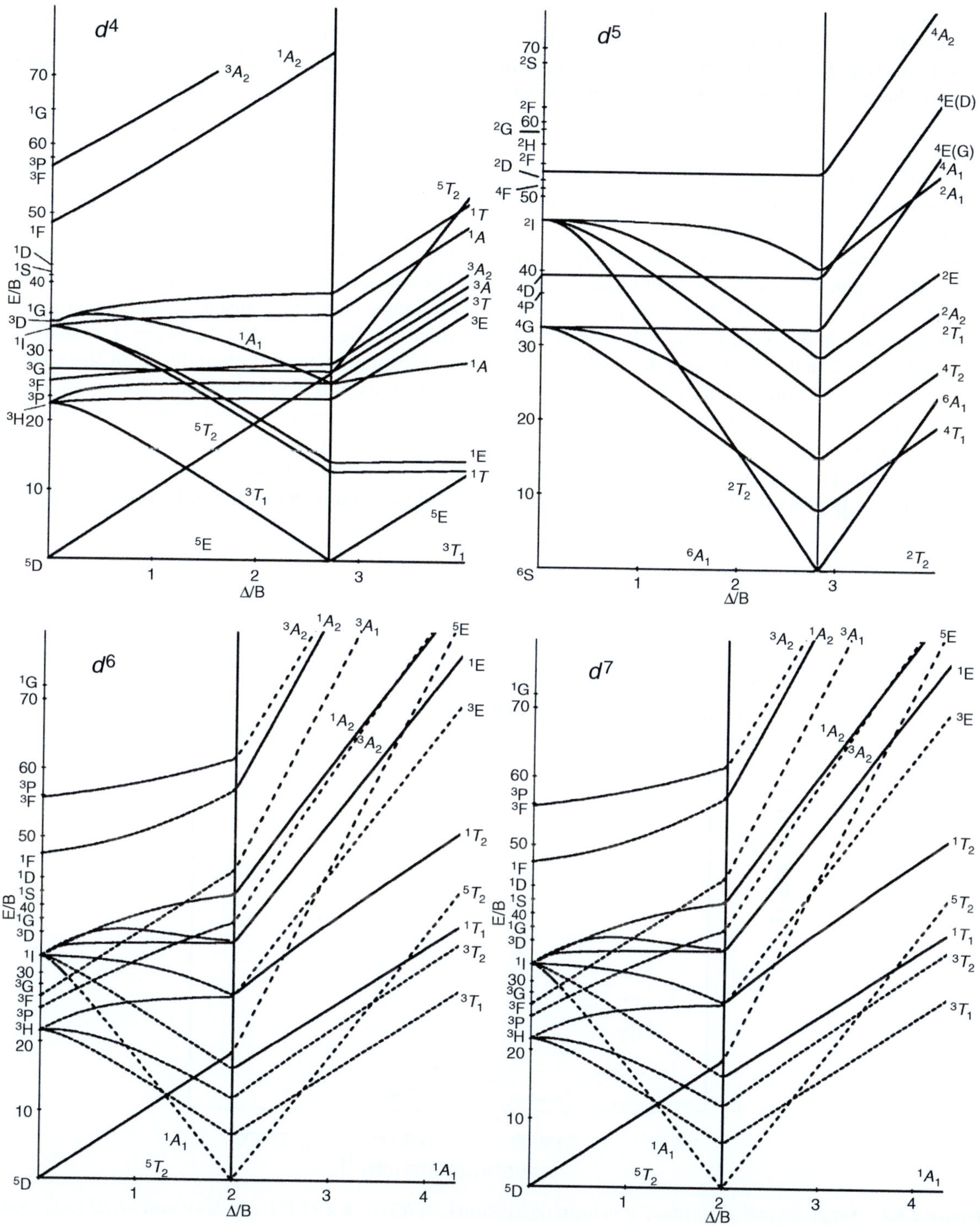

Figure 13.6 Tanabe–Sugano diagrams for transition metal ions of d^4–d^7 electronic configurations.

Due to the fact that the spin-crossover phenomenon occurs for a relatively narrow range of ligand field strengths, the visible absorption spectra of different SCO complexes with the same metal ion and the same donor atoms (e.g. $Fe^{II}N_6$) are expected to be fairly similar. In reality, however, the colour of the compounds can be rather different due to the occurrence of various charge transfer transitions in the visible range. In particular, metal–ligand charge transfer (MLCT) bands are often observed for ligands with extended π electron systems. Charge transfer transitions involving molecular orbitals located primarily on the metal ion will be inevitably altered by its spin state. Since these transitions are usually very intense when compared to the relatively weak (Laporte forbidden) *d–d* ligand-field transitions they can completely dominate the perceptible colour of the sample.

As an example Figure 13.7 shows single crystal optical absorption spectra in the visible and near-infrared (NIR) wavelength ranges of the spin-crossover complex $[Fe(bbtr)_3](ClO_4)_2$ (where bbtr stands for 1,4-di(1,2,3-triazol-1-yl)).[16] In the high temperature (HS) state the sample is nearly transparent and only a weak, broad absorption is observed in the NIR around 830 nm, which can be assigned to the ${}^5T_2 \rightarrow {}^5E$ transition. On the other hand, in the low temperature (LS) state the sample displays a purple colour due to an absorption band centred at ~590 nm, which can be assigned to the ${}^1A_1 \rightarrow {}^1T_1$ transition. The molar extinction coefficients for these metal-centred transitions are rather low: they span from a few to a few tens of $dm^3 mol^{-1} cm^{-1}$. In the LS state, however, one can also observe the onset of a very strong absorption below ca. 430 nm, which is an indication of the possible occurrence of intense singlet charge transfer transition(s) in the UV region. Since the absorbance is proportional to the concentration of the colour centres, when normalised appropriately,

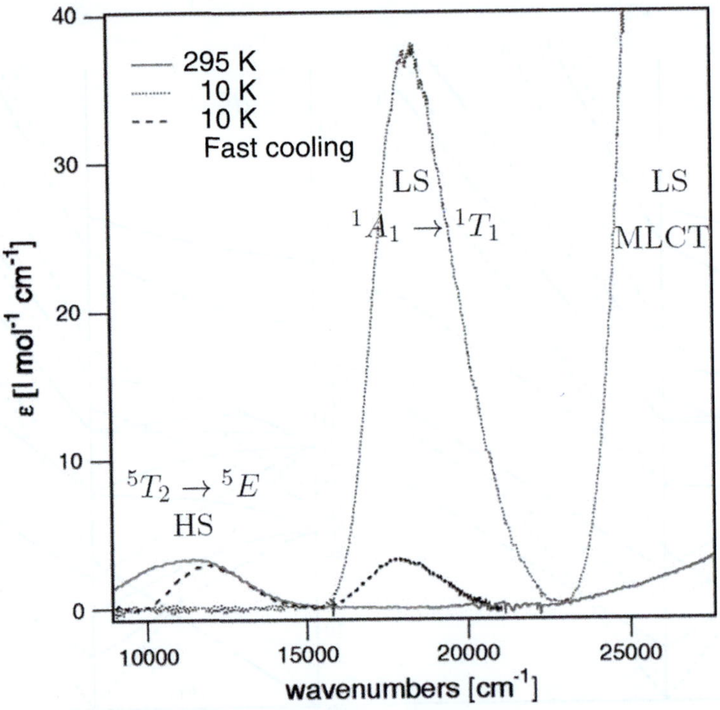

Figure 13.7 *Single crystal absorption spectrum of $[Fe(bbtr)_3](ClO_4)_2$ at 295 K (solid line) and at 10 K on slow cooling (narrow dash) and after temperature quenching (wide dash). Reprinted from [16], with permission from Elsevier.*

these absorption spectra directly yield the temperature dependence of the molar fraction of iron(II) ions in the LS or HS states.

13.4 Materials with Combined Spin-Crossover and Luminescent Functionalities

13.4.1 General Considerations

For efficient coupling between luminescence and SCO properties, the *spectral overlap* of the two functional entities and its consequences for quenching must be considered. As stated previously, the quenching energy transfer mechanisms depend strongly on the *relative proximity of the two centres*. Different synthetic strategies have been employed to combine the two properties. In this review we will divide them into two families.

The first type of synthetic strategy aims to *combine the luminophore and SCO centres into a single entity*, for example through the use of luminescent ligands or counterions to form luminescent SCO complexes and compounds. In these cases the relative location of each component is usually observable, often through structural analysis techniques and hence energy transfer mechanisms may be rationalised. Furthermore, in these cases the distance between the centres may be to some extent rationally controlled through varying the length of the chemical bridging unit. It is however important to state that chemically changing the SCO molecule to include a luminescent moiety risks changing, or indeed losing, one or both properties in an unpredictable manner. Using such an approach there should be a stoichiometric proportion of luminophore and SCO centres, possibly facilitating quantitative analysis of the spin-transition process using the fluorescence technique. However, an aspect usually forgotten while analysing this type of compound is the self-quenching effect. This particular approach may lead to materials that have a high concentration of luminescent centres (i.e. comparable quantity of SCO centres and luminophores) that can permanently quench the luminescent response of the system at all temperatures, even if the conditions for obtaining both properties at a molecular level are respected. Even though the concept of having one luminophore that probes one SCO centre is an ideal situation for monitoring the spin-transition in a material, in the solid state this situation may in fact be *detrimental* to the overall luminescent response. For these reasons, attempting to design combined molecules with specific spectral properties from individual components is difficult and is largely of interest only for fundamental mechanistic studies. It seems difficult to consider this type of approach for the development of practical applications.

Other synthetic approaches concern *doping luminescent entities (metal ions, molecules, etc.) into a SCO material*, at various concentrations, into more or less uncertain positions within the lattice. In fact, in this case the luminophore is used as a structural probe, which is a well-known approach in several fields of science. In these materials, a structural description of the interaction between SCO centres and luminophores is more difficult to obtain and therefore the mechanistic elucidation of energy transfer mechanisms may be rather more complex. Also, one should not overlook the fact that due to its relatively small proportion, the luminophore will probe only a part of the system and it is not guaranteed that the results inferred describe the properties of the ensemble of the material. On the other hand, if the dopant is present in sufficiently low concentrations, the risk of adversely interfering with the SCO properties is reduced. This two component system has the further advantage that each component is usually independent and so individual modification can be possible. For example, one can fine-tune the SCO properties of certain complexes using well-known methods. In many cases, the optical properties of the SCO complex will not change much and therefore it will be possible to work with the same dopant, which is obviously a huge benefit. On the whole, this synthetic strategy appears thus better adapted for practical developments and perhaps less for mechanistic studies.

We must stress that the frontier between these two main approaches is not always well-defined. For example, if one considers a case in which all counterions are luminescent it would constitute the first approach. And if a very low percentage of nonluminescent counterions were replaced with an ionic luminophore, it would clearly represent an example of doping. However, at intermediate proportions of luminophore the distinction is not obvious and perhaps may be less helpful. There will also be several additional (general) requirements for the use of these materials in application (*vide infra*), which include *low photobleaching* of the luminophore so as to ensure the system may be cycled repeatedly, sufficiently high intensity of the emission (at least in one spin state), and low sensitivity of the luminophore to the stimuli that are used to change the spin state of the system.

13.4.2 Examples of Luminescent Spin-Crossover Compounds (Ligands, Counterions)

There are several reports of systems in which the luminescent entity and the SCO centre are part of the same molecule, either with the luminophore being a part of the ligand system surrounding the SCO metal centre,[20–23] or forming part of a mixed metal *d-f* molecular material.[24] There are also conceptually related complexes in which the luminophore acts as a counterion to the cationic SCO molecule.[25]

Fabbrizzi *et al.*[20] reported a system in which a fluorescent fragment (naphthalene) was substituted onto the macrocyclic tetradentate 1,4,8,11-tetraazacyclo-tetradecane ligand to form the neutral ligand **L1**. This ligand coordinates to Ni^{2+} to form complexes of general formula $[Ni(L1)X_2]$, where X is a monoionic anion. The result is either low-spin square-planar complexes (in the case of weakly coordinating anions) or high-spin octahedral complexes (when anions coordinate to the Ni in the axial positions). In solution the spin state at the Ni^{2+} centre changes through the reversible coordination of solvent molecules (S) in axial positions of the molecules according to the following equilibrium:

$[Ni^{II}(L1)S_2]^{2+}$	$\rightleftharpoons$	$[Ni^{II}(L1)]^{2+} + 2S$
Octahedral		Square planar
Low temp.		High temp.
High spin		Low spin

The position of this equilibrium is dependent on the temperature of the system, and unlike in most SCO materials (in which the SCO is not accompanied by a change in the coordination number), higher temperatures favour the low spin, square planar state. Between 300 and 338 K, the fluorescence intensity of the naphthalene molecule increases with increasing LS proportion by a factor of approximately 2.5 (Fig. 13.8), a result of greater quenching efficiency in the HS state. The quenching mechanism in this material was proposed to be energy transfer via double electron exchange, which involves redistribution of electrons between the molecular orbitals of the photoexcited fluorophore and the *d*-orbitals of the metal centre. In this particular system, the temperature-dependence of the spin state equilibrium may be only observed in free molecules in solution where solvent mobility is high and so it is not strictly comparable to more traditional SCO systems. However, from a wider point of view, this study provides an important proof of concept that the spin state of the material can be reported by fluorescent ligands bound to the SCO centre.

During the 1990s, Piguet *et al.*[26, 27] combined a series of rare earth metals with transition metal complexes, some of which also displayed SCO properties. These works were the first to introduce the prospect of employing iron(II) as a versatile acceptor for modulating the luminescence of lanthanide ions in heterodimetallic *d-f* complexes via FRET. Due to the distinctive *d–d* and MLCT absorption bands of iron(II) in the two spin states and its spectral overlap with the emission of Eu(III), it was expected that the resonant energy transfer Eu(III)→Fe(II) would depend strongly on the spin state of the iron centre. As a result, the lanthanide ion could be employed as the 'reporter' of the spin state in iron-based SCO molecules. For example this was attempted

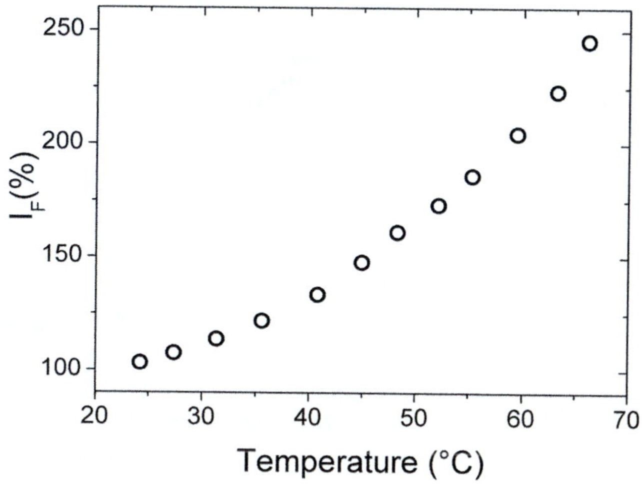

Figure 13.8 *Fluorescence intensity for* $[Ni^{II}(L1)](ClO_4)_2$ *as a function of temperature. Reproduced from [17] with permission of The Royal Society of Chemistry.*

using a heterodimetallic complex (HHH)-$[LnFe(L)_3]^{5+}$, which showed SCO around room temperature.[24] However, even at high temperatures, where the highest possible fraction of $Fe(II)_{HS}$ (>0.5) was observed, an extremely efficient nonradiative energy transfer between Eu(III) and the Fe(II) centres completely quenched the luminescent response of the lanthanide. Thus, in order to study the energy transfer processes between Eu(III) and Fe(II) in the HS state, they attached a sterically hindering methyl group to the ligand so as to prevent the contraction of the Fe–N bonds necessary for the formation of $Fe(II)_{LS}$ species. This was done in such a manner as to avoid affecting the coordination sphere of Eu(III). Even when all iron centres were in the HS state, only a weak luminescence signal could be detected above 580 nm. A drastic decrease of the luminescence lifetime of the Eu(III) ions was observed for the $Fe(II)_{HS}$ complex ($\tau = 0.277$ ms) when compared to the Zn analogue ($\tau = 2.63$ ms). Taking into account the blue shifted MLCT absorption band of the Fe present in both spin configurations (with different oscillator strengths) within that specific coordination environment and the calculated intermetallic distance $\approx$0.9 nm, it was concluded that there was no orbitally mediated interaction (Dexter energy transfer) between the metals. As a result, the energy transfer mechanism was attributed to a Förster resonance type that relies strongly on the spectral overlap of the sensitiser and the acceptor.

Hasegawa *et al.*[21] reported the temperature-dependant fluorescent behaviour (in methanol/ethanol solution) of two Fe(II) SCO complexes with derivatives of the fluorescent bzp ligand (bzp = 2,6-bis(benzimidazole-2′-yl)pyridine), $[Fe(bzp\text{-}H_2)_2]\cdot 2ClO_4$ and $[Fe(bzp\text{-}H)_2]$. In this study, the electronic absorption spectra of both bzp and $[Fe(bzp\text{-}H)_2]$ were assigned using molecular orbital calculations, while the fluorescent spectra of the two iron complexes were used in the assignment of electronic transitions. The origins of the broad bands observed in the emission spectra of the two metal complexes were also assigned. Both complexes were fluorescent in both the HS and LS states, with emission intensity increasing on cooling. The proportion of HS and LS states was estimated from fluorescence spectra using the Boltzman equation at various temperatures and was in good agreement with other detection methods. Importantly, they showed that following the fluorescent properties as a function of temperature may be able to provide quantitative information regarding the fraction of each spin state in Fe(II) materials.

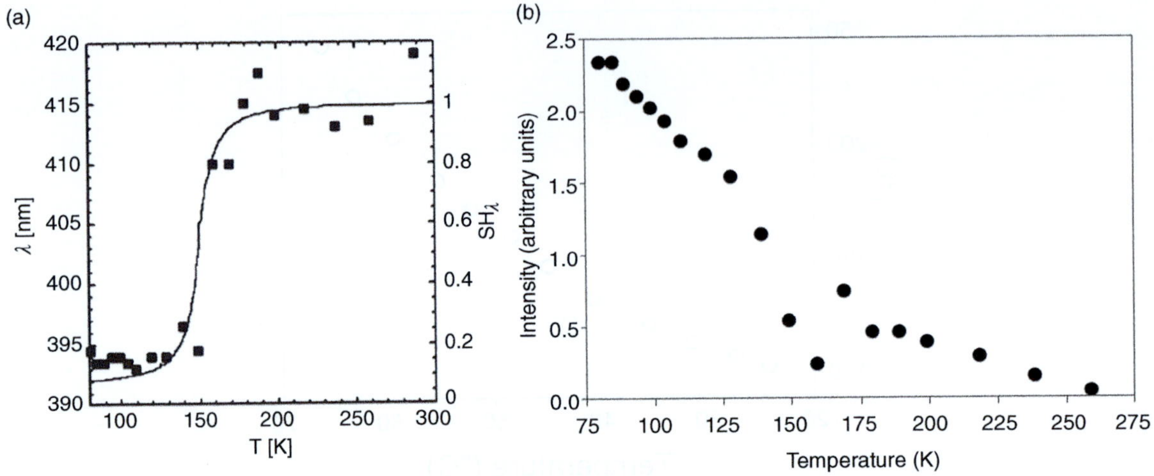

Figure 13.9 *(a) Variation of wavelength of the luminescence emission maximum (λ_{max}) as a function of temperature for $[Fe_2(L2)_5(NCS)_4]_3{\cdot}(MeOH)_4$. Reprinted with permission from [20]. Copyright 2011, American Chemical Society. (b) Luminescence intensity at 414 nm (350 nm excitation) as a function of temperature for the same compound, estimated from Figure S3a of ref. 23.*

Garcia *et al.*[23] substituted a functional moiety at the 4-position of the well-known 1,2,4-triazole ligand system to form N-salicylidene-4-amino-1,2,4-triazole (L2), which shows strong emission in the solid state. Complexation of this ligand to $Fe(NCS)_2$ salt yields the crystalline dinuclear material $[Fe_2(L2)_5(NCS)_4]_3{\cdot}(MeOH)_4$. This system shows a relatively abrupt one-step SCO with $T_c = 150$ K and ca. 6% residual HS fraction at low temperature. Here, the wavelength of the emission maximum, rather than intensity, was followed as a function of temperature and a striking correlation with the SCO curve determined from SQUID measurements was observed (as shown in Fig. 13.9a), demonstrating the high sensitivity of this probe to changes in its environment as a means for monitoring spin state. This approach however is limited by the experimental requirement for high resolution spectra, which in turn requires a high signal-to-noise ratio, something that will of course decrease significantly when moving from the bulk crystalline scale of this study towards nanomaterials. Indeed, when the fluorescence intensity rather than λ_{max} is considered, the results become less clear, as demonstrated in Figure 13.9b, which shows the variation in intensity at constant wavelength (414 nm) after excitation $\lambda_{ex} = 350$ nm. While there is an observed discontinuity in the temperature range of the SCO, the significant thermal dependence in the region in which no transition occurs may make analysis of SCO properties difficult in the absence of magnetic data. In this case, the increasing intensity of the luminescence below T_c was attributed by the authors to the stiffening of the lattice.

In a study by Ruben *et al.*[19] a fluorescent pyrene moiety was incorporated into the bpp ligand, either separated by a long, flexible organic group (4-(2,6-di(1H-pyrazol-1-yl)pyridin-4-yl)benzyl-4-(pyren-1-yl)butanoate (L3)), or directly attached via an sp^3 C-C bond (2,6-di(1H-pyrazol-1-yl)-4-(pyren-1-yl)pyridine (L4)). Complexation of these ligands to Fe^{II} metal centres afforded materials with very different SCO properties. As a consequence of accommodating the supramolecular π-π stacking interactions of the pyrene ligands in $[Fe(L4)_2](ClO_4)_2$, the iron coordination sphere is highly distorted and thus does not undergo SCO, remaining in the HS state across the investigated temperature range. By contrast, the two complexes

$[Fe(L3)_2](ClO_4)_2$ and $[Fe(L3)_2](BF_4)_2 \cdot CH_3CN \cdot H_2O$ both show spin-crossover behaviour without hysteresis. $[Fe(L3)_2](BF_4)_2 \cdot CH_3CN \cdot H_2O$ is fluorescent across the investigated temperature range (17–295 K), and the emission spectrum at 17 K shows features that can be attributed to monomer and excimer pyrene fluorescence. Despite this, no obvious correlation between the SCO and fluorescence properties could be established in this material, with the temperature dependent quenching of the emission being attributed to an intrinsic feature of the pyrene and linker moiety by reference to related non-SCO compounds. As no structural information is available, it is not clear to what extent the pyrene and the SCO centre interact. This study demonstrates the importance of chemical environment and crystal packing effects in determining the luminescent and SCO properties and highlights what may be one of the most significant problems of combining these properties in a single molecule – it is possible that one, or indeed both, of the properties will be lost in the mixed compound. Hence rational design of such hybrid molecules with specific properties (a particular T_c for example) is extremely difficult.

Kimizuka and co-workers[25] synthesised linear iron 1,2,4-triazole complexes with a long lipophilic chain substituted at the 4 position of the triazole ligand. The organic chromophores 9,10-dimethoxyanthracene-2-sulphonate (Anth-SO_3^-) and 1-pyrenesulphonate (Py-SO_3^-) are electrostatically accumulated on the surface of the iron triazole chains, acting as counterions in the two different samples. Combining the luminophore and SCO centre through ionic interaction rather than chemically binding them into a single molecule means the components can be replaced individually to tune the properties. In both cases, the complex forms fibrous nanostructures which undergo SCO in toluene solution. While there was no obvious connection between the SCO and luminescence intensity in the case of Anth-SO_3^-, the complex in the LS state quenches the fluorescence of the Py-SO_3^- anion via energy transfer processes. This is attributed to significant overlap between the emission spectrum of the luminophore and the MLCT band of the SCO molecule in the LS state. Quenching efficiency diminishes in the HS state and hence luminescence intensity increases with the HS proportion. Above 30 °C (in the HS state) the luminescence intensity is again reduced (Fig. 13.10), which was attributed to dynamic quenching of the luminophore by the HS Fe centres, occurring as a result of greater contact between the two units on thermal disintegration of the fibrous structures.

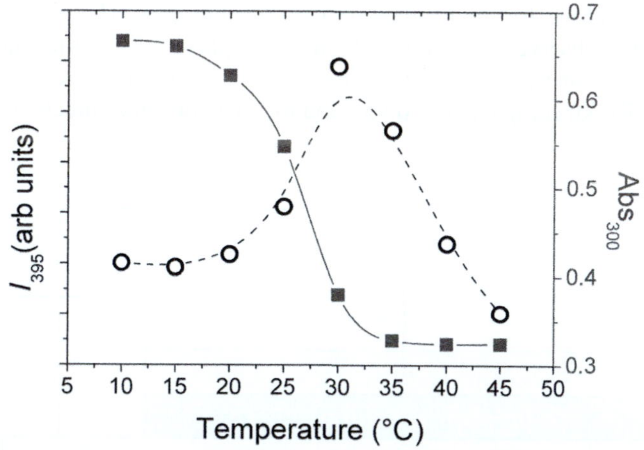

Figure 13.10 *Temperature dependence of emission (λ_{ex} = 380 nm, open circles) and absorption (at 300 nm, filled squares) intensity for the tris(4-(3-dodecyloxypropyl)-1,2,4-triazole)iron(II) nano-fibres with Py-SO_3^- anions in toluene solution as given in ref. 25.*

13.4.3 Luminescent Doping

We have seen that combining SCO and luminescence entities within a single molecule can result in unpredictable or uncontrollable properties of both processes. A conceptually different approach concerns coupling SCO and luminescence properties not within one molecule, but rather through physical mixing of two chemically separate entities via the doping or decoration of a SCO material with a luminophore. This field includes the co-crystallisation of structurally similar molecules to afford SCO-luminescent bulk materials,[28] synthesis of nano-objects and thin films doped with trace quantities of luminophore[29–31] and the elaboration of nano-particles with a SCO core decorated by a luminescent shell.[32] The benefit of this approach is that the individual components may be more readily replaced and hence it may facilitate tuning of individual properties in the hybrid material. However, in such systems, there is not necessarily a stoichiometric proportion of each component, and the relative locations of each are not usually known. Thus the issues of quantitative measurements and elucidation of quenching mechanisms may become rather more complex.

Matsuda *et al.*[33, 34] developed a simple OLED device, shown schematically in Figure 13.11, in which the electroluminescent emission from a thin film of chlorophyll *a* was quenched by the LS state of the SCO complex $[Fe(dpp)_2](BF)_4$, allowing for on/off switching of the device output as a function of temperature.

Electroluminescence differs from the photoluminescence that has been described previously in that it is the result of an electrical input, rather than light. Electron-hole pairs are generated in OLEDs through a charge injection process, and subsequently recombine with electrons producing radiative luminescent emission. The quenching process in this system was reported *not* to be the result of energy transfer from an excited chlorophyll *a* state to the Fe complex, resulting in radiationless decay. Instead it was speculated that the excited state of the luminophore that is responsible for the electroluminescence above T_c (and in devices without the SCO complex) does not form in the presence of the LS iron complex due to a difference in the efficiency of the charge injection process.

Co-crystallisation was used by Halcrow and co-workers[28] as a tool for doping the SCO material $[Fe(bpp)_2][BF_4]_2$ (bpp = 2,6-di(pyrazol-1-yl)pyridine) with the structurally related fluorescent complex $[Ru(terpy)_2][BF_4]_2$ (terpy = 2,2′:6′,2″-terpyridine). The complementary supramolecular interactions shown in the crystal structures of both pure materials – the so-called terpyridine embrace motif – as well as the overall structural similarity allows for the formation of solid solutions of $[Fe(bpp)_2]_x[Ru(terpy)_2]_{1-x}[BF_4]_2$. Virtually complete SCO was observed in all Fe containing complexes, with little change in T_c. An expected result of the reduction in Fe concentration in the lattice is that the small hysteresis shown by the pure Fe complex disappears with 5% Ru complex ($x = 0.95$), and the SCO becomes much less abrupt as x decreases.

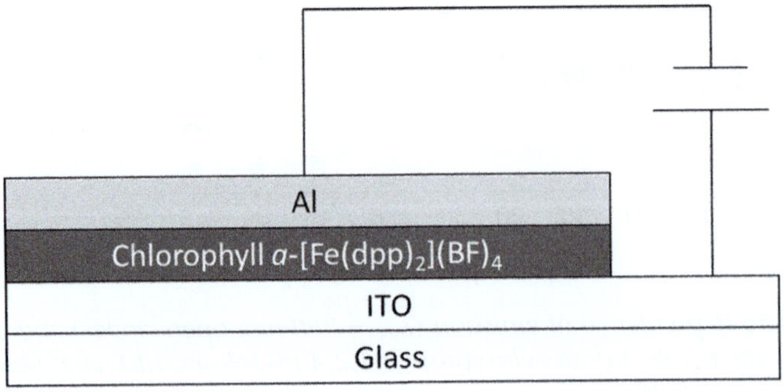

Figure 13.11 *Schematic view of the OLED device constructed in refs 33 and 34.*

Crystalline $[Ru(terpy)_2][BF_4]_2$ has a strong fluorescence emission at 77 K, although this property is much less pronounced in the mixed $[Fe(bpp)_2]_x[Ru(terpy)_2]_{1-x}[BF_4]_2$ species. When $x = 0.47$ the fluorescence intensity at 77 K is only 7% of that observed in the pure Ru complex, and when $x = 0.75$ no obvious emission was observed at all. Thus the material of composition $[Fe(bpp)_2]_{0.47}[Ru(terpy)_2]_{0.53}[BF_4]_2$ has been shown to have both SCO and very weak fluorescence properties, albeit within different temperature ranges. One advantage of this approach is that the relative location of the luminophore and SCO elements can be reliably inferred in these co-crystals, which is not necessarily true for other studies involving doping of the luminophore into the SCO lattice. In principle, either of the individual complexes could be exchanged for others with similar structural features, and in this way the properties of either component may be tuned to some extent. It should be noted that directed co-crystallisation of molecular components is far from routine due to the abundance of weak interactions that pervade the lattice and serve to direct the crystal packing in a largely unpredictable manner. However, it is clear from this study that supramolecular complementarity is another useful tool to consider in the development of multifunctional materials.

Bousseksou *et al.*[29, 30] combined luminescence and SCO properties in nano-materials by doping a very low proportion (between 10^{-4} and 3%) of rhodamine-110 (Rh-110) into small (diameter = 10 ± 6 nm) nano-particles of $[[Fe(NH_2trz)_3](OTs)_2]_n$, dispersed in a surfactant-mediated reverse-micelle solution. It was possible to observe SCO in these particles in the suspension between 288 and 318 K by following the thermal variation of the emission intensity at 540 nm ($\lambda_{excitation} = 475$ nm), as shown in Figure 13.12b. The values of $T_c\uparrow$ and $T_c\downarrow$ are in good agreement with those obtained from absorption measurements of the undoped SCO material. It was suggested that the quenching mechanism is based on a radiative energy transfer regime, based on the assumption that the Rh-110 dopant accumulates in some manner within the reverse micelles. The increase in intensity of the luminescence can be easily assigned to the disappearance of the singlet absorption band of the complex while the $^1A \rightarrow {}^5T$ spin-crossover phenomenon takes place. However, the opposite effect could be observed by changing the luminophore for one with good spectral overlap with the $[[Fe(NH_2trz)_3](OTs)_2]_n$ HS absorption band in the near-IR. One should also notice that

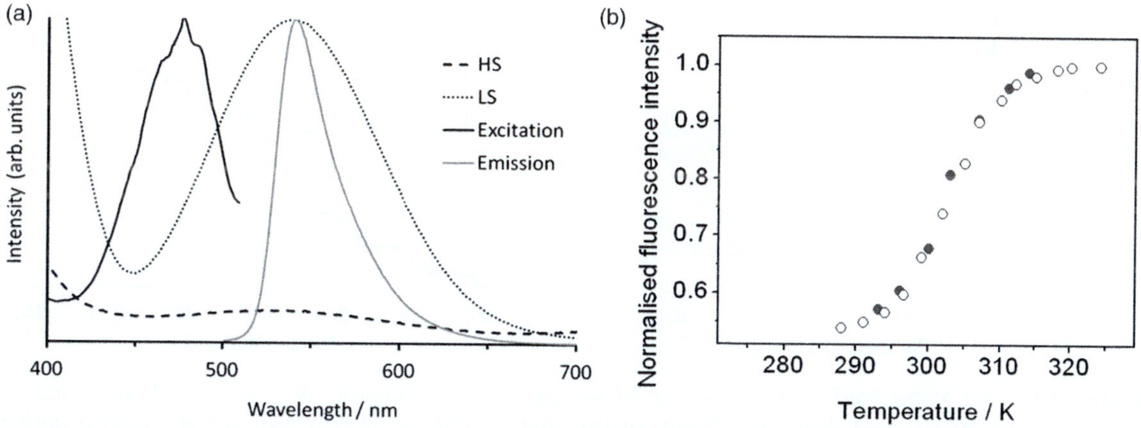

Figure 13.12 *(a) Fluorescence excitation (540 nm) and emission (475 nm) spectra of a rhodamine-110 doped $[Fe(NH_2trz)_3](OTs)_2$ nano-particle suspension at room temperature. The dotted and dashed lines show the absorption spectra of the particle suspension in the LS and HS state, respectively. (b) Thermal variation of the fluorescence intensity at 540 nm in the heating (open symbols) and cooling (closed symbols) modes for a rhodamine-110 doped $[Fe(NH_2trz)_3](NO_3)_2$ nano-particle suspension. Reproduced from [26] with permission of The Royal Society of Chemistry.*

unlike the majority of studies involving luminescent materials where thermal quenching causes a reduction in intensity with increasing temperature, the emission intensity of this dopant remains nearly constant with temperature in the range of interest (i.e. around room temperature). Hence the thermal SCO phenomenon is very easily observed. Independent modification of the two components was also possible in this system. For the first time, the tuning of the SCO properties was achieved by chemically altering the properties (transition temperature, cooperativity, etc.) of the SCO nano-particles, independently of the fluorescent probe. The tosylate counterion was replaced with nitrate ions to afford even smaller (ca. 3 nm) doped particles of $[Fe(NH_2trz)_3](NO_3)_2$ with a lower T_c than in the larger particles and without significant hysteretic effects.[29]

In a related effort to improve the nonradiative energy transfer interaction, the cationic Rh-110 luminophore was replaced with the anionic alizarine red, with the aim of replacing a fraction of the OTs counterions, and hence reducing the separation between SCO and luminescent centres.[30] With Rh-110 dopant the luminescent intensity approximately doubled on LS→HS SCO, while in the case of alizarine red the increase was by a factor of three, revealing the efficacy of this approach.

These ideas were extended beyond solution to the solid state by Bousseksou and Molnár *et al.*[31] to include fluorescent detection of SCO in thin films and single, isolated nano-dots of $[[Fe^{II}(hptrz)_3](OTs)_2]_n$ doped with acridine orange. In this system, the $^1A_1 \rightarrow {}^1T_1$ absorption band of the iron complex overlaps perfectly with the emission spectrum of acridine orange. The thermal response of the thin films and nano-objects was followed at 550 nm after excitation at 450 nm, as shown in Figure 13.13. Homogenous thin films of different thicknesses (between 60 and 200 nm) were prepared by spin coating a solution of the iron tosylate salt, heptyltriazole ligand and 0.1 mol% acridine orange. An increase in luminescence intensity between 293 and 323 K was observed for an 85 nm thick film, and was fully reproducible for three successive thermal cycles, demonstrating the absence of photobleaching in this regime. Again, despite the lack of definitive

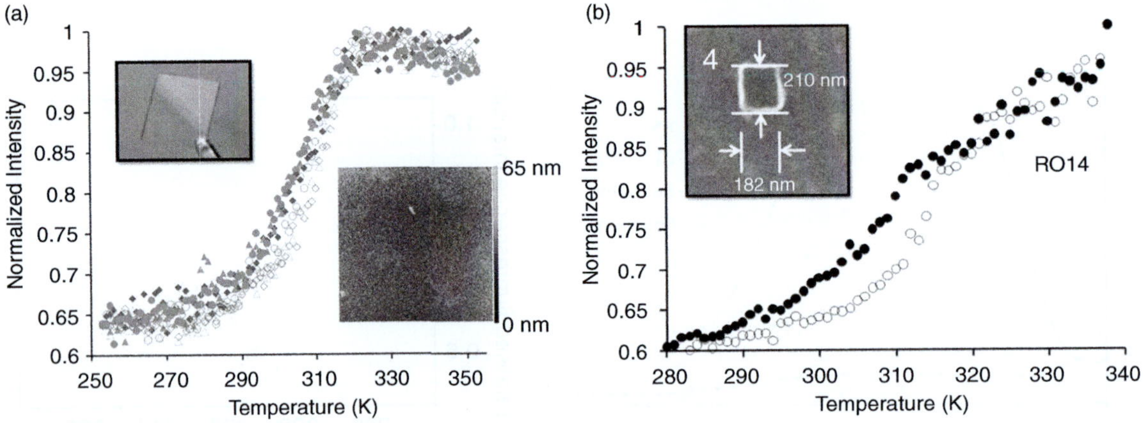

Figure 13.13 (a) Normalised thermal variation of the luminescence intensity at 550 nm with an excitation at 450 nm in a spin coated thin film (85 nm) of $[[Fe^{II}(hptrz)_3](OTs)_2]_n$ doped with acridine orange during three consecutive thermal cycles ($dT/dt = 2$ K/min; open and closed symbols for heating and cooling modes, respectively). Insets: photograph of a spin coated SCO thin film on a 1 cm^2 quartz substrate and AFM image (34×34 μm^2) of its surface. (b) Normalised thermal variation of the luminescence at 550 nm for a nano-dot of $[[Fe^{II}(hptrz)_3](OTs)_2]_n$ doped with acridine orange. Open and closed symbols indicate heating and cooling modes, respectively. The inset shows the magnified SEM image of the nano-dot. Reproduced from [28] with permission of The Royal Society of Chemistry.

proof regarding the location of the luminophore dopant, the quenching mechanism was largely attributed to nonradiative transfer between the acridine orange and the Fe^{II} ions of the SCO complex. Structured arrays of solvent- and surfactant-free nano-dots with average dimensions of ca. 150 nm height and 200 nm lateral size composed of the same doped compound were also created by a soft lithographic technique. The thermal luminescent modulation resulting from the SCO was observed for individual isolated dots, free from the effects of ensemble averaging. Furthermore, the SCO behaviour of each dot could be associated with its own geometrical features as determined by SEM, an important step in the drive toward correlating size reduction effects with SCO properties in these materials.

These fluorescent spin-crossover thin films and nano-objects provide very interesting scope for thermometry and thermal imaging purposes. When compared to conventional IR thermography, fluorescence thermometry can provide better sensitivity (especially at low temperatures) and higher spatial resolution. Thus it may find applications in various fields of science and technology where accurate temperature measurements are required at reduced scales (microelectonic circuit testing, lab-on-chip devices, and so on).[35] The combination of luminescence with SCO is an appealing approach because it gives higher flexibility in the design of the thermometer through the separation of the sensor (SCO material) and signal transducer (fluorophore) functions.[29] Figure 13.14 shows a proof-of-concept experiment where a thin film of the compound $[Fe^{II}(hptrz)_3](OTs)_2$ doped with Rhodamine 110 has been spin coated on a passivated gold microwire. The reduced cross-section of the wire allows for a local confinement of the heating along its axis when passing an electrical current through the device (Joule effect).[35] As temperature increases, the spin state of the system will change locally from LS to HS and as a consequence the luminescence intensity from the heated area will increase. With precise knowledge of the relationship between the luminescence intensity and the temperature, a given change in intensity can be converted to temperature. In Figure 13.14a the bottom panel shows the fluorescent image around the wire when no current is applied. The middle and top panels show the same measurement, but for increasing currents of 9 mA and 15 mA. The strong confinement of the temperature increase in the vicinity of the wire is clearly perceptible. As shown in Figure 13.14b the measured temperature distribution profile along the wire is reasonably flat, with a steep change of the temperature at the ends of the wire.[36]

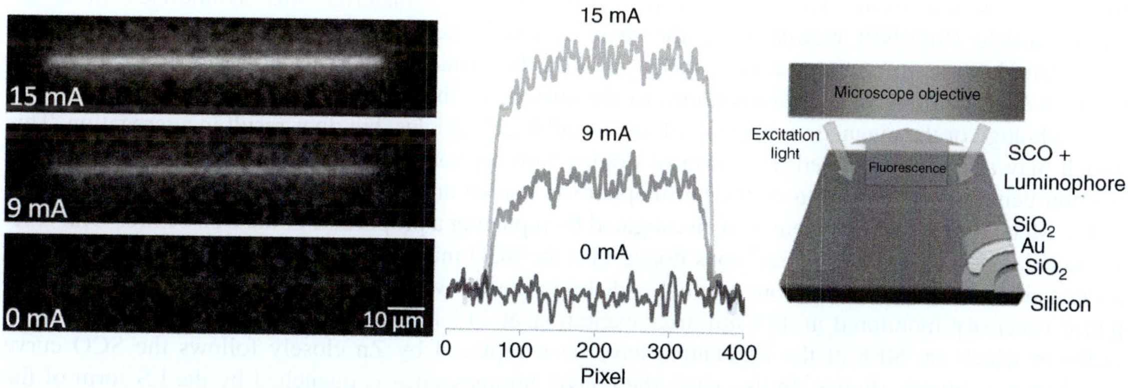

Figure 13.14 (a) Fluorescence microscopy images (excitation 450 nm, emission 550 nm) of a gold nanowire (80 μm length, 1 μm width, 50 nm thickness) covered by a thin film of the spin-crossover compound $[Fe(hptrz)_3](OTs)_2$ doped with Rhodamine 110. During the fluorescence measurements the nano-wires were heated by Joule effect: the current intensity is indicated for each image. (b) Cross-section of the fluorescence images along the nano-wires. (c) Schematic picture of the device and the measurement setup.

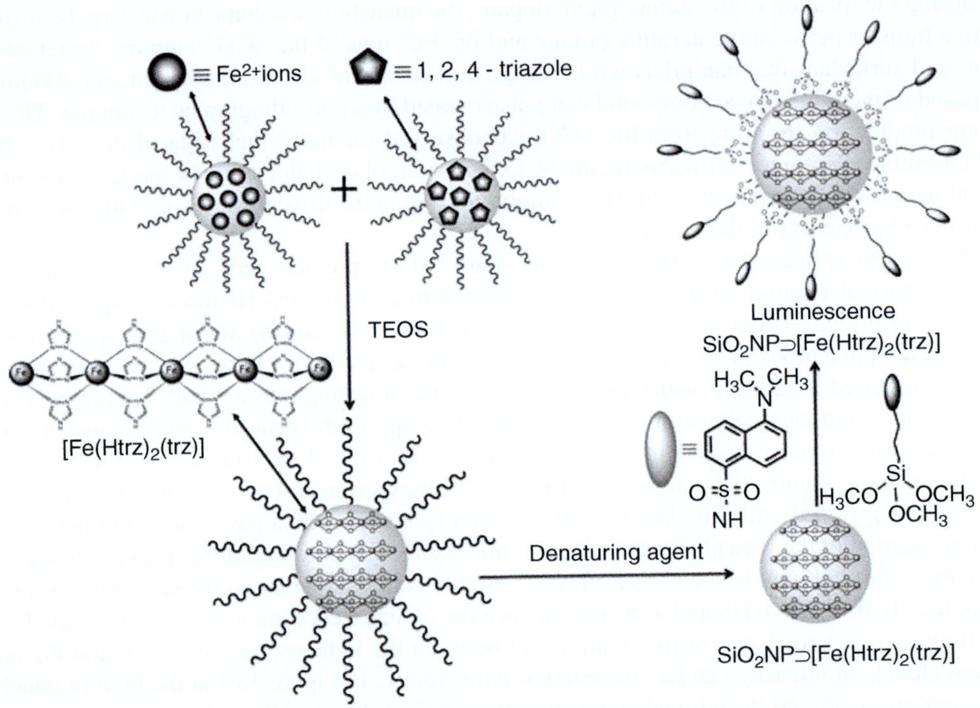

Figure 13.15 *Synthetic route and schematic representation of the SCO/luminescent SiO_2 nano-particles. Reproduced from [29] with permission from John Wiley and Sons, 2011.*

A different approach to the combination of luminescence with nano-scale SCO materials was presented by Herrera and Colacio *et al.*,[32] in which a core (SCO)–shell (luminescence) nano-particle architecture was employed. The $[[Fe(HTrz)_2(Trz)](BF_4)]_n$ SCO material was synthesised in a silica nano-particle (SiO_2NP) matrix using the reverse micelle technique and the luminophore, dansyl (3-(dansylamido)propyltrimethoxysilane), was grafted on the surface of the particles, shown schematically in Figure 13.15. Grafting of the luminophore to the surface of the particles does not significantly affect their morphology or the magnetic properties of an ensemble of particles, but does result in aggregation. This approach may also be considered as a form of doping luminophores into the SCO material, the essential distinction being that the location of the luminophore is known and confined to the surface of the hybrid particles. Metal dilution effects were also investigated by replacing a proportion of the iron centres with zinc. As expected, as the proportion of Zn^{II} ions doped into the SCO material increases, the transition is shifted to lower temperatures, becomes more gradual and the hysteresis width narrows. The thermal luminescent response (intensity monitored at 495 nm after excitation at 315 nm) of an ensemble of these core-shell particles in which ca. 50% of the Fe centres have been replaced by Zn closely follows the SCO curve obtained from magnetic studies. In this case, the dansyl luminescence is quenched by the LS form of the Fe complex in the core of the particle via an energy transfer mechanism, resulting from a degree of spectral overlap between the dansyl luminescence and the absorption band of the LS ion. Again, this two component system should allow for the modification of the SCO material so as to tune the sensing properties. However, no fluorescence results are presented concerning the undiluted particles and hence this issue remains to be fully explored.

13.5 Concluding Remarks

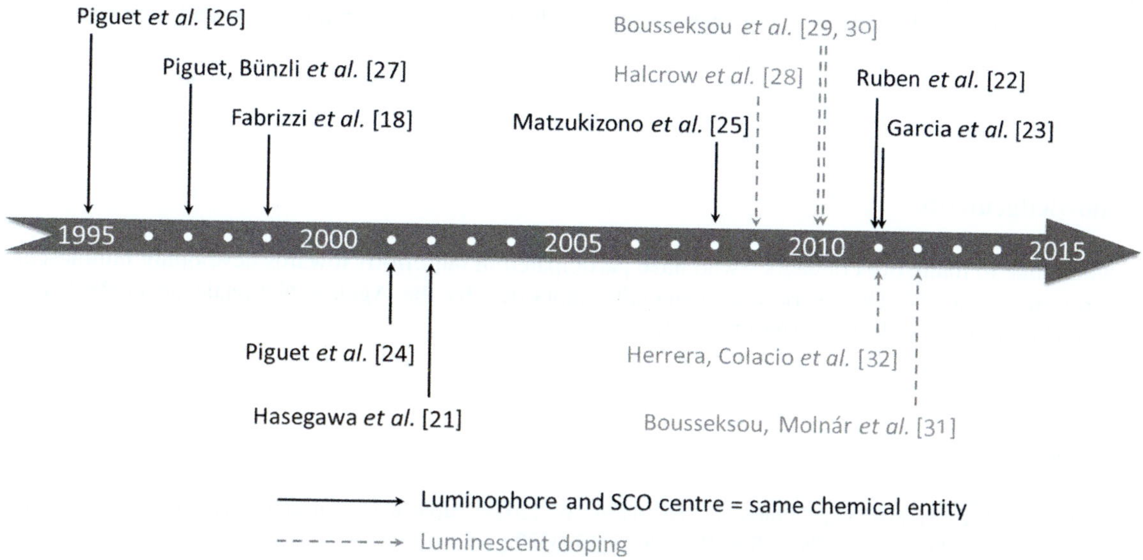

Scheme 13.1 *Timeline for the development of fluorescent SCO materials.*

Luminescence detection is a promising alternative to traditional methods for studying the spin-crossover phenomenon from materials at the bulk scale to individual nano-objects and even down to a single molecule. However, due to the high sensitivity to the environment of both SCO and luminescent centres, their combination into one material is a challenging task and may result in the loss of one or even both properties.

During the last 12 years, essentially two different approaches have been adopted for the design of these novel materials: combination of the SCO centre and luminophore into a single molecule and luminescent doping. In the first case, uniting both entities into the same molecule provides clear advantages for rationalising the structural relationship between the two components. Furthermore, it could allow not only the elucidation, but also the quantification of the interactions between the luminophore and the SCO centre. However, this approach may suffer from self-quenching problems. Furthermore, designing each chemical unit may include several compatibility challenges in order to preserve the spin-transition and the luminescence, and combining them often requires complex synthetic strategies. In the second case, the doping of a SCO material with low concentrations of a luminescent agent may have lower risks of deleteriously affecting its original SCO properties. There are fewer constraints for the choice of the luminophore and it will be basically restricted to an appropriate spectral overlap between the emission and the characteristic SCO absorption bands of the material to study. Thus, the characteristics of the spin transition (T_C, cooperativity, etc.) may be tuned independently of the luminophore giving a unique flexibility to the approach. This can be a clear advantage for technological applications. Nonetheless, the uncertainty of the luminophore location in the lattice introduces additional difficulties in understanding the energy transfer mechanisms involved in these hybrid materials and providing quantitative information regarding the SCO phenomenon.

Clearly this field of research holds many possibilities for future research. First of all, there is a need for in depth mechanistic studies of existing systems, including luminescence lifetime and quantum yield measurements as a function of the spin fractions. Then, new systems could be developed based on advanced ligand engineering and supramolecular chemistry. For example, luminescent doping in a hetero-bimetallic (Fe–M) SCO complex, where a small fraction of the non-luminescent M metal centre is replaced with a luminescent metal, may provide a quite ideal system from different points of view. Completely new strategies may be also envisaged, such as using metal ions, which display both luminescence and spin-crossover. Manganese complexes may be good candidates. Finally, there are also promising application possibilities for combined luminescence SCO systems, including tunable photonic devices as well as sensors.

Acknowledgements

We would like to thank our colleagues who have participated in our efforts towards developing luminescent spin-crossover systems. This work was financially supported by the Agence Nationale de la Recherche (project Thermospin, ANR-10-NANO-012-02).

References

1. Bousseksou, A., Molnár, G., Salmon, L., Nicolazzi, W. (2011) Molecular spin crossover phenomenon: recent achievements and prospects. Chem. Soc. Rev., 40: 3313–3335.
2. Moerner, W. E., Fromm, D. P. (2003) Methods of single-molecule fluorescence spectroscopy and microscopy. Rev. Sci. Instrum., 74: 3597–3619.
3. Marfunin, A. S. (1979) Spectroscopy, Luminescence and Radiation Centers in Minerals. Springer-Verlag, Berlin, Germany, p. 352.
4. Blasse, G., Grabmaier, B. C. (1994) Luminescent Materials. Springer-Verlag, Berlin, Germany, p. 232.
5. Lakowicz, J. R. (2006) Principles of Fluorescence Spectroscopy. Springer, New York, USA, p. 954.
6. Gaft, M., Reisfeld, R., Panczer, G. (2005) Modern Luminescence Spectroscopy of Minerals and Materials. Spinger, Berlin, Germany, p. 356.
7. Valeur, B. (2002) Molecular Fluorescence: Principles and Applications. Wiley VCH, Weinheim, Germany, p. 402.
8. Powel, R. C., Blasse, G. (1980) Energy transfer in concentrated systems. Struct. Bonding (Berlin), 42: 43–96.
9. Gütlich, P., Goodwin, H. A. (Eds.) (2004) Spin Crossover in Transition Metal Compounds I-III. Top. Curr. Chem., vols. 233–235. Springer, Berlin / Heidelberg, Germany.
10. Braslavsky, S. E. (2007) Glossary of terms used in photochemistry, 3rd Edn (IUPAC recommendations 2006). Pure Appl. Chem.,79: 293–465.
11. Figgis, B. N. (1964) Introduction to Ligand Fields. Wiley-Interscience, New York, USA, p. 362.
12. Förster, T. (1946). Energiewanderung und Fluoreszenz. Naturwissenshaften, 6: 166–175.
13. Förster, T. (1948). Zwischenmolekulare Energiewanderung und Fluoreszenz. Ann. Phys., 2: 55–75.
14. Förster, T. (1993). Intermolecular energy migration and fluorescence. in: Mielczarek, E. V., Greenbaum, E., Knox, R. S. (Eds) Biological Physics. American Institute of Physics, New York, USA, pp. 148–160.
15. Dexter, D. (1953) A theory of sensitized luminescence in solids. J. Chem. Phys., 21: 836–850.
16. Dexter, D. L., Knox, R. S. (1965). Excitons. Interscience, New York, USA, p. 139.
17. Clegg, R. M. (2009) Förster resonance energy transfer – FRET. What it is, why do it, and how it's done, in: Gadella, T. W. J. (Ed.) FRET and FLIM techniques. Laboratory techniques in biochemistry and molecular biology, vol. 33. Elsevier, Amsterdam, Netherlands, ch. 1, pp. 1–57.
18. Hauser, A. (2004) Ligand field theoretical considerations, in: Gütlich, P., Goodwin, H. A. (Eds) Spin Crossover in Transition Metal Compounds I. Top. Curr. Chem., 233: 49–58.
19. Krivokapic, I., Enachescu, C., Bronisz, R., Hauser, A. (2008) The interaction between the spin transition and a crystallographic phase transition in the spin-crossover compound $[Fe(bbtr)_3](ClO_4)_2$: nucleation, formation of domains and fluctuations. Inorg. Chim. Acta, 361: 3616–3622.

20. Engeser, M., Fabbrizzi, L., Licchelli, M., Sacchi, D. (1999) A fluorescent molecular thermometer based on the nickel(II) high-spin/low-spin interconversion. Chem. Commun., 1191–1192.
21. Hasegawa, M., Renz, F., Hara, T., Kikuchi, Y., Fukud, Y., Okubo, J., Hoshi, T., Linert, W. (2002) Fluorescence spectra of Fe(II) spin crossover complexes with 2,6-bis(benzimidazole-2′-yl)pyridine. Chem. Phys., 277: 21–30.
22. González-Prieto, R., Fleury, B., Schramm, F., Zoppellaro, G., Chandrasekar, R., Fuhr, O., Lebedkin, S., Kappesa, M., Ruben, M. (2011) Tuning the spin-transition properties of pyrene-decorated 2,6-bispyrazolylpyridine based Fe(II) complexes. Dalton Trans., 40: 7564–7570.
23. Garcia, Y., Robert, F., Naik, A. D., Zhou, G., Tinant, B., Robeyns, K., Michotte, S., Piraux, L. (2011) Spin transition charted in a fluorophore-tagged thermochromic dinuclear iron(II) complex. J. Am. Chem. Soc., 133: 15850–15853.
24. Edder, C., Piguet, C., Bünzli, J.-C. G., Hopfgartner, G. (2001) High-spin iron(II) as a semitransparent partner for tuning europium(III) luminescence in heterodimetallic d–f complexes. Chem. Eur. J., 7: 3014–3024.
25. Matzukizono, H., Kuroiwa, K., Kimizuka, N. (2008) Self-assembly-directed spin conversion of iron(II) 1,2,4-triazole complexes in solution and their effect on photorelaxation processes of fluorescent counter ions. Chem. Lett., 37: 446–447.
26. Piguet, C., Rivara-Minten, E., Hopfgartner, G., Bünzli, J.-C. G. (1995) Molecular magnetism and iron(II) spin-state equilibrium as structural probes in heterodinuclear d-f complexes. Helv. Chim. Acta, 78: 1651–1672.
27. Piguet, C., Rivara-Minten, E., Bernardinelli, G., Bünzli, J.-C. G., Hopfgartner, G. (1997) Non-covalent lanthanide podates with predetermined physicochemical properties: iron(II) spin-state equilibria in self-assembled heterodinuclear d–f supramolecular complexes. J. Chem. Soc. Dalton Trans., 421–433.
28. Tovee, C. A., Kilner, C. A., Thomas, J. A., Halcrow, M. A. (2009) Co-crystallising two functional complex molecules in a terpyridine embrace lattice. Cryst. Eng. Comm., 11: 2069–2077.
29. Salmon, L., Molnár, G., Zitouni, D., Quintero, C., Bergaud, C., Micheau, J.-C., Bousseksou, A. (2010) A novel approach for fluorescent thermometry and thermal imaging purposes using spin crossover nanoparticles. J. Mater. Chem., 20: 5499–5503.
30. Quintero, C., Molnár, G., Salmon, L., Tokarev, A., Bergaud, C., Bousseksou, A. (2010) Design of fluorescent spin-crossover nanoparticles for thermometry applications. 16th International workshop on thermal investigations of ICs and systems (THERMINIC), Barcelona, Spain, pp. 1–5.
31. Quintero, C. M., Gural'skiy, I. A., Salmon, L., Molnár, G., Bergaud, C., Bousseksou, A. (2012) Soft lithographic patterning of spin crossover complexes. Part 1: fluorescent detection of the spin transition in single nano-objects. J. Mater. Chem., 22: 3745–3751.
32. Titos-Padilla, S., Herrera, J. M., Chen, X.-W., Delgado, J. J., Colacio, E. (2011) Bifunctional hybrid SiO_2 nanoparticles showing synergy between core spin crossover and shell luminescence properties. Angew. Chem. Int. Ed., 50: 3290–3293.
33. Matsuda, M., Isozaki, H., Tajima, H. (2008) Electroluminescence quenching caused by a spin-crossover transition. Chem. Lett., 37: 374–375.
34. Matsuda, M., Isozaki, H., Tajima, H. (2008) Reproducible on–off switching of the light emission from the electroluminescent device containing a spin crossover complex. Thin Solid Films, 517: 1465–1467.
35. Löw, P., Kim, B., Takama, N., Bergaud, C. (2008) High-spatial-resolution surface-temperature mapping using fluorescent thermometry. Small, 4: 908–914.
36. Molnár, G., Gural'skiy, I. A., Salmon, L., Nicolazzi, W., Quintero, C. M., Akou, A., Abdul-Kader, K., Félix, G., Mahfoud, T., Bergaud, C., Bartual, C., Thibault, C., Vieu, C., Bousseksou, A. (2012) Bistable photonic nanostructures based on molecular spin crossover complexes. Proceedings of SPIE, Photonic Crystal Materials and Devices X, vol. 8425, p. 842513.

14

Nanoparticles, Thin Films and Surface Patterns from Spin-Crossover Materials and Electrical Spin State Control

Paulo Nuno Martinho, Cyril Rajnak and Mario Ruben

Institute of Nanotechnology, Karlsruhe Institute of Technology, Germany

14.1 Introduction

Since the report of the first example of a spin-crossover (SCO) complex in 1931 by Cambi and co-workers[1] and the vision of Kahn and co-workers in 1996 towards the application of SCO compounds in data processing,[2] a long and impressive literature on SCO has developed with the discovery of several new examples of SCO compounds, the explanation of different types of SCO profiles, and the modification of SCO examples to increase cooperativity and to direct the application to materials science. Scientists have developed SCO networks,[3–5] SCO frameworks,[6–8] SCO gels,[9–11] SCO liquid crystals,[12–14] SCO nanoparticles and nanocrystals, SCO thin films and have also applied patterning techniques to fabricate SCO devices. More recently electrical control of SCO has been achieved and this work reviews research made so far with concern to nanoparticles, thin films, surface patterns and devices from SCO materials.

SCO complexes are a particular class of compounds of $3d^4$ to $3d^7$ transition metals ions in an octahedral (O_h) field, which can interchange between two electronic states, high spin (HS) and low spin (LS), by application of an external perturbation such as temperature, pressure, light or magnetic field.[15] This property can be either present in solution or solid state and while in solution the system is diluted and the process essentially molecular, in solid state the process may involve elastic interactions conferring different types and shapes of transition curves. SCO was first reported by Cambi and co-workers when they observed unusual magnetism in iron(III) derivatives of various dithiocarbamates which resulted in the recognition of the interconversion of two spin states as a result of variation in temperature.[1, 16, 17] Despite this discovery in the early 1930s a boom

Spin-Crossover Materials: Properties and Applications, First Edition. Edited by Malcolm A. Halcrow.

in SCO research has only occurred in the last 15 years with the miniaturisation of electronic devices and the approach of the superparamagnetic limit for magnetic materials.

14.2 Nanoparticles and Nanocrystals

A key factor in the application of SCO systems into materials development is the control of the size of the object that displays hysteresis.[18] For production of functional devices, the use of nanometre-range SCO materials is required and, combined with this, their magnetic and cooperative behaviour should be retained at low dimensions. SCO nanoparticles have been produced, based mainly on two distinct families of compounds: the $[Fe(R\text{-}trz)_3]X_2$ family (trz = 1,2,4-triazole) and the $[Fe(pz)][M(CN)_4]$ family (pz = pyrazole). The results obtained for SCO nanoparticles derived from $[Fe(R\text{-}trz)_3]X_2$ (Fig. 14.1) have been very interesting.

14.2.1 Reverse Micelle (Microemulsion) Technique

In this technique, the resultant particle volume is defined by the size of the water droplet within the emulsion. The most effective way to control the water droplet size is to vary the relative ratio of the water-to-oil, such that smaller droplets are formed when the amount of oil is increased.[19, 20]

Diverse techniques have been used to produce SCO nanoparticles but undoubtedly the most widely used has been the reverse micelle method, Figure 14.1. This section reviews work done to produce SCO nanoparticles using reverse micelle, dispersion and sol-gel techniques.

14.2.1.1 Triazole Derivatives

Coronado *et al.* pioneered the preparation of SCO nanoparticles using the reverse micelle type technique, which used either a water-in-oil or ethanol-in-oil microemulsion to grow SCO nanoparticles of $[Fe(Htrz)_2(trz)](BF_4)$ which were purple in colour and with an average size of around 15 nm.[21] Evaporation of the solvent produced a solid that could be suspended in *n*-octane yielding a bright transparent suspension. The thermochromic behaviour of such suspensions was investigated, with the nanoparticles presenting a colour change associated with both HS (pink) and LS (colourless) states. Transmission electron microscopy (TEM) images of isolated particles showed regular sized spherical nanoparticles, Figure 14.2a. Magnetic

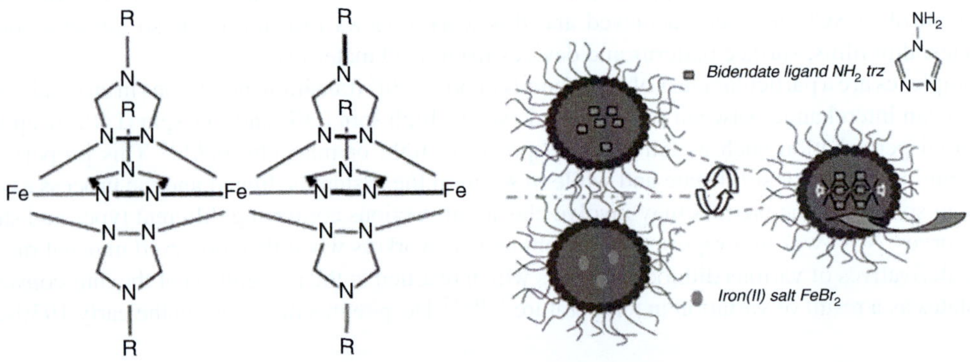

Figure 14.1 Iron(II) triazole-nanoparticles. (Left) representation of main family of compounds ($[Fe(R\text{-}trz)_3]X_2$) used to produce SCO nanoparticles; (right) synthetic route used to produce nanoparticles of $[Fe(NH_2\text{-}trz)_3]$. Reproduced from [22] with permission of The Royal Society of Chemistry.

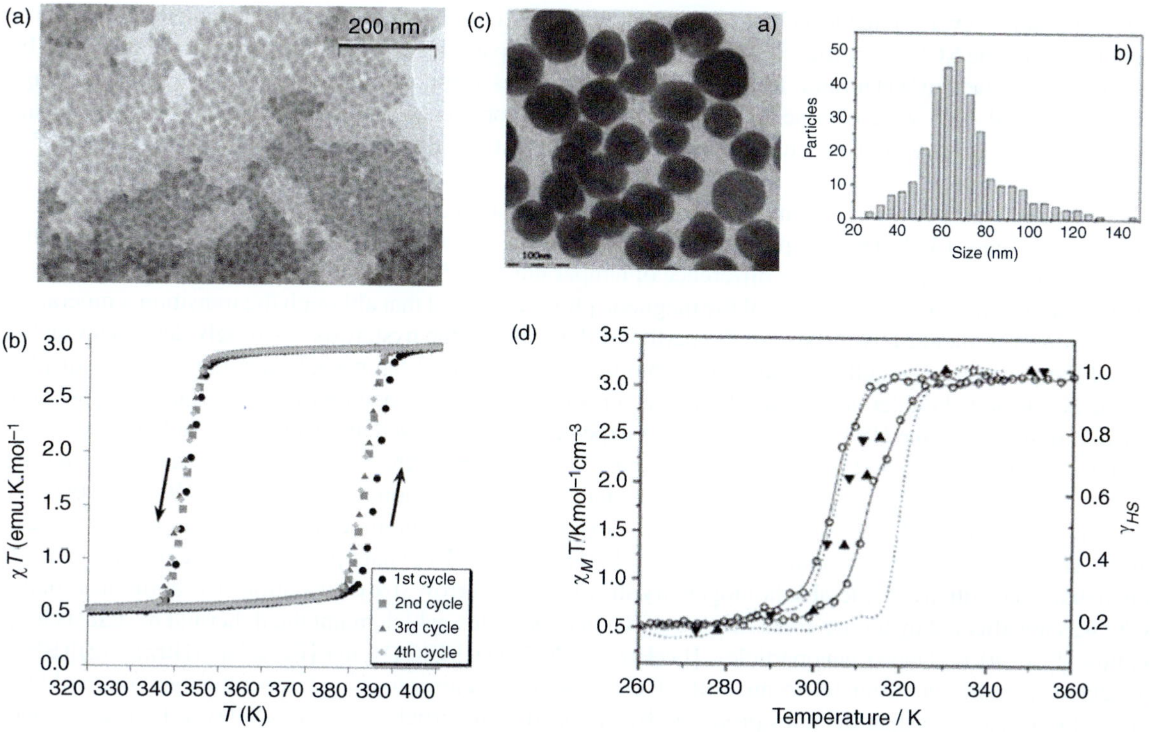

Figure 14.2 *SCO nanoparticles of ([Fe(R-trz)*$_3$*]X*$_2$*: (a) TEM image of [Fe(Htrz)*$_2$*(trz)](BF*$_4$*) nanoparticles; (b) magnetic thermal hysteresis of [Fe(Htrz)*$_2$*(trz)](BF*$_4$*) nanoparticles with 1st cycle represented by black circles, 2nd cycle represented by gray squares, 3rd cycle represented by gray triangles and 4th cycle represented by pale gray diamonds;*[21] *(c) TEM image of [Fe(NH*$_2$*trz)*$_3$*](Br)*$_2$*·3H*$_2$*O·0.03(surfactant) nanoparticles with statistic distribution based TEM images on ca. 300 particles, showing average size 69 nm, minimum and maximum sizes, 26 and 147 nm and disparity 19 nm; (d) Magnetic plot of [Fe(NH*$_2$*trz)*$_3$*](Br*$_2$*)·3H*$_2$*O·0.03(surfactant) for nanoparticles (-) and for macroscopic particles obtained from a typical synthesis (–). Reprinted with permission from [21]. Copyright © 2007 WILEY-VCH Verlag GmbH & Co. KGaA, and [22] with permission from The Royal Society of Chemistry.*

investigation of the isolated nanoparticles showed that both the transition temperature ($T_{1/2}\uparrow = 386$ K and $T_{1/2}\downarrow = 343$ K) and the large 43 K wide hysteresis window were preserved when compared to the original nonengineered powder samples, Figure 14.2b.

In a parallel development Létard *et al.* prepared nanoparticles of a related compound where the triazole is appended with an amino function and the bromide salt is used, [Fe(NH$_2$-trz)$_3$](Br)$_2$·3H$_2$O.0.03(surfactant = Lauropal).[22] In contrast to the nanoparticles previously described, their hysteresis window is preserved only for large particles (>50 nm). It is interesting that these authors used a nonionic surfactant in contrast to sodium dioctyl sulfosuccinate used by Coronado and co-workers. TEM images of isolated nanoparticles dispersed in an ethanol solution, showed a good distribution of spherical particles with an average size of 69 nm, Figure 14.2c.

Magnetic measurements for a collection of nanoparticles in both cooling and warming modes showed an 8 K wide hysteretic SCO system ($T_{1/2}\uparrow = 311$ K and $T_{1/2}\downarrow = 303$ K). When compared to the macroscopic particles (bulk sample) it was possible to observe that the width of the hysteresis window and consequently the degree of cooperativity decreases for the nanoparticles, Figure 14.2d. In a follow up, Létard and co-workers applied the same reverse micelle technique to the previously studied system and analysed the influence of the particle size on the SCO properties.[23] In a controlled fashion, particles with different sizes were produced by changing the surfactant/water ratio. TEM images of the particles as function of the surfactant/water percentage ranging from 50–85% were obtained, Figure 14.3a. The authors concluded that homogeneity increases and particle size decreases with increasing proportions of surfactant, yielding 30 nm size nanoparticles when 85% of surfactant was used.

The magnetic properties in cooling and warming modes as a function of surfactant/water percentage were determined, Figure 14.3b. The abruptness of the transition was quantified using a parameter introduced by the authors (ΔT_{80}), corresponding to the difference of temperature at which 80 and 20% of the iron(II) metal ions undergo spin-transition. Examination of the magnetic plots concluded that although the transition temperature is not strongly influenced by the size of the particles, the width of the hysteresis is strongly dependent on the size of the particles. The authors also pointed out that irrespective of the particle size the materials display a large absorption shoulder at around 520 nm, as a result of the *d–d* transition of the LS state ($^1A_{1g}\rightarrow{}^1T_{1g}$). The *d–d* transition of the HS state ($^5T_{2g}\rightarrow{}^5E_g$) is expected to be observed at lower energy at 830 nm.

Recently, Coronado and Galán-Mascarós have studied the effect of chemical and size tuning on the thermal hysteresis of bistable SCO nanoparticles. Nanoparticles of $[Fe(Htrz)_2(trz)](BF_4)$, $[Fe(Htrz)_{3-x}(NH_2trz)_x](ClO_4)_2$ and $[Fe_{0.8}Zn_{0.2}(Htrz)_2(trz)](BF_4)$ were prepared using the reverse micelle technique. The authors observed that mean size of the $[Fe(Htrz)_2(trz)](BF_4)$ nanoparticles can be tuned down to 6 nm in diameter, with a considerable homogeneity shown by the narrow size distribution. The abruptness of the SCO was not affected by tuning down the size of the nanoparticles, which maintained thermal hysteresis loop widths of 30–40 K. Doped nanoparticles, $[Fe(Htrz)_{3-x}(NH_2trz)_x](ClO_4)_2$ and $[Fe_{0.8}Zn_{0.2}(Htrz)_2(trz)](BF_4)$, presented sizes ranging from 10–15 nm with SCO temperatures closer to room temperature and thermal hysteresis loops narrower than the ones produced by undoped nanoparticles (15–20 K). TEM, Figure 14.4a–c, and atomic force microscopy (AFM), Figure 14.4d and Figure 14.4e, after deposition on gold or silicon surfaces were used to characterise produced nanoparticles.[24]

Bousseksou *et al.* also contributed considerably to advance the research on SCO nanoparticles by preparing nanoparticles of $[Fe(NH_2trz)_3](tos)_2$ (tos = tosylate) doped with fluorescent Rhodamine 110 (Rh110), Figure 14.5.[25] The authors choose Rh110 as fluorescent marker because this exhibits only a very weak thermal extinction of its luminescence around room temperature. Its emission spectrum and the low energy tail of its excitation spectrum both overlap the $^1A_{1g}\rightarrow{}^1T_{1g}$ absorption band of the iron(II)-triazole complex. Finally, Rh110 also exhibits relatively weak photobleaching, and its properties within reverse micelles had been previously explored by others.[26]

Undoped nanoparticles showed reversible thermochromic behaviour, with a violet suspension turning colourless when heated to 50 °C which indicated possible SCO behaviour, Figure 14.5a. Fluorescence of 10(6) nm nanoparticles of $[Fe(NH_2trz)_3](tos)_2$ doped with 3% Rh110, and 3 nm nanoparticles of $[Fe(NH_2trz)_3](NO_3)_2$ doped with 0.01% Rh110, were investigated. The thermal variation of the fluorescence intensity at 540 nm showed that for $[Fe(NH_2trz)_3](tos)_2$ doped with 3% of Rh110, the transition temperatures are 301 and 306 K in the heating and cooling modes, respectively. In contrast to what would be expected, the authors observed that the emission intensity increases with the temperature. In the case of $[Fe(NH_2trz)_3](NO_3)_2$ nanoparticles doped with 0.01% of Rh110, the absence of hysteresis allowed the authors to develop a thermometer for real-time thermometry, Figure 14.5c.

In parallel to the developments on fluorescent SCO nanoparticles, Bousseksou and co-workers also investigated the cooperative SCO phenomena in $[Fe(NH_2trz)_3](tos)_2$ nanoparticles.[27] Particles with 3–4 nm size were obtained using a nonionic surfactant to avoid anion exchange. TEM, high-resolution transmission electron microscopy (HRTEM), dynamic light scattering (DLS), UV-vis spectroscopy and magnetic measurements were used to characterise dried particles and particles in suspension. TEM images of nanoparticles of $[Fe(NH_2trz)_3](tos)_2$ show similar, spherical, nonaggregated nanoparticles with a narrow size distribution.

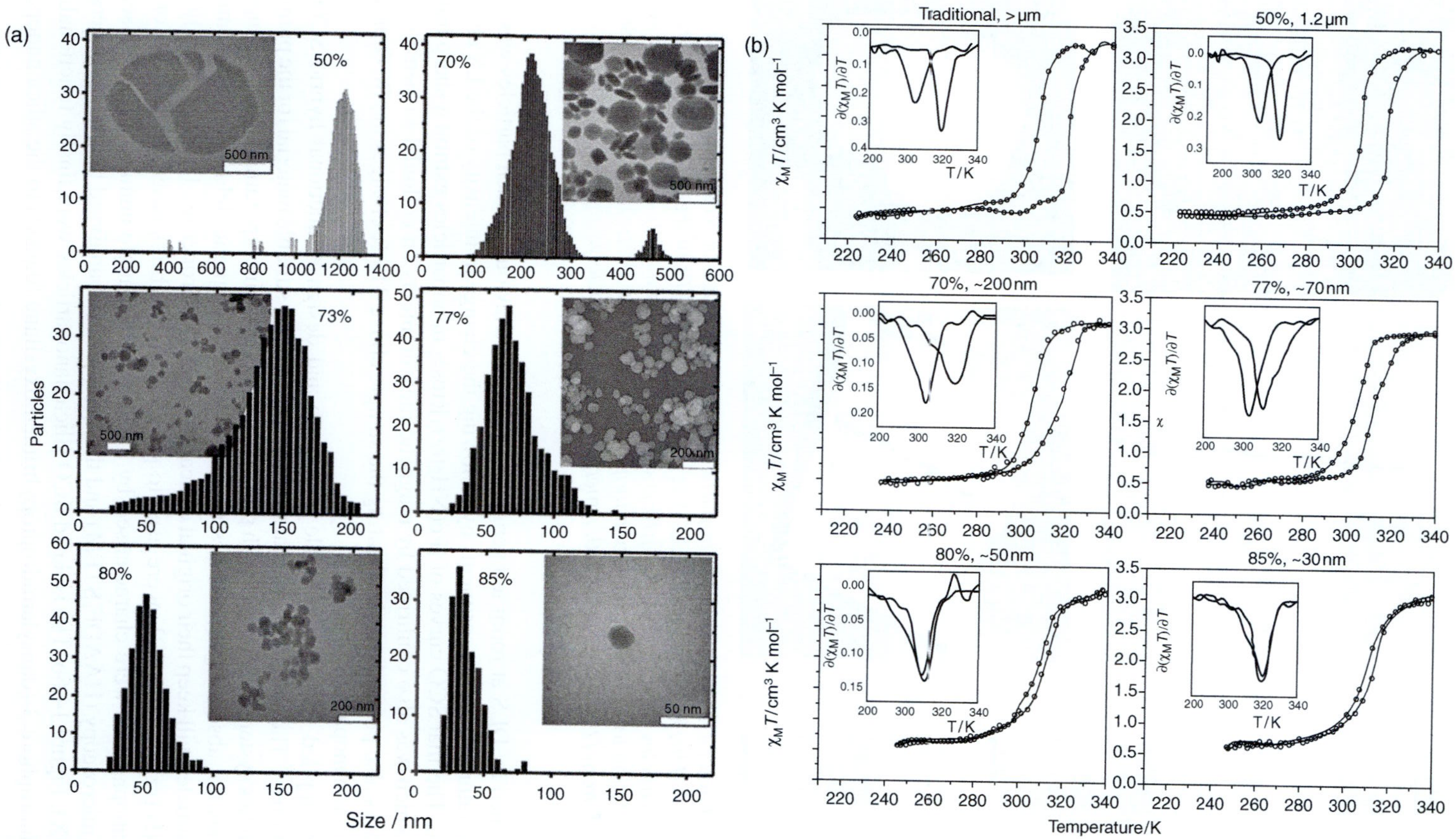

Figure 14.3 *Size distribution of [Fe(NH$_2$trz)$_3$]Br$_2$ nanoparticles: (a) Number of particles versus particle sizes from TEM images of nanoparticles prepared by a reverse micelle method as function of the surfactant/water percentage. Statistic distribution evaluated according TEM images on around 300 particles; (b) magnetic plots of [Fe(NH$_2$trz)$_3$]Br$_2$·y(surfactant) for the macroscopic particles prepared on traditional way and for nanoparticles depending of the surfactant/water ratio. Reprinted with permission from [23]. Copyright 2009 WILEY-VCH Verlag GmbH & Co. KGaA, Weinheim.*

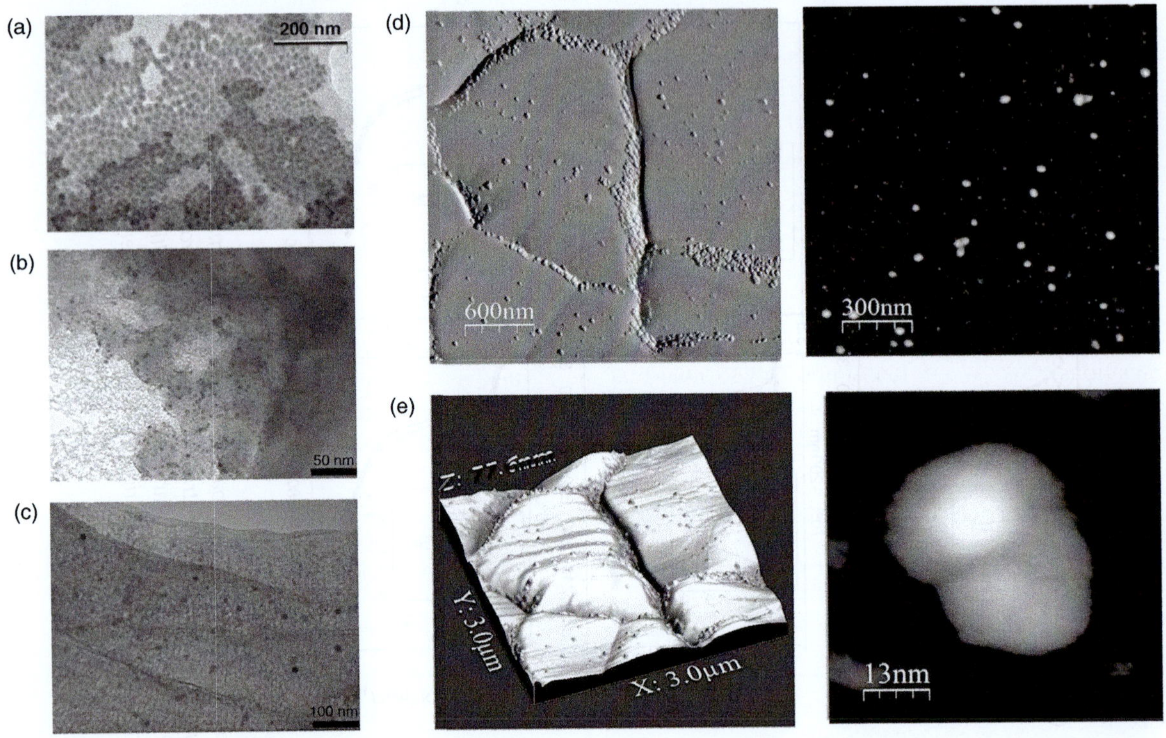

Figure 14.4 *Structural characterisation of SCO nanoparticles: (a) TEM image of NPs of [Fe(Htrz)$_2$(trz)](BF$_4$); (b) TEM image of NPs of [Fe(Htrz)$_{3-x}$(NH$_2$trz)$_x$](ClO$_4$)$_2$; (c) TEM image of NPs of [Fe$_{0.8}$Zn$_{0.2}$(Htrz)$_2$(trz)](BF$_4$); (d) AFM images of nanoparticles of [Fe(Htrz)$_2$(trz)](BF$_4$) on a gold substrate (left) and on a silicon substrate (right); (e) AFM high resolution topography image of nanoparticles of [Fe(Htrz)$_2$(trz)](BF$_4$) on a gold substrate. Reprinted with permission from [24]. Copyright 2010, American Chemical Society.*

This was further confirmed by DLS at room temperature. The SCO behaviour of such particles was investigated by variable temperature UV-vis experiments by following the change in intensity of the LS absorption peak centred at 540 nm. Thermal SCO curves of [Fe(NH$_2$trz)$_3$](tos)$_2$ nanoparticles exhibit a sharp transition around $T_{1/2} = 295$ K, which was also confirmed by magnetic measurements on colloidal suspensions of the same sample. The authors also pointed out that heating the sample above 313 K changes the SCO properties of the [Fe(NH$_2$trz)$_3$](tos)$_2$ nanoparticles.

Colacio and Herrera applied the same principle to prepare and design bifunctional hybrid SCO silica nanoparticles, Figure 14.6a.[28] The authors stated that silica is a particularly suitable material for the preparation of SCO silica nanoparticles systems because its high porosity allows for the incorporation of SCO compounds. Moreover, as silica does not absorb light and does not interfere with magnetic fields the SCO compounds inside the silica nanoparticles will keep their original optical and magnetic properties. [[Fe(Htrz)$_2$(trz)](BF$_4$)]$_n$ and doped [[Fe$_{1-x}$Zn$_x$(HTrz)$_2$(Trz)](BF$_4$)]$_n$ were used to produce three hybrids based on the ratio of Fe/Zn used. The resultant nanoparticles were characterised by TEM and high-angle annular dark-field-scanning transmission electron microscopy (HAADF-STEM) and their composition analysed by energy dispersive X-ray spectroscopy (EDX), Figure 14.6b. The isolated SCO silica nanoparticles were finally functionalised by grafting the organic fluorophore 3-(dansylamido)propyltrimethoxysilane (dansyl) to the silica nanoparticles surface.

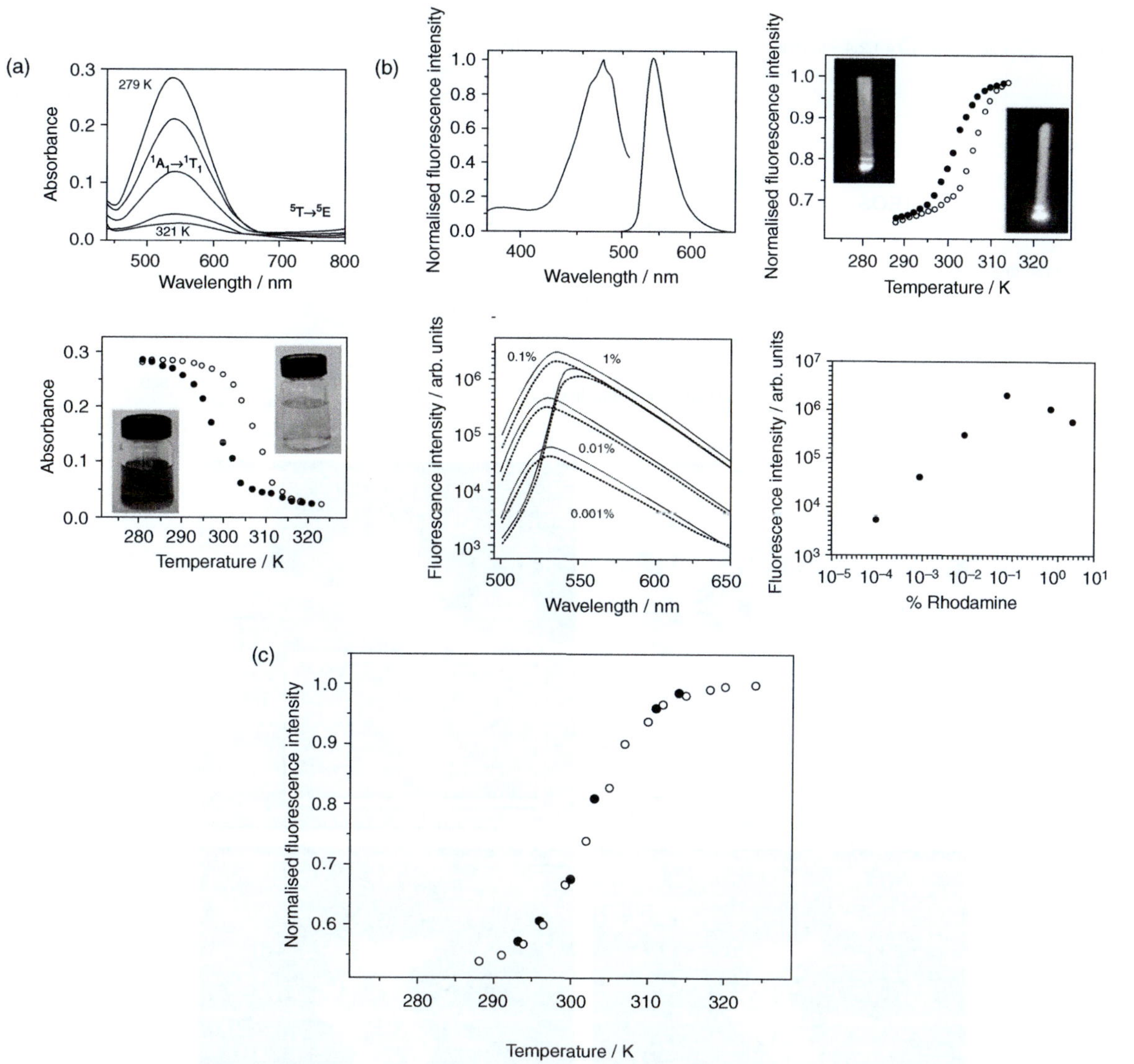

Figure 14.5 *Optical properties of $Fe(NH_2trz)_3(tos)_2$ nanoparticle suspensions: (a) (top) absorption spectra of an octane suspension of $Fe(NH_2trz)_3(tos)_2$ nanoparticles at selected temperatures in the cooling mode; (bottom) thermal variation of the absorbance of the suspension at 540 nm in the heating (open symbols) and cooling (closed symbols) modes, showing photographs of the sample at 295 K (violet) and 320 K (transparent); (b) (top-left) fluorescence excitation ($\lambda_{emission} = 540$ nm) and emission ($\lambda_{excitation} = 475$ nm) spectra of a rhodamine-110 doped (3%) $Fe(NH_2trz)_3(tos)_2$ nanoparticle suspension at room temperature; (top-right) thermal variation of the emission intensity at 540 nm in the heating (open symbols) and cooling (closed symbols) modes, showing photographs of a sample (0.001%) under white light excitation at 295 K and 320 K; (bottom-left) emission spectra of suspensions with diverse rhodamine concentrations. For each concentration the spectra were recorded successively at 288 K (solid line), 318 K (dashed line) and 288 K (dotted line); (bottom-right) fluorescence intensity maxima as a function of the rhodamine concentration (288 K); (c) thermal variation of the fluorescence intensity at 540 nm in the heating (open symbols) and cooling (closed symbols) modes for a 3 nm rhodamine-110 doped (0.01%) $[Fe(NH_2trz)_3](NO_3)_2$ nanoparticle suspension. Reproduced from [25] with permission of The Royal Society of Chemistry.*

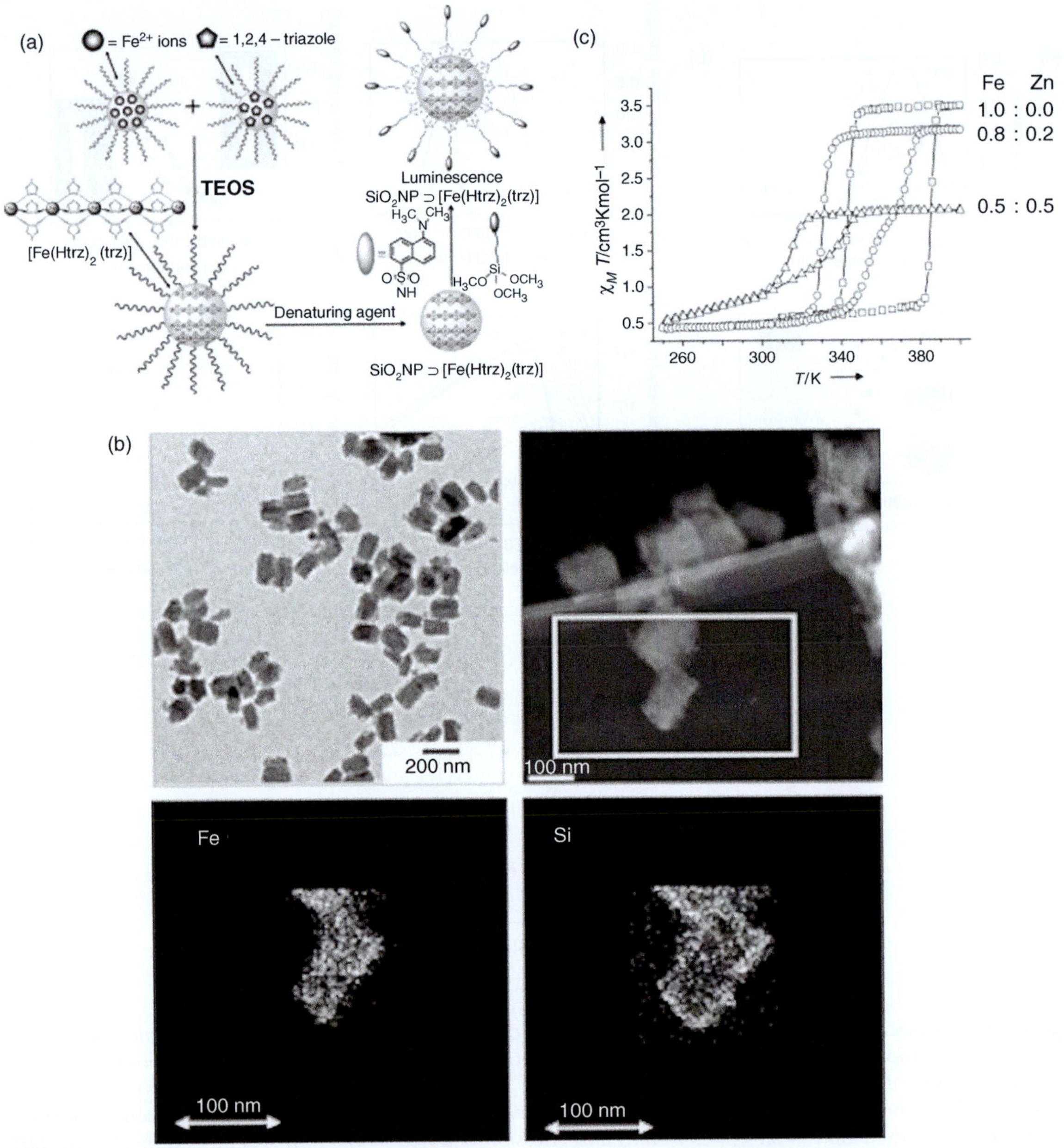

Figure 14.6 *SCO luminescent SiO_2 nanoparticles: (a) synthetic route; (b) TEM (top left) and high angle annular dark field-scanning transmission electron microscopy (HAADF-STEM) (top right) images of nanostructures of $[[Fe_{1-x}Zn_x(HTrz)_2(Trz)](BF_4)]_n$. EDX compositional maps of Fe (bottom left) and Si (bottom right) collected from the square area indicated on the HAADF-STEM image; (c) magnetic plots for different Fe/Zn ratios. Reprinted with permission from [28]. Copyright 2011 WILEY-VCH Verlag GmbH & Co. KGaA.*

The size of the particles produced depends on the ratio of Fe/Zn used with average dimensions ranging from side length and width of 147 ± 11.7 nm and 92.1 ± 12.1 nm when no Zn is used to 188.9 ± 37.4 nm length and 28.8 ± 7.1 nm width when the ratio of Fe/Zn is 0.5:0.5. Silica nanoparticles grafted with dansyl have shown no significant difference in dimension when compared to the ungrafted nanoparticles.

Magnetic measurements on the three different silica nanoparticles showed that SCO behaviour is not lost when these are incorporated in silica or even when the system is diluted with Zn, Figure 14.6c. For the undoped system, a complete and abrupt SCO was observed with $T_{1/2}\downarrow = 345$ K and $T_{1/2}\uparrow = 385$ K and a hysteresis loop of 40 K, which resembles the magnetic profile of the bulk sample.[29] For silica nanoparticles doped with Fe/Zn 0.8:0.2, SCO becomes less abrupt and shifts to lower temperatures with $T_{1/2}\downarrow = 330$ K and, in the warming mode, a stepped transition with $T_{1/2}\uparrow = 353$ K and $T_{1/2}\uparrow = 374$ K. Finally, for the system doped with Fe/Zn 0.5:0.5, the transition becomes rather smooth with $T_{1/2}\downarrow = 316$ K and $T_{1/2}\uparrow = 342$ K. Grafting the surface of SCO silica nanoparticles did not alter the magnetic profile of the nanoparticles. The authors also observed that the luminescence properties of the fluorophores grafted on the surface of the SCO silica nanoparticles can be adjusted by the spin state of the SCO polymer.

A very interesting approach has been described by Pillet and co-workers, who recently reported the preparation of nanoparticles of $[Fe(Htrz)_2(trz)](BF_4)$ using the reverse micelle technique and reverse hexagonal liquid crystalline phases.[30] The latter permitted the fabrication of polydispersive 1D nanoparticles which were characterised by variable temperature Raman spectroscopy. 1D nanoparticles have been shown to retain their SCO behaviour displaying abrupt SCO above room temperature with a hysteresis loop of 27 K.

14.2.1.2 *Hofmann-Like Derivatives*

SCO nanoparticles produced from Hofmann-like frameworks have also been prepared, Figure 14.7a.[31] Catala and Mallah also used the reverse micelle technique which consisted of mixing a reverse microemulsion of $Fe(BF_4)_2{\cdot}6H_2O$ and pyrazine with a second microemulsion containing $K_2Pt(CN)_4$. This was followed by precipitation with *p*-nitrobenzylpyridine in the presence of the surfactant sodium bis(2-ethylhexyl sulfosuccinate). HAADF-STEM showed that the size of the particles is dependent on the concentration of metal ions used with sizes varying from 7.7 nm for higher concentrations to 40 nm for lower concentrations of metal ions, Figure 14.7b. In the case of nanoparticles of Hofmann-like frameworks the SCO profile of the nanoparticles differs from the bulk powder, occurring at a lower temperature for the nanoparticle samples, Figure 14.7c.

Matrix-dependent cooperativity in anisotropic SCO nanoparticles of $Fe(pyrazine)Pt(CN)_4$ was investigated by Catala and Mallah.[32] The authors tried to prove that the environment of the nanoparticles plays a crucial role in the SCO process and that the cooperativity can be tuned by the matrix surrounding the nanoparticles. TEM images showed mainly square-like particles with 10 × 10 nm. Matrix effects on the SCO and cooperativity of the system have been studied using two coating agents: a calixarene based ligand bearing 8 pyridine functions; and an inorganic polymer silica with thin and thick SiO_2 shells. Magnetic studies have shown that for particles with a thin silica shell, a hysteresis loop of 15 K is observed with transition temperatures close to room temperature. For calix-covered nanoparticles, only a very small hysteresis loop of 2 K is observed and no hysteresis was shown for the diluted composites whether in silica or in PVP.

SCO nanoparticles and nanocrystals of a 2D coordination polymer derived from $[Fe(3\text{-}Fpy)_2M(CN)_4]$ (M = Ni, Pd, Pt and 3-Fpy = 3-fluoropyridine) have also been prepared using the coating polymer poly(vinylpyrrolidone) (PVP), which was added to an aqueous solution of the starting materials and subsequent mixtures of these.[33] The bulk material, nanoparticles and nanocrystals have very distinct magnetic profiles. Their SCO properties are drastically influenced by the dimensions of the crystal or the particle. Square-like nanocrystals of Ni, Pt, and Pd derivatives with dimensions 400 × 400 × 30 nm, undergo a first order spin-transition. Earlier, other techniques, such as a variation of the reverse micelle synthesis, have been

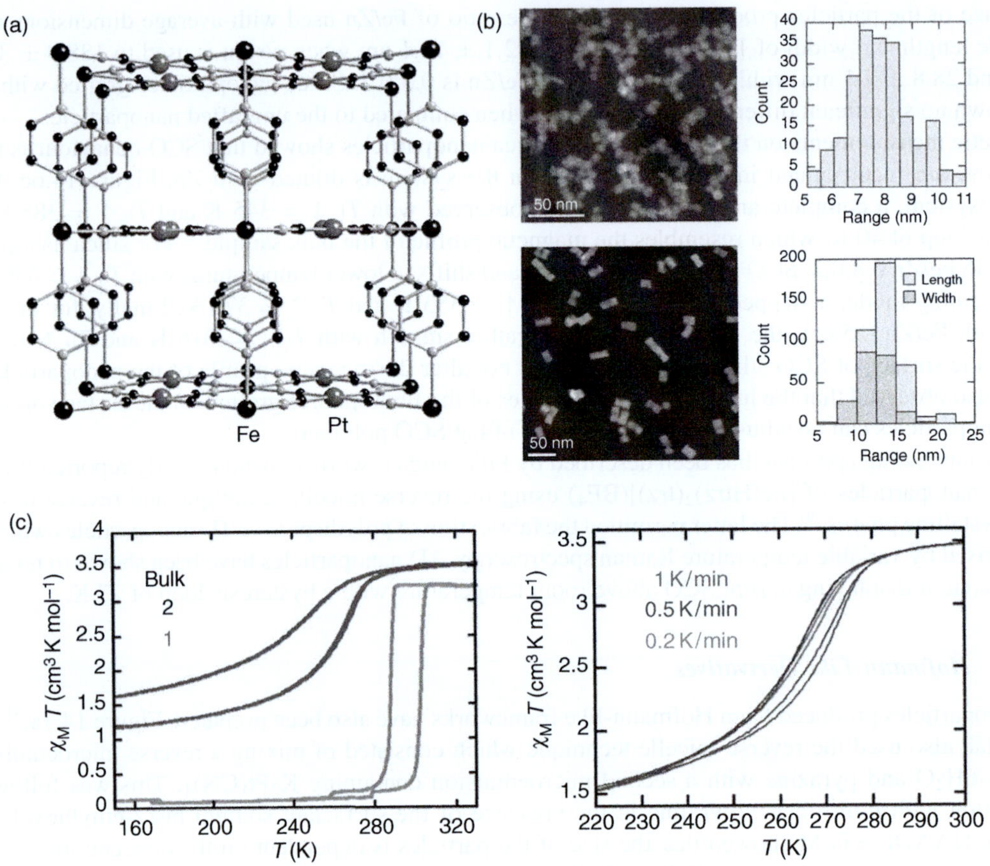

*Figure 14.7 SCO nanostructures derived from Hoffmann-like objects: (a) representation of a 3D network of Fe(pyrazine)[Pt(CN)$_4$];[8] (b) HAADF-STEM imaging and size distribution of particles (top, scale-bar 50 nm and bottom, scale-bar 50 nm); (c) magnetic plots of the objects obtained from two different concentrations: **1**, 0.14 mol L^{-1} and **2**, 0.08 mol L^{-1} and bulk compound at a sweep rate of 1 K min^{-1} (left) and for 0.08 mol L^{-1} at three different sweep rates (right). Reprinted with permission from [31]. Copyright 2008, American Chemical Society.*

used to give surfactant-free nanocrystals of [Fe(pz)Pt(CN)$_4$] displaying size-dependent SCO with hysteresis.[34] Ultra-small monodisperse nanoparticles of the 3D SCO coordination polymer [Fe(pyrazine)[Ni(CN)$_4$]] obtained using the biopolymer chitosan as matrix were also reported with these particles retaining the cooperative SCO behaviour.[35]

14.2.1.3 Pyridyl Derivatives

In 2010 Létard *et al.* investigated light-induced storage of information in nanoparticles.[20] Nanoparticles of [Fe(NCS)$_2$(bpe)$_2$] [bpe = 1,2-bis(4′-pyridyl)ethane] were prepared by the water-in-oil reverse micelle technique, varying the amount of surfactant used, Figure 14.8a. The reactant concentration was kept constant (0.05 M) and the mass of the nonionic polyoxyethylenic surfactant Ifralan D0205 was varied between [Fe(NCS)$_2$(bpe)$_2$]-25%, [Fe(NCS)$_2$(bpe)$_2$]-78%, [Fe(NCS)$_2$(bpe)$_2$]-85%. The consequences of particle size reduction on light-induced and thermally induced SCO properties have been investigated. TEM images of

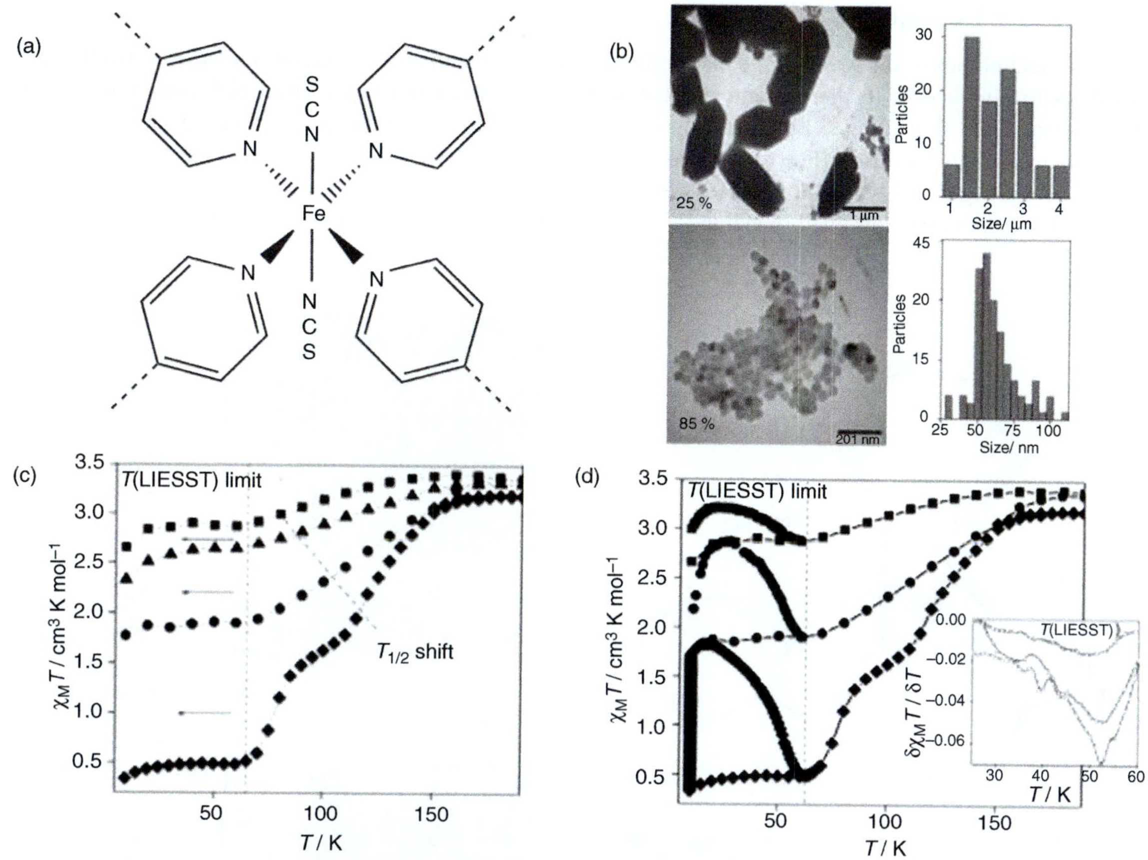

Figure 14.8 *Nanostructures from [Fe(NCS)$_2$(bpe)$_2$]: (a) structure of the compound used to produce nanoparticles; (b) (top) TEM images when 25% of surfactant was used (1–4 μm) and (bottom) when 85% of surfactant was used (30–80 nm) with respective size distribution histograms; (c) magnetic plot for bulk [Fe(NCS)2(bpe)$_2$] (♦, $T_{1/2}$ = 80/133 K), [Fe(NCS)$_2$(bpe)$_2$]-(25%-surfactant) (•, $T_{1/2}$ = 120 K), [Fe(NCS)$_2$(bpe)$_2$]-(78%-surfactant) (▲, $T_{1/2}$ = 110 K), [Fe(NCS)2(bpe)$_2$]-(85%-surfactant) (■, $T_{1/2}$ = 90 K); (d) magnetic plot for the thermally induced and light-induced SCO of bulk [Fe(NCS)$_2$(bpe)$_2$] (macroscale particles, ♦), [Fe(NCS)$_2$(bpe)$_2$]-(25%-surfactant) (microscale particles, •) and [Fe(NCS)$_2$(bpe)2]-(85%-surfactant) (nanoscale particles, ■). Reprinted with permission from [20]. Copyright 2010 WILEY-VCH Verlag GmbH & Co. KGaA.*

the particles showed that their size is dependent on the amount of surfactant used. This is not surprising, as it had been previously shown. For lower amounts of surfactant the particles show rod-like features with sizes ranging 1–4 μm. The size can be tuned down to 30–80 nm when 85% of surfactant is used, Figure 14.8b.

Unlike the bulk material, the magnetic profile of the nanoparticles shows a slow one-step SCO for all amounts of surfactant used with $T_{1/2}$ shifting to lower temperatures with increasing of the amount of surfactant, Figure 14.8c.

Light-induced excited spin state trapping effect (LIESST) was investigated by sample irradiation at 10 K with light working at $\lambda = 530.2$ nm. A light-induced excitation of the LS state to a metastable HS state was observed with the degree of conversion being dependent on the amount of surfactant used and consequently the size of the particles, Figure 14.8d.

14.2.2 Sol-Gel Techniques

Alternative techniques to the reverse micelle method to prepare nanoparticles have been used by Boillot and co-workers to give access to thermo and photoswitchable SCO nanoparticles of an Fe^{II} complex trapped in transparent silica thin films.[36] For that the authors used the sol-gel technique to fabricate nanoparticles of discrete molecules based on $[Fe((mepy)_3tren)](PF_6)_2$ ($(mepy)_3tren$ = tris(4-[(6-Me)-2-pyridyl]-3-aza-3-butenyl)amine), Figure 14.9a. Nanoparticles were produced by mixing the crystalline solid in a previously aged solution containing tetramethoxysilane (TMOS), dimethyldimethoxysilane (DMDMS), acetone and

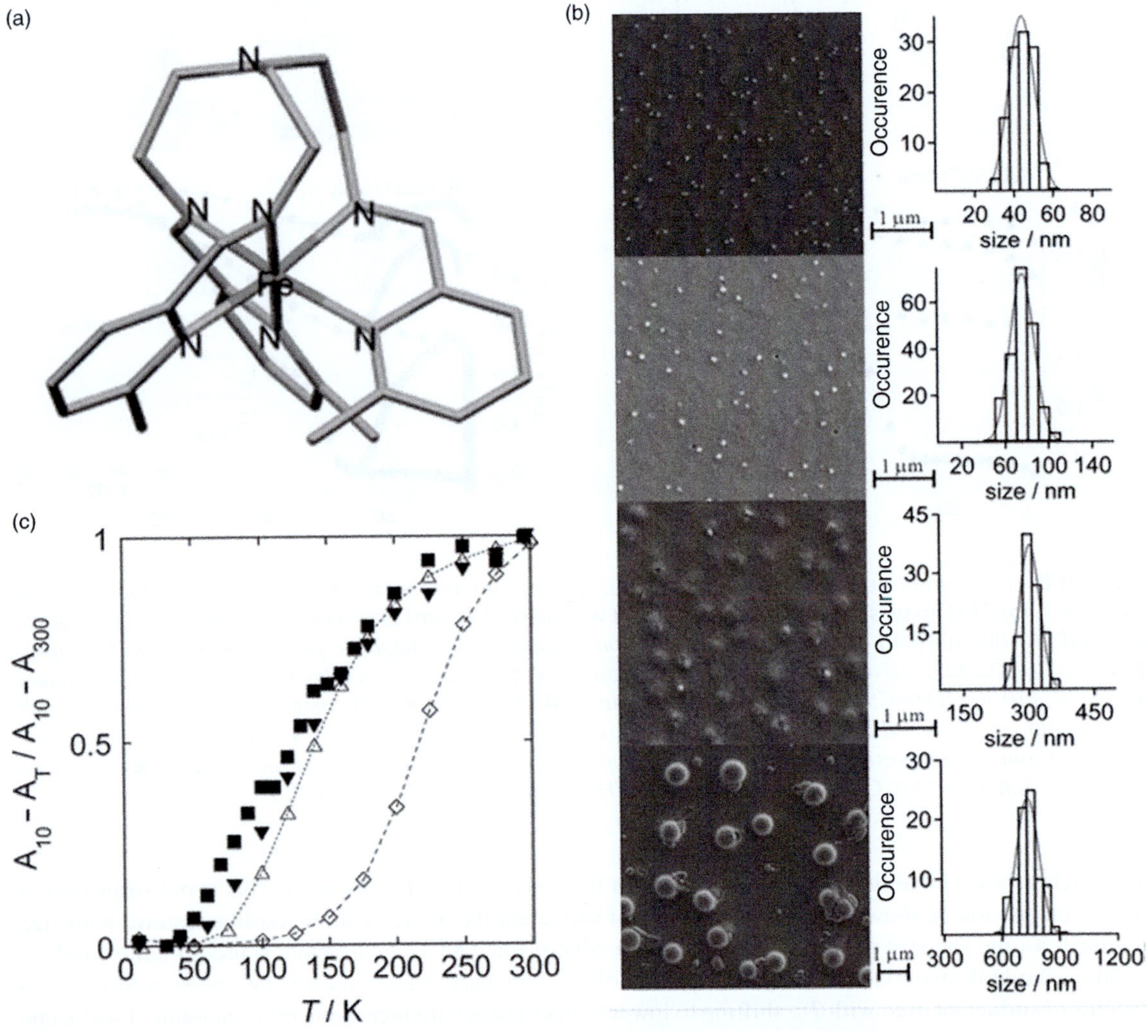

Figure 14.9 *Sol-gel technique to produce SCO nanoparticles: (a) molecular formula; (b) SEM images of $[Fe((mepy)_3tren)](PF_6)_2$ nanoparticles embedded in silica thin films; (c) relative area variation of MLCT absorption extracted from the Vis absorption data for 47 nm nanoparticles (▼), 730 nm nanoparticles (■), precipitated powder (Δ) and crystalline powder (◇). The ratio $(A_{10} - A_T)/(A_{10} - A_{300})$ corresponds to the fraction of SCO compounds in the HS state. Reproduced from [36] with permission of The Royal Society of Chemistry.*

H_2O. The obtained solution was then spin-coated on a glass slide where rapid evaporation of solvents quenches the growth of the particles in the silica thin film. Nanoparticles with different dimensions were obtained depending on the aging time of the sol, giving particles with dimensions of 730 (± 80), 300 (± 40), 72 (± 16) to 47 (± 10) nm after 1 h, 2, 10 and 15 days, respectively, Figure 14.9b.

Variable-temperature magnetic measurements were carried out both for the microcrystalline and the precipitated powders. Although the nanoparticles show thermochromic behaviour, magnetic measurements on thin films of nanoparticles prepared by spin coating were inconclusive and the authors used variable temperature UV-vis to probe the spin-transition in the nanoparticles, Figure 14.9c. These have shown that the nanoparticles have a magnetic behaviour very similar to that of the precipitated powders with $T_{1/2} = 130 \pm 5$ K resulting in decreased cooperativity when compared to the crystalline sample. The authors also observed that the magnetic profile of the nanoparticles is size independent. Like the precipitated powders, the nanoparticles also show LIESST effect.

Faulmann *et al.* prepared nanoparticles of the SCO complex $[Fe(Htrz)_2(trz)](BF_4)$ in xerogel transparent composite films.[37] The dispersion and consecutive inclusion of the iron complex in a silica matrix prepared from TMOS or tetraethoxysilane (TEOS) afforded monoliths or films with a violet colour at room temperature, which turns white above 380 K. Particles with sizes ranging from 1 to 5 nm were imaged by TEM and their composition was confirm by EDX. Magnetic measurements showed that for films of nanoparticles in TMOS the transition temperatures shifted towards higher temperatures with the hysteresis loops also getting narrower with the number of cycles like the bulk sample. The SCO of crushed thin films and as prepared thin films was further confirmed by variable temperature Raman.

14.3 Thin Films

The engineering of SCO molecules towards processable better: information processing devices is the main goal of applied SCO research. One of the most active routes pursued has been the deposition of such SCO active molecules on surfaces. For this, several techniques have been used (Langmuir–Blodgett (LB) deposition, surface-assisted molecular self-assembly, dip coating/drop casting, spin coating, complexes imbedded into matrixes), although only a few of these have been successful. One technique which has been most used to fabricate thin films of SCO active molecules is LB deposition. This and the other surface deposition techniques are reviewed below.

14.3.1 Langmuir–Blodgett Deposition

A single layer of molecules on a liquid–air interface is termed a *Langmuir monolayer*, and after transfer, is termed an *LB film*. It may be possible to transfer the monolayer to a surface if the surface substrate is immersed into and emmersed carefully from the trough. More than one layer may be transferred by successive immersions and emersions, although this is not always possible. Upon LB layer formation on a surface three different film architectures can result. Y-type multilayers are most common and X-type and Z-type films are rare.[38]

The first attempts to form a SCO LB film were made by Kahn and Ruaudel-Texier using an amphiphilic derivative of the well-known $[Fe(phen)_2(SCN)_2]$ complex, Figure 14.10a.[39, 40]

The LB film of this complex was transferred onto calcium fluoride and quartz substrates accomplishing a successful transfer of up to 200 layers, Figure 14.10b. The type of film was inconclusive and of type X or Z and has been shown by variable temperature IR to exhibit thermal SCO with $T_{1/2} = 260$ K and a hysteresis loop of ca. 4 K, Figure 14.10c.[40]

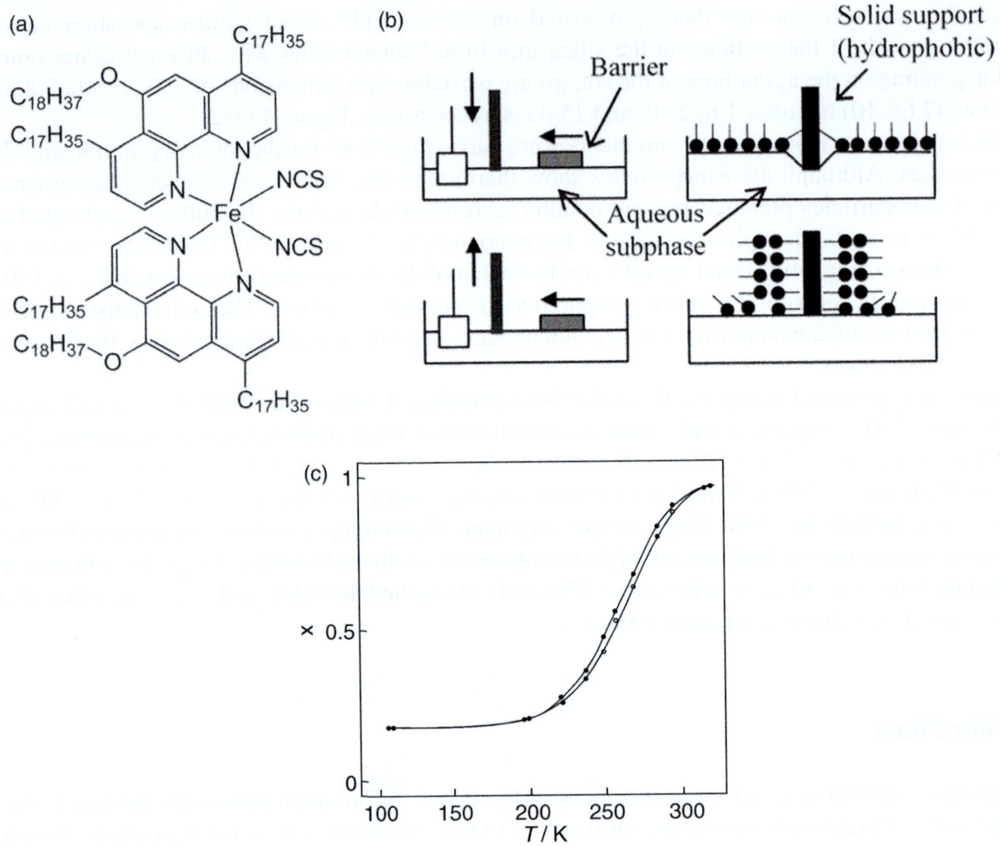

Figure 14.10 *Production of SCO LB films: (a) molecular structure of [Fe(phen)$_2$(SCN)$_2$] derivative;*[39, 40] *(b) Langmuir–Blodgett deposition on a surface starting with a hydrophobic substrate;*[38] *(c) temperature dependence of molar fraction of HS species in both cooling and warming modes based on vibrational modes changes followed by IR. Reprinted from [38] with permission from American Chemical Society 2004 and from [40] with permission of The Royal Society of Chemistry.*

Later in 1998, Mingotaud and co-workers studied the formation of Langmuir and LB films on a derivatised [Fe(bpy)$_2$(SCN)$_2$] with bpy = 2,2′-bipyridine. Langmuir films were found to be very unstable when formed at the air–water interface. The problem was overcome when a mixed liquid phase was used. The stability of the Langmuir films increased considerably when a water:DMF mixture was used as liquid phase, and a further increase in stability was achieved when a mixture of water:DMF:KNCS was used. Y-type LB multi-layered films were transferred onto a calcium fluoride substrate and characterised by variable temperature IR spectroscopy. The SCO showed a quite distinctive profile if the magnetic measurements on the film were performed below 340 K (gradual and incomplete with $T_{1/2} = 292 \pm 7$ K), or by heating it above 340 K. The latter transition is more abrupt and complete than the former cycle, but is irreversible.[41, 42]

A third Fe(II) LB film was obtained using a small variation of the bpy ligand described in the previous example. Langmuir films obtained using an aqueous solution of KSCN as subphase and complexes with a ligand derived from bpy with R groups $(CH_2)_{12}$-$(CF_2)_5$-CH_3 and Me were transferred onto a Mylar substrate.

Multilayered (>1200 layers) LB films were also shown to display thermal SCO with hysteresis. Furthermore, the complex was shown to present the LIESST effect, showing a more efficient LIESST process when compared to the powdered sample.[43–45]

Armand *et al.* used the same approach to fabricate LB films of the alkylated coordination polymer $[Fe(C_{18}trz)_3]X_2$ (X = perchlorate, triflate, tosylate).[46] The attempts were not successful due to the instability of the complexes at the air–water interface.[46] This is not surprising, as LB films of SCO complexes have often exhibited assembly problems, due to chemical instability of the iron complexes at the air–water interface and consequent hydrolysis of the coordination bonds, or oxidation of the metal centres. Attempts to minimise the hydrolysis have included the use of a DMF-water mixture as the subphase instead of pure water, and the use of semifluorinated chains to avoid organic solvents. A very interesting approach by Roubeau towards stabilisation of the Langmuir films at the air–water interface was the use of aqueous solutions of metal salts as a subphase. Solutions of $Co(NO_3)_2$, $Ni(NO_3)_2$ and $Zn(NO_3)_2$ all proved successful. These salt solutions permitted closer packing and formation of stable Langmuir monolayers, which were suitable for the fabrication of up to 1100 LB multilayers.[47] When this technique was applied to SCO complexes, partial thermal SCO in $[Fe(C_{18}trz)_3](NO_3)_2$ polymers was detected by magnetic measurements of 400–1100 LB multilayers deposited from a subphase containing $Fe(NO_3)_2$. For subphases containing other metal salts, no SCO was observed in LB films of $[Fe(C_{18}trz)_3](NO_3)_2$ polymers.[47, 48]

A different and more recent approach was developed by Kurth and co-workers where they fabricated LB multilayers using a polyelectrolyte-amphiphile complex (PAC). PAC were formed by self-assembly of a metallo-supramolecular polyelectrolyte (MEPE) based on the ditopic bis-terpyridine 1,4-bis(2,2′:6′,2″-terpyridine-4′-yl)benzene and dihexadecyl phosphate (DHP). LB films of PAC with 11–15 layers have been prepared and structurally characterised.[49, 50] Unlike the bulk material, the SCO of the PAC LB films is incomplete but reversible.[51, 52]

Brooker and Albrecht have used the LB technique to prepare Langmuir films of discrete Fe(II) complexes which have been shown to be very unstable.[53] More recently they have shown that functionalisation of the 3,5-di(2-pyridyl)-4*H*-1,2,4-triazole (dpt) ligand with a C_{16} aliphatic chain forms a SCO Fe(II) complex, $[Fe(C_{16}dpt)_2(NCS)_2]$, with $T_{1/2} = 290$ K. This Fe(II) complex was also found to form stable Langmuir monolayers at the air–water interface but no attempts to form LB films were reported.[54]

Recently Albrecht and Morgan extrapolated these same principles to Fe(III), Co(III) and Mn(III) amphiphiles. The authors reported the formation of stable Langmuir films with redox stable Fe(III) amphiphilic complexes derived from $[Fe(sal)_2trien]^+$ (sal = salicylaldehyde, trien = triethylenetetramine), Figure 14.11.[55] Although these complexes do not show SCO in the solid state, the longer C_{18} complex shows an abrupt SCO in solution.[56] Complexes with alkyl chains with a length ranging from six to eighteen carbons were tested and their ability to form Langmuir films was investigated. The authors observed that only the longer C_{18} complexes formed suitable Langmuir films for surface transfer. The transfer process was limited to four layers and the spectroscopic characterisation was inconclusive with respect to occurrence of LB film SCO.

The extension of similar amphiphilic systems to Mn(III) was accomplished by functionalisation of a ligand know to promote SCO with Mn(III) ions.[57] It was observed that in these ions the SCO is sensitive to the length and position of the alkyl chains with all spin transitions gentle and incomplete. It was also observed that the complexes with longer alkyl chains (C_{12} and C_{18}) formed densely packed Langmuir monolayers with intermolecular contacts occurring at earlier stages of compression for the C_{18}-functionalised complexes. Stability tests using the C_{18}-functionalised complexes have shown that only the complex bearing *cisoid* alkyl chains were able to form reasonably stable Langmuir films over extended periods of time. Attempts to transfer Langmuir films were not successful resulting in only partial transfer due to significant desorption of the material during downstrokes.

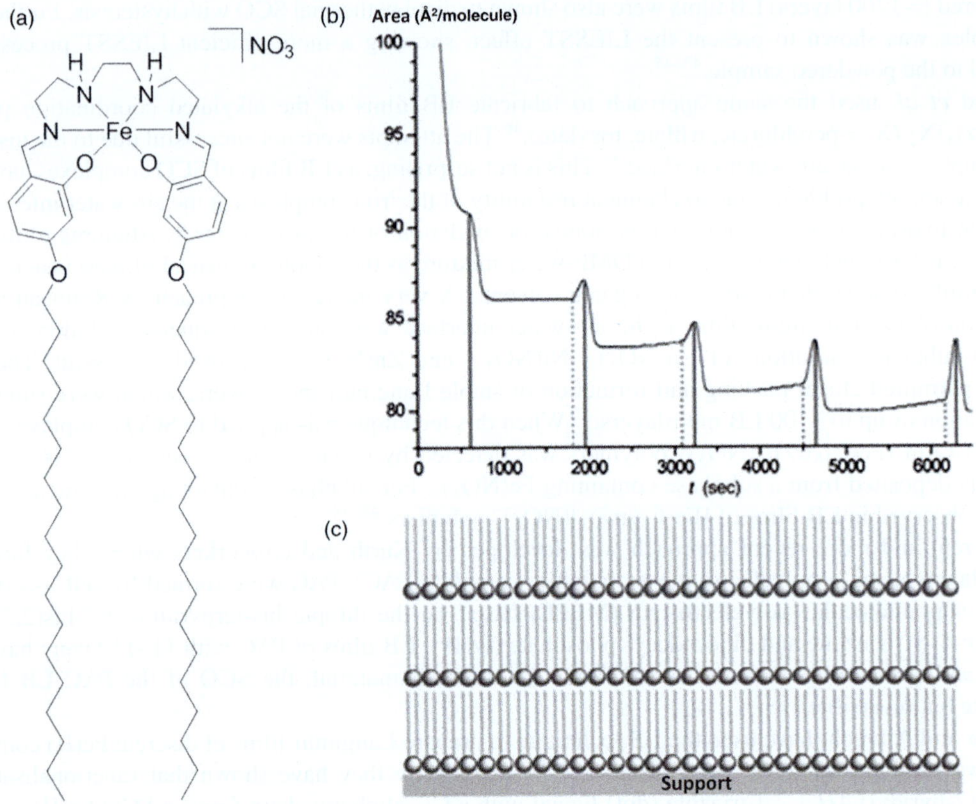

Figure 14.11 Production of [Fe(R-sal)$_2$trien]$^+$ LB films: (a) C_{18} amphiphilic compound used to fabricate LB films; (b) Langmuir layers of [Fe(C_{18}-sal)$_2$trien]NO_3, transferred onto glass support; (c) representation of type of LB film formed. Reproduced from [55] with permission of The Royal Society of Chemistry.

14.3.2 Surface-Assisted Molecular Self-assembly

A more successful approach than the LB technique is the multilayer sequential assembly (MSA) technique. This technique consists of the deposition/assembly of sequential layers by coordination bonds.[58, 59] Thus, Bousseksou and co-workers have prepared thin films with multilayers of the Hofmann-type SCO compound, [Fe(pz)M$(CN)_4$] where M = Ni, Pd, or Pt using the MSA method by epitaxial growth on gold surfaces, Figure 14.12a.[60]

The thin films formed by MSA retained the SCO characteristic of the bulk material. The SCO in the multilayers shows a 25 K wide hysteresis window centred around 310 K but the transition is less abrupt than in the powdered sample, Figure 14.12b and 14.12c.

14.3.3 Diverse Techniques

Spin coating is a widely used and very versatile technique with several advantages. It is simple, cost effective, produces a low amount of waste, is rapid and highly reproducible and produces homogeneous films on a substrate. Spin coating is widely used in fabrication of inorganic nanostructure self-assemblies but major drawbacks are the limitation by the solvent and the lack of order after evaporation of the solvent.[61]

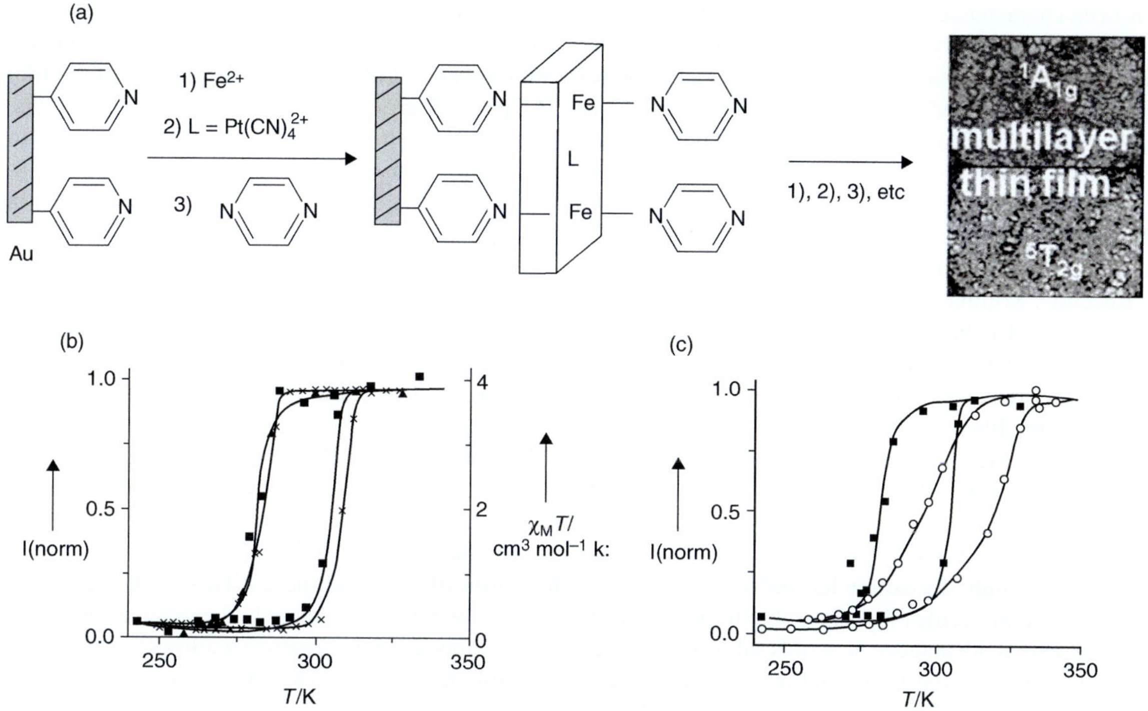

Figure 14.12 *Surface confined Hofmann-type SCO assemblies: (a) representation of SCO thin film formation of [Fe(pz)Pt(CN)$_4$].; (b) temperature dependence of the χ_mT product (×) and the normalised Raman intensity ratio (■) for [Fe(pyrazine)Pt(CN)$_4$] powder on cooling and heating modes. (c) Temperature dependence of the normalised Raman intensity ratio for [Fe(pyrazine)Pt(CN)$_4$] powder (■) and film (○) samples on cooling and heating modes. Reprinted with permission from [60]. Copyright 2006 WILEY-VCH Verlag GmbH & Co. KGaA.*

The spin coating technique has also been applied to SCO complexes and the first report on attempts to fabricate a SCO thin-film using this technique was by Matsuda and Tajima.[62] [Fe(dpp)$_2$](BF$_4$)$_2$ (dpp = 2,6-di(pyrazolyl)pyridine) was chosen, which shows an abrupt spin-transition centred at 259 K with a 3 K hysteresis loop in the solid state, and a gradual transition in acetone solution. Smooth 30 nm thick films were prepared on a glass substrate from a saturated acetonitrile solution of the complex. The SCO property of the prepared film was investigated by variable temperature UV-vis spectroscopy and SQUID magnetometry and shows that these films show an abrupt and reversible SCO at ca. 260 K. Conductivity measurements performed on films sandwiched by indium-thin oxide (ITO) and aluminium over the SCO transition temperature have shown that there is a small change in resistivity with change in spin state.

As a follow up, the authors utilised the same SCO molecule and spin coating method to embed a SCO complex into a light-emitting layer of an EL device of chlorophyll *a*. By doping chlorophyll *a* with a SCO complex the authors achieved a drastic change of the EL intensity accompanying the SCO of [Fe(dpp)$_2$](BF$_4$)$_2$.[63]

A different approach has been developed by Kojima and co-workers, in which a Nafion film was doped with the SCO complex [Fe(Htrz)$_3$].[64] The properties of thermal spin-transition as well as photoinduced SCO on the film have been reported. The Fe(II)-doped Nafion film was prepared by immersion of a Nafion film in an aqueous solution of an iron salt followed by immersion in a methanolic solution of 1,2,4-1H-triazole. The magnetic profile of the film, which shows a SCO around $T = 250$ K with a small hysteresis loop of about 5 K, has been characterised in a separate report by Kojima *et al.* published around the same time.[65] Photoinduced SCO has been observed at low temperatures with an efficiency of conversion of 70% and decay dynamics suggesting that stochastic nucleation dominates the relaxation process. Later, the SCO behaviour of the film was further confirmed by Mößbauer spectroscopy.[66] The *P-T* phase diagram for the photoexcited steady state of the Nafion-[Fe(Htrz)$_3$] film has also been determined[67] and the same investigations have been extended to the parent system [Fe(NH$_2$-trz)$_3$]-Nafion. The prepared film shows a smoother SCO centred at 198 K.[68] To finalise the extensive work developed on these Nafion-[Fe(R-trz)$_3$] films, Kojima and co-workers reported in 2005 the preparation of transparent SCO Nafion-[Fe(Htrz)$_3$] film showing thermochromic behaviour changing from transparent (HS) to purple (LS).[69]

Recently, Kojima *et al.* demonstrated that the same principle can be applied to other SCO systems.[70] The elected candidate between the broad range of examples of Fe(II) SCO complexes has been $[Fe^{II}(diAMsar)]^{2+}$ (diAMsar = 1,8-diaminosarcophagine) where, as in the powdered samples, [Fe^{II}(diAMsar)]-Nafion films show a thermochromic dependence with the *p*H which is correlated with the SCO property of either the powder or the films.

Boillot reported poly(methyl methacrylate) (PMMA) polymeric films doped with the Fe(II) SCO complex [Fe(stpy)$_4$(NCSe)$_2$] (stpy = styrylpyridine).[71] Films were prepared from a mixture of PMMA with increasing amounts (2–20 wt %) of the [Fe(stpy)$_4$(NCSe)$_2$] complex. Transparent and semi-rigid PMMA doped thin films showed SCO at $T = 135 \pm 25$ K, which is comparable to that of the crystalline sample ($T_{1/2} = 163$ K).

Finally, vacuum deposition has been used to produce thin films of SCO complexes. This method can only be applied to molecules that are stable to decomposition under the ultra-high vacuum deposition conditions and has several advantages: it avoids dilution of the complexes embedded in a matrix; it allows exact control of the thickness of the films; and it can be applied to patterning techniques. Despite all these advantages, until now only two examples have been reported.

The first example was reported by Shi and Beaurepaire.[72] The authors described high-quality thin films of [Fe(phen)$_2$(SCN)$_2$] deposited on silicon or glass substrates by evaporation under high vacuum. However, the film morphology was shown to be sensitive to air and became coarse under ambient conditions. The composition of the film and confirmation of its integrity was demonstrated by X-ray photoelectron spectroscopy (XPS) on a 280 nm thick [Fe(phen)$_2$(SCN)$_2$] film. Optical transmission measurements were used to investigate the electronic structure of the films and the authors claimed that for films ranging from 7 to 530 nm the electronic structure of these remains practically unaltered. Moreover the magnetic profile of a 280 nm thick film was characterised by SQUID magnetometer measurements, showing that thicker films retain the SCO property with $T_{1/2} = 175$ K.

A more recent report by Quandt and Tuczek exploits the formation of thin films by high vacuum deposition method of SCO complexes exhibiting a LIESST effect.[73] The complexes studied were [Fe(H$_2$Bpz$_2$)$_2$(phen)] (H$_2$Bpz$_2$ = bis(pyrazolyl)borate) and [Fe(H$_2$Bpz$_2$)$_2$(bipy)] (bipy = 2,2′-bipyridine). In the solid state, [Fe(H$_2$Bpz$_2$)$_2$(bipy)] shows a SCO centred at $T_{1/2} = 159.5$ K and a $T_{LIESST} = 52$ K, while [Fe(H$_2$Bpz$_2$)$_2$(phen)] shows a SCO centred at $T_{1/2}\uparrow = 165.0$ K, $T_{1/2}\downarrow = 162.7$ K and a $T_{LIESST} = 44$ K.[7] Thin films were deposited on ITO-coated glass or polymer tape and the thicknesses of the films were determined by a profilometer, their surface quality was investigated by AFM. A thin film of [Fe(H$_2$Bpz$_2$)$_2$(phen)] was prepared with a thickness of 480 nm and a rms roughness of 3.15 nm and a second thin film of [Fe(H$_2$Bpz$_2$)$_2$(bipy)] was also prepared with a thickness of 410 nm and rms roughness of 3.7 nm. The thermal and light-induced spin-transitions of the thin films were monitored by UV-vis absorption spectroscopy.

From the UV-vis spectra, the authors deduced the spin-transition temperatures for each of the thin films. For thin films of [Fe(H$_2$Bpz$_2$)$_2$(phen)] the $T_{1/2} = 155$ K ($\pm$ 1 K) which is comparable with the transition temperature of the solid sample ($T_{1/2}\uparrow = 165.0$ K, $T_{1/2}\downarrow = 162.7$ K). For thin films of [Fe(H$_2$Bpz$_2$)$_2$(bipy)] the $T_{1/2} = 166$ K ($\pm$ 1 K) which is also close to the transition temperature of the solid sample ($T_{1/2} = 159.5$ K).

LIESST-effect studies on thin films of both complexes were performed by irradiation of the film with a 525 nm wavelength light. For thin films of [$Fe(H_2Bpz_2)_2$(phen)], $T_{LIESST} = 44$ K with an efficiency of conversion to metastable HS of 83%. For films of [$Fe(H_2Bpz_2)_2$(bipy)], $T_{LIESST} = 51$ K with an efficiency of conversion to metastable HS of 85%.

14.4 Surface Patterns

The search for practical applications within the field of SCO either by insertion of active molecules into devices or by constructing devices out of SCO materials have brought together chemists and material scientists. In a different but complementary approach to those described above, Ruben and Cavallini, Molnár and Bousseksou, and finally Vieu, have investigated the possibility of depositing SCO compounds on surfaces through patterning techniques.

14.4.1 Surface Patterns of Spin-Crossover

Ruben *et al.* used discrete molecules to reproduce the pattern of a CD through lithographic techniques. For this the authors considered that the information generated in a SCO switching event occurs within the nanometric regime, thus the interfacing of the molecular switching units with the microscaled device environment would be of crucial importance. Considering this, the elected candidate for patterning was the neutral SCO complex *cis*-bis(thiocyanato)bis(1,10-phenanthroline)iron(II) which shows a thermal spin-transition centred at 176K, Figure 14.13a.[75, 76]

The techniques used to deposit the SCO molecules on surfaces were Micro Injection Molding in Capillaries (MIMIC) to fabricate micrometric stripes and Lithographically Controlled Wetting (LCW) to pattern submicrometric and nanostructures, Figure 14.13b. It has been observed that the SCO compound crystallises into micro- and nanowires or dots. Such patterns were characterised by AFM, polarised optical microscopy, grazing incidence X-ray diffraction (GIXD) and Raman spectroscopy.

AFM studies of structures of the SCO compound printed by MIMIC on silicon revealed the formation of strip-like structures (micrometric stripes) 1 μm in width, similar to the features of a stamp, Figure 14.13c (left). Optical images obtained by polarised microscopy indicated behaviour typical for optically anisotropic materials exhibiting birefringence, Figure 14.13c (centre and right). In particular, the microstripes appeared homogeneously coloured indicating that their thickness is almost constant over the entire stripe but changes between different stripes. Based on these observations it was deduced that confined deposition by MIMIC induced a coherent, long-range order along the length of the stripes. To obtain features in the nanometre regime the LCW technique was used. Structures with a resolution of approximately 160 nm were achieved, by estimation of the full width at half-maximum; below this limit the stamp features were not perfectly replicated.

The structure of nanopatterned films was investigated by GIXD. Azimuthal scans for both 010 and 001 reflections, forbidden for the original structure, were measured, indicating that the molecular arrangement (with loss of symmetry elements) is slightly different from that in the bulk, Figure 14.13d. Finally, the authors aimed to pattern SCO representing information storage. For this, films of [$Fe(phen)_2(SCN)_2$] were moulded by LCW into a logic pattern replicating a recorded compact disk. The imaging of the patterned film by dark-field optical microscopy and AFM revealed an ordered array of crystallites of [$Fe(phen)_2(SCN)_2$]. The SCO was probed by variable temperature Raman performed on freshly prepared films, Figure 14.13e. Freshly prepared films exhibited Raman spectra very similar to those of the powder retaining the SCO property.[75] This property was found to be unstable over time due to aging of the micro/nano architectures.

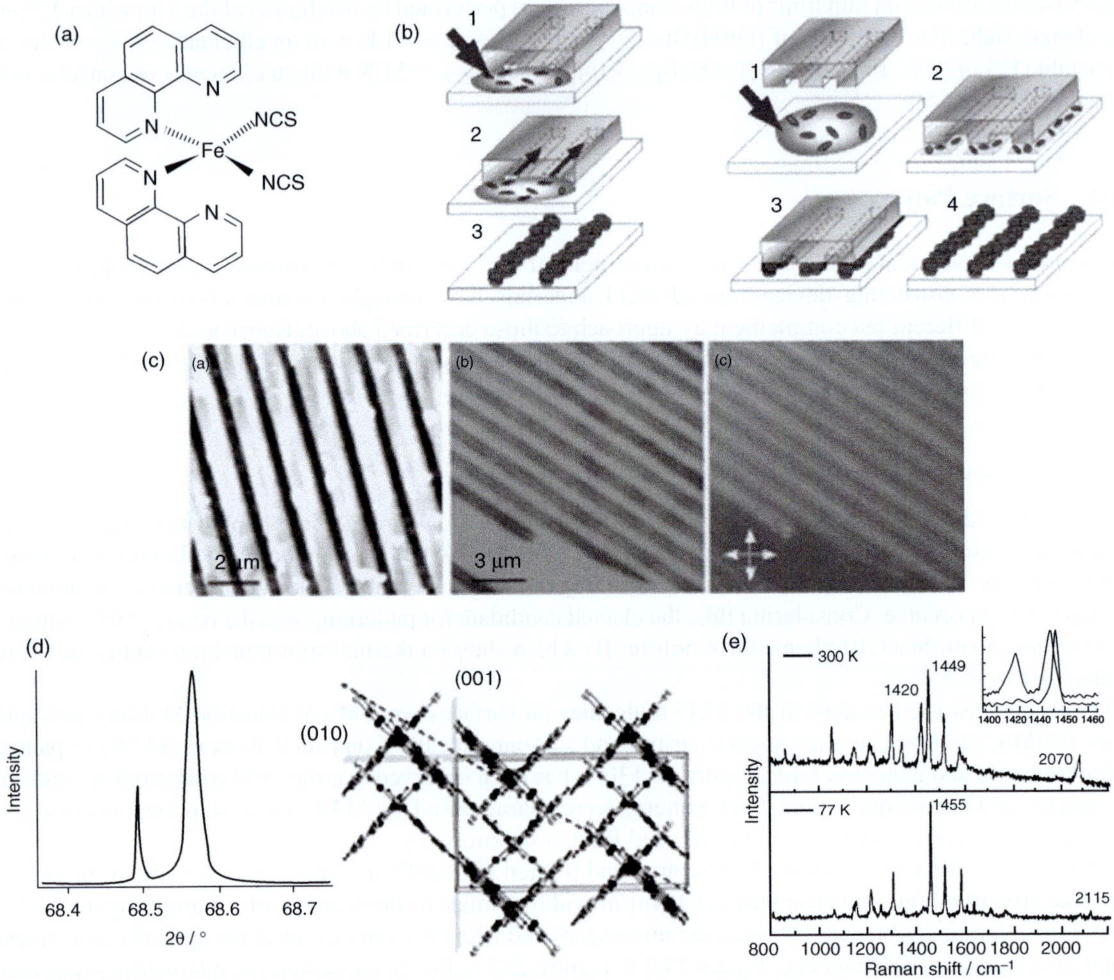

Figure 14.13 Stamping of SCO nanostructures: (a) molecular structure of compound used to create patterns; (b) representations of the processes used to patterning SCO molecules; (c) microstripes of [Fe(phen)$_2$(SCN)$_2$] fabricated by MIMIC on a silicon surface: (left) AFM topography (z scale is 50 nm), (middle) optical micrographs recorded with unpolarised light, (right) micrographs recorded with crossed polars oriented along the axes of the image; (d) GIXD azimuthal scan for the (010) of microstripes printed by MIMIC and respective top view of the 3D structure of [Fe(phen)$_2$(SCN)$_2$] in the microstripes, as oriented in one domain. The red arrow indicates the orientation of the microstripes; (e) Raman spectra of freshly prepared films, at 300 and 77 K with SCO evident from the vibrational modes at 1449, 1420 and 2070 cm^{-1}. Reprinted with permission from [75]. Copyright 2008 WILEY-VCH Verlag GmbH & Co. KGaA, and [76] American Chemical Society. Copyright 1964.

More recently, the same authors built on the experience obtained with the previous system, to fabricate thin deposits and patterns of a room-temperature-switchable 1D SCO compound which crystallises to form an infinite 1-D chain exhibiting a SCO at $T = 286$ K.[77] With the aim of controlling the spatial distribution and the morphology of thin deposits, the authors used the LCW technique. The thin films and patterns prepared were characterised by AFM, polarised optical microscopy, XRD and Raman spectroscopy. Patterns of the

SCO compound prepared by LCW revealed the formation of stripe-like structures of 1 μm width, observed by optical microscopy and AFM.

It was found after a detailed AFM analysis of the morphology of the stripes, that the patterns show that each stripe is formed by rod-like crystallites. The SCO properties of these stripes were investigated by Raman spectroscopy. Raman experiments showed that the compound retains its SCO nature when the solvent shrinks slowly; interestingly, when fast shrinkage was employed, the thin deposit lost the ability to undergo SCO.

In 2007, Bousseksou and Molnár reported on a process for nano and microscale assembly of the 3-D SCO coordination polymer $Fe(pyrazine)[Pt(CN)_4]$ which displays a very strong cooperative spin-transition with a ca. 25 K wide hysteresis loop centred around room-temperature.[78] These nano and microscale assemblies were produced using a combination of top-down (lift-off) and bottom-up (MSA) methods, employing a gold-coated silicon surface covered by a conventional polymethylmethacrylate (PMMA) EBL resist. Square-shaped patterns of different sizes and densities were written by a focused electron beam. After resist development, the substrates were dipped into a solution of 4,4′-azopyridine to prepare an anchoring layer for the thin film. Patterns of sizes ranging from 50 nm to 2 μm were created and imaged by SEM.

Limiting factors present during the process were desorption and re-adsorption accompanied by the formation of crystallites which originated 3-D crystallites with substantial roughness at large thickness. To probe the spin state of the micro and nanopatterned assemblies, Raman spectroscopy was used on patterns of 2 μm, 500 nm and 200 nm. Raman spectra of the larger nanopatterns allowed the identification of a thermal spin-transition where the vibrational mode at 1232 cm^{-1} changes dramatically in relative intensity. For smaller sizes the identification of the change in relative intensities was much more difficult, with arrays of 200 nm nanopatterns giving Raman spectra at both temperatures that were very similar, but sufficient for the authors to claim a spin-transition at the nanometric regime.

Later, Bousseksou produced 3-D coordination polymers derived from $[Fe^{II}(azpy)(M^{II}(CN)_4)]\cdot nH_2O$ (azpy = 4,4′-azopyridine; M = Ni, Pd, and Pt), to fabricate nanopatterned thin films on gold substrates. Micro- and nanometric patterns were obtained using a PMMA mask as a physical barrier for the assembly of the multilayers on the substrate, Figure 14.14a, b and c.[79] The authors have characterised the SCO characteristic of both bulk materials and nanopatterned materials by Raman and concluded that nanopatterned thin films on gold substrates retained their SCO characteristics, Figure 14.14d and e.

Recently, Bousseksou and Salmon used the previously described methods to fabricate micro- and nanopatterned thin layers of the compound $[Fe(bpac)(M(CN)_4)]$ (bpac = bis(4-pyridyl)acetylene; M = Ni, Pd, and Pt) with pattern sizes of 2000, 200 and 30 nm.[80] The authors investigated the uptake of guest solvent molecules by the porous patterns and observed that the SCO behaviour is different and reversible when inclusion of pyridine occurred.

Finally, Vieu *et al.* reported their work on soft lithographic patterning of SCO nanoparticles using micro-transfer moulding (μTM) to pattern $[Fe(NH_2trz)_3](tos)_2$ (tos = tosylate, NH_2trz = 4-amino-1,2,4-triazole). This SCO compound is known to exhibit a relatively abrupt thermal transition between the HS and LS forms around room temperature. To form such patterns of SCO nanoparticles, a drop of the nanoparticle suspension was placed on the patterned surface of a PDMS stamp and the excess solution was removed by a stream of nitrogen until the PDMS stamp was completely dry. Finally, the filled stamp was placed in contact with a substrate and peeled away.[81] Using this method the authors fabricated micrometric features of 5 μm meshes that were perfectly ordered with no defects. The resolution limit of the patterned stripes was further moved into the nanometer regime by using μTM. Structures with a resolution of approximately 300 nm could be achieved with good thickness homogeneity and very few defects. Despite the difficulties in assigning the SCO transition of the patterned nanoparticles because the spectrum of the microstructures is dominated by the strong Raman scattering of the NaAOT surfactant used during the fabrication process of the nanoparticles, the authors claimed to have prepared micro- and nanopatterns of SCO nanoparticles retaining their spin switching characteristic after fabrication of such materials.

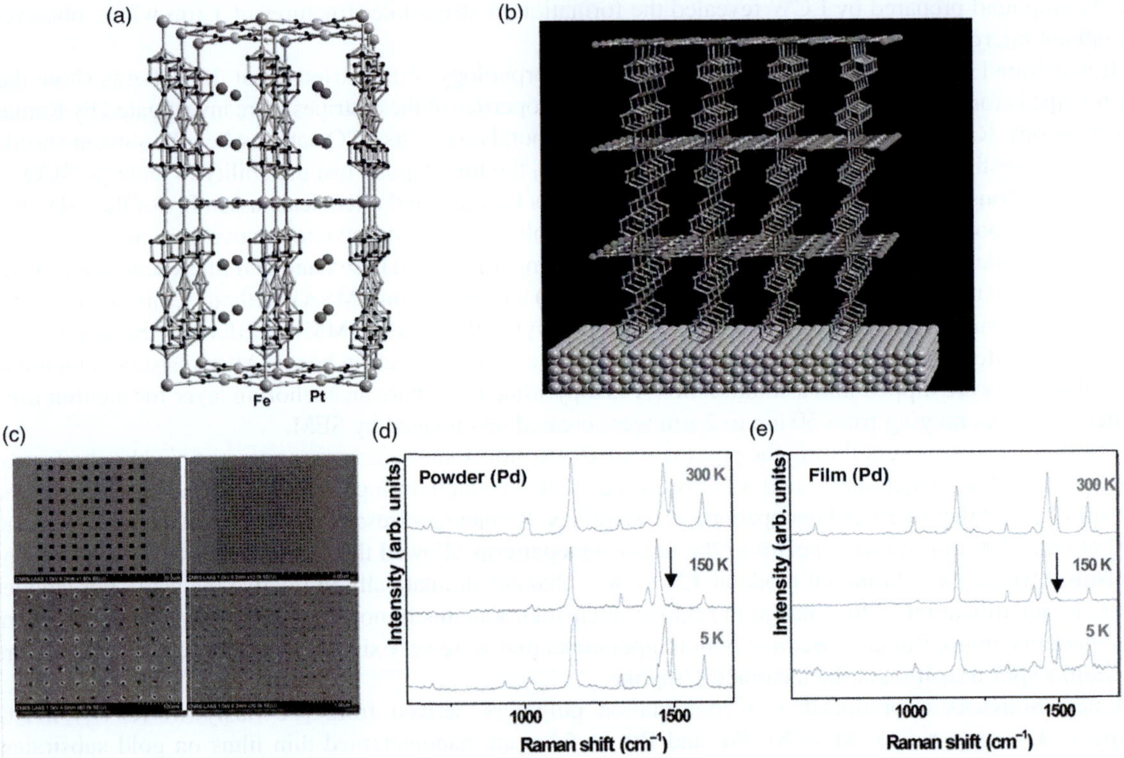

Figure 14.14 *Templating of SCO nanostructures: (a) 3-D coordination polymer of $[Fe^{II}(azpy)(M^{II}(CN)_4)]\cdot nH_2O$; (b) scheme of deposition of 3-D coordination polymer on gold surface; (c) SEM images of patterns of $Fe(azpy)[Pd(CN)_4]$ obtained following 10 deposition cycles for different shapes with distinct sizes between 2 μm and 30 nm; (d) variable temperature Raman spectra of bulk material indicating SCO; (e) variable temperature Raman spectra of nanopatterns indicating SCO. Reproduced from [79] with permission from American Chemical Society © 2008.*

14.5 Electrical Spin State Control

Fabrication of typical data storage devices is performed using a *top-down* approach where the size of discrete magnetic domains is constantly reduced in dimensions.[82] The size of discrete magnetic domains can be reduced until the super paramagnetic limit is reached, where ambient temperature is sufficient to invert the magnetisation of such domains leading to highly unstable devices.[83, 84] To overcome this problem scientists have been developing strategies using diverse and imaginative solutions such as antiferromagnetic coupling media recording,[85] perpendicular recording[86] and lithographically patterned media recording,[87] while the assembly of magnetic molecules into nanowires has been reported producing a huge advance in the area of information storage.[88, 89]

In a complementary *bottom-up* approach, chemists have applied supramolecular principles to fabricate molecular machines, motors and switches, and devices for information storage and processing represent a particularly attractive field in this research area. For example, reversible data storage may become possible on the nanoscale by imprinting information via a change of magnetisation. Reversibility of this process requires materials with variable magnetisation states that can be addressed selectively and without perturbation of

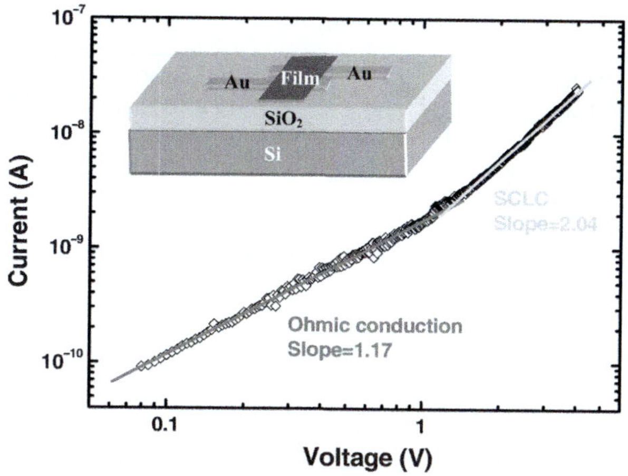

Figure 14.15 *Electrical properties of a SCO film: current–voltage characteristics for a 240 nm thick film of [Fe(phen)$_2$(SCN)$_2$] evaporated on a silicon substrate at 300 K and linear fits to the data. Reprinted with permission from [72], Copyright 2009, American Institute of Physics.*

the entire environment. Research has been developed where bistable magnetic states are addressed mainly by application of temperature or light. For electronic purposes, it is desirable to apply an electric field to switch between two states which would permit fabrication of molecular devices where the magnetic states are controlled electrically.

Studies on attempts to electrically control the spin states on SCO complexes are quite recent and the first studies on devices have been reported by Shi and Beaurepaire, Figure 14.15.[72] A device was constructed using a 240 nm thick film of [Fe(phen)$_2$(SCN)$_2$] evaporated on a silicon substrate with gold electrodes. The current–voltage (*I-V*) characteristics at room temperature of the device were determined, and at low voltage the ln(*I*)-ln(*V*) characteristic is linear with a slope of 1.17 suggesting Ohmic conduction. Above 1.4 V, the ln(*I*)-ln(*V*) characteristic remained linear but with a slope of 2.04, indicating a space charge-limited current (SCLC) regime. The authors extracted a mobility of 6.53×10^{-6} cm^2/Vs in the space charge-limited regime, comparable to that found in typical small organic molecules.

Developments of Coronado's work with SCO nanoparticles reported in 2007 have been recently achieved, where Coronado and van der Zant described devices built from a single SCO nanoparticle.[90] Nanoparticles based on a core of the Fe(II) coordination polymer [Fe(trz)$_3$](BF$_4$)$_2$ (trz = triazole) and a surfactant shell around 11 nm in diameter were placed between two nanometer-spaced electrodes. Such particles have been characterised in detail and show a cooperative spin-transition from Fe(II) low spin, $S = 0$, to high spin, $S = 2$, exhibiting a thermal hysteresis loop above room temperature with a width of ca. 40 K.[21, 24]

Current–voltage (*I–V*) experiments at 10 K led to the observation that for the 100-nm-wide electrode, equidistant step-like features and peaks in the corresponding derivative (d*I*/d*V*), are reminiscent of a Coulomb staircase. Based on these observations, the authors concluded that the transport mechanism is due to sequential single-electron tunnelling through an asymmetric double-barrier junction. The nanoparticle core is isolated from the electrodes by tunnel barriers defined by the surfactant layer where, electrons first tunnel onto the particle and remain there for some time before tunnelling off again. Coulomb steps are equidistant and therefore indicate that transport occurs through a single particle.

The current–voltage characteristics of the devices measured between 300–400 K show that there was a change in conductance in the temperature region of the SCO. Moreover, a hysteresis loop was observed in the current–voltage characteristic of the devices, although this was only observed in 30% of the devices

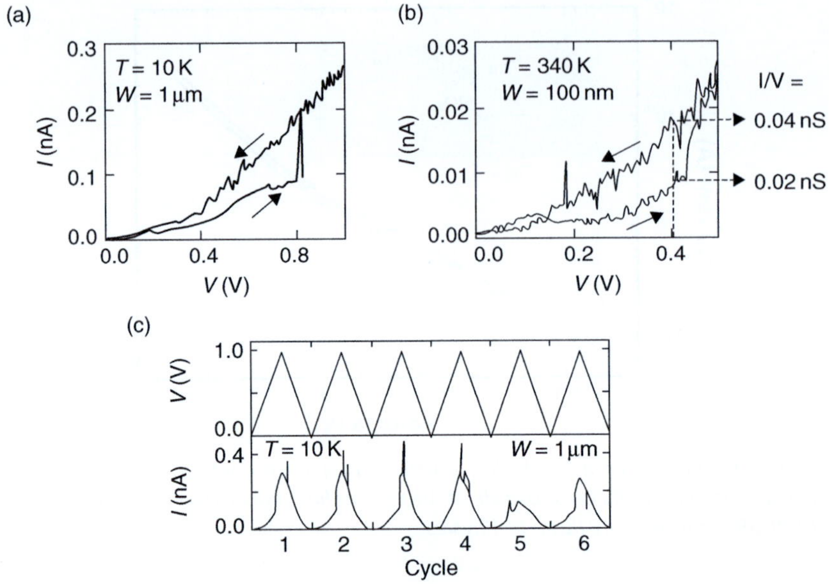

Figure 14.16 *Device of [Fe(trz)$_3$](BF$_4$)$_2$SCO nanoparticles: (a) Current–voltage characteristics of a 1-μm-wide electrode device at low temperature; (b) current–voltage characteristic of a 100-nm-wide electrode device at 340 K. The conductance (I/V) at 0.4 V coincides with the values observed in the thermal hysteresis of the same device, showing that the bistability can be probed with an applied voltage; (c) repetitive cycling of the voltage between 0 and 1 V at low temperature (10 K) of the device. Reprinted with permission from [90]. Copyright 2011 WILEY-VCH Verlag GmbH & Co. KGaA.*

tested. The authors found that by keeping the temperature constant but changing the applied bias voltage, a transition between the low- and the high-conductance state occurs, being the effect more pronounced at low temperature, Figure 14.16.

The device whose conductance is represented in Figure 14.16a was cycled 6 times and it was observed that the lifetime of the high-conductance states varies. Factors such as electric field polarisation, intersystem crossing through excited states and local resistive heating are possible explanations for this observation.

More recently, Ruben and van der Zant reported on the electrical control of SCO in a single molecule.[91] The approach used by the authors was different from that previously reported, where SCO is triggered by charging the molecule with two electrons. This was predicted by theoretical calculations and confirmed experimentally by a split of zero bias (Kondo) resonance in the current–voltage characteristics. Theoretical calculations suggest that the spin-transition should manifest itself in an electron transport experiment by the specific structure of its magnetic excitations and experiments were carried out with a three terminal device and the low-energy excitation probed by inelastic tunnelling spectroscopy. Transport experiments revealed a double-quantum-dot behaviour in combination with a split Kondo peak, Figure 14.17. This experiment opens the realm of active spin state modulation in a molecule; the smallest possible SCO switching unit.

Finally, Bousseksou reports on electrical properties and the nonvolatile memory effect of a SCO complex.[92] The authors built a nonvolatile memory device based on changes of the dielectric permittivity allowing the information to be stored in the high- and low-capacitance states. The compound elected for these studies was [Fe(HB(pz)$_3$)$_2$] where HB(pz)$_3$ = hydrotris(1-pyrazolyl)borate. This compound is known to undergo a thermal spin transition in the 300–450 K temperature range with the first heating cycle completely different from successive thermal cycles. Such a difference has been attributed to a crystallographic transformation.

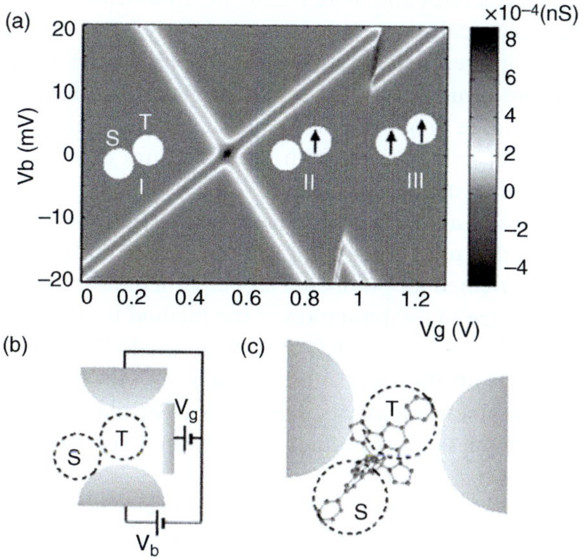

Figure 14.17 (a) Simulation of the differential conductance vs. source-drain (V_b) and gate (Vg) voltages calculated for $T = 4.2$ K; (b) schematic representation of the double-dot system. T: transfer dot, S: spectator dot; (c) schematic representation of molecule-electrode contacts realising the double-dot system. Reproduced from [91] with permission from American Physical Society © 2011.

Thin films of the compound, of ca. 200 nm thickness, were deposited on gold interdigitated microelectrodes by thermal evaporation and the composition of the films was confirmed by Raman spectroscopy.

The resulting thin film had a densely covered surface with a granular morphology composed of nanocrystals, and was used for AC conductivity measurements as a function of frequency and temperature. Isothermal frequency measurements revealed a change in conductivity with a frequency dependence more pronounced than in the powdered sample, which the authors attributed to a higher degree of disorder of the thin films. I-V measurements were performed on a thin film, at room temperature, before and after heating the film to 408 K, showing that these differ considerably. A complementary experiment was carried out by changing the applied electric bias from 1 to 10 V. An increase in current was observed upon changing the applied bias from 1 to 2 V, but over time the current drops to low conductivity values. A further increase in voltage showed a small increase in current, which remained in the low conductivity region. Based on these observations the authors claimed the construction of a nonvolatile read-only memory (ROM) using a SCO compound.

14.6 Conclusion

SCO is a fascinating interdisciplinary area of research which has attracted experimentalists and theoreticians from chemistry, physics, materials science and engineering. SCO research emerged 80 years ago and a long and impressive literature on SCO has been developed with the discovery of several new examples of SCO compounds, the explanation of different types of SCO profiles and the modification of SCO examples to increase cooperativity and to direct the application to materials science. The development of materials from SCO molecules has attracted scientists and has been a very active area in recent years. These include the fabrication of SCO nanoparticles and nanocrystals, the fabrication of SCO thin films, the use of patterning techniques to fabricate SCO nanostructures and more recently the fabrication of devices using SCO materials where spin states are addressed electrically.

SCO nanoparticles are of great interest and have shown to display interesting properties. Reports on SCO nanoparticles and nanocrystals are limited to a few research groups and it will be important in the future to investigate physical properties of a single particle and size effects on physical properties. Also, new examples of nanoparticles and other nano-objects with higher control of size and spatial orientation of the SCO molecules in the nano-objects would be desirable.

Fabrication of SCO thin films was a very active area in former years with work initiated by Kahn and built on by others. Unfortunately, due to the lack of predictability and control this area has been neglected. Undoubtedly, the most interesting method for producing regular thin films is the ultra-high vacuum deposition of thin films. This lacks examples due to the limitation of the method to molecules that are stable under the deposition conditions. New examples of neutral SCO molecules would help to advance this field.

Patterning of SCO molecules permits the fabrication of materials with different sizes and shapes which is very appealing for materials development and device fabrication. The major drawback of this technique is its cost and processing times, and without new cost-effective hardware to produce nanopatterned materials, this will not be pushed forward.

Finally, devices where spin states are addressed electrically, in particular at the single molecule limit, are the most interesting and important of the advances in SCO research towards application of SCO materials. This would permit the fabrication of cost-effective hybrid assemblies, where SCO molecules or objects are integrated into electronic circuits acting as active switching units in spintronic devices.[93]

References

1. Cambi, L., Szegö, L. (1931) Über die magnetische Susceptibilität der komplexen Verbindungen. Ber. Dtsch. Chem. Ges. Teil B, 64: 2591–2598.
2. Kahn, O., Codjovi, E., Garcia, Y., van Koningsbruggen, P. J., Lapouyade, R., Sommier, L. (1996) Spin-transition molecular materials for display and data processing. ACS Symp. Ser., 644: 298–310.
3. Erenburg, S. B., Bausk, N. V., Lavrenova, L. G., Varnek, V. A., Mazalov, L. N. (1997) Relation between electronic and spatial structure and spin-transition parameters in chain-like Fe(II) compounds. Solid State Ionics, 101: 571–577.
4. Garcia, Y., van Koningsbruggen, P. J., Lapouyade, R., Fournès, L., Rabardel, L., Kahn, O., Ksenofontov, V., Levchenko, G., Gütlich, P. (1998) Influences of temperature, pressure, and lattice solvents on the spin transition regime of the polymeric compound $[Fe(hyetrz)_3]A_2{\cdot}3H_2O$ (hyetrz = 4-(2′-hydroxyethyl)-1,2,4-triazole and $A^- =$ 3-nitrophenylsulfonate). Chem Mater, 10: 2426–2433.
5. Garcia, Y., Niel, V., Muñoz, M. C., Real, J. A.(2004) Spin crossover in 1D, 2D and 3D polymeric Fe(II) networks, in: Gütlich, P., Goodwin, H. A. (Eds) Spin Crossover in Transition Metal Compounds I. Top. Curr. Chem., 233: 229–257.
6. Ruben, M., Ziener, U., Lehn, J.-M., Ksenofontov, V., Gütlich, P., Vaughan, G. B. M. (2004) Hierarchical self-assembly of supramolecular spintronic modules into 1D- and 2D-architectures with emergence of magnetic properties. Chem. Eur. J., 11: 94–100.
7. Kitazawa, T., Gomi, Y., Takahashi, M., Takeda, M., Enomoto, M., Miyazaki, A., Enoki, T. (1996) Spin-crossover behaviour of the coordination polymer $Fe^{II}(C_5H_5N)_2Ni^{II}(CN)_4$. J. Mater. Chem., 6: 119–121.
8. Niel, V., Martinez-Agudo, J.M., Muñoz, M. C., Gaspar, A. B., Real, J. A. (2001) Cooperative spin crossover behavior in cyanide-bridged Fe(II)-M(II) bimetallic 3D Hofmann-like networks (M = Ni, Pd, and Pt). Inorg. Chem., 40: 3838–3839.
9. Roubeau, O., Colin, A., Schmitt, W., Clérac, R. (2004) Thermoreversible gels as magneto-optical switches. Angew. Chem. Int. Ed., 43: 3283–3286.

10. Grondin, P., Roubeau, O., Castro, M., Saadaoui, H., Colin, A., Clérac, R. (2010) Multifunctional gels from polymeric spin-crossover metallo-gelators. Langmuir, 26: 5184–5195.
11. Fujigaya, T., Jiang, D. L., Aida, T. (2007) Spin-crossover physical gels: a quick thermoreversible response assisted by dynamic self-organization. Chem. Asian J., 2:106–113.
12. Galyametdinov, Y., Ksenofontov, V., Prosvirin, A., Ovchinnikov, I., Ivanova, G., Gütlich, P., Haase, W. (2001) First example of coexistence of thermal spin transition and liquid-crystal properties. Angew. Chem. Int. Ed., 40: 4269–4271.
13. Bodenthin, Y., Schwarz, G., Tomkowicz, Z., Lommel, M., Geue, T., Haase, W., Mohwald, H., Pietsch, U., Kurth, D. G. (2009) Spin-crossover phenomena in extended multi-component metallo-supramolecular assemblies. Coord. Chem. Rev., 253: 2414–2422.
14. Gaspar, A.B., Seredyuk, M., Gütlich, P. (2009) Spin crossover in metallomesogens. Coord. Chem. Rev., 253: 2399–2413.
15. Gütlich, P., Goodwin, H. A. (2004) Spin crossover–an overall perspective, in: Gütlich, P., Goodwin, H. A. (Eds) Spin Crossover in Transition Metal Compounds I. Top. Curr. Chem., 233: 1–47.
16. Cambi, L., Szegö, L. (1933) Über die magnetische Susceptibilität der komplexen Verbindungen (II. Mitteil.). Ber. Dtsch. Chem. Ges. Teil B, 66 (1933) 656–661.
17. Cambi, L., Malatesta, L. (1937) Magnetismus und Polymorphie innerer Komplexsalze: Eisensalze der Dithiocarbamidsäuren. Ber. Dtsch. Chem. Ges. Teil B, 70: 2067–2078.
18. Létard, J.-F., Guionneau, P., Goux-Capes, L. (2004) Towards spin crossover applications, in: Gütlich, P., Goodwin, H. A. (Eds) Spin Crossover in Transition Metal Compounds III. Top. Curr. Chem., 235: 221–249.
19. Catala, L., Volatron, F., Brinzei, D., Mallah, T. (2009) Functional coordination nanoparticles. Inorg. Chem., 48: 3360–3370.
20. Neville, S. M., Etrillard, C., Asthana, S., Létard, J.-F. (2010) Light-induced stored information in nanoparticles. Eur. J. Inorg. Chem., 282–288.
21. Coronado, E., Galán-Mascarós, J. R., Monrabal-Capilla, M., García-Martínez, J., Pardo-Ibáñez, P. (2007) Bistable spin-crossover nanoparticles showing magnetic thermal hysteresis near room temperature. Adv. Mater., 19: 1359–1362.
22. Forestier, T., Mornet, S., Daro, N., Nishihara, T., Mouri, S., Tanaka, K., Fouché, O., Freysz, E., Létard, J.-F. (2008) Nanoparticles of iron(II) spin-crossover. Chem. Commun., 4327–4329.
23. Forestier, T., Kaiba, A., Pechev, S., Denux, D., Guionneau, P., Etrillard, C., Daro, N., reysz, E., Létard, J.-F. (2009) Nanoparticles of $[Fe(NH_2\text{-}trz)_3]Br_2{\cdot}3H_2O$ (NH_2-trz = 2-amino-1,2,4-triazole) prepared by the reverse micelle technique: influence of particle and coherent domain sizes on spin-crossover properties. Chem. Eur. J., 15: 6122–6130.
24. Galán-Mascarós, J.R., Coronado, E., Forment-Aliaga, A., Monrabal-Capilla, M., Pinilla-Cienfuegos, E., Ceolin, M. (2010) Tuning size and thermal hysteresis in bistable spin crossover nanoparticles. Inorg. Chem., 49: 5706–5714.
25. Salmon, L., Molnár, G., Zitouni, D., Quintero, C., Bergaud, C., Micheau, J.-C., Bousseksou, A. (2010) A novel approach for fluorescent thermometry and thermal imaging purposes using spin crossover nanoparticles. J. Mater. Chem., 20: 5499–5503.
26. Dutt, G. B. (2008) Fluorescence anisotropy of ionic probes in AOT reverse micelles: influence of water droplet size and electrostatic interactions on probe dynamics. J. Phys. Chem. B, 112: 7220–7226.
27. Tokarev, A., Salmon, L., Guari, Y., Nicolazzi, W., Molnár, G., Bousseksou, A. (2010) Cooperative spin crossover phenomena in $[Fe(NH_2trz)_3](tosylate)_2$ nanoparticles. Chem. Commun., 46: 8011–8013.
28. Titos-Padilla, S., Herrera, J. M., Chen, X.-W., Delgado, J. J., Colacio, E. (2011) Bifunctional hybrid SiO_2 nanoparticles showing synergy between core spin crossover and shell luminescence properties. Angew. Chem. Int. Ed., 50: 3290–3293.
29. Kröber, J., Audière, J.-P., Claude, R., Codjovi, E., Kahn, O., Haasnoot, J. G., Grolière, F., Jay, C., Bousseksou, A. (1994) Spin transitions and thermal hystereses in the molecular – based materials $[Fe(Htrz)_2(trz)](BF_4)$ and $[Fe(Htrz)_3](BF_4)_2{\cdot}H_2O$ (Htrz = 1,2,4-4*H*-triazole; trz = 1,2,4- triazolato) Chem. Mater., 6: 1404–1412.
30. Mader, D., Pillet, S., Carteret, C., Stébé, M.-J., Blin, J.-L. (2011) Confined growth of spin crossover nanoparticles in surfactant-based matrices: enhancing shape anisotropy. J. Dispers. Sci. Tech., 32: 1771–1779.
31. Volatron, F., Catala, L., Rivière, E., Gloter, A., Stéphan, O., Mallah, T. (2008) Spin-crossover coordination nanoparticles. Inorg. Chem., 47: 6584–6586.

32. Raza, Y., Volatron, F., Moldovan, S., Ersen, O., Huc, V., Martini, C., Brisset, F., Gloter, A., Stéphan, O., Bousseksou, A., Catala, L., Mallah, T. (2011) Matrix-dependent cooperativity in spin crossover Fe(pyrazine)Pt$(CN)_4$ nanoparticles. Chem Commun, 47: 11501–11503.
33. Martínez, V., Boldog, I., Gaspar, A. B., Ksenofontov, V., Bhattacharjee, A., Gütlich, P., Real, J. A. (2010) Spin crossover phenomenon in nanocrystals and nanoparticles of [Fe(3-Fpy)$_2$M(CN)$_4$] (M^{II} = Ni, Pd, Pt) two-dimensional coordination polymers. Chem. Mater., 22: 4271–4281.
34. Boldog, I., Gaspar, A. B., Martínez, V., Pardo-Ibañez, P., Ksenofontov, V., Bhattacharjee, A., Gütlich, P., Real, J. A. (2008) Spin-crossover nanocrystals with magnetic, optical, and structural bistability near room temperature. Angew. Chem. Int. Ed., 47: 6433–6437.
35. Larionova, J., Salmon, L., Guarl, Y., Tokarev, A., Molvinger, K., Molnár, G., Bousseksou, A. (2008) Towards the ultimate size limit of the memory effect in spin-crossover solids. Angew. Chem. Int. Ed., 47: 8236–8240.
36. Tissot, A., Bardeau, J.-F., Rivière, E., Brisset, F., Boillot, M.-L. (2010) Thermo- and photoswitchable spin-crossover nanoparticles of an iron(II) complex trapped in transparent silica thin films. Dalton Trans., 39: 7806–7812.
37. Faulmann, C., Chahine, J., Malfant, I., de Caro, D., Cormary, B., Valade, L. (2011) A facile route for the preparation of nanoparticles of the spin-crossover complex [Fe(Htrz)$_2$(trz)](BF_4) in xerogel transparent composite films. Dalton Trans., 40: 2480–2485.
38. Talham, D. R. (2004) Conducting and magnetic Langmuir–Blodgett films. Chem. Rev., 104: 5479–5502.
39. Ruaudel-Teixier, A., Barraud, A., Coronel, P., Kahn, O. (1988) Spin transition in a magnetic Langmuir-Blodgett film. Thin Solid Films, 160: 107–115.
40. Coronel, P., Barraud, A., Claude, R., Kahn, O., Ruaudel-Teixier, A., Zarembowitch, J. (1989) Spin transition in a Langmuir–Blodgett film. J. Chem. Soc. Chem. Commun., 193–194.
41. Soyer, H., Mingotaud, C., Boillot, M.-L., Delhaès, P. (1998) Spin-crossover complex stabilized on a formamide/water subphase. Thin Solid Films, 329: 435–438.
42. Soyer, H., Mingotaud, C., Boillot, M.-L., Delhaès, P. (1998) Spin crossover of a Langmuir-Blodgett film based on an amphiphilic iron(II) complex. Langmuir, 14: 5890–5895.
43. Soyer, H., Dupart, E., Gómez-García, C. J., Mingotaud, C., Delhaès, P. (1999) First magnetic observation of a spin crossover in a Langmuir-Blodgett film. Adv. Mater., 11: 382–384.
44. Létard, J.-F., Nguyen, O., Soyer, H., Mingotaud, C., Delhaès, P., Kahn, O. (1999) First evidence of the LIESST effect in a Langmuir-Blodgett film. Inorg. Chem., 38: 3020–3021.
45. Soyer, H., Dupart, E., Mingotaud, C., Gomez-Garcia, C. J., Delhaès, P. (2000) Spin crossover phenomenon of a semi-fluorinated iron (II) complex organized in a Langmuir-Blodgett film. Colloids Surf. A, 171: 275–282.
46. Armand, F., Badoux, C., Bonville, P., Ruaudel-Teixier, A., Kahn, O. (1995) Langmuir-Blodgett films of spin transition iron(II) metalloorganic polymers. 1. Iron(II) complexes of octadecyl-1,2,4-triazole. Langmuir, 11: 3467–3472.
47. Roubeau, O., Agricole, B., Clérac, R., Ravaine, S. (2004) Triazole-based magnetic Langmuir–Blodgett films: paramagnetic to spin-crossover behavior. J. Phys. Chem. B, 108: 15110–15116.
48. Roubeau, O., Natividad, E., Agricole, B., Ravaine, S. (2007) Formation, structure, and morphology of triazole-based Langmuir–Blodgett films. Langmuir, 23: 3110–3117.
49. Kurth, D. G., Lehmann, P., Schütte, M. (2000) A route to hierarchical materials based on complexes of metallo-supramolecular polyelectrolytes and amphiphiles. Proc. Natl. Acad. Sci. USA, 97: 5704–5707.
50. Lehmann, P., Kurth, D. G., Brezesinski, G., Symietz, C. (2001) Structural analysis of a metallosupramolecular polyelectrolyte-amphiphile complex at the air/water interface. Chem. Eur. J., 7: 1646–1651.
51. Bodenthin, Y., Pietsch, U., Möhwald, H., Kurth, D. G. (2005) Inducing spin crossover in metallo-supramolecular polyelectrolytes through an amphiphilic phase transition. J. Am. Chem. Soc., 12: 3110–3114.
52. Bodenthin, Y., Schwarz, G., Tomkowicz, Z., Nefedov, A., Lommel, M., Möhwald, H., Haase, W., Kurth, D. G., Pietsch, U. (2007) Structure-driven remanent high-spin state in metallosupramolecular assemblies. Phys. Rev. B, 76: 64422/1–064422/6.
53. White, N. G., Feltham, H. L. C., Gandolfi, C., Albrecht, M., Brooker, S. (2010) Towards Langmuir–Blodgett films of magnetically interesting materials: solution equilibria in amphiphilic iron(II) complexes of a triazole-containing ligand. Dalton Trans., 39: 3751–3758.

54. Kitchen, J. A., White, N. G., Gandolfi, C., Albrecht, M., Jameson, G. N. L., Tallon, J. L., Brooker, S. (2010) Room-temperature spin crossover and Langmuir–Blodgett film formation of an iron(II) triazole complex featuring a long alkyl chain substituent: the tail that wags the dog. Chem. Commun., 46: 6464–6466.
55. Gandolfi, C., Miyashita, N., Kurth, D. G., Martinho, P. N., Morgan, G. G., Albrecht, M. (2010) Organization of spin- and redox-labile metal centers into Langmuir and Langmuir–Blodgett films. Dalton Trans., 39: 4508–4516.
56. Gandolfi, C., Moitzi, C., Schurtenberger, P., Morgan, G. G., Albrecht, M. (2008) Improved cooperativity of spin-labile iron(III) centers by self-assembly in solution. J. Am. Chem. Soc., 130: 14434–14435.
57. Morgan, G. G., Murnaghan, K. D., Müller-Bunz, H., Mckee, V., Harding, C. J. (2006) A manganese(III) complex that exhibits spin crossover triggered by geometric tuning. Angew. Chem. Int. Ed., 45: 7192–7195.
58. Bell, C. M., Keller, S. W., Lynch, V. M., Mallouk, T. E. (1993) New solids and surfaces, via coordination chemistry. Mater. Chem. Phys., 35: 225–232.
59. Bell, C. M., Arendt, M. F., Gomez, L., Schmehl, R. H., Mallouk, T. E. (1994) Growth of lamellar Hofmann clathrate films by sequential ligand exchange reactions: assembling a coordination solid one layer at a time. J. Am. Chem. Soc., 116: 8374–8375.
60. Cobo, S., Molnár, G., Real, J. A., Bousseksou, A. (2006) Multilayer sequential assembly of thin films that display room-temperature spin crossover with hysteresis. Angew. Chem. Int. Ed., 45: 5786–5789.
61. Wang, Y., Zhou, W. (2010) A review on inorganic nanostructure self-assembly. J. Nanosci. Nanotech., 10: 1563–1583.
62. Matsuda, M., Tajima, H. (2007) Thin film of a spin crossover complex $[Fe(dpp)_2](BF_4)_2$. Chem. Lett., 36: 700–701.
63. Matsuda, M., Isozaki, H., Tajima, H. (2008) Electroluminescence quenching caused by a spin-crossover transition. Chem Lett, 37: 374–375.
64. Liu, X. J., Moritomo, Y., Nakamura, A., Hirao, T., Toyazaki, S., Kojima, N. (2001) Photoinduced phase transition and relaxation behavior in a spin-crossover Fe(II) complex Nafion-$[Fe(Htrz)_3]$ film. J. Phys. Soc. Jpn., 70: 2521–2524.
65. Kojima, N., Toyazaki, S., Itoi, M., Ono, Y., Aoki, W., Kobayashi, Y., Seto, M., Yokoyama, T. (2002) Search on multi-functional properties of spin-crossover system. Mol. Cryst. Liq. Cryst., 376: 567–574.
66. Nakamoto, A., Ono, Y., Kojima, N., Matsumura, D., Yokoyama, T., Liu, X. J., Moritomo, Y. (2003) Spin transition and its photo-induced effect in spin crossover complex film based on $[Fe(II)(trz)_3]$. Synth. Met., 137:1219–1220.
67. Liu, X. J., Moritomo, Y., Kawamoto, T., Nakamoto, A., Kojima, N. (2003) Dynamical phase transition in a spin-crossover complex. J. Phys. Soc. Jpn., 72: 1615–1618.
68. Nakamoto, A., Ono, Y., Kojima, N., Matsumura, D., Yokoyama, T. (2003) Spin crossover complex film, $[Fe^{II}(H\text{-}trz)_3]$-Nafion, with a spin transition around room temperature. Chem. Lett., 32: 336–337.
69. Nakamoto, A., Kojima, N., Liu, X. J., Moritomo, Y., Nakamura, A. (2005) Demonstration of the thermally induced high spin–low spin transition for a transparent spin crossover complex film $[Fe(II)(H\text{-}trz)_3]$-Nafion (trz = triazole). Polyhedron, 24: 2909–2912.
70. Kamebuchi, H., Jo, T., Shimizu, H., Okazawa, A., Enomoto, M., Kojima, N. (2011) Development of pH-sensitive spin-crossover iron(II) complex films, $[Fe^{II}(diAMsar)]$-Nafion: manipulation of the spin state by proton concentration. Chem. Lett., 40: 888–889.
71. Boillot, M.-L., Pillet, S., Tissot, A., Rivière, E., Claiser, N., Lecomte, C. (2009) Ligand-driven light-induced spin change activity and bidirectional photomagnetism of styrylpyridine iron(II) complexes in polymeric media. Inorg. Chem., 48: 4729–4736.
72. Shi, S., Schmerber, G., Arabski, J., Beaufrand, J. B., Kim, D. J., Boukari, S., Bowen, M., Kemp, N. T., Viart, N., Rogez, G., Beaurepaire, E., Aubriet, H., Petersen, J., Becker, C., Ruch, D. (2009) Study of molecular spin-crossover complex $Fe(phen)_2(NCS)_2$ thin films. Appl. Phys. Lett., 95: 043303/1–043303/3.
73. Naggert, H., Bannwarth, A., Chemnitz, S., von Hofe, T., Quandt, E., Tuczek, F. (2011) First observation of light-induced spin change in vacuum deposited thin films of iron spin crossover complexes. Dalton Trans., 40: 6364–6366.
74. Moliner, N., Salmon, L., Capes, L., Muñoz, M. C., Létard, J.-F., Bousseksou, A., Tuchagues, J.-P., McGarvey, J. J., Dennis, A. C., Castro, M., Burriel, R., Real, J. A. (2002) Thermal and optical switching of molecular spin states in the $\{[FeL[H_2B(pz)_2]_2\}$ spin-crossover system (L = bpy, phen). J. Phys. Chem. B, 106: 4276–4283.
75. Cavallini, M., Bergenti, I., Milita, S., Ruani, G., Salitros, I., Qu, Z. R., Chandrasekar, R., Ruben, M. (2008) Micro- and nanopatterning of spin-transition compounds into logical structures. Angew. Chem. Int. Ed., 47: 8596–8600.
76. Baker, W. A. jr., Bobonich, H. M. (1964) Magnetic properties of some high-spin complexes of iron(II). Inorg. Chem., 3: 1184–1188.

77. Cavallini, M., Bergenti, I., Milita, S., Kengne, J. C., Gentili, D., Ruani, G., Salitros, I., Meded, V., Ruben, M. (2011) Thin deposits and patterning of room-temperature-switchable one-dimensional spin-crossover compounds. Langmuir, 27: 4076–4081.
78. Molnár, G., Cobo, S., Real, J. A., Carcenac, F., Daran, E., Vieu, C., Bousseksou, A. (2007) A combined top-down/bottom-up approach for the nanoscale patterning of spin-crossover coordination polymers. Adv. Mater., 19: 2163–2167.
79. Agustí, G., Cobo, S., Gaspar, A. B., Molnár, G., Moussa, N.O., Szilágyi, P. A., Pálfi, V., Vieu, C., Muñoz, M. C., Real, J. A., Bousseksou, A. (2008) Thermal and light-induced spin crossover phenomena in new 3D Hofmann-like microporous metalorganic frameworks produced as bulk materials and nanopatterned thin films. Chem. Mater., 20: 6721–6732.
80. Bartual-Murgui, C., Akou, A., Salmon, L., Molnár, G., Thibault, C., Real, J. A., Bousseksou, A. (2011) Guest effect on nanopatterned spin-crossover thin films. Small, 7: 3385–3391.
81. Thibault, C., Molnár, G., Salmon, L., Bousseksou, A., Vieu, C. (2010) Soft lithographic patterning of spin crossover nanoparticles. Langmuir, 26: 1557–1560.
82. Dietzel, A. (2003) Hard disk drives, in: Waser, R. (Ed.) Nanoelectronics and Information Technology. Wiley-VCH, Berlin, p. 617–631.
83. Charap, S. H., Lu, P. L., He, Y. J. (1997) Thermal stability of recorded information at high densities. IEEE Trans. Magn., 33: 978–983.
84. Martín, J. I., Nogués, J., Liu, K., Vicent, J. L., Schuller, I. K. (2003) Ordered magnetic nanostructures: fabrication and properties. J. Magn. Magn. Mater., 256: 449–501.
85. Skumryev, V., Stoyanov, S., Zhang, Y., Hadjipanayis, G., Givord, D., Nogués, J. (2003) Beating the superparamagnetic limit with exchange bias. Nature, 423: 850–853.
86. Iwasaki, S., Takemura, K. (1975) An analysis for the circular mode of magnetization in short wavelength recording. IEEE Trans. Magn., 11: 1173–1175.
87. Terris, B., Thomson, T., Hu, G. (2007) Patterned media for future magnetic data storage. Microsystem Technol., 13: 189–196.
88. Thomas, L., Hayashi, M., Jiang, X., Moriya, R., Rettner, C., Parkin, S. (2007) Resonant amplification of magnetic domain-wall motion by a train of current pulses. Science, 315 (2007) 1553–1556.
89. Thomas, L., Hayashi, M., Jiang, X., Moriya, R., Rettner, C., Parkin, S. S. P. (2006) Oscillatory dependence of current-driven magnetic domain wall motion on current pulse length. Nature, 443: 197–200.
90. Prins, F., Monrabal-Capilla, M., Osorio, E. A., Coronado, E., van der Zant, H. S. J. (2011) Room-temperature electrical addressing of a bistable spin-crossover molecular system. Adv. Mater., 23: 1545–1549.
91. Meded, V., Bagrets, A., Fink, K., Chandrasekar, R., Ruben, M., Evers, F., Bernand-Mantel, A., Seldenthuis, J. S., Beukman, A., van der Zant, H. S. J. (2011) Electrical control over the Fe(II) spin crossover in a single molecule: theory and experiment. Phys. Rev. B, 8: 245415/1–245415/13.
92. Mahfoud, T., Molnár, G., Cobo, S., Salmon, L., Thibault, C., Vieu, C., Demont, P., Bousseksou, A. (2011) Electrical properties and non-volatile memory effect of the [Fe(HB(pz)3)2] spin crossover complex integrated in a microelectrode device. Appl. Phys. Lett., 99: 053307/1–053307/3.
93. Osorio, E., Bjornholm, T., Lehn, J.-M., Ruben, M., van der Zant, H. S. J. (2008) Single-molecule transport in three-terminal devices. J. Phys.: Condens. Matter, 20: 374121/1–374121/14.

15

Ultrafast Studies of the Light-Induced Spin Change in Fe(II)-Polypyridine Complexes

Majed Chergui

Ecole Polytechnique Fédérale de Lausanne, Laboratoire de Spectroscopie Ultrarapide (LSU), Switzerland

15.1 Introduction

The phenomenon of spin-crossover (SCO), which implies a thermally induced transition from a low spin (LS) ground state to a high spin (HS) excited state (or the reverse) in metal-based molecular complexes has been demonstrated for derivatives of ions with d^4, d^5, d^6 and d^7 configurations and is observed for all these in complexes of the first transition series.[1] SCO processes can also be triggered by pressure or irradiation by visible light. Research on the latter phenomenon witnessed a major development after McGarvey *et al.*[2, 3] discovered that for a number of iron(II) and iron(III) polypyridine complexes in solution, the HS state could be populated efficiently at the expense of the LS state by pulsed laser excitation. Near ambient temperatures these light-induced HS states have lifetimes ranging from nanoseconds to microseconds, which are governed by the HS to LS relaxation kinetics. Following this discovery, Decurtins *et al.*[4] showed that at cryogenic temperatures the HS/LS relaxation slows down such that under continuous irradiation, the iron(II) spin-crossover systems can be trapped in the HS state, for which the expression 'light-induced excited state spin trapping (LIESST)' was coined. The discovery of LIESST triggered several years of intense research not only with regard to the mechanism of trapping in the HS state of the system but also into the chemical and physical parameters governing the lifetimes of the low temperature metastable HS states.[5, 6] The fact that LIESST can be observed in crystalline solids raised interest in the study of cooperative effects. The interest in spin trapping is not just academic but is also stirred by the potential applications in optical writing/magnetic reading for magnetic data storage.[7, 8] Iron(II) polypyridine complexes are also considered as potential (and cheaper) candidates for dye-sensitized solar cells,[9–12] which have thus far been most successful with ruthenium-based complexes.[13] Finally, Fe(II) polypyridine complexes appear to have several photophysical properties in common with iron porphyrins, which play an important role in respiration and oxygen transport.

Spin-Crossover Materials: Properties and Applications, First Edition. Edited by Malcolm A. Halcrow.

Several studies have been carried out to characterize the spectroscopic, magnetic, structural and dynamic properties of Fe(II) complexes with different ligands, with the aim of identifying the parameters that are crucial to efficient SCO and LIESST switching. In particular, the aim was to identify what controls the lifetime of the lowest excited quintet state, which differs by several orders of magnitude depending on the nature of the ligand, and to identify the route leading from the initial excited state to the lowest quintet state in the SCO process. These studies were nicely reviewed in two volumes published in 2004 that were dedicated to SCO complexes.[1] Since then, there has been an upsurge of ultrafast studies aimed at describing in detail the mechanisms whereby the excited singlet states populates the quintet state on ultrashort timescales.[14, 15] In parallel, several time-resolved structural studies using X-ray diffraction[16–19] and X-ray absorption spectroscopy[15, 20–25] were carried out, aimed at getting information about the electronic and molecular structure of the short lived excited quintet state, or the intermediates leading to it from the initial excited singlet state. Finally, this intense experimental activity has been accompanied by several new theoretical studies that deliver insight into the energetics and the couplings between excited states.[26–32]

Previously, steady-state spectroscopic and X-ray studies on SCO complexes showed several electronic and structural properties common to nearly all Fe(II) complexes.[33–35] This is of importance as properties that are determined for one complex could, to a certain extent, be generalized to the rest of the family of iron(II)-based complexes. The present chapter deals with studies of isolated Fe(II) complexes, mostly in solution. Understanding of the photoinduced dynamics in isolated complexes is of importance for the description of the SCO in crystals, where it is more likely to find applications. In particular, there has been some discussion about the photoswitching of spin of entire domains in such crystals,[19, 36, 37] and it is therefore crucial to understand what occurs at the molecular level. Excellent reviews have appeared that present the results of steady-state and time-resolved spectroscopic studies of the SCO process.[6, 14, 38] This chapter focuses on the recent developments based on novel ultrafast optical and X-ray techniques which have led to new insight into the ultrafast SCO phenomenon.

15.2 Properties of Fe(II) Complexes

We first recall a number of properties that are common to all Fe(II)-based SCO complexes.

15.2.1 Electronic Structure

Iron(II) consists of six electrons in the 3*d* shell. In perfectly octahedral O_h coordination, the five *nd* orbitals of a transition metal ion are split into a subset of three orbitals, namely d_{xy}, d_{yz} and d_{zx}, which are the basis of the irreducible representation t_{2g}, and a subset of two orbitals, namely d_{z^2} and $d_{x^2-y^2}$, which are the basis of the irreducible representation e_g (see Fig. 15.1a). The t_{2g} orbitals are basically nonbonding and are therefore at lower energy than the antibonding e_g orbitals. The splitting between the two sets is referred to as ligand field splitting and is symbolized by the parameter of the ligand field strength. The latter depends upon both the particular set of ligands and the given metal ion. Because the spin-crossover involves transferring electrons from the t_{2g} subshell to the e_g subshell, the Fe–N bond length increases as a result.

Figure 15.1b shows the potential energy level diagram of $[Fe(bpy)_3]^{2+}$ along the Fe–N coordinate.[30] The $^1A_1(t_{2g}{}^6)$ low spin ground state, the lowest excited $^5T_2(t_{2g}{}^4e_g{}^2)$ high spin electronic state, as well as the higher excited singlet, triplet and quintet ligand-field (LF, also called metal-centred, MC) states are shown. In addition, the manifold of low lying singlet and triplet metal–ligand charge transfer (MLCT) states for ligands with extended π-systems is shown as a shaded region. This diagram can be generalized to the family of Fe(II) polypyridine molecular complexes, in terms of the ordering of states but not their absolute energies. This can be gathered from the absorption spectra of the various complexes,[6] and from quantum-chemical calculations

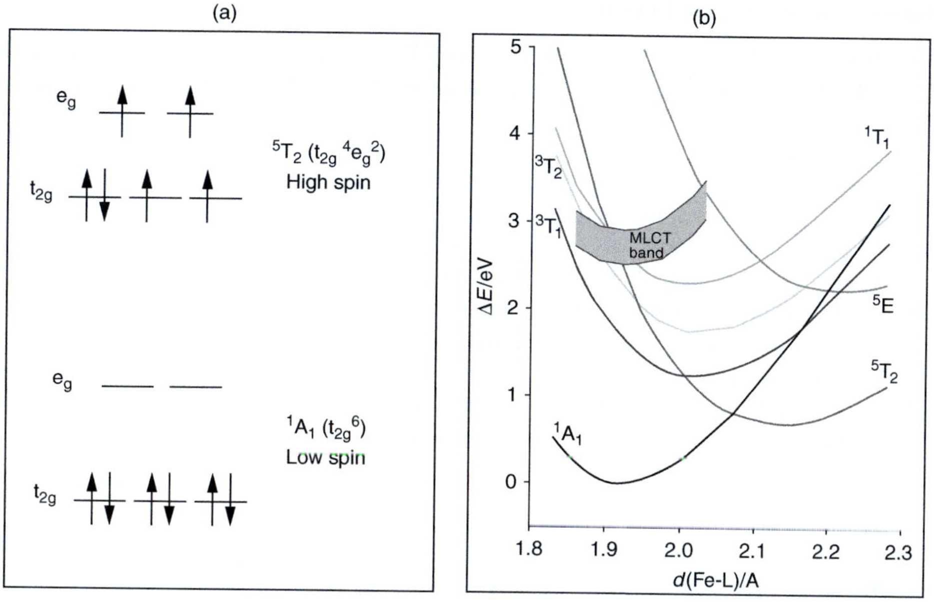

Figure 15.1 (a) Occupancy of the ligand-field split d-orbitals in the low and high spin states; (b) potential curves of $[Fe(bpy)_3]^{2+}$ as a function of the Fe-ligand coordinate. The manifold of metal-to-ligand charge transfer (MLCT) states is shown as a grey shaded area. Reprinted with permission from [30]. Copyright 2011 WILEY-VCH Verlag GmbH & Co. KGaA.

on complexes such as $[Fe(bpy)_3]^{2+}$,[27, 30, 31, 39] $[Fe(ptz)_6]^{2+}$ (ptz = 1-propyltetrazole),[26] $[Fe(tz)_6]^{2+}$ (tz = 1*H*-tetrazole),[28] and $[Fe(pap)_2]^+$ (pap = *N*-2-pyridylmethylidene-2-hydroxyphenylaminato).[40]

15.2.2 Molecular Structure

Per electron that is promoted from the t_{2g} to the e_g orbital, the metal–ligand bond length increases by as much as 0.1 Å.[41] In the low spin $^1A_1(t_{2g}{}^6)$ state, the bond length is centred at 1.96 ± 0.04 Å. In the high spin $^5T_2(t_{2g}{}^4e_g{}^2)$ state, the bond length increases by 0.20 Å typically.[28–31, 39] Obviously these parameters change somewhat for different ligand types. They have been determined by static X-ray diffraction and X-ray absorption spectroscopy studies of complexes on the LS state and on the metastable quintet states.[34, 42, 43] We will later see that the same structural change from LS to HS is also verified for systems having short-lived HS, so that the bond length increase is not the parameter that determines the HS lifetime.

For the series of states $^{1,3}T_{1,2}(t_{2g}{}^5e_g{}^1)$, the bond length is expected to be intermediate between the LS and HS state, and for the $^5E(t_{2g}{}^3e_g{}^3)$ state it is expected to be larger than that of the $^5T_2$ state, as confirmed by quantum-chemical calculations.[27–31, 39] In the case of the $^{1,3}T$ states, their similar bond distances are also corroborated to a certain extent by the width of the absorption bands from the ground state to these states, which are observed at low temperatures.[6]

15.2.3 Vibrational Modes

Another property that is common to all Fe(II)-based complexes is their Fe–N bond stretch frequency, which lies around 390–430 cm^{-1} in the LS state and around 210–250 cm^{-1} in the HS state,[33, 44] supporting the idea of a softer potential due to the more antibonding character in the HS state.

15.2.4 Kinetics of Ground State Recovery

The conversion from the HS quintet state to the LS ground state has been investigated in detail by several groups using steady state spectroscopy of low temperature crystals, and these studies have been reviewed in detail by Hauser in ref. 6. The HS to LS state relaxation was interpreted invoking a multiphonon relaxation in an energy gap law kinetics.

Time-resolved studies of the light-induced SCO started in the 1980s using picosecond and sub-picosecond laser techniques.[6, 14, 38, 45–50] Most of these studies, which concentrated on the determination of the LS ground state recovery in over 10 different Fe(II) polypyridine complexes, are nicely reviewed in ref. 38. It was found that the lowest excited (HS) quintet state relaxes nonradiatively to the LS ground state with times ranging from ~0.65 ns to ~150 ns in room temperature solutions.[35] This was established by correlating the timescale of the ground state bleach recovery with the decay time of the quintet excited state absorption, which was found to lie in the 300 nm region.[6, 48, 51]

While the LS ground state recovery seems a well understood process, that leading from the initial excited state to the lowest quintet state has raised several questions regarding the nonradiative cascade leading to the population of the HS state. Here we mainly concentrate on this cascade and on the structural changes occurring in the HS state.

15.3 From the Singlet to the Quintet State

We concentrate on the excitation into the manifold of singlet metal-to-ligand-charge-transfer (^{1}MLCT) states, which act as doorway states to the $\Delta S = 2$ spin change. Based on ideas from organic physical chemistry, initial models describing the photophysics of SCO tended towards the picture that the intermediate $^1T_1$, $^3T_2$ and $^3T_1$ MC states (Fig. 15.1) mediate the relaxation process in $\Delta S = 1$ steps. Vibrational relaxation should then occur on the timescale of the relevant vibrations of the molecule (typically picoseconds), while the electronic processes involved in the SCO are internal conversion (IC) and intersystem crossing (ISC), with the former being much faster than the latter, due to its conservation of spin angular momentum.

However, several results showed that this picture may not be valid in the case of metal-complexes, and in particular Fe(II) ones. Back in the early 1980s, Sutin and co-workers[45] and Netzel and co-workers[52] published a picosecond time-resolved absorption study of the excited-state dynamics of $[Fe(bpy)_3]^{2+}$ in which they reported a unity quantum yield for formation of the lowest energy HS excited $^5T_2$ state following ^{1}MLCT excitation. This was later supported by a study by Hauser, who reported a near unity quantum efficiency for $[Fe(2\text{-}pic)_3]^{2+}$ in the solid state at 11 K.[6, 53]

These unity quantum yields are difficult to reconcile with the picture of a relaxation cascade via the intermediate $^1T_1$, $^3T_2$ and $^3T_1$ states, since these are expected to have significant electronic coupling to the ground state. Moreover, in the case of the singlet states, relaxation back to the ground state should be more favourable than cascading towards higher spin states. Furthermore, as mentioned in Section 15.2.4, return to the ground state proceeds exclusively through the $^5T_2$ state. One reason to nevertheless invoke the intermediate state is based on Hauser's low temperature single-crystal study of the spectroscopy of $[Fe(ptz)_6]^{2+}$,[54] where he observed the formation of the $^5T_2$ state with near unity quantum yield following excitation into all four MC states (i.e. $^1T_2$, $^1T_1$, $^3T_2$ and $^3T_1$).

Coming back to time-resolved studies, Hendrickson and co-workers carried out sub-picosecond transient absorption studies of $[Fe(tpen)]^{2+}$ in solution. Although they could not observe a spectroscopic signature of the quintet state, they concluded that it would be formed within the time resolution of the experiment (<700 fs).[47, 48] Similar results were found for several other LS Fe(II) complexes, including $[Fe(bpy)_3]^{2+}$, suggesting that such ultrafast formation of the $^5T_2$ state is a general feature of the photophysics of Fe(II) polypyridine complexes.

These initial studies were complemented by McCusker and co-workers using $[Fe(tren(py)_3)]^{2+}$ ($tren(py)_3$ = tris(2-pyridylmethyliminoethyl)amine),[10] where they were able to resolve dynamic spectral shifts associated with the formation of the HS state, which would be formed in ~300 fs. They also observed additional dynamics attributed to vibrational relaxation in the $^5T_2$ state with a time constant of 8 ± 2 ps. Information concerning dynamics associated with the initially formed MLCT state was also unravelled by monitoring spectral changes in regions where the difference spectra for the MC and MLCT states are of opposite sign. They thus concluded that depopulation of the MLCT manifold occurs in 80 ± 20 fs.

The model that was proposed to explain these observations invoked a distinct kinetic process corresponding to conversion from the MLCT to MC manifolds. Beyond that, the model ruled out any intermediate electronic state other than $^5T_2$ as being involved in the photoinduced SCO for Fe(II) complexes. In this picture there is extensive mixing among all of the excited ligand-field (MC) states that lie between the ^{1}MLCT and $^5T_2$ states. However, the picture of the relaxation within the ligand-field manifold was not so clear and it was suggested that it may probably involve a 'barrier less potential slope in which the wave function continuously evolves in its character from something approximating a singlet to that of the $^5T_2$ state'.[38] The problem with the latter idea is that the MC states seem to be parallel to each other, in the sense that they all have the same equilibrium distance and probably the same curvature along the Fe–N coordinate.[6, 27, 28, 30] It also does not explain why the preference would be to feed population into the quintet state at the expense of the ground state. In order to clearly address the issue of the conversion from the MLCT states to the HS states, it is necessary to have means to probe the departure, the intermediate steps (if any) and the arrival of population into the quintet state. Over the past few years there has been an upsurge of studies in ultrafast spectroscopies that aim at specifically addressing this issue.

15.3.1 Departing from the MCLT Manifold

Contrary to their ruthenium analogues, Fe(II)-polypyridine complexes do not exhibit a long lived ^{3}MLCT phosphorescence, due to the fact that the ligand-field states lie at lower energies than the MLCT states.

Broadband femtosecond fluorescence up-conversion was first used on the study of $[Ru(bpy)_3]^{2+}$ by the Lausanne group who could observe both the very short lived ^{1}MLCT for the first time but also the long lived ^{3}MLCT.[55] Decay of the ^{1}MLCT fluorescence (≤30 fs) was reflected in the rise of the ^{3}MLCT emission. The ability to observe the ultrafast depopulation of the ^{1}MLCT by fluorescence up-conversion prompted the same group to investigate the case of $[Fe(bpy)_3]^{2+}$. Figure 15.2a shows a typical 2D time-energy plot of the fluorescence of aqueous $[Fe(bpy)_3]^{2+}$ upon excitation at 400 nm.[56] The time-dependent emission spectra, shown in Figure 15.2b, are obtained by taking slices at fixed time delay, while cuts at fixed emission wavelength provide the kinetic traces shown in Figure 15.2c. Strikingly (and just as for all Ru complexes[55, 57]), the emission centred around 600 nm is already visible at $t = 0$ with a mirror-image like profile with respect to the absorption band (here, in the 500–650 nm region). This behaviour was recently reinvestigated, and showed no dependence on the pump wavelength.[57] The ultrafast relaxation processes occurring in the manifold of ^{1}MLCT states that lead to a mirror-like image (with respect to the absorption band) fluorescence were estimated to occur in <10 fs and were attributed to intramolecular vibrational redistribution (IVR) and internal conversion (IC) processes. The fluorescence appears cold with respect to the high frequency modes of the molecules (which show up as a progression in the absorption band).

Coming back to Figure 15.2a, a weak emission is observed at $\lambda \geq 650$ nm and $t \geq 100$ fs. This observation is confirmed in Figure 15.2b, where the spectrum at $t = 100$ fs exhibits two weak bands of almost identical intensity, the remnant of the fluorescence centred at ~600 nm, the other at ~660 nm. Just as for $[Ru(bpy)_3]^{2+}$,[55]

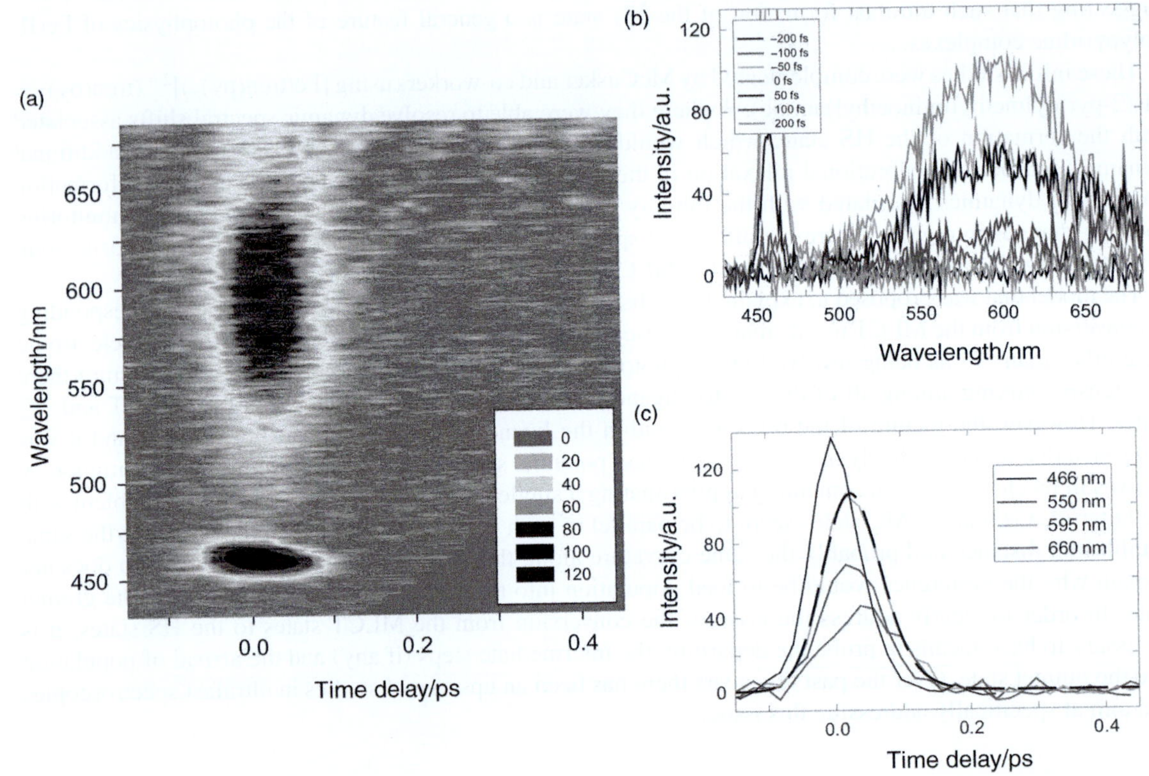

Figure 15.2 *(a) Two-dimensional plot of the femtosecond fluorescence spectra of aqueous* $[Fe(bpy)_3]^{2+}$ *upon excitation at 400 nm. The signal at 466 nm corresponds to a Raman line of* H_2O*. (b) Fluorescence spectra at different time delays. (c) Time traces at different wavelengths. Time delays are taken with respect to the water Raman line. The trace at 595 nm is shown together with its fit function. Reprinted with permission from [56]. Copyright 2008, American Chemical Society.*

the comparison of Raman and emission time traces allow us to extract an exponential decay of 30 ± 10 fs for the ^{1}MLCT fluorescence. The weaker emission at ∼660 nm was found to decay in ∼130 fs and is assigned to the ^{3}MLCT state. Indeed, it lies in the wavelength range where the ^{3}MLCT phosphorescence is expected to occur. These results represent the first observation of fluorescence and phosphorescence of Fe(II) SCO complexes. Interestingly, the decay of the ^{3}MLCT state in $[Fe(bpy)_3]^{2+}$ is comparable to the value of 80 ± 20 fs reported for $[Fe(tren(py)_3)]^{2+}$.[10]

The decay of the ^{3}MLCT state of $[Fe(bpy)_3]^{2+}$ was confirmed by transient absorption measurements. Figure 15.3a shows representative spectra recorded at various time delays within the initial 2 ps after photoexcitation. A short-lived excited state absorption (positive signal) appears in the 340–440 nm region, while in the 560–640 nm range it is present even at longer time delays. The former is a clear signature of the MLCT state, since it corresponds to the absorption of the reduced bpy ligand, as established from spectroelectrochemical studies of Fe(II) polypyridine complexes,[58] and from studies of $[Ru(bpy)_3]^{2+}$ whose ^{3}MLCT state is long-lived.[59] In <200 fs, the short wavelength (< 400 nm) excited state absorption is replaced by a negative signal, which we identify as the ground state bleach (GSB) signal, caused by depletion of the ground state due to the photoexcitation by the pump pulse. This bleach signal is dominant in the 450–550 nm spectral range

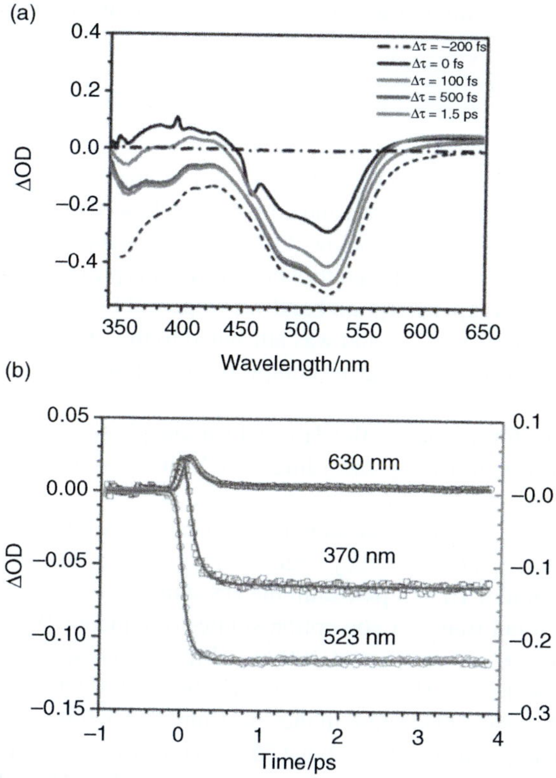

Figure 15.3 *(a) Transient absorption spectra of aqueous $[Fe(bpy)_3]^{2+}$ upon 400 nm excitation. The spectral evolution in the first 1.5 ps after photoexcitation ($t = 0$) is shown. The peak around 400 nm at $t = 0$ is the residual scattered light contribution from the pump pulse. The strong negative absorption feature present in the early spectra around 450–460 nm is the impulsively-induced Raman signal of H_2O. The inverted ground state absorption spectrum is displayed (dashed line) for comparison. (b) Kinetic traces of aqueous $[Fe(bpy)_3]^{2+}$ in a 5 ps time window at three selected wavelengths, together with their fit functions using a global fit model (see ref. 56 for details). Reprinted with permission from [56]. Copyright 2008, American Chemical Society.*

and it fully reflects the ground state absorption. Kinetic traces at three characteristic wavelengths (370 nm, 523 nm and 630 nm) are shown in Figure 15.3b. It can be seen that the positive absorption signals at 370 and 630 nm appear within the experimental temporal resolution of 140 fs but behave differently, as already noted in Figure 15.3a. The 630 nm trace is due to the strong absorption of aqueous electrons, which are produced in small quantities by multiphoton excitation, as established from pump power dependence studies.[60] The analysis and modelling of these TA data[56] delivers three global time constants: 115 ± 10 fs, 960 ± 100 fs and 665 ± 35 ps, that fit simultaneously all the kinetic traces in the spectral range of interest. The shortest component reflects the lifetime of the ^{3}MLCT state, in agreement with the above fluorescence up-conversion measurements, while the longest component reflects the ground state recovery by the decay of the $^5T_2$ state. The intermediate component could not be assigned to any given photoinduced species in a straightforward way, and is either due to the solvated electron or to a 1 ps component of the relaxation within the quintet state, which will be presented below. Finally, neither information about a channel via the intermediate $^{1,3}T$ states, nor about the arrival in the quintet states, could be retrieved from these data, either because these states are spectroscopically silent in the region of our probe wavelength range and/or they are not populated in the subsequent nonradiative cascade.

15.3.2 Arrival into the HS State

Of all the ligand-field states lying lower than the MLCT manifold, the $^5T_2$ state is the only one for which an excited state absorption was known. Hendrickson and co-workers showed that for the complexes $[Fe(tpen)]^{2+}$ and $[Fe(tptn)]^{2+}$, a quintet absorption occurs below 330 nm.[61] The same was established by Hauser in his studies of the absorption spectrum of low temperature crystals of $[Fe(ptz)_6]^{2+}$ entirely photoconverted to the high spin state.[6] Another lower energy absorption band was reported by both groups[6, 62] in the 800–900 nm region but is very weak ($\varepsilon = 5$ lmol^{-1}cm^{-1}), and was attributed to the $^5T_2$–^5E transition between MC states. Finally the UV HS absorption also shows up in complexes that have a high spin ground state, such as $[Fe(tren(6\text{-}CH_3\text{-}py)_3)]^{2+}$.[51]

Smeigh *et al.*[51] probed the formation of the HS state upon photoexcitation of the ^{1}MLCT state of $[Fe(tren(py)_3)]^{2+}$ complex in acetonitrile by recording its UV absorption in the region of 285 to 325 nm. From these results, they estimated that the ^5T state is populated in <250 fs, which was their time resolution. They complemented this result by fs stimulated Raman scattering studies where, by following the CN stretch frequency of the ligand as a function of time, they deduced a population time of the HS state of 190 ± 50 fs, alongside a vibrational relaxation of 10 ± 3 ps within the HS state.

The Lausanne group carried out transient absorption studies of aqueous $[Fe(bpy)_3]^{2+}$ using a white light continuum in the UV, with a cross-correlation of 100 fs.[63] Figure 15.4 shows a representative selection of UV TA spectra at different time delays. The positive region between 300 nm and ~330 nm corresponds to an excited state absorption (ESA), while elsewhere in the spectrum a bleach signal is dominant, which matches quite well the ground state absorption spectrum above 350 nm (see Fig. 15.3). The ESA is due to the quintet state, as evidenced by its decay time of 650–700 ps (see bottom inset, Fig. 15.4), but the kinetics also exhibits a short component of 2–3 ps. By normalizing the absorption spectrum to the TA spectrum above 350 nm, which consists only of the ground state bleach (GSB) signal, we could retrieve the absorption spectrum of

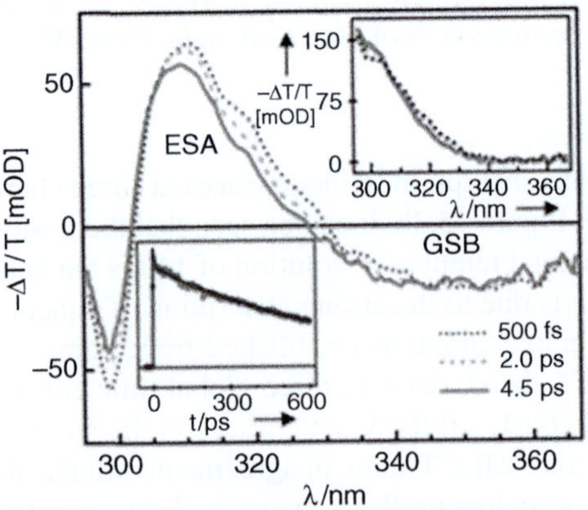

Figure 15.4 *UV transient absorption spectra at different time delays upon excitation of aqueous $[Fe(bpy)_3]^{2+}$ at 530 nm. The lower inset shows a long-time kinetic trace recorded at 315 nm. The upper inset shows the excited state absorption spectra at different time delays, obtained by subtracting the ground state spectrum from the transient spectra. Reprinted with permission from [63]. Copyright 2008 WILEY-VCH Verlag GmbH & Co. KGaA.*

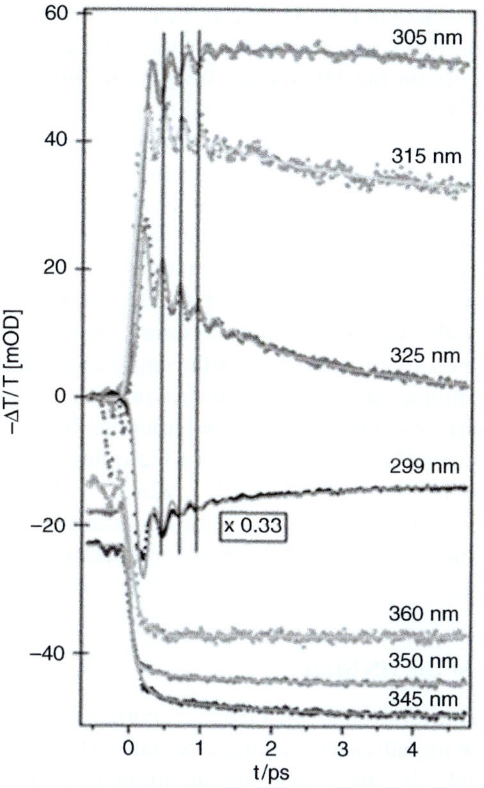

Figure 15.5 *Kinetic traces at different probe wavelengths covering the ESA (305, 315 and 325 nm) and the bleach regions (299, 345, 350 and 360 nm). Solid lines represent the fits with a bimodal relaxation kinetics and an oscillatory contribution. The vertical lines show the phase shift of $\sim\pi$ of the oscillations between the blue-most and the red-most probe wavelengths in the ESA region. For clarity, time-traces at $\lambda > 340$ nm have been vertically shifted and the 299 nm one is scaled to $^1/_3$. Reprinted with permission from [63]. Copyright 2008 WILEY-VCH Verlag GmbH & Co. KGaA.*

the quintet state at different time delays (top inset, Fig. 15.4). The retrieved spectra agree well with those of stable and metastable HS Fe(II) complexes.[6, 47, 49, 51]

Kinetic traces between 300 to 360 nm over the first 5 ps are shown in Figure 15.5. The traces in the ESA region are all characterized by a fast rise followed by an oscillatory pattern and a longer decaying component. The oscillations at the blue and red edges of the ESA (300 and 325 nm) exhibit a phase shift close to π (vertical lines), which is typical of a vibrational wave packet oscillating between the turning points of the potential surface on which it evolves. These traces were analysed in detail in ref. 63, with the experimental data described in terms of MLCT ESA and ground state bleach (GSB) contributions being instantaneously excited and decaying with 130 fs and 650 ps, respectively, while the quintet state ESA rises in 130 fs. Bimodal dynamics with 1.1 ± 0.17 ps and 3.4 ± 1.2 ps time-constants were necessary to obtain optimal fits at short times, along with an oscillatory pattern with a period of 254 ± 2 fs, independent of probe wavelength. However, its damping constants varied between $\sim$400 to $\sim$800 fs, depending on the probe wavelength. The results show good agreement with the experimental traces over the whole spectral range (Fig. 15.5). The 1.1 and 3.4 ps components describe a narrowing of the quintet state absorption band due to vibrational cooling, as actually evidenced by increase of its maximum and the presence of an isobestic point around 305 nm

(Fig. 15.4, top inset). The 1.1 ps component may well be associated with the 0.96 ps component derived from the global fit of the visible TA data (Fig. 15.3), in which case the latter cannot be due to the solvated electron.

A vibrational wave packet occurs on the HS surface because it only appears in the region of the ESA region, and is missing in the GSB traces above 325 nm and in the visible region.[56] The reaction coordinate of the photoexcited LS–HS spin change is the Fe–N bond stretch (Fig. 15.1), whose vibrational frequency is ~220 cm^{-1} (~150 fs) in the HS state. Hence, the observed wave packet is not directly excited by the laser pump pulse, but by the ultrafast elongation of the Fe–N bonds, which excites bending modes due an impulsive change of N–Fe–N angle. Based on DFT calculations[64] we attribute the observed oscillation to a combination of chelate ring modes, benzene cycle bending modes and N–Fe–N bending modes. The indirect excitation of wave packets of low frequency modes via laser excited (not necessarily in the form of wave packets) higher frequency ones is rather common in many systems, including metalloporphyrins[65] and organic molecules.[66] The invariance of the oscillation period over the entire range of probe wavelengths suggests that the wave packet is created in a harmonic part of the HS surface on which it evolves, that is near its minimum. Alternatively, it is the potential surface itself which is very harmonic, but this seems unlikely.

Most notable in relation to the population mechanism of the quintet state is the fact that the rise of the quintet absorption corresponds to the decay of the ^{3}MLCT, which implies that relaxation into the $^5T_2$ state bypasses intermediate MC states. This conclusion was further confirmed in the case of $[Fe(phen)_3]^{2+}$ in acetonitrile by Tribollet *et al.*[67] who measured a ^{3}MLCT→$^5T_2$ relaxation time of 220 fs.

15.3.3 Vibrational relaxation of the HS State

A direct ^{3}MLCT to $^5T_2$ transition implies that ~1.3 eV[27, 56], corresponding to the energy difference of the relative minima, is stored as vibrational energy in the HS state. The damping times of the wave packet represent a lower limit of its dephasing time since, as mentioned above, they also depend on the probe window. The dephasing causes the coherent (oscillatory) wave packet dynamics to turn into an incoherent kinetic regime (exponential). Given that the largest dephasing time we found is ~800 fs, we associate the 1.1 ps decay component to the dephasing of the 130 cm^{-1} mode. Concerning the 3.4 ps decay, we associate it to the relaxation time of the (incoherently) excited Fe–N stretch mode in the quintet state. Since the ^{1,3}MLCT emission is already vibrationally cold at very early times, at least as far as the high frequency Franck–Condon modes are concerned,[57] upon crossing into the quintet state only the Fe–N stretch and the N–Fe–N chelate modes are excited, which are responsible for the bimodal relaxation regime.

The issue of vibrational energy storage and relaxation in the quintet state of Fe(II) polypyridine complexes in solution is important for the understanding of the spin photoswitching in crystals. It has mainly been studied by time-resolved resonance Raman spectroscopy.[35, 38] Recently McCusker, Mathies and co-workers used fs stimulated Raman scattering to study the vibrational relaxation in the HS state of $[Fe(tren(py)_3)]^{2+}$ in acetonitrile,[51] and reported a bimodal time evolution of the high frequency C–N stretch mode with time constants of 190 ± 50 fs and 10 ± 3 ps. A previous TA study had inferred a relaxation time of 8 ± 3 ps for $[Fe(tren(py)_3)]^{2+}$.[10] The 10 ps component was attributed to vibrational cooling, while the 190 fs component was associated with the structural change from LS to HS. Wolf *et al.*[68] reported on a sub-ps IR study of $[Fe(btpa)]^{2+}$ and $[Fe(b(bdpa))]^{2+}$ in the 1000–1100 cm^{-1} region, and found relaxation times of up to 13 ps. Thus, contrary to the case of the electronic relaxation, vibrational relaxation in the HS state depends on the complex, which is logical since the vibrational modes change. Nevertheless, the dissipation of excess vibrational energy is still not fully understood, nor is the role of the solvent clearly described. It is likely that in crystals, the vibrational cooling is slower, but the amount of energy dissipated per molecule would be sufficient to switch unexcited molecules to the HS state, simply by heating.

15.4 Ultrafast X-Ray Studies

Probing the structure of the HS state in SCO complexes using X-ray diffraction (XRD) or X-ray absorption spectroscopy (XAS) has been carried out on crystals whose HS state is long-lived and can be populated by pressure, light or temperature.[18, 34, 43, 69–72] In all the cases studied the Fe–N bond elongation from the LS to the HS state amounted to ~0.2 Å, as a result of populating the antibonding $d(e_g)$ orbitals. The derivation of the structure of short lived HS states is not possible by quasi-steady state XRD or XAS methods, which need to be pushed into the short time domain in order to capture the short lived HS state. In recent years, picosecond XRD[19] and XAS[73–75] have been developed, which enable the probing of transient structures with the 50–100 ps resolution of synchrotrons. XAS in particular offers the advantage of being chemically selective since one tunes to the core transitions of a specific atom, and of providing information about the electronic structure of the intermediate via the pre-edge absorption bands at the K-edges (which are due to transitions from the 1*s* core orbital) in the hard X-ray range, or the edge bands at the L-edges in the soft X-ray range. In particular, the $L_{2,3}$ edges are due to transitions from the $2p_{1/2,3/2}$ core orbitals and are dipole allowed to the 3*d* valence orbitals, which are exactly those involved in the SCO process (Fig. 15.1). In addition, molecular structure can be extracted from analysis of the multiple scattering XANES (X-ray near edge absorption structure) features, and the single scattering EXAFS (extended X-ray absorption fine structure). Finally, XAS can be implemented in any type of medium, in particular in solutions, in which most of the short-lived SCO complexes have been identified.[38, 46–49]

As previously mentioned, for Fe(II) complexes with long-lived metastable HS states, quasi-static X-ray diffraction and X-ray absorption spectroscopy studies determined the Fe–N bond elongation to be ~0.2 Å with respect to the ground state, independent of the ligands.[34] For short-lived HS states with ns and sub-ns lifetimes, one may ask if the bond elongation is also similar. Quantum-chemical calculations indeed predicted this to be the case for various Fe(II) complexes.[26–30] In order to confirm it, Khalil *et al.*[20] determined the structure of $[Fe(tren(py)_3)]^{2+}$ in acetonitrile solution, whose HS lifetime is 60 ns, using picosecond XAS,[76] and found an elongation of 0.21 ± 0.03 Å. In the case of $[Fe(bpy)_3]^{2+}$, the HS lifetime is 650 ps and with the 70 ps time resolution of the Swiss Light Source synchrotron, the transient HS structure was determined by the Lausanne group.

Figure 15.6a shows the static K-edge XAS spectrum of a 25 mM aqueous $[Fe(bpy)_3]^{2+}$complex. It is characterized by a number of XANES features, which are displayed in the inset and have already been discussed for similar ferrous polypyridine complexes.[70, 77] Suffices it to say that all have been shown in steady-state XAS studies to undergo significant changes upon light-, pressure- or temperature-induced spin-transitions.[42, 43, 70–72, 77] It is important to note that the B-band is a multiple scattering (i.e. a structural) XANES feature, as was clearly shown in ref. 70. Most of these changes have been attributed to changes of metal–ligand and intra-ligand bond distances and angles, with the Fe–N bond being the dominant contributor. Additional changes in the high energy (EXAFS) region were also clearly observed, which likewise point to a significant Fe–N bond change.[42, 43, 72, 78, 79]

Figure 15.6b shows the transient difference XAS spectrum measured 50 ps after laser excitation. All the aforementioned changes are indeed occurring as a result of laser excitation and transient population of the $^5T_2$ state. That the latter is responsible for these changes is confirmed by the inset showing the temporal evolution of the absorption changes at the B-feature (7126 eV), which maps the lifetime of the quintet state measured by optical laser spectroscopy.[56] The structural analysis of the excited state based on the transient XANES and EXAFS spectra deliver the same Fe–N bond elongation of $\Delta R_{Fe\text{–}N} = 0.20$ Å. Thus, along with the results of Khalil *et al.* for a complex with a 60 ns HS lifetime,[20] and of the literature for metastable HS states,[34] our result on the shortest-lived HS complex shows that the Fe–N bond elongation in the HS state is indeed nearly the same for all Fe(II) polypyridine complexes. This is therefore not the parameter that controls the quintet

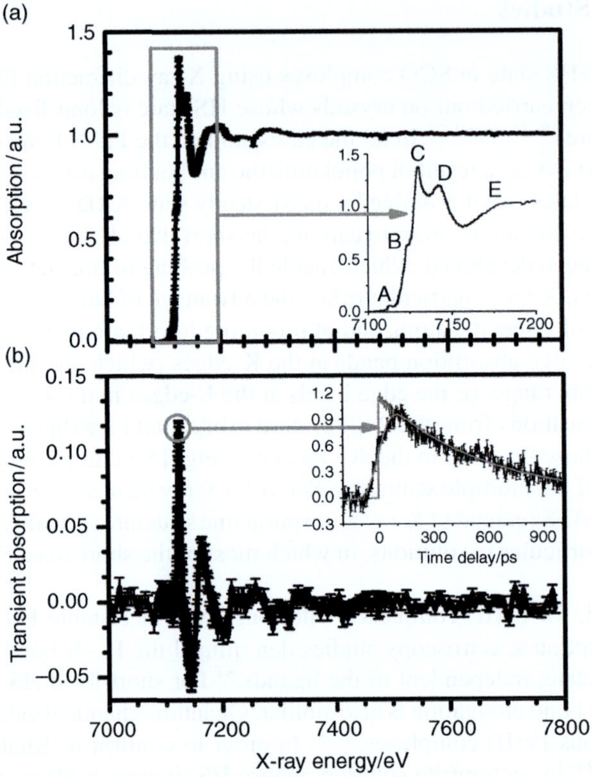

Figure 15.6 *(a) K-edge X-ray absorption spectrum of the ground LS state of aqueous* $[Fe(bpy)_3]^{2+}$. *The inset zooms into the first 100 eV of the spectrum and displays the details of the absorption edge and the nearest-lying XANES resonances (labelled from A–D). The onset of the EXAFS range is represented by the E feature. (b) Transient difference spectrum, recorded 50 ps after the laser pump pulse, with its error bars. The inset displays the kinetics traces of the transient X-ray absorption signal recorded at 7126 eV (feature B – black line) as compared to the optical signal recorded in transmission at 523 nm (grey line).*

state lifetime. Rather, the adiabatic energy and the couplings between low spin and high spin state are the crucial parameters. Indeed, of all Fe(II) SCO complexes, $[Fe(bpy)_3]^{2+}$ has the highest lying quintet state.

In order to follow in real-time the change of structure from the ^{1}MLCT state to the $^5T_2$ state, we carried out a femtosecond XANES study of $[Fe(bpy)_3]^{2+}$ exploiting the fact that the multiple-scattering B-feature is sensitive to the Fe–N bond elongation, as seen in Figure 15.6b. This sensitivity to the Fe–N bond length is supported by simulations based on multiple scattering theories.[23] Thus, because the $^{1,3}T$ states all lie at an intermediate distance between the LS and HS states (Fig. 15.1), and considering that they have no known spectroscopic signatures in the optical domain, one should in principle be able to identify them in the XANES spectrum by the amplitude of the B-feature if they are at all involved in the relaxation cascade. However, this necessitates tuneable femtosecond hard X-ray pulses since the population time of the quintet state is ~130 fs (see Section 15.3.1). Using the pulses generated by the slicing scheme at the Swiss Light Source,[80] the Lausanne team carried out an optical pump/X-ray probe experiment following the evolution of the B-feature as a function of pump-probe delay. Figure 15.7 presents the time scan obtained at the B-feature (7.122 keV) showing that the signal stabilizes from about 300 fs up to the scan limit of 10 ps (see inset), which is evidence

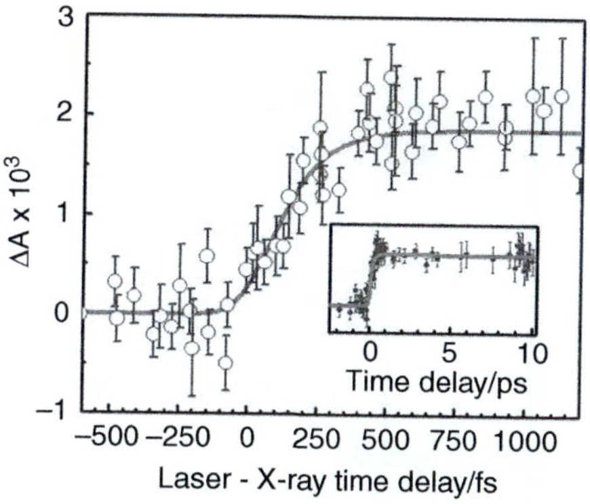

Figure 15.7 *Femtosecond XANES results: time scan of the signal (circles) at the B-feature (Fig. 15.6), as a function of laser pump/X-ray probe time delay after excitation of aqueous* $[Fe(bpy)_3]^{2+}$ *at 400 nm. The inset shows a long time scan up to 10 ps time delay. The solid line is the simulated signal assuming a simple 4-step kinetic model* $^1A_1 \rightarrow {}^1MLCT \rightarrow {}^3MLCT \rightarrow {}^5T$ *to describe the spin conversion process. See ref. 23 for details.*

that the system is already in the HS state. Further evidence is the energy scan at 300 fs, which reproduces the transient spectrum at 50 ps time resolution (see Fig. 2 of ref. 23). The fit in Figure 15.7 shows that the quintet state is reached in 150 ± 50 fs, confirming the above result using broadband UV TA (Section 15.3.2), which implies that the ultrafast spin conversion is a simple three-step ^{1}MLCT–^{3}MLCT–$^5T_2$ cascade that bypasses the intermediate $^{1,3}T$ states.

Later, Huse *et al.* carried out ps[24] and fs[25] studies at the Fe $L_{2,3}$ edges of the ultrafast SCO of $[Fe(tren(py)_3)]^{2+}$ in solution. These experiments are particularly challenging as the Fe L edges lie in the 700 eV region, which means that one needs to work under vacuum. They achieved this *tour de force* using liquid cells equipped with soft X-ray transparent silicon nitride windows. While their results confirmed the scenario already presented in refs 15 and 23, they do represent a precedent in demonstrating the possibility of carrying out ultrafast XAS at the Fe L-edges, which provides more details on the electronic structure than the K-edge, which is more adequate for molecular structure. Another tool that delivers much insight into the electronic structure of species is X-ray emission spectroscopy (XES). Indeed, while XAS probe the density of unoccupied states, XES probes that of unoccupied states and is a valuable tool to extract information about the spin structure of the system. In a proof of principle experiment, Vanko *et al.*[81] recorded, with a resolution of 70 ps, the changes of the $K_{\alpha 1}$ XES spectrum of Fe upon photoexcitation of aqueous $[Fe(bpy)_3]^{2+}$.

15.5 Summary and Outlook

The combination of cutting-edge ultrafast optical and X-ray spectroscopic tools allows one to retrieve the complete photocycle of Fe^{II} SCO complexes in solution. Here we focused on the model compound $[Fe(bpy)_3]^{2+}$ in water, and the various steps involved in its photocycle are shown in Figure 15.8. The first event is an ultrafast sequence of IVR/IC in the ^{1}MLCT state and ISC to the ^{3}MLCT state, followed by a second ISC into the quintet state, leaving the latter with a significant excess of vibrational energy. The HS undergoes vibrational relaxation in a bimodal fashion, which we believe to be due to the relaxation of the Fe–N elongation mode and of the bending and chelate modes of bpy. The latter are coherently excited by the impulsive Fe–N bond elongation, giving rise to wave packet oscillations. Finally, the return to the ground state occurs on a much slower timescale of 650 ps. The main conclusion here is that population of the HS

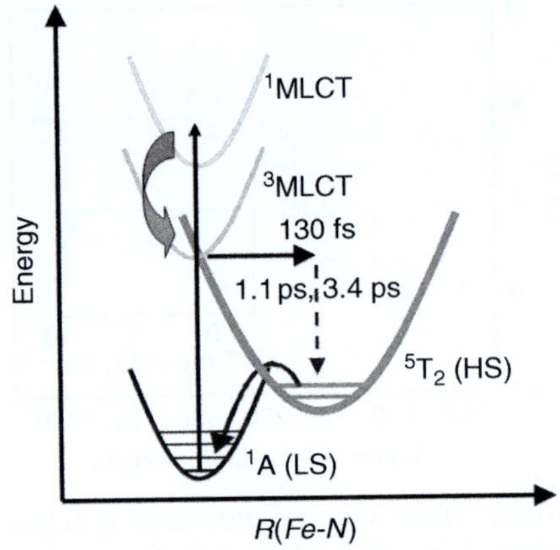

Figure 15.8 *The full photocycle of the ultrafast light-induced spin-crossover process in [Fe(bpy)$_3$]$^{2+}$ in water at room temperature.*

state proceeds directly and on an ultrafast timescale (<200 fs) from the ^{1,3}MLCT manifold, bypassing the intermediate MC states. This mechanism is most probably identical in the other Fe(II) SCO complexes so far studied by ultrafast spectroscopy.[25, 51, 67]

The direct population of the HS state from the MLCT manifold resolves a number of issues that were debated in the literature. The exclusion of the $^{1,3}T$ states solves the problem of multiple ultrafast ISC steps, which have to occur on an ultrafast timescale between states that are quasi-parallel with respect to the Fe–N coordinate (i.e. with poor vibrational overlap, Fig. 15.1). In addition, the unit quantum efficiency of the SCO process from the ^{1}MLCT state into the quintet state now makes sense in the context of excluding any leakage back to the ground state through the bypassing of the $^{1,3}T$ states. It is also interesting to note that the $^1A_1 \rightarrow ^{1,3}$MLCT$\rightarrow ^5T_2$ mechanism we unravel is quite close to the mechanism proposed by Hendrickson and co-workers in 1993, except that they did not include the ^{3}MLCT state in their cascade.[47] However, the timescale of the ^{1}MLCT$\rightarrow ^3$MLCT ISC,[56] the presence of a weak ^{3}MLCT absorption on the red edge of the ^{1}MLCT absorption, and the fact that the latter has an absorption coefficient that is one order of magnitude smaller than that of organic dyes with pure singlet states, suggest that the singlet and triplet MLCT states are strongly mixed. Surprisingly, the initially proposed relaxation mechanism was later replaced by a mechanism invoking the intermediate $^{1,3}T$ states that would be strongly mixed. The latter statement does not seem to be supported by the low temperature steady-state spectroscopic studies,[6] which point to a clear classification of the LF states according to their spin quantum number.

The proposed mechanism is also in line with the steady-state spectroscopic studies at cryogenic temperatures, which showed that excitation into the $^{1,3}T$ states leads to population of the $^5T_2$ state with a quantum efficiency of ~80%.[6] However, for excitation of the ^{1}MLCT state, the relaxation process to the HS state was reported to occur with 100% efficiency at both 10 K[82] and at room temperature.[45, 52] This high efficiency suggested a direct decay of the ^{1}MLCT state to the HS state, since including the intermediate $^{1,3}T$ states can only decrease the conversion efficiency. The timescale of ~150 fs for the MLCT$\rightarrow ^5T_2$ conversion corresponds to about two oscillations of the Fe–N stretch vibration,[33] suggesting a non-Born–Oppenheimer process. Excitation of the MC states also populates the HS state with high efficiency,[6] but the pathways and their timescales still need to be determined.

Given the common electronic, structural and ultrafast kinetic properties of Fe(II) complexes, we believe that the ultrafast spin relaxation in [Fe(bpy)$_3$]$^{2+}$ is representative for the entire family of such complexes.

However, the vibrational relaxation kinetics differs from complex to complex, just as the well-studied HS to LS relaxation back to the ground state.

There are, however, still some outstanding issues that need to be solved in order to fully describe the ultrafast light-induced SCO process. First, it is still not clear why the MC quintet state is exclusively populated from the MLCT manifold of states, while MC triplet and singlet states are avoided although they cross the MLCT states near the same point as the quintet state. Theoretical work has attempted to answer this question. Chang *et al.*[83] proposed a cascade with an intermediate step involving a ^{5}MLCT state, instead of the direct conversion of the ^{3}MLCT to the final HS state. This means that the complete SCO takes place in the MLCT manifold on a very short timescale. Once the system is in the quintet spin coupling, there is no way back through the triplet and singlet coupled ligand field states. The weakness of this model is that experimental evidence for a ^{5}MLCT state is lacking. Indeed, the absorption spectrum of the metastable HS state has been recorded in low temperature crystals and it shows only the red-infrared band due to the ^5T–^5E transition between MC states,[6] and the UV band similar to that reported in Figure 15.4. de Graaf and Sousa[31] also rule out a role for the ^{5}MLCT state, as their estimated energy places it too high to account for the cascade.

The understanding of the exact mechanism leading to population of the HS state is not just of relevance for simple Fe(II) polypyridine complexes. It has been reported with similar timescales in mono- and homodinuclear and heterodinuclear six-coordinate $[Fe\text{-}M(PMK)_3]^{2+}$ (M = Fe, Zn; PMK= bis-bidentate ligand 2-(pyridylmethyl)ketazine) complexes.[84] In addition, ultrafast spin $\Delta S = 2$ changes have been reported in metalloporphyrins[85] upon excitation of the Soret band and quite remarkably, they occur on the same timescale (~150 fs) as reported here.

Secondly, the excitation of the ^{1,3}MC states has also been found to populate the ^5T state with high efficiency (~80%).[6] This is quite surprising as these states also cross the ground state potential surface and in some case they share the same spin with it. It therefore seems surprising that the nonradiative channels funnel the population preferentially into the HS state. This behaviour needs to be clarified by time-resolved studies that would selectively excite the ^{1,3}MC states.

Lastly, there is growing evidence of a solvent effect on the electronic structure of the dissolved molecules,[86] in addition to its effects on their relaxation dynamics.[87–89] For the ultrafast SCO process, only one theoretical study has so far addressed this issue specifically,[32] showing the involvement of the solvent in the process of HS state formation. A recent ultrafast experimental study on a diplatinum complex by van der Veen *et al.*[90] showed a clear solvent dependence of the ISC rate.

Future studies should address the above questions with a multiple-method approach as presented here. With the advent of novel tools such as ultrafast soft X-ray absorption spectroscopy at the L-edge, time-resolved X-ray emission spectroscopy and ultrafast photoelectron spectroscopy of liquid jets,[91] there is no doubt that new insight will be reached.

Acknowledgements

This work was funded by the Swiss National Science Foundation (FNRS), via contracts 110464, 107956, 620-066145 and 105239, and by the Staatssekretariat für Bildung und Forschungcontract COST D35 C06.0016. Great thanks to all my co-workers who contributed to this work: W. Gawelda, C. Consani, M. Prémont-Schwarz, V.-T. Pham, A. ElNahhas, R. M. van der Veen, C. Bressler, C.J. Milne, T. Penfold and F. van Mourik. We also acknowledge the great help of the team at the MicroXAS beamline of the Swiss Light Source (PSI, Villigen): S. Johnson, P. Beaud, D. Grolimund, C. N. Borca, G. Ingold, R. Abela. Thanks also to M. Benfatto (Rome) for theoretical support and A. Hauser (Geneva) for many fruitful discussions and for samples.

References

1. Gütlich, P., Goodwin, H. A. (2004) Spin crossover – an overall perspective, in: Gütlich, P., Goodwin, H. A. (Eds) Spin Crossover in Transition Metal Compounds I. Top. Curr. Chem., 233: 1–47.
2. McGarvey, J. J., Lawthers, I., Heremans, K., Toftlund, H. (1984) Spin-state relaxation dynamics in iron(II) complexes: solvent on the activation and reaction and volumes for the $^1A \rightleftharpoons {}^5T$ interconversion. J. Chem. Soc. Chem. Commun., 1575–1576.
3. Lawthers, I., McGarvey, J. J. (1984) Spin-state relaxation dynamics in iron(III) complexes: photochemical perturbation of the $^2T \rightleftharpoons {}^6A$ spin equilibrium by pulsed-laser irradiation in the ligand-to-metal charge-transfer absorption band. J. Am. Chem. Soc., 106: 4280–4282.
4. Decurtins, S., Gutlich, P., Hasselbach, K. M., Hauser, A., Spiering, H. (1985) Light-induced excited-spin-state trapping in iron(II) spin-crossover systems. Optical spectroscopic and magnetic susceptibility study. Inorg. Chem., 24: 2174–2178.
5. Hauser, A. (2004) Ligand field theoretical considerations, in: Gütlich, P., Goodwin, H. A. (Eds) Spin Crossover in Transition Metal Compounds I. Top. Curr. Chem., 233: 49–58.
6. Hauser, A. (2004) Light-induced spin crossover and the high-spin→low-spin relaxation, in: Gütlich, P., Goodwin, H. A. (Eds) Spin Crossover in Transition Metal Compounds II. Top. Curr. Chem., 234: 155–198.
7. Létard, J.-F., Guionneau, P., Goux-Capes, L. (2004) Towards spin crossover applications, in: Gütlich, P., Goodwin, H. A. (Eds) Spin Crossover in Transition Metal Compounds III. Top. Curr. Chem., 235: 221–249.
8. Bousseksou, A., Molnár, G., Salmon, L., Nicolazzi, W. (2011) Molecular spin crossover phenomenon: recent achievements and prospects. Chem. Soc. Rev., 40: 3313–3335.
9. Ferrere, S., Gregg, B. A. (1998) Photosensitization of TiO_2 by $[Fe^{II}(2,2'\text{-bipyridine-4,4'-dicarboxylic acid})_2(CN)_2]$: band selective electron injection from ultra-short-lived excited states. J. Am. Chem. Soc., 120: 843–844.
10. Monat, J. E., McCusker, J. K. (2000) Femtosecond excited-state dynamics of an iron(ii) polypyridyl solar cell sensitizer model. J. Am. Chem. Soc., 122: 4092–4097.
11. Ferrere, S. (2002) New photosensitizers based upon $[Fe^{II}(L)_2(CN)_2]$ and $[Fe^{II}L_3]$, where L is substituted 2,2′-bipyridine. Inorg. Chim. Acta, 329: 79–92.
12. Xia, H.-L., Ardo, S., Sarjeant, A. A. N., Huang, S. X., Meyer, G. J. (2009) Photodriven spin change of Fe(II) benzimidazole compounds anchored to nanocrystalline TiO_2 thin films. Langmuir, 25: 13641–13652.
13. Grätzel, M. (2009) Recent advances in sensitized mesoscopic solar cells. Acc. Chem. Res., 42: 1788–1798.
14. Juban, E. A., Smeigh, A. L., Monat, J. E., McCusker, J. K. (2006) Ultrafast dynamics of ligand-field excited states. Coord. Chem. Rev., 250: 1783–1791.
15. Cannizzo, A., Milne, C. J., Consani, C., Gawelda, W., Bressler, C., van Mourik, F., Chergui, M. (2010) Light-induced spin crossover in Fe(II)-based complexes: the full photocycle unraveled by ultrafast optical and X-ray spectroscopies. Coord. Chem. Rev., 254: 2677–2686.
16. Glijer, D., Hébert, J., Trzop, E., Collet, E., Toupet, L., Cailleau, H., Matouzenko, G. S., Lazar, H. Z., Létard, J.-F., Koshihara, S., Buron-Le Cointe, M. (2008) Photoinduced phenomena and structural analysis associated with the spin-state switching in the $[Fe^{II}(DPEA)(NCS)_2]$ complex. Phys. Rev. B, 78: 134112/1–134112/9.
17. Moisan, N., Servol, M., Lorenc, M., Tissot, A., Boillot, M.-L., Cailleau, H., Koshihara, S., Collet, E. (2008) Towards ultrafast spin-state switching in the solid state. C. R. Chimie, 11: 1235–1240.
18. Ichiyanagi, K., Hébert, J., Toupet, L., Cailleau, H., Guionneau, P., Létard, J.-F., Collet, E. (2006) Nature and mechanism of the photoinduced spin transition in $[Fe(PM\text{-}BiA)_2(NCS)_2]$. Phys. Rev. B, 73: 060408/1–060408/4.
19. Cailleau, H., Lorenc, M., Guérin, L., Servol, M., Collet, E., Buron-Le Cointe, M. (2010) Structural dynamics of photoinduced molecular switching in the solid state. Acta Cryst. Sect. A, 66: 189–197.
20. Khalil, M., Marcus, M. A., Smeigh, A. L., McCusker, J. K., Chong, H. H. W., Schoenlein, R. W. (2006) Picosecond X-ray absorption spectroscopy of a photoinduced iron(ii) spin crossover reaction in solution. J. Phys. Chem. A, 110: 38–44.
21. Gawelda, W., Pham, V. T., Benfatto, M., Zaushitsyn, Y., Kaiser, M., Grolimund, D., Johnson, S. L., Abela, R., Hauser, A., Bressler, C., Chergui, M. (2007) Structural determination of a short-lived excited iron(II) complex by picosecond x-ray absorption spectroscopy. Phys. Rev. Lett., 98: 057401/1–057401/4.

22. Gawelda, W., Pham, V. T., Van der Veen, R. M., Grolimund, D., Abela, R., Chergui, M., Bressler, C. (2009) Structural analysis of ultrafast extended X-ray absorption fine structure with subpicometer spatial resolution: application to spin crossover complexes J. Chem. Phys., 130: 124520/1–124520/9.
23. Bressler, C., Milne, C., Pham, V. T., El Nahhas, A., van der Veen, R. M., Gawelda, W., Johnson, S., Beaud, P., Grolimund, D., Kaiser, M., Borca, C. N., Ingold, G., Abela, R., Chergui, M. (2009) Femtosecond XANES study of the light-induced spin crossover dynamics in an iron(II) complex. Science, 323: 489–492.
24. Huse, N., Kim, T. K., Jamula, L., McCusker, J. K., de Groot, F. M. F., Schoenlein, R. W. (2010) Photo-induced spin-state conversion in solvated transition metal complexes probed via time-resolved soft x-ray spectroscopy. J. Am. Chem. Soc., 132: 6809–6816.
25. Huse, N., Cho, H., Hong, K., Jamula, L., de Groot, F. M. F., Kim, T. K., McCusker, J. K., Schoenlein, R. W. (2011) Femtosecond soft X-ray spectroscopy of solvated transition-metal complexes: deciphering the interplay of electronic and structural dynamics. J. Phys. Chem. Lett., 2: 880–884.
26. Rodriguez, J. H. (2005) Ground- and excited-state electronic structure of an iron-containing molecular spin photo-switch. J. Chem. Phys., 123: 094709/1–094709/6.
27. Daku, L. M. L., Vargas, A., Hauser, A., Fouqueau, A., Casida, M. E. (2005) Assessment of density functionals for the high-spin/low-spin energy difference in the low-spin iron(II) tris(2,2′-bipyridine) complex. ChemPhysChem, 6: 1393–1410.
28. Ordejon, B., de Graaf, C., Sousa, C. (2008) Light-induced excited-state spin trapping in tetrazole-based spin crossover systems. J. Am. Chem. Soc., 130: 13961–13968.
29. Suaud, N., Bonnet, M. L., Boilleau, C., Labeguerie, P., Guihery, N. (2009) Light-induced excited spin state trapping: ab initio study of the physics at the molecular level. J. Am. Chem. Soc., 131: 715–722.
30. de Graaf, C., Sousa, C. (2010) Study of the light-induced spin crossover process of the $[Fe^{II}(bpy)_3]^{2+}$ complex. Chem. Eur. J. 16: 4550–4556.
31. De Graaf, C., Sousa, C. (2011) On the role of the metal-to-ligand charge transfer states in the light-induced spin crossover in $Fe^{II}(bpy)_3$. Int. J. Quantum Chem., 111: 3385–3393.
32. Daku, L. M. L., Hauser, A. (2010) Ab initio molecular dynamics study of an aqueous solution of $[Fe(bpy)_3](Cl)_2$ in the low-spin and in the high-spin states. J. Phys. Chem. Lett., 1: 1830–1835.
33. Tuchagues, J. P., Bousseksou, A., Molnar, G., McGarvey, J. J., Varret, F. (2004) The role of molecular vibrations in the spin crossover phenomenon, in: Gütlich, P., Goodwin, H. A. (Eds) Spin Crossover in Transition Metal Compounds III. Top. Curr. Chem., 235: 85–103.
34. Guionneau, P., Marchivie, M., Bravic, G., Létard, J.-F., Chasseau, D. (2004) Structural aspects of spin crossover. Example of the $[Fe^{II}L_n(NCS)_2]$ complexes, in: Gütlich, P., Goodwin, H. A. (Eds) Spin Crossover in Transition Metal Compounds II. Top. Curr. Chem., 234: 97–128.
35. Brady, C., Callaghan, P. L., Ciunik, Z., Coates, C. G., Døssing, A., Hazell, A., McGarvey, J. J., Schenker, S., Toftlund, H., Trautwein, A. X., Winkler, H., Wolny, J. A. (2004) Molecular structure and vibrational spectra of spin-crossover complexes in solution and colloidal media: resonance Raman and time-resolved resonance Raman studies. Inorg. Chem., 43: 4289–4299.
36. Huby, N., Guérin, L., Collet, E., Toupet, L., Ameline, J. C., Cailleau, H., Roisnel, T., Tayagaki, T., Tanaka, K. (2004) Photoinduced spin transition probed by x-ray diffraction. Phys. Rev. B, 69: 020101/1–020101/4.
37. Lorenc, M., Hébert, J., Moisan, N., Trzop, E., Servol, M., Buron-Le Cointe, M., Cailleau, H., Boillot, M.-L., Pontecorvo, E., Wulff, M., Koshihara, S., Collet, E. (2009) Successive dynamical steps of photoinduced switching of a molecular Fe(III) spin-crossover material by time-resolved X-ray diffraction. Phys. Rev. Lett., 103: 028301/1–028301/4.
38. Brady, C., McGarvey, J. J., McCusker, J. K., Toftlund, H., Hendrickson, D. N. (2004) Time-resolved relaxation studies of spin crossover systems in solution, in: Gütlich, P., Goodwin, H. A. (Eds) Spin Crossover in Transition Metal Compounds III. Top. Curr. Chem., 235: 1–22.
39. Daku, L. M. L., Vargas, A., Hauser, A. (2008) Private communication.
40. Ando, H., Nakao, Y., Sato, H., Sakaki, S. J. (2007) Theoretical study of low-spin, high-spin, and intermediate-spin states of $[Fe^{III}(pap)_2]^+$ (pap = N-2-pyridylmethylidene-2-hydroxyphenylaminato). Mechanism of light-induced excited spin state trapping. Phys. Chem. A, 111: 5515–5522.

41. Figgis, B. N., Hitchman, M. A. (2000) Ligand Field Theory and its Applications. Wiley-VCH, New York, p. 376.
42. Okamoto, K., Nagai, K., Miyawaki, J., Kondoh, H., Ohta, T. (2003) XAFS study on the photoinduced spin transition of $[Fe(2\text{-}pic)_3]Cl_2{\cdot}C_2H_5OH$. Chem. Phys. Lett., 371: 707–712.
43. Oyanagi, H., Tayagaki, T., Tanaka, K. (2004) Photo-induced phase transitions probed by X-ray absorption spectroscopy: Fe(II) spin crossover complex. J. Phys. Chem. Solids., 65: 1485–1489.
44. Ronayne, K. L., Paulsen, H., Höfer, A., Dennis, A. C., Wolny, J. A., Chumakov, A. I., Schünemann, V., Winkler, H., Spiering, H., Bousseksou, A., Gütlich, P., Trautwein, A. X., McGarvey, J. J. (2006) Vibrational spectrum of the spin crossover complex $[Fe(phen)_2(NCS)_2]$ studied by IR and Raman spectroscopy, nuclear inelastic scattering and DFT calculations. Phys. Chem. Chem. Phys., 8: 4685–4693.
45. Creutz, C., Chou, M., Netzel, T. L., Okumura, M., Sutin, N. (1980) Lifetimes, spectra, and quenching of the excited states of polypyridine complexes of iron(II), ruthenium(II), and osmium(II). J. Am. Chem. Soc., 102: 1309–1319.
46. McCusker, J. K., Toftlund, H., Rheingold, A. L., Hendrickson, D. N. (1993) Ligand conformational changes affecting $^5T_2{\rightarrow}^1A_1$ intersystem crossing in a ferrous complex. J. Am. Chem. Soc., 115: 1797–1804.
47. McCusker, J. K., Walda, K. N., Dunn, R. C., Simon, J. D., Magde, D., Hendrickson, D. N. (1992) Sub-picosecond $\Delta S = 2$ intersystem crossing in low-spin ferrous complexes. J. Am. Chem. Soc., 114: 6919–6920.
48. McCusker, J. K., Walda, K. N., Dunn, R. C., Simon, J. D., Magde, D., Hendrickson, D. N. (1993) Subpicosecond $^1MLCT{\rightarrow}^5T_2$ intersystem crossing of low-spin polypyridyl ferrous complexes. J. Am. Chem. Soc., 115: 298–307.
49. McCusker, J. K., Rheingold, A. L., Hendrickson, D. N. (1996) Variable-temperature studies of laser-initiated $^5T_2{\rightarrow}^1A_1$ intersystem crossing in spin-crossover complexes: empirical correlations between activation parameters and ligand structure in a series of polypyridyl ferrous complexes. Inorg. Chem., 35:, 2100–2112.
50. McCusker, J. K. (2003) Femtosecond absorption spectroscopy of transition metal charge-transfer complexes. Acc. Chem. Res., 36:, 876–887.
51. Smeigh, A. L., Creelman, M., Mathies, R. A., McCusker, J. K. (2008) Femtosecond time-resolved optical and raman spectroscopy of photoinduced spin crossover: temporal resolution of low-to-high spin optical switching. J. Am. Chem. Soc., 130: 14105–14107.
52. Bergkamp, M. A., Chang, C. K., Netzel, T. L. (1983) Quantum yield determinations for subnanosecond-lived excited states and photoproducts: applications to inorganic complexes and photosynthetic models. J. Phys. Chem., 87: 4441–4446.
53. Enachescu, C., Oetliker, U., Hauser, A. (2002) Photoexcitation in the spin-crossover compound $[Fe(pic)_3]Cl_2{\cdot}EtOH$ (pic = 2-picolylamine). J. Phys. Chem. B, 106: 9540–9545.
54. Hauser, A. (1991) Intersystem crossing in the $[Fe(ptz)_6](BF_4)_2$ spin crossover system (ptz=1-propyltetrazole). J. Chem. Phys., 94: 2741–2748.
55. Cannizzo, A., van Mourik, F., Gawelda, W., Zgrablic, G., Bressler, C., Chergui, M. (2006) Broadband femtosecond fluorescence spectroscopy of $[Ru(bpy)_3]^{2+}$. Angew. Chem. Int. Ed., 45: 3174–3176.
56. Gawelda, W., Cannizzo, A., Pham, V. T., van Mourik, F., Bressler, C., Chergui, M. (2007) Ultrafast Nonadiabatic Dynamics of $[Fe^{II}(bpy)_3]^{2+}$ in Solution. J. Am. Chem. Soc., 129: 8199–8206.
57. Bräm, O., Messina, F., El-Zohry, A., Cannizzo, A., Chergui, M. (2012) Polychromatic femtosecond fluorescence studies of metal–polypyridine complexes in solution. Chem. Phys., 393: 51–57.
58. Braterman, P. S., Song, J. I., Peacock, R. D. (1992) Electronic absorption spectra of the iron(II) complexes of 2,2′-bipyridine, 2,2′-bipyrimidine, 1,10-phenanthroline, and 2,2′:6′,2″-terpyridine and their reduction products. Inorg. Chem., 31: 555–559.
59. Tarnovsky, A., Gawelda, W., Johnson, M., Bressler, C., Chergui, M. (2006) Photexcitation of aqueous ruthenium(II)-tris-(2,2′-bipyridine) with high-intensity femtosecond laser pulses. J. Phys. Chem. B, 110: 26497–26505.
60. Gawelda, W., Johnson, M., de Groot, F. M. F, Abela, R., Bressler, C., Chergui, M. (2006) Electronic and molecular structure of photoexcited $[Ru^{II}(bpy)_3]^{2+}$ probed by picosecond x-ray absorption spectroscopy. J. Am. Chem. Soc., 128: 5001–5009.
61. Chang, H. R., Mccusker, J. K., Toftlund, H., Wilson, S. R., Trautwein, A. X., Winkler, H., Hendrickson, D. N. (1990) [Tetrakis(2-pyridylmethyl)ethylenediamine]iron(II) perchlorate, the first rapidly interconverting ferrous spin-crossover complex. J. Am. Chem. Soc., 112: 6814–6827.
62. Xie, C. L., Hendrickson, D. N. (1987) Mechanism of spin-state interconversion in ferrous spin-crossover complexes: direct evidence for quantum mechanical tunneling. J. Am. Chem. Soc., 109: 6981–6988.

63. Consani, C., Prémont-Schwarz, M., Elnahhas, A., Bressler, C., van Mourik, F., Cannizzo, A., Chergui, M. (2009) Vibrational coherences and relaxation in the high-spin state of aqueous $[Fe^{II}(bpy)_3]^{2+}$. Angew. Chem. Int. Ed., 48: 7184–7187.
64. Alexander, B. D., Dines, T. J., Longhurst, R. W. (2008) DFT calculations of the structures and vibrational spectra of the $[Fe(bpy)_3]^{2+}$ and $[Ru(bpy)_3]^{2+}$ complexes. Chem. Phys., 352: 19–27.
65. Rosca, F., Kumar, A. T. N., Ionascu, D., Ye, X., Demidov, A. A., Sjodin, T., Wharton, D., Barrick, D., Sligar, S. G., Yonetani, T., Champion, P. M. (2002) Investigations of anharmonic low-frequency oscillations in heme proteins. J. Phys. Chem. A, 106: 3540–3552.
66. Zgrablic, G., Haacke, S., Chergui, M. (2007) Vibrational coherences of the protonated Schiff base of all-trans retinal in solution. Chem. Phys.,338: 168–174.
67. Tribollet, J., Galle, G., Jonusauskas, G., Deldicque, D., Tondusson, M., Létard, J.-F., Freysz, E. (2011) Transient absorption spectroscopy of the iron(II) $[Fe(phen)_3]^{2+}$ complex: Study of the non-radiative relaxation of an isolated iron(II) complex. Chem. Phys. Lett., 513: 42–47.
68. Wolf, M. M. N., Groß, R., Schumann, C., Wolny, J. A., Schünemann, V., Døssing, A., Paulsen, H., McGarvey, J. J., Diller, R. (2008) Sub-picosecond time resolved infrared spectroscopy of high-spin state formation in Fe(II) spin crossover complexes. Phys. Chem. Chem. Phys., 10: 4264–4273.
69. Hannay, C., HubinFranskin, M. J., Grandjean, F., Briois, V., Itié, J.-P., Polian, A., Trofimenko, S., Long, G. J. (1997) X-ray absorption spectroscopic study of the temperature and pressure dependence of the electronic spin states in several iron(II) and cobalt(ii) tris(pyrazolyl)borate complexes. Inorg. Chem., 36: 5580–5588.
70. Briois, V., Sainctavit, P., Long, G. J., Grandjean, F. (2001) Importance of photoelectron multiple scattering in the iron K-edge X-ray absorption spectra of spin-crossover complexes: full multiple scattering calculations for several iron(II) trispyrazolylborate and trispyrazolylmethane complexes. Inorg. Chem., 40: 912–918.
71. Boillot, M.-L., Zarembowitch, J., Itié, J.-P., Polian, A., Bourdet, E., Haasnoot, J. G. (2002) Pressure-induced spin-state crossovers at room temperature in iron(II) complexes: comparative analysis; a XANES investigation of some new transitions. New J. Chem., 26: 313–322.
72. Oyanagi, H., Tayagaki, T., Tanaka, K. (2006) Synchrotron radiation study of photo-induced spin-crossover transitions: microscopic origin of nonlinear phase transition. J. Luminesc., 119: 361–369.
73. Bressler, C., Abela, R., Chergui, M. (2008) Exploiting EXAFS and XANES for time-resolved molecular structures in liquids. Z. Kristallogr. 223: 307–321.
74. Bressler, C., Chergui, M. (2010) Molecular structural dynamics probed by ultrafast X-ray absorption spectroscopy. Ann. Rev. Phys. Chem., 61: 263–282.
75. Chergui, M. (2010) Picosecond and femtosecond X-ray absorption spectroscopy of molecular systems. Acta Cryst. Sect. A, 66: 229–239.
76. Saes, M., Bressler, C., van Mourik, F., Gawelda, W., Kaiser, M., Chergui, M., Bressler, C., Grolimund, D., Abela, R., Glover, T. E., Heimann, P. A., Schoenlein, R. W., Johnson, S. L., Lindenberg, A. M., Falcone, R. W. (2004) A setup for ultrafast time-resolved x-ray absorption spectroscopy. Rev. Sci. Instrum., 75: 24–30.
77. Briois, V., Moulin, C. C. D., Sainctavit, P., Brouder, C., Flank, A. M. (1995) Full multiple scattering and crystal field multiplet calculations performed on the spin transition $Fe^{II}(phen)_2(NCS)_2$ complex at the iron K and $L_{2,3}$ X-ray absorption edges. J. Am. Chem. Soc., 117: 1019–1026.
78. Yokoyama, T., Murakami, Y., Kiguchi, M., Komatsu, T., Kojima, N. (1998) Spin-crossover phase transition of a chain Fe(II) complex studied by X-ray-absorption fine-structure spectroscopy. Phys. Rev. B, 58: 14238–14244.
79. Boča, R., Vrbová, M., Werner, R., Haase, W. (2000) Spin crossover in iron(II) tris(2-(2-pyridyl)benzimidazole) complex monitored by the variable temperature EXAFS. Chem. Phys. Lett., 328: 188–196.
80. Beaud, P., Johnson, S. L., Streun, A., Abela, R., Abramsohn, D., Grolimund, D., Krasniqi, F., Schmidt, T., Schlott, V., Ingold, G. (2007) Spatiotemporal stability of a femtosecond hard–X-ray undulator source studied by control of coherent optical phonons. Phys. Rev. Lett., 99: 174801/1–174801/4.
81. Vankó, G., Glatzel, P., Pham, V. T., Abela, R., Grolimund, D., Borca, C. N., Johnson, S. L., Milne, C. J., Bressler, C. (2010) Picosecond time-resolved X-ray emission spectroscopy: ultrafast spin-state determination in an iron complex. Angew. Chem. Int. Ed., 49: 5910–5912.
82. Enachescu, C., Oetliker, U., Hauser, A. (2003) The quantum efficiency of the photo-excitation in a Fe (II) spin-crossover compound. J. Opt. Adv. Mater., 5: 267–272.

83. Chang, J., Fedro, A. J., van Veenendaal, M. (2010) Ultrafast cascading theory of intersystem crossings in transition-metal complexes. Phys. Rev. B, 82: 075124/1–075124/5.
84. Lapini, A., Foggi, P., Bussotti, L., Righini, R., Dei, A. (2008) Relaxation dynamics in three polypyridyl iron(II)-based complexes probed by nanosecond and sub-picosecond transient absorption spectroscopy. Inorg. Chim. Acta, 361: 3937–3943.
85. Champion, P. M., Rosca, F., Ionascu, D., Cao, W. X., Ye, X. (2004) Rapid timescale processes and the role of electronic surface coupling in the photolysis of diatomic ligands from heme proteins. Faraday Discuss., 127: 123–135.
86. Aziz, E. F., Rittmann-Frank, M. H., Lange, K. M., Bonhommeau, S., Chergui, M. (2010) Charge transfer to solvent identified using dark channel fluorescence-yield L-edge spectroscopy. Nature Chem., 2: 853–857.
87. Moret, M. E., Tapavicza, E., Guidoni, L., Röhrig, U. F., Sulpizi, M., Tavernelli, I., Rothlisberger, U. (2005) Quantum mechanical/molecular mechanical (QM/MM) Car-Parrinello simulations in excited states. Chimia, 59: 493–498.
88. Moret, M. E., Tavernelli, I., Rothlisberger, U. (2009) Combined QM/MM and classical molecular dynamics study of $[Ru(bpy)_3]^{2+}$ in water. J. Phys. Chem. B, 113: 7737–7744.
89. Moret, M. E., Tavernelli, I., Chergui, M., Rothlisberger, U. (2010) Electron localization dynamics in the triplet excited state of $[Ru(bpy)_3]^{2+}$ in aqueous solution. Chem. Eur. J., 16: 5889–5894.
90. van der Veen, R. M., Cannizzo, A., van Mourik, F., Vlček, A. jr., Chergui, M. (2011) Vibrational relaxation and intersystem crossing of binuclear metal complexes in solution. J. Am. Chem. Soc., 133: 305–315.
91. Link, O., Lugovoy, E., Siefermann, K., Liu, Y., Faubel, M., Abel, B. (2009) Ultrafast electronic spectroscopy for chemical analysis near liquid water interfaces: concepts and applications. Appl. Phys. A Mater. Sci. Process., 96: 117–135.

16

Real-Time Observation of Spin-Transitions by Optical Microscopy

François Varret[1], Christian Chong[1], Ahmed Slimani[1], Damien Garrot[1], Yann Garcia[2] and Anil D. Naik[2]

[1]*GEMAC Université de Versailles, France*
[2]*Institute of Condensed Matter and Nanosciences, MOST – Inorganic Chemistry, Université Catholique de Louvain, Belgium*

16.1 Introduction

Optical microscope investigations of spin-transition single crystals[1] have been actively developed since 2008.[2–11] A crucial milestone in these earlier investigations was the observation of macroscopic domains with well-defined transformation fronts.[3,7] In high-quality crystals, the usual situation was that of a single macroscopic domain irreversibly growing through the crystal, even at constant temperature.[6,9,11] The propagation velocity of the domain boundaries was unexpectedly low (~ some μm/s), that is, easily measurable by standard camera techniques. We show here in a further example[12] that the velocity can reach some hundred μm/s. The physical features revealed by optical microscopy of spin-transition crystals are consistent with well-established observations on (first-order) phase transitions in organic crystals.[13,14] That is, they evidence a nucleation and growth process for the spin transition. We therefore reached the crucial conclusion that *the spin-transition obeys a multiscale, stress-driven, nucleation and growth process.*[10] In the present case of switchable molecular solids, due to their thermo- and photochromic properties, optical microscopy can be used as a local colorimetric tool.[2,6] On one hand, the change in optical density (OD,[16] see the next section) associated with a specific absorption band enables local measurements of the HS fraction of the system. On the other hand, comparison of the ODs for the various colours provides information on the scattering effects associated with the presence of stresses or defects, generated by the volume change during the transition.[6] The use of cross-polarizer geometry may reveal the onset of a symmetry change concomitant to the spin change,[7] thus allowing an investigation of the interplay between the

Spin-Crossover Materials: Properties and Applications, First Edition. Edited by Malcolm A. Halcrow.

spin and structural changes, which will be presented here. Cross-polarizer geometry may also reveal the presence of internal stresses.[6] Lastly, the experimental image data were often organized in the form of movies, which cannot be reproduced here, but are available in the supplementary materials of refs 3, 8, 9, 10 and 11.

An important physical issue for the propagation of the transformation front of the thermally-induced transition is the nucleation and growth of Like-Spin Domains (LSDs). The available experiments[15] have led to the general agreement that these hypothetic LSDs should be far below the micron size. Such a small size is out of reach of the present (far field) optical microscopy, and will not be discussed here. Another crucial concern is the ageing of crystals under repeated thermal switching which generally generates opacity, cracks, self-cleavage and other defects. These mechanical effects are obviously generated by the volume change at the transition, and they are clearly enhanced in the case of large crystals. The brittleness of the crystals is a major obstacle for a real physical investigation of the transition which has to be based on 'single-shot' experiments performed on separate fresh crystals. Encouragingly, there is now a growing availability of robust crystals,[3, 12] based on the idea that a partial (or stepped) transition should generate a lesser volume change. The most recent example will be briefly reported here.

16.2 Experimental Aspects

16.2.1 Single Crystals

For obtaining good images in transmission geometry, which is exclusively considered here, single crystals with well defined facets are needed. The region of the crystal between facets perpendicular to the microscope axis will provide quantitative data through the analysis of the optical densities (ODs) defined as the ratio $\log_{10}(I_{\text{incident}}/I_{\text{transmitted}})$ (OD is proportional to Absorbance, $A = \ln$ (*I*incident/*I*transmitted). With a colour camera, red, green and blue intensity data are obtained independently, and their incident values are easily determined from the bright field background. ODs add up along the beam path through the crystal and for this reason are suited for measuring concentration variations, such as the high spin fraction, n_{HS}. On the other hand the lateral faces, as well as cracks and defects, are seen as dark regions due to shade effects, which are sometimes associated with brighter zones due to extra reflections on faces or interfaces. A suitable crystal thickness should result in OD values in the typical range of 0.1–2. This corresponds to ~10–200 μm for the case of weakly absorbing crystals, which are mostly considered here, but should be smaller in the case of more strongly absorbing crystals. We show in Figure 16.1 some typical crystals, whose detailed formulations will be given in the following sections.

16.2.2 The Sample Cell

In previous work[6] we introduced the use of a helium-tight sample cell aiming to provide an efficient heat transfer to the heat exchanger of the cryostat. Indeed, excellent temperature control and homogeneity are needed for an accurate investigation of the first-order transition. In the absence of this sample cell, large temperature deviations were reported in our early work,[2] where we discounted the common use of silver lacquer which might have generated stresses on the fragile and sensitive crystals. Also, the temperature ramp should remain very slow (typically ~0.2 K/min) so as to minimize the temperature gradients associated with the thermal flux through the crystal. A good example of a phase transition investigated by optical absorption[16, 17] was recently reported by the Hauser group at Genève.

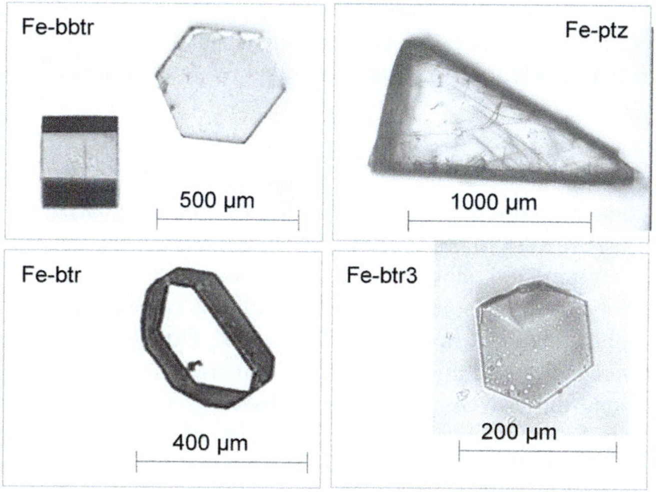

Figure 16.1 *Typical crystals investigated in the present report (at room temperature, before thermal cycling). Their complete formulae will be given in text.* (See the plate section for a colour version of this figure).

16.2.3 Cryostat, Objective, Camera

The optical microscope facility used at Rennes University consisted of a Leica Laborlux 12 pol microscope, a Sony Digital Hyper HAD SSC-DC38P camera and an Oxford Instruments Microstat helium flow cryostat. The recent set up at Versailles University[12] consists of a Nikon Eclipse LV100 microscope, a Dalsa Falcon 1.4M100 colour camera (10 bits, 100 fps) and an Oxford Instruments Microstat-HiResII helium flow cryostat.

The important parameters of the objective are the magnification, the numeric aperture and the working distance. So far the Rennes and Versailles setups have respectively used $\times 7$ and $\times 20$ objectives with working distances >12 mm, numeric apertures of 0.12 and 0.4, and (in-transmission) resolution limits of ~ 2.5 and 1.4 μm. The sensitive matrix of the camera is located in the plane of the image generated by the objective, with an inter-pixel distance (L) shorter than the product (ℓg) resolution limit $\times$ the magnification of the objective. This criterion was better fulfilled by the Versailles equipment ($\ell g = 16$ μm, L = 7 μm) than the Rennes ($\ell g =$ 16 μm, $L = 20$ μm). In colour devices, each pixel consists of a mini-matrix (typically 2×2) of detectors sensitive to red, green and blue light; or, to red, green-red, green-blue and blue in the case of more recent cameras. In all cases the image data are extracted as a set of three matrices, associated with the red, green, blue (R, G, B) components of the intensity of light. Such coding can be made in TIFF or JPEG format and the images are conveniently treated by the IMAGE J software. TIFF images make possible all kinds of numerical treatments with the MATLAB package. The camera data can also be saved in the form of AVI movies which require less memory space and can be converted into JPEG images using the VirtualDub software.

16.2.4 Setting of the Cryostat

In the transmission geometry, the cryostat is located between the condenser and the objective of the microscope. At Rennes, the cryostat was merely attached to the moving stage of the microscope, and the centering of the crystal was often lost. At Versailles we followed a different design which ensured a robust and easy control of the centering of the crystal: the cryostat is fixed on a solid flange attached to the experimental table by four strong legs, which leave room beneath for the lower part of the microscope, which is sitting on a micrometric XY support attached to the table. The XY support also allows us to draw the microscope backwards when the cryostat is to be opened. The experimental table is heavy, and is standing on flexible plots to minimize as much as possible the vibrations generated by a rather noisy ambience.

16.2.5 Observation Modes

We used a transmission geometry in bright field with and without crossed polarizers. In bright field geometry, the measured ODs are linearly related to the concentrations in the different absorbing species (HS and LS, for instance). This way the thermo- and photoinduced transitions can be investigated as well.[1, 2] Knowledge of the absorption spectra in the different spin states can help, but is not mandatory, since the OD scales can be normalized using the values obtained in the saturated states. The absorption coefficients derived from optical microscopy data (once the crystal thickness has been measured) are average values over the spectral bands of the RGB pixels, and these average values are obviously reduced with respect to the true peak values. Eventual scattering of light, which would lift the baseline of the absorption spectrum, is expected to contribute almost equally to the RGB measured ODs. Between crossed polarizers, a perfect crystal free from stresses and birefringence gives a dark image, as shown in Figure 16.2, and any transmitted intensity will reveal the presence of either birefringence or scattering effects. Birefringence effects are characterized by an angular dependence of the transmitted intensities upon simultaneous rotation of the polarizers; they may originate from a low-symmetry structure or from mechanical stresses generated by the volume effects at the transition. Scattering or depolarization effects might reveal the presence of defects or microstructures. Dark-field transmission images[8] are also obtained with a specific objective and condenser set which eliminates the central part of the light beam, and they provide an increased contrast for visualizing defects and stresses. So far, we have made little use of the reflection mode. For this mode, the lightening beam is transmitted through the objective, leading to a significant warming of the crystal that we detected by a downward shift of the measured transition temperatures. This effect was not observed in transmission geometry, because we used a lower lightening intensity (with the same sensitivity range) of the detector.

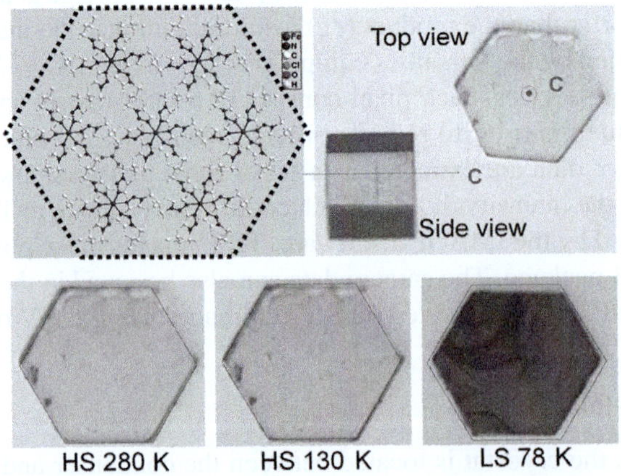

Figure 16.2 *Bright-field transmission images of typical Fe-bbtr crystals with different orientations. The crystal in top view (hexagonal contour) was ∼100 μm thick; the crystal in side view was ∼165 μm. The low temperature image (LS state) is characterized by the dark purple colour (absorption by the LS* d-d *band ∼590 nm), by the visible shrinkage of the sample and by the presence of major damages generated by the volume effects during the transition. After ref. 7.* (See the plate section for a colour version of this figure).

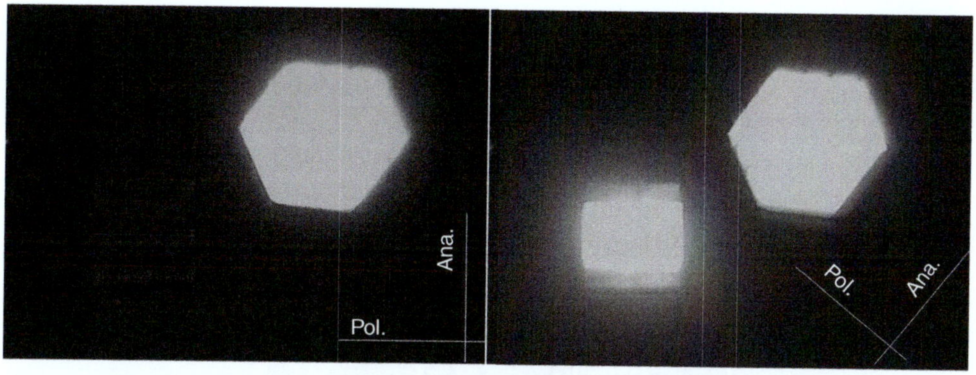

Figure 16.3 *The set of Fe-bbtr crystals, observed at room temperature between crossed polarizers. Adapted from ref. 7.* (See the plate section for a colour version of this figure).

We present in Figures 16.2 and 16.3 some typical images obtained with high quality crystals of $[Fe(bbtr)_3](ClO_4)_2$ (bbtr = 1,4-di(1,2,3-triazol-1-yl)butane),[18] here abbreviated Fe(bbtr), prepared by R. Bronisz at Wroclaw University (see also[17, 18]). The structure of the compound is $P\bar{3}$ at room temperature and in the vicinity of the transition temperature it turns into a parented disordered structure which was not fully characterized. This structural transformation is reminiscent of the $R\bar{3} \leftrightarrow R\bar{3}dis$ transition reported for Fe-ptz.[19] In the microscope sample holder, the crystals were oriented according to their different *facies*, for providing 'top' or 'side' views, that is, with their 3-fold axis parallel or perpendicular to the microscope axis, respectively. They could also be cleaved due to the 2D polymeric character of the packing, in such a way that both views could be obtained using separate parts of the same crystal.

The side view crystal exhibits a perfect extinction (birefringence effect), see Figure 16.3, which characterizes the high optical quality of the crystal. In the top view, a similar extinction was not obtained, whatever the orientation of the analyser. Between crossed polarizers, the transmitted intensity remained constant when the crystal was rotated. These features are typical for scattering effects. The anisotropy of the present scattering effects seems consistent with the likely presence of defects in the packing of the polymeric planes.

16.3 Selected Investigations

16.3.1 The Interplay Between Structure and Spin Transformations: $[Fe(bbtr)_3](ClO_4)_2$

We now report on an intriguing question, which is the possible interplay between structure and spin transformations, on the example of the 2D spin-transition polymer Fe-bbtr already presented above. Previous studies by optical absorption using Fe-bbtr single crystals[17] revealed a complex behaviour, for which the temperature of the HS $\rightarrow$ LS transformation was strongly perturbed by the temperature at which a previous thermal annealing was performed. These effects were related to the promotion or hindrance of the structural transformation $P\bar{3} \rightarrow$ "$P\bar{3}dis$" in the HS phase. We provide here an additional piece of information, which is the possible triggering of the structure transformation by the spin state switching (Fig. 16.4). This result was obtained serendipitously,[7] as follows: in order to minimize the severe damage created by mechanical stresses, we prepared two thin pieces cleaved from the same large crystal (thanks Loïc Toupet!), and let them slowly cool down overnight. We reached 125 K and both crystals remained transparent. The following day, we pursued the cooling of the crystal at a faster rate (~2 K/min), and started measuring the crystals between crossed polarizers. At 120 K we observed that one crystal was displaying structural twins, while the second

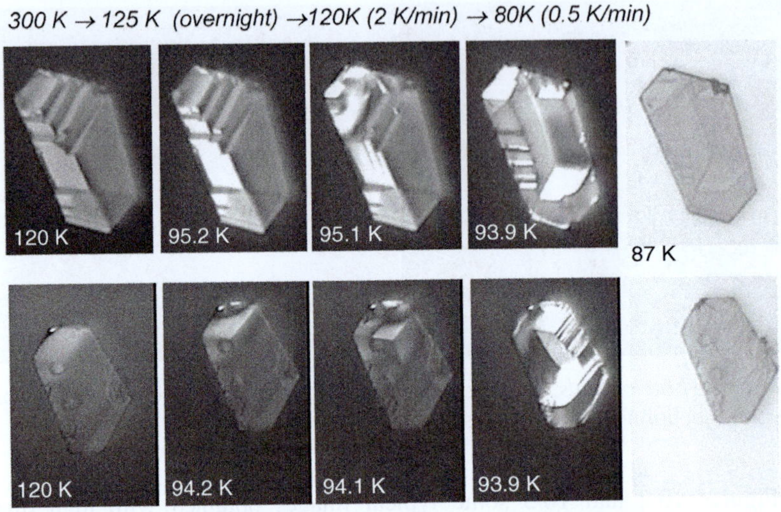

Figure 16.4 *Cross-polarizer images of two pieces (~370 μm long, ~60 μm thick), cleaved from the same Fe-bbtr crystal, simultaneously measured on cooling. The development of the LS phase is revealed by the onset of the pink colour in all geometries (with and without crossed polarizers). The development of the structural transition is revealed by the facetted patterning in the cross-polarizer images. The different impacts of the spin-transition, albeit starting from the same top corner of both crystals, is discussed in the text. Some of the structural domain walls are also detectable on the bright-field images in the LS state (right images). Adapted from ref. 7.* (See the plate section for a colour version of this figure).

remained almost unaffected. This interesting situation was obviously generated by chance, presumably due to a minor difference between the crystals. Only one of the crystals, during the time of the experiment, possessed the single germ needed for nucleation and growth of the twinning process. Then we carefully followed the spin-transition by the simultaneous observation of both crystals, on cooling at 0.5 K/min. The spin-transition was detected through the onset of a pink colour, at different temperatures, ~95.2 and 94.2 K for the two crystals, and it induced different effects. For the crystal in the "$P\bar{3}dis$" state (top images), the propagation of the spin transformation induced a complete reorganization of the structure pattern. For the second crystal (bottom line), the onset of the spin-transition triggered the structure transformation. For both crystals the course of the transformation front resulted in visible scattering effects which can be assigned to the presence of transient micro-structures or defects. We attribute the efficiency of the spin-transition for triggering the structure transition, and the absence of the reverse effect, to the larger volume change generated by the spin transformation.

16.3.2 Colorimetric Investigation of $[Fe(ptz)_6](BF_4)_2$

The following example deals with a textbook example of spin-transitions, $[Fe(ptz)_6](BF_4)_2$ (ptz = 1-propyl-tetrazole), here abbreviated Fe-ptz. A typical illustration of the thermo- and photochromic properties of the system is shown in Figure 16.5. As already described, the structural properties of Fe-ptz are related to those of Fe-bbtr, with a $R\bar{3} \leftrightarrow R\bar{3}dis$ transformation associated with the thermal spin-transition. However, we did not observe any structural patterning in the case of Fe-ptz. It can be inferred that structural domains in Fe-ptz are sub-micronic in size, a conclusion which agrees with the previous observation by A. Hauser of an overall increase in the baseline signal of the low temperature optical absorption spectra in the low

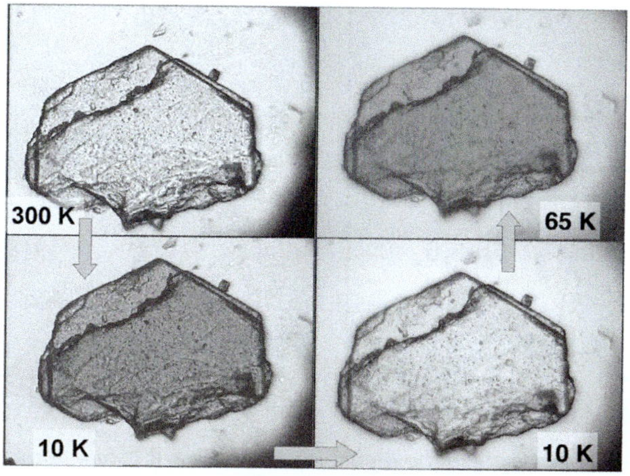

Figure 16.5 *Thermo- and photochromic properties of Fe-ptz. It is worth noting that the photoexcitation and subsequent relaxation processes did not induce any patterning of the Fe-ptz images, in contrast to most of the thermal transitions reported here. Adapted from ref. 7.* (See the plate section for a colour version of this figure).

temperature structural state.[20] Another difference is that Fe-ptz is a true molecular compound (no covalent bonds between molecular units) instead of a 2D polymer. For this reason Fe-ptz is very brittle. Also, the high temperature structure of Fe-ptz could be retained by rapid cooling (>10 K/min) through the thermal transition range, while this was not possible in the case of Fe-bbtr. At the thermal transition, we reported in our early microscopy investigation[3] the presence of a rather irregular "forehead" (transformation front) which propagated at ~1 μm/s. This was only obtained at the $LS - R\bar{3}dis \rightarrow HS - R\bar{3}$ transition, using crossed polarizer geometry. Images in the absence of crossed polarizers did not yield a defined transformation front line, which now seems rather puzzling, but might be due to the noisy character of the OD maps. Indeed the crystals were not that beautiful, and here we prefer to focus on the more conclusive colorimetric study. This study was based on average OD values, determined over a large area of the crystal. The efficiency of the colorimetric method in the present case is due to the location of the metal–ligand charge transfer band which does not overlap the narrow *d-d* bands. In addition, these *d-d* bands are located in such a way[21] that absorption effects only impact the green pixels (Fig. 16.6). In such a fortunate situation, any detected value of the red or blue ODs can be assigned to scattering effects. Therefore the measurement of the RGB ODs can be used for monitoring the electronic and structural transformations which interplay at the thermal transition, with their differing kinetics.

We show in Figure 16.7 the results obtained by the colorimetric method with a Fe-ptz crystal initially quenched in the$LS - R\bar{3}dis$ state. The initial $LS - R\bar{3}$ state is characterized by a large green OD (typical for the *LS* state) and negligible scattering of light. Upon temperature increase, the electronic transformation exhibits first a tentative onset (slight decrease in green OD) due to the approach of the spin-transition in the $R\bar{3}dis$ state. This is expected to be a continuous transition according to low temperature relaxation studies.[22] The course of this spin transformation is interrupted by a simultaneous increase in all ODs at ~118 K due to scattering effects, which peaks at ~125 K and reveals the onset of the structure transition, $R\bar{3} \rightarrow R\bar{3}dis$. Upon further heating, all ODs exhibit a rapid decrease slightly above 130 K, due to the linked spin/structure transition $LS - R\bar{3}dis \rightarrow HS - R\bar{3}$. This establishes that the scattering of light is correlated to the presence of the disordered phase. Further thermal investigation of the system evidenced the hysteretic $HS - R\bar{3} \leftrightarrow LS - R\bar{3}dis$ linked transformation. It can be concluded that the first-order character of the

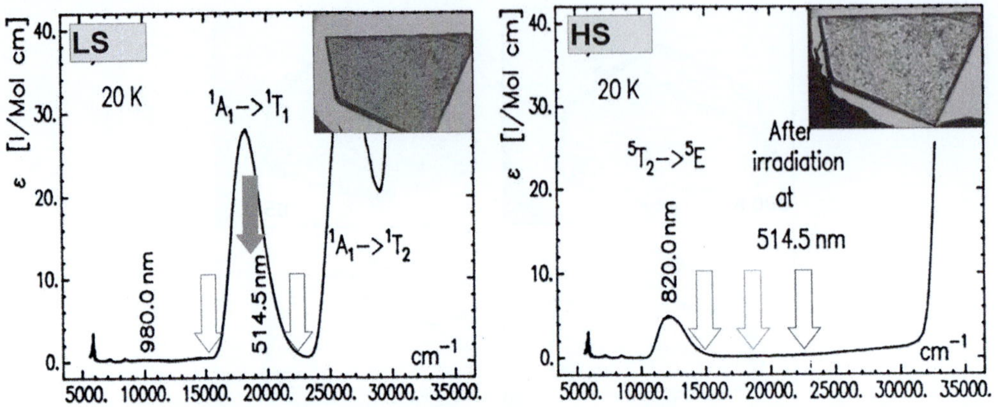

Figure 16.6 *Absorption spectra of Fe-ptz in the high temperature HS-R$\bar{3}$ state and in the low temperature LS-R$\bar{3}$ quenched state. Adapted from ref. 21.* (See the plate section for a colour version of this figure).

joined transition is due to the synergy between electronic and structural transformations. The incomplete character of the initial $R\bar{3} \rightarrow R\bar{3}dis$ transformation illustrates the kinetic nature of all these transformations. A deeper insight into the interplay between the electronic and structural transformations (not reported here) was obtained by analysing the correlation plots of the different ODs. The desirable investigation of the spatio-temporal properties of this linked spin/structure transition, using high quality crystals, seems hardly feasible due to the brittleness of the crystals, which undergo severe damage during the first switching of the spin state.

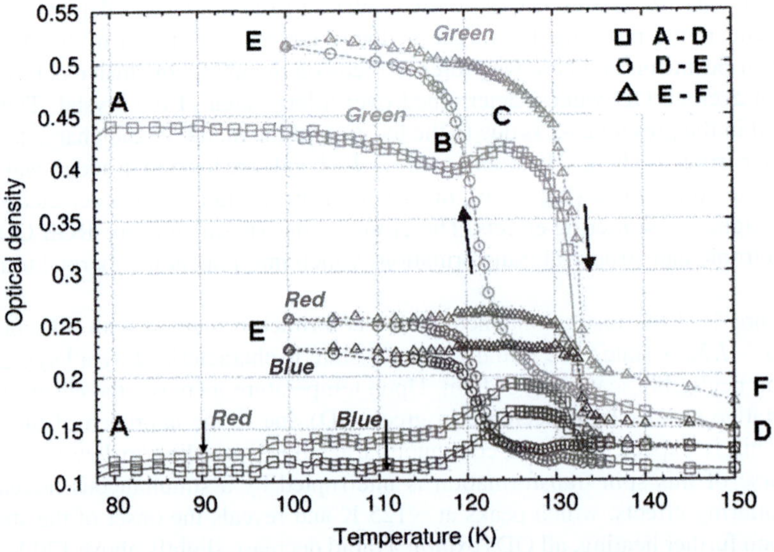

Figure 16.7 *The RGB ODs of a Fe-ptz single crystal during successive transformations at ± 0.1 K/min: [A–D] the irreversible process from quenched LS – R$\bar{3}$ to HS – R$\bar{3}$; [D-E-F] the reversible joined transition HS – R$\bar{3}$ ↔ LS – R$\bar{3}$dis. After ref. 6.* (See the plate section for a colour version of this figure).

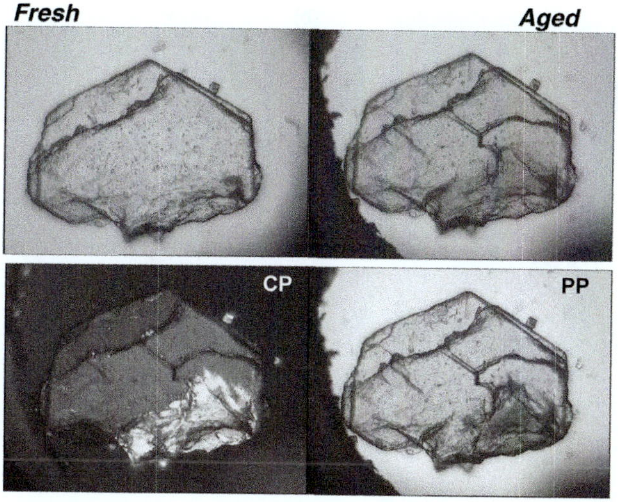

Figure 16.8 *Images of a Fe-ptz crystal at 10 K after "saturated" photoexcitation. The area resistant to photoexcitation, in the aged crystal, is characterized by the purple colour in the top right image. Stresses are evidenced by the colour-changing areas in the bottom images (CP, PP = crossed, parallel polarizers). The area contaminated by the R$\bar{3}$dis phase is evidenced by scattering effects (CP image). After ref. 6.* (See the plate section for a colour version of this figure).

Another intriguing problem was the ageing of the Fe-ptz crystals during photoexcitation experiments. Indeed we observed in previous experiments on crystals in the quenched $R\bar{3}$ state[23, 24] that after several switching events, the photoresponse of the crystal sample was not accurately reproducible. This ageing process was explained by the microscope investigation reported in Figure 16.8. It evidences the irreversible growth of the stable structural phase, $R\bar{3}dis$ under the repeated effect of mechanical stresses generated by the switching process. Although this disordered phase has a more facile photoexcitation than the ordered phase, we observed that the net efficiency of the system was decreased. This is presumably due to the stresses generated by the microstructure associated with the growth of this second phase.

16.3.3 The Transformation Front in $[Fe(btr)_2(NCS)_2]{\cdot}H_2O$ Crystals

$[Fe(btr)_2(NCS)_2]{\cdot}H_2O$ (btr=4,4′-bis-1,2,4-bis-triazole) here abbreviated Fe-btr is a 2D spin-crossover coordination polymer[25] due to the bridging role of btr ligands which build up a covalent square planar lattice of spin-crossover units. This compound provided one of the first examples of well-defined transformation fronts.[5] We first observed "batch crystals"[5, 11] prepared years ago by J. Haasnoot at Leiden, and more recently we were able to investigate high quality crystals[10] prepared by S. Pillet at Nancy, and reached the crucial conclusion that the spin-transition obeys a multiscale, stress-driven, nucleation and growth process.

The images recorded during the on-cooling transition of fresh crystals always showed a well defined frontline with almost regular shape. Conversely, at the on-heating transition, the patterns were very complex, due to the presence of defects generated by the on-cooling transition, which in turn impact the propagation of the on-heating front. We therefore focus on the on-cooling transition which typically requires one minute or more to proceed through the whole crystal. This made possible the recording of successive high-definition images with the Rennes microscope and in this way we accessed the spatio-temporal properties of the transition for the first time. In Figure 16.9 we show the fresh crystal at room temperature (a) and an image

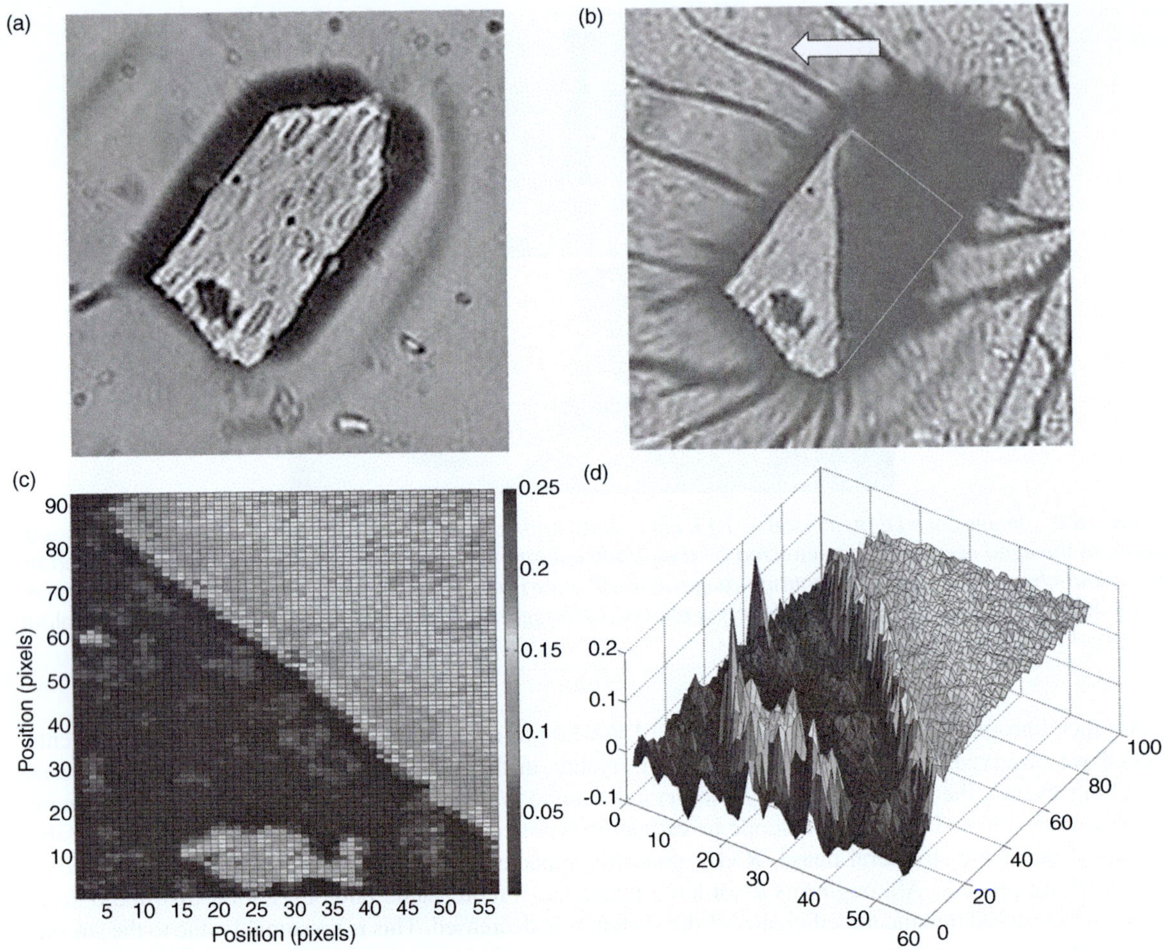

Figure 16.9 *A high-quality Fe-btr crystal, at room temperature (a) and in the course of the thermal transition, at 102.5 K (b).The crystal was embedded in oil so as to prevent dehydration, and the oil film underwent several cracks by ~120 K. The bright and dark areas respectively correspond to HS, LS phases. On the bottom line we mapped the red ODs derived from image (b) inside the rectangular frame; they are plotted in projection (c) or using a pseudo 3D representation (d). The propagation sense of the frontline is shown by an arrow. Adapted from ref. 10 (including video files).* (See the plate section for a colour version of this figure).

midway through the spin-transition (b). The transformation started from the top corner of the crystal and after a while it reached a steady progression regime with a frontline almost straight and constant in orientation, as shown in (b). Images (c) (d) are false-colour maps of the local values of the red OD derived from image (b).

Information relevant to the propagation velocity (~2.2 μm/s) was straightforwardly derived from the successive positions of the frontline.[10] In addition, more detailed information on spatio-temporal effects was derived from the OD profiles, that is, the spatial variation of the ODs along a reference line parallel to the propagation direction; and from the kinetic responses of the individual pixels. We focus here on the spatial information which can be derived from a careful inspection of a typical image, with suitable processing, such as shown in Figure 16.9 (c, d):

(i) the frontline is a smeared out over a few pixels (say ~10 μm);
(ii) on the rear side of the frontline, the ODs peak so as to form a visible ridge;
(iii) behind this ridge the ODs are transiently depressed;
(iv) the fluctuations in the final LS state are far smaller than those of the initial HS state.

The smearing of the frontline is consistent with the reported interpretation of the kinetic data in terms of a nucleation and growth process based on the diffusion of sub-micronic domains. The complex changes to the ODs through the frontline were assigned to purely optical effects due to the variation of the refractive index under the effect of mechanical stresses, resulting in shade and mirage effects. The large static fluctuations ahead of the frontline were due to static impurities or defects, and their decrease behind the frontline was tentatively attributed to the onset of scattering effects which blur out the image of these impurities.

After thermal cycling, the images of all Fe-btr crystals showed the presence of severe damage, see Figure 16.10. On the top line: cracking lines subsequent to the thermal cycling of the crystal shown in Figure 16.9; the orientation of a large crack can be correlated to the orientation of the frontline. On the bottom line: the damage undergone by another crystal cooled at a slower rate (0.05 K/min); the crystal underwent a stepped transition at a lower temperature, with larger propagation velocities. Once returned to room temperature, the crystal displayed incredible features: large volume expansion and visible opalescence. The role of the rheological properties of the material is certainly crucial for the onset of this amazing behaviour. No doubt the rheological properties of the crystal are also deeply involved in the propagation of the nucleation and growth process.

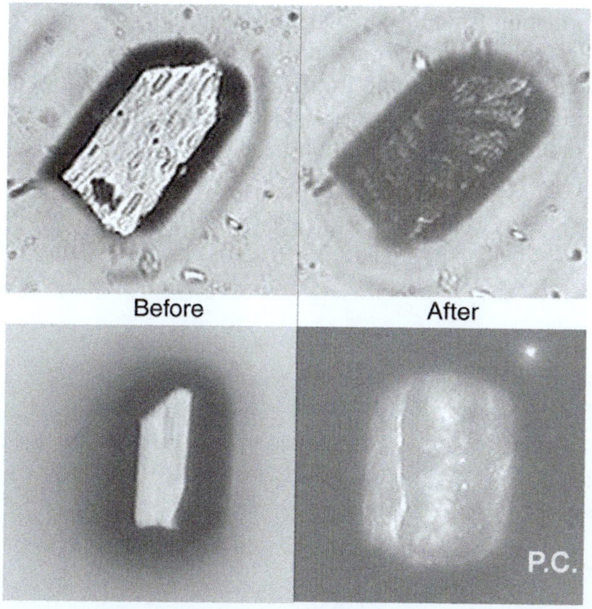

Figure 16.10 *Various damages of Fe-btr crystals, generated by the mechanical stresses associated with the spin transition. Adapted from ref. 10.* (See the plate section for a colour version of this figure).

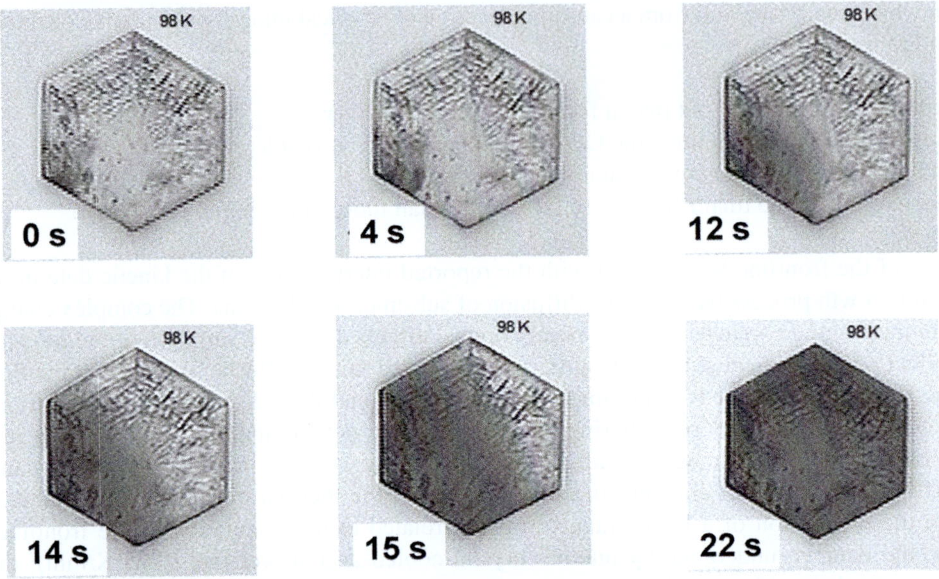

Figure 16.11 *Selected images of a top view crystal during its first HS→LS transition, at 98 K after cooling at 0.2 K/min. Crystal thickness (135–190 μm) was determined from a side view of the same crystal. After ref. 9 (including video files).* (See the plate section for a colour version of this figure).

16.3.4 The Evolution of the Frontline in $[Fe(bbtr)_3](ClO_4)_2$ Crystals

We now focus on the transformation front of the 2D spin-transition polymer Fe-bbtr, see Figures 16.11 and 16.12 which respectively illustrate the transformation of top view and side view crystals. The top view, Figure 16.11, shows a rapid propagation starting from a corner at the thin side of the crystal. (The thin side was determined thanks to a side view of the same crystal, shown in Table 1 "crystal C" of ref. 9.) The HS→LS transition took place with a rather smooth propagation front, whose shape was circular at the beginning but

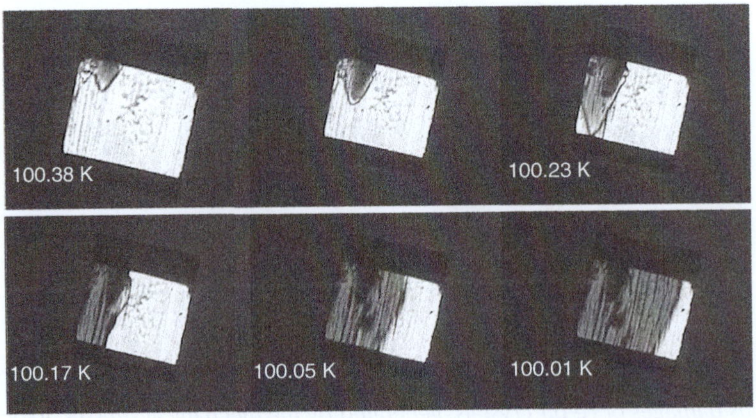

Figure 16.12 *Selected images of a side-view crystal during its first HS→LS transition, by cooling at 0.2 K/min. After ref. 9 (including video files).* (See the plate section for a colour version of this figure).

evolved subsequently under the influence of the edges of the crystal. The propagation velocity of the front was determined to be ~9 μm/s in the main course of the transformation but was sizably larger at the end of the process. The side view, Figure 16.12, showed a rather confused start, followed by a well-defined linear transformation front, indicating an LS/HS interface roughly in a plane parallel to the microscope axis. The propagation velocity was determined ~2 μm/s. The orientation of the transformation front, at ~77° from the crystallographic *c* axis, is in excellent agreement with the ratio of the parallel and perpendicular components, respectively determined from the side and top views: $v(//c) / v(\perp c) \sim 2/9 \sim 1/\tan 77° = 0.23$. Such a large anisotropy of the propagation velocity obviously implies the anisotropic character of the elastic (or visco-elastic) properties of the compound, in agreement with its 2D polymeric nature.

16.3.5 An Example of a Robust Crystal: $[Fe(btr)_3](ClO_4)_2$

$[Fe(btr)_3](ClO_4)_2$ here abbreviated Fe-btr3 is a 3D compound which displays a 2-step transition.[26] A beautiful single crystal (~10 μm thick) was grown at Louvain. Fortunately, the combination of its partial spin-transition and 3D covalent character resulted in an ideal robustness against repeated switching, which was very useful for a detailed study based on a unique crystal. The coordination sphere of the Fe(II) atom is similar to that in Fe-ptz and Fe-btbr, and affords similar optical properties, with a metal–ligand charge transfer band shifted to the UV range. Indeed, the crystal is colourless at room temperature and turns progressively to light pink in the HS–LS and LS phases, as illustrated in Figure 16.13, where we plotted the measured green OD of a thin crystal as a function of temperature. These data are in excellent agreement with the literature magnetic data, revealing first on cooling a gradual conversion and subsequently a hysteretic transition.[28]

The colour changes displayed by this thin crystal were hardly visible, as illustrated in the low OD values reported in Figure 16.13, and required recourse to image processing. The first task consisted of locating the transition temperatures, by investigating the complete temperature range [170–200 K]. At 0.2 K/min, the complete scan films required 150 min recording, leading to 9000 images at 1 frame per second (fps). These images at the transition could be found by visual inspection through a dichotomy strategy. The transition

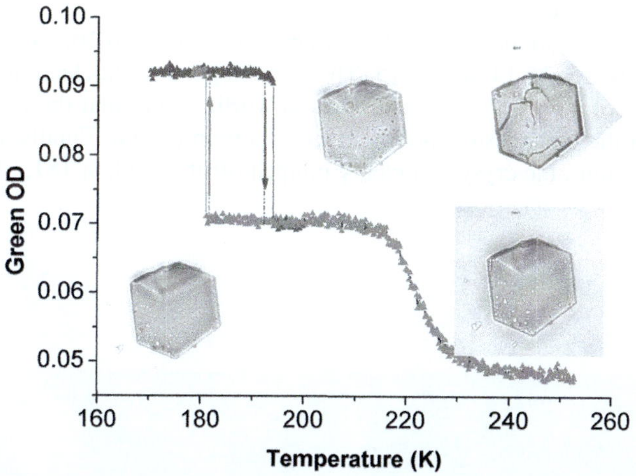

Figure 16.13 *Fe-btr3: the thermal variation of the green OD, illustrating the 2-step transformation of the system. The high temperature step occurs progressively in the range [220–230 K] (scan rate 1 K/min). The low temperature step (scan rate 0.2 K/min) shows first-order character with hysteresis ~10–12 K wide. Top right: view of the aged crystal (after ~10 thermal cycles).* (See the plate section for a colour version of this figure).

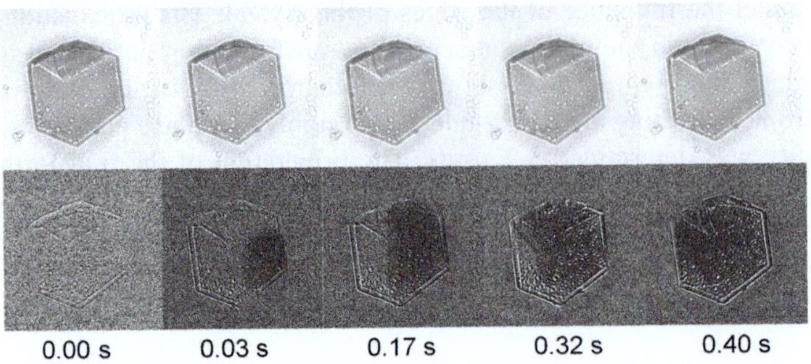

Figure 16.14 Fe-btr3: selected images of the low-temperature step, on cooling at 0.2 K/min, T = 181.6 K. The processed images (bottom line) evidence the progression of the transformation front, at an average velocity ~500 μm/s. (See the plate section for a colour version of this figure).

occurred within less than 1 s, that is, between two successive images. Further films were then recorded in narrower temperature ranges, with increased time resolutions: 10 fps and finally 100 fps, always at the same cooling rate 0.2 K/min. This stepped procedure was imposed by the available space of the buffer memory of the camera. For saving memory space, the object field (image size) of the camera was also restricted as much as possible. We made videos by selecting "active" sections of these films, and finally could observe the frontline and its propagation within less than one second through the crystal. However the frontlines were not yet visible on the raw images, and we processed the images by subtracting to the intensities of each of them the intensities of the first one. These "difference" images were rather noisy, but provided a clear visualization of the frontline, as shown in Figures 16.14 and 16.15. The evolution of the frontline, in both heating and cooling modes, revealed front shapes similar to those observed on the fresh Fe-bbtr crystal, that ideally are: nucleation at one corner of the crystal; initial growth with a ~ circular frontline; continued growth with progressive decrease of the curvature of the frontline; and a final stage with a reversed curvature, finishing at the opposite side of the crystal. The present observations confirm the importance of the edge effects on the front shape and propagation process.

Coming briefly back to the continuous step of the transition, Figure 16.16, we did not observe any clear patterning, in contrast to Figures 16.14 and 16.15. There, homogeneous nucleation starts everywhere in the crystal, due to the absence of a first-order transition. It is worth noting that the present crystal had a large extended defect, visible at the top, presumably the trace of a twin crystal. Finally, during an additional on-cooling experiment (0.2 K/min), the crystal suddenly jumped off the field, at 181.3 K under the effect of the

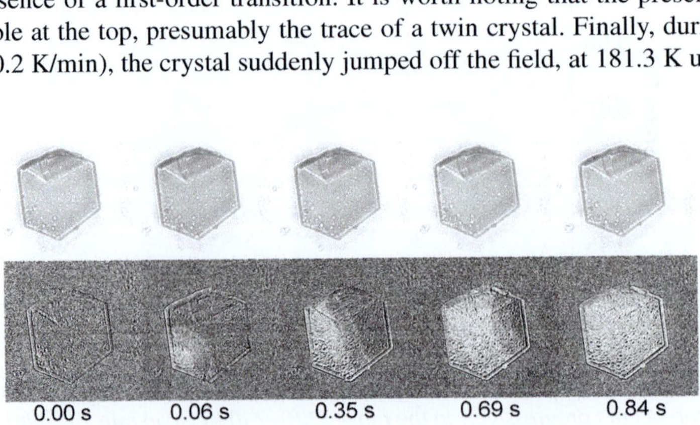

Figure 16.15 Fe-btr4: same caption as Figure 16.14, on heating at 0.2 K/min, T = 192.2 K, at average velocity ~250 μm/s. (See the plate section for a colour version of this figure).

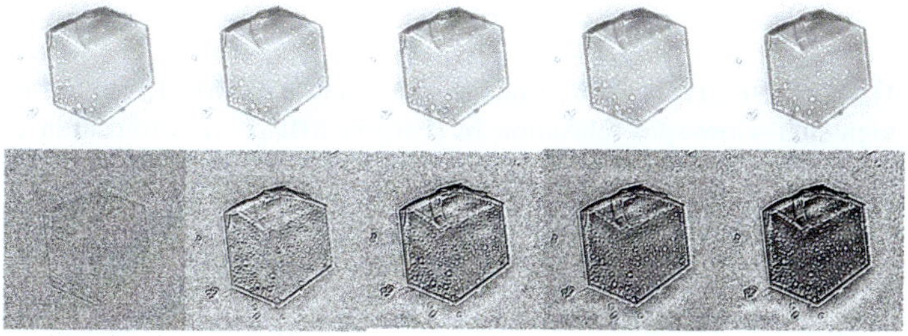

Figure 16.16 *Fe-btr3: selected images of the high-temperature step; raw images (top line) and processed images (bottom), see text.* (See the plate section for a colour version of this figure).

transition. This rough event generated the first visible damage in the crystal, but subsequent switching (on heating) did not visibly increase the damage, at variance from the other crystals studied previously. The proven robustness of the Fe-btr$_3$ system is presumably due to its 3D covalent structure. The present investigation should be completed by the investigation of higher quality and thicker crystals, for a better analysis of the frontline and an accurate determination of its propagation velocity.

16.4 Conclusions and Prospects

Optical Microscopy (OM) has clearly thrown light on the nucleation and growth process involved in the spin-transition. However the present conclusions are obviously limited by the resolution limit of the microscope, which cannot be further improved to a very sizeable degree. The desired observation of the sub-micronic like-spin domains (LSD) which are generated by the irreversible nucleation and growth (NG) process is out of reach of OM, which however makes possible monitoring at μm resolution the propagation of the conditions suitable for the nm-range NG process. Observation of LSD patterns will recourse to near-field microscopies such as AFM, SNOM, STM and TEM, which may be very slow techniques. Therefore an important prerequisite of future near-field investigations is the freezing of the NG process. We are presently working on this way by trying to freeze the frontline observed by optical microscopy. Other challenges for OM are the detailed analysis of the mechanical stresses and their location with respect to the frontline, the 3D analysis of the transformation, and the observation of transitions triggered by external means. A crucial question for which OM is perfectly suited is to understand the damage undergone by the crystals under repeated switching. The theoretical side of the problem, which couples the local electronic Hamiltonian and the rheological properties of the material, is fascinating. From the experimental viewpoint, it seems that the size of the crystals plays a major role, which deserves to be investigated in more detail for different systems. In particular, the expected robustness of nanoparticles should be confirmed by specific studies using electron microscopy or near-field techniques.

Acknowledgements

The authors thank Guillaume Bouchez, Kamel Boukheddaden, Epiphane Codjovi, Stéphane Denise, Frédéric Mourgès (Versailles), Jean-Claude Ameline, Hervé Cailleau, Eric Collet, Jelena Jeftic, Loïc Toupet (Rennes), Sébastien Pillet (Nancy), Jaap Haasnoot (Leiden), Andreas Hauser (Genève), Robert Bronisz (Wroclaw).

References

1. Jeftic, J., Romstedt, H., Varret, F., Hauser, A., Roubeau, O., Matsarski, M., Rivera, J. P. (1999) Patterns during photoexcitation and high-spin→low-spin relaxation in $[Fe(ptz)_6](BF_4)_2$ spin transition crystal. Mol. Cryst. Liq. Cryst., 335: 1223–1232
2. Goujon, A., Varret, F., Boukheddaden, K., Chong, C., Jeftic, J., Garcia, Y., Naik, A. D., Ameline, J.-C., Collet, E. (2008) An optical microscope study of photo-switching and relaxation in single crystals of the spin transitions solid $[Fe(ptz)_6](BF_4)_2$ with image processing. Inorg. Chim. Acta, 361: 4055–4064.
3. Bonnet, S., Molnár, G., Costa, J. S., Siegler, M. A., Spek, A. L., Bousseksou, A., Fu, W.-T., Gamez, P., Reedijk, J. (2009) Influence of sample preparation, temperature, light, and pressure on the two-step spin crossover mononuclear compound $[Fe(bapbpy)(NCS)_2]$. Chem. Mater., 21: 1123–1136.
4. Varret, F., Chong, C., Goujon, A., Boukheddaden, K. (2009) Light-induced phase separation (LIPS) in $[Fe(ptz)_6](BF_4)_2$ spin-crossover single crystals: experimental data revisited through optical microscope investigation. J. Phys. Conf. Ser., 148: 012036/1–012036/5.
5. Varret, F. (2009) Voyage au coeur de la transition de spin, $[Fe(btr)_2(NCS)_2]\cdot H_2O$. Oral presentation at CNRS GDR-MCM meeting, Dourdan, France.
6. Chong, C., Mishra, H., Boukheddaden, K., Denise, S., Bouchez, G., Varret, F., Garcia, Y., Collet, E., Ameline, J.-C. (2010) Electronic and structural aspects of spin transitions observed by optical microscopy. The case of $[Fe(ptz)_6](BF_4)_2$. J. Phys. Chem. B, 114: 1975–1984
7. Chong, C. (2010) Observation et modélisation d'effets d'auto-organisation dans les solides thermo et photo-commutables. PhD thesis , University of Versailles.
8. Bedoui, S., Molnár, G., Bonnet, S., Quintero, C., Shepherd, H. J., Nicolazzi, W., Salmon, L., Bousseksou, A. (2010) Raman spectroscopic and optical imaging of high spin/low spin domains in a spin crossover complex. Chem. Phys. Lett., 499: 94–99
9. Chong, C., Slimani, A., Varret, F., Boukheddaden, K., Collet, E., Ameline, J.-C., Bronisz, R., Hauser, A. (2011) The kinetic features of a thermal spin transition characterized by optical microscopy on the example of $[Fe(bbtr)_3](ClO_4)_2$ single crystals: size effect and mechanical instability. Chem. Phys. Lett., 504: 29–33.
10. Slimani, A., Varret, F., Boukheddaden, K., Chong, C., Mishra, H., Haasnoot, J., Pillet, S. (2011) Visualization and quantitative analysis of spatiotemporal behavior in a first-order thermal spin transition: a stress-driven multiscale process. Phys. Rev. B, 84: 094442/1–094442/8.
11. Varret, F., Slimani, A., Boukheddaden, K., Chong, C., Mishra, H., Collet, E., Haasnoot, J., Pillet, S. (2011) The propagation of the thermal spin transition of $[Fe(btr)_2(NCS)_2]\cdot H_2O$ single crystals, observed by optical microscopy. New J. Chem., 35: 2333–2340.
12. Slimani, A. PhD Thesis, University of Versailles, in preparation.
13. Mnyukh, Y. V., Petropavlov, N. N. (1972) Polymorphic transitions in molecular crystals -1. Orientations of lattices and interfaces. J. Phys. Chem. Solids, 33: 2079–2087.
14. Mnyukh, Y. V. (2001) Fundamentals of Solid-State Phase Transitions, Ferromagnetism and Ferroelectricity. 1st Books, Fairfield, USA, p. 376.
15. Molnár, G., Bousseksou, A., Zwick, A., McGarvey, J. (2003) The spin-crossover phenomenon in the solid state: do domains play a role? A micro-Raman study. Chem. Phys. Lett., 367: 593–598.
16. Krivokapic, I., Enachescu, C., Bronisz, R., Hauser, A. (2008) The interaction between the spin transition and a crystallographic phase transition in the spin-crossover compound $[Fe(bbtr)_3](ClO_4)_2$: nucleation, formation of domains and fluctuations. Inorg. Chim. Acta, 361: 3616–3622.
17. Krivokapic, I., Enachescu, C., Bronisz, R., Hauser, A. (2008) Spin transition and relaxation dynamics coupled to a crystallographic phase transition in a polymeric iron(II) spin-crossover system. Chem. Phys. Lett., 455: 192–196.
18. Bronisz, R. (2005) 1,4-di(1,2,3-triazol-1-yl)butane as building block for the preparation of the iron(II) spin-crossover 2D coordination polymer. Inorg. Chem., 44: 4463–4365.
19. Kusz, J., Gütlich, P., Spiering, H. (2004) Structural investigations of tetrazole complexes of Fe(II), in: Gütlich, P., Goodwin, H. A. (Eds) Spin Crossover in Transition Metal Compounds II. Top. Curr. Chem., 234: 129–153.
20. Hauser, A. Private communication, based on earlier unpublished results (1986) A quantitative analysis similar to that in ref. 17 remains to be done.

21. Hauser, A. (1991) Intersystem crossing in the $[Fe(ptz)_6](BF_4)_2$ spin crossover system (ptz = 1-propyltetrazole). J. Chem. Phys., 94 : 2741–2748.
22. Jeftic, J., Romstedt, H., Hauser, A. (1996) Interplay between the spin transition and the crystallographic phase transition in the Fe(II) spin-crossover system $[Zn_{1-x}Fe_x(ptz)_6](BF_4)_2$ (x = 0.1, 1; ptz = 1-propyltetrazole). J. Phys. Chem. Solids 57: 1743-1750.
23. Varret, F., Boukheddaden, K., Chong, C., Goujon, A., Gillon, B., Jeftic, J., Hauser, A. (2007) Light-induced phase separation in the $[Fe(ptz)_6](BF_4)_2$ spin-crossover single crystal. Eur. Phys. Lett., 77: 30007/1–30007/6.
24. Varret, F., Boukheddaden, K., Goujon, A., Gillon, B., McIntyre, G. J. (2008) Light-induced phase separation (LIPS) into like-spin phases observed by Laue neutron diffraction on a single crystal of $[Fe(ptz)_6](BF_4)_2$. Z. Kristallogr., 22: 250–258.
25. Vreugdenhil, W., van Diemen, J. H., de Graaf, R. A, Haasnoot, J. G., van der Kraan, A. M., Kahn, O., Zarembowitch, J. (1990) High-spin ↔ low-spin transition in $[Fe(NCS)_2(4,4\text{-bis-}1,2,4\text{-triazole})_2](H_2O)$. X-ray crystal structure and magnetic, Mössbauer and EPR properties. Polyhedron, 9: 2971–2979.
26. Garcia, Y., Kahn, O., Rabardel, L., Chansou, B., Salmon, L., Tuchagues, J.-P. (1999) Two-step spin conversion for the three-dimensional compound tris(4,4′-bis-1,2,4-triazole)iron(II) diperchlorate. Inorg. Chem., 38: 4663–4670.

17

Theoretical Prediction of Spin-Crossover at the Molecular Level

Robert J. Deeth, Christopher M. Handley and Benjamin J. Houghton

Inorganic Computational Chemistry Group, Department of Chemistry, University of Warwick, UK

17.1 Introduction

The spin state of a transition metal (TM) complex is an electronic property. It is therefore not surprising that most theoretical work has employed quantum mechanics (QM) although other models are also available. In this chapter, we will explore the application of computational chemistry in its broadest sense to the calculation of the spin state energetics of coordination complexes. Many applications of theory have not been directed to spin crossover (SCO) *per se* although clearly any method which can accurately predict the relative energies of different spin states could be applied to SCO.

The chapter is arranged roughly along historical lines beginning with early crystal field theory (CFT) and valence bond (VB) approaches which provided at least a qualitative description of spin states. We then move through *ab initio* quantum mechanics and density functional theory (DFT) before returning to more empirical methods.

17.2 Beginnings: Valence Bond and Ligand Field Theories

Baker and Bobonich reported the strong temperature dependence of the magnetic moments of $[Fe(phen)_2(NCS)_2]$, $[Fe(bipy)_2(NCS)_2]$ and $[Fe(phen)_2(NCSe)_2]$ in 1964[1] but were unable to provide a satisfactory explanation of this behaviour. Three years later, König and Madeja correctly formulated the behaviour as due to a thermal spin equilibrium.[2] They also attempted to justify their measurements theoretically although the methods then available for treating transition metal systems were very limited.

It was well known for certain metal centres that the spin state could depend on the structure of the complex and the nature of the ligands. However, the dominant theoretical models of the day, namely valence bond

Spin-Crossover Materials: Properties and Applications, First Edition. Edited by Malcolm A. Halcrow.

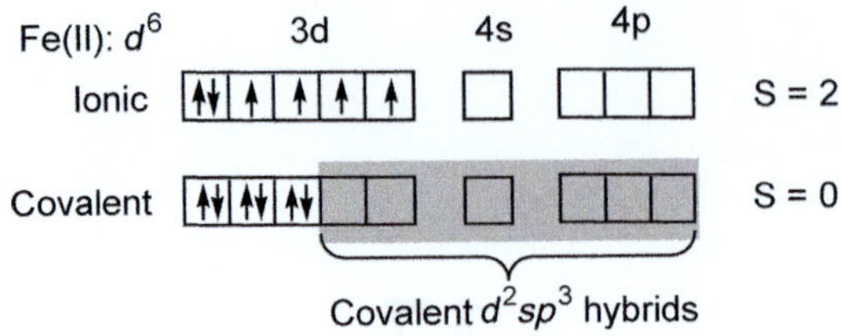

Figure 17.1 Illustration of Pauling's magnetic criterion of bond type. For ionic/ion dipole bonding (top) the electronic configuration of the free metal ion is adopted leading to a high spin configuration. For covalent bonding (bottom) two of the d orbitals are required for forming hybrid orbitals and the d electrons reside in the others giving a low spin state.

(VB) theory and ligand field theory (LFT) gave quite different rationalisations. In 1931, Pauling employed VB concepts to develop for metal complexes the notion of a 'magnetic criterion of bond type' (Fig. 17.1).[3] The difference in spin states between high spin $[Fe(OH_2)_6]^{2+}$ and low spin $[Fe(CN)_6]^{4-}$ was deemed due to the change from ionic/ion-dipole bonding in the former to covalent bonding in the latter. Thus, since the Fe-OH_2 interaction was considered ion-dipole, $[Fe(OH_2)_6]^{2+}$ should have the spin state of the parent free d^6 Fe(II) ion – that is $S = 2$. In contrast, the six covalent Fe–CN bonds of $[Fe(CN)_6]^{4-}$ require a d^2sp^3 hybridisation state on the metal centre to accommodate the twelve bonding electrons from the ligands. This leaves the three remaining *d* orbitals to accommodate the six Fe *d* electrons leading to the diamagnetic $S = 0$ state.

In contrast, LFT describes a very different role for the d orbitals. LFT, or rather the crystal field theory model from which it was derived, was used by van Vleck in 1935[4] to rationalise the variable magnetic moments for TM complexes. In CFT, the only explicit orbitals are the d functions and the spin state is determined by the energetic balance between *d-d* inter-electron repulsion and, in octahedral symmetry, the $t_{2g} \rightarrow e_g$ promotion energy, Δ_{oct} (Fig. 17.2). For a d^6 system, inter-electron repulsion favours the maximum number of unpaired spins which requires two d electrons in the e_g orbitals. Hence, compared to the $t_{2g}{}^6$ state, the $t_{2g}{}^4 e_g{}^2$ configuration is higher in energy by $2\Delta_{oct}$ but is stabilised by two lots of spin-pairing energy, *E*(pair). Thus, in the weak-field case of $[Fe(OH_2)_6]^{2+}$, spin pairing dominates resulting in a high spin $^5T_{2g}$ state while for strong-field ligands such as cyanide, Δ_{oct} dominates resulting in a low spin $^1A_{1g}$ state. However, the essential premise of the crystal field model is that the *d* orbitals are not themselves directly involved in metal–ligand bonding. This is also the case for LFT. The failure of CFT to provide a *quantitative* model

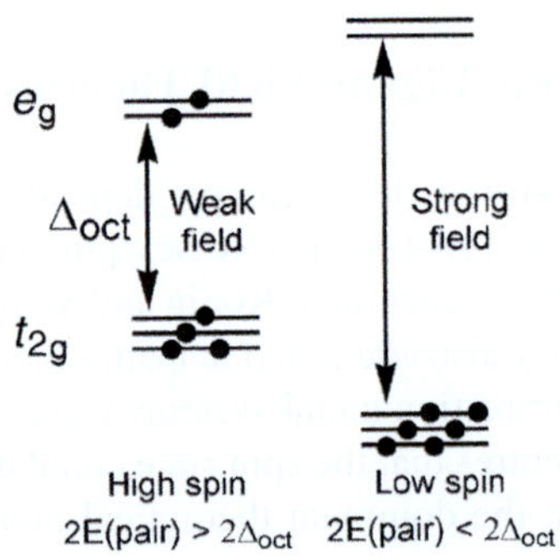

Figure 17.2 Illustration of the LFT description of spin state energetics for d^6 complexes. The high spin $^5T_{2g}$ state on the left is dominated by spin pairing while the low spin $^1A_{1g}$ state on the right is dominated by Δ_{oct}.

for d orbital splitting is resolved in the ligand field picture by abandoning any attempt to compute values for Δ_{oct} from first principles electrostatics. Instead, these quantities are treated as free variables with values determined by fitting to experimental data.

In contrast to Pauling's VB treatment which is only qualitative, LFT provides in principle a quantitative, albeit parameterised, method of computing the magnetic behaviour of metal complexes. König and Madeja[2] invoke the van Vleck equation[5] which expresses the temperature dependence of the total magnetic moment in terms of the contributions from the low spin $^1A_{1g}$ ground state and the high spin $^5T_{2g}$ state. The latter is presumed to lie at some energy, ΔE, above the ground state. While the agreement between theory and experiment is not perfect, the authors conclude that ΔE should be of the order of a few hundred wavenumbers – that is in a thermally accessible regime.

A more complete LF picture of the energy levels of d^6 complexes is provided by the relevant Tanabe–Sugano diagram (Fig. 17.3).[6] Tanabe–Sugano diagrams display the relative energies of all the Russell–Saunders

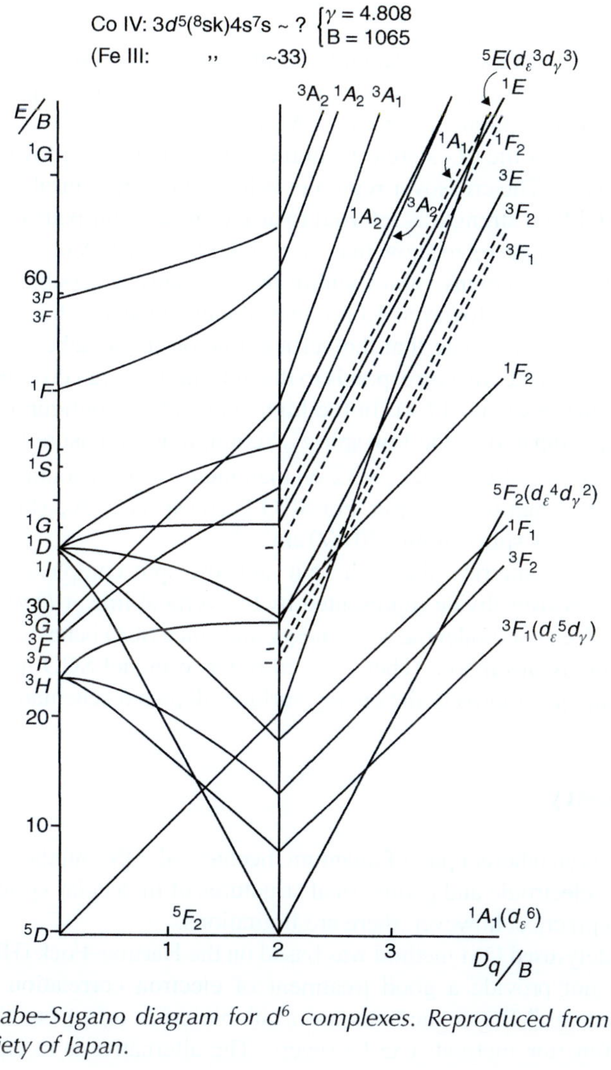

Figure 17.3 Original Tanabe–Sugano diagram for d^6 complexes. Reproduced from [6] with permission from Journal of the Physical Society of Japan.

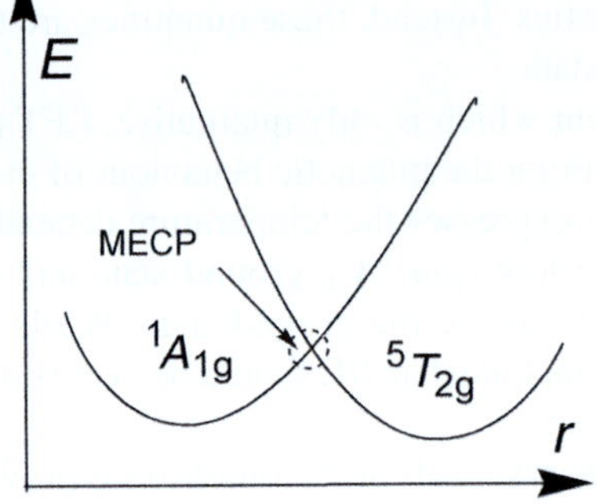

Figure 17.4 The minimum energy crossing point intersection between the low spin and high spin potential-energy surfaces corresponds to the spin-crossover point in the d^6 Tanabe–Sugano diagram.

multiplet terms arising from a given d configuration as a function of the crystal-field splitting parameter Δ_{oct} or, equivalently, Dq (where $\Delta_{oct} = 10Dq$). As the magnitude of the crystal-field splitting increases (relative to the Racah interelectronic repulsion parameter, B), the $^1A_1$ term arising from the free-ion ^1I term rapidly falls in energy and at $Dq/B = 2$ crosses the $^5T_2$ state ($^5F_2$ in the older notation employed by Tanabe and Sugano) to become the new ground state. The crossover region is indicated by the central vertical line in the diagram and the corresponding crystal field strength is referred to as the 'mean spin-pairing energy', π.[7]

The Tanabe–Sugano diagram appears to show that 'if, at a field strength Δ close to π, the separation of the two states of different multiplicity attains values within the thermally accessible range, equilibria between these states are expected to occur'.[2] However, König and Madeja appear to have misinterpreted the earlier analysis of Ewald *et al.*[7] who clearly state that the vertical line on the Tanabe–Sugano diagram refers to a nonequilibrium geometry which actually corresponds to the minimum energy crossing point between the $^1A_{1g}$ and $^5T_{2g}$ potential energy surfaces (Fig. 17.4). In the high spin $t_{2g}{}^4e_g{}^2$ configuration, the extra, strongly σ antibonding e_g d electrons, compared to the low spin $t_{2g}{}^6$ configuration, ensure that the Fe–L bond lengths in high spin complexes are appreciably longer (~0.2 Å) than their low spin counterparts. Hence, Δ_{oct} for the high spin case is much smaller than for low spin. For SCO Fe(II) species, Δ_{oct}(high spin) is estimated to be around 12 000 cm^{-1} and Δ_{oct}(low spin) around 20 000 cm^{-1}.[8]

Thus, at their *equilibrium* geometries, the high spin and low spin complexes are rather well removed from the ligand field spin crossover divide represented by the vertical line at $Dq/B = 2$ in Figure 17.3. The problem with LFT is that it does not model the total energy and since SCO behaviour is inherently a two-state problem, LFT is fundamentally unsuitable. Therefore, in order to model SCO more effectively, we need a method which is able to map quantitatively the energy surfaces depicted schematically in Figure 17.4.

17.3 Quantum Chemistry

The total molecular energy is an inherent part of quantum mechanical (QM) methods which therefore provide, in principle, a model of the electronic and geometrical structures of molecular systems and can be applied to SCO in TM complexes. In practice, however, there are limitations.

Historically, the first widely-used QM method was based on the Hartree–Fock (HF) approximation. Unfortunately, HF theory does not provide a good treatment of electron correlation and is a relatively poor approximation for TM systems.[9] Better methods are available (*vide infra*) but the computational cost of accurate, correlated wavefunction methods can be severe. The alternative is to employ Density Functional

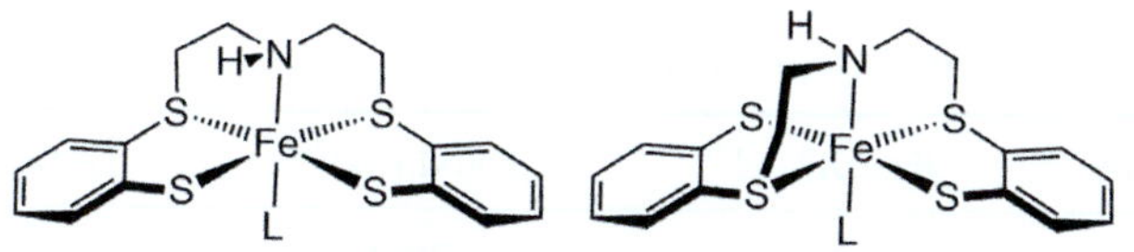

Figure 17.5 Structural diagrams of complexes used by Reiher to explore the amount of exact HF exchange in the hybrid B3LYP functional necessary to reproduce the experimental spin states.

Theory (DFT). DFT has revolutionised our ability to carry out good quality calculations on relatively large systems and there has been no shortage of applications of DFT to spin state energetics of TM complexes in general and spin-crossover systems in particular. However, rather than attempt to provide a comprehensive review, we will focus on a selected but illustrative subset.

A feature of DFT is that there are very many functionals to choose from. This becomes problematic for SCO systems since different functionals can give wildly different spin state energy differences. However, this apparently aberrant behaviour masks the fact that qualitatively, most functionals give the correct trend. Hence, for example, we showed[10] for the 'pure' (i.e. nonhybrid) functionals LDA, PW91, BLYP, PBE, revPBE and RPBE that, irrespective of the functional, the stability of the low spin state varies as $[Fe(CN)_6]^{3-/4-}$ < $[Fe(py)_6]^{2+/3+}$ < $[FeCl_6]^{3-/4-}$. The cyanide complexes are always computed to be low spin and the chlorides always high spin while pyridine systems are predicted to be low spin for the LDA, PW91, BLYP, PBE functionals and high spin for revPBE and RPBE. Thus, considering a wider set of complexes, the 'correct' spin-crossover point can be obtained by systematically correcting the computed spin state energies. For example, the RPBE functional tends to overstabilise the low spin state by $\sim$ 13 kcal mol^{-1}. By adding this correction to the low spin energies, the experiment may be reproduced.

In general, pure functionals tend to overly stabilise the low spin state. Conversely, the extremely popular hybrid B3LYP functional tends to stabilise high spin.[11] Reiher *et al.*[12] traced this to the amount of 'exact' HF exchange inherent in Becke's three-parameter exchange functional and showed a linear relationship between the amount of HF exchange and the spin state energy difference. By comparing computed spin state energy differences for the series of complexes shown in Figure 17.5 to experimental data, they suggest that the 20% HF exchange found in 'normal' B3LYP should be reduced to ~15% to give B3LYP* which delivers good spin state energetics for iron complexes.

The problem here, of course, is that this modification of B3LYP is not universal and different complexes require different amounts of exact HF exchange. Alternatively, different hybrid functionals may require a different amount of HF exchange such as the 25% suggested for PBE0.[13, 14] Harvey argues that the 15% admixture in B3LYP* 'should not be relied upon, and the development of new functionals is [. . .] highly desirable'.[11] Indeed, by analysing vertical excitation energies, Hughes and Freisner[15] find that B3LYP* can lead to much larger spin state energy errors than B3LYP.

Several other functionals have been suggested. Ghosh[16] favours OLYP and Swart[17] OPBE, both of which employ Handy's OPTX exchange functional,[18] while Neese[19] dismisses OPBE and suggests that a double-hybrid functional such as Grimme's B2PLYP, in conjunction with large and flexible basis sets (def2-QZVPP) 'is able to provide qualitatively correct results of spin-state energetics' although the error margins seem quite large: a calculated high spin minus low spin energy difference between 0 and 6 kcal mol^{-1} is deemed indicative of possible SCO behaviour.

So which method is correct? The answer, as is so often the case in DFT, is all of them and none. By its very nature, DFT cannot be systematically improved so that each functional is the product of some kind of fitting process. In general, TM systems do not feature in the fitting dataset – for example Becke's hybrid exchange[20]

Table 17.1 Comparison of low spin/high spin energy differences ($E_{LS} - E_{HS}$, kcal mol^{-1}) for different theoretical methods.

Complex	$[Fe(OH_2)_6]^{2+}$	$[Fe(NH_3)_6]^{2+}$	$[Fe(bipy)_3]^{2+}$
CASPT2	**46.6**	**20.3**	**−13.2**
HF	81.1	72.8	70.2
B3LYP	33.1	14.1	0.6
OPBE	**48.6**	**19.5**	**−13.3**

and OPTX[18] are based on main group elements only – so, to some extent, their ability to treat TM systems effectively is fortuitous. Moreover, given the variability in the performance of different functionals, selecting one which happens to correlate with some experimental observations is also potentially fraught. One cannot guarantee that a DFT protocol designed for a particular type of complex will be applicable to others.

In contrast, wavefunction methods are capable of systematic improvement and hence benchmarking functionals against the results of high-level wavefunction calculations appears to be a less arbitrary way of assessing the formers' performance. Perhaps the most accurate wavefunction approach currently available for TM complexes is CASPT2.[21] The complete active space (CAS) part takes care of the static correlation (i.e. the near-degeneracy effects commonly encountered in open shell TM complexes) while the second-order perturbation theory (PT2) approach deals with the dynamic correlation which is important for relatively heavy TM centres. Swart[17] has compared Hartree–Fock, B3LYP and OPBE spin state energy differences (not corrected for zero point energy) with published CASPT2 data for $[Fe(OH_2)_6]^{2+}$, $[Fe(NH_3)_6]^{2+}$ and $[Fe(bipy)_3]^{2+}$ (Table 17.1). The agreement for OPBE is remarkable and appears to be quite consistent for Fe complexes. However, it would be dangerous to assume that OPBE is a universal functional for *all* spin state energetics. There are many outstanding issues.

First, the spin state energies must to some extent depend on the structure. CASPT2 is too expensive a method for geometry optimisation so the 'benchmark' results do not refer to an actual CASPT2 stationary point. However, the expectation is that the total energy is a fairly weak function of the geometry.

Second, while the benchmark comparison can refer to gas phase calculations, comparisons for 'real' systems invariably involve condensed-phase data (usually X-ray crystal structures). The calculations presumably need to include the effect of the environment if they purport to reproduce condensed phase data such as, for example, the spin state implied by a bulk magnetic susceptibility measurement. The most convenient way to include 'environmental effects' is via a continuum solvation model such as COSMO and we have shown the importance of condensed phase effects on the geometries of TM species.[22] In general, solvation will favour low spin. The low spin structure has shorter metal–ligand bond lengths leading to a smaller complex and the Born model predicts that the solvation energy is proportional to the inverse of the ionic radius. Thus, a smaller complex is associated with an increased (more negative) solvation energy. The effect is of the order of 4–6 kcal mol^{-1}.

Third, by default, QM methods optimise to the lowest local minimum on the potential energy surface. If the structure in one spin state is known and there is no change in conformation with SCO, the QM estimate of the spin state energy difference can be quite accurate. However, if the structure is unknown, then, in general, some kind of conformational search may be required to establish the most likely structure. Depending on the system, this could imply generating many thousand possible starting points, optimising each one and placing them in energy order. As in drug discovery, QM methods of whatever flavour, including DFT, rapidly become too expensive.

Fourth, many common functionals – B3LYP and OPBE amongst them – do not include the dispersion force, which can be significant, while CASPT2 certainly does. Dispersion is commonly incorporated into DFT via an empirical, molecular mechanics correction[23] and its inclusion to give a 'better' calculation will affect the calculated spin state energy difference such that any correlation based on reproduction of experimental data may change. Clearly, while great progress has been made, SCO remains difficult and subtle and the performance of quantum mechanical methods needs to be continually reviewed.

17.4 Empirical Methods

Although often billed as 'first principles' or '*ab initio*', quantum methods in general, and DFT in particular, often contain healthy numbers of empirical parameters. In general, this is to improve the accuracy rather than the speed of a calculation. Thus, while it may be desirable to use QM, it may not be feasible if the number of separate calculations becomes very large. As alluded to in the previous section, comprehensive conformational searching, screening of large libraries of complexes, or molecular dynamics may require many hundreds of thousands or even millions of individual computations which can make QM simply unaffordable. In this section, we will explore methods which use parameters to make the computations significantly faster.

17.4.1 Semi-Empirical MO Theory

Semi-empirical MO theory employs minimal, valence atomic orbital basis sets, and makes some fairly severe theoretical approximations. The latter are usually based on the neglect of diatomic differential overlap (NDDO) approximation which allows for a number of difficult integrals to be treated as parameters. These simplifications speed up the calculation by several orders of magnitude compared to 'full' QM. However, this acceleration can be at the expense of accuracy. The situation is much worse for TM systems than 'organic' systems composed of light elements since good semi-empirical MO parameters for metal centres are difficult to generate. Although Clark and co-workers have made great strides to develop new AM1* parameters for a range of metal centres, in general, the accuracy is somewhat patchy and for Fe in particular, they do not address the issue of spin states.[24] Ball *et al.*[25], have for technetium complexes, reparameterised the semi-empirical PM3(tm) Hamiltonian using a Genetic Algorithm and claim accuracy comparable with DFT, although individual errors can be large ($>10\,\text{kcal mol}^{-1}$). Given the sensitivity of SCO, it seems unlikely that semi-empirical MO theory will ever be accurate enough.

17.4.2 Ligand Field Molecular Mechanics

Finally in this section, we turn to classical methods based on molecular mechanics (MM). It may seem somewhat incongruous given the previous comments concerning the accuracy of the parameters in semi-empirical MO theory to even consider a method which is completely parametric. However, as will be shown, this feature of MM can be turned to our advantage.

In MM, the total strain energy is expressed as a function of various contributions. In its simplest 'diagonal' form (Eq. (17.1)), these correspond to bond stretching, angle bending, torsional twisting and nonbonding interactions such as electrostatics and van der Waals interactions. In Eq. (17.1), the k are appropriate force constants, θ are bond angles, τ are torsion angles, n is the torsional periodicity parameter, ϕ is the torsion offset, ρ are partial atomic charge, ε is the dielectric constant, A and B are Lennard–Jones vdW parameters and the summations run over bonded atom pairs (ij), angle triples (ijk) and torsional quadruples ($ijkl$). The nonbonded terms are summed over the distances, d_{ij}, between unique atom pairs excluding bonded pairs and

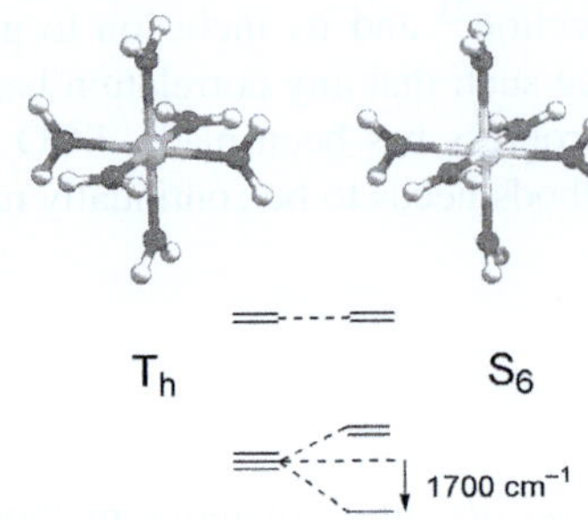

Figure 17.6 *The Jahn–Teller effect in high spin d^6 [Fe(OH$_2$)$_6$]$^{2+}$ where a rotation about the Fe–O bonds raises the ground state orbital degeneracy without affecting the Fe–O bond lengths.*

the atoms at either end of an angle triple. For the atoms at the ends of a torsion quadruple, the non-bonded term may be omitted or scaled.

$$E_{\text{tot}} = \sum_{i,j} k_{ij}(r_{ij} - r_{0,ij})^2 + \sum_{i,j,k} k_{ijk}(\theta_{ijk} - \theta_{0,ijk})^2 + \sum_{i,j,k,l} k_{ijkl}[1 + \cos(n_{ijkl}\tau - \phi_{ijkl})] + \left[\sum_{i<j} \frac{\rho_i \rho_j}{\varepsilon d_{ij}} + \sum_{i<j}\left(\frac{A_{ij}}{d_{ij}^{12}} - \frac{B_{ij}}{d_{ij}^{6}}\right)\right] \tag{17.1}$$

The mathematical form of each term is quite simple and therefore very quick to compute. However, a large number of parameters is required to describe the so-called force field (FF). Nevertheless, MM is an extremely popular and powerful method for handling very large systems and the fact that it is based on empirical parameters does not preclude good accuracy. However, conventional MM is not well suited to TM systems in general[26] and SCO in particular.

The harmonic expression for the angle-bending potential in Eq. (17.1) does not easily lend itself to the higher coordination numbers encountered in TM chemistry where there may be more than one value for a given L–M–L angle – for example 90° and 180° for octahedral complexes. Electronic effects such as Jahn–Teller distortions are difficult to incorporate in a general way, although the tetragonal elongations typically observed for six-coordinate Cu(II) complexes can be incorporated by adding Lagrangian constraints.[27] Jahn–Teller effects are quite widespread in coordination complexes and can be energetically important in high spin Fe(II) species but not necessarily easy to spot.[28] For example, $[Fe(OH_2)_6]^{2+}$ can relieve the ground state orbital degeneracy of the high-symmetry T_h form by rotations around the Fe–O bonds to generate an S_6 geometry (Fig. 17.6) which has the same Fe–O bond lengths but is computed to be about 5 kcal mol^{-1} lower in energy.

Conventional MM would also struggle with SCO for Fe(II) species since the large difference in Fe–L bond lengths for the two spin states would require two sets of FF parameters each with its own (hypothetical) reference state. Hence, the absolute total spin state energy difference would be meaningless, although variations from complex to complex might be more reliable.

Our response is to recognise that SCO is a quintessential ligand field problem. Thus, we have augmented conventional MM with a generalised calculation of the ligand field stabilisation energy (LFSE) which enables us to design a single set of FF parameters capable of describing the structure and energy in both spin states. Thus, ligand field molecular mechanics (LFMM)[29, 30] can be designed to reproduce experimental and/or

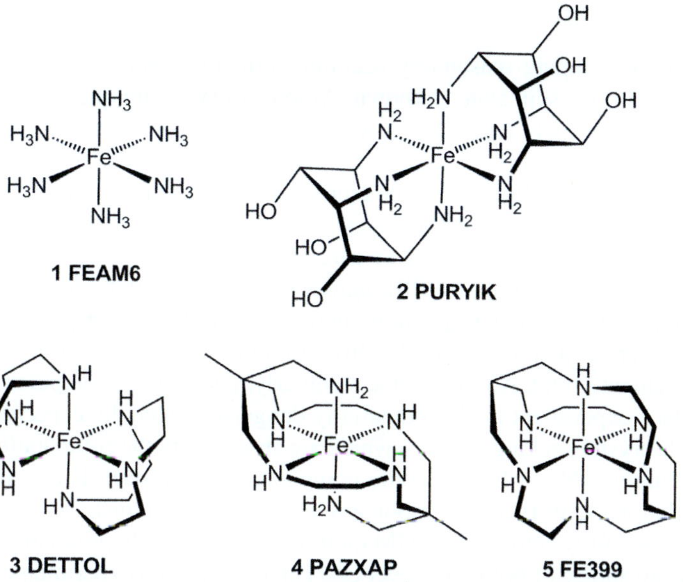

Figure 17.7 Fe-am(m)ine complexes used to develop the LFMM FF.

calculated data to good accuracy. Given our recent development of a multi-objective optimisation algorithm based on machine learning,[31] we can now quickly and automatically optimise LFMM FF parameters for Fe-amine complexes (Fig. 17.7) which reproduce both the DFT OPBE structures and spin state energy differences, the latter to about 0.2 kcal mol^{-1} (Fig. 17.8).

Thus, LFMM can be designed to reproduce accurate QM estimates of spin state energetics and since it is many orders of magnitude faster even than semi-empirical MO theory, LFMM can be used to design new Fe(II) complexes with spin state energy differences near zero and which therefore may be able to support SCO.[32]

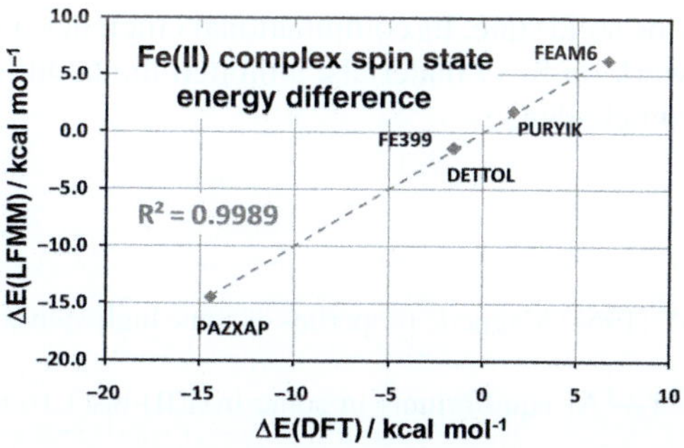

Figure 17.8 Fit between DFT and LFMM spin state energy differences. LFMM parameters derived from multi-objective evolutionary algorithm optimisation.[31] DFT calculations used the OPBE functional, TZP basis sets and the COSMO solvation model.

17.5 Conclusions

Currently, it is possible to compute reasonably accurate spin state energetics for isolated TM complexes especially if the structure in one spin state is known. Amongst QM techniques, several DFT protocols have been suggested and tested especially for Fe(II) systems with nitrogen donors. In particular, the 'pure' OPBE and double hybrid B2PLYP functionals are claimed to deliver good results, although reasonably large basis sets of at least triple-ζ plus polarisation should be employed. Tuning the amount of Hartree–Fock exchange in B3LYP, while possible, receives less support.

Of course, the performance of DFT needs to be carefully monitored since any apparent agreement must, to some extent, be fortuitous. For example, OPBE agrees very nicely with high-level CASPT2 wavefunction results yet the latter implicitly includes dispersion energies while the former provides at best only a partial treatment. The possibility arises that 'improving' OPBE by adding a Grimme-style dispersion correction might actually make the agreement with CASPT2 worse. Similarly, justifying a computational method according to its ability to reproduce experiment is also potentially dangerous since the experimental data may include influences –for example crystal packing effects – which are absent from the computations. Again, the apparent agreement may be fortuitous. Nevertheless, if the computational approach is consistently good, we might accept a happy cancellation of errors and overlook any formal shortcomings.

In that spirit, the authors' fully empirical LFMM method is a particularly promising approach for 'materials discovery'. It can take advantage of both high level QM calculations and experimental data to develop highly accurate, tailored force fields which deliver results fast enough to enable conformational searching and dynamics simulations. However, all the approaches described in this chapter suffer from one fairly significant shortcoming. That is, they focus on the spin state energetics of isolated complexes. Thus, even though isotropic condensed phase effects can be included via some continuum solvation model, the crystalline environment is likely to be inherently anisotropic. These 'crystal packing' effects can easily be of the order of several kcal mol^{-1} such that even if the isolated molecule estimate is exactly correct, the subsequent behaviour in the solid state is subject to significant perturbation. This is similar to the problem faced in drug discovery where the strength of the drug–target interaction is only one part of a multifaceted problem. Thus, as recognised by us and others[15],[19,32] the best we can currently hope for from 'single molecule' studies is to find likely species spanning a range of spin state energy differences sufficiently close to zero for them to be at least worthy of further investigation – say, a spin state energy difference of 3–6 kcal mol^{-1}. This target energy range is sufficiently large that further refinements of DFT protocols seem unwarranted, at least for Fe systems where extensive testing and validation has been carried out.

As for the future, the undoubted success of DFT and LFMM to capture the spin state energies of isolated complexes needs to be taken into the solid state. Its computational efficiency and high accuracy makes LFMM a promising avenue for further work on SCO materials, although the additional complexities of solid state modelling present many significant challenges.

References

1. Baker, W. A. jr., Bobonich, H. M. (1964) Magnetic properties of some high-spin complexes of iron(II). Inorg. Chem., 3: 1184–1188.
2. König, E., Madeja, K. (1967) $^5T_2$–$^1A_1$ equilibriums in some iron(II)-bis(1,10-phenanthroline) complexes. Inorg. Chem., 6: 48–55.
3. Pauling, L. (1931) The nature of the chemical bond. Application of results obtained from the quantum mechanics and from a theory of paramagnetic susceptibility to the structure of molecules. J. Am. Chem. Soc., 53: 1367–1400.

4. Van Vleck, J. H. (1935) Valence strength and the magnetism of complex salts. J. Chem. Phys., 3: 807–813.
5. Van Vleck, J. H. (1932) The Theory of Electric and Magnetic Susceptibilities. Oxford University Press, Oxford, UK, p. 384.
6. Tanabe, Y., Sugano, S. (1954) On the absorption spectra of complex ions. II. J. Phys. Soc. Jpn., 9: 766–779.
7. Ewald, A. H., Martin, R. L., Ross, I. G., White, A. H. (1964) Anomalous behaviour at the $^6A_1$–$^2T_2$ crossover in iron (III) complexes. Proc. R. Soc. London Ser. A, 280: 235–257.
8. Gütlich, P., Hauser, A., Spiering, H. (1994) Thermal and optical switching of iron(II) complexes. Angew. Chem. Int. Ed., 33: 2024–2054.
9. Szabo, A., Ostlund, N. S. (1989) Modern Quantum Chemistry: Introduction to Advanced Electronic Structure Theory. McGraw-Hill, New York, USA, p. 466.
10. Deeth, R. J., Fey, N. (2004) The performance of nonhybrid density functionals for calculating the structures and spin states of Fe(II) and Fe(III) complexes. J. Comp. Chem., 25: 1840–1848.
11. Harvey, J. N. (2004) DFT computation of relative spin-state energetics of transition metal compounds. Struct. Bonding (Berlin), 112: 151–183.
12. Reiher, M., Salomon, O., Hess, B. A. (2001) Reparameterization of hybrid functionals based on energy differences of states of different multiplicity. Theor. Chem. Acc., 107: 48–55.
13. Fouqueau, A., Casida, M. E., Daku, L. M. L., Hauser, A., Neese, F. (2005) Comparison of density functionals for energy and structural differences between the high-[$^5T_{2g}$:$(t_{2g})^4(e_g)^2$] and low-[$^1A_{1g}$:$(t_{2g})^6(e_g)^0$] spin states of iron(II) coordination compounds. II. More functionals and the hexaminoferrous cation, [$Fe(NH_3)_6]^{2+}$. J. Chem. Phys., 122: 044110/1–044110/13.
14. Fouqueau, A., Mer, S., Casida, M. E., Daku, L. M. L., Hauser, A., Mineva, T., Neese, F. (2004) Comparison of density functionals for energy and structural differences between the high-[$^5T_{2g}$:$(t_{2g})^4(e_g)^2$] and low-[$^1A_{1g}$:$(t_{2g})^6(e_g)^0$] spin states of the hexaquoferrous cation [$Fe(H_2O)_6]^{2+}$. J. Chem. Phys., 120: 9473–9486.
15. Hughes, T. F., Friesner, R. A. (2011) Correcting systematic errors in DFT spin-splitting energetics for transition metal complexes. J. Chem. Theory Comput., 7: 19–32.
16. Conradie, J., Ghosh, A. (2007) DFT calculations on the spin-crossover complex Fe(salen)(NO): a quest for the best functional. J. Phys. Chem. B, 111: 12621–12624.
17. Swart, M. (2008) Accurate spin-state energies for iron complexes. J. Chem. Theory Comput., 4: 2057–2066.
18. Handy, N. C., Cohen, A. J. (2001) Left-right correlation energy. Mol. Phys., 99: 403–412.
19. Ye, S. F., Neese, F.(2010) Accurate modeling of spin-state energetics in spin-crossover systems with modern density functional theory. Inorg. Chem., 49: 772–774.
20. Becke, A. D. (1988) Density-functional exchange-energy approximation with correct asymptotic behavior. Phys. Rev. A, 38: 3098–3100.
21. Andersson, K., Roos, B. O. (1992) Excitation energies in the nickel atom studies with the complete active space SCF method and second-order perturbation theory. Chem. Phys. Lett., 191: 507–514.
22. Hocking, R. K., Deeth, R. J., Hambley, T. W. (2007) DFT study of the systematic variations in metal–ligand bond lengths of coordination complexes: the crucial role of the condensed phase. Inorg. Chem., 46: 8238–8244.
23. Grimme, S. (2004) Accurate description of van der Waals complexes by density functional theory including empirical corrections. J. Comp. Chem., 25: 1463–1473.
24. Kayi, H., Clark, T. (2010) AM1* parameters for manganese and iron. J. Mol. Model., 16: 1109–1126.
25. Ball, D. M., Buda, C., Gillespie, A. M., White, D. P., Cundari, T. R. (2002) Can semiempirical quantum mechanics be used to predict the spin state of transition metal complexes? An application of de novo prediction. Inorg. Chem., 41: 152–156.
26. Deeth, R. J., Anastasi, A., Diedrich, C., Randell, K. (2009) Molecular modelling for transition metal complexes: dealing with d-electron effects. Coord. Chem. Rev., 253: 795–816.
27. Comba, P., Zimmer, M. (1994) Molecular mechanics and the Jahn-Teller effect. Inorg. Chem., 33: 5368–5369.
28. Deeth, R. J., Anastasi, A., Randell, K. (2009) Ligand field torque: a π-type electronic driving force for determining ligand rotational preferences. Dalton Trans., 6007–6012.
29. Deeth, R. J. (2010) Molecular modelling for systems containing transition metal centres, in: Comba, P. (Ed.), Structure and Function, Springer, Dordrecht, Netherlands. Ch. 2, pp. 21–51.

30. Deeth, R. J. (2001) The ligand field molecular mechanics model and the stereoelectronic effects of d and s electrons. Coord. Chem. Rev., 212: 11–34.
31. Handley, C. M., Deeth, R. J. (2012) A multi-objective approach to force field optimization: structures and spin state energetics of d^6 Fe(II) complexes. J. Chem. Theory Comput., 8: 194–202.
32. Deeth, R. J., Anastasi, A. E., Wilcockson, M. J. (2010) An in silico design tool for Fe(II) spin crossover and light-induced excited spin state-trapped complexes. J. Am. Chem. Soc., 132: 6876–6877.

18

Theoretical Descriptions of Spin-Transitions in Bulk Lattices

Cristian Enachescu[1], Masamichi Nishino[2] and Seiji Miyashita[3,4]

[1]*Department of Physics, Faculty of Physics, Alexandru Ioan Cuza University, Romania*
[2]*Computational Materials Science Centre, National Institute for Materials Science, Japan*
[3]*Department of Physics, Graduate School of Science, The University of Tokyo, Japan*
[4]*CREST, JST, Japan*

18.1 Introduction

As spin-crossover (SC) materials are of interest not only for their potential applications as future memories or sensors,[1–3] but also as archetypes to test general theories for phase transitions, their theoretical descriptions have fascinated researchers over the last few decades. Experimental observation of abrupt thermal transitions, sometimes accompanied by hysteresis, and the sigmoidal shape of HS–LS relaxation curves, made clear that, in addition to the large vibrational entropy difference between the two states, the cooperative nature of many SC compounds must be taken into account. To explain these phenomena, several models with a large number of variants have therefore been elaborated, starting from a simple reproduction of the thermal equilibrium involving a Boltzmann distribution between the vibronic manifolds of the high spin (HS) and low spin (LS) states, which is only valid for noninteracting compounds, to complex models taking into account distortions, anisotropy effects or lattice strains.

In theoretical analyses of SC, following initial phenomenological mean-field approaches,[4–8] Ising-like models[9–12] and their extensions[13–16] or phenomenological elastic models[17–20] have been intensively studied. These approaches have been able to reproduce successfully several of the characteristic features of SC compounds, such as: the occurrence of thermal hysteresis and its dependence on the strength of intermolecular interactions[5, 10, 12, 13], the sigmoidal shape of relaxation curves[4, 10, 16], or light induced bistability.[6–8, 10]

The use of Ising models to study SC complexes started with the pioneering work of Wajnflasz and Pick, which introduced intermolecular interactions through an Ising-like Hamiltonian and was the first to predict

Spin-Crossover Materials: Properties and Applications, First Edition. Edited by Malcolm A. Halcrow.

the existence of a first order transition associated with the spin transition.[9] In these models, the two states of molecules have different degeneracies and a lower energy for the LS state. By associating a fictitious spin to each molecule ($\sigma_i = \pm 1$ for HS(LS) state),[9] and defining the short-range interactions J, the energy difference D and the degeneracy ratio g between the two states, the prototype Hamiltonian is given by $H_{WP} = -J\sum_{i,j}\sigma_i\sigma_j + \frac{1}{2}\sum_i (D - k_BT\ln g)\sigma_i$, where k_B is the Boltzmann constant and T the temperature. Here the interactions between SC units are not of magnetic origin, but result from the change in bond length accompanying every spin change. The discovery of 2-step transitions[21] revealed the importance of short-range interactions, and the existence of correlations which are not allowed by long range interactions alone. Boussekou *et al.* proposed an Ising-like model with short-range antiferro interaction in addition to the ferro interaction, which gives the condition of 2-step transitions, solved in mean-field approximation.[13] Subsequent extensive studies have successfully treated the dynamic[10] and static properties[11] of SC compounds. Later on, more complex Ising-like models as $H = H_{WP} - G\sum_i \sigma_i \langle\sigma\rangle$ that simultaneously take into account short and long range interactions have been developed.[14, 15] Using mean-field theory of H_{WP}, it was possible to obtain the full classification of ordering processes and to reveal the existence of a metastable phase at low temperature,[12] which was observed in rapid quenching experiments,[22] light induced excited spin state trapping[5, 23] or light induced phase collapse phenomena.[24] Monte Carlo simulations considering short range correlations with a nonrandom distribution of HS and LS complexes have been used to treat 2-step transitions or complex thermal transitions accurately.[14] Recent developments of Ising models include their use for nanoparticulate systems using open boundaries with forced HS molecules at the edge[25] or embedded in polymers.[26, 27]

In parallel to the development of studies of Ising-like models and their extension, studies of phenomenological elastic models have been carried out for SC phenomena. Following an initial phenomenological treatment of interactions proposed by Slichter and Drickamer,[28] Spiering pointed out that the elastic interactions arise from stresses due to changes of the volume, shape and elasticity of the lattice during transition, introducing the concept of 'image pressure' in his model of lattice expansion and elastic interactions, which is treated in a mean-field approach.[17–19] The total interaction is the sum between an infinite-range contribution originating in lattice strains,[17] as first described by Onishi and Sugano,[20] and a direct contribution from pair interaction.[18] Later on, it was stated that the interaction constant can be explained by elasticity theory in a proportion of around 80%.[29] In a related approach, in his elastic model based on ligand field theory within a mean-field approach, Kambara discussed the effect of coupling between *d*-electrons and lattice strains.[30]

In the atom–phonon coupling model,[31] SC compounds are modelled analogous to the Ising-like model, but neighbouring molecules are linked by springs, whose elastic constant has three values depending on the spins of the closest molecules. While Ising-like models can reproduce several features of SC behaviour, the physical meaning of the interaction parameters is not clear as it does not take into account the distortions created by the different sizes of molecules, which are precisely the origin of cooperative interactions. On the other side, while conventional elastic models revealed the importance of elasticity for SC phenomena, due to lack of treatment of local distortion and global deformation simultaneously (Section 18.2), they described only partially the relations between elastic interactions and cooperativity. Properties such as volume expansion, pressure effects, cluster spreading during the transition, the dependence of SC phenomena on crystal shape, microscopic mechanisms of SC phenomena and so forth, cannot be correctly treated by the above mentioned approaches. Hence, in order to study these aspects, qualitatively new models and methods based on elastic interactions were necessary.

In recent years, a new class of cooperative SC models has been developed. Based on the so-called ball and spring concept, this new theoretical description of SC materials implies that elastic interactions arise from lattice distortion due to the difference of the molecular sizes between the LS and HS states in higher dimensions (2D and 3D).[32–45] As in 1D systems, the elastic interactions are basically expressed by Ising-type interactions,[46] but the higher dimensionality is essential to reproduce the cooperative features of the system.

In the first part of this chapter we present the main results obtained in the frame of elastic models, solved by using either the Molecular Dynamic approach or the Monte Carlo Metropolis method, while the second part is devoted to the mechanoelastic model applied both to the bulk and to nanoparticles.

18.2 Elastic Interaction Models for Spin-Crossover Systems

Previous studies of elastic interactions beyond mean-field approaches considered either the lattice distortion around a fixed lattice point, or the independent elongation effect in each direction (x, y, z).[47, 48] However, due to the drastic simplifications involved, they lead to a false representation of the relation between the local distortion and global deformation. Therefore, in order to properly represent the change of lattice shape due to the volume (size) change of each molecule, the volume change of each molecule and the distance between molecules must both be treated correctly. Unlike past studies, the present approach incorporates the mechanism of the global and inhomogeneous deformation of the system, which is responsible for the feature of the long-range interaction of elasticity. Combining this kind of model with molecular dynamics (MD) and Monte Carlo (MC) methods, has allowed various new aspects of SC phenomena to be clarified.[32–34, 37, 39, 43, 45] In this section we show several important features of these new models.

A schematic picture of the model for SC is given in Figure 18.1. In Figure 18.1a, an intramolecular potential is shown as a function of the molecular radius r, where the growth of r from the ideal radius of the LS state (r_{LS}) is defined by $x = r - r_{\mathrm{LS}}$ and r_{HS} is the ideal radius of the HS state. Figure 18.1b depicts a picture of the lattice distortion due to the different sizes of the LS and HS molecules. In the following we present both approaches based on MC and MD methods.

For Monte Carlo (MC) methods the following Hamiltonian is adopted:[33, 34, 45]

$$H = {}^{1}\!/_{2}\sum_{i}(D - k_B T \ln g)\sigma_i + \sum_{\langle i,j\rangle} V_{ij}^{\mathrm{inter}}(\vec{X}_i, \vec{X}_j, r_i, r_j) \tag{18.1}$$

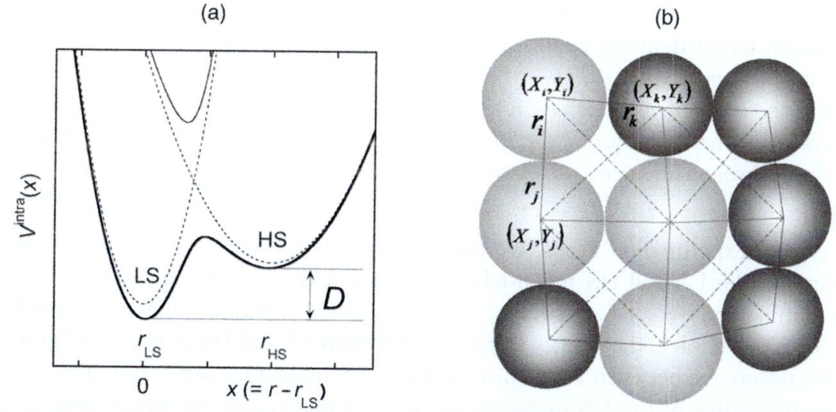

Figure 18.1 (a) Bistable intramolecular potential (solid curve). The dotted curves are the LS and HS potentials without mixing. (b) SC model of elastic interactions, caused by lattice distortions originated from the different molecular sizes. Dark and light grey denotes LS and HS molecules, respectively. Solid lines depict nearest-neighbour interactions and broken lines weak next-nearest-neighbour interactions. Note that molecular positions change.

where the first term is similar to Ising Hamiltonian, with $V_g = -{}^1/_2 k_B T \ln g \sum_{i=1}^{N} \sigma_i$ standing for the entropy effect and the second term corresponds to the intermolecular elastic interactions (see Fig. 18.1b), with $\vec{X}_i = (X_i, Y_i)$ representing the coordinate of the centre of mass for the ith molecule and r_i is the radius of the molecule. To detail the second term of Hamiltonian (1) the following elastic interactions are considered as harmonic potentials:

$$\sum_{\langle i,j \rangle} V_{ij}^{\text{inter}} = \frac{k_1}{2} \sum_{i,j} \left[X_{ij} - (r_i + r_j) \right]^2 + \frac{k_2}{2} \sum_{i,k} \left[X_{ij} - \sqrt{2}(r_i + r_k) \right]^2 \tag{18.2}$$

where $X_{ij} = |\vec{X}_i - \vec{X}_j|$ is the distance between the ith and jth sites, and k_1 and k_2 are the elastic constants between nearest-neighbours (i, j) and next-nearest-neighbours (i, k). We set the ratio of the elastic constants $k_1/k_2 = 10$, where a small value of k_2 is necessary to maintain the square lattice structure, and the ratio of the radii $r_{HS}/r_{LS} = 1.1$.

The following Hamiltonian for molecular dynamics (MD) methods is considered:[32, 37, 39, 43]

$$\hat{H} = K + V \tag{18.3}$$

where:

$$K = \sum_{i=1}^{N} \frac{p_i^2}{2m} + \sum_{i=1}^{N} \frac{\vec{P}_i^2}{2M} \tag{18.4}$$

$$V = \sum_{i=1}^{N} V_i^{\text{intra}}(r_i) + \sum_{\langle i,j \rangle} V_{ij}^{\text{inter}}(\vec{X}_i, \vec{X}_j, r_i, r_j) \tag{18.5}$$

In the kinetic energy term (K), $\vec{P}_i$ and p_i represent the conjugate momenta of $\vec{X}_i$ and r_i respectively for the ith molecule, and M and m are the corresponding masses. The mass M corresponds to that of the molecule, which is usually much larger than the central ion (Fe) mass, and the mass m for the motion of the radius corresponds to that of ligands, which move when the molecular size changes. Both these are of the same order of magnitude, and we took $M = m = 1$ for simplicity. Figure 18.1a shows the intramolecular potential $V_i^{\text{intra}}(x_i)$ which gives symmetric mode for the ith molecule. The intramolecular LS potential $y = ax^2$ and HS one $y = b\,(x - c)^2 + d$(broken lines) are mixed by the off-diagonal element J, which causes the intramolecular adiabatic potentials: $V(x) = \frac{A}{2}\left(d + b(c - x)^2 + ax^2 - \sqrt{4J^2 + (d + b(c - x)^2 - ax^2)^2}\right)$.

According to typical experimental cases, we adopt a realistic ratio $\omega_{LS} / \omega_{HS} = 2.0$ for the frequency between the intramolecular LS and HS potentials.[49] Because $a \propto {\omega_{LS}}^2$ and $b \propto {\omega_{HS}}^2$, we take $a = 4$ and $b = 1$. For other parameters, we set $A = 10$, $c = 1$, $d = 0.2$ and $J = 0.3$ to obtain a reasonable shape of the mixed potential. With this choice, the energy difference between LS and HS is $D \approx 2.47$ in the present units. In this notation the difference between the radii of HS and LS molecules are 1 and thus, for intermolecular interaction, we set the radius of the HS molecule as $r_{HS} = 9$ and $r_{LS} = 10$, which gives a large r_{HS} / r_{LS} ratio than in most SC compounds (about 5%).[50] However, this choice does not change qualitative features of the study.

In contrast to the MC formulation, the entropy term (V_g) cannot be easily introduced because the density of state is uniquely determined by the given potentials (Eq. (18.5)). Therefore, a formulation to construct the entropy term for MD methods is needed. In typical SC materials the entropy difference is much larger than expected from the simple relation $\Delta S = Nk_B \ln(\omega_{LS} / \omega_{HS})$. In order to realize the larger entropy difference, one introduces additional oscillators u_{0i} with a frequency depending on the spin state of the ith molecule, and adds the following Hamiltonian to Eq. (18.3).

$$H_g = \sum_{i=1}^{N} \frac{p_{0i}^2}{2m_0} + \sum_{i=1}^{N} \frac{1}{2} m_0 \bar{\omega}_{0i}^2 u_{0i}^2 \tag{18.6}$$

where the inverse of the frequency is proportional to the density of states of the oscillator. These additional degrees of freedom may correspond to the motion of ligand molecules. The frequency ϖ_{0i} is given by:

$$\frac{1}{\bar{\omega}_{0i}} = \frac{\alpha(x_i)^2}{\bar{\omega}_{0LS}} + \frac{\beta(x_i)^2}{\bar{\omega}_{0HS}}, \text{ where } \frac{\bar{\omega}_{0LS}}{\bar{\omega}_{0HS}} = \prod_k \frac{\omega_k^{LS}}{\omega_k^{HS}} \times \frac{2S_{HS}+1}{2S_{LS}+1} \tag{18.7}$$

Here $\alpha(x)^2 = \left|\langle LS(x) \mid \Psi_g(x)\rangle\right|^2$ and $\beta(x)^2 = \left|\langle HS(x) \mid \Psi_g(x)\rangle\right|^2$ denote the LS and HS components for the ground state $\left|\Psi_g\right\rangle = \alpha\,|LS\rangle + \beta\,|HS\rangle$, respectively. We set $\bar{\omega}_{0LS}/\bar{\omega}_{0HS} = 100$ and $m_0 = 1.0$. Considering ω_{LS} / $\omega_{HS} = 2$ for the intramolecular potential $V_i^{\text{intra}}(r_i)$, the ratio of the total densities of states is $g = 200$.

We adopt the intermolecular potential $V_{ij}^{\text{inter}}(X_i, X_j, r_i, r_j)$ in the following form:

$$\sum_{\langle i,j\rangle} V_{ij}^{\text{inter}} = \sum_{i,j} f_{NN}(d_{ij}) + \sum_{i,k} f_{NNN}(d_{ik} - \Delta d) \tag{18.8}$$

with $f(u) = C[\exp(a'(u - u_0)) + \exp(-b'(u - u_0))]$, a function which has a minimum at $u = 0$, and d_{ij} given by $d_{ij} = X_{ij} - (r_i + r_j)$ where X_{ij} is the distance between ith and jth molecules and r_i and r_j are their radii. The nearest-neighbour (i,j) and next-nearest-neighbour (i, k) interactions, denoted by f_{NN} and f_{NNN}, respectively, are taken into account. We set $a' = 0.5$ and $b' = 1.0$ for f_{NN} and $a' = 0.1$ and $b' = 0.2$ for f_{NNN} (weaker interactions). For the next-nearest-neighbour pair an adjustment is considered: $\Delta d = 2(\sqrt{2} - 1)\bar{r} = (\sqrt{2} - 1)(r_{LS} + r_{HS})$.

In the following we adopt a harmonic potential for MC study and an exponential type potential for MD study for the choice of intermolecular potentials. However, these choices are interchangeable and others such as Lenard–Johns potentials do not change the qualitative features in the present study; the essence of the results can be reproduced with either method.

18.2.1 Thermal Expansion of Volume and Pressure-Induced Transitions

First, we present the thermal expansion of the volume under a constant pressure in SC systems using MD methods.[37] The thermal expansion has been previously observed in experiments,[50, 51] and a linear dependence of the volume expansion with temperature was detected. To control the temperature we use the Nosé–Hoover thermostat[52, 53] and to control the pressure p the Andersen formalism is applied.[54] We set the following parameters: $m = 1.0$, $M = 1.0$, $m_0 = 1.0$, $M_0 = 0.2$. Here the normalized volume is defined as $V_n \equiv (V - V_{LS})/(V_{HS} - V_{LS})$, where V_{LS} and V_{HS} are the volumes for perfect crystals of the LS and HS states, and the HS fraction is defined as $n_{HS} \equiv \left\langle \frac{1}{N} \sum_{i=1}^{N} \beta(x_i)^2 \right\rangle$, where $\langle\rangle$ means the statistical average.

We analyse the dependency of the normalized volume V_n and of n_{HS} on the temperature, as functions of the strength of the interaction (C) in Eq. (18.8). Here, the system was warmed from $T = 0.1$ to 5.6 in increments of 0.1 under a pressure $p = 0.001$, and then re-cooled to the initial temperature $T = 0.1$. At each temperature, 2×10^5 MD steps were discarded as transient time, and the subsequent 10^5 MD steps were used to measure physical quantities with the MD time step $\Delta t = 0.01$. In our parameter choice,[43] the energy unit

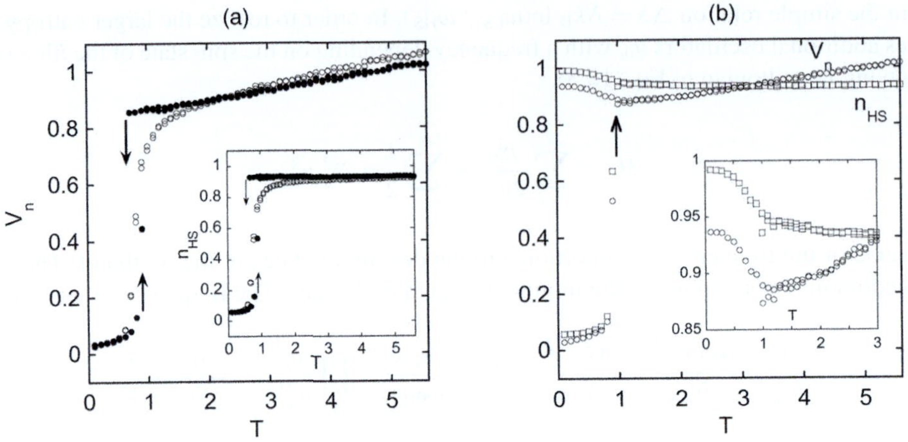

Figure 18.2 *(a)Temperature dependences of V_n under pressure $p = 0.001$. Open and closed circles denote the dependences for $C = 5$ (weak interaction) and $C = 12$ (intermediate interaction), respectively. In the inset temperature dependences of n_{HS} are shown. (b) Temperature dependences of V_n (circles) and n_{HS} (squares) for $C = 15$ (strong interaction) under the pressure $p = 0.001$. Temperature dependence around the minimum of V_n is focused on in the inset.*

is of the order of 100 K and the SC transition point ($T_{SC} \approx 0.8$) is consistent with experimental ones ($T_{SC} \approx$ 100–300 K).

In Figure 18.2, the dependences of n_{HS} and V_n on T are depicted for weak ($C = 5$) and intermediate ($C = 12$) intermolecular interaction cases. The former case shows a gradual (continuous) SC transition and the latter case shows a first-order SC transition. In the warming process, n_{HS} increases and approaches a saturated value close to unity, while V_n increases at high temperatures and finally exceeds unity. It is worth noting that $V_n(V)$ has a linear dependence on T at high temperatures. This linear dependence has often been reproduced in analyses of $V-T$ curves in experiments using phenomenological lattice expansion models.[50, 51] The thermal expansion dV_n / dT in the weaker interaction case is larger at high temperatures ($2 < T$).

Figure 18.2b presents the strong intermolecular interaction case ($C = 15$), showing the characteristic features of the n_{HS} and V_n dependencies on T, which both drastically change with C. While in weaker interaction cases (Fig. 18.2a), n_{HS} decrease continuously in the cooling process, for larger C (Fig. 18.2b) n_{HS} increases during the cooling process when the state is in the HS phase, although n_{HS} is almost constant at high temperatures. In such cases, the HS metastable state at low temperatures has a larger n_{HS} at lower T. This is similar to the strong-interaction case of the short-range Ising-like model, and may be a general feature in bistable systems with large entropy difference.

Concerning the T dependence of V_n, it can be seen that for smaller C V_n also decreases in the cooling process, while for larger C, in contrast to n_{HS}, V_n decreases during the cooling process at high temperatures, passes through a minimum value, and then increases again at lower T. This suggests that a metastable HS state exists at low temperatures. Due to the small thermal fluctuations, the state of the molecules tends to be around the minimum point of the metastable HS intramolecular potential ($x = 1$). When the temperature increases, the state of the molecules can cross the barrier to the LS state and some HS molecules change to LS molecules. Consequently, the volume is reduced and the derivative $\frac{\partial V}{\partial T}$ is negative, which leads to negative thermal expansion. In conclusion the present MD formulation reproduces the nontrivial T dependences of the total volume and of n_{HS}. As negative and zero thermal expansions[55] are one of the topics of molecular solids, this observation will promote research into the unusual behaviour of thermal expansion.

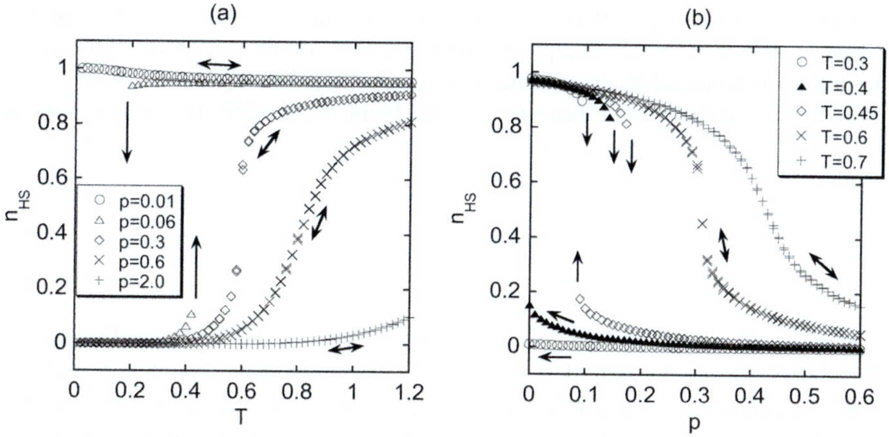

Figure 18.3 *(a) Temperature dependence of the HS fraction n_{HS} with $g = 20$, $D = 1$, $k_1 = 50$ for several values of applied pressure. (b) Pressure dependence of the HS fraction n_{HS} with $g = 20$, $D = 1$, $k_1 = 50$ for several values of temperature.*

We also studied the relaxation curve of n_{HS} from the metastable HS state to the LS state for various T at a constant pressure ($p = 0.001$) and for various p at a constant temperature ($T = 0.15$). It was found that relaxations occur with sigmoidal curves.[37] When T is higher, the decay time becomes shorter due to large thermal fluctuations. When p is larger, it also becomes shorter because the LS state is more favoured.

In the following, we discuss the pressure effect on $n_{HS}(T)$ by MC simulations,[34] in the frame of NPT-ensemble Metropolis algorithm. For this purpose, we performed 10^4 MC steps for transient steps and 10^4 MC steps to measure the physical quantities and considered a system size of $N = 16 \times 16 \times 16$, and set the parameters as $g = 20$, $D = 1$, and $k_1 = 50$ to study thermal properties. In Figure 18.3a, we depict $n_{HS}(T)$ in a cooling and heating process. The initial state corresponds to the equilibrium state at a high temperature ($T = 1.2$); subsequently the system is cooled to $T = 0.02$ and then reheated to the initial temperature in steps of $T = 0.02$.

The equilibrium state at $T = 1.2$ depends on the pressure: it is HS for small p and LS for higher pressure (e.g. $p = 2.0$). However, at very low pressures, $p = 0.01$, the complex is always in the HS state because the metastable HS state is strong even at low temperatures. In contrast, at a high pressure ($p = 2.0$) the complex is always in the LS state. At $p = 0.06$, the metastability of the HS state at low temperatures becomes weaker and the complex shows hysteresis curves. At $p = 0.6$, the hysteresis completely disappears and a gradual change is observed. A transient case between a first order transition and a gradual change can be noticed at $p = 0.3$.

In order to study the pressure dependency of n_{HS} at various temperatures, the HS state is set as the initial state, and the pressure is first increased to $p = 0.6$ and then decreased. In Figure 18.3b, the pressure dependence of n_{HS} is given for several temperatures. In the case of $T = 0.3$, a transition from the HS to the LS phases takes place at $p = 0.1$. As the pressure is reduced, this pressure-induced LS phase does not return to the HS phase. At $T = 0.45$, a pressure-induced hysteresis is observed. The transition is smooth at $T = 0.7$, while at $T = 0.6$ a transient case between a first order transition and a gradual change can be observed. The present observations indicate that pressure also plays an important role as a control parameter for the SC transition.

18.2.2 Long-Range Interactions and Nucleation Features

In this subsection we focus on the configuration of spin states of the elastic model in the bulk. We use the MC method for a 2D simple square lattice and for a 3D simple cubic lattice with periodic boundary conditions

(PBC).[33] Here the pressure is set to $p = 0$. First we study the critical properties and spin configurations at the critical temperature T_C, which is located at a point where $n_{HS} = {}^1\!/_2$ on the coexistence line given by $T = D/(k_B \ln g)$. We investigate the temperature dependence of n_{HS}, adopting the magnetization $\langle M \rangle = \langle \sum \sigma_i \rangle$ instead of n_{HS} for convenience along the coexistence line. For the Ising-like model, the critical properties are those of the corresponding ferromagnetic Ising model. The spontaneous magnetization m_S and the susceptibility per spin χ are obtained from the relation:

$$\frac{\langle M^2 \rangle}{N^2} = m_S^2 + k_B T \frac{\chi}{N} \tag{18.9}$$

where $N = L^d$ is the total number of sites, and $\chi = \langle M^2 \rangle / N k_B T$ is the susceptibility per site above the critical point. We studied the temperature dependences of $\langle M^2 \rangle / N^2$ and χ^{-1} for the 2D and 3D models and found that the critical exponents of β and γ are $\beta = 0.5$ and $\gamma = 1$, where $\langle m_S^2 \rangle / N^2 \propto (T_C - T)^{2\beta}$ and $\chi^{-1} \propto (T_C - T)^{\gamma}$. This set of critical exponents corresponds to those of the mean-field universality class. We also estimated the critical temperature by analysis of the Binder cumulant and found that it agrees well with that of the infinite range (Husimi–Temperley) model for both 2D and 3D systems. Therefore the universality class of the elastic model is different from that of Ising-like models.

In Figure 18.4a, we depict a snapshot of a spin configuration at the critical point (T_c). In Figure 18.4b the time dependence of the magnetization at the end of a hysteresis loop near the spinodal point is given for the parameters $g = 20$, $k_1 = 40$. Here we decreased the temperature gradually from the HS phase in the two-dimensional model with $D = 0.4$ ($< D_c \sim 0.6$), where $D_c = k_B T_c \ln g$. The temperature was decreased from $T = 0.1056$ in steps of $\Delta T = 0.0008$ every 40 000 MCS. A rapid change of the magnetization takes place at $T = 0.1040$ between 8×10^4 and 1.2×10^4 MCS. Open circles are given in the magnetization curve at 96 000, 101 000, 102 000 and 103 000 MCS, in which $M = 0.70$, 0.41, –0.14 and –0.90, respectively. The corresponding configurations for these four points are shown in Figure 18.4c. We find that there are no large domain structures at either the critical point or in the relaxation process from the metastable state. Therefore, one may state that the usual critical behaviour associated with two-phase coexistence is suppressed in the present model. In contrast to short-range interaction models, in which the phase change occurs through

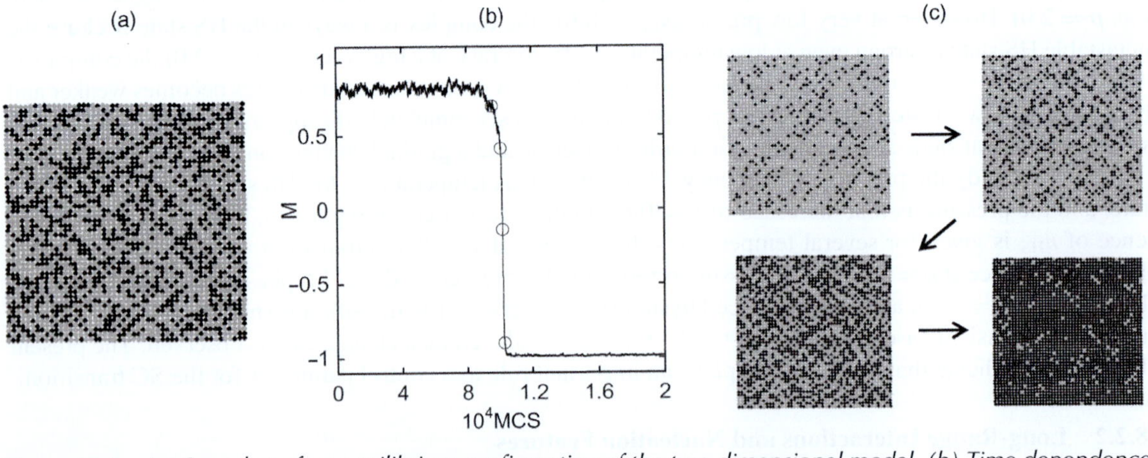

Figure 18.4 (a) Snapshot of an equilibrium configuration of the two-dimensional model. (b) Time dependence of the magnetization at the end of a hysteresis loop. ($g = 20$, $D = 0.4$, $k_1 = 40$). (c) Spin configurations at the end of a hysteresis loop.

growth of compact critical droplets of the bulk equilibrium phase, the present system remains macroscopically uniform during the whole transformation process. This is consistent with the accepted picture of spinodal nucleation in systems with long-range interactions.[45]

The elastic interaction among the molecules with different volumes induces effective long-range interactions. However, in contrast to the infinite-range model, the system is still aware of the shape of the lattice. Because the range of the interaction is infinite, the boundary condition plays a significant role in the dynamics of the bulk. In the following, we present the boundary effect, focusing on nucleation dynamics of the system with open boundary conditions (OBC).[39] Using the MD method, we observe relaxation from the metastable HS phase to the stable LS phase at the edge of a relatively large hysteresis loop at low temperatures. The system consists of $N = L^2 = 100 \times 100$ molecules. The transitions occur at around $T = 1.0$ in the heating process and at around $T = 0.4$ in the cooling process (Fig. 18.5a). The snapshots of the configuration during

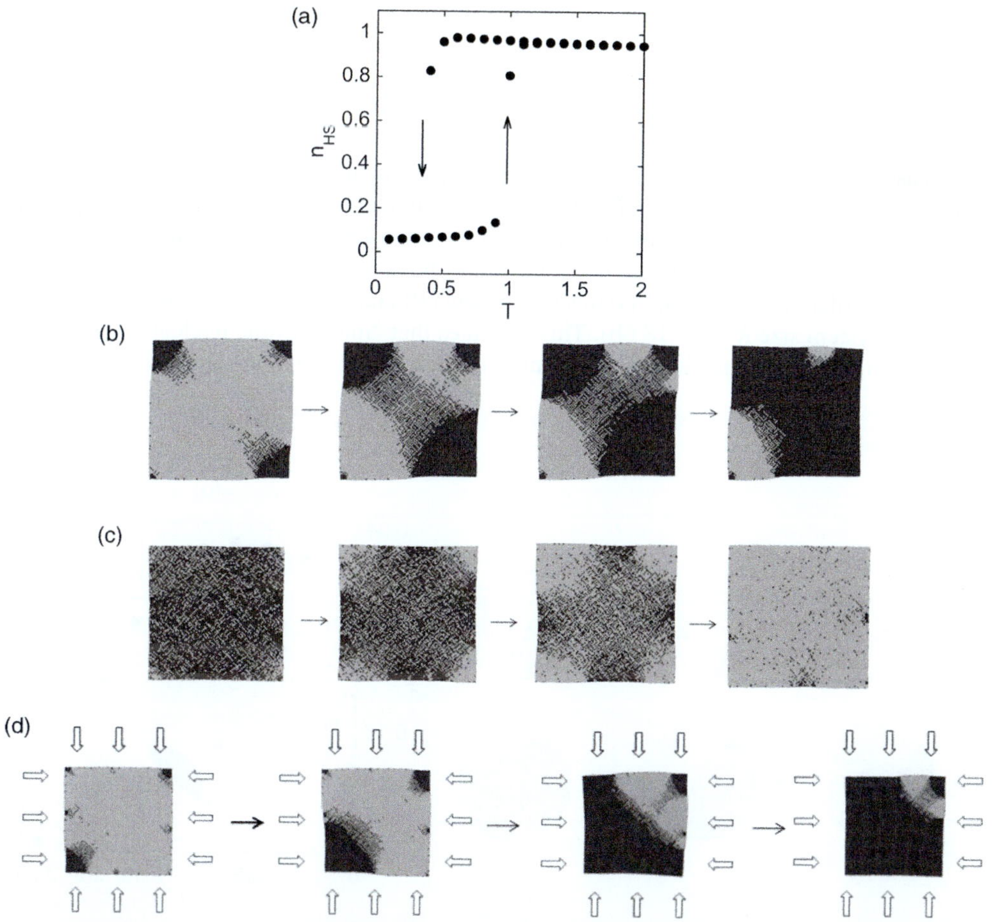

Figure 18.5 (a) Thermal hysteresis loop of high spin fraction with OBC ($L = 100$, $C = 20$). (b) Snapshots of configurations in the left branch of the hysteresis loop during the cooling process from the HS to LS phase (dark and light grey denotes LS and HS phases, respectively) and (c) those in the right branch of the hysteresis loop during the heating process from LS to HS phase. (d) Corner clustering of LS phase when applying pressure ($f_p = 0.1$) at the boundary.

the HS→LS relaxation in Figure 18.5b show the spreading of domains from corners during the transition from the HS to LS phase. LS domains grow and finally combine to extend to the whole system. Domain growth occurs in the diagonal directions.

The evolution of clusters contrasts with Ising-type models with short-range interactions, where the nucleation occurs from the inner part or the sides in large systems. In order to clarify this difference, we have studied the energy dependence on the clustering pattern, finding that growth from sides and bulk is not acceptable in the elastic model.

We next investigate the process in heating. Snapshots of transient states from the LS to HS phase in the heating are given in Figure 18.5(c). Local clusters of HS molecules around the corners appear. However, in contrast to the case of the process from the HS to LS phase (Fig. 18.5(b)), a large homogeneous region appears as is observed with PBC. The feature of corner clustering is also observed when the pressure is applied to the elastic system with OBC. We apply a constant pressure to a metastable HS phase at $T = 0.18$, in which the force per site $f_p = 0.1$ is applied in the direction of x, $-x$, y, or $-y$ at the boundary molecules. After the constant pressure is applied at the boundary, as shown in Figure 18.5(d) clusters grow from the corners and the domain growth is accelerated.

In contrast to the case of periodic boundary conditions, where uniform configurations are maintained during the switching process, we found that domain structures appear with OBC and a macroscopic inhomogeneity is produced.

As another case for awareness of the lattice shape we present the relaxation from the metastable HS state to the LS state at a low temperature ($T = 0.2$) in approximately circular crystals on a square lattice.[43] Snapshots of the configuration during the course of a relaxation event are depicted in Figure 18.6a. The radius of the system is defined by R. In detailed analyses of relaxation we found that the domain formation exhibits geometric similarity for circular crystals of any size (R) and that the size of the critical nucleus r_C is proportional to the system size R (Fig. 18.6b). This indicates that "macroscopic nucleation" can be realized. It is also worth noting that single nucleation is essential in such elastically long-range systems, in contrast to

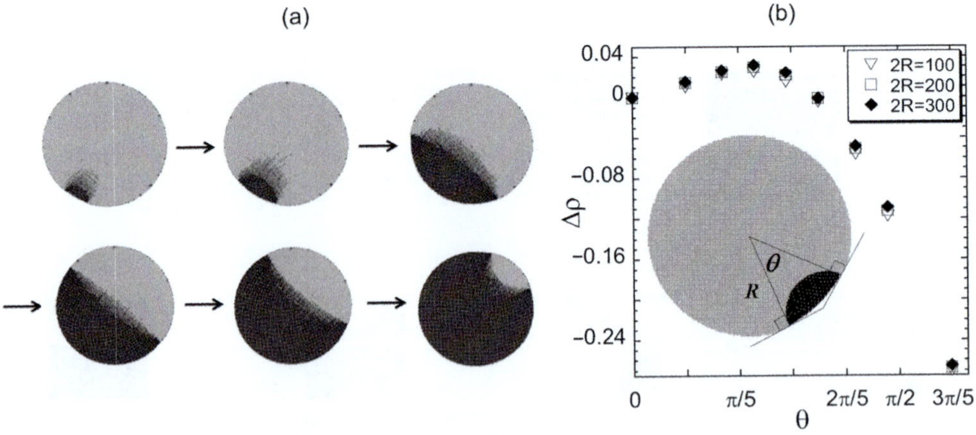

Figure 18.6 *(a) Snapshots of the configuration during relaxation from the HS phase for $2R = 200$ molecules. (b) The excess energy density Δ_ρ as a function of θ for the system sizes $2R = 100, 200, 300$. The inset is the definition of θ and domain region is given for θ using the contact angle of $\pi/2$. For all R, the values of $n_{HS} = 0.98, 0.97, 0.95, 0.92, 0.9, 0.86, 0.83$ are given at $\theta/n = 0.1, 0.17, 0.23, 0.29, 0.35, 0.42, 0.48$ respectively. After $\theta/n = 0.48$, Δ_ρ decreases monotonically until $\theta = \pi / 2$ (LS phase).*

the multi-droplet nucleation mechanism which is universal in short-range interaction systems of large sizes. The single nucleation picture agrees with recent experiments of domain formation of SC compounds.[56]

Traditionally, the nucleation process is considered a microscopic process (at most nanometer), implying the formation of a microscopic nucleus of the stable phase. The size of a critical nucleus is determined by the microscopic competition between the surface and bulk free energies of a microscopic cluster. To be precise, this situation is realized in short-range interaction systems, where the separation of energy between the bulk and surface is allowed. Nucleation theories have been based on this idea, and so far only microscopic nucleation is known. The mechanism shown here is different from that of conventional nucleation theory, introducing a new concept of "macroscopic nucleation" for systems with long-range interactions. The recognition of this mechanism should give important insights for all systems in which local structural changes cause a distortion of the lattice.

18.3 Mechano-Elastic Model

The main idea of the mechano-elastic model[38, 40–42, 44] is similar to that of elastic model presented in previous section but the solution method is different. As there, the SC molecules are represented as rigid spheres, situated in open boundary lattices and interacting by way of connecting springs. In the mechano-elastic model, the equilibrium positions of molecules for a given spin state configuration are calculated by the equation of motion, while the Monte Carlo method is used to determine the switching molecules. The change in volume of a switching molecule produces an instantaneous elastic force in its neighbouring springs and consequently determines the shift in position, initially of its closest neighbours and then of all other molecules in the system. Therefore, the volume and the shape of the whole system changes as each molecule switches. By this procedure, the effect of one single switch propagates progressively through the whole sample; even if the interactions considered here are short range by their definition, their effect manifests, certainly to a smaller extent, at long-range distances. Consequently, the elastic constant of the springs stands for both short-range and long-range parameters in the classical Ising-like model applied to SC complexes.

We shall discuss here the mechano-elastic model for two-dimensional hexagonal lattices with open boundary conditions (Fig. 18.7), which was first introduced for the study of HS–LS relaxation processes, using

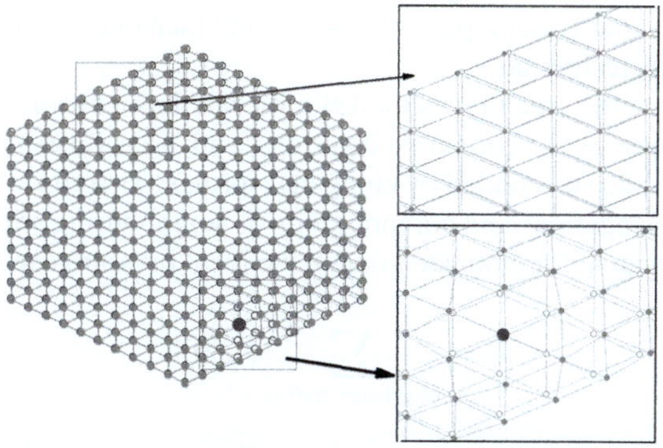

Figure 18.7 *(a) The hexagonal network of particles connected with springs. In the initial state all the molecules are in the LS state (open circles). When a molecule switches its state to HS (dark discs), all the other molecules (small grey discs) change their positions. The hexagon then becomes distorted: the distortions are bigger in the vicinity of switched molecule (short range effect), but they are present even in remote parts of the system (long range effect).*

Arrhenius dynamics.[38] It was subsequently adapted for the study of thermal[44] and light induced hysteresis,[40] for the investigation of the clusters evolution during the transition[41] and for the analysis of the role of impurities in diluted SC complexes.[42]

The classical mechano-elastic model implies finding out-of-equilibrium positions of all molecules in the system for given spin state configurations. Whenever a molecule changes its state, all the molecules in the system move to find their new equilibrium positions, with the property that the vector sum of elastic forces acting on all molecules in the system is zero. As it is difficult to solve the equilibrium equations for all the molecules in the system, we consider instead that every molecule has a damped oscillatory-type motion and iteratively solve the following system of coupled differential equations for all molecules in the system.

$$\begin{cases} m\dfrac{d^2 y_i}{dt^2} = F_{xi} - \mu\dfrac{dx_i}{dt} \\ m\dfrac{d^2 x_i}{dt^2} = F_{yi} - \mu\dfrac{dy_i}{dt} \end{cases} \tag{18.10}$$

until the maximum change in position of any molecule between two consecutive steps is smaller than a given threshold value. The following notations have been used: x_i, y_i are the Cartesian coordinates of molecule i, μ is the damping constant and F_{xi}, F_{yi} are the components of the instantaneous force $\vec{F_i}$ acting on molecule i given by the sum of the forces from the neighbouring springs:

$$\begin{aligned} F_{i,x} &= \sum_{\text{neighbor springs}} k\delta r_{ij,x} \\ F_{i,y} &= \sum_{\text{neighbor springs}} k\delta r_{ij,y}, \end{aligned} \tag{18.11}$$

where k is the spring constant, and $\delta\vec{r}_{ij}$ is the deviation of the neighbour spring j from its unstressed value.

Once the consequences of the switching of one molecule on the lattice have been clarified, we need to decide when individual switching occurs. In a first approach, the switch is influenced by the local pressure, as the average force per unit area acting on a given molecule, while in a second technique one examines the total energy of the system before and after a possible switch and then decides if the switch takes place or not using a Monte Carlo Metropolis technique.

The first approach is based on the idea that as the LS molecule has a smaller volume, it is expected to be favoured when its neighbour springs are compressed springs, and, vice versa, the HS state is expected to be favoured for the case of a molecule surrounded by elongated springs. Therefore, if $\Sigma\ \delta r_{ij}$ is the algebraic sum at mechanical equilibrium of elongations and compressions of neighbouring springs (δr_{ij} is taken as positive for compressed springs and negative for elongated ones), the local pressure can be defined as:

$$p_i = \sum_{\text{neighbours springs}} k \cdot \delta r_{ij} \tag{18.12}$$

The local pressure will be equal for all molecules in the system only for perfectly hexagonal shaped systems (this corresponds to fully HS or LS systems); for intermediate states, the local pressures will be different for every molecule in the system. This difference in local pressure favours the formation of clusters and can lead to avalanche phenomenon.[38, 41] Then, the individual HS→LS and LS→HS transition probabilities depend on the temperature, on the intrinsic features of the SC molecules and on the interactions between molecules, represented here by way of the local pressure p_i, according to:

$$\begin{aligned} P^i_{\mathrm{HS}\to\mathrm{LS}} &= \frac{1}{\tau}\exp\left(-\frac{E_A - \kappa p_i}{k_B T}\right) \\ P^i_{\mathrm{LS}\to\mathrm{HS}} &= \frac{1}{\tau}\exp\left(-\frac{D - k_B T \ln g}{k_B T}\right)\exp\left(-\frac{E_A + \kappa p_i}{k_B T}\right), \end{aligned} \tag{18.13}$$

where E_A is the effective activation energy corresponding to a fully HS state, D is the enthalpy difference for non-interacting molecules. τ a constant scaling factor, chosen so that the above probabilities are well below unity at any temperature, and κ a constant that establishes the extent to which the local pressure influences the relaxation.

In the second approach, the following Hamiltonian is considered:

$$H = \frac{1}{2}\sum_i (D - k_B T \ln g)\sigma_i + \frac{k}{2}\sum_{i,j} dx_{ij}^2 \tag{18.14}$$

where the first term corresponds to the classical Hamiltonian used in the Ising-like system and the second term stands for the elastic energy W_{elastic} calculated as the sum of energies for all the springs in the system. Accordingly, the Metropolis probability is written:

$$P = \exp\left(-\frac{(D - k_B T \ln g)\cdot\left(\sigma_i^{\mathrm{final}} - \sigma_i^{\mathrm{initial}}\right) + \kappa\cdot\left(W_{\mathrm{elastic}}^{\mathrm{final}} - W_{\mathrm{elastic}}^{\mathrm{initial}}\right)}{k_B T}\right) \tag{18.15}$$

where the final elastic energy corresponds to the new positions of molecules calculated considering that the switch has been produced.

In this way, the switching probabilities depend on the energies (which are proportional to the square of all deformations) instead of local pressure, and are proportional to the local deformations. In both approaches, the switching of individual molecules is decided by comparing the above probabilities with a random number between 0 and 1. Only if this number is lower than the calculated probability is the molecule allowed to change its state. A Monte Carlo step (MCS) is completed when all the molecules in the system have been interrogated once on average. After every MCS, the new mechanical equilibrium positions of all molecules are calculated. The two approaches show some differences concerning the evolution of clusters of the same spin state molecules, but the key results are the same;[57] therefore here we shall present here the results obtained using the first approach.

In the simulations presented below, a 10% variation between the radius of molecules in the HS and the LS state has been assumed. The values of all the other parameters, as specified in the Figure captions, do not correspond to any specific SC solid, but are in line with those experimentally found for several classes of SC complexes or used in previous simulations.

The modelling of HS–LS relaxations is one of the most attractive phenomena for SC theoreticians. Even before the discovery of light induced conversion of the stable LS at low temperatures (LIESST effect),[58] a theoretical model based on a tunnelling process[59] was elaborated that predicted HS–LS relaxation. This model has been followed by several others, with different complexity and accuracy, from simple phenomenological treatments in a mean-field approximation[4, 60] to dynamical Ising-like models.[10] In order to reproduce relaxation curves in the frame of the mechano-elastic model, we consider that in the initial state of the system

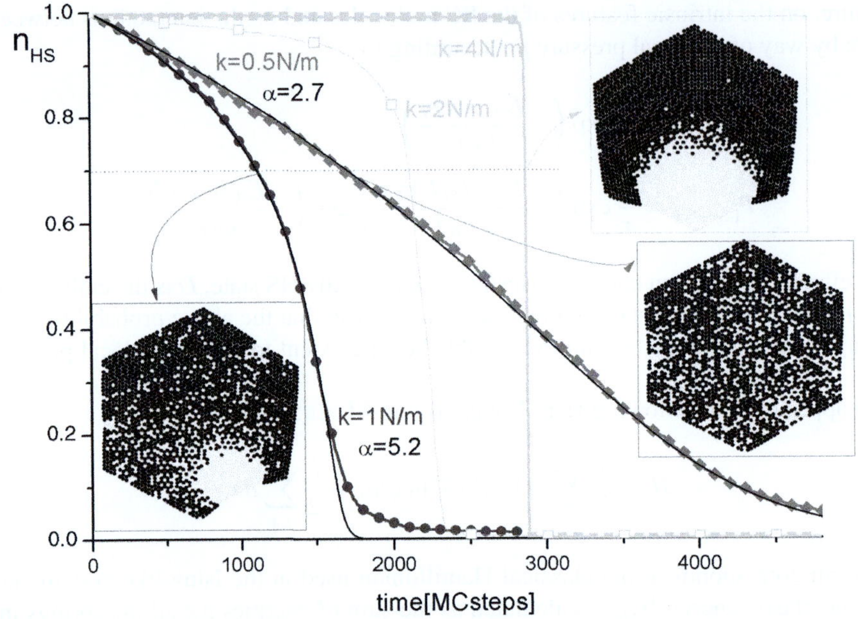

Figure 18.8 *Relaxation curves for different spring constant at 60 K obtained within the mechano-elastic model (line + symbols) and the best fit in the frame of the classical mean-field model (black lines). The other parameters used in simulations are* $E_a = 400$ *K,* $k_B = 1$, $\kappa = 1000$, $g = 1096$, $\Delta H = 1100$ *K. We notice the change in the relaxation regime by comparing the shape of relaxation curves and the cluster evolution inside the system.*

all the molecules are in their metastable HS states. Therefore, in absence of an external pressure, all springs are unstressed and no local pressure acts on molecules; if an external pressure is applied to the network, all the springs are compressed with the same factor.

By the procedure described above, considering both probabilities (Eq. (18.13)), one may calculate relaxation curves at various temperatures or external pressures and the same dependence of the relaxation time as in experimental data will be found.[38] In Figure 18.8 we present relaxation curves obtained for different spring constants. One can notice that depending on the value of the elastic spring constant, not only does the shape of the relaxation curves vary, but also the propagation of changes at the molecular level inside the system. For weak interactions, as in the case of solutions or diluted crystals, the relaxation is close to a single exponential, no clusters are formed and the molecules flip almost independently of each other. For moderate values of spring constant, the relaxation curve is sigmoidal, as expected for a cooperative process with the energy barrier decreasing progressively with HS fraction while several small clusters spread slowly, starting from edge or corners. In fact, the clusters spread from edges only at very low temperatures, for which the local pressure cannot compensate the large value of $D - k_B T \ln g$ term. However, for very big systems, we expect that the clustering can start only from corners, regardless of the temperature, as the stress produced around edge domains would prevent their development. The relaxation curves in these two situations can be reproduced using the simple mean-field approximation, considering a rate constant for the HS–LS relaxation which depends exponentially on the HS fraction following the law $k_{HL} = k_{HL}(n_{HS} = 1) \cdot \exp(\alpha \cdot (1 - n_{HS}))$, with α being the cooperativity parameter.[4] For strongly interacting systems, one single cluster starting from one of six corners develops faster throughout the sample, leading to deviations from

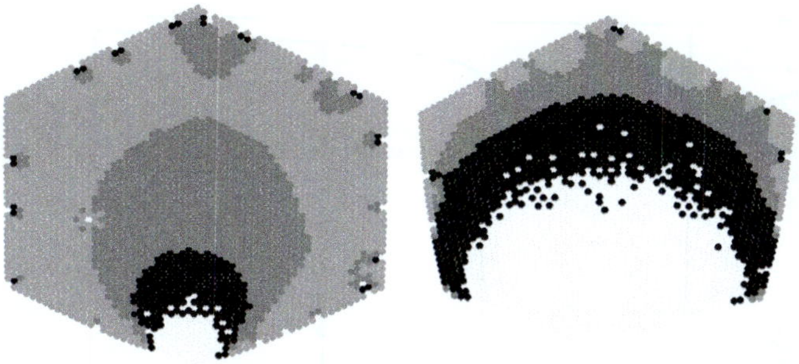

Figure 18.9 Snapshots of a system with high interactions ($k = 4N / m$) taken during relaxation at $n_{HS} = 0.95$ and $n_{HS} = 0.5$ illustrating the concept of local pressure. Colour code: LS molecules (●): HS molecules with higher local pressure (●): HS molecules with average local pressure (●): HS molecules with low local pressure (●).

mean-field behaviour, nonrandom distributions of HS and LS molecules, nucleation and growth phenomena. In this case, when several close molecules change their state to LS, they will trigger their neighbours to flip and will produce an infinite avalanche phenomenon with a discontinuous jump in the HS fraction. The evolution of clusters starting from corners is so fast that very little time remains for allowing fluctuations inside the sample.

In real SC systems, cooperativity, and consequently hysteresis, are due to the large volume difference between HS and LS molecules. In general, cooperativity effects manifest in cluster formation inside the lattice as the conversion proceeds. To understand why this simple model exhibits cooperativity and in particular why the clusters usually build up from the corners, we have calculated the local pressure for all molecules in the system at different moments during the transition. As stated above, when a molecule switches, the local pressure acting on all others molecules changes. In Figure 18.9 we depict the local pressure diagram for two situations during the HS–LS transition: a small fraction of LS molecules (left) and an intermediate fraction of LS molecules (right). At the very beginning of the transition, the HS molecules switch randomly, as the local pressure does stay approximately the same for all the molecules in the system. After a while, the switch of several HS molecules determines the rearrangement of all the molecules in the system and consequently the change of the local pressures. The plot in Figure 18.9 (left) reveals the reason for cluster propagation from corners: the local pressure for HS molecules situated near the newly switched LS molecules near corners dramatically increases, while the pressure for more distant HS molecules increases less or stays approximately constant. This behaviour can be hardly intuitively understood, as one may naturally think that the local pressure in the vicinity of freshly switched LS molecules should be smaller due to the initial compression of neighboring springs. The explanation of this effect is given by the competition between the local and global volume changes, which results in decreasing distances between most molecules and their neghbours, as the edge molecules are pulled inside the lattice while the HS–LS transition proceeds. According to Eq. (18.13) the HS molecules subjected to a higher local pressure will have a higher probability of switching. Therefore, the spread of clusters starts from one corner in the case of strong interactions, while clusters start from several corners for moderate interactions,[41] as can be seen in Figure 18.9.

The thermal transition can be easily simulated within the mechano-elastic model by varying the temperature with a constant sweeping rate and then applying the probabilities (Eq. (18.13)) in order to determine the HS fraction. In this way, it is possible to study the influence of temperature sweep rates, interaction strength,

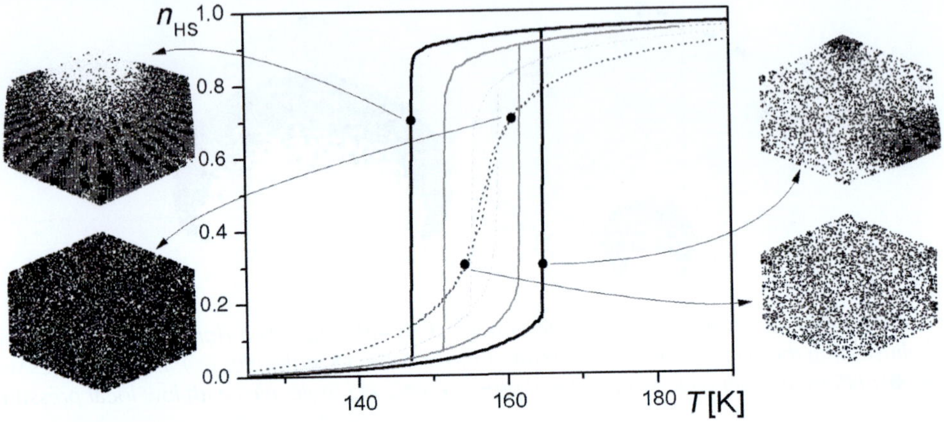

Figure 18.10 *Thermal hysteresis obtained for systems with different interactions (from outside to inside hysteresis loops: k = 1.5 N / m, 1.25N / m, 1N / m, 0.75N / m) and snapshots of the system during both transitions for extreme hysteresis loops for* $n_{HS} = 0.7$ *and* $n_{HS} = 0.3$.

external pressure and sample size on hysteresis shape and width.[44] Examples of thermal hysteresis for two different spring constants are presented in Figure 18.10 (middle), together with snapshots of the system for the same proportion of switched molecules (30%) for both transitions HS→LS (left) and LS→HS (right side). As expected, the width and shape of the thermal hysteresis change with the spring constant, becoming larger and more abrupt for increasing spring constants. Strong interactions, expressed by high spring constants, result in the formation of clusters, which spread starting from corners; this feature is more obvious in the case of the downward transition because of smaller thermal fluctuations. Lower elastic constants result in a rather smoother and gradual transition with almost random switching inside the sample. The same behaviour can be observed equally for systems with impurities, if these exceed a threshold value, which separates the avalanche and mean-field regimes.[44]

The open boundary conditions intrinsic to the mechano-elastic model make it appropriate for the study of the thermal behaviour of SC nanoparticles. As shown elsewhere in this book, SC nanoparticles show different behaviour compared to bulk samples. In the nanoparticle case the thermal transition, sometimes incomplete, is smoother and shifted towards lower temperatures with narrower or even no hysteresis. By simply varying the system size, the experimental data have been partially reproduced: only the less steep transition and the smaller hysteresis which vanishes for a critical size could be reproduced. Therefore, in order to account for all experimental features we consider instead samples embedded in a polymer surrounding environment.[44] For practical reasons, the polymer is represented as a rigid shell of nonswitching molecules which interact with the molecules situated at the edge of the lattice by way of springs, with a given elastic constant k_{pol}. In this way, if the polymer is not flexible, the reduction of the overall volume of the nanoparticles while the transition proceeds induces an increasing negative pressure at the edge of the system, originating in pulling forces from the polymer network. With this premise, the hysteresis not only shifts to lower temperatures, for moderate interactions between edge molecules and polymer, but also the transition is no longer complete with a residual HS fraction if these interactions are higher (Fig. 18.11). A preliminary study showed also that even the switching probability of edge molecules is lowered, due to the negative pressures induced by the polymer molecules, clusters also tend to form from corners, as in the case of open boundary systems, due to the propagation of this negative pressure throughout the entire sample.[44]

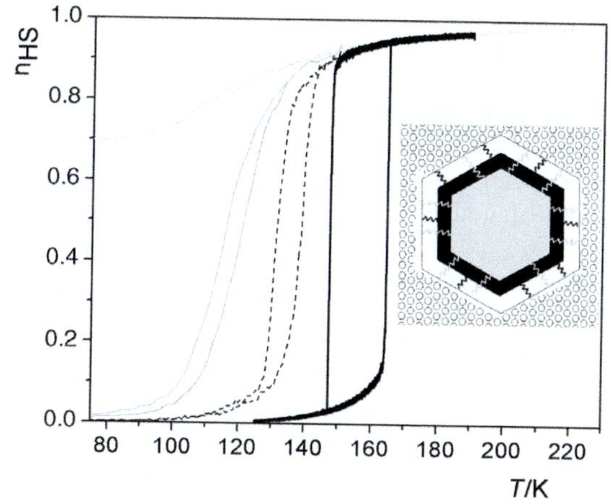

Figure 18.11 *Thermal transition for bulk system (thick full line) and for nanoparticles (2791 molecule system) in polymer with different interactions with the environment (dashed line, $k_{pol} = 0.03N / m$; thin line, $k_{pol} = 0.06N / m$; dotted line, $k_{pol} = 0.2N / m$) (inset) Hexagonal shaped nanoparticles in full HS (black) and full LS states (grey) in polymer environment (open circles). If the polymer is not flexible, then the length of springs connecting edge molecules with polymer increase as the HS–LS conversion proceeds.*

18.4 Conclusions

Though different in method, both the molecular dynamics and Monte Carlo approaches and mechano-elastic approaches lead to similar conclusions and are able to describe accurately the different aspects of spin-transition using simple and realistic assumptions. In particular, open boundary conditions allow the reproduction of the spreading of clusters from corners as universal phenomena in systems with elastic interactions, in accordance with the experimental spatio-temporal properties of SC crystals recently revealed by means of optical microscopy,[61, 62] showing the propagation of a phase boundary. These models are suitable not only for the study of SC systems but also for other systems in which local structural changes determine the modification of the volume of the whole lattice, such as Jahn–Teller[63] or martensitic systems.[64]

References

1. Kahn, O., Martinez, C. J. (1998) Spin-transition polymers: From molecular materials toward memory devices. Science, 279: 44–48.
2. Létard, J.-F., Guionneau, P., Goux-Capes, L. (2004) Towards spin crossover applications, in: Gütlich, P., Goodwin, H. A. (Eds) Spin Crossover in Transition Metal Compounds III. Top. Curr. Chem., 235: 221–249.
3. Matsuda, T., Isozaki, H.,Tajima, H. (2008) Reproducible on-off switching of the light emission from the electroluminescent device containing a spin crossover complex. Thin Solid Films, 517: 1465–1467.
4. Hauser, A. (1992) Cooperative effects on the HS→LS relaxation in the $[Fe(ptz)_6](BF_4)_2$ spin-crossover system. Chem. Phys. Lett., 192: 65–70.
5. Gütlich, P., Hauser, A., Spiering, H. (1994) Thermal and optical switching of iron(II) complexes. Angew. Chem. Int. Ed., 33: 2024–2054.
6. Hinek, R., Spiering, H., Gütlich, P., Hauser, A. (1996) The $[Fe(etz)_6](BF_4)_2$ spin-crossover system - part two: hysteresis in the LIESST regime. Chem. Eur. J., 2: 1435–1439.

7. Desaix, A., Roubeau, O., Jeftic, J., Haasnoot, J. G., Boukheddaden, K., Codjovi, E., Linares, J., Nogues, M., Varret, F. (1998) Light-induced bistability in spin transition solids leading to thermal and optical hysteresis. Eur. Phys. J. B, 6: 183–193.
8. Varret, F., Boukheddaden, K., Codjovi, E., Enachescu, C., Linarès, J. (2004) On the competition between relaxation and photoexcitations in spin crossover solids under continuous irradiation, in: Gütlich, P., Goodwin, H. A. (Eds) Spin Crossover in Transition Metal Compounds II. Top. Curr. Chem., 234: 199–229.
9. Wajnflasz, J., Pick, R. (1971) Low-spin-high-spin transitions in Fe^{2+} complexes. J. Phys. Coll. (France), Cl-91 – Cl-92.
10. Boukheddaden, K., Shteto, I., Hôo, B., Varret, F. (2000) Dynamical model for spin-crossover solids. I. Relaxation effects in the mean-field approach. Phys. Rev. B, 62: 14796–14805.
11. Hôo, B., Boukheddaden, K., Varret, F. (2000) Two-variable macroscopic model for spin-crossover solids: Static and dynamic effects of the correlations. Eur. Phys. J. B, 17: 449–457.
12. Miyashita, S., Honishi, Y., Tokoro, H., Nishino, M., Boukheddaden, K., Varret, F. (2005) Structures of metastable states in phase transitions with a high-spin low-spin degree of freedom. Prog. Theor. Phys., 114: 719–735.
13. Bousseksou, A., Nasser, J., Linares, J., Boukheddaden, K., Varret, F. (1992) Ising-like model for the two-step spin-crossover. J. Phys. 1 (France), 2: 1381–1403.
14. Kohlhaas, T., Spiering, H., Gütlich, P. (1997) Monte Carlo study of the two-step spin transition in $[Fe_xZn_{1-x}(2\text{-}pic)_3]Cl_2\cdot EtOH$. Z. Phys. B, 102: 455–459.
15. Varret, F., Salunke, S. A., Boukheddaden, K., Bousseksou, A., Codjovi, E., Enachescu, C., Linares, J. (2003) The Ising-like model applied to switchable inorganic solids: discussion of the static properties. C. R. Chimie, 6: 385–393.
16. Nishino, M., Boukheddaden, K., Miyashita, S., Varret, F. (2003) Arrhenius Monte Carlo study of two-step spin crossover: equilibrium and relaxation paths. Phys. Rev. B, 68: 224402/1 – 224402/10.
17. Spiering, H., Meissner, E., Köppen, H., Müller, E. W., Gütlich, P. (1982) The effect of the lattice expansion on high spin $\rightleftharpoons$ low spin transitions. Chem. Phys., 68: 65–71.
18. Spiering, H., Willenbacher, N. (1989) Elastic interaction of high-spin and low-spin complex molecules in spin-crossover compounds. II. J. Phys. Condens. Matter, 1: 10089–10105.
19. Spiering, H. (2004) Elastic interaction in spin crossover compounds, in: Gütlich, P., Goodwin, H. A. (Eds) Spin Crossover in Transition Metal Compounds III. Top. Curr. Chem., 235: 171–195.
20. Onishi, S., Sugano, S. (1981) Strain interaction effects on the high-spin-low-spin transition of transition-metal compounds. J. Phys. C, 14: 39–55.
21. Real, J. A., Bolvin, H., Bousseksou, A., Dworkin, A., Kahn, O., Varret, F., Zarembowitch, J. (1992) Two-step spin crossover in the new dinuclear compound $[Fe(bt)(NCS)_2]_2bpym$, with bt = 2,2′-bi-2-thiazoline and bpym = 2,2′-bipyrimidine: experimental investigation and theoretical approach. J. Am. Chem. Soc., 114: 4650–4658.
22. Tokoro, H., Miyashita, S., Hashimoto, K., Ohkoshi, S. (2006) Huge thermal hysteresis loop and a hidden stable phase in a charge-transfer phase transition of $Rb_{0.64}Mn[Fe(CN)_6]_{0.88}\cdot 1.7H_2O$. Phys. Rev. B, 73: 172415/1 – 172415/4.
23. Letard, J.-F. (2006) Photomagnetism of iron(II) spin crossover complexes - the T(LIESST) approach. J. Mater. Chem., 16: 2550–2559.
24. Tokoro, H., Ohkoshi, S. (2008) Phase collapse caused by blue-light irradiation in a cyanobridged coordination polymer. Appl. Phys. Lett., 93: 021906/1 – 021906/3.
25. Muraoka, A., Boukheddaden, K., Linarès, J., Varret, F. (2011) Two-dimensional Ising-like model with specific edge effects for spin-crossover nanoparticles: A Monte Carlo study. Phys. Rev. B, 84: 054119/1 – 054119/7.
26. Dobrinescu, A., Enachescu, C., Stancu, A. (2009) Ising-like model study of size dependence relaxation in spin crossover complexes. J. Magn. Magn. Mater., 321: 4132–4138.
27. Atitoaie, A., Enachescu, C., Tanasa, R. (2012) Size dependent thermal hysteresis in spin crossover nanoparticles reflected within a Monte Carlo based Ising-like model. J. Magn. Magn. Mater., 324: 1596–1600.
28. Slichter, C. P., Drickamer, H. G. (1972) Pressure-induced electronic changes in compounds of iron. J. Chem. Phys., 56: 2142–2160.
29. Jung, J., Schmitt, G., Wiehl, L., Hauser, A., Knorr, K., Spiering, H., Gütlich, P. (1996) The cooperative spin transition in $[Fe_xZn_{1-x}(ptz)_6](BF_4)_2$: II. Structural properties and calculation of the elastic interaction. Z. Phys. B, 100: 523–534.

30. Kambara, T. (1981) Theory of high-spin ⇌ low-spin transitions in transition metal compounds induced by cooperative molecular distortions and lattice strains. J. Chem. Phys., 74: 4557–4565.
31. Nasser, J. A., Topçu, S., Chassagne, L., Wakim, M., Bennali, B., Linares, J., Alayli, Y. (2011) Two-dimensional atom-phonon coupling model for spin conversion: role of metastable states. Eur. Phys. J. B, 83: 115–132.
32. Nishino, M., Boukheddaden, K., Konishi, Y., Miyashita, S. (2007) Simple two-dimensional model for the elastic origin of cooperativity among spin states of spin-crossover complexes. Phys. Rev. Lett., 98: 247203/1 – 247203/4.
33. Miyashita, S., Konishi, Y., Nishino, M., Tokoro, H., Rikvold, P. A. (2008) Realization of the mean-field universality class in spin-crossover materials. Phys. Rev. B, 77: 014105/1 – 014105/12.
34. Konishi, Y., Tokoro, H., Nishino, M., Miyashita, S. (2008) Monte Carlo simulation of pressure-induced phase transitions in spin-crossover materials. Phys. Rev. Lett., 100: 067206/1 – 067206/4.
35. Stoleriu, L., Enachecsu, C., Stancu, A., Hauser, A. (2008) Elastic model for complex hysteretic processes in molecular magnets. IEEE Trans. Magn., 44: 3052–3055.
36. Nicolazzi, W., Pillet, S., Lecomte, C. (2008) Two-variable anharmonic model for spin-crossover solids: a like-spin domains interpretation. Phys. Rev. B, 78: 174401/1 – 174401/12.
37. Nishino, M., Boukheddaden, K., Miyashita, S. (2009) Molecular dynamics study of thermal expansion and compression in spin-crossover solids using a microscopic model of elastic interactions. Phys. Rev. B, 79: 012409/1 – 012409/4.
38. Enachescu, C., Stoleriu, L., Stancu, A., Hauser, A. (2009) Model for elastic relaxation phenomena in finite 2D hexagonal molecular lattices. Phys. Rev. Lett., 102: 257204/1 – 257204/4.
39. Nishino, M., Enachescu, C., Miyashita, S., Boukheddaden, K., Varret, F. (2010) Intrinsic effects of the boundary condition on switching processes in effective long-range interactions originating from local structural change. Phys. Rev. B, 82: 020409/1 – 020409/4.
40. Enachescu, C., Stoleriu, L., Stancu, A., Hauser, A. (2010) Competition between photoexcitation and relaxation in spin-crossover complexes in the frame of a mechanoelastic model. Phys. Rev. B, 82: 104114/1 – 104114/8.
41. Enachescu, C., Nishino, M., Miyashita, S., Stoleriu, L., Stancu, A., Hauser, A. (2010) Cluster evolution in spin crossover systems observed in the frame of a mechano-elastic model. Europhys. Lett. 91, 27003/1 – 27003/6.
42. Enachescu, C., Stoleriu, L., Stancu, A., Hauser, A. (2011) Study of the relaxation in diluted spin crossover molecular magnets in the framework of the mechano-elastic model. J. Appl. Phys., 109: 07B111/1 – 07B111/3.
43. Nishino, M., Enachescu, C., Miyashita, S., Rikvold, P. A., Boukheddaden, K., Varret, F. (2011) Macroscopic nucleation phenomena in continuum media with long-range interactions. Sci. Rep., 1: 162/1 – 162/5.
44. Stoleriu, L., Chakraborty, P., Hauser, A., Stancu, A., Enachescu, C. (2011) Thermal hysteresis in spin-crossover compounds studied within the mechanoelastic model and its potential application to nanoparticles. Phys. Rev. B, 84: 134102/1 – 134102/9.
45. Miyashita, S., Rikvold, P. A., Mori, T., Konishi, Y., Nishino, M., Tokoro, H. (2009) Threshold phenomena under photoexcitation of spin-crossover materials with cooperativity due to elastic interactions. Phys. Rev. B, 80: 064414/1 – 064414/6.
46. Boukheddaden, K., Miyashita, S., Nishino, M. (2007) Elastic interaction among transition metals in one-dimensional spin-crossover solids. Phys. Rev. B, 75: 094112/1 – 094112/10.
47. Sakai, O., Ogawa, T., Koshino, K. (2002) Monte Carlo simulation for the photo-induced low-spin to high-spin state transitions-difference and similarity with the thermal transition. J. Phys. Soc. Jpn., 71: 978–988.
48. Sakai, O., Ishii, M., Ogawa, T., Koshino, K. (2002) Monte Carlo simulation for the macroscopic phase separation in photoinduced low-spin to high-spin state transitions. J. Phys. Soc. Jpn., 71: 2052–2060.
49. Tuchagues, J.-P., Bousseksou, A., Molnar, G., McGarvey, J. J., Varret, F. (2004) The role of molecular vibrations in the spin crossover phenomenon, in: Gütlich, P., Goodwin, H. A. (Eds) Spin Crossover in Transition Metal Compounds III. Top. Curr. Chem., 235: 85–103.
50. Meissner, E., Köppen, H., Spiering, H., Gütlich, P. (1983) The effect of low pressure on a high-spin—low-spin transition. Chem. Phys. Lett., 95: 163–166.
51. Wiehl, L., Kiel, G., Köhler, C. P., Spiering, H., Gütlich, P. (1986) Structure determination and investigation of the high-spin ↔ low-spin transition of $[Fe(pic)_3]Br_2 \cdot EtOH$. Inorg. Chem., 25: 1565–1571.

52. Nosé, S. (1984) A unified formulation of the constant temperature molecular dynamics methods. J. Chem. Phys., 81: 511–519.
53. Hoover, W. G. (1985) Canonical dynamics: equilibrium phase-space distributions. Phys. Rev. A, 31: 1695–1697.
54. Andersen, H. C. (1980) Molecular dynamics simulations at constant pressure and/or temperature. J. Chem. Phys., 72: 2384–2393.
55. Matsuda, T., Tokoro, H., Hashimoto, K., Ohkoshi, S. (2006) Charge-transfer phase transition and zero thermal expansion in caesium manganese hexacyanoferrates. Dalton Trans., 5046–5050.
56. Slimani, A., Varret, F., Boukheddaden, K., Chong, C., Mishra, H., Haasnoot, J., Pillet, S. (2011) Visualization and quantitative analysis of spatiotemporal behavior in a first-order thermal spin transition: a stress-driven multiscale process. Phys. Rev. B, 84: 094442/1 – 094442/8.
57. Enachescu, C., Nishino, M., Miyashita, S., Stancu, A., Stoleriu, L. (2012) Monte Carlo Metropolis study of cluster evolution in spin-crossover solids within the framework of a mechanoelastic model. Phys. Rev. B, 86: 054114/1–054114/7.
58. Decurtins, S., Gütlich, P., Köhler, C. P., Spiering, H., Hauser, A. (1984) Light-induced excited spin state trapping in a transition-metal complex: the hexa-1-propyltetrazole-iron (II) tetrafluoroborate spin-crossover system. Chem. Phys. Lett., 105: 1–4.
59. Buhks, E., Navon, G., Bixon, M., Jortner, J. (1980) Spin conversion processes in solutions. J. Am. Chem. Soc., 102: 2918–2923.
60. Hauser, A. (2004) Light-induced spin crossover and the high-spin→low-spin relaxation, in: Gütlich, P., Goodwin, H. A. (Eds) Spin Crossover in Transition Metal Compounds II. Top. Curr. Chem., 234: 155–198.
61. Chong, C., Mishra, H., Boukheddaden, K., Denise, S., Bouchez, G., Collet, E., Ameline, J.-C., Naik, A. D., Garcia, Y., Varret, F. (2010) Electronic and structural aspects of spin transitions observed by optical microscopy. The case of $[Fe(ptz)_6](BF_4)_2$. J. Phys. Chem. B, 114: 1975–1984.
62. Chong, C., Slimani, A., Varret, F., Boukheddaden, K., Collet, E., Ameline, J.-C., Bronisz, R., Hauser, A. (2011) The stepwise temporal evolution of the thermal spin transition in $[Fe(bbtr)_3](ClO_4)_2$ single crystals explained by optical microscopy in terms of inhomogeneous stresses. Chem. Phys. Lett., 504: 29–33.
63. Englman, R. (1972) The Jahn-Teller Effect in Molecules and Crystals. Wiley-Blackwell, Oxford, UK, p. 350.
64. Porter, D. A., Easterling, K. E., Sherif, M. (2009) Phase Transformations in Metals and Alloys, 3rd edn. CRC Press, Boca Raton FL, USA, p. 520.

19

Optimizing the Stability of Trapped Metastable Spin States

Jean-François Létard, Guillaume Chastanet, Philippe Guionneau and Cedric Desplanches

CNRS, Université de Bordeaux, ICMCB, France

19.1 Introduction

In the domain of transition metal complexes, there are several possibilities for generating light-induced switches accompanied by drastic modifications of magnetic, volume and/or optical properties.[1–4] Among the best known and extensively studied systems, we can cite metal-centered thermal spin-transition (observed in iron(II) SCO materials), valence tautomerism (catecholate complexes of Co(II)), metal-to-metal charge transfer (Prussian Blue analogs), metal-to-ligand charge transfer (nitroprusside complexes), and finally, all ligand modifications that induce a change of spin state, such as isomerization (stilbenoid complexes) and open/close (diarylethene type ligands) processes.

The objective of this chapter is to focus on the Light-Induced Excited Spin-State Trapping (abbreviated as LIESST) encountered in iron(II) spin-crossover complexes,[5] which is of major interest for the design of optical switches,[6] and to reflect on ten years of work since we introduced the *T*(LIESST) procedure.[7] This systematically measures the limiting temperature above which a photomagnetic effect in a material is erased, by warming the sample from 10 K at a rate of 0.3 K min^{-1}. Until now, any applications based on the LIESST effect have been prohibited because the lifetimes of the photomagnetic states are sufficiently long only at low temperatures. Nevertheless, it seems that certain factors can affect the stability of the photoinduced HS state and the *T*(LIESST) limit has been increased progressively.

This chapter is organized into four subsections. Section 19.2 briefly recalls some background considerations on the Light-Induced Excited Spin-State Trapping (LIESST) effect and on the various strategies used to compare the photomagnetic properties of iron(II) SCO materials, including the ‘inverse energy gap’ law introduced by Hauser which represents the first guideline allowing some prediction of the lifetime of a photoinduced high spin state. This will lead us to introduce the *T*(LIESST) procedure used to compare

Spin-Crossover Materials: Properties and Applications, First Edition. Edited by Malcolm A. Halcrow.

the photomagnetic properties of SCO materials. Section 19.3 reports the linear relationship found between *T*(LIESST) and the thermal spin-transition ($T_{1/2}$) temperature, and the work done over the past ten years to identify the parameters affecting the T_0 factor. Some comments will be also made about the photomagnetic properties of iron(III) SCO analogs. Section 19.4 will consider *T*(LIESST) measurements of polynuclear iron(II) SCO materials, and in particular the case of binuclear SCO complexes with antiferromagnetic coupling. Finally, Section 19.5 reports the simulation of the *T*(LIESST) curves for a mono- and a binuclear SCO material. Some predictions are also discussed about the relaxation decay expected for highly polynuclear compounds.

19.2 Light-Induced Excited Spin-State Trapping (LIESST) Effect

19.2.1 LIESST Effect

The first evidence of the light sensitive electronic changes in an iron(II) SCO material was reported by McGarvey *et al.*[8] in solution and then by Decurtins *et al.*[5] in solid state. These authors demonstrated the possibility of converting a low spin (LS) state into a metastable high spin (HS) state at low temperatures ($\leq$50 K) by using a green light irradiation. This phenomenon, called Light-Induced Excited Spin-State Trapping (abbreviated as LIESST), was extended later by Hauser, who showed that red light switches the system back to the LS state (reverse-LIESST).[9]

The mechanism of LIESST is explained in Figure 19.1. Irradiating the sample at low temperature, in LS state, with the 514.5 nm line of an argon laser, induces the spin-allowed $^1A_1 \cdot {}^1T_n$ transition. The excited singlet state, characterized by a short lifetime, quickly decays back to the $^1A_1$ ground state. An alternative decay path is also possible due to spin orbit coupling, which enables an intersystem crossing step with $\Delta S = 1$ to the triplet states $^3T_1$ or $^3T_2$. These in turn decay via a second intersystem crossing either to the $^1A_1$ ground state or to the metastable $^5T_2$ state. The system remains trapped in the HS state with a very long lifetime if the temperature is low enough, so that the energy barrier between the HS and LS potential surfaces, which are well separated by the large difference in metal–ligand bond length, cannot be overcome thermally.

Today, the LIESST effect has been observed in a large number of iron(II) systems.[10–13] From all these examples, we have learned that due to the large increase in metal–ligand bond lengths during the light-induced population of the HS state at low temperatures, the role of the elastic interactions are particularly important for the interpretation of the photomagnetic data. The resulting elastic interactions may be pictured as an internal pressure which is proportional to the concentration of the LS species. In noncooperative systems where the intermolecular interaction in the solid is weak, the HS→LS relaxation is single exponential (or stretched exponential) whereas in strongly cooperative systems the curves follow a sigmoidal behaviour due to a self-accelerating process.[10–13]

At present, most of the photomagnetic studies on iron(II) complexes involve the LIESST and reverse-LIESST processes with the ultimate goal of obtaining suitable materials for technical applications such as displays, information storage and sensors.[6] In regard to these prospects, the main deficiency of the LIESST effect is its temperature domain. The lifetime of the photoinduced HS state is usually long at very low temperatures, for example weeks at 20 K for $[Fe(ptz)_6](BF_4)_2$ (ptz = 1-propyltetrazole),[14] but above 50 K the relaxation process becomes thermally activated and in a few seconds the stored light induced information vanishes. To get past this limitation, it has been recently indicated that in some iron(II) molecular materials the light-induced phenomenon can be produced in the centre of a thermal hysteresis.[15–17] But the integration into a genuine device for data storage will impose defining an iron(II) spin-crossover material with a thermal hysteresis of around 100 K (i.e. with typical working temperatures from 253 K up to 353 K), a value not yet

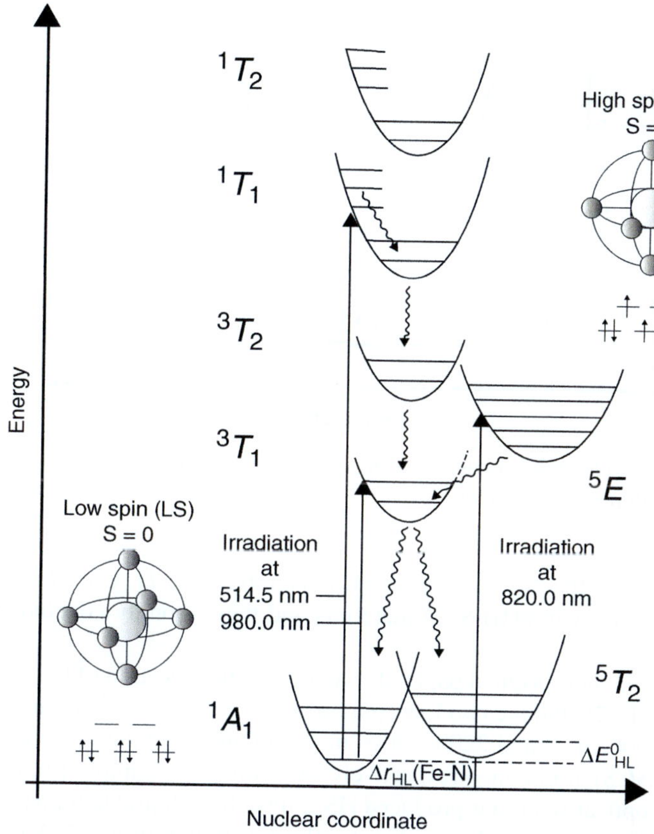

Figure 19.1 *Potential wells for the $^1A_1$ ground state and the thermally accessible $^5T_2$ state, as well as the higher excited states of a d^6 spin-crossover system. Arrows indicate the mechanism of LIESST and reverse LIESST. Adapted from [12], with permission from Elsevier.*

reached. An alternative strategy would be to identify the factors accountable for stabilizing the light-induced metastable state at low temperatures (i.e. outside the thermal hysteresis). Along with that it is interesting to note that some iron(II) SCO complexes have been described in the literature with unusual long lifetimes up to 100 K.[18–22] This indicates that some factors are able to stabilize the photoinduced HS state.

19.2.2 Variable Temperature Fourier Transform Infrared Spectroscopy (VTFTIR)

The first discussion of the stability of the photoinduced HS state was had by Herber *et al.* using Variable Temperature Fourier Transform Infrared Spectroscopy (VTFTIR).[23, 24] In agreement with the studies of Baker and Long,[25] Teodorescu[26, 27] and Maddock *et al.*,[28] Herber *et al.* used the fact that the infrared signatures of the HS and LS forms of $[Fe(phen)_2(NCS)_2]$ (phen = 1,10-phenanthroline) are sufficiently different to identify the state under light irradiation.[23, 24] For example, the room temperature HS form, and the LS form stable at liquid nitrogen temperature, can be readily distinguished by their characteristic infrared signatures, both in the ν_{CN} region (~2100 cm^{-1}) and in the spectral range 900–600 cm^{-1}. Based on this, these authors defined a limiting temperature T_2 above which the light-induced phenomenon is no longer observable, and

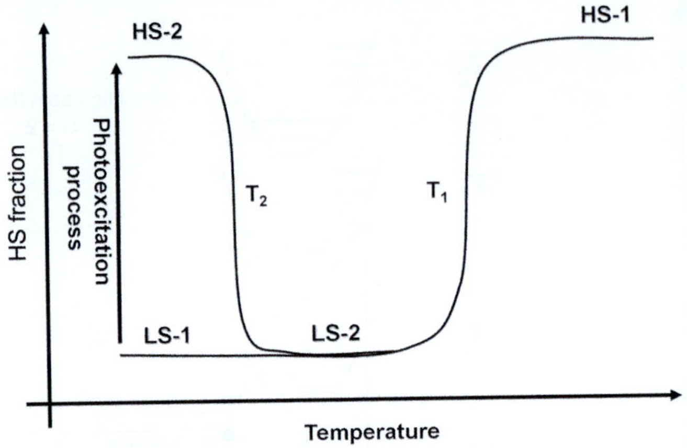

Figure 19.2 Schematic diagram for the successive conversion of an iron(II) complex from the high spin (HS) to the low spin (LS) forms. Adapted with permission from [23]. Copyright 1986, American Chemical Society.

attempted some correlations with the thermal spin-transition temperature T_1. Unfortunately, this comparison was only made on the basis of six iron(II) SCO materials and they wrongly concluded that T_2 and T_1 are lowered concomittantly.[24]

Interestingly, Herber *et al.* in their original paper also addressed the question of the nature of two high spin forms, HS-1 and HS-2 (Fig. 19.2), and the two low spin forms LS-1 and LS-2.[23] From the comparison of the IR spectra in both the 900–600 cm^{-1} and 2200–1900 cm^{-1} regions, these authors concluded that LS-1 and LS-2 forms of $[Fe(phen)_2(NCS)_2]$, obtained respectively by cooling the sample from room temperature to ~80 K and by warming the optical-pumping-produced HS-2 form from liquid helium temperature to ~ 80 K, were completely identical. This agreed with previous conclusions drawn from ^{57}Fe Mössbauer spectroscopy by Gütlich *et al.*[29] Concerning the comparison of HS-1 and HS-2 (Fig. 19.2), a small but significant difference in the infrared absorption maxima at 300 K and at 8 K was noticed, suggesting that the high spin form of $[Fe(phen)_2(NCS)_2]$ at room temperature (HS-1), and the high spin form of this complex generated by the LIESST process near liquid helium temperature (HS-2), are not completely identical. This conclusion was also supported by the ^{57}Fe Mössbauer data of Gütlich *et al.* who note a significant difference in the isomer shift and quadrupole splitting between HS-1 and HS-2.[29] Later, Guionneau *et al.* reported the X-ray structure of $[Fe(phen)_2(NCS)_2]$,[30] and definitively indicated that while the distortion parameter Σ[31] was identical for HS-1 and HS-2, the Fe–N bond lengths are slightly different.

19.2.3 The Low-Energy Gap

In 1991, Hauser introduced the first real guideline allowing some prediction of the lifetime of a photoinduced HS state.[32–35] By carefully investigating the dynamics of the LIESST effect in different diluted SCO materials, this author noticed a strong correlation between the lifetime of the LIESST state extrapolated at $T \to 0$ (i.e. in the tunneling region[32–35]), expressed as $\ln k_{HL}(T \to 0)$, and the thermal spin-crossover temperature $T_{1/2}$ (Fig. 19.3). In other words, Hauser demonstrates that the lifetime of the metastable LIESST state is inversely proportional to $T_{1/2}$ (assumed to vary as ΔE^0_{HL}, see Figure 19.1). This relation is today known as the inverse-energy-gap law.

This approach is based on the nonadiabatic multiphonon relaxation theory,[36] where the low-temperature tunneling rate depends on both the Huang–Rhys factor S, (Eqs (19.1) and (19.2)), which is a measure of the horizontal displacement of the potential wells of the HS and LS states, and the energy gap between the lowest

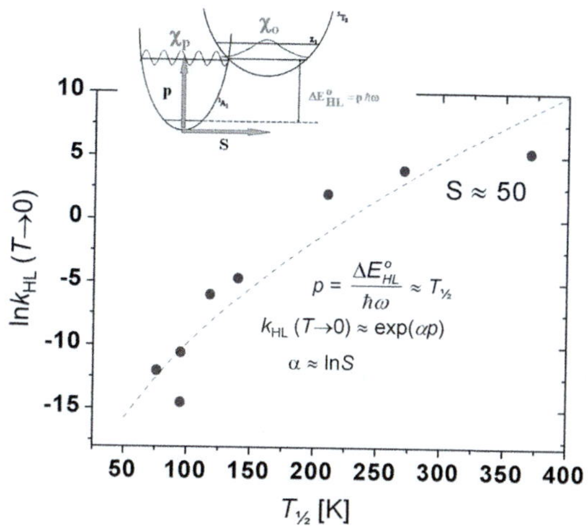

Figure 19.3 Schematic diagram for the successive conversion of an iron(II) complex from the high spin (HS) to the low spin (LS) forms. Adapted from [12], with permission from Elsevier., and with permission from [33], Copyright 2009, American Institute of Physics.

vibronic levels of the HS and LS states. More precisely β_{HL} is the electronic tunneling matrix element, $\hbar\omega$ is the breathing vibration energy of the FeN_6 core, $p = \Delta E^0{}_{HL} / \hbar\omega$ is the reduced energy gap and S is the Huang–Rhys factor:

$$k_o^{LT} = \frac{2\pi}{\hbar^2\omega}(\beta_{HL})^2\Omega_{LS}\left(\frac{S^p e^{-S}}{p!}\right) \tag{19.1}$$

$$S = \frac{1}{2}f\frac{\Delta Q^2}{\hbar\omega} \text{ with } \Delta Q = \sqrt{6}\Delta r_{HL} \tag{19.2}$$

One difficulty of the inverse-energy-gap law lies in properly determining the $T\rightarrow 0$ lifetime, particularly when the stabilization of the photoinduced HS state in the tunnelling region is very effective. For this reason, in 1999 in partnership with the industrial Motorola Research Center, we proposed comparing various SCO materials by systematically measuring the T(LIESST) value,[7, 37] that is the limit temperature above which the light- induced magnetic HS information is erased when the temperature is increased from 10 K at a rate of 0.3 K min^{-1}. The idea was to define a procedure allowing a rapid comparison of the light-induced property of a given material.

19.3 The *T*(LIESST) Approach: The Case of Mononuclear Compounds

19.3.1 Principle of the *T*(LIESST) Measurement

Figure 19.4 shows a typical measurement of the *T*(LIESST) temperature for a mononuclear SCO complex, $[Fe(PM\text{-}BiA)_2(NCS)_2]$ (PM-BiA = *N*-2′-pyridylmethylene-4-aminobiphenyl).[7, 38] The magnetic response is expressed as the $\chi_M T$ product (χ_M stands for the molecular magnetic susceptibility and T the temperature). Light irradiation is applied to the material at 10 K. At this temperature, the system is usually in a LS form, and population of the paramagnetic HS state through the LIESST phenomenon results in a dramatic increase of the magnetic signal. The irradiation is only stopped when saturation of the signal is reached, that is when

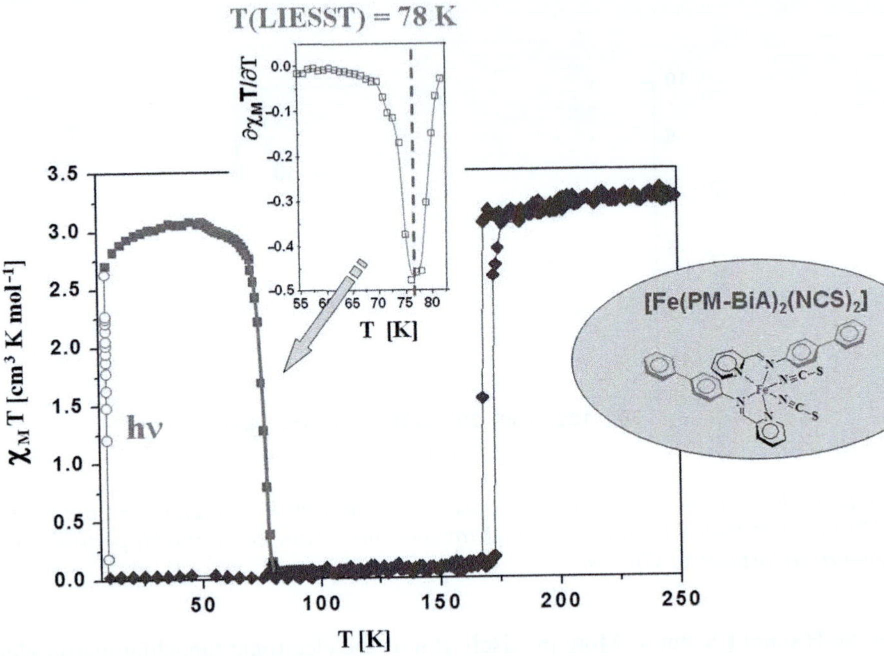

Figure 19.4 T(LIESST) experiment recorded for the mononuclear $[Fe(PM\text{-}BiA)_2(NCS)_2]$ complex with (•) the temperature dependence of $\chi_M T$ recorded in the cooling and warming modes without irradiation, (○) the change recorded during 1 h of irradiation at 10 K, and (■) the behavior recorded during the warming mode (0.3 K min^{-1}) when the light irradiation was switched off. The insert graph shows the derivative $\partial(\chi_M T)/\partial T$ plot as function of the temperature. The solid lines represent the simulation obtained according to Eq. (19.10) discussed in Section 19.5.

for a given laser power of 5 mW cm^2 an equilibrium is operating between the population and relaxation of the excited state. The temperature is then slowly increased at a rate of 0.3 K min^{-1} and the magnetic behavior recorded. Typically, the magnetic response of the light-induced HS state remains almost constant up to 40–50 K. Only a slight increase of $\chi_M T$ is observed between ca. 10–20 K which can be attributed to zero-field splitting of the HS state in a nonideal octahedral environment. Up to 40–50 K, the relaxation process is governed by the nonadiabatic tunneling regime through the barrier, and the experimental timescale is negligible with regard to the lifetime of the photoinduced HS state.[32–36] This behavior contrasts with the region above 50 K, where the system reaches the thermally activated relaxation regime, which is more strictly speaking tunneling from thermally populated vibrational levels of the HS state. In this temperature range, $\chi_M T$ drastically decreases and rapidly recovers its initial value. The minimum of the $\partial \chi_M T/\partial T$ *vs.* T curve determines the *T*(LIESST) temperature.[7, 37]

Techniques like Mössbauer, infrared, optical reflectivity and Raman spectroscopies could be used to detect the occurrence of a LIESST effect, but we made the choice to record the photomagnetic property of SCO complex with a SQUID magnetometer coupled to an optical source through an optical fiber.[7, 37] Hereafter, we list some of the advantages and difficulties in following the photomagnetic response of a material.

First, let us recall that photomagnetism defines a change of magnetic response between initial and final states induced by an optical perturbation. For iron(II) SCO materials, the LS state is diamagnetic ($S = 0$, $t_{2g}{}^6 e_g{}^0$) and the HS state paramagnetic ($S = 2$, $t_{2g}{}^4 e_g{}^2$). The LIESST effect in a mononuclear SCO compound results in an increase of the magnetic signal from 0 to 3.5 cm^3 K mol^{-1}. An advantage of the SQUID technique is its sensitivity: a signal can be detected down to 10^{-6}–10^{-7} emu. The quantity of the sample is then typically around 0.1 mg. The use of such a small quantity is important, particularly when the sample is strongly colored to avoid light penetration problems in a bulk material. Another advantage of the photomagnetic technique is that the sample, in powder or crystal forms, is easily deposited on a commercial transparent adhesive tape placed close to the edge of an optical fiber installed in the rod sample holder. The temperature is then easily controlled and the experiment is rapid. In the *T*(LIESST) procedure, the time associated with the three extractions used to detect the magnetic signal is 60 s and the time to reach the next temperature by steps of 1 K is 120 s.

But of course the investigation of the photomagnetism requires some effort. One of them is to properly eliminate oxygen gas by purging the SQUID cavity; otherwise a magnetic peak is recorded in the 50 K region. Another difficulty, which can be solved when the experimenter acquires some expertise, is to properly center the signal in the SQUID cavity. Both LS state and commercial transparent adhesive tape are diamagnetic, and then the resulting magnetic signal (expressed in emu) is negative, while the HS state is paramagnetic with a positive emu response. The magnetic signals in the initial and final states are then of opposite sign, and during the photoexcitation and relaxation processes the signal irremediably passes through the zero emu position. Any centering procedure is thus prohibited. A third difficulty concerns light excitation into a bulk material. As indicated previously a thin layer of sample is prepared, but in some cases the opacity of the sample is strong enough to prevent light penetration. Photoexcitation is thus incomplete, and the sample can be considered biphasic with a top-phase at the surface saturated in its photoexcitation, and a bottom phase practically unexcited. Such inhomogeneity of photoexcitation has to be taken into account when the studied material is cooperative, as the relaxation activation energy is a function of the photoexcited fraction. Optimization of light excitation requires determination of the absorption spectra of the two states involved. For a deeply colored compound, it has been shown that the photoexcitation is better in the shoulder of an absorption band than at the maximum.[38] In our case, photomagnetic measurements were performed using a mixed gas Krypton/Argon laser and/or a series of laser diodes.[7, 37] The use of such laser sources allows one to precisely select the wavelength, in contrast to a halogen (or xenon) lamp fitted with an optical filter, where a range of wavelengths is produced meaning that some opposite electronic transitions can conversely be excited. Regarding the power of the light excitation, the intensity was usually adjusted to 5 mW cm^{-2}, and we verified that when the light was switched off the magnetic response did not jump. Such an artificial effect is, in fact, a way to evidence that a sample experiences a heating effect. The final difficulty of a photomagnetic measurement is properly estimating the weight of the sample. Over time we have learned that the best way is to calculate the weight by comparing the thermal spin-crossover curve obtained with a more accurately weighed sample of the same material.

Of course, the *T*(LIESST) procedure can be criticized as strictly speaking it does not correspond to an intrinsic critical parameter. The determination of a limiting *T*(LIESST) temperature is kinetically dependent on the measurement. Figure 19.5 reports the influence of the warming rate on the *T*(LIESST) value. As expected, the faster the warming rate, the higher the *T*(LIESST) temperature. This explains why we have imposed a 'strict' protocol, that is irradiation at 10 K and a warming rate of 0.3 K min^{-1}, for comparison of the photomagnetic properties of SCO materials. Under such conditions, the *T*(LIESST) value can be regarded as a relevant parameter. The great advantage of such an approach is that in a few hours the potentiality of a material to retain the optical information is evaluated. Furthermore, the *T*(LIESST) temperature is a macroscopic value which intrinsically combines both the temperature independent region and the thermally activated region, as well as the existence of cooperativity or heterogeneity of photoexcitation. In other words, an analysis of the *T*(LIESST) temperatures integrates all the factors affecting the lifetime of the photoinduced HS state.

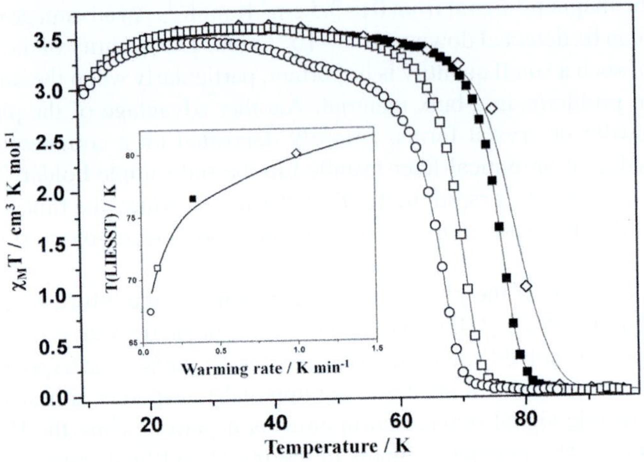

Figure 19.5 *T(LIESST) experiments recorded for the mononuclear [Fe(PM-BiA)$_2$(NCS)$_2$] complex at the warming rate (◊) 1 K min^{-1}, (■) 0.33 K min^{-1}, (□) 0.1 K min^{-1} and (○) 0.05 K min^{-1}. The inset graph reports the change of the T(LIESST) temperature as function of the warming rate. The solid line represents the simulation obtained according to Eq. 19.10 by changing the warming rate.*

19.3.2 The *T*(LIESST) Database

The first work based on a comparison of *T*(LIESST) temperatures was made in 1999.[39] In this study, the magnetic and the photomagnetic properties of 22 SCO compounds were determined (Fig. 19.6, samples **1–22**), including the thermal spin-transition $T_{1/2}$ as well as the *T*(LIESST) temperatures. When all these data were plotted in a graph defined by *T*(LIESST) vs. $T_{1/2}$, we first observed a large distribution of points. By considering the chemical formulae of the different samples, however, we realized that the physical properties of the complexes followed some general tendencies. Most of the investigated SCO complexes in that case are members of the [FeL$_2$(NCX)$_2$] family, with L an aromatic unit and NCX a thiocyanato (X = S) or a selenocyanato (X = Se) anionic ligand. The originality of our idea was to define a couple when two complexes possessed the same aromatic L unit ligand but involved either a S or Se unit. By doing this, several couples (S and Se) were found to follow the same tendency (samples **1–2**, **7–8**, **9–10**, **11–12**, **13–14**) and two virtual lines were proposed with a general equation, $T(\text{LIESST}) = T_0 - 0.3T_{1/2}$. The extrapolated T_0 values at $T \rightarrow 0$ were respectively 100 K and 120 K.[39]

Three years later, this analysis was confirmed by additional work in which magnetic and photomagnetic studies of eight new SCO materials were investigated (Fig. 19.6, samples **23–30**).[40] The idea was to compare a series of iron(II) metal complexes possessing various $T_{1/2}$ and *T*(LIESST) values without changing the FeN$_6$ coordination sphere. The selected family was [Fe(bpp)$_2$]X$_2$·*n*H$_2$O (bpp = 2,6-bis(pyrazol-3-yl)pyridine), which is well-known for exhibiting various thermal spin-crossover behaviors depending on the nature of the anion and the degree of hydration, and for involving an iron(II) metal center surrounded by six donor nitrogen atoms with two ligand molecules coordinated in an octahedral meridional plane.[41–44] Moreover, in some cases unusual long lifetimes have been found for the metastable HS state generated both by thermal trapping and by the LIESST effect.[18, 19, 45] Consequently, we systematically determined the thermal spin-transition and the limit temperature of the LIESST phenomenon of all the series of hydrated and dehydrated [Fe(bpp)$_2$]X$_2$·*n*H$_2$O complexes, with X = PF$_6^-$, NCS$^-$, NCSe$^-$, BF$_4^-$, Br$^-$ and I$^-$.[40] Interestingly, the [Fe(bpp)$_2$]X$_2$·*n*H$_2$O data on the *T*(LIESST)/$T_{1/2}$ plot were distributed in such a way that the previously established $T_0 - 0.3T_{1/2}$ relation

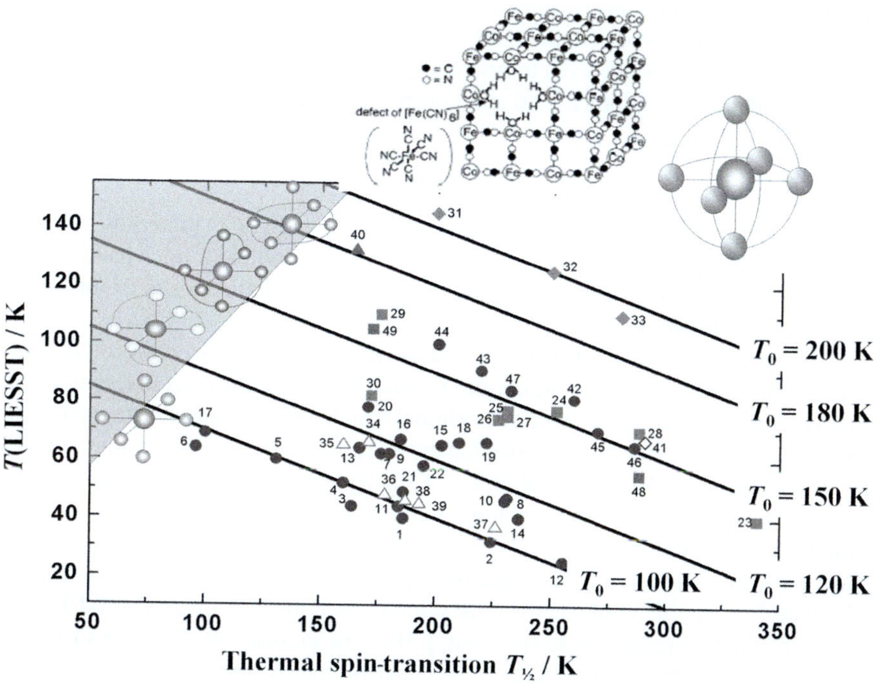

Figure 19.6 Variation of T(LIESST) vs. $T_{1/2}$ for spin-crossover compounds.[48] Data for the 120 and 100 T_0 lines are reported in ref. 39. The 150 T_0 line is the result of the linear regression obtained with the $[Fe(bpp)_2]X_2{\cdot}nH_2O$ family[40] and the 200 T_0 line is based on the work of Hashimoto.[46] The region in gray represents the situation where the T(LIESST) temperature is inferior or at least equal to $T_{1/2}$ temperature.

remained valid, and a third line with a T_0 value of 150 K was proposed (Fig. 19.6, samples **23–30**).[40] What is remarkable is that for the first time the 150 T_0 line defines some SCO complexes with a $T_{1/2}$ value around room temperature and a measurable *T*(LIESST) value (sample **28**). Such a result was very exciting, as it represented the first step towards the design of switchable SCO materials useable in a genuine device.[6]

In parallel, Hashimoto provided new evidence for the validly of the $T(\text{LIESST}) = T_0 - 0.3T_{1/2}$ relationship.[46] The selected family was the Co–Fe Prussian Blue analogs, $[Na_xCo_yFe(CN)_6]{\cdot}zH_2O$ with atomic Co/Fe composition ratios of 1.37, 1.32 and 1.26. In such complexes, a change of spin states may be induced (i) by a temperature gradient, from room temperature $Fe^{III}(t_{2g}{}^5e_g{}^0$, LS, $S = {}^1/_2)$–CN–$Co^{II}(t_{2g}{}^5e_g{}^2$, HS, $S = {}^3/_2)$ to low temperature $Fe^{II}(t_{2g}{}^6e_g{}^0$, LS, $S = 0)$–CN–$Co^{III}(t_{2g}{}^6e_g{}^0$, LS, $S = 0)$; and (ii) by a light irradiation at 5 K with the population of the metastable $Fe^{III}(t_{2g}{}^5e_g{}^0$, LS, $S = {}^1/_2)$–CN–$Co^{II}(t_{2g}{}^5e_g{}^2$, HS, $S = {}^3/_2)$ state.[47] The authors consequently determined the $T_{1/2}$ and *T*(LIESST) temperatures for each Co/Fe composition and observed the presence of a new T_0 line (Fig. 19.6, samples **31–33**) parallel to the third previous one.[46] The T_0 value was estimated at 200 K. Of course, on the one hand it is clear that the physical process involved in these Prussian Blues analogs is not a pure spin-crossover phenomenon, but a Charge-Transfer-Induced Spin-transition (CTIST) effect,[47] and any overall comparison has to be made cautiously. But on the other hand, this indicates that the *T*(LIESST) relation ($= T_0 - 0.3T_{1/2}$) is probably more general. With that in mind, let us mention that the acronym LIESST refers to an excited spin state trapped by light. This has been intensively used for describing the light-induced phenomenon of SCO material, but basically the definition is not limiting.

In our sense, the T(LIESST) temperature is linked to the thermally activated region and represents a general approach to follow the relaxation process occurring between two potential wells separated by an activation energy. This can explain why the T(LIESST) temperatures, associated to the CTIST effect of Prussian Blue complexes, can be also added to the Figure 19.6 and compared to the purely SCO materials.[48, 49]

19.3.3 Parameters Affecting the T_0 Factor

The T(LIESST) database today compiles the properties of more than 80 SCO materials and a general relation has been evidenced $T(\text{LIESST}) = T_0 - 0.3T_{1/2}$.[48, 49] Until now five parallel T_0 lines have been obtained with, respectively, values of 100, 120, 150, 180 and 200 K and one of the challenges is to further increase the T_0 value for reaching room temperature.

A first general tendency, noticeable for all the T_0 lines (Fig. 19.6), is a decrease of T(LIESST) with the increase of $T_{1/2}$, reflecting the fact that the higher the thermal spin-transition occurs, the less photomagnetic information remains. This behavior agrees with the original conclusion of Hauser with the inverse-energy-gap law;[32–35] the HS→LS relaxation is a direct dependence of ΔE^0_{HL} energy gap (i.e. the enthalpy factor, ΔH, at $T \rightarrow 0$), which is typically smaller than ΔH at $T_{1/2}$. Consequently, if we assume in a first approach that the entropy factor, ΔS, is almost constant for all SCO compounds, the dynamics of the relaxation and thus T(LIESST) become a direct function of $T_{1/2}$.

Concerning now the way to affect T_0, it is sometimes proposed that the cooperativity factor stabilizes the photoinduced HS state.[50, 51] In this context, we have effectively noticed for the [Fe(PM-L)$_2$(NCX)$_2$] family that for an almost constant $T_{1/2}$ temperature of about 170 K, an increase of T(LIESST) from 31 K, for the gradual SCO [Fe(PM-AzA)$_2$(NCS)$_2$] (PM-AzA = *N*-2′-pyridylmethylene-4-(phenylazo)aniline; compound **11**), to 78 K for the abrupt SCO [Fe(PM-BiA)$_2$(NCS)$_2$] compound (**20**), and we have proposed a general relation between the T(LIESST) temperature and the cooperativity factor C.[50] But it should be admitted that it is not so easy to properly disconnect the cooperativity factor from the effects of local deformation (such as a twist of the FeN_6 core or an elongation of the Fe–N bond length) and the change in electronic distribution through modification of the aromatic substituent.[31] The X-ray structures of thermodynamic stable states (LS and HS) and metastable state (photoinduced HS) really have to be systematically solved. Unfortunately, until now only a few articles have reported a careful comparison of all these structures and it is not yet possible to obtain an overall view of the effect of the cooperativity.

If we consider the influence of the cooperativity on the HS→LS relaxations, it is well-known from the work of Hauser *et al.*[10–13, 32–35] that the activation energy barrier is a function of the HS (or LS) fraction because of an internal pressure created by the large change in the metal–ligand bond lengths between the HS and the LS states. As a consequence, the activation energy barrier for a given HS fraction is reduced when the cooperativity is increased and thus, as recently demonstrated through theoretical simulations, T(LIESST) is slightly reduced.[38, 49] In fact, the influence of the cooperativity on T(LIESST) appears to be very weak. This conclusion is also supported by experimental work performed on the [Fe(bpp)$_2$]$X_2 \cdot nH_2O$ family (Fig. 19.6, samples **23–30**).[40] The photomagnetic property of these series was found to follow the same $T_0 = 150$ K line whatever the cooperativity, from an incomplete spin conversion to a gradual or an abrupt spin-transition without or with thermal hysteresis. Moreover, we have also prepared some [Fe(bpp)$_2$]-Nafion films characterized by a gradual thermal spin-transition.[52] This absence of cooperativity was reflected by investigation of the kinetics of the photoinduced HS state which followed a stretched exponential model as usually recorded for a noncooperative system.[11] But what is interesting in regard to the question of the cooperativity is that the T(LIESST)/$T_{1/2}$ couple of the [Fe(bpp)$_2$]-Nafion film[52] is once more in the vicinity of the $T_0 = 150$ K line as for the other [Fe(bpp)$_2$]$X_2 \cdot nH_2O$ analogs prepared in powder form and displaying cooperative SCO behavior.[40] All this demonstrates that the cooperativity factor is not the key factor for stabilizing the photo-induced HS state.

By looking at the data from the $[Fe(bpp)_2]X_2{\cdot}nH_2O$ family (Fig. 19.6, samples **23–30**) we also notice that whatever the nature of the anion X, the T(LIESST)/$T_{1/2}$ properties follow the same T_0 line.[40] In other words, the final T_0 value appears to be insensitive to any change outside the inner coordination sphere. Basically, in a coordination sphere, the first important factor which has to be taken into account is the electronic contribution of the ligand. In this specific context, the work performed on the $[Fe(bpp)_2]X_2{\cdot}nH_2O$ family gives some interesting information. Goodwin *et al.* showed that the thermal SCO properties of the $[Fe(bpp)_2]X_2{\cdot}nH_2O$ series are strongly affected by the degree of hydration, because the hydrated salts participate in an extended hydrogen-bonded network involving the uncoordinated NH groups of the pyrazole moieties, the anion and the water molecules.[45] Consequently, any change in hydration affects the electronic density on the ligand and modifies the thermal spin-crossover behavior. But interestingly the photomagnetic properties of the hydrated and dehydrated compounds follow the same T_0 line, suggesting that the influence of electronic factors on the T_0 factor is negligible. This idea is also supported by the results found on the $[Fe(PM\text{-}L)_2(NCX)_2]$ family (Fig. 19.6, samples **1–22**), where for a given aromatic ligand (L) the T(LIESST)/$T_{1/2}$ values of thiocyanato (X = S) and selenocyanato (Se) complexes follow the same T_0 line, independently of the electronic modification accompanying the S/Se substitution.[39]

In parallel to this electronic contribution of the ligand, which appears to be negligible, it may be anticipated that if we consider the single configurational coordinate (SCC) model,[36] a horizontal displacement of the potential wells between the LS and HS state is able to affect the final T_0 value. Indirect information has been obtained by investigating some iron(II) spin-crossover compounds containing phosphorus atoms in the coordination sphere. In the specific case of $[Fe(dppen)_2Cl_2].2(CH_3)_2CO$ (dppen = *cis*-1,2-bis(diphenylphosphino)ethene), the single-crystal structures determined in HS and LS states showed a large Δr_{HL}(Fe–P) variation (0.3 Å)[53] and the relaxation kinetics of the HS→LS conversion occurring after the LIESST effect were found to be slower than for classical FeN_6 complexes.[20] But, by looking at other iron(II) phosphorus complexes, we realized that the T(LIESST)/$T_{1/2}$ data (Fig. 19.6, samples **34–39**) remained close to the $T_0 = 100$ K and 120 K lines and the latitude to further increase T_0 by tuning the iron(II) bond length is not so large.[54, 55]

We then focused our attention on the distortion of the FeN_6 octahedron. By using the few X-ray data available on mononuclear iron(II) complexes of general formula $[FeL_2(NCS)_2]$, we obtained an interesting result; T(LIESST) shows a linear correlation with the distortion of the metal coordination sphere (Fig. 19.7).[56] The stronger the distortion of the FeN_6 octahedron the higher the value of T(LIESST), and the lower the $T_{1/2}$ value. Based on Figure 19.7b, the LS→HS photoconversion is only possible at temperatures lower than both $T_{1/2}$ and T(LIESST), in region 3 of Figure 19.7. In regions 1 and 2, the thermal HS and LS states are respectively stable. The crossing point between the two lines corresponds to the equality between T(LIESST) and $T_{1/2}$, and defines the highest T(LIESST) value obtainable for this series of complex (125 K).[56] Incidentally, this helps to predict that the highest T(LIESST) value for these bidentate iron(II) complexes will be obtained with a $\Delta\Theta$ value of around 145°.

Based on this we have decided to compare different T(LIESST)/$T_{1/2}$ values via the chemical nature of the ligand involved into the inner coordination sphere.[48, 49] For this we arbitrarily separate the iron(II) complexes involving six independent monodentate ligands (FeL_6 coordination sphere), with those possessing three bidentate ligands (FeL_3), and those involving only two tridentate ligands (FeL_2). Interestingly, in each three cases (FeL_6, FeL_3 and FeL_2) the highest T(LIESST) values are found on a different T_0 line. More precisely, the highest T(LIESST) temperature of FeL_6 complexes corresponds to the well-known $[Fe(ptz)_6](BF_4)_2$, situated close to the $T_0 = 100$ K line (sample **5**, Figure 19.6); for FeL_3 systems it is $[Fe(PM\text{-}BiA)_2(NCS)_2]$ (78 K, sample **20**, Figure 19.6), situated on the $T_0 = 120$ K line; and for FeL_2 compounds, it is $[Fe(bpp)_2](BF_4)_2$ obeying at the $T_0 = 150$ K line (sample **29**, Figure 19.6). All these results confirm the assumption that the T(LIESST) and the T_0 values are mainly dependent on the inner coordination sphere and not on cooperativity or other surrounding parameters.[48, 49]

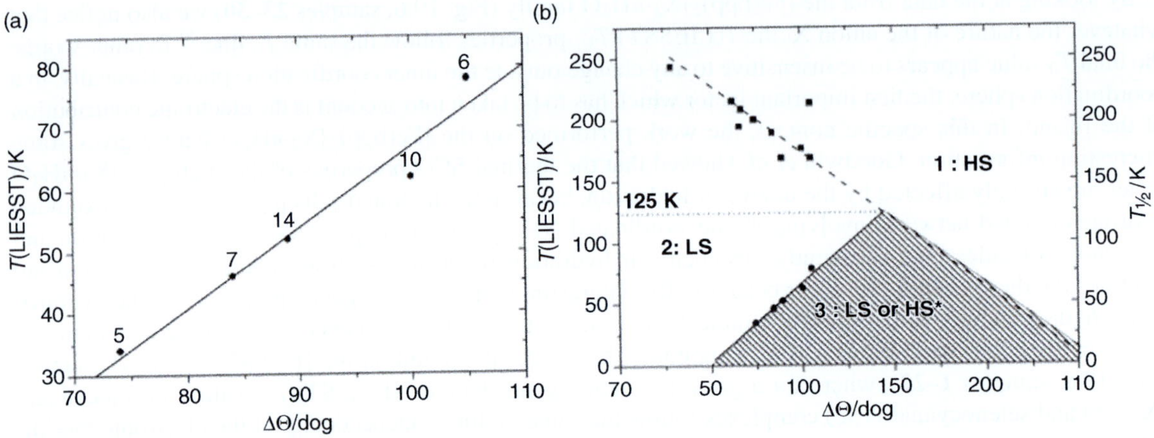

Figure 19.7 *For the [$FeL_n(NCS)_2$] complexes (a) plot of the variation at the spin-crossover of the metal environment distortion, ΔΘ, vs. the T(LIESST) values, and (b) variation of T(LIESST) (circle) and $T_{1/2}$ (square) as a function of the structural parameter ΔΘ. Three zones are defined as a function of the possible spin state. Zone 3 corresponds to the region where the photoinduced LS to HS conversion is possible. The maximum T(LIESST) value for this series is therefore shown. HS* represents the photoinduced HS state (adapted from ref. 56).*

As a continuation to this approach, it is interesting to note that the [$Fe(L_{222}(N_3O_2))(CN)_2$]·$H_2O$, which is a macrocyclic iron(II) spin-crossover compound that can be regarded as a model of a FeL_4 system, possesses a *T*(LIESST) temperature of 130 K.[22, 57, 58] That is exactly where the *T*(LIESST)/$T_{1/2}$ point behaves on a virtual T_0 = 180 K line (Fig. 19.6, sample **40**).[49] Moreover, we also recently reported the photomagnetic properties of [$Fe(L_{222}N_5)(CN)_2$]·H_2O ($L_{222}N_5$ = (2,13-dimethyl-3,6,9,12,18-pentaazabicyclo[12.3.1]octadeca-1(18),2,12,14,16-pentaene).[59] This complex was found at low spin (LS) up to 400 K and we demonstrated at 10 K the existence of a LIESST effect with a long lived photoinduced HS state. The *T*(LIESST) limit was estimated at 105 K.[59] All this demonstrates that by playing with the inner coordination sphere, the chemist is able to stabilize the lifetime of the photoinduced HS state.

Conversely, we have observed that the introduction of flexibility into the inner coordination sphere drastically reduces the T_0 value.[60, 61] For instance, if we consider the peculiar case of SCO materials with rigid tridentate ligands, the *T*(LIESST)/$T_{1/2}$ points lie on the virtual T_0 = 150 K line (Fig. 19.6). But the introduction of freedom in the tridentate ligands, through the incorporation of an unsaturated carbon atom between the aromatic rings, such as in the bpypz unit, drastically reduces *T*(LIESST) towards the T_0 = 100 K line.[60, 61] This emphasizes the concept that the vibrational aspect, the stiffness and/or the distortion of the inner coordination sphere are the key factors for stabilizing the metastability of the light-induced HS state.[48, 49]

19.3.4 The *T*(LIESST) Approach to Fe(III) Metal Complexes

In comparison to the large number *T*(LIESST) measurements performed on Fe(II) metal complexes, very few studies have been dedicated to Fe(III) mononuclear materials. In fact, while the first LIESST experiment on an iron(II) metal complex was reported in 1984,[5] the first example on an Fe(III) system was only described in 2000 by Hayami *et al.*[51] Before that, iron(III) molecular materials were considered to display LIESST properties with fast decay;[62, 63] typically with a low temperature tunneling rate $k_{HL}(T \rightarrow 0)$ higher than 10^2 s^{-1} and a lifetime lower than 10^{-2} s, which would prevent any photomagnetic measurement in a SQUID cavity. The reason for such fast decay is linked to the smaller change in the metal–ligand distance between the LS and the HS states for Fe(III).[32–35] Indeed, in terms of a nonadiabatic multiphonon process, the main difference between Fe(III) and Fe(II) complexes is the metal–ligand bond length Δr_{HL}, which is about 0.12 Å for Fe(III) complexes instead of 0.16–0.21 Å for Fe(II) complexes.

The first atypical long-lived photoinduced excited spin state of an Fe(III) complex was found for $[Fe(pap)_2]ClO_4{\cdot}H_2O$.[51] This system was characterized by an abrupt thermal spin-transition ($T_{1/2}\uparrow = 180$ K; $T_{1/2}\downarrow = 165$ K) with hysteresis loop of 15 K, and the T(LIESST) temperature was estimated at 105 K. The existence of a long lived photoinduced HS state was attributed by the authors to the existence of strong intermolecular π–π interactions.[51] Later, in 2002, a new salt of the same $[Fe(pap)_2]^+$ complex, with the PF_6^- counteranion was described with an abrupt SCO transition at $T_{1/2} = 288$ K without thermal hysteresis and T(LIESST) = 55 K.[64] A few years later, Hayami *et al.* described the properties of $[Fe(pap)_2]BF_4{\cdot}H_2O$, $[Fe(CH_3\text{-}pap)_2]ClO_4{\cdot}H_2O$ (CH_3-pap = methylated ligand) and two metal diluted complexes $[Fe_{0.5}Al_{0.5}(pap)_2]ClO_4{\cdot}CH_3OH$ and $[Fe_{0.25}Al_{0.75}(pap)_2]ClO_4{\cdot}CH_3OH$.[65] All these systems exhibit long photoinduced lifetimes, and the T(LIESST) values were systematically recorded. The authors propose a direct relation between T(LIESST) and the cooperativity factor C defined by the Slichter and Drickamer model ($= \Gamma/(2RT_{1/2})$): T(LIESST) $= 42.6C + 29.2$.[65]

In 2009, two new examples of the LIESST effect in iron(III) systems, $[Fe(qsal)_2]NCS$ and $[Fe(qsal)_2]NCSe$, were reported.[66] These two compounds displayed SCO transitions with wide hysteresis: for $[Fe(qsal)_2]NCS$ $T_{1/2}\downarrow = 205$ K and $T_{1/2}\uparrow = 289$ K; and, for $[Fe(qsal)_2]NCSe$ $T_{1/2}\downarrow = 212$ K and $T_{1/2}\uparrow = 282$ K. The LIESST properties were systematically determined by irradiating the sample at 800–1000 nm but unfortunately the level of photoexcitation was low at approximately 8%. Later, the $[Fe(qsal)_2]^+$ architecture was used in $[Fe(qsal)_2][Ni(dmit)_2]{\cdot}2CH_3CN$ to combine spin-crossover and conductivity properties in the same material.[67] An incomplete photoexcitation was described, but T(LIESST) was estimated to be around 46 K. Parallel to that another compound, $[Fe(qsal)_2][Pd(dmit)]_5{\cdot}$acetone, has been described as combining LIESST and conductivity.[68] The thermal spin-transition occurs between 200 and 250 K but the LIESST properties generated by a YAG laser (532 nm) were very small ($\chi_M T$ increases by about 0.1 cm^3 K mol^{-1} at 2 K) and vanish at about 40 K.[68]

More recently, another polyfunctional compound presenting a LIESST effect has been reported resulting from the insertion of a $[Fe(saltrien)]^+$ complex into a 2D oxalate network $[Fe(saltrien)][MnCr(ox)_3]{\cdot}$

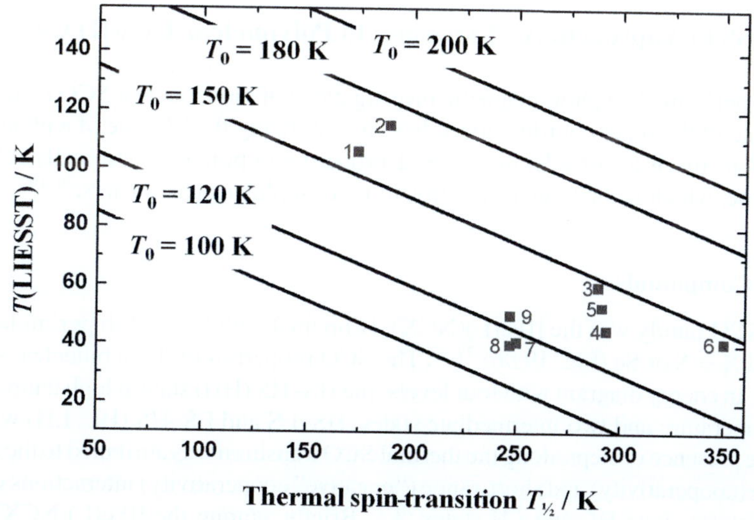

Figure 19.8 *Variation of T(LIESST) vs. $T_{1/2}$ for iron(III) SCO compounds reported in Table 19.1.*

Table 19.1 *T(LIESST) and $T_{1/2}$ values of iron(III) SCO compounds.*

Label	Formula	$T_{1/2}$ / K	T(LIESST) / K	Comment	Ref.
1	$[Fe(pap)_2]ClO_4{\cdot}H_2O$	174	106		51
2	$[Fe(pap)_2]BF_4{\cdot}H_2O$	189	115		64
3	$[Fe(pap)_2]PF_6{\cdot}CH_3OH$	289	60		64
4	$[Fe(CH_3\text{-}pap)_2]ClO_4{\cdot}H_2O$	293	45		65
5	$[Fe_{0.5}Al_{0.5}(pap)_2]ClO_4{\cdot}CH_3OH$	291	53		65
6	$[Fe_{0.25}Al_{0.75}(pap)_2]ClO_4{\cdot}CH_3OH$	350	41		65
7	$[Fe(saltrien)][MnCr(ox)_3]{\cdot}CH_2Cl_2$	250	41		69
8	$[Fe(qsal)_2]NCS$	247	40	T(LIESST) approx., weak photoconversion	66
9	$[Fe(qsal)_2]NCSe$	247	50	T(LIESST) approx., weak photoconversion	66
10	$[Fe(qsal)_2][Ni(dmit)_2]{\cdot}2CH_3CN$	–	46	stepped thermal spin conversion	67
11	$[Fe(qnal)_2][Pd(dmit)]_5{\cdot}$acetone	220	–	very weak photoconversion	68

CH_2Cl_2.[69] Thanks to the oxalate network, this compound combines magnetic ordering with a Curie temperature T_C of 5.6 K, with a gradual spin-transition at about 200 K. The light-induced effect at low temperature on this compound was recorded using red light (647 nm) and the *T*(LIESST) value was 41 K.[69]

In conclusion, over the last few years the number of iron(III) metal ions displaying photoinduction with long lifetimes and measurable *T*(LIESST) values has increased. Figure 19.8 and Table 19.3 compile the *T*(LIESST)/$T_{1/2}$ properties for these materials. Up to now, the number of examples has remained too low to allow any discussion about the nature of the inner coordination sphere. Hayami *et al.* have proposed to attribute this atypical stabilization of the photoinduced lifetime to strong cooperativity, obtained through π–π stacking.[51] Other factors are also certainly important and the use of a classical single coordinate configuration model may not be so appropriate.

19.4 The *T*(LIESST) Approach: An Extension to Polynuclear Iron(II) Complexes

The intensive works performed on photomagnetic investigations of mononuclear SCO materials have demonstrated the importance of the inner coordination sphere in stabilizing the lifetime of a photoexcited HS state. In this section, our objective is to consider the LIESST properties of polynuclear iron(II) SCO materials, from the simplest archetype, which is the binuclear entity, to more sophisticated clusters.[70–105]

19.4.1 Binuclear Compounds

The first binuclear SCO family was the $[[Fe(L)(NCX)_2]_2(bpym)]$ with L = 2,2′-bipyrimidine (bpym) or 2,2′-bithiazoline (bt) and X = S or Se (Fig. 19.9b).[72, 73] The SCO properties of these binuclear systems have been interpreted by using an energy diagram with four levels: the HS–HS (HH) state at high temperature, the LS–LS (LL) one at low temperature and two intermediate states, HS–LS and LS–HS (HL, LH) which are generally degenerate.[70–73] The presence of steps along the thermal SCO transition was attributed to the antagonistic effect between long-range (cooperativity) and short-range ("negative" cooperativity) interactions with a stabilization energy (*W*) of the intermediate HL and LH states.[70, 71] Briefly, among the $[[Fe(L)(NCX)_2]_2(bpym)]$ series the two iron(II) ions of $[[Fe(bpym)(NCS)_2]_2(bpym)]$ (**b1**) remain HS whatever the temperature and couple

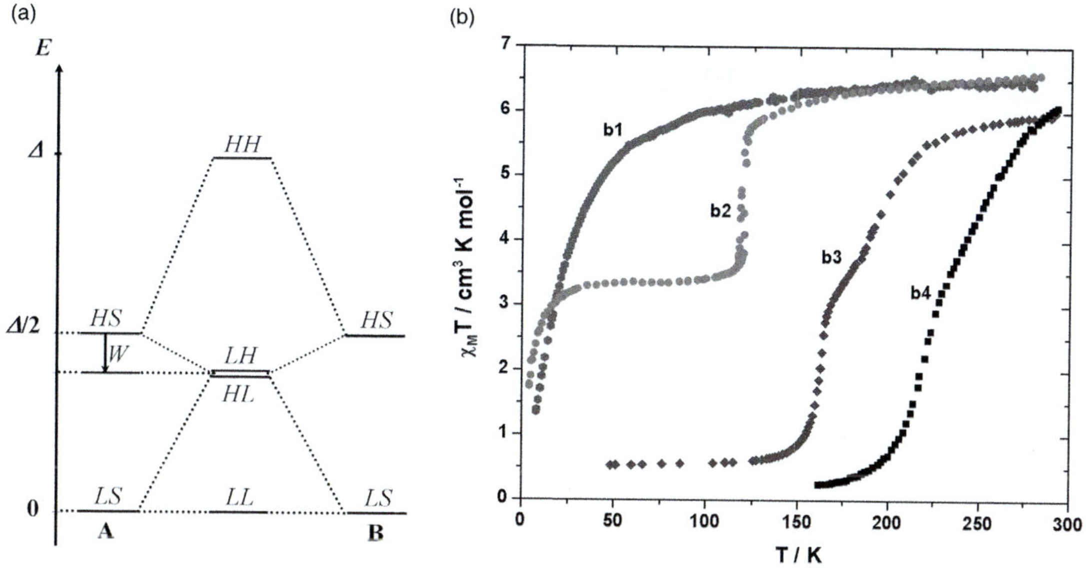

*Figure 19.9 (a) Energy diagram of the binuclear systems and (b) thermal dependence of the magnetic susceptibility time the temperature (χ_MT) of the binuclear compounds **b1–b4** (addapted from refs. 70 and 72).*

antiferromagnetically through the bpym bridge ($J = -4.1$ cm^{-1} in the zero-field Hamiltonian $\hat{\mathbf{H}} = -JS_A{\cdot}S_B$), giving a $S = 0$ ground pair HH state at low temperature.[72] In contrast, [[Fe(bpym)$(NCSe)_2$]$_2$bpym] (**b2**) exhibits a thermal half-SCO at low temperature.[71] The ground state is the paramagnetic HL pair and the spin-crossover has been attributed to the HH↔HL equilibrium. Finally, [[Fe(bt)$(NCS)_2$]$_2$bpym] (**b3**) and [[Fe(bt)$(NCSe)_2$]$_2$bpym] (**b4**) show almost complete thermal SCO with a 2-step character, attributed to the equilibria HH↔LH↔LL.[70, 73]

Concerning their photomagnetic properties, let us first consider [[Fe(bpym)$(NCSe)_2$]$_2$bpym] **b2**, which exhibits SCO at only one metal center between the HH and HL states. Irradiation below 17 K in the red region (647 nm) induced a decrease of the magnetic response, whereas above this temperature an increase was observed (Fig. 19.10a).[74–77] Usually in a mononuclear system, a decrease of the magnetic signal under irradiation is attributed to occurrence of the reverse-LIESST process.[9] In a binuclear case, however, this situation can be complicated by the presence of intramolecular magnetic interactions between the two metal ions. A typical example of intramolecular antiferromagnetic exchange interaction between two HS iron(II) ions is given by the compound **b1**. The shape of the magnetic curve (Fig. 19.10a) reflects progressive population of the nonmagnetic $S = 0$ ground spin state and the $S = 1, 2, 3$ and 4 low-lying excited spin states. More precisely, the analysis can be done according to the spin Hamiltonian $\hat{\mathbf{H}} = -JS_A.S_B + g\beta(S_A + S_B)H$, with $J = -4.1$ cm^{-1} and $g = 2.13$ as the best parameters (Eq. (19.3)).

$$\chi_M T = \frac{Ng^2\beta^2}{3k_B}\,\frac{6e^{J/k_BT} + 30e^{3J/k_BT} + 84e^{6J/k_BT} + 180e^{10J/k_BT}}{1 + 3e^{J/k_BT} + 5e^{3J/k_BT} + 7e^{6J/k_BT} + 9e^{10J/k_BT}} \tag{19.3}$$

Under irradiation at low temperature, the HL ground state of **b2** is converted to the HH metastable state, explaining why the photostationary limits reached the antiferromagnetic value of compound **b1**. Figure 19.10b completes this finding by comparing the *T*(LIESST) curve recorded for **b2** and the magnetic response of **b1**.

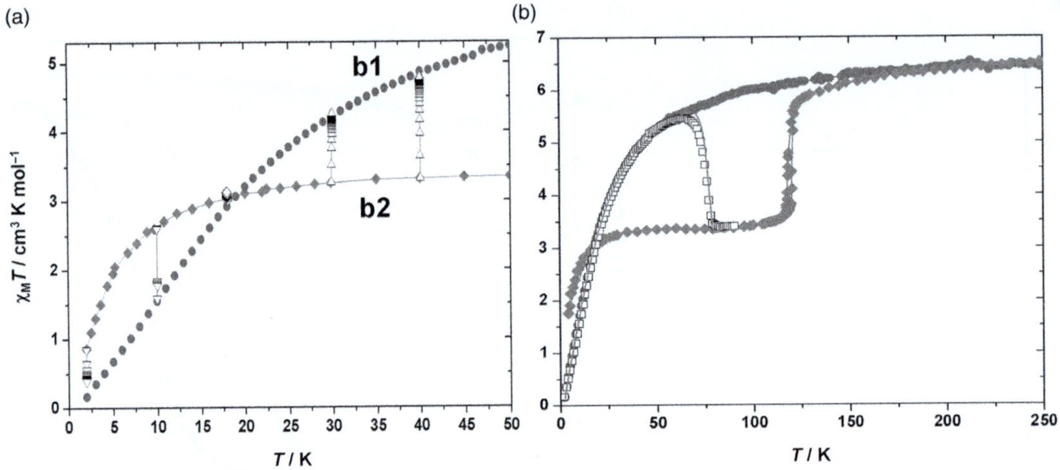

Figure 19.10 *Thermal dependence of the magnetic susceptibility times the temperature ($\chi_M T$) after irradiation at 647 nm (◇) of the binuclear **b2** compare to the magnetic behavior of **b1** (•): (a) Influence of irradiation at 647 nm (∇ and Δ) and (b) T(LIESST) experiment and simulation as discussed in the text (addapted from ref. 75).*

Interestingly, the *T*(LIESST) curve was found to perfectly follow the shape of the antiferromagnetic curve of **b1**. The only deviation appears above 55 K where the relaxation process becomes thermally activated.

The peculiar behavior found at 17 K where the magnetic response remains similar before and after light irradiation (Fig. 19.10a) is interesting. Classically, this can be understood as an absence of photoreaction. In fact, the reality is totally different; we are faced with two different spin states. Before irradiation, the system behaves as a HL mixed-spin pair, resulting in a $S = 2$ spin state with a $\chi_M T$ value at 17 K of 3 cm^3 K mol^{-1}. After irradiation, the two iron(II) ions are HS and coupled antiferromagnetically with a thermal population of the low-lying spin states, S = 1–4, located at the energies $E(S) = -JS(S+1)/_2$ above the ground state $S = 0$. Assuming a J factor of –4.1 cm^{-1} and a g value of 2.13,[71] at 17 K the distribution of the HS–HS pair is 16% $S = 0$, 35% $S = 1$, 29% $S = 2$, 15% $S = 3$ and 5% $S = 4$. The resulting $\chi_M T$ value after light irradiation is then 3 cm^3 K mol^{-1}, the same as the HL mixed-spin pair before irradiation. This first binuclear example demonstrates that their *T*(LIESST) shape strongly differs from a mononuclear SCO complex (Fig. 19.4).

This is also confirmed by looking the photomagnetic properties of [[Fe(bt)$(NCS)_2$]$_2$bpym] (**b3**) and [[Fe(bt)$(NCSe)_2$]$_2$bpym] (**b4**), characterized by a complete thermal SCO of the two metal ions (with a 2-step character attributed to the HH↔LH↔LL transitions). The shape of the *T*(LIESST) curves was also very close to the antiferromagnetic curve of **b1** (Fig. 19.11a).[74–77] The $\chi_M T$ product was found to dramatically increase up to 4.4 cm^3 K mol^{-1} at 44 K for **b3** and up to 2.8 cm^3 K mol^{-1} at 31 K for **b4**. Above these maxima, the magnetic signal rapidly dropped and recovered the initial diamagnetic response. This finding was consistent with the quantitative transformation of the LL ground state under irradiation into the HH metastable state.[74–77] This mechanism was further confirmed by Raman spectroscopy.[78–81]

To complete this view, recently a study was performed on six binuclear complexes **b6–b10** based on the bpypz ligand (bpypz = 3,5-bis-(pyridine-2-yl)pyrazolate) with various substituted pyridines as co-ligand.[82] Interestingly, none of these displays a step in their thermal SCO properties[83–86] and all of them show "bell-shape" *T*(LIESST) curves comparable to the antiferromagnetic behavior of the reference compound for this family, **b5**.[82]

Other binuclear compounds have been obtained.[87–89] The compound [$Fe_2(ddpp)_2(NCS)_4$].4CH_2Cl_2 (**b13**) exhibits complete spin-crossover with two separate steps with an incomplete photoconversion. The shape of its *T*(LIESST) curve was obviously close to a mononuclear complex.[87] Binuclear complexes

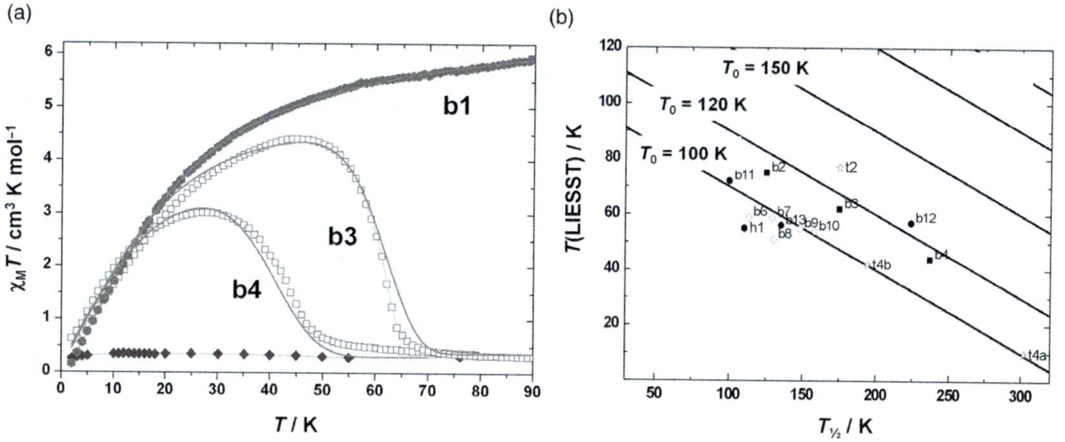

Figure 19.11 (a) Thermal dependence of the magnetic susceptibility times the temperature ($\chi_M T$) after irradiation at 647 nm (□) of the binuclear **b3** and **b4** compare to the magnetic behavior of **b1** (●). Straight lines stand for the T(LIESST) simulation as discussed in the text. (b) Variation of the T(LIESST) vs. $T_{1/2}$ for polynuclear spin crossover compounds (Table 9.2). (■) stands for the $[[Fe(L)(NCX)_2]_2(bpym)]$ family **b2–b4**; (◊) corresponds to the $[[Fe(NCS)(py\text{-}X)]_2(bpypz)_2]$ family **b5–b10**; (●) represents compounds **b11–b13**. (☆) are related to the tetranuclear **t2** and **t4** compounds. (⬬) corresponds to the hexanuclear **h1**.

$[[Fe(3\text{-}bpp)(NCS)_2]_2(4,4'\text{-}bpy)]$ (**b11**)[88] and $[Fe_2(PMAT)_2](BF_4)_4$.DMF ((PMAT = 4-amino-3,5-bis{[(2-pyridylmethyl)amino]methyl}-4H-1,2,4-triazole; **b12**),[89] exhibit incomplete spin-crossover transitions between the HH state and the intermediate HL state. At low temperatures, the HL state is quantitatively photoconverted into the HH state and the *T*(LIESST) values were respectively 72 (**b11**) and 57 K (**b12**).

In conclusion, the number of binuclear complexes with *T*(LIESST)/$T_{1/2}$ properties is still limited (Fig. 19.11b and Table 19.2). From the series **b2–b4**, the *T*(LIESST) values were found to increase as **b2** (75 K) > **b3** (62 K) > **b4** (44 K), that is in the reverse order to the SCO temperature $T_{1/2}$: **b2** (125 K) < **b3** (175 K) <

Table 19.2 T(LIESST) and $T_{1/2}$ values of iron(II) binuclear SCO compounds.

Label	Formula[a]	$T_{1/2}$ / K	T(LIESST) / K	Ref.
b1	$[[Fe(bpym)(NCS)_2]_2(bpym)]$	HS	–	73
b2	$[[Fe(bpym)(NCSe)_2]_2(bpym)]$	125	75	75–77
b3	$[[Fe(bt)(NCS)_2]_2(bpym)]$	175	62	75–77
b4	$[[Fe(bt)(NCSe)_2]_2(bpym)]$	237	44	75–77
b5	$[[Fe(NCS)_2(3,5\text{-}(CH_3)_2\text{-}py)]_2(bpypz)_2]$	HS	–	82
b6	$[[Fe(NCS)_2(4\text{-}(CH_3)\text{-}py)]_2(bpypz)_2]$	106	59	82
b7	$[[Fe(NCS)_2(py)]_2(bpypz)_2]$	129	59	82
b8	$[[Fe(NCS)_2(3\text{-}(CH_3)\text{-}py)]_2(bpypz)_2]$	130	51	82
b9	$[[Fe(NCS)_2(3\text{-}Cl\text{-}py)]_2(bpypz)_2]$	148	55	82
b10	$[[Fe(NCS)_2(3\text{-}Br\text{-}py)]_2(bpypz)_2]$	158	54	82
b11	$[[Fe(3\text{-}bpp)(NCS)_2]_2(4,4'\text{-}bpy)]$	100	72	88
b12	$[Fe_2(PMAT)_2](BF_4)_4$.DMF	224	57	89
b13	$[Fe_2(ddpp)_2(NCS)_4]$.$4CH_2Cl_2$	90 & 175	56	87

[a] bpym = 2,2′-bipyrimidine; bt = 2,2′-bithiazoline; bpypz = 3,5-bis-(pyridine-2-yl)pyrazolate; py = pyridine; 3-bpp = 2,6-bis(3-pyrazolyl)pyridine; 4,4′-bpy = 4,4′-bipyrine; PMAT = 4-amino-3,5-bis{[(2-pyridylmethyl) amino]methyl}-4H-1,2,4-triazole; ddpp = 2,5-di(2′,2″-dipyridylamino)pyridine.

b4 (237 K). This agrees with the "inverse–energy-gap law" that correlates high $T_{1/2}$ to fast relaxation rates.[32–35] The *T*(LIESST) vs. $T_{1/2}$ graph for (**b2–b4**) indicates that they belong to the T_0 = 120 K line (Fig. 19.11b), which is consistent with the bidentate nature of the bpym and/or bt ligands.

If we consider now the [[Fe(NCS)(py-X)]$_2$(bpypz)$_2$] family (**b5–b10**; Fig. 19.11b), their *T*(LIESST) values are distributed around the T_0 = 100 K line.[82] For an equivalent $T_{1/2}$ temperature, stabilization of the photoinduced HS state is less efficient for this family than for the [[Fe(L)(NCX)$_2$]$_2$(bpym)] series. Such a difference can be attributed to the existence of axial unidentate ligands, which can be a source of flexibility for the inner coordination sphere, as previously mentioned for the mononuclear study. In line with that, the binuclear complex [Fe$_2$(PMAT)$_2$](BF$_4$)$_4$.DMF (**b12**), possessing an apical ligand donor connected to the PMAT bridge belongs to the T_0 = 120 K line.[89] This emphasizes the concept that the vibrational aspect and stiffness of the inner coordination sphere are the key factors for stabilizing the light-induced HS state. This also suggests that in regard to the *T*(LIESST) limit binuclear and mononuclear systems behave similarly. The main difference concerns the shape of the *T*(LIESST) curve when the two iron(II) metal centers are magnetically coupled.

19.4.2 Trinuclear/Tetranuclear Complexes

If we consider the trinuclear architecture, most of the reported examples are based on triazole ligands.[90–96] These mostly exhibit incomplete thermal spin-crossover transition on the central iron(II) ion between the HHH↔HLH states, and up to now no *T*(LIESST) measurement has been reported.

Concerning tetranuclear [Fe$_4$] complexes, the basic architecture is a square topology. A first family concerns [2 × 2] grid-like cationic structures which have been initially represented by the [Fe$_4$L$_4$] SCO material where L is a bis-tridentate ligand.[97] A second family is the cyanide-bridged tetranuclear squares, [Fe$_4$(CN)$_4$L$_4$L$'_x$] with L a bidentate ligand and L′ a bi-or tetra-dentate ligand.[98–100]

Historically, the first square SCO complex belonging to the [2 × 2] grid-like cationic [Fe$_4$L$_4$] structure was reported in 2000.[97] The [Fe$_4$L$_4$](ClO$_4$)$_8$ (L = 4,6-bis-(2′,2″-bipyridyl-6′-yl)-2-phenylpyrimidine) (**t1**) undergoes a gradual SCO transition between three iron(II) metal centers, that is HHHL↔HHLL↔HLLL, and its LIESST properties were investigated using Mössbauer spectroscopy. The photoinduced state was observed to be stable up to 50 K.[97] Later on several substitutions of ligands and/or counteranions were performed on this framework.[101–103] The only example combining SCO and LIESST properties was the [Fe$_4$L$_4$](BF$_4$)$_4$ compound (**t2**) described by Sato *et al.*[103] This tetranuclear system displays an incomplete thermal SCO regime with only one iron(II) metal ion changing from HHHH to HHHL. The *T*(LIESST) curve recorded after irradiation at 785 nm shows a "bell-shape" comparable with the [Fe$_4$L$_4$](PF$_6$)$_4$ compound, characteristic of antiferromagnetic exchange between the four iron(II) ions in the HHHH state (Fig. 19.12a). The *T*(LIESST) value was estimated to 77 K for $T_{1/2}$ = 175 K.[103]

Concerning now cyanide-bridged tetranuclear squares [Fe$_4$(CN)$_4$L$_4$L$'_x$], a first example is given by the [Fe$_4$(CN)$_4$(bpy)$_4$(tpa)$_2$](PF$_6$)$_2$ (bpy = 2,2′-bipyridine and tpa = *tris*(2-pyridylmethyl)amine), a Prussian Blue analog, which undergoes a thermally induced 2-step spin conversion attributed to a HLHL↔HHHL↔LLLL process.[98] At low temperature, irradiation of the LLLL state with red light gives rise to a conversion towards the LLLH state.[99] An alternative tetranuclear square is based on dicyanamide bridging ligand, [Fe$_4$(N(CN)$_2$)$_4$(tpa)$_4$](BF$_4$)$_4$ (**t3**).[104] This cluster undergoes an incomplete 2-step HHHH↔HHHL↔HHLL SCO at 302 K and 194 K. Irradiation at 5 K (λ = 457 nm) induced a photoexcitation of about 50% and *T*(LIESST) was estimated at 42 K (Fig. 19.12b).[104]

The number of tetranuclear complexes on the *T*(LIESST)/$T_{1/2}$ database (Fig. 19.11) is consequently too low to be consistent. Nevertheless, interestingly the most rigid architecture based on a unique ligand **t2** is placed on an upper line (T_0 = 120 K line), while the **t3** complexes built from the flexible dicyanamide lie on the T_0 = 100 K line.

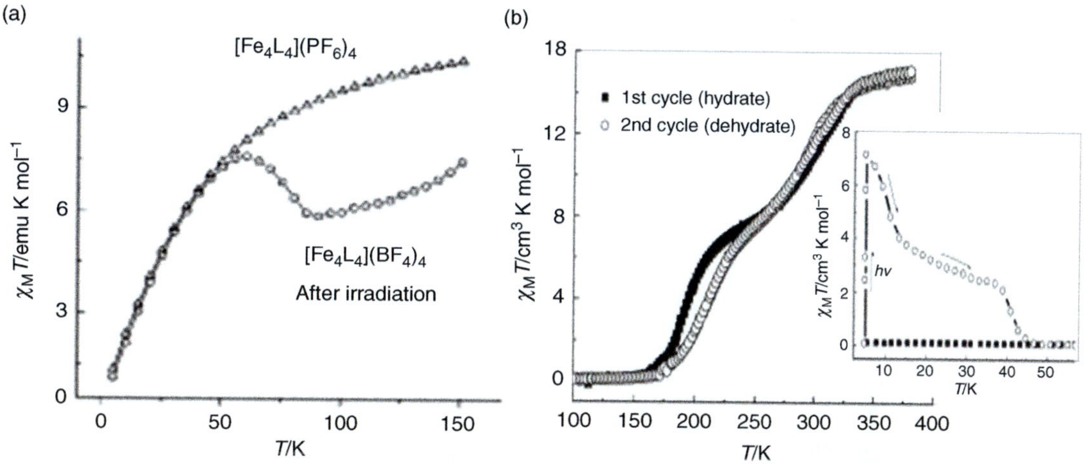

Figure 19.12 (a) Photomagnetic properties of $[Fe_4L_4](BF_4)_4$ compound (**t2**) compared to the $[Fe_4L_4](PF_6)_4$ behavior linked to an antiferromagnetic exchange between the four iron(II) ions in the HHHH state.[103] (b) Magnetic product vs. temperature for the tetranuclear $[Fe_4(N(CN)_2)_4(tpa)_4](BF_4)_4$ (**t3**) system.[104] The inset graph reports the thermal dependence of the magnetic signal after light irradiation generated at 5 K.[104]

19.4.3 Hexanuclear Complexes

A further step towards the design of SCO clusters has been recently achieved with the elaboration of a hexanuclear complex.[105] This cluster exhibits a nanoball shape 3 nm in size, built from a preformed metalloligand $[Cu^I(Tp^{4\text{-}py})(CH_3CN)]$ ($Tp^{4\text{-}py}$ = *tris*{3-(4′-pyridyl)-pyrazol-1-yl}hydroborate) which possesses three terminal pyridine ligands. Coordination with iron(II) ions gave the architecture $[[Cu^I(Tp^{4\text{-}py})(CH_3CN)]_8[Fe^{II}(NCS)_2]_{10/3}[Fe^{II}(NCS)(CH_3CN)]_{8/3}](ClO_4)_{8/3}$ (**h1**) (Fig. 19.13).[105] Hexanuclear **h1** shows an incomplete thermal spin-crossover regime with $T_{1/2}$ = 125 K and an important HS residue at low temperature. LIESST properties generated by irradiating at 514 nm gave a *T*(LIESST) shape very reminiscent to the one recorded for a mononuclear complex. The absence of any magnetic exchange within the cluster can be easily attributed to the fact that the paramagnetic iron(II) metal centers are isolated by diamagnetic Cu^I ions. The *T*(LIESST) value of **h1** was 55 K, close to the T_0 = 100 K line (Fig. 19.11), revealing a quite flexible complex.[105]

19.4.4 Polymeric Complexes

Following this systematic study of the relation between the dimensionality and the stability of the LIESST properties, it is interesting to consider the peculiar case of polymeric iron(II) SCO materials. Nevertheless, even if the number of polymeric SCO materials with 1D, 2D and 3D architectures is very large,[106–109] only a very few of them exhibit LIESST properties, and the number of reported *T*(LIESST) values is consequently extremely limited.

For instance, some $[Fe(trz)_3]A_2$ chains were irradiated in Nafion films but the lifetimes appear extremely short and *T*(LIESST) measurements could not be extracted.[110, 111] Other 2D and 3D SCO architectures of the Hofmann-like family have been shown to display some photoinduced effect in the center of the thermal hysteresis[17] but studies at low temperature are extremely rare.[112, 113] Other 2D SCO architectures based on the *bis*-triazole (btr) ligand, known for displaying LIESST properties at low temperature,[114, 115] have been mostly investigated for the existence of Light-Induced Thermal Hysteresis (LITH effect[7]).

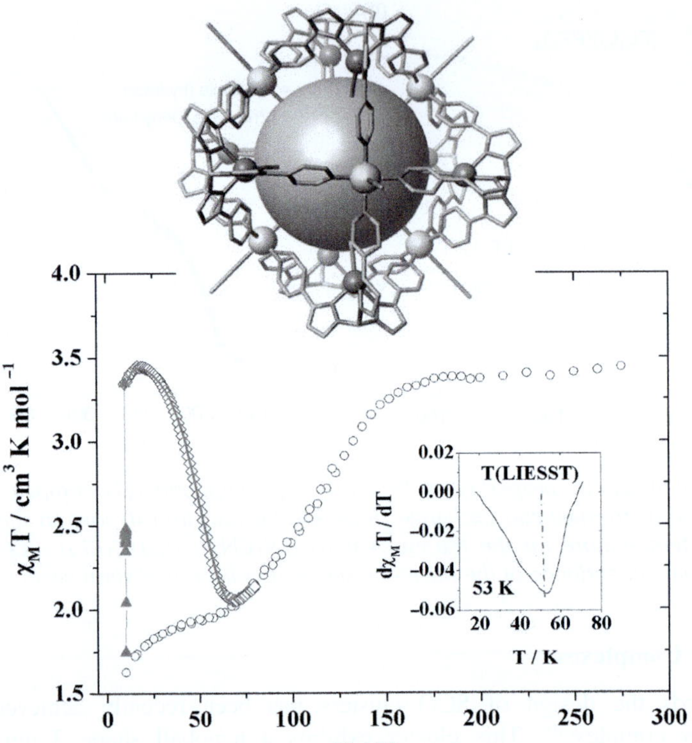

Figure 19.13 *Crystal structure of the* ***h1*** *nanoball (the sphere is included to highlight the nanoball shape) with magnetic and photomagnetic properties.*

One interesting study was based on the spin-crossover coordination polymers ([Fe(n-ditz)$_3$](ClO$_4$)$_2$]), where n-ditz is 1,n-bis(tetrazol-1-yl)alkane with $n = 4$–9.[116] The ditetrazole bridging ligands provide octahedral symmetry at the iron(II) centers, while allowing the distance between iron(II) centers to be varied. These polymers have therefore been investigated for the effects of spacer length on their thermal and light-induced spin-transition behavior. It was demonstrated that the increase in the number of carbon atoms in the spacer (n) raises the thermal spin-crossover temperature and decreases the stability of the light-induced metastable state, generated through the LIESST effect by irradiating the sample at 530 nm. Remarkably, however, the parity of the spacer also has an effect, enabling the complexes to be divided into two subseries depending on whether the bridging ligand possesses an even or an odd number of carbon atoms (Fig. 19.14).[116]

19.4.5 Nanoparticles

Today many research groups are working intensively to design nanomaterials with the ultimate goal of studying the relation between the thermal SCO regime (in particular the cooperativity and the existence of hysteresis) and the size and/or the shape of the particles. Concerning the relation with LIESST properties, the only reported example is based on the coordination polymer [Fe(NCS)$_2$(bpe)$_2$] [bpe = 1,2-*bis*(4-pyridyl)ethane] prepared by the water-in-oil reverse micelle method.[117] Comparison of the magnetic properties with particle sizes reducing from 2 μm to 55 nm reveals that the spin-transition becomes more gradual and incomplete, and concomitantly the transition temperature $T_{1/2}$ decreases. Nevertheless, it was demonstrated that T(LIESST) remains almost constant for all particle sizes, despite the diminution of $T_{1/2}$.[117] This tendency confirms the

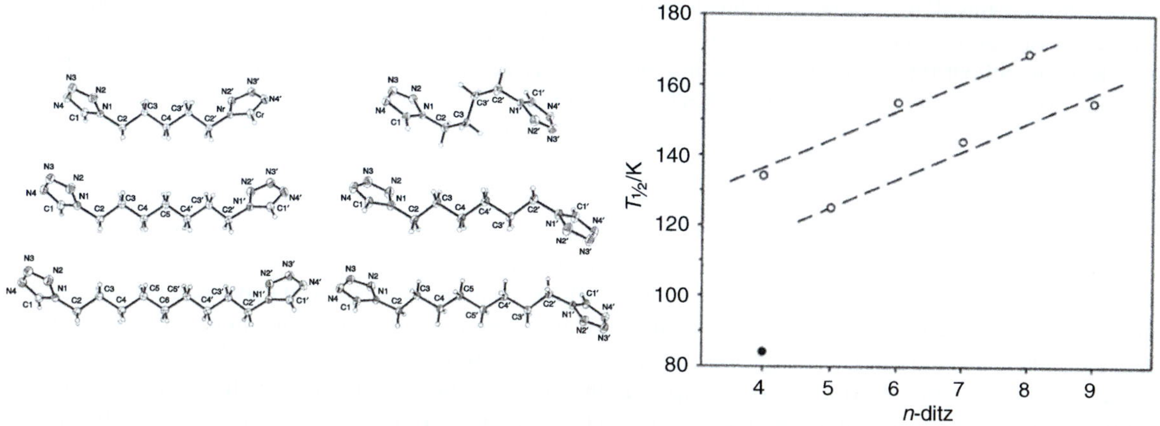

Figure 19.14 *Structural views and $T_{1/2}$ properties as a function of the number of carbons (n) in the spacer of the ditetrazole ligands. (•) = $T_{1/2}$ of the lower temperature spin-transition of [Fe(4-ditz)$_3$](ClO$_4$)$_2$. The parallel dashed lines are drawn to clearly show the separation of the data points into two subseries, even and odd, but have no further significance.*

current hypothesis that the stability of the photoinduced HS state is mainly governed by the inner coordination sphere. In other words, since the local coordination sphere around the iron(II) metal ion does not change with particle size, it is reasonable to record the same *T*(LIESST) value. Of course, future studies will have to confirm such view.

19.5 Simulation and Extrapolation of a *T*(LIESST) Experiment

19.5.1 Simulation of *T*(LIESST) Curve

As indicated, the *T*(LIESST) measurement is a macroscopic approach which compiles the influence of intrinsic and kinetic parameters. Nevertheless, if all these factors are taken into account a simulation of the *T*(LIESST) curve can be carried out.[38, 49] From study of the kinetics at different temperatures, the intrinsic parameters governing the relaxation process can be determined; the activation energy E_a, the pre-exponential factor k_∞, the acceleration factor connected to cooperativity (α), and the relaxation rate at T approaching absolute zero $k_{HL}(T\to0)$. The treatment of the kinetics is done with a single exponential (or a stretched exponential) for noncooperative systems, and a sigmoidal treatment for cooperative SCO (Eqs (19.4) and (19.5)). The E_a and k_∞ parameters are calculated from the straight line found on the plot of $\ln k_{HL}$ *vs.* $1/T$ (Eq. (19.6)) in the high temperature region. Concerning now the estimation of $k_{HL}(T\to0)$, one of the difficulties is that the relaxation process in the quantum tunneling regime is very slow and proper measurement of the rate constant is not realistic. One way to estimate $k_{HL}(T\to0)$ is to use the kinetics recorded at the lowest temperature, where a full relaxation occurs on the timescale investigated (typically in the range of 12 hours), and to consider that as an upper limit.

$$\frac{d\gamma_{HS}}{dt} = -k_{HL}(T, \gamma_{HS})\gamma_{HS} \tag{19.4}$$

$$k_{HL}(T, \gamma_{HS}) = k_{HL}(T)\exp[\alpha(T)(1-\gamma_{HS})] \tag{19.5}$$

$$k_{HL}(T) = k_\infty \exp\left(-\frac{E_a}{k_B T}\right) \tag{19.6}$$

Taking into account all these kinetic parameters, *T*(LIESST) can be simulated if we are able to reproduce the time and temperature dependencies. In a measurement of a *T*(LIESST) curve, the sample was first irradiated at 10 K, then without further irradiation the temperature was slowly warmed at the rate of approximately 0.3 K min^{-1}. The temperature during a *T*(LIESST) measurement, in a SQUID cavity, is rigorously changed in 1 K steps. At each temperature T_i, the time for the signal acquisition is 60 s and the time to reach the next temperature is 120 s. In reality, it is difficult to reproduce relaxation connected to the stabilization of the temperature. In a first approximation, we neglect this subtle effect and the relaxation is calculated at each temperature T_i during the global time of 180 s. The γ_{HS} fraction obtained after 180 s of relaxation at T_i is used as started value for the next temperature, T_{i+1}; that is $(\gamma_{HS})_{t=180}^{T=T_i} = (\gamma_{HS})_{t=0}^{T=T_{i+1}}$.

Another important point in the simulation of the *T*(LIESST) curve is that such an experiment combines the relaxation of both the tunneling and the thermally activated regions. Rigorously, this imposes the use of the theory of nonadiabatic multiphonon processes in the strong vibronic coupling limit.[32–35] We have assumed that the evolution of the HS fraction is a simple summation of these two effects.[38, 49] In the absence of any significant cooperativity, the time dependence of the HS fraction at temperature T_i is given by Eq. (19.7). For a cooperative compound exhibiting sigmoidal relaxation curves, as encountered, Eq. (19.8) can be used.

$$\left(\frac{\partial \gamma_{HS}}{\partial t}\right)_{Ti} = -\gamma_{HS}\,\{k_{HL}\,(T \to 0) + k_\infty \exp(-E_a/k_B T_i\,)\} \tag{19.7}$$

$$\left(\frac{\partial \gamma_{HS}}{\partial t}\right)_{Ti} = -\,\gamma_{HS}\,\{k_{HL}(T \to 0) + k_\infty \exp(-E_a/k_B T_i\,)\}\,\exp[\alpha(T_i)(1-\gamma_{HS})] \tag{19.8}$$

For a mononuclear SCO complex, the *T*(LIESST) curve is finally calculated by taking into account the anisotropy of the HS iron(II) ion in an octahedral surrounding. This phenomenon, called zero-field splitting, is associated with spin-orbit coupling between the ground state and the excited state in a zero applied magnetic field. For an iron(II) ion in HS configuration, the $S = 2$ ground state is split into three levels. The magnetic contribution of each state is determined by their energy separation, *D*, and their thermal population. For a powder sample, the $(\chi_M T)_{ZFS}$ product associated with zero-field splitting and the $\chi_M T$ of the *T*(LIESST) curve are given by Eqs (19.9) and (19.10) respectively.

$$(\chi_M T_i)_{ZFS} = \frac{\chi_M T_{i//} + 2\chi_M T_{i\perp}}{3} \tag{19.9}$$

$$\text{with } \chi_M T_{i//} = \frac{3}{4} g^2 \frac{e^{-D/k_B T_i} + 4e^{-4D/k_B T_i}}{1 + 2e^{-D/k_B T_i} + 2e^{-4D/k_B T_i}}$$

$$\text{and } \chi_M T_{i\perp} = \frac{1}{4} g^2 \frac{k_B T_i}{D} \frac{9 - 7e^{-D/k_B T_i} - 2e^{-4D/k_B T_i}}{1 + 2e^{-D/k_B T_i} + 2e^{-4D/k_B T_i}}$$

$$\chi_M T_i = (\chi_M T_i)_{ZFS}\,\gamma_{HS}(t, T_i) \tag{19.10}$$

Figure 19.4 shows the calculated *T*(LIESST) curve for a mononuclear example. The $k_{HL}(T\to 0)$, k_∞, E_a, and $\alpha(T_i)$ parameters used in the simulation are those obtained from the experimental kinetics. The agreement with the experimental data is very good. The shape and the capacity of the compound to retain the light-induced HS information, estimated through the determination of *T*(LIESST), are well described. This demonstrates the validity of our fitting procedure as well as the estimation of the various kinetic parameters, at least in this temperature range. Moreover, by changing the warming rate we have also reproduced the experimental changes recorded in the *T*(LIESST) temperatures (Fig. 19.5).

In the binuclear case, the main difference with the simulation of *T*(LIESST) curves of mononuclear systems lies in the fact that it is not the zero-field splitting of the iron(II) ion that was taken into account (except for compound **b2**) but antiferromagnetic exchange. The *T*(LIESST) curves have been fitted using Eq. (19.11) for compound **b2** and Eq. (19.12) for compounds **b3**, **b4** and **b6–b10**.[76] Figure 19.11 reports the calculated *T*(LIESST) curves of **b2–4** using kinetic parameters obtained in the thermal activated region. Interestingly, the agreement with experimental data is quite good for all the compounds. The shape and the capacity of these binuclear compounds to retain the light-induced HS information estimated through the determination of the *T*(LIESST) are well described.

$$\chi_M T = \gamma_{HS}(t, T).[(\chi_M T)_{AF} - (\chi_M T)_{HL}] + (\chi_M T)_{HL} \tag{19.11}$$

$$\chi_M T = (\chi_M T)_{AF} \cdot \chi_{HS}(t, T) \tag{19.12}$$

19.5.2 Simulation and Extrapolation

Based on the good agreement between experiments and simulation, it becomes possible to extract the influence of each components of the master equation on the *T*(LIESST) value. Figure 19.15a shows the link between the cooperativity and the *T*(LIESST) value. The decrease of the activation energy term, E_a^* (attributed to cooperativity) from 300 to 0 cm^{-1} is minor, just a slight increase of 6 K for the *T*(LIESST) limit. Figure 19.15b shows the influence of the tunneling relaxation constant, $k_{HL}(T \rightarrow 0)$. For a given E_a and $k_{HL}(T \rightarrow \infty)$ values, a change of $k_{HL}(T \rightarrow 0)$ from 10^{-10} to $10^{-3.6}$ s^{-1} only slightly affects the *T*(LIESST) limit. Moreover, for $k_{HL}(T \rightarrow 0) > 10^{-3.6}$ s^{-1}, the $\chi_M T$ vs. *T* curve no longer displays an inflexion point and *T*(LIESST) cannot be defined. Figure 19.15c reports the evolution of *T*(LIESST) according to $k_{HL}(T \rightarrow \infty)$. For a given activation energy, a decrease of $k_{HL}(T \rightarrow \infty)$ from 2×10^7 s^{-1} to 2×10^4 s^{-1} leads to an important increase in *T*(LIESST).

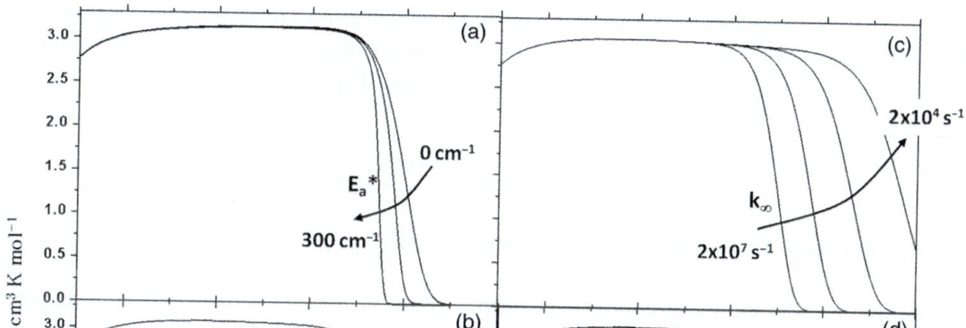

Figure 19.15 *Influence of E_a^* (a), $k_{HL}(T \rightarrow 0)$ (b), $k_{HL}(T \rightarrow \infty)$ (c) and E_a (d) on the T(LIESST) curve.*

Finally, Figure 19.15d illustrates the effect of E_a on the *T*(LIESST) value. The influence of the activation energy on *T*(LIESST) appears here to also be very important. For example, *T*(LIESST) varies over 85 K for E_a ranging from 200 cm^{-1} to 1100 cm^{-1}.

It follows from this study that the variation of T(LIESST) mainly arises from the E_a and $k_{HL}(T\rightarrow\infty)$ values. The role of $E_a{}^*$ and $k_{HL}(T\rightarrow 0)$ appears negligible. Therefore, to establish an analytical relationship between *T*(LIESST), E_a and $k_{HL}(T\rightarrow\infty)$, it seems reasonable in a first approximation to neglect $E_a{}^*$ and $k_{HL}(T\rightarrow 0)$. Under these conditions, the master equation can be written as Eq. (19.13).

$$\frac{d\gamma_{HS}}{dt} = -\gamma_{HS}\left\{k_{HL}(T\rightarrow\infty)\exp\left[-\frac{E_a}{k_B T}\right]\right\} \qquad (19.13)$$

Since the *T*(LIESST) measurement is carried out at constant heating rate, if we consider θ to be constant, the temperature *T* becomes a linear function of time *t* as $T_{i+1} = T_i + \theta t$. This translation leads to Eq. (19.14).

$$\frac{d\gamma_{HS}}{dT} = -\frac{\gamma_{HS}}{\theta}\left\{k_{HL}(T\rightarrow\infty)\exp\left[-\frac{E_a}{k_B T}\right]\right\} \qquad (19.14)$$

The second order derivative of this equation upon *T* gives rise to Eq. (19.15).

$$\frac{d^2\gamma_{HS}}{dT^2} = -\frac{1}{\theta}\left[\frac{d\gamma_{HS}}{dT}\left(k_{HL}(T\rightarrow\infty)\exp\left(-\frac{E_a}{k_B T}\right)\right) + \gamma_{HS}\frac{d}{dT}\left(k_{HL}(T\rightarrow\infty)\exp\left(-\frac{E_a}{k_B T}\right)\right)\right]$$
$$\frac{d^2\gamma_{HS}}{dT^2} = -\frac{1}{\theta}\left[\frac{d\gamma_{HS}}{dT}\left(k_{HL}(T\rightarrow\infty)\exp\left(-\frac{E_a}{k_B T}\right)\right) + \gamma_{HS}\frac{k_{HL}(T\rightarrow\infty)E_a}{k_B T^2}\exp\left(-\frac{E_a}{k_B T}\right)\right] \qquad (19.15)$$

This second order derivative becomes zero at $T = T$(LIESST), which allows us to extract a relationship between the *T*(LIESST), E_a and $k_{HL}(T\rightarrow\infty)$ (Eq. (19.16)). It is then possible to plot the evolution of *T*(LIESST) as function of the activation energy for different $k_{HL}(T\rightarrow\infty)$ values (Fig. 19.16).

$$k_{HL}(T\rightarrow\infty) = \frac{\theta E_a}{k_B T(LIESST)^2}\exp\left(\frac{E_a}{k_B T(LIESST)}\right) \qquad (19.16)$$

The data reported in Figure 19.16 leads to the following remarks:

(i) It is interesting to note that several (E_a, k_∞) couples give the same *T*(LIESST) value. In other words, *T*(LIESST) does not change if the two parameters change in the same way.
(ii) Such simulation and extrapolation define for the first time the ideal conditions which should shift the *T*(LIESST) limit towards room temperature; that is $k_{HL}(T\rightarrow\infty)$ should be between 10^{-2} and 10 s^{-1} and associated with an energy barrier of between 900 and 2000 cm^{-1}.

Hauser demonstrated that, on the basis of the SCC approach,[36] values of $k_{HL}(T\rightarrow\infty)$ are predicted in the range of 10^6–10^9 s^{-1}.[10–13, 32–35] In SCC theory, the Fe–Ligand distance is taken as the reaction coordinate with a single average vibrational mode, each spin state being represented by a harmonic potential well. These two wells are separated horizontally by Δr_{HL} and vertically by an energy $\Delta E^0{}_{HL}$ (Fig. 19.17a). Nevertheless, as discussed in this chapter, some iron materials are characterized by large distortion of the coordination sphere upon HS↔LS crossover. In these cases, the SCC model may become insufficient. It therefore becomes necessary to consider a system with several vibrational modes to correctly describe HS→LS

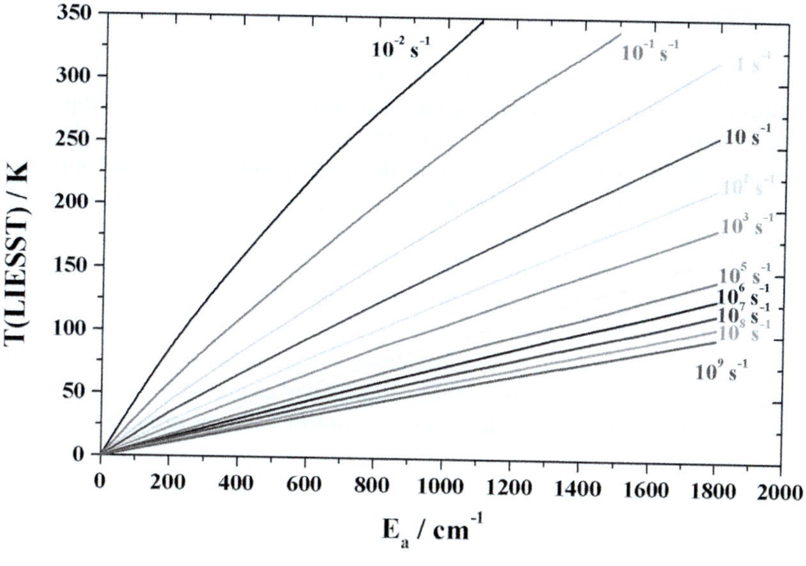

Figure 19.16 Correlation between T(LIESST, E_a and $k_{HL}(T \to \infty)$.

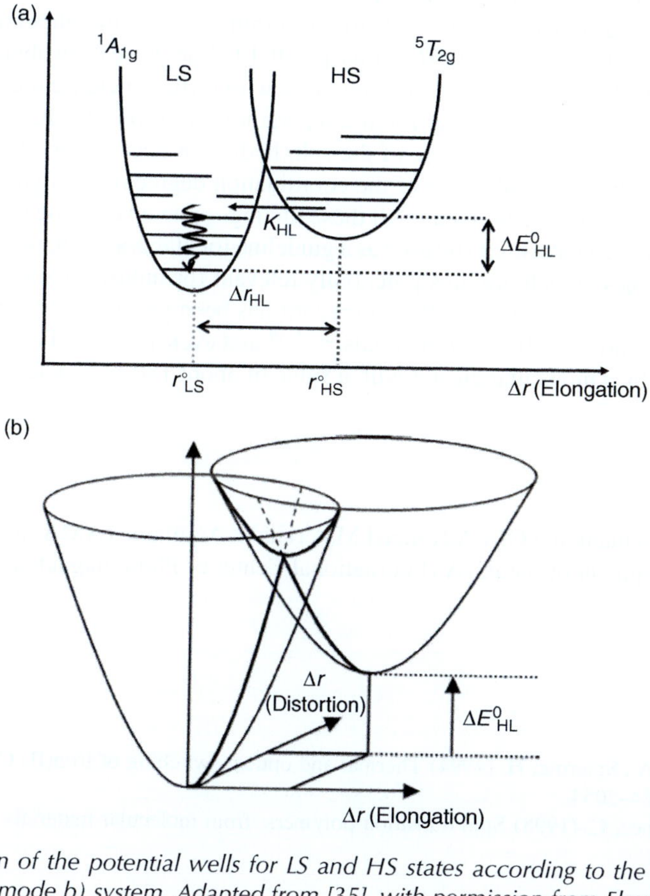

Figure 19.17 Illustration of the potential wells for LS and HS states according to the Fe–ligand distance in a single mode a) and dual-mode b) system. Adapted from [35], with permission from Elsevier.

relaxation. Such a case has been reported to correctly describe the relaxation process of $[Fe(TRIM)_2]Cl_2$ (TRIM = 4-(4-imidazolylmethyl)-2-(2-imidazolylmethyl)imidazole).[60] In this complex two modes were reported respectively at 220 cm^{-1} and 20 cm^{-1}. In 2006, Hauser *et al.* discussed the concept of a model based on two vibrational modes, one being relative to the Fe–Ligand distance and the other relative to the distortion (Fig. 19.17b).[35] These authors conclude that for SCO complexes with anisotropic expansion of the coordination sphere the single mode model is clearly inadequate.[35] The additional large difference in the axial Fe–N bond length results in a significant increase in the barrier between the two states, and thus slows down relaxation at low temperatures by several orders of magnitude. From a practical point of view this opens a new avenue to search for complexes with a slow relaxation at higher temperatures than was thought possible based on the single mode model. Today based on our simulation and extrapolation it appears that crucial parameters are both $k_{HL}(T\rightarrow\infty)$ and E_a values. We have to extensively pursue both our synthetic and fundamental reasearch efforts to further undertand how to control and optimize such parameters.

19.6 Conclusions

Extensive progress has been made over the last 40 years on spin-crossover materials, particularly in the light-induced field. Some years ago we introduced the idea of using *T*(LIESST) to determine the stability of a photoinduced HS state. Based on that the *T*(LIESST)/$T_{1/2}$ values of a large number of spin-crossover materials have been investigated and a general relation governing most of the photomagnetic properties has been evidenced, namely $T(\text{LIESST}) = T_0 - 0.3T_{1/2}$. Five parallel T_0 lines have been obtained with, respectively, values of 100, 120, 150, 180 and 200 K. The challenge is now to further increase the T_0 value and to find an operative molecular light-induced complex at room temperature. From our database, it seems that changes outside the sphere (cooperativity factor, nature of the salt, hydration degree) are relatively unimportant for tuning the T_0 factor. In contrast, the influence of the coordination degree of the ligand involved in the inner coordination sphere appears to be predominant. All these findings open up exciting prospects and we will see in the future if the *T*(LIESST) database can be used as a guideline for elaborating new photomagnetic materials with predicted long lifetimes, which are thus potentially relevant for industrial applications. Particularly, if we consider that in the last 10 years the *T*(LIESST) record has been progressively shifted from 60 K in the late 1990s to 130 K for a pure iron(II) SCO materials[22, 57–59] and even 150 K in Prussian Blue analogs.[46, 118] The record today is held by a molecular cluster with a value of about 180 K.[119] Clearly, the *T*(LIESST) limit has not yet been reached.

Acknowledgements

The authors would like to thank the GIS-Advanced Materials in Aquitaine (AMA) and the Aquitaine Region for supporting the development of the ICPA (International Center of Photomagnetism in Aquitaine) platform at the ICMCB.

References

1. Gütlich, P., Hauser, A., Spiering, H. (1994) Thermal and optical switching of iron(II) Complexes. Angew. Chem. Int. Ed. Engl., 33:2024–2054.
2. Kahn, O., Jay Martinez, C. (1998) Spin transition polymers: from molecular materials towards memory devices. Science, 279: 44–48.

3. Gütlich, P., Garcia, Y., Woike, T. (2001) Photoswitchable coordination compounds. Coord. Chem. Rev., 219–221: 839–879.
4. Gütlich, P., GoodwinH. A. (Eds) (2004) Spin Crossover in Transition Metal Compounds I-III. Top. Curr. Chem., vols. 233–235. Springer, Berlin / Heidelberg, Germany.
5. Decurtins, S., Gütlich, P., Köhler, C. P., Spiering, H., Hauser, A. (1984) Light-induced excited spin state trapping in a transition metal complex: the hexa-1-propyltetrazole-iron(II) tetrafluoroborate spin-crossover system. Chem. Phys. Lett., 105: 1–4.
6. Létard, J.-F., Guionneau, P., Goux-Capes, L. (2004) Towards spin crossover applications, in: Gütlich, P., Goodwin, H. A. (Eds) Spin Crossover in Transition Metal Compounds III. Top. Curr. Chem., 235: 221– 249.
7. Létard, J.-F., Guionneau, P., Rabardel, L., Howard, J. A. K., Goeta, A. E., Chasseau, D., Kahn, O. (1998) Structural, magnetic, and photomagnetic studies of a mononuclear iron(II) derivative exhibiting an exceptionally abrupt spin transition. light-induced thermal hysteresis phenomenon. Inorg. Chem., 37: 4432–4441.
8. McGarvey, J. J., Lawthers, I. (1982) Photochemically-induced perturbation of the $^{1}A \rightleftharpoons {}^{5}T$ equilibrium in Fe^{II} complexes by pulsed laser irradiation in the metal-to-ligand charge-transfer absorption band. J. Chem. Soc. Chem. Comm., 16: 906–907.
9. Hauser, A. (1986) Reversibility of light-induced excited spin state trapping in the $Fe(ptz)_6(BF_4)_2$, and the $Zn_{1-x}Fe_x(ptz)_6(BF_4)_2$ spin-crossover systems. Chem. Phys. Lett., 124: 543–548.
10. Hauser, A. (2004) Light-induced spin crossover and the high-spin→low-spin relaxation, in: Gütlich, P., Goodwin, H. A. (Eds) Spin Crossover in Transition Metal Compounds II. Top. Curr. Chem., 234: 155–198.
11. Hauser, A., Adler, J., Gütlich, P. (1988) Light-induced excited spin state trapping (LIESST) in $[Fe(2\text{-mephen})_3]^{2+}$ embedded in polymer matrices. Chem. Phys. Lett., 152: 468–472.
12. Hauser, A. (1992) Cooperative effects on the HS→LS relaxation in the $[Fe(ptz)_6](BF_4)_2$ spin-crossover system. Chem. Phys. Lett. 192: 65–70.
13. Hauser, A., Jeftić, J., Romstedt, H., Hinek, R., Spiering, H. (1999) Cooperative phenomena and light-induced bistability in iron(II) spin-crossover compounds. Coord. Chem. Rev., 190-192: 471–491.
14. Hauser, A., Gütlich, P., Spiering, H. (1986) High-spin→low-spin relaxation kinetics and cooperative effects in the hexakis(1-propyltetrazole)iron bis(tetrafluoroborate) and $[Zn_{1-x}Fe_x(ptz)_6](BF_4)_2$ (ptz = 1-propyltetrazole) spin-crossover systems. Inorg. Chem., 25: 4245–4248.
15. Shimamoto, N., Ohkoshi, S., Sato, O., Hashimoto, K. (2002) One-shot-laser-pulse-induced cooperative charge transfer accompanied by spin transition in a Co-Fe Prussian blue analog at room temperature. Chem. Lett., 31: 486–487.
16. Freysz, E., Montant, S., Létard, S., Létard, J.-F. (2004) Single laser pulse induces spin state transition within the hysteresis loop of an iron compound. Chem. Phys. Lett., 394: 318–323.
17. Bonhommeau, S., Molnár, G., Galet, A., Zwick, A., Real, J. A., McGarvey, J. J., Bousseksou, A. (2005) One shot laser pulse induced reversible spin transition in the spin-crossover complex $[Fe(C_4H_4N_2)\{Pt(CN)_4\}]$ at room temperature. Angew. Chem. Int. Ed., 44: 4069–4073.
18. Buchen, T., Gütlich, P., Goodwin, H. A. (1994) Non-first-order kinetics of the high spin-low spin relaxation in $[Fe(bpp)_2](BF_4)_2$ after LIESST and thermal spin trapping. Inorg. Chem., 33: 4573–4576.
19. Buchen, T., Gütlich, P., Sugiyarto, K. H., Goodwin, H. A. (1996) High-spin→low-spin relaxation in $[Fe(bpp)_2](CF_3SO_3)_2{\cdot}H_2O$ after LIESST and thermal spin-state trapping – dynamics of spin transition *vs.* dynamics of phase transition. Chem. Eur. J., 2: 1134–1138.
20. Wu, C. H., Jung, J., Gantzel, P. K., Gütlich, P., Hendrickson, D. N. (1997) LIESST effect studies of iron(II) spin-crossover complexes with phosphine ligands: relaxation kinetics and effects of solvent molecules. Inorg. Chem., 36: 5339–5347.
21. Renz, F., Oshio, H., Ksenofontov, V., Waldeck, M., Spiering, H., Gütlich, P. (2000) Strong field iron(II) complex converted by light into a long-lived high-spin state. Angew. Chem. Int. Ed., 39: 3699–3700.
22. Hayami, S., Gu, Z. Z., Einaga, Y., Kobayashi, Y., Ishikawa, Y., Yamada, Y., Fujishima, A., Sato, O. (2001) A novel LIESST iron(II) complex exhibiting a high relaxation temperature. Inorg. Chem., 40: 3240–3242.
23. Herber, R., Casson, L. M. (1986) Light-induced excited-spin-state trapping: evidence from variable temperature Fourier transform measurements. Inorg. Chem., 25: 847–852.

24. Herber, R. (1987) VTFTIR and light-induced excited spin state trapping in bis(2,2′-bipyridine)bis(thiocyanato)iron and related spin-crossover compounds. Inorg. Chem., 26: 173–178.
25. Baker, W. A., Long, G. (1965) Infrared spectra of some magnetically anomalous iron(II) complexes. Chem. Commun., 368–369.
26. Spacu, P., Teodorescu, M., Ciomartan, D. (1967) Magnetic properties and structure of some complex amines of iron(II). Rev. Roum. Chim., 12: 145–149.
27. Teodorescu, M. (1976) The structure of dark-red compound dithiocyanatobis(phenanthroline)iron(II) monohydrate. Rev. Roum. Chim., 21: 1031–1036.
28. Savage, S., Jia-Long, Z., Maddock, A. G. (1985) The red form of bis(1,10-phenanthroline)dithiocyanatoiron(II). J Chem. Soc. Dalton Trans., 991–996.
29. Ganguli, P., Gütlich, P., Müller, E. W., Irler, W. (1981) Further studies on the spin cross-over phenomenon in di-isothiocyanatobis(1,10-phenanthroline)iron(II). J. Chem. Soc. Dalton. Trans., 441–446.
30. Marchivie, M., Guionneau, P., Howard, J. A. K., Chastanet, G., Létard, J.-F., Goeta, A. E., Chasseau, D. (2002) Structural characterization of a photoinduced molecular switch. J. Am. Chem. Soc., 124: 194–195.
31. Guionneau, P., Marchivie, M., Bravic, G., Létard, J.-F., Chasseau, D. (2004) Structural aspects of spin crossover. Example of the $[Fe^{II}L_n(NCS)_2]$ complexes, in: Gütlich, P., Goodwin, H. A. (Eds) Spin Crossover in Transition Metal Compounds II. Top. Curr. Chem., 234: 97–128.
32. Hauser, A. (1991) Intersystem crossing in Fe(II) coordination compounds. Coord. Chem. Rev., 111: 275–290.
33. Hauser, A., Vef, A., Adler, P. (1991) Intersystem crossing dynamics in Fe(II) coordination compounds. J. Chem. Phys., 95: 8710–8717.
34. Hauser, A. (1995) Intersystem crossing in iron(II) coordination compounds: a model process between classical and quantum mechanical behaviour. Comments Inorg. Chem., 17: 17–40.
35. Hauser, A., Enachescu, C., Daku, M. L, Vargas, A., Amstutz, N. (2006) Low-temperature lifetimes of metastable high-spin states in spin-crossover and in low-spin iron(II) compounds: the rule and exceptions to the rule. Coord. Chem. Rev., 250: 1642–1652.
36. Buhks, E., Navon, G., Bixon, M., Jortner, J. (1980) Spin conversion processes in solutions. J. Am. Chem. Soc., 102: 2918–2923.
37. Capes, L. (2000) Ph.D. Thesis CNRS-MOTOROLA, Bordeaux, France.
38. Létard, J.-F., Chastanet, G., Nguyen, O., Marcen, S., Marchivie, M., Guionneau, P., Chasseau, D., Gütlich, P. (2003) Spin crossover properties of the $[Fe(PM\text{-}BiA)_2(NCS)_2]$ complex – phases I and II. Monatsh. Chem., 134: 165–182.
39. Létard, J.-F., Capes, L., Chastanet, G., Moliner, N., Létard, S., Real, J. A., Kahn, O. (1999) Critical temperature of the LIESST effect in iron(II) spin crossover compounds. Chem. Phys. Lett., 313: 115–120.
40. Marcen, S., Lecren, L., Capes, L., Goodwin, H. A., Létard, J.-F. (2002) Critical temperature of the LIESST effect in a series of hydrated and anhydrous complex salts $[Fe(bpp)_2]X_2$. Chem. Phys. Lett., 358: 87–95.
41. Sugiyarto, K. H., Goodwin, H. A. (1988) Coordination of pyridine-substituted pyrazolesand their influence on the spin state of iron(II). Aust. J. Chem., 41: 1645–1663.
42. Sugiyarto, K. H., Craig, D. C., Rae, A. D., Goodwin, H. A. (1994) Structural, magnetic and Mössbauer spectral studies of salts of bis[2,6-bis(pyrazol-3-yl)pyridine]iron(II) – a spin crossover system. Aust. J. Chem., 47: 869–890.
43. Sugiyarto, K. H., Weitzner, K., Craig, D. C., Goodwin, H. A. (1997) Structural, magnetic and Mössbauer studies of bis(2,6-bis(pyrazol-3-yl)pyridine)iron(II) triflate and its hydrates. Aust. J. Chem., 50: 869–873.
44. Sugiyarto, K. H., Scuddler, M. L., Craig, D. C., Goodwin, H. A. (2000) Electronic and structural properties of the spin crossover systems bis(2,6-bis(pyrazol-3-yl)pyridine)iron(II) thiocyanate and selenocyanate. Aust. J. Chem., 53: 755–765.
45. Goodwin, H. A., Sugiyarto, K. H. (1987) Lattice trapping of metastable quintet state bis(2,6-bis(pyrazol-3-yl)pyridine)iron(II) bis(tetrafluoroborate), a spin crossover system, and kinetics of the quintet-singlet transformation. Chem. Phys. Lett., 139: 470–474.
46. Shimamoto, N., Ohkoshi, S. S., Sato, O., Hashimoto, K. (2002) Control of charge-transfer-induced spin transition temperature on cobalt–iron Prussian blue analogues. Inorg. Chem., 41: 678–684.
47. Sato, O., Iyoda, T., Fujishima, A., Hashimoto, K. (1996) Photoinduced magnetization of a cobalt-iron cyanide. Science, 272: 704–705.

48. Létard, J.-F., Guionneau, P., Nguyen, O., Sanchez Costa, J., Marcén, S., Chastanet, G., Marchivie, M., Goux-Capes, L. (2005) A guideline to the design of molecular-based materials with long-lived photomagnetic lifetimes. Chem. Eur. J., 11: 4582–4589.
49. Létard, J.-F. (2006) Photomagnetism of iron(II) spin crossover complexes – the *T*(LIESST) approach. J. Mater. Chem., 16: 2550–2559.
50. Capes, L., Létard, J.-F., Kahn, O. (2000) Photomagnetic properties in a series of spin crossover compounds $[Fe(PM\text{-}L)_2(NCX)_2]$ (X = S, Se) with substituted 2′-pyridylmethylene-4-amino ligands. Chem. Eur. J., 6: 2246–2255.
51. Hayami, S., Gu, Z.-Z., Shiro, M., Einaga, Y., Fujishima, A., Sato, O. (2000) First observation of light-induced excited spin state trapping for an iron(III) complex. J. Am. Chem. Soc., 122: 7126–7127.
52. Létard, J.-F., Costa, J.S., Marcen, S., Carbonera, C., Desplanches, C., Kobayashi, A., Daro, N., Guionneau, P., Ader, J. P. (2005) Does cooperativity influence the lifetime of the photo-induced HS state? J. Phys. Conf. Ser., 21: 23–29.
53. Cecconi, F., Di Vaira, M., Midollini, S., Orlandini, A., Sacconi, L. (1981) Singlet ⇌ quintet spin transitions of iron(II) complexes with a P_4Cl_2 donor set. X-ray structures of the compound $FeCl_2(Ph_2PCH{=}CHPPh_2)_2$ and of its acetone solvate at 130 and 295 K. Inorg. Chem., 20: 3423–3430.
54. Rosa, P., Debay, A., Capes, L., Chastanet, G., Bousseksou, A., Le Floch, P., Létard, J.-F. (2004) Heat- and light-induced spin transition of an iron(II) polymer containing the 1,2,4,5-tetrakis(diphenylphosphanyl)benzene ligand. Eur. J. Inorg. Chem., 3017–3019.
55. Marcen, S. (2003) Ph.D. Thesis, CNRS, Bordeaux, France.
56. Marchivie, M., Guionneau, P., Létard, J.-F., Chasseau, D. (2005) Photo-induced spin-transition: the role of the iron(II) environment distortion. Acta Cryst. Sect. B, 61: 25–28.
57. Costa, J. S. (2005) Ph.D. Thesis, CNRS, Bordeaux, France.
58. Costa, J. S., Guionneau, P., Létard, J.-F. (2005) Photomagnetic properties of the $[Fe(L_{222}(N_3O_2))(CN)_2]{\cdot}H_2O$ complex: a fascinating example of multi-metastability. J Phys Conf. Ser., 21: 67–72.
59. Costa, J. S., Balde, C., Carbonera, C., Denux, D., Wattiaux, A., Desplanches, C., Ader, J. P., Gütlich, P., Létard, J.-F. (2007) Photomagnetic properties of an iron(II) low-spin complex with an unusually long-lived metastable LIESST state. Inorg. Chem., 46: 4114–4119.
60. Bonhommeau, S., Bréfuel, N., Pálfi, V. K., Molnár, G., Zwick, A., Salmon, L., Tuchagues, J.-P., Costa, J. S., Létard, J.-F., Paulsen, H., Bousseksou, A. (2005) High-spin to low-spin relaxation kinetics in the $[Fe(TRIM)_2]Cl_2$ complex. Phys. Chem. Chem. Phys., 7: 2909–2914.
61. Enachescu, C., Linares, J., Varret, F., Boukheddaden, K., Codjovi, E., Salunke, S. G., Mukherjee, R. (2004) Nonexponential relaxation of the metastable state of the spin-crossover system $[Fe(L)_2](ClO_4)_2{\cdot}H_2O$ [L = 2,6-Bis(pyrazol-1-ylmethyl)pyridine]. Inorg. Chem., 43: 4880–4888.
62. Schenker, S., Hauser, A. (1994) Low-temperature tunneling in the spin-state relaxation dynamics of iron(III) spin-crossover compounds. J. Am. Chem. Soc., 116: 5497–5498.
63. Schenker, S., Hauser, A., Dyson, R. (1996) Intersystem crossing dynamics in the iron(III) spin-crossover compounds $[Fe(acpa)_2]PF_6$ and $[Fe(Sal_2tr)]PF_6$. Inorg. Chem., 35: 4676–4682.
64. Juhasz, G., Hayami, S., Sato, O., Maeda, Y. (2002) Photo-induced spin transition for iron(III) compounds with π–π interactions. Chem. Phys. Lett., 364: 164–170.
65. Hayami, S., Kawahara, T., Maeda, Y., Inoue, K., Sato, O. (2005) Influence of the intermolecular interaction on photo-induced spin-transition of $[Fe(R\text{-}pap)_2]X$. J. Rad. Nucl. Chem., 266: 521–525.
66. Hayami, S., Hiki, K., Kawahara, T., Maeda, Y., Urakami, D., Inoue, K., Ohama, M., Kawata, S., Sato, O. (2009) Photo-induced spin transition of iron(III) compounds with π–π intermolecular interactions. Chem. Eur. J., 15: 3497–3508.
67. Takahashi, K., Cui, H.-B., Kobayashi, H., Einaga, Y., Sato, O. (2005) The light-induced excited spin state trapping effect on $Ni(dmit)_2$ salt with an Fe(III) spin-crossover cation: $[Fe(qsal)_2][Ni(dmit)_2]{\cdot}2CH_3CN$. Chem. Lett., 34: 1240–1241.
68. Takahashi, K., Cui, H.-B., Okano, Y., Kobayashi, H., Mori, H., Tajima, H., Einaga, Y., Sato, O. (2009) Evidence of the chemical uniaxial strain effect on electrical conductivity in the spin-crossover conducting molecular system: $[Fe^{III}(qnal)_2][Pd(dmit)_2]_5{\cdot}$acetone. J. Am. Chem. Soc., 130: 6688–6689.

69. Clemente-León, M., Coronado, E., López-Jordà, M., Desplanches, C., Asthana, S., Wang, H., Létard, J.-F. (2011) A hybrid magnet with coexistence of ferromagnetism and photoinduced Fe(III) spin-crossover. Chem. Sci., 2: 1121–1127.
70. Real, J. A., Bolvin, H., Bousseksou, A., Dworkin, A., Kahn, O., Varret, F., Zarembowitch, J. (1992) Two-step spin crossover in the new dinuclear compound $[Fe(bt)(NCS)_2]_2bpym$, with bt = 2,2′-bi-2-thiazoline and bpym = 2,2′-bipyrimidine: experimental investigation and theoretical approach. J. Am. Chem. Soc., 114: 4650–4658.
71. Linares, J., Nasser, J., Bousseksou, A., Boukkheddaden, K., Varret, F. (1995) Monte Carlo simulations of spin-crossover transitions using the two level model. II: Binuclear case. J. Magn. Magn. Mater., 140: 1503–1504.
72. Real, J. A., Castro, I., Bousseksou, A., Verdaguer, M., Burriel, R., Castro, M., Linares, J., Varret, F. (1997) Spin crossover in the 2,2′-bipyrimidine- (bpym-) bridged iron(II) complexes $[Fe(L)(NCX)_2]_2(bpym)$ (L = 2,2′-bithiazoline (bt) and bpym; X = S, Se). X-ray absorption spectroscopy, magnetic susceptibility, calorimetric, and Mössbauer Spectroscopy Studies. Inorg. Chem., 36: 455–464.
73. Real, J. A., Zarembowitch, J., Kahn, O., Solans, X. (1987) Magnetic interaction and spin transition in iron(II) dinuclear compounds. Crystal structure of (μ-2,2′-bipyrimidine)bis[(2,2′-bipyrimidine)bis(thiocyanato)iron(II)]. Inorg. Chem., 26: 2939–2943.
74. Létard, J.-F., Real, J. A., Moliner, N., Gaspar, A. B., Capes, L., Cador, O., Kahn, O. (1999) Light induced excited pair spin state in an iron(II) binuclear spin-crossover compound. J. Am. Chem. Soc., 121: 10630–10631.
75. Chastanet, G., Gaspar, A. B., Real, J. A., Létard, J.-F. (2001) Photo-switching spin pairs – synergy between LIESST effect and magnetic interaction in an iron(II) binuclear spin-crossover compound. Chem. Commun., 819–820.
76. Chastanet, G., Carbonera, C., Mingotaud, C., Létard, J.-F. (2004) Atypical photomagnetic properties in a series of binuclear iron(II) spin crossover complexes. J. Mater. Chem., 14: 3516–3523.
77. Chastanet, G. (2002) Ph.D. Thesis, CNRS, Bordeaux, France.
78. Ould Moussa, N., Molnár, G., Bonhommeau, S., Zwick, A., Mouri, S., Tanaka, K., Real, J. A., Bousseksou, A. (2005) Selective photoswitching of the binuclear spin crossover compound $\{[Fe(bt)(NCS)_2]_2(bpm)\}$ into two distinct macroscopic phases. Phys. Rev. Lett., 94: 107205/1–107205/4.
79. Ould Moussa, N., Trzop, E., Mouri, S., Zein, S., Molnár, G., Gaspar, A. B., Collet, E., Buron Le Cointe, M., Real, J. A., Borshch, S., Tanaka, K., Cailleau, H., Bousseksou, A. (2007) Wavelength selective light-induced magnetic effects in the binuclear spin crossover compound $\{[Fe(bt)(NCS)_2]_2(bpym)\}$. Phys. Rev. B, 75: 054101/1–054101/8.
80. Ould Moussa, N., Mouri, S., Molnár, G., Tanaka, K., Real, J. A., Bousseksou, A. (2008) On the photomagnetic properties of the binuclear spin crossover complexes $\{[Fe(bt)(NCSe)_2]_2(bpym)\}$ and $\{[Fe(bpym)(NCSe)_2]_2(bpym)\}$. J. Inorg. Organomet. Polym., 18: 195–200.
81. Mouri, S., Ould Moussa, N., Molnár, G., Real, J. A., Gaspar, A. B., Bousseksou, A., Tanaka, K. (2008) Light induced excited spin state trapping in the binuclear spin crossover compound $[Fe(bpym)(NCS)_2]_2(bpym)$ exhibiting a high-spin ground state. Chem. Phys. Lett., 456: 215–219.
82. Létard, J.-F., Carbonera, C., Real, J. A., Kawata, S., Kaizaki, S. (2009) Photomagnetism of a series of dinuclear iron(II) complexes. Chem. Eur. J., 15: 4146–4155.
83. Nakano, K., Suemura, N., Kawata, S., Fuyuhiro, A., Yagi, T., Nasu, S., Morimoto, S., Kaizaki, S. (2004) Magnetic behavior and Mössbauer spectra of spin-crossover pyrazolate bridged dinuclear diiron(II) complexes: X-ray structures of high-spin and low-spin $[\{Fe(NCBH_3)(py)\}_2(\mu\text{-bpypz})_2]$. Dalton Trans., 7: 982–988.
84. Nakano, K., Kawata, S., Fuyuhiro, A., Yagi, T., Yoneda, K., Nasu, S., Morimoto, S., Kaizaki, S. (2004) Direct two-step spin-crossover through [HS–HS]... [LS–LS] at the plateau in dinuclear diiron(II) complex $[\{Fe(NCBH_3)(4phpy)\}_2(\mu\text{-bpypz})_2]$. Chem. Commun., 2892–2893.
85. Nakano, K., Suemura, N., Yoneda, K., Kawata, S., Kaizaki, S. (2005) Substituent effect of the coordinated pyridine in a series of pyrazolato bridged dinuclear diiron(II) complexes on the spin-crossover behavior. Dalton Trans., 740–743.
86. Yoneda, K., Adachi, K., Hayami, S., Maeda, Y., Katada, M., Fuyuhiro, A., Kawata, S., Kaizaki, S. (2006) A steep one-step [HS–HS] to [LS–LS] spin transition in a 4,4′-bipyridine linked one-dimensional coordination polymer constructed from a pyrazolato bridged Fe(II) dimer. Chem. Commun., 45–47.
87. Amoore, J. J. M., Neville, S. M., Moubaraki, B., Iremonger, S. S., Murray, K. S., Létard, J.-F., Kepert, C. J. (2010) Thermal- and light-induced spin crossover in a guest-dependent dinuclear iron(II) system. Chem. Eur. J., 16: 1973–1982.

88. Fedaoui, D., Bouhadja, Y., Kaiba, A., Guionneau, P., Létard, J.-.F, Rosa, P. (2008) Complexation of 2,6-bis(3-pyrazolyl)pyridine-bis(thiocyanato)iron(II) with a bridging 4,4′-bipyridine: a new example of a dinuclear spin crossover complex. Eur. J. Inorg. Chem., 1022–1026.
89. Bhattacharjee, A., Ksenofontov, V., Kitchen, J. A., White, N. G., Brooker, S., Gütlich, P. (2008) Effect of pressure and light on the spin transition behavior of the dinuclear iron(II) compound $[Fe^{II}_2(PMAT)_2](BF_4)_4 \cdot DMF$. Appl. Phys. Lett., 92: 174104/1–174104/3.
90. Vos, G., Le Fêbre, R. A., de Graaff, R. A. G., Haasnoot, J. G., Reedijk, J. (1983) Unique high-spin–low-spin transition of the central ion in a linear, trinuclear iron(II) triazole compound. J. Am. Chem. Soc., 105: 1682–1683.
91. Thomann, M., Kahn, O., Guilhem, J., Varret, F. (1994) Spin conversion *vs.* antiferromagnetic interaction in iron(i1) trinuclear species. crystal structures and magnetic properties of $[Fe_3(p\text{-MeOptrz})_8(H_2O)_4](BF_4)_6$ and $[Fe_3(p\text{-MeOptrz})_6(H_2O)_6](tos)_6$ [*p*-MeOptrz = 4-(*p*-methoxyphenyl)-1,2,4-triazole, tos = tosylate]. Inorg. Chem., 33: 6029–6037.
92. Kolnaar, J. J. A., van Dijk, G., Kooijman, H., Spek, A. L., Ksenofontov, V. G., Gütlich, P., Haasnoot, J. G., Reedijk, J. (1997) Synthesis, structure, magnetic behavior, and Mössbauer spectroscopy of two new iron(II) spin-transition compounds with the ligand 4-isopropyl-1,2,4-triazole. X-ray structure of $[Fe_3(4\text{-isopropyl-1,2,4-triazole})_6(H_2O)_6](tosylate)_6 \cdot 2H_2O$. Inorg. Chem., 36: 2433–2440.
93. Garcia, Y., Guionneau, P., Bravic, G., Chasseau, D., Howard, J. A. K., Kahn, O., Ksenofontov, V., Reiman, S., Gütlich, P. (2000) Synthesis, crystal structure, magnetic properties and ^{57}Fe Mössbauer spectroscopy of the new trinuclear $[Fe_3(4\text{-}(2'\text{-hydroxyethyl})\text{-1,2,4-triazole})_6(H_2O)_6](CF_3SO_3)_6$ spin crossover compound. Eur. J. Inorg. Chem., 7: 1531–1538.
94. Psomas, G., Bréfuel, N., Dahan, F., Tuchagues, J.-P. (2004) An unprecedented trinuclear structure involving two high-spin and one spin-crossover iron(II) centers. Inorg. Chem., 43: 4590–4594.
95. Savard, D., Cook, C., Enright, G. D., Korobkov, I., Burchell, T. J., Murugesu, M. (2011) Gradual spin crossover behaviour in a linear trinuclear Fe^{II} complex. Cryst Eng Comm, 13: 5190–5197.
96. Schneider, C. J., Moubaraki, B., Cashion, J. D., Turner, D. R., Leita, B. A., Batten, S. R., Murray, K. S. (2011) Spin crossover in di-, tri- and tetranuclear, mixed-ligand tris(pyrazolyl)methane iron(II) complexes. Dalton Trans., 40: 6939–6951.
97. Breuning, E., Ruben, M., Lehn, J.-M., Renz, F., Garcia, Y., Ksenofontov, V., Gütlich, P., Wegelius, E., Rissanen, K. (2000) Spin crossover in a supramolecular Fe_4^{II} [2 × 2] grid triggered by temperature, pressure, and light. Angew. Chem. Int. Ed., 39: 2504–2507.
98. Nihei, M., Ui, M., Yokota, M., Han, L., Maeda, A., Kishida, H., Okamoto, H., Oshio, H. (2005) Two-step spin conversion in a cyanide-bridged ferrous square. Angew. Chem. Int. Ed., 44: 6484–6487.
99. Nishihara, T., Nihei, M., Oshio, H., Tanaka, K. (2009) The light-induced spin transition of tetranuclear spincrossover complex $[Fe_4(CN)_4(bpy)_4(tpa)_2](PF_6)_4$. J. Phys. Conf. Ser., 148: 012033/1–012033/4.
100. Newton, G. N., Nihei, M., Oshio, H. (2011) Cyanide-bridged molecular squares – the building units of Prussian blue. Eur. J. Inorg. Chem., 3031–3042.
101. Ruben, M., Breuning, E., Lehn, J.-M., Ksenofontov, V., Renz, F., Gütlich, P., Vaughan, G. B. M. (2003) Supramolecular spintronic devices: spin transitions and magnetostructural correlations in $[Fe_4^{II}L_4]^{8+}$ [2 × 2]-grid-type complexes. Chem. Eur. J., 9: 4422–4429.
102. Schneider, B., Demeshko, S., Dechert, S., Meyer, F. (2010) A double-switching multistable Fe_4 grid complex with stepwise spin-crossover and redox transitions. Angew. Chem. Int. Ed., 49: 9274–9277.
103. Wu, D.-Y., Sato, O., Einaga, Y., Duan, C.-Y. (2009) A spin-crossover cluster of iron(II) exhibiting a mixed-spin structure and synergy between spin transition and magnetic interaction. Angew. Chem. Int. Ed., 48: 1475–1478.
104. Wei, R.-J., Huo, Q., Tao, J., Huang, R.-B., Zheng, L.-S. (2011) Spin-crossover Fe^{II}_4 squares: two-step complete spin transition andreversible single-crystal-to-single-crystal transformation. Angew. Chem. Int. Ed., 50: 8940–8943.
105. Duriska, M. B., Neville, S. M., Moubaraki, B., Cashion, J. D., Halder, G. J., Chapman, K. W., Baldé, C., Létard, J.-F., Murray, K. S., Kepert, C. J., Batten, S. R. (2009) A nanoscale molecular switch triggered by thermal, light, and guest perturbation. Angew. Chem. Int. Ed., 48: 2549–2552.
106. Garcia, Y., Niel, V., Muñoz, M. C., Real, J. A. (2004) Spin crossover in 1D, 2D and 3D polymeric Fe(II) networks, in: Gütlich, P., Goodwin, H. A. (Eds) Spin Crossover in Transition Metal Compounds I. Top. Curr. Chem., 233: 229–257.

107. Real, J. A., Gaspar, A. B., Niel, V., Muñoz, M. C. (2003) Communication between iron(II) building blocks in cooperative spin transition phenomena. Coord. Chem. Rev., 236: 121–141.
108. Muñoz, M. C., Real, J. A. (2011) Thermo-, piezo-, photo- and chemo-switchable spin crossover iron(II)-metallocyanate based coordination polymers. Coord. Chem. Rev., 255: 2068–2093.
109. Aromí, G., Barrios, L. A., Roubeau, O., Gamez, P. (2011) Triazoles and tetrazoles: prime ligands to generate remarkable coordination materials. Coord. Chem. Rev., 255: 485–546.
110. Moritomo, Y., Isobe, Y., Liu, X. J., Kawamoto, T., Nakamoto, A., Kojima, N., Kato, K., Takata, M. (2004) Dynamical phase transition under photo-excitation in a spin-crossover complex. J. Luminesc., 108: 229–232.
111. Nakamoto, A., Ono, Y., Kojima, N., Matsumura, D., Yokoyama, T., Liu, X. J., Moritomo, Y. (2003) Spin transition and its photo-induced effect in spin crossover complex film based on $[Fe(II)(trz)_3]$. Synth. Met., 137: 1219–1220.
112. Niel, V., Thompson, A. L., Goeta, A. E., Enachescu, C., Hauser, A., Galet, A., Muñoz, M. C., Real, J. A. (2005) Thermal- and photoinduced spin-state switching in an unprecedented three-dimensional bimetallic coordination polymer. Chem. Eur. J., 11: 2047–2060.
113. Agustí, G., Cobo, S., Gaspar, A. B., Molnár, G., Ould Moussa, N., Szilágyi, P. A., Pálfi, V., Vieu, C., Muñoz, M. C., Real, J. A., Bousseksou, A. (2008) Thermal and light-induced spin crossover phenomena in new 3D Hofmann-like microporous metalorganic frameworks produced as bulk materials and nanopatterned thin films. Chem. Mater., 20: 6721–6732.
114. Haasnoot, J. G. (2000) Mononuclear, oligonuclear and polynuclear metal coordination compounds with 1,2,4-triazole derivatives as ligands. Coord. Chem. Rev., 200–202: 131–185.
115. Desaix, A., Roubeau, O., Jeftić, J., Haasnoot, J. G., Boukheddaden, K., Codjovi, E., Linarès, J., Noguès, M., Varret, F. (1998) Light-induced bistability in spin transition solids leading to thermal and optical hysteresis. Eur. Phys. J. B, 6: 183–193.
116. Absmeier, A., Bartel, M., Carbonera, C., Jameson, G. N. L., Weinberger, P., Caneschi, A., Mereiter, K., Létard, J.-F., Linert, W. (2006) Both spacer length and parity influence the thermal and light-induced properties of iron(II) α,ω-bis(tetrazole-1-yl)alkane coordination polymers. Chem. Eur. J., 12: 2235–2243.
117. Neville, S. M., Etrillard, C., Asthana, S., Létard, J.-F. (2010) Light-induced stored information in nanoparticles. Eur. J. Inorg. Chem., 282–288.
118. Lebris, R., Mathonière, C., Létard, J.-F (2006) Cooperative relaxation of the metastable states in the photomagnetic octacyanotungstate $Cs^{I}[\{Co^{II}(3\text{-CN-py})_2\}\{W^{V}(CN)_8\}]\cdot H_2O$. Chem. Phys. Lett., 426: 380–386.
119. Li, D., Clérac, R., Roubeau, O., Harté, E., Mathonière, C., Le Bris, R., Holmes, S. M. (2008) Magnetic and optical bistability driven by thermally and photoinduced intramolecular electron transfer in a molecular cobalt−iron Prussian blue analogue. J. Am. Chem. Soc., 130: 252–258.

20

Piezo- and Photo-Crystallography Applied to Spin-Crossover Materials

Philippe Guionneau[1] and Eric Collet[2]

[1] *CNRS, Université de Bordeaux, France*
[2] *Institut de Physique de Rennes, Université de Rennes 1-CNRS, France*

Dedicated to the memory of Andrés E. Goeta.

20.1 Introduction

Spin-crossover (SCO) materials are multistable systems which can be switched between different electronic states by using external stimuli. Almost 80 years ago, it was initially discovered that temperature is an efficient stimulus.[1] The first studies of pressure-induced SCO in solutions[2] or on solids[3] then followed about four decades later. The study of light-induced SCO started in the early 1980s with observations on solutions[4] and the discovery of the LIESST phenomenon.[5] In such systems where electronic and structural degrees of freedom are strongly coupled, crystallographic analysis undoubtedly plays a key role not only for understanding the SCO phenomenon itself but also for the subsequent rational design of SCO materials suitable for applications. The present chapter reviews recent piezo- and photo-crystallographic approaches to SCO materials highlighting the important role of structural features in the SCO process.

20.2 Spin-Crossover and Piezo-Crystallography

The SCO phenomenon is highly pressure-sensitive. Pressure-induced SCO has been observed either in solutions[2] or on solids[3, 6] and was theoretically explained four decades ago.[7] As a general matter, this phenomenon represents one of numerous potential effects of external pressure on the electronic configurations

Spin-Crossover Materials: Properties and Applications, First Edition. Edited by Malcolm A. Halcrow.

of coordination compounds, which has been well reviewed by Butler *et al.*[8] Note that, in this regard, the transformation from Fe(III) to Fe(II) is also a pressure-induced electronic effect of notable interest in the field of SCO materials, in which Fe ions play a key role. Such a possible transformation had been reported in initial investigations.[7] Pressure-induced SCO has implications for many fields as diverse as geology, biology or physics and chemistry of synthetic materials. The phenomenon was proposed early on to be relevant to the properties of the Earth's mantle[9, 10] and it is presently receiving close attention from the geochemistry and geophysics communities.[11, 12] In biology, the influence of high pressure upon biomolecules is well documented[13] and the role of SCO in haemoglobin has been thoroughly studied,[14] as has the influence of pressure on the spin state of molecules of biological interest.[15] In the research field of laboratory developed materials, piezo-SCO attracts attention not only due to its intricate fundamental aspects[16] but also because it opens up a large number of potential applications, including piezo-chromism.[17]

20.2.1 Pressure-Induced SCO: Expectation Versus Observation

The application of pressure on a SCO material is likely to favour the low spin, LS, state since the corresponding volume of the ion is smaller. Consequently, the expected basic modifications of the features of thermal SCO are as follows: with increasing pressure (i) *the temperature of SCO increases*, (ii) *the SCO becomes more gradual* and (iii) *the LS residue increases*.

These features are confirmed by considerations based on thermodynamics and/or theoretical models, which are well reviewed elsewhere.[7, 18–23] These considerations also envisage interesting phenomena, such as: (iv) *the hysteresis width decreases with increasing pressure*, (v) $T_{1/2}$ *varies linearly with pressure*, (vi) *the pressure hysteresis width is linearly linked to the thermal hysteresis width*, and (vii) *the pressure effect is reversible*.

The very first experimental observations of pressure-induced SCO, which were mainly based on Mössbauer studies, already confirmed features (i)–(iii).[6] Since then, in the last four decades, a large variety of techniques has been successively used to investigate pressure-induced SCO, including Mössbauer,[25] infrared and optical spectroscopies,[9, 26] reflectivity,[27] X-ray absorption,[28, 29] magnetometry,[30] X-ray diffraction,[31] Raman,[32] neutron diffraction[33] and inelastic X-ray scattering.[34] The interesting point is that while many investigations support expectations (i)–(vii) (Fig. 20.1), many also offer counter-examples. For instance, Mössbauer studies on $[Fe(phy)_2]X_2$ ($X = BF_4$ or ClO_4) confirm features (i)–(iii), (v) and (vii) but contradict (iv), since the hysteresis appears larger under pressure.[35]Conversely, (iv) is supported by numerous other studies based,

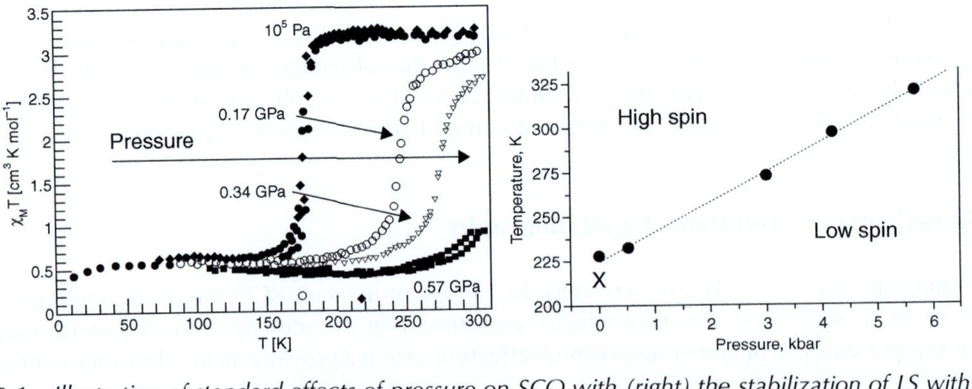

Figure 20.1 Illustration of standard effects of pressure on SCO with (right) the stabilization of LS with increasing pressure for $[Fe(phen)_2(NCS)_2]$ and (left) an example of linear variation of $T_{1/2}$ with pressure in $[Fe(sal_2$-trien)]$[Ni(dmit)_2]$. Modified from refs 16 and 24 respectively.

for instance, on optical measurements[27, 36] or magnetic measurements.[37] Elsewhere, since the reference case of $[Fe(phen)_2(NCS)_2]$,[6] expectation (v) has been confirmed in numerous investigations (Fig. 20.1)[19, 24, 38–41] but nonlinear behaviour has been reported as well in complexes that nonetheless clearly fulfil expectations (i)–(iii).[16, 37] In the same manner, while the reversibility of pressure effects, expectation (vii), is often reported,[42, 43] some nonreversible effects have also been observed.[24, 28]

All attempts to rationalize the behaviour of SCO materials under pressure using comparative analysis of data finally came to the conclusion that the large diversity of situations is almost unpredictable. For instance, analysis of X-ray absorption spectra of eight iron mononuclear or polymeric complexes showed that about half of the investigated materials fulfilled the above expectations, but the other half showed clear differences, notably on (v).[44] Similar conclusions were reached from magnetic measurements on five chemically very similar complexes belonging to the $[Fe(PM\text{-}L)_2(NCS)_2]$ family, showing that each of expectations (i)–(vii) can be fulfilled, or not, depending on the nature of the L part of the ligand.[30] Furthermore, in some cases, the application of pressure on an SCO material can even favour the HS state. This nonintuitive behaviour was observed in early Mössbauer studies in mononuclear phenanthroline[25] or poly(1-pyrazol)borate[42] iron(II) complexes, and was also reported later for the two-dimensional compound $[Fe(btr)_2(NCS)_2]\cdot H_2O$ based on magnetic measurements.[45] The list of discrepancies between expectation and observation found in the literature is long and will certainly increase with the number of high-pressure investigations.

In fact the above contradictions are only apparent. The decrease of the metal-coordination sphere volume from HS to LS is not necessarily – and is probably never – propagated in an isotropic mode all along the material. It is quite common to observe large anisotropic changes in the crystal packing that can correspond to the increase of one unit-cell parameter, as the crystal transforms from HS to LS. In other words, structural modifications under pressure at the molecular scale, at the crystal packing scale or at the microscopic level can in some cases conversely lead to a better compressibility of the HS state. That is, it is the whole compressibility of the real material in a given spin state that must be taken into account. The observed contradictions from the initial expectations (i)–(vii) are in most cases probably the result of antagonistic pressure-induced effects between the SCO phenomenon and structural modifications, including pressure-induced phase transitions. The prediction of the effect of pressure on genuine SCO materials is consequently almost impossible and experimental investigations under pressure are absolutely required. Note that in almost all the contradictory cases mentioned above, although no structural information under pressure was available, structural modifications under pressure were argued to explain the observed divergence from expectations (i)–(vii).

In addition, other questions requiring structural data under pressure have been raised by contradictory experimental observations. This is the case for the possible disappearance or occurrence of steps in the course of the SCO under pressure,[46, 47] as well as potential pressure-induced polymorphism in SCO materials.[33, 43] All the above comments emphasize the need for crystallography under high pressure, that is piezo-crystallography, in the field of SCO.

20.2.2 Piezo-Crystallography and SCO: Investigations

In general, piezo-crystallographic investigations involve a confinement of the sample in a pressure cell during the data collection. X-ray diffraction pressure cells in use are mainly Diamond Anvil Cells, namely *DAC*, that were designed about 50 years ago[48] and are being continuously improved. A *DAC* is schematically made of two diamond anvils that press a gasket enclosing the pressure cavity. The latter contains the sample, the pressure sensor and a pressure-transmitting fluid (Fig. 20.2). The pressure range and the hydrostatic character of the pressure inside the cavity depend on the nature of the *DAC* components. The most common kind of *DAC* is the *Merril–Basset* design[49] while others can also be used, such as the *Ahsbahs* type.[50] Thoroughly documented

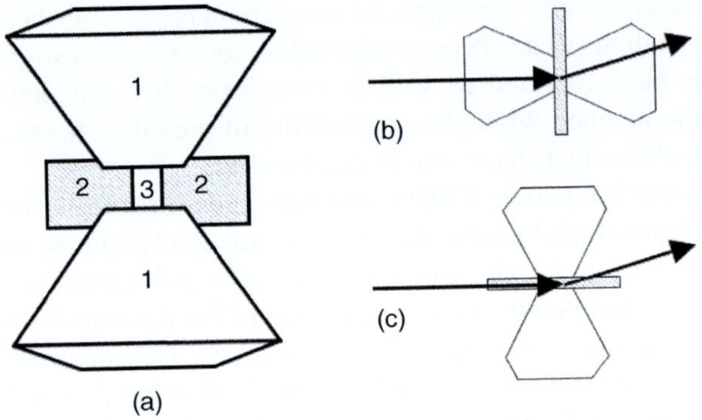

Figure 20.2 Schematic views of (a) a Diamond Anvils Cell, DAC, and its components – (1) diamonds, (2) gasket and (3) the pressure cavity and of (b) the Merril–Basset and (c) the Ahsbahs geometries, arrows figure the X-ray beam.

reviews have been recently published and must be consulted for further technical and historical details about *DAC*s and other available pressure cells like those used in neutron diffraction experiments.[51, 52] Note that single-crystal X-ray diffraction pressure cells that are simpler than *DAC*s have also been designed, although their use is still in development.[53, 54] Nowadays, access to relatively simple X-ray diffraction laboratory *DAC* investigations – notably thanks to the adaptation to two-dimensional detector experiments –[55] opens up high-pressure crystallography to a larger range of research fields. Accordingly, a promising future is expected for piezo-crystallography.[51, 56] SCO materials are concerned with this dynamic since almost all high-pressure structural data on SCO materials published so far derive from *DAC* experiments.

At the time of writing, the number of piezo-crystallography studies of SCO materials is rather limited but the situation is likely to change rapidly for the reasons stated above. Full high-pressure X-ray diffraction structural investigations, including crystal structure determinations, are currently very few and concern five mononuclear Fe(II)[31, 41, 57, 58] and one Mn(III)[59] complexes as well as one dinuclear Fe(II) compound.[60] The comprehensive pioneering work performed on the SCO compounds [Fe(phen)$_2$(NCS)$_2$] and [Fe(btz)$_2$(NCS)$_2$] showed a reference case of fulfilled expectations of the effects of high pressure on the features of SCO (see above) and proposed the first study of the corresponding structural effects.[31] The similarity of the structural modifications of the metal-coordination sphere during thermal- and pressure-induced SCO was then proved. Note that structure–property relationships for thermal SCO are reported elsewhere.[61–63] For example, the shortening of the metal–ligand bond lengths during the HS to LS conversion and the distortion of the metal-coordination sphere associated with SCO were shown to be equal whatever the stimulus. Incidentally this result was later extended to light-induced SCO through photo-crystallography investigations on the same compound.[64] In contrast, significant differences between the low temperature and the high pressure modifications of the crystal packing were reported. In addition, the different compressibilities of the HS and LS states were already quoted as an important feature for investigation, notably through the determination of the bulk modulus derived from the unit cell parameter variations. These results were later confirmed and completed by comparison of the low temperature and high pressure structural features of SCO compounds.[57, 59] It was then demonstrated that even in the absence of a phase transition, the crystal packing can be modified in different ways during the course of piezo- or thermal-SCO. A recent study of the dinuclear complex [[Fe(bpp)(NCS)$_2$]$_2$4,4′-bipy]·2MeOH, provides an example of the structural origin of the difference between the thermal- and piezo- SCO features.[60] In this material, the piezo-SCO corresponds to a complete, gradual SCO over a range of 20 kbar while the thermal-SCO corresponds to a partial, abrupt SCO below 120 K. This discrepancy has been attributed to a purely structural phase transition occurring at 161 K that is disfavoured

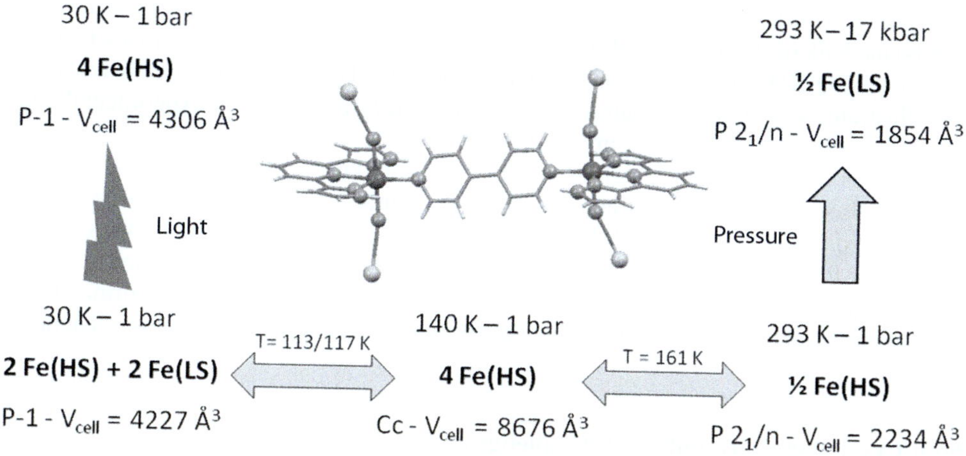

Figure 20.3 *Temperature–pressure–light phase diagram of [[Fe(bpp)(NCS)$_2$]$_2$4,4′-bipy]·2MeOH derived from X-ray diffraction investigations showing the structural differences associated to the thermal- and the piezo-SCO. Thermodynamic conditions for each phase are reported as well as the unit-cell volume (V_{cell}), the space group and, in bold, the number of iron atoms in the asymmetric unit and the corresponding spin state. This figure has been drawn from a compilation of published data.*[60, 65, 66]

by high pressure (Fig. 20.3). In contrast, a structural study of [Fe(bapbpy)(NCS)$_2$] has revealed a similar behaviour between thermal- and piezo-SCO.[41] This mononuclear compound undergoes an unusual stepped first-order transition between HS and LS at high pressure or at low temperature. The step in fact corresponds to the formation of an intermediate phase characterized by a tripling of the crystallographic *c*-axis that occurs at low temperature and at high pressure. Elsewhere, although piezo-crystallography was the only possible approach to determine the exact nature of the pressure-induced step in this case, the value of combining Raman spectroscopy and X-ray diffraction experiments under pressure has been also demonstrated through the determination of the temperature–pressure phase diagram.

Last, high-pressure single-crystal X-ray diffraction on the iron(II) complex [Fe(dpp)$_2$(NCS)$_2$]·py has revealed an antagonism between the structural modifications due to high pressure on one part and SCO on the other.[58] The shape of the complex can be schematically represented as scissor-like. Thermal-SCO corresponds to a closing while the application of high pressure results in opening of the scissors. As a consequence of this scissor-like mechanism, the material shows a Negative Linear Compression, that is the material expands in one direction, and the SCO is suppressed, under pressure. This piezo-crystallography study has shown the antagonistic structural effects of SCO and high pressure in this compound, which was already known for its thermal behaviour.[67] Hence, it has revealed the relevance of this SCO compound to the field of NLC materials. In general, as illustrated in this last example, structural high-pressure investigations are likely to open new potentialities of SCO materials.

The sole determination of the unit-cell parameter pressure-dependence can also provide interesting information on SCO materials. For instance, high-pressure neutron diffraction on two [Fe(PM-L)$_2$(NCS)$_2$] complexes has confirmed pressure-induced SCO at room temperature above 0.6 GPa in [Fe(PM-AzA)$_2$(NCS)$_2$], and has revealed a pressure-induced transition between two well known polymorphs of [Fe(PM-BiA)$_2$(NCS)$_2$] at 0.75 GPa.[33] The latter result appeared consistent with the change in magnetic behaviour observed above 0.7 GPa from high-pressure SQUID measurements,[30] but is contradicted by more recent diffuse reflectivity measurements under high pressure.[68] The latter confirmed the pressure-induced phase transition, but rules out its assignment as the previously described polymorphism. The origin of this contradiction is still unclear, but the differences between the techniques must be borne in mind. Neutron diffraction and SQUID measurements essentially probe the bulk of the material while diffuse reflectivity probes the surface. Elsewhere, hydrostatic conditions inside the SQUID, and the neutron diffraction and optical pressure cells, are obviously not similar. This example illustrates why it is important to compare data taken in equivalent experimental conditions in general and in particular for high-pressure investigations.

Elsewhere, neutron diffraction under high pressure on the well known $[Fe(ptz)_6](BF_4)_2$ complex demonstrated for the first time that the spin-conversion and the structural order–disorder phase transition from $R\bar{3}$ to $P\bar{1}$ are in fact two distinct phenomena:[69, 70] at high pressure the first-order symmetry breaking is followed by spin-crossover at lower temperature. Apart from revealing the pressure–temperature phase diagram of this material, already identified as very intricate in isobaric conditions,[71] this study confirmed the interest in getting high pressure structural data at low temperatures, which is one of the remaining challenges in the field.

20.2.3 Piezo-Crystallography and SCO: Challenges

Considering the investigation of external pressure effects on SCO materials and notably through piezo-crystallography, the actual challenges can be summarized in three points: first, adding more data by using *DAC*s on a large diversity of SCO systems; second, expanding the investigated field to a wider definition of the concept of pressure; and third, applying pressure to obtain new crystalline materials. The first point was addressed a long time ago.[16, 30] It is currently on course and a rapid growth of available data is predicted. However, further experimental challenges remain, including getting reliable structural data from combined high-pressure, low-temperature and light-irradiation experiments. The interest in these investigations has been highlighted recently by magnetic measurements under combined high-pressure and light-irradiation conditions.[72] This challenge should also include experimental measurements of bulk moduli for SCO materials, which can simply be achieved by careful unit-cell parameter measurements. Access to these values represents an actual lack in the SCO field, even though compressibility was pointed out as a crucial point in the early days of high-pressure SCO materials studies.[7] Towards this aim, high-pressure powder X-ray diffraction may be used, taking advantage of all the recent developments.[73]

The second point concerns a multitude of unexplored routes in the field of SCO. It includes the effects of "small" pressures, <1000 bar, which are not covered by *DAC*s but are probably of key importance for applications; negative pressures also deserve to be explored. The investigation of dynamic pressure, that is. shocks, on SCO materials represents another open field, being almost unexplored in the literature even though mechanical-stress effects have been already mentioned.[74] In addition, dynamical pressure effects generated by pulsed laser irradiation are discussed in Section 20.3.5 with time-resolved X-ray studies. Note that "One-Finger-Push" piezo-chromism has already been reported in the field of inorganic materials,[75] and molecular SCO materials may be good candidates for presenting similar properties. The third point means that high pressure should be employed as a tool to obtain new SCO materials, as it is commonly used in other research fields like pharmaceutical compounds[76] or functional small molecules.[77] This can be achieved through pressure-induced polymorphism; that is, by obtaining new structural phases of already known compounds, or by crystallizing new materials that cannot be obtained at ambient pressure. Without doubt, piezo-crystallography will play a key role in all these coming challenges for the SCO phenomenon.

20.3 Crystallography of Photoexcited SCO Materials

Photo-crystallography is a new frontier in the field of structural science,[78–80] promising direct access to structural rearrangements driven by laser irradiation toward transient, permanent or photo-steady states. In materials, real photoinduced transitions can take place,[81] representing a fascinating perspective to manipulate the macroscopic physical properties of a material with light, as is the case for many SCO systems. These out-of-equilibrium phenomena are different in nature from processes induced at thermal equilibrium by external fields, and in some cases light irradiation allows new states to be reached that are unattainable at thermal equilibrium, as demonstrated in SCO compounds.[82, 83] Photoinduced phase transition processes can also involve phenomena such as phase nucleation and hysteretic behaviour. It is of fundamental interest to understand the mechanisms driving such photoswitching at the macroscopic and microscopic scales, in order to design future materials.

X-ray diffraction gives access to key structural (and electronic) information, as well as to the nature and mechanisms of photoinduced transitions from the molecular to the macroscopic scales. Slow time-resolved experiments can reveal the kinetics of phase separation processes, for example.[84–88] Recent progress in ultrafast X-ray science[89] has delivered a hitherto elusive temporal dimension to atomic and molecular motions in condensed matter. It is now possible to record "molecular movies" during the transformation of matter initiated by the light pulse. Laser pump and X-ray probe techniques now provide an outstanding opportunity for the direct observation of photoinduced processes in SCO systems.[90]

The following review, based on recent investigations, illustrates that photocrystallography experiments are not merely data collections under laser irradiation. Accordingly, it is shown that through varying different external parameters (such as temperature, laser irradiation wavelength or power) and careful investigation of the structural signatures, key information on the different processes can be obtained. For instance, it has been observed that for some systems the nature of the photoinduced state can be triggered by choosing the appropriate excitation wavelength, or even that transitions between several excited states may occur. Furthermore, extension of this work into the 100 ps time domain[91, 92] using a time-resolved technique around synchrotron sources is presented.

20.3.1 Photo-Crystallography of SCO: Probing the Change of Molecular Structure

The photoswitching observed in many SCO materials[93] is a prototype of bistability and cooperative transformation in molecular solids. Diffraction techniques make it possible to understand how electronic and structural degrees of freedom are interconnected, and more importantly to understand how photoinduced states can be stabilized.[94] The recent construction of He cryostats with coupled laser excitation has allowed the development of photo-crystallography of SCO compounds.[64, 65, 71, 84, 94–100]

A nice illustration of the information one can obtain from detailed analysis of the molecular structures of photoinduced states is provided by the comparison of the structural changes in the two polymorphs of $[Fe(PM\text{-}BiA)_2(NCS)_2]$, denoted BIA I and BIA II,[94] whose *T*(LIESST) values are quite different (78 K for BIA I and 35 K for BIA II). The evolution of structural parameters at the intramolecular level during thermal conversion, and after photoexcitation at low temperature, are shown in Figure 20.4. In addition to the well-known average Fe–N bond length (Fig. 20.4a), the anisotropic deformation of the coordination sphere around the Fe atom is taken into account via the deformation parameters ζ (Fig. 20.4b) and Σ (Fig. 20.4c). The Σ parameter is the sum of the absolute deviation of the 12 *cis* N–Fe–N angles from 90° and the ζ parameter is the sum of the absolute deviation of the 6 Fe–N bond-lengths from the average <Fe–N> value. Both compounds show the typical increase of 0.2 Å of <Fe–N> between the pure thermal and photoinduced LS and HS states.

The key point here is that the octahedral deformation is stronger in BIA I than in BIA II. In a two-mode description of the site potential, including Fe–N elongation and deformation coordinates, a larger deformation results in an increase of the energy barrier between the HS and LS states (Fig. 20.4d). This deformation of the octahedron results from intermolecular interactions, which are also stronger in BIA I than in BIA II, and are therefore responsible for the stabilization of the photoinduced HS state.

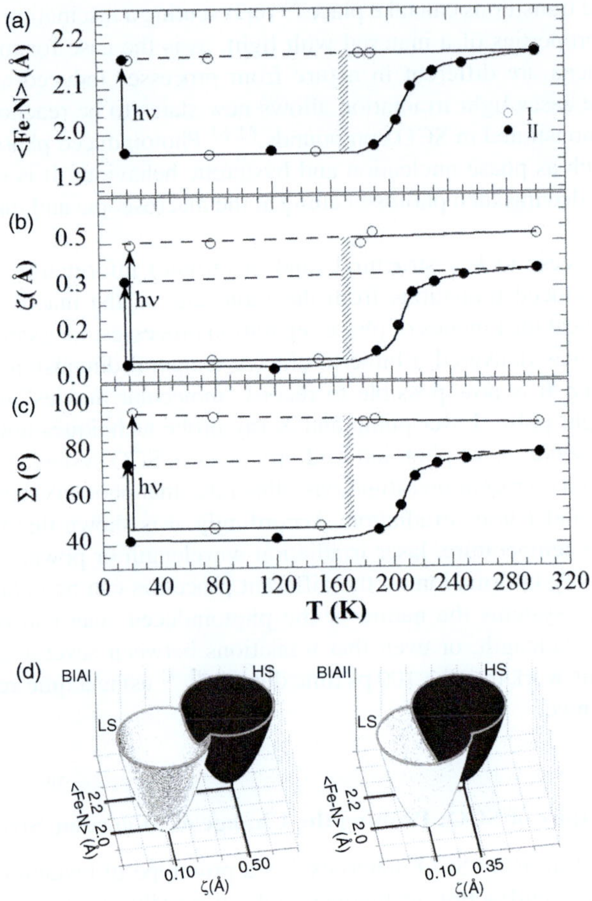

Figure 20.4 *Temperature dependence of (a) the average Fe–N lengths (Å), (b) the parameter* ζ *(Å) and (c) the parameter* Σ*. (d) Two-mode description of on-site molecular potentials for BIA I and BIA II as a function of the average <Fe–N> bond length (Å) and of the length distortion parameter. Modified from ref. 94.*

High-resolution X-ray diffraction measurements make it possible to track electronic redistribution in SCO compounds. Experimental analysis of the change of electron density between the ground low spin (LS) state and the metastable thermally quenched high spin (HS) state was performed for $[Fe(btr)_2(NCS)_2]\cdot H_2O$.[95] The electronic populations for the LS state are characteristic of Fe(II) ($3d^6$) with an almost $t_{2g}{}^6e_g{}^0$ electron configuration, whereas in the metastable HS state the $3d$ orbitals are more uniformly populated.

20.3.2 Light-Induced Broken Symmetry: Reaching New States by Laser Excitation

The SCO phenomenon occurs in most cases without symmetry breaking: only the respective population of HS and LS states changes. Besides, for a few systems undergoing thermally-induced stepped transition, the intermediate phases may result from long-range ordering of HS and LS states.[83, 101–106] Very recently, it was demonstrated in the SCO materials $[Fe^{II}H_2L^{2\text{-}Me}][XF_6]_2$, with X = P or As[82, 83] that light can drive symmetry breaking. Photo-crystallography experiments indicate that in the metastable photoinduced HS state (PIHS)

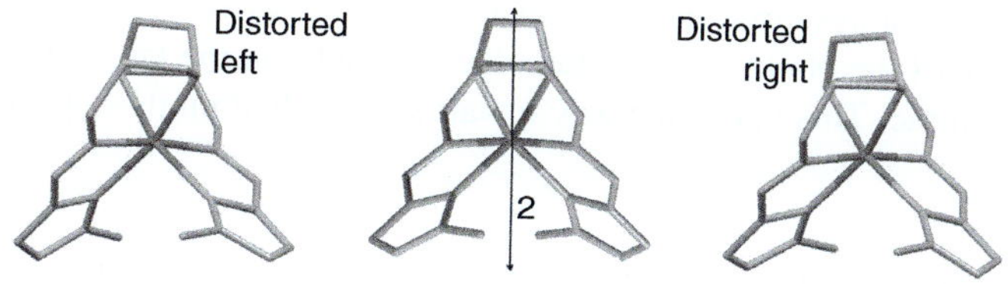

Figure 20.5 Schematic view of the $[Fe^{II}H_2L^{2\text{-}Me}]$ cation in the regular geometry with 2-fold axis, or the distorted geometry as in the PIHS state.

the crystal structure differs from the one in the HS state at room temperature (Fig. 20.5). Above 250 K, the cations are located on 2-fold axis. However, some structural multi-stability exists at low temperature and some cations may lose the 2-fold axis to reach a distorted state. A long range order of the distorted and undistorted HS states takes place only in the PIHS state, resulting in triplication of the unit cell, which can only be probed by diffraction techniques. Among the different competing false ground states existing in these multistable systems, some of them can only be reached under nonequilibrium conditions by light irradiation, as it is the case for the symmetry broken HS phase. This is one of the important aspects of the research developed in the emerging field of photoinduced phase transitions.

20.3.3 Photoswitching between Different Excited States

In addition, by using different laser-excitation wavelengths it is possible to generate different states. For binuclear systems made of two sites which may be HS and/or LS, the nature of the photoinduced state from a LS–LS state can be selectively triggered by light excitation wavelength toward HS–HS or HS–LS. In SCO systems, photoinduced states have lifetimes from several hours up to days at very low temperature. It is therefore possible to switch this photoinduced state by re-irradiating it at a different wavelength, as was achieved in the binuclear compound $[[Fe(bt)(NCS)_2]_2(bpym)]$.[96] Photo-crystallography can track the structural signatures of the LS–LS, HS–HS and HS–LS states in terms of their molecular structure, as well as the unit cell volume (Fig. 20.6). Excitation at 808 nm of the LS–LS state generates the HS–HS state. Excitation of this photoinduced HS–HS state at 1310 nm induces volume and bond-length contractions, characteristic of the HS–LS state. Conversely, once the HS–LS state is reached (after excitation at 1310 nm), further excitation of this state by irradiation at 808 nm results in volume and bond expansion, characteristic of a return to the

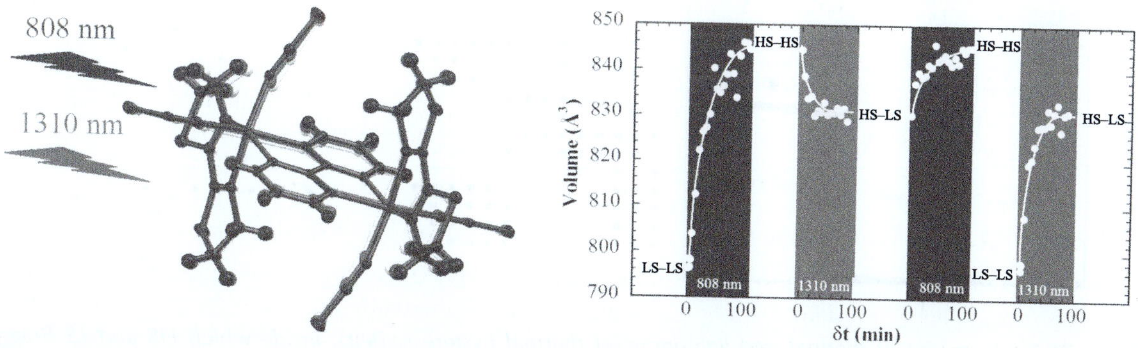

Figure 20.6 Evolution of the molecular structure and volume between the LS–LS (green) and photoinduced HS–LS (red) and HS–HS (blue) as a function of the laser wavelength. Modified from ref. 96.

HS–HS state. For such binuclear systems, photoinduced transitions between excited states can genuinely be photocontrolled. The nature of the photoinduced state depends only on the laser irradiation wavelength, not on the initial state of the sample. This opens the way towards controlled multi-stepped molecular switches.

20.3.4 Slow Phase Nucleation Dynamical Process and Hysteretic Behaviour

Furthermore, when considering the physical properties of a material, several orders of magnitude in spatial dimension have to be taken into account, compared to descriptions at the molecular level given above in Sections 20.3.2 and 20.3.3. In addition, when molecules behave in a more or less independent way, or when molecules in the crystal are coupled to a greater or lesser extent, different types of macroscopic response arise. The mechanisms of the photoinduced transformation and relaxation processes, associated with the recovery of thermal equilibrium state once the laser irradiation is switched off, are of fundamental interest. It has been shown that for weakly cooperative systems, with a homogeneous distribution of photoexcited molecules in the crystal, a gradual evolution of the Bragg peaks occurs when the system transforms.[85, 94] For more cooperative systems, the behaviour is different and a phase separation (spatially separated phases) process takes place, as evidenced by X-ray diffraction for the first time in the $[Fe(pic)_3]^{2+}$ system.[84] This phase separation process was described as indicative of Avrami behaviour, that is growth of the HS phase inside the bulk LS phase (or vice versa). Detailed investigation of the phase nucleation of such Avrami behaviour was carried out on several other compounds[86, 88, 106, 107] and direct imaging of the mechanism was reported by optical microscopy techniques.[108] As a result of the macroscopic coexistence of LS and HS domains, Bragg reflections split during the relaxation process (Fig 20.7). Because of the significant difference in their lattice parameters, the positions of Bragg reflections are modified in the diffraction pattern between the HS and LS states, and the coexistence of the two phases is associated with the coexistence of the peaks. This is observed around different types of nucleation mechanisms due to dynamic growth of macroscopic phases inside the thermal hysteresis, as well as hysteresis resulting from photoinduced effects (LIOH and LITH).[87]

Note that phase separation is a universal phenomenon observed in materials (not only SCO) undergoing first-order phase transitions. In SCO, the propagation dynamics are governed by the (elastic) energy cost of the interface between the phases, and therefore crystalline defects can play a major role in phase propagation, as illustrated by optical microscopy.[109] Photo-crystallography by means of X-ray diffraction has therefore demonstrated that similar phase separation takes place inside LITH and LIOH (Fig. 20.7).

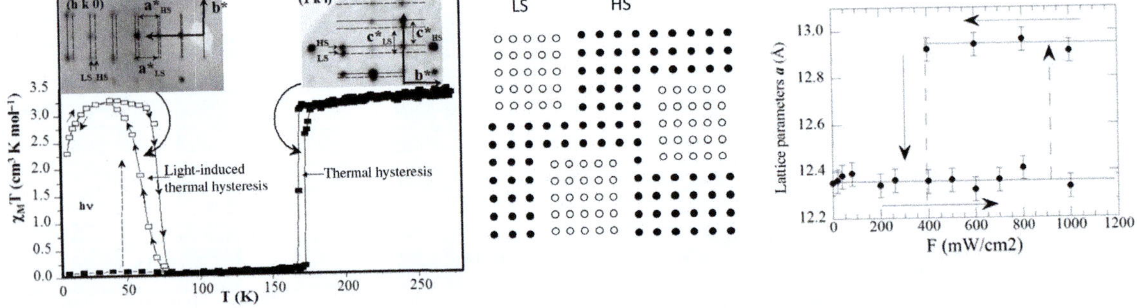

Figure 20.7 Examples of thermal and light-induced thermal hysteresis (left), inside which HS and LS Bragg peaks' coexistence results from phase separation schematically represented (middle) as well as light-induced optical hysteresis detected by evolution of lattice parameter (right). Modified from ref. 87.

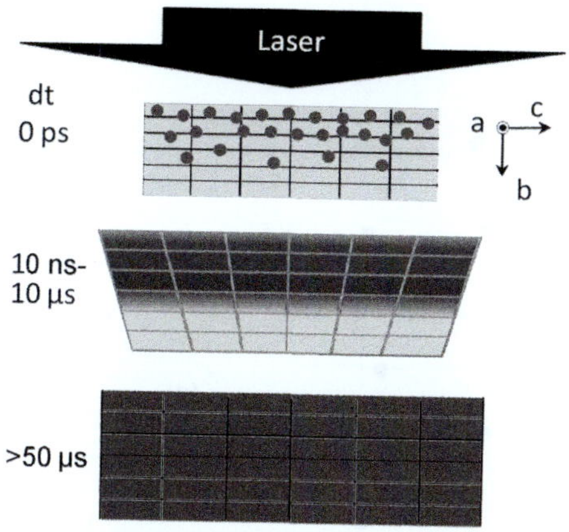

Figure 20.8 Schematic drawing of the multistep response to femtosecond laser excitation: HS molecules (black circles) generated by laser pulse in the cold (grey) lattice with mainly LS molecules, warm lattice (dark) expansion on 10 ns, thermal stabilization of HS population within μs.

20.3.5 Ultrafast Time-Resolved Crystallography of SCO Photoswitching Dynamics

Nowadays, one of the key questions in bistable materials is "what is the ultimate timescale for switching the system?" Pulsed lasers, with short ns[110, 111] or ultrashort fs[91] duration, are widely regarded as the most likely source for achieving this goal. Depending on the time resolution of the experiment, different processes can be investigated, spanning intrinsic atomic or molecular dynamics (100 fs) to macroscopic propagation or diffusion effects (μs). When the phenomena under investigation are fast, or when the excited state is transient, more sophisticated techniques are required. These are based on pump-probe methods, in which a laser flash drives a photoswitching process and a probe pulse (X-ray or electrons) visualizes the excited structure after laser excitation.[89] X-ray pulses at synchrotrons are of the order of 100 ps,[90, 112, 113] and detailed structural analyses of photoexcited structures have been performed with 100 fs or 100 ps time resolutions in SCO systems as diverse as molecules in solution[114, 115] or single crystals.[92, 116]

Recent reports, combining ultrafast optical and X-ray diffraction techniques,[91] revealed the spin state photoswitching dynamics in macroscopic crystals. These studies were performed in the solid Fe(III) compound [(TPA)Fe(TCC)]PF_6, for which a femtosecond laser flash can drive LS-to-HS photoswitching.[90, 116] The existence of four steps (Fig. 20.8) reflects a sequence of physical processes, hidden in the time domain from molecular transformation, cell deformation to macroscopic thermal crystal switching.

Step 1: Local LMCT to HS relaxation cascade, characterized by <Fe–N> elongation on a 300 fs timescale (Fig 20.9a and b). This local transformation changes the average crystal structure since the average structure factor $<F_{hkl}>$ is the weighted contribution of the HS and LS ones:

$$< \mathrm{F}_{hkl}(dt) >= \mathrm{X}_{\mathrm{HS}}(dt)\mathrm{F}_{hkl}^{\mathrm{HS}} + [1 - \mathrm{X}_{\mathrm{HS}}]\mathrm{F}_{hkl}^{\mathrm{LS}} \tag{20.1}$$

This structural change is also observed in the photodifference maps obtained by comparing the structure factor before and after laser excitation ($F_{\mathrm{obs}}{}^{dt>0} - F_{\mathrm{obs}}{}^{-1\mathrm{ns}}$). In the system studied here, this results in a shift of electron density from the atom Fe1 towards O2 (Fig. 20.9c). This fs molecular photoswitching steps occurs in a constant unit-cell volume.

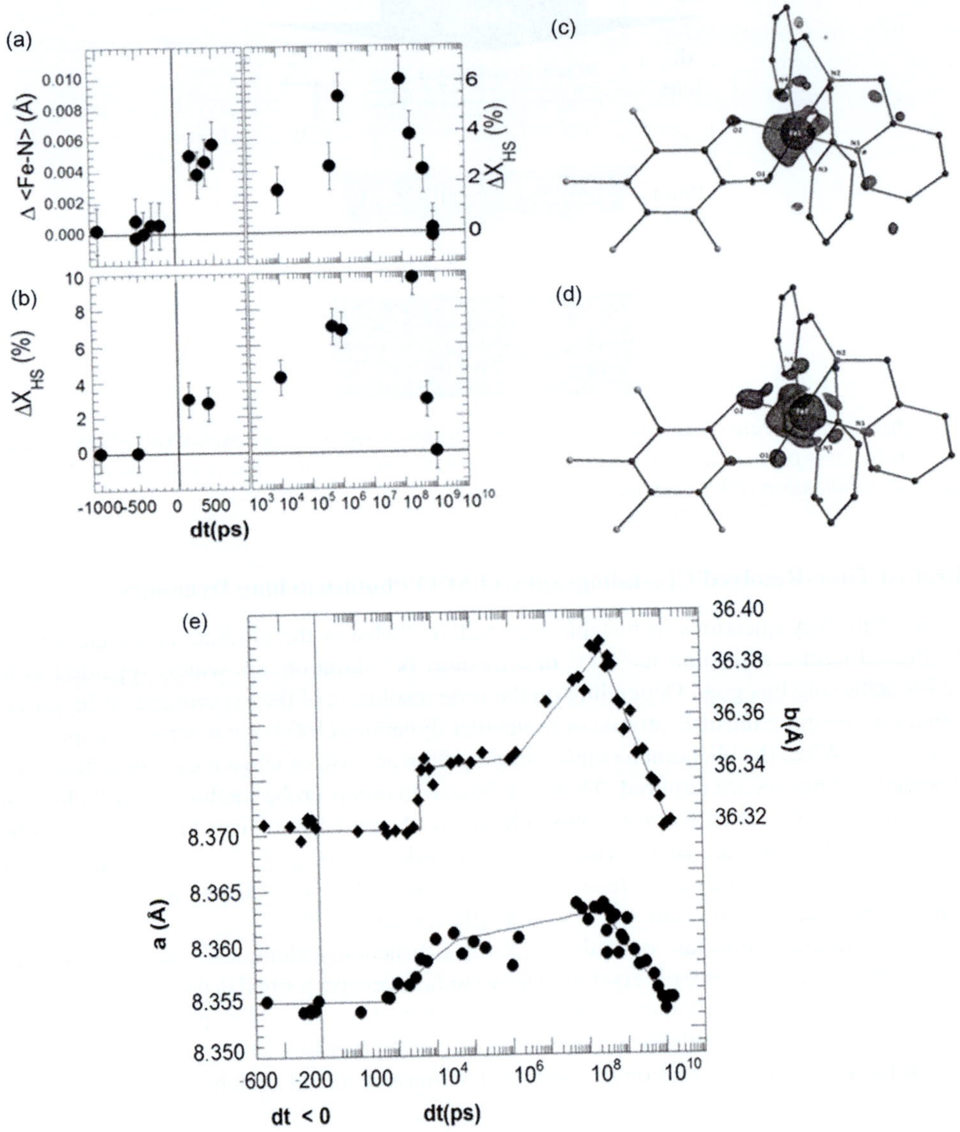

Figure 20.9 Examples of time delay dependence of <Fe–N> (a), X_{HS} (b), photodifference maps at 500 ps (c) data and 5 μs data (d) and lattice parameters (e). Modified from ref. 90.

Step 2: The structural reorganization associated with the molecular swelling and lattice heating induces crystal expansion within 1–50 ns. Because of the anisotropic sample shape, volume expansion is anisotropic and faster in the thicker dimension since it is governed by propagation effects (shock wave and heat diffusivity) (Fig. 20.9e).

Step 3: The temperature increase in the crystal corresponds to a higher fraction of HS species, thermally populated in the 100 ns–100 μs time window. This thermal-switching is again characterized by an additional increase of the <Fe–N> bond length (Fig. 20.9a) and by a larger structural shift (Fig. 20.9d). Since conversion related to heat diffusion effects is more rapidly activated when the size is small, the excitation of crystals of anisotropic shape induces anisotropic deformations in this time domain. This can be observed in deformation of the Bragg peaks in the reciprocal lattice.[92]

Step 4: Recovery to the thermal equilibrium state, limited by heat exchange with the sample environment. Typically this occurs on the ms timescale, with homogeneous relaxation in this noncooperative material.

Time-resolved diffraction is a very nice tool for directly monitoring structural dynamics of materials, during transformations induced by ultrashort laser pulses. The structural information obtained is crucial to establish the physical foundations of how to control macroscopic photoswitching in materials, notably in SCO systems. A key feature is that dynamics follow a complex pathway from the molecular to the material scales, through a sequence of processes. SCO crystals are constituted of subsystems that are different in nature with a macroscopic number of degrees of freedom (electrons, spins, phonons, molecular configuration, unit cell deformation etc.). Not only is the pathway indirect, the nature of dynamical processes along the pathway depends on the timescale.[116] This dictates which degrees of freedom are involved in the subsequent dynamics or kinetics, and which are frozen or statistically averaged. Unlike independent photochemical processes in solution, the solid medium is an active participant in light-induced switching, as illustrated here by the successive switching of silent molecules in the material.

Ultrafast time-resolved diffraction studies are fundamentally different from those of conventional diffraction studies of long-lived photoinduced HS states. Ultrafast diffraction probes different physical quantities on their intrinsic timescales, shedding new light on successive processes driving macroscopic switching in a functionalized material. Finally, regarding the transformation process at the molecular level, femtosecond photo-crystallography experiments on the next generation of X-ray sources will ultimately allow the investigation of the fs pathway already detected by fs optical spectroscopy for the real photoswitch process.[92]

Acknowledgements

Many co-workers, students and friends have helped in obtaining experimental data and developing the ideas outlined in this article. In particular, P.G. would like to thank H.J. Shepherd, G. Molnár and P. Rosa for high-pressure studies and J.-F. Létard for a long-standing collaboration on SCO systems. EC additionally thanks H. Cailleau, M. Buron, M. Lorenc, M. Servol, L. Toupet, J.-F. Létard, M.-L. Boillot, M. Wulff, N. Bréfuel, J.-P. Tuchagues and S. Koshihara for their contributions.

List of Abbreviations

AzA	4-(phenylazo)aniline
BiA	4-(aminobiphenyl)aniline
bapbpy	N-(6-(6-(pyridin-2-ylamino)pyridin-2-yl)pyridin-2-yl)pyridin-2-amine
bipy	4,4′-bipy = 4,4′-bipyridine
bpp	2,6-bis(pyrazol-3-yl)pyridine
btr	4,4′-bis-1,2,4-triazole
btz	2,2′-bi-4,5-dihydrothiazine
DAC	diamond anvil cell
dmit	2-thioxo-1,3-dithiole-4,5-dithiolato
dpp	dipyrido[3,2-a:2′3′-c]phenazine
Fepic	$[Fe(2\text{-picolylamine})_3],Cl_2,EtOH$
$Fe^{II}H_2L_2{}^{-Me}$	bis[((2-methylimidazol-4-yl)methylidene)-3-aminopropyl]ethylenediamine
F_{hkl}	Structure factor

fs	Femtosecond = 10^{-15}s
HS	high spin
LIESST	Light-Induced Excited Spin-State Trapping
LIOH	Light-Induced Optical Hysteresis
LITH	Light-Induced Thermal Hysteresis
LMCT	Ligand-metal charge transfer
LS	low spin
ns	nanosecond = 10^{-9}s
phen	1,10-phenantroline
phy	1,10-phenantroline-2-carbaldehyde phenylhydrazone
ps	Picosecond = 10^{-12}s
ptz	1-propyltetrazole
py	pyridine
PM	N-(2′-pyridylmethylene)
SCO	Spin-crossover
TCC	3,4,5,6-tetrachlorocatecholate
TPA	tris(2-pyridylmethyl)amine
T(LIESST)	temperature above which HS state thermally relaxes (heating rate of 0.3 $K.mn^{-1}$)
X_{HS}	High spin fraction
Σ	distortion parameter of the cis ϕ angles of the FeN_6 octahedron: $\sum = \sum_{i=1}^{12} \lvert 90 - \varphi_i \rvert$
ζ	distortion parameter of the Fe-N bond lengths of FeN_6 octahedron: $\zeta = \sum_{i=1}^{6} \lvert Fe - N_i - < Fe - N > \rvert$

References

1. Cambi, L., Szego, L. (1931) Über die magnetische Susceptibilität der komplexen Verbindungen. Ber. Drsch. Chem. Ges. Teil B, 64: 2591–2598.
2. Ewald, A. H., Martin, R. L., Sinn, E., White, A. H. (1969) Electronic equilibrium between the $^6A_1$ and $^2T_2$ states in iron(III) dithio chelates. Inorg. Chem., 8: 1837–1846.
3. Bargeron, C. B., Avinor, M., Drickamer, H. G. (1971) Effect of pressure on the spin state of iron(II) in manganese(IV) sulfide. Inorg. Chem., 10: 1338–1339.
4. McGarvey, J. J., Lawthers, I. (1982) Photochemically-induced perturbation of the $^1A \rightleftharpoons {}^5T$ equilibrium in Fe^{11} complexes by pulsed laser irradiation in the metal-to-ligand charge-transfer absorption band. J. Chem. Soc. Chem. Comm., 906–907.
5. Decurtins, S., Gütlich, P., Kölher, C. P., Spiering, H., Hauser, A. (1984) Light-induced excited spin state trapping in a transition-metal complex: the hexa-1-propyltetrazole-iron (II) tetrafluoroborate spin-crossover system. Chem. Phys. Lett., 105: 1–4.
6. Fisher, D. C., Drickamer, H. G. (1971) Effect of pressure on the spin state of iron in ferrous phenanthroline compounds. J. Chem. Phys., 54: 4825–4837.
7. Slichter, C. P., Drickamer, H. G. (1972) Pressure-induced electronic changes in compounds of iron. J. Chem. Phys., 56: 2142–2160.
8. Grey, J. K., Butler, I. S. (2001) Effects of high external pressures on the electronic spectra of coordination compounds. Coord. Chem. Rev., 219–221: 713–759.
9. Ferraro, J. R., Basile, L. J. (1979) High spin-low spin crossover in Co(II) complexes with tris(2-diphenyl phosphinoethyl)amine (NP_3) by application of high external pressures. Inorg. Chim. Acta, 35: 317–318.

10. Ohnishi, S. (1978) A theory of the pressure-induced high-spin—low-spin transition of transition-metal oxides. Phys. Earth Planet. Int., 17: 130–139.
11. McGammon, C., Kantor, I., Narygina, O., Rouquette, J., Ponkratz, U., Sergueev, I., Mezouar, M., Prakapenka, V., Dubrovinsky, L. (2008) Stable intermediate-spin ferrous iron in lower-mantle perovskite. Nature Geosci., 1: 684–687.
12. Fortes, A. D., Suard, E., Knight, K. S. (2011) Negative linear compressibility and massive anisotropic thermal expansion in methanol monohydrate. Science, 311: 742–746.
13. Weber, G., Drickamer, H. G. (1983) The effect of high pressure upon proteins and other biomolecules. Q. Rev. Biophys., 16: 89–112.
14. Perutz, M. F., Kilmartin, J. V., Nagai, K., Szabo, A., Simon, S. R. (1976) Influence of globin structures on the state of the heme: ferrous low spin derivatives. Biochemistry, 15: 378–387.
15. Di Primo, C., Deprez, E., Hoa, G. H. B., Douzou, P. (1995) Antagonistic effects of hydrostatic pressure and osmotic pressure on cytochrome P-450cam spin transition. Biophys. J., 68: 2056–2061.
16. Gütlich, P., Ksenofontov, V., Gaspar, A. B. (2005) Pressure effect studies on spin crossover systems. Coord. Chem. Rev., 249: 1811–1829.
17. Létard, J.-F., Guionneau, P., Goux-Capes, L. (2004) Towards spin crossover applications, in: Gütlich, P., Goodwin, H. A. (Eds.) Spin crossover in transition metal compounds III. Top. Curr. Chem., 235: 221–249.
18. Ksenofontov, V., Gaspar, A. B., Gütlich, P. (2004) Pressure effect studies on spin crossover and valence tautomeric systems, in: Gütlich, P., Goodwin, H. A. (Eds) Spin Crossover in Transition Metal Compounds III. Top. Curr. Chem., 235: 23–64
19. Gütlich, P., Gaspar, A. B., Ksenofontov, V., Garcia, Y. (2004) Pressure effect studies in molecular magnetism. J. Phys. Condens. Matter, 16: S1087–S1108.
20. Spiering, H. (2004) Elastic interaction in spin-crossover compounds, in: Gütlich, P., Goodwin, H. A. (Eds) Spin Crossover in Transition Metal Compounds III. Top. Curr. Chem., 235: 171–195.
21. Bonhommeau, S., Molnár, G., Goiran, M., Boukheddaden, K., Bousseksou, A. (2006) Theoretical study of the effect of an intense and pulsed magnetic field and a pressure pulse on the spin crossover complex $Fe(phen)_2(NCS)_2$. Phys. Rev. B, 74: 064424/1–064424/8.
22. Bousseksou, A., Molnár, G., Salmon, L., Nicolazzi, W. (2011) Molecular spin crossover phenomenon: recent achievements and prospects. Chem. Soc. Rev., 40: 3313–3335.
23. Stoleriu, L., Chakraborty, P., Hauser, A., Stancu, A., Enachescu, C. (2011) Thermal hysteresis in spin-crossover compounds studied within the mechanoelastic model and its potential application to nanoparticles. Phys. Rev. B, 84: 134102/1–134102/9.
24. Szilágyi, P. A., Dorbes, S., Molnár, G., Real, J. A., Homonnay, Z., Faulmann, C., Bousseksou, A. (2008) Temperature and pressure effects on the spin state of ferric ions in the $[Fe(sal_2\text{-trien})][Ni(dmit)_2]$ spin crossover complex. J. Phys. Chem. Solids, 69: 2681–2686.
25. Bargeron, C. B., Drickamer, H. G. (1971) Effect of pressure on the electronic structure of complexes of ferrous iron with substituted phenanthrolines. J. Chem. Phys., 7: 3471–3482.
26. Jeftic, J., Romstedt, H., Hauser, A. (1996) The interplay between the spin transition and the crystallographic phase transition in the Fe(II) spin-crossover system $[Zn_{1-x}Fe_x(ptz)_6](BF_4)_2$ ($x = 0.1, 1$; ptz = 1-propyltetrazole). J. Phys. Chem. Solids, 57: 1743–1750.
27. Jeftic, J., Menéndez, N., Wack, A., Codjovi, E., Linarès, J., Goujon, A., Hamel, G., Klotz, S., Syfosse, G., Varret, F. (1999) Helium gas pressure apparatus with optical reflectivity detection. Meas. Sci. Technol., 10: 1059–1064.
28. Roux, C., Zarembowitch, J., Itié, J. P., Polian, A., Verdaguer, M. (1996) Pressure-induced spin-state crossovers in six-coordinate $Fe^{II}L_nL'_m(NCS)_2$ complexes with $L = L'$ and $L \neq L'$: a XANES investigation. Inorg. Chem., 35: 574–580.
29. Hannay, C., Hubin-Franskin, M. J., Grandjean, F., Briois, V., Itié, J. P., Polian, A., Trofimenko, S., Long, G. J. (1997) X-ray absorption spectroscopic study of the temperature and pressure dependence of the electronic spin states in several iron(II) and cobalt(II) tris(pyrazolyl)borate complexes. Inorg. Chem., 36: 5580–5588.
30. Ksenofontov, V., Levchenko, G., Spiering, H., Gütlich, P., Létard, J.-F., Bouhedja, Y., Kahn, O. (1998) Spin crossover behavior under pressure of $Fe(PM\text{-}L)_2(NCS)_2$ compounds with substituted 2′-pyridylmethylene 4-anilino ligands. Chem. Phys. Lett., 294: 545–553.

31. Granier, T., Gallois, B., Gaultier, J., Real, J. A., Zarembowitch, J. (1993) High-pressure single-crystal X-ray diffraction study of two spin-crossover iron(II) complexes: $Fe(Phen)_2(NCS)_2$ and $Fe(Btz)_2(NCS)_2$. Inorg. Chem., 32: 5305–5312.
32. Molnár, G., Niel, V., Real, J. A., Dubrovinsky, L., Bousseksou, A., McGarvey, J. J. (2003) Raman spectroscopic dtusy of pressure effects on the spin-crossover coordination polymers Fe(pyrazine)$[M(CN)_4]\cdot 2H_2O$ (M = Ni, Pd, Pt). First observation of a piezo-hysteresis loop at room temperature. J. Phys. Chem. B, 107: 3149–3155.
33. Legrand, V., Le Gac, F., Guionneau, P., Létard, J.-F. (2008) Neutron powder diffraction studies of two spin transition Fe^{II} complexes under pressure. J. Appl. Cryst., 41: 637–640.
34. Antonangeli, D., Siebert, J., Aracne, C. M., Farber, D. L., Bosak, A., Hoesch, M., Krisch, M., Ryerson, F. J., Fiquet, G., Badro, J. (2011) Spin crossover in ferropericlase at high pressure: a seismologically transparent transition? Science, 331: 64–67.
35. König, E., Ritter, G., Waigel, J., Goodwin, H. A. (1985) The effect of pressure on the thermal hysteresis of the first-order spin transition in bis(1,10-phenanthroline-2-carbaldehyde phenylhydrazone) iron (II) complexes. J. Chem. Phys., 83: 3055–3061.
36. Codjovi, E., Menéndez, N., Jeftic, J., Varret, F. (2001) Pressure and temperature hysteresis in the spin-transition solid $Fe(btr)_2(NCS)_2.H_2O$, pure and diluted in Ni matrix. C. R. Chimie, 4: 181–188.
37. Galet, A., Gaspar, A. B., Agustí, G., Muñoz, M. C., Real, J. A. (2007) Pressure effect studies on the 3D spin crossover system: $\{Fe(3CN\text{-}py)_2[M(CN)_2]_2\}\cdot nH_2O$ ($n \leq 2/3$, M = Ag(I), Au(I)). Chem. Phys. Lett., 434: 68–72.
38. Romstedt, H., Hauser, A., Spiering, H. (1998) High-spin → low-spin relaxation in the two-step spincrossover compound $[Fe(pic)_3]Cl_2EtOH$ (pic = 2-picolylamine). J. Phys. Chem. Solids, 59: 265–275.
39. Gaspar, A. B., Muñoz, M. C., Moliner, N., Ksenofontov, V., Levchenko, G., Gütlich, P., Real, J. A. (2003) Polymorphism and pressure driven thermal spin crossover phenomenon in $[Fe(abpt)_2(NCX)_2]$ (X = S, and Se): synthesis, structure and magnetic properties. Monatsh. Chem., 134: 285–294.
40. Li, B., Wei, R. J., Tao, J., Huang, R.-B., Zheng, L.-S. (2010) Pressure effects on a spin-crossover monomeric compound $[Fe(pmea)(SCN)_2]$ (pmea = bis[(2-pyridyl)methyl]-2-(2-pyridyl)ethylamine). Inorg. Chem., 49: 745–751.
41. Shepherd, H. J., Bonnet, S., Guionneau, P., Bedoui, S., Garbarino, G., Nicolazzi, W., Bousseksou, A., Molnár, G. (2011) Pressure-induced two-step spin transition with structural symmetry breaking: X-ray diffraction, magnetic, and Raman studies. Phys. Rev. B, 84: 144107/1–144107/9.
42. Long, G. J., Hutchinson, B. B. (1987) Spin equilibrium in iron(II) poly(1-pyrazolyl)borate complexes: low-temperature and high-pressure Mössbauer spectral studies. Inorg. Chem., 26: 608–613.
43. Konno, M., Mikami-Kido, M. (1991) Temperature- or pressure-induced structure changes of a spin crossover Fe(II) complex; $[Fe(bpy)_2(NCS)_2]$. Bull. Chem. Soc. Jpn., 64: 339–345.
44. Boillot, M.-L., Zarembovitch, J., Itié, J.-P., Polian, A., Bourdet, E., Haasnoot, J. G. (2002) Pressure-induced spin-state crossovers at room temperature in iron(II) complexes: comparative analysis; a XANES investigation of some new transitions. New J. Chem., 26: 313–322.
45. Garcia, Y., Ksenofontov, V., Levchenko, G., Schmitt, G., Gütlich, P. (2000) Pressure-induced high spin state in $[Fe(btr)_2(NCS)_2]\cdot H_2O$ (btr = 4,4′-bis-1,2,4-triazole). J. Phys. Chem. B, 104: 5045–5048.
46. Molnár, G., Guillon, T., Moussa, N. O., Rechignat, L., Kitazawa, T., Nardone, M., Bousseksou, A. (2006) Two-step spin-crossover phenomenon under high pressure in the coordination polymer Fe(3-methylpyridine)$_2[Ni(CN)_4]$. Chem. Phys. Lett., 423: 152–156.
47. Bonnet, S., Molnár, G., Costa, J. S., Siegler, M. A., Spek, A. L., Bousseksou, A., Fu, W.-T., Gamez, P., Reedijk, J. (2009) Influence of sample preparation, temperature, light, and pressure on the two-step spin crossover mononuclear compound $[Fe(bapbpy)(NCS)_2]$. Chem. Mater., 21: 1123–1136.
48. Weir, C. E., Lippincott, E. R., Van Valkenburg, A., Bunting, E. N. (1959) Infrared studies in the 1-micron to 15-micron region to 30000 atmosphere. J. Res. Natl. Bur. Stand. Sect. A, 63: 55–62.
49. Merrill, L., Basset, W. A. (1974) Miniature diamond anvil pressure cell for single crystal X-ray diffraction studies. Rev. Sci. Instrum., 45: 290–294.
50. Ahsbahs, H. (1984) Diamond-anvil high pressure cell for improved single-crystal X-ray diffraction measurements. Rev. Sci. Instrum., 55: 99–102.
51. Katrusiak, A. (2008) High-pressure crystallography. Acta Cryst. Sect. A, 64: 135–148.

52. Basset, W. A. (2009) Diamond anvil cell, 50th birthday. High Press. Res., 29: 163–186.
53. Tilton, R. F. (1988) A fixture for X-ray crystallographic studies of biomolecules under high gas pressure. J. Appl. Cryst., 21: 4–10.
54. Yufit, D. S., Howard, J. A. K. (2005) Simple pressure cell for single-crystal X-ray crystallography. J. Appl. Cryst., 38: 583–586.
55. Dawson, A., Allan, D. R., Parsons, S., Rulf, M. (2004) Use of a CCD diffractometer in crystal structure determinations at high pressure. J. Appl. Cryst., 37: 410–416.
56. Boldyreva, E. V. (2008) High-pressure diffraction studies of molecular organic solids. A personal view. Acta Cryst. Sect. A, 64: 218–231,
57. Guionneau, P., Brigouleix, C., Barrans, Y., Goeta, A. E., Létard, J.-F., Howard, J. A. K., Gaultier, J., Chasseau, D. (2001) High pressure and very low temperature effects on the crystal structures of some iron(II) complexes. C. R. Chimie, 4: 161–171.
58. Shepherd, H. J., Palamarciuc, T., Rosa, P., Guionneau, P., Molnár, G., Létard, J.-F., Bousseksou, A. (2012). Antagonism between extreme negative linear compression and spin crossover in $[Fe(dpp)_2(NCS)_2]\cdot py$. Angew. Chem. Int. Ed., 51: 3910–3914.
59. Guionneau, P., Marchivie, M., Garcia, Y., Howard, J. A. K., Chasseau, D. (2005) Spin crossover in $[Mn^{II(}pyrol)_3tren]$ probed by high-pressure and low-temperature X-ray diffraction. Phys. Rev. B, 72: 214408/1–214408/8.
60. Shepherd, H. J., Rosa, P., Vendier, L., Casati, N., Létard, J.-F., Bousseksou, A., Guionneau, P., Molnár, G. (2012) High-pressure spin-crossover in a dinuclear Fe(II) complex. Phys. Chem. Chem. Phys., 14: 5265–5271.
61. König, E. (1987) Structural changes accompanying continuous and discontinuous spin-state transitions Prog. Inorg. Chem., 35: 527–622.
62. Guionneau, P., Marchivie, M., Bravic, G., Létard, J.-F., Chasseau, D. (2004) Structural aspects of spin crossover. example of the $[Fe^{II}L_n(NCS)_2]$ complexes in: Gütlich, P., Goodwin, H. A. (Eds) Spin Crossover in Transition Metal Compounds II. Top. Curr. Chem. 234: 97–128.
63. Halcrow, M. A. (2011) Structure:function relationships in molecular spin-crossover complexes. Chem. Soc. Rev., 40: 4119–4142.
64. Marchivie, M., Guionneau, P., Howard, J. A. K., Chastanet, G., Létard, J.-F., Goeta, A. E., Chasseau, D. (2002) Structural characterization of a photoinduced molecular switch. J. Am. Chem. Soc., 124: 194–195.
65. Kaiba, A., Shepherd, H. J., Fedaouik, D., Rosa, P., Goeta, A. E., Rebbani, N., Létard, J.-F., Guionneau, P. (2010) Crystallographic elucidation of purely structural, thermal and light-induced spin transitions in an iron(ii) binuclear complex. Dalton Trans., 39: 2910–2918.
66. Fedaoui, D., Bouhadja, Y., Kaiba, A., Guionneau, P., Létard, J.-F., Rosa, P. (2008) Complexation of 2,6-bis(3-pyrazolyl)pyridine–bis(thiocyanato)iron(II) with a bridging 4,4_-bipyridine: a new example of a dinuclear spin crossover complex. Eur. J. Inorg. Chem., 1022–1026.
67. Kusz, J., Zubko, M., Fitch, A., Gütlich, P. (2011) Isostructural phase transition in the spin crossover compound $[Fe(dpp)_2(NCS)_2]py$. Z. Kristallogr., 226: 576–584.
68. Rotaru, A., Varret, F., Codjovi, E., Boukheddaden, K., Linares, J., Stancu, A., Guionneau, P., Létard, J.-F. (2008) Hydrostatic pressure investigation of the spin crossover compound $[Fe(PM\text{-}BiA)_2(NCS)_2]$ polymorph I using reflectance detection. J. Appl. Phys., 106: 053515/1–053515/5.
69. Lemée-Cailleau, M. H., Ecolivet, C., Ouladdiafa, B., Moussa, F., Létard, J.-F. (2009) Multi-phase spin crossover in $Fe(ptz)_6(BF_4)_2$. Physica B, 404: 379–381.
70. Lemée-Cailleau, M. H., Ecolivet, C., Ouladdiafa, B., Moussa, F., Jeftic, J., Létard, J.-F. (2007) Intermediate ferroelastic phase of the photo-sensible spin-crossover system $[Fe(ptz)_6](BF_4)_2$. J. Magn. Magn. Mater., 310: 1792–1793.
71. Kusz, J., Zubko, M., Neder, R. B., Gütlich, P. (2012) Structural phase transition to disorder low-temperature phase in $[Fe(ptz)_6](BF_4)_2$ spin-crossover compounds. Acta Cryst. Sect. B, 68: 40–56.
72. Sugahara, A., Moriya, K., Enomoto, M., Okazawa, A., Kojima, N. (2011) Study on the spin-crossover transition in [Fe(*cis-/trans*-stpy)$_4$(X)$_2$] (stpy: styrylpyridine, X: NCS, $NCBH_3$) under high pressure toward ligand-driven light-induced spin change. Polyhedron, 30: 3127–3130.
73. Paszkowicz, W. (2002) High-pressure powder X-ray diffraction at the turn of the century. Nucl. Instr. Methods in Phys. Res. Sect. B, 198: 142–182.

74. Tsuchiya, N., Tsukamoto, A., Ohshita, T., Isobe, T., Senna, M., Yoshioka, N., Inoue, H. (2001) Stress-induced ligand field distribution and consequent multi-mode spin crossover in $Fe^{II}(phen)_2(NCS)_2$ and $Fe^{II}[HB(pz)_3]_2$. Solid State Sci., 3: 705–714.
75. Gaudon, M., Deniard, P., Demourgues, A., Thiry, A.-E., Carbonera, C., LeNestour, A., Largeteau, A., Létard, J.-F., Jobic, S. (2007) Unprecedented "one-finger-push"-induced phase transition with a drastic color change in an inorganic material. Adv. Mater., 19: 3517–3519.
76. Fabbiani, F. P. A., Pulham, C. R. (2006) High-pressure studies of pharmaceutical compounds and energetic materials. Chem. Soc. Rev., 35: 932–942.
77. Oswald, I. D. H., Allan, D. R., Motherwell, W. D. S., Parsons, S. (2005) Structures of the monofluoro- and monochlorophenols at low temperature and high pressure. Acta Cryst. Sect. B, 61: 69–79.
78. Coppens, P., Vorontsov, I. I., Graber, T., Gembicky, M., Kovalevsky, A. Y. (2005) The structure of short-lived excited states of molecular complexes by time-resolved X-ray diffraction. Acta Cryst. Sect. A, 61: 162–172.
79. Collet, E., Buron-Le Cointe, M., Cailleau, H. (2006) X-ray diffraction investigation of the nature and the mechanism of photoinduced phase transition in molecular materials. J. Phys. Soc. Jpn., 75: 011002/1–011002/9.
80. Woike, T., Schaniel, D. (2008) Photocrystallography. Z. Kristallogr., 223: 235–369.
81. Buron, M., Collet, E. (2005) Second international conference on photoinduced phase transition; cooperative, non-linear and functional properties. J. Phys.: Conf. Ser., 21: doi:10.1088/1742-6596/21/1/E01.
82. Bréfuel, N., Watanabe, H., Toupet, L., Come, J., Matsumoto, N., Collet, E., Tanaka, K., Tuchagues, J.-P. (2009) Concerted spin crossover and symmetry breaking yield three thermally and one light-induced crystallographic phases of a molecular material. Angew. Chem. Int. Ed., 48: 9304–9307.
83. Bréfuel, N., Collet, E., Watanabe, H., Kojima, M., tsumoto, N., Toupet, L., Tanaka, K., Tuchagues, J.-P. (2010) Nanoscale self-hosting of molecular spin-states in the intermediate phase of a spin-crossover material. Chem. Eur. J., 16: 14060–14068.
84. Huby, N., Guérin, L., Collet, E., Ameline, J.-C., Cailleau, H., Roisnel, T., Tayagaki, T., Tanaka, K. (2004) Photoinduced spin transition probed by X-ray diffraction. Phys. Rev. B, 69: 020101/1–020101/4.
85. Goujon, A., Gillon, B., Debede, A., Cousson, A., Gukasov, A., Jeftic, J., McIntyre, G. J., Varret, F. (2006) Neutron Laue diffraction on the spin crossover crystal $[Fe(1\text{-}n\text{-propyltetrazole})_6](BF_4)_2$ showing continuous photoinduced transformation. Phys. Rev. B, 73: 104413/1–104413/5.
86. Pillet, S., Legrand, V., Souhassou, M., Lecomte, C. (2006) Kinetics of light-induced first-order phase transformation in molecular solids $Fe(btr)_2(NCS)_2{\cdot}H_2O$. Phys. Rev. B, 74: 140101/1–140101/4.
87. Ichiyanagi, K., Hébert, J., Toupet, L., Cailleau, H., Guionneau, P., Létard, J.-F., Collet, E. (2006) Nature and mechanism of the photoinduced spin transition in $[Fe(PM\text{-}BiA)_2(NCS)_2]$. Phys. Rev. B, 73: 060408/1–060408/4.
88. Nicolazzi, W., Pillet, S. (2012) Structural aspects of the relaxation process in spin crossover solids: Phase separation, mapping of lattice strain, and domain wall structure. Phys. Rev. B, 85: 094101/1–094101/13.
89. Collet, E. (Ed.) (2010) Dynamical structural science. Acta Cryst. Sect. A, 66: 133–280.
90. Collet, E., Lorenc, M., Cammarata, M., Guérin, L., Servol, M., Tissot, A., Boillot, M.-L., Cailleau, H., Buron-Le Cointe, M. (2012) 100 Picosecond diffraction catches structural transients of laser-pulse triggered switching in a spin-crossover crystal. Chem. Eur. J., 18: 2051–2055.
91. Lorenc, M., Hébert, J., Moisan, N., Trzop, E., Servol, M., Buron-Le Cointe, M., Cailleau, H., Boillot, M.-L., Pontecorvo, E., Wulff, M., Koshihara, S., Collet, E. (2009) Successive dynamical steps of photoinduced switching of a molecular Fe(III) spin-crossover material by time-resolved X-ray diffraction. Phys. Rev. Lett., 103: 028301/1–028301/4.
92. Collet, E., Moisan, N., Baldé, C., Bertoni, R., Trzop, E., Laulhé, C., Lorenc, M., Servol, M., Cailleau, H., Tissot, A., Boillot, M.-L., Graber, T., Henning, R., Coppens, P., Buron-Le Cointe, M. (2012) Ultrafast spin-state photoswitching in a crystal and slower consecutive processes investigated by femtosecond optical spectroscopy and picosecond X-ray diffraction. Phys. Chem. Chem. Phys., 14: 6192–6199.
93. Gütlich, P., Goodwin, H. A. (Eds) (2004) Spin Crossover in Transition Metal Compounds I-III. Top. Curr. Chem., vols. 233–235. Springer, Berlin / Heidelberg, Germany.
94. Buron-Le Cointe, M., Hébert, J., Baldé, C., Moisan, N., Toupet, L., Guionneau, P., Létard, J.-F., Freysz, E., Cailleau, H., Collet, E. (2012) Intermolecular control of thermoswitching and photoswitching phenomena in two spin-crossover polymorphs. Phys. Rev. B, 85: 064114/1–064114/9.

95. Legrand, V., Pillet, S., Souhassou, M., Lugan, N., Lecomte, C. (2006) Extension of the experimental electron density analysis to metastable states: a case example of the spin crossover complex $Fe(btr)_2(NCS)_2{\cdot}H_2O$. J. Am. Chem. Soc., 128: 13921–13931.
96. Trzop, E., Buron-Le Cointe, M., Cailleau, H., Toupet, L., Molnár, G., Bousseksou, A., Gaspar, A. B., Real, J. A., Collet, E. (2007) Structural investigation of the photoinduced spin conversion in the dinuclear compound $[Fe(bt)(NCS)_2]_2(bpym)\}$: toward controlled multi-stepped molecular switches. J. Appl. Cryst., 40: 158–164.
97. Agustí, G., Thompson, A. L., Gaspar, A. B., Muñoz, M. C., Goeta, A. E., Rodríguez-Velamazán, J. R., Castro, M., Burriel, R., Real, J. A. (2008) Thermal, pressure and light induced spin transition in the two-dimensional coordination polymer $\{Fe(pmd)_2[Cu(CN)_2]_2\}$. Dalton Trans., 642–649.
98. Glijer, D., Hébert, J., Trzop, E., Collet, E., Toupet, L., Cailleau, H., Matouzenko, G. S., Lazar, H. Z., Létard, J.-F., Koshihara, S., Buron-Le Cointe, M. (2008) Photoinduced phenomena and structural analysis associated with the spin state switching in the $[Fe^{II}(DPEA)(NCS)_2]$ complex. Phys. Rev. B,78: 134112/1–134112/9.
99. Pillet, S., Legrand, V., Weber, H.-P., Souhassou, M., Létard, J.-F., Guionneau, P., Lecomte, C. (2008) Out-of-equilibrium charge density distribution of spin crossover complexes from steady-state photocrystallographic measurements: experimental methodology and results. Z. Kristallogr., 223: 235–249.
100. Dupouy, G., Marchivie, M., Triki, S., Sala-Pala, J., Gómez-García, C. J., Pillet, S., Lecomte, C., Létard, J.-F. (2009) Photoinduced HS state in the first spin-crossover chain containing a cyanocarbanion as bridging ligand. Chem. Commun., 3404–3406.
101. Boinnard, D., Bousseksou, A., Dworkin, A., Savariault, J.-M., Varret, F., Tuchagues, J. P. (1994) Two-step spin conversion of $[Fe^{II}(5\text{-}NO_2\text{-sal-N}(1,4,7,10))]$: 292, 153, and 103 K X-ray crystal and molecular structure, infrared, magnetic, Mössbauer, calorimetric, and theoretical studies. Inorg. Chem., 33: 271–281.
102. Bonnet, S., Siegler, M. A., Costa, J. S, Molnár, G., Bousseksou, A., Spek, A. L., Gamez, P., Reedijk, J. (2008) A two-step spin crossover mononuclear iron(II) complex with a [HS–LS–LS] intermediate phase Chem. Commun., 5619–5621.
103. Chernyshov, D., Hostettler, M., Törnroos, K. W., Bürgi, H.-B. (2003) Ordering phenomena and phase transitions in a spin-crossover compound—uncovering the nature of the intermediate phase of $[Fe(2\text{-pic})_3]Cl_2{\cdot}EtOH$. Angew. Chem. Int. Ed., 42: 3825–3830.
104. Neville, S. M., Leita, B. A., Halder, G. J., Kepert, C. J., Moubaraki, B., Létard, J.-F., Murray, K. S. (2008) Understanding the two-step spin-transition phenomenon in iron(II) 1D chain materials. Chem. Eur. J., 14: 10123–10133.
105. Yamada, M., Ooidemizu, M., Ikuta, Y., Osa, S., Matsumoto, N., Iijima, S., Kojima, M., Dahan, F., Tuchagues, J.-P. (2003) Interlayer interaction of two-dimensional layered spin crossover complexes $[Fe^{II}H_3L^{Me}][Fe^{II}LMe]X$ ($X^- = ClO_4^-$, BF_4^-, PF_6^{--}, AsF_6^-, and SbF_6^-; H_3L^{Me} = tris[2-(((2-methylimidazol-4-yl)methylidene)amino)ethyl]amine). Inorg. Chem., 42: 8406–8416.
106. Yamada, M., Hagiwara, H., Torigoe, H., Matsumoto, N., Kojima, M., Dahan, F., Tuchagues, J.-P., Re, N., Iijima, S. (2006) A variety of spin-crossover behaviors depending on the counter anion: two-dimensional complexes constructed by $NH{\cdots}Cl^-$ Hydrogen Bonds, $[Fe^{II}H_3L^{Me}]$ Cl·X ($X = PF_6^-$, AsF_6^-, SbF_6^-, $CF_3SO_3^-$; H_3L^{Me} = Tris[2-{[(2-methylimidazol-4-yl)methylidene] amino}ethyl]amine). Chem. Eur. J., 12: 4536–4549.
107. Nicolazzi, W., Pillet, S., Lecomte, C. (2009) Photoinduced phase separation in spin-crossover materials: numerical simulation of a dynamic photocrystallographic experiment. Phys. Rev. B, 80: 132102/1–132102/4.
108. Chong, C., Slimani, A., Varret, F., Boukheddaden, K., Collet, E., Ameline, J.-C., Bronisz, R., Hauser, A. (2011) The stepwise temporal evolution of the thermal spin transition in $[Fe(bbtr)_3](ClO_4)_2$ single crystals explained by optical microscopy in terms of inhomogeneous stresses. Chem. Phys. Lett., 504: 29–33.
109. Varret, F., Slimani, A., Boukheddaden, K., Chong, C., Mishra, H., Collet, E., Haasnoot, J., Pillet, S. (2011) The propagation of the thermal spin transition of $[Fe(btr)_2(NCS)_2].H_2O$ single crystals, observed by optical microscopy. New J. Chem., 35: 2333–2340.
110. Cobo, S., Ostrovskii, D., Bonhommeau, S., Vendier, L., Molnár, G., Salmon, L., Tanaka, K., Bousseksou, A. (2008) Single-laser-shot-induced complete bidirectional spin transition at room temperature in single crystals of $(Fe^{II}(pyrazine)(Pt(CN)_4))$. J. Am. Chem. Soc., 130: 9019–9024.
111. Galle, G., Degert, J., Mauriac, C., Etrillard, C., Létard, J.-F., Freysz, E. (2010) Nanosecond study of spin state transition induced by a single nanosecond laser shot on $[Fe(NH_2)trz)_3]$ compounds inside and outside their thermal hysteresis loops. Chem. Phys. Lett. 500: 18–22.

112. Collet, E., Lemée-Cailleau, M. H., Buron-Le Cointe, M., Cailleau, H., Wulff, M., Luty, T., Koshihara, S.-Y., Meyer, M., Toupet, L., Rabiller, P., Techert, S. (2003) Laser-induced ferroelectric structural order in an organic charge-transfer crystal. Science, 300: 612–615.
113. Nozawa, S., Adachi, S., Takahashi, J., Tazaki, R., Guérin, L., Daimon, M., Tomita, A., Sato, T., Chollet, M., Collet, E., Cailleau, H., Yamamoto, S., Tsuchiya, K., Shioya, T., Sasaki, H., Mori, T., Ichiyanagi, K., Sawa, H., Kawata, H., Koshihara, S. (2007) Developing 100 ps-resolved X-ray structural analysis capabilities on beamline NW14A at the Photon Factory Advanced Ring. J. Synchrotron Rad., 14: 313–319.
114. Bressler, C., Milne, C., Pham, V. T., El Nahhas, A., van der Veen, R. M., Gawelda, W., Johnson, S., Beaud, P., Grolimund, D., Kaiser, M., Borca, C. N., Ingold, G., Abela, R., Chergui, M. (2009) Femtosecond XANES study of the light-induced spin crossover dynamics in an iron(II) complex. Science, 323: 489–492.
115. Chergui, M. (2010) Picosecond and femtosecond X-ray absorption spectroscopy of molecular systems. Acta Cryst. Sect. A, 66: 229–239.
116. Cailleau, H., Lorenc, M., Guérin, L., Servol, M., Collet, E., Buron-Le Cointe, M. (2010) Structural dynamics of photoinduced molecular switching in the solid state. Acta Cryst. Sect. A, 66: 189–197.

21
Spin-Transitions in Metal Oxides

Jean-Pascal Rueff

Synchrotron SOLEIL, France

21.1 Introduction

Metal oxides (MOs) are among the most widespread materials in nature, frequently found in minerals, meteorites, or bacteria. Yet their fascinating properties are still far from being fully understood. Let's consider the example of magnetite (Fe_3O_4), the most magnetic of all the naturally occurring minerals on Earth. Magnetite is seemingly a simple iron oxide, known since the Iron Age for its ability to attract small pieces of iron. But its magnetic structure and physical properties are still very much debated. More recently, MO materials have attracted much attention largely due to the large diversity of their physical properties and the potential of controlling them via various dopant levels, as well as strain in multilayer structures or via external parameters such as temperature, bias, or pressure. On the other hand, oxides exhibit an exceptionally broad range of functionalities: they can be insulators, semiconductors, metals, superconductors, heavy fermions systems, ferromagnets, antiferromagnets, ferroelectrics, piezoelectric, or multiferroics combining these properties. Because of the low lattice mismatch between many of them, oxides also offer the possibility of being combined into heterostructures controlled at the atomic scale. This allows definition of meta-materials with enhanced or even wholly new properties and can exploit the potentially remarkable properties of their interfaces to create a new generation of optical, electronic, and spintronic devices.

Among this swarming inventory of properties, we will focus in this chapter on perhaps one of the most fundamental ones: the spin state. Magnetism in MOs is provided by the polarization of the 3*d* states on the metal site. Because crystal electric field (CEF) strength in 3*d* elements is strong compared to both spin orbit (SO) coupling and the Zeeman interaction, the orbital moment is usually considered as 'quenched', so that their total magnetic moment is to a large extent governed by the spin moment only (Table 21.1). Spin not only affects MO magnetic properties but also, as discussed in the following, structural stabilization and electronic or transport properties.

Spin-Crossover Materials: Properties and Applications, First Edition. Edited by Malcolm A. Halcrow.

Table 21.1 *Magnetic moments of 3d ions (from ref. 1). The magnetic moment is mainly given by the spin moment.*

Example of magnetic ions	Electronic configuration	Ground state term symbol	Calc. total magnetic moment $g[J(J+1)]^{1/2}$	Calc. spin magnetic moment $2[S(S+1)]^{1/2}$	Exp. magnetic moment
Ti^{3+}	$3d^1$	$^2D_{3/2}$	1.55	1.73	1.8
V^{3+}	$3d^2$	$^3F_2$	1.63	2.83	2.8
Cr^{3+}	$3d^3$	$^4F_{3/2}$	0	3.87	3.8
Mn^{3+}	$3d^4$	$^5D_0$	0.77	4.90	4.9
Fe^{3+}	$3d^5$	$^6S_{5/2}$	5.92	5.92	5.9
Fe^{2+}	$3d^6$	$^5D_4$	6.70	4.90	5.4
Co^{2+}	$3d^7$	$^4F_{9/2}$	6.63	3.87	4.8
Ni^{2+}	$3d^8$	$^3F_4$	5.59	2.83	3.2
Cu^{2+}	$3d^9$	$^2D_{5/2}$	3.55	1.73	1.9

21.1.1 CEF Approach to Spin State Stability

The prime action of the CEF is to lift the degeneracy of the ground state so that the 3d orbitals are split into symmetry-dependent components. The simplistic case of a metal ion in an octahedral symmetry is illustrated in Figure 21.1, where $10Dq$ is the crystal field parameter. The five d orbitals are split in two groups of orbitals with e_g and t_{2g} symmetry respectively. The former point towards the ligand sites, which tends to enhance the Coulomb interaction whereas the latter point towards the octahedral edge centers, thus gaining in energy.

The stability in energy of the orbitals depends on the crystal field strength. This will subsequently affect the magnetic properties of the 3d ion. Figure 21.2 shows the case of a $3d^5$ ion in an octahedral symmetry. Depending on the relative strength of the CEF, the ground state configuration will switch from high moment ('high spin' state) to low moment ('low spin' state). This simple picture is the prime rationale for magnetic collapse transitions in metal-oxides under high pressure discussed below, as pressure predominantly modifies the inter-atomic distances, and therefore the CEF strength.

21.1.2 Stoner Criterion for Itinerant Magnetism

The atomic approach finds its limits when considering metal oxides. Although spin retains its atomic aspect in the presence of oxygen ligands – X-ray spectroscopy of MO is reasonably well accounted for by atomic

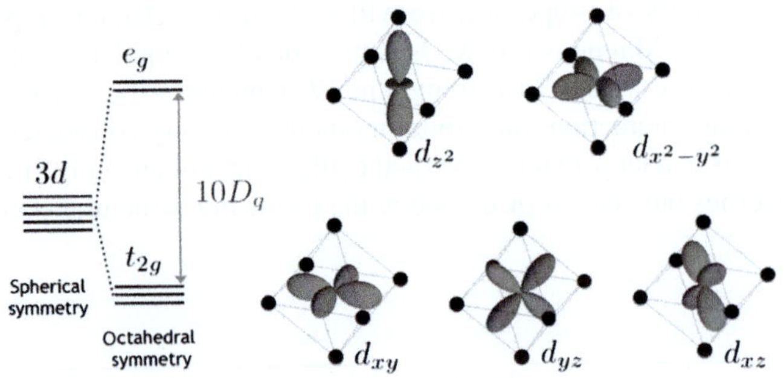

Figure 21.1 *3d orbitals in octahedral symmetry.*

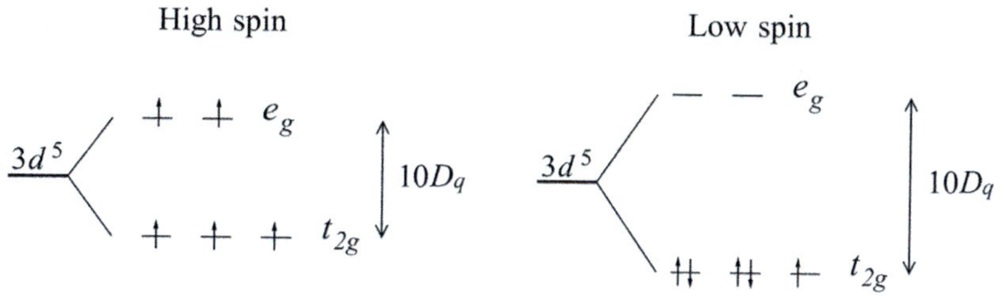

Figure 21.2 *Stabilization of the high spin (S = 5/2) or low spin (S = 1/2) state of a 3d5 metal ion in octahedral symmetry depending on the crystal field strength 10 Dq. Reproduced from [2] with permission from American Physical Society. Copyright 2010.*

multiplet approaches including crystal field effects and metal-oxygen charge transfer – solid state effects cannot be disregarded when dealing with the magnetic stability of the metal ion.

The Stoner criterion is a useful approach when dealing with magnetic- and spin-transitions, from the point of view of itinerant magnetism. It stipulates that a system will exhibit a spontaneous magnetic moment whenever the following condition is valid:

$$I \cdot N(E_F) \geq 1$$

I is the strength of the exchange interaction and $N(E_F)$ the density of states at the Fermi energy. The Stoner criterion is especially useful to describe band-like behavior of electrons, such as in monoxides under high pressure conditions where the d electron band is sufficiently broadened to be considered itinerant.

21.1.3 Probes of the Spin-Transitions

Spin-transitions can be observed through various techniques that can tackle either directly or indirectly the spin properties. In this first category, the average magnetic moment can be retrieved by magnetic susceptibility, while X-ray magnetic circular dichroism (XMCD) at the metal $L_{2,3}$ edges provides a handle on both the local spin and orbital magnetic moment through the so-called 'sum rules'. XMCD at the metal K-edge retains this sensitivity to magnetism (although the sum rules do not apply) with the advantage of being a high energy probe with bulk sensitivity. Mössbauer spectroscopy has also been widely used for its unique sensitivity to the spin state, magnetic order, and chemical environment in isotopically substituted metal ion compounds. Magnetic neutron diffraction is best suited to the determination of the magnetic structure in single crystals. On the other hand, indirect phenomena associated with the spin-transitions can serve to disclose spin-transitions. For instance, changes in *d–d* crystal field excitations can be determined by optical or infrared spectroscopy while any spin-driven volume anomaly is accessible by standard structural probes such as diffraction and X-ray absorption fine structure (XAFS) analysis.

In this chapter, we will focus on the use of resonant inelastic X-ray scattering (RIXS). This technique is considered to be a powerful probe of the electronic properties of materials, but has recently attracted great interest for its ability to observe in detail the electronic properties of materials, and more specifically to monitor spin-transitions. RIXS is a spectroscopic technique with the same orbital and chemical selectivity as XMCD. As an all photon technique, RIXS when applied in the hard X-ray range ($E > 5$ keV), turns out to be highly suitable for investigating the electronic properties of materials in difficult sample environments such as in a high-pressure cell. This particular topic has been the subject of a recent review article that depicts the strength of the technique in detail.[2] We nevertheless recall in the following section the basic ideas about RIXS and its applications to spin properties before discussing the results in MOs.

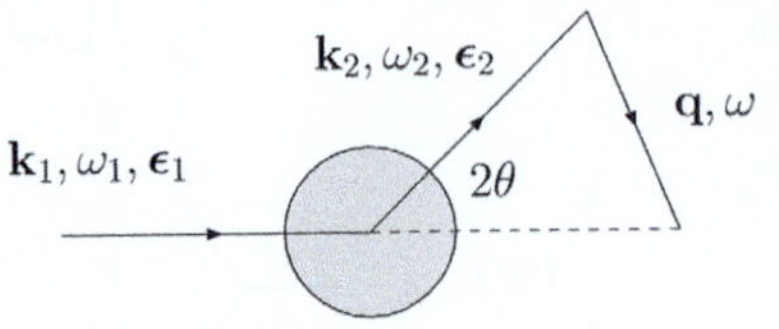

Figure 21.3 General scattering process. Reproduced from [2] with permission from American Physical Society. Copyright 2010.

21.2 RIXS: A Probe of the 3*d* Electronic Properties

21.2.1 Overview of the RIXS Process

RIXS is primarily a scattering process where an incident photon beam interacts with the electron assembly and a secondary photon is emitted (c.f. Fig. 21.3). During the scattering event, the photons exchange part of their energy and momentum with the electrons so that elementary excitations of the system can be probed along with their dispersion.

When the energy of the incident photons is tuned close to an absorption edge, RIXS is better described as a coherent, resonantly enhanced, combined absorption and emission process: the incident photon prompts transition of core electrons to empty states before radiative decay takes place, leaving the system either with a valence-hole or a core-hole (alternatively, the system may return to its ground state with a null energy transfer, and no core-hole left). Such core-to-core or core-to-valence RIXS transitions are described by the Kramers–Heisenberg (KH) formula:

$$\frac{d^2\sigma}{d\omega_1 d\omega_2} = \sum_f \left| \sum_n \frac{\langle f\, |T_2| n\rangle \langle n\, |T_1| g\rangle}{E_g - E_n + \hbar\omega_1 - i\frac{\Gamma_n}{2}} \right|^2 \times \frac{\Gamma_f/2\pi}{(E_g - E_f + \hbar\omega_1 - \hbar\omega_2)^2 + \frac{\Gamma_f^2}{4}} \tag{21.1}$$

where $\hbar\omega_1$ and $\hbar\omega_2$ are the incident and emitted photon energies, and T_1 and T_2 are the transition operators from the ground state g to the intermediate states n, and from the intermediate state to the final state, f respectively. The denominator in the first term ensures resonant behavior when the incident photon energy equals the transition energy of the core electrons $\hbar\omega_1$. Lifetime broadening effects in the intermediate and final states are introduced via the gamma coefficient in the numerator and the Lorentzian function respectively.

21.2.2 X-Ray Emission as a Probe for the Spin State

An extreme case of the general RIXS process occurs when the incident photon energy exceeds the ionization potential of the considered core electrons. In contrast to the general resonant regime where the KH formula applies, the RIXS process reduces to a pure fluorescence decay channel.

In 3*d* elements, this X-ray emission spectroscopy (XES) can be efficiently used as a probe of the metal spin state when performed around the metal K-edge. Among the different decay channels that follow the creation of a 1*s* core-hole, the Kα ($2p \rightarrow 1s$) and Kβ ($3p \rightarrow 1s$) lines are of special interest as they involved a *p*-symmetry core-hole in the XES final state. Final state interactions between the 2*p* or 3*p* core-hole and the spin-polarized 3*d* band yield the sensitivity to the spin magnetic moment. A closer look at the Kβ emission line is shown in Figure 21.4 in a $3d^5$ ion. The emission line is split into several multiplet states whose energy splitting is determined by both Coulomb and exchange interactions, producing several satellite features near the main line. As the 3*d* moment is changed, so will the energy splitting and intensity of these satellite features.[3]

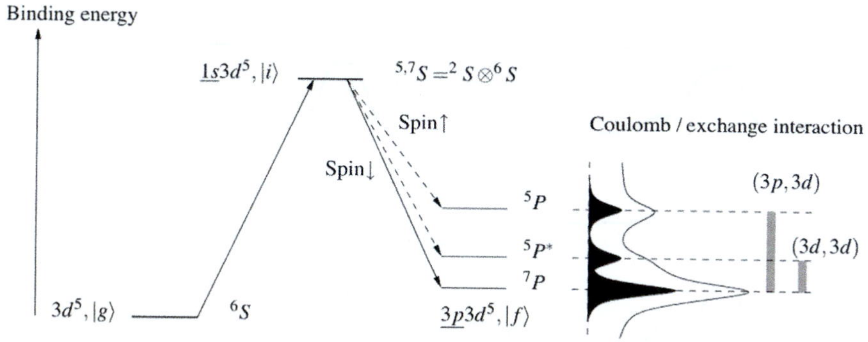

Figure 21.4 *Multiplet view of the Kβ emission line in a 3d⁵ ion. Reproduced from [2] with permission from American Physical Society. Copyright 2010.*

This sensitivity to the spin state is pictured in Figure 21.5 for the whole series of $3d$ ions in O_h symmetry. The spectra for both Kα and Kβ emission lines are calculated within the ligand field multiplet approach using the CTM4XAS code.[4] Different magnetic ground states are imposed by changing the crystal field parameter $10Dq$ from low field to high field values, thereby mimicking a high spin to low spin-transition. The spectra exhibit a net sensitivity to the local spin state through a decrease of the satellite intensity and a shift of the main line. In contrast to magnetic susceptibility, Mössbauer spectroscopy, or XMCD, which all require a magnetic field to align the moments, XES at the Kα and Kβ lines is a local probe which is not constrained by long range magnetic order and therefore can be operated in the absence of an external field in both paramagnetic and antiferromagnetic phases. The technique has been successfully applied to monitor spin-transitions in metal oxides induced by high-pressure conditions or temperature changes (Section 21.3).

21.2.3 Direct View of the 3*d*: Pre-Edge Features at the Metal K-Edge by RIXS and PFY-XAS

In contrast to XES, RIXS may be used to directly probe the $3d$ states either at the $L_{2,3}$ ($2p \rightarrow 3d$) edges of the metal, or via quadrupolar excited transitions at the K-edge ($1s \rightarrow 3d$). We will focus more specifically on the latter technique, which provides both magnetic sensitivity and bulk-like sensitivity. The former aspect has been amply discussed in the recent literature (c.f. the review article)[5] but its range of applicability is hampered by the RIXS surface sensitivity of soft X-rays.

Quadrupolar transitions show up in the metal XAS K-edge as features on the low energy side of the white line. Because these pre-edge features are of weak intensity, they are normally overwhelmed by the white line tail that extends towards low energy. One of the benefits of RIXS is to overcome this limitation by taking advantage of different broadening effects in the intermediate (Γ_n) and final state (Γ_f). As the core-hole left in the final state is shallower than that created in the intermediate state, one may safely consider that $\Gamma_f \ll \Gamma_n$. The total broadening Γ_{tot} to be considered for the $1s2p$-RIXS approach is thus defined by:

$$\frac{1}{\Gamma_{tot}} = \sqrt{\frac{1}{\Gamma_n^2} + \frac{1}{\Gamma_f^2}} \sim \frac{1}{\Gamma_f} \tag{21.2}$$

Figure 21.6 shows a simulation of a $1s2p$-RIXS process (panel (a)) in a $3d^4$ ion in both high spin and low spin configurations. The results are represented as an intensity map function of incident and emitted energies in panel (d). Besides the sharpening effect, the simulations prove the high sensitivity of the $1s2p$-RIXS approach to the spin state. Note that by projecting the map along the emitted energy (panel (b)), one

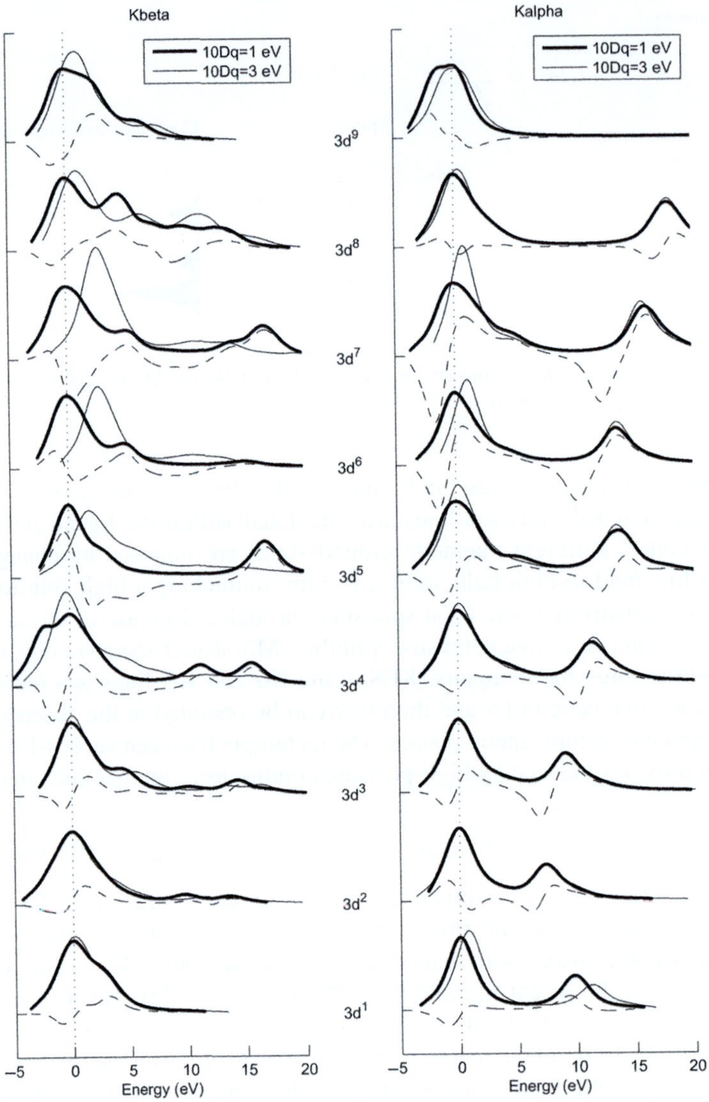

*Figure 21.5 Overview of the Kα and Kβ lines the 3*d *series calculated with the CTM4XAS code in the low field (10* Dq = *1 eV) and high field (10* Dq = *3 eV) cases. Difference spectra are also shown. We used a Lorentzian broadening of 1 eV, and a Gaussian broadening of 0.5 eV to emphasize the spectral fine structure. The Slater integrals are reduced by 80% to account for covalent effects.*

retrieves the Kα emission XES line (c.f. Fig. 21.6), while a projection along the incident energy axis (panel (c)) provides a view of the quadrupolar component of the metal K-edge.

This intrinsic sharpening effect allows RIXS to overcome the natural core-hole lifetime broadening effects of standard XAS, providing spectral features with enhanced resolving power. The measurement – in the so-called partial fluorescence yield (PFY) mode as opposed to the total fluorescence yield (TFY) mode – consists of monitoring the scattered intensity at the Kα channel while sweeping the incident energy across

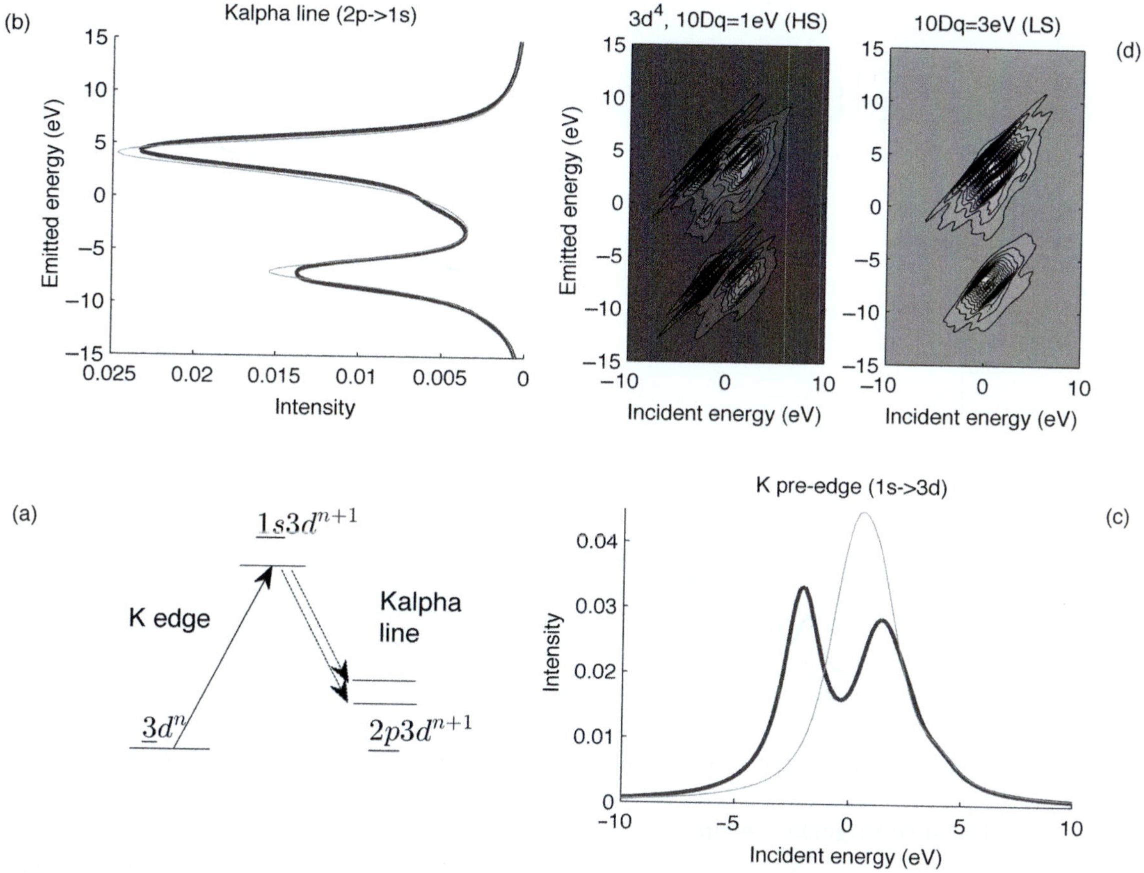

Figure 21.6 *Simulated 1s2p RIXS in $3d^4$ ion in low field (10 Dq = 1 eV) and high field (10 Dq = 3 eV).*

the absorption. As an illustration of the method strength for MO applications, Figure 21.7 compares both standard TFY XAS and 1*s*2*p*-RIXS PFY spectra simultaneously measured from V_2O_3.[6] The resolving power is clearly enhanced in the RIXS spectrum, yielding highly resolved pre-edge features, and therefore the *d* density of states.

An exciting extension of this method was recently proposed in order to extract magnetic information out of the pre-edge feature.[7] Similarly to XMCD, the measurements are carried out with circular polarized light. By inversing the helicity of the incident photons, it is possible to retrieve a dichroic signal in the pre-edge region of comparable intensity to that obtained at the $L_{2,3}$ edge. The signal shows clear signatures of the 3*d* magnetism, hence sum rules could be applied.

21.3 Experimental Results

We now turn to reviewing some remarkable results about spin-transitions in MOs as observed by RIXS. We will emphasize a selection of results of broad importance at the expense of being exhaustive, concentrating on our own research work that we know best.

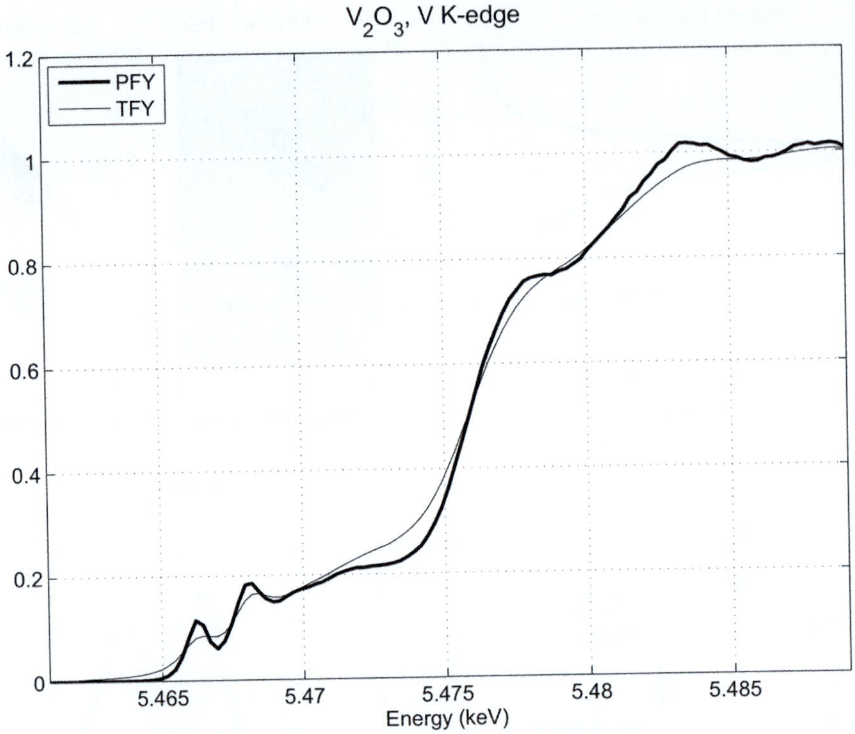

Figure 21.7 RIXS high-resolution XAS spectrum (PFY-XAS) at the K-edge in V_2O_3 (adapted from ref. 6).

21.3.1 High Pressure Magnetic Collapse

Monoxides are the simplest possible oxides among the vast MO family. Yet, their importance was recognized early on as model systems of strongly correlated insulators. The partially filed d band is split by an on-site Coulomb interaction U, that energetically drains free carriers away from the Fermi energy, opening a gap at E_F. The correct description of the d electron behavior involves also the O p-ligand states that notably

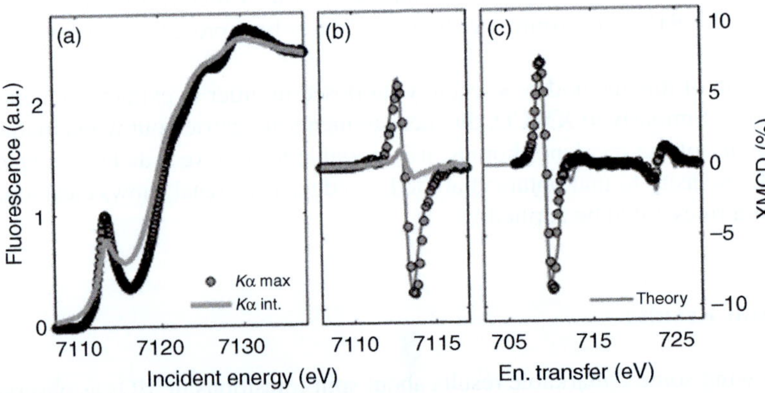

Figure 21.8 Dichroism in the pre-edge region, observed in magnetite. Reproduced from [7] with permission from American Psychical Society. Copyright 2010.

interact with the d-electrons through electron charge transfer. Thus, the insulating gap in the monoxides is of p-d character, defined by the charge transfer energy Δ and not purely by U as in other strongly correlated materials.

As it turns out, monoxides form antiferromagnetically ordered cubic structures with large Néel temperatures. The magnetic stability upon gap closure has been questioned within the Stoner model.[8] The main outcome of this exploratory work is the prediction of simultaneous metal-insulator and magnetic collapse transitions at very high pressure, in the megabar range.

The spin state of a series of monoxides, MnO, FeO, CoO, and NiO was monitored under high pressure conditions by XES spectroscopy.[9, 10] The main results are depicted in Figure 21.9 for the two extreme pressures. We observe a clear change of the satellite intensity at high pressure that is reminiscent of a high spin to low spin-transition, in all the monoxides except for NiO ($3d$),[8] where such a transition is not expected. Notice it was not possible here to determine possible energy shifts of the main emission lines because of the experimental difficulties related to the high-pressure setup.

Using multiplet calculations as a guideline (see right panel in Fig. 21.9), the spin-transition in the monoxides is understood as resulting from the combined effects of an increase in the crystal-field strength $10Dq$ and

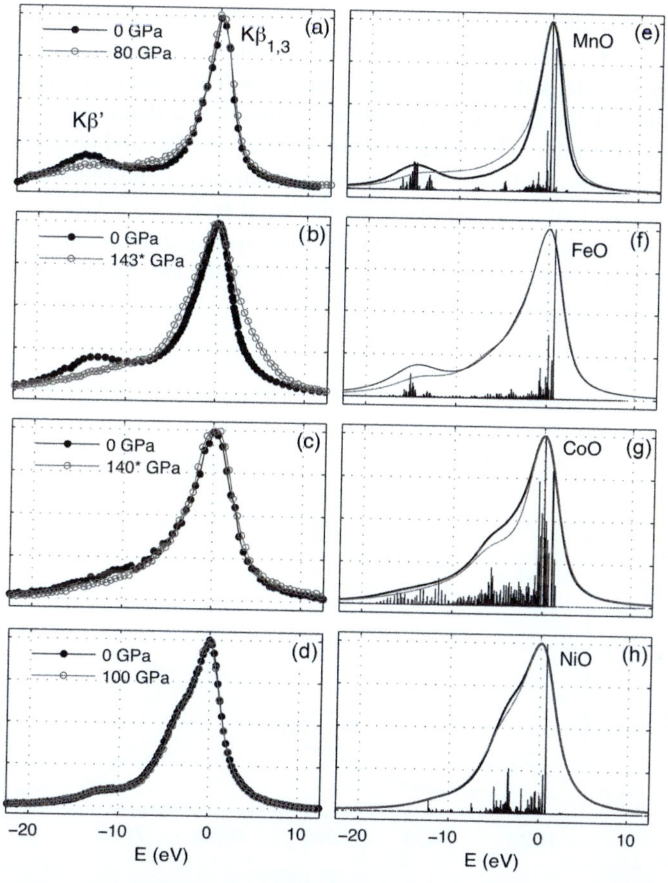

Figure 21.9 *Magnetic collapse transition in monoxides under pressure; (left) experimental XES spectra at low and high pressure; (right) calculation with the ligand field multiplet approach. Reproduced from [9] with permission from American Psychical Society. Copyright 2007.*

a broadening of the O-2*p* bandwidth, along with an increased covalent contribution from the hybridization to the ligand field at high pressures. The reinforcement of crystal field strength is seen as the driving force towards the LS state.

21.3.2 Application to Geophysics

Metal oxides are commonly found in minerals. Two of the most important rock phases of the lower Earth mantle, namely (Mg,Fe)SiO_3 magnesium silicate perovskite and (Mg,Fe)O ferropericlase, belong to this category. The knowledge of the spin state in these materials under realistic conditions of pressure and temperature is primordial to determine phase partitioning and heat transport properties of the Earth's interior.

Recent measurements were carried out in both rock phases by X-ray emission spectroscopy. As an example, Figure 21.10 shows the Kβ emission spectra in (Mg,Fe)O under high pressure.[11] The decrease of the satellite intensity with increasing pressure indicates a magnetic transition from a high spin to low spin state around 60–70 GPa, corresponding to depths of 2000 km in Earth's lower mantle. This transition implies that the partition coefficient of iron between ferropericlase and magnesium silicate perovskite may increase by several orders of magnitude, depleting the perovskite phase of its iron. In the latter, two successive spin-transitions are found around 70 at 120 GPa. They are understood as transitions from high spin ($S = 2$) to intermediate spin ($S = 1$) then low spin ($S = 0$) states[12] as pressure increases. The change of magnetic state may provoke an increase of transparency of the mantle in the infrared region at a depth region consistent with the D" layer above the core-mantle boundary. The measurements on perovskite (and post-perovskite) minerals were further extended at high pressure and high temperature, in physical conditions that better match those of the Earth's interior. The spin state at high pressure was found to be remarkably stable up 3000 K with a $S = 1$ intermediate spin configuration (c.f. Fig. 21.11), demonstrating that radiative thermal conductivity and iron partitioning in the lowermost mantle is rather controlled by the structural transition from perovskite to post-perovskite, rather than the electronic transition of Fe^{2+}.[12]

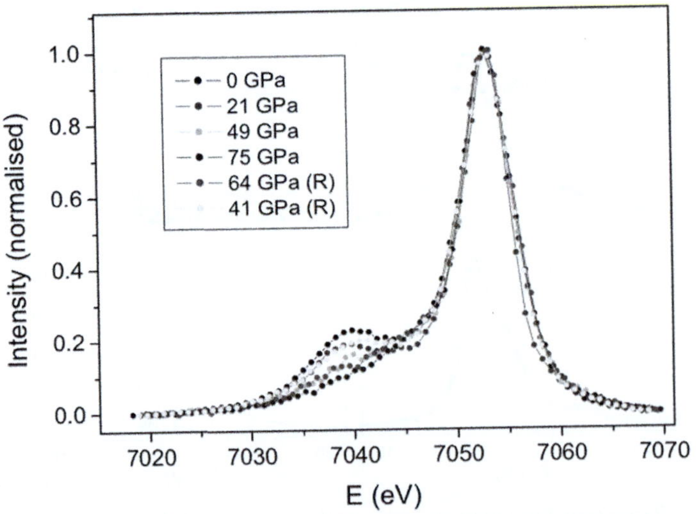

Figure 21.10 Kβ emission from (Mg,Fe)O ferropericlase. Reproduced from [10] with permission from AAAS. Copyright 2003.

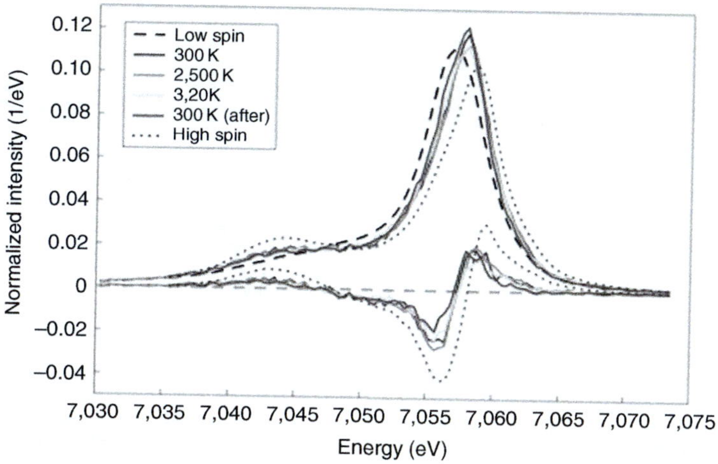

Figure 21.11 Kβ emission in (Mg,Fe)SiO3 perovskite at high pressure and high temperature (from ref. 12). The spectra are consistent with a S = 1 intermediate spin state. Reproduced from [12] with permission from Nature Publishing Group.

21.3.3 Occurrence of Intermediate Spin State in Cobaltates

Spin state transitions also play a crucial role in the understanding of electronic and magnetic properties of cobaltates. Spin entropy, for instance, was invoked as a possible cause of the outstanding thermoelectric power of the $Na_xCo_2O_4$ or $La_{1-x}Sr_xCoO_3$ families, about ten times larger than in typical metals.[14] In these materials, the cobalt ions are distributed between Co^{4+} and Co^{3+} oxidation states. Whereas Co^{3+} adopts a nonmagnetic $S = 0$ low spin configuration, Co^{4+} is believed to retain a magnetic moment with an $S = {}^1/_2$, low spin state. In this picture, it is the hole hopping between Co^{4+} and Co^{3+} sites that conveys spin entropy, providing the mechanism for thermoelectricity. The strength of the thermoelectric power depends on the interplay between degenerate spin configurations. Both Co^{4+} and Co^{3+} can adopt three spin states (respectively low spin, intermediate spin or high spin) depending on the occupancy of the high-lying e_g orbitals (with 0, 1, or 2 electrons respectively).[14] The different spin states are close in energy so that different spin states can in principle be simultaneously occupied. Large values of thermoelectric power are found when both ions are in their low spin configuration with an equal population, as observed experimentally. A similar reasoning applies to the $La_{1-x}Sr_xCoO_3$ families that show interesting coupling between electronic and spin-transitions.

The parent compound $LaCoO_3$ is a charge transfer insulator at low temperature, with the Co ions in a low spin $S = 0$ $t_{2g}{}^6$ configuration. Two successive, sluggish magnetic transitions are observed with increasing temperature. The first occurs at 100 K and is associated with a spin state transition towards an intermediate or high spin state with progressive occupation of the Co e_g orbitals. The second transition leads to a metallic state around 500 K and could be due to a gradual delocalization of the e_g electrons. The spin state in $LaCoO_3$ was measured with X-ray emission spectroscopy, using the Kβ sensitivity to the Co spin state (c.f. Fig. 21.12(a)).[16] The growing intensity in the satellite region with temperature is interpreted as a signature of a transition towards a nonzero spin state. Careful analysis of the data (black line in Fig. 21.12(b)) shows that the transition is best described using an orbitally nondegenerate intermediate spin configuration at medium temperature and a high spin state at high temperature.

The stability of the spin state was also verified under high pressure in both the parent and Sr-doped compounds.[16, 17] In both cases, pressure drives the system towards a low spin state. In the Sr-doped material, the spin state is found to be stable against temperature in contrast to the parent compound. These results also

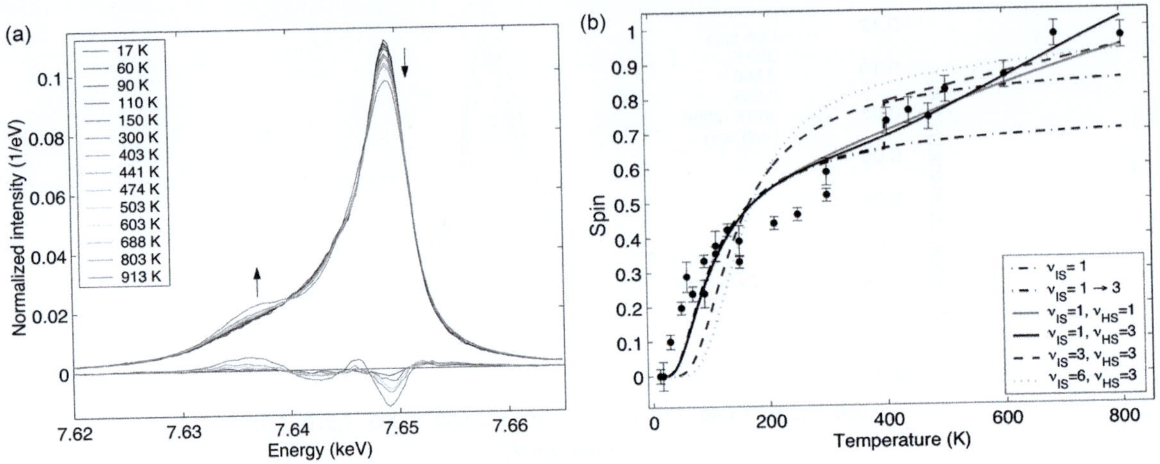

Figure 21.12 *(a) Kβ emission in $LaCoO_3$ as a function of temperature; (b) Temperature dependent spin state – the best fit yields transition towards a nondegenerate three-state model. Reproduced from [15] with permission from American Psychical Society. Copyright 2007.*

provide microscopic evidence that the pressure-induced metal-insulator transition in $La_{1-x}Sr_xCoO_3$ is driven by a simultaneous spin state transition.

21.3.4 Photoexcited Spin-Transition in Crossover Compounds and ps Dynamics

'Spin-crossover' complexes are coordination compounds of transition-metal ions with medium ligand field strength and an electronic configuration ranging between $3d^4$ and $3d^7$. In these systems, the spin state of the molecules can be switched back and forth between low spin and high spin states by triggering a redistribution of the $3d$ electrons among the t_{2g} and e_g orbitals either through temperature or pressure change, irradiation with light, or application of a high magnetic field.

As spin state can be easily controlled, spin-crossover complexes are foreseen for data storage applications and display devices. They also make a perfect testing ground for X-ray spectroscopic techniques. The symmetry of the metal site and ground state spin configuration can be tuned by chemistry engineering. In a recent study, several of these spin-crossovers complexes with Fe^{3+}, Fe^{2+}, and Co^{2+} were investigated by RIXS.[17] Figure 21.13 summarizes the results obtained for a Fe^{2+} complex during the low spin ($S = 0$) to high spin ($S = 2$) temperature- induced transition. We observed clear signature of the spin-transition in the different emission lines (panel a), in the 1*s*2*p*-RIXS map and Fe K pre-edge region (panel b).

A more spectacular result was obtained by ultra-fast time-resolved spectroscopy. The spin-transition was induced under laser photoexcitation of a Fe^{2+} low spin molecular complex and spectral changes were monitored as a function of the time delay between the laser pump and the X-ray probe. A 10% conversion of the spin state towards a high spin state was found after a 60 ps time delay (c.f. Fig. 21.14).

21.4 Conclusions and Perspectives

Spin state is directly involved in many important features of transition metal oxides, illustrating the strong coupling in these materials between magnetic, structural, and electronic degrees of freedom. The spin state affects structural stability, especially when the system is subjected to high-pressure conditions. This has

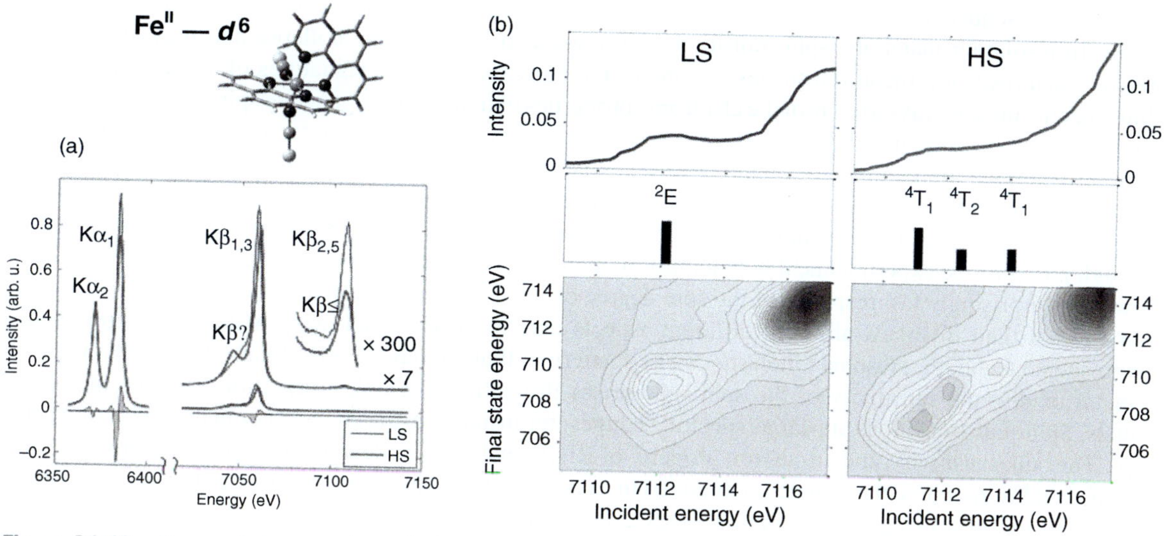

Figure 21.13 Spectral changes through the temperature induced spin-transition in a Fe^{2+} complex: (a) Emission line, (b) 1s2p-RIXS map and Fe K-pre edge. Reprinted with permission from [17]. Copyright 2006, American Chemical Society.

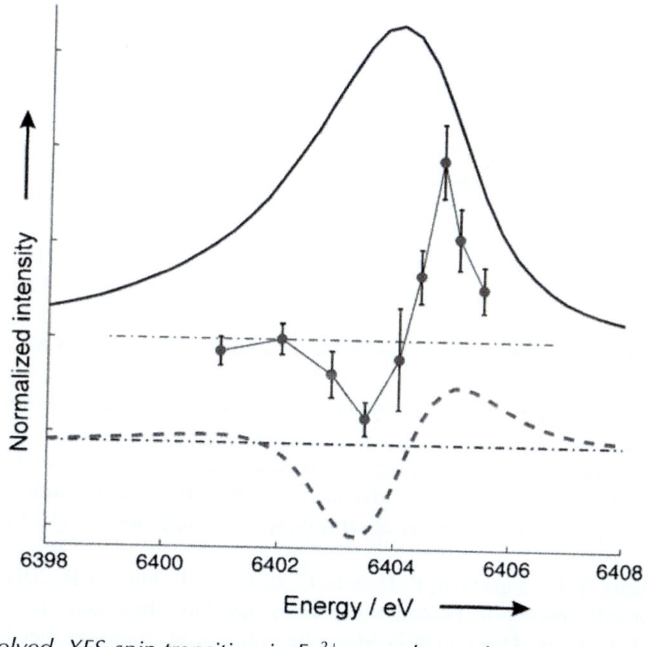

Figure 21.14 Time resolved -XES spin-transition in Fe^{2+} complex at the $K\alpha_1$ emission (solid line). Circles are spectral difference between XES measured at 60 ps after laser excitation. The expected effect at 10% excitation yield of the high spin state is shown as a dashed line. Reprinted with permission from [18]. Copyright 2010 WILEY-VCH Verlag GmbH & Co. KGaA.

immediate consequences for geophysics since major components of the Earth's interior are magnetic metal oxides. In insulating materials, spin-transitions are associated with volume collapse and may lead to gap closure. Such insulator–metal transitions are one of the most fundamental signatures of strong correlations while opening up new ways to control the electronic properties of materials. Degenerate spin states are also held responsible for the exceptional thermoelectric power of metal oxides through the 'spin entropy' contribution. In these materials, the complexity of the spin states is revealed in the strong temperature dependence of the magnetic and electronic properties. Finally, metal oxide complexes offer an ideal playground for spin-transitions as spin state can be altered either by pressure or temperature, or under laser photoexcitation. Through pump-probe experiments, the dynamics of spin state transitions were investigated in the ps regime, showing a surprisingly fast response of the spin degree of freedom to the excitation.

In spite of their diversity, all these different aspects may be investigated using a simple, yet powerful, spectroscopic technique, resonant inelastic X-ray scattering, that we have mostly focused on in this review. In metal oxides, the sensitivity of the spin state arises through a core-hole interaction with the magnetic orbitals. Spin state transitions appear as spectral changes, for instance in the emission or in the RIXS intensity map'. The bulk sensitivity, and all-photon property of RIXS, makes it an ideal probe of spin-transitions under extreme conditions where the sample is contained in a pressure cell. Other types of spin excitations (magnons) may also be revealed using high resolution RIXS, complementing the interest in this technique for correlated magnetic metal oxides.[20] As a core-hole spectroscopy, RIXS also allows ultra-fast spin-related phenomena to be observed in the ps, or fs timescale either through the 'core-hole clock' process or, as discussed, by pump-probe approaches. There is no doubt that the construction of new synchrotron fs sources of high brilliance coupled to RIXS techniques will offer new opportunities for investigating ultra-fast spin-transition processes, renewing the picture of spin states in metal oxides.

References

1. Kittel, C. (2004) Introduction to Solid State Physics, 8th edn. John Wiley & Sons, New York, USA, p. 680.
2. Rueff, J.-P., Shukla, A. (2010) Inelastic x-ray scattering by electronic excitations under high pressure. Rev. Mod. Phys., 82: 847–896.
3. Peng, G., de Groot, F. M. F., Hämäläinen, K., Moore, J. A., Wang, X., Grush, M. M., Hastings, J. B., Siddons, D. P., Armstrong, W. H. (1994) High-resolution manganese x-ray fluorescence spectroscopy. Oxidation-state and spin-state sensitivity. J. Am. Chem. Soc., 116: 2914–2920.
4. Stavitski, E., de Groot, F. M. F. (2010) The CTM4XAS program for EELS and XAS spectral shape analysis of transition metal L edges. Micron, 41: 687–694.
5. Kotani, A., Shin, S. (2001) Resonant inelastic x-ray scattering spectra for electrons in solids. Rev. Mod. Phys., 73: 203–246.
6. Rodolakis, F., Rueff, J.-P., Sikora, M., Alliot, I., Itié, J.-P., Baudelet, F., Ravy, S., Wzietek, P., Hansmann, P., Toschi, A., Haverkort, M. W., Sangiovanni, G., Held, K., Metcalf, P., Marsi, M. (2011) Evolution of the electronic structure of a Mott system across its phase diagram: x-ray absorption spectroscopy study of $(V_{1-x}Cr_x)_2O_3$. Phys. Rev. B, 84: 245113/1 – 245113/10.
7. Sikora, M., Juhin, A., Weng, T.-C., Sainctavit, P., Detlefs, C., de Groot, F., Glatzel, P. (2010) Strong *K*-edge magnetic circular dichroism observed in photon-in–photon-out spectroscopy. Phys. Rev. Lett., 105: 037202/1 – 037202/4.
8. Cohen, R. E., Mazin, I. I., Isaak, D. G. (1997) Magnetic collapse in transition metal oxides at high pressure: implications for the Earth. Science, 275: 654–657.
9. Rueff, J.-P., Mattila, A., Badro, J., Vankó, G., Shukla, A. (2005) Electronic properties of transition-metal oxides under high pressure revealed by x-ray emission spectroscopy. J. Phys. Condens. Matter, 17: S717.
10. Mattila, A., Rueff, J.-P., Badro, J., Vankó, G., Shukla, A. (2007) Metal-ligand interplay in strongly correlated oxides: a parametrized phase diagram for pressure-induced spin-transitions. Phys. Rev. Lett., 98: 196404/1 – 196404/4.

11. Badro, J., Fiquet, G., Guyot, F., Rueff, J.-P., Struzhkin, V. V., Vankó, G., Monaco, G. (2003) Iron partitioning in earth's mantle: toward a deep lower mantle discontinuity. Science, 300: 789–791.
12. Badro, J., Rueff, J.-P., Vankó, G., Monaco, G., Fiquet, G., Guyot, F. (2004) Electronic transitions in perovskite: possible nonconvecting layers in the lower mantle. Science, 305: 383–386.
13. Lin, J.-F., Watson, H., Vankó, G., Alp, E. E., Prakapenka, V. B., Dera, P., Struzhkin, V. V., Kubo, A., Zhao, J., McCammon, C., Evans, W. J. (2008) Intermediate-spin ferrous iron in lowermost mantle post-perovskite and perovskite. Nature Geosci., 1: 688–691.
14. Wang, Y., Rogado, N. S., Cava, R. J., Ong, N. P. (2003) Spin entropy as the likely source of enhanced thermopower in $Na_xCo_2O_4$. Nature, 423: 425–428.
15. Koshibae, W., Tsutsui, K., Maekawa, S. (2000) Thermopower in cobalt oxides. Phys. Rev. B, 62: 6869–6872.
16. Vankó, G., Rueff, J.-P., Mattila, A., Németh, Z., Shukla, A. (2006) Temperature- and pressure-induced spin-state transitions in $LaCoO_3$. Phys. Rev. B, 73: 024424/1 – 024424/9.
17. Lengsdorf, R., Rueff, J.-P., Vankó, G., Lorenz, T., Tjeng, L. H., Abd-Elmeguid, M. M. (2007) Spin-state-driven metal-insulator transition in $(La,Sr)CoO_3$ under high-pressure. Phys. Rev. B, 75: 180401/1 – 180401/4.
18. Vankó, G., Neisius, T., Molnár, G., Renz, F., Kárpáti, S., Shukla, A., de Groot, F. M. F., (2006) Probing the 3d spin momentum with x-ray emission spectroscopy: the case of molecular-spin-transitions. J. Phys. Chem. B, 110: 11647–11653.
19. Vankó, G., Glatzel, P., Pham, V.-T., Abela, R., Grolimund, D., Borca, C. N., Johnson, S. L., Milne, C. J., Bressler, C. (2010) Picosecond time-resolved x-ray emission spectroscopy: ultrafast spin-state determination in an iron complex. Angew. Chem. Int. Ed., 49: 5910–5912.
20. Le Tacon, M., Ghiringhelli, G., Chaloupka, J., Moretti Sala, M., Hinkov, V., Haverkort, M. W., Minola, M., Bakr, M., Zhou, K. J., Blanco-Canosa, S., Monney, C., Song, Y. T., Sun, G. L., Lin, C. T., De Luca, G. M., Salluzzo, M., Khaliullin, G., Schmitt, T., Braicovich, L., Keimer, B. (2011) Intense paramagnon excitations in a large family of high-temperature superconductors. Nature Phys., 7: 725–730.

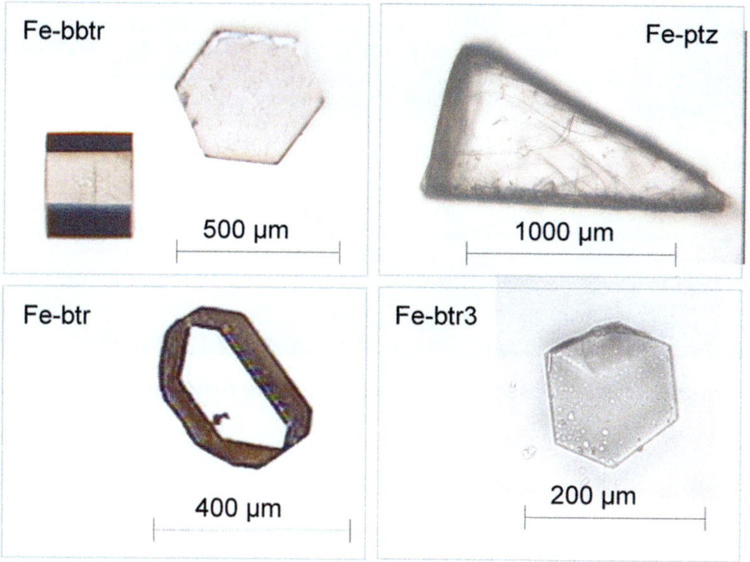

Plate 16.1 *Typical crystals investigated in the present report (at room temperature, before thermal cycling). Their complete formulae will be given in text.*

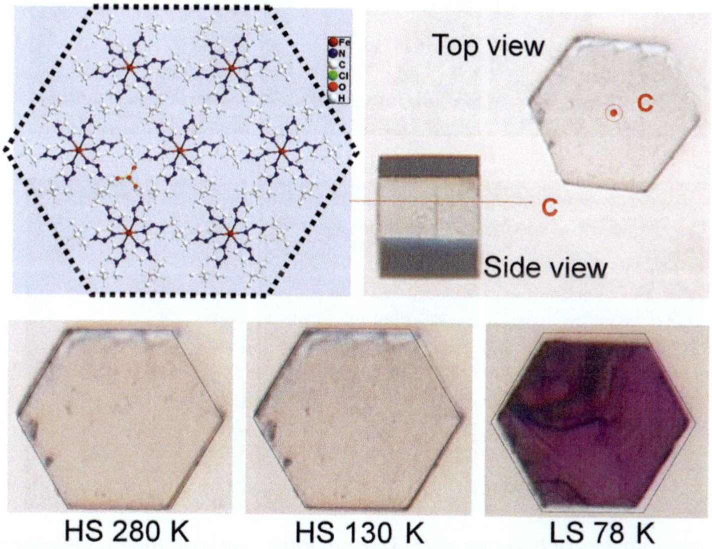

Plate 16.2 *Bright-field transmission images of typical Fe-bbtr crystals with different orientations. The crystal in top view (hexagonal contour) was ~100 μm thick; the crystal in side view was ~165 μm. The low temperature image (LS state) is characterized by the dark purple colour (absorption by the LS* d-d *band ~590 nm), by the visible shrinkage of the sample and by the presence of major damages generated by the volume effects during the transition. After ref. 7.*

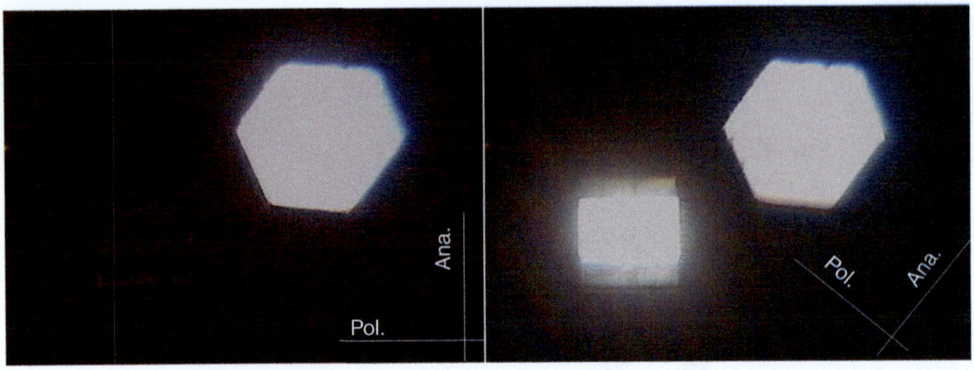

Plate 16.3 *The set of Fe-bbtr crystals, observed at room temperature between crossed polarizers. Adapted from ref. 7.*

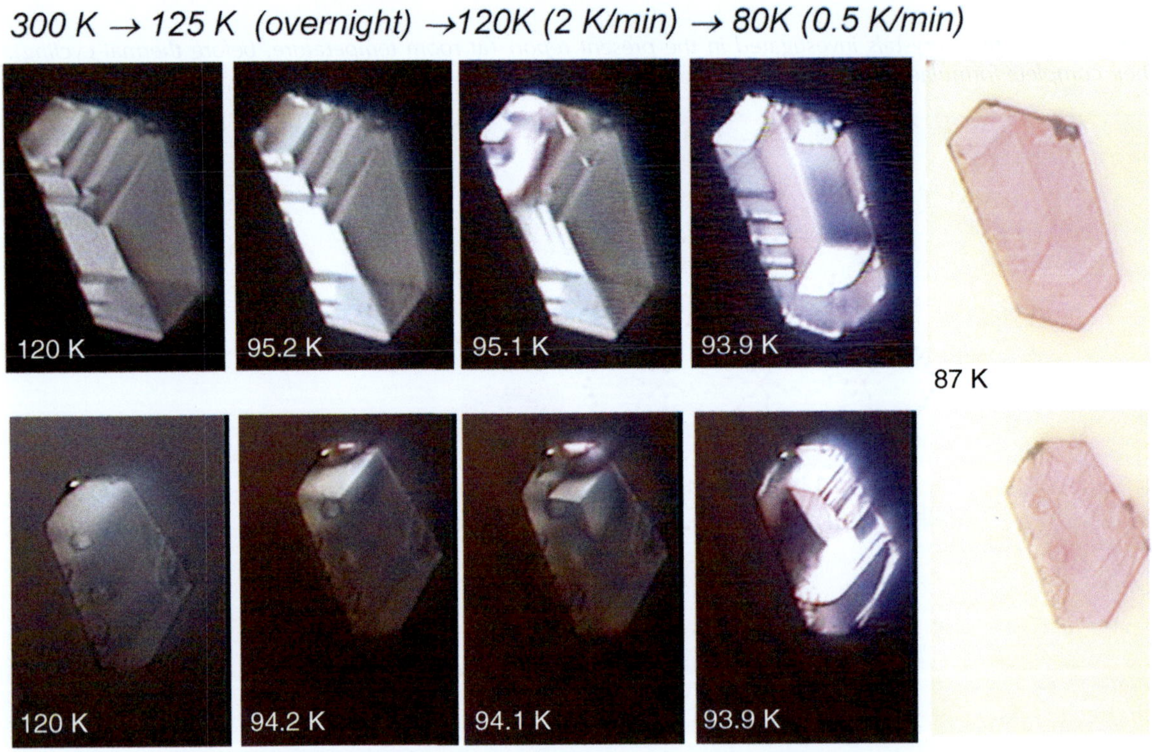

Plate 16.4 *Cross-polarizer images of two pieces (~370 μm long, ~60 μm thick), cleaved from the same Fe-bbtr crystal, simultaneously measured on cooling. The development of the LS phase is revealed by the onset of the pink colour in all geometries (with and without crossed polarizers). The development of the structural transition is revealed by the facetted patterning in the cross-polarizer images. The different impacts of the spin-transition, albeit starting from the same top corner of both crystals, is discussed in the text. Some of the structural domain walls are also detectable on the bright-field images in the LS state (right images). Adapted from ref. 7.*

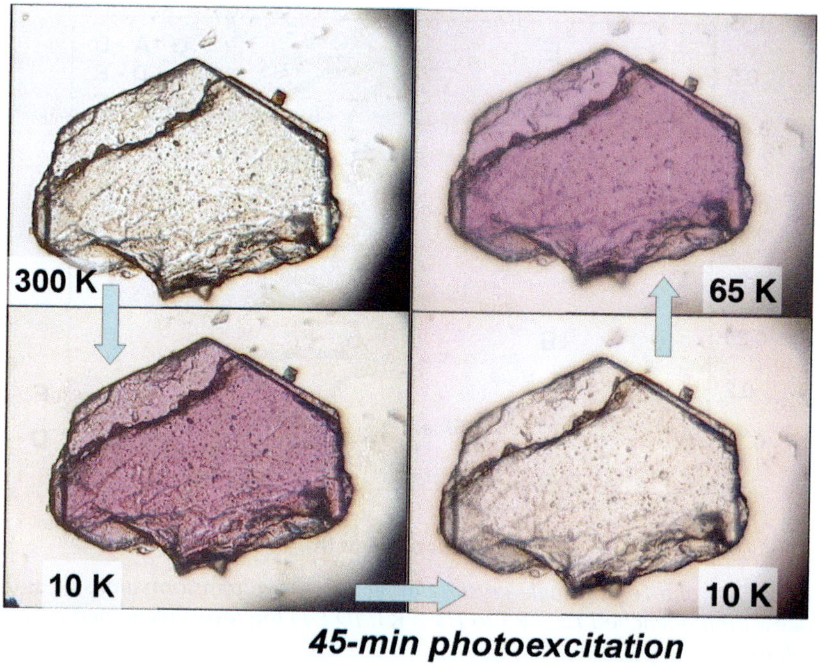

Plate 16.5 *Thermo- and photochromic properties of Fe-ptz. It is worth noting that the photoexcitation and subsequent relaxation processes did not induce any patterning of the Fe-ptz images, in contrast to most of the thermal transitions reported here. Adapted from ref. 7.*

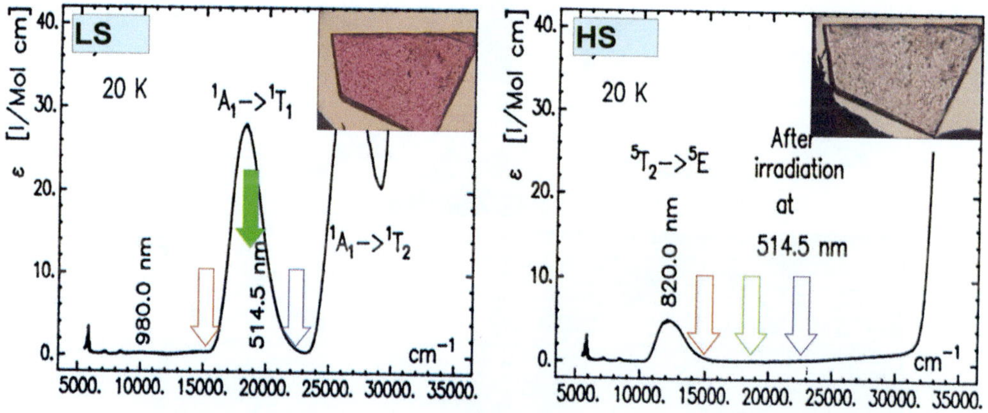

Plate 16.6 *Absorption spectra of Fe-ptz in the high temperature HS-R$\bar{3}$ state and in the low temperature LS-R$\bar{3}$ quenched state. Adapted from ref. 21.*

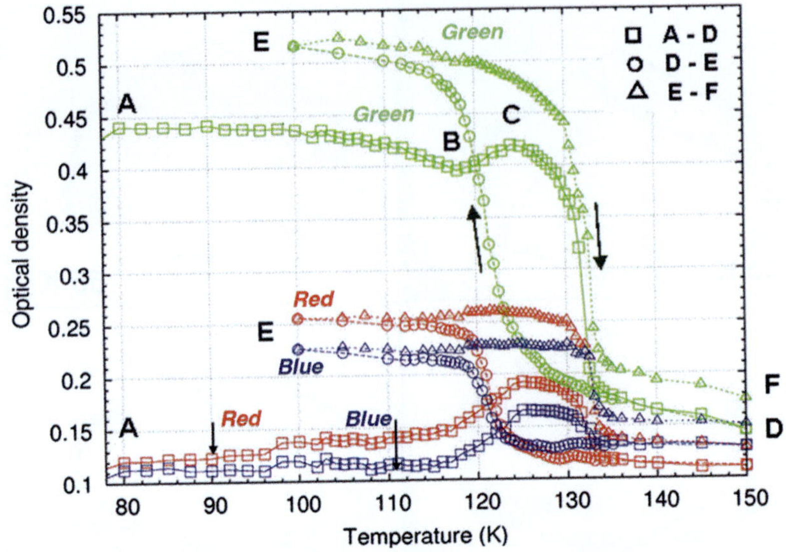

Plate 16.7 The RGB ODs of a Fe-ptz single crystal during successive transformations at ± 0.1 K/min: [A–D] the irreversible process from quenched $LS - R\bar{3}$ to $HS - R\bar{3}$; [D-E-F] the reversible joined transition $HS - R\bar{3} \leftrightarrow LS - R\bar{3}dis$. After ref. 6.

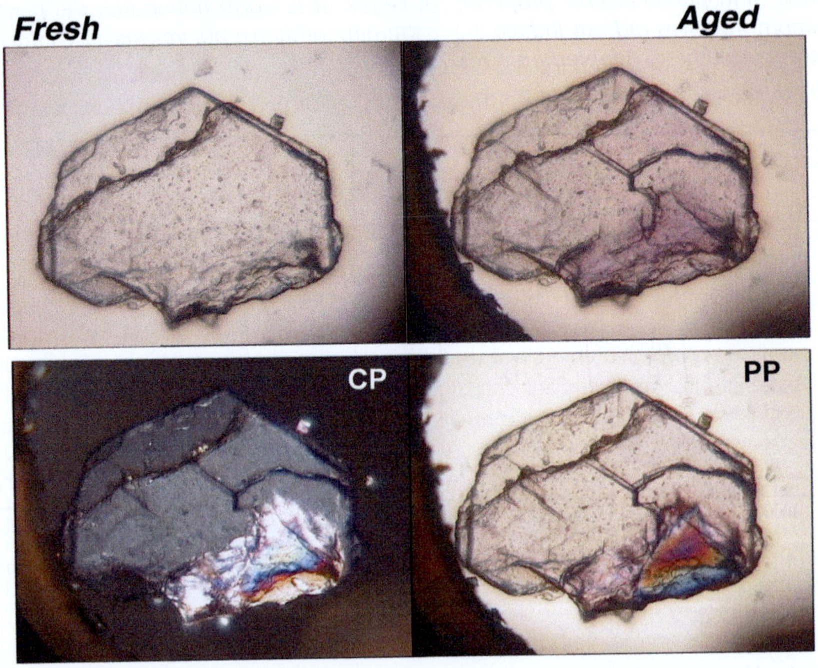

Plate 16.8 *Images of a Fe-ptz crystal at 10 K after "saturated" photoexcitation. The area resistant to photoexcitation, in the aged crystal, is characterized by the purple colour in the top right image. Stresses are evidenced by the colour-changing areas in the bottom images (CP, PP = crossed, parallel polarizers). The area contaminated by the* $R\bar{3}dis$ *phase is evidenced by scattering effects (CP image). After ref. 6.*

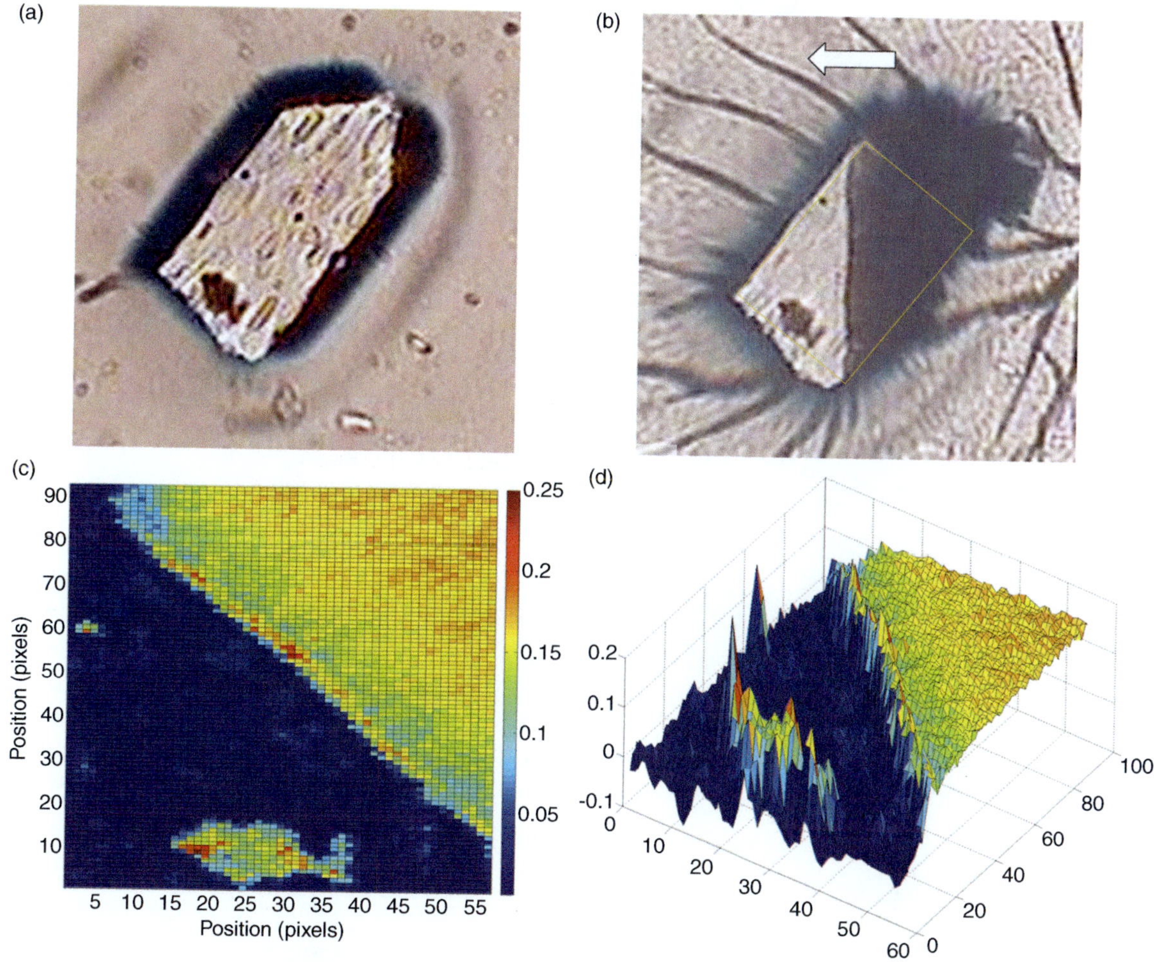

Plate 16.9 *A high-quality Fe-btr crystal, at room temperature (a) and in the course of the thermal transition, at 102.5 K (b).The crystal was embedded in oil so as to prevent dehydration, and the oil film underwent several cracks by ~120 K. The bright and dark areas respectively correspond to HS, LS phases. On the bottom line we mapped the red ODs derived from image (b) inside the rectangular frame; they are plotted in projection (c) or using a pseudo 3D representation (d). The propagation sense of the frontline is shown by an arrow. Adapted from ref. 10 (including video files).*

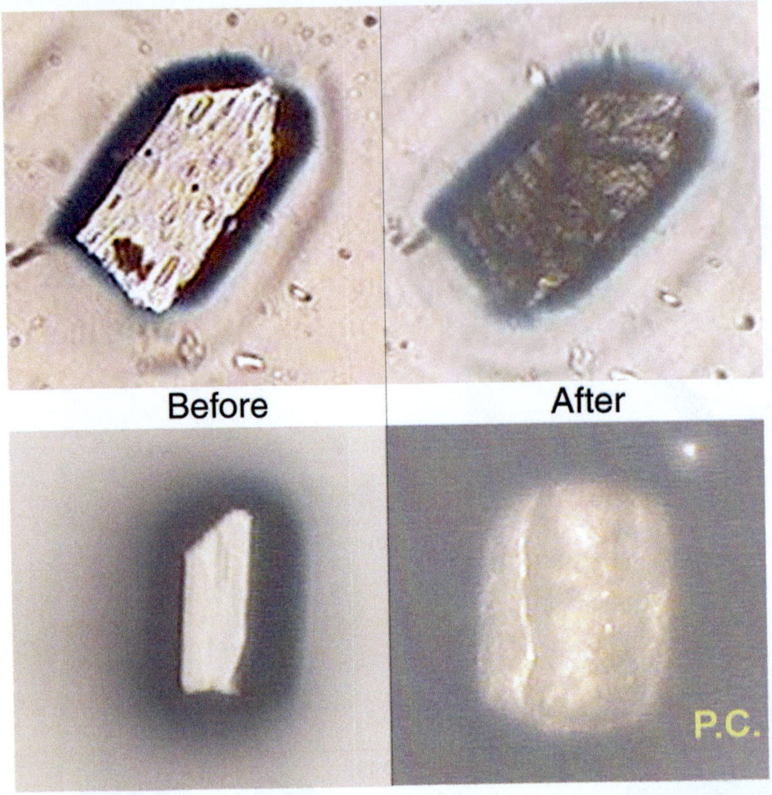

Plate 16.10 *Various damages of Fe-btr crystals, generated by the mechanical stresses associated with the spin transition. Adapted from ref. 10.*

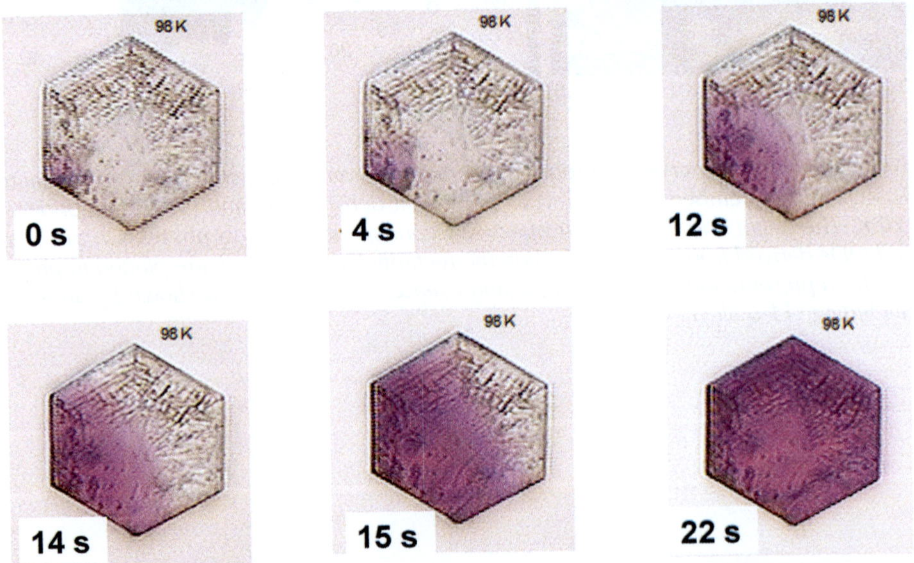

Plate 16.11 *Selected images of a top view crystal during its first HS→LS transition, at 98 K after cooling at 0.2 K/min. Crystal thickness (135–190 μm) was determined from a side view of the same crystal. After ref. 9 (including video files).*

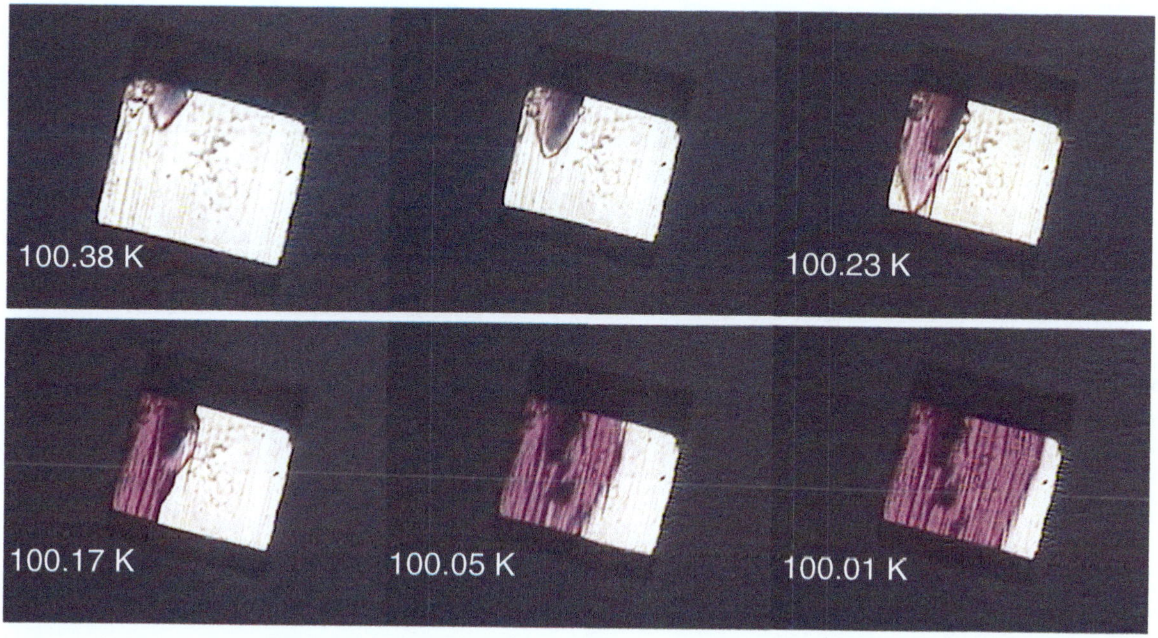

Plate 16.12 *Selected images of a side-view crystal during its first HS→LS transition, by cooling at 0.2 K/min. After ref. 9 (including video files).*

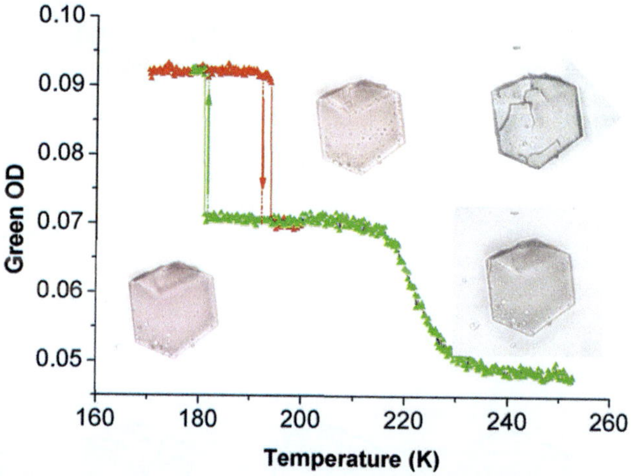

Plate 16.13 *Fe-btr3: the thermal variation of the green OD, illustrating the 2-step transformation of the system. The high temperature step occurs progressively in the range [220–230 K] (scan rate 1 K/min). The low temperature step (scan rate 0.2 K/min) shows first-order character with hysteresis ~10–12 K wide. Top right: view of the aged crystal (after ~10 thermal cycles).*

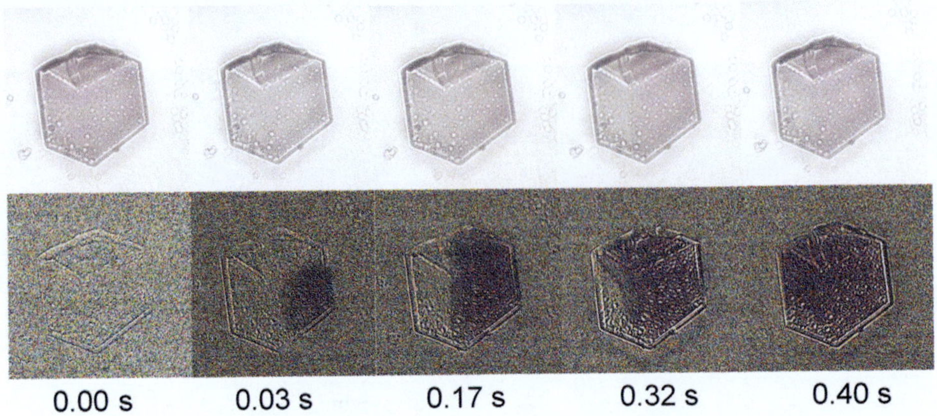

Plate 16.14 *Fe-btr3: selected images of the low-temperature step, on cooling at 0.2 K/min, T = 181.6 K. The processed images (bottom line) evidence the progression of the transformation front, at an average velocity ~500 µm/s.*

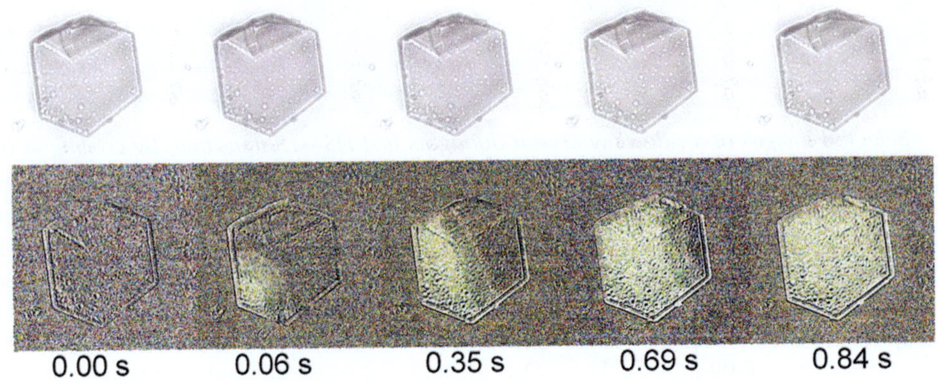

Plate 16.15 *Fe-btr4: same caption as Figure 16.14, on heating at 0.2 K/min, T = 192.2 K, at average velocity ~250 µm/s.*

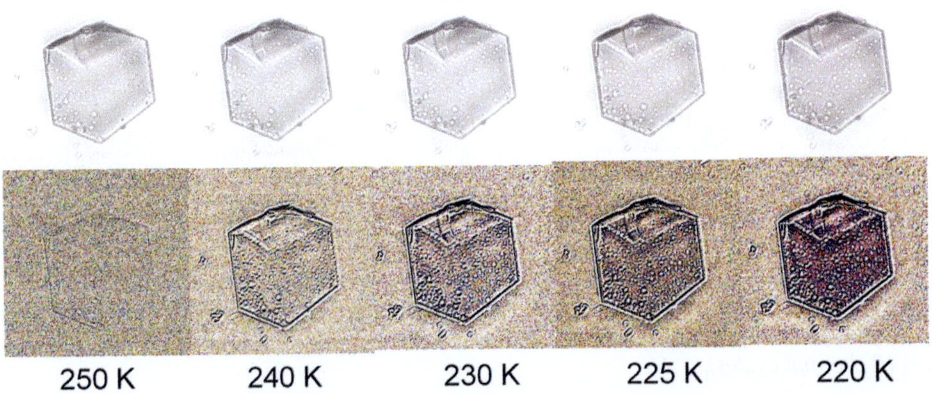

Plate 16.16 *Fe-btr3: selected images of the high-temperature step; raw images (top line) and processed images (bottom), see text.*

Index

Spin-Crossover Materials: Properties and Applications, First Edition. Edited by Malcolm A. Halcrow.

Printed and bound by CPI Group (UK) Ltd, Croydon, CR0 4YY
25/09/2026
14969413-0001